$\Phi(z) = P(Z \le z)$

z	0.00	0.01	0.02	0.03	0.04	0.05	0.06	0.07	0.08	0.09
0.0	0.5000	0.5040	0.5080	0.5120	0.5160	0.5199	0.5239	0.5279	0.5319	0.5359
0.1	0.5398	0.5438	0.5478	0.5517	0.5557	0.5596	0.5636	0.5675	0.5714	0.5753
0.2	0.5793	0.5832	0.5871	0.5910	0.5948	0.5987	0.6026	0.6064	0.6103	0.6141
0.3	0.6179	0.6217	0.6255	0.6293	0.6331	0.6368	0.6406	0.6443	0.6480	0.6517
0.4	0.6554	0.6591	0.6628	0.6664	0.6700	0.6736	0.6772	0.6808	0.6844	0.6879
0.5	0.6915	0.6950	0.6985	0.7019	0.7054	0.7088	0.7123	0.7157	0.7190	0.7224
0.6	0.7257	0.7291	0.7324	0.7357	0.7389	0.7422	0.7454	0.7486	0.7517	0.7549
0.7	0.7580	0.7611	0.7642	0.7673	0.7704	0.7734	0.7764	0.7794	0.7823	0.7852
0.8	0.7881	0.7910	0.7939	0.7967	0.7995	0.8023	0.8051	0.8078	0.8106	0.8133
0.9	0.8159	0.8186	0.8212	0.8238	0.8264	0.8289	0.8315	0.8340	0.8365	0.8389
1.0	0.8413	0.8438	0.8461	0.8485	0.8508	0.8531	0.8554	0.8577	0.8599	0.8621
1.1	0.8643	0.8665	0.8686	0.8708	0.8729	0.8749	0.8770	0.8790	0.8810	0.8830
1.2	0.8849	0.8869	0.8888	0.8907	0.8925	0.8944	0.8962	0.8980	0.8997	0.9015
1.3	0.9032	0.9049	0.9066	0.9082	0.9099	0.9115	0.9131	0.9147	0.9162	0.9177
1.4	0.9192	0.9207	0.9222	0.9236	0.9251	0.9265	0.9278	0.9292	0.9306	0.9319
1.5	0.9332	0.9345	0.9357	0.9370	0.9382	0.9394	0.9406	0.9418	0.9429	0.9441
1.6	0.9452	0.9463	0.9474	0.9484	0.9495	0.9505	0.9515	0.9525	0.9535	0.9545
1.7	0.9554	0.9564	0.9573	0.9582	0.9591	0.9599	0.9608	0.9616	0.9625	0.9633
1.8	0.9641	0.9649	0.9656	0.9664	0.9671	0.9678	0.9686	0.9693	0.9699	0.9706
1.9	0.9713	0.9719	0.9726	0.9732	0.9738	0.9744	0.9750	0.9756	0.9761	0.9767
2.0	0.9772	0.9778	0.9783	0.9788	0.9793	0.9798	0.9803	0.9808	0.9812	0.9817
2.1	0.9821	0.9826	0.9830	0.9834	0.9838	0.9842	0.9846	0.9850	0.9854	0.9857
2.2	0.9861	0.9864	0.9868	0.9871	0.9875	0.9878	0.9881	0.9884	0.9887	0.9890
2.3	0.9893	0.9896	0.9898	0.9901	0.9904	0.9906	0.9909	0.9911	0.9913	0.9916
2.4	0.9918	0.9920	0.9922	0.9925	0.9927	0.9929	0.9931	0.9932	0.9934	0.9936
2.5	0.9938	0.9940	0.9941	0.9943	0.9945	0.9946	0.9948	0.9949	0.9951	0.9952
2.6	0.9953	0.9955	0.9956	0.9957	0.9959	0.9960	0.9961	0.9962	0.9963	0.9964
2.7	0.9965	0.9966	0.9967	0.9968	0.9969	0.9970	0.9971	0.9972	0.9973	0.9974
2.8	0.9974	0.9975	0.9976	0.9977	0.9977	0.9978	0.9979	0.9979	0.9980	0.9981
2.9	0.9981	0.9982	0.9982	0.9983	0.9984	0.9984	0.9985	0.9985	0.9986	0.9986
3.0	0.9987	0.9987	0.9987	0.9988	0.9988	0.9989	0.9989	0.9989	0.9990	0.9990
3.1	0.9990	0.9991	0.9991	0.9991	0.9992	0.9992	0.9992	0.9992	0.9993	0.9993
3.2	0.9993	0.9993	0.9994	0.9994	0.9994	0.9994	0.9994	0.9995	0.9995	0.9995
3.3	0.9995	0.9995	0.9995	0.9996	0.9996	0.9996	0.9996	0.9996	0.9996	0.9997
3.4	0.9997	0.9997	0.9997	0.9997	0.9997	0.9997	0.9997	0.9997	0.9997	0.9998

FIFTH EDITION

Probability and Statistics for Engineering and the Sciences

FIFTH EDITION

Probability and Statistics for Engineering and the Sciences

Jay L. Devore

California Polytechnic State University

Duxbury
Thomson Learning™

Australia • Canada • Mexico • Singapore • Spain • United Kingdom • United States

Sponsoring Editor: *Carolyn Crockett*
Marketing Team: *Tom Ziolkowski/Beth Kroenke*
Editorial Assistants: *Kimberly Raburn/Ann Day*
Production Editor: *Tessa McGlasson Avila*
Production Service: *Susan L. Reiland*
Manuscript Editor: *Christine Levesque*
Interior Design: *John Edeen*

Cover Design: *Laurie Albrecht*
Interior Illustration: *Lori Heckelman*
Print Buyer: *Vena Dyer*
Typesetting: *Graphic World, Inc.*
Cover Printing: *Phoenix Color Corporation*
Printing and Binding: *R. R. Donnelley/Crawfordsville*

For more information about this or any other Duxbury products, contact:
DUXBURY
511 Forest Lodge Road
Pacific Grove, CA 93950 USA
www.duxbury.com
1-800-423-0563 (Thomson Learning Academic Resource Center)

For permission to use material from this work, contact us by
Web: www.thomsonrights.com
fax: 1-800-730-2215
phone: 1-800-730-2214

Printed in the United States of America

10   9   8   7

Library of Congress Cataloging-in-Publication Data
Devore, Jay L.
  Probability and statistics for engineering and the sciences / Jay L. Devore—5th ed.
  Includes bibliographical references and index.
  ISBN 0-534-37281-3
    1. Probabilities. 2. Mathematical statistics. I. Title.
QA273 .D46 1999
519.5—dc21   99-046944

To my wife, Carol:
Your dedication to teaching
is a continuing inspiration to me.

To my daughters, Allison and Teresa:
The great pride I take in your
accomplishments knows no bounds.

Contents

Preface

Purpose

The use of probability models and statistical methods for analyzing data has become common practice in virtually all scientific disciplines. This book attempts to provide a comprehensive introduction to those models and methods most likely to be encountered and used by students in their careers in engineering and the natural sciences. Although the examples and exercises have been designed with scientists and engineers in mind, most of the methods covered are basic to statistical analyses in many other disciplines, so that students of business and the social sciences will also profit from reading the book.

Approach

Students in a statistics course designed to serve other majors may be initially skeptical of the value and relevance of the subject matter, but my experience is that students *can* be turned on to statistics by the use of good examples and exercises that blend their everyday experiences with their scientific interests. Consequently, I have worked hard to find examples of real, rather than artificial, data—data that someone thought was worth collecting and analyzing. Many of the methods presented, especially in the later chapters on statistical inference, are illustrated by analyzing data taken from a published source, and many of the exercises also involve working with such data. Sometimes the reader may be unfamiliar with the context of a particular problem (as indeed I often was), but I have found that students are more attracted by real problems with a somewhat strange context than by patently artificial problems in a familiar setting.

Mathematical Level

The exposition is relatively modest in terms of mathematical development. Substantial use of the calculus is made only in Chapter 4 and parts of Chapters 5 and 6. In particular, with the exception of an occasional remark or aside, calculus appears in the inference part of the book only in the second section of Chapter 6. Matrix algebra is not used at all. Thus almost all the exposition should be accessible to those whose mathematical background includes one semester or two quarters of differential and integral calculus.

Content

Chapter 1 begins with some basic concepts and terminology—population, sample, descriptive and inferential statistics, enumerative versus analytic studies, and so on—and continues with a survey of important graphical and numerical descriptive methods. A rather traditional development of probability is given in Chapter 2, followed by probability distributions of discrete and continuous random variables in Chapters 3 and 4, respectively. Joint distributions and their properties are discussed in the first part of Chapter 5. The latter part of this chapter introduces statistics and their sampling distributions, which form the bridge between probability and inference. The next three chapters cover point estimation, statistical intervals, and hypothesis testing based on a single sample. Methods of inference involving two independent samples and paired data are presented in Chapter 9. The analysis of variance is the subject of Chapters 10 and 11 (single-factor and multifactor, respectively). Regression makes its initial appearance in Chapter 12 (the simple linear regression model and correlation) and returns for an extensive encore in Chapter 13. The last three chapters develop chi-squared methods, distribution-free (nonparametric) procedures, and techniques from statistical quality control.

Helping Students Learn

Although the book's mathematical level should give most science and engineering students little difficulty, working toward an understanding of the concepts and gaining an appreciation for the logical development of the methodology may sometimes require substantial effort. To help students gain such an understanding and appreciation, I have provided numerous exercises ranging in difficulty from many that involve routine application of text material to some that ask the reader to extend concepts discussed in the text to somewhat new situations. There are many more exercises than most instructors would want to assign during any particular course, but I recommend that students be required to work a substantial number of them; in a problem-solving discipline, active involvement of this sort is the surest way to identify and close the gaps in understanding that inevitably arise. Answers to odd-numbered exercises appear in the answer section at the back of the text. In addition, a Student Solutions Manual, consisting of worked-out solutions to virtually all the odd-numbered exercises, is available.

New for This Edition

- The first section of Chapter 1 has been rewritten to emphasize from the outset that variation is the source from which all statistical methodology flows. The techniques from exploratory and descriptive statistics introduced in this chapter are utilized to a greater extent than before in the inferential part of the book.
- The material on sampling distributions in Chapter 5 has been reorganized to convey more clearly the central idea on which inferential methods are based: The value of any statistic (quantity calculated from sample data) will in general vary when sample after sample is selected from the same population.
- One-sided confidence and prediction intervals are now featured in Chapter 7 along with their two-sided counterparts. A new confidence interval for a population propor-

tion (the Agresti–Coull "score" interval) is included. Normal tolerance intervals, previously relegated to an exercise, are now discussed in more detail in the text itself, and a table of tolerance critical values for one- and two-sided intervals is included.

- There is increased emphasis on P-values for testing hypotheses. The appendix now contains a table of t curve tail areas, so that a statement such as P-value $\approx .017$, rather than just $.01 < P$-value $< .025$, can be made. A new chi-squared table also allows for more precise P-value information for chi-squared tests, and a more detailed F table does the same thing for F tests.

- Notation in the regression chapters has been streamlined, allowing for the use of more concise formulas. There is now a short subsection in Chapter 13 on logistic regression.

- Finally, numerous examples have been updated, and many new exercises have supplemented or replaced those from previous editions.

Recommended Coverage

There is enough material in this book for a year-long course. Anyone teaching a course of shorter duration will have to be selective in the choice of topics to be included. At Cal Poly, we teach a two-quarter sequence, meeting four hours per week. During the first ten weeks we cover much of the material in Chapters 1–7 (going lightly over joint distributions and the details of estimation by maximum likelihood and the method of moments). The second quarter begins with hypothesis testing and moves on to two-sample inferences, ANOVA, regression, and selections from the chi-squared, distribution-free, and quality control chapters. Coverage of material in a one-semester course would obviously have to be somewhat more restrictive. There is, of course, never enough time to teach students all that we would like them to know!

Acknowledgments

My colleagues here at Cal Poly have provided me with invaluable support and encouragement over the years. I am also grateful to the many users of previous editions who have made suggestions for improvement (and pointed out occasional errors). A note of thanks goes to Julie Seely and Beth Eltinge for their work on the Student Solutions Manual.

I gratefully acknowledge the plentiful feedback provided by the following reviewers of this and previous editions: Robert L. Armacost, University of Central Florida; Douglas M. Bates, University of Wisconsin–Madison; David M. Cresap, University of Portland; Don E. Deal, University of Houston; Charles E. Donaghey, University of Houston; Mark Duva, University of Virginia; Nasser S. Fard, Northeastern University; Celso Grebogi, University of Maryland; James J. Halavin, Rochester Institute of Technology; Wei-Min Huang, Lehigh University; Stephen Kokoska, Colgate University; Arnold R. Miller, University of Denver; Don Ridgeway, North Carolina State University; Larry J. Ringer, Texas A&M University; Richard M. Soland, The George Washington University; Clifford Spiegelman, Texas A&M University; Jery Stedinger, Cornell University; David Steinberg, Tel Aviv University; G. Geoffrey Vining, University of Florida; Bhutan Wadhwa, Cleveland State University; and Michael G. Zabetakis, University of Pittsburgh.

I very much appreciate the editorial and production services provided by Susan Reiland, Christine Levesque, and Lori Heckelman (I can't conceive of a more effective and congenial production coordinator than Susan). The staff at Duxbury and Brooks/Cole has as usual been extremely supportive—thanks in particular to Carolyn Crockett, Curt Hinrichs, Tessa Avila, Seema Atwal, and Kimberly Raburn. I wish also to commend the sales representatives of Thomson Learning for their hard work over the years in getting the word out about earlier editions of this book as well as other books I have written. Finally, words cannot adequately express my gratitude toward my wife, Carol, for her support of my writing efforts over the course of the last 25 years.

Jay L. Devore

FIFTH EDITION

Probability and Statistics for Engineering and the Sciences

1

Overview and Descriptive Statistics

Introduction

Statistical concepts and methods are not only useful but indeed often indispensable in understanding the world around us. They provide ways of gaining new insights into the behavior of many phenomena that you will encounter in your chosen field of specialization in engineering or science.

The discipline of statistics teaches us how to make intelligent judgments and informed decisions in the presence of uncertainty and variation. Without uncertainty or variation, there would be little need for statistical methods or statisticians. If every component of a particular type had exactly the same lifetime, if all resistors produced by a certain manufacturer had the same resistance value, if pH determinations for soil specimens from a particular locale gave identical results, and so on, then a single observation would reveal all desired information.

An interesting manifestation of variation arises in the course of performing emissions testing on motor vehicles. The expense and time requirements of the Federal Test Procedure (FTP) preclude its widespread use in vehicle inspection programs. As a result, many agencies have developed less costly and quicker tests, which it is hoped replicate FTP results. According to the article "Motor Vehicle

Emissions Variability" (*J. of the Air and Waste Mgmt. Assoc.,* 1996: 667–675), the acceptance of the FTP as a gold standard has led to the widespread belief that repeated measurements on the same vehicle would yield identical (or nearly identical) results. The authors of the article applied the FTP to seven vehicles characterized as "high emitters." Here are the results for one such vehicle:

HC (gm/mile)	13.8	18.3	32.2	32.5
CO (gm/mile)	118	149	232	236

The substantial variation in both the HC and CO measurements casts considerable doubt on conventional wisdom and makes it much more difficult to make precise assessments about emissions levels.

How can statistical techniques be used to gather information and draw conclusions? Suppose, for example, that a materials engineer has developed a coating for retarding corrosion in metal pipe under specified circumstances. If this coating is applied to different segments of pipe, variation in environmental conditions and in the segments themselves will result in more substantial corrosion on some segments than on others. Methods of statistical analysis could be used on data from such an experiment to decide whether the *average* amount of corrosion exceeds an upper specification limit of some sort or to predict how much corrosion will occur on a single piece of pipe.

Alternatively, suppose the engineer has developed the coating in the belief that it will be superior to the currently used coating. A comparative experiment could be carried out to investigate this issue by applying the current coating to some segments of pipe and the new coating to other segments. This must be done with care lest the wrong conclusion emerge. For example, perhaps the average amount of corrosion is identical for the two coatings. However, the new coating may be applied to segments that have superior ability to resist corrosion and under less stressful environmental conditions compared to the segments and conditions for the current coating. The investigator would then likely observe a difference between the two coatings attributable not to the coatings themselves but just to extraneous variation. Statistics offers not only methods for analyzing the results of experiments once they have been carried out, but also suggestions for how experiments can be performed in an efficient manner to mitigate the effects of variation and have a better chance of producing correct conclusions.

1.1 | Populations, Samples, and Processes

Engineers and scientists are constantly exposed to collections of facts, or **data,** both in their professional capacities and in everyday activities. The discipline of statistics provides methods for organizing and summarizing data, and for drawing conclusions based on information contained in the data.

An investigation will typically focus on a well-defined collection of objects constituting a **population** of interest. In one study, the population might consist of all gelatin capsules of a particular type produced during a specified period. Another investigation might involve the population consisting of all individuals who received a B.S. in engineering during the most recent academic year. When desired information is available for all objects in the population, we have what is called a **census.** Constraints on time, money, and other scarce resources usually make a census impractical or infeasible. Instead, a subset of the population—a **sample**—is selected in some prescribed manner. Thus we might obtain a sample of bearings from a particular production run as a basis for investigating whether bearings are conforming to manufacturing specifications, or we might select a sample of last year's engineering graduates to obtain feedback about the quality of the engineering curricula.

We are usually interested only in certain characteristics of the objects in a population: the number of flaws on the surface of each casing, the thickness of each capsule wall, the gender of an engineering graduate, the age at which the individual graduated, and so on. A characteristic may be categorical, such as gender or type of malfunction, or it may be numerical in nature. In the former case, the *value* of the characteristic is a category (e.g., female or insufficient solder), whereas in the latter case, the value is a number (e.g., age = 23 years or diameter = .502 cm). A **variable** is any characteristic whose value may change from one object to another in the population. We shall initially denote variables by lowercase letters from the end of our alphabet. Examples include

x = gender of a graduating engineer

y = number of major defects on a newly manufactured automobile

z = braking distance of an automobile under specified conditions

Data results from making observations either on a single variable or simultaneously on two or more variables. A **univariate** data set consists of observations on a single variable. For example, we might determine the type of transmission, automatic (A) or manual (M), on each of ten automobiles recently purchased at a certain dealership, resulting in the categorical data set

M   A   A   A   M   A   A   M   A   A

The following sample of lifetimes (hours) of brand D batteries put to a certain use is a numerical univariate data set:

5.6   5.1   6.2   6.0   5.8   6.5   5.8   5.5

We have **bivariate** data when observations are made on each of two variables. Our data set might consist of a (height, weight) pair for each basketball player on a team, with the first observation as (72, 168), the second as (75, 212), and so on. If an engineer determines the value of both x = component lifetime and y = reason for component failure, the resulting data set is bivariate with one variable numerical and the other categorical. **Multivariate** data arises when observations are made on more than two variables. For example, a research physician might determine the systolic blood pressure, diastolic blood pressure, and serum cholesterol level for each patient participating in a study. Each observation would be a triple of numbers, such as (120, 80, 146). In many multivariate data sets, some variables are numerical and others are categorical. Thus the annual automobile issue of *Consumer Reports* gives values of such variables as type of vehicle (small, sporty, compact, mid-size, large), city fuel efficiency (mpg), highway fuel efficiency (mpg), drive train type (rear wheel, front wheel, four wheel), and so on.

Branches of Statistics

An investigator who has collected data may wish simply to summarize and describe important features of the data. This entails using methods from **descriptive statistics.** Some of these methods are graphical in nature; the construction of histograms, boxplots, and scatter plots are primary examples. Other descriptive methods involve calculation of numerical summary measures, such as means, standard deviations, and correlation coefficients. The wide availability of statistical computer software packages has made these tasks much easier to carry out than they used to be. Computers are much more efficient than human beings at calculation and the creation of pictures (once they have received appropriate instructions from the user!). This means that the investigator doesn't have to expend much effort on "grunt work" and will have more time to study the data and extract important messages. Throughout this book, we will present output from various packages such as MINITAB, SAS, and S-Plus.

Example 1.1   The tragedy that befell the space shuttle *Challenger* and its astronauts in 1986 led to a number of studies to investigate the reasons for mission failure. Attention quickly focused on the behavior of the rocket engine's O-rings. Here is data consisting of observations on x = O-ring temperature (°F) for each test firing or actual launch of the shuttle rocket engine (*Presidential Commission on the Space Shuttle Challenger Accident,* Vol. 1, 1986: 129–131).

84	49	61	40	83	67	45	66	70	69	80	58
68	60	67	72	73	70	57	63	70	78	52	67
53	67	75	61	70	81	76	79	75	76	58	31

Without any organization, it is very difficult to get a sense of what a typical or representative temperature might be, whether the values are highly concentrated about a typical value or quite spread out, whether there are any gaps in the data, what percentage of the values are in the 60's, and so on. Figure 1.1 shows what is called a *stem-and-leaf display* of the data, as well as a *histogram.* Shortly, we will discuss construction and interpretation of these pictorial summaries; for the moment, we hope you see how they begin to tell us how the values of temperature are distributed along the measurement scale. Some of these launches/firings were successful and others resulted in failure. In Chapter 13, we will consider whether temperature had a bearing on the likelihood of a successful launch.

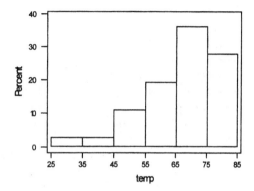

```
Stem-and-leaf of temp  N = 36
Leaf Unit = 1.0
   1    3   1
   1    3
   2    4   0
   4    4   59
   6    5   23
   9    5   788
  13    6   0113
  (7)   6   6777789
  16    7   000023
  10    7   556689
   4    8   0134
```

Figure 1.1 A MINITAB stem-and-leaf display and histogram of the O-ring temperature data ■

Having obtained a sample from a population, an investigator would frequently like to use sample information to draw some type of conclusion (make an inference of some sort) about the population. That is, the sample is a means to an end rather than an end in itself. Techniques for generalizing from a sample to a population are gathered within the branch of our discipline called **inferential statistics.**

Example 1.2   Material strength investigations provide a rich area of application for statistical methods. The article "Effects of Aggregates and Microfillers on the Flexural Properties of Concrete" (*Magazine of Concrete Research,* 1997: 81–98) reported on a study of strength properties of high-performance concrete obtained by using superplasticizers and certain binders. The compressive strength of such concrete had previously been investigated, but not much was known about flexural strength (a measure of ability to resist failure in bending). The accompanying data on flexural strength (in MegaPascal, MPa, where 1 Pa (Pascal) = 1.45×10^{-4} psi) appeared in the article cited:

5.9	7.2	7.3	6.3	8.1	6.8	7.0	7.6	6.8	6.5	7.0	6.3	7.9	9.0
8.2	8.7	7.8	9.7	7.4	7.7	9.7	7.8	7.7	11.6	11.3	11.8	10.7	

Suppose we want an *estimate* of the average value of flexural strength for all beams that could be made in this way (if we conceptualize a population of all such beams, we are trying to estimate the population mean). It can be shown that, with a high degree of confidence, the population mean strength is between 7.48 MPa and 8.80 MPa; we call this a *confidence interval* or *interval estimate*. Alternatively, this data could be used to predict the flexural strength of a *single* beam of this type. With a high degree of confidence, the strength of a single such beam will exceed 7.35 MPa; the number 7.35 is called a *lower prediction bound*. ■

The main focus of this book is on presenting and illustrating methods of inferential statistics that are useful in scientific work. The most important types of inferential procedures—point estimation, hypothesis testing, and estimation by confidence intervals—are introduced in Chapters 6–8 and then used in more complicated settings in Chapters 9–16. The remainder of this chapter presents methods from descriptive statistics that are most used in the development of inference.

Chapters 2–5 present material from the discipline of probability. This material ultimately forms a bridge between the descriptive and inferential techniques and leads to a better understanding of how inferential procedures are developed and used, how statistical conclusions can be translated into everyday language and interpreted, and when and where pitfalls can occur in applying the methods. Probability and statistics both deal with questions involving populations and samples, but do so in an "inverse manner" to one another.

In a probability problem, properties of the population under study are assumed known (e.g., in a numerical population, some specified distribution of the population values may be assumed), and questions regarding a sample taken from the population are posed and answered. In a statistics problem, characteristics of a sample are available to the experimenter, and this information enables the experimenter to draw conclusions about the population. The relationship between the two disciplines can be summarized by saying that probability reasons from the population to the sample (deductive reasoning), whereas inferential statistics reasons from the sample to the population (inductive reasoning). This is illustrated in Figure 1.2.

Figure 1.2 The relationship between probability and inferential statistics

Before we can understand what a particular sample can tell us about the population, we should first understand the uncertainty associated with taking a sample from a given population. This is why we study probability before statistics.

As an example of the contrasting focus of probability and inferential statistics, consider drivers' use of manual lap belts in cars equipped with automatic shoulder belt systems. (The article "Automobile Seat Belts: Usage Patterns in Automatic Belt Systems," *Human Factors,* 1998: 126–135, summarizes usage data.) In probability, we might assume that 50% of all drivers of cars equipped in this way in a certain metropolitan area regularly use their lap belt (an assumption about the population), so we might ask, "How likely is it that a sample of 100 such drivers will include at least 70 who regularly use their lap belt?" or "How many of the drivers in a sample of size 100 can we expect to regularly use their lap belt?" On the other hand, in inferential statistics, we have sample information available; for example, a sample of 100 drivers of such cars revealed that 65 regularly use their lap belt. We might then ask, "Does this provide substantial evidence for concluding that more than 50% of all such drivers in this area regularly use their lap belt?" In this latter scenario, we are attempting to use sample information to answer a question about the structure of the entire population from which the sample was selected.

In the lap belt example, the population is well defined and concrete: all drivers of cars equipped in a certain way in a particular metropolitan area. In Example 1.1, however, a sample of O-ring temperatures is available, but it is from a population that does not actually exist. Instead, it is convenient to think of the population as consisting of all possible temperature measurements that might be made under similar experimental conditions. Such a population is referred to as a **conceptual** or **hypothetical population.** There are a number of problem situations in which we fit questions into the framework of inferential statistics by conceptualizing a population.

Enumerative Versus Analytic Studies

W. E. Deming, a very influential American statistician who was a moving force in Japan's quality revolution during the 1950s and 1960s, introduced the distinction between *enumerative studies* and *analytic studies.* In the former, interest is focused on a finite, identifiable, unchanging collection of individuals or objects that make up a population. A *sampling frame*—that is, a listing of the individuals or objects to be sampled—is either available to an investigator or else can be constructed. For example, the frame might consist of all signatures on a petition to qualify a certain initiative for the ballot in an upcoming election; a sample is usually selected to ascertain whether the number of *valid* signatures exceeds a specified value. As another example, the frame may contain serial numbers of all furnaces manufactured by a particular company during a certain time period; a sample may be selected to infer something about the average lifetime of these units. The use of inferential methods to be developed in this book is reasonably noncontroversial in such settings (though statisticians may still argue over which particular methods should be used).

An analytic study is broadly defined as one that is not enumerative in nature. Such studies are often carried out with the objective of improving a future product by taking action on a process of some sort (e.g., recalibrating equipment or adjusting the level of some input such as the amount of a catalyst). Data can often be obtained only on an existing process, one that may differ in important respects from the future process. There is thus no sampling frame listing the individuals or objects of interest. For example, a sample of five turbines with a new design may be experimentally manufactured and tested to investigate efficiency. These five could be viewed as a sample from the conceptual population of all prototypes that could be manufactured under similar conditions, but *not* necessarily as representative of the population of units manufactured once regular production gets underway. Methods for using sample information to draw conclusions about future production units may be problematic. Someone with expertise in the area of turbine design and engineering (or whatever other subject area is relevant) should be called upon to judge whether such extrapolation is sensible. A good exposition of these issues is contained in the article "Assumptions for Statistical Inference" by Gerald Hahn and William Meeker (*The American Statistician,* 1993: 1–11).

Collecting Data

Statistics deals not only with the organization and analysis of data once it has been collected but also with the development of techniques for collecting the data. If data is not properly collected, an investigator may not be able to answer the questions under consideration with a reasonable degree of confidence. One common problem is that the target population—the one about which conclusions are to be drawn— may be different from the population actually sampled. For example, advertisers would like various kinds of information about the television-viewing habits of potential customers. The most systematic information of this sort comes from placing monitoring devices in a small number of homes across the United States. It has been conjectured that placement of such devices in and of itself alters viewing behavior, so that characteristics of the sample may be different from those of the target population.

When data collection entails selecting individuals or objects from a frame, the simplest method for ensuring a representative selection is to take a *simple random sample.* This is one for which any particular subset of the specified size (e.g., a sample of size 100) has the same chance of being selected. For example, if the frame consists of 1,000,000 serial numbers, the numbers 1, 2, . . . , up to 1,000,000 could be placed on identical slips of paper. After placing these slips in a box and thoroughly mixing, slips could be drawn one by one until the requisite sample size has been obtained. Alternatively (and much to be preferred), a table of random numbers or a computer's random number generator could be employed.

Sometimes alternative sampling methods can be used to make the selection process easier, to obtain extra information, or to increase the degree of confidence in conclusions. One such method, *stratified sampling,* entails separating the population units into nonoverlapping groups and taking a sample from each one. For example, a manufacturer of VCRs might want information about customer satisfaction for units produced during the previous year. If three different models were manufactured and sold, a separate sample could be selected from each of the three corresponding strata. This would result in information on all three models and ensure that no one model was over- or underrepresented in the entire sample.

Frequently a "convenience" sample is obtained by selecting individuals or objects without systematic randomization. As an example, a collection of bricks may be stacked in such a way that it is extremely difficult for those in the center to be selected. If the bricks on the top and sides of the stack were somehow different from the others, resulting sample data would not be representative of the population. Often an investigator will assume that such a convenience sample approximates a random sample, in which case a statistician's repertoire of inferential methods can be used; however, this is a judgment call. Most of the methods discussed herein are based on a variation of simple random sampling described in Chapter 5.

Engineers and scientists often collect data by carrying out some sort of designed experiment. This may involve deciding how to allocate several different treatments (such as fertilizers or coatings for corrosion protection) to the various experimental units (plots of land or pieces of pipe). Alternatively, an investigator may systematically vary the levels or categories of certain factors (e.g., pressure or type of insulating material) and observe the effect on some response variable (such as yield from a production process).

Example 1.3   An article in the *New York Times* (Jan. 27, 1987) reported that heart attack risk could be reduced by taking aspirin. This conclusion was based on a designed experiment involving both a control group of individuals who took a placebo having the appearance of aspirin but known to be inert and a treatment group who took aspirin according to a specified regimen. Subjects were randomly assigned to the groups to protect against any biases and so that probability-based methods could be used to analyze the data. Of the 11,034 individuals in the control group, 189 subsequently experienced heart attacks, whereas only 104 of the 11,037 in the aspirin group had a heart attack. The incidence rate of heart attacks in the treatment group was only about half that in the control group. One possible explanation for this result is chance variation—that aspirin really doesn't have the desired effect and the observed difference is just typical variation in the same way that tossing two identical coins would usually produce different numbers of heads. However, in this case, inferential methods suggest that chance variation by itself cannot adequately explain the magnitude of the observed difference.   ■

Example 1.4   An engineer wishes to investigate the effects of both adhesive type and conductor material on bond strength when mounting an integrated circuit (IC) on a certain substrate. Two adhesive types and two conductor materials are under consideration. Two observations are made for each adhesive-type/conductor-material combination, resulting in the accompanying data:

Adhesive Type	Conductor Material	Observed Bond Strength	Average
1	1	82, 77	79.5
1	2	75, 87	81.0
2	1	84, 80	82.0
2	2	78, 90	84.0

The resulting average bond strengths are pictured in Figure 1.3. It appears that adhesive type 2 improves bond strength as compared with type 1 by about the same amount whichever one of the conducting materials is used, with the 2, 2 combination being best. Inferential methods can again be used to judge whether these effects are real or simply due to chance variation.

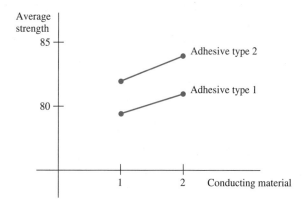

Figure 1.3   Average bond strengths in Example 1.4

Suppose additionally that there are two cure times under consideration and also two types of IC post coating. There are then $2 \cdot 2 \cdot 2 \cdot 2 = 16$ combinations of these four factors, and our engineer may not have enough resources to make even a single observation for each of these combinations. In Chapter 11, we will see how the careful selection of a fraction of these possibilities will usually yield the desired information. ▪

Exercises | Section 1.1 (1–9)

1. Give one possible sample of size 4 from each of the following populations:
   a. All daily newspapers published in the United States
   b. All companies listed on the New York Stock Exchange
   c. All students at your college or university
   d. All grade point averages of students at your college or university

2. For each of the following hypothetical populations, give a plausible sample of size 4.
   a. All distances that might result when you throw a football
   b. Page lengths of books published 5 years from now
   c. All possible earthquake-strength measurements (Richter scale) that might be recorded in California during the next year
   d. All possible yields (in grams) from a certain chemical reaction carried out in a laboratory

3. Consider the population consisting of all VCRs of a certain brand and model, and focus on whether a VCR needs service while under warranty.
   a. Pose several probability questions based on selecting a sample of 100 such VCRs.
   b. What inferential statistics question might be answered by determining the number of such VCRs in a sample of size 100 that need warranty service?

4. a. Give three different examples of concrete populations and three different examples of hypothetical populations.
   b. For one each of your concrete and your hypothetical populations, give an example of a probability question and an example of an inferential statistics question.

5. Many universities and colleges have instituted supplemental instruction (SI) programs, in which a student facilitator meets regularly with a small group of students enrolled in the course to promote discussion of course material and enhance subject mastery. Suppose that students in a large statistics course (what else?) are randomly divided into a control group that will not participate in SI and a treatment group that will participate. At the end of the term, each student's total score in the course is determined.
   a. Are the scores from the SI group a sample from an existing population? If so, what is it? If not, what is the relevant conceptual population?
   b. What do you think is the advantage of randomly dividing the students into the two groups rather than letting each student choose which group to join?
   c. Why didn't the investigators put all students in the treatment group? *Note:* The article "Supplemental Instruction: An Effective Component of Student Affairs Programming" (*J. of College Student Devel.,* 1997: 577–586) discusses the analysis of data from several SI programs.

6. The California State University (CSU) system consists of 23 campuses, from San Diego State in the south to Humboldt State near the Oregon border. A CSU administrator wishes to make an inference about the average distance between the hometowns of students and their campuses. Describe and discuss several different sampling methods that might be employed. Would this be an enumerative or an analytic study? Explain your reasoning.

7. A certain city divides naturally into ten district neighborhoods. How might a real estate appraiser select a sample of single-family homes that could be used as a basis for developing an equation to predict appraised value from characteristics such as age, size, number of bathrooms, distance to the nearest school, and so on? Is the study enumerative or analytic?

8. The amount of flow through a solenoid valve in an automobile's pollution-control system is an impor-

tant characteristic. An experiment was carried out to study how flow rate depended on three factors: armature length, spring load, and bobbin depth. Two different levels (low and high) of each factor were chosen, and a single observation on flow was made for each combination of levels.
a. The resulting data set consisted of how many observations?
b. Is this an enumerative or analytic study? Explain your reasoning.

9. In a famous experiment carried out in 1882, Michelson and Newcomb obtained 66 observations on the time it took for light to travel between two locations in Washington, D.C. A few of the measurements (coded in a certain manner) were 31, 23, 32, 36, −2, 26, 27, and 31.
a. Why are these measurements not identical?
b. Is this an enumerative study? Why or why not?

1.2 | Pictorial and Tabular Methods in Descriptive Statistics

Descriptive statistics can be divided into two general subject areas. In this section, we will discuss the first of these areas—representing a data set using visual techniques. In Sections 1.3 and 1.4, we will develop some numerical summary measures for data sets. Many visual techniques may already be familiar to you: frequency tables, tally sheets, histograms, pie charts, bar graphs, scatter diagrams, and the like. Here we focus on a selected few of these techniques that are most useful and relevant to probability and inferential statistics.

Notation

Some general notation will make it easier to apply our methods and formulas to a wide variety of practical problems. The number of observations in a single sample will often be denoted by n, so that $n = 4$ for the sample of universities {Stanford, Iowa State, Wyoming, Rochester} and also for the sample of pH measurements {6.3, 6.2, 5.9, 6.5}. If two samples are simultaneously under consideration, either m and n or n_1 and n_2 can be used to denote the numbers of observations. Thus, if {29.7, 31.6, 30.9} and {28.7, 29.5, 29.4, 30.3} are thermal-efficiency measurements for two different types of diesel engines, then $m = 3$ and $n = 4$.

Given a data set consisting of n observations on some variable x, the individual observations will be denoted by $x_1, x_2, x_3, \ldots, x_n$ (though any other letter could be used in place of x). The subscript bears no relation to the magnitude of a particular observation, so that x_1 will not in general be the smallest observation in the set, nor will x_n typically be the largest. In many applications, x_1 will be the first observation gathered by the experimenter, x_2 the second, and so on. The ith observation in the data set will be denoted by x_i.

Stem-and-Leaf Displays

Suppose we have a numerical data set $x_1, x_2, \ldots, x_n$ for which each x_i consists of at least two digits. A quick way to obtain an informative visual representation of the data set is to construct a *stem-and-leaf display*.

Steps for Constructing a Stem-and-Leaf Display

1. Select one or more leading digits for the stem values. The trailing digits become the leaves.

2. List possible stem values in a vertical column.

3. Record the leaf for every observation beside the corresponding stem value.

4. Indicate the units for stems and leaves someplace in the display.

Thus, if the data set consists of exam scores, each between 0 and 100, the score of 83 would have a stem of 8 and a leaf of 3. For a data set of automobile-fuel efficiencies (mpg), all between 8.1 and 47.8, we could use the tens digit as the stem, so 32.6 would then have a leaf of 2.6. In general, a display based on between 5 and 20 stems is recommended.

Example 1.5   The use of alcohol by college students is of great concern not only to those in the academic community but also, because of potential health and safety consequences, to society at large. The article "Health and Behavioral Consequences of Binge Drinking in College" (*J. of the Amer. Med. Assoc.*, 1994: 1672–1677) reported on a comprehensive study of heavy drinking on campuses across the United States. A binge episode was defined as five or more drinks in a row for males and four or more for females. Figure 1.4 shows a stem-and-leaf display of 140 values of x = the percentage of undergraduate students who are binge drinkers. (These values were not given in the cited article, but our display agrees with a picture of the data that did appear.)

The first leaf on the stem 2 row is 1, which tells us that 21% of the students at one of the colleges in the sample were binge drinkers. Without the identification of stem digits and leaf digits on the display, we wouldn't know whether the stem 2, leaf 1 observation should be read as 21%, 2.1%, or .21%.

```
0 | 4
1 | 1345678889
2 | 1223456666777889999               Stem: tens digit
3 | 01122333445556666777778888899999  Leaf: ones digit
4 | 11122222334444455666666677788888999
5 | 0011122222334556666667777888899
6 | 01111244455666778
```

Figure 1.4 Stem-and-leaf display for percentage binge drinkers at each of 140 colleges

When creating a display by hand, ordering the leaves from smallest to largest on each line can be time-consuming, and this ordering usually contributes little if any extra information. Suppose the observations had been listed in alphabetical order by school name, as

16%   33%   64%   37%   31%   . . .

Then placing these values on the display in this order would result in the stem 1 row having 6 as its first leaf, and the beginning of the stem 3 row would be

3 | 371 . . .

The display suggests that a typical or representative value is in the stem 4 row, perhaps in the mid-40% range. The observations are not highly concentrated about this typical value, as would be the case if all values were between 20% and 49%. The display rises to a single peak as we move downward, and then declines; there are no gaps in the display. The shape of the display is not perfectly symmetric, but instead appears to stretch out a bit more in the direction of low leaves than in the direction of high leaves. Lastly, there are no observations that are unusually far from the bulk of the data (no *outliers*), as would be the case if one of the 26% values had instead been 86%. The most surprising feature of this data is that, at most colleges in the sample, at least one-quarter of the students are binge drinkers. The problem of heavy drinking on campuses is much more pervasive than many had suspected. ■

A stem-and-leaf display conveys information about the following aspects of the data:

- identification of a typical or representative value
- extent of spread about the typical value
- presence of any gaps in the data
- extent of symmetry in the distribution of values
- number and location of peaks
- presence of any outlying values

Example 1.6 Figure 1.5 presents stem-and-leaf displays for a random sample of yardages of golf courses that have been designated by *Golf Magazine* as among the most challenging in the United States. Among the sample of 40 yardages, the shortest course is 6433 yards long, and the longest is 7280 yards. The yardages appear to be distributed in a roughly uniform fashion over the range of values in the sample. Notice that a stem choice here of either a single digit (6 or 7) or three digits (643, . . ., 728) would yield an uninformative display, the first because of too few stems and the latter because of too many.

```
Stem: Thousands and hundreds digits
64 | 35  64  33  70    Leaf: Tens and ones digits
65 | 26  27  06  83                        Stem-and-leaf of yardage  N = 40
66 | 05  94  14                            Leaf Unit = 10
67 | 90  70  00  98  70  45  13                4      64   3367
68 | 90  70  73  50                            8      65   0228
69 | 00  27  36  04                           11      66   019
70 | 51  05  11  40  50  22                    18      67   0147799
71 | 31  69  68  05  13  65                   (4)     68   5779
72 | 80  09                                    18      69   0023
                                               14      70   012455
                                                8      71   013666
                                                2      72   08
              (a)                                         (b)
```

Figure 1.5 Stem-and-leaf displays of golf course yardages: (a) two-digit leaves; (b) display from MINITAB with truncated one-digit leaves ■

A stem-and-leaf display does not show the order in which observations were obtained, possibly hiding important information about the mechanism that generated the data. For example, during a particular time period, the width of a slot cut in a certain part by a milling machine may tend to increase relative to the target value, indicating an "out-of-control" process. A *digidot plot* combines a picture of the observations over time with a stem-and-leaf display.

Example 1.7   Each observation in the stem-and-leaf display of Figure 1.6 is the value of U.S. beer production (millions of barrels) for a different quarter during the period 1975–1982. The display uses *repeated stems*; for example, the 4L row is for observations with a "low" leaf—0, 1, 2, 3, or 4—and observations with high leaves are placed in the 4H row. The *time-series plot* on the right shows both an increasing trend over time and also higher production in the second and third quarters of any given year than in the other two quarters (a *seasonal* effect).

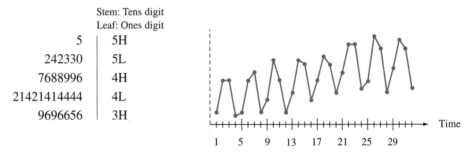

Figure 1.6   A digidot plot for U.S. beer production ■

Dotplots

A dotplot is an attractive summary of numerical data when the data set is reasonably small or there are relatively few distinct data values. Each observation is represented by a dot above the corresponding location on a horizontal measurement scale. When a value occurs more than once, there is a dot for each occurrence, and these dots are stacked vertically. As with a stem-and-leaf display, a dotplot gives information about location, spread, extremes, and gaps.

Example 1.8   Figure 1.7 shows a dotplot for the O-ring temperature data introduced in Example 1.1 in the previous section. A representative temperature value is one in the mid-60's (°F), and there is quite a bit of spread about the center. The data stretches out more at the lower end than at the upper end, and the smallest observation, 31, can fairly be described as an outlier.

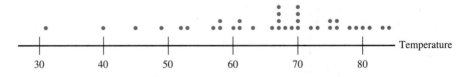

Figure 1.7   A dotplot of the O-ring temperature data (°F) ■

If the data set discussed in Example 1.8 had consisted of 50 or 100 temperature observations, each recorded to a tenth of a degree, it would have been much more cumbersome to construct a dotplot. Our next technique is well suited to such situations.

Histograms

Some numerical data is obtained by counting to determine the value of a variable (the number of traffic citations a person received during the last year, the number of persons arriving for service during a particular period), whereas other data is obtained by taking measurements (weight of an individual, reaction time to a particular stimulus). The prescription for drawing a histogram is generally different for these two cases.

DEFINITION

A variable is **discrete** if its set of possible values either is finite or else can be listed in an infinite sequence (one in which there is a first number, a second number, and so on). A variable is **continuous** if its possible values consist of an entire interval on the number line.

A discrete variable x almost always results from counting, in which case possible values are 0, 1, 2, 3, . . . or some subset of these integers. Continuous variables arise from making measurements. For example, if x is the pH of a chemical substance, then in theory x could be any number between 0 and 14: 7.0, 7.03, 7.032, etc. Of course, in practice there are limitations on the degree of accuracy of any measuring instrument, so we may not be able to determine pH, reaction time, height, and concentration to an arbitrarily large number of decimal places. However, from the point of view of creating mathematical models for distributions of data, it is helpful to imagine an entire continuum of possible values.

Consider data consisting of observations on a discrete variable x. The **frequency** of any particular x value is the number of times that value occurs in the data set. The **relative frequency** of a value is the fraction or proportion of time the value occurs:

$$\text{relative frequency of a value} = \frac{\text{number of times the value occurs}}{\text{number of observations in the data set}}$$

Suppose, for example, that our data set consists of 200 observations on x = the number of major defects on a new car of a certain type. If 70 of these x values are 1, then

frequency of the x value 1:   70

relative frequency of the x value 1: $\dfrac{70}{200} = .35$

Multiplying a relative frequency by 100 gives a percentage; in the defect example, 35% of the cars in the sample had just one major defect. The relative frequencies, or percentages, are usually of more interest than the frequencies themselves. In theory, the relative frequencies should sum to 1, but in practice the sum may differ slightly from 1 because of rounding. A **frequency distribution** is a tabulation of the frequencies and/or relative frequencies.

> ### Constructing a Histogram for Discrete Data
>
> First, determine the frequency and relative frequency of each x value. Then mark possible x values on a horizontal scale. Above each value, draw a rectangle whose height is the relative frequency (or alternatively, the frequency) of that value.

This construction ensures that the *area* of each rectangle is proportional to the relative frequency of the value. Thus if the relative frequencies of $x = 1$ and $x = 5$ are .35 and .07, respectively, then the area of the rectangle above 1 is five times the area of the rectangle above 5.

Example 1.9   How unusual is a no-hitter or a one-hitter in a major league baseball game, and how frequently does a team get more than 10, 15, or even 20 hits? Table 1.1 is a frequency distribution for the number of hits per team per game for all nine-inning games that were played between 1989 and 1993.

Table 1.1   Frequency Distribution for Hits in Nine-Inning Games

Hits/Game	Number of Games	Relative Frequency	Hits/Game	Number of Games	Relative Frequency
0	20	.0010	14	569	.0294
1	72	.0037	15	393	.0203
2	209	.0108	16	253	.0131
3	527	.0272	17	171	.0088
4	1048	.0541	18	97	.0050
5	1457	.0752	19	53	.0027
6	1988	.1026	20	31	.0016
7	2256	.1164	21	19	.0010
8	2403	.1240	22	13	.0007
9	2256	.1164	23	5	.0003
10	1967	.1015	24	1	.0001
11	1509	.0779	25	0	.0000
12	1230	.0635	26	1	.0001
13	834	.0430	27	1	.0001
				19,383	1.0005

The corresponding histogram in Figure 1.8 rises rather smoothly to a single peak and then declines. The histogram extends a bit more on the right (toward large values) than it does on the left—a slight "positive skew."

Either from the tabulated information or from the histogram itself, we can determine the following:

$$\text{proportion of games with at most two hits} = \begin{matrix}\text{relative}\\\text{frequency}\\\text{for }x=0\end{matrix} + \begin{matrix}\text{relative}\\\text{frequency}\\\text{for }x=1\end{matrix} + \begin{matrix}\text{relative}\\\text{frequency}\\\text{for }x=2\end{matrix}$$

$$= .0010 + .0037 + .0108 = .0155$$

Similarly,

$$\begin{array}{c}\text{proportion of games with}\\\text{between 5 and 10 hits (inclusive)}\end{array} = .0752 + .1026 + \cdots + .1015 = .6361$$

That is, roughly 64% of all these games resulted in between 5 and 10 (inclusive) hits.

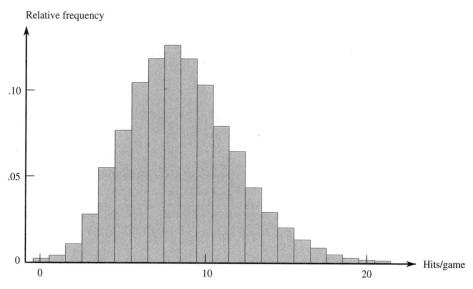

Figure 1.8 Histogram of number of hits per nine-inning game   ■

Constructing a histogram for continuous data (measurements) entails subdividing the measurement axis into a suitable number of **class intervals** or **classes,** such that each observation is contained in exactly one class. Suppose, for example, that we have 50 observations on $x =$ fuel efficiency of an automobile (mpg), the smallest of which is 27.8 and the largest of which is 31.4. Then we could use the class boundaries 27.5, 28.0, 28.5, . . . , and 31.5 as shown here:

One potential difficulty is that occasionally an observation lies on a class boundary so therefore does not lie in exactly one interval, for example, 29.0. One way to deal with this problem is to use boundaries like 27.55, 28.05, . . . , 31.55. Adding a hundredths digit to the class boundaries prevents observations from falling on the resulting boundaries.

Another way to deal with this problem is to use the classes 27.5–<28.0, 28.0–<28.5, ..., 31.0–<31.5. Then 29.0 falls in the class 29.0–<29.5 rather than in the class 28.5–<29.0. In other words, with this convention, an observation on a boundary is placed in the interval to the right of the boundary. This is how MINITAB constructs a histogram.

Constructing a Histogram for Continuous Data: Equal Class Widths

Determine the frequency and relative frequency for each class. Mark the class boundaries on a horizontal measurement axis. Above each class interval, draw a rectangle whose height is the corresponding relative frequency (or frequency).

Example 1.10 Power companies need information about customer usage to obtain accurate forecasts of demands. Investigators from Wisconsin Power and Light determined energy consumption (BTUs) during a particular period for a sample of 90 gas-heated homes. An adjusted consumption value was calculated as follows:

$$\text{adjusted consumption} = \frac{\text{consumption}}{(\text{weather, in degree days})(\text{house area})}$$

This resulted in the accompanying data (part of the stored data set FURNACE.MTW available in MINITAB), which we have ordered from smallest to largest).

2.97	4.00	5.20	5.56	5.94	5.98	6.35	6.62	6.72	6.78
6.80	6.85	6.94	7.15	7.16	7.23	7.29	7.62	7.62	7.69
7.73	7.87	7.93	8.00	8.26	8.29	8.37	8.47	8.54	8.58
8.61	8.67	8.69	8.81	9.07	9.27	9.37	9.43	9.52	9.58
9.60	9.76	9.82	9.83	9.83	9.84	9.96	10.04	10.21	10.28
10.28	10.30	10.35	10.36	10.40	10.49	10.50	10.64	10.95	11.09
11.12	11.21	11.29	11.43	11.62	11.70	11.70	12.16	12.19	12.28
12.31	12.62	12.69	12.71	12.91	12.92	13.11	13.38	13.42	13.43
13.47	13.60	13.96	14.24	14.35	15.12	15.24	16.06	16.90	18.26

We let MINITAB select the class intervals. The most striking feature of the histogram in Figure 1.9 is its resemblance to a bell-shaped (and therefore symmetric) curve, with the point of symmetry roughly at 10.

Class	1–<3	3–<5	5–<7	7–<9	9–<11	11–<13	13–<15	15–<17	17–<19
Frequency	1	1	11	21	25	17	9	4	1
Relative frequency	.011	.011	.122	.233	.278	.189	.100	.044	.011

From the histogram,

proportion of observations less than 9 $\approx .01 + .01 + .12 + .23 = .37$ (exact value $= \dfrac{34}{90} = .378$)

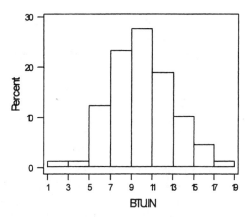

Figure 1.9   Histogram of the energy consumption data from Example 1.10

The relative frequency for the 9–<11 class is about .27, so we estimate that roughly half of this, or .135, is between 9 and 10. Thus

$$\text{proportion of observations less than 10} \approx .37 + .135 = .505 \text{ (slightly more than 50\%)}$$

The exact value of this proportion is 47/90 = .522. ∎

There are no hard-and-fast rules concerning either the number of classes or the choice of classes themselves. Between 5 and 20 classes will be satisfactory for most data sets. Generally, the larger the number of observations in a data set, the more classes should be used. A reasonable rule of thumb is

$$\text{number of classes} \approx \sqrt{\text{number of observations}}$$

Equal-width classes may not be a sensible choice if a data set contains a few outliers. Figure 1.10 (page 20) shows a dot diagram of such a data set. Using a small number of equal-width classes results in almost all observations falling in just one or two of the classes. If a large number of equal-width classes are used, many classes will have zero frequency. A sound choice is to use a few wider intervals near extreme observations and narrower intervals in the region of high concentration.

**Constructing a Histogram for Continuous Data:
Unequal Class Widths**

After determining frequencies and relative frequencies, calculate the height of each rectangle using the formula

$$\text{rectangle height} = \frac{\text{relative frequency of the class}}{\text{class width}}$$

The resulting rectangle heights are usually called *densities,* and the vertical scale is the **density scale.** This prescription will also work when class widths are equal.

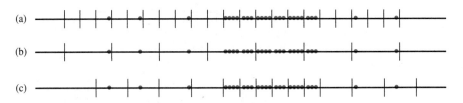

Figure 1.10   Selecting class intervals when there are outliers: (a) many short equal-width intervals; (b) a few wide equal-width intervals; (c) unequal-width intervals

Example 1.11   Corrosion of reinforcing steel is a serious problem in concrete structures located in environments affected by severe weather conditions. For this reason, researchers have been investigating the use of reinforcing bars made of composite material. One study was carried out to develop guidelines for bonding glass-fiber-reinforced plastic rebars to concrete ("Design Recommendations for Bond of GFRP Rebars to Concrete," *J. of Structural Engr.*, 1996: 247–254). Consider the following 48 observations on measured bond strength:

11.5	12.1	9.9	9.3	7.8	6.2	6.6	7.0	13.4	17.1	9.3	5.6
5.7	5.4	5.2	5.1	4.9	10.7	15.2	8.5	4.2	4.0	3.9	3.8
3.6	3.4	20.6	25.5	13.8	12.6	13.1	8.9	8.2	10.7	14.2	7.6
5.2	5.5	5.1	5.0	5.2	4.8	4.1	3.8	3.7	3.6	3.6	3.6

Class	2–<4	4–<6	6–<8	8–<12	12–<20	20–<30
Frequency	9	15	5	9	8	2
Relative frequency	.1875	.3125	.1042	.1875	.1667	.0417
Density	.094	.156	.052	.047	.021	.004

The resulting histogram appears in Figure 1.11. The right or upper tail stretches out much farther than does the left or lower tail—a substantial departure from symmetry.

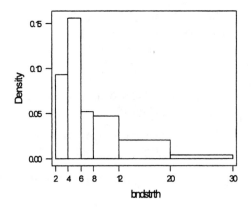

Figure 1.11   A MINITAB density histogram for the bond strength data of Example 1.11

When class widths are unequal, not using a density scale will give a picture with distorted areas. For equal-class widths, the divisor is the same in each density calculation, and the extra arithmetic simply results in a rescaling of the vertical axis (that is, the histogram using relative frequency and the one using density will have exactly the same appearance). A density histogram does have one interesting property. Multiplying both sides of the formula for density by the class width gives

$$\text{relative frequency} = (\text{class width})(\text{density}) = (\text{rectangle width})(\text{rectangle height})$$
$$= \text{rectangle area}$$

That is, *the area of each rectangle is the relative frequency of the corresponding class.* Furthermore, since the sum of relative frequencies must be 1.0 (except for roundoff), *the total area of all rectangles in a density histogram is* 1. It is always possible to draw a histogram so that the area equals the relative frequency (this is true also for a histogram of discrete data—just use the density scale). This property will play an important role in creating models for distributions in Chapter 4.

Histogram Shapes

Histograms come in a variety of shapes. A **unimodal** histogram is one that rises to a single peak and then declines. A **bimodal** histogram has two different peaks. Bimodality occurs when the data set consists of observations on two different kinds of individuals or objects. For example, a histogram of heights of college students would show one peak at a representative female height and another at a typical male height. A histogram with more than two peaks is said to be **multimodal.** Of course, the number of peaks will sometimes depend on the choice of class intervals, particularly with a small number of observations. A histogram is **symmetric** if the left half is a mirror image of the right half. A unimodal histogram is **positively skewed** if the right or upper tail is stretched out compared with the left or lower tail and **negatively skewed** if the stretching is to the left. Figure 1.12 shows "smoothed" histograms, obtained by superimposing a smooth curve on the rectangles, that illustrate the various possibilities.

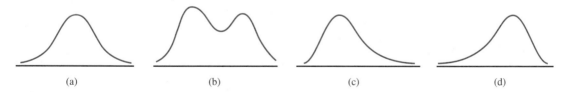

(a)   (b)   (c)   (d)

Figure 1.12 Smoothed histograms: (a) symmetric unimodal; (b) bimodal; (c) positively skewed; and (d) negatively skewed

Qualitative Data

Both a frequency distribution and a histogram can be constructed when the data set is *qualitative* (categorical) in nature. In some cases, there will be a natural ordering of classes—for example, freshmen, sophomores, juniors, seniors, graduate students—whereas in other cases the order will be arbitrary—for example, Catholic, Jewish, Protestant, and the like. With such categorical data, the intervals above which rectangles are constructed should have equal width.

Example 1.12   Each member of a sample of 120 individuals owning motorcycles was asked for the name of the manufacturer of his or her bike. The frequency distribution for the resulting data is given in Table 1.2 and the histogram is shown in Figure 1.13.

Table 1.2   Frequency distribution for motorcycle data

Manufacturer	Frequency	Relative Frequency
1. Honda	41	.34
2. Yamaha	27	.23
3. Kawasaki	20	.17
4. Suzuki	18	.15
5. Harley-Davidson	3	.03
6. Other	11	.09
	120	1.01

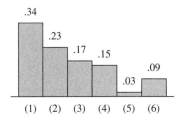

Figure 1.13   Histogram for motorcycle data ■

Multivariate Data

The techniques presented so far have been exclusively for situations in which each observation in a data set is either a single number or a single category. Often, however, the data is *multivariate* in nature. That is, if we obtain a sample of individuals or objects and on each one we make two or more measurements, then each "observation" would consist of several measurements on one individual or object. The sample is bivariate if each observation consists of two measurements or responses, so that the data set can be represented as $(x_1, y_1), \ldots, (x_n, y_n)$. For example, x might refer to engine size and y to miles per gallon, or x might refer to brand of calculator owned and y to academic major. In Chapters 11–14, we will analyze multivariate data sets of this sort, so we will postpone a detailed discussion until that time.

Exercises | Section 1.2 (10–32)

10. Consider the strength data for beams given in Example 1.2.
    a. Construct a stem-and-leaf display of the data. What appears to be a representative strength value? Do the observations appear to be highly concentrated about the representative value or rather spread out?
    b. Does the display appear to be reasonably symmetric about a representative value, or would you describe its shape in some other way?
    c. Do there appear to be any outlying strength values?
    d. What proportion of strength observations in this sample exceed 10 MPa?

11. The following data on motor octane ratings for various gasoline blends is taken from an article in *Technometrics* (Vol. 19, p. 425), a journal devoted to statistical applications in the physical sciences and engineering:

88.5	87.7	83.4	86.7	87.5	91.5	88.6	100.3
95.6	93.3	94.7	91.1	91.0	94.2	87.8	89.9
88.3	87.6	84.3	86.7	88.2	90.8	88.3	98.8
94.2	92.7	93.2	91.0	90.3	93.4	88.5	90.1
89.2	88.3	85.3	87.9	88.6	90.9	89.0	96.1
93.3	91.8	92.3	90.4	90.1	93.0	88.7	89.9
89.8	89.6	87.4	88.9	91.2	89.3	94.4	92.7
91.8	91.6	90.4	91.1	92.6	89.8	90.6	91.1
90.4	89.3	89.7	90.3	91.6	90.5	93.7	92.7
92.2	92.2	91.2	91.0	92.2	90.0	90.7	

    Construct a stem-and-leaf display of this data. Why is it relatively easy to identify a representative octane value? Does the display reveal any interesting features of the data?

12. The accompanying data set consists of observations on shower-flow rate (L/min) for a sample of $n = 129$ houses in Perth, Australia ("An Application of Bayes Methodology to the Analysis of Diary Records in a Water Use Study," *J. Amer. Stat. Assoc.*, 1987: 705–711):

4.6	12.3	7.1	7.0	4.0	9.2	6.7	6.9	11.5	5.1
11.2	10.5	14.3	8.0	8.8	6.4	5.1	5.6	9.6	7.5
7.5	6.2	5.8	2.3	3.4	10.4	9.8	6.6	3.7	6.4
8.3	6.5	7.6	9.3	9.2	7.3	5.0	6.3	13.8	6.2
5.4	4.8	7.5	6.0	6.9	10.8	7.5	6.6	5.0	3.3
7.6	3.9	11.9	2.2	15.0	7.2	6.1	15.3	18.9	7.2
5.4	5.5	4.3	9.0	12.7	11.3	7.4	5.0	3.5	8.2

8.4	7.3	10.3	11.9	6.0	5.6	9.5	9.3	10.4	9.7
5.1	6.7	10.2	6.2	8.4	7.0	4.8	5.6	10.5	14.6
10.8	15.5	7.5	6.4	3.4	5.5	6.6	5.9	15.0	9.6
7.8	7.0	6.9	4.1	3.6	11.9	3.7	5.7	6.8	11.3
9.3	9.6	10.4	9.3	6.9	9.8	9.1	10.6	4.5	6.2
8.3	3.2	4.9	5.0	6.0	8.2	6.3	3.8	6.0	

    a. Construct a stem-and-leaf display of the data.
    b. What is a typical, or representative, flow rate?
    c. Does the display appear to be highly concentrated or spread out?
    d. Does the distribution of values appear to be reasonably symmetric? If not, how would you describe the departure from symmetry?
    e. Would you describe any observation as being far from the rest of the data (an outlier)?

13. A *Consumer Reports* article on peanut butter (Sept. 1990) reported the following scores for various brands:

    Creamy 56 44 62 36 39 53 50 65 45 40
    56 68 41 30 40 50 56 30 22
    Crunchy 62 53 75 42 47 40 34 62 52
    50 34 42 36 75 80 47 56 62

    Construct a *comparative* stem-and-leaf display by listing stems in the middle of your page and then displaying the creamy leaves out to the right and the crunchy leaves out to the left. Describe similarities and differences for the two types.

14. The article cited in Example 1.2 also gave the accompanying strength observations for cylinders:

6.1	5.8	7.8	7.1	7.2	9.2	6.6	8.3	7.0	8.3
7.8	8.1	7.4	8.5	8.9	9.8	9.7	14.1	12.6	11.2

    a. Construct a comparative stem-and-leaf display (see the previous exercise) of the beam and cylinder data, then answer the questions in parts (b)–(d) of Exercise 10 for the observations on cylinders.
    b. In what ways are the two sides of the display similar? Are there any obvious differences between the beam observations and the cylinder observations?
    c. Construct a dotplot of the cylinder data.

15. Every score in the following batch of exam scores is in the 60's, 70's, 80's, or 90's. A stem-and-leaf display with only the four stems 6, 7, 8, and 9

would not give a very detailed description of the distribution of scores. In such situations, it is desirable to use repeated stems. Here we could repeat the stem 6 twice, using 6L for scores in the low 60's (leaves 0, 1, 2, 3, and 4) and 6H for scores in the high 60's (leaves 5, 6, 7, 8, and 9). Similarly, the other stems can be repeated twice to obtain a display consisting of eight rows. Construct such a display for the given scores. What feature of the data is highlighted by this display?

74 89 80 93 64 67 72 70 66 85 89 81 81
71 74 82 85 63 72 81 81 95 84 81 80 70
69 66 60 83 85 98 84 68 90 82 69 72 87
88

16. The accompanying specific gravity values for various wood types used in construction appeared in the article "Bolted Connection Design Values Based on European Yield Model" (*J. of Structural Engr.*, 1993: 2169–2186):

.31 .35 .36 .36 .37 .38 .40 .40 .40
.41 .41 .42 .42 .42 .42 .42 .43 .44
.45 .46 .46 .47 .48 .48 .48 .51 .54
.54 .55 .58 .62 .66 .66 .67 .68 .75

Construct a stem-and-leaf display using repeated stems (see the previous exercise), and comment on any interesting features of the display.

17. Temperature transducers of a certain type are shipped in batches of 50. A sample of 60 batches was selected, and the number of transducers in each batch not conforming to design specifications was determined, resulting in the following data:

2 1 2 4 0 1 3 2 0 5 3 3 1 3 2 4 7 0 2 3
0 4 2 1 3 1 1 3 4 1 2 3 2 2 8 4 5 1 3 1
5 0 2 3 2 1 0 6 4 2 1 6 0 3 3 3 6 1 2 3

   a. Determine frequencies and relative frequencies for the observed values of x = number of nonconforming transducers in a batch.
   b. What proportion of batches in the sample have at most five nonconforming transducers? What proportion have fewer than five? What proportion have at least five nonconforming units?
   c. Draw a histogram of the data using relative frequency on the vertical scale, and comment on its features.

18. In a study of author productivity ("Lotka's Test," *Collection Mgmt.*, 1982: 111–118), a large number of authors were classified according to the number of articles they had published during a certain period. The results were presented in the accompanying frequency distribution:

Number of papers	1	2	3	4	5	6	7	8
Frequency	784	204	127	50	33	28	19	19

Number of papers	9	10	11	12	13	14	15	16	17
Frequency	6	7	6	7	4	4	5	3	3

   a. Construct a histogram corresponding to this frequency distribution. What is the most interesting feature of the shape of the distribution?
   b. What proportion of these authors published at least five papers? At least ten papers? More than ten papers?
   c. Suppose the five 15's, three 16's, and three 17's had been lumped into a single category displayed as "≥15." Would you be able to draw a histogram? Explain.
   d. Suppose that instead of the values 15, 16, and 17 being listed separately, they had been combined into a 15–17 category with frequency 11. Would you be able to draw a histogram? Explain.

19. The number of contaminating particles on a silicon wafer prior to a certain rinsing process was determined for each wafer in a sample of size 100, resulting in the following frequencies:

Number of particles	0	1	2	3	4	5	6	7
Frequency	1	2	3	12	11	15	18	10

Number of particles	8	9	10	11	12	13	14
Frequency	12	4	5	3	1	2	1

   a. What proportion of the sampled wafers had at least one particle? At least five particles?
   b. What proportion of the sampled wafers had between five and ten particles, inclusive? Strictly between five and ten particles?
   c. Draw a histogram using relative frequency on the vertical axis. How would you describe the shape of the histogram?

20. The article "Determination of Most Representative Subdivision" (*J. of Energy Engr.*, 1993: 43–55) gave data on various characteristics of subdivisions that could be used in deciding whether to provide electrical power using overhead lines or under-

ground lines. Here are the values of the variable x = total length of streets within a subdivision:

1280	5320	4390	2100	1240	3060	4770
1050	360	3330	3380	340	1000	960
1320	530	3350	540	3870	1250	2400
960	1120	2120	450	2250	2320	2400
3150	5700	5220	500	1850	2460	5850
2700	2730	1670	100	5770	3150	1890
510	240	396	1419	2109		

a. Construct a stem-and-leaf display using the thousands digit as the stem and the hundreds digit as the leaf, and comment on the various features of the display.

b. Construct a histogram using class boundaries 0, 1000, 2000, 3000, 4000, 5000, and 6000. What proportion of subdivisions have total length less than 2000? Between 2000 and 4000? How would you describe the shape of the histogram?

21. The article cited in Exercise 20 also gave the following values of the variables y = number of culs-de-sac and z = number of intersections:

y 1 0 1 0 0 2 0 1 1 1 2 1 0 0 1 1 0 1 1
z 1 8 6 1 1 5 3 0 0 4 4 0 0 1 2 1 4 0 4

y 1 1 0 0 0 1 1 2 0 1 2 2 1 1 0 2 1 1 0
z 0 3 0 1 1 0 1 3 2 4 6 6 0 1 1 8 3 3 5

y 1 5 0 3 0 1 1 0 0
z 0 5 2 3 1 0 0 0 3

a. Construct a histogram for the y data. What proportion of these subdivisions had no culs-de-sac? At least one cul-de-sac?

b. Construct a histogram for the z data. What proportion of these subdivisions had at most five intersections? Fewer than five intersections?

22. How does the speed of a runner vary over the course of a marathon (a distance of 42.195 km)? Consider determining both the time to run the first 5 km and the time to run between the 35-km and 40-km points, and then subtracting the former time from the latter time. A positive value of this difference corresponds to a runner slowing down toward the end of the race. The accompanying histogram is based on times of runners who participated in several different Japanese marathons ("Factors Affecting Runners' Marathon Performance," *Chance*, Fall, 1993: 24–30).

What are some interesting features of this histogram? What is a typical difference value? Roughly what proportion of the runners ran the late distance more quickly than the early distance?

Histogram for Exercise 22

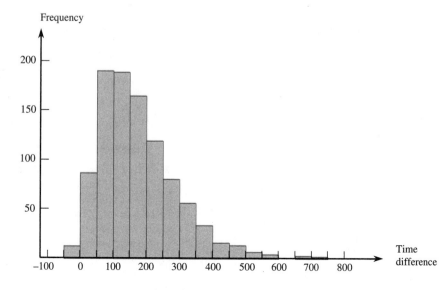

23. In a study of warp breakage during the weaving of fabric (*Technometrics*, 1982: 63), 100 specimens of yarn were tested. The number of cycles of strain to breakage was determined for each yarn specimen, resulting in the following data:

```
 86 146 251 653   98 249 400 292 131 169
175 176  76 264   15 364 195 262  88 264
157 220  42 321  180 198  38  20  61 121
282 224 149 180  325 250 196  90 229 166
 38 337  65 151  341  40  40 135 597 246
211 180  93 315  353 571 124 279  81 186
497 182 423 185  229 400 338 290 398  71
246 185 188 568   55  55  61 244  20 284
393 396 203 829  239 236 286 194 277 143
198 264 105 203  124 137 135 350 193 188
```

a. Construct a relative frequency histogram based on the class intervals 0–<100, 100–<200, . . . , and comment on features of the histogram.

b. Construct a histogram based on the following class intervals: 0–<50, 50–<100, 100–<150, 150–<200, 200–<300, 300–<400, 400–<500, 500–<600, and 600–<900.

c. If weaving specifications require a breaking strength of at least 100 cycles, what proportion of the yarn specimens in this sample would be considered satisfactory?

24. The accompanying data set consists of observations on shear strength (lb) of ultrasonic spot welds made on a certain type of alclad sheet. Construct a relative frequency histogram based on ten equal-width classes with boundaries 4000, 4200, ... [The histogram will agree with the one in "Comparison of Properties of Joints Prepared by Ultrasonic Welding and Other Means" (*J. of Aircraft*, 1983: 552–556).] Comment on its features.

```
5434 4948 4521 4570 4990 5702 5241
5112 5015 4659 4806 4637 5670 4381
4820 5043 4886 4599 5288 5299 4848
5378 5260 5055 5828 5218 4859 4780
5027 5008 4609 4772 5133 5095 4618
4848 5089 5518 5333 5164 5342 5069
4755 4925 5001 4803 4951 5679 5256
5207 5621 4918 5138 4786 4500 5461
5049 4974 4592 4173 5296 4965 5170
4740 5173 4568 5653 5078 4900 4968
5248 5245 4723 5275 5419 5205 4452
5227 5555 5388 5498 4681 5076 4774
4931 4493 5309 5582 4308 4823 4417
5364 5640 5069 5188 5764 5273 5042
5189 4986
```

25. A transformation of data values by means of some mathematical function, such as $\sqrt{x}$ or $1/x$, can often yield a set of numbers that has "nicer" statistical properties than the original data. In particular, it may be possible to find a function for which the histogram of transformed values is more symmetric (or, even better, more like a bell-shaped curve) than the original data. As an example, the article "Time Lapse Cinematographic Analysis of Beryllium–Lung Fibroblast Interactions" (*Environ. Research*, 1983: 34–43) reported the results of experiments designed to study the behavior of certain individual cells that had been exposed to beryllium. An important characteristic of such an individual cell is its interdivision time (IDT). IDTs were determined for a large number of cells both in exposed (treatment) and unexposed (control) conditions. The authors of the article used a logarithmic transformation, that is, transformed value = log(original value). Consider the following representative IDT data:

IDT	$\log_{10}$(IDT)	IDT	$\log_{10}$(IDT)	IDT	$\log_{10}$(IDT)
28.1	1.45	60.1	1.78	21.0	1.32
31.2	1.49	23.7	1.37	22.3	1.35
13.7	1.14	18.6	1.27	15.5	1.19
46.0	1.66	21.4	1.33	36.3	1.56
25.8	1.41	26.6	1.42	19.1	1.28
16.8	1.23	26.2	1.42	38.4	1.58
34.8	1.54	32.0	1.51	72.8	1.86
62.3	1.79	43.5	1.64	48.9	1.69
28.0	1.45	17.4	1.24	21.4	1.33
17.9	1.25	38.8	1.59	20.7	1.32
19.5	1.29	30.6	1.49	57.3	1.76
21.1	1.32	55.6	1.75	40.9	1.61
31.9	1.50	25.5	1.41		
28.9	1.46	52.1	1.72		

Use class intervals 10–<20, 20–<30, . . . to construct a histogram of the original data. Use intervals 1.1–<1.2, 1.2–<1.3, . . . to do the same for the transformed data. What is the effect of the transformation?

26. The clearness index was determined for the skies over Baghdad for each of the 365 days during a particular year ("Contribution to the Study of the Solar Radiation Climate of the Baghdad Environment," *Solar Energy*, 1990: 7–12). The accompanying table gives the results.

Class	Frequency
.15–<.25	8
.25–<.35	14
.35–<.45	28
.45–<.50	24
.50–<.55	39
.55–<.60	51
.60–<.65	106
.65–<.70	84
.70–<.75	11

a. Determine relative frequencies and draw the corresponding histogram.

b. Cloudy days are those with a clearness index smaller than .35. What percentage of the days were cloudy?

c. Clear days are those for which the index is at least .65. What percentage of the days were clear?

27. The concentration of suspended solids in river water is an important environmental characteristic. The article "Water Quality in Agricultural Watershed: Impact of Riparian Vegetation During Base Flow" (*Water Resources Bull.*, 1981: 233–239) reported on concentration (in parts per million, or ppm) for several different rivers. Suppose the following 50 observations had been obtained for a particular river:

55.8	60.9	37.0	91.3	65.8
42.3	33.8	60.6	76.0	69.0
45.9	39.1	35.5	56.0	44.6
71.7	61.2	61.5	47.2	74.5
83.2	40.0	31.7	36.7	62.3
47.3	94.6	56.3	30.0	68.2
75.3	71.4	65.2	52.6	58.2
48.0	61.8	78.8	39.8	65.0
60.7	77.1	59.1	49.5	69.3
69.8	64.9	27.1	87.1	66.3

a. Construct a stem-and-leaf display.

b. Why can you not base a frequency distribution on the class intervals 0–10, 10–20, 20–30, 30–40, . . . , 90–100?

c. Construct a frequency distribution using class intervals 20–<30, 30–<40, . . . , 90–<100. (The resulting distribution agrees with that for one of the rivers discussed in the article.)

d. What proportion of the concentration observations were less than 50? At least 60?

28. Construct a digidot plot for the accompanying time series. The data is monthly and was obtained during the period 1985–1989. Each value is average solar radiation in the band 385–530 nm as a percentage of total radiation ("Global Energy in the Different Spectral Bands at Dhahran, Saudi Arabia," *J. Solar Energy Engr.*, 1991: 290–294). Comment on any interesting features of the data:

20.9	19.6	20.4	20.3	20.8	20.6	20.5	20.4
19.9	19.8	19.5	20.2	16.5	18.3	18.7	19.6
20.0	20.0	19.5	19.6	19.1	18.8	18.3	17.6
17.2	17.8	18.7	19.0	19.0	18.6	18.8	19.0
18.5	18.3	17.5	16.9	17.0	17.8	18.1	18.8
18.9	18.9	19.1	18.8	18.4	17.8	17.0	16.8
17.9	18.4	19.0	19.4	19.7	19.5	19.5	19.5
19.0	18.7	18.1	17.9				

29. Consider the following data on type of health complaint (J = joint swelling, F = fatigue, B = back pain, M = muscle weakness, C = coughing, N = nose running/irritation, O = other) made by tree planters. Obtain frequencies and relative frequencies for the various categories, and draw a histogram. (The data is consistent with percentages given in the article "Physiological Effects of Work Stress and Pesticide Exposure in Tree Planting by British Columbia Silviculture Workers," *Ergonomics*, 1993: 951–961):

```
O  O  N  J  C  F  B  B  F  O  J  O  O  M
O  F  F  O  O  N  O  N  J  F  J  B  O  C
J  O  J  J  F  N  O  B  M  O  J  M  O  B
O  F  J  O  O  B  N  C  O  O  O  M  B  F
J  O  F  N
```

30. A **Pareto diagram** is a variation of a histogram for categorical data resulting from a quality control study. Each category represents a different type of product nonconformity or production problem. The categories are ordered so that the one with the largest frequency appears on the far left, then the category with the second largest frequency, and so on. Suppose the following information on nonconformities in circuit packs is obtained: failed component, 126; incorrect component, 210; insufficient solder, 67; excess solder, 54; missing component, 131. Construct a Pareto diagram.

31. The **cumulative frequency** and cumulative relative frequency for a particular class interval are the sum of frequencies and relative frequencies, respectively, for that interval and all intervals lying below

it. If, for example, there are four intervals with frequencies 9, 16, 13, and 12, then the cumulative frequencies are 9, 25, 38, and 50, and the cumulative relative frequencies are .18, .50, .76, and 1.00. Compute the cumulative frequencies and cumulative relative frequencies for the data of Exercise 24.

32. Fire load (MJ/m²) is the heat energy that could be released per square meter of floor area by combustion of contents and the structure itself. The article "Fire Loads in Office Buildings" (*J. of Structural Engr.,* 1997: 365–368) gave the following cumulative percentages (read from a graph) for fire loads in a sample of 388 rooms:

Value	0	150	300	450	600
Cumulative %	0	19.3	37.6	62.7	77.5

Value	750	900	1050	1200	1350
Cumulative %	87.2	93.8	95.7	98.6	99.1

Value	1500	1650	1800	1950
Cumulative %	99.5	99.6	99.8	100.0

a. Construct a relative frequency histogram and comment on interesting features.

b. What proportion of fire loads are less than 600? At least 1200?

c. What proportion of the loads are between 600 and 1200?

1.3 | Measures of Location

Visual summaries of data are excellent tools for obtaining preliminary impressions and insights. More formal data analysis often requires the calculation and interpretation of numerical summary measures. That is, from the data we try to extract several summarizing numbers—numbers that might serve to characterize the data set and convey some of its salient features. Our primary concern will be with numerical data; some comments regarding categorical data appear at the end of the section.

Suppose, then, that our data set is of the form $x_1, x_2, \ldots, x_n$, where each x_i is a number. What features of such a set of numbers are of most interest and deserve emphasis? One important characteristic of a set of numbers is its location, and in particular its center. This section presents methods for describing the location of a data set, whereas in Section 1.4 we will turn to methods for measuring the variability in a set of numbers.

The Mean

For a given set of numbers $x_1, x_2, \ldots, x_n$, the most familiar and useful measure of the center is the *mean,* or arithmetic average of the set. Because we will almost always think of the x_i's as constituting a sample, we will often refer to the arithmetic average as the *sample mean* and denote it by $\bar{x}$.

DEFINITION

The **sample mean** $\bar{x}$ of observations $x_1, x_2, \ldots, x_n$ is given by

$$\bar{x} = \frac{x_1 + x_2 + \cdots + x_n}{n} = \frac{\sum_{i=1}^{n} x_i}{n}$$

The numerator of $\bar{x}$ can be written more informally as $\sum x_i$, where the summation is over all sample observations.

For reporting $\bar{x}$, we recommend using decimal accuracy of one digit more than the accuracy of the x_i's. Thus if observations are stopping distances with $x_1 = 125$, $x_2 = 131$, and so on, we might have $\bar{x} = 127.3$ ft.

Example 1.13 Caustic stress corrosion cracking of iron and steel has been studied because of failures around rivets in steel boilers and failures of steam rotors. Consider the accompanying observations on $x = $ crack length (μm) as a result of constant load stress corrosion tests on smooth bar tensile specimens for a fixed length of time. (The data is consistent with a histogram and summary quantities from the article "On the Role of Phosphorus in the Caustic Stress Corrosion Cracking of Low Alloy Steels," *Corrosion Science*, 1989: 53–68.)

$$x_1 = 16.1 \quad x_2 = 9.6 \quad x_3 = 24.9 \quad x_4 = 20.4 \quad x_5 = 12.7 \quad x_6 = 21.2 \quad x_7 = 30.2$$
$$x_8 = 25.8 \quad x_9 = 18.5 \quad x_{10} = 10.3 \quad x_{11} = 25.3 \quad x_{12} = 14.0 \quad x_{13} = 27.1 \quad x_{14} = 45.0$$
$$x_{15} = 23.3 \quad x_{16} = 24.2 \quad x_{17} = 14.6 \quad x_{18} = 8.9 \quad x_{19} = 32.4 \quad x_{20} = 11.8 \quad x_{21} = 28.5$$

Figure 1.14 shows a stem-and-leaf display of the data; a crack length in the low 20's appears to be "typical."

```
0H │ 96  89
1L │ 27  03  40  46  18
1H │ 61  85
2L │ 49  04  12  33  42          Stem: tens digit
2H │ 58  53  71  85              Leaf: one and tenths digit
3L │ 02  24
3H │
4L │
4H │ 50
```

Figure 1.14   A stem-and-leaf display of the crack-length data

With $\sum x_i = 444.8$, the sample mean is

$$\bar{x} = \frac{444.8}{21} = 21.18$$

a value consistent with information conveyed by the stem-and-leaf display. ■

A physical interpretation of $\bar{x}$ demonstrates how it measures the location (center) of a sample. Think of drawing and scaling a horizontal measurement axis, and then represent each sample observation by a 1-lb weight placed at the corresponding point on the axis. The only point at which a fulcrum can be placed to balance the system of weights is the point corresponding to the value of $\bar{x}$ (see Figure 1.15).

Just as $\bar{x}$ represents the average value of the observations in a sample, the average of all values in the population can be calculated. This average is called the **population mean**

Figure 1.15   The mean as the balance point for a system of weights

and is denoted by the Greek letter μ. When there are N values in the population (a finite population), then μ = (sum of the N population values)/N. In Chapters 3 and 4, we will give a more general definition for μ that applies to both finite and (conceptually) infinite populations. Just as $\bar{x}$ is an interesting and important measure of sample location, μ is an interesting and important (often the most important) characteristic of a population. In the chapters on statistical inference, we will present methods based on the sample mean for drawing conclusions about a population mean. For example, we might use the sample mean $\bar{x} = 21.18$ computed in Example 1.13 as a point estimate (a single number which is our "best" guess) of μ = the true average crack length for all specimens treated as described.

The mean suffers from one deficiency that makes it an inappropriate measure of center under some circumstances: Its value can be greatly affected by the presence of even a single outlier (unusually large or small observation). In Example 1.13, the value $x_{14} = 45.0$ is obviously an outlier. Without this observation, $\bar{x} = 399.8/20 = 19.99$; the outlier increases the mean by more than 1 μm. If the 45.0 μm observation were replaced by the catastrophic value 295.0 μm, a really extreme outlier, then $\bar{x} = 694.8/21 = 33.09$, which is larger than all but one of the observations!

A sample of incomes often produces a few such outlying values (those lucky few who earn astronomical incomes), and the use of average income as a measure of location will often be misleading. Such examples suggest that we look for a measure that is less sensitive to outlying values than $\bar{x}$, and we will momentarily propose one. However, although $\bar{x}$ does have this potential defect, it is still the most widely used measure, largely because there are many populations for which an extreme outlier in the sample would be highly unlikely. When sampling from such a population (a normal or bell-shaped population being the most important example), the sample mean will tend to be stable and quite representative of the sample.

The Median

The word *median* is synonymous with "middle," and the sample median is indeed the middle value when the observations are ordered from smallest to largest. When the observations are denoted by $x_1, \ldots, x_n$, we will use the symbol $\tilde{x}$ to represent the sample median.

DEFINITION

The **sample median** is obtained by first ordering the n observations from smallest to largest (with any repeated values included, so that every sample observation appears in the ordered list). Then,

$$\tilde{x} = \begin{cases} \text{The single middle value if } n \text{ is odd} & = \left(\dfrac{n+1}{2}\right)^{\text{th}} \text{ ordered value} \\[2em] \text{The average of the two middle values if } n \text{ is even} & = \text{average of } \left(\dfrac{n}{2}\right)^{\text{th}} \text{ and } \left(\dfrac{n}{2}+1\right)^{\text{th}} \text{ ordered values} \end{cases}$$

Example 1.14   The risk of developing iron deficiency is especially high during pregnancy. The problem with detecting such deficiency is that some methods for determining iron status can be affected by the state of pregnancy itself. Consider the following data on transferrin receptor concentration for a sample of women with laboratory evidence of overt iron-deficiency anemia ("Serum Transferrin Receptor for the Detection of Iron Deficiency in Pregnancy," *Amer. J. of Clinical Nutrition,* 1991: 1077–1081):

$$x_1 = 15.2 \quad x_2 = 9.3 \quad x_3 = 7.6 \quad x_4 = 11.9 \quad x_5 = 10.4 \quad x_6 = 9.7$$
$$x_7 = 20.4 \quad x_8 = 9.4 \quad x_9 = 11.5 \quad x_{10} = 16.2 \quad x_{11} = 9.4 \quad x_{12} = 8.3$$

The list of ordered values is

7.6   8.3   9.3   9.4   9.4   9.7   10.4   11.5   11.9   15.2   16.2   20.4

Since $n = 12$ is even, we average the $n/2 =$ sixth- and seventh-ordered values:

$$\text{sample median} = \frac{9.7 + 10.4}{2} = 10.05$$

Notice that if the largest observation, 20.4, had not appeared in the sample, the resulting sample median for the $n = 11$ observations would have been the single middle value, 9.7 (the $(n + 1)/2 =$ sixth-ordered value). The sample mean is $\bar{x} = \sum x_i/n = 139.3/12 = 11.61$, which is somewhat larger than the median because of the outliers, 15.2, 16.2, and 20.4. ∎

The data set in Example 1.14 illustrates an important property of $\tilde{x}$ in contrast to $\bar{x}$: The sample median is very insensitive to a number of extremely small or extremely large data values. If, for example, we increased the two largest x_i's from 16.2 and 20.4 to 26.2 and 30.4, respectively, $\tilde{x}$ would be unaffected. Thus, in the treatment of outlying data values, $\bar{x}$ and $\tilde{x}$ are at opposite ends of a spectrum: $\bar{x}$ is sensitive to even one such value, whereas $\tilde{x}$ is insensitive to a large number of outlying values.

Because the large values in the sample of Example 1.14 affect $\bar{x}$ more than $\tilde{x}$, $\tilde{x} < \bar{x}$ for that data. Whereas $\bar{x}$ and $\tilde{x}$ both provide a measure for the center of a data set, they will not in general be equal because they focus on different aspects of the sample.

Analogous to $\tilde{x}$ as the middle value in the sample, there is a middle value in the population, the **population median,** denoted by $\tilde{\mu}$. As with $\bar{x}$ and μ, we can think of using the sample median $\tilde{x}$ to make an inference about $\tilde{\mu}$. In Example 1.14, we might use $\tilde{x} = 10.05$ as an estimate of the median concentration in the entire population from which the sample was selected. A median is often used to describe income or salary data (because it is not greatly influenced by a few large salaries), so if the median salary for a sample of engineers were $\tilde{x} = \$56,416$, we might use this as a basis for concluding that the median salary for all engineers exceeds $50,000.

The population mean μ and median $\tilde{\mu}$ will not generally be equal to one another. If the population distribution is either positively or negatively skewed, as pictured in Figure 1.16 (page 32), then $\mu \neq \tilde{\mu}$. When this is the case, in making inferences we must first decide which of the two population characteristics is of greater interest and then proceed accordingly.

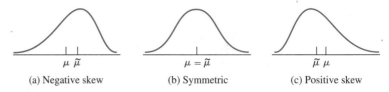

μ $\tilde{\mu}$	$\mu = \tilde{\mu}$	$\tilde{\mu}$ μ
(a) Negative skew	(b) Symmetric	(c) Positive skew

Figure 1.16   Three different shapes for a population distribution

Other Measures of Location: Quartiles, Percentiles, and Trimmed Means

The median (population or sample) divides the data set into two parts of equal size. To obtain finer measures of location, we could divide the data into more than two such parts. Roughly speaking, quartiles divide the data set into four equal parts, with the observations above the third quartile constituting the upper quarter of the data set, the second quartile being identical to the median, and the first quartile separating the lower quarter from the upper three-quarters. Similarly, a data set (sample or population) can be divided into 100 equal parts using percentiles; the 99th percentile separates the highest 1% from the bottom 99%, and so on. Unless the number of observations is a multiple of 100, care must be exercised in obtaining percentiles. We will use percentiles in Chapter 4 in connection with certain models for infinite populations, so we will postpone discussion until that point.

The sample mean and sample median are influenced by outlying values in a very different manner—the mean greatly and the median not at all. Since extreme behavior of either type might be undesirable, we briefly consider alternative measures that are neither as sensitive as $\bar{x}$ nor as insensitive as $\tilde{x}$. To motivate these alternatives, note that $\bar{x}$ and $\tilde{x}$ are at opposite extremes of the same "family" of measures. After the data set is ordered, $\tilde{x}$ is computed by throwing away as many values on each end as one can without eliminating everything (leaving just one or two middle values) and averaging what is left, whereas to compute $\bar{x}$ one throws away nothing before averaging. To paraphrase, the mean involves trimming 0% from each end of the sample, whereas for the median the maximum possible amount is trimmed from each end. A **trimmed mean** is a compromise between $\bar{x}$ and $\tilde{x}$. A 10% trimmed mean, for example, would be computed by eliminating the smallest 10% and the largest 10% of the sample and then averaging what is left over.

Example 1.15   Consider the following 20 observations, ordered from smallest to largest, each one representing the lifetime (in hours) of a certain type of incandescent lamp:

612	1016
623	1022
666	1029
744	1058
883	1085
898	1088
964	1122
970	1135
983	1197
1003	1201

The average of all 20 observations is $\bar{x} = 965.0$, and $\tilde{x} = 1009.5$. The 10% trimmed mean is obtained by deleting the smallest two observations (612 and 623) and the largest two (1197 and 1201) and then averaging the remaining 16 to obtain $\bar{x}_{tr(10)} = 979.1$. The effect of trimming here is to produce a "central value" that is somewhat above the mean ($\bar{x}$ is pulled down by a few small lifetimes) and yet considerably below the median. Similarly, the 20% trimmed mean averages the middle 12 values to obtain $\bar{x}_{tr(20)} = 999.9$, even closer to the median. (See Figure 1.17.)

Figure 1.17   Dotplot of lifetimes (in hours) of incandescent lamps   ■

Generally speaking, using a trimmed mean with a moderate trimming proportion (10% or 20%) will yield a measure that is neither as sensitive to outliers as the mean (since any small number of outliers will be deleted before averaging) nor as insensitive as the median. For this reason, trimmed means have merited increasing attention from statisticians for both descriptive and inferential purposes. More will be said about trimmed means when point estimation is discussed in Chapter 6. As a final point, if the trimming proportion is denoted by α and $n\alpha$ is not an integer, then it is not obvious how the $100\alpha\%$ trimmed mean should be computed. For example, if $\alpha = .10$ (10%) and $n = 22$, then $n\alpha = (22)(.10) = 2.2$, and we cannot trim 2.2 observations from each end of the ordered sample. In this case, the 10% trimmed mean would be obtained by first trimming two observations from each end and calculating $\bar{x}_{tr}$, then trimming three and calculating $\bar{x}_{tr}$, and finally interpolating between the two values to obtain $\bar{x}_{tr(10)}$.

Categorical Data and Sample Proportions

When the data is categorical, a frequency distribution or relative frequency distribution provides an effective tabular summary of the data. The natural numerical summary quantities in this situation are the individual frequencies and the relative frequencies. For example, if a survey of individuals who own stereo receivers is undertaken to study brand preference, then each individual in the sample would identify the brand of receiver that he or she owned, from which we could count the number owning Sony, Marantz, Pioneer, and so on. Consider sampling a dichotomous population—one that consists of only two categories (such as voted or did not vote in the last election, does or does not own a stereo receiver, etc.). If we let x denote the number in the sample falling in category 1, then the number in category 2 is $n - x$. The relative frequency or *sample proportion* in category 1 is x/n and the sample proportion in category 2 is $1 - x/n$. Let's denote a response that falls in category 1 by a 1 and a response that falls in category 2 by a 0. A sample size of $n = 10$ might then yield the responses, 1, 1, 0, 1, 1, 1, 0, 0, 1, 1. The sample mean for this numerical sample is (since number of 1's $= x = 7$)

$$\frac{x_1 + \cdots + x_n}{n} = \frac{1 + 1 + 0 + \cdots + 1 + 1}{10} = \frac{7}{10} = \frac{x}{n} = \text{sample proportion}$$

This result can be generalized and summarized as follows: *If in a categorical data situation we focus attention on a particular category and code the sample results so that a 1 is recorded for an individual in the category and a 0 for an individual not in the category, then the sample proportion of individuals in the category is the sample mean of the sequence of 1's and 0's.* Thus, a sample mean can be used to summarize the results of a categorical sample. These remarks also apply to situations in which categories are defined by grouping values in a numerical sample or population (e.g., we might be interested in knowing whether individuals have owned their present automobile for at least 5 years, rather than studying the exact length of ownership).

Analogous to the sample proportion x/n of individuals falling in a particular category, let p represent the proportion of individuals in the entire population falling in the category. As with x/n, p is a quantity between 0 and 1. While x/n is a sample characteristic, p is a characteristic of the population; the relationship between the two parallels the relationship between $\tilde{x}$ and $\tilde{\mu}$ and between $\bar{x}$ and μ. In particular, we will subsequently use x/n to make inferences about p. If, for example, a sample of 100 car owners reveals that 22 owned their car at least 5 years, then we might use $22/100 = .22$ as a point estimate of the proportion of all owners who have owned their car at least 5 years. We will study the properties of x/n as an estimator of p and see how x/n can be used to answer other inferential questions. With k categories ($k > 2$), we can use the k sample proportions to answer questions about the population proportions $p_1, \ldots, p_k$.

Exercises | Section 1.3 (33–43)

33. The article "The Pedaling Technique of Elite Endurance Cyclists" (*Int. J. of Sport Biomechanics,* 1991: 29–53) reported the accompanying data on single-leg power at a high workload:

244 191 160 187 180 176 174
205 211 183 211 180 194 200

a. Calculate and interpret the sample mean and median.

b. Suppose that the first observation had been 204 rather than 244. How would the mean and median change?

c. Calculate a trimmed mean by eliminating the smallest and largest sample observations. What is the corresponding trimming percentage?

d. The article also reported values of single-leg power for a low workload. The sample mean for $n = 13$ observations was $\bar{x} = 119.8$ (actually 119.7692), and the 14th observation, somewhat of an outlier, was 159. What is the value of $\bar{x}$ for the entire sample?

34. Consider the following observations on shear strength (MPa) of a joint bonded in a particular man-

ner (from a graph in the article "Diffusion of Silicon Nitride to Austenitic Stainless Steel without Interlayers," *Metallurgical Trans.*, 1993: 1835–1843):

22.2 40.4 16.4 73.7 36.6 109.9
30.0 4.4 33.1 66.7 81.5

a. Determine the value of the sample mean.

b. Determine the value of the sample median. Why is it so different from the mean?

c. Calculate a trimmed mean by deleting the smallest and largest observations. What is the corresponding trimming percentage? How does the value of this $\bar{x}_{tr}$ compare to the mean and median?

35. The minimum injection pressure (psi) for injection molding specimens of high amylose corn was determined for eight different specimens (higher pressure corresponds to greater processing difficulty), resulting in the following observations (from "Thermoplastic Starch Blends with a Polyethylene-Co-Vinyl Alcohol: Processability and Physical Properties," *Polymer Engr. and Science,* 1994: 17–23):

15.0 13.0 18.0 14.5 12.0 11.0 8.9 8.0

a. Determine the values of the sample mean, sample median, and 12.5% trimmed mean, and compare these values.

b. By how much could the smallest sample observation, currently 8.0, be increased without affecting the value of the sample median?

c. Suppose we want the values of the sample mean and median when the observations are expressed in kilograms per square inch (ksi) rather than psi. Is it necessary to reexpress each observation in ksi, or can the values calculated in part (a) be used directly? *Hint:* 1 kg = 2.2 lb.

36. A sample of 26 offshore oil workers took part in a simulated escape exercise, resulting in the accompanying data on time (sec) to complete the escape ("Oxygen Consumption and Ventilation During Escape from an Offshore Platform," *Ergonomics,* 1997: 281–292):

389 356 359 363 375 424 325 394 402
373 373 370 364 366 364 325 339 393
392 369 374 359 356 403 334 397

a. Construct a stem-and-leaf display of the data. How does it suggest that the sample mean and median will compare?

b. Calculate the values of the sample mean and median. *Hint:* $\Sigma x_i = 9638$.

c. By how much could the largest time, currently 424, be increased without affecting the value of the sample median? By how much could this value be decreased without affecting the value of the sample median?

d. What are the values of $\bar{x}$ and $\tilde{x}$ when the observations are reexpressed in minutes?

37. The article "Snow Cover and Temperature Relationships in North America and Eurasia" (*J. Climate and Applied Meteorology,* 1983: 460–469) used statistical techniques to relate the amount of snow cover on each continent to average continental temperature. Data presented there included the following ten observations on October snow cover for Eurasia during the years 1970–1979 (in million km^2):

6.5 12.0 14.9 10.0 10.7 7.9 21.9 12.5 14.5 9.2

What would you report as a representative, or typical, value of October snow cover for this period, and what prompted your choice?

38. Blood pressure values are often reported to the nearest 5 mmHg (100, 105, 110, etc.). Suppose the actual blood pressure values for nine randomly selected individuals are

118.6 127.4 138.4 130.0 113.7 122.0 108.3
131.5 133.2

a. What is the median of the *reported* blood pressure values?

b. Suppose the blood pressure of the second individual is 127.6 rather than 127.4 (a small change in a single value). How does this affect the median of the reported values? What does this say about the sensitivity of the median to rounding or grouping in the data?

39. The propagation of fatigue cracks in various aircraft parts has been the subject of extensive study in recent years. The accompanying data consists of propagation lives (flight hours/10^4) to reach a given crack size in fastener holes intended for use in military aircraft ("Statistical Crack Propagation in Fastener Holes under Spectrum Loading," *J. Aircraft,* 1983: 1028–1032):

.736 .863 .865 .913 .915 .937 .983 1.007
1.011 1.064 1.109 1.132 1.140 1.153 1.253 1.394

a. Compute the values of the sample mean and median.

b. By how much could the largest sample observation be decreased without affecting the value of the median?

40. Compute the sample median, 25% trimmed mean, 10% trimmed mean, and sample mean for the concentration data given in Exercise 27, and compare these measures (a stem-and-leaf display will aid in ordering the observations).

41. A sample of $n = 10$ automobiles was selected, and each was subjected to a 5-mph crash test. Denoting a car with no visible damage by S (for success) and a car with such damage by F, results were as follows:

S S F S S S F F S S

a. What is the value of the sample proportion of successes x/n?

b. Replace each S with a 1 and each F with a 0. Then calculate $\bar{x}$ for this numerically coded sample. How does $\bar{x}$ compare to x/n?

c. Suppose it is decided to include 15 more cars in the experiment. How many of these would have to be S's to give $x/n = .80$ for the entire sample of 25 cars?

42. a. If a constant c is added to each x_i in a sample, yielding $y_i = x_i + c$, how do the sample mean and median of the y_i's relate to the mean and median of the x_i's? Verify your conjectures.

b. If each x_i is multiplied by a constant c, yielding $y_i = cx_i$, answer the question of part (a). Again, verify your conjectures.

43. An experiment to study the lifetime (in hours) for a certain type of component involved putting ten components into operation and observing them for 100 hours. Eight of the components failed during that period, and those lifetimes were recorded. Denote the lifetimes of the two components still functioning after 100 hours by 100+. The resulting sample observations were

48 79 100+ 35 92 86 57 100+ 17 29

Which of the measures of center discussed in this section can be calculated, and what are the values of those measures? (*Note:* The data from this experiment is said to be "censored on the right.")

1.4 | Measures of Variability

Reporting a measure of center gives only partial information about a data set or distribution. Different samples or populations may have identical measures of center yet differ from one another in other important ways. Figure 1.18 shows dotplots of three samples with the same mean and median, yet the extent of spread about the center is different for all three samples. The first sample has the largest amount of variability, the third has the smallest amount, and the second is intermediate to the other two in this respect.

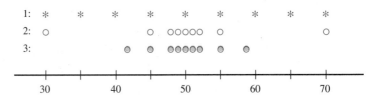

Figure 1.18 Samples with identical measures of center but different amounts of variability

Measures of Variability for Sample Data

The simplest measure of variability in a sample is the **range,** which is the difference between the largest and smallest sample values. Notice that the value of the range for sample 1 in Figure 1.18 is much larger than it is for sample 3, reflecting more variability in the first sample than in the third. A defect of the range, though, is that it depends on only the two most extreme observations and disregards the positions of the remaining $n - 2$ values. Samples 1 and 2 in Figure 1.18 have identical ranges, yet when we take into account the observations between the two extremes, there is much less variability or dispersion in the second sample than in the first.

Our primary measures of variability involve the **deviations from the mean,** $x_1 - \bar{x}$, $x_2 - \bar{x}, \ldots, x_n - \bar{x}$. That is, the deviations from the mean are obtained by subtracting $\bar{x}$ from each of the n sample observations. A deviation will be positive if the observation is larger than the mean (to the right of the mean on the measurement axis) and negative if the observation is smaller than the mean. If all the deviations are small in magnitude, then all x_i's are close to the mean and there is little variability. On the other hand, if some of the deviations are large in magnitude, then some x_i's lie far from $\bar{x}$, suggesting a greater amount of variability. A simple way to combine the deviations into a single quantity is to average them (sum them and divide by n). Unfortunately, there is a major problem with this suggestion:

$$\text{sum of deviations} = \sum_{i=1}^{n} (x_i - \bar{x}) = 0$$

so that the average deviation is always zero. The verification uses several standard rules of summation and the fact that $\sum \bar{x} = \bar{x} + \bar{x} + \cdots + \bar{x} = n\bar{x}$:

$$\sum(x_i - \bar{x}) = \sum x_i - \sum \bar{x} = \sum x_i - n\bar{x} = \sum x_i - n\left(\frac{1}{n}\sum x_i\right) = 0$$

How can we change the deviations to nonnegative quantities so the positive and negative deviations do not counteract one another when they are combined? One possibility is to work with the absolute values of the deviations and calculate the average absolute deviation $\sum |x_i - \bar{x}|/n$. Because the absolute value operation leads to a number of theoretical difficulties, consider instead the squared deviations $(x_1 - \bar{x})^2, (x_2 - \bar{x})^2, \ldots, (x_n - \bar{x})^2$. We might now use the average squared deviation $\sum(x_i - \bar{x})^2/n$, but for several reasons we will divide the sum of squared deviations by $n - 1$ rather than n.

DEFINITION

The **sample variance,** denoted by s^2, is given by

$$s^2 = \frac{\sum(x_i - \bar{x})^2}{n - 1} = \frac{S_{xx}}{n - 1}$$

The **sample standard deviation,** denoted by s, is the (positive) square root of the variance:

$$s = \sqrt{s^2}$$

The unit for s is the same as the unit for each of the x_i's. If, for example, the observations are fuel efficiencies in miles per gallon, then we might have $s = 2.0$ mpg. A rough interpretation of the sample standard deviation is that it is the size of a typical or representative deviation from the sample mean within the given sample. Thus if $s = 2.0$ mpg, then some x_i's in the sample are closer than 2.0 to $\bar{x}$, whereas others are farther away; 2.0 is a representative (or "standard") deviation from the mean fuel efficiency. If $s = 3.0$ for a second sample of cars of another type, a typical deviation in this sample is roughly one and one half times what it is in the first sample, an indication of more variability in the second sample.

Example 1.16 Strength is an important characteristic of materials used in prefabricated housing. Each of $n = 11$ prefabricated plate elements was subjected to a severe stress test and the maximum width (mm) of the resulting cracks was recorded. The given data (Table 1.3)

Table 1.3 Data for Example 1.16

x_i	$x_i - \bar{x}$	$(x_i - \bar{x})^2$
.684	−.9841	.9685
2.540	.8719	.7602
.924	−.7441	.5537
3.130	1.4619	2.1372
1.038	−.6301	.3970
.598	−1.0701	1.1451
.483	−1.1851	1.4045
3.520	1.8519	3.4295
1.285	−.3831	.1468
2.650	.9819	.9641
1.497	−.1711	.0293
$\sum x_i = 18.349$	$\sum(x_i - \bar{x}) = -.0001$	$S_{xx} = \sum(x_i - \bar{x})^2 = 11.9359$

$$\bar{x} = \frac{18.349}{11} = 1.6681$$

appeared in the article "Prefabricated Ferrocement Ribbed Elements for Low-Cost Housing" (*J. Ferrocement,* 1984: 347–364). Effects of rounding account for the sum of deviations not being exactly zero. The numerator of s^2 is 11.9359; therefore $s^2 = 11.9359/(11 - 1) = 11.9359/10 = 1.19359$ and $s = \sqrt{1.19359} = 1.0925$ mm. ∎

Motivation for s^2

To explain why s^2 rather than the average squared deviation is used to measure variability, note first that whereas s^2 measures sample variability, there is a measure of variability in the population called the population variance. We will use σ^2 (the square of the lowercase Greek letter sigma) to denote the population variance and σ to denote the population standard deviation (the square root of σ^2). When the population is finite and consists of N values,

$$\sigma^2 = \sum_{i=1}^{N} (x_i - \mu)^2/N$$

which is the average of all squared deviations from the population mean (for the population, the divisor is N and not $N - 1$). More general definitions of σ^2 appear in Chapters 3 and 4.

Just as $\bar{x}$ will be used to make inferences about the population mean μ, we should define the sample variance so that it can be used to make inferences about σ^2. Now note that σ^2 involves squared deviations about the population mean μ. If we actually knew the value of μ, then we could define the sample variance as the average squared deviation of the sample x_i's about μ. However, the value of μ is almost never known, so the sum of squared deviations about $\bar{x}$ must be used. But *the x_i's tend to be closer to their average $\bar{x}$ than to the population average μ, so to compensate for this the divisor $n - 1$ is used rather than n.* In other words, if we used a divisor n in the sample variance, then the resulting quantity would tend to underestimate σ^2

(produce estimated values that are too small on the average), whereas dividing by the slightly smaller $n - 1$ corrects this underestimating.

It is customary to refer to s^2 as being based on $n - 1$ **degrees of freedom** (df). This terminology results from the fact that although s^2 is based on the n quantities $x_1 - \bar{x}, x_2 - \bar{x}, \ldots, x_n - \bar{x}$, these sum to 0, so specifying the values of any $n - 1$ of the quantities determines the remaining value. For example, if $n = 4$ and $x_1 - \bar{x} = 8$, $x_2 - \bar{x} = -6$, and $x_4 - \bar{x} = -4$, then automatically we have $x_3 - \bar{x} = 2$, so only three of the four values of $x_i - \bar{x}$ are freely determined (3 df).

A Computing Formula for s^2

Computing and squaring the deviations can be tedious, especially if enough decimal accuracy is being used in $\bar{x}$ to guard against the effects of rounding. An alternative formula for the numerator of s^2 circumvents the need for all the subtraction necessary to obtain the deviations. The formula involves both $\left(\sum x_i\right)^2$, summing and then squaring, and $\sum x_i^2$, squaring and then summing.

> An alternative expression for the numerator of s^2 is
> $$S_{xx} = \sum (x_i - \bar{x})^2 = \sum x_i^2 - \frac{\left(\sum x_i\right)^2}{n}$$

Proof   Because $\bar{x} = \sum x_i / n$, $n\bar{x}^2 = \left(\sum x_i\right)^2 / n$. Then,

$$\sum (x_i - \bar{x})^2 = \sum (x_i^2 - 2\bar{x} \cdot x_i + \bar{x}^2) = \sum x_i^2 - 2\bar{x} \sum x_i + \sum (\bar{x})^2$$
$$= \sum x_i^2 - 2\bar{x} \cdot n\bar{x} + n(\bar{x})^2 = \sum x_i^2 - n(\bar{x})^2$$

Example 1.17   The amount of light reflectance by leaves has been used for various purposes, including evaluation of turf color, estimation of nitrogen status, and measurement of biomass. The article "Leaf Reflectance–Nitrogen–Chlorophyll Relations in Buffel-Grass" (*Photogrammetric Engr. and Remote Sensing*, 1985: 463–466) gave the following observations, obtained using spectrophotogrammetry, on leaf reflectance under specified experimental conditions.

Observation	x_i	x_i^2	Observation	x_i	x_i^2
1	15.2	231.04	9	12.7	161.29
2	16.8	282.24	10	15.8	249.64
3	12.6	158.76	11	19.2	368.64
4	13.2	174.24	12	12.7	161.29
5	12.8	163.84	13	15.6	243.36
6	13.8	190.44	14	13.5	182.25
7	16.3	265.69	15	12.9	166.41
8	13.0	169.00			
				$\sum x_i = 216.1$	$\sum x_i^2 = 3168.13$

The computational formula now gives

$$S_{xx} = \sum x_i^2 - \frac{(\sum x_i)^2}{n} = 3168.13 - \frac{(216.1)^2}{15}$$

$$= 3168.13 - 3113.28 = 54.85$$

from which $s^2 = S_{xx}/(n - 1) = 54.85/14 = 3.92$ and $s = 1.98$.   ■

The shortcut method can yield values of s^2 and s that differ from the values computed using the definitions. These differences are due to effects of rounding and will not be important in most samples. To minimize the effects of rounding when using the shortcut formula, particularly when there is little variability in the data, intermediate calculations should be done using several more significant digits than are to be retained in the final answer. Because the numerator of s^2 is the sum of nonnegative quantities (squared deviations), s^2 is guaranteed to be nonnegative. Yet if the shortcut method is used, particularly with data having little variability, a slight numerical error can result in a negative numerator ($\sum x_i^2$ smaller than $(\sum x_i)^2/n$). If your value of s^2 is negative, you have made a computational error.

Several other properties of s^2 can sometimes be used to increase computational efficiency. These are summarized in the following proposition.

PROPOSITION

Let $x_1, x_2, \ldots, x_n$ be a sample and c be any nonzero constant.

1. If $y_1 = x_1 + c$, $y_2 = x_2 + c$, $\ldots$, $y_n = x_n + c$, then $s_y^2 = s_x^2$, and
2. If $y_1 = cx_1, \ldots, y_n = cx_n$, then $s_y^2 = c^2 s_x^2$, $s_y = |c| s_x$,

where s_x^2 is the sample variance of the x's and s_y^2 is the sample variance of the y's.

In words, Result 1 says that if a constant c is added to (or subtracted from) each data value, the variance is unchanged. This is intuitive, since adding or subtracting c shifts the location of the data set but leaves distances between data values unchanged. According to Result 2, multiplication of each x_i by c results in s^2 being multiplied by a factor of c^2. These properties can be proved by noting in Result 1 that $\bar{y} = \bar{x} + c$ and in Result 2 that $\bar{y} = c\bar{x}$.

Boxplots

Stem-and-leaf displays and histograms convey rather general impressions about a data set, whereas a single summary such as the mean or standard deviation focuses on just one aspect of the data. In recent years, a pictorial summary called a *boxplot* has been used successfully to describe several of a data set's most prominent features. These features include (1) center, (2) spread, (3) the extent and nature of any departure from symmetry, and (4) identification of "outliers," observations that lie unusually far from the main body of the data. Because even a single outlier can drastically affect the value of some numerical summaries (such as $\bar{x}$ and s), a boxplot is based on measures that are "resistant" to the presence of a few outliers—the median and a measure of spread called the *fourth spread.*

DEFINITION

After the *n* observations in a data set are ordered from smallest to largest, the **lower fourth** and **upper fourth** are given by

$$\frac{\text{lower}}{\text{fourth}} = \begin{cases} \text{median of the smallest } n/2 \text{ observations,} & n \text{ even} \\ \text{median of the smallest } (n + 1)/2 \text{ observations,} & n \text{ odd} \end{cases}$$

$$\frac{\text{upper}}{\text{fourth}} = \begin{cases} \text{median of the largest } n/2 \text{ observations,} & n \text{ even} \\ \text{median of the largest } (n + 1)/2 \text{ observations,} & n \text{ odd} \end{cases}$$

That is, the lower (upper) fourth is the median of the smallest (largest) half of the data, where the median $\tilde{x}$ is included in both halves if *n* is odd. A measure of spread that is resistant to outliers is the **fourth spread** f_s, given by

$$f_s = \text{upper fourth} - \text{lower fourth}$$

Roughly speaking, the fourth spread is unaffected by the positions of those observations in the smallest 25% or the largest 25% of the data.

The simplest boxplot is based on the following five-number summary:

smallest x_i lower fourth   median   upper fourth   largest x_i

First, draw a horizontal measurement scale. Then place a rectangle above this axis; the left edge of the rectangle is at the lower fourth, and the right edge is at the upper fourth (so box width $= f_s$). Place a vertical line segment or some other symbol inside the rectangle at the location of the median; the position of the median symbol relative to the two edges conveys information about skewness in the middle 50% of the data. Finally, draw "whiskers" out from either end of the rectangle to the smallest and largest observations. A boxplot with a vertical orientation can also be drawn by making obvious modifications in the construction process.

Example 1.18 Ultrasound was used to gather the accompanying corrosion data on the thickness of the floor plate of an aboveground tank used to store crude oil ("Statistical Analysis of UT Corrosion Data from Floor Plates of a Crude Oil Aboveground Storage Tank," *Materials Eval.*, 1994: 846–849); each observation is the largest pit depth in the plate, expressed in milli-in.

40 52 55 60 70 75 85 85 90 90 92 94 94 95 98 100 115 125 125

The five-number summary is as follows:

smallest $x_i = 40$ lower fourth $= 72.5$ $\tilde{x} = 90$ upper fourth $= 96.5$
largest $x_i = 125$

Figure 1.19 (page 42) shows the resulting boxplot. The right edge of the box is much closer to the median than is the left edge, indicating a very substantial skew in the middle half of the data. The box width (f_s) is also reasonably large relative to the range of the data (distance between the tips of the whiskers).

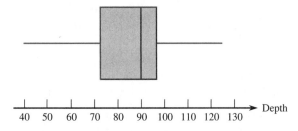

Figure 1.19 A boxplot of the corrosion data

Figure 1.20 shows MINITAB output from a request to describe the corrosion data. The trimmed mean is the average of the 17 observations that remain after the largest and smallest values are deleted (trimming percentage ≈ 5%). Q1 and Q3 are the lower and upper quartiles; these are similar to the fourths but are calculated in a slightly different manner. SE Mean is $s/\sqrt{n}$; this will be an important quantity in our subsequent work concerning inferences about μ.

Variable	N	Mean	Median	TrMean	StDev	SE Mean
depth	19	86.32	90.00	86.76	23.32	5.35

Variable	Minimum	Maximum	Q1	Q3
depth	40.00	125.00	70.00	98.00

Figure 1.20 MINITAB description of the pit-depth data ■

Boxplots that Show Outliers

A boxplot can be embellished to indicate explicitly the presence of outliers.

DEFINITION

> Any observation farther than $1.5f_s$ from the closest fourth is an **outlier.** An outlier is **extreme** if it is more than $3f_s$ from the nearest fourth, and it is **mild** otherwise.

Many inferential procedures are based on the assumption that the sample came from a normal distribution. Even a single extreme outlier in the sample warns the investigator that such procedures should not be used, and the presence of several mild outliers conveys the same message.

Let's now modify our previous construction of a boxplot by drawing a whisker out from each end of the box to the smallest and largest observations that are *not* outliers. Each mild outlier is represented by a closed circle and each extreme outlier by an open circle. Some statistical computer packages do not distinguish between mild and extreme outliers.

Example 1.19 The effects of partial discharges on the degradation of insulation cavity material have important implications for the lifetimes of high-voltage components. Consider the following sample of $n = 25$ pulse widths from slow discharges in a cylindrical cavity made of polyethylene. (This data is consistent with a histogram of 250 observations in the article "Assessment of Dielectric Degradation by Ultrawide-band PD Detection," *IEEE Trans. on Dielectrics and Elec. Insul.,* 1995: 744–760.) The article's author notes the impact of a wide variety of statistical tools on the interpretation of discharge data.

$$5.3 \quad 8.2 \quad 13.8 \quad 74.1 \quad 85.3 \quad 88.0 \quad 90.2 \quad 91.5 \quad 92.4 \quad 92.9 \quad 93.6 \quad 94.3 \quad 94.8$$
$$94.9 \quad 95.5 \quad 95.8 \quad 95.9 \quad 96.6 \quad 96.7 \quad 98.1 \quad 99.0 \quad 101.4 \quad 103.7 \quad 106.0 \quad 113.5$$

Relevant quantities are

$$\tilde{x} = 94.8 \quad \text{lower fourth} = 90.2 \quad \text{upper fourth} = 96.7$$
$$f_s = 6.5 \quad 1.5f_s = 9.75 \quad 3f_s = 19.50$$

Thus any observation smaller than $90.2 - 9.75 = 80.45$ or larger than $96.7 + 9.75 = 106.45$ is an outlier. There is one outlier at the upper end of the sample, and four outliers are at the lower end. Because $90.2 - 19.5 = 70.7$, the three observations 5.3, 8.2, and 13.8 are extreme outliers; the other two outliers are mild. The whiskers extend out to 85.3 and 106.0, the most extreme observations that are not outliers. The resulting boxplot is in Figure 1.21. There is a great deal of negative skewness in the middle half of the sample as well as in the entire sample.

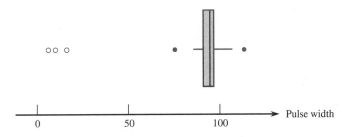

Figure 1.21 A boxplot of the pulse width data showing mild and extreme outliers ■

Comparative Boxplots

A comparative or side-by-side boxplot is a very effective way of revealing similarities and differences between two or more data sets consisting of observations on the same variable.

Example 1.20 In recent years, some evidence suggests that high indoor radon concentration may be linked to the development of childhood cancers, but many health professionals remain unconvinced. A recent article ("Indoor Radon and Childhood Cancer," *The Lancet,* 1991: 1537–1538) presented the accompanying data on radon concentration (Bq/m³) in two different samples of houses. The first sample consisted of houses in which a child diagnosed with cancer had been residing. Houses in the second sample had no recorded cases of childhood cancer. Figure 1.22 (page 44) presents a stem-and-leaf display of the data.

```
         1. Cancer                              2. No cancer

              9683795 │ 0 │ 95768397678993
86071815066815233150 │ 1 │ 12271713114
             12302731 │ 2 │ 99494191
                 8349 │ 3 │ 839
                    5 │ 4 │
                    7 │ 5 │ 55
                      │ 6 │
                      │ 7 │              Stem: Tens digit
             HI: 210  │ 8 │ 5            Leaf: Ones digit
```

Figure 1.22   Stem-and-leaf display for Example 1.20

Numerical summary quantities are as follows:

	$\bar{x}$	$\tilde{x}$	s	f_s
Cancer	22.8	16.0	31.7	11.0
No cancer	19.2	12.0	17.0	18.0

The values of both the mean and median suggest that the cancer sample is centered somewhat to the right of the no-cancer sample on the measurement scale. The mean, however, exaggerates the magnitude of this shift, largely because of the observation 210 in the cancer sample. The values of s suggest more variability in the cancer sample than in the no-cancer sample, but this impression is contradicted by the fourth spreads. Again, the observation 210, an extreme outlier, is the culprit. Figure 1.23 shows a comparative boxplot from the S-Plus computer package. The no-cancer box is stretched out compared with the cancer box ($f_s = 18$ vs. $f_s = 11$), and the positions of the median lines in the two boxes show much more skewness in the middle half of the no-cancer sample than the cancer sample. Outliers are represented by horizontal line segments, and there is no distinction between mild and extreme outliers.

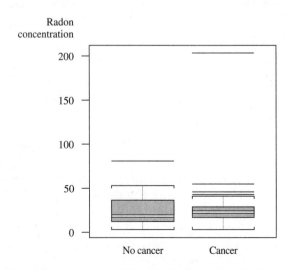

Figure 1.23   A boxplot of the data in Example 1.20, from S-Plus

Exercises | Section 1.4 (44–61)

44. The article "Oxygen Consumption During Fire Suppression: Error of Heart Rate Estimation" (*Ergonomics*, 1991: 1469–1474) reported the following data on oxygen consumption (mL/kg/min) for a sample of ten firefighters performing a fire-suppression simulation:

29.5 49.3 30.6 28.2 28.0 26.3 33.9 29.4 23.5 31.6

Compute the following:
a. The sample range
b. The sample variance s^2 from the definition (i.e., by first computing deviations, then squaring them, etc.)
c. The sample standard deviation
d. s^2 using the shortcut method

45. The value of Young's modulus (GPa) was determined for cast plates consisting of certain intermetallic substrates, resulting in the following sample observations ("Strength and Modulus of a Molybdenum-Coated Ti-25Al-10Nb-3U-1Mo Intermetallic," *J. of Materials Engr. and Performance*, 1997: 46–50):

116.4 115.9 114.6 115.2 115.8

a. Calculate $\bar{x}$ and the deviations from the mean.
b. Use the deviations calculated in part (a) to obtain the sample variance and the sample standard deviation.
c. Calculate s^2 by using the computational formula for the numerator S_{xx}.
d. Subtract 100 from each observation to obtain a sample of transformed values. Now calculate the sample variance of these transformed values, and compare it to s^2 for the original data.

46. The accompanying observations on stabilized viscosity (cP) for specimens of a certain grade of asphalt with 18% rubber added are from the article "Viscosity Characteristics of Rubber-Modified Asphalts" (*J. of Materials in Civil Engr.*, 1996: 153–156):

2781 2900 3013 2856 2888

a. What are the values of the sample mean and sample median?
b. Calculate the sample variance using the computational formula. *Hint:* First subtract a convenient number from each observation.

47. Calculate and interpret the values of the sample mean and sample standard deviation for the following observations on fracture strength (MPa, read from a graph in "Heat-Resistant Active Brazing of Silicon Nitride: Mechanical Evaluation of Braze Joints," *Welding J.*, August, 1997).

87 93 96 98 105 114 128 131 142 168

48. Exercise 36 in Section 1.3 presented a sample of 26 escape times for oil workers in a simulated escape exercise. Calculate and interpret the sample standard deviation. *Hint:* $\sum x_i = 9638$ and $\sum x_i^2 = 3,587,566$.

49. A study of the relationship between age and various visual functions (such as acuity and depth perception) reported the following observations on area of scleral lamina (mm^2) from human optic nerve heads ("Morphometry of Nerve Fiber Bundle Pores in the Optic Nerve Head of the Human," *Experimental Eye Research*, 1988: 559–568):

2.75 2.62 2.74 3.85 2.34 2.74 3.93 4.21 3.88
4.33 3.46 4.52 2.43 3.65 2.78 3.56 3.01

a. Calculate $\sum x_i$ and $\sum x_i^2$.
b. Use the values calculated in part (a) to compute the sample variance s^2 and then the sample standard deviation s.

50. The accompanying data on bearing load-life (million revs.) for bearings of a certain type when subjected to a 9.56 kN load appeared in the article "The Load-Life Relationship for M50 Bearings with Silicon Nitride Ceramic Balls" (*Lubric. Engr.*, 1984: 153–159):

14.5 25.6 52.4 66.3 69.3 69.8 76.2

a. Calculate the values of the sample mean and median. What do their values relative to one another tell you about the sample?
b. Compute the values of the sample variance and standard deviation.

51. The article "A Thin-Film Oxygen Uptake Test for the Evaluation of Automotive Crankcase Lubricants" (*Lubric. Engr.*, 1984: 75–83) reported the following data on oxidation-induction time (min) for various commercial oils:

87 103 130 160 180 195 132 145 211 105 145
153 152 138 87 99 93 119 129

a. Calculate the sample variance and standard deviation.

b. If the observations were reexpressed in hours, what would be the resulting values of the sample variance and sample standard deviation? Answer without actually performing the reexpression.

52. The first four deviations from the mean in a sample of $n = 5$ reaction times were .3, .9, 1.0, and 1.3. What is the fifth deviation from the mean? Give a sample for which these are the five deviations from the mean.

53. Reconsider the data on area of scleral lamina given in Exercise 49.
a. Determine the lower and upper fourths.
b. Calculate the value of the fourth spread.
c. If the two largest sample values, 4.33 and 4.52, had instead been 5.33 and 5.52, how would this affect f_s? Explain.
d. By how much could the observation 2.34 be increased without affecting f_s? Explain.
e. If an 18th observation, $x_{18} = 4.60$, is added to the sample, what is f_s?

54. Reconsider the accompanying shear strength observations (MPa) introduced in Exercise 34 of this chapter:

22.2   40.4   16.4   73.7   36.6   109.9
30.0   4.4   33.1   66.7   81.5

a. What are the values of the fourths, and what is the value of f_s?
b. Construct a boxplot based on the five-number summary, and comment on its features.
c. How large or small does an observation have to be to qualify as an outlier? As an extreme outlier?
d. By how much could the largest observation be decreased without affecting f_s?

55. Here is a stem-and-leaf display of the escape time data introduced in Exercise 36 of this chapter.

32	55
33	49
34	
35	6699
36	34469
37	03345
38	9
39	2347
40	23
41	
42	4

a. Determine the value of the fourth spread.
b. Are there any outliers in the sample? Any extreme outliers?

c. Construct a boxplot and comment on its features.
d. By how much could the largest observation, currently 424, be decreased without affecting the value of the fourth spread?

56. The amount of aluminum contamination (ppm) in plastic of a certain type was determined for a sample of 26 plastic specimens, resulting in the following data ("The Lognormal Distribution for Modeling Quality Data when the Mean Is Near Zero," *J. of Quality Technology,* 1990: 105–110):

30	30	60	63	70	79	87	90	101
102	115	118	119	119	120	125	140	145
172	182	183	191	222	244	291	511	

Construct a boxplot that shows outliers, and comment on its features.

57. A sample of 20 glass bottles of a particular type was selected, and the internal pressure strength of each bottle was determined. Consider the following partial sample information:

median = 202.2   lower fourth = 196.0
upper fourth = 216.8

Three smallest observations 125.8   188.1   193.7
Three largest observations   221.3   230.5   250.2

a. Are there any outliers in the sample? Any extreme outliers?
b. Construct a boxplot that shows outliers, and comment on any interesting features.

58. A company utilizes two different machines to manufacture parts of a certain type. During a single shift, a sample of $n = 20$ parts produced by each machine is obtained, and the value of a particular critical dimension for each part is determined. The comparative boxplot at the top of page 47 is constructed from the resulting data. Compare and contrast the two samples.

59. Blood cocaine concentration (mg/L) was determined both for a sample of individuals who had died from cocaine-induced excited delirium (ED) and for a sample of those who had died from a cocaine overdose without excited delirium; survival time for people in both groups was at most 6 hours. The accompanying data was read from a comparative boxplot in the article "Fatal Excited Delirium Following Cocaine Use" (*J. of Forensic Sciences,* 1997: 25–31).

ED   0 0 0 0 .1 .1 .1 .1 .2 .2 .3 .3
          .3 .4 .5 .7 .8 1.0 1.5 2.7 2.8
          3.5 4.0 8.9 9.2 11.7 21.0

Comparative boxplot for Exercise 58

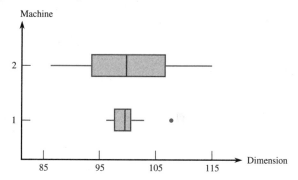

Non-ED   0 0 0 0 0 .1 .1 .1 .1 .2 .2 .2
.3 .3 .3 .4 .5 .5 .6 .8 .9 1.0
1.2 1.4 1.5 1.7 2.0 3.2 3.5 4.1
4.3 4.8 5.0 5.6 5.9 6.0 6.4 7.9
8.3 8.7 9.1 9.6 9.9 11.0 11.5
12.2 12.7 14.0 16.6 17.8

a. Determine the medians, fourths, and fourth spreads for the two samples.
b. Are there any outliers in either sample? Any extreme outliers?
c. Construct a comparative boxplot, and use it as a basis for comparing and contrasting the ED and non-ED samples.

60. Observations on burst strength (lb/in²) were obtained both for test nozzle closure welds and for production cannister nozzle welds ("Proper Proce-

dures Are the Key to Welding Radioactive Waste Cannisters," *Welding J.,* Aug. 1997: 61–67).

Test	7200	6100	7300	7300	8000	7400
	7300	7300	8000	6700	8300	
Cannister	5250	5625	5900	5900	5700	6050
	5800	6000	5875	6100	5850	6600

Construct a comparative boxplot and comment on interesting features (the cited article did not include such a picture, but the authors commented that they had looked at one).

61. The accompanying comparative boxplot of gasoline vapor coefficients for vehicles in Detroit appeared in the article "Receptor Modeling Approach to VOC Emission Inventory Validation," (*J. of Envir. Engr.,* 1995: 483–490). Discuss any interesting features.

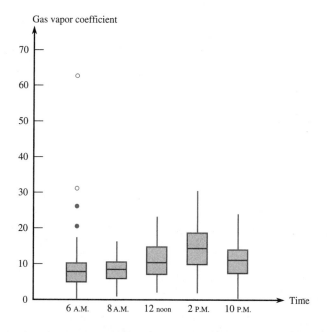

Supplementary Exercises (62–81)

62. Consider the following information on ultimate tensile strength (lb/in) for a sample of $n = 4$ hard zirconium copper wire specimens (from "Characterization Methods for Fine Copper Wire," *Wire J. Intl.*, Aug., 1997: 74–80).

$\bar{x} = 76,831 \quad s = 180 \quad$ smallest $x_i = 76,683$
largest $x_i = 77,048$

Determine the values of the two middle sample observations (and don't do it by successive guessing!).

63. Three different C_2F_6 flow rates (SCCM) were considered in an experiment to investigate the effect of flow rate on the uniformity (%) of the etch on a silicon wafer used in the manufacture of integrated circuits, resulting in the following data:

Flow rate
125	2.6	2.7	3.0	3.2	3.8	4.6
160	3.6	4.2	4.2	4.6	4.9	5.0
200	2.9	3.4	3.5	4.1	4.6	5.1

Compare and contrast the uniformity observations resulting from these three different flow rates.

64. The amount of radiation received at a greenhouse plays an important role in determining the rate of photosynthesis. The accompanying observations on incoming solar radiation were read from a graph in the article "Radiation Components over Bare and Planted Soils in a Greenhouse" (*Solar Energy*, 1990: 1011–1016).

6.3	6.4	7.7	8.4	8.5	8.8	8.9
9.0	9.1	10.0	10.1	10.2	10.6	10.6
10.7	10.7	10.8	10.9	11.1	11.2	11.2
11.4	11.9	11.9	12.2	13.1		

Use some of the methods discussed in this chapter to describe and summarize this data.

65. The following data on HC and CO emissions for one particular vehicle was given in the chapter introduction.

HC (gm/mi)	13.8	18.3	32.2	32.5
CO (gm/mi)	118	149	232	236

a. Compute the sample standard deviations for the HC and CO observations. Does the widespread belief appear to be justified?

b. The *sample coefficient of variation* $s/\bar{x}$ (or $100 s/\bar{x}$) assesses the extent of variability relative to the mean. Values of this coefficient for several different data sets can be compared to determine which data sets exhibit more or less variation. Carry out such a comparison for the given data.

66. The accompanying frequency distribution of fracture strength (MPa) observations for ceramic bars fired in a particular kiln appeared in the article "Evaluating Tunnel Kiln Performance" (*Amer. Ceramic Soc. Bull.*, Aug. 1997: 59–63).

Class	81– <83	83– <85	85– <87	87– <89	89– <91
Frequency	6	7	17	30	43

Class	91– <93	93– <95	95– <97	97– <99
Frequency	28	22	13	3

a. Construct a histogram based on relative frequencies, and comment on any interesting features.

b. What proportion of the strength observations are at least 85? Less than 95?

c. Roughly what proportion of the observations are less than 90?

67. Fifteen air samples from a certain region were obtained, and for each one the carbon monoxide concentration was determined. The results (in ppm) were

9.3	10.7	8.5	9.6	12.2	15.6	9.2	10.5
9.0	13.2	11.0	8.8	13.7	12.1	9.8	

Using the interpolation method suggested in Section 1.3, compute the 10% trimmed mean.

68. a. For what value of c is the quantity $\Sigma(x_i - c)^2$ minimized? (*Hint:* Take the derivative with respect to c, set equal to 0, and solve.)

b. Using the result of part (a), which of the two quantities $\Sigma(x_i - \bar{x})^2$ and $\Sigma(x_i - \mu)^2$ will be smaller than the other (assuming that $\bar{x} \neq \mu$)?

69. a. Let a and b be constants and let $y_i = ax_i + b$ for $i = 1, 2, \ldots, n$. What are the relationships between $\bar{x}$ and $\bar{y}$ and between s_x^2 and s_y^2?

b. A sample of temperatures for initiating a certain chemical reaction yielded a sample average (°C) of 87.3 and a sample standard deviation of 1.04. What are the sample average and standard deviation measured in °F? (*Hint:* $F = \frac{9}{5} C + 32$.)

70. The percentage of juice lost after thawing for 19 different strawberry varieties appeared in the article "Evaluation of Strawberry Cultivars with Different Degrees of Resistance to Red Stele" (*Fruit Varieties J.,* 1991: 12–17):

```
46  51  44  50  33  46  60  41  55  46
53  53  42  44  50  54  46  41  48
```

a. Are there any observations that are mild outliers? Extreme outliers?

b. Construct a boxplot and comment.

71. The article "Can We Really Walk Straight?" (*Amer. J. of Physical Anthropology,* 1992: 19–27) reported on an experiment in which each of 20 healthy men was asked to walk as straight as possible to a target 60 m away at normal speed. Consider the following observations on cadence (number of strides per second):

```
.95  .85  .92   .95  .93   .86  1.00  .92  .85  .81
.78  .93  .93  1.05  .93  1.06  1.06  .96  .81  .96
```

Use the methods developed in this chapter to summarize the data; include an interpretation or discussion wherever appropriate. (*Note:* The author of the article used a rather sophisticated statistical analysis to conclude that people cannot walk in a straight line and suggested several explanations for this.)

72. The **mode** of a numerical data set is the value that occurs most frequently in the set.

a. Determine the mode for the cadence data given in Exercise 71.

b. For a categorical sample, how would you define the modal category?

73. Specimens of three different types of rope wire were selected, and the fatigue limit (MPa) was determined for each specimen, resulting in the accompanying data.

Type 1	350	350	350	358	370	370	370	371
	371	372	372	384	391	391	392	
Type 2	350	354	359	363	365	368	369	371
	373	374	376	380	383	388	392	
Type 3	350	361	362	364	364	365	366	371
	377	377	377	379	380	380	392	

a. Construct a comparative boxplot, and comment on similarities and differences.

b. Construct a comparative dotplot (a dotplot for each sample with a common scale). Comment on similarities and differences.

c. Does the comparative boxplot of part (a) give an informative assessment of similarities and differences? Explain your reasoning.

74. The three measures of center introduced in this chapter are the mean, median, and trimmed mean. Two additional measures of center that are occasionally used are the *midrange,* which is the average of the smallest and largest observations, and the *midfourth,* which is the average of the two fourths. Which of these five measures of center are resistant to the effects of outliers and which are not? Explain your reasoning.

75. Consider the following data on active repair time (hours) for a sample of $n = 46$ airborne communications receivers:

```
.2   .3   .5   .5   .5   .6   .6    .7    .7    .7  .8   .8
.8  1.0  1.0  1.0  1.0  1.1  1.3   1.5   1.5   1.5 1.5  2.0
2.0  2.2  2.5  2.7  3.0  3.0  3.3   3.3   4.0   4.0 4.5  4.7
5.0  5.4  5.4  7.0  7.5  8.8  9.0  10.3  22.0  24.5
```

Construct the following:

a. A stem-and-leaf display in which the two largest values are displayed separately in a row labeled HI

b. A histogram based on six class intervals with 0 as the lower limit of the first interval and interval widths of 2, 2, 2, 4, 10, and 10, respectively

76. Consider a sample $x_1, x_2, \ldots, x_n$ and suppose that the values of $\bar{x}$, s^2, and s have been calculated.

a. Let $y_i = x_i - \bar{x}$ for $i = 1, \ldots, n$. How do the values of s^2 and s for the y_i's compare to the corresponding values for the x_i's? Explain.

b. Let $z_i = (x_i - \bar{x})/s$ for $i = 1, \ldots, n$. What are the values of the sample variance and sample standard deviation for the z_i's?

77. Let $\bar{x}_n$ and s_n^2 denote the sample mean and variance for the sample $x_1, \ldots, x_n$ and let $\bar{x}_{n+1}$ and s_{n+1}^2 denote these quantities when an additional observation x_{n+1} is added to the sample.

a. Show how $\bar{x}_{n+1}$ can be computed from $\bar{x}_n$ and x_{n+1}.

b. Show that

$$ns_{n+1}^2 = (n-1)s_n^2 + \frac{n}{n+1}(x_{n+1} - \bar{x}_n)^2$$

so that s_{n+1}^2 can be computed from x_{n+1}, $\bar{x}_n$, and s_n^2.

c. Suppose that a sample of 15 strands of drapery yarn has resulted in a sample mean thread elongation of 12.58 mm and a sample standard deviation of .512 mm. A 16th strand results in an elongation value of 11.8. What are the values of the sample mean and sample standard deviation for all 16 elongation observations?

78. Lengths of bus routes for any particular transit system will typically vary from one route to another. The article "Planning of City Bus Routes" (*J. of the Institution of Engineers,* 1995: 211–215) gives the following information on lengths (km) for one particular system:

Length	6–<8	8–<10	10–<12	12–<14	14–<16
Frequency	6	23	30	35	32
Length	16–<18	18–<20	20–<22	22–<24	24–<26
Frequency	48	42	40	28	27
Length	26–<28	28–<30	30–<35	35–<40	40–<45
Frequency	26	14	27	11	2

a. Draw a histogram corresponding to these frequencies.

b. What proportion of these route lengths are less than 20? What proportion of these routes have lengths of at least 30?

c. Roughly what is the value of the 90th percentile of the route length distribution?

d. Roughly what is the median route length?

79. A study carried out to investigate the distribution of total braking time (reaction time plus accelerator-to-brake movement time, in ms) during real driving conditions at 60 km/hr gave the following summary information on the distribution of times ("A Field Study on Braking Responses during Driving," *Ergonomics,* 1995: 1903–1910):

mean = 535   median = 500   mode = 500
sd = 96   minimum = 220   maximum = 925

5th percentile = 400   10th percentile = 430
90th percentile = 640   95th percentile = 720

What can you conclude about the shape of a histogram of this data? Explain your reasoning.

80. The sample data $x_1, x_2, \ldots, x_n$ sometimes represents a **time series,** where $x_t =$ the observed value of a response variable x at time t. Often the observed series shows a great deal of random varia-

tion, which makes it difficult to study longer-term behavior. In such situations, it is desirable to produce a smoothed version of the series. One technique for doing so involves **exponential smoothing.** The value of a smoothing constant α is chosen $(0 < \alpha < 1)$. Then with $\bar{x}_t =$ smoothed value at time t, we set $\bar{x}_1 = x_1$, and for $t = 2, 3, \ldots, n$, $\bar{x}_t = \alpha x_t + (1 - \alpha)\bar{x}_{t-1}$.

a. Consider the following time series in which $x_t =$ temperature (°F) of effluent at a sewage treatment plant on day t: 47, 54, 53, 50, 46, 46, 47, 50, 51, 50, 46, 52, 50, 50. Plot each x_t against t on a two-dimensional coordinate system (a time-series plot). Does there appear to be any pattern?

b. Calculate the $\bar{x}_t$'s using $\alpha = .1$. Repeat using $\alpha = .5$. Which value of α gives a smoother $\bar{x}_t$ series?

c. Substitute $\bar{x}_{t-1} = \alpha x_{t-1} + (1 - \alpha)\bar{x}_{t-2}$ on the right-hand side of the expression for $\bar{x}_t$, then substitute $\bar{x}_{t-2}$ in terms of x_{t-2} and $\bar{x}_{t-3}$, and so on. On how many of the values $x_t, x_{t-1}, \ldots, x_1$ does $\bar{x}_t$ depend? What happens to the coefficient on x_{t-k} as k increases?

d. Refer to part (c). If t is large, how sensitive is $\bar{x}_t$ to the initialization $\bar{x}_1 = x_1$? Explain.

(*Note:* A relevant reference is the article "Simple Statistics for Interpreting Environmental Data," *Water Pollution Control Fed. J.,* 1981: 167–175.)

81. Consider numerical observations $x_1, \ldots, x_n$. It is frequently of interest to know whether the x_i's are (at least approximately) symmetrically distributed about some value. If n is at least moderately large, the extent of symmetry can be assessed from a stem-and-leaf display or histogram. However, if n is not very large, such pictures are not particularly informative. Consider the following alternative. Let y_1 denote the smallest x_i, y_2 the second smallest x_i, and so on. Then plot the following pairs as points on a two-dimensional coordinate system: $(y_n - \tilde{x}, \tilde{x} - y_1)$, $(y_{n-1} - \tilde{x}, \tilde{x} - y_2)$, $(y_{n-2} - \tilde{x}, \tilde{x} - y_3), \ldots$. There are $n/2$ points when n is even and $(n - 1)/2$ when n is odd.

a. What does this plot look like when there is perfect symmetry in the data? What does it look like when observations stretch out more above the median than below it (a long upper tail)?

b. The accompanying data on rainfall (acre-feet) from 26 seeded clouds is taken from the article "A Bayesian Analysis of a Multiplicative Treatment Effect in Weather Modification" (*Technometrics,* 1975: 161–166). Construct the plot and comment on the extent of symmetry or nature of departure from symmetry.

4.1	7.7	17.5	31.4	32.7	40.6	92.4
115.3	118.3	119.0	129.6	198.6	200.7	242.5
255.0	274.7	274.7	302.8	334.1	430.0	489.1
703.4	978.0	1656.0	1697.8	2745.6		

Bibliography

Chambers, John, William Cleveland, Beat Kleiner, and Paul Tukey, *Graphical Methods for Data Analysis,* Brooks/Cole, Pacific Grove, CA (1983). A highly recommended presentation of both older and more recent graphical and pictorial methodology in statistics.

Devore, Jay, and Roxy Peck, *Statistics: The Exploration and Analysis of Data* (3rd ed.), Duxbury Press, Pacific Grove, CA (1997). The first few chapters give a very nonmathematical survey of methods for describing and summarizing data.

Freedman, David, Robert Pisani, Roger Purves, and Ani Adhikari, *Statistics* (2nd ed.), Norton, New York (1991). An excellent, very nonmathematical survey of basic statistical reasoning and methodology.

Hoaglin, David, Frederick Mosteller, and John Tukey, *Understanding Robust and Exploratory Data Analysis,* Wiley, New York (1983). Discusses why, as well as how, exploratory methods should be employed; it is good on details of stem-and-leaf displays and boxplots.

Hoaglin, David, and Paul Velleman, *Applications, Basics, and Computing of Exploratory Data Analysis,* Duxbury Press, Boston (1980). A good discussion of some basic exploratory methods.

Moore, David, *Statistics: Concepts and Controversies* (4th ed.), Freeman, San Francisco (1997). An extremely readable and entertaining paperback that contains an intuitive discussion of problems connected with sampling and designed experiments.

Tanur, Judith, et al. (eds.), *Statistics: A Guide to the Unknown* (3rd ed.), Duxbury Press, Belmont, CA (1988). Contains many short nontechnical articles describing various applications of statistics.

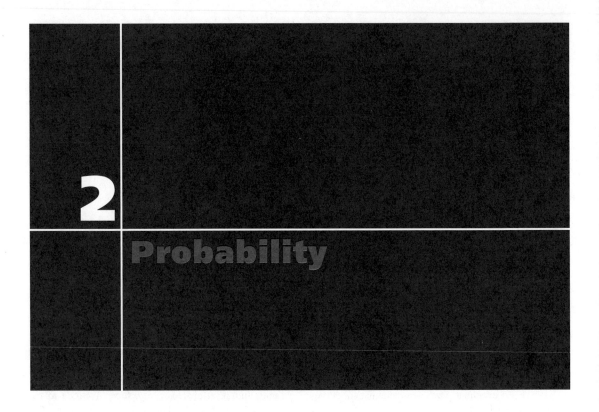

2

Probability

Introduction

The term **probability** refers to the study of randomness and uncertainty. In any situation in which one of a number of possible outcomes may occur, the theory of probability provides methods for quantifying the chances, or likelihoods, associated with the various outcomes. The language of probability is constantly used in an informal manner in both written and spoken contexts. Examples include such statements as "It is likely that the Dow-Jones average will increase by the end of the year," "There is a 50–50 chance that the incumbent will seek reelection," "There will probably be at least one section of that course offered next year," "The odds favor a quick settlement of the strike," and "It is expected that at least 20,000 concert tickets will be sold." In this chapter, we introduce some elementary probability concepts, indicate how probabilities can be interpreted, and show how the rules of probability can be applied to compute the probabilities of many interesting events. The methodology of probability will then permit us to express in precise language such informal statements as those given above.

The study of probability as a branch of mathematics goes back over 300 years, where it had its genesis in connection with questions involving games of chance.

Many books are devoted exclusively to probability, but our objective here is to cover only that part of the subject that has the most direct bearing on problems of statistical inference.

2.1 | Sample Spaces and Events

An **experiment** is any action or process that generates observations. Although the word *experiment* generally suggests a planned or carefully controlled laboratory testing situation, we use it here in a much wider sense. Thus, experiments that may be of interest include tossing a coin once or several times, selecting a card or cards from a deck, weighing a loaf of bread, ascertaining the commuting time from home to work on a particular morning, obtaining blood types from a group of individuals, or measuring the compressive strengths of different steel beams.

The Sample Space of an Experiment

DEFINITION

The **sample space** of an experiment, denoted by $\mathscr{S}$, is the set of all possible outcomes of that experiment.

Example 2.1   The simplest experiment to which probability applies is one with two possible outcomes. One such experiment consists of examining a single fuse to see whether it is defective. The sample space for this experiment can be abbreviated as $\mathscr{S} = \{N, D\}$, where N represents not defective, D represents defective, and the braces are used to enclose the elements of a set. Another such experiment would involve tossing a thumbtack and noting whether it landed point up or point down, with sample space $\mathscr{S} = \{U, D\}$, and yet another would consist of observing the sex of the next child born at the local hospital, with $\mathscr{S} = \{M, F\}$.   ∎

Example 2.2   If we examine three fuses in sequence and note the result of each examination, then an outcome for the entire experiment is any sequence of N's and D's of length 3, so

$$\mathscr{S} = \{NNN, NND, NDN, NDD, DNN, DND, DDN, DDD\}$$

If we had tossed a thumbtack three times, the sample space would be obtained by replacing N by U in $\mathscr{S}$ above, with a similar notational change yielding the sample space for the experiment in which the sexes of three newborn children are observed.   ∎

Example 2.3   Two gas stations are located at a certain intersection. Each one has six gas pumps. Consider the experiment in which the number of pumps in use at a particular time of day is determined for each of the stations. An experimental outcome specifies how many pumps are in use at the first station and how many are in use at the second one. One possible outcome is (2, 2), another is (4, 1), and yet another is (1, 4). The 49 outcomes in $\mathscr{S}$ are displayed in the accompanying table. The sample space for the experiment in which a six-sided die is thrown twice results from deleting the 0 row and 0 column from the table, giving 36 outcomes.

		Second Station						
		0	**1**	**2**	**3**	**4**	**5**	**6**
	0	(0, 0)	(0, 1)	(0, 2)	(0, 3)	(0, 4)	(0, 5)	(0, 6)
	1	(1, 0)	(1, 1)	(1, 2)	(1, 3)	(1, 4)	(1, 5)	(1, 6)
	2	(2, 0)	(2, 1)	(2, 2)	(2, 3)	(2, 4)	(2, 5)	(2, 6)
First Station	**3**	(3, 0)	(3, 1)	(3, 2)	(3, 3)	(3, 4)	(3, 5)	(3, 6)
	4	(4, 0)	(4, 1)	(4, 2)	(4, 3)	(4, 4)	(4, 5)	(4, 6)
	5	(5, 0)	(5, 1)	(5, 2)	(5, 3)	(5, 4)	(5, 5)	(5, 6)
	6	(6, 0)	(6, 1)	(6, 2)	(6, 3)	(6, 4)	(6, 5)	(6, 6)

∎

Example 2.4   If a new type-D flashlight battery has a voltage that is outside certain limits, that battery is characterized as a failure (*F*); if the battery has a voltage within the prescribed limits, it is a success (*S*). Suppose an experiment consists of testing each battery as it comes off an assembly line until we first observe a success. Although it may not be very likely, a possible outcome of this experiment is that the first 10 (or 100 or 1000 or . . .) are *F*'s and the next one is an *S*. That is, for any positive integer *n*, we may have to examine *n* batteries before seeing the first *S*. The sample space is $\mathscr{S} = \{S, FS, FFS, FFFS, \ldots\}$, which contains an infinite number of possible outcomes. The same abbreviated form of the sample space is appropriate for an experiment in which, starting at a specified time, the sex of each newborn infant is recorded until the birth of a male is observed.   ∎

Events

In our study of probability, we will be interested not only in the individual outcomes of $\mathscr{S}$ but also in any collection of outcomes from $\mathscr{S}$.

DEFINITION

An **event** is any collection (subset) of outcomes contained in the sample space $\mathscr{S}$. An event is said to be **simple** if it consists of exactly one outcome and **compound** if it consists of more than one outcome.

When an experiment is performed, a particular event *A* is said to occur if the resulting experimental outcome is contained in *A*. In general, exactly one simple event will occur, but many compound events will occur simultaneously.

Example 2.5   Consider an experiment in which each of three automobiles taking a particular freeway exit turns left (*L*) or right (*R*) at the end of the exit ramp. The eight possible outcomes that comprise the sample space are *LLL, RLL, LRL, LLR, LRR, RLR, RRL,* and *RRR*. Thus, there are eight simple events, among which are $E_1 = \{LLL\}$ and $E_5 = \{LRR\}$. Some compound events include

$A = \{RLL, LRL, LLR\}$ = the event that exactly one of the three cars turns right

$B = \{LLL, RLL, LRL, LLR\}$ = the event that at most one of the cars turns right

$C = \{LLL, RRR\}$ = the event that all three cars turn in the same direction

Suppose that when the experiment is performed, the outcome is LLL. Then the simple event E_1 has occurred and so also have the events B and C (but not A).   ■

Example 2.6
(Example 2.3 continued)

When the number of pumps in use at each of two six-pump gas stations is observed, there are 49 possible outcomes, so there are 49 simple events: $E_1 = \{(0, 0)\}$, $E_2 = \{(0, 1)\}, \ldots, E_{49} = \{(6, 6)\}$. Examples of compound events are

$A = \{(0, 0), (1, 1), (2, 2), (3, 3), (4, 4), (5, 5), (6, 6)\}$ = the event that the number of pumps in use is the same for both stations

$B = \{(0, 4), (1, 3), (2, 2), (3, 1), (4, 0)\}$ = the event that the total number of pumps in use is four

$C = \{(0, 0), (0, 1), (1, 0), (1, 1)\}$ = the event that at most one pump is in use at each station   ■

Example 2.7
(Example 2.4 continued)

The sample space for the battery examination experiment contains an infinite number of outcomes, so there are an infinite number of simple events. Compound events include

$A = \{S, FS, FFS\}$ = the event that at most three batteries are examined

$E = \{FS, FFFS, FFFFFS, \ldots\}$ = the event that an even number of batteries are examined   ■

Some Relations from Set Theory

An event is nothing but a set, so that relationships and results from elementary set theory can be used to study events. The following concepts from set theory will be used to construct new events from given events.

DEFINITION

> **1.** The **union** of two events A and B, denoted by $A \cup B$ and read "A or B," is the event consisting of all outcomes that are *either in A or in B or in both events* (so that the union includes outcomes for which both A and B occur as well as outcomes for which exactly one occurs).
> **2.** The **intersection** of two events A and B, denoted by $A \cap B$ and read "A and B," is the event consisting of all outcomes that are in *both A and B*.
> **3.** The **complement** of an event A, denoted by A', is the set of all outcomes in $\mathcal{S}$ that are not contained in A.

Example 2.8
(Example 2.3 continued)

For the experiment in which the number of pumps in use at a single six-pump gas station is observed, let $A = \{0, 1, 2, 3, 4\}$, $B = \{3, 4, 5, 6\}$, and $C = \{1, 3, 5\}$. Then

$$A \cup B = \{0, 1, 2, 3, 4, 5, 6\} = \mathcal{S}, \quad A \cup C = \{0, 1, 2, 3, 4, 5\},$$
$$A \cap B = \{3, 4\}, \quad A \cap C = \{1, 3\}, \quad A' = \{5, 6\}, \quad \{A \cup C\}' = \{6\}. \quad ■$$

Example 2.9
(Example 2.4
continued)

In the battery experiment, define A, B, and C by

$$A = \{S, FS, FFS\}$$
$$B = \{S, FFS, FFFFS\}$$

and

$$C = \{FS, FFFS, FFFFFS, \ldots\}$$

Then

$$A \cup B = \{S, FS, FFS, FFFFS\}$$
$$A \cap B = \{S, FFS\}$$
$$A' = \{FFFS, FFFFS, FFFFFS, \ldots\}$$

and

$$C' = \{S, FFS, FFFFS, \ldots\} = \{\text{an odd number of batteries are examined}\} \quad \blacksquare$$

Sometimes A and B have no outcomes in common, so that the intersection of A and B contains no outcomes.

DEFINITION

> When A and B have no outcomes in common, they are said to be **mutually exclusive** or **disjoint** events.

Example 2.10   A small city has three automobile dealerships: a GM dealer selling Chevrolets, Pontiacs, and Buicks; a Ford dealer selling Fords and Mercurys; and a Chrysler dealer selling Plymouths and Chryslers. If an experiment consists of observing the brand of the next car sold, then the events $A = \{\text{Chevrolet, Pontiac, Buick}\}$ and $B = \{\text{Ford, Mercury}\}$ are mutually exclusive because the next car sold cannot be both a GM product and a Ford product. $\quad \blacksquare$

The operations of union and intersection can be extended to more than two events. For any three events A, B, and C, the event $A \cup B \cup C$ is the set of outcomes contained in at least one of the three events, whereas $A \cap B \cap C$ is the set of outcomes contained in all three events. Given events $A_1, A_2, A_3, \ldots$, these events are said to be mutually exclusive (or pairwise disjoint) if no two events have any outcomes in common.

A pictorial representation of events and manipulations with events is obtained by using Venn diagrams. To construct a Venn diagram, draw a rectangle whose interior will represent the sample space $\mathscr{S}$. Then any event A is represented as the interior of a closed curve (often a circle) contained in $\mathscr{S}$. Figure 2.1 shows examples of Venn diagrams.

(a) Venn diagram of events A and B

(b) Shaded region is A ∩ B

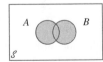

(c) Shaded region is A ∪ B

(d) Shaded region is A′

(e) Mutually exclusive events

Figure 2.1   Venn diagrams

Exercises | Section 2.1 (1–10)

1. Four universities—1, 2, 3, and 4—are participating in a holiday basketball tournament. In the first round, 1 will play 2 and 3 will play 4. Then the two winners will play for the championship, and the two losers will also play. One possible outcome can be denoted by 1324 (1 beats 2 and 3 beats 4 in first-round games, and then 1 beats 3 and 2 beats 4).
   a. List all outcomes in $\mathcal{S}$.
   b. Let A denote the event that 1 wins the tournament. List outcomes in A.
   c. Let B denote the event that 2 gets into the championship game. List outcomes in B.
   d. What are the outcomes in $A \cup B$ and in $A \cap B$? What are the outcomes in A'?

2. Suppose that vehicles taking a particular freeway exit can turn right (R), turn left (L), or go straight (S). Consider observing the direction for each of three successive vehicles.
   a. List all outcomes in the event A that all three vehicles go in the same direction.
   b. List all outcomes in the event B that all three vehicles take different directions.
   c. List all outcomes in the event C that exactly two of the three vehicles turn right.
   d. List all outcomes in the event D that exactly two vehicles go in the same direction.
   e. List outcomes in D', $C \cup D$, and $C \cap D$.

3. Three components are connected to form a system as shown in the accompanying diagram. Because the components in the 2–3 subsystem are connected in parallel, that subsystem will function if at least one of the two individual components functions. For the entire system to function, component 1 must function and so must the 2–3 subsystem.

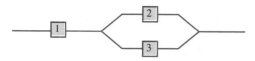

   The experiment consists of determining the condition of each component [S (success) for a functioning component and F (failure) for a nonfunctioning component].
   a. What outcomes are contained in the event A that exactly two out of the three components function?
   b. What outcomes are contained in the event B that at least two of the components function?
   c. What outcomes are contained in the event C that the system functions?
   d. List outcomes in C', $A \cup C$, $A \cap C$, $B \cup C$, and $B \cap C$.

4. Each of a sample of four home mortgages is classified as fixed rate (F) or variable rate (V).
   a. What are the 16 outcomes in $\mathcal{S}$?
   b. Which outcomes are in the event that exactly three of the selected mortgages are fixed rate?
   c. Which outcomes are in the event that all four mortgages are of the same type?
   d. Which outcomes are in the event that at most one of the four is a variable-rate mortgage?
   e. What is the union of the events in parts (c) and (d), and what is the intersection of these two events?
   f. What are the union and intersection of the two events in parts (b) and (c)?

5. A family consisting of three persons—A, B, and C—belongs to a medical clinic that always has a doctor at each of stations 1, 2, and 3. During a certain week, each member of the family visits the clinic once and is assigned at random to a station. The experiment consists of recording the station number for each member. One outcome is (1, 2, 1) for A to station 1, B to station 2, and C to station 1.
   a. List the 27 outcomes in the sample space.
   b. List all outcomes in the event that all three members go to the same station.
   c. List all outcomes in the event that all members go to different stations.
   d. List all outcomes in the event that no one goes to station 2.

6. A college library has five copies of a certain text on reserve. Two copies (1 and 2) are first printings, and the other three (3, 4, and 5) are second printings. A student examines these books in random order, stopping only when a second printing has been selected. One possible outcome is 5, and another is 213.
   a. List the outcomes in $\mathcal{S}$.
   b. Let A denote the event that exactly one book must be examined. What outcomes are in A?
   c. Let B be the event that book 5 is the one selected. What outcomes are in B?
   d. Let C be the event that book 1 is not examined. What outcomes are in C?

7. An academic department has just completed voting by secret ballot for a department head. The ballot box contains four slips with votes for candidate A and three slips with votes for candidate B. Suppose these slips are removed from the box one by one.
   a. List all possible outcomes.
   b. Suppose a running tally is kept as slips are removed. For what outcomes does A remain ahead of B throughout the tally?

8. An engineering construction firm is currently working on power plants at three different sites. Let A_i denote the event that the plant at site i is completed by the contract date. Use the operations of union, intersection, and complementation to describe each of the following events in terms of A_1, A_2, and A_3, draw a Venn diagram, and shade the region corresponding to each one.
   a. At least one plant is completed by the contract date.
   b. All plants are completed by the contract date.
   c. Only the plant at site 1 is completed by the contract date.
   d. Exactly one plant is completed by the contract date.
   e. Either the plant at site 1 or both of the other two plants are completed by the contract date.

9. Use Venn diagrams to verify the following two relationships for any events A and B (these are called De Morgan's laws):
   a. $(A \cup B)' = A' \cap B'$
   b. $(A \cap B)' = A' \cup B'$

10. a. In Example 2.10, identify three events that are mutually exclusive.
    b. Suppose there is no outcome common to all three of the events A, B, and C. Are these three events necessarily mutually exclusive? If your answer is yes, explain why; if your answer is no, give a counterexample using the experiment of Example 2.10.

2.2 Axioms, Interpretations, and Properties of Probability

Given an experiment and a sample space $\mathscr{S}$, *the objective of probability is to assign to each event A a number P(A), called the probability of the event A, which will give a precise measure of the chance that A will occur.* To ensure that the probability assignments will be consistent with our intuitive notions of probability, all assignments should satisfy the following axioms (basic properties) of probability.

AXIOM 1	For any event A, $P(A) \geq 0$.
AXIOM 2	$P(\mathscr{S}) = 1$.
AXIOM 3	**a.** If $A_1, A_2, \ldots, A_k$ is a finite collection of mutually exclusive events, then

$$P(A_1 \cup A_2 \cup \cdots \cup A_k) = \sum_{i=1}^{k} P(A_i)$$

b. If $A_1, A_2, A_3, \ldots$ is an infinite collection of mutually exclusive events, then

$$P(A_1 \cup A_2 \cup A_3 \cup \cdots) = \sum_{i=1}^{\infty} P(A_i)$$

Axiom 1 reflects the intuitive notion that the chance of A occurring should be at least 0, so that negative probabilities are not allowed. The sample space is by definition an event that must occur when the experiment is performed ($\mathscr{S}$ contains all possible outcomes), so Axiom 2 says that the maximum possible probability of 1 is assigned to $\mathscr{S}$.

The third axiom formalizes the idea that if we wish the probability that at least one of a number of events will occur and no two of the events can occur simultaneously, then the chance of at least one occurring is the sum of the chances of the individual events.

Example 2.11 In the experiment in which a single coin is tossed, the sample space is $\mathscr{S} = \{H, T\}$. The axioms specify $P(\mathscr{S}) = 1$, so to complete the probability assignment, it remains only to determine $P(H)$ and $P(T)$. Since H and T are disjoint events and $H \cup T = \mathscr{S}$, Axiom 3 implies that

$$1 = P(\mathscr{S}) = P(H) + P(T)$$

This implies that $P(T) = 1 - P(H)$. Thus, the only freedom allowed by the axioms in this experiment is the probability assigned to H. One possible assignment of probabilities is $P(H) = .5$, $P(T) = .5$, whereas another possible assignment is $P(H) = .75$, $P(T) = .25$. In fact, letting p represent any fixed number between 0 and 1, $P(H) = p$ and $P(T) = 1 - p$ is an assignment consistent with the axioms. ∎

Example 2.12 Consider the experiment in Example 2.4, in which batteries coming off an assembly line are tested one by one until one having a voltage within prescribed limits is found. The simple events are $E_1 = \{S\}$, $E_2 = \{FS\}$, $E_3 = \{FFS\}$, $E_4 = \{FFFS\}$, ... Suppose the probability of any particular battery being satisfactory is .99. Then it can be shown that $P(E_1) = .99$, $P(E_2) = (.01)(.99)$, $P(E_3) = (.01)^2(.99)$, ... is an assignment of probabilities to the simple events that satisfies the axioms. In particular, because the E_i's are disjoint and $\mathscr{S} = E_1 \cup E_2 \cup E_3 \cup \dots$, it must be the case that

$$1 = P(\mathscr{S}) = P(E_1) + P(E_2) + P(E_3) + \cdots$$
$$= .99[1 + .01 + (.01)^2 + (.01)^3 + \cdots]$$

The validity of this equality is a consequence of a mathematical result concerning the sum of a geometric series.

However, another legitimate (according to the axioms) probability assignment of the same "geometric" type is obtained by replacing .99 by any other number p between 0 and 1 (and .01 by $1 - p$). Thus, there are an infinite number of legitimate probability assignments of this type. ∎

Interpreting Probability

Examples 2.11 and 2.12 show that the axioms do not completely determine an assignment of probabilities to events. The axioms serve only to rule out assignments inconsistent with our intuitive notions of probability. In the coin-tossing experiment of Example 2.11, two particular assignments were suggested. The appropriate or correct assignment depends on the manner in which the experiment is carried out and also on one's interpretation of probability. The interpretation that is most frequently used and most easily understood is based on the notion of relative frequencies.

Consider an experiment that can be repeatedly performed in an identical and independent fashion, and let A be an event consisting of a fixed set of outcomes of the experiment. Simple examples of such repeatable experiments include the coin-tossing and die-tossing experiments previously discussed. If the experiment is performed n times, on some of the replications the event A will occur (the outcome will be in the set A), and on others, A will not occur. Let $n(A)$ denote the number of replications on which A does occur. Then the ratio $n(A)/n$ is called the *relative frequency* of occurrence of the event A in the sequence of n replications. Empirical evidence, based on the results of many of these sequences of repeatable experiments, indicates that as n grows large, the relative frequency $n(A)/n$ stabilizes, as pictured in Figure 2.2. That is, as n gets arbitrarily large, the relative frequency approaches a limiting value we refer to as the *limiting relative frequency* of the event A. The objective interpretation of probability identifies this limiting relative frequency with $P(A)$.

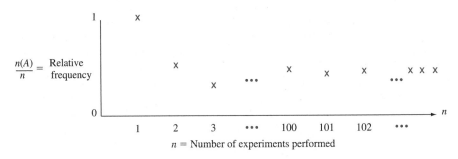

Figure 2.2   Stabilization of relative frequency

   If probabilities are assigned to events in accordance with their limiting relative frequencies, then we can interpret a statement such as "The probability of that coin landing with the head facing up when it is tossed is .5" to mean that in a large number of such tosses, a head will appear on approximately half the tosses and a tail on the other half.

   This relative frequency interpretation of probability is referred to as an objective interpretation because it rests on a property of the experiment rather than on any particular individual concerned with the experiment. For example, two different observers of a sequence of coin tosses should both use the same probability assignments since the observers have nothing to do with limiting relative frequency. In practice, this interpretation is not as objective as it might seem, since the limiting relative frequency of an event will not be known. Thus, we will have to assign probabilities based on our beliefs about the limiting relative frequency of events under study. Fortunately, there are many experiments for which there will be a consensus with respect to probability assignments. When we speak of a fair coin, we shall mean $P(H) = P(T) = .5$, and a fair die is one for which limiting relative frequencies of the six outcomes are all $\frac{1}{6}$, suggesting probability assignments $P(\{1\}) = \cdots = P(\{6\}) = \frac{1}{6}$.

   Because the objective interpretation of probability is based on the notion of limiting frequency, its applicability is limited to experimental situations that are repeatable. Yet the language of probability is often used in connection with situations that are inherently unrepeatable. Examples include: "The chances are good for a peace agreement"; "It is likely that our company will be awarded the contract"; and "Because their best quarterback is injured, I expect them to score no more than 10 points against us." In such situations we would like, as before, to assign numerical probabilities to various outcomes and events (e.g., the probability is .9 that we will get the contract). We must therefore adopt an alternative interpretation of these probabilities. Because different observers may have different prior information and opinions concerning such experimental situations, probability assignments may now differ from individual to individual. Interpretations in such situations are thus referred to as *subjective*. The book by Robert Winkler listed in the chapter references gives a very readable survey of several subjective interpretations.

Properties of Probability

PROPOSITION

> For any event A, $P(A) = 1 - P(A')$.

Proof
In Axiom 3a, let $k = 2$, $A_1 = A$, and $A_2 = A'$. Since by definition of A', $A \cup A' = \mathcal{S}$ while A and A' are disjoint, $1 = P(\mathcal{S}) = P(A \cup A') = P(A) + P(A')$, from which the desired result follows. ∎

This proposition is surprisingly useful because there are many situations in which $P(A')$ is more easily obtained by direct methods than is $P(A)$.

Example 2.13 Consider a system of five identical components connected in series, as illustrated in Figure 2.3.

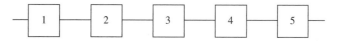

Figure 2.3   A system of five components connected in series

Denote a component that fails by F and one that doesn't fail by S (for success). Let A be the event that the *system* fails. For A to occur, at least one of the individual components must fail. Outcomes in A include $SSFSS$ (1, 2, 4, and 5 all work, but 3 does not), $FFSSS$, and so on. There are in fact 31 different outcomes in A. However, A', the event that the system works, consists of the single outcome $SSSSS$. We will see in Section 2.5 that if 90% of all these components do not fail and different components fail independently of one another, then $P(A') = P(SSSSS) = .9^5 = .59$. Thus, $P(A) = 1 - .59 = .41$; so among a large number of such systems, roughly 41% will fail. ∎

In general, the foregoing proposition is useful when the event of interest can be expressed as "at least . . . ," since then the complement "less than . . ." is often easier to work with (in some problems, "more than . . ." is easier to deal with than "at most . . ."). When you are having difficulty calculating $P(A)$ directly, think of determining $P(A')$.

PROPOSITION

> If A and B are mutually exclusive, then $P(A \cap B) = 0$.

Proof

Because $A \cap B$ contains no outcomes, $(A \cap B)' = \mathscr{S}$. Thus, we have that $1 = P[A \cap B)'] = 1 - P(A \cap B)$, which implies $P(A \cap B) = 1 - 1 = 0$.   ■

When events A and B are mutually exclusive, Axiom 3 gives $P(A \cup B) = P(A) + P(B)$. When A and B are not mutually exclusive, the probability of the union is obtained from the following result.

PROPOSITION

For any two events A and B,

$$P(A \cup B) = P(A) + P(B) - P(A \cap B)$$

Notice that the proposition is valid even if A and B are mutually exclusive, since then $P(A \cap B) = 0$. The key idea is that, in adding $P(A)$ and $P(B)$, the probability of the intersection $A \cap B$ is actually counted twice, so $P(A \cap B)$ must be subtracted out.

Proof

Note first that $A \cup B = A \cup (B \cap A')$, as illustrated in Figure 2.4. Since A and $(B \cap A')$ are mutually exclusive, $P(A \cup B) = P(A) + P(B \cap A')$. But $B = (B \cap A) \cup (B \cap A')$ (the union of that part of B in A and that part of B not in A), with $(B \cap A)$ and $(B \cap A')$ mutually exclusive, so that $P(B) = P(B \cap A) + P(B \cap A')$. Combining these results gives

$$P(A \cup B) = P(A) + P(B \cap A') = P(A) + [P(B) - P(A \cap B)]$$
$$= P(A) + P(B) - P(A \cap B)$$

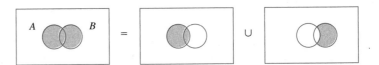

Figure 2.4 $A \cup B = A \cup (B \cap A')$ ■

Example 2.14   In a certain residential suburb, 60% of all households subscribe to the metropolitan newspaper published in a nearby city, 80% subscribe to the local afternoon paper, and 50% of all households subscribe to both papers. If a household is selected at random, what is the probability that it subscribes to (1) at least one of the two newspapers and (2) exactly one of the two newspapers?

With $A = \{$subscribes to the metropolitan paper$\}$ and $B = \{$subscribes to the local paper$\}$, the given information implies that $P(A) = .6$, $P(B) = .8$, and $P(A \cap B) = .5$. The previous proposition then applies to give

$P($subscribes to at least one of the two newspapers$)$

$$= P(A \cup B) = P(A) + P(B) - P(A \cap B) = .6 + .8 - .5 = .9$$

The event that a household subscribes only to the local paper can be written as $A' \cap B$ [(not metropolitan) and local]. Now Figure 2.4 implies that

$$.9 = P(A \cup B) = P(A) + P(A' \cap B) = .6 + P(A' \cap B)$$

from which $P(A' \cap B) = .3$. Similarly, $P(A \cap B') = P(A \cup B) - P(B) = .1$. This is all illustrated in Figure 2.5, from which we see that

$$P(\text{exactly one}) = P(A \cap B') + P(A' \cap B) = .1 + .3 = .4$$

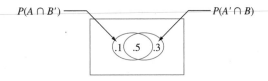

Figure 2.5 Probabilities for Example 2.14 ∎

The probability of a union of more than two events can be computed analogously. For three events A, B, and C, the result is

$$P(A \cup B \cup C) = P(A) + P(B) + P(C) - P(A \cap B) - P(A \cap C)$$
$$- P(B \cap C) + P(A \cap B \cap C)$$

This can be seen by examining a Venn diagram of $A \cup B \cup C$, which is shown in Figure 2.6. When $P(A)$, $P(B)$, and $P(C)$ are added, certain intersections are counted twice, so they must be subtracted out, but this results in $P(A \cap B \cap C)$ being subtracted once too often.

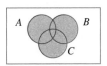

Figure 2.6 $A \cup B \cup C$

Determining Probabilities Systematically

When the number of possible outcomes (simple events) is large, there will be many compound events. A simple way to determine probabilities for these events that avoids violating the axioms and derived properties is to first determine probabilities $P(E_i)$ for all simple events. These should satisfy $P(E_i) \geq 0$ and $\sum_{\text{all } i} P(E_i) = 1$. Then the probability of any compound event A is computed by adding together the $P(E_i)$'s for all E_i's in A:

$$P(A) = \sum_{\text{all } E_i\text{'s in } A} P(E_i)$$

Example 2.15 Denote the six elementary events $\{1\}, \ldots, \{6\}$ associated with tossing a six-sided die once by $E_1, \ldots, E_6$. If the die is constructed so that any of the three even outcomes is twice as likely to occur as any of the three odd outcomes, then an appropriate assignment of probabilities to elementary events is $P(E_1) = P(E_3) = P(E_5) = \frac{1}{9}$, $P(E_2) = P(E_4) = P(E_6) = \frac{2}{9}$. Then for the event $A = \{\text{outcome is even}\} = E_2 \cup E_4 \cup E_6$, $P(A) = P(E_2) + P(E_4) + P(E_6) = \frac{6}{9} = \frac{2}{3}$; for $B = \{\text{outcome} \leq 3\} = E_1 \cup E_2 \cup E_3$, $P(B) = \frac{1}{9} + \frac{2}{9} + \frac{1}{9} = \frac{4}{9}$. ∎

Equally Likely Outcomes

In many experiments consisting of N outcomes, it is reasonable to assign equal probabilities to all N simple events. These include such obvious examples as tossing a fair coin or fair die once or twice (or any fixed number of times), or selecting one or several cards from a well-shuffled deck of 52. With $p = P(E_i)$ for every i,

$$1 = \sum_{i=1}^{N} P(E_i) = \sum_{i=1}^{N} p = p \cdot N \qquad \text{so } p = \frac{1}{N}$$

That is, if there are N possible outcomes, then the probability assigned to each is $1/N$.

Now consider an event A, with $N(A)$ denoting the number of outcomes contained in A. Then

$$P(A) = \sum_{E_i \text{ in } A} P(E_i) = \sum_{E_i \text{ in } A} \frac{1}{N} = \frac{N(A)}{N}$$

Once we have counted the number N of outcomes in the sample space, to compute the probability of any event we must count the number of outcomes contained in that event and take the ratio of the two numbers. Thus, when outcomes are equally likely, computing probabilities reduces to counting.

Example 2.16 When two dice are rolled separately, there are $N = 36$ outcomes (delete the first row and column from the table in Example 2.3). If both the dice are fair, all 36 outcomes are equally likely, so $P(E_i) = \frac{1}{36}$. Then the event $A = \{\text{sum of two numbers} = 7\}$ consists of the six outcomes (1, 6), (2, 5), (3, 4), (4, 3), (5, 2), and (6, 1), so

$$P(A) = \frac{N(A)}{N} = \frac{6}{36} = \frac{1}{6}$$ ∎

Exercises | Section 2.2 (11–28)

11. A mutual fund company offers its customers several different funds: a money-market fund, three different bond funds (short, intermediate, and long-term), two stock funds (moderate and high-risk), and a balanced fund. Among customers who own shares in just one fund, the percentages of customers in the different funds are as follows:

Money-market	20%	High-risk stock	18%
Short bond	15%	Moderate-risk	
Intermediate		stock	25%
bond	10%	Balanced	7%
Long bond	5%		

A customer who owns shares in just one fund is randomly selected.

a. What is the probability that the selected individual owns shares in the balanced fund?

b. What is the probability that the individual owns shares in a bond fund?

c. What is the probability that the selected individual does not own shares in a stock fund?

12. Consider randomly selecting a student at a certain university, and let A denote the event that the selected individual has a Visa credit card and B be the analogous event for a MasterCard. Suppose that $P(A) = .5$, $P(B) = .4$, and $P(A \cap B) = .25$.

a. Compute the probability that the selected individual has at least one of the two types of cards (that is, the probability of the event $A \cup B$).

b. What is the probability that the selected individual has neither type of card?

c. Describe, in terms of A and B, the event that the selected student has a Visa card but not a MasterCard, and then calculate the probability of this event.

13. A computer consulting firm presently has bids out on three projects. Let $A_i = \{\text{awarded project } i\}$, for $i = 1, 2, 3$, and suppose that $P(A_1) = .22$, $P(A_2) = .25$, $P(A_3) = .28$, $P(A_1 \cap A_2) = .11$, $P(A_1 \cap A_3) = .05$, $P(A_2 \cap A_3) = .07$, $P(A_1 \cap A_2 \cap A_3) = .01$. Express in words each of the following events, and compute the probability of each event.

a. $A_1 \cup A_2$

b. $A_1' \cap A_2'$ [*Hint:* $(A_1 \cup A_2)' = A_1' \cap A_2'$]

c. $A_1 \cup A_2 \cup A_3$ **d.** $A_1' \cap A_2' \cap A_3'$

e. $A_1' \cap A_2' \cap A_3$ **f.** $(A_1' \cap A_2') \cup A_3$

14. A utility company offers a lifeline rate to any household whose electricity usage falls below 240 kWh during a particular month. Let A denote the event that a randomly selected household in a certain community does not exceed the lifeline usage during January, and let B be the analogous event for the month of July (A and B refer to the same household). Suppose $P(A) = .8$, $P(B) = .7$, and $P(A \cup B) = .9$. Compute the following:

a. $P(A \cap B)$.

b. The probability that the lifeline usage amount is exceeded in exactly one of the two months. Describe this event in terms of A and B.

15. Consider the type of clothes dryer (gas or electric) purchased by each of five different customers at a certain store.

a. If the probability that at most one of these purchases an electric dryer is .087, what is the probability that at least two purchase an electric dryer?

b. If $P(\text{all five purchase gas}) = .0768$ and $P(\text{all five purchase electric}) = .0102$, what is the probability that at least one of each type is purchased?

16. An individual is presented with three different glasses of cola, labeled C, D, and P. He is asked to taste all three and then list them in order of preference. Suppose the same cola has actually been put into all three glasses.

a. What are the simple events in this ranking experiment, and what probability would you assign to each one?

b. What is the probability that C is ranked first?

c. What is the probability that C is ranked first and D is ranked last?

17. Let A denote the event that the next item checked out at a public library is a nonfiction book, and let B be the event that the next item checked out is a work of fiction. Suppose that $P(A) = .35$ and $P(B) = .50$.

a. Why is it not the case that $P(A) + P(B) = 1$?

b. Calculate $P(A')$.

c. Calculate $P(A \cup B)$.

d. Calculate $P(A' \cap B')$.

18. A box contains four 40-W bulbs, five 60-W bulbs, and six 75-W bulbs. If bulbs are selected one by one in random order, what is the probability that at least two bulbs must be selected to obtain one that is rated 75 W?

19. Human visual inspection of solder joints on printed circuit boards can be very subjective. Part of the problem stems from the numerous types of solder defects (e.g., pad nonwetting, knee visibility, voids) and even the degree to which a joint possesses one or more of these defects. Consequently, even highly trained inspectors can disagree on the disposition of a particular joint. In one batch of 10,000 joints, inspector A found 724 that were judged defective, inspector B found 751 such joints, and 1159 of the joints were judged defective by at least one of the inspectors. Suppose that one of the 10,000 joints is randomly selected.

a. What is the probability that the selected joint was judged to be defective by neither of the two inspectors?

b. What is the probability that the selected joint was judged to be defective by inspector B but not by inspector A?

20. A large company offers its employees two different health insurance plans and two different dental insurance plans. Plan 1 of each type is relatively inexpensive, but restricts the choice of providers, whereas plan 2 is more expensive but more flexible. The accompanying table gives the percentages of employees who have chosen the various plans:

Health Plan	Dental Plan	
	1	2
1	25%	16%
2	22%	37%

Suppose that an employee is randomly selected and both the health plan and dental plan chosen by the selected employee are determined.

a. What are the four simple events?

b. What is the probability that the selected employee has chosen the more restrictive plan of each type?

c. What is the probability that the employee has chosen the more flexible dental plan?

21. An insurance company offers four different deductible levels—none, low, medium, and high—for its homeowner's policyholders and three different levels—low, medium, and high—for its automobile policyholders. The accompanying table gives proportions for the various categories of policyholders who have both types of insurance. For example, the proportion of individuals with both low homeowner's deductible and low auto deductible is .06 (6% of all such individuals).

Auto	Homeowner's			
	N	L	M	H
L	.04	.06	.05	.03
M	.07	.10	.20	.10
H	.02	.03	.15	.15

Suppose an individual having both types of policies is randomly selected.

a. What is the probability that the individual has a medium auto deductible and a high homeowner's deductible?

b. What is the probability that the individual has a low auto deductible? A low homeowner's deductible?

c. What is the probability that the individual is in the same category for both auto and homeowner's deductibles?

d. Based on your answer in part (c), what is the probability that the two categories are different?

e. What is the probability that the individual has at least one low deductible level?

f. Using the answer in part (e), what is the probability that neither deductible level is low?

22. The route used by a certain motorist in commuting to work contains two intersections with traffic signals. The probability that he must stop at the first signal is .4, the analogous probability for the second signal is .5, and the probability that he must stop at at least one of the two signals is .6. What is the probability that he must stop

a. At both signals?

b. At the first signal but not at the second one?

c. At exactly one signal?

23. A library has five copies of a certain text, of which copies 1 and 2 are first printings and copies 3, 4, and 5 are second printings. Two copies are to be randomly selected to be placed on 2-hour reserve (implying 10 equally likely outcomes).

a. What is the probability that both selected copies are first printings?

b. What is the probability that both selected copies are second printings?

c. What is the probability that at least one selected copy is a first printing?

d. What is the probability that the selected copies are different printings?

24. Use the axioms to show that if one event A is contained in another event B (i.e., A is a subset of B), then $P(A) \le P(B)$. [*Hint:* For such A and B, A and $B \cap A'$ are disjoint and $B = A \cup (B \cap A')$, as can be seen from a Venn diagram.] For general A and B, what does this imply about the relationship among $P(A \cap B)$, $P(A)$, and $P(A \cup B)$?

25. The three major options on a certain type of new car are an automatic transmission (A), a sunroof (B), and a stereo with compact disc player (C). If 70% of all purchasers request A, 80% request B, 75% request C, 85% request A or B, 90% request A or C, 95% request B or C, and 98% request A or B or C, compute the probabilities of the following events. [*Hint:* "A or B" is the event that at least one of the two options is requested; try drawing a Venn diagram and labeling all regions.]

a. The next purchaser will request at least one of the three options.

b. The next purchaser will select none of the three options.

c. The next purchaser will request only an automatic transmission and not either of the other two options.

d. The next purchaser will select exactly one of these three options.

26. A certain system can experience three different types of defects. Let A_i ($i = 1, 2, 3$) denote the event that the system has a defect of type i. Suppose that

$P(A_1) = .12 \quad P(A_2) = .07 \quad P(A_3) = .05$
$P(A_1 \cup A_2) = .13 \quad P(A_1 \cup A_3) = .14$
$P(A_2 \cup A_3) = .10 \quad P(A_1 \cap A_2 \cap A_3) = .01$

  a. What is the probability that the system does not have a type 1 defect?

  b. What is the probability that the system has both type 1 and type 2 defects?

  c. What is the probability that the system has both type 1 and type 2 defects but not a type 3 defect?

  d. What is the probability that the system has at most two of these defects?

27. An academic department with five faculty members—Anderson, Box, Cox, Cramer, and Fisher—must select two of its members to serve on a personnel review committee. Because the work will be time-consuming, no one is anxious to serve, so it is decided that the representative will be selected by putting five slips of paper in a box, mixing them, and selecting two.

  a. What is the probability that both Anderson and Box will be selected? (*Hint:* List the equally likely outcomes.)

  b. What is the probability that at least one of the two members whose name begins with C is selected?

  c. If the five faculty members have taught for 3, 6, 7, 10, and 14 years, respectively, at the university, what is the probability that the two chosen representatives have at least 15 years' teaching experience at the university?

28. In Exercise 5, suppose that any incoming individual is equally likely to be assigned to any of the three stations irrespective of where other individuals have been assigned. What is the probability that

  a. All three family members are assigned to the same station?

  b. At most two family members are assigned to the same station?

  c. Every family member is assigned to a different station?

2.3 | Counting Techniques

When the various outcomes of an experiment are equally likely (the same probability is assigned to each simple event), then the task of computing probabilities reduces to counting. In particular, if N is the number of outcomes in a sample space and $N(A)$ is the number of outcomes contained in an event A, then

$$P(A) = \frac{N(A)}{N} \tag{2.1}$$

If a list of the outcomes is available or easy to construct and N is small, then the numerator and denominator of Equation (2.1) can be obtained without the benefit of any general counting principles.

    There are, however, many experiments for which the effort involved in constructing such a list is prohibitive because N is quite large. By exploiting some general counting rules, it is possible to compute probabilities of the form (2.1) without a listing of outcomes. These rules are also useful in many problems involving outcomes that are not equally likely. Several of the rules developed here will be used in studying probability distributions in the next chapter.

The Product Rule for Ordered Pairs

Our first counting rule applies to any situation in which a set (event) consists of ordered pairs of objects and we wish to count the number of such pairs. By an ordered pair, we mean that, if O_1 and O_2 are objects, then the pair (O_1, O_2) is different from the pair (O_2, O_1). For example, if an individual selects one airline for a trip from Los Angeles to Chicago and (after transacting business in Chicago) a second one for continuing on to New York, one possibility is (American, United), another is (United, American), and still another is (United, United).

PROPOSITION

If the first element or object of an ordered pair can be selected in n_1 ways, and for each of these n_1 ways the second element of the pair can be selected in n_2 ways, then the number of pairs is $n_1 n_2$.

Example 2.17   A homeowner doing some remodeling requires the services of both a plumbing contractor and an electrical contractor. If there are 12 plumbing contractors and 9 electrical contractors available in the area, in how many ways can the contractors be chosen? If we denote the plumbers by $P_1, \ldots, P_{12}$ and the electricians by $Q_1, \ldots, Q_9$, then we wish the number of pairs of the form (P_i, Q_j). With $n_1 = 12$ and $n_2 = 9$, the product rule yields $N = (12)(9) = 108$ possible ways of choosing the two types of contractors.   ∎

In Example 2.17, the choice of the second element of the pair did not depend on which first element was chosen or occurred. As long as there is the same number of choices of the second element for each first element, the product rule is valid even when the set of possible second elements depends on the first element.

Example 2.18   A family has just moved to a new city and requires the services of both an obstetrician and a pediatrician. There are two easily accessible medical clinics, each having two obstetricians and three pediatricians. The family will obtain maximum health insurance benefits by joining a clinic and selecting both doctors from that clinic. In how many ways can this be done? Denote the obstetricians by O_1, O_2, O_3, and O_4 and the pediatricians by $P_1, \ldots, P_6$. Then we wish the number of pairs (O_i, P_j) for which O_i and P_j are associated with the same clinic. Because there are four obstetricians, $n_1 = 4$, and for each there are three choices of pediatrician, so $n_2 = 3$. Applying the product rule gives $N = n_1 n_2 = 12$ possible choices.   ∎

Tree Diagrams

In problems in which the product rule can be applied, a configuration called a *tree diagram* can be used to represent pictorially all the possibilities. The tree diagram associated with Example 2.18 appears in Figure 2.7. Starting from a point on the left side of the diagram, for each possible first element of a pair a straight-line segment emanates rightward. Each of these lines is referred to as a first-generation branch. Now for any given first-generation branch we construct another line segment emanating from the tip of the branch for each possible choice of a second element of the pair. Each such line segment is a second-generation branch. Because there are four obstetricians, there are four first-generation branches, and three pediatricians for each obstetrician yields three second-generation branches emanating from each first-generation branch.

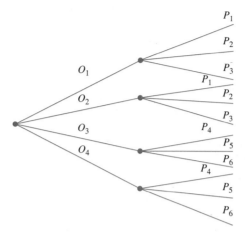

Figure 2.7 Tree diagram for Example 2.18

In the general case, there are n_1 first-generation branches, and for each first-generation branch there are n_2 second-generation branches. The total number of second-generation branches is therefore $n_1 n_2$. Since each second-generation branch corresponds to exactly one possible pair (choosing a first element and then a second puts us at the end of exactly one second-generation branch), there are $n_1 n_2$ pairs, so the product rule is verified.

The construction of a tree diagram does not depend on having the same number of second-generation branches emanating from each first-generation branch. If the second clinic had four pediatricians, then there would be only three branches emanating from two of the first-generation branches and four emanating from each of the other two first-generation branches. A tree diagram can thus be used to represent pictorially experiments other than those to which the product rule applies.

A More General Product Rule

If a six-sided die is tossed five times in succession rather than just twice, then each possible outcome is an ordered collection of five numbers such as (1, 3, 1, 2, 4) or (6, 5, 2, 2, 2). We will call an ordered collection of k objects a *k-tuple* (so a pair is a 2-tuple and a triple is a 3-tuple). Each outcome of the die-tossing experiment is then a 5-tuple.

Product Rule for *k*-Tuples

Suppose a set consists of ordered collections of k elements (k-tuples) and that there are n_1 possible choices for the first element; for each choice of the first element, there are n_2 possible choices of the second element; . . .; for each possible choice of the first $k - 1$ elements, there are n_k choices of the kth element. Then there are $n_1 n_2 \cdot \cdots \cdot n_k$ possible k-tuples.

This more general rule can also be illustrated by a tree diagram; simply construct a more elaborate diagram by adding third-generation branches emanating from the tip of each second generation, then fourth-generation branches, and so on, until finally kth-generation branches are added.

Example 2.19
(Example 2.17 continued)

Suppose the home remodeling job involves first purchasing several kitchen appliances. They will all be purchased from the same dealer, and there are five dealers in the area. With the dealers denoted by $D_1, \ldots, D_5$, there are $N = n_1 n_2 n_3 = (5)(12)(9) = 540$ 3-tuples of the form (D_i, P_j, Q_k), so there are 540 ways to choose first an appliance dealer, then a plumbing contractor, and finally an electrical contractor. ■

Example 2.20
(Example 2.18 continued)

If each clinic has both three specialists in internal medicine and two general surgeons, there are $n_1 n_2 n_3 n_4 = (4)(3)(3)(2) = 72$ ways to select one doctor of each type such that all doctors practice at the same clinic. ■

Permutations

So far the successive elements of a k-tuple were selected from entirely different sets (e.g., appliance dealers, then plumbers, and finally electricians). In several tosses of a die, the set from which successive elements are chosen is always $\{1, 2, 3, 4, 5, 6\}$, but the choices are made "with replacement" so that the same element can appear more than once. We now consider a fixed set consisting of n distinct elements and suppose that a k-tuple is formed by selecting successively from this set *without replacement* so that an element can appear in at most one of the k positions.

DEFINITION

> Any ordered sequence of k objects taken from a set of n distinct objects is called a **permutation** of size k of the objects. The number of permutations of size k that can be constructed from the n objects is denoted by $P_{k,n}$.

The number of permutations of size k is obtained immediately from the general product rule. The first element can be chosen in n ways, for each of these n ways the second element can be chosen in $n - 1$ ways, and so on; finally, for each way of choosing the first $k - 1$ elements, the kth element can be chosen in $n - (k - 1) = n - k + 1$ ways, so

$$P_{k,n} = n(n - 1)(n - 2) \cdot \cdots \cdot (n - k + 2)(n - k + 1)$$

Example 2.21

There are eight teaching assistants available for grading papers in a particular course. The first exam consists of four questions, and the professor wishes to select a different assistant to grade each question (only one assistant per question). In how many ways can assistants be chosen to grade the exam? Here $n =$ the number of assistants $= 8$ and $k =$ the number of questions $= 4$. The number of different grading assignments is then $P_{4,8} = (8)(7)(6)(5) = 1680$. ■

The use of factorial notation allows $P_{k,n}$ to be expressed more compactly.

DEFINITION

For any positive integer m, $m!$ is read "m factorial" and is defined by $m! = m(m - 1) \cdot \cdots \cdot (2)(1)$. Also, $0! = 1$.

Using factorial notation yields

$$P_{k,n} = n(n - 1) \cdot \cdots \cdot (n - k + 1)$$

$$= \frac{n(n - 1) \cdot \cdots \cdot (n - k + 1)(n - k)(n - k - 1) \cdot \cdots \cdot (2)(1)}{(n - k)(n - k - 1) \cdot \cdots \cdot (2)(1)}$$

which becomes

$$P_{k,n} = \frac{n!}{(n - k)!}$$

For example, $P_{3,9} = 9!/(9 - 3)! = 9!/6! = 9 \cdot 8 \cdot 7 \cdot 6!/6! = 9 \cdot 8 \cdot 7$. Note also that because $0! = 1$, $P_{n,n} = n!/(n - n)! = n!/0! = n!/1 = n!$, as it should.

Combinations

There are many counting problems in which one is given a set of n distinct objects and wishes to count the number of *unordered* subsets of size k. For example, in bridge it is only the 13 cards in a hand and not the order in which they are dealt that is important; in the formation of a committee, the order in which committee members are listed is frequently unimportant.

DEFINITION

Given a set of n distinct objects, any unordered subset of size k of the objects is called a **combination.** The number of combinations of size k that can be formed from n distinct objects will be denoted by $\binom{n}{k}$. (This notation is more common in probability than $C_{k,n}$, which would be analogous to notation for permutations.)

The number of combinations of size k from a particular set is smaller than the number of permutations because, when order is disregarded, a number of permutations correspond to the same combination. Consider, for example, the set $\{A, B, C, D, E\}$ consisting of five elements. We know that there are $5!/(5 - 3)! = 60$ permutations of size 3. There are six permutations of size 3 consisting of the elements A, B, and C since these three can be ordered $3 \cdot 2 \cdot 1 = 3! = 6$ ways: (A, B, C), (A, C, B), (B, A, C), (B, C, A), (C, A, B), and (C, B, A). These six permutations are equivalent to the single combination $\{A, B, C\}$. Similarly, for any other combination of size 3, there are $3!$ permutations, each obtained by ordering the three objects. Thus,

$$60 = P_{3,5} = \binom{5}{3} \cdot 3!; \quad \text{so} \quad \binom{5}{3} = \frac{60}{3!} = 10$$

These ten combinations are

$$\{A, B, C\} \{A, B, D\} \{A, B, E\} \{A, C, D\} \{A, C, E\} \{A, D, E\}, \{B, C, D\}$$
$$\{B, C, E\} \{B, D, E\} \{C, D, E\}$$

When there are n distinct objects, any permutation of size k is obtained by ordering the k unordered objects of a combination in one of $k!$ ways, so the number of permutations is the product of $k!$ and the number of combinations. This gives

$$\binom{n}{k} = \frac{P_{k,n}}{k!} = \frac{n!}{k!(n-k)!}$$

Notice that $\binom{n}{n} = 1$ and $\binom{n}{0} = 1$ since there is only one way to choose a set of (all) n elements or of no elements, and $\binom{n}{1} = n$ since there are n subsets of size 1.

Example 2.22   A bridge hand consists of any 13 cards selected from a 52-card deck without regard to order. There are $\binom{52}{13} = 52!/13!39!$ different bridge hands, which works out to approximately 635 billion. Since there are 13 cards in each suit, the number of hands consisting entirely of clubs and/or spades (no red cards) is $\binom{26}{13} = 26!/13!13! = 10,400,600$. One of these $\binom{26}{13}$ hands consists entirely of spades, and one consists entirely of clubs, so there are $[\binom{26}{13} - 2]$ hands that consist entirely of clubs and spades with both suits represented in the hand. Suppose a bridge hand is dealt from a well-shuffled deck (i.e., 13 cards are randomly selected from among the 52 possibilities) and let

$A = \{$the hand consists entirely of spades and clubs with both suits represented$\}$

$B = \{$the hand consists of exactly two suits$\}$

The $N = \binom{52}{13}$ possible outcomes are equally likely, so

$$P(A) = \frac{N(A)}{N} = \frac{\binom{26}{13} - 2}{\binom{52}{13}} = .0000164$$

Since there are $\binom{4}{2} = 6$ combinations consisting of two suits, of which spades and clubs is one such combination,

$$P(B) = \frac{6\left[\binom{26}{13} - 2\right]}{\binom{52}{13}} = .0000983$$

That is, a hand consisting entirely of cards from exactly two of the four suits will occur roughly once in every 10,000 hands. If you play bridge only once a month, it is likely that you will never be dealt such a hand. ■

Example 2.23   A rental car service facility has 10 foreign cars and 15 domestic cars waiting to be serviced on a particular Saturday morning. Because there are so few mechanics working on Saturday, only 6 can be serviced. If the 6 are chosen at random, what is the probability that 3 of the cars selected are domestic and the other 3 are foreign?

Let $D_3 = \{$exactly 3 of the 6 cars chosen are domestic$\}$. Assuming that any particular set of 6 cars is as likely to be chosen as is any other set of 6, we have equally likely outcomes, so $P(D_3) = N(D_3)/N$, where N is the number of ways of choosing 6 cars from the 25 and $N(D_3)$ is the number of ways of choosing 3 domestic cars and 3 foreign cars. Thus $N = \binom{25}{6}$. To obtain $N(D_3)$, think of first choosing 3 of the 15 domestic cars and then 3 of the foreign cars. There are $\binom{15}{3}$ ways of choosing the 3 domestic cars, and there are $\binom{10}{3}$ ways of choosing the 3 foreign cars; $N(D_3)$ is now the product of these two numbers (visualize a tree diagram—we are really using a product rule argument here), so

$$P(D_3) = \frac{N(D_3)}{N} = \frac{\binom{15}{3}\binom{10}{3}}{\binom{25}{6}} = \frac{\frac{15!}{3!12!} \cdot \frac{10!}{3!7!}}{\frac{25!}{6!19!}} = .3083$$

Let $D_4 = \{$exactly 4 of the 6 cars chosen are domestic$\}$ and define D_5 and D_6 in an analogous manner. Then the probability that at least 3 domestic cars are selected is

$$P(D_3 \cup D_4 \cup D_5 \cup D_6) = P(D_3) + P(D_4) + P(D_5) + P(D_6)$$
$$= \frac{\binom{15}{3}\binom{10}{3}}{\binom{25}{6}} + \frac{\binom{15}{4}\binom{10}{2}}{\binom{25}{6}} + \frac{\binom{15}{5}\binom{10}{1}}{\binom{25}{6}} + \frac{\binom{15}{6}\binom{10}{0}}{\binom{25}{6}} = .8530$$

This is also the probability that at most 3 foreign cars are selected. ■

Exercises | Section 2.3 (29–44)

29. The Student Engineers Council at a certain college has one student representative from each of the five engineering majors (civil, electrical, industrial, materials, and mechanical). In how many ways can
   a. Both a council president and a vice president be selected?
   b. A president, a vice president, and a secretary be selected?
   c. Two members be selected for the President's Council?

30. A real estate agent is showing homes to a prospective buyer. There are ten homes in the desired price range listed in the area. The buyer has time to visit only three of them.
   a. In how many ways could the three homes be chosen if the order of visiting is considered?
   b. In how many ways could the three homes be chosen if the order is disregarded?
   c. If four of the homes are new and six have previously been occupied and if the three homes to

visit are randomly chosen, what is the probability that all three are new? (The same answer results regardless of whether order is considered.)

31. a. Beethoven wrote 9 symphonies and Mozart wrote 27 piano concertos. If a university radio station announcer wishes to play first a Beethoven symphony and then a Mozart concerto, in how many ways can this be done?
   b. The station manager decides that on each successive night (7 days per week), a Beethoven symphony will be played, followed by a Mozart piano concerto, followed by a Schubert string quartet (of which there are 15). For roughly how many years could this policy be continued before exactly the same program would have to be repeated?

32. A chain of stereo stores is offering a special price on a complete set of components (receiver, compact disc player, speakers, cassette deck). A purchaser is offered a choice of manufacturer for each component:

Receiver: Kenwood, Onkyo, Pioneer, Sony, Sherwood

Compact disc player: Onkyo, Pioneer, Sony, Technics

Speakers: Boston, Infinity, Polk

Cassette deck: Onkyo, Sony, Teac, Technics

A switchboard display in the store allows a customer to hook together any selection of components (consisting of one of each type). Use the product rules to answer the following questions.

a. In how many ways can one component of each type be selected?

b. In how many ways can components be selected if both the receiver and the compact disc player are to be Sony?

c. In how many ways can components be selected if none is to be Sony?

d. In how many ways can a selection be made if at least one Sony component is to be included?

e. If someone flips switches on the selection in a completely random fashion, what is the probability that the system selected contains at least one Sony component? Exactly one Sony component?

33. Shortly after being put into service, some buses manufactured by a certain company have developed cracks on the underside of the main frame. Suppose a particular city has 20 of these buses, and cracks have actually appeared in 8 of them.

a. How many ways are there to select a sample of 5 buses from the 20 for a thorough inspection?

b. In how many ways can a sample of 5 buses contain exactly 4 with visible cracks?

c. If a sample of 5 buses is chosen at random, what is the probability that exactly 4 of the 5 will have visible cracks?

d. If buses are selected as in part (c), what is the probability that at least 4 of those selected will have visible cracks?

34. A production facility employs 20 workers on the day shift, 15 workers on the swing shift, and 10 workers on the graveyard shift. A quality control consultant is to select 6 of these workers for in-depth interviews. Suppose the selection is made in such a way that any particular group of 6 workers has the same chance of being selected as does any other group (drawing 6 slips without replacement from among 45).

a. How many selections result in all 6 workers coming from the day shift? What is the probability that all 6 selected workers will be from the day shift?

b. What is the probability that all 6 selected workers will be from the same shift?

c. What is the probability that at least two different shifts will be represented among the selected workers?

d. What is the probability that at least one of the shifts will be unrepresented in the sample of workers?

35. An academic department with five faculty members narrowed its choice for department head to either candidate *A* or candidate *B*. Each member then voted on a slip of paper for one of the candidates. Suppose there are actually three votes for *A* and two for *B*. If the slips are selected for tallying in random order, what is the probability that *A* remains ahead of *B* throughout the vote count (e.g., this event occurs if the selected ordering is *AABAB*, but not for *ABBAA*)?

36. An experimenter is studying the effects of temperature, pressure, and type of catalyst on yield from a certain chemical reaction. Three different temperatures, four different pressures, and five different catalysts are under consideration.

a. If any particular experimental run involves the use of a single temperature, pressure, and catalyst, how many experimental runs are possible?

b. How many experimental runs are there that involve use of the lowest temperature and two lowest pressures?

37. Refer to Exercise 36 and suppose that five different experimental runs are to be made on the first day of experimentation. If the five are randomly selected from among all the possibilities, so that any group of five has the same probability of selection, what is the probability that a different catalyst is used on each run?

38. A box in a certain supply room contains four 40-W lightbulbs, five 60-W bulbs, and six 75-W bulbs. Suppose that three bulbs are randomly selected.

a. What is the probability that exactly two of the selected bulbs are rated 75 W?

b. What is the probability that all three of the selected bulbs have the same rating?

c. What is the probability that one bulb of each type is selected?

d. Suppose now that bulbs are to be selected one by one until a 75-W bulb is found. What is the probability that it is necessary to examine at least six bulbs?

39. Fifteen telephones have just been received at an authorized service center. Five of these telephones are cellular, five are cordless, and the other five are corded phones. Suppose that these components are randomly allocated the numbers 1, 2, . . . , 15 to establish the order in which they will be serviced.
   a. What is the probability that all the cordless phones are among the first ten to be serviced?
   b. What is the probability that after servicing ten of these phones, phones of only two of the three types remain to be serviced?
   c. What is the probability that two phones of each type are among the first six serviced?

40. Three molecules of type A, three of type B, three of type C, and three of type D are to be linked together to form a chain molecule. One such chain molecule is *ABCDABCDABCD*, and another is *BCDDAAABDBCC*.
   a. How many such chain molecules are there? (*Hint:* If the three A's were distinguishable from one another—A_1, A_2, A_3—and the B's, C's, and D's were also, how many molecules would there be? How is this number reduced when the subscripts are removed from the A's?)
   b. Suppose a chain molecule of the type described is randomly selected. What is the probability that all three molecules of each type end up next to one another (such as in *BBBAAADDDCCC*)?

41. A mathematics professor wishes to schedule an appointment with each of her eight teaching assistants, four men and four women, to discuss her calculus course. Suppose all possible orderings of appointments are equally likely to be selected.
   a. What is the probability that at least one female assistant is among the first three with whom the professor meets?
   b. What is the probability that after the first five appointments she has met with all female assistants?
   c. Suppose the professor has the same eight assistants the following semester and again schedules appointments without regard to the ordering during the first semester. What is the probability that the orderings of appointments are different?

42. Three married couples have purchased theater tickets and are seated in a row consisting of just six seats. If they take their seats in a completely random fashion (random order), what is the probability that Jim and Paula (husband and wife) sit in the two seats on the far left? What is the probability that Jim and Paula end up sitting next to one another? What is the probability that at least one of the wives ends up sitting next to her husband?

43. In five-card poker, a straight consists of five cards with adjacent denominations (e.g., 9 of clubs, 10 of hearts, jack of hearts, queen of spades, and king of clubs). Assuming that aces can be high or low, if you are dealt a five-card hand, what is the probability that it will be a straight with high card 10? What is the probability that it will be a straight? What is the probability that it will be a straight flush (all cards in the same suit)?

44. Show that $\binom{n}{k} = \binom{n}{n-k}$. Give an interpretation involving subsets.

2.4 | Conditional Probability

The probabilities assigned to various events depend on what is known about the experimental situation when the assignment is made. Subsequent to the initial assignment, partial information about or relevant to the outcome of the experiment may become available, and this information may cause us to revise some of our probability assignments. For a particular event A, we have used P(A) to represent the probability assigned to A; we now think of P(A) as the original or unconditional probability of the event A.

In this section, we examine how the information "an event B has occurred" affects the probability assigned to A. For example, A might refer to an individual having a particular disease in the presence of certain symptoms. If a blood test is performed on the individual and the result is negative (B = negative blood test), then the probability of having the disease will change (it should decrease, but not usually to zero, since blood tests are not infallible). We will use the notation $P(A \mid B)$ to represent the **conditional probability of A given that the event B has occurred.**

Example 2.24   Complex components are assembled in a plant that uses two different assembly lines, A and A'. Line A uses older equipment than A', so it is somewhat slower and less reliable. Suppose on a given day line A has assembled 8 components, of which 2 have been identified as defective (B) and 6 as nondefective (B'), whereas A' has produced 1 defective and 9 nondefective components. This information is summarized in the accompanying table.

		Condition	
		B	B'
Line	A	2	6
	A'	1	9

Unaware of this information, the sales manager randomly selects 1 of these 18 components for a demonstration. Prior to the demonstration

$$P(\text{line } A \text{ component selected}) = P(A) = \frac{N(A)}{N} = \frac{8}{18} = .44$$

However, if the chosen component turns out to be defective, then the event B has occurred, so the component must have been 1 of the 3 in the B column of the table. Since these 3 components are equally likely among themselves after B has occurred,

$$P(A \mid B) = \frac{2}{3} = \frac{\frac{2}{18}}{\frac{3}{18}} = \frac{P(A \cap B)}{P(B)} \tag{2.2}$$

In Equation (2.2), the conditional probability is expressed as a ratio of unconditional probabilities: The numerator is the probability of the intersection of the two events, whereas the denominator is the probability of the conditioning event B. A Venn diagram illuminates this relationship (Figure 2.8).

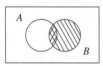

Figure 2.8   Motivating the definition of conditional probability

Given that B has occurred, the relevant sample space is no longer $\mathcal{S}$ but consists of outcomes in B; A has occurred if and only if one of the outcomes in the intersection occurred, so that the conditional probability of A given B is proportional to $P(A \cap B)$. The proportionality constant $1/P(B)$ is used to ensure that the probability $P(B \mid B)$ of the new sample space B equals 1.

The Definition of Conditional Probability

Example 2.24 demonstrates that when outcomes are equally likely, computation of conditional probabilities can be based on intuition. When experiments are more complicated, though, intuition may fail us, so we want to have a general definition of conditional probability that will yield intuitive answers in simple problems. The Venn diagram and Equation (2.2) suggest the appropriate definition.

DEFINITION	For any two events A and B with $P(B) > 0$, the **conditional probability of A given that B has occurred** is defined by $$P(A \mid B) = \frac{P(A \cap B)}{P(B)} \qquad (2.3)$$

Example 2.25 Suppose that of all individuals buying a certain personal computer, 60% include a word processing program in their purchase, 40% include a spreadsheet program, and 30% include both types of programs. Consider randomly selecting a purchaser and let $A = $ {word processing program included} and $B = $ {spreadsheet program included}. Then $P(A) = .60$, $P(B) = .40$, and $P(\text{both included}) = P(A \cap B) = .30$. Given that the selected individual included a spreadsheet program, the probability that a word processing program was also included is

$$P(A \mid B) = \frac{P(A \cap B)}{P(B)} = \frac{.30}{.40} = .75$$

That is, of all those purchasing a spreadsheet program, 75% purchased a word processing program. Similarly,

$$P(\text{spreadsheet} \mid \text{word processing}) = P(B \mid A) = \frac{P(A \cap B)}{P(A)} = \frac{.30}{.60} = .50$$

Notice that $P(A \mid B) \neq P(A)$ and $P(B \mid A) \neq P(B)$. ■

Example 2.26 A news magazine publishes three columns entitled "Art" (A), "Books" (B), and "Cinema" (C). Reading habits of a randomly selected reader with respect to these columns are

Read regularly	A	B	C	$A \cap B$	$A \cap C$	$B \cap C$	$A \cap B \cap C$
Probability	.14	.23	.37	.08	.09	.13	.05

(See Figure 2.9.)

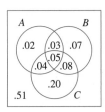

Figure 2.9 Venn diagram for Example 2.26

We thus have

$$P(A \mid B) = \frac{P(A \cap B)}{P(B)} = \frac{.08}{.23} = .348$$

$$P(A \mid B \cup C) = \frac{P(A \cap (B \cup C))}{P(B \cup C)} = \frac{.04 + .05 + .03}{.47} = \frac{.12}{.47} = .255$$

$$P(A \mid \text{reads at least one}) = P(A \mid A \cup B \cup C) = \frac{P(A \cap (A \cup B \cup C))}{P(A \cup B \cup C)}$$

$$= \frac{P(A)}{P(A \cup B \cup C)} = \frac{.14}{.49} = .286$$

and

$$P(A \cup B \mid C) = \frac{P((A \cup B) \cap C)}{P(C)} = \frac{.04 + .05 + .08}{.37} = .459$$ ∎

The Multiplication Rule for $P(A \cap B)$

The definition of conditional probability yields the following result, obtained by multiplying both sides of Equation (2.3) by $P(B)$.

The Multiplication Rule

$$P(A \cap B) = P(A \mid B) \cdot P(B)$$

This rule is important because it is often the case that $P(A \cap B)$ is desired, whereas both $P(B)$ and $P(A \mid B)$ can be specified from the problem description. Consideration of $P(B \mid A)$ gives $P(A \cap B) = P(B \mid A) \cdot P(A)$.

Example 2.27 Four individuals have responded to a request by a blood bank for blood donations. None of them has donated before, so their blood types are unknown. Suppose only type A+ is desired and only one of the four actually has this type. If the potential donors are selected in random order for typing, what is the probability that at least three individuals must be typed to obtain the desired type?

Making the identification $B = \{\text{first type not A+}\}$ and $A = \{\text{second type not A+}\}$, $P(B) = \frac{3}{4}$. Given that the first type is not A+, two of the three individuals left are not A+, so $P(A \mid B) = \frac{2}{3}$. The multiplication rule now gives

$$P(\text{at least three individuals are typed}) = P(A \cap B)$$

$$= P(A \mid B) \cdot P(B)$$

$$= \frac{2}{3} \cdot \frac{3}{4} = \frac{6}{12}$$

$$= .5$$ ∎

The multiplication rule is most useful when the experiment consists of several stages in succession. The conditioning event B then describes the outcome of the first stage and A the outcome of the second, so that $P(A \mid B)$—conditioning on what occurs first—will often be known. The rule is easily extended to experiments involving more than two stages. For example,

$$P(A_1 \cap A_2 \cap A_3) = P(A_3 \mid A_1 \cap A_2) \cdot P(A_1 \cap A_2)$$

$$= P(A_3 \mid A_1 \cap A_2) \cdot P(A_2 \mid A_1) \cdot P(A_1) \qquad (2.4)$$

where A_1 occurs first, followed by A_2, and finally A_3.

Example 2.28 For the blood typing experiment of Example 2.27,

$$P(\text{third type is A+}) = P(\text{third is} \mid \text{first isn't} \cap \text{second isn't})$$
$$\cdot P(\text{second isn't} \mid \text{first isn't}) \cdot P(\text{first isn't})$$
$$= \frac{1}{2} \cdot \frac{2}{3} \cdot \frac{3}{4} = \frac{1}{4} = .25 \qquad ∎$$

When the experiment of interest consists of a sequence of several stages, it is convenient to represent these with a tree diagram. Once we have an appropriate tree diagram, probabilities and conditional probabilities can be entered on the various branches; this will make repeated use of the multiplication rule quite straightforward.

Example 2.29 A chain of video stores sells three different brands of VCRs. Of its VCR sales, 50% are brand 1 (the least expensive), 30% are brand 2, and 20% are brand 3. Each manufacturer offers a 1-year warranty on parts and labor. It is known that 25% of brand 1's VCRs require warranty repair work, whereas the corresponding percentages for brands 2 and 3 are 20% and 10%, respectively.

1. What is the probability that a randomly selected purchaser has bought a brand 1 VCR that will need repair while under warranty?

2. What is the probability that a randomly selected purchaser has a VCR that will need repair while under warranty?

3. If a customer returns to the store with a VCR that needs warranty repair work, what is the probability that it is a brand 1 VCR? A brand 2 VCR? A brand 3 VCR?

The first stage of the problem involves a customer selecting one of the three brands of VCR. Let $A_i = \{\text{brand } i \text{ is purchased}\}$, for $i = 1, 2$, and 3. Then $P(A_1) = .50$, $P(A_2) = .30$, and $P(A_3) = .20$. Once a brand of VCR is selected, the second stage involves observing whether the selected VCR needs warranty repair. With $B = \{\text{needs repair}\}$ and $B' = \{\text{doesn't need repair}\}$, the given information implies that $P(B \mid A_1) = .25$, $P(B \mid A_2) = .20$, and $P(B \mid A_3) = .10$.

The tree diagram representing this experimental situation is shown in Figure 2.10. The initial branches correspond to different brands of VCRs; there are two second-generation branches emanating from the tip of each initial branch, one for "needs repair" and the other for "doesn't need repair." The probability $P(A_i)$ appears on the ith initial branch, whereas the conditional probabilities $P(B \mid A_i)$ and $P(B' \mid A_i)$ appear on the second-generation branches. To the right of each second-generation branch corresponding to the occurrence of B, we display the product of probabilities on the branches leading out to that point. This is simply the multiplication rule in action. The answer to the question posed in 1 is thus $P(A_1 \cap B) = P(B|A_1) \cdot P(A_1) = .125$. The answer to question 2 is

$$P(B) = P[(\text{brand 1 and repair) or (brand 2 and repair) or (brand 3 and repair)}]$$

$$= P(A_1 \cap B) + P(A_2 \cap B) + P(A_3 \cap B)$$

$$= .125 + .060 + .020 = .205$$

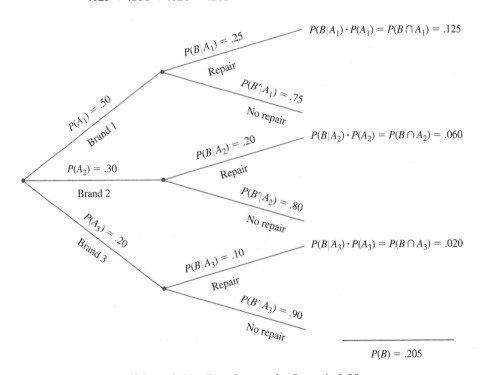

Figure 2.10   Tree diagram for Example 2.29

Finally,

$$P(A_1 | B) = \frac{P(A_1 \cap B)}{P(B)} = \frac{.125}{.205} = .61$$

$$P(A_2 | B) = \frac{P(A_2 \cap B)}{P(B)} = \frac{.060}{.205} = .29$$

and

$$P(A_3 | B) = 1 - P(A_1 | B) - P(A_2 | B) = .10$$

Notice that the initial or *prior probability* of brand 1 is .50, whereas once it is known that the selected VCR needed repair, the *posterior probability* of brand 1 increases to .61. This is because brand 1 VCRs are more likely to need warranty repair than are the other brands. The posterior probability of brand 3 is $P(A_3 | B) = .10$, which is much less than the prior probability $P(A_3) = .20$.   ∎

Bayes' Theorem

The computation of a posterior probability $P(A_j | B)$ from given prior probabilities $P(A_i)$ and conditional probabilities $P(B | A_i)$ occupies a central position in elementary probability. The general rule for such computations, which is really just a simple application of the multiplication rule, goes back to Reverend Thomas Bayes, who lived in the eighteenth century. To state it we first need another result. Recall that events $A_1, \ldots, A_k$ are mutually exclusive if no two have any common outcomes. The events are *exhaustive* if one A_i must occur, so that $A_1 \cup \cdots \cup A_k = \mathcal{S}$.

The Law of Total Probability

Let $A_1, \ldots, A_k$ be mutually exclusive and exhaustive events. Then for any other event B,

$$P(B) = P(B | A_1)P(A_1) + \cdots + P(B | A_k)P(A_k) \qquad (2.5)$$
$$= \sum_{i=1}^{k} P(B | A_i)P(A_i)$$

Proof

Because the A_i's are mutually exclusive and exhaustive, if B occurs it must be in conjunction with exactly one of the A_i's. That is, $B = (A_1$ and $B)$ or $\ldots$ or $(A_k$ and $B) = (A_1 \cap B) \cup \cdots \cup (A_k \cap B)$, where the events $(A_i \cap B)$ are mutually exclusive. This "partitioning of B" is illustrated in Figure 2.11. Thus,

$$P(B) = \sum_{i=1}^{k} P(A_i \cap B) = \sum_{i=1}^{k} P(B | A_i)P(A_i)$$

as desired.

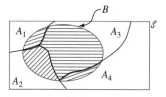

Figure 2.11   Partition of B by mutually exclusive and exhaustive A_i's   ∎

An example of the use of Equation (2.5) appeared in answering Question 1 of Example 2.29, where $A_1 = \{$brand 1$\}$, $A_2 = \{$brand 2$\}$, $A_3 = \{$brand 3$\}$, and $B = \{$repair$\}$.

Bayes' Theorem

Let $A_1, A_2, \ldots, A_k$ be a collection of k mutually exclusive and exhaustive events with $P(A_i) > 0$ for $i = 1, \ldots, k$. Then for any other event B for which $P(B) > 0$

$$P(A_j | B) = \frac{P(A_j \cap B)}{P(B)} = \frac{P(B | A_j)P(A_j)}{\sum_{i=1}^{k} P(B | A_i) \cdot P(A_i)} \qquad j = 1, \ldots, k \qquad (2.6)$$

The transition from the second to the third expression in (2.6) rests on using the multiplication rule in the numerator and the law of total probability in the denominator.

The proliferation of events and subscripts in (2.6) can be a bit intimidating to probability newcomers. As long as there are relatively few events in the partition, a tree diagram (as in Example 2.29) can be used as a basis for calculating posterior probabilities without ever referring explicitly to Bayes' theorem.

Example 2.30 *Incidence of a rare disease.* Only 1 in 1000 adults is afflicted with a rare disease for which a diagnostic test has been developed. The test is such that, when an individual actually has the disease, a positive result will occur 99% of the time, whereas an individual without the disease will show a positive test result only 2% of the time. If a randomly selected individual is tested and the result is positive, what is the probability that the individual has the disease?

To use Bayes' theorem, let $A_1 = \{$individual has the disease$\}$, $A_2 = \{$individual does not have the disease$\}$, and $B = \{$positive test result$\}$. Then $P(A_1) = .001$, $P(A_2) = .999$, $P(B | A_1) = .99$, and $P(B | A_2) = .02$. The tree diagram for this problem is in Figure 2.12.

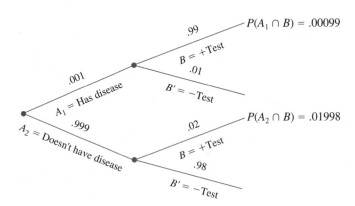

Figure 2.12 Tree diagram for the rare-disease problem

Next to each branch corresponding to a positive test result, the multiplication rule yields the recorded probabilities. Therefore, $P(B) = .00099 + .01998 = .02097$, from which we have

$$P(A_1 \mid B) = \frac{P(A_1 \cap B)}{P(B)} = \frac{.00099}{.02097} = .047$$

This result seems counterintuitive; the diagnostic test appears so accurate we expect someone with a positive test result to be highly likely to have the disease, whereas the computed conditional probability is only .047. The reason for this seemingly paradoxical result is that, because the disease is rare and the test only moderately reliable, most positive test results arise from errors rather than from diseased individuals. The probability of having the disease has increased by a multiplicative factor of 47 (from prior .001 to posterior .047); but to get a further increase in the posterior probability, a diagnostic test with much smaller error rates is needed. If the disease were not so rare (e.g., 25% incidence in the population), then the error rates for the present test would provide good diagnoses. ■

Exercises | Section 2.4 (45–67)

45. A certain sports car comes equipped with either an automatic or a manual transmission, and the car is available in one of four colors. Relevant probabilities for various combinations of transmission type and color are given in the accompanying table.

		Color		
		White	**Blue**	**Black**	**Red**
Transmission	**A**	.15	.10	.10	.10
Type	**M**	.15	.05	.15	.20

Let A = {automatic transmission}, B = {black}, and C = {white}.
  a. Calculate $P(A)$, $P(B)$, and $P(A \cap B)$.
  b. Calculate both $P(A \mid B)$ and $P(B \mid A)$, and explain in context what each of these probabilities represents.
  c. Calculate and interpret $P(A \mid C)$ and $P(A \mid C')$.

46. Suppose an individual is randomly selected from the population of all adult males living in the United States. Let A be the event that the selected individual is over 6 ft in height, and let B be the event that the selected individual is a professional basketball player. Which do you think is larger, $P(A \mid B)$ or $P(B \mid A)$? Why?

47. Return to the credit card scenario of Exercise 12 (Section 2.2), where A = {Visa}, B = {Master-Card}, $P(A)$ = .5, $P(B)$ = .4, and $P(A \cap B)$ = .25.

Calculate and interpret each of the following probabilities (a Venn diagram might help).
  a. $P(B \mid A)$ **b.** $P(B' \mid A)$
  c. $P(A \mid B)$ **d.** $P(A' \mid B)$
  e. Given that the selected individual has at least one card, what is the probability that he or she has a Visa card?

48. Reconsider the system defect situation described in Exercise 26 (Section 2.2).
  a. Given that the system has a type 1 defect, what is the probability that it has a type 2 defect?
  b. Given that the system has a type 1 defect, what is the probability that it has all three types of defects?
  c. Given that the system has at least one type of defect, what is the probability that it has exactly one type of defect?
  d. Given that the system has both of the first two types of defects, what is the probability that it does not have the third type of defect?

49. If two bulbs are randomly selected from the box of lightbulbs described in Exercise 38 (Section 2.3) and at least one of them is found to be rated 75 W, what is the probability that both of them are 75-W bulbs? Given that at least one of the two selected is not rated 75 W, what is the probability that both selected bulbs have the same rating?

50. A department store sells sport shirts in three sizes (small, medium, and large), three patterns (plaid, print, and stripe), and two sleeve lengths (long and short). The accompanying tables give the proportions of shirts sold in the various category combinations.

Short-sleeved

Size	Pattern		
	Pl	Pr	St
S	.04	.02	.05
M	.08	.07	.12
L	.03	.07	.08

Long-sleeved

Size	Pattern		
	Pl	Pr	St
S	.03	.02	.03
M	.10	.05	.07
L	.04	.02	.08

a. What is the probability that the next shirt sold is a medium, long-sleeved, print shirt?

b. What is the probability that the next shirt sold is a medium print shirt?

c. What is the probability that the next shirt sold is a short-sleeved shirt? A long-sleeved shirt?

d. What is the probability that the size of the next shirt sold is medium? That the pattern of the next shirt sold is a print?

e. Given that the shirt just sold was a short-sleeved plaid, what is the probability that its size was medium?

f. Given that the shirt just sold was a medium plaid, what is the probability that it was short-sleeved? Long-sleeved?

51. One box contains six red balls and four green balls, and a second box contains seven red balls and three green balls. A ball is randomly chosen from the first box and placed in the second box. Then a ball is randomly selected from the second box and placed in the first box.

a. What is the probability that a red ball is selected from the first box and a red ball is selected from the second box?

b. At the conclusion of the selection process, what is the probability that the numbers of red and green balls in the first box are identical to the numbers at the beginning?

52. A system consists of two identical pumps, #1 and #2. If one pump fails, the system will still operate. However, because of the added strain, the extra remaining pump is now more likely to fail than was originally the case. That is, $r = P(\#2 \text{ fails} \mid \#1 \text{ fails}) > P(\#2 \text{ fails}) = q$. If at least one pump fails by the end of the pump design life in 7% of all systems and both pumps fail during that period in only 1%, what is the probability that pump #1 will fail during the pump design life?

53. A certain shop repairs both audio and video components. Let A denote the event that the next component brought in for repair is an audio component, and let B be the event that the next component is a compact disc player (so the event B is contained in A). Suppose that $P(A) = .6$ and $P(B) = .05$. What is $P(B \mid A)$?

54. In Exercise 13, $A_i = \{$awarded project $i\}$, for $i = 1$, 2, 3. Use the probabilities given there to compute the following probabilities:
a. $P(A_2 \mid A_1)$
b. $P(A_2 \cap A_3 \mid A_1)$
c. $P(A_2 \cup A_3 \mid A_1)$
d. $P(A_1 \cap A_2 \cap A_3 \mid A_1 \cup A_2 \cup A_3)$. Express in words the probability you have calculated.

55. In Exercise 42, six people (three married couples) chose seats at random in a row consisting of six seats.
a. Use the multiplication rule to compute the probability that Jim and Paula sit together on the far left (event A) and that John and Mary Lou (husband and wife) sit together in the middle (event B).
b. Given that John and Mary Lou sit together in the middle, what is the probability that the two other husbands sit next to their wives?
c. Given that John and Mary Lou sit together, what is the probability that all husbands sit next to their wives?

56. If $P(B \mid A) > P(B)$, show that $P(B' \mid A) < P(B')$. *Hint:* Add $P(B' \mid A)$ to both sides of the given inequality and then use the result of Exercise 57.

57. For any events A and B with $P(B) > 0$, show that $P(A \mid B) + P(A' \mid B) = 1$.

58. Show that for any three events A, B, and C with $P(C) > 0$, $P(A \cup B \mid C) = P(A \mid C) + P(B \mid C) - P(A \cap B \mid C)$.

59. At a certain gas station, 40% of the customers use regular unleaded gas (A_1), 35% use extra unleaded gas (A_2), and 25% use premium unleaded gas (A_3).

Of those customers using regular gas, only 30% fill their tanks (event B). Of those customers using extra gas, 60% fill their tanks, whereas of those using premium, 50% fill their tanks.

a. What is the probability that the next customer will request extra unleaded gas and fill the tank $(A_2 \cap B)$?

b. What is the probability that the next customer fills the tank?

c. If the next customer fills the tank, what is the probability that regular gas is requested? Extra gas? Premium gas?

60. Seventy percent of the light aircraft that disappear while in flight in a certain country are subsequently discovered. Of the aircraft that are discovered, 60% have an emergency locator, whereas 90% of the aircraft not discovered do not have such a locator. Suppose a light aircraft has disappeared.

a. If it has an emergency locator, what is the probability that it will not be discovered?

b. If it does not have an emergency locator, what is the probability that it will be discovered?

61. Components of a certain type are shipped to a supplier in batches of ten. Suppose that 50% of all such batches contain no defective components, 30% contain one defective component, and 20% contain two defective components. Two components from a batch are randomly selected and tested. What are the probabilities associated with 0, 1, and 2 defective components being in the batch under each of the following conditions?

a. Neither tested component is defective.

b. One of the two tested components is defective. (*Hint:* Draw a tree diagram with three first-generation branches for the three different types of batches.)

62. A company that manufactures video cameras produces a basic model and a deluxe model. Over the past year, 40% of the cameras sold have been of the basic model. Of those buying the basic model, 30% purchase an extended warranty, whereas 50% of all deluxe purchasers do so. If you learn that a randomly selected purchaser has an extended warranty, how likely is it that he or she has a basic model?

63. For customers purchasing a full set of tires at a particular tire store, consider the events

A = {tires purchased were made in the United States}

B = {purchaser has tires balanced immediately}

C = {purchaser requests front-end alignment}

along with A', B', and C'. Assume the following unconditional and conditional probabilities:

$P(A) = .75$ $P(B|A) = .9$ $P(B|A') = .8$

$P(C|A \cap B) = .8$ $P(C|A \cap B') = .6$

$P(C|A' \cap B) = .7$ $P(C|A' \cap B') = .3$

a. Construct a tree diagram consisting of first-, second-, and third-generation branches and place an event label and appropriate probability next to each branch.

b. Compute $P(A \cap B \cap C)$.

c. Compute $P(B \cap C)$.

d. Compute $P(C)$.

e. Compute $P(A|B \cap C)$, the probability of a purchase of U.S. tires given that both balancing and an alignment were requested.

64. In Example 2.30, suppose that the incidence rate for the disease is 1 in 25 rather than 1 in 1000. What then is the probability of a positive test result? Given that the test result is positive, what is the probability that the individual has the disease? Given a negative test result, what is the probability that the individual does not have the disease?

65. At a large university, in the never-ending quest for a satisfactory textbook, the Statistics Department has tried a different text during each of the last three quarters. During the fall quarter, 500 students used the text by Professor Mean; during the winter quarter, 300 students used the text by Professor Median; and during the spring quarter, 200 students used the text by Professor Mode. A survey at the end of each quarter showed that 200 students were satisfied with Mean's book, 150 were satisfied with Median's book, and 160 were satisfied with Mode's book. If a student who took statistics during one of these quarters is selected at random and admits to having been satisfied with the text, is the student most likely to have used the book by Mean, Median, or Mode? Who is the least likely author? (*Hint:* Draw a tree diagram or use Bayes' theorem.)

66. A friend who works in a big city owns two cars, one small and one large. Three-quarters of the time he drives the small car to work, and one-quarter of the time he takes the large car. If he takes the small car, he usually has little trouble parking, and so is at work on time with probability .9. If he takes the large car, he is on time to work with probability .6.

Given that he was on time on a particular morning, what is the probability that he drove the small car?

67. In Exercise 59, consider the following additional information on credit card usage:

70% of all regular fill-up customers use a credit card.

50% of all regular non-fill-up customers use a credit card.

60% of all extra fill-up customers use a credit card.

50% of all extra non-fill-up customers use a credit card.

50% of all premium fill-up customers use a credit card.

40% of all premium non-fill-up customers use a credit card.

Compute the probability of each of the following events for the next customer to arrive (a tree diagram might help).
a. {extra and fill-up and credit card}
b. {premium and non-fill-up and credit card}
c. {premium and credit card}
d. {fill-up and credit card}
e. {credit card}
f. If the next customer uses a credit card, what is the probability that premium was requested?

2.5 | Independence

The definition of conditional probability enables us to revise the probability $P(A)$ originally assigned to A when we are subsequently informed that another event B has occurred; the new probability of A is $P(A|B)$. In our examples, it was frequently the case that $P(A|B)$ was unequal to the unconditional probability $P(A)$, indicating that the information "B has occurred" resulted in a change in the chance of A occurring. There are other situations, though, in which the chance that A will occur or has occurred is not affected by knowledge that B has occurred, so that $P(A|B) = P(A)$. It is then natural to think of A and B as independent events, meaning that the occurrence or nonoccurrence of one event has no bearing on the chance that the other will occur.

DEFINITION

> Two events A and B are **independent** if $P(A|B) = P(A)$ and are **dependent** otherwise.

The definition of independence might seem "unsymmetric" because we do not demand that $P(B|A) = P(B)$ also. However, using the definition of conditional probability and the multiplication rule,

$$P(B|A) = \frac{P(A \cap B)}{P(A)} = \frac{P(A|B)P(B)}{P(A)} \tag{2.7}$$

The right-hand side of Equation (2.7) is $P(B)$ if and only if $P(A|B) = P(A)$ (independence), so the equality in the definition implies the other equality (and vice versa). It is also straightforward to show that if A and B are independent, then so are the following pairs of events: (1) A' and B, (2) A and B', and (3) A' and B'.

Example 2.31 Consider tossing a fair six-sided die once and define events $A = \{2, 4, 6\}$, $B = \{1, 2, 3\}$, and $C = \{1, 2, 3, 4\}$. We then have $P(A) = \frac{1}{2}$, $P(A|B) = \frac{1}{3}$, and $P(A|C) = \frac{1}{2}$. That is, events A and B are dependent, whereas events A and C are independent. Intuitively, if such a die is tossed and we are informed that the outcome was 1, 2, 3, or 4 (*C* has occurred), then the probability that *A* occurred is $\frac{1}{2}$, as it originally was, since two of the four relevant outcomes are even and the outcomes are still equally likely. ∎

Example 2.32 Let *A* and *B* be any two mutually exclusive events with $P(A) > 0$. For example, for a randomly chosen automobile, let *A* = {the car has four cylinders} and *B* = {the car has six cylinders}. Since the events are mutually exclusive, if *B* occurs, then *A* cannot possibly have occurred, so $P(A|B) = 0 \neq P(A)$. The message here is that *if two events are mutually exclusive, they cannot be independent.* When *A* and *B* are mutually exclusive, the information that *A* occurred says something about *B* (it cannot have occurred), so independence is precluded. ∎

$P(A \cap B)$ When Events Are Independent

Frequently the nature of an experiment suggests that two events *A* and *B* should be assumed independent. This is the case, for example, if a manufacturer receives a circuit board from each of two different suppliers, each board is tested on arrival, and *A* = {first is defective} and *B* = {second is defective}. If $P(A) = .1$, it should also be the case that $P(A|B) = .1$; knowing the condition of the second board shouldn't provide information about the condition of the first. Our next result shows how to compute $P(A \cap B)$ when the events are independent.

PROPOSITION

> *A* and *B* are independent if and only if
>
> $$P(A \cap B) = P(A) \cdot P(B) \tag{2.8}$$

To paraphrase the proposition, *A* and *B* are independent events iff* the probability that they both occur ($A \cap B$) is the product of the two individual probabilities. The verification is as follows:

$$P(A \cap B) = P(A|B) \cdot P(B) = P(A) \cdot P(B) \tag{2.9}$$

where the second equality in Equation (2.9) is valid iff *A* and *B* are independent. Because of the equivalence of independence with Equation (2.8), the latter can be used as a definition of independence.

Example 2.33 It is known that 30% of a certain company's washing machines require service while under warranty, whereas only 10% of its dryers need such service. If someone purchases both a washer and a dryer made by this company, what is the probability that both machines need warranty service?

Let *A* denote the event that the washer needs service while under warranty, and let *B* be defined analogously for the dryer. Then $P(A) = .30$ and $P(B) = .10$. Assuming that the two machines function independently of one another, the desired probability is

$$P(A \cap B) = P(A) \cdot P(B) = (.30)(.10) = .03$$

---

*iff is an abbreviation for "if and only if."

The probability that neither machine needs service is

$$P(A' \cap B') = P(A') \cdot P(B') = (.70)(.90) = .63$$ ■

Example 2.34 Each day, Monday through Friday, a batch of components sent by a first supplier arrives at a certain inspection facility. Two days a week, a batch also arrives from a second supplier. Eighty percent of all supplier 1's batches pass inspection, and 90% of supplier 2's do likewise. What is the probability that, on a randomly selected day, two batches pass inspection? We will answer this assuming that, on days when two batches are tested, whether the first batch passes is independent of whether the second batch does so. Figure 2.13 displays the relevant information.

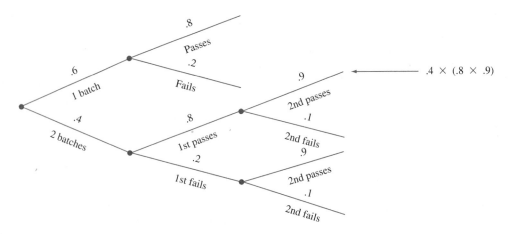

Figure 2.13 Tree diagram for Example 2.34

$$P(\text{two pass}) = P(\text{two received} \cap \text{both pass})$$
$$= P(\text{both pass} \mid \text{two received}) \cdot P(\text{two received})$$
$$= [(.8)(.9)](.4) = .288$$ ■

Independence of More Than Two Events

The notion of independence of two events can be extended to collections of more than two events. Although it is possible to extend the definition for two independent events by working in terms of conditional and unconditional probabilities, it is more direct and less cumbersome to proceed along the lines of the last proposition.

DEFINITION

Events $A_1, \ldots, A_n$ are **mutually independent** if for every k ($k = 2, 3, \ldots, n$) and every subset of indices $i_1, i_2, \ldots, i_k$,

$$P(A_{i_1} \cap A_{i_2} \cap \cdots \cap A_{i_k}) = P(A_{i_1}) \cdot P(A_{i_2}) \cdot \cdots \cdot P(A_{i_k}).$$

To paraphrase the definition, the events are mutually independent if the probability of the intersection of any subset of the n events is equal to the product of the individual probabilities. As was the case with two events, we frequently specify at the outset of a problem the independence of certain events. The definition can then be used to calculate the probability of an intersection.

Example 2.35 A system consists of four components, as illustrated in Figure 2.14. The entire system will work if either the 1–2 subsystem works or if the 3–4 subsystem works (since the two subsystems are connected in parallel). Since the two components in each subsystem are connected in series, a subsystem will work only if both its components work. If components work or fail independently of one another and if each works with probability .9, what is the probability that the entire system will work (the system reliability coefficient)? Letting A_i ($i = 1, 2, 3, 4$) be the event that the ith component works, the A_i's are mutually independent. The event that the 1–2 subsystem works is $A_1 \cap A_2$, and similarly, $A_3 \cap A_4$ denotes the event that the 3–4 subsystem works. The event that the entire system works is $(A_1 \cap A_2) \cap (A_3 \cap A_4)$, so

$$P[(A_1 \cap A_2) \cup (A_3 \cap A_4)]$$
$$= P(A_1 \cap A_2) + P(A_3 \cap A_4) - P[(A_1 \cap A_2) \cap (A_3 \cap A_4)]$$
$$= P(A_1) \cdot P(A_2) + P(A_3) \cdot P(A_4) - P(A_1) \cdot P(A_2) \cdot P(A_3) \cdot P(A_4)$$
$$= (.9)(.9) + (.9)(.9) - (.9)(.9)(.9)(.9) = .9639$$

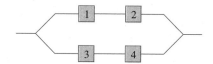

Figure 2.14 The system of components for Example 2.35

Letting $x = P(A_i)$, for $i = 1, 2, 3, 4$, what value of x would yield a system reliability of .99? Proceeding analogously, $P(\text{system works}) = x^2 + x^2 - x^4 = .99$ or $y^2 - 2y + .99 = 0$ where $y = x^2$. Solving this quadratic equation gives $y = .9$, so $x = \sqrt{.9} \approx .95$ (recall that $\approx$ means approximately equal). To achieve a system reliability of .99 without introducing more components or changing the system configuration, the reliability of each component would have to be increased from .9 to .95. ■

Exercises | Section 2.5 (68–85)

68. Reconsider the credit card scenario of Exercise 47 (Section 2.4), and show that A and B are dependent first by using the definition of independence and then by verifying that the multiplication property does not hold.

69. An executive on a business trip must rent a car in each of two different cities. Let A denote the event that the executive is offered a free upgrade in the first city and B represent the analogous event for the second city. Suppose that $P(A) = .2$, $P(B) = .3$, and that A and B are independent events.

a. If the executive is not offered a free upgrade in the first city, what is the probability of not getting a free upgrade in the second city? Explain your reasoning.

b. What is the probability that the executive is offered a free upgrade in at least one of the two cities?

c. If the executive is offered a free upgrade in at least one of the two cities, what is the probability that such an offer was made only in the first city?

70. In Exercise 13, is any A_i independent of any other A_j? Answer using the multiplication property for independent events.

71. If A and B are independent events, show that A' and B are also independent. [*Hint:* First establish a relationship between $P(A' \cap B)$, $P(B)$, and $P(A \cap B)$.]

72. Suppose that the proportions of blood phenotypes in a particular population are as follows:

A	B	AB	O
.42	.10	.04	.44

Assuming that the phenotypes of two randomly selected individuals are independent of one another, what is the probability that both phenotypes are O? What is the probability that the phenotypes of two randomly selected individuals match?

73. One of the assumptions underlying the theory of control charting (see Chapter 16) is that successive plotted points are independent of one another. Each plotted point can signal either that a manufacturing process is operating correctly or that there is some sort of malfunction. Even when a process is running correctly, there is a small probability that a particular point will signal a problem with the process. Suppose that this probability is .01. What is the probability that at least one of 10 successive points indicates a problem when in fact the process is operating correctly? Answer this question for 25 successive points.

74. The probability that a grader will make a marking error on any particular question of a multiple-choice exam is .1. If there are ten questions and questions are marked independently, what is the probability that no errors are made? That at least one error is made? If there are n questions and the probability of a marking error is p rather than .1, give expressions for these two probabilities.

75. An aircraft seam requires 25 rivets. The seam will have to be reworked if any of these rivets is defective. Suppose rivets are defective independently of one another, each with the same probability.

a. If 14% of all seams need reworking, what is the probability that a rivet is defective?

b. How small should the probability of a defective rivet be to ensure that only 10% of all seams need reworking?

76. A boiler has five identical relief valves. The probability that any particular valve will open on demand is .95. Assuming independent operation of the valves, calculate P(at least one valve opens) and P(at least one valve fails to open).

77. Two pumps connected in parallel fail independently of one another on any given day. The probability that only the older pump will fail is .10, and the probability that only the newer pump will fail is .05. What is the probability that the pumping system will fail on any given day (which happens if both pumps fail)?

78. Consider the system of components connected as in the accompanying picture. Components 1 and 2 are connected in parallel, so that subsystem works iff either 1 or 2 works; since 3 and 4 are connected in series, that subsystem works iff both 3 and 4 work. If components work independently of one another and P(component works) = .9, calculate P(system works).

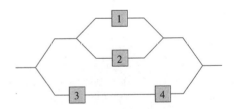

79. Components arriving at a distributor are checked for defects by two different inspectors (each component is checked by both inspectors). The first inspector detects 90% of all defectives that are present, and the second inspector does likewise. At least one inspector does not detect a defect on 20% of all defective components. What is the probability that

a. A defective component will be detected only by the first inspector? By exactly one of the two inspectors?

b. All three defective components in a batch escape detection by both inspectors (assuming inspections of different components are independent of one another)?

80. Sixty percent of all vehicles examined at a certain emissions inspection station pass the inspection. Assuming that successive vehicles pass or fail independently of one another, calculate the following probabilities.
    a. P(all of the next three vehicles inspected pass)
    b. P(at least one of the next three inspected fails)
    c. P(exactly one of the next three inspected passes)
    d. P(at most one of the next three vehicles inspected passes)
    e. Given that at least one of the next three vehicles passes inspection, what is the probability that all three pass (a conditional probability)?

81. A quality control inspector is inspecting newly produced items for faults. The inspector searches an item for faults in a series of independent fixations, each of a fixed duration. Given that a flaw is actually present, let p denote the probability that the flaw is detected during any one fixation (this model is discussed in "Human Performance in Sampling Inspection," *Human Factors,* 1979: 99–105).
    a. Assuming that an item has a flaw, what is the probability that it is detected by the end of the second fixation (once a flaw has been detected, the sequence of fixations terminates)? (*Hint:* {flaw detected in at most two fixations} = {flaw detected on the first fixation} ∪ {flaw undetected on the first and detected on the second}.)
    b. Give an expression for the probability that a flaw will be detected by the end of the nth fixation.
    c. If when a flaw has not been detected in three fixations, the item is passed, what is the probability that a flawed item will pass inspection?
    d. Suppose 10% of all items contain a flaw [P(randomly chosen item is flawed) = .1]. With the assumption of part (c), what is the probability that a randomly chosen item will pass inspection (it will automatically pass if it is not flawed, but could also pass if it is flawed)?
    e. Given that an item has passed inspection (no flaws in three fixations), what is the probability that it is actually flawed? Calculate for $p = .5$.

82. a. A lumber company has just taken delivery on a lot of 10,000 2 × 4 boards. Suppose that 20% of these boards (2000) are actually too green to be used in first-quality construction. Two boards are selected at random, one after the other. Let

A = {the first board is green} and B = {the second board is green}. Compute $P(A)$, $P(B)$, and $P(A \cap B)$ (a tree diagram might help). Are A and B independent?
    b. With A and B independent and $P(A) = P(B) = .2$, what is $P(A \cap B)$? How much difference is there between this answer and $P(A \cap B)$ in part (a)? For purposes of calculating $P(A \cap B)$, can we assume that A and B of part (a) are independent to obtain essentially the correct probability?
    c. Suppose the lot consists of ten boards, of which two are green. Does the assumption of independence now yield approximately the correct answer for $P(A \cap B)$? What is the critical difference between the situation here and that of part (a)? When do you think that an independence assumption would be valid in obtaining an approximately correct answer to $P(A \cap B)$?

83. Refer to the assumptions stated in Exercise 78 and answer the question posed there for the system in the accompanying picture. How would the probability change if this were a subsystem connected in parallel to the subsystem pictured in Example 2.35?

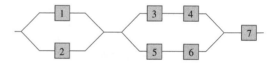

84. Professor Stan der Deviation can take one of two routes on his way home from work. On the first route, there are four railroad crossings. The probability that he will be stopped by a train at any particular one of the crossings is .1, and trains operate independently at the four crossings. The other route is longer but there are only two crossings, independent of one another, with the same stoppage probability for each as on the first route. On a particular day, Professor Deviation has a meeting scheduled at home for a certain time. Whichever route he takes, he calculates that he will be late if he is stopped by trains at at least half the crossings encountered.
    a. Which route should he take to minimize the probability of being late to the meeting?
    b. If he tosses a fair coin to decide on a route and he is late, what is the probability that he took the four-crossing route?

85. Suppose identical tags are placed on both the left ear and the right ear of a fox. The fox is then let loose for

a period of time. Consider the two events $C_1 = \{$left ear tag is lost$\}$ and $C_2 = \{$right ear tag is lost$\}$. Let $\pi = P(C_1) = P(C_2)$ and assume C_1 and C_2 are independent events. Derive an expression (involving π) for the probability that exactly one tag is lost given

that at most one is lost ("Ear Tag Loss in Red Foxes," *J. Wildlife Mgmt.*, 1976: 164–167). (*Hint:* Draw a tree diagram in which the two initial branches refer to whether the left ear tag was lost.)

Supplementary Exercises (86–111)

86. A small manufacturing company will start operating a night shift. There are 20 machinists employed by the company.
  a. If a night crew consists of 3 machinists, how many different crews are possible?
  b. If the machinists are ranked 1, 2, . . . , 20 in order of competence, how many of these crews would not have the best machinist?
  c. How many of the crews would have at least 1 of the 10 best machinists?
  d. If one of these crews is selected at random to work on a particular night, what is the probability that the best machinist will not work that night?

87. A certain type of refrigerator comes in three sizes, each of which is available with or without an icemaker. The accompanying table gives the relevant probabilities for a randomly selected purchaser:

		Size		
		17.5 ft³	**21 ft³**	**24.5 ft³**
Icemaker	**Yes**	.12	.17	.24
	No	.18	.13	.16

  a. What is the probability that the purchaser selects a 21-ft³ refrigerator? A refrigerator with an icemaker?
  b. Given that the purchaser selected a model with an icemaker, what is the probability that it was a 21-ft³ model? How does this compare to the unconditional probability of selecting a 21-ft³ model?
  c. Given that the purchaser selected a model with at least 20 ft³ of space, what is the probability that the selected refrigerator did not have an icemaker?

88. An employee of the records office at a certain university currently has ten forms on his desk awaiting

processing. Six of these are withdrawal petitions and the other four are course substitution requests.
  a. If he randomly selects six of these forms to give to a subordinate, what is the probability that only one of the two types of forms remains on his desk?
  b. Suppose he has time to process only four of these forms before leaving for the day. If these four are randomly selected one by one, what is the probability that each succeeding form is of a different type from its predecessor?

89. Suppose I fly to New York on airline X and return to Los Angeles on airline Y. Let $A = \{$X loses my baggage$\}$ and $B = \{$Y loses my baggage$\}$. If A and B are independent events with $P(A) > P(B)$, $P(A \cap B) = .0002$, and $P(A \cup B) = .03$, determine $P(A)$ and $P(B)$.

90. A transmitter is sending a message by using a binary code, namely, a sequence of 0's and 1's. Each transmitted bit (0 or 1) must pass through three relays to reach the receiver. At each relay, the probability is .20 that the bit sent will be different from the bit received (a reversal). Assume that the relays operate independently of one another.

Transmitter → Relay 1 → Relay 2 → Relay 3
                                                   → Receiver
  a. If a 1 is sent from the transmitter, what is the probability that a 1 is sent by all three relays?
  b. If a 1 is sent from the transmitter, what is the probability that a 1 is received by the receiver? (*Hint:* The eight experimental outcomes can be displayed on a tree diagram with three generations of branches, one generation for each relay.)
  c. Suppose 70% of all bits sent from the transmitter are 1's. If a 1 is received by the receiver, what is the probability that a 1 was sent?

91. Individual A has a circle of five close friends (B, C, D, E, and F). A has heard a certain rumor from outside the circle and has invited the five friends to a party to circulate the rumor. To begin, A selects one of the five at random and tells the rumor to the chosen individual. That individual then selects at random one of the four remaining individuals and repeats the rumor. Continuing, a new individual is selected from those not already having heard the rumor by the individual who has just heard it, until everyone has been told.

a. What is the probability that the rumor is repeated in the order B, C, D, E, and F?

b. What is the probability that F is the third person at the party to be told the rumor?

c. What is the probability that F is the last person to hear the rumor?

92. Refer to Exercise 91. If at each stage the person who currently "has" the rumor does not know who has already heard it and selects the next recipient at random from all five possible individuals, what is the probability that F has still not heard the rumor after it has been told ten times at the party?

93. An automobile insurance company classifies each driver as a good risk (A_1), a medium risk (A_2), or a poor risk (A_3). Of those currently insured, 30% are good risks, 50% are medium risks, and 20% are poor risks. In any given year, the probability that a driver will have at least one citation is .1 for a good risk, .3 for a medium risk, and .5 for a poor risk. If a randomly selected driver insured by this company has at least one citation during the next year, what is the probability that the driver was actually a good risk? A medium risk?

94. In Exercise 93, suppose that each insuree has a 3-year policy. If citations are independent from year to year and a randomly selected driver reports no citations during the 3 years, what is the probability that the driver was actually a good risk?

95. A chemical engineer is interested in determining whether a certain trace impurity is present in a product. An experiment has a probability of .80 of detecting the impurity if it is present. The probability of not detecting the impurity if it is absent is .90. The prior probabilities of the impurity being present and being absent are .40 and .60, respectively. Three separate experiments result in only two detections. What is the posterior probability that the impurity is present?

96. Each contestant on a quiz show is asked to specify one of six possible categories from which questions will be asked. Suppose P(contestant requests category i) $= \frac{1}{6}$ and successive contestants choose their categories independently of one another. If there are three contestants on each show and all three contestants on a particular show select different categories, what is the probability that exactly one has selected category 1?

97. Fasteners used in aircraft manufacturing are slightly crimped so that they lock enough to avoid loosening during vibration. Suppose that 95% of all fasteners pass an initial inspection. Of the 5% that fail, 20% are so seriously defective that they must be scrapped. The remaining fasteners are sent to a recrimping operation, where 40% cannot be salvaged and are discarded. The other 60% of these fasteners are corrected by the recrimping process and subsequently pass inspection.

a. What is the probability that a randomly selected incoming fastener will pass inspection either initially or after recrimping?

b. Given that a fastener passed inspection, what is the probability that it passed the initial inspection and did not need recrimping?

98. One percent of all individuals in a certain population are carriers of a particular disease. A diagnostic test for this disease has a 90% detection rate for carriers and a 5% detection rate for noncarriers. Suppose the test is applied independently to two different blood samples from the same randomly selected individual.

a. What is the probability that both tests yield the same result?

b. If both tests are positive, what is the probability that the selected individual is a carrier?

99. A system consists of two components. The probability that the second component functions in a satisfactory manner during its design life is .9, the probability that at least one of the two components does so is .96, and the probability that both components do so is .75. Given that the first component functions in a satisfactory manner throughout its design life, what is the probability that the second one does also?

100. A certain company sends 40% of its overnight mail parcels via express mail service E_1. Of these parcels, 2% arrive after the guaranteed delivery time (denote by L the event "late delivery"). If a record of an overnight mailing is randomly selected from the company's file, what is the probability that the parcel went via E_1 and was late?

101. Refer to Exercise 100. Suppose that 50% of the overnight parcels are sent via express mail service E_2 and the remaining 10% are sent via E_3. Of those sent via E_2, only 1% arrive late, whereas 5% of the parcels handled by E_3 arrive late.
 a. What is the probability that a randomly selected parcel arrived late?
 b. If a randomly selected parcel has arrived on time, what is the probability that it was not sent via E_1?

102. A company uses three different assembly lines— A_1, A_2, and A_3—to manufacture a particular component. Of those manufactured by line A_1, 5% need rework to remedy a defect, whereas 8% of A_2's components need rework and 10% of A_3's need rework. Suppose that 50% of all components are produced by line A_1, 30% are produced by line A_2, and 20% come from line A_3. If a randomly selected component needs rework, what is the probability that it came from line A_1? From line A_2? From line A_3?

103. Disregarding the possibility of a February 29 birthday, suppose a randomly selected individual is equally likely to have been born on any one of the other 365 days.
 a. If ten people are randomly selected, what is the probability that all have different birthdays? That at least two have the same birthday?
 b. With k replacing ten in part (a), what is the smallest k for which there is at least a 50–50 chance that two or more people will have the same birthday?
 c. If ten people are randomly selected, what is the probability that either at least two have the same birthday or at least two have the same last three digits of their Social Security numbers? [*Note:* The article "Methods for Studying Coincidences" (F. Mosteller and P. Diaconis, *J. Amer. Stat. Assoc.*, 1989: 853–861) discusses problems of this type.]

104. One method used to distinguish between granitic (G) and basaltic (B) rocks is to examine a portion of the infrared spectrum of the sun's energy reflected from the rock surface. Let R_1, R_2, and R_3 denote measured spectrum intensities at three different wavelengths; typically, for granite $R_1 < R_2 < R_3$, whereas for basalt $R_3 < R_1 < R_2$. When measurements are made remotely (using aircraft), various orderings of the R_i's may arise whether the rock is basalt or granite. Flights over regions of known composition have yielded the following information:

	Granite	Basalt
$R_1 < R_2 < R_3$	60%	10%
$R_1 < R_3 < R_2$	25%	20%
$R_3 < R_1 < R_2$	15%	70%

Suppose that for a randomly selected rock in a certain region, $P(\text{granite}) = .25$ and $P(\text{basalt}) = .75$.
 a. Show that $P(\text{granite} \mid R_1 < R_2 < R_3) > P(\text{basalt} \mid R_1 < R_2 < R_3)$. If measurements yielded $R_1 < R_2 < R_3$, would you classify the rock as granite or basalt?
 b. If measurements yielded $R_1 < R_3 < R_2$, how would you classify the rock? Answer the same question for $R_3 < R_1 < R_2$.
 c. Using the classification rules indicated in parts (a) and (b), when selecting a rock from this region, what is the probability of an erroneous classification? [*Hint:* Either G could be classified as B or B as G, and $P(B)$ and $P(G)$ are known.]
 d. If $P(\text{granite}) = p$ rather than .25, are there values of p (other than 1) for which one would always classify a rock as granite?

105. A subject is allowed a sequence of glimpses to detect a target. Let $G_i = \{$the target is detected on the ith glimpse$\}$, with $p_i = P(G_i)$. Suppose the G_i's are independent events and write an expression for the probability that the target has been detected by the end of the nth glimpse. (*Note:* This model is discussed in "Predicting Aircraft Detectability," *Human Factors*, 1979: 277–291.)

106. In a Little League baseball game, team A's pitcher throws a strike 50% of the time and a ball 50% of the time, successive pitches are independent of one another, and the pitcher never hits a batter. Knowing this, team B's manager has instructed the first batter not to swing at anything. Calculate the probability that
 a. The batter walks on the fourth pitch.
 b. The batter walks on the sixth pitch (so two of the first five must be strikes), using a counting argument or constructing a tree diagram.
 c. The batter walks.
 d. The first batter up scores while no one is out (assuming that each batter pursues a no-swing strategy).

107. Four engineers, A, B, C, and D, have been scheduled for job interviews at 10 A.M. on Friday, January 13, at Random Sampling, Inc. The personnel manager has scheduled the four for interview rooms 1, 2, 3, and 4, respectively. However, the manager's secretary does not know this, so assigns them to the four rooms in a completely random fashion (what else!). What is the probability that
   a. All four end up in the correct rooms?
   b. None of the four ends up in the correct room?

108. A particular airline has 10 A.M. flights from Chicago to New York, Atlanta, and Los Angeles. Let A denote the event that the New York flight is full and define events B and C analogously for the other two flights. Suppose $P(A) = .6$, $P(B) = .5$, $P(C) = .4$, and the three events are independent. What is the probability that
   a. All three flights are full? That at least one flight is not full?
   b. Only the New York flight is full? That exactly one of the three flights is full?

109. A personnel manager is to interview four candidates for a job. These are ranked 1, 2, 3, and 4 in order of preference and will be interviewed in random order. However, at the conclusion of each interview, the manager will know only how the current candidate compares to those previously interviewed. For example, the interview order 3, 4, 1, 2 generates no information after the first interview, shows that the second candidate is worse than the first, and that the third is better than the first two. However, the order 3, 4, 2, 1 would generate the same information after each of the first three interviews. The manager wants to hire the best candidate, but must make an irrevocable hire/no hire decision after each interview. Consider the following strategy: Automatically reject the first s candidates and then hire the first subsequent candidate who is best among those already interviewed (if no such candidate appears, the last one interviewed is hired).

For example, with $s = 2$, the order 3, 4, 1, 2 would result in the best being hired, whereas the order 3, 1, 2, 4 would not. Of the four possible s values (0, 1, 2, and 3), which one maximizes P(best is hired)? (*Hint:* Write out the 24 equally likely interview orderings: $s = 0$ means that the first candidate is automatically hired.)

110. Consider four independent events A_1, A_2, A_3, and A_4 and let $p_i = P(A_i)$ for $i = 1, 2, 3, 4$. Express the probability that at least one of these four events occurs in terms of the p_i's and do the same for the probability that at least two of the events occur.

111. A box contains the following four slips of paper, each having exactly the same dimensions: (1) win prize 1; (2) win prize 2; (3) win prize 3; (4) win prizes 1, 2, and 3. One slip will be randomly selected. Let $A_1 = \{$win prize 1$\}$, $A_2 = \{$win prize 2$\}$, and $A_3 = \{$win prize 3$\}$. Show that A_1 and A_2 are independent, that A_1 and A_3 are independent, and that A_2 and A_3 are also independent (this is *pairwise* independence). However, show that $P(A_1 \cap A_2 \cap A_3) \neq P(A_1) \cdot P(A_2) \cdot P(A_3)$, so the three events are *not* mutually independent.

Bibliography

Durrett, Richard, *The Essentials of Probability,* Duxbury Press, Belmont, CA, 1993. A concise presentation at a slightly higher level than this text.

Mosteller, Frederick, Robert Rourke, and George Thomas, *Probability with Statistical Applications* (2nd ed.), Addison-Wesley, Reading, MA, 1970. A very good precalculus introduction to probability, with many entertaining examples; especially good on counting rules and their application.

Olkin, Ingram, Cyrus Derman, and Leon Gleser (2nd ed.), *Probability Models and Applications,* Macmillan, New York, 1994. A comprehensive introduction to probability, written at a slightly higher mathematical level than this text but containing many good examples.

Ross, Sheldon, *A First Course in Probability* (5th ed.), Macmillan, New York, 1998. Rather tightly written and more mathematically sophisticated than this text, but contains a wealth of interesting examples and exercises.

Winkler, Robert, *Introduction to Bayesian Inference and Decision,* Holt, Rinehart & Winston, New York, 1972. A very good introduction to subjective probability.

3

Discrete Random Variables and Probability Distributions

Introduction

Whether an experiment yields qualitative or quantitative outcomes, methods of statistical analysis require that we focus on certain numerical aspects of the data (such as a sample proportion x/n, mean $\bar{x}$, or standard deviation s). The concept of a random variable allows us to pass from the experimental outcomes themselves to a numerical function of the outcomes. There are two fundamentally different types of random variables—discrete random variables and continuous random variables. In this chapter, we examine the basic properties and discuss the most important examples of discrete variables, and we will study continuous variables in Chapter 4.

3.1 | Random Variables

In any experiment, there are numerous characteristics that can be observed or measured, but in most cases an experimenter will focus on some specific aspect or aspects of a sample. For example, in a study of commuting patterns in a metropolitan area, each individual in a sample might be asked about commuting distance and the number of people commuting in the same vehicle, but not about IQ, income, family size, and other such characteristics. Alternatively, a researcher may test a sample of components and record only the number that have failed within 1000 hours, rather than recording the individual failure times.

In general, each outcome of an experiment can be associated with a number by specifying a rule of association (e.g., the number among the sample of ten components that fail to last 1000 hours or the total weight of baggage for a sample of 25 airline passengers). Such a rule of association is called a **random variable**—a variable because different numerical values are possible and random because the observed value depends on which of the possible experimental outcomes results (Figure 3.1).

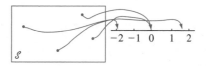

Figure 3.1 A random variable

DEFINITION

> For a given sample space $\mathscr{S}$ of some experiment, a **random variable** is any rule that associates a number with each outcome in $\mathscr{S}$.

We will often use the abbreviation rv in place of random variable. Random variables are customarily denoted by uppercase letters, such as X and Y, near the end of our alphabet. In contrast to our previous use of a lowercase letter, such as x, to denote a variable, we will now use lowercase letters to represent some particular value of the corresponding random variable. The notation $X(s) = x$ means that x is the value associated with the outcome s by the rv X.

Example 3.1 When a student attempts to log on to a computer time-sharing system, either all ports are busy (F), in which case the student will fail to obtain access, or else there is at least one port free (S), in which case the student will be successful in accessing the system. With $\mathscr{S} = \{S, F\}$, define an rv X by

$$X(S) = 1 \qquad X(F) = 0$$

The rv X indicates whether (1) or not (0) the student can log on. ∎

In Example 3.1, the rv X was specified by explicitly listing each element of $\mathscr{S}$ and the associated number. If $\mathscr{S}$ contains more than a few outcomes, such a listing is tedious, but it can frequently be avoided.

Example 3.2 Consider the experiment in which a telephone number in a certain area code is dialed using a random number dialer (such devices are used extensively by polling organizations), and define an rv Y by

$$Y = \begin{cases} 1 & \text{if the selected number is unlisted} \\ 0 & \text{if the selected number is listed in the directory} \end{cases}$$

For example, if 5282966 appears in the telephone directory, then $Y(5282966) = 0$, whereas $Y(7727350) = 1$ tells us that the number 7727350 is unlisted. A word description of this sort is more economical than a complete listing, so we will use such a description whenever possible.   ■

In Examples 3.1 and 3.2, the only possible values of the random variable were 0 and 1. Such a random variable arises frequently enough to be given a special name, after the individual who first studied it.

DEFINITION

Any random variable whose only possible values are 0 and 1 is called a **Bernoulli random variable.**

We will often want to define and study several different random variables from the same sample space.

Example 3.3   Example 2.3 described an experiment in which the number of pumps in use at each of two gas stations was determined. Define rv's X, Y, and U by

$X = $ the total number of pumps in use at the two stations

$Y = $ the difference between the number of pumps in use at station 1 and the number in use at station 2

$U = $ the maximum of the numbers of pumps in use at the two stations

If this experiment is performed and $s = (2, 3)$ results, then $X((2, 3)) = 2 + 3 = 5$, so we say that the observed value of X was $x = 5$. Similarly, the observed value of Y would be $y = 2 - 3 = -1$, and the observed value of U would be $u = \max(2, 3) = 3$.   ■

Each of the random variables of Examples 3.1–3.3 can assume only a finite number of possible values. This need not be the case.

Example 3.4   In Example 2.4, we considered the experiment in which batteries coming off an assembly line were examined until a good one (S) was obtained. The sample space was $\mathscr{S} = \{S, FS, FFS, \ldots\}$. Define an rv X by

$X = $ the number of batteries examined before the experiment terminates

Then $X(S) = 1$, $X(FS) = 2$, $X(FFS) = 3, \ldots, X(FFFFFFS) = 7$, and so on. Any positive integer is a possible value of X, so the set of possible values is infinite.   ■

Example 3.5   Suppose that in some random fashion, a location (latitude and longitude) in the continental United States is selected. Define an rv Y by

$Y = $ the height above sea level at the selected location

For example, if the selected location were (39°50′N, 98°35′W), then we might have $Y((39°50′N, 98°35′W)) = 1748.26$ ft. The largest possible value of Y is 14,494 (Mt. Whitney), and the smallest possible value is −282 (Death Valley). The set of all possible values of Y is the set of all numbers in the interval between -282 and $14,494$—that is,

$$\{y: y \text{ is a number}, -282 \leq y \leq 14,494\}$$

and there are an infinite number of numbers in this interval.   ∎

Two Types of Random Variables

In Section 1.2, a distinction was made between two different types of numerical variables, those that are *discrete* and those that are *continuous*. This same distinction carries over to random variables.

DEFINITION

A **discrete** random variable is an rv whose possible values either constitute a finite set or else can be listed in an infinite sequence in which there is a first element, a second element, and so on. A random variable is **continuous** if its set of possible values consists of an entire interval on the number line.

Although any interval on the number line contains an infinite number of numbers, it can be shown that there is no way to create an infinite listing of all these values—there are just too many of them.

Example 3.6   All random variables of Examples 3.1–3.4 are discrete. As another example, suppose we select married couples at random and do a blood test on each person until we find a husband and wife who both have the same Rh factor. With $X =$ the number of blood tests to be performed, possible values of X are $D = \{2, 4, 6, 8, \ldots\}$. Since the possible values have been listed in sequence, X is a discrete rv.   ∎

To study basic properties of discrete rv's, only the tools of discrete mathematics—summation and differences—are required. The study of continuous variables requires the continuous mathematics of the calculus—integrals and derivatives. This is the reason for making the distinction between the two types of variables.

Exercises | Section 3.1 (1–10)

1. Three automobiles are selected at random, and each is categorized as having a diesel (*S*) or nondiesel (*F*) engine (so outcomes are *SSS, SSF,* etc.). If $X =$ the number of cars among the three with diesel engines, list each outcome in $\mathcal{S}$ and its associated X value.

2. Give three examples of Bernoulli rv's (other than those in the text).

3. Using the experiment in Example 3.3, define two or more random variables and list the possible values of each.

4. Let $X =$ the number of nonzero digits in a randomly selected zip code. What are the possible values of X? Give three possible outcomes and their associated X values.

5. If the sample space $\mathscr{S}$ is an infinite set, does this necessarily imply that any rv X defined from $\mathscr{S}$ will have an infinite set of possible values? If yes, say why. If no, give an example.

6. Starting at a fixed time, each car entering an intersection is observed to see whether it turns left (L), right (R), or goes straight ahead (A). The experiment terminates as soon as a car is observed to turn left. Let X = the number of cars observed. What are possible X values? List five outcomes and their associated X values.

7. For each random variable defined here, describe the set of possible values for the variable and state whether the variable is discrete.
  a. X = the number of unbroken eggs in a randomly chosen standard egg carton
  b. Y = the number of students on a class list for a particular course who are absent on the first day of classes
  c. U = the number of times a duffer has to swing at a golf ball before hitting it
  d. X = the length of a randomly selected rattlesnake
  e. Z = the amount of royalties earned from the sale of a first edition of 10,000 textbooks
  f. Y = the pH of a randomly chosen soil sample
  g. X = the tension (psi) at which a randomly selected tennis racket has been strung
  h. X = the total number of coin tosses required for three individuals to obtain a match (*HHH* or *TTT*)

8. Each time a component is tested, the trial is a success (S) or failure (F). Suppose the component is tested repeatedly until a success occurs on three *consecutive* trials. Let Y denote the number of trials necessary to achieve this. List all outcomes corresponding to the five smallest possible values of Y and state which Y value is associated with each one.

9. Consider an experiment in which an individual named Claudius is located at the point 0 in the accompanying diagram.

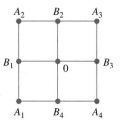

Using an appropriate randomization device (such as a tetrahedral die, one having four sides), Claudius first moves to one of the four locations B_1, B_2, B_3, B_4. Once at one of these locations, another randomization device is used to decide whether Claudius next returns to 0 or next visits one of the other two adjacent points. The experiment then continues in this fashion; after each move, another move to one of the (new) adjacent points is determined by tossing an appropriate die or coin.
  a. Let X = the number of moves that Claudius makes before first returning to 0. What are possible values of X? Is X discrete or continuous?
  b. If moves are allowed also along the diagonal paths connecting 0 to A_1, A_2, A_3, and A_4, respectively, answer the questions in part (a).

10. The number of pumps in use at both a six-pump station and a four-pump station will be determined. Give the possible values for each of the following random variables.
  a. T = the total number of pumps in use
  b. X = the difference between the numbers in use at stations 1 and 2
  c. U = the maximum number of pumps in use at either station
  d. Z = the number of stations having exactly two pumps in use

---

3.2 Probability Distributions for Discrete Random Variables

When probabilities are assigned to various outcomes in $\mathscr{S}$, these in turn determine probabilities associated with the values of any particular rv X. The *probability distribution of X* says how the total probability of 1 is distributed among (allocated to) the various possible X values.

Example 3.7 Six lots of components are ready to be shipped by a certain supplier. The number of defective components in each lot is as follows:

Lot	1 2 3 4 5 6
Number of defectives	0 2 0 1 2 0

One of these lots is to be randomly selected for shipment to a particular customer. Let X be the number of defectives in the selected lot. The three possible X values are 0, 1, and 2. Of the six equally likely simple events, three result in $X = 0$, one in $X = 1$, and the other two in $X = 2$. Let $p(0)$ denote the probability that $X = 0$ and $p(1)$ and $p(2)$ represent the probabilities of the other two possible values of X. Then

$$p(0) = P(X = 0) = P(\text{lot 1 or 3 or 6 is sent}) = \frac{3}{6} = .500$$

$$p(1) = P(X = 1) = P(\text{lot 4 is sent}) = \frac{1}{6} = .167$$

$$p(2) = P(X = 2) = P(\text{lot 2 or 5 is sent}) = \frac{2}{6} = .333$$

That is, a probability of .500 is distributed to the X value 0, a probability of .167 is placed on the X value 1, and the remaining probability, .333, is associated with the X value 2. The values of X along with their probabilities collectively specify the probability distribution or *probability mass function of X*. If this experiment were repeated over and over again, in the long run $X = 0$ would occur one-half of the time, $X = 1$ one-sixth of the time, and $X = 2$ one-third of the time. ■

DEFINITION

> The **probability distribution** or **probability mass function** (pmf) of a discrete rv is defined for every number x by $p(x) = P(X = x) = P(\text{all } s \in \mathcal{S}: X(s) = x)$.*

In words, for every possible value x of the random variable, the pmf specifies the probability of observing that value when the experiment is performed. The conditions $p(x) \geq 0$ and $\sum_{\text{all possible } x} p(x) = 1$ are required of any pmf.

Example 3.8 Suppose we go to a university bookstore during the first week of classes and observe whether the next person buying a computer buys a laptop or a desktop model. Let

$$X = \begin{cases} 1 & \text{if the customer purchases a laptop computer} \\ 0 & \text{if the customer purchases a desktop computer} \end{cases}$$

If 20% of all purchasers during that week select a laptop, the pmf for X is

$$p(0) = P(X = 0) = P(\text{next customer purchases a desktop model}) = .8$$

$$p(1) = P(X = 1) = P(\text{next customer purchases a laptop model}) = .2$$

$$p(x) = P(X = x) = 0 \quad \text{for } x \neq 0 \text{ or } 1$$

---

*$P(X = x)$ is read "the probability that the rv X assumes the value x." For example, $P(X = 2)$ denotes the probability that the resulting X value is 2.

An equivalent description is

$$p(x) = \begin{cases} .8 & \text{if } x = 0 \\ .2 & \text{if } x = 1 \\ 0 & \text{if } x \neq 0 \text{ or } 1 \end{cases}$$

Figure 3.2 is a picture of this pmf, called a *line graph*.

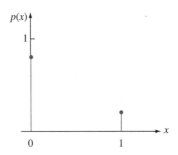

Figure 3.2   The line graph for the pmf in Example 3.8

Example 3.9   Consider a group of five potential blood donors—A, B, C, D, and E—of whom only A and B have type O+ blood. Five blood samples, one from each individual, will be typed in random order until an O+ individual is identified. Let the rv Y = the number of typings necessary to identify an O+ individual. Then the pmf of Y is

$$p(1) = P(Y = 1) = P(A \text{ or } B \text{ typed first}) = \frac{2}{5} = .4$$

$$p(2) = P(Y = 2) = P(C, D, \text{ or } E \text{ first, and then } A \text{ or } B)$$

$$= P(C, D, \text{ or } E \text{ first}) \cdot P(A \text{ or } B \text{ next} \,|\, C, D, \text{ or } E \text{ first}) = \frac{3}{5} \cdot \frac{2}{4} = .3$$

$$p(3) = P(Y = 3) = P(C, D, \text{ or } E \text{ first and second, and then } A \text{ or } B)$$

$$= \left(\frac{3}{5}\right)\left(\frac{2}{4}\right)\left(\frac{2}{3}\right) = .2$$

$$p(4) = P(Y = 4) = P(C, D, \text{ and } E \text{ all done first}) = \left(\frac{3}{5}\right)\left(\frac{2}{4}\right)\left(\frac{1}{3}\right) = .1$$

$$p(y) = 0 \quad \text{if } y \neq 1, 2, 3, 4$$

The pmf can be presented nicely in tabular form:

y	1	2	3	4
$p(y)$	.4	.3	.2	.1

where any y value not listed receives zero probability. This pmf can also be displayed in a line graph (Figure 3.3).

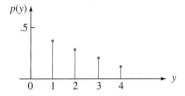

Figure 3.3 The line graph for the pmf in Example 3.9   ■

The name "probability mass function" is suggested by a model used in physics for a system of "point masses." In this model, masses are distributed at various locations x along a one-dimensional axis. Our pmf describes how the total probability mass of 1 is distributed at various points along the axis of possible values of the random variable (where and how much mass at each x).

Another useful pictorial representation of a pmf, called a **probability histogram,** is similar to histograms discussed in Chapter 1. Above each y with $p(y) > 0$, construct a rectangle centered at y. The height of each rectangle is proportional to $p(y)$, and the base is the same for all rectangles. When possible values are equally spaced, the base is frequently chosen as the distance between successive y values (though it could be smaller). Figure 3.4 shows two probability histograms.

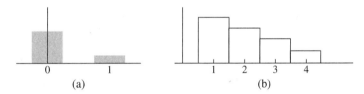

Figure 3.4 Probability histograms: (a) Example 3.8; (b) Example 3.9

A Parameter of a Probability Distribution

In Example 3.8, we had $p(0) = .8$ and $p(1) = .2$ because 20% of all purchasers selected a laptop computer. At another bookstore, it may be the case that $p(0) = .9$ and $p(1) = .1$. More generally, the pmf of any Bernoulli rv can be expressed in the form $p(1) = \alpha$ and $p(0) = 1 - \alpha$, where $0 < \alpha < 1$. Because the pmf depends on the particular value of α, we often write $p(x; \alpha)$ rather than just $p(x)$:

$$p(x; \alpha) = \begin{cases} 1 - \alpha & \text{if } x = 0 \\ \alpha & \text{if } x = 1 \\ 0 & \text{otherwise} \end{cases} \qquad (3.1)$$

Then each choice of α in Expression (3.1) yields a different pmf.

DEFINITION

Suppose $p(x)$ depends on a quantity that can be assigned any one of a number of possible values, with each different value determining a different probability distribution. Such a quantity is called a **parameter** of the distribution. The collection of all probability distributions for different values of the parameter is called a **family** of probability distributions.

The quantity α in Expression (3.1) is a parameter. Each different number α between 0 and 1 determines a different member of a family of distributions; two such members are

$$p(x; .6) = \begin{cases} .4 & \text{if } x = 0 \\ .6 & \text{if } x = 1 \\ 0 & \text{otherwise} \end{cases} \quad \text{and} \quad p(x; .5) = \begin{cases} .5 & \text{if } x = 0 \\ .5 & \text{if } x = 1 \\ 0 & \text{otherwise} \end{cases}$$

Every probability distribution for a Bernoulli rv has the form of Expression (3.1), so it is called the *family of Bernoulli distributions.*

Example 3.10   Starting at a fixed time, we observe the gender of each newborn child at a certain hospital until a boy (B) is born. Let $p = P(B)$, assume that successive births are independent, and define the rv X by X = number of births observed. Then

$$p(1) = P(X = 1) = P(B) = p$$
$$p(2) = P(X = 2) = P(GB) = P(G) \cdot P(B) = (1 - p)p$$

and

$$p(3) = P(X = 3) = P(GGB) = P(G) \cdot P(G) \cdot P(B) = (1 - p)^2 p$$

Continuing in this way, a general formula emerges:

$$p(x) = \begin{cases} (1 - p)^{x-1}p & x = 1, 2, 3, \ldots \\ 0 & \text{otherwise} \end{cases} \tag{3.2}$$

The quantity p in Expression (3.2) represents a number between 0 and 1 and is a parameter of the probability distribution. In the gender example, $p = .51$ might be appropriate, but if we were looking for the first child with Rh-positive blood, then we might let $p = .85$.   ∎

The Cumulative Distribution Function

For some fixed value x, we often wish to compute the probability that the observed value of X will be at most x. For example, the pmf in Example 3.7 was

$$p(x) = \begin{cases} .500 & x = 0 \\ .167 & x = 1 \\ .333 & x = 2 \\ 0 & \text{otherwise} \end{cases}$$

The probability that X is at most 1 is then

$$P(X \leq 1) = p(0) + p(1) = .500 + .167 = .667$$

In this example, $X \leq 1.5$ iff $X \leq 1$, so $P(X \leq 1.5) = P(X \leq 1) = .667$. Similarly, $P(X \leq 0) = P(X = 0) = .5$, and $P(X \leq .75) = .5$ also. Since 0 is the smallest possible value of X, $P(X \leq -1.7) = 0$, $P(X \leq -.0001) = 0$, and so on. The largest possible X value is 2, so $P(X \leq 2) = 1$, and if x is any number larger than 2, $P(X \leq x) = 1$; that is, $P(X \leq 5) = 1$, $P(X \leq 10.23) = 1$, and so on. Notice that $P(X < 1) = .5 \neq P(X \leq 1)$, since the probability of the X value 1 is included in the latter probability but not in the former. When X is a discrete random variable and x is a possible value of X, $P(X < x) < P(X \leq x)$.

DEFINITION

The **cumulative distribution function** (cdf) $F(x)$ of a discrete rv variable X with pmf $p(x)$ is defined for every number x by

$$F(x) = P(X \leq x) = \sum_{y:\, y \leq x} p(y)$$

For any number x, $F(x)$ is the probability that the observed value of X will be at most x.

Example 3.11 The pmf of Y for Example 3.9 was

y	1	2	3	4
$p(y)$	.4	.3	.2	.1

We first determine $F(y)$ for each value in the set $\{1, 2, 3, 4\}$ of possible values:

$$F(1) = P(Y \leq 1) = P(Y = 1) = p(1) = .4$$
$$F(2) = P(Y \leq 2) = P(Y = 1 \text{ or } 2) = p(1) + p(2) = .7$$
$$F(3) = P(Y \leq 3) = P(Y = 1 \text{ or } 2 \text{ or } 3) = p(1) + p(2) + p(3) = .9$$
$$F(4) = P(Y \leq 4) = P(Y = 1 \text{ or } 2 \text{ or } 3 \text{ or } 4) = 1$$

Now for any other number y, $F(y)$ will equal the value of F at the closest possible value of Y to the left of y. For example, $F(2.7) = P(Y \leq 2.7) = P(Y \leq 2) = .7$, and $F(3.999) = F(3) = .9$. The cdf is thus

$$F(y) = \begin{cases} 0 & \text{if } y < 1 \\ .4 & \text{if } 1 \leq y < 2 \\ .7 & \text{if } 2 \leq y < 3 \\ .9 & \text{if } 3 \leq y < 4 \\ 1 & \text{if } 4 \leq y \end{cases}$$

A graph of $F(y)$ is shown in Figure 3.5.

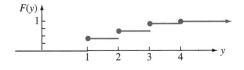

Figure 3.5 A graph of the cdf of Example 3.11

For X a discrete rv, the graph of $F(x)$ will have a jump at every possible value of X and will be flat between possible values. Such a graph is called a **step function.**

Example 3.12 In Example 3.10, any positive integer was a possible X value, and the pmf was

$$p(x) = \begin{cases} (1-p)^{x-1}p & x = 1, 2, 3, \ldots \\ 0 & \text{otherwise} \end{cases}$$

For any positive integer x,

$$F(x) = \sum_{y \leq x} p(y) = \sum_{y=1}^{x} (1-p)^{y-1} p = p \sum_{y=0}^{x-1} (1-p)^y \tag{3.4}$$

To evaluate this sum, we use the fact that the partial sum of a geometric series is

$$\sum_{y=0}^{k} a^y = \frac{1 - a^{k+1}}{1 - a}$$

Using this in Equation (3.4), with $a = 1 - p$ and $k = x - 1$, gives

$$F(x) = p \cdot \frac{1 - (1-p)^x}{1 - (1-p)} = 1 - (1-p)^x \quad x \text{ a positive integer}$$

Since F is constant in between positive integers,

$$F(x) = \begin{cases} 0 & x < 1 \\ 1 - (1-p)^{[x]} & x \geq 1 \end{cases} \tag{3.5}$$

where $[x]$ is the largest integer $\leq x$ (e.g., $[2.7] = 2$). Thus, if $p = .51$ as in the birth example, then the probability of having to examine at most five births to see the first boy is $F(5) = 1 - (.49)^5 = 1 - .0282 = .9718$, whereas $F(10) \approx 1.0000$. This cdf is graphed in Figure 3.6.

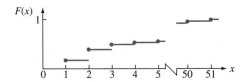

Figure 3.6 A graph of $F(x)$ for Example 3.12

In our examples thus far, the cdf has been derived from the pmf. It is possible to reverse this procedure and obtain the pmf from the cdf whenever the latter function is available. Suppose, for example, that X represents the number of defective components in a shipment consisting of six components, so that possible X values are $0, 1, \ldots, 6$. Then

$$\begin{aligned} p(3) &= P(X = 3) \\ &= [p(0) + p(1) + p(2) + p(3)] - [p(0) + p(1) + p(2)] \\ &= P(X \leq 3) - P(X \leq 2) \\ &= F(3) - F(2) \end{aligned}$$

More generally, the probability that X falls in a specified interval is easily obtained from the cdf. For example,

$$P(2 \leq X \leq 4) = p(2) + p(3) + p(4)$$
$$= [p(0) + \cdots + p(4)] - [p(0) + p(1)]$$
$$= P(X \leq 4) - P(X \leq 1)$$
$$= F(4) - F(1)$$

Notice that $P(2 \leq X \leq 4) \neq F(4) - F(2)$. This is because the X value 2 is included in $2 \leq X \leq 4$, so we do not want to subtract out its probability. However, $P(2 < X \leq 4) = F(4) - F(2)$ because $X = 2$ is not included in the interval $2 < X \leq 4$.

PROPOSITION

For any two numbers a and b with $a \leq b$,

$$P(a \leq X \leq b) = F(b) - F(a-)$$

where "$a-$" represents the largest possible X value that is strictly less than a. In particular, if the only possible values are integers and if a and b are integers, then

$$P(a \leq X \leq b) = P(X = a \text{ or } a + 1 \text{ or } \ldots \text{ or } b)$$
$$= F(b) - F(a - 1)$$

Taking $a = b$ yields $P(X = a) = F(a) - F(a - 1)$ in this case.

Example 3.13 Let X = the number of days of sick leave taken by a randomly selected employee of a large company during a particular year. If the maximum number of allowable sick days per year is 14, possible values of X are 0, 1, . . . , 14. With $F(0) = .58$, $F(1) = .72$, $F(2) = .76$, $F(3) = .81$, $F(4) = .88$, and $F(5) = .94$,

$$P(2 \leq X \leq 5) = P(X = 2, 3, 4, \text{ or } 5) = F(5) - F(1) = .22$$

and

$$P(X = 3) = F(3) - F(2) = .05 \qquad \blacksquare$$

The reason for subtracting $F(a-)$ rather than $F(a)$ is that we want to include $P(X = a)$; $F(b) - F(a)$ gives $P(a < X \leq b)$. This proposition will be used extensively when computing binomial and Poisson probabilities in Sections 3.4 and 3.6.

Another View of Probability Mass Functions

It is often helpful to think of a pmf as specifying a mathematical model for a discrete population.

Example 3.14 Consider selecting at random a student who is among the 15,000 registered for the current term at Mega University. Let X = the number of courses for which the selected student is registered and suppose that X has the pmf shown at the top of page 108.

x	1	2	3	4	5	6	7
$p(x)$	.01	.03	.13	.25	.39	.17	.02

One way to view this situation is to think of the population as consisting of 15,000 individuals, each having his or her own X value; the proportion with each X value is given by $p(x)$. An alternative viewpoint is to forget about the students and think of the population itself as consisting of the X values: There are some 1's in the population, some 2's, ..., and finally some 7's. The population then consists of the numbers 1, 2, ..., 7 (so is discrete), and $p(x)$ gives a model for the distribution of population values. ■

Once we have such a mathematical model for a population, we will use it to compute values of population characteristics (such as the mean μ) and make inferences about such characteristics.

Exercises | Section 3.2 (11–27)

11. An automobile service facility specializing in engine tune-ups knows that 45% of all tune-ups are done on four-cylinder automobiles, 40% on six-cylinder automobiles, and 15% on eight-cylinder automobiles. Let X = the number of cylinders on the next car to be tuned.
    a. What is the pmf of X?
    b. Draw both a line graph and a probability histogram for the pmf of part (a).

12. Let X = the number of tires on a randomly selected automobile that are underinflated.
    a. Which of the following three $p(x)$ functions is a legitimate pmf for X, and why are the other two not allowed?

x	0	1	2	3	4
$p(x)$	.3	.2	.1	.05	.05
$p(x)$	.4	.1	.1	.1	.3
$p(x)$	.4	.1	.2	.1	.3

   b. For the legitimate pmf of part (a), compute $P(2 \leq X \leq 4)$, $P(X \leq 2)$, and $P(X \neq 0)$.
   c. If $p(x) = c \cdot (5 - x)$ for $x = 0, 1, \ldots, 4$, what is the value of c? [*Hint:* $\sum_{x=0}^{4} p(x) = 1$.]

13. A mail-order computer business has six telephone lines. Let X denote the number of lines in use at a specified time. Suppose the pmf of X is as given in the accompanying table.

x	0	1	2	3	4	5	6
$p(x)$	.10	.15	.20	.25	.20	.06	.04

Calculate the probability of each of the following events.
   a. {at most 3 lines are in use}
   b. {fewer than 3 lines are in use}
   c. {at least 3 lines are in use}
   d. {between 2 and 5 lines, inclusive, are in use}
   e. {between 2 and 4 lines, inclusive, are not in use}
   f. {at least 4 lines are not in use}

14. A contractor is required by a county planning department to submit one, two, three, four, or five forms (depending on the nature of the project) in applying for a building permit. Let Y = the number of forms required of the next applicant. The probability that y forms are required is known to be proportional to y—that is, $p(y) = ky$ for $y = 1, \ldots, 5$.
    a. What is the value of k? [*Hint:* $\sum_{y=1}^{5} p(y) = 1$.]
    b. What is the probability that at most three forms are required?
    c. What is the probability that between two and four forms (inclusive) are required?
    d. Could $p(y) = y^2/50$ for $y = 1, \ldots, 5$ be the pmf of Y?

15. Many manufacturers have quality control programs that include inspection of incoming materials for defects. Suppose a computer manufacturer

receives computer boards in lots of five. Two boards are selected from each lot for inspection. We can represent possible outcomes of the selection process by pairs. For example, the pair (1, 2) represents the selection of boards 1 and 2 for inspection.

a. List the ten different possible outcomes.

b. Suppose that boards 1 and 2 are the only defective boards in a lot of five. Two boards are to be chosen at random. Define X to be the number of defective boards observed among those inspected. Find the probability distribution of X.

c. Let $F(x)$ denote the cdf of X. First determine $F(0) = P(X \le 0)$, $F(1)$, and $F(2)$ and then obtain $F(x)$ for all other x.

16. Some parts of California are particularly earthquake-prone. Suppose that in one such area, 30% of all homeowners are insured against earthquake damage. Four homeowners are to be selected at random; let X denote the number among the four who have earthquake insurance.

a. Find the probability distribution of X. [*Hint:* Let S denote a homeowner who has insurance and F one who does not. Then one possible outcome is *SFSS*, with probability (.3)(.7)(.3)(.3) and associated X value 3. There are 15 other outcomes.]

b. Draw the corresponding probability histogram.

c. What is the most likely value for X?

d. What is the probability that at least two of the four selected have earthquake insurance?

17. A new battery's voltage may be acceptable (*A*) or unacceptable (*U*). A certain flashlight requires two batteries, so batteries will be independently selected and tested until two acceptable ones have been found. Suppose that 90% of all batteries have acceptable voltages. Let Y denote the number of batteries that must be tested.

a. What is $p(2)$, that is $P(Y = 2)$?

b. What is $p(3)$? (*Hint:* There are two different outcomes that result in $Y = 3$.)

c. To have $Y = 5$, what must be true of the fifth battery selected? List the four outcomes for which $Y = 5$ and then determine $p(5)$.

d. Use the pattern in your answers for parts (a)–(c) to obtain a general formula for $p(y)$.

18. Two fair six-sided dice are tossed independently. Let M = the maximum of the two tosses (so $M(1, 5) = 5$, $M(3, 3) = 3$, etc.).

a. What is the pmf of M? [*Hint:* First determine $p(1)$, then $p(2)$, and so on.]

b. Determine the cdf of M and graph it.

19. In Example 3.9, suppose there are only four potential blood donors, of whom only one has type O+ blood. Compute the pmf of Y.

20. A library subscribes to two different weekly news magazines, each of which is supposed to arrive in Wednesday's mail. In actuality, each one may arrive on Wednesday, Thursday, Friday, or Saturday. Suppose the two arrive independently of one another, and for each one $P(\text{Wed.}) = .3$, $P(\text{Thurs.}) = .4$, $P(\text{Fri.}) = .2$, and $P(\text{Sat.}) = .1$. Let Y = the number of days beyond Wednesday that it takes for both magazines to arrive (so possible Y values are 0, 1, 2, or 3). Compute the pmf of Y. [*Hint:* There are 16 possible outcomes; $Y(W, W) = 0$, $Y(F, Th) = 2$, and so on.]

21. Refer to Exercise 13, and calculate and graph the cdf $F(x)$. Then use it to calculate the probabilities of the events given in parts (a)–(d) of that problem.

22. A consumer organization that evaluates new automobiles customarily reports the number of major defects on each car examined. Let X denote the number of major defects on a randomly selected car of a certain type. The cdf of X is as follows:

$$F(x) = \begin{cases} 0 & x < 0 \\ .06 & 0 \le x < 1 \\ .19 & 1 \le x < 2 \\ .39 & 2 \le x < 3 \\ .67 & 3 \le x < 4 \\ .92 & 4 \le x < 5 \\ .97 & 5 \le x < 6 \\ 1 & 6 < x \end{cases}$$

Calculate the following probabilities directly from the cdf.

a. $p(2)$, that is, $P(X = 2)$

b. $P(X > 3)$

c. $P(2 \le X \le 5)$

d. $P(2 < X < 5)$

23. An insurance company offers its policyholders a number of different premium payment options. For a randomly selected policyholder, let X = the

number of months between successive payments. The cdf of X is as follows:

$$F(x) = \begin{cases} 0 & x < 1 \\ .30 & 1 \le x < 3 \\ .40 & 3 \le x < 4 \\ .45 & 4 \le x < 6 \\ .60 & 6 \le x < 12 \\ 1 & 12 \le x \end{cases}$$

a. What is the pmf of X?
b. Using just the cdf, compute $P(3 \le X \le 6)$ and $P(4 \le X)$.

24. In Example 3.10, let $Y =$ the number of girls born before the experiment terminates. With $p = P(B)$ and $1 - p = P(G)$, what is the pmf of Y? (*Hint:* First list the possible values of Y, starting with the smallest, and proceed until you see a general formula.)

25. Alvie Singer lives at 0 in the accompanying diagram and has four friends who live at A, B, C, and D. One day Alvie decides to go visiting, so he tosses a fair coin twice to decide which of the four to visit. Once at a friend's house, he will either return home or else proceed to one of the two adjacent houses (such as 0, A, or C when at B), with each of the three possibilities having probability $\frac{1}{3}$. In this way, Alvie continues to visit friends until he returns home.

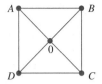

a. Let $X =$ the number of times that Alvie visits a friend. Derive the pmf of X.
b. Let $Y =$ the number of straight-line segments that Alvie traverses (including those leading to and from 0). What is the pmf of Y?
c. Suppose that female friends live at A and C and male friends at B and D. If $Z =$ the number of visits to female friends, what is the pmf of Z?

26. After all students have left the classroom, a statistics professor notices that four copies of the text were left under desks. At the beginning of the next lecture, the professor distributes the four books in a completely random fashion to each of the four students (1, 2, 3, and 4) who claim to have left books. One possible outcome is that 1 receives 2's book, 2 receives 4's book, 3 receives his or her own book, and 4 receives 1's book. This outcome can be abbreviated as (2, 4, 3, 1).
a. List the other 23 possible outcomes.
b. Let X denote the number of students who receive their own book. Determine the pmf of X.

27. Show that the cdf $F(x)$ is a nondecreasing function; that is, $x_1 < x_2$ implies that $F(x_1) \le F(x_2)$. Under what condition will $F(x_1) = F(x_2)$?

3.3 | Expected Values of Discrete Random Variables

In Example 3.14, we considered a university having 15,000 students and let $X =$ the number of courses for which a randomly selected student is registered. The pmf of X follows. Since $p(1) = .01$, we know that $(.01) \cdot (15,000) = 150$ of the students are registered for one course, and similarly for the other x values.

x	1	2	3	4	5	6	7	
$p(x)$	.01	.03	.13	.25	.39	.17	.02	(3.6)
Number registered	150	450	1950	3750	5850	2550	300	

To compute the average number of courses per student, or the average value of X in the population, we should compute the total number of courses and divide by the total number of students. Since each of 150 students is taking one course, these 150 contribute 150 courses to the total. Similarly, 450 students contribute 2(450) courses, and so on. The population average value of X is then

$$\frac{1(150) + 2(450) + 3(1950) + \cdots + 7(300)}{15,000} = 4.57 \tag{3.7}$$

Since $150/15,000 = .01 = p(1)$, $450/15,000 = .03 = p(2)$, and so on, an alternative expression for (3.7) is

$$1 \cdot p(1) + 2 \cdot p(2) + \cdots + 7 \cdot p(7) \tag{3.8}$$

Expression (3.8) shows that to compute the population average value of X, we need only the possible values of X along with their probabilities (proportions). In particular, the population size is irrelevant as long as the pmf is given by (3.6). The average or mean value of X is then a *weighted* average of the possible values $1, \ldots, 7$, where the weights are the probabilities of those values.

The Expected Value of *X*

DEFINITION

> Let X be a discrete rv with set of possible values D and pmf $p(x)$. The **expected value** or **mean value** of X, denoted by $E(X)$ or μ_X, is
>
> $$E(X) = \mu_X = \sum_{x \in D} x \cdot p(x)$$

When it is clear to which X the expected value refers, μ rather than μ_X is often used.

Example 3.15 For the pmf in (3.6),

$$\begin{aligned}
\mu &= 1 \cdot p(1) + 2 \cdot p(2) + \cdots + 7 \cdot p(7) \\
&= (1)(.01) + 2(.03) + \cdots + (7)(.02) \\
&= .01 + .06 + .39 + 1.00 + 1.95 + 1.02 + .14 = 4.57
\end{aligned}$$

If we think of the population as consisting of the X values $1, 2, \ldots, 7$, then $\mu = 4.57$ is the population mean. In the sequel, we will often refer to μ as the *population mean* rather than the mean of X in the population. ∎

In Example 3.15, the expected value μ was 4.57, which is not a possible value of X. The word *expected* should be interpreted with caution because one would not expect to see an X value of 4.57 when a single student is selected.

Example 3.16 Just after birth, each newborn child is rated on a scale called the Apgar scale. The possible ratings are $0, 1, \ldots, 10$, with the child's rating determined by color, muscle tone, respiratory effort, heartbeat, and reflex irritability (the best possible score is 10). Let X

be the Apgar score of a randomly selected child born at a certain hospital during the next year and suppose that the pmf of X is

x	0	1	2	3	4	5	6	7	8	9	10
$p(x)$	.002	.001	.002	.005	.02	.04	.18	.37	.25	.12	.01

Then the mean value of X is

$$E(X) = \mu = 0(.002) + 1(.001) + 2(.002)$$
$$+ \cdots + 8(.25) + 9(.12) + 10(.01)$$
$$= 7.15$$

Again, μ is not a possible value of the variable X. Also, because the variable refers to a future child, there is no concrete existing population to which μ refers. Instead, we think of the pmf as a model for a conceptual population consisting of the values 0, 1, 2, . . . , 10. The mean value of this conceptual population is then $\mu = 7.15$. ∎

Example 3.17 Let X be a Bernoulli rv with pmf

$$p(x) = \begin{cases} 1 - p & x = 0 \\ p & x = 1 \\ 0 & x \neq 0, 1 \end{cases}$$

Then $E(X) = 0 \cdot p(0) + 1 \cdot p(1) = 0(1 - p) + 1(p) = p$. That is, the expected value of X is just the probability that X takes on the value 1. If we conceptualize a population consisting of 0's in proportion $1 - p$ and 1's in proportion p, then the population average is $\mu = p$. ∎

Example 3.18 The general form for the pmf of X = number of children born up to and including the first boy is

$$p(x) = \begin{cases} p(1 - p)^{x-1} & x = 1, 2, 3, \ldots \\ 0 & \text{otherwise} \end{cases}$$

From the definition,

$$E(X) = \sum_D x \cdot p(x) = \sum_{x=1}^{\infty} xp(1 - p)^{x-1} = p \sum_{x=1}^{\infty} \left[-\frac{d}{dp} (1 - p)^x \right] \tag{3.9}$$

If we interchange the order of taking the derivative and the summation, the sum is that of a geometric series. After the sum is computed, the derivative is taken, and the final result is $E(X) = 1/p$. If p is near 1, we expect to see a boy very soon, whereas if p is near 0, we expect many births before the first boy. For $p = .5$, $E(X) = 2$. ∎

There is another frequently used interpretation of μ. Consider the pmf

$$p(x) = \begin{cases} (.5) \cdot (.5)^{x-1} & \text{if } x = 1, 2, 3, \ldots \\ 0 & \text{otherwise} \end{cases}$$

This is the pmf of X = the number of tosses of a fair coin necessary to obtain the first H (a special case of Example 3.18). Suppose we observe a value x from this pmf (toss a coin until an H appears), then observe independently another value (keep tossing), then another, and so on. If after observing a very large number of x values, we average them, the resulting sample average will be very near to $\mu = 2$. That is, μ can be interpreted as the long-run average observed value of X when the experiment is performed repeatedly.

Example 3.19 Let X have pmf

$$p(x) = \begin{cases} k/x^2 & x = 1, 2, 3, \ldots \\ 0 & \text{otherwise} \end{cases}$$

where k is chosen so that $\sum_{x=1}^{\infty} (k/x^2) = 1$. (In a mathematics course on infinite series, it is shown that $\sum_{x=1}^{\infty} (1/x^2) < \infty$, which implies that such a k exists, but its exact value need not concern us.) The expected value of X is

$$\mu = E(X) = \sum_{x=1}^{\infty} x \cdot \frac{k}{x^2} = k \sum_{x=1}^{\infty} \frac{1}{x} \tag{3.10}$$

The sum on the right of Equation (3.10) is the famous harmonic series of mathematics and can be shown to equal ∞. $E(X)$ is not finite here because $p(x)$ does not decrease sufficiently fast as x increases; statisticians say that the probability distribution of X has "a heavy tail." If a sequence of X values is chosen using this distribution, the sample average will not settle down to some finite number but will tend to grow without bound.

Statisticians use the phrase "heavy tails" in connection with any distribution having a large amount of probability far from μ (so heavy tails do not require $\mu = \infty$). Such heavy tails make it difficult to make inferences about μ. ∎

The Expected Value of a Function

Often we will be interested in the expected value of some function $h(X)$ rather than X itself.

Example 3.20 Suppose a bookstore purchases ten copies of a book at $6.00 each, to sell at $12.00 with the understanding that at the end of a 3-month period any unsold copies can be redeemed for $2.00. If X = the number of copies sold, then net revenue = $h(X) = 12X + 2(10 - X) - 60 = 10X - 40$. ∎

An easy way of computing the expected value of $h(X)$ is suggested by the following example.

Example 3.21 Let X = the number of cylinders in the engine of the next car to be tuned up at a certain facility. The cost of a tune-up is related to X by $h(X) = 20 + 3X + .5X^2$. Since X is a random variable, so is $h(X)$; denote this latter rv by Y. The pmf's of X and Y are as follows:

x	4	6	8		y	40	56	76
$p(x)$	.5	.3	.2		$p(y)$	.5	.3	.2

With D^* denoting possible values of Y,

$$E(Y) = E[h(X)] = \sum_{D^*} y \cdot p(y) \tag{3.11}$$

$$= (40)(.5) + (56)(.3) + (76)(.2)$$

$$= h(4) \cdot (.5) + h(6) \cdot (.3) + h(8) \cdot (.2)$$

$$= \sum_{D} h(x) \cdot p(x)$$

According to Equation (3.11), it was not necessary to determine the pmf of Y to obtain $E(Y)$; instead, the desired expected value is a weighted average of the possible $h(x)$ (rather than x) values. ∎

PROPOSITION

> If the rv X has set of possible values D and pmf $p(x)$, then the expected value of any function $h(X)$, denoted by $E[h(X)]$ or $\mu_{h(X)}$, is computed by
>
> $$E[h(X)] = \sum_{D} h(x) \cdot p(x)$$

According to this proposition, $E[h(X)]$ is computed in the same way that $E(X)$ itself is, except that $h(x)$ is substituted in place of x.

Example 3.22 A computer store has purchased three computers of a certain type at $500 apiece. It will sell them for $1000 apiece. The manufacturer has agreed to repurchase any computers still unsold after a specified period at $200 apiece. Let X denote the number of computers sold and suppose that $p(0) = .1$, $p(1) = .2$, $p(2) = .3$, and $p(3) = .4$. With $h(X)$ denoting the profit associated with selling X units, the given information implies that $h(X) = $ revenue $-$ cost $= 1000X + 200(3 - X) - 1500 = 800X - 900$. The expected profit is then

$$E[h(X)] = h(0) \cdot p(0) + h(1) \cdot p(1) + h(2) \cdot p(2) + h(3) \cdot p(3)$$

$$= (-900)(.1) + (-100)(.2) + (700)(.3) + (1500)(.4)$$

$$= \$700$$

∎

Rules of Expected Value

The $h(X)$ function of interest is quite frequently a linear function $aX + b$. In this case, $E[h(X)]$ is easily computed from $E(X)$.

PROPOSITION

> $E(aX + b) = a \cdot E(X) + b$
>
> (or, using alternative notation, $\mu_{aX+b} = a \cdot \mu_X + b$).

To paraphrase, the expected value of a linear function equals the linear function evaluated at the expected value $E(X)$. Since $h(X)$ in Example 3.22 is linear and $E(X) = 2$, $E[h(X)] = 800(2) - 900 = \$700$ as before.

Proof

$$E(aX + b) = \sum_D (ax + b) \cdot p(x) = a\sum_D x \cdot p(x) + b\sum_D p(x)$$
$$= aE(X) + b \qquad \blacksquare$$

Two special cases of the proposition yield two important rules of expected value.

---

1. For any constant a, $E(aX) = a \cdot E(X)$ (take $b = 0$).   (3.12)
2. For any constant b, $E(X + b) = E(X) + b$ (take $a = 1$).

Multiplication of X by a constant a changes the unit of measurement (from dollars to cents, where $a = 100$, inches to cm, where $a = 2.54$, etc.). Rule 1 says that the expected value in the new units equals the expected value in the old units multiplied by the conversion factor a. Similarly, if a constant b is added to each possible value of X, then the expected value will be shifted by that same constant amount.

The Variance of X

The expected value of X measures where the probability distribution is centered. Using the physical analogy of placing point mass $p(x)$ at the value x on a one-dimensional axis, if the axis were then supported by a fulcrum placed at μ, there would be no tendency for the axis to tilt. This is illustrated for two different distributions in Figure 3.7.

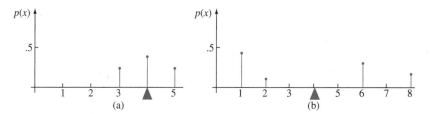

Figure 3.7   Two different probability distributions with $\mu = 4$

Although both distributions pictured in Figure 3.7 have the same center μ, the distribution of Figure 3.7(b) has greater spread or variability or dispersion than does that of Figure 3.7(a). We will use the variance of X to measure the amount of variability in (the distribution of) X, just as s^2 was used in Chapter 1 to measure variability in a sample.

DEFINITION

Let X have pmf $p(x)$ and expected value μ. Then the **variance** of X, denoted by $V(X)$ or σ_X^2, or just σ^2, is

$$V(X) = \sum_D (x - \mu)^2 \cdot p(x) = E[(X - \mu)^2]$$

The **standard deviation** (SD) of X is

$$\sigma_X = \sqrt{\sigma_X^2}$$

The quantity $h(X) = (X - \mu)^2$ is the squared deviation of X from its mean, and σ^2 is the expected squared deviation. If most of the probability distribution is close to μ, then σ^2 will be relatively small. However, if there are x values far from μ that have large $p(x)$, then σ^2 will be quite large.

Example 3.23 If X is the number of cylinders on the next car to be tuned at a service facility, with pmf as given in Example 3.21 [$p(4) = .5$, $p(6) = .3$, $p(8) = .2$, from which $\mu = 5.4$], then

$$V(X) = \sigma^2 = \sum_{x=4}^{8} (x - 5.4)^2 \cdot p(x)$$

$$= (4 - 5.4)^2(.5) + (6 - 5.4)^2(.3) + (8 - 5.4)^2(.2) = 2.44$$

The standard deviation of X is $\sigma = \sqrt{2.44} = 1.562$. ∎

When the pmf $p(x)$ specifies a mathematical model for the distribution of population values, both σ^2 and σ measure the spread of values in the population; σ^2 is the population variance, and σ is the population standard deviation.

A Shortcut Formula for σ^2

The number of arithmetic operations necessary to compute σ^2 can be reduced by using an alternative computing formula.

PROPOSITION

$$V(X) = \sigma^2 = \left[\sum_{D} x^2 \cdot p(x) \right] - \mu^2 = E(X^2) - [E(X)]^2$$

In using this formula, $E(X^2)$ is computed first without any subtraction; then $E(X)$ is computed, squared, and subtracted (once) from $E(X^2)$.

Example 3.24 The pmf of the number of cylinders X on the next car to be tuned at a certain facility was given in Example 3.23 as $p(4) = .5$, $p(6) = .3$, and $p(8) = .2$, from which $\mu = 5.4$ and

$$E(X^2) = (4^2)(.5) + (6^2)(.3) + (8^2)(.2) = 31.6$$

Thus $\sigma^2 = 31.6 - (5.4)^2 = 2.44$ as in Example 3.23. ∎

Proof of the Shortcut Formula

Expand $(x - \mu)^2$ in the definition of σ^2 to obtain $x^2 - 2\mu x + \mu^2$ and then carry Σ through to each of the three terms:

$$\sigma^2 = \sum_{D} x^2 \cdot p(x) - 2\mu \cdot \sum_{D} x \cdot p(x) + \mu^2 \sum_{D} p(x)$$

$$= E(X^2) - 2\mu \cdot \mu + \mu^2 = E(X^2) - \mu^2$$ ∎

Rules of Variance

The variance of $h(X)$ is the expected value of the squared difference between $h(X)$ and its expected value:

$$V[h(X)] = \sigma_{h(X)}^2 = \sum_{D} \{h(x) - E[h(X)]\}^2 \cdot p(x) \tag{3.13}$$

When $h(X)$ is a linear function, $V[h(X)]$ is easily related to $V(X)$.

PROPOSITION

$$V(aX + b) = \sigma^2_{aX+b} = a^2 \cdot \sigma^2_X \quad \text{and} \quad \sigma_{aX+b} = |a| \cdot \sigma_X$$

This result says that the addition of the constant b does not affect the variance, which is intuitive, because the addition of b changes the location (mean value) but not the spread of values. In particular,

1. $\sigma^2_{aX} = a^2 \cdot \sigma^2_X, \sigma_{aX} = |a| \cdot \sigma_X$
2. $\sigma^2_{X+b} = \sigma^2_X$ (3.14)

The reason for the absolute value in σ_{aX} is that a may be negative, whereas a standard deviation cannot be negative; a^2 results when a is brought outside the term being squared in Equation (3.13).

Example 3.25   In the computer sales problem of Example 3.22, $E(X) = 2$ and

$$E(X^2) = (0)^2(.1) + (1)^2(.2) + (2)^2(.3) + (3)^2(.4) = 5$$

so $V(X) = 5 - (2)^2 = 1$. The profit function $h(X) = 800X - 900$ then has variance $(800)^2 \cdot V(X) = (640{,}000)(1) = 640{,}000$ and standard deviation 800. ∎

Exercises | Section 3.3 (28–43)

28. The pmf for $X =$ the number of major defects on a randomly selected appliance of a certain type is

x	0	1	2	3	4
$p(x)$	.08	.15	.45	.27	.05

Compute the following:
a. $E(X)$
b. $V(X)$ directly from the definition
c. The standard deviation of X
d. $V(X)$ using the shortcut formula

29. An individual who has automobile insurance from a certain company is randomly selected. Let Y be the number of moving violations for which the individual was cited during the last 3 years. The pmf of Y is

y	0	1	2	3
$p(y)$	.60	.25	.10	.05

a. Compute $E(Y)$.
b. Suppose an individual with Y violations incurs a surcharge of $\$100Y^2$. Calculate the expected amount of the surcharge.

30. Refer to Exercise 29 and calculate $V(Y)$ and σ_Y. Then determine the probability that Y is within 1 standard deviation of its mean value.

31. An appliance dealer sells three different models of upright freezers having 13.5, 15.9, and 19.1 cubic feet of storage space, respectively. Let $X =$ the amount of storage space purchased by the next customer to buy a freezer. Suppose that X has pmf

x	13.5	15.9	19.1
$p(x)$	.2	.5	.3

a. Compute $E(X)$, $E(X^2)$, and $V(X)$.

b. If the price of a freezer having capacity X cubic feet is $25X - 8.5$, what is the expected price paid by the next customer to buy a freezer?

c. What is the variance of the price $25X - 8.5$ paid by the next customer?

d. Suppose that although the rated capacity of a freezer is X, the actual capacity is $h(X) = X - .01X^2$. What is the expected actual capacity of the freezer purchased by the next customer?

32. Let X be a Bernoulli rv with pmf as in Example 3.17.
a. Compute $E(X^2)$.
b. Show that $V(X) = p(1 - p)$.
c. Compute $E(X^{79})$.

33. Suppose that the number of plants of a particular type found in a rectangular region (called a quadrat by ecologists) in a certain geographic area is an rv X with pmf

$$p(x) = \begin{cases} c/x^3 & x = 1, 2, 3, \ldots \\ 0 & \text{otherwise} \end{cases}$$

Is $E(X)$ finite? Justify your answer (this is another distribution that statisticians would call heavy-tailed).

34. A small drugstore orders copies of a news magazine for its magazine rack each week. Let X = demand for the magazine, with pmf

x	1	2	3	4	5	6
$p(x)$	$\frac{1}{15}$	$\frac{2}{15}$	$\frac{3}{15}$	$\frac{4}{15}$	$\frac{3}{15}$	$\frac{2}{15}$

Suppose the store owner actually pays $1.00 for each copy of the news magazine and the price to customers is $2.00. If magazines left at the end of the week have no salvage value, is it better to order three or four copies of the magazine? (*Hint:* For both three and four copies ordered, express net revenue as a function of demand X and then compute the expected revenue.)

35. Let X be the damage incurred (in $) in a certain type of accident during a given year. Possible X values are 0, 1000, 5000, and 10000, with probabilities .8, .1, .08, and .02, respectively. A particular company offers a $500 deductible policy. If the company wishes its expected profit to be $100, what premium amount should it charge?

36. The n candidates for a job have been ranked 1, 2, 3, $\ldots$, n. Let X = the rank of a randomly selected candidate, so that X has pmf

$$p(x) = \begin{cases} 1/n & x = 1, 2, 3, \ldots, n \\ 0 & \text{otherwise} \end{cases}$$

(this is called the discrete uniform distribution). Compute $E(X)$ and $V(X)$ using the shortcut formula. [*Hint:* The sum of the first n positive integers is $n(n + 1)/2$, whereas the sum of their squares is $n(n + 1)(2n + 1)/6$.]

37. Let X = the outcome when a fair die is rolled once. If before the die is rolled you are offered either $(1/3.5)$ dollars or $h(X) = 1/X$ dollars, would you accept the guaranteed amount or would you gamble? [*Note:* It is not generally true that $1/E(X) = E(1/X)$.]

38. A chemical supply company currently has in stock 100 lb of a certain chemical, which it sells to customers in 5-lb lots. Let X = the number of lots ordered by a randomly chosen customer, and suppose that X has pmf

x	1	2	3	4
$p(x)$	.2	.4	.3	.1

Compute $E(X)$ and $V(X)$. Then compute the expected number of pounds left after the next customer's order is shipped, and the variance of the number of pounds left. (*Hint:* The number of pounds left is a linear function of X.)

39. a. Draw a line graph of the pmf of X in Exercise 34. Then determine the pmf of $-X$ and draw its line graph. From these two pictures, what can you say about $V(X)$ and $V(-X)$?
b. Use the proposition involving $V(aX + b)$ to establish a general relationship between $V(X)$ and $V(-X)$.

40. Use the definition in Expression (3.13) to prove that $V(aX + b) = a^2 \cdot \sigma_X^2$. [*Hint:* With $h(X) = aX + b$, $E[h(X)] = a\mu + b$ where $\mu = E(X)$.]

41. Suppose $E(X) = 5$ and $E[X(X - 1)] = 27.5$. What is
a. $E(X^2)$? [*Hint:* $E[X(X - 1)] = E[X^2 - X] = E(X^2) - E(X)$.]
b. $V(X)$?
c. The general relationship among the quantities $E(X)$, $E[X(X - 1)]$, and $V(X)$?

42. Write a general rule for $E(X - c)$ where c is a constant. What happens when you let $c = \mu$, the expected value of X?

43. A result called **Chebyshev's inequality** states that for any probability distribution of an rv X and any number k that is at least 1, $P(|X - \mu| \geq k\sigma) \leq 1/k^2$. In words, the probability that the value of X lies at least k standard deviations from its mean is at most $1/k^2$.

a. What is the value of the upper bound for $k = 2$? $k = 3$? $k = 4$? $k = 5$? $k = 10$?

b. Compute μ and σ for the distribution of Exercise 13. Then evaluate $P(|X - \mu| \geq k\sigma)$ for the values of k given in part (a). What does this suggest about the upper bound relative to the corresponding probability?

c. Let X have three possible values, -1, 0, and 1, with probabilities $\frac{1}{18}$, $\frac{8}{9}$, and $\frac{1}{18}$, respectively. What is $P(|X - \mu| \geq 3\sigma)$, and how does it compare to the corresponding bound?

d. Give a distribution for which $P(|X - \mu| \geq 5\sigma) = .04$.

3.4 | The Binomial Probability Distribution

There are many experiments that conform either exactly or approximately to the following list of requirements:

1. The experiment consists of a sequence of n trials, where n is fixed in advance of the experiment.

2. The trials are identical, and each trial can result in one of the same two possible outcomes, which we denote by success (S) or failure (F).

3. The trials are independent, so that the outcome on any particular trial does not influence the outcome on any other trial.

4. The probability of success is constant from trial to trial; we denote this probability by p.

DEFINITION

> An experiment for which Conditions 1–4 are satisfied is called a **binomial experiment.**

Example 3.26 The same coin is tossed successively and independently n times. We arbitrarily use S to denote the outcome H (heads) and F to denote the outcome T (tails). Then this experiment satisfies Conditions 1–4. Tossing a thumbtack n times, with S = point up and F = point down, also results in a binomial experiment. ∎

Many experiments involve a sequence of independent trials for which there are more than two possible outcomes on any one trial. A binomial experiment can then be created by dividing the possible outcomes into two groups.

Example 3.27 The color of pea seeds is determined by a single genetic locus. If the two alleles at this locus are AA or Aa (the genotype), then the pea will be yellow (the phenotype), and if the allele is aa, the pea will be green. Suppose we pair off 20 Aa seeds and cross the two seeds in each of the ten pairs to obtain ten new genotypes. Call each new genotype a success S if it is aa and a failure otherwise. Then with this identification of S and F, the experiment is binomial with $n = 10$ and $p = P(\text{aa genotype})$. If each member of the pair is equally likely to contribute a or A, then $p = P(\text{a}) \cdot P(\text{a}) = \left(\frac{1}{2}\right)\left(\frac{1}{2}\right) = \frac{1}{4}$. ∎

Example 3.28 Suppose a certain city has 50 licensed restaurants, of which 15 currently have at least one serious health code violation and the other 35 have no serious violations. There are five inspectors, each of whom will inspect one restaurant during the coming week. The name of each restaurant is written on a different slip of paper, and after the slips are thoroughly mixed, each inspector in turn draws one of the slips *without replacement*. Label the *i*th trial as a success if the *i*th restaurant selected ($i = 1, \ldots, 5$) has no serious violations. Then

$$P(S \text{ on first trial}) = \frac{35}{50} = .70$$

and

$$P(S \text{ on second trial}) = P(SS) + P(FS) = P(\text{second } S \mid \text{first } S) \, P(\text{first } S) \\ + P(\text{second } S \mid \text{first } F) \, P(\text{first } F)$$

$$= \frac{34}{49} \cdot \frac{35}{50} + \frac{35}{49} \cdot \frac{15}{50} = \frac{35}{50} \left(\frac{34}{49} + \frac{15}{49} \right) = \frac{35}{50} = .70$$

Similarly, it can be shown that $P(S \text{ on } i\text{th trial}) = .70$ for $i = 3, 4, 5$. However,

$$P(S \text{ on fifth trial} \mid SSSS) = \frac{31}{46} = .67$$

whereas

$$P(S \text{ on fifth trial} \mid FFFF) = \frac{35}{46} = .76$$

so the experiment is not binomial because the trials are not independent. In general, if sampling is without replacement, the experiment will not yield independent trials. If each slip had been replaced after being drawn, then trials would have been independent, but this might have resulted in the same restaurant being inspected by more than one inspector. ∎

Example 3.29 Suppose a certain state has 500,000 licensed drivers, of whom 400,000 are insured. A sample of ten drivers is chosen without replacement. The *i*th trial is labeled *S* if the *i*th driver chosen is insured. Although this situation would seem identical to that of Example 3.28, the important difference is that the size of the population being sampled is very large relative to the sample size. In this case

$$P(S \text{ on } 2 \mid S \text{ on } 1) = \frac{399,999}{499,999} = .80000$$

and

$$P(S \text{ on } 10 \mid S \text{ on first } 9) = \frac{399,991}{499,991} = .799996 \approx .80000$$

These calculations suggest that, although the trials are not exactly independent, the conditional probabilities differ so slightly from one another that for practical purposes the trials can be regarded as independent with constant $P(S) = .8$. Thus, to a very good approximation, the experiment is binomial with $n = 10$ and $p = .8$. ∎

We will use the following rule of thumb in deciding whether a "without-replacement" experiment can be treated as a binomial experiment.

RULE | Suppose each trial of an experiment can result in S or F, but the sampling is without replacement from a population of size N. If the sample size (number of trials) n is at most 5% of the population size, the experiment can be analyzed as though it were exactly a binomial experiment.

By "analyzed," we mean that probabilities based on the binomial experiment assumptions will be quite close to the actual "without-replacement" probabilities, which are typically more difficult to calculate. In Example 3.28, $n/N = 5/50 = .1 > .05$, so the binomial experiment is not a good approximation, but in Example 3.29, $n/N = 10/500,000 < .05$.

The Binomial Random Variable and Distribution

In most binomial experiments, it is the total number of S's, rather than knowledge of exactly which trials yielded S's, that is of interest.

DEFINITION | Given a binomial experiment consisting of n trials, the **binomial random variable** X associated with this experiment is defined as

$$X = \text{the number of } S\text{'s among the } n \text{ trials}$$

Suppose, for example, that $n = 3$. Then there are eight possible outcomes for the experiment:

SSS SSF SFS SFF FSS FSF FFS FFF

From the definition of X, $X(SSF) = 2$, $X(SFF) = 1$, and so on. Possible values for X in an n-trial experiment are $x = 0, 1, 2, \ldots, n$. We will often write $X \sim \text{Bin}(n, p)$ to indicate that X is a binomial rv based on n trials with success probability p.

NOTATION | Because the pmf of a binomial rv X depends on the two parameters n and p, we denote the pmf by $b(x; n, p)$.

To gain insight into $b(x; n, p)$ for general n, consider the case $n = 4$ for which each outcome, its probability, and corresponding x value are listed in Table 3.1 (page 122). For example,

$$
\begin{aligned}
P(SSFS) &= P(S) \cdot P(S) \cdot P(F) \cdot P(S) \quad \text{(independent trials)} \\
&= p \cdot p \cdot (1 - p) \cdot p \quad \text{(constant } P(S)\text{)} \\
&= p^3 \cdot (1 - p)
\end{aligned}
$$

Table 3.1 Outcomes and probabilities for a binomial experiment with four trials

Outcome	x	Probability	Outcome	x	Probability
SSSS	4	p^4	FSSS	3	$p^3(1-p)$
SSSF	3	$p^3(1-p)$	FSSF	2	$p^2(1-p)^2$
SSFS	3	$p^3(1-p)$	FSFS	2	$p^2(1-p)^2$
SSFF	2	$p^2(1-p)^2$	FSFF	1	$p(1-p)^3$
SFSS	3	$p^3(1-p)$	FFSS	2	$p^2(1-p)^2$
SFSF	2	$p^2(1-p)^2$	FFSF	1	$p(1-p)^3$
SFFS	2	$p^2(1-p)^2$	FFFS	1	$p(1-p)^3$
SFFF	1	$p(1-p)^3$	FFFF	0	$(1-p)^4$

In this special case, we wish $b(x; 4, p)$ for $x = 0, 1, 2, 3,$ and 4. For $b(3; 4, p)$, we identify which of the 16 outcomes yield an x value of 3 and add up the probabilities associated with each such outcome:

$$b(3; 4, p) = P(FSSS) + P(SFSS) + P(SSFS) + P(SSSF) = 4p^3(1-p)$$

Since the number of outcomes with $x = 3$ is four and each of the four has probability $p^3(1-p)$ (the order of S's and F's is not important, but only the number of S's),

$$b(3; 4, p) = \left\{ \begin{matrix} \text{number of outcomes} \\ \text{with } X = 3 \end{matrix} \right\} \cdot \left\{ \begin{matrix} \text{probability of any particular} \\ \text{outcome with } X = 3 \end{matrix} \right\}$$

Similarly, $b(2; 4, p) = 6p^2(1-p)^2$, which is also the product of the number of outcomes with $X = 2$ and the probability of any such outcome.

In general,

$$b(x; n, p) = \left\{ \begin{matrix} \text{number of sequences of} \\ \text{length } n \text{ consisting of } x \text{ } S\text{'s} \end{matrix} \right\} \cdot \left\{ \begin{matrix} \text{probability of any} \\ \text{particular such sequence} \end{matrix} \right\}$$

Since the ordering of S's and F's is not important, the second factor in the previous equation is $p^x(1-p)^{n-x}$ (e.g., the first x trials resulting in S and the last $n - x$ resulting in F). The first factor is the number of ways of choosing x of the n trials to be S's—that is, the number of combinations of size x that can be constructed from n distinct objects (trials here).

THEOREM

$$b(x; n, p) = \begin{cases} \dbinom{n}{x} p^x(1-p)^{n-x} & x = 0, 1, 2, \ldots n \\ 0 & \text{otherwise} \end{cases}$$

Example 3.30 Each of six randomly selected cola drinkers is given a glass containing cola S and one containing cola F. The glasses are identical in appearance except for a code on the bottom to identify the cola. Suppose there is actually no tendency among cola drinkers to prefer one cola to the other. Then $p = P$(a selected individual prefers S) $= .5$, so with $X =$ the number among the six who prefer S, $X \sim$ Bin(6, .5).

Thus,

$$P(X = 3) = b(3; 6, .5) = \binom{6}{3}(.5)^3(.5)^3 = 20(.5)^6 = .313$$

The probability that at least three prefer S is

$$P(3 \le X) = \sum_{x=3}^{6} b(x; 6, .5) = \sum_{x=3}^{6} \binom{6}{x}(.5)^x(.5)^{6-x} = .656$$

and the probability that at most one prefers S is

$$P(X \le 1) = \sum_{x=0}^{1} b(x; 6, .5) = .109$$ ■

Using Binomial Tables

Even for a relatively small value of n, the computation of binomial probabilities can be tedious. Appendix Table A.1 tabulates the cdf $F(x) = P(X \le x)$ for $n = 5, 10, 15, 20, 25$ in combination with selected values of p. Various other probabilities can then be calculated using the proposition on cdf's from Section 3.2.

NOTATION

> For $X \sim$ Bin(n, p), the cdf will be denoted by
>
> $$P(X \le x) = B(x; n, p) = \sum_{y=0}^{x} b(y; n, p) \qquad x = 0, 1, \ldots, n$$

Example 3.31 Suppose that 20% of all copies of a particular textbook fail a certain binding strength test. Let X denote the number among 15 randomly selected copies that fail the test. Then X has a binomial distribution with $n = 15$ and $p = .2$.

1. The probability that at most 8 fail the test is

$$P(X \le 8) = \sum_{y=0}^{8} b(y; 15, .2) = B(8; 15, .2)$$

which is the entry in the $x = 8$ row and the $p = .2$ column of the $n = 15$ binomial table. From Appendix Table A.1, the probability is $B(8; 15, .2) = .999$.

2. The probability that exactly 8 fail is

$$P(X = 8) = P(X \le 8) - P(X \le 7) = B(8; 15, .2) - B(7; 15, .2)$$

which is the difference between two consecutive entries in the $p = .2$ column. The result is $.999 - .996 = .003$.

3. The probability that at least 8 fail is

$$P(X \geq 8) = 1 - P(X \leq 7) = 1 - B(7; 15, .2)$$

$$= 1 - \begin{pmatrix} \text{entry in } x = 7 \\ \text{row of } p = .2 \text{ column} \end{pmatrix}$$

$$= 1 - .996 = .004$$

4. Finally, the probability that between 4 and 7, inclusive, fail is

$$P(4 \leq X \leq 7) = P(X = 4, 5, 6, \text{ or } 7) = P(X \leq 7) - P(X \leq 3)$$

$$= B(7; 15, .2) - B(3; 15, .2) = .996 - .648 = .348$$

Notice that this latter probability is the difference between entries in the $x = 7$ and $x = 3$ rows, *not* the $x = 7$ and $x = 4$ rows.   ■

Example 3.32   An electronics manufacturer claims that at most 10% of its power supply units need service during the warranty period. To investigate this claim, technicians at a testing laboratory purchase 20 units and subject each one to accelerated testing to simulate use during the warranty period. Let p denote the probability that a power supply unit needs repair during the period (the proportion of all such units that need repair). The laboratory technicians must decide whether the data resulting from the experiment supports the claim that $p \leq .10$. Let X denote the number among the 20 sampled that need repair, so $X \sim \text{Bin}(20, p)$. Consider the decision rule

> Reject the claim that $p \leq .10$ in favor of the conclusion that $p > .10$ if $x \geq 5$ (where x is the observed value of X), and consider the claim plausible if $x \leq 4$.

The probability that the claim is rejected when $p = .10$ (an incorrect conclusion) is

$$P(X \geq 5 \text{ when } p = .10) = 1 - B(4; 20, .1) = 1 - .957 = .043$$

The probability that the claim is not rejected when $p = .20$ (a different type of incorrect conclusion) is

$$P(X \leq 4 \text{ when } p = .2) = B(4; 20, .2) = .630$$

The first probability is rather small, but the second is intolerably large. When $p = .20$, so that the manufacturer has grossly understated the percentage of units that need service, and the stated decision rule is used, 63% of all samples will result in the manufacturer's claim being judged plausible!

One might think that the probability of this second type of erroneous conclusion could be made smaller by changing the cutoff value 5 in the decision rule to something else. However, although replacing 5 by a smaller number would yield a probability smaller than .630, the other probability would then increase. The only way to make both "error probabilities" small is to base the decision rule on an experiment involving many more units.   ■

Note that a table entry of 0 signifies only that a probability is 0 to three significant digits, for all entries in the table are actually positive. Much more extensive tables of binomial probabilities have been published. Alternatively, statistical computer packages such as MINITAB will generate either $b(x; n, p)$ or $B(x; n, p)$ for any specified values of n and p. In Chapter 4, we will present a method for obtaining quick and accurate approximations to binomial probabilities when n is large.

The Mean and Variance of *X*

For $n = 1$, the binomial distribution becomes the Bernoulli distribution. From Example 3.17, the mean value of a Bernoulli variable is $\mu = p$, so the expected number of S's on any single trial is p. Since a binomial experiment consists of n trials, intuition suggests that for $X \sim \text{Bin}(n, p)$, $E(X) = np$, the product of the number of trials and the probability of success on a single trial. The expression for $V(X)$ is not so intuitive.

PROPOSITION

> If $X \sim \text{Bin}(n, p)$, then $E(X) = np$, $V(X) = np(1 - p) = npq$, and $\sigma_X = \sqrt{npq}$ (where $q = 1 - p$).

Thus, calculating the mean and variance of a binomial rv does not necessitate evaluating summations. The proof of the result for $E(X)$ is sketched in Exercise 58.

Example 3.33 If 75% of all purchases at a certain store are made with a credit card and X is the number among ten randomly selected purchases made with a credit card, then $X \sim \text{Bin}(10, .75)$. Thus $E(X) = np = (10)(.75) = 7.5$, $V(X) = npq = 10(.75)(.25) = 1.875$, and $\sigma = \sqrt{1.875}$. Again, even though X can take on only integer values, $E(X)$ need not be an integer. If we perform a large number of independent binomial experiments, each with $n = 10$ trials and $p = .75$, then the average number of S's per experiment will be close to 7.5. ∎

Exercises | Section 3.4 (44–61)

44. Compute the following binomial probabilities directly from the formula for $b(x; n, p)$.
  a. $b(3; 8, .6)$
  b. $b(5; 8, .6)$
  c. $P(3 \leq X \leq 5)$ when $n = 8$ and $p = .6$
  d. $P(1 \leq X)$ when $n = 12$ and $p = .1$

45. Use Appendix Table A.1 to obtain the following probabilities:
  a. $B(4; 10, .3)$
  b. $b(4; 10, .3)$
  c. $b(6; 10, .7)$
  d. $P(2 \leq X \leq 4)$ when $X \sim \text{Bin}(10, .3)$
  e. $P(2 \leq X)$ when $X \sim \text{Bin}(10, .3)$
  f. $P(X \leq 1)$ when $X \sim \text{Bin}(10, .7)$
  g. $P(2 < X < 6)$ when $X \sim \text{Bin}(10, .3)$

46. When circuit boards used in the manufacture of compact disc players are tested, the long-run percentage of defectives is 5%. Let $X =$ the number of defective boards in a random sample of size $n = 25$, so $X \sim \text{Bin}(25, .05)$.
  a. Determine $P(X \leq 2)$.
  b. Determine $P(X \geq 5)$.
  c. Determine $P(1 \leq X \leq 4)$.
  d. What is the probability that none of the 25 boards are defective?
  e. Calculate the expected value and standard deviation of X.

47. A company that produces fine crystal knows from experience that 10% of its goblets have cosmetic flaws and must be classified as "seconds."

a. Among six randomly selected goblets, how likely is it that only one is a second?

b. Among six randomly selected goblets, what is the probability that at least two are seconds?

c. If goblets are examined one by one, what is the probability that at most five must be selected to find four that are not seconds?

48. Suppose that only 20% of all drivers come to a complete stop at an intersection having flashing red lights in all directions when no other cars are visible. What is the probability that, of 20 randomly chosen drivers coming to an intersection under these conditions,

a. At most 6 will come to a complete stop?

b. Exactly 6 will come to a complete stop?

c. At least 6 will come to a complete stop?

d. How many of the next 20 drivers do you expect to come to a complete stop?

49. Exercise 29 (Section 3.3) gave the pmf of Y, the number of traffic citations for a randomly selected individual insured by a particular company. What is the probability that among 15 randomly chosen such individuals

a. At least 10 have no citations?

b. More than half have at least one citation?

c. The number that have at least one citation is between 5 and 10, inclusive?*

50. A particular type of tennis racket comes in a midsize version and an oversize version. Sixty percent of all customers at a certain store want the oversize version.

a. Among ten randomly selected customers who want this type of racket, what is the probability that at least six want the oversize version?

b. Among ten randomly selected customers, what is the probability that the number who want the oversize version is within 1 standard deviation of the mean value?

c. The store currently has seven rackets of each version. What is the probability that all of the next ten customers who want this racket can get the version they want from current stock?

51. Twenty percent of all telephones of a certain type are submitted for service while under warranty. Of these, 60% can be repaired whereas the other 40% must be replaced with new units. If a company purchases ten of these telephones, what is the probability that exactly two will end up being replaced under warranty?

52. A very large batch of components has arrived at a distributor. The batch can be characterized as acceptable only if the proportion of defective components is at most .10. The distributor decides to randomly select 10 components and to accept the batch only if the number of defective components in the sample is at most 2.

a. What is the probability that the batch will be accepted when the actual proportion of defectives is .01? .05? .10? .20? .25?

b. Let p denote the actual proportion of defectives in the batch. A graph of P(lot is accepted) as a function of p with p on the horizontal axis and P(lot is accepted) on the vertical axis is called the *operating characteristic curve* for the acceptance sampling plan. Use the results of part (a) to sketch this curve for $0 \leq p \leq 1$.

c. Repeat parts (a) and (b) with "1" replacing "2" in the acceptance sampling plan.

d. Repeat parts (a) and (b) with "15" replacing "10" in the acceptance sampling plan.

e. Which of the three sampling plans, that of part (a), (c), or (d), appears most satisfactory, and why?

53. An ordinance requiring that a smoke detector be installed in all previously constructed houses has been in effect in a particular city for one year. The fire department is concerned that many houses remain without detectors. Let p = the true proportion of such houses having detectors and suppose that a random sample of 25 homes is inspected. If the sample strongly indicates that fewer than 80% of all houses have a detector, the fire department will campaign for a mandatory inspection program. Because of the costliness of the program, the department prefers not to call for such inspections unless sample evidence strongly argues for their necessity. Let X denote the number of homes with detectors among the 25 sampled. Consider rejecting the claim that $p \geq .8$ if $x \leq 15$, where x is the observed value of X.

a. What is the probability that the claim is rejected when the actual value of p is .8?

b. What is the probability of not rejecting the claim when $p = .7$? When $p = .6$?

c. How do the "error probabilities" of parts (a) and (b) change if the value 15 in the decision rule is replaced by 14?

54. A toll bridge charges $1.00 for passenger cars and $2.50 for other vehicles. Suppose that during daytime hours, 60% of all vehicles are passenger cars.

---

*"Between a and b, inclusive" is equivalent to $(a \leq X \leq b)$.

If 25 vehicles cross the bridge during a particular daytime period, what is the resulting expected toll revenue? [*Hint:* Let $X =$ the number of passenger cars; then the toll revenue $h(X)$ is a linear function of X.]

55. A student who is trying to write a paper for a course has a choice of two topics, A and B. If topic A is chosen, the student will order two books through interlibrary loan, whereas if topic B is chosen, the student will order four books. The student believes that a good paper necessitates receiving and using at least half the books ordered for either topic chosen. If the probability that a book ordered through interlibrary loan actually arrives in time is .9 and books arrive independently of one another, which topic should the student choose to maximize the probability of writing a good paper? What if the arrival probability is only .5 instead of .9?

56. a. For fixed n, are there values of p ($0 \le p \le 1$) for which $V(X) = 0$? Explain why this is so.
  b. For what value of p is $V(X)$ maximized? [*Hint:* Either graph $V(X)$ as a function of p or else take a derivative.]

57. a. Show that $b(x; n, 1 - p) = b(n - x; n, p)$.
  b. Show that $B(x; n, 1 - p) = 1 - B(n - x - 1; n, p)$. [*Hint:* At most x S's is equivalent to at least $(n - x)$ F's.]
  c. What do parts (a) and (b) imply about the necessity of including values of p greater than .5 in Appendix Table A.1?

58. Show that $E(X) = np$ when X is a binomial random variable. [*Hint:* First express $E(X)$ as a sum with lower limit $x = 1$. Then factor out np, let $y = x - 1$ so that the sum is from $y = 0$ to $y = n - 1$, and show that the sum equals 1.]

59. Customers at a gas station pay with a credit card (*A*), debit card (*B*), or cash (*C*). Assume that successive customers make independent choices, with $P(A) = .5$, $P(B) = .2$, and $P(C) = .3$.
  a. Among the next 100 customers, what are the mean and variance of the number who pay with a debit card? Explain your reasoning.
  b. Answer part (a) for the number among the 100 who don't pay with cash.

60. An airport limousine can accommodate up to four passengers on any one trip. The company will accept a maximum of six reservations for a trip, and a passenger must have a reservation. From previous records, 20% of all those making reservations do not appear for the trip. Answer the following questions, assuming independence wherever appropriate.
  a. If six reservations are made, what is the probability that at least one individual with a reservation cannot be accommodated on the trip?
  b. If six reservations are made, what is the expected number of available places when the limousine departs?
  c. Suppose the probability distribution of the number of reservations made is given in the accompanying table.

Number of reservations	3	4	5	6
Probability	.1	.2	.3	.4

Let X denote the number of passengers on a randomly selected trip. Obtain the probability mass function of X.

61. Refer to Chebyshev's inequality given in Exercise 43. Calculate $P(|X - \mu| \ge k\sigma)$ for $k = 2$ and $k = 3$ when $X \sim \text{Bin}(20, .5)$, and compare to the corresponding upper bound. Repeat for $X \sim \text{Bin}(20, .75)$.

3.5 Hypergeometric and Negative Binomial Distributions

The hypergeometric and negative binomial distributions are both closely related to the binomial distribution. Whereas the binomial distribution is the approximate probability model for sampling without replacement from a finite dichotomous (*S–F*) population, the hypergeometric distribution is the exact probability model for the number of *S*'s in the sample. The binomial rv X is the number of *S*'s when the number n of trials is fixed, whereas the negative binomial distribution arises from fixing the number of S's and letting the number of trials be random.

The Hypergeometric Distribution

The assumptions leading to the hypergeometric distribution are as follows:

1. The population or set to be sampled consists of N individuals, objects, or elements (a *finite* population).

2. Each individual can be characterized as a success (S) or a failure (F), and there are M successes in the population.

3. A sample of n individuals is selected without replacement in such a way that each subset of size n is equally likely to be chosen.

The random variable of interest is X = the number of S's in the sample. The probability distribution of X depends on the parameters n, M, and N, so we wish to obtain $P(X = x) = h(x; n, M, N)$.

Example 3.34   An undergraduate library has 20 copies of a certain introductory economics text, of which 8 are first printings and 12 are second printings (containing corrections of some minor errors that appeared in the first printing). The course instructor has requested that 5 copies be put on 2-hour reserve. If the copies are selected in a completely random fashion, so that every subset of size 5 has the same probability of being selected, what is the probability that x ($x = 0, 1, 2, 3, 4,$ or 5) of those selected are second printings?

In this example, the population size is $N = 20$, the sample size is $n = 5$, and the number of S's (second printing $= S$) and F's in the population are $M = 12$ and $N - M = 8$, respectively. Consider the value $x = 2$. Because all outcomes (each one consisting of five particular books) are equally likely,

$$P(X = 2) = h(2; 5, 12, 20) = \frac{\text{number of outcomes having } X = 2}{\text{number of possible outcomes}}$$

The number of possible outcomes in the experiment is the number of ways of selecting 5 from the 20 objects without regard to order—that is, $\binom{20}{5}$. To count the number of outcomes having $X = 2$, note that there are $\binom{12}{2}$ ways of selecting 2 of the second printings, and for each such way there are $\binom{8}{3}$ ways of selecting the 3 first printings to fill out the sample. The product rule from Chapter 2 then gives $\binom{12}{2}\binom{8}{3}$ as the number of outcomes with $X = 2$, so

$$h(2; 5, 12, 20) = \frac{\binom{12}{2}\binom{8}{3}}{\binom{20}{5}} = \frac{77}{323} = .238 \qquad \blacksquare$$

In general, if the sample size n is smaller than the number of successes in the population (M), then the largest possible X value is n. However, if $M < n$ (e.g., a sample size of 25 and only 15 successes in the population), then X can be at most M. Similarly, whenever the number of population failures ($N - M$) exceeds the sample size, the smallest possible X value is 0 (since all sampled individuals might then be failures).

However, if $N - M < n$, the smallest possible X value is $n - (N - M)$. Summarizing, the possible values of X satisfy the restriction $\max(0, n - (N - M)) \leq x \leq \min(n, M)$. An argument parallel to that of the previous example gives the pmf of X.

PROPOSITION

> If X is the number of S's in a completely random sample of size n drawn from a population consisting of M S's and $(N - M)$ F's, then the probability distribution of X, called the **hypergeometric distribution,** is given by
>
> $$P(X = x) = h(x; n, M, N) = \frac{\dbinom{M}{x}\dbinom{N - M}{n - x}}{\dbinom{N}{n}} \tag{3.15}$$
>
> for x an integer satisfying $\max(0, n - N + M) \leq x \leq \min(n, M)$.

In Example 3.34, $n = 5$, $M = 12$, and $N = 20$, so $h(x; 5, 12, 20)$ for $x = 0, 1, 2, 3, 4, 5$ can be obtained by substituting these numbers into Equation (3.15).

Example 3.35 Five individuals from an animal population thought to be near extinction in a certain region have been caught, tagged, and released to mix into the population. After they have had an opportunity to mix, a random sample of 10 of these animals is selected. Let X = the number of tagged animals in the second sample. If there are actually 25 animals of this type in the region, what is the probability that (a) $X = 2$? (b) $X \leq 2$?

The parameter values are $n = 10$, $M = 5$ (5 tagged animals in the population), and $N = 25$, so

$$h(x; 10, 5, 25) = \frac{\dbinom{5}{x}\dbinom{20}{10 - x}}{\dbinom{25}{10}} \qquad x = 0, 1, 2, 3, 4, 5$$

For part (a),

$$P(X = 2) = h(2; 10, 5, 25) = \frac{\dbinom{5}{2}\dbinom{20}{8}}{\dbinom{25}{10}} = .385$$

For part (b),

$$P(X \leq 2) = P(X = 0, 1, \text{ or } 2) = \sum_{x=0}^{2} h(x; 10, 5, 25)$$

$$= .057 + .257 + .385 = .699 \qquad \blacksquare$$

Comprehensive tables of the hypergeometric distribution are available, but because the distribution has three parameters, these tables require much more space than tables for the binomial distribution.

The Mean and Variance of X

As in the binomial case, there are simple expressions for $E(X)$ and $V(X)$ for hypergeometric rv's.

PROPOSITION

The mean and variance of the hypergeometric rv X having pmf $h(x; n, M, N)$ are

$$E(X) = n \cdot \frac{M}{N} \qquad V(X) = \left(\frac{N - n}{N - 1}\right) \cdot n \cdot \frac{M}{N} \cdot \left(1 - \frac{M}{N}\right)$$

The ratio M/N is the proportion of S's in the population. If we replace M/N by p in $E(X)$ and $V(X)$, we get

$$E(X) = np$$

$$V(X) = \left(\frac{N - n}{N - 1}\right) \cdot np(1 - p) \qquad (3.16)$$

Expression (3.16) shows that the means of the binomial and hypergeometric rv's are equal, whereas the variances of the two rv's differ by the factor $(N - n)/(N - 1)$, often called the **finite population correction factor.** This factor is less than 1, so the hypergeometric variable has smaller variance than does the binomial rv. The correction factor can be written $(1 - n/N)/(1 - 1/N)$, which is approximately 1 when n is small relative to N.

Example 3.36
(Example 3.35 continued)

In the animal-tagging example, $n = 10$, $M = 5$, and $N = 25$, so $p = \frac{5}{25} = .2$ and

$$E(X) = 10(.2) = 2$$

$$V(X) = \frac{15}{24}(10)(.2)(.8) = (.625)(1.6) = 1$$

If the sampling was carried out with replacement, $V(X) = 1.6$.

Suppose the population size N is not actually known, so the value x is observed and we wish to estimate N. It is reasonable to equate the observed sample proportion of S's, x/n, with the population proportion, M/N, giving the estimate

$$\hat{N} = \frac{M \cdot n}{x}$$

If $M = 100$, $n = 40$, and $x = 16$, then $\hat{N} = 250$. ∎

Approximating Hypergeometric Probabilities

Our general rule of thumb in Section 3.4 stated that if sampling was without replacement but n/N was at most .05, then the binomial distribution could be used to compute approximate probabilities involving the number of S's in the sample. A more precise statement is as follows: Let the population size, N, and number of population S's, M, get large with the ratio M/N approaching p. Then $h(x; n, M, N)$ approaches $b(x; n, p)$;

so for n/N small, the two are approximately equal provided that p is not too near either 0 or 1. This is the rationale for our rule of thumb.

The Negative Binomial Distribution

The negative binomial rv and distribution are based on an experiment satisfying the following conditions:

1. The experiment consists of a sequence of independent trials.

2. Each trial can result in either a success (S) or a failure (F).

3. The probability of success is constant from trial to trial, so $P(S \text{ on trial } i) = p$ for $i = 1, 2, 3 \ldots$.

4. The experiment continues (trials are performed) until a total of r successes have been observed, where r is a specified positive integer.

The random variable of interest is $X =$ the number of failures that precede the rth success; X is called a **negative binomial random variable** because, in contrast to the binomial rv, the number of successes is fixed and the number of trials is random.

Possible values of X are $0, 1, 2, \ldots$. Let $nb(x; r, p)$ denote the pmf of X. The event $\{X = x\}$ is equivalent to $\{r - 1 \ S$'s in the first $(x + r - 1)$ trials and an S on the $(x + r)$th trial$\}$ (e.g., if $r = 5$ and $x = 10$, then there must be four S's in the first 14 trials and the trial 15 must be an S). Since trials are independent,

$$nb(x; r, p) = P(X = x)$$
$$= P(r - 1 \ S\text{'s on the first } x + r - 1 \text{ trials}) \cdot P(S) \qquad (3.17)$$

The first probability on the far right of Expression (3.17) is the binomial probability

$$\binom{x + r - 1}{r - 1} p^{r-1}(1 - p)^x \qquad \text{where } P(S) = p$$

PROPOSITION

The pmf of the negative binomial rv X with parameters $r =$ number of S's and $p = P(S)$ is

$$nb(x; r, p) = \binom{x + r - 1}{r - 1} p^r(1 - p)^x \qquad x = 0, 1, 2, \ldots$$

Example 3.37 A pediatrician wishes to recruit 5 couples, each of whom is expecting their first child, to participate in a new natural childbirth regimen. Let $p = P(\text{a randomly selected couple agrees to participate})$. If $p = .2$, what is the probability that 15 couples must be asked before 5 are found who agree to participate? That is, with $S = \{\text{agrees to participate}\}$, what is the probability that 10 F's occur before the fifth S? Substituting $r = 5$, $p = .2$, and $x = 10$ into $nb(x; r, p)$ gives

$$nb(10; 5, .2) = \binom{14}{4}(.2)^5(.8)^{10} = .034$$

The probability that at most 10 F's are observed (at most 15 couples are asked) is

$$P(X \leq 10) = \sum_{x=0}^{10} nb(x; 5, .2) = (.2)^5 \sum_{x=0}^{10} \binom{x+4}{4}(.8)^x = .164 \qquad \blacksquare$$

In some sources, the negative binomial rv is taken to be the number of trials $X + r$ rather than the number of failures.

In the special case $r = 1$, the pmf is

$$nb(x; 1, p) = (1 - p)^x p \qquad x = 0, 1, 2, \ldots \qquad (3.18)$$

In Example 3.10, we derived the pmf for the number of trials necessary to obtain the first S, and the pmf there is similar to Expression (3.18). Both $X =$ number of F's and $Y =$ number of trials ($= 1 + X$) are referred to in the literature as **geometric random variables,** and the pmf in Expression (3.18) is called the **geometric distribution.**

In Example 3.18, the expected number of trials until the first S was shown to be $1/p$, so that the expected number of F's until the first S is $(1/p) - 1 = (1 - p)/p$. Intuitively, we would expect to see $r \cdot (1 - p)/p$ F's before the rth S, and this is indeed $E(X)$. There is also a simple formula for $V(X)$.

PROPOSITION

> If X is a negative binomial rv with pmf $nb(x; r, p)$, then
>
> $$E(X) = \frac{r(1 - p)}{p} \qquad V(X) = \frac{r(1 - p)}{p^2}$$

Finally, by expanding the binomial coefficient in front of $p^r(1 - p)^x$ and doing some cancellation, it can be seen that $nb(x; r, p)$ is well defined even when r is not an integer. This *generalized negative binomial distribution* has been found to fit observed data quite well in a wide variety of applications.

Exercises | Section 3.5 (62–72)

62. A shipment of 15 concrete cylinders has been received by a contractor, 5 for a small project and the other 10 for a larger project. Suppose that 6 of the 15 have a crushing strength below the specified minimum. If the 5 for the smaller project are randomly selected from the 15 and $X =$ the number among the 5 that have a below minimum crushing strength, then X has a hypergeometric distribution with parameters $n = 5$, $M = 6$, and $N = 15$. Compute the following:
   a. $P(X = 2)$
   b. $P(X \leq 2)$
   c. $P(X \geq 2)$
   d. $E(X)$ and $V(X)$

63. Each of 12 refrigerators of a certain type has been returned to a distributor because of the presence of a high-pitched oscillating noise when the refrigerator is running. Suppose that 5 of these 12 have defective compressors and the other 7 have less serious problems. If they are examined in random order, let $X =$ the number among the first 6 examined that have a defective compressor. Compute the following:
   a. $P(X = 1)$ **b.** $P(X \geq 4)$ **c.** $P(1 \leq X \leq 3)$

64. An instructor who taught two sections of engineering statistics last term, the first with 20 students and the second with 30, decided to assign a term project. After all projects had been turned in, the instructor randomly ordered them before grading. Consider the first 15 graded projects.

    a. What is the probability that exactly 10 of these are from the second section?

    b. What is the probability that at least 10 of these are from the second section?

    c. What is the probability that at least 10 of these are from the same section?

    d. What are the mean value and standard deviation of the number among these 15 that are from the second section?

    e. What are the mean value and standard deviation of the number of projects not among these first 15 that are from the second section?

65. A geologist has collected 10 specimens of basaltic rock and 10 specimens of granite. The geologist instructs a laboratory assistant to randomly select 15 of the specimens for analysis.

    a. What is the pmf of the number of granite specimens selected for analysis?

    b. What is the probability that all specimens of one of the two types of rock are selected for analysis?

    c. What is the probability that the number of granite specimens selected for analysis is within 1 standard deviation of its mean value?

66. A personnel director interviewing 11 senior engineers for four job openings has scheduled six interviews for the first day and five for the second day of interviewing. Assume that the candidates are interviewed in random order.

    a. What is the probability that x of the top four candidates are interviewed on the first day?

    b. How many of the top four candidates can be expected to be interviewed on the first day?

67. Twenty pairs of individuals playing in a bridge tournament have been seeded $1, \ldots, 20$. In the first part of the tournament, the 20 are randomly divided into 10 east–west pairs and 10 north–south pairs.

    a. What is the probability that x of the top 10 pairs end up playing east–west?

    b. What is the probability that all of the top five pairs end up playing the same direction?

    c. If there are $2n$ pairs, what is the pmf of $X =$ the number among the top n pairs who end up playing east–west? What are $E(X)$ and $V(X)$?

68. A second-stage smog alert has been called in a certain area of Los Angeles County in which there are 50 industrial firms. An inspector will visit 10 randomly selected firms to check for violations of regulations.

    a. If 15 of the firms are actually violating at least one regulation, what is the pmf of the number of firms visited by the inspector that are in violation of at least one regulation?

    b. If there are 500 firms in the area, of which 150 are in violation, approximate the pmf of part (a) by a simpler pmf.

    c. For $X =$ the number among the 10 visited that are in violation, compute $E(X)$ and $V(X)$ both for the exact pmf and the approximating pmf in part (b).

69. Suppose that $p = P(\text{male birth}) = .5$. A couple wishes to have exactly two female children in their family. They will have children until this condition is fulfilled.

    a. What is the probability that the family has x male children?

    b. What is the probability that the family has four children?

    c. What is the probability that the family has at most four children?

    d. How many male children would you expect this family to have? How many children would you expect this family to have?

70. A family decides to have children until it has three children of the same gender. Assuming $P(B) = P(G) = .5$, what is the pmf of $X =$ the number of children in the family?

71. Three brothers and their wives decide to have children until each family has two female children. What is the pmf of $X =$ the total number of male children born to the brothers? What is $E(X)$, and how does it compare to the expected number of male children born to each brother?

72. Individual A has a red die and B has a green die (both fair). If they each roll until they obtain five "doubles" (1–$1, \ldots, 6$–6), what is the pmf of $X =$ the total number of times a die is rolled? What are $E(X)$ and $V(X)$?

3.6 | The Poisson Probability Distribution

The binomial, hypergeometric, and negative binomial distributions were all derived by starting with an experiment consisting of trials or draws and applying the laws of probability to various outcomes of the experiment. There is no simple experiment on which the Poisson distribution is based, though we will shortly describe how it can be obtained by certain limiting operations.

DEFINITION

A random variable X is said to have a **Poisson distribution** if the pmf of X is

$$p(x; \lambda) = \frac{e^{-\lambda}\lambda^x}{x!} \qquad x = 0, 1, 2, \ldots$$

for some $\lambda > 0$.

The value of λ is frequently a rate per unit time or per unit area. The letter e in $p(x; \lambda)$ represents the base of the natural logarithm system; its numerical value is approximately 2.71828. Because λ must be positive, $p(x; \lambda) > 0$ for all possible x values. The fact that $\sum_{x=0}^{\infty} p(x; \lambda) = 1$ is a consequence of the Maclaurin infinite series expansion of e^{λ}, which appears in most calculus texts:

$$e^{\lambda} = 1 + \lambda + \frac{\lambda^2}{2!} + \frac{\lambda^3}{3!} + \cdots = \sum_{x=0}^{\infty} \frac{\lambda^x}{x!} \tag{3.19}$$

If the two extreme terms in Expression (3.19) are multiplied by $e^{-\lambda}$ and then $e^{-\lambda}$ is placed inside the summation, the result is

$$1 = \sum_{x=0}^{\infty} e^{-\lambda} \frac{\lambda^x}{x!}$$

which shows that $p(x; \lambda)$ fulfills the second condition necessary for specifying a pmf.

Example 3.38 Let X denote the number of creatures of a particular type captured in a trap during a given time period. Suppose that X has a Poisson distribution with $\lambda = 4.5$, so on average traps will contain 4.5 creatures. [The article "Dispersal Dynamics of the Bivalve *Gemma Gemma* in a Patchy Environment (*Ecological Monographs*, 1995: 1–20) suggests this model; the bivalve *Gemma gemma* is a small clam.] The probability that a trap contains exactly five creatures is

$$P(X = 5) = \frac{e^{-4.5}(4.5)^5}{5!} = .1708$$

The probability that a trap has at most five creatures is

$$P(X \le 5) = \sum_{x=0}^{5} \frac{e^{-4.5}(4.5)^x}{x!} = e^{-4.5}\left[1 + 4.5 + \frac{(4.5)^2}{2!} + \cdots + \frac{(4.5)^5}{5!}\right] = .7029 \quad \blacksquare$$

The Poisson Distribution as a Limit

The rationale for using the Poisson distribution in many situations is provided by the following proposition.

PROPOSITION

Suppose that in the binomial pmf $b(x; n, p)$, we let $n \rightarrow \infty$ and $p \rightarrow 0$ in such a way that np approaches a value $\lambda > 0$. Then $b(x; n, p) \rightarrow p(x; \lambda)$.

According to this proposition, *in any binomial experiment in which n is large and p is small, $b(x; n, p) \approx p(x; \lambda)$ where $\lambda = np$*. As a rule of thumb, this approximation can safely be applied if $n \geq 100$, $p \leq .01$, and $np \leq 20$.

Example 3.39 If a publisher of nontechnical books takes great pains to ensure that its books are free of typographical errors, so that the probability of any given page containing at least one such error is .005 and errors are independent from page to page, what is the probability that one of its 400-page novels will contain exactly one page with errors? At most three pages with errors?

With S denoting a page containing at least one error and F an error-free page, the number X of pages containing at least one error is a binomial rv with $n = 400$ and $p = .005$, so $np = 2$. We wish

$$P(X = 1) = b(1; 400, .005) \approx p(1; 2) = \frac{e^{-2}(2)^1}{1!} = .271$$

Similarly,

$$P(X \leq 3) \approx \sum_{x=0}^{3} p(x; 2) = \sum_{x=0}^{3} e^{-2} \frac{2^x}{x!} = .135 + .271 + .271 + .180$$

$$= .857 \qquad \blacksquare$$

Appendix Table A.2 exhibits the cdf $F(x; \lambda)$ for $\lambda = .1, .2, \ldots, 1, 2, \ldots, 10, 15$, and 20. For example, if $\lambda = 2$, then $P(X \leq 3) = F(3; 2) = .857$ as in Example 3.39, whereas $P(X = 3) = F(3; 2) - F(2; 2) = .180$. Alternatively, many statistical computer packages will generate $p(x; \lambda)$ and $F(x; \lambda)$ upon request.

The Mean and Variance of *X*

Since $b(x; n, p) \rightarrow p(x; \lambda)$ as $n \rightarrow \infty$, $p \rightarrow 0$, $np \rightarrow \lambda$, the mean and variance of a binomial variable should approach those of a Poisson variable. These limits are $np \rightarrow \lambda$ and $np(1 - p) \rightarrow \lambda$.

PROPOSITION

If X has a Poisson distribution with parameter λ, then $E(X) = V(X) = \lambda$.

These results can also be derived directly from the definitions of mean and variance.

Example 3.40
(Example 3.38
continued)

Both the expected number of creatures trapped and the variance of the number trapped equal 4.5, and $\sigma_X = \sqrt{\lambda} = \sqrt{4.5} = 2.12$. ∎

The Poisson Process

A very important application of the Poisson distribution arises in connection with the occurrence of events of a particular type over time. As an example, suppose that starting from a time point that we label $t = 0$, we are interested in counting the number of radioactive pulses recorded by a Geiger counter. We make the following assumptions about the way in which pulses occur:

1. There exists a parameter $\alpha > 0$ such that for any short time interval of length Δt, the probability that exactly one pulse is received is $\alpha \cdot \Delta t + o(\Delta t)$.*

2. The probability of more than one pulse being received during Δt is $o(\Delta t)$ [which, along with Assumption 1, implies that the probability of no pulses during Δt is $1 - \alpha \cdot \Delta t - o(\Delta t)$].

3. The number of pulses received during the time interval Δt is independent of the number received prior to this time interval.

Informally, Assumption 1 says that, for a short interval of time, the probability of receiving a single pulse is approximately proportional to the length of the time interval, where α is the constant of proportionality. Now let $P_k(t)$ denote the probability that k pulses will be received by the counter during any particular time interval of length t.

PROPOSITION

$P_k(t) = e^{-\alpha t} \cdot (\alpha t)^k / k!$, so that the number of pulses during a time interval of length t is a Poisson rv with parameter $\lambda = \alpha t$. The expected number of pulses during any such time interval is then αt, so the expected number during a unit interval of time is α.

Example 3.41

Suppose pulses arrive at the counter at an average rate of six per minute, so that $\alpha = 6$. To find the probability that in a .5-min interval at least one pulse is received, note that the number of pulses in such an interval has a Poisson distribution with parameter $\alpha t = 6(.5) = 3$ (.5 min is used because α is expressed as a rate per minute). Then with $X = $ the number of pulses received in the 30-sec interval,

$$P(1 \leq X) = 1 - P(X = 0) = 1 - \frac{e^{-3}(3)^0}{0!} = .950$$ ∎

If in Assumptions 1–3 we replace "pulse" by "event," then the number of events occurring during a fixed time interval of length t has a Poisson distribution with parameter αt. Any process that has this distribution is called a **Poisson process,** and α is

---

*A quantity is $o(\Delta t)$ (read "little o of delta t") if, as Δt approaches 0, so does $o(\Delta t)/\Delta t$. That is, $o(\Delta t)$ is even more negligible than Δt itself. The quantity $(\Delta t)^2$ has this property, but $\sin(\Delta t)$ does not.

called the *rate of the process*. Other examples of situations giving rise to a Poisson process include monitoring the status of a computer system over time, with breakdowns constituting the events of interest, recording the number of accidents in an industrial facility over time, answering calls at a telephone switchboard, and observing the number of cosmic-ray showers from a particular observatory over time.

Instead of observing events over time, consider observing events of some type that occur in a two- or three-dimensional region. For example, we might select on a map a certain region R of a forest, go to that region, and count the number of trees. Each tree would represent an event occurring at a particular point in space. Under assumptions similar to 1–3, it can be shown that the number of events occurring in a region R has a Poisson distribution with parameter $\alpha \cdot a(R)$, where $a(R)$ is the area of R. The quantity α is the expected number of events per unit area or volume.

Exercises | Section 3.6 (73–87)

73. Let X, the number of flaws on the surface of a randomly selected boiler of a certain type, have a Poisson distribution with parameter $\lambda = 5$. Use Appendix Table A.2 to compute the following probabilities.
  a. $P(X \leq 8)$ **b.** $P(X = 8)$ **c.** $P(9 \leq X)$
  d. $P(5 \leq X \leq 8)$ **e.** $P(5 < X < 8)$

74. Suppose the number X of tornadoes observed in a particular region during a 1-year period has a Poisson distribution with $\lambda = 8$.
  a. Compute $P(X \leq 5)$.
  b. Compute $P(6 \leq X \leq 9)$.
  c. Compute $P(10 \leq X)$.
  d. How many tornadoes can be expected to be observed during the 1-year period? What is the standard deviation of the number of observed tornadoes?

75. Suppose that the number of drivers who travel between a particular origin and destination during a designated time period has a Poisson distribution with parameter $\lambda = 20$ (suggested in the article "Dynamic Ride Sharing: Theory and Practice," *J. of Transp. Engr.,* 1997: 308–312). What is the probability that the number of drivers will
  a. Be at most 10?
  b. Exceed 20?
  c. Be between 10 and 20, inclusive? Be strictly between 10 and 20?
  d. Exceed the mean number by more than two standard deviations?

76. Consider writing onto a computer disk and then sending it through a certifier that counts the number of missing pulses. Suppose this number X has a Poisson distribution with parameter $\lambda = .2$. (Suggested in "Average Sample Number for Semi-Curtailed Sampling Using the Poisson Distribution," *J. Quality Technology,* 1983: 126–129.)
  a. What is the probability that a disk has exactly one missing pulse?
  b. What is the probability that a disk has at least two missing pulses?
  c. If two disks are independently selected, what is the probability that neither contains a missing pulse?

77. An article in the *Los Angeles Times* (Dec. 3, 1993) reports that 1 in 200 people carry the defective gene that causes inherited colon cancer. In a sample of 1000 individuals, what is the approximate distribution of the number who carry this gene? Use this distribution to calculate the approximate probability that
  a. Between 5 and 8 (inclusive) carry the gene.
  b. At least 8 carry the gene.

78. A notice is sent to all owners of a certain type of automobile, asking them to bring their cars to a dealer to check for the presence of a particular manufacturing defect. Suppose that only .05% of such cars have the defect. Consider a random sample of 10,000 cars.
  a. What are the expected value and standard deviation of the number of cars in the sample that have the defect?
  b. What is the (approximate) probability that more than 10 sampled cars have the defect?

c. What is the (approximate) probability that no sampled cars have the defect?

79. Suppose small aircraft arrive at a certain airport according to a Poisson process with rate $\alpha = 8$ per hour, so that the number of arrivals during a time period of t hours is a Poisson rv with parameter $\lambda = 8t$.

   a. What is the probability that exactly 5 small aircraft arrive during a 1-hour period? At least 5? At least 10?

   b. What are the expected value and standard deviation of the number of small aircraft that arrive during a 90-min period?

   c. What is the probability that at least 20 small aircraft arrive during a $2\frac{1}{2}$-hour period? That at most 10 arrive during this period?

80. The number of tickets issued by a meter reader for parking-meter violations can be modeled by a Poisson process with a rate parameter of five per hour.

   a. What is the probability that exactly four tickets are given out during a particular hour?

   b. What is the probability that at least four tickets are given out during a particular hour?

   c. How many tickets do you expect to be given during a 45-min period?

81. The number of requests for assistance received by a towing service is a Poisson process with rate $\alpha = 4$ per hour.

   a. Compute the probability that exactly ten requests are received during a particular 2-hour period.

   b. If the operators of the towing service take a 30-min break for lunch, what is the probability that they do not miss any calls for assistance?

   c. How many calls would you expect during their break?

82. In proof testing of circuit boards, the probability that any particular diode will fail is .01. Suppose a circuit board contains 200 diodes.

   a. How many diodes would you expect to fail, and what is the standard deviation of the number that are expected to fail?

   b. What is the (approximate) probability that at least four diodes will fail on a randomly selected board?

   c. If five boards are shipped to a particular customer, how likely is it that at least four of them will work properly? (A board works properly only if all its diodes work.)

83. The article "Reliability-Based Service-Life Assessment of Aging Concrete Structures" (*J. Structural Engr.,* 1993: 1600–1621) suggests that a Poisson process can be used to represent the occurrence of structural loads over time. Suppose the mean time between occurrences of loads is .5 year.

   a. How many loads can be expected to occur during a 2-year period?

   b. What is the probability that more than five loads occur during a 2-year period?

   c. How long must a time period be so that the probability of no loads occurring during that period is at most .1?

84. Let X have a Poisson distribution with parameter λ. Show that $E(X) = \lambda$ directly from the definition of expected value. (*Hint:* The first term in the sum equals 0, and then x can be canceled. Now factor out λ and show that what is left sums to 1.)

85. Suppose that trees are distributed in a forest according to a two-dimensional Poisson process with parameter α, the expected number of trees per acre, equal to 80.

   a. What is the probability that in a certain quarter-acre plot, there will be at most 16 trees?

   b. If the forest covers 85,000 acres, what is the expected number of trees in the forest?

   c. Suppose you select a point in the forest and construct a circle of radius .1 mile. Let $X = $ the number of trees within that circular region. What is the pmf of X? (*Hint:* 1 sq mile = 640 acres.)

86. Automobiles arrive at a vehicle equipment inspection station according to a Poisson process with rate $\alpha = 10$ per hour. Suppose that with probability .5 an arriving vehicle will have no equipment violations.

   a. What is the probability that exactly ten arrive during the hour and all ten have no violations?

   b. For any fixed $y \geq 10$, what is the probability that y arrive during the hour, of which ten have no violations?

   c. What is the probability that ten "no-violation" cars arrive during the next hour? [*Hint:* Sum the probabilities in part (b) from $y = 10$ to ∞.]

87. a. In a Poisson process, what has to happen in both the time interval $(0, t)$ and the interval $(t, t + \Delta t)$ so that no events occur in the entire interval $(0, t + \Delta t)$? Use this and Assumptions 1–3 to write a relationship between $P_0(t + \Delta t)$ and $P_0(t)$.

b. Use the result of part (a) to write an expression for the difference $P_0(t + \Delta t) - P_0(t)$. Then divide by Δt and let $\Delta t \to 0$ to obtain an equation involving $(d/dt)P_0(t)$, the derivative of $P_0(t)$ with respect to t.

c. Verify that $P_0(t) = e^{-\alpha t}$ satisfies the equation of part (b).

d. It can be shown in a manner similar to parts (a) and (b) that the $P_k(t)$'s must satisfy the system of differential equations

$$\frac{d}{dt} P_k(t) = \alpha P_{k-1}(t) - \alpha P_k(t)$$

$$k = 1, 2, 3, \ldots$$

Verify that $P_k(t) = e^{-\alpha t}(\alpha t)^k/k!$ satisfies the system. (This is actually the only solution.)

Supplementary Exercises (88–114)

88. Consider a deck consisting of seven cards, marked $1, 2, \ldots, 7$. Three of these cards are selected at random. Define an rv W by $W = $ the sum of the resulting numbers, and compute the pmf of W. Then compute μ and σ^2. [*Hint:* Consider outcomes as unordered, so that $(1, 3, 7)$ and $(3, 1, 7)$ are not different outcomes. Then there are 35 outcomes, and they can be listed. (This type of rv actually arises in connection with a hypothesis test called Wilcoxon's rank-sum test, in which there is an x sample and a y sample and W is the sum of the ranks of the x's in the combined sample.)]

89. After shuffling a deck of 52 cards, a dealer deals out 5. Let $X = $ the number of suits represented in the five-card hand.

a. Show that the pmf of X is

x	1	2	3	4
$p(x)$	.002	.146	.588	.264

[*Hint:* $p(1) = 4P(\text{all are spades})$, $p(2) = 6P(\text{only spades and hearts with at least one of each})$, and $p(4) = 4P(2 \text{ spades} \cap \text{one of each other suit}).$]

b. Compute μ, σ^2, and σ.

90. The negative binomial rv X was defined as the number of F's preceding the rth S. Let $Y = $ the number of trials necessary to obtain the rth S. In the same manner in which the pmf of X was derived, derive the pmf of Y.

91. Of all customers purchasing automatic garage-door openers, 75% purchase a chain-driven model. Let $X = $ the number among the next 15 purchasers who select the chain-driven model.

a. What is the pmf of X?

b. Compute $P(X \leq 10)$.

c. Compute $P(8 \leq X \leq 12)$.

d. Compute μ and σ^2.

e. If the store currently has in stock 10 chain-driven models and 8 shaft-driven models, what is the probability that the requests of these 15 customers can all be met from existing stock?

92. A friend recently planned a camping trip. He had two flashlights, one that required a single 6-V battery and another that used two size-D batteries. He had previously packed two 6-V and four size-D batteries in his camper. Suppose the probability that any particular battery works is p and that batteries work or fail independently of one another. Our friend wants to take just one flashlight. For what values of p should he take the 6-V flashlight?

93. A *k-out-of-n system* is one that will function if and only if at least k of the n individual components in the system function. If individual components function independently of one another, each with probability .9, what is the probability that a 3-out-of-5 system functions?

94. A manufacturer of flashlight batteries wishes to control the quality of its product by rejecting any lot in which the proportion of batteries having unacceptable voltage appears to be too high. To this end, out of each large lot (10,000 batteries), 25 will be selected and tested. If at least 5 of these generate an unacceptable voltage, the entire lot will be rejected. What is the probability that a lot will be rejected if

a. 5% of the batteries in the lot have unacceptable voltages?

b. 10% of the batteries in the lot have unacceptable voltages?

c. 20% of the batteries in the lot have unacceptable voltages?

d. What would happen to the probabilities in parts (a)–(c) if the critical rejection number were increased from 5 to 6?

95. Of the people passing through an airport metal detector, .5% activate it; let X = the number among a randomly selected group of 500 who activate the detector.

a. What is the (approximate) pmf of X?

b. Compute $P(X = 5)$.

c. Compute $P(5 \leq X)$.

96. An educational consulting firm is trying to decide whether high school students who have never before used a hand-held calculator can solve a certain type of problem more easily with a calculator that uses reverse Polish logic or one that does not use this logic. A sample of 25 students is selected and allowed to practice on both calculators. Then each student is asked to work one problem on the reverse Polish calculator and a similar problem on the other. Let $p = P(S)$ where S indicates that a student worked the problem more quickly using reverse Polish logic than without and let X = number of S's.

a. If $p = .5$, what is $P(7 \leq X \leq 18)$?

b. If $p = .8$, what is $P(7 \leq X \leq 18)$?

c. If the claim that $p = .5$ is to be rejected when either $X \leq 7$ or $X \geq 18$, what is the probability of rejecting the claim when it is actually correct?

d. If the decision to reject the claim $p = .5$ is made as in part (c), what is the probability that the claim is not rejected when $p = .6$? When $p = .8$?

e. What decision rule would you choose for rejecting the claim $p = .5$ if you wanted the probability in part (c) to be at most .01?

97. Consider a disease whose presence can be identified by carrying out a blood test. Let p denote the probability that a randomly selected individual has the disease. Suppose n individuals are independently selected for testing. One way to proceed is to carry out a separate test on each of the n blood samples. A potentially more economical approach, group testing, was introduced during World War II to identify syphilitic men among army inductees. First, take a part of each blood sample, combine these specimens, and carry out a single test. If no one has the disease, the result will be negative, and only the one test is required. If at least one individ-

ual is diseased, the test on the combined sample will yield a positive result, in which case the n individual tests are then carried out. If $p = .1$ and $n = 3$, what is the expected number of tests using this procedure? What is the expected number when $n = 5$? [The article "Random Multiple-Access Communication and Group Testing" (*IEEE Trans. on Commun.*, 1984: 769–774) applied these ideas to a communication system in which the dichotomy was active/idle user rather than diseased/nondiseased.]

98. Let p_1 denote the probability that any particular code symbol is erroneously transmitted through a communication system. Assume that on different symbols, errors occur independently of one another. Suppose also that with probability p_2 an erroneous symbol is corrected upon receipt. Let X denote the number of correct symbols in a message block consisting of n symbols (after the correction process has ended). What is the probability distribution of X?

99. The purchaser of a power-generating unit requires c consecutive successful start-ups before the unit will be accepted. Assume that the outcomes of individual start-ups are independent of one another. Let p denote the probability that any particular start-up is successful. The random variable of interest is X = the number of start-ups that must be made prior to acceptance. Give the pmf of X for the case $c = 2$. If $p = .9$, what is $P(X \leq 8)$? [*Hint:* For $x \geq 5$, express $p(x)$ "recursively" in terms of the pmf evaluated at the smaller values $x - 3$, $x - 4, \ldots, 2$.] (This problem was suggested by the article "Evaluation of a Start-Up Demonstration Test," *J. Quality Technology,* 1983: 103–106.)

100. A plan for an executive travelers' club has been developed by an airline on the premise that 10% of its current customers would qualify for membership.

a. Assuming the validity of this premise, among 25 randomly selected current customers, what is the probability that between 2 and 6 (inclusive) qualify for membership?

b. Again assuming the validity of the premise, what are the expected number of customers who qualify and the standard deviation of the number who qualify in a random sample of 100 current customers?

c. Let X denote the number in a random sample of 25 current customers who qualify for membership. Consider rejecting the company's premise in favor of the claim that $p > .10$ if $x \geq 7$. What is the probability that the company's premise is rejected when it is actually valid?

d. Refer to the decision rule introduced in part (c). What is the probability that the company's premise is not rejected even though $p = .20$ (i.e., 20% qualify)?

101. Forty percent of seeds from maize (modern-day corn) ears carry single spikelets, and the other 60% carry paired spikelets. A seed with single spikelets will produce an ear with single spikelets 29% of the time, whereas a seed with paired spikelets will produce an ear with single spikelets 26% of the time. Consider randomly selecting ten seeds.

a. What is the probability that exactly five of these seeds carry a single spikelet and produce an ear with a single spikelet?

b. What is the probability that exactly five of the ears produced by these seeds have single spikelets? What is the probability that at most five ears have single spikelets?

102. A trial has just resulted in a hung jury because eight members of the jury were in favor of a guilty verdict and the other four were for acquittal. If the jurors leave the jury room in random order and each of the first four leaving the room is accosted by a reporter in quest of an interview, what is the pmf of X = the number of jurors favoring acquittal among those interviewed? How many of those favoring acquittal do you expect to be interviewed?

103. A telephone company employs five information operators who receive requests for information independently of one another, each according to a Poisson process with rate $\alpha = 2$ per minute.

a. What is the probability that during a given 1-min period, the first operator receives no requests?

b. What is the probability that during a given 1-min period, exactly four of the five operators receive no requests?

c. Write an expression for the probability that during a given 1-min period, all of the operators receive exactly the same number of requests.

104. Grasshoppers are distributed at random in a large field according to a Poisson distribution with parameter $\alpha = 2$ per square yard. How large should the radius R of a circular sampling region be taken so that the probability of finding at least one in the region equals .99?

105. A newsstand has ordered five copies of a certain issue of a photography magazine. Let X = the number of individuals who come in to purchase this magazine. If X has a Poisson distribution with parameter $\lambda = 4$, what is the expected number of copies that are sold?

106. Individuals A and B begin to play a sequence of chess games. Let S = {A wins a game}, and suppose that outcomes of successive games are independent with $P(S) = p$ and $P(F) = 1 - p$ (they never draw). They will play until one of them wins ten games. Let X = the number of games played (with possible values 10, 11, . . . , 19).

a. For $x = 10, 11, \ldots , 19$, obtain an expression for $p(x) = P(X = x)$.

b. If a draw is possible, with $p = P(S)$, $q = P(F)$, $1 - p - q = P(\text{draw})$, what are the possible values of X? What is $P(20 \leq X)$? [*Hint:* $P(20 \leq X) = 1 - P(X < 20).$]

107. A test for the presence of a certain disease has probability .20 of giving a false-positive reading (indicating that an individual has the disease when this is not the case) and probability .10 of giving a false-negative result. Suppose that ten individuals are tested, five of whom have the disease and five of whom do not. Let X = the number of positive readings that result.

a. Does X have a binomial distribution? Explain your reasoning.

b. What is the probability that exactly three of the ten test results are positive?

108. The generalized negative binomial pmf is given by $nb(x; r, p) = k(r, x) \cdot p^r(1 - p)^x$

$$x = 0, 1, 2, \ldots$$

Let X, the number of plants of a certain species found in a particular region, have this distribution with $p = .3$ and $r = 2.5$. What is $P(X = 4)$? What is the probability that at least one plant is found?

109. Define a function $p(x; \lambda, \mu)$ by

$p(x; \lambda, \mu)$

$$= \begin{cases} \dfrac{1}{2}e^{-\lambda}\dfrac{\lambda^x}{x!} + \dfrac{1}{2}e^{-\mu}\dfrac{\mu^x}{x!} & x = 0, 1, 2, \ldots \\ 0 & \text{otherwise} \end{cases}$$

a. Show that $p(x; \lambda, \mu)$ satisfies the two conditions necessary for specifying a pmf, [*Note:* If a firm employs two typists, one of whom makes typographical errors at the rate of λ per page and the other at rate μ per page and they each do half the firm's typing, then $p(x; \lambda, \mu)$ is the pmf of $X =$ the number of errors on a randomly chosen page.]

b. If the first typist (rate λ) types 60% of all pages, what is the pmf of X of part (a)?

c. What is $E(X)$ for $p(x; \lambda, \mu)$ given by the displayed expression?

d. What is σ^2 for $p(x; \lambda, \mu)$ given by that expression?

110. The *mode* of a discrete random variable X with pmf $p(x)$ is that value x^* for which $p(x)$ is largest (the most probable x value).

a. Let $X \sim \text{Bin}(n, p)$. By considering the ratio $b(x + 1; n, p)/b(x; n, p)$, show that $b(x; n, p)$ increases with x as long as $x < np - (1 - p)$. Conclude that the mode x^* is the integer satisfying $(n + 1)p - 1 \le x^* \le (n + 1)p$.

b. Show that if X has a Poisson distribution with parameter λ, the mode is the largest integer less than λ. If λ is an integer, show that both $\lambda - 1$ and λ are modes.

111. A computer disk storage device has ten concentric tracks, numbered 1, 2, . . . , 10 from outermost to innermost, and a single access arm. Let $p_i =$ the probability that any particular request for data will take the arm to track i ($i = 1, . . . , 10$). Assume that the tracks accessed in successive seeks are independent. Let $X =$ the number of tracks over which the access arm passes during two successive requests (excluding the track that the arm has just left, so possible X values are $x = 0$, 1, . . . , 9). Compute the pmf of X. [*Hint:* P(the arm is now on track i and $X = j$) $= P(X = j \mid$ arm now on i) $\cdot p_i$. After the conditional probability is written in terms of $p_1, . . . , p_{10}$, by the law of total probability, the desired probability is obtained by summing over i.]

112. If X is a hypergeometric rv, show directly from the definition that $E(X) = nM/N$ (consider only the case $n < M$). [*Hint:* Factor nM/N out of the sum for $E(X)$, and show that the terms inside the sum are of the form $h(y; n - 1, M - 1, N - 1)$, where $y = x - 1$.]

113. Use the fact that

$$\sum_{\text{all } x} (x - \mu)^2 p(x) \ge \sum_{x:\, |x - \mu|\, \ge k\sigma} (x - \mu)^2 p(x)$$

to prove Chebyshev's inequality given in Exercise 43.

114. The simple Poisson process of Section 3.6 is characterized by a constant rate α at which events occur per unit time. A generalization of this is to suppose that the probability of exactly one event occurring in the interval $[t, t + \Delta t]$ is $\alpha(t) \cdot \Delta t + o(\Delta t)$. It can then be shown that the number of events occurring during an interval $[t_1, t_2]$ has a Poisson distribution with parameter

$$\lambda = \int_{t_1}^{t_2} \alpha(t)\, dt$$

The occurrence of events over time in this situation is called a *nonhomogeneous Poisson process.* The article "Inference Based on Retrospective Ascertainment," *J. Amer. Stat. Assoc.,* 1989: 360–372, considers the intensity function

$$\alpha(t) = e^{a + bt}$$

as appropriate for events involving transmission of HIV (the AIDS virus) via blood transfusions. Suppose that $a = 2$ and $b = .6$ (close to values suggested in the paper), with time in years.

a. What is the expected number of events in the interval [0, 4]? In [2, 6]?

b. What is the probability that at most 15 events occur in the interval [0, .9907]?

Bibliography

Johnson, Norman, Samuel Kotz, and Adrienne Kemp, *Discrete Univariate Distributions,* Wiley, New York, 1992. An encyclopedia of information on discrete distributions.

Olkin, Ingram, Cyrus Derman, and Leon Gleser, *Probability Models and Applications* (2nd ed.), Macmillan, New York, 1994. Contains an in-depth discussion of both general properties of discrete and continuous distributions and results for specific distributions.

Ross, Sheldon, *Introduction to Probability Models* (5th ed.), Academic Press, New York, 1993. A good source of material on the Poisson process and generalizations and a nice introduction to other topics in applied probability.

4

Continuous Random Variables and Probability Distributions

Introduction

As mentioned at the beginning of Chapter 3, not all random variables are discrete. In this chapter, we study the second general type of random variable that arises in many applied problems. Sections 4.1 and 4.2 present the basic definitions and properties of continuous random variables and their probability distributions. In Section 4.3, we study in detail the normal random variable and distribution, unquestionably the most important and useful in probability and statistics. Sections 4.4 and 4.5 discuss some other continuous distributions that are often used in applied work. In Section 4.6, we introduce a method for assessing whether given sample data is consistent with a specified distribution.

4.1 Continuous Random Variables and Probability Density Functions

A discrete random variable (rv) is one whose possible values either constitute a finite set or else can be listed in an infinite sequence (a list in which there is a first element, a second element, etc.). A random variable whose set of possible values is an entire interval of numbers is not discrete.

Continuous Random Variables

<table>
<tr>
<td>DEFINITION</td>
<td>A random variable X is said to be **continuous** if its set of possible values is an entire interval of numbers—that is, if for some $A < B$, any number x between A and B is possible.</td>
</tr>
</table>

Example 4.1 If in the study of the ecology of a lake, we make depth measurements at randomly chosen locations, then $X =$ the depth at such a location is a continuous rv. Here A is the minimum depth in the region being sampled, and B is the maximum depth. ∎

Example 4.2 If a chemical compound is randomly selected and its pH X is determined, then X is a continuous rv because any pH value between 0 and 14 is possible. If more is known about the compound selected for analysis, then the set of possible values might be a subinterval of [0, 14] such as $5.5 \leq x \leq 6.5$, but X would still be continuous. ∎

If the measurement scale of X can be subdivided to any extent desired, then the variable is continuous; if it cannot, the variable is discrete. For example, if the variable is height or length, then it can be measured in kilometers, meters, centimeters, millimeters, and so on, so the variable is continuous. If, however, $X =$ the billing amount on a randomly chosen monthly natural gas statement, then the smallest unit of measurement is cents, so any value of X is a multiple of $.01 and X is a discrete variable.

One might argue that although in principle variables such as height, weight, and temperature are continuous, in practice the limitations of our measuring instruments restrict us to a discrete (though sometimes very finely subdivided) world. However, continuous models often approximate real-world situations very well, and continuous mathematics (the calculus) is frequently easier to work with than mathematics of discrete variables and distributions.

Probability Distributions for Continuous Variables

Suppose the variable X of interest is the depth of a lake at a randomly chosen point on the surface. Let $M =$ the maximum depth (in meters), so that any number in the interval [0, M] is a possible value of X. If we "discretize" X by measuring depth to the nearest meter, then possible values are nonnegative integers less than or equal to M. The resulting discrete distribution of depth can be pictured using a probability histogram. If we draw the histogram so that the area of the rectangle above any possible integer k is the proportion of the lake whose depth is (to the nearest meter) k, then the total area of all rectangles is 1. A possible histogram appears in Figure 4.1(a).

If depth is measured to the nearest centimeter and the same measurement axis as in Figure 4.1(a) is used, each rectangle in the resulting probability histogram is much narrower, though the total area of all rectangles is still 1. A possible histogram is pictured in Figure 4.1(b); it has a much smoother appearance than the histogram in Figure 4.1(a). If we continue in this way to measure depth more and more finely, the resulting sequence of histograms approaches a smooth curve, such as is pictured in Figure 4.1(c). Because for each histogram the total area of all rectangles equals 1, the total area under the smooth curve is also 1. The probability that the depth at a randomly chosen point is

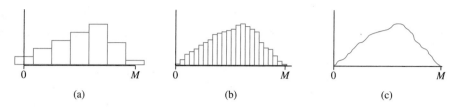

Figure 4.1 (a) Probability histogram of depth measured to the nearest meter; (b) probability histogram of depth measured to the nearest centimeter; (c) a limit of a sequence of discrete histograms

between a and b is just the area under the smooth curve between a and b. It is exactly a smooth curve of the type pictured in Figure 4.1(c) that specifies a continuous probability distribution.

DEFINITION

Let X be a continuous rv. Then a **probability distribution** or **probability density function** (pdf) of X is a function $f(x)$ such that for any two numbers a and b with $a \leq b$,

$$P(a \leq X \leq b) = \int_a^b f(x)dx$$

That is, the probability that X takes on a value in the interval $[a, b]$ is the area under the graph of the density function, as illustrated in Figure 4.2. The graph of $f(x)$ is often referred to as the *density curve*.

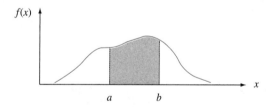

Figure 4.2 $P(a \leq X \leq b)$ = the area under the density curve between a and b

For $f(x)$ to be a legitimate pdf, it must satisfy the following two conditions:

1. $f(x) \geq 0$ for all x

2. $\int_{-\infty}^{\infty} f(x)dx$ = area under the entire graph of $f(x)$
$\qquad\qquad = 1$

Example 4.3

Suppose I take a bus to work, and that every 5 minutes a bus arrives at my stop. Because of variation in the time that I leave my house, I don't always arrive at the bus stop at the same time, so my waiting time X for the next bus is a continuous rv. The set of possible values of X is the interval $[0, 5]$. One possible pdf for X is

$$f(x) = \begin{cases} \dfrac{1}{5} & 0 \leq x \leq 5 \\ 0 & \text{otherwise} \end{cases}$$

The pdf $f(x)$ is graphed in Figure 4.3. Clearly $f(x) \geq 0$, and the total area under the graph is $5 \cdot \left(\frac{1}{5}\right) = 1$. The probability that I wait between 1 and 3 minutes is

$$P(1 \leq X \leq 3) = \int_1^3 f(x)dx = \int_1^3 \frac{1}{5}\, dx = \frac{x}{5}\Big|_{x=1}^{x=3} = \frac{2}{5}$$

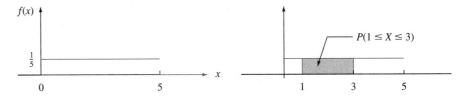

Figure 4.3   The pdf of Example 4.3

Similarly, $P(2 \leq X \leq 4) = \frac{2}{5}$. The probability that I wait at least 4 minutes is

$$P(4 \leq X) = \int_4^\infty f(x)dx = \int_4^5 \frac{1}{5}dx + \int_5^\infty 0\, dx = \frac{x}{5}\Big|_{x=4}^{x=5} = \frac{1}{5} \qquad \blacksquare$$

Because whenever $0 \leq a \leq b \leq 5$ in Example 4.3, $P(a \leq X \leq b)$ depends only on the length $b - a$ of the interval, X is said to have a uniform distribution.

<table>
<tr><td>DEFINITION</td><td>

A continuous rv X is said to have a **uniform distribution** on the interval $[A, B]$ if the pdf of X is

$$f(x; A, B) = \begin{cases} \dfrac{1}{B - A} & A \leq x \leq B \\ 0 & \text{otherwise} \end{cases}$$

</td></tr>
</table>

The graph of any uniform pdf looks like the graph in Figure 4.3 except that the interval of positive density is $[A, B]$ rather than $[0, 5]$.

In the discrete case, a probability mass function (pmf) tells us how little "blobs" of probability mass of various magnitudes are distributed along the measurement axis. In the continuous case, probability density is "smeared" in a continuous fashion along the interval of possible values. When density is smeared uniformly over the interval, a uniform pdf, as in Figure 4.3, results.

When X is a discrete rv, each possible value is assigned positive probability, but this is not the case when X is continuous.

<table>
<tr><td>PROPOSITION</td><td>

If X is a continuous rv, then for any number c, $P(X = c) = 0$. Furthermore, for any two numbers a and b with $a < b$,

$$\begin{aligned} P(a \leq X \leq b) &= P(a < X \leq b) \\ &= P(a \leq X < b) = P(a < X < b) \end{aligned} \qquad (4.1)$$

</td></tr>
</table>

In words, the probability assigned to any particular value is zero, and the probability of an interval does not depend on whether either of its endpoints is included. These properties follow from the facts that the area under the graph of $f(x)$ and above the single value c is zero and that the area under the graph above an interval is unaffected by exclusion or inclusion of the endpoints of the interval. If X is discrete and there is positive probability mass at both $X = a$ and $X = b$, then all four probabilities in Expression (4.1) will differ from one another.

The fact that a continuous distribution assigns probability zero to each value has a physical analog. Consider a solid circular rod with cross-sectional area $= 1$ in.^{2}. Place the rod alongside a measurement axis and suppose that the density of the rod at any point x is given by the value $f(x)$ of a density function. Then if the rod is sliced at points a and b and this segment is removed, the amount of mass removed is $\int_a^b f(x)\,dx$; if the rod is sliced just at the point c, no mass is removed. Mass is assigned to interval segments of the rod but not to individual points.

Example 4.4 "Time headway" in traffic flow is the elapsed time between the time that one car finishes passing a fixed point and the instant that the next car begins to pass that point. Let $X =$ the time headway for two randomly chosen consecutive cars on a freeway during a period of heavy flow. The following pdf of X is essentially the one suggested in "The Statistical Properties of Freeway Traffic" (*Transp. Research,* vol. 11: 221–228):

$$f(x) = \begin{cases} .15e^{-.15(x-.5)} & x \geq .5 \\ 0 & \text{otherwise} \end{cases}$$

The graph of $f(x)$ is given in Figure 4.4; there is no density associated with headway times less than .5, and headway density decreases rapidly (exponentially fast) as x increases from .5. Clearly, $f(x) \geq 0$; to show that $\int_{-\infty}^{\infty} f(x)dx = 1$, we use the calculus result $\int_a^{\infty} e^{-kx}\,dx = (1/k)e^{-k \cdot a}$. Then

$$\int_{-\infty}^{\infty} f(x)dx = \int_{.5}^{\infty} .15e^{-.15(x-.5)}\,dx = .15e^{.075}\int_{.5}^{\infty} e^{-.15x}dx$$

$$= .15e^{.075} \cdot \frac{1}{.15}e^{-(.15)(.5)} = 1$$

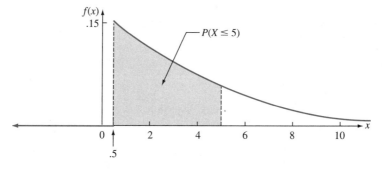

Figure 4.4 The density curve for headway time in Example 4.4

The probability that headway time is at most 5 sec is

$$P(X \le 5) = \int_{-\infty}^{5} f(x)dx = \int_{.5}^{5} .15e^{-.15(x-.5)}dx$$

$$= .15e^{.075} \int_{.5}^{5} e^{-.15x} dx = .15e^{.075} \cdot -\frac{1}{.15}e^{-.15x}\Big|_{x=.5}^{x=5}$$

$$= e^{.075}(-e^{-.75} + e^{-.075}) = 1.078(-.472 + .928) = .491$$

$$= P(\text{less than 5 sec}) = P(X < 5)$$

■

Unlike discrete distributions such as the binomial, hypergeometric, and negative binomial, the distribution of any given continuous rv cannot usually be derived using simple probabilistic arguments. Instead, one must make a judicious choice of pdf based on prior knowledge and available data. Fortunately, there are some general families of pdf's that have been found to fit well in a wide variety of experimental situations; several of these are discussed later in the chapter.

Just as in the discrete case, it is often helpful to think of the population of interest as consisting of X values rather than individuals or objects. The pdf is then a model for the distribution of values in this numerical population, and from this model various population characteristics (such as the mean) can be calculated.

Exercises Section 4.1 (1–10)

1. Let X denote the amount of time for which a book on 2-hour reserve at a college library is checked out by a randomly selected student and suppose that X has density function

$$f(x) = \begin{cases} .5x & 0 \le x \le 2 \\ 0 & \text{otherwise} \end{cases}$$

   Calculate the following probabilities:
   a. $P(X \le 1)$
   b. $P(.5 \le X \le 1.5)$
   c. $P(1.5 < X)$

2. Suppose the reaction temperature X (in °C) in a certain chemical process has a uniform distribution with $A = -5$ and $B = 5$.
   a. Compute $P(X < 0)$.
   b. Compute $P(-2.5 < X < 2.5)$.
   c. Compute $P(-2 \le X \le 3)$.
   d. For k satisfying $-5 < k < k + 4 < 5$, compute $P(k < X < k + 4)$.

3. Suppose the error involved in making a certain measurement is a continuous rv X with pdf

$$f(x) = \begin{cases} .09375(4 - x^2) & -2 \le x \le 2 \\ 0 & \text{otherwise} \end{cases}$$

   a. Sketch the graph of $f(x)$.
   b. Compute $P(X > 0)$.
   c. Compute $P(-1 < X < 1)$.
   d. Compute $P(X < -.5 \text{ or } X > .5)$.

4. Let X denote the vibratory stress (psi) on a wind turbine blade at a particular wind speed in a wind tunnel. The article "Blade Fatigue Life Assessment with Application to VAWTS" (*J. Solar Energy Engr.*, 1982: 107–111) proposes the Rayleigh distribution, with pdf

$$f(x; \theta) = \begin{cases} \dfrac{x}{\theta^2} \cdot e^{-x^2/(2\theta^2)} & x > 0 \\ 0 & \text{otherwise} \end{cases}$$

   as a model for the X distribution.
   a. Verify that $f(x; \theta)$ is a legitimate pdf.
   b. Suppose $\theta = 100$ (a value suggested by a graph in the article). What is the probability that X is at most 200? Less than 200? At least 200?
   c. What is the probability that X is between 100 and 200 (again assuming $\theta = 100$)?
   d. Give an expression for $P(X \le x)$.

5. A college professor never finishes his lecture before the bell rings to end the period and always fin-

ishes his lectures within 2 min after the bell rings. Let X = the time that elapses between the bell and the end of the lecture and suppose the pdf of X is

$$f(x) = \begin{cases} kx^2 & 0 \leq x \leq 2 \\ 0 & \text{otherwise} \end{cases}$$

a. Find the value of k. [*Hint:* Total area under the graph of $f(x)$ is 1.]
b. What is the probability that the lecture ends within 1 min of the bell ringing?
c. What is the probability that the lecture continues beyond the bell for between 60 and 90 sec?
d. What is the probability that the lecture continues for at least 90 sec beyond the bell?

6. The actual tracking weight of a stereo cartridge that is set to track at 3 g on a particular changer can be regarded as a continuous rv X with pdf

$$f(x) = \begin{cases} k[1 - (x - 3)^2] & 2 \leq x \leq 4 \\ 0 & \text{otherwise} \end{cases}$$

a. Sketch the graph of $f(x)$.
b. Find the value of k.
c. What is the probability that the actual tracking weight is greater than the prescribed weight?
d. What is the probability that the actual weight is within .25 g of the prescribed weight?
e. What is the probability that the actual weight differs from the prescribed weight by more than .5 g?

7. The time X (min) for a lab assistant to prepare the equipment for a certain experiment is believed to have a uniform distribution with $A = 25$ and $B = 35$.
a. Write the pdf of X and sketch its graph.
b. What is the probability that preparation time exceeds 33 min?
c. What is the probability that preparation time is within 2 min of the mean time? [*Hint:* Identify μ from the graph of $f(x)$.]
d. For any a such that $25 < a < a + 2 < 35$, what is the probability that preparation time is between a and $a + 2$ min?

8. In commuting to work, I must first get on a bus near my house and then transfer to a second bus. If the waiting time (in minutes) at each stop has a uniform distribution with $A = 0$ and $B = 5$, then it can be shown that my total waiting time Y has the pdf

$$f(y) = \begin{cases} \dfrac{1}{25}y & 0 \leq y < 5 \\ \dfrac{2}{5} - \dfrac{1}{25}y & 5 \leq y \leq 10 \\ 0 & y < 0 \text{ or } y > 10 \end{cases}$$

a. Sketch a graph of the pdf of Y.
b. Verify that $\int_{-\infty}^{\infty} f(y)dy = 1$.
c. What is the probability that total waiting time is at most 3 min?
d. What is the probability that total waiting time is at most 8 min?
e. What is the probability that total waiting time is between 3 and 8 min?
f. What is the probability that total waiting time is either less than 2 min or more than 6 min?

9. Consider again the pdf of X = time headway given in Example 4.4. What is the probability that time headway is
a. At most 6 sec?
b. More than 6 sec? At least 6 sec?
c. Between 5 and 6 sec?

10. A family of pdf's that has been used to approximate the distribution of income, city population size, and size of firms is the Pareto family. The family has two parameters, k and θ, both > 0, and the pdf is

$$f(x; k, \theta) = \begin{cases} \dfrac{k \cdot \theta^k}{x^{k+1}} & x \geq \theta \\ 0 & x < \theta \end{cases}$$

a. Sketch the graph of $f(x; k, \theta)$.
b. Verify that the total area under the graph equals 1.
c. If the rv X has pdf $f(x; k, \theta)$, for any fixed $b > \theta$, obtain an expression for $P(X \leq b)$.
d. For $\theta < a < b$, obtain an expression for the probability $P(a \leq X \leq b)$.

4.2 | Cumulative Distribution Functions and Expected Values

Several of the most important concepts introduced in the study of discrete distributions also play an important role for continuous distributions. Definitions analogous to those in Chapter 3 involve replacing summation by integration.

The Cumulative Distribution Function

The cumulative distribution function (cdf) $F(x)$ for a discrete rv X gives, for any specified number x, the probability $P(X \le x)$. It is obtained by summing the pmf $p(y)$ over all possible values y satisfying $y \le x$. The cdf of a continuous rv gives the same probabilities $P(X \le x)$ and is obtained by integrating the pdf $f(y)$ between the limits $-\infty$ and x.

DEFINITION

The **cumulative distribution function** $F(x)$ for a continuous rv X is defined for every number x by

$$F(x) = P(X \le x) = \int_{-\infty}^{x} f(y)dy$$

For each x, $F(x)$ is the area under the density curve to the left of x. This is illustrated in Figure 4.5, where $F(x)$ increases smoothly as x increases.

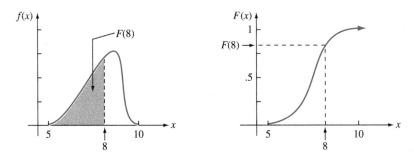

Figure 4.5   A pdf and associated cdf

Example 4.5   Let X have a uniform distribution on $[A, B]$. The density function is shown in Figure 4.6. For $x < A$, $F(x) = 0$, since there is no area under the graph of the density function to the left of such an x. For $x \ge B$, $F(x) = 1$, since all the area is accumulated to the left of such an x. Finally, for $A \le x \le B$,

$$F(x) = \int_{-\infty}^{x} f(y)dy = \int_{A}^{x} \frac{1}{B-A}\,dy = \frac{1}{B-A} \cdot y \Big|_{y=A}^{y=x} = \frac{x-A}{B-A}$$

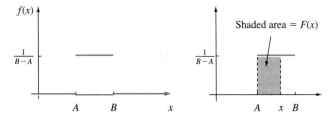

Figure 4.6   The pdf for a uniform distribution

The entire cdf is

$$F(x) = \begin{cases} 0 & x < A \\ \dfrac{x - A}{B - A} & A \le x < B \\ 1 & x \ge B \end{cases}$$

The graph of this cdf appears in Figure 4.7.

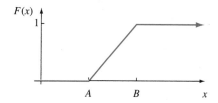

Figure 4.7   The cdf for a uniform distribution   ∎

Using *F(x)* to Compute Probabilities

The importance of the cdf here, just as for discrete rv's, is that probabilities of various intervals can be computed from a formula for or table of $F(x)$.

PROPOSITION

Let X be a continuous rv with pdf $f(x)$ and cdf $F(x)$. Then for any number a,

$$P(X > a) = 1 - F(a)$$

and for any two numbers a and b with $a < b$,

$$P(a \le X \le b) = F(b) - F(a)$$

Figure 4.8 illustrates the second part of this proposition; the desired probability is the shaded area under the density curve between a and b, and it equals the difference between the two shaded cumulative areas.

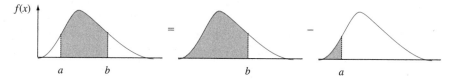

Figure 4.8   Computing $P(a \le X \le b)$ from cumulative probabilities

Example 4.6   Suppose the pdf of the magnitude X of a dynamic load on a bridge (in newtons) is given by

$$f(x) = \begin{cases} \dfrac{1}{8} + \dfrac{3}{8}x & 0 \le x \le 2 \\ 0 & \text{otherwise} \end{cases}$$

For any number x between 0 and 2,

$$F(x) = \int_{-\infty}^{x} f(y)dy = \int_{0}^{x} \left(\frac{1}{8} + \frac{3}{8} y \right) dy = \frac{x}{8} + \frac{3}{16} x^2$$

Thus

$$F(x) = \begin{cases} 0 & x < 0 \\ \dfrac{x}{8} + \dfrac{3}{16} x^2 & 0 \le x \le 2 \\ 1 & 2 < x \end{cases}$$

The graphs of $f(x)$ and $F(x)$ are shown in Figure 4.9. The probability that the load is between 1 and 1.5 is

$$P(1 \le X \le 1.5) = F(1.5) - F(1)$$
$$= \left[\frac{1}{8}(1.5) + \frac{3}{16}(1.5)^2 \right] - \left[\frac{1}{8}(1) + \frac{3}{16}(1)^2 \right]$$
$$= \frac{19}{64} = .297$$

The probability that the load exceeds 1 is

$$P(X > 1) = 1 - P(X \le 1) = 1 - F(1) = 1 - \left[\frac{1}{8}(1) + \frac{3}{16}(1)^2 \right]$$
$$= \frac{11}{16} = .688$$

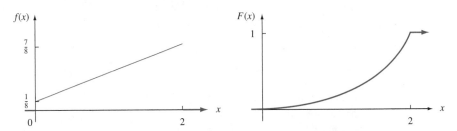

Figure 4.9 The pdf and cdf for Example 4.6   ∎

Once the cdf has been obtained, any probability involving X can easily be calculated without any further integration.

Obtaining f(x) from F(x)

For X discrete, the pmf is obtained from the cdf by taking the difference between two $F(x)$ values. The continuous analog of a difference is a derivative. The following result is a consequence of the Fundamental Theorem of Calculus.

PROPOSITION

If X is a continuous rv with pdf $f(x)$ and cdf $F(x)$, then at every x at which the derivative $F'(x)$ exists, $F'(x) = f(x)$.

Example 4.7
(Example 4.5
continued)

When X has a uniform distribution, $F(x)$ is differentiable except at $x = A$ and $x = B$, where the graph of $F(x)$ has sharp corners. Since $F(x) = 0$ for $x < A$ and $F(x) = 1$ for $x > B$, $F'(x) = 0 = f(x)$ for such x. For $A < x < B$,

$$F'(x) = \frac{d}{dx}\left(\frac{x - A}{B - A}\right) = \frac{1}{B - A} = f(x)$$

∎

Percentiles of a Continuous Distribution

When we say that an individual's test score was at the 85th percentile of the population, we mean that 85% of all population scores were below that score and 15% were above. Similarly, the 40th percentile is the score that exceeds 40% of all scores and is exceeded by 60% of all scores.

<table>
<tr><td>DEFINITION</td><td>Let p be a number between 0 and 1. The **(100p)th percentile** of the distribution of a continuous rv X, denoted by $\eta(p)$, is defined by

$$p = F(\eta(p)) = \int_{-\infty}^{\eta(p)} f(y)dy \qquad (4.2)$$</td></tr>
</table>

According to Expression (4.2), $\eta(p)$ is that value on the measurement axis such that $100p\%$ of the area under the graph of $f(x)$ lies to the left of $\eta(p)$ and $100(1 - p)\%$ lies to the right. Thus $\eta(.75)$, the 75th percentile, is such that the area under the graph of $f(x)$ to the left of $\eta(.75)$ is .75. Figure 4.10 illustrates the definition.

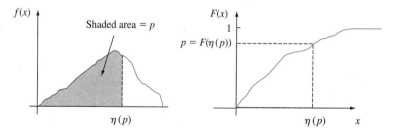

Figure 4.10 The (100p)th percentile of a continuous distribution

Example 4.8

The distribution of the amount of gravel (in tons) sold by a particular construction supply company in a given week is a continuous rv X with pdf

$$f(x) = \begin{cases} \frac{3}{2}(1 - x^2) & 0 \le x \le 1 \\ 0 & \text{otherwise} \end{cases}$$

The cdf of sales is then, for $0 < x < 1$,

$$F(x) = \int_0^x \frac{3}{2}(1 - y^2)dy = \frac{3}{2}\left(y - \frac{y^3}{3}\right)\Bigg|_{y=0}^{y=x} = \frac{3}{2}\left(x - \frac{x^3}{3}\right)$$

The graphs of both $f(x)$ and $F(x)$ appear in Figure 4.11. The $(100p)$th percentile of this distribution satisfies the equation

$$p = F(\eta(p)) = \frac{3}{2}\left[\eta(p) - \frac{(\eta(p))^3}{3}\right]$$

that is,

$$(\eta(p))^3 - 3\eta(p) + 2p = 0$$

For the 50th percentile, $p = .5$, and the equation to be solved is $\eta^3 - 3\eta + 1 = 0$; the solution is $\eta = \eta(.5) = .347$. If the distribution remains the same from week to week, then in the long run 50% of all weeks will result in sales of less than .347 ton and 50% in more than .347 ton.

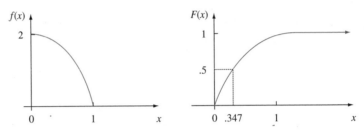

Figure 4.11   The pdf and cdf for Example 4.8

DEFINITION

> The **median** of a continuous distribution, denoted by $\tilde{\mu}$, is the 50th percentile, so $\tilde{\mu}$ satisfies $.5 = F(\tilde{\mu})$. That is, half the area under the density curve is to the left of $\tilde{\mu}$ and half is to the right of $\tilde{\mu}$.

A continuous distribution whose pdf is **symmetric**—which means that the graph of the pdf to the left of some point is a mirror image of the graph to the right of that point— has median $\tilde{\mu}$ equal to the point of symmetry, since half the area under the curve lies to either side of this point. Figure 4.12 gives several examples. The amount of error in a measurement of a physical quantity is often assumed to have a symmetric distribution.

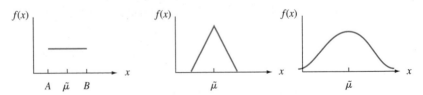

Figure 4.12   Medians of symmetric distributions

Expected Values for Continuous Random Variables

For a discrete random variable X, $E(X)$ was obtained by summing $x \cdot p(x)$ over possible X values. Here we replace summation by integration and the pmf by the pdf to get a continuous weighted average.

DEFINITION

The **expected** or **mean value** of a continuous rv X with pdf $f(x)$ is

$$\mu_X = E(X) = \int_{-\infty}^{\infty} x \cdot f(x)dx$$

Example 4.9
(Example 4.8
continued)

The pdf of weekly gravel sales X was

$$f(x) = \begin{cases} \dfrac{3}{2}(1 - x^2) & 0 \le x \le 1 \\ 0 & \text{otherwise} \end{cases}$$

so

$$E(X) = \int_{-\infty}^{\infty} x \cdot f(x)dx = \int_0^1 x \cdot \frac{3}{2}(1 - x^2)dx$$

$$= \frac{3}{2}\int_0^1 (x - x^3)dx = \frac{3}{2}\left(\frac{x^2}{2} - \frac{x^4}{4}\right)\Bigg|_{x=0}^{x=1} = \frac{3}{8} \qquad \blacksquare$$

When the pdf $f(x)$ specifies a model for the distribution of values in a numerical population, then μ is the population mean, which is the most frequently used measure of population location or center.

Often we wish to compute the expected value of some function $h(X)$ of the rv X. If we think of $h(X)$ as a new rv Y, techniques from mathematical statistics can be used to derive the pdf of Y, and $E(Y)$ can be computed from the definition. Fortunately, as in the discrete case, there is an easier way to compute $E[h(X)]$.

PROPOSITION

If X is a continuous rv with pdf $f(x)$ and $h(X)$ is any function of X, then

$$E[h(X)] = \mu_{h(X)} = \int_{-\infty}^{\infty} h(x) \cdot f(x)dx$$

Example 4.10
Two species are competing in a region for control of a limited amount of a certain resource. Let X = the proportion of the resource controlled by species 1 and suppose X has pdf

$$f(x) = \begin{cases} 1 & 0 \le x \le 1 \\ 0 & \text{otherwise} \end{cases}$$

which is a uniform distribution on [0, 1]. (In her book *Ecological Diversity*, E. C. Pielou calls this the "broken-stick" model for resource allocation, since it is analogous to breaking a stick at a randomly chosen point.) Then the species that controls the majority of this resource controls the amount

$$h(X) = \max(X, 1 - X) = \begin{cases} 1 - X & \text{if } 0 \le X < \dfrac{1}{2} \\ X & \text{if } \dfrac{1}{2} \le X \le 1 \end{cases}$$

The expected amount controlled by the species having majority control is then

$$E[h(X)] = \int_{-\infty}^{\infty} \max(x, 1 - x) \cdot f(x)dx = \int_{0}^{1} \max(x, 1 - x) \cdot 1 \, dx$$

$$= \int_{0}^{1/2} (1 - x) \cdot 1 \, dx + \int_{1/2}^{1} x \cdot 1 \, dx = \frac{3}{4}$$ ■

For $h(X)$ a linear function, $E[h(X)] = E(aX + b) = aE(X) + b$.

The Variance of a Continuous Random Variable

DEFINITION

The **variance** of a continuous random variable X with pdf $f(x)$ and mean value μ is

$$\sigma_X^2 = V(X) = \int_{-\infty}^{\infty} (x - \mu)^2 \cdot f(x) \, dx = E[(X - \mu)^2]$$

The **standard deviation** (SD) of X is $\sigma_X = \sqrt{V(X)}$.

As in the discrete case, σ_X^2 is the expected or average squared deviation about the mean μ and gives a measure of how much spread there is in the distribution or population of x values. The easiest way to compute σ^2 is to again use a shortcut formula.

PROPOSITION

$$V(X) = E(X^2) - [E(X)]^2$$

Example 4.11
(Example 4.9
continued)

For $X =$ weekly gravel sales, we computed $E(X) = \frac{3}{8}$. Since

$$E(X^2) = \int_{-\infty}^{\infty} x^2 \cdot f(x)dx = \int_{0}^{1} x^2 \cdot \frac{3}{2} (1 - x^2)dx$$

$$= \int_{0}^{1} \frac{3}{2}(x^2 - x^4)dx = \frac{1}{5}$$

$$V(X) = \frac{1}{5} - \left(\frac{3}{8}\right)^2 = \frac{19}{320} = .059 \qquad \text{and} \qquad \sigma_X = .244$$ ■

When $h(X)$ is a linear function and $V(X) = \sigma^2$, $V[h(X)] = V(aX + b) = a^2 \cdot \sigma^2$ and $\sigma_{aX+b} = |a| \cdot \sigma$.

Exercises | Section 4.2 (11–25)

11. The cdf of checkout duration X as described in Exercise 1 is

$$F(x) = \begin{cases} 0 & x < 0 \\ \dfrac{x^2}{4} & 0 \le x < 2 \\ 1 & 2 \le x \end{cases}$$

Use this to compute the following:
a. $P(X \le 1)$
b. $P(.5 \le X \le 1)$
c. $P(X > .5)$
d. The median checkout duration $\tilde{\mu}$ [solve $.5 = F(\tilde{\mu})$]
e. $F'(x)$ to obtain the density function $f(x)$

12. The cdf for X ($=$ measurement error) of Exercise 3 is

$$F(x) = \begin{cases} 0 & x < -2 \\ \dfrac{1}{2} + \dfrac{3}{32}\left(4x - \dfrac{x^3}{3}\right) & -2 \le x < 2 \\ 1 & 2 \le x \end{cases}$$

a. Compute $P(X < 0)$.
b. Compute $P(-1 < X < 1)$.
c. Compute $P(.5 < X)$.
d. Verify that $f(x)$ is as given in Exercise 3 by obtaining $F'(x)$.
e. Verify that $\tilde{\mu} = 0$.

13. Let X denote checkout time duration with pdf given in Exercise 1.
a. Compute $E(X)$.
b. Compute $V(X)$ and σ_X.
c. If the borrower is charged an amount $h(X) = X^2$ when checkout duration is X, compute the expected charge $E[h(X)]$.

14. The article "Modeling Sediment and Water Column Interactions for Hydrophobic Pollutants" (*Water Research,* 1984: 1169–1174) suggests the uniform distribution on the interval (7.5, 20) as a model for depth (cm) of the bioturbation layer in sediment in a certain region.
a. What are the mean and variance of depth?
b. What is the cdf of depth?

c. What is the probability that observed depth is at most 10? Between 10 and 15?
d. What is the probability that the observed depth is within 1 standard deviation of the mean value? Within 2 standard deviations?

15. Let X denote the amount of space occupied by an article placed in a 1-ft^3 packing container. The pdf of X is

$$f(x) = \begin{cases} 90x^8(1 - x) & 0 < x < 1 \\ 0 & \text{otherwise} \end{cases}$$

a. Graph the pdf. Then obtain the cdf of X and graph it.
b. What is $P(X \le .5)$ [i.e., $F(.5)$]?
c. Using part (a), what is $P(.25 < X \le .5)$? What is $P(.25 \le X \le .5)$?
d. What is the 75th percentile of the distribution?
e. Compute $E(X)$ and σ_X.
f. What is the probability that X is within 1 standard deviation of its mean value?

16. Answer parts (a)–(f) of Exercise 15 with $X =$ lecture time past the bell given in Exercise 5.

17. Consider the pdf of $X =$ actual tracking weight given in Exercise 6.
a. Obtain and graph the cdf of X.
b. From the graph of $f(x)$, what is $\tilde{\mu}$?
c. Compute $E(X)$ and $V(X)$.

18. Let X have a uniform distribution on the interval $[A, B]$.
a. Obtain an expression for the $(100p)$th percentile.
b. Compute $E(X)$, $V(X)$, and σ_X.
c. For n a positive integer, compute $E(X^n)$.

19. Let X be a continuous rv with cdf

$$F(x) = \begin{cases} 0 & x \le 0 \\ \dfrac{x}{4}\left[1 + \ln\left(\dfrac{4}{x}\right)\right] & 0 < x \le 4 \\ 1 & x > 4 \end{cases}$$

[This type of cdf is suggested in the article "Variability in Measured Bedload-Transport Rates" (*Water Resources Bull.,* 1985: 39–48) as a model for a certain hydrologic variable.] What is

a. $P(X \le 1)$?
b. $P(1 \le X \le 3)$?
c. The pdf of X?

20. Consider the pdf for total waiting time Y for two buses

$$f(y) = \begin{cases} \dfrac{1}{25} y & 0 \le y < 5 \\ \dfrac{2}{5} - \dfrac{1}{25} y & 5 \le y \le 10 \\ 0 & \text{otherwise} \end{cases}$$

introduced in Exercise 8.
a. Compute and sketch the cdf of Y. [*Hint:* Consider separately $0 \le y < 5$ and $5 \le y \le 10$ in computing $F(y)$. A graph of the pdf should be helpful.]
b. Obtain an expression for the $(100p)$th percentile. (*Hint:* Consider separately $0 < p < .5$ and $.5 < p < 1$.)
c. Compute $E(Y)$ and $V(Y)$. How do these compare with the expected waiting time and variance for a single bus when the time is uniformly distributed on $[0, 5]$?

21. An ecologist wishes to mark off a circular sampling region having radius 10 m. However, the radius of the resulting region is actually a random variable R with pdf

$$f(r) = \begin{cases} \dfrac{3}{4}[1 - (10 - r)^2] & 9 \le r \le 11 \\ 0 & \text{otherwise} \end{cases}$$

What is the expected area of the resulting circular region?

22. The weekly demand for propane gas (in 1000's of gallons) from a particular facility is an rv X with pdf

$$f(x) = \begin{cases} 2\left(1 - \dfrac{1}{x^2}\right) & 1 \le x \le 2 \\ 0 & \text{otherwise} \end{cases}$$

a. Compute the cdf of X.
b. Obtain an expression for the $(100p)$th percentile. What is the value of $\tilde{\mu}$?
c. Compute $E(X)$ and $V(X)$.
d. If 1.5 thousand gallons is in stock at the beginning of the week and no new supply is due in during the week, how much of the 1.5 thousand gallons is expected to be left at the end of the week? *Hint:* Let $h(x)$ = amount left when demand = x.

23. If the temperature at which a certain compound melts is a random variable with mean value 120°C and standard deviation 2°C, what are the mean temperature and standard deviation measured in °F? (*Hint:* °F = 1.8°C + 32.)

24. Let X have the Pareto pdf

$$f(x; k, \theta) = \begin{cases} \dfrac{k \cdot \theta^k}{x^{k+1}} & x \ge \theta \\ 0 & x < \theta \end{cases}$$

introduced in Exercise 10.
a. If $k > 1$, compute $E(X)$.
b. What can you say about $E(X)$ if $k = 1$?
c. If $k > 2$, show that $V(X) = k\theta^2(k-1)^{-2}(k-2)^{-1}$.
d. If $k = 2$, what can you say about $V(X)$?
e. What conditions on k are necessary to ensure that $E(X^n)$ is finite?

25. Let X be the temperature in °C at which a certain chemical reaction takes place, and let Y be the temperature in °F (so $Y = 1.8X + 32$).
a. If the median of the X distribution is $\tilde{\mu}$, show that $1.8\tilde{\mu} + 32$ is the median of the Y distribution.
b. How is the 90th percentile of the Y distribution related to the 90th percentile of the X distribution? Verify your conjecture.
c. More generally, if $Y = aX + b$, how is any particular percentile of the Y distribution related to the corresponding percentile of the X distribution?

4.3 | The Normal Distribution

The normal distribution is the most important one in all of probability and statistics. Many numerical populations have distributions that can be fit very closely by an appropriate normal curve. Examples include heights, weights, and other physical characteristics (the famous 1903 *Biometrika* article "On the Laws of Inheritance in Man" discussed many examples of this sort), measurement errors in scientific experiments, anthropometric measurements on fossils, reaction times in psychological experiments, measurements of intelligence and aptitude, scores on various tests, and numerous economic measures and indicators. Even when the underlying distribution is discrete, the normal curve often gives an excellent approximation. In addition, even when individual variables themselves are not normally distributed, sums and averages of the variables will under suitable conditions have approximately a normal distribution; this is the content of the Central Limit Theorem discussed in the next chapter.

DEFINITION

> A continuous rv X is said to have a **normal distribution with parameters μ and σ (or μ and σ^2),** where $-\infty < \mu < \infty$ and $0 < \sigma$, if the pdf of X is
>
> $$f(x; \mu, \sigma) = \frac{1}{\sqrt{2\pi}\,\sigma}\, e^{-(x-\mu)^2/(2\sigma^2)} \qquad -\infty < x < \infty \qquad (4.3)$$

Again e denotes the base of the natural logarithm system and equals approximately 2.71828, and π represents the familiar mathematical constant with approximate value 3.14159. The statement that X is normally distributed with parameters μ and σ^2 is often abbreviated $X \sim N(\mu, \sigma^2)$.

Clearly $f(x; \mu, \sigma) \geq 0$ for any number x, but techniques from multivariate calculus must be used to show that $\int_{-\infty}^{\infty} f(x; \mu, \sigma)\, dx = 1$. It can be shown that $E(X) = \mu$ and $V(X) = \sigma^2$, so the parameters are the mean and the standard deviation of X. Figure 4.13 presents graphs of $f(x; \mu, \sigma)$ for several different (μ, σ) pairs. Each graph is symmetric about μ and bell-shaped, so the center of the bell (point of symmetry) is both the mean of the distribution and the median. The value of σ is the distance from μ to the inflection points of the curve (the points at which the curve changes from turning downward to turning upward). Large values of σ yield graphs that are quite spread out about μ, whereas small values of σ yield graphs with a high peak above μ and most of the area under the graph quite close to μ. Thus, a large σ implies that a value of X far from μ may well be observed, whereas such a value is quite unlikely when σ is small.

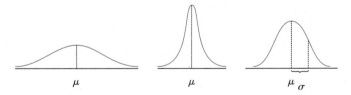

Figure 4.13   Graphs of normal pdf's

The Standard Normal Distribution

To compute $P(a \leq X \leq b)$ when X is a normal rv with parameters μ and σ, we must evaluate

$$\int_a^b \frac{1}{\sqrt{2\pi}\,\sigma}\, e^{-(x-\mu)^2/(2\sigma^2)}\, dx \qquad (4.4)$$

None of the standard integration techniques can be used to evaluate Expression (4.4). Instead, for $\mu = 0$ and $\sigma = 1$, Expression (4.4) has been numerically evaluated and tabulated for certain values of a and b. This table can also be used to compute probabilities for any other values of μ and σ under consideration.

DEFINITION

> The normal distribution with parameter values $\mu = 0$ and $\sigma = 1$ is called a **standard normal distribution.** A random variable that has a standard normal distribution is called a **standard normal random variable** and will be denoted by Z. The pdf of Z is
>
> $$f(z; 0, 1) = \frac{1}{\sqrt{2\pi}} e^{-z^2/2} \qquad -\infty < z < \infty$$
>
> The cdf of Z is $P(Z \leq z) = \int_{-\infty}^{z} f(y; 0, 1)dy$, which we will denote by $\Phi(z)$.

The standard normal distribution does not frequently serve as a model for a naturally arising population. Instead, it is a reference distribution from which information about other normal distributions can be obtained. Appendix Table A.3 gives $\Phi(z) = P(Z \leq z)$, the area under the graph of the standard normal pdf to the left of z, for $z = -3.49, -3.48, \ldots, 3.48, 3.49$. Figure 4.14 illustrates the type of cumulative area (probability) tabulated in Table A.3. From this table, various other probabilities involving Z can be calculated.

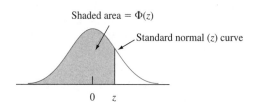

Shaded area $= \Phi(z)$

Standard normal (z) curve

$0 \quad z$

Figure 4.14   Standard normal cumulative areas tabulated in Appendix Table A.3

Example 4.12   Compute the following probabilities: (a) $P(Z \leq 1.25)$,   (b) $P(Z > 1.25)$, (c) $P(Z \leq -1.25)$, and (d) $P(-.38 \leq Z \leq 1.25)$.

a. $P(Z \leq 1.25) = \Phi(1.25)$, a probability that is tabulated in Appendix Table A.3 at the intersection of the row marked 1.2 and the column marked .05. The number there is .8944, so $P(Z \leq 1.25) = .8944$. See Figure 4.15(a).

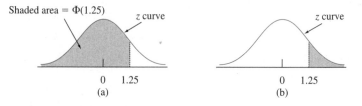

Shaded area $= \Phi(1.25)$

z curve

z curve

$0 \quad 1.25$
(a)

$0 \quad 1.25$
(b)

Figure 4.15   Normal curve areas (probabilities) for Example 4.12

b. $P(Z > 1.25) = 1 - P(Z \le 1.25) = 1 - \Phi(1.25)$, the area under the standard normal curve to the right of 1.25 (an upper-tail area). Since $\Phi(1.25) = .8944$, $P(Z > 1.25) = .1056$. Since Z is a continuous rv, $P(Z \ge 1.25)$ also equals .1056. See Figure 4.15(b).

c. $P(Z \le -1.25) = \Phi(-1.25)$, a lower-tail area. Directly from Appendix Table A.3, $\Phi(-1.25) = .1056$. By symmetry of the normal curve, this is the same answer as in part (b).

d. $P(-.38 \le Z \le 1.25)$ is the area under the standard normal curve above the interval whose left endpoint is $-.38$ and whose right endpoint is 1.25. From Section 4.2, if X is a continuous rv with cdf $F(x)$, then $P(a \le X \le b) = F(b) - F(a)$. This gives $P(-.38 \le Z \le 1.25) = \Phi(1.25) - \Phi(-.38) = .8944 - .3520 = .5424$. (See Figure 4.16.)

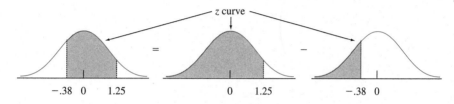

Figure 4.16 $P(-.38 \le Z \le 1.25)$ as the difference between two cumulative areas ■

Percentiles of the Standard Normal Distribution

For any p between 0 and 1, Appendix Table A.3 can be used to obtain the $(100p)$th percentile of the standard normal distribution.

Example 4.13   The 99th percentile of the standard normal distribution is that value on the axis such that the area under the curve to the left of the value is .9900. Now Appendix Table A.3 gives for fixed z the area under the standard normal curve to the left of z, whereas here we have the area and want the value of z. This is the "inverse" problem to $P(Z \le z) = ?$ so the table is used in an inverse fashion: Find in the middle of the table .9900; the row and column in which it lies identify the 99th z percentile. Here .9901 lies in the row marked 2.3 and column marked .03, so the 99th percentile is (approximately) $z = 2.33$. (See Figure 4.17.) By symmetry, the first percentile is the negative of the 99th percentile, so equals -2.33 (1% lies below the first and above the 99th). (See Figure 4.18.)

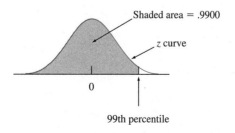

Figure 4.17   Finding the 99th percentile

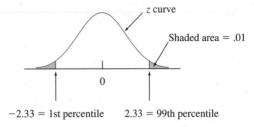

Figure 4.18 The relationship between the 1st and 99th percentiles ■

In general, the ($100p$)th percentile is identified by the row and column of Appendix Table A.3 in which the entry p is found (e.g., the 67th percentile is obtained by finding .6700 in the body of the table, which gives $z = .44$). If p does not appear, the number closest to it is often used, although linear interpolation gives a more accurate answer. For example, to find the 95th percentile, we look for .9500 inside the table. Although .9500 does not appear, both .9495 and .9505 do, corresponding to $z = 1.64$ and 1.65, respectively. Since .9500 is halfway between the two probabilities that do appear, we will use 1.645 as the 95th percentile and -1.645 as the 5th percentile.

z_α Notation

In statistical inference, we will need the values on the measurement axis that capture certain small tail areas under the standard normal curve.

Notation

z_α will denote the value on the measurement axis for which α of the area under the z curve lies to the right of z_α. (See Figure 4.19.)

For example, $z_{.10}$ captures upper-tail area .10 and $z_{.01}$ captures upper-tail area .01.

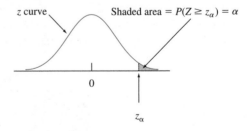

Figure 4.19 z_α notation illustrated

Since α of the area under the standard normal curve lies to the right of z_α, $1 - \alpha$ of the area lies to the left of z_α. Thus, z_α is the $100(1 - \alpha)$th percentile of the standard normal distribution. Further, by symmetry the area under the standard normal curve to the left of $-z_\alpha$ is also α. The z_α's are usually referred to as **z critical values.** Table 4.1 lists the most useful standard normal percentiles and z_α values.

Table 4.1 Standard normal percentiles and critical values

Percentile	90	95	97.5	99	99.5	99.9	99.95
α (tail area)	.1	.05	.025	.01	.005	.001	.0005
$z_\alpha = 100(1 - \alpha)$th percentile	1.28	1.645	1.96	2.33	2.58	3.08	3.27

Example 4.14 $z_{.05}$ is the $100(1 - .05)$th $= 95$th percentile of the standard normal distribution, so $z_{.05} = 1.645$. The area under the standard normal curve to the left of $-z_{.05}$ is also .05. (See Figure 4.20.)

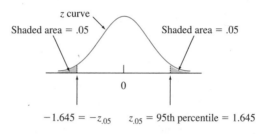

z curve

Shaded area = .05   Shaded area = .05

0

$-1.645 = -z_{.05}$ $z_{.05} = $ 95th percentile $= 1.645$

Figure 4.20   Finding $z_{.05}$

Nonstandard Normal Distributions

When $X \sim N(\mu, \sigma^2)$, probabilities involving X are computed by "standardizing." The **standardized variable** is $(X - \mu)/\sigma$. Subtracting μ shifts the mean from μ to zero, and then dividing by σ scales the variable so that the standard deviation is 1 rather than σ.

PROPOSITION

If X has a normal distribution with mean μ and standard deviation σ, then

$$Z = \frac{X - \mu}{\sigma}$$

has a standard normal distribution. Thus

$$P(a \le X \le b) = P\left(\frac{a - \mu}{\sigma} \le Z \le \frac{b - \mu}{\sigma}\right)$$

$$= \Phi\left(\frac{b - \mu}{\sigma}\right) - \Phi\left(\frac{a - \mu}{\sigma}\right)$$

$$P(X \le a) = \Phi\left(\frac{a - \mu}{\sigma}\right) \qquad P(X \ge b) = 1 - \Phi\left(\frac{b - \mu}{\sigma}\right)$$

The key idea of the proposition is that by standardizing, any probability involving X can be expressed as a probability involving a standard normal rv Z, so that Appendix

Table A.3 can be used. This is illustrated in Figure 4.21. The proposition can be proved by writing the cdf of $Z = (X - \mu)/\sigma$ as

$$P(Z \le z) = P(X \le \sigma z + \mu) = \int_{-\infty}^{\sigma z + \mu} f(x;\ \mu,\ \sigma)dx$$

Using a result from calculus, this integral can be differentiated with respect to z to yield the desired pdf $f(z:\ 0,\ 1)$.

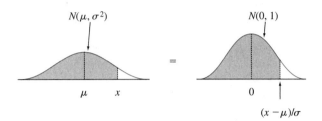

$N(\mu, \sigma^2)$ $N(0, 1)$

μ x 0

$(x - \mu)/\sigma$

Figure 4.21 Equality of nonstandard and standard normal curve areas

Example 4.15 The time that it takes a driver to react to the brake lights on a decelerating vehicle is critical in helping to avoid rear-end collisions. The article "Fast-Rise Brake Lamp as a Collision-Prevention Device" (*Ergonomics,* 1993: 391–395) suggests that reaction time for an in-traffic response to a brake signal from standard brake lights can be modeled with a normal distribution having mean value 1.25 sec and standard deviation of .46 sec. What is the probability that reaction time is between 1.00 sec and 1.75 sec? If we let X denote reaction time, then standardizing gives

$$1.00 \le X \le 1.75$$

if and only if

$$\frac{1.00 - 1.25}{.46} \le \frac{X - 1.25}{.46} \le \frac{1.75 - 1.25}{.46}$$

Thus,

$$P(1.00 \le X \le 1.75) = P\left(\frac{1.00 - 1.25}{.46} \le Z \le \frac{1.75 - 1.25}{.46}\right)$$

$$= P(-.54 \le Z \le 1.09) = \Phi(1.09) - \Phi(-.54)$$

$$= .8621 - .2946 = .5675$$

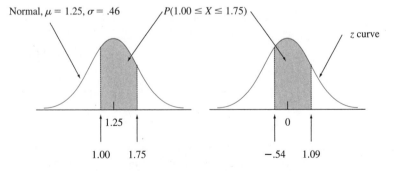

Normal, $\mu = 1.25, \sigma = .46$ $P(1.00 \le X \le 1.75)$ z curve

1.25 0

1.00 1.75 $-.54$ 1.09

Figure 4.22 Normal curves for Example 4.15

This is illustrated in Figure 4.22. Similarly, if we view 2 sec as a critically long reaction time, the probability that actual reaction time will exceed this value is

$$P(X > 2) = P\left(Z > \frac{2 - 1.25}{.46}\right) = P(Z > 1.63) = 1 - \Phi(1.63) = .0516$$ ∎

Standardizing amounts to nothing more than calculating a distance from the mean value and then reexpressing the distance as some number of standard deviations. For example, if $\mu = 100$ and $\sigma = 15$, then $x = 130$ corresponds to $z = (130 - 100)/15 = 30/15 = 2.00$. That is, 130 is 2 standard deviations above (to the right of) the mean value. Similarly, standardizing 85 gives $(85 - 100)/15 = -1.00$, so 85 is 1 standard deviation below the mean. The z table applies to *any* normal distribution provided that we think in terms of number of standard deviations away from the mean value.

Example 4.16 The breakdown voltage of a randomly chosen diode of a particular type is known to be normally distributed. What is the probability that a diode's breakdown voltage is within 1 standard deviation of its mean value? This question can be answered without knowing either μ or σ, as long as the distribution is known to be normal; in other words, the answer is the same for *any* normal distribution:

$P(X$ is within 1 standard deviation of its mean$) = P(\mu - \sigma \le X \le \mu + \sigma)$

$$= P\left(\frac{\mu - \sigma - \mu}{\sigma} \le Z \le \frac{\mu + \sigma - \mu}{\sigma}\right)$$

$$= P(-1.00 \le Z \le 1.00)$$

$$= \Phi(1.00) - \Phi(-1.00) = .6826$$

The probability that X is observed to be within 2 standard deviations is $P(-2.00 \le Z \le 2.00) = .9544$ and within 3 standard deviations is $P(-3.00 \le Z \le 3.00) = .9974$. ∎

The results of Example 4.16 are often reported in percentage form and referred to as the *empirical rule* (because empirical evidence has shown that histograms of real data can very frequently be approximated by normal curves).

If the population distribution of a variable is (approximately) normal, then

1. Roughly 68% of the values are within 1 SD of the mean.
2. Roughly 95% of the values are within 2 SDs of the mean.
3. Roughly 99.7% of the values are within 3 SDs of the mean.

It is indeed unusual to observe a value from a normal population that is much farther than 2 standard deviations from μ. These results will be important in the development of hypothesis-testing procedures in later chapters.

Percentiles of an Arbitrary Normal Distribution

The $(100p)$th percentile of a normal distribution with mean μ and standard deviation σ is easily related to the $(100p)$th percentile of the standard normal distribution.

PROPOSITION

$$\begin{array}{c} (100p)\text{th percentile} \\ \text{for normal } (\mu, \sigma) \end{array} = \mu + \left[\begin{array}{c} (100p)\text{th for} \\ \text{standard normal} \end{array}\right] \cdot \sigma$$

Another way of saying this is that if z is the desired percentile for the standard normal distribution, then the desired percentile for the normal (μ, σ) distribution is z standard deviations from μ.

Example 4.17   The amount of distilled water dispensed by a certain machine is normally distributed with mean value 64 oz and standard deviation .78 oz. What container size c will ensure that overflow occurs only .5% of the time? If X denotes the amount dispensed, the desired condition is that $P(X > c) = .005$, or, equivalently, that $P(X \le c) = .995$. Thus c is the 99.5th percentile of the normal distribution with $\mu = 64$ and $\sigma = .78$. The 99.5th percentile of the standard normal distribution is 2.58, so

$$c = \eta(.995) = 64 + (2.58)(.78) = 64 + 2.0 = 66 \text{ oz}$$

This is illustrated in Figure 4.23.

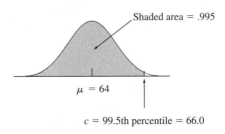

Shaded area = .995

$\mu = 64$

$c = 99.5\text{th percentile} = 66.0$

Figure 4.23   Distribution of amount dispensed for Example 4.17   ■

The Normal Distribution and Discrete Populations

The normal distribution is often used as an approximation to the distribution of values in a discrete population. In such situations, extra care must be taken to ensure that probabilities are computed in an accurate manner.

Example 4.18   IQ in a particular population (as measured by a standard test) is known to be approximately normally distributed with $\mu = 100$ and $\sigma = 15$. What is the probability that a randomly selected individual has an IQ of at least 125? Letting $X =$ the IQ of a randomly chosen person, we wish $P(X \ge 125)$. The temptation here is to standardize $X \ge 125$ immediately as in previous examples. However, the IQ population is actually discrete, since IQs are integer-valued, so the normal curve is an approximation to a discrete probability histogram, as pictured in Figure 4.24.

The rectangles of the histogram are *centered* at integers, so IQs of at least 125 correspond to rectangles beginning at 124.5, as shaded in Figure 4.24. Thus, we really want $P(X \ge 124.5)$, which can now be standardized to yield $P(Z \ge 1.63) = .0516$. If we had standardized $X \ge 125$, we would have obtained $P(Z \ge 1.67) = .0475$. The difference is not great, but the answer .0516 is more accurate. Similarly, $P(X = 125)$ would be approximated by the area between 124.5 and 125.5, since the area under the normal curve above the single value 125 is zero.

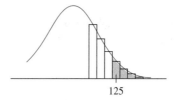

125

Figure 4.24 A normal approximation to a discrete distribution ■

The correction for discreteness of the underlying distribution in Example 4.18 is often called a **continuity correction.** It is useful in the following application of the normal distribution to the computation of binomial probabilities.

The Normal Approximation to the Binomial Distribution

Recall that the mean value and standard deviation of a binomial random variable X are $\mu_X = np$ and $\sigma_X = \sqrt{npq}$, respectively. Figure 4.25 displays a binomial probability histogram for the binomial distribution with $n = 20$, $p = .6$ [so $\mu = 20(.6) = 12$ and $\sigma = \sqrt{20(.6)(.4)} = 2.19$]. A normal curve with mean value and standard deviation equal to the corresponding values for the binomial distribution has been superimposed on the probability histogram. Although the probability histogram is a bit skewed (because $p \neq .5$), the normal curve gives a very good approximation, especially in the middle part of the picture. The area of any rectangle (probability of any particular X value) except those in the extreme tails can be accurately approximated by the corresponding normal curve area. For example, $P(X = 10) = B(10; 20, .6) - B(9; 20, .6) = .117$, whereas the area under the normal curve between 9.5 and 10.5 is $P(-1.14 \leq Z \leq -.68) = .1212$.

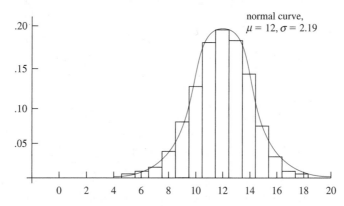

Figure 4.25 Binomial probability histogram for $n = 20$, $p = .6$ with normal approximation curve superimposed

More generally, as long as the binomial probability histogram is not too skewed, binomial probabilities can be well approximated by normal curve areas. It is then customary to say that X has approximately a normal distribution.

PROPOSITION

Let X be a binomial rv based on n trials with success probability p. Then if the binomial probability histogram is not too skewed, X has approximately a normal distribution with $\mu = np$ and $\sigma = \sqrt{npq}$. In particular, for $x = $ a possible value of X,

$$P(X \leq x) = B(x; n, p) \approx \left(\begin{array}{c} \text{area under the normal curve} \\ \text{to the left of } x + .5 \end{array} \right)$$

$$= \Phi\left(\frac{x + .5 - np}{\sqrt{npq}} \right)$$

In practice, the approximation is adequate provided that both $np \geq 10$ and $nq \geq 10$.

If either $np < 10$ or $nq < 10$, the binomial distribution is too skewed for the (symmetric) normal curve to give accurate approximations.

Example 4.19 Suppose that 25% of all licensed drivers in a particular state do not have insurance. Let X be the number of uninsured drivers in a random sample of size 50 (somewhat perversely, a success is an uninsured driver), so that $p = .25$. Then $\mu = 12.5$ and $\sigma = 3.06$. Since $np = 50(.25) = 12.5 \geq 10$ and $nq = 37.5 \geq 10$, the approximation can safely be applied:

$$P(X \leq 10) = B(10; 50, .25) \approx \Phi\left(\frac{10 + .5 - 12.5}{3.06} \right)$$

$$= \Phi(-.65) = .2578$$

Similarly, the probability that between 5 and 15 (inclusive) of the selected drivers are uninsured is

$$P(5 \leq X \leq 15) = B(15; 50, .25) - B(4; 50, .25)$$

$$\approx \Phi\left(\frac{15.5 - 12.5}{3.06} \right) - \Phi\left(\frac{4.5 - 12.5}{3.06} \right) = .8320$$

The exact probabilities are .2622 and .8348, respectively, so the approximations are quite good. In the last calculation, the probability $P(5 \leq X \leq 15)$ is being approximated by the area under the normal curve between 4.5 and 15.5—the continuity correction is used for both the upper and lower limits. ∎

When the objective of our investigation is to make an inference about a population proportion p, interest will focus on the sample proportion of successes X/n rather than on X itself. Because this proportion is just X multiplied by the constant $1/n$, it will also have approximately a normal distribution (with mean $\mu = p$ and standard deviation

$\sigma = \sqrt{pq/n}$) provided that both $np \geq 10$ and $nq \geq 10$. This normal approximation is the basis for several large-sample inferential procedures to be discussed in later chapters.

It is quite difficult to give a direct proof of the validity of this normal approximation (the first one goes back over 150 years to Laplace). In the next chapter, we'll show that it is a consequence of an important general result called the Central Limit Theorem.

Exercises | Section 4.3 (26–52)

26. Let Z be a standard normal random variable and calculate the following probabilities, drawing pictures wherever appropriate.
  a. $P(0 \leq Z \leq 2.17)$
  b. $P(0 \leq Z \leq 1)$
  c. $P(-2.50 \leq Z \leq 0)$
  d. $P(-2.50 \leq Z \leq 2.50)$
  e. $P(Z \leq 1.37)$
  f. $P(-1.75 \leq Z)$
  g. $P(-1.50 \leq Z \leq 2.00)$
  h. $P(1.37 \leq Z \leq 2.50)$
  i. $P(1.50 \leq Z)$
  j. $P(|Z| \leq 2.50)$

27. In each case, determine the value of the constant c that makes the probability statement correct.
  a. $\Phi(c) = .9838$
  b. $P(0 \leq Z \leq c) = .291$
  c. $P(c \leq Z) = .121$
  d. $P(-c \leq Z \leq c) = .668$
  e. $P(c \leq |Z|) = .016$

28. Find the following percentiles for the standard normal distribution. Interpolate where appropriate.
  a. 91st
  b. 9th
  c. 75th
  d. 25th
  e. 6th

29. Determine z_α for the following:
  a. $\alpha = .0055$
  b. $\alpha = .09$
  c. $\alpha = .663$

30. If X is a normal rv with mean 80 and standard deviation 10, compute the following probabilities by standardizing.
  a. $P(X \leq 100)$
  b. $P(X \leq 80)$
  c. $P(65 \leq X \leq 100)$
  d. $P(70 \leq X)$

  e. $P(85 \leq X \leq 95)$
  f. $P(|X - 80| \leq 10)$

31. Suppose the force acting on a column that helps to support a building is normally distributed with mean 15.0 kips and standard deviation 1.25 kips. What is the probability that the force
  a. Is at most 17 kips?
  b. Is between 10 and 12 kips?
  c. Differs from 15.0 kips by at most 2 standard deviations?

32. The article "Reliability of Domestic-Waste Biofilm Reactors" (*J. of Envir. Engr.*, 1995: 785–790) suggests that substrate concentration (mg/cm³) of influent to a reactor is normally distributed with $\mu = .30$ and $\sigma = .06$.
  a. What is the probability that the concentration exceeds .25?
  b. What is the probability that the concentration is at most .10?
  c. How would you characterize the largest 5% of all concentration values?

33. Suppose the diameter at breast height (in.) of trees of a certain type is normally distributed with $\mu = 8.8$ and $\sigma = 2.8$, as suggested in the article "Simulating a Harvester-Forwarder Softwood Thinning" (*Forest Products J.*, May 1997: 36–41).
  a. What is the probability that the diameter of a randomly selected tree will be at least 10 in.? Will exceed 10 in.?
  b. What is the probability that the diameter of a randomly selected tree will exceed 20 in.?
  c. What is the probability that the diameter of a randomly selected tree will be between 5 and 10 in.?
  d. What value c is such that the interval $(8.8 - c, 8.8 + c)$ includes 98% of all diameter values?

34. There are two machines available for cutting corks intended for use in wine bottles. The first produces corks with diameters that are normally distributed with mean 3 cm and standard deviation .1 cm. The second machine produces corks with diameters that have a normal distribution with mean 3.04 cm and standard deviation .02 cm. Acceptable corks have diameters between 2.9 cm and 3.1 cm. Which machine is more likely to produce an acceptable cork?

35. a. If a normal distribution has $\mu = 25$ and $\sigma = 5$, what is the 91st percentile of the distribution?
    b. What is the 6th percentile of the distribution?
    c. The width of a line etched on an integrated circuit chip is normally distributed with mean 3.000 μm and standard deviation .150. What width value separates the widest 10% of all such lines from the other 90 percent?

36. The article "Monte Carlo Simulation—Tool for Better Understanding of LRFD" (*J. Structural Engr.*, 1993: 1586–1599) suggests that yield strength (ksi) for A36 steel grade is normally distributed with $\mu = 43$ and $\sigma = 4.5$.
    a. What is the probability that yield strength is at most 40? Greater than 60?
    b. What yield strength value separates the strongest 75% from the others?

37. The automatic opening device of a military cargo parachute has been designed to open when the parachute is 200 m above the ground. Suppose opening altitude actually has a normal distribution with mean value 200 m and standard deviation 30 m. Equipment damage will occur if the parachute opens at an altitude of less than 100 m. What is the probability that there is equipment damage to the payload of at least one of five independently dropped parachutes?

38. The temperature reading from a thermocouple placed in a constant-temperature medium is normally distributed with mean μ, the actual temperature of the medium, and standard deviation σ. What would the value of σ have to be to ensure that 95% of all readings are within .1° of μ?

39. The distribution of resistance for resistors of a certain type is known to be normal, with 10% of all resistors having a resistance exceeding 10.256 ohms, and 5% having a resistance smaller than 9.671 ohms. What are the mean value and standard deviation of the resistance distribution?

40. If bearing diameter is normally distributed, what is the probability that the diameter of a randomly selected bearing is
    a. Within 1.5 SDs of its mean value?
    b. Farther than 2.5 SDs from its mean value?
    c. Between 1 and 2 SDs from its mean value?

41. The air pressure in a randomly selected tire put on a certain model new car is normally distributed with mean value 31 psi and standard deviation .2 psi.
    a. What is the probability that the pressure for a randomly selected tire exceeds 30.5 psi?
    b. What is the probability that the pressure for a randomly selected tire is between 30.5 and 31.5 psi? Between 30 and 32 psi?
    c. Suppose a tire is classed as underinflated if its pressure is less than 30.4 psi. What is the probability that at least one of the four tires on a car is underinflated? (*Hint:* If $A = \{$at least 1 tire is underinflated$\}$, what is the complement of A?)

42. The Rockwell hardness of a metal is determined by impressing a hardened point into the surface of the metal and then measuring the depth of penetration of the point. Suppose the Rockwell hardness of a particular alloy is normally distributed with mean 70 and standard deviation 3. (Rockwell hardness is measured on a continuous scale.)
    a. If a specimen is acceptable only if its hardness is between 67 and 75, what is the probability that a randomly chosen specimen has an acceptable hardness?
    b. If the acceptable range of hardness is $(70 - c, 70 + c)$, for what value of c would 95% of all specimens have acceptable hardness?
    c. If the acceptable range is as in part (a) and the hardness of each of ten randomly selected specimens is independently determined, what is the expected number of acceptable specimens among the ten?
    d. What is the probability that at most eight of ten independently selected specimens have a hardness of less than 73.84? (*Hint:* $Y =$ the number among the ten specimens with hardness less than 73.84 is a binomial variable; what is p?)

43. The weight distribution of parcels sent in a certain manner is normal with mean value 10 lb and standard deviation 2 lb. The parcel service wishes to establish a weight value c beyond which there will be a surcharge. What value of c is such that 99% of all parcels are at least 1 lb under the surcharge weight?

44. Suppose Appendix Table A.3 contained $\Phi(z)$ only for $z \geq 0$. Explain how you could still compute
    a. $P(-1.72 \leq Z \leq -.55)$.
    b. $P(-1.72 \leq Z \leq .55)$.

Is it necessary to table $\Phi(z)$ for z negative? What property of the standard normal curve justifies your answer?

45. Chebyshev's inequality, introduced in Exercise 43 (Chapter 3), is valid for continuous as well as discrete distributions. It states that for any number k satisfying $k \geq 1$, $P(|X - \mu| \geq k\sigma) \leq 1/k^2$ (see Exercise 43 in Chapter 3 for an interpretation). Obtain this probability in the case of a normal distribution for $k = 1$, 2, and 3 and compare to the upper bound.

46. Let X denote the number of flaws along a 100-m reel of magnetic tape (an integer-valued variable). Suppose X has approximately a normal distribution with $\mu = 25$ and $\sigma = 5$. Use the continuity correction to calculate the probability that the number of flaws is

  a. Between 20 and 30, inclusive.

  b. At most 30. Less than 30.

47. Let X have a binomial distribution with parameters $n = 25$ and p. Calculate each of the following probabilities using the normal approximation (with the continuity correction) for the cases $p = .5$, .6, and .8 and compare to the exact probabilities calculated from Appendix Table A.1.

  a. $P(15 \leq X \leq 20)$

  b. $P(X \leq 15)$

  c. $P(20 \leq X)$

48. Suppose that 10% of all steel shafts produced by a certain process are nonconforming but can be reworked (rather than having to be scrapped). Consider a random sample of 200 shafts, and let X denote the number among these that are nonconforming and can be reworked. What is the (approximate) probability that X is

  a. At most 30?

  b. Less than 30?

  c. Between 15 and 25 (inclusive)?

49. Suppose only 40% of all drivers in a certain state regularly wear a seatbelt. A random sample of 500 drivers is selected. What is the probability that

  a. Between 180 and 230 (inclusive) of the drivers in the sample regularly wear a seatbelt?

  b. Fewer than 175 of those in the sample regularly wear a seatbelt? Fewer than 150?

50. Show that the relationship between a general normal percentile and the corresponding z percentile is as stated in this section.

51. a. Show that if X has a normal distribution with parameters μ and σ, then $Y = aX + b$ (a linear function of X) also has a normal distribution. What are the parameters of the distribution of Y [i.e., $E(Y)$ and $V(Y)$]? [*Hint:* Write the cdf of Y, $P(Y \leq y)$, as an integral involving the pdf of X, and then differentiate with respect to y to get the pdf of Y.]

  b. If when measured in °C, temperature is normally distributed with mean 115 and standard deviation 2, what can be said about the distribution of temperature measured in °F?

52. There is no nice formula for the standard normal cdf $\Phi(z)$, but several good approximations have been published in articles. The following is from "Approximations for Hand Calculators Using Small Integer Coefficients" (*Mathematics of Computation*, 1977: 214–222). For $0 \leq z \leq 5.5$,

$$P(Z \geq z) = 1 - \Phi(z)$$

$$\approx .5 \exp \left\{ -\left[\frac{(83z + 351)z + 562}{703/z + 165} \right] \right\}$$

The relative error of this approximation is less than .042%. Use this to calculate approximations to the following probabilities and compare whenever possible to the probabilities obtained from Appendix Table A.3.

  a. $P(Z \geq 1)$

  b. $P(Z < -3)$

  c. $P(-4 < Z < 4)$

  d. $P(Z > 5)$

4.4 | The Gamma Distribution and Its Relatives

The graph of any normal pdf is bell-shaped and thus symmetric. There are many practical situations in which the variable of interest to the experimenter might have a skewed distribution. A family of pdf's that yields a wide variety of skewed distributional shapes is the gamma family. To define the family of gamma distributions, we first need to introduce a function that plays an important role in many branches of mathematics.

DEFINITION

> For $\alpha > 0$, the **gamma function** $\Gamma(\alpha)$ is defined by
>
> $$\Gamma(\alpha) = \int_0^\infty x^{\alpha-1}e^{-x}dx \qquad (4.5)$$

The most important properties of the gamma function are the following:

1. For any $\alpha > 1$, $\Gamma(\alpha) = (\alpha - 1) \cdot \Gamma(\alpha - 1)$ [via integration by parts]

2. For any positive integer, n, $\Gamma(n) = (n - 1)!$

3. $\Gamma\left(\frac{1}{2}\right) = \sqrt{\pi}$

By Expression (4.5), if we let

$$f(x; \alpha) = \begin{cases} \dfrac{x^{\alpha-1}e^{-x}}{\Gamma(\alpha)} & x \geq 0 \\ 0 & \text{otherwise} \end{cases} \qquad (4.6)$$

then $f(x; \alpha) \geq 0$ and $\int_0^\infty f(x; \alpha)dx = \Gamma(\alpha)/\Gamma(\alpha) = 1$, so $f(x; \alpha)$ satisfies the two basic properties of a pdf.

The Family of Gamma Distributions

DEFINITION

> A continuous random variable X is said to have a **gamma distribution** if the pdf of X is
>
> $$f(x; \alpha, \beta) = \begin{cases} \dfrac{1}{\beta^\alpha\Gamma(\alpha)} x^{\alpha-1}e^{-x/\beta} & x \geq 0 \\ 0 & \text{otherwise} \end{cases} \qquad (4.7)$$
>
> where the parameters α and β satisfy $\alpha > 0$, $\beta > 0$. The **standard gamma distribution** has $\beta = 1$, so the pdf of a standard gamma rv is given by (4.6).

Figure 4.26a illustrates the graphs of the gamma pdf $f(x; \alpha, \beta)$ (4.7) for several (α, β) pairs, whereas Figure 4.26(b) presents graphs of the standard gamma pdf. For the standard pdf, when $\alpha \leq 1$, $f(x; \alpha)$ is strictly decreasing as x increases from 0; when $\alpha > 1$, $f(x; \alpha)$ rises from 0 at $x = 0$ to a maximum and then decreases. The parameter β in (4.7) is called the *scale parameter* because values other than 1 either stretch or compress the pdf in the x direction.

$E(X)$ and $E(X^2)$ can be obtained from a reasonably straightforward integration, and then $V(X) = E(X^2) - [E(X)]^2$.

PROPOSITION

> The mean and variance of a random variable X having the gamma distribution $f(x; \alpha, \beta)$ are
>
> $$E(X) = \mu = \alpha\beta \qquad V(X) = \sigma^2 = \alpha\beta^2$$

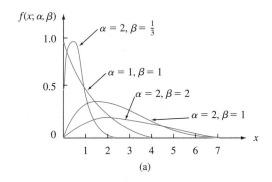

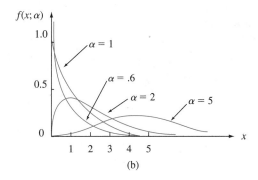

Figure 4.26 (a) Gamma density curves; (b) standard gamma density curves

Computing Probabilities from the Gamma Distribution

When X is a standard gamma rv, the cdf of X,

$$F(x; \alpha) = \int_0^x \frac{y^{\alpha-1}e^{-y}}{\Gamma(\alpha)}\, dy \qquad x > 0 \tag{4.8}$$

is called the **incomplete gamma function** [sometimes the incomplete gamma function refers to Expression (4.8) without the denominator $\Gamma(\alpha)$ in the integrand]. There are extensive tables of $F(x; \alpha)$ available; in Appendix Table A.4, we present a small tabulation for $\alpha = 1, 2, \ldots, 10$ and $x = 1, 2, \ldots, 15$.

Example 4.20 Suppose the reaction time X of a randomly selected individual to a certain stimulus has a standard gamma distribution with $\alpha = 2$ sec. Since

$$P(a \le X \le b) = F(b) - F(a)$$

when X is continuous,

$$P(3 \le X \le 5) = F(5; 2) - F(3; 2) = .960 - .801 = .159$$

The probability that the reaction time is more than 4 sec is

$$P(X > 4) = 1 - P(X \le 4) = 1 - F(4; 2) = 1 - .908 = .092 \qquad \blacksquare$$

The incomplete gamma function can also be used to compute probabilities involving nonstandard gamma distributions.

PROPOSITION

> Let X have a gamma distribution with parameters α and β. Then for any $x > 0$, the cdf of X is given by
>
> $$P(X \leq x) = F(x; \alpha, \beta) = F\left(\frac{x}{\beta}; \alpha\right)$$
>
> where $F(\cdot; \alpha)$ is the incomplete gamma function.*

Example 4.21 Suppose the survival time X in weeks of a randomly selected male mouse exposed to 240 rads of gamma radiation has a gamma distribution with $\alpha = 8$ and $\beta = 15$. (Data in *Survival Distributions: Reliability Applications in the Biomedical Services,* by A. J. Gross and V. Clark, suggests $\alpha \approx 8.5$ and $\beta \approx 13.3$.) The expected survival time is $E(X) = (8)(15) = 120$ weeks, whereas $V(X) = (8)(15)^2 = 1800$ and $\sigma_X = \sqrt{1800} = 42.43$ weeks. The probability that a mouse survives between 60 and 120 weeks is

$$
\begin{aligned}
P(60 \leq X \leq 120) &= P(X \leq 120) - P(X \leq 60) \\
&= F(120/15; 8) - F(60/15; 8) \\
&= F(8; 8) - F(4; 8) = .547 - .051 = .496
\end{aligned}
$$

The probability that a mouse survives at least 30 weeks is

$$
\begin{aligned}
P(X \geq 30) &= 1 - P(X < 30) = 1 - P(X \leq 30) \\
&= 1 - F(30/15; 8) = .999
\end{aligned}
$$ ∎

The Exponential Distribution

The family of exponential distributions provides probability models that are very widely used in engineering and science disciplines.

DEFINITION

> X is said to have an **exponential distribution** if the pdf of X is
>
> $$f(x; \lambda) = \begin{cases} \lambda e^{-\lambda x} & x \geq 0 \\ 0 & \text{otherwise} \end{cases} \qquad \text{where } \lambda > 0 \qquad (4.9)$$

The exponential pdf is a special case of the general gamma pdf Expression (4.7) in which $\alpha = 1$ and β has, been replaced by $1/\lambda$ [some authors use the form $(1/\beta)e^{-x/\beta}$]. The mean and variance of X are then

$$\mu = \alpha\beta = \frac{1}{\lambda} \qquad \sigma^2 = \alpha\beta^2 = \frac{1}{\lambda^2}$$

---

*MINITAB and other statistical packages calculate $F(x; \alpha, \beta)$ once values of x, α, and β are specified.

Both the mean and standard deviation of the exponential distribution equal $1/\lambda$. Graphs of several exponential pdf's are in Figure 4.27.

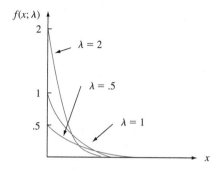

Figure 4.27 Exponential density curves

Unlike the general gamma pdf, the exponential pdf can be easily integrated. In particular, the cdf of X is

$$F(x; \lambda) = \begin{cases} 0 & x < 0 \\ 1 - e^{-\lambda x} & x \geq 0 \end{cases}$$

Example 4.22 Suppose the response time X at a certain on-line computer terminal (the elapsed time between the end of a user's inquiry and the beginning of the system's response to that inquiry) has an exponential distribution with expected response time equal to 5 sec. Then $E(X) = 1/\lambda = 5$, so $\lambda = .2$. The probability that the response time is at most 10 sec is

$$P(X \leq 10) = F(10; .2) = 1 - e^{-(.2)(10)} = 1 - e^{-2} = 1 - .135 = .865$$

The probability that response time is between 5 and 10 sec is

$$P(5 \leq X \leq 10) = F(10; .2) - F(5; .2)$$
$$= (1 - e^{-2}) - (1 - e^{-1}) = .233 \qquad \blacksquare$$

Applications of the Exponential Distribution

The exponential distribution is frequently used as a model for the distribution of times between the occurrence of successive events, such as customers arriving at a service facility or calls coming in to a switchboard. The reason for this is that the exponential distribution is closely related to the Poisson process discussed in Chapter 3.

PROPOSITION

Suppose that the number of events occurring in any time interval of length t has a Poisson distribution with parameter αt (where α, the rate of the event process, is the expected number of events occurring in 1 unit of time) and that numbers of occurrences in nonoverlapping intervals are independent of one another. Then the distribution of elapsed time between the occurrence of two successive events is exponential with parameter $\lambda = \alpha$.

Although a complete proof is beyond the scope of the text, the result is easily verified for the time X_1 until the first event occurs:

$$P(X_1 \leq t) = 1 - P(X_1 > t) = 1 - P[\text{no events in } (0, t)]$$

$$= 1 - \frac{e^{-\alpha t} \cdot (\alpha t)^0}{0!} = 1 - e^{-\alpha t}$$

which is exactly the cdf of the exponential distribution.

Example 4.23 Suppose that calls are received at a 24-hour "suicide hotline" according to a Poisson process with rate $\alpha = .5$ call per day. Then the number of days X between successive calls has an exponential distribution with parameter value .5, so the probability that more than 2 days elapse between calls is

$$P(X > 2) = 1 - P(X \leq 2) = 1 - F(2; .5) = e^{-(.5)(2)} = .368$$

The expected time between successive calls is $1/.5 = 2$ days. ■

Another important application of the exponential distribution is to model the distribution of component lifetime. A partial reason for the popularity of such applications is the **"memoryless" property** of the exponential distribution. Suppose component lifetime is exponentially distributed with parameter λ. After putting the component into service, we leave for a period of t_0 hours and then return to find the component still working; what now is the probability that it lasts at least an additional t hours? In symbols, we wish $P(X \geq t + t_0 \,|\, X \geq t_0)$. By the definition of conditional probability,

$$P(X \geq t + t_0 \,|\, X \geq t_0) = \frac{P[(X \geq t + t_0) \cap (X \geq t_0)]}{P(X \geq t_0)}$$

But the event $X \geq t_0$ in the numerator is redundant, since both events can occur if and only if $X \geq t + t_0$. Therefore,

$$P(X \geq t + t_0 \,|\, X \geq t_0) = \frac{P(X \geq t + t_0)}{P(X \geq t_0)} = \frac{1 - F(t + t_0; \lambda)}{1 - F(t_0; \lambda)} = e^{-\lambda t}$$

This conditional probability is identical to the original probability $P(X \geq t)$ that the component lasted t hours. Thus, *the distribution of additional lifetime is exactly the same as the original distribution of lifetime,* so at each point in time the component shows no effect of wear. In other words, the distribution of remaining lifetime is independent of current age.

Although the memoryless property can be justified at least approximately in many applied problems, in other situations components deteriorate with age or occasionally improve with age (at least up to a certain point). More general lifetime models are then furnished by the gamma, Weibull, or lognormal distributions (the latter two are discussed in the next section).

The Chi-Squared Distribution

DEFINITION

Let ν be a positive integer. Then a random variable X is said to have a **chi-squared distribution** with parameter ν if the pdf of X is the gamma density with $\alpha = \nu/2$ and $\beta = 2$. The pdf of a chi-squared rv is thus

$$f(x; \nu) = \begin{cases} \dfrac{1}{2^{\nu/2}\Gamma(\nu/2)} x^{(\nu/2)-1} e^{-x/2} & x \geq 0 \\ 0 & x < 0 \end{cases} \tag{4.10}$$

The parameter ν is called the **number of degrees of freedom** (df) of X. The symbol χ^2 is often used in place of "chi-squared."

The chi-squared distribution is important because it is the basis for a number of procedures in statistical inference. The reason for this is that chi-squared distributions are intimately related to normal distributions (see Exercise 65). We will discuss the chi-squared distribution in more detail in the chapters on inference.

Exercises | Section 4.4 (53–65)

53. Evaluate the following:
  a. $\Gamma(6)$
  b. $\Gamma(5/2)$
  c. $F(4; 5)$ (the incomplete gamma function)
  d. $F(5; 4)$
  e. $F(0; 4)$

54. Let X have a standard gamma distribution with $\alpha = 7$. Evaluate the following:
  a. $P(X \leq 5)$
  b. $P(X < 5)$
  c. $P(X > 8)$
  d. $P(3 \leq X \leq 8)$
  e. $P(3 < X < 8)$
  f. $P(X < 4 \text{ or } X > 6)$

55. Suppose the time (in hours) taken by a homeowner to mow his lawn is an rv X having a gamma distribution with parameters $\alpha = 2$ and $\beta = \frac{1}{2}$. What is the probability that it takes:
  a. At most 1 hour to mow the lawn?
  b. At least 2 hours to mow the lawn?
  c. Between .5 and 1.5 hours to mow the lawn?

56. Suppose the time spent by a randomly selected student who uses a terminal connected to a local time-sharing computer facility has a gamma distribution with mean 20 min and variance 80 min².
  a. What are the values of α and β?
  b. What is the probability that a student uses the terminal for at most 24 min?
  c. What is the probability that a student spends between 20 and 40 min using the terminal?

57. Suppose that when a transistor of a certain type is subjected to an accelerated life test, the lifetime X (in weeks) has a gamma distribution with mean 24 weeks and standard deviation 12 weeks.
  a. What is the probability that a transistor will last between 12 and 24 weeks?
  b. What is the probability that a transistor will last at most 24 weeks? Is the median of the lifetime distribution less than 24? Why or why not?
  c. What is the 99th percentile of the lifetime distribution?
  d. Suppose the test will actually be terminated after t weeks. What value of t is such that only .5% of all transistors would still be operating at termination?

58. Let $X =$ the time between two successive arrivals at the drive-up window of a local bank. If X has an exponential distribution with $\lambda = 1$ (which is identical to a standard gamma distribution with $\alpha = 1$), compute the following:
  a. The expected time between two successive arrivals
  b. The standard deviation of the time between successive arrivals
  c. $P(X \leq 4)$
  d. $P(2 \leq X \leq 5)$

59. Let X denote the distance (m) that an animal moves from its birth site to the first territorial vacancy it encounters. Suppose that for banner-tailed kangaroo rats, X has an exponential distribution with parameter $\lambda = .01386$ (as suggested in the article "Competition and Dispersal from Multiple Nests," *Ecology,* 1997: 873–883).
  a. What is the probability that the distance is at most 100 m? At most 200 m? Between 100 and 200 m?

b. What is the probability that distance exceeds the mean distance by more than 2 standard deviations?

c. What is the value of the median distance?

60. Extensive experience with fans of a certain type used in diesel engines has suggested that the exponential distribution provides a good model for time until failure. Suppose the mean time until failure is 25,000 hours. What is the probability that

a. A randomly selected fan will last at least 20,000 hours? At most 30,000 hours? Between 20,000 and 30,000 hours?

b. The lifetime of a fan exceeds the mean value by more than 2 standard deviations? More than 3 standard deviations?

61. The special case of the gamma distribution in which α is a positive integer n is called an Erlang distribution. If we replace β by $1/\lambda$ in Expression (4.7), the Erlang pdf is

$$f(x; \lambda, n) = \begin{cases} \dfrac{\lambda(\lambda x)^{n-1} e^{-\lambda x}}{(n-1)!} & x \geq 0 \\ 0 & x < 0 \end{cases}$$

It can be shown that if the times between successive events are independent, each with an exponential distribution with parameter λ, then the total time X that elapses before all of the next n events occur has pdf $f(x; \lambda, n)$.

a. What is the expected value of X? If the time (in minutes) between arrivals of successive customers is exponentially distributed with $\lambda = .5$, how much time can be expected to elapse before the tenth customer arrives?

b. If customer interarrival time is exponentially distributed with $\lambda = .5$, what is the probability that the tenth customer (after the one who has just arrived) will arrive within the next 30 min?

c. The event $\{X \leq t\}$ occurs iff at least n events occur in the next t units of time. Use the fact that the number of events occurring in an interval of length t has a Poisson distribution with parameter λt to write an expression (involving Poisson probabilities) for the Erlang cdf $F(t; \lambda, n) = P(X \leq t)$.

62. A system consists of five identical components connected in series as shown:

As soon as one component fails, the entire system will fail. Suppose each component has a lifetime that is exponentially distributed with $\lambda = .01$ and that components fail independently of one another. Define events $A_i = \{i\text{th component lasts at least } t \text{ hours}\}$, $i = 1, \ldots, 5$, so that the A_i's are independent events. Let $X =$ the time at which the system fails—that is, the shortest (minimum) lifetime among the five components.

a. The event $\{X \geq t\}$ is equivalent to what event involving $A_1, \ldots, A_5$?

b. Using the independence of the A_i's, compute $P(X \geq t)$. Then obtain $F(t) = P(X \leq t)$ and the pdf of X. What type of distribution does X have?

c. Suppose there are n components, each having exponential lifetime with parameter λ. What type of distribution does X have?

63. If X has an exponential distribution with parameter λ, derive a general expression for the $(100p)$th percentile of the distribution. Then specialize to obtain the median.

64. Show that $\int_0^\infty f(x; \alpha, \beta)dx = 1$ for the gamma pdf of (4.7). [*Hint:* Let $y = x/\beta$ to obtain an integrand like that of Expression (4.6).]

65. a. The event $\{X^2 \leq y\}$ is equivalent to what event involving X itself?

b. If X has a standard normal distribution, use part (a) to write the integral that equals $P(X^2 \leq y)$. Then differentiate this with respect to y to obtain the pdf of X^2 [the square of a $N(0, 1)$ variable]. Finally, show that X^2 has a chi-squared distribution with $\nu = 1$ df [see (4.10)]. (*Hint:* Use the following identity.)

$$\frac{d}{dy}\left\{\int_{a(y)}^{b(y)} f(x)dx\right\} = f[b(y)] \cdot b'(y) - f[a(y)] \cdot a'(y)$$

4.5 | Other Continuous Distributions

The normal, gamma (including exponential), and uniform families of distributions provide a wide variety of probability models for continuous variables, but there are many practical situations in which no member of these families fits a set of observed data very well. Statisticians and other investigators have developed other families of distributions that are often appropriate in practice.

The Weibull Distribution

The family of Weibull distributions was introduced by the Swedish physicist Waloddi Weibull in 1939; his 1951 article "A Statistical Distribution Function of Wide Applicability" (*J. Applied Mechanics,* vol. 18: 293–297) discusses a number of applications.

DEFINITION

A random variable X is said to have a **Weibull distribution** with parameters α and β ($\alpha > 0$, $\beta > 0$) if the pdf of X is

$$f(x; \alpha, \beta) = \begin{cases} \dfrac{\alpha}{\beta^\alpha} x^{\alpha-1} e^{-(x/\beta)^\alpha} & x \geq 0 \\ 0 & x < 0 \end{cases} \tag{4.11}$$

In some situations, there are theoretical justifications for the appropriateness of the Weibull distribution, but in many applications $f(x; \alpha, \beta)$ simply provides a good fit to observed data for particular values of α and β. When $\alpha = 1$, the pdf reduces to the exponential distribution (with $\lambda = 1/\beta$), so the exponential distribution is a special case of both the gamma and Weibull distributions. However, there are gamma distributions that are not Weibull distributions and vice versa, so one family is not a subset of the other. Both α and β can be varied to obtain a number of different distributional shapes, as illustrated in Figure 4.28. β is a scale parameter, so different values stretch or compress the graph in the x direction.

Integrating to obtain $E(X)$ and $E(X^2)$ yields

$$\mu = \beta\Gamma\left(1 + \frac{1}{\alpha}\right) \quad \sigma^2 = \beta^2\left\{\Gamma\left(1 + \frac{2}{\alpha}\right) - \left[\Gamma\left(1 + \frac{1}{\alpha}\right)\right]^2\right\}$$

The computation of μ and σ^2 thus necessitates using the gamma function.

The integration $\int_0^x f(y; \alpha, \beta)dy$ is easily carried out to obtain the cdf of X.

PROPOSITION

The cdf of a Weibull rv having parameters α and β is

$$F(x; \alpha, \beta) = \begin{cases} 0 & x < 0 \\ 1 - e^{-(x/\beta)^\alpha} & x \geq 0 \end{cases} \tag{4.12}$$

Example 4.24 Let X = the ultimate tensile strength (ksi) at $-200°$F of a type of steel that exhibits "cold brittleness" at low temperatures. Suppose X has a Weibull distribution with parameters $\alpha = 20$ and $\beta = 100$. Then

$$P(X \leq 105) = F(105; 20, 100) = 1 - e^{-(105/100)^{20}} = 1 - .070 = .930$$

and

$$P(98 \leq X \leq 102) = F(102; 20, 100) - F(98; 20, 100)$$
$$= e^{-(.98)^{20}} - e^{-(1.02)^{20}} = .513 - .226 = .287 \quad \blacksquare$$

Frequently, in practical situations, a Weibull model may be reasonable except that the smallest possible X value may be some value γ not assumed to be zero (this would also apply to a gamma model). The quantity γ can then be regarded as a third parameter of the distribution, which is what Weibull did in his original work. For, say, $\gamma = 3$, all curves in Figure 4.28 would be shifted 3 units to the right. This is equivalent to saying that $X - \gamma$ has the pdf (4.11), so that the cdf of X is obtained by replacing x in (4.12) by $x - \gamma$.

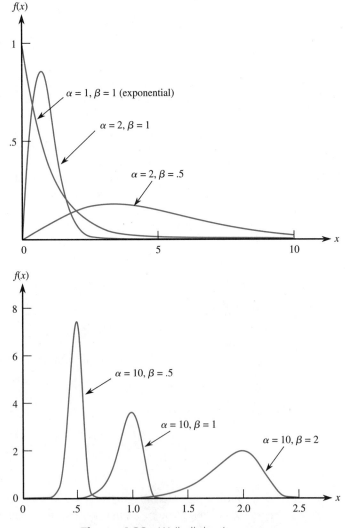

Figure 4.28 Weibull density curves

Example 4.25 Let X = the corrosion weight loss for a small square magnesium alloy plate immersed for 7 days in an inhibited aqueous 20% solution of $MgBr_2$. Suppose the minimum possible weight loss is $\gamma = 3$ and that the excess $X - 3$ over this minimum has a Weibull distribution with $\alpha = 2$ and $\beta = 4$. (This example was considered in "Practical Applications of the Weibull Distribution," *Industrial Quality Control*, Aug. 1964: 71–78; values for α and β were taken to be 1.8 and 3.67, respectively, though a slightly different choice of parameters was used in the article.) The cdf of X is then

$$F(x; \alpha, \beta, \gamma) = F(x; 2, 4, 3) = \begin{cases} 0 & x < 3 \\ 1 - e^{-[(x-3)/4]^2} & x \geq 3 \end{cases}$$

Therefore,

$$P(X > 3.5) = 1 - F(3.5; 2, 4, 3) = e^{-.0156} = .985$$

and

$$P(7 \leq X \leq 9) = 1 - e^{-2.25} - (1 - e^{-1}) = .895 - .632 = .263 \qquad \blacksquare$$

The Lognormal Distribution

DEFINITION

A nonnegative rv X is said to have a **lognormal distribution** if the rv $Y = \ln(X)$ has a normal distribution. The resulting pdf of a lognormal rv when $\ln(X)$ is normally distributed with parameters μ and σ is

$$f(x; \mu, \sigma) = \begin{cases} \dfrac{1}{\sqrt{2\pi}\,\sigma x} e^{-[\ln(x)-\mu]^2/(2\sigma^2)} & x \geq 0 \\ 0 & x < 0 \end{cases}$$

Be careful here; μ and σ are not the mean and standard deviation of X but of $\ln(X)$. The mean and variance of X can be shown to be

$$E(X) = e^{\mu + \sigma^2/2} \qquad V(X) = e^{2\mu + \sigma^2} \cdot (e^{\sigma^2} - 1)$$

In Chapter 5, we will present a theoretical justification for this distribution in connection with the Central Limit Theorem, but as with other distributions, the lognormal can be used as a model even in the absence of such justification. Figure 4.29 (page 182) illustrates the graphs of the lognormal pdf; although a normal curve is symmetric, a lognormal curve has a positive skew.

Because $\ln(X)$ has a normal distribution, the cdf of X can be expressed in terms of the cdf $\Phi(z)$ of a standard normal rv Z. For $x \geq 0$,

$$F(x; \mu, \sigma) = P(X \leq x) = P[\ln(X) \leq \ln(x)]$$
$$= P\left(Z \leq \frac{\ln(x) - \mu}{\sigma}\right) = \Phi\left(\frac{\ln(x) - \mu}{\sigma}\right) \qquad (4.13)$$

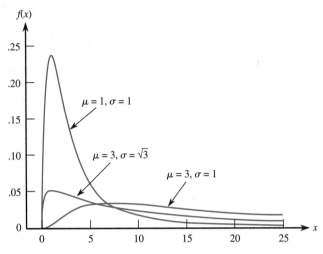

Figure 4.29 Lognormal density curves

Example 4.26 The lognormal distribution is frequently used as a model for various material properties. The article "Reliability of Wood Joist Floor Systems with Creep" (*J. of Structural Engr.*, 1995: 946–954) suggests that the lognormal distribution with $\mu = .375$ and $\sigma = .25$ is a plausible model for $X =$ the modulus of elasticity (MOE, in 10^6 psi) of wood joist floor systems constructed from #2 grade hem-fir. The mean value and variance of MOE are

$$E(X) = e^{.375 + (.25)^2/2} = e^{.40625} = 1.50$$

$$V(X) = e^{.8125}(e^{.0625} - 1) = .1453$$

The probability that MOE is between 1 and 2 is

$$P(1 \leq X \leq 2) = P(\ln(1) \leq \ln(X) \leq \ln(2))$$
$$= P(0 \leq \ln(X) \leq .693)$$
$$= P\left(\frac{0 - .375}{.25} \leq Z \leq \frac{.693 - .375}{.25}\right)$$
$$= \Phi(1.27) - \Phi(-1.50) = .8312$$

What value c is such that only 1% of all systems have an MOE exceeding c? We wish the c for which

$$.99 = P(X \leq c) = P\left(Z \leq \frac{\ln(c) - .375}{.25}\right)$$

from which $(\ln(c) - .375)/.25 = 2.33$ and $c = 2.605$. Thus 2.605 is the 99th percentile of the MOE distribution. ∎

The Beta Distribution

All families of continuous distributions discussed so far except for the uniform distribution have positive density over an infinite interval (though typically the density function decreases rapidly to zero beyond a few standard deviations from the mean). The beta distribution provides positive density only for X in an interval of finite length.

DEFINITION

A random variable X is said to have a **beta distribution** with parameters α, β (both positive), A, and B if the pdf of X is

$$f(x;\ \alpha,\ \beta,\ A,\ B) = \begin{cases} \dfrac{1}{B-A} \cdot \dfrac{\Gamma(\alpha+\beta)}{\Gamma(\alpha) \cdot \Gamma(\beta)} \left(\dfrac{x-A}{B-A}\right)^{\alpha-1} \left(\dfrac{B-x}{B-A}\right)^{\beta-1} & A \le x \le B \\ 0 & \text{otherwise} \end{cases}$$

The case $A = 0$, $B = 1$ gives the **standard beta distribution.**

Figure 4.30 illustrates several standard beta pdf's. Graphs of the general pdf are similar, except they are shifted and then stretched or compressed to fit over $[A, B]$. Unless α and β are integers, integration of the pdf to calculate probabilities is difficult, so a table of the incomplete beta function is generally used. The mean and variance of X are

$$\mu = A + (B - A) \cdot \frac{\alpha}{\alpha + \beta} \qquad \sigma^2 = \frac{(B - A)^2 \alpha \beta}{(\alpha + \beta)^2(\alpha + \beta + 1)}$$

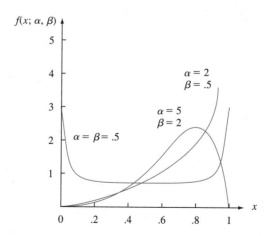

Figure 4.30 Standard beta density curves

Example 4.27 Project managers often use a method labeled PERT—for program evaluation and review technique—to coordinate the various activities making up a large project. (One successful application was in the construction of the *Apollo* spacecraft.) A standard assumption in PERT analysis is that the time necessary to complete any particular activity once it has been started has a beta distribution with $A = $ the optimistic time (if everything goes well) and $B = $ the pessimistic time (if everything goes badly).

Suppose that in constructing a single-family house, the time X (in days) necessary for laying the foundation has a beta distribution with $A = 2$, $B = 5$, $\alpha = 2$, and $\beta = 3$. Then $\alpha/(\alpha + \beta) = .4$, so $E(X) = 2 + (3)(.4) = 3.2$. For these values of α and β, the pdf of X is a simple polynomial function. The probability that it takes at most 3 days to lay the foundation is

$$P(X \leq 3) = \int_2^3 \frac{1}{3} \cdot \frac{4!}{1!2!} \left(\frac{x-2}{3}\right) \left(\frac{5-x}{3}\right)^2 dx$$

$$= \frac{4}{27} \int_2^3 (x-2)(5-x)^2 \, dx = \frac{4}{27} \cdot \frac{11}{4} = \frac{11}{27} = .407 \qquad \blacksquare$$

The standard beta distribution is commonly used to model variation in the proportion or percentage of a quantity occurring in different samples, such as the proportion of a 24-hour day that an individual is asleep or the proportion of a certain element in a chemical compound.

Exercises | Section 4.5 (66–78)

66. The lifetime X (in hundreds of hours) of a certain type of vacuum tube has a Weibull distribution with parameters $\alpha = 2$ and $\beta = 3$. Compute the following:
  a. $E(X)$ and $V(X)$
  b. $P(X \leq 6)$
  c. $P(1.5 \leq X \leq 6)$

(This Weibull distribution is suggested as a model for time in service in "On the Assessment of Equipment Reliability: Trading Data Collection Costs for Precision," *J. Engr. Manuf.*, 1991: 105–109).

67. The authors of the article "A Probabilistic Insulation Life Model for Combined Thermal-Electrical Stresses" (*IEEE Trans. on Elect. Insulation*, 1985: 519–522) state that "the Weibull distribution is widely used in statistical problems relating to aging of solid insulating materials subjected to aging and stress." They propose the use of the distribution as a model for time (in hours) to failure of solid insulating specimens subjected to AC voltage. The values of the parameters depend on the voltage and temperature; suppose $\alpha = 2.5$ and $\beta = 200$ (values suggested by data in the article).
  a. What is the probability that a specimen's lifetime is at most 200? Less than 200? More than 300?
  b. What is the probability that a specimen's lifetime is between 100 and 200?
  c. What value is such that exactly 50% of all specimens have lifetimes exceeding that value?

68. Let X = the time (in 10^{-1} weeks) from shipment of a defective product until the customer returns the product. Suppose that the minimum return time is $\gamma = 3.5$ and that the excess $X - 3.5$ over the minimum has a Weibull distribution with parameters $\alpha = 2$ and $\beta = 1.5$ (see the *Industrial Quality Control* article referenced in Example 4.25).
  a. What is the cdf of X?
  b. What are the expected return time and variance of return time? [*Hint:* First obtain $E(X - 3.5)$ and $V(X - 3.5)$.]
  c. Compute $P(X > 5)$.
  d. Compute $P(5 \leq X \leq 8)$.

69. Let X have a Weibull distribution with the pdf from Expression (4.11). Verify that $\mu = \beta \Gamma(1 + 1/\alpha)$. (*Hint:* In the integral for $E(X)$, make the change of variable $y = (x/\beta)^{\alpha}$, so that $x = \beta y^{1/\alpha}$.)

70. a. In Exercise 66, what is the median lifetime of such tubes? [*Hint:* Use Expression (4.12).]
  b. In Exercise 68, what is the median return time?
  c. If X has a Weibull distribution with the cdf from Expression (4.12), obtain a general expression for the $(100p)$th percentile of the distribution.
  d. In Exercise 68, the company wants to refuse to accept returns after t weeks. For what value of t will only 10% of all returns be refused?

71. Let X = the hourly median power (in decibels) of received radio signals transmitted between two cities. The authors of the article "Families of Distributions for Hourly Median Power and Instantaneous Power of Received Radio Signals" (*J. Research National Bureau of Standards*, vol. 67D, 1963: 753–762) argue that the lognormal distribution provides a reasonable probability model for X. If the parameter values are $\mu = 3.5$ and $\sigma = 1.2$, calculate the following:

a. The mean value and standard deviation of received power

b. The probability that received power is between 50 and 250 dB

c. The probability that X is less than its mean value. Why is this probability not .5?

72. a. Use Equation (4.13) to write a formula for the median $\tilde{\mu}$ of the lognormal distribution. What is the median for the power distribution of Exercise 71?

b. Recalling that z_α is our notation for the $100(1 - \alpha)$ percentile of the standard normal distribution, write an expression for the $100(1 - \alpha)$ percentile of the lognormal distribution. In Exercise 71, what value will received power exceed only 5% of the time?

73. A theoretical justification based on a certain material failure mechanism underlies the assumption that ductile strength X of a material has a lognormal distribution. Suppose the parameters are $\mu = 5$ and $\sigma = .1$.

a. Compute $E(X)$ and $V(X)$.

b. Compute $P(X > 120)$.

c. Compute $P(110 \leq X \leq 130)$.

d. What is the value of median ductile strength?

e. If ten different samples of an alloy steel of this type were subjected to a strength test, how many would you expect to have strength at least 120?

f. If the smallest 5% of strength values were unacceptable, what would the minimum acceptable strength be?

74. The article "The Statistics of Phytotoxic Air Pollutants" (*J. Royal Stat. Soc.*, 1989: 183–198) suggests the lognormal distribution as a model for SO_2 concentration above a certain forest. Suppose the parameter values are $\mu = 1.9$ and $\sigma = .9$.

a. What are the mean value and standard deviation of concentration?

b. What is the probability that concentration is at most 10? Between 5 and 10?

75. What condition on α and β is necessary for the standard beta pdf to be symmetric?

76. Suppose the proportion X of surface area in a randomly selected quadrat that is covered by a certain plant has a standard beta distribution with $\alpha = 5$ and $\beta = 2$.

a. Compute $E(X)$ and $V(X)$.

b. Compute $P(X \leq .2)$.

c. Compute $P(.2 \leq X \leq .4)$.

d. What is the expected proportion of the sampling region not covered by the plant?

77. Let X have a standard beta density with parameters α and β.

a. Verify the formula for $E(X)$ given in the section.

b. Compute $E[(1 - X)^m]$. If X represents the proportion of a substance consisting of a particular ingredient, what is the expected proportion that does not consist of this ingredient?

78. Stress is applied to a 20-in. steel bar that is clamped in a fixed position at each end. Let $Y =$ the distance from the left end at which the bar snaps. Suppose $Y/20$ has a standard beta distribution with $E(Y) = 10$ and $V(Y) = \frac{100}{7}$.

a. What are the parameters of the relevant standard beta distribution?

b. Compute $P(8 \leq Y \leq 12)$.

c. Compute the probability that the bar snaps more than 2 in. from where you expect it to.

4.6 | Probability Plots

An investigator will often have obtained a numerical sample $x_1, x_2, \ldots, x_n$ and wish to know whether it is plausible that it came from a population distribution of some particular type (e.g., from a normal distribution). For one thing, many formal procedures from statistical inference are based on the assumption that the population distribution is of a specified type. The use of such a procedure is inappropriate if the actual underlying probability distribution differs greatly from the assumed type. Additionally, understanding the underlying distribution can sometimes give insight into the physical mechanisms involved in generating the data. An effective way to check a distributional assumption is to construct what is called a **probability plot.** The essence of such a plot is that if the distribution on which the plot is based is correct, the points in the plot will fall close to a straight line. If the actual distribution is quite different from the one used to construct the plot, the points should depart substantially from a linear pattern.

Sample Percentiles

The details involved in constructing probability plots differ a bit from source to source. The basis for our construction is a comparison between percentiles of the sample data and the corresponding percentiles of the distribution under consideration. Recall that the $(100p)$th percentile of a continuous distribution with cdf $F(\cdot)$ is the number $\eta(p)$ that satisfies $F(\eta(p)) = p$. That is, $\eta(p)$ is the number on the measurement scale such that the area under the density curve to the left of $\eta(p)$ is p. Thus, the 50th percentile $\eta(.5)$ satisfies $F(\eta(.5)) = .5$, and the 90th percentile satisfies $F(\eta(.9)) = .9$. Consider as an example the standard normal distribution, for which we have denoted the cdf by $\Phi(\cdot)$. From Appendix Table A.3, we find the 20th percentile by locating the row and column in which .2000 (or a number as close to it as possible) appears inside the table. Since .2005 appears at the intersection of the $-.8$ row and the .04 column, the 20th percentile is approximately $-.84$. Similarly, the 25th percentile of the standard normal distribution is (using linear interpolation) approximately $-.675$.

Roughly speaking, sample percentiles are defined in the same way that percentiles of a population distribution are defined. The fiftieth-sample percentile should separate the smallest 50% of the sample from the largest 50%, the 90th percentile should be such that 90% of the sample lies below that value and 10% lies above, and so on. Unfortunately, we run into problems when we actually try to compute the sample percentiles for a particular sample of n observations. If, for example, $n = 10$, we can split off 20% of these values or 30% of the data, but there is no value that will split off exactly 23% of these ten observations. To proceed further, we need an operational definition of sample percentiles (this is one place where different people do slightly different things). Recall that when n is odd, the sample median or 50th sample percentile is the middle value in the ordered list, for example, the sixth largest value when $n = 11$. This amounts to regarding the middle observation as being half in the lower half of the data and half in the upper half. Similarly, suppose $n = 10$. Then if we call the third smallest value the 25th percentile, we are regarding that value as being half in the lower group (consisting of the two smallest observations) and half in the upper group (the seven largest observations). This leads to the following general definition of sample percentiles.

DEFINITION

> Order the n-sample observations from smallest to largest. Then the ith smallest observation in the list is taken to be the $[100(i - .5)/n]$**th sample percentile.**

Once the percentage values $100(i - .5)/n$ $(i = 1, 2, \ldots, n)$ have been calculated, sample percentiles corresponding to intermediate percentages can be obtained by linear interpolation. For example, if $n = 10$, the percentages corresponding to the ordered sample observations are $100(1 - .5)/10 = 5\%$, $100(2 - .5)/10 = 15\%$, 25%, ... , and $100(10 - .5)/10 = 95\%$. The 10th percentile is then halfway between the 5th percentile (smallest sample observation) and the 15th percentile (second smallest observation). For our purposes such interpolation is not necessary because a probability plot will be based only on the percentages $100(i - .5)/n$ corresponding to the n sample observations.

A Probability Plot

Suppose now that for percentages $100(i - .5)/n$ ($i = 1, \ldots, n$) the percentiles are determined for a specified population distribution whose plausibility is being investigated. If the sample was actually selected from the specified distribution, the sample percentiles (ordered sample observations) should be reasonably close to the corresponding population distribution percentiles. That is, for $i = 1, 2, \ldots, n$ there should be reasonable agreement between the ith smallest sample observation and the $[100(i - .5)/n]$th percentile for the specified distribution. Consider the (population percentile, sample percentile) pairs—that is, the pairs

$$\begin{pmatrix} [100(i - .5)/n]\text{th percentile} & i\text{th smallest sample} \\ \text{of the distribution} & , & \text{observation} \end{pmatrix}$$

for $i = 1, \ldots, n$. Each such pair can be plotted as a point on a two-dimensional coordinate system. If the sample percentiles are close to the corresponding population distribution percentiles, the first number in each pair will be roughly equal to the second number. The plotted points will then fall close to a 45° line. Substantial deviations of the plotted points from a 45° line cast doubt on the assumption that the distribution under consideration is the correct one.

Example 4.28 The value of a certain physical constant is known to an experimenter. The experimenter makes $n = 10$ independent measurements of this value using a particular measurement device and records the resulting measurement errors (error = observed value − true value). These observations appear in the accompanying table.

Percentage	5	15	25	35	45
z percentile	−1.645	−1.037	−.675	−.385	−.126
Sample observation	−1.91	−1.25	−.75	−.53	.20

Percentage	55	65	75	85	95
z percentile	.126	.385	.675	1.037	1.645
Sample observation	.35	.72	.87	1.40	1.56

Is it plausible that the random variable *measurement error* has a standard normal distribution? The needed standard normal (z) percentiles are also displayed in the table. Thus, the points in the probability plot are $(-1.645, -1.91)$, $(-1.037, -1.25)$, ... , and $(1.645, 1.56)$. Figure 4.31 (page 188) shows the resulting plot. Although the points deviate a bit from the 45° line, the predominant impression is that this line fits the points very well. The plot suggests that the standard normal distribution is a reasonable probability model for measurement error.

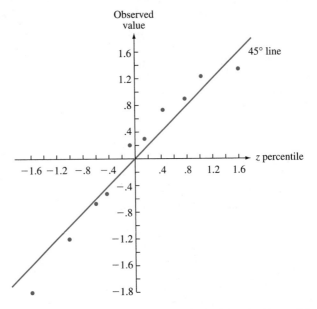

Figure 4.31 Plots of pairs (z percentile, observed value) for the data of Example 4.28: first sample

Figure 4.32 shows a plot of pairs (z percentile, observation) for a second sample of ten observations. The 45° line gives a good fit to the middle part of the sample but not to the extremes. The plot has a well-defined S-shaped appearance. The two small-

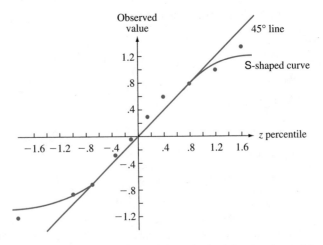

Figure 4.32 Plots of pairs (z percentile, observed value) for the data of Example 4.28: second sample

est sample observations are considerably larger than the corresponding z percentiles (the points on the far left of the plot are well above the 45° line). Similarly, the two largest sample observations are much smaller than the associated z percentiles. This plot indicates that the standard normal distribution would not be a plausible choice for the probability model that gave rise to these observed measurement errors. ∎

An investigator is typically not interested in knowing just whether a specified probability distribution, such as the standard normal distribution (normal with $\mu = 0$ and $\sigma = 1$) or the exponential distribution with $\lambda = .1$, is a plausible model for the population distribution from which the sample was selected. Instead, the investigator will want to know whether *some* member of a family of probability distributions specifies a plausible model—the family of normal distributions, the family of exponential distributions, the family of Weibull distributions, and so on. The values of the parameters of a distribution are usually not specified at the outset. If the family of Weibull distributions is under consideration as a model for lifetime data, the issue is whether there are *any* values of the parameters α and β for which the corresponding Weibull distribution gives a good fit to the data. Fortunately, it is almost always the case that just one probability plot will suffice for assessing the plausibility of an entire family. If the plot deviates substantially from a straight line, no member of the family is plausible. When the plot is quite straight, further work is necessary to estimate values of the parameters (e.g., find values for μ and σ) that yield the most reasonable distribution of the specified type.

Let's focus on a plot for checking normality. Such a plot can be very useful in applied work because many formal statistical procedures are appropriate (give accurate inferences) only when the population distribution is at least approximately normal. These procedures should generally not be used if the normal probability plot shows a very pronounced departure from linearity. The key to constructing an omnibus normal probability plot is the relationship between standard normal (z) percentiles and those for any other normal distribution:

$$\begin{array}{c} \text{percentile for a normal} \\ (\mu, \sigma) \text{ distribution} \end{array} = \mu + \sigma \cdot (\text{corresponding } z \text{ percentile})$$

Consider first the case, $\mu = 0$. Then if each observation is exactly equal to the corresponding normal percentile for a particular value of σ, the pairs ($\sigma \cdot [z$ percentile], observation) fall on a 45° line, which has slope 1. This implies that the pairs (z percentile, observation) fall on a line passing through (0, 0) (i.e., one with y-intercept 0) but having slope σ rather than 1. The effect of a nonzero value of μ is simply to change the y-intercept from 0 to μ.

A plot of the n pairs

([$100(i - .5)/n$]th z percentile, ith smallest observation)

on a two-dimensional coordinate system is called a **normal probability plot.** If the sample observations are in fact drawn from a normal distribution with mean value μ and standard deviation σ, the points should fall close to a straight line with slope σ and intercept μ. Thus, a plot for which the points fall close to some straight line suggests that the assumption of a normal population distribution is plausible.

Example 4.29   The accompanying sample consisting of $n = 20$ observations on dielectric breakdown voltage of a piece of epoxy resin appeared in the article "Maximum Likelihood Estimation in the 3-Parameter Weibull Distribution (*IEEE Trans. on Dielectrics and Elec. Insul.*, 1996: 43–55). The values of $(i - .5)/n$ for which z percentiles are needed are $(1 - .5)/20 = .025$, $(2 - .5)/20 = .075$, ... , and $.975$.

Observation	24.46	25.61	26.25	26.42	26.66	27.15	27.31	27.54	27.74	27.94
z percentile	−1.96	−1.44	−1.15	−.93	−.76	−.60	−.45	−.32	−.19	−.06

Observation	27.98	28.04	28.28	28.49	28.50	28.87	29.11	29.13	29.50	30.88
z percentile	.06	.19	.32	.45	.60	.76	.93	1.15	1.44	1.96

Figure 4.33 shows the resulting normal probability plot. The pattern in the plot is quite straight, indicating it is plausible that the population distribution of dielectric breakdown voltage is normal.

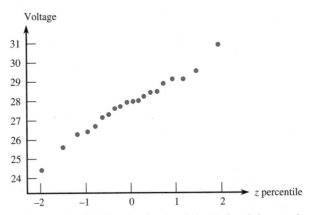

Figure 4.33   Normal probability plot for the dielectric breakdown voltage sample   ∎

There is an alternative version of a normal probability plot in which the z percentile axis is replaced by a nonlinear probability axis. The scaling on this axis is constructed so that plotted points should again fall close to a line when the sampled distribution is normal. Figure 4.34 shows such a plot from MINITAB for the breakdown voltage data of Example 4.29.

A nonnormal population distribution can often be placed in one of the following three categories:

1. It is symmetric and has "lighter tails" than does a normal distribution; that is, the density curve declines more rapidly out in the tails than does a normal curve.

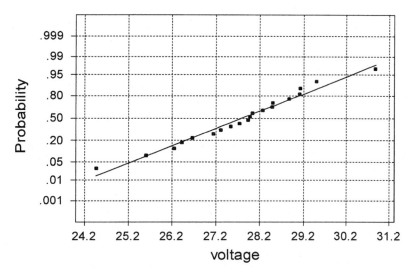

Figure 4.34 Normal probability plot of the breakdown voltage data from MINITAB

2. It is symmetric and heavy-tailed compared to a normal distribution.

3. It is skewed.

A uniform distribution is light-tailed, since its density function drops to zero outside a finite interval. The density function $f(x) = 1/[\pi(1 + x^2)]$ for $-\infty < x < \infty$ is one example of a heavy-tailed distribution, since $1/(1 + x^2)$ declines much less rapidly than does $e^{-x^2/2}$. Lognormal and Weibull distributions are among those that are skewed. When the points in a normal probability plot do not adhere to a straight line, the pattern will frequently suggest that the population distribution is in a particular one of these three categories.

When the distribution from which the sample is selected is light-tailed, the largest and smallest observations are usually not as extreme as would be expected from a normal random sample. Visualize a straight line drawn through the middle part of the plot; points on the far right tend to be below the line (observed value $< z$ percentile), whereas points on the left end of the plot tend to fall above the straight line (observed value $> z$ percentile). The result is an S-shaped pattern of the type pictured in Figure 4.32.

A sample from a heavy-tailed distribution also tends to produce an S-shaped plot. However, in contrast to the light-tailed case, the left end of the plot curves downward (observed $< z$ percentile), as shown in Figure 4.35(a) (page 192). If the underlying distribution is positively skewed (a short left tail and a long right tail), the smallest sample observations will be larger than expected from a normal sample and so will the largest observations. In this case, points on both ends of the plot will fall above a straight line through the middle part, yielding a curved pattern as illustrated in Figure 4.35(b). A sample from a lognormal distribution will usually produce such a pattern. A plot of (z percentile, $\ln(x)$) pairs should then resemble a straight line.

Even when the population distribution is normal, the sample percentiles will not coincide exactly with the theoretical percentiles because of sampling variability. How much can the points in the probability plot deviate from a straight-line pattern before

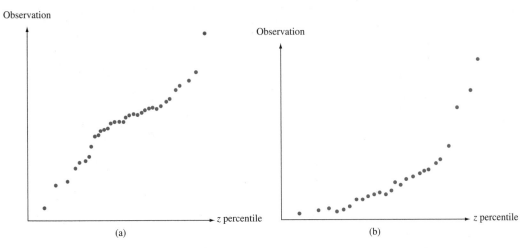

Figure 4.35 Probability plots that suggest a nonnormal distribution: (a) a plot consistent with a heavy-tailed distribution; (b) a plot consistent with a positively skewed distribution

the assumption of population normality is no longer plausible? This is not an easy question to answer. Generally speaking, a small sample from a normal distribution is more likely to yield a plot with a nonlinear pattern than is a large sample. The book *Fitting Equations to Data* (see the Chapter 13 bibliography) presents the results of a simulation study in which numerous samples of different sizes were selected from normal distributions. The authors concluded that there is typically greater variation in the appearance of the probability plot for sample sizes smaller than 30, and only for much larger sample sizes does a linear pattern generally predominate. When a plot is based on a small sample size, only a very substantial departure from linearity should be taken as conclusive evidence of nonnormality. A similar comment applies to probability plots for checking the plausibility of other types of distributions.

Beyond Normality

Consider a family of probability distributions involving two parameters, θ_1 and θ_2, and let $F(x; \theta_1, \theta_2)$ denote the corresponding cdf's. The family of normal distributions is one such family, with $\theta_1 = \mu$, $\theta_2 = \sigma$, and $F(x; \mu, \sigma) = \Phi[(x - \mu)/\sigma]$. Another example is the Weibull family, with $\theta_1 = \alpha$, $\theta_2 = \beta$, and

$$F(x; \alpha, \beta) = 1 - e^{-(x/\beta)^\alpha}$$

Still another family of this type is the gamma family, for which the cdf is an integral involving the incomplete gamma function that cannot be expressed in any simpler form.

The parameters θ_1 and θ_2 are said to be **location** and **scale parameters,** respectively, if $F(x; \theta_1, \theta_2)$ is a function of $(x - \theta_1)/\theta_2$. The parameters μ and σ of the normal family are location and scale parameters, respectively. Changing μ shifts the location of the bell-shaped density curve to the right or left, and changing σ amounts to stretching or compressing the measurement scale (the scale on the horizontal axis when the density function is graphed). Another example is given by the cdf

$$F(x; \theta_1, \theta_2) = 1 - e^{-e^{(x - \theta_1)/\theta_2}} \qquad -\infty < x < \infty$$

A random variable with this cdf is said to have an *extreme value distribution.* It is used in applications involving component lifetime and material strength.

Although the form of the extreme value cdf might at first glance suggest that θ_1 is the point of symmetry for the density function, and therefore the mean and median, this is not the case. Instead, $P(X \leq \theta_1) = F(\theta_1; \theta_1, \theta_2) = 1 - e^{-1} = .632$, and the density function $f(x; \theta_1, \theta_2) = F'(x; \theta_1, \theta_2)$ is negatively skewed (a long lower tail). Similarly, the scale parameter θ_2 is not the standard deviation ($\mu = \theta_1 - .5772\theta_2$ and $\sigma = 1.283\theta_2$). However, changing the value of θ_1 does change the location of the density curve, whereas a change in θ_2 rescales the measurement axis.

The parameter β of the Weibull distribution is a scale parameter, but α is not a location parameter. The parameter α is usually referred to as a **shape parameter.** A similar comment applies to the parameters α and β of the gamma distribution. In the usual form, the density function for any member of either the gamma or Weibull distribution is positive for $x > 0$ and zero otherwise. A location parameter can be introduced as a third parameter γ (we did this for the Weibull distribution) to shift the density function so that it is positive if $x > \gamma$ and zero otherwise.

When the family under consideration has only location and scale parameters, the issue of whether any member of the family is a plausible population distribution can be addressed via a single easily constructed probability plot. One first obtains the percentiles of the *standard distribution,* the one with $\theta_1 = 0$ and $\theta_2 = 1$, for percentages $100(i - .5)/n$ ($i = 1, \ldots, n$). The n (standardized percentile, observation) pairs give the points in the plot. This is of course exactly what we did to obtain an omnibus normal probability plot. Somewhat surprisingly, this methodology can be applied to yield an omnibus Weibull probability plot. The key result is that if X has a Weibull distribution with shape parameter α and scale parameter β, then the transformed variable $\ln(X)$ has an extreme value distribution with location parameter $\theta_1 = \ln(\beta)$ and scale parameter α. Thus a plot of the (extreme value standardized percentile, $\ln(x)$) pairs that shows a strong linear pattern provides support for choosing the Weibull distribution as a population model.

Example 4.30   The accompanying observations are on lifetime (in hours) of power apparatus insulation when thermal and electrical stress acceleration were fixed at particular values ("On the Estimation of Life of Power Apparatus Insulation Under Combined Electrical and Thermal Stress," *IEEE Trans. on Electrical Insulation,* 1985: 70–78). A Weibull probability plot necessitates first computing the 5th, 15th, . . . , and 95th percentiles of the standard extreme value distribution. The $(100p)$th percentile $\eta(p)$ satisfies

$$p = F(\eta(p)) = 1 - e^{-e^{\eta(p)}}$$

from which $\eta(p) = \ln[-\ln(1 - p)]$.

Percentile	−2.97	−1.82	−1.25	−.84	−.51
x	282	501	741	851	1072
ln(x)	5.64	6.22	6.61	6.75	6.98

Percentile	−.23	.05	.33	.64	1.10
x	1122	1202	1585	1905	2138
ln(x)	7.02	7.09	7.37	7.55	7.67

The pairs (−2.97, 5.64), (−1.82, 6.22), . . . , (1.10, 7.67) are plotted as points in Figure 4.36. The straightness of the plot argues strongly for using the Weibull distribution as a model for insulation life, a conclusion also reached by the author of the cited article.

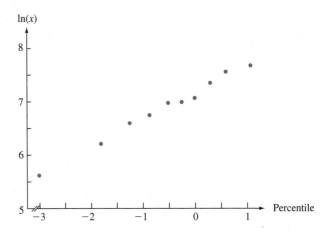

Figure 4.36 A Weibull probability plot of the insulation lifetime data

The gamma distribution is an example of a family involving a shape parameter for which there is no transformation $h(\cdot)$ such that $h(X)$ has a distribution that depends only on location and scale parameters. Construction of a probability plot necessitates first estimating the shape parameter from sample data (some methods for doing this are described in Chapter 6). Sometimes an investigator wishes to know whether the transformed variable X^{θ} has a normal distribution for some value of θ (by convention, $\theta = 0$ is identified with the logarithmic transformation, in which case X has a lognormal distribution). The book *Graphical Methods for Data Analysis,* listed in the Chapter 1 bibliography, discusses this type of problem as well as other refinements of probability plotting.

Exercises | Section 4.6 (79–87)

79. The accompanying normal probability plot was constructed from a sample of 30 readings on tension for mesh screens behind the surface of video display tubes used in computer monitors. Does it appear plausible that the tension distribution is normal?

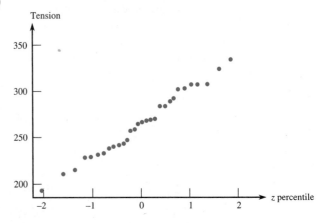

80. Consider the following ten observations on bearing lifetime (in hours):

152.7   172.0   172.5   173.3   193.0
204.7   216.5   234.9   262.6   422.6

Construct a normal probability plot and comment on the plausibility of the normal distribution as a model for bearing lifetime (data from "Modified Moment Estimation for the Three-Parameter Lognormal Distribution," *J. Quality Technology,* 1985: 92–99).

81. Construct a normal probability plot for the following sample of observations on coating thickness for low-viscosity paint ("Achieving a Target Value for a Manufacturing Process: A Case Study," *J. of Quality Technology,* 1992: 22–26). Would you feel comfortable estimating population mean thickness using a method that assumed a normal population distribution?

.83   .88   .88   1.04   1.09   1.12   1.29   1.31
1.48   1.49   1.59   1.62   1.65   1.71   1.76   1.83

82. The article "A Probabilistic Model of Fracture in Concrete and Size Effects on Fracture Toughness" (*Magazine of Concrete Res.,* 1996: 311–320) gives arguments for why the distribution of fracture toughness in concrete specimens should have a Weibull distribution, and presents several histograms of data that appear well fit by superimposed Weibull curves. Consider the following sample of size $n = 18$ observations on toughness for high-strength concrete (consistent with one of the histograms); values of $p_i = (i - .5)/18$ are also given.

Observation	.47	.58	.65	.69	.72	.74
p_i	.0278	.0833	.1389	.1944	.2500	.3056
Observation	.77	.79	.80	.81	.82	.84
p_i	.3611	.4167	.4722	.5278	.5833	.6389
Observation	.86	.89	.91	.95	1.01	1.04
p_i	.6944	.7500	.8056	.8611	.9167	.9722

Construct a Weibull probability plot and comment.

83. Construct a normal probability plot for the fatigue-crack propagation data given in Exercise 39 (Chapter 1). Does it appear plausible that propagation life has a normal distribution? Explain.

84. The article "The Load-Life Relationship for M50 Bearings with Silicon Nitride Ceramic Balls" (*Lu-*brication Engr.,* 1984: 153–159) reports the accompanying data on bearing load life (million revs.) for bearings tested at a 6.45 kN load.

47.1   68.1   68.1   90.8   103.6   106.0   115.0
126.0   146.6   229.0   240.0   240.0   278.0   278.0
289.0   289.0   367.0   385.9   392.0   505.0

a. Construct a normal probability plot. Is normality plausible?
b. Construct a Weibull probability plot. Is the Weibull distribution family plausible?

85. Construct a probability plot that will allow you to assess the plausibility of the lognormal distribution as a model for the rainfall data of Exercise 81 (Chapter 1).

86. The accompanying observations are precipitation values during March over a 30-year period in Minneapolis–St. Paul.

.77   1.20   3.00   1.62   2.81   2.48
1.74   .47   3.09   1.31   1.87   .96
.81   1.43   1.51   .32   1.18   1.89
1.20   3.37   2.10   .59   1.35   .90
1.95   2.20   .52   .81   4.75   2.05

a. Construct and interpret a normal probability plot for this data set.
b. Calculate the square root of each value and then construct a normal probability plot based on this transformed data. Does it seem plausible that the square root of precipitation is normally distributed?
c. Repeat part (b) after transforming by cube roots.

87. The following failure time observations (1000's of hours) resulted from accelerated life testing of 16 integrated circuit chips of a certain type.

82.8   11.6   359.5   502.5   307.8   179.7
242.0   26.5   244.8   304.3   379.1   212.6
229.9   558.9   366.7   204.8

Use the corresponding percentiles of the exponential distribution with $\lambda = 1$ to construct a probability plot. Then explain why the plot assesses the plausibility of the sample having been generated from *any* exponential distribution.

Supplementary Exercises (88–114)

88. Let X = the time it takes a read/write head to locate a desired record on a computer disk memory device once the head has been positioned over the correct track. If the disks rotate once every 25 millisec, a reasonable assumption is that X is uniformly distributed on the interval [0, 25].
  a. Compute $P(10 \leq X \leq 20)$.
  b. Compute $P(X \geq 10)$.
  c. Obtain the cdf $F(X)$.
  d. Compute $E(X)$ and σ_X.

89. A 12-in. bar that is clamped at both ends is to be subjected to an increasing amount of stress until it snaps. Let Y = the distance from the left end at which the break occurs. Suppose Y has pdf

$$f(y) = \begin{cases} \left(\dfrac{1}{24}\right)y\left(1 - \dfrac{y}{12}\right) & 0 \leq y \leq 12 \\ 0 & \text{otherwise} \end{cases}$$

Compute the following:
  a. The cdf of Y, and graph it.
  b. $P(Y \leq 4)$, $P(Y > 6)$, and $P(4 \leq Y \leq 6)$.
  c. $E(Y)$, $E(Y^2)$, and $V(Y)$.
  d. The probability that the break point occurs more than 2 in. from the expected break point
  e. The expected length of the shorter segment when the break occurs

90. Let X denote the time to failure (in years) of a certain hydraulic component. Suppose the pdf of X is $f(x) = 32/(x + 4)^3$ for $x > 0$.
  a. Verify that $f(x)$ is a legitimate pdf.
  b. Determine the cdf.
  c. Use the result of part (b) to calculate the probability that time to failure is between 2 and 5 years.
  d. What is the expected time to failure?
  e. If the component has a salvage value equal to $100/(4 + x)$ when its time to failure is x, what is the expected salvage value?

91. The completion time X for a certain task has cdf $F(x)$ given by

$$\begin{cases} 0 & x < 0 \\ \dfrac{x^3}{3} & 0 \leq x < 1 \\ 1 - \dfrac{1}{2}\left(\dfrac{7}{3} - x\right)\left(\dfrac{7}{4} - \dfrac{3}{4}x\right) & 1 \leq x \leq \dfrac{7}{3} \\ 1 & x \geq \dfrac{7}{3} \end{cases}$$

  a. Obtain the pdf $f(x)$ and sketch its graph.
  b. Compute $P(.5 \leq X \leq 2)$.
  c. Compute $E(X)$.

92. The breakdown voltage of a randomly chosen diode of a certain type is known to be normally distributed with mean value 40 V and standard deviation 1.5 V.
  a. What is the probability that the voltage of a single diode is between 39 and 42?
  b. What value is such that only 15% of all diodes have voltages exceeding that value?
  c. If four diodes are independently selected, what is the probability that at least one has a voltage exceeding 42?

93. The article "Computer Assisted Net Weight Control" (*Quality Progress*, 1983: 22–25) suggests a normal distribution with mean 137.2 oz and standard deviation 1.6 oz. for the actual contents of jars of a certain type. The stated contents was 135 oz.
  a. What is the probability that a single jar contains more than the stated contents?
  b. Among ten randomly selected jars, what is the probability that at least eight contain more than the stated contents?
  c. Assuming that the mean remains at 137.2, to what value would the standard deviation have to be changed so that 95% of all jars contain more than the stated contents?

94. When circuit boards used in the manufacture of compact disc players are tested, the long-run percentage of defectives is 5%. Suppose that a batch of 250 boards has been received and that the condition of any particular board is independent of that of any other board.
  a. What is the approximate probability that at least 10% of the boards in the batch are defective?
  b. What is the approximate probability that there are exactly 10 defectives in the batch?

95. The article "Characterization of Room Temperature Daming in Aluminum-Indium Alloys" (*Metallurgical Trans.*, 1993: 1611–1619) suggests that Al matrix grain size (μm) for an alloy consisting of 2% indium could be modeled with a normal distribution with a mean value 96 and standard deviation 14.
  a. What is the probability that grain size exceeds 100?
  b. What is the probability that grain size is between 50 and 75?

c. What interval (a, b) includes the central 90% of all grain sizes (so that 5% are below a and 5% are above b)?

96. The reaction time (in seconds) to a certain stimulus is a continuous random variable with pdf

$$f(x) = \begin{cases} \dfrac{3}{2} \cdot \dfrac{1}{x^2} & 1 \le x \le 3 \\ 0 & \text{otherwise} \end{cases}$$

a. Obtain the cdf.

b. What is the probability that reaction time is at most 2.5 sec? Between 1.5 and 2.5 sec?

c. Compute the expected reaction time.

d. Compute the standard deviation of reaction time.

e. If an individual takes more than 1.5 sec to react, a light comes on and stays on either until one further second has elapsed or until the person reacts (whichever happens first). Determine the expected amount of time that the light remains lit. [*Hint:* Let $h(X) =$ the time that the light is on as a function of reaction time X.]

97. Let X denote the temperature at which a certain chemical reaction takes place. Suppose that X has pdf

$$f(x) = \begin{cases} \dfrac{1}{9}(4 - x^2) & -1 \le x \le 2 \\ 0 & \text{otherwise} \end{cases}$$

a. Sketch the graph of $f(x)$.

b. Determine the cdf and sketch it.

c. Is 0 the median temperature at which the reaction takes place? If not, is the median temperature smaller or larger than 0?

d. Suppose this reaction is independently carried out once in each of ten different labs and that the pdf of reaction time in each lab is as given. Let $Y =$ the number among the ten labs at which the temperature exceeds 1. What kind of distribution does Y have? (Give the name and values of any parameters.)

98. The article "Determination of the MTF of Positive Photoresists Using the Monte Carlo Method" (*Photographic Sci. and Engr.,* 1983: 254–260) proposes the exponential distribution with parameter $\lambda = .93$ as a model for the distribution of a photon's free path length (μm) under certain circumstances. Suppose this is the correct model.

a. What is the expected path length, and what is the standard deviation of path length?

b. What is the probability that path length exceeds 3.0? What is the probability that path length is between 1.0 and 3.0?

c. What value is exceeded by only 10% of all path lengths?

99. The article "The Prediction of Corrosion by Statistical Analysis of Corrosion Profiles" (*Corrosion Science,* 1985: 305–315) suggests the following cdf for the depth X of the deepest pit in an experiment involving the exposure of carbon manganese steel to acidified seawater.

$$F(x; \alpha, \beta) = e^{-e^{-(x-\alpha)/\beta}} \qquad -\infty < x < \infty$$

The authors propose the values $\alpha = 150$ and $\beta = 90$. Assume this to be the correct model.

a. What is the probability that the depth of the deepest pit is at most 150? At most 300? Between 150 and 300?

b. Below what value will the depth of the maximum pit be observed in 90% of all such experiments?

c. What is the density function of X?

d. The density function can be shown to be unimodal (a single peak). Above what value on the measurement axis does this peak occur? (This value is the mode.)

e. It can be shown that $E(X) \approx .5772\beta + \alpha$. What is the mean for the given values of α and β, and how does it compare to the median and mode? Sketch the graph of the density function. (*Note:* This is called the *largest extreme value distribution.*)

100. A component has lifetime X that is exponentially distributed with parameter λ.

a. If the cost of operation per unit time is c, what is the expected cost of operating this component over its lifetime?

b. Instead of a constant cost rate c as in part (a), suppose the cost rate is $c(1 - .5e^{ax})$ with $a < 0$, so that the cost per unit time is less than c when the component is new and gets more expensive as the component ages. Now compute the expected cost of operation over the lifetime of the component.

101. The *mode* of a continuous distribution is the value x^* that maximizes $f(x)$.

a. What is the mode of a normal distribution with parameters μ and σ?

b. Does the uniform distribution with parameters A and B have a single mode? Why or why not?

c. What is the mode of an exponential distribution with parameter λ? (Draw a picture.)

d. If X has a gamma distribution with parameters α and β, and $\alpha > 1$, find the mode. [*Hint:*

$\ln[f(x)]$ will be maximized iff $f(x)$ is, and it may be simpler to take the derivative of $\ln[f(x)]$.]

e. What is the mode of a chi-squared distribution having ν degrees of freedom?

102. The article "Error Distribution in Navigation" (*J. Institute of Navigation,* 1971: 429–442) suggests that the frequency distribution of positive errors (magnitudes of errors) is well approximated by an exponential distribution. Let $X =$ the lateral position error (nautical miles), which can be either negative or positive. Suppose the pdf of X is

$$f(x) = (.1)e^{-2|x|} \quad -\infty < x < \infty$$

a. Sketch a graph of $f(x)$ and verify that $f(x)$ is a legitimate pdf (show that it integrates to 1).

b. Obtain the cdf of X and sketch it.

c. Compute $P(X \leq 0)$, $P(X \leq 2)$, $P(-1 \leq X \leq 2)$, and the probability that an error of more than 2 miles is made.

103. In some systems, a customer is allocated to one of two service facilities. If the service time for a customer served by facility i has an exponential distribution with parameter λ_i ($i = 1, 2$) and p is the proportion of all customers served by facility 1, then the pdf of $X =$ the service time of a randomly selected customer is

$f(x; \lambda_1, \lambda_2, p)$

$$= \begin{cases} p\lambda_1 e^{-\lambda_1 x} + (1-p)\lambda_2 e^{-\lambda_2 x} & x \geq 0 \\ 0 & \text{otherwise} \end{cases}$$

This is often called the hyperexponential or mixed exponential distribution. As an example, many computer systems process some programs using both a fast in-core compiler (FORTRAN, WATFIV) and a slower compiler. This distribution is also proposed as a model for rainfall amount in "Modeling Monsoon Affected Rainfall of Pakistan by Point Processes" (*J. Water Resources Planning and Mgmnt.,* 1992: 671–688).

a. Verify that $f(x; \lambda_1, \lambda_2, p)$ is indeed a pdf.

b. What is the cdf $F(x; \lambda_1, \lambda_2, p)$?

c. If X has $f(x; \lambda_1, \lambda_2, p)$ as its pdf, what is $E(X)$?

d. Using the fact that $E(X^2) = 2/\lambda^2$ when X has an exponential distribution with parameter λ, compute $E(X^2)$ when X has pdf $f(x; \lambda_1, \lambda_2, p)$. Then compute $V(X)$.

e. The coefficient of variation of a random variable (or distribution) is $CV = \sigma/\mu$. What is CV for an exponential rv? What can you say about

the value of CV when X has a hyperexponential distribution?

f. What is CV for an Erlang distribution with parameters λ and n as defined in Exercise 61? (*Note:* In applied work, the sample CV is used to decide which of the three distributions might be appropriate.)

104. Suppose a particular state allows individuals filing tax returns to itemize deductions only if the total of all itemized deductions is at least \$5000. Let X (in 1000's of dollars) be the total of itemized deductions on a randomly chosen form. Assume that X has the pdf

$$f(x; \alpha) = \begin{cases} k/x^\alpha & x \geq 5 \\ 0 & \text{otherwise} \end{cases}$$

a. Find the value of k. What restriction on α is necessary?

b. What is the cdf of X?

c. What is the expected total deduction on a randomly chosen form? What restriction on α is necessary for $E(X)$ to be finite?

d. Show that $\ln(X/5)$ has an exponential distribution with parameter $\alpha - 1$.

105. Let I_i be the input current to a transistor and I_0 be the output current. Then the current gain is proportional to $\ln(I_0/I_i)$. Suppose the constant of proportionality is 1 (which amounts to choosing a particular unit of measurement), so that current gain $= X = \ln(I_0/I_i)$. Assume X is normally distributed with $\mu = 1$ and $\sigma = .05$.

a. What type of distribution does the ratio I_0/I_i have?

b. What is the probability that the output current is more than twice the input current?

c. What are the expected value and variance of the ratio of output to input current?

106. The article "Response of SiC_f/Si_3N_4 Composites Under Static and Cyclic Loading—An Experimental and Statistical Analysis" (*J. of Engr. Materials and Technology,* 1997: 186–193) suggests that tensile strength (MPa) of composites under specified conditions can be modeled by a Weibull distribution with $\alpha = 9$ and $\beta = 180$.

a. Sketch a graph of the density function.

b. What is the probability that the strength of a randomly selected specimen will exceed 175? Will be between 150 and 175?

c. If two randomly selected specimens are chosen and their strengths are independent of one another, what is the probability that at least one has a strength between 150 and 175?

d. What strength value separates the weakest 10% of all specimens from the remaining 90%?

107. Let Z have a standard normal distribution and define a new rv Y by $Y = \sigma Z + \mu$. Show that Y has a normal distribution with parameters μ and σ. (*Hint:* $Y \leq y$ iff $Z \leq$? Use this to find the cdf of Y and then differentiate it with respect to y.)

108. a. Suppose the lifetime X of a component, when measured in hours, has a gamma distribution with parameters α and β. Let $Y =$ the lifetime measured in minutes. Derive the pdf of Y. (*Hint:* $Y \leq y$ iff $X \leq y/60$. Use this to obtain the cdf of Y and then differentiate to obtain the pdf.)

b. If X has a gamma distribution with parameters α and β, what is the probability distribution of $Y = cX$?

109. In Exercises 107 and 108, as well as many other situations, one has the pdf $f(x)$ of X and wishes to know the pdf of $Y = h(X)$. Assume that $h(\cdot)$ is an invertible function, so that $y = h(x)$ can be solved for x to yield $x = k(y)$. Then it can be shown that the pdf of Y is

$$g(y) = f[k(y)] \cdot |k'(y)|$$

a. If X has a uniform distribution with $A = 0$ and $B = 1$, derive the pdf of $Y = -\ln(X)$.

b. Work Exercise 107, using this result.

c. Work Exercise 108(b), using this result.

110. Let X denote the lifetime of a component, with $f(x)$ and $F(x)$ the pdf and cdf of X. The probability that the component fails in the interval $(x, x + \Delta x)$ is approximately $f(x) \cdot \Delta x$. The conditional probability that it fails in $(x, x + \Delta x)$ given that it has lasted at least x is $f(x) \cdot \Delta x/[1 - F(x)]$. Dividing this by Δx produces the **failure rate function**:

$$r(x) = \frac{f(x)}{1 - F(x)}$$

An increasing failure rate function indicates that older components are increasingly likely to wear out, whereas a decreasing failure rate is evidence of increasing reliability with age. In practice, a "bathtub-shaped" failure is often assumed.

a. If X is exponentially distributed, what is $r(x)$?

b. If X has a Weibull distribution with parameters α and β, what is $r(x)$? For what parameter values will $r(x)$ be increasing? For what parameter values will $r(x)$ decrease with x?

c. Since $r(x) = -(d/dx)\ln[1 - F(x)]$, $\ln[1 - F(x)] = -\int r(x)dx$. Suppose

$$r(x) = \begin{cases} \alpha\left(1 - \dfrac{x}{\beta}\right) & 0 \leq x \leq \beta \\ 0 & \text{otherwise} \end{cases}$$

so that if a component lasts β hours, it will last forever (while seemingly unreasonable, this model can be used to study just "initial wearout"). What are the cdf and pdf of X?

111. Let U have a uniform distribution on the interval [0, 1]. Then observed values having this distribution can be obtained from a computer's random number generator. Let $X = -(1/\lambda) \ln(1 - U)$.

a. Show that X has an exponential distribution with parameter λ. [*Hint:* The cdf of X is $F(x) = P(X \leq x)$; $X \leq x$ is equivalent to $U \leq$?]

b. How would you use part (a) and a random number generator to obtain observed values from an exponential distribution with parameter $\lambda = 10$?

112. Consider an rv X with mean μ and standard deviation σ, and let $g(X)$ be a specified function of X. The first-order Taylor series approximation to $g(X)$ in the neighborhood of μ is

$$g(X) \approx g(\mu) + g'(\mu) \cdot (X - \mu)$$

The right-hand side of this equation is a linear function of X. If the distribution of X is concentrated in an interval over which $g(\cdot)$ is approximately linear [e.g., $\sqrt{x}$ is approximately linear in (1, 2)], then the equation yields approximations to $E(g(X))$ and $V(g(X))$.

a. Give expressions for these approximations. (*Hint:* Use rules of expected value and variance for a linear function $aX + b$.)

b. If the voltage v across a medium is fixed but current I is random, then resistance will also be a random variable related to I by $R = v/I$. If $\mu_I = 20$ and $\sigma_I = .5$, calculate approximations to μ_R and σ_R.

113. A function $g(x)$ is *convex* if the chord connecting any two points on the function's graph lies above the graph. When $g(x)$ is differentiable, an equivalent condition is that for every x, the tangent line at x lies entirely on or below the graph. (See the figures on page 200.) How does $g(\mu) = g(E(X))$ compare to $E(g(X))$? [*Hint:* The equation of the tangent line at $x = \mu$ is $y = g(\mu) + g'(\mu) \cdot (x - \mu)$. Use the condition of convexity, substitute X for x, and take

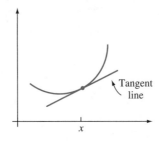

expected values. *Note:* Unless $g(x)$ is linear, the resulting inequality (usually called Jensen's inequality) is strict ($<$ rather than $\leq$); it is valid for both continuous and discrete rv's.]

114. Let X have a Weibull distribution with parameters $\alpha = 2$ and β. Show that $Y = 2X^2/\beta^2$ has a chi-squared distribution with $\nu = 2$. [*Hint:* The cdf of Y is $P(Y \leq y)$; express this probability in the form $P(X \leq g(y))$, use the fact that X has cdf of the form in Expression (4.12), and differentiate with respect to y to obtain the pdf of Y.]

Bibliography

Bury, Karl, *Statistical Distributions in Engineering,* Cambridge Univ. Press, Cambridge, England, 1999. A readable and informative survey of distributions and their properties.

Johnson, Norman, Samuel Kotz, and N. Balakrishnan *Continuous Univariate Distributions,* vols. 1–2, Wiley, New York, 1994. These two volumes together present an exhaustive survey of various continuous distributions.

Nelson, Wayne, *Applied Life Data Analysis,* Wiley, New York, 1982. Gives a comprehensive discussion of distributions and methods that are used in the analysis of lifetime data.

Olkin, Ingram, Cyrus Derman, and Leon Gleser, *Probability Models and Applications* (2nd ed.), Macmillan, New York, 1994. Good coverage of general properties and specific distributions.

5

Joint Probability Distributions and Random Samples

Introduction

In Chapters 3 and 4, we studied probability models for a single random variable. Many problems in probability and statistics lead to models involving several random variables simultaneously. In this chapter, we first discuss probability models for the joint behavior of several random variables, putting special emphasis on the case in which the variables are independent of one another. We then study expected values of functions of several random variables, including covariance and correlation as measures of the degree of association between two variables.

The last three sections of the chapter consider functions of n random variables $X_1, X_2, \ldots, X_n$, focusing especially on their average $(X_1 + \cdots + X_n)/n$. We call any such function, itself a random variable, a *statistic*. Results from probability are used to obtain information about the distribution of a statistic. The premier result of this type is the Central Limit Theorem (CLT), the basis for many inferential procedures involving large sample sizes.

5.1 | Jointly Distributed Random Variables

There are many experimental situations in which more than one random variable (rv) will be of interest to an investigator. We shall first consider joint probability distributions for two discrete rv's, then for two continuous variables, and finally for more than two variables.

The Joint Probability Mass Function for Two Discrete Random Variables

The probability mass function (pmf) of a single discrete rv X specifies how much probability mass is placed on each possible X value. The joint pmf of two discrete rv's X and Y describes how much probability mass is placed on each possible pair of values (x, y).

DEFINITION

Let X and Y be two discrete rv's defined on the sample space $\mathcal{S}$ of an experiment. The **joint probability mass function** $p(x, y)$ is defined for each pair of numbers (x, y) by

$$p(x, y) = P(X = x \text{ and } Y = y)$$

Let A be any set consisting of pairs of (x, y) values. Then the probability $P[(X, Y) \in A]$ is obtained by summing the joint pmf over pairs in A:

$$P[(X, Y) \in A] = \sum \sum_{(x, y) \in A} p(x, y)$$

Example 5.1   A large insurance agency services a number of customers who have purchased both a homeowner's policy and an automobile policy from the agency. For each type of policy, a deductible amount must be specified. For an automobile policy, the choices are $100 and $250, whereas for a homeowner's policy the choices are 0, $100, and $200. Suppose an individual with both types of policy is selected at random from the agency's files. Let X = the deductible amount on the auto policy and Y = the deductible amount on the homeowner's policy. Possible (X, Y) pairs are then (100, 0), (100, 100), (100, 200), (250, 0), (250, 100), and (250, 200); the joint pmf specifies the probability associated with each one of these pairs, with any other pair having probability zero. Suppose the joint pmf is given in the accompanying **joint probability table:**

$p(x, y)$		0	y 100	200
x	100	.20	.10	.20
	250	.05	.15	.30

Then $p(100, 100) = P(X = 100 \text{ and } Y = 100) = P(\$100 \text{ deductible on both policies}) = .10$. The probability $P(Y \geq 100)$ is computed by summing probabilities of all (x, y) pairs for which $y \geq 100$:

$$P(Y \geq 100) = p(100, 100) + p(250, 100) + p(100, 200) + p(250, 200)$$
$$= .75 \qquad \blacksquare$$

A function $p(x, y)$ can be used as a joint pmf provided that $p(x, y) \geq 0$ for all x and y and $\sum_x \sum_y p(x, y) = 1$.

The pmf of one of the variables alone is obtained by summing $p(x, y)$ over values of the other variable. The result is called a *marginal pmf* because when the $p(x, y)$ values appear in a rectangular table, the sums are just marginal (row or column) totals.

DEFINITION	The **marginal probability mass functions** of X and of Y, denoted by $p_X(x)$ and $p_Y(y)$, respectively, are given by $$p_X(x) = \sum_y p(x, y) \qquad p_Y(y) = \sum_x p(x, y)$$

Thus, to obtain the marginal pmf of X evaluated at, say, $x = 100$, the probabilities $p(100, y)$ are added over all possible y values. Doing this for each possible X value gives the marginal pmf of X alone (without reference to Y). From the marginal pmf's, probabilities of events involving only X or only Y can be computed.

Example 5.2
(Example 5.1 continued)

The possible X values are $x = 100$ and $x = 250$, so computing row totals in the joint probability table yields

$$p_X(100) = p(100, 0) + p(100, 100) + p(100, 200) = .50$$

and

$$p_X(250) = p(250, 0) + p(250, 100) + p(250, 200) = .50$$

The marginal pmf of X is then

$$p_X(x) = \begin{cases} .5 & x = 100, 250 \\ 0 & \text{otherwise} \end{cases}$$

Similarly, the marginal pmf of Y is obtained from column totals as

$$p_Y(y) = \begin{cases} .25 & y = 0, 100 \\ .50 & y = 200 \\ 0 & \text{otherwise} \end{cases}$$

so $P(Y \geq 100) = p_Y(100) + p_Y(200) = .75$ as before.   ■

The Joint Probability Density Function for Two Continuous Random Variables

The probability that the observed value of a continuous rv X lies in a one-dimensional set A (such as an interval) is obtained by integrating the pdf $f(x)$ over the set A. Similarly, the probability that the pair (X, Y) of continuous rv's falls in a two-dimensional set A (such as a rectangle) is obtained by integrating a function called the *joint density function*.

DEFINITION

> Let X and Y be continuous rv's. Then $f(x, y)$ is the **joint probability density function** for X and Y if for any two-dimensional set A
>
> $$P[(X, Y) \in A] = \int \int_A f(x, y)dx\, dy$$
>
> In particular, if A is the two-dimensional rectangle $\{(x, y): a \le x \le b, c \le y \le d\}$, then
>
> $$P[(X, Y) \in A] = P(a \le X \le b, c \le Y \le d) = \int_a^b \int_c^d f(x, y)dy\, dx$$

For $f(x, y)$ to be a candidate for a joint pdf, it must satisfy $f(x, y) \ge 0$ and $\int_{-\infty}^\infty \int_{-\infty}^\infty f(x, y)dx\, dy = 1$. We can think of $f(x, y)$ as specifying a surface at height $f(x, y)$ above the point (x, y) in a three-dimensional coordinate system. Then $P[(X, Y) \in A]$ is the volume underneath this surface and above the region A, analogous to the area under a curve in the one-dimensional case. This is illustrated in Figure 5.1.

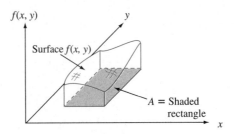

Figure 5.1 $P[(X, Y) \in A]$ = volume under density surface above A

Example 5.3

A bank operates both a drive-up facility and a walk-up window. On a randomly selected day, let X = the proportion of time that the drive-up facility is in use (at least one customer is being served or waiting to be served) and Y = the proportion of time that the walk-up window is in use. Then the set of possible values for (X, Y) is the rectangle $D = \{(x, y): 0 \le x \le 1, 0 \le y \le 1\}$. Suppose the joint pdf of (X, Y) is given by

$$f(x, y) = \begin{cases} \dfrac{6}{5}(x + y^2) & 0 \le x \le 1, 0 \le y \le 1 \\ 0 & \text{otherwise} \end{cases}$$

To verify that this is a legitimate pdf, note that $f(x, y) \ge 0$ and

$$\int_{-\infty}^\infty \int_{-\infty}^\infty f(x, y)dx\, dy = \int_0^1 \int_0^1 \frac{6}{5}(x + y^2)dx\, dy$$

$$= \int_0^1 \int_0^1 \frac{6}{5}x\, dx\, dy + \int_0^1 \int_0^1 \frac{6}{5}y^2\, dx\, dy$$

$$= \int_0^1 \frac{6}{5}x\, dx + \int_0^1 \frac{6}{5}y^2\, dy = \frac{6}{10} + \frac{6}{15} = 1$$

The probability that neither facility is busy more than one-quarter of the time is

$$P\left(0 \leq X \leq \frac{1}{4}, 0 \leq Y \leq \frac{1}{4}\right) = \int_0^{1/4} \int_0^{1/4} \frac{6}{5}(x + y^2) dx\, dy$$

$$= \frac{6}{5} \int_0^{1/4} \int_0^{1/4} x\, dx\, dy + \frac{6}{5} \int_0^{1/4} \int_0^{1/4} y^2\, dx\, dy$$

$$= \frac{6}{20} \cdot \frac{x^2}{2} \Big|_{x=0}^{x=1/4} + \frac{6}{20} \cdot \frac{y^3}{3} \Big|_{y=0}^{y=1/4} = \frac{7}{640}$$

$$= .0109 \qquad \blacksquare$$

As with joint pmf's, from the joint pdf of X and Y, each of the two marginal density functions can be computed.

DEFINITION

> The **marginal probability density functions** of X and Y, denoted by $f_X(x)$ and $f_Y(y)$, respectively, are given by
>
> $$f_X(x) = \int_{-\infty}^{\infty} f(x, y) dy \qquad \text{for } -\infty < x < \infty$$
>
> $$f_Y(y) = \int_{-\infty}^{\infty} f(x, y) dx \qquad \text{for } -\infty < y < \infty$$

Example 5.4
(Example 5.3 continued)

The marginal pdf of X, which gives the probability distribution of busy time for the drive-up facility without reference to the walk-up window, is

$$f_X(x) = \int_{-\infty}^{\infty} f(x, y) dy = \int_0^1 \frac{6}{5}(x + y^2) dy = \frac{6}{5}x + \frac{2}{5}$$

for $0 \leq x \leq 1$ and 0 otherwise. The marginal pdf of Y is

$$f_Y(y) = \begin{cases} \dfrac{6}{5} y^2 + \dfrac{3}{5} & 0 \leq y \leq 1 \\ 0 & \text{otherwise} \end{cases}$$

Then

$$P\left(\frac{1}{4} \leq Y \leq \frac{3}{4}\right) = \int_{1/4}^{3/4} f_Y(y) dy = \frac{37}{80} = .4625 \qquad \blacksquare$$

In Example 5.3, the region of positive joint density was a rectangle, which made computation of the marginal pdf's relatively easy. Consider now an example in which the region of positive density is a more complicated figure.

Example 5.5

A nut company markets cans of deluxe mixed nuts containing almonds, cashews, and peanuts. Suppose the net weight of each can is exactly 1 lb, but the weight contribution of each type of nut is random. Because the three weights sum to 1, a joint probability model for any two gives all necessary information about the weight of the third type.

Let X = the weight of almonds in a selected can and Y = the weight of cashews. Then the region of positive density is $D = \{(x, y): 0 \leq x \leq 1, 0 \leq y \leq 1, x + y \leq 1\}$, the shaded region pictured in Figure 5.2.

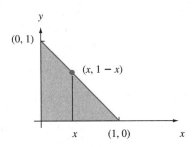

Figure 5.2   Region of positive density for Example 5.5

Now let the joint pdf for (X, Y) be

$$f(x, y) = \begin{cases} 24xy & 0 \leq x \leq 1, 0 \leq y \leq 1, x + y \leq 1 \\ 0 & \text{otherwise} \end{cases}$$

For any fixed x, $f(x, y)$ increases with y; for fixed y, $f(x, y)$ increases with x. This is appropriate because the word *deluxe* implies that most of the can should consist of almonds and cashews rather than peanuts, so that the density function should be large near the upper boundary and small near the origin. The surface determined by $f(x, y)$ slopes upward from zero as (x, y) moves away from either axis.

Clearly $f(x, y) \geq 0$. To verify the second condition on a joint pdf, recall that a double integral is computed as an iterated integral by holding one variable fixed (such as x as in Figure 5.2), integrating over values of the other variable lying along the straight line passing through the value of the fixed variable, and finally integrating over all possible values of the fixed variable. Thus,

$$\int_{-\infty}^{\infty}\int_{-\infty}^{\infty} f(x, y)dy\, dx = \int\int_D f(x, y)dy\, dx = \int_0^1 \left\{ \int_0^{1-x} 24xy\, dy \right\} dx$$

$$= \int_0^1 24x \left\{ \frac{y^2}{2} \Big|_{y=0}^{y=1-x} \right\} dx = \int_0^1 12x(1-x)^2\, dx = 1$$

To compute the probability that the two types of nuts together make up at most 50% of the can, let $A = \{(x, y): 0 \leq x \leq 1, 0 \leq y \leq 1, \text{and } x + y \leq .5\}$, as shown in Figure 5.3. Then

$$P((X, Y) \in A) = \int\int_A f(x, y)dx\, dy = \int_0^{.5}\int_0^{.5-x} 24xy\, dy\, dx = .0625$$

The marginal pdf for almonds is obtained by holding X fixed at x and integrating $f(x, y)$ along the vertical line through x:

$$f_X(x) = \int_{-\infty}^{\infty} f(x, y)dy = \begin{cases} \int_0^{1-x} 24xy\, dy = 12x(1-x)^2 & 0 \leq x \leq 1 \\ 0 & \text{otherwise} \end{cases}$$

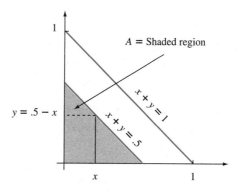

Figure 5.3 Computing $P[(X, Y) \in A]$ for Example 5.5

By symmetry of $f(x, y)$ and the region D, the marginal pdf of Y is obtained by replacing x and X in $f_X(x)$ by y and Y, respectively. ∎

Independent Random Variables

In many situations, information about the observed value of one of the two variables X and Y gives information about the value of the other variable. In Example 5.1, the marginal probability of X at $x = 250$ was .5, as was the probability that $X = 100$. If, however, we are told that the selected individual had $Y = 0$, then $X = 100$ is four times as likely as $X = 250$. Thus, there is a dependence between the two variables.

In Chapter 2 we pointed out that one way of defining independence of two events is to say that A and B are independent if $P(A \cap B) = P(A) \cdot P(B)$. We now use an analogous definition for the independence of two rv's.

DEFINITION

Two random variables X and Y are said to be **independent** if for every pair of x and y values,

$$p(x, y) = p_X(x) \cdot p_Y(y) \qquad \text{when } X \text{ and } Y \text{ are discrete}$$

or (5.1)

$$f(x, y) = f_X(x) \cdot f_Y(y) \qquad \text{when } X \text{ and } Y \text{ are continuous}$$

If (5.1) is not satisfied for all (x, y), then X and Y are said to be **dependent.**

The definition says that two variables are independent if their joint pmf or pdf is the product of the two marginal pmf's or pdf's.

Example 5.6 In the insurance situation of Examples 5.1 and 5.2,

$$p(100, 100) = .10 \neq (.5)(.25) = p_X(100) \cdot p_Y(100)$$

so X and Y are not independent. Independence of X and Y requires that *every* entry in the joint probability table be the product of the corresponding row and column marginal probabilities. ∎

Example 5.7
(Example 5.5
continued)

Because $f(x, y)$ has the form of a product, X and Y would appear to be independent. However, although $f_X(\frac{3}{4}) = f_Y(\frac{3}{4}) = \frac{9}{16}$, $f(\frac{3}{4}, \frac{3}{4}) = 0 \neq \frac{9}{16} \cdot \frac{9}{16}$, so the variables are not in fact independent. To be independent, $f(x, y)$ must have the form $g(x) \cdot h(y)$ *and* the region of positive density must be a rectangle whose sides are parallel to the coordinate axes.   ∎

Independence of two random variables is most useful when the description of the experiment under study tells us that X and Y have no effect on one another. Then once the marginal pmf's or pdf's have been specified, the joint pmf or pdf is simply the product of the two marginal functions. It follows that

$$P(a \leq X \leq b, c \leq Y \leq d) = P(a \leq X \leq b) \cdot P(c \leq Y \leq d)$$

Example 5.8

Suppose that the lifetimes of two components are independent of one another and that the first lifetime, X_1, has an exponential distribution with parameter λ_1 whereas the second, X_2, has an exponential distribution with parameter λ_2. Then the joint pdf is

$$f(x_1, x_2) = f_{X_1}(x_1) \cdot f_{X_2}(x_2)$$

$$= \begin{cases} \lambda_1 e^{-\lambda_1 x_1} \cdot \lambda_2 e^{-\lambda_2 x_2} = \lambda_1 \lambda_2 e^{-\lambda_1 x_1 - \lambda_2 x_2} & x_1 > 0, x_2 > 0 \\ 0 & \text{otherwise} \end{cases}$$

Let $\lambda_1 = 1/1000$ and $\lambda_2 = 1/1200$, so that the expected lifetimes are 1000 hours and 1200 hours, respectively. The probability that both component lifetimes are at least 1500 hours is

$$P(1500 \leq X_1, 1500 \leq X_2) = P(1500 \leq X_1) \cdot P(1500 \leq X_2)$$

$$= e^{-\lambda_1(1500)} \cdot e^{-\lambda_2(1500)}$$

$$= (.2231)(.2865) = .0639$$ ∎

More Than Two Random Variables

To model the joint behavior of more than two random variables, we extend the concept of a joint distribution of two variables.

DEFINITION

If $X_1, X_2, \ldots, X_n$ are all discrete random variables, the joint pmf of the variables is the function

$$p(x_1, x_2, \ldots, x_n) = P(X_1 = x_1, X_2 = x_2, \ldots, X_n = x_n)$$

If the variables are continuous, the joint pdf of $X_1, \ldots, X_n$ is the function $f(x_1, x_2, \ldots, x_n)$ such that for any n intervals $[a_1, b_1], \ldots, [a_n, b_n]$,

$$P(a_1 \leq X_1 \leq b_1, \ldots, a_n \leq X_n \leq b_n) = \int_{a_1}^{b_1} \cdots \int_{a_n}^{b_n} f(x_1, \ldots, x_n) dx_n \cdots dx_1$$

Example 5.9

In a binomial experiment, each trial could result in one of only two possible outcomes. Consider now an experiment consisting of n independent and identical trials, in which each trial can result in any one of r possible outcomes. Let $p_i = P($outcome i on any particular trial$)$ and define random variables by $X_i =$ the number of trials resulting in outcome i $(i = 1, \ldots, r)$. Such an experiment is called a **multinomial experiment,** and the joint pmf of $X_1, \ldots, X_r$ is called the **multinomial distribution.** By using a counting argument analogous to the one used in deriving the binomial distribution, the joint pmf of $X_1, \ldots, X_r$ can be shown to be

$p(x_1, \ldots, x_r)$
$$= \begin{cases} \dfrac{n!}{(x_1!)(x_2!) \cdot \cdots \cdot (x_r!)} \, p_1^{x_1} \cdot \cdots \cdot p_r^{x_r} & x_i = 0, 1, 2, \ldots, \text{ with } x_1 + \cdots + x_r = n \\ 0 & \text{otherwise} \end{cases}$$

As an example, if the allele of each of ten independently obtained pea sections is determined and $p_1 = P(AA), p_2 = P(Aa), p_3 = P(aa), X_1 =$ number of AA's, $X_2 =$ number of Aa's, and $X_3 =$ number of aa's, then

$$p(x_1, x_2, x_3) = \dfrac{10!}{(x_1!)(x_2!)(x_3!)} \, p_1^{x_1} p_2^{x_2} p_3^{x_3} \qquad x_i = 0, 1, \ldots \quad \text{and } x_1 + x_2 + x_3 = 10$$

If $p_1 = p_3 = .25, p_2 = .5$, then

$$P(X_1 = 2, X_2 = 5, X_3 = 3) = p(2, 5, 3)$$
$$= \dfrac{10!}{2! \, 5! \, 3!} (.25)^2 (.5)^5 (.25)^3 = .0769$$

The case $r = 2$ gives the binomial distribution, with $X_1 =$ number of successes and $X_2 = n - X_1 =$ number of failures. ∎

Example 5.10 When a certain method is used to collect a fixed volume of rock samples in a region, there are four resulting rock types. Let X_1, X_2, and X_3 denote the proportion by volume of rock types 1, 2, and 3 in a randomly selected sample (the proportion of rock type 4 is $1 - X_1 - X_2 - X_3$, so a variable X_4 would be redundant). If the joint pdf of X_1, X_2, X_3 is

$$f(x_1, x_2, x_3) = \begin{cases} kx_1x_2(1 - x_3) & 0 \le x_1 \le 1, 0 \le x_2 \le 1, 0 \le x_3 \le 1, x_1 + x_2 + x_3 \le 1 \\ 0 & \text{otherwise} \end{cases}$$

then k is determined by

$$1 = \int_{-\infty}^{\infty} \int_{-\infty}^{\infty} \int_{-\infty}^{\infty} f(x_1, x_2, x_3) dx_3 \, dx_2 \, dx_1$$
$$= \int_0^1 \left\{ \int_0^{1-x_1} \left[\int_0^{1-x_1-x_2} kx_1x_2(1 - x_3) dx_3 \right] dx_2 \right\} dx_1$$

This iterated integral has value $k/144$, so $k = 144$. The probability that rocks of types 1 and 2 together account for at most 50% of the sample is

$$P(X_1 + X_2 \le .5) = \underset{\substack{0 \le x_i \le 1 \text{ for } i = 1, 2, 3 \\ x_1 + x_2 + x_3 \le 1, \, x_1 + x_2 \le .5}}{\iiint} f(x_1, x_2, x_3) dx_3 \, dx_2 \, dx_1$$
$$= \int_0^{.5} \left\{ \int_0^{.5-x_1} \left[\int_0^{1-x_1-x_2} 144x_1x_2(1 - x_3) dx_3 \right] dx_2 \right\} dx_1$$
$$= .6066$$

■

The notion of independence of more than two random variables is similar to the notion of independence of more than two events.

DEFINITION

> The random variables $X_1, X_2, \ldots, X_n$ are said to be **independent** if for *every* subset $X_{i_1}, X_{i_2}, \ldots, X_{i_k}$ of the variables (each pair, each triple, and so on), the joint pmf or pdf of the subset is equal to the product of the marginal pmf's or pdf's.

Thus, if the variables are independent with $n = 4$, then the joint pmf or pdf of any two variables is the product of the two marginals, and similarly for any three variables and all four variables together. Most important, once we are told that n variables are independent, then the joint pmf or pdf is the product of the n marginals.

Example 5.11 If $X_1, \ldots, X_n$ represent the lifetimes of n components, the components operate independently of one another, and each lifetime is exponentially distributed with parameter λ, then

$$f(x_1, x_2, \ldots, x_n) = (\lambda e^{-\lambda x_1}) \cdot (\lambda e^{-\lambda x_2}) \cdot \cdots \cdot (\lambda e^{-\lambda x_n})$$

$$= \begin{cases} \lambda^n e^{-\lambda \Sigma x_i} & x_1 \geq 0, x_2 \geq 0, \ldots, x_n \geq 0 \\ 0 & \text{otherwise} \end{cases}$$

If these n components constitute a system that will fail as soon as a single component fails, then the probability that the system lasts past time t is

$$P(X_1 > t, \ldots, X_n > t) = \int_t^\infty \cdots \int_t^\infty f(x_1, \ldots, x_n) dx_1 \ldots dx_n$$

$$= \left(\int_t^\infty \lambda e^{-\lambda x_1} dx_1 \right) \cdots \left(\int_t^\infty \lambda e^{-\lambda x_n} dx_n \right)$$

$$= (e^{-\lambda t})^n = e^{-n\lambda t}$$

Therefore,

$$P(\text{system lifetime} \leq t) = 1 - e^{-n\lambda t} \qquad \text{for } t \geq 0$$

which shows that *system* lifetime has an exponential distribution with parameter $n\lambda$; the expected value of system lifetime is $1/n\lambda$. ∎

In many experimental situations to be considered in this book, independence is a reasonable assumption, so that specifying the joint distribution reduces to deciding on appropriate marginal distributions.

Conditional Distributions

Suppose X = the number of major defects on a randomly selected new automobile and Y = the number of minor defects on that same auto. If we learn that the number of major defects on the selected car is one, what now is the probability that the car has at most three minor defects—that is, what is $P(Y \leq 3 \mid X = 1)$? Similarly, if X and Y denote the lifetimes of two components in a system and it happens that $X = 100$, what is the probability that $Y \geq 200$, and what is the expected lifetime of the second component "conditional on" this value of X? Questions of this sort can be answered by studying conditional probability distributions.

DEFINITION

Let X and Y be two continuous rv's with joint pdf $f(x, y)$ and marginal X pdf $f_X(x)$. Then for any X value x for which $f_X(x) > 0$, the **conditional probability density function of Y given that $X = x$** is

$$f_{Y|X}(y\,|\,x) = \frac{f(x, y)}{f_X(x)} \qquad -\infty < y < \infty$$

If X and Y are discrete, replacing pdf's by pmf's in this definition gives the **conditional probability mass function of Y when $X = x$.**

Notice that the definition of $f_{Y|X}(y\,|\,x)$ parallels that of $P(B\,|\,A)$, the conditional probability that B will occur, given that A has occurred. Once the conditional pdf or pmf has been determined, questions of the type posed at the outset of this subsection can be answered by integrating or summing over an appropriate set of Y values.

Example 5.12   Reconsider the situation of Examples 5.3 and 5.4 involving $X =$ the proportion of time that a bank's drive-up facility is busy and $Y =$ the analogous proportion for the walk-up window. The conditional pdf of Y given that $X = .8$ is

$$f_{Y|X}(y\,|\,.8) = \frac{f(.8, y)}{f_X(.8)} = \frac{1.2(.8 + y^2)}{1.2(.8) + .4} = \frac{1}{34}(24 + 30y^2) \qquad 0 < y < 1$$

The probability that the walk-in facility is busy at most half the time given that $X = .8$ is then

$$P(Y \le .5\,|\,X = .8) = \int_{-\infty}^{.5} f_{Y|X}(y\,|\,.8)dy = \int_0^{.5} \frac{1}{34}(24 + 30y^2)dy = .390$$

Using the marginal pdf of Y gives $P(Y \le .5) = .350$. Also $E(Y) = .6$, whereas the expected proportion of time that the walk-in facility is busy given that $X = .8$ (a *conditional* expectation) is

$$E(Y\,|\,X = .8) = \int_{-\infty}^{\infty} y \cdot f_{Y|X}(y\,|\,.8)dy = \frac{1}{34}\int_0^1 y(24 + 30y^2)dy = .574 \qquad \blacksquare$$

Exercises | Section 5.1 (1–21)

1. A service station has both self-service and full-service islands. On each island, there is a single regular unleaded pump with two hoses. Let X denote the number of hoses being used on the self-service island at a particular time, and let Y denote the number of hoses on the full-service island in use at that time. The joint pmf of X and Y appears in the accompanying tabulation.

			y	
$p(x, y)$		0	1	2
	0	.10	.04	.02
x	1	.08	.20	.06
	2	.06	.14	.30

a. What is $P(X = 1$ and $Y = 1)$?
b. Compute $P(X \le 1$ and $Y \le 1)$.

c. Give a word description of the event $\{X \neq 0 \text{ and } Y \neq 0\}$ and compute the probability of this event.

d. Compute the marginal pmf of X and of Y. Using $p_X(x)$, what is $P(X \leq 1)$?

e. Are X and Y independent rv's? Explain.

2. When an automobile is stopped by a roving safety patrol, each tire is checked for tire wear, and each headlight is checked to see whether it is properly aimed. Let X denote the number of headlights that need adjustment, and let Y denote the number of defective tires.

a. If X and Y are independent with $p_X(0) = .5$, $p_X(1) = .3$, $p_X(2) = .2$, and $p_Y(0) = .6$, $p_Y(1) = .1$, $p_Y(2) = p_Y(3) = .05$, $p_Y(4) = .2$, display the joint pmf of (X, Y) in a joint probability table.

b. Compute $P(X \leq 1 \text{ and } Y \leq 1)$ from the joint probability table and verify that it equals the product $P(X \leq 1) \cdot P(Y \leq 1)$.

c. What is $P(X + Y = 0)$ (the probability of no violations)?

d. Compute $P(X + Y \leq 1)$.

3. A certain market has both an express checkout line and a superexpress checkout line. Let X_1 denote the number of customers in line at the express checkout at a particular time of day and let X_2 denote the number of customers in line at the superexpress checkout at the same time. Suppose the joint pmf of X_1 and X_2 is as given in the accompanying table.

		x_2		
	0	1	2	3
0	.08	.07	.04	.00
1	.06	.15	.05	.04
x_1 2	.05	.04	.10	.06
3	.00	.03	.04	.07
4	.00	.01	.05	.06

a. What is $P(X_1 = 1, X_2 = 1)$, that is, the probability that there is exactly one customer in each line?

b. What is $P(X_1 = X_2)$, that is, the probability that the numbers of customers in the two lines are identical?

c. Let A denote the event that there are at least two more customers in one line than in the other line. Express A in terms of X_1 and X_2, and calculate the probability of this event.

d. What is the probability that the total number of customers in the two lines is exactly four? At least four?

4. Return to the situation described in Exercise 3.

a. Determine the marginal pmf of X_1 and then calculate the expected number of customers in line at the express checkout.

b. Determine the marginal pmf of X_2.

c. By inspection of the probabilities $P(X_1 = 4)$, $P(X_2 = 0)$, and $P(X_1 = 4, X_2 = 0)$, are X_1 and X_2 independent random variables? Explain.

5. The number of customers waiting for gift-wrap service at a department store is an rv X with possible values 0, 1, 2, 3, 4 and corresponding probabilities .1, .2, .3, .25, .15. A randomly selected customer will have 1, 2, or 3 packages for wrapping with probabilities .6, .3, and .1, respectively. Let $Y =$ the total number of packages to be wrapped for the customers waiting in line (assume that the number of packages submitted by one customer is independent of the number submitted by any other customer).

a. Determine $P(X = 3, Y = 3)$, i.e., $p(3, 3)$.

b. Determine $p(4, 11)$.

6. Let X denote the number of brand X VCRs sold during a particular week by a certain store. The pmf of X is

x	0	1	2	3	4
$p_X(x)$	.1	.2	.3	.25	.15

Sixty percent of all customers who purchase brand X VCRs also buy an extended warranty. Let Y denote the number of purchasers during this week who buy an extended warranty.

a. What is $P(X = 4, Y = 2)$? [*Hint:* This probability equals $P(Y = 2 \mid X = 4) \cdot P(X = 4)$; now think of the four purchases as four trials of a binomial experiment, with success on a trial corresponding to buying an extended warranty.]

b. Calculate $P(X = Y)$.

c. Determine the joint pmf of X and Y and then the marginal pmf of Y.

7. The joint probability distribution of the number X of cars and the number Y of buses per signal cycle at a proposed left turn lane is displayed in the accompanying joint probability table.

$p(x, y)$			y	
		0	1	2
	0	.025	.015	.010
	1	.050	.030	.020
	2	.125	.075	.050
x	3	.150	.090	.060
	4	.100	.060	.040
	5	.050	.030	.020

a. What is the probability that there is exactly one car and exactly one bus during a cycle?

b. What is the probability that there is at most one car and at most one bus during a cycle?

c. What is the probability that there is exactly one car during a cycle? Exactly one bus?

d. Suppose the left turn lane is to have a capacity of five cars and one bus is equivalent to three cars. What is the probability of an overflow during a cycle?

e. Are X and Y independent rv's? Explain.

8. A stockroom currently has 30 components of a certain type, of which 8 were provided by supplier 1, 10 by supplier 2, and 12 by supplier 3. Six of these are to be randomly selected for a particular assembly. Let X = the number of supplier 1's components selected, Y = the number of supplier 2's components selected, and p(x, y) denote the joint pmf of X and Y.

a. What is p(3, 2)? [*Hint:* Each sample of size six is equally likely to be selected. Therefore, p(3, 2) = (number of outcomes with X = 3 and Y = 2)/(total number of outcomes). Now use the product rule for counting to obtain the numerator and denominator.]

b. Using the logic of part (a), obtain p(x, y). (This can be thought of as a multivariate hypergeometric distribution—sampling without replacement from a finite population consisting of more than two categories.)

9. Each front tire on a particular type of vehicle is supposed to be filled to a pressure of 26 psi. Suppose the actual air pressure in each tire is a random variable—X for the right tire and Y for the left tire, with joint pdf

$$f(x, y) = \begin{cases} K(x^2 + y^2) & 20 \leq x \leq 30, 20 \leq y \leq 30 \\ 0 & \text{otherwise} \end{cases}$$

a. What is the value of K?

b. What is the probability that both tires are underfilled?

c. What is the probability that the difference in air pressure between the two tires is at most 2 psi?

d. Determine the (marginal) distribution of air pressure in the right tire alone.

e. Are X and Y independent rv's?

10. Annie and Alvie have agreed to meet between 5:00 P.M. and 6:00 P.M. for dinner at a local health-food restaurant. Let X = Annie's arrival time and Y = Alvie's arrival time. Suppose X and Y are in-

dependent with each uniformly distributed on the interval [5, 6].

a. What is the joint pdf of X and Y?

b. What is the probability that they both arrive between 5:15 and 5:45?

c. If the first one to arrive will wait only 10 min before leaving to eat elsewhere, what is the probability that they have dinner at the health-food restaurant? [*Hint:* The event of interest is A = $\{(x, y): |x - y| \leq \frac{1}{6}\}$.]

11. Two different professors have just submitted final exams for duplication. Let X denote the number of typographical errors on the first professor's exam and Y denote the number of such errors on the second exam. Suppose X has a Poisson distribution with parameter λ, Y has a Poisson distribution with parameter θ, and X and Y are independent.

a. What is the joint pmf of X and Y?

b. What is the probability that at most one error is made on both exams combined?

c. Obtain a general expression for the probability that the total number of errors in the two exams is m (where m is a nonnegative integer). [*Hint:* A = $\{(x, y): x + y = m\}$ = $\{(m, 0), (m - 1, 1), \ldots, (1, m - 1), (0, m)\}$. Now sum the joint pmf over (x, y) ∈ A and use the binomial theorem, which says that

$$\sum_{k=0}^{m} \binom{m}{k} a^k b^{m-k} = (a + b)^m$$

for any a, b.]

12. Two components of a minicomputer have the following joint pdf for their useful lifetimes X and Y:

$$f(x, y) = \begin{cases} xe^{-x(1+y)} & x \geq 0 \text{ and } y \geq 0 \\ 0 & \text{otherwise} \end{cases}$$

a. What is the probability that the lifetime X of the first component exceeds 3?

b. What are the marginal pdf's of X and Y? Are the two lifetimes independent? Explain.

c. What is the probability that the lifetime of at least one component exceeds 3?

13. You have two lightbulbs for a particular lamp. Let X = the lifetime of the first bulb and Y = the lifetime of the second bulb (both in 1000's of hours). Suppose that X and Y are independent and that each has an exponential distribution with parameter λ = 1.

a. What is the joint pdf of X and Y?

b. What is the probability that each bulb lasts at most 1000 hours (i.e., X ≤ 1 and Y ≤ 1)?

c. What is the probability that the total lifetime of the two bulbs is at most 2? [*Hint:* Draw a picture of the region $A = \{(x, y): x \geq 0, y \geq 0, x + y \leq 2\}$ before integrating.]

d. What is the probability that the total lifetime is between 1 and 2?

14. Suppose that you have ten lightbulbs, that the lifetime of each is independent of all the other lifetimes, and that each lifetime has an exponential distribution with parameter λ.

a. What is the probability that all ten bulbs fail before time t?

b. What is the probability that exactly k of the ten bulbs fail before time t?

c. Suppose that nine of the bulbs have lifetimes that are exponentially distributed with parameter λ and that the remaining bulb has a lifetime that is exponentially distributed with parameter θ (it is made by another manufacturer). What is the probability that exactly five of the ten bulbs fail before time t?

15. Consider a system consisting of three components as pictured. The system will continue to function as long as the first component functions and either component 2 or component 3 functions. Let X_1, X_2, and X_3 denote the lifetimes of components 1, 2, and 3, respectively. Suppose the X_i's are independent of one another and each X_i has an exponential distribution with parameter λ.

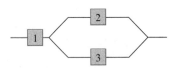

a. Let Y denote the system lifetime. Obtain the cumulative distribution function of Y and differentiate to obtain the pdf. [*Hint:* $F(y) = P(Y \leq y)$; express the event $\{Y \leq y\}$ in terms of unions and/or intersections of the three events $\{X_1 \leq y\}$, $\{X_2 \leq y\}$, and $\{X_3 \leq y\}$.]

b. Compute the expected system lifetime.

16. a. For $f(x_1, x_2, x_3)$ as given in Example 5.10, compute the **joint marginal density function** of X_1 and X_3 alone (by integrating over x_2).

b. What is the probability that rocks of types 1 and 3 together make up at most 50% of the sample? [*Hint:* Use the result of part (a).]

c. Compute the marginal pdf of X_1 alone. [*Hint:* Use the result of part (a).]

17. An ecologist wishes to select a point inside a circular sampling region according to a uniform distribution (in practice this could be done by first selecting a direction and then a distance from the center in that direction). Let $X =$ the x coordinate of the point selected and $Y =$ the y coordinate of the point selected. If the circle is centered at $(0, 0)$ and has radius R, then the joint pdf of X and Y is

$$f(x, y) = \begin{cases} \dfrac{1}{\pi R^2} & x^2 + y^2 \leq R^2 \\ 0 & \text{otherwise} \end{cases}$$

a. What is the probability that the selected point is within $R/2$ of the center of the circular region? [*Hint:* Draw a picture of the region of positive density D. Because $f(x, y)$ is constant on D, computing a probability reduces to computing an area.]

b. What is the probability that both X and Y differ from 0 by at most $R/2$?

c. Answer part (b) for $R/\sqrt{2}$ replacing $R/2$.

d. What is the marginal pdf of X? Of Y? Are X and Y independent?

18. Refer to Exercise 1 and answer the following questions.

a. Given that $X = 1$, determine the conditional pmf of Y—i.e., $p_{Y|X}(0 | 1)$, $p_{Y|X}(1 | 1)$, and $p_{Y|X}(2 | 1)$.

b. Given that two hoses are in use at the self-service island, what is the conditional pmf of the number of hoses in use on the full-service island?

c. Use the result of part (b) to calculate the conditional probability $P(Y \leq 1 | X = 2)$.

d. Given that two hoses are in use at the full-service island, what is the conditional pmf of the number in use at the self-service island?

19. The joint pdf of pressures for right and left front tires is given in Exercise 9.

a. Determine the conditional pdf of Y given that $X = x$ and the conditional pdf of X given that $Y = y$.

b. If the pressure in the right tire is found to be 22 psi, what is the probability that the left tire has a pressure of at least 25 psi? Compare this to $P(Y \geq 25)$.

c. If the pressure in the right tire is found to be 22 psi, what is the expected pressure in the left tire, and what is the standard deviation of pressure in this tire?

20. Let X_1, X_2, and X_3 be the lifetimes of components 1, 2, and 3 in a three-component system.

    a. How would you define the conditional pdf of X_3 given that $X_1 = x_1$ and $X_2 = x_2$?

    b. How would you define the conditional joint pdf of X_2 and X_3 given that $X_1 = x_1$?

21. What condition on $f_{Y|X}(y\,|\,x)$ is equivalent to the independence of X and Y?

5.2 | Expected Values, Covariance, and Correlation

We previously saw that any function $h(X)$ of a single rv X is itself a random variable. However, to compute $E[h(X)]$, it was not necessary to obtain the probability distribution of $h(X)$; instead, $E[h(X)]$ was computed as a weighted average of $h(x)$ values, where the weight function was the pmf $p(x)$ or pdf $f(x)$ of X. A similar result holds for a function $h(X, Y)$ of two jointly distributed random variables.

PROPOSITION

Let X and Y be jointly distributed rv's with pmf $p(x, y)$ or pdf $f(x, y)$ according to whether the variables are discrete or continuous. Then the expected value of a function $h(X, Y)$, denoted by $E[h(X, Y)]$ or $\mu_{h(X, Y)}$, is given by

$$E[h(X, Y)] = \begin{cases} \displaystyle\sum_x \sum_y h(x, y) \cdot p(x, y) & \text{if } X \text{ and } Y \text{ are discrete} \\ \displaystyle\int_{-\infty}^{\infty} \int_{-\infty}^{\infty} h(x, y) \cdot f(x, y)dx\, dy & \text{if } X \text{ and } Y \text{ are continuous} \end{cases}$$

Example 5.13

Five friends have purchased tickets to a certain concert. If the tickets are for seats 1–5 in a particular row and the tickets are randomly distributed among the five, what is the expected number of seats separating any particular two of the five? Let X and Y denote the seat numbers of the first and second individuals, respectively. Possible (X, Y) pairs are $\{(1, 2), (1, 3), \ldots, (5, 4)\}$, and the joint pmf of (X, Y) is

$$p(x, y) = \begin{cases} \dfrac{1}{20} & x = 1, \ldots, 5; y = 1, \ldots, 5; x \neq y \\ 0 & \text{otherwise} \end{cases}$$

The number of seats separating the two individuals is $h(X, Y) = |X - Y| - 1$. The accompanying table gives $h(x, y)$ for each possible (x, y) pair.

$h(x, y)$		x 1	2	3	4	5
	1	—	0	1	2	3
	2	0	—	0	1	2
y	3	1	0	—	0	1
	4	2	1	0	—	0
	5	3	2	1	0	—

Thus,

$$E[h(X, Y)] = \sum\sum_{(x, y)} h(x, y) \cdot p(x, y) = \sum_{x=1}^{5} \sum_{\substack{y=1 \\ x \ne y}}^{5} (|x - y| - 1) \cdot \frac{1}{20} = 1 \qquad \blacksquare$$

Example 5.14 In Example 5.5, the joint pdf of the amount X of almonds and amount Y of cashews in a 1-lb can of nuts was

$$f(x, y) = \begin{cases} 24xy & 0 \le x \le 1, 0 \le y \le 1, x + y \le 1 \\ 0 & \text{otherwise} \end{cases}$$

If 1 lb of almonds costs the company $1.00, 1 lb of cashews costs $1.50, and 1 lb of peanuts costs $.50, then the total cost of the contents of a can is

$$h(X, Y) = (1)X + (1.5)Y + (.5)(1 - X - Y) = .5 + .5X + Y$$

(since $1 - X - Y$ of the weight consists of peanuts). The expected total cost is

$$E[h(X, Y)] = \int_{-\infty}^{\infty}\int_{-\infty}^{\infty} h(x, y) \cdot f(x, y)dx\, dy$$

$$= \int_{0}^{1}\int_{0}^{1-x} (.5 + .5x + y) \cdot 24xy\, dy\, dx = \$1.10 \qquad \blacksquare$$

The method of computing the expected value of a function $h(X_1, \ldots, X_n)$ of n random variables is similar to that for two random variables. If the X_i's are discrete, $E[h(X_1, \ldots, X_n)]$ is an n-dimensional sum; if the X_i's are continuous, it is an n-dimensional integral.

Covariance

When two random variables X and Y are not independent, it is frequently of interest to assess how strongly they are related to one another.

DEFINITION

The **covariance** between two rv's X and Y is

$$\text{Cov}(X, Y) = E[(X - \mu_X)(Y - \mu_Y)]$$

$$= \begin{cases} \sum_x \sum_y (x - \mu_X)(y - \mu_Y)p(x, y) & X, Y \text{ discrete} \\ \int_{-\infty}^{\infty}\int_{-\infty}^{\infty} (x - \mu_X)(y - \mu_Y)f(x, y)dx\, dy & X, Y \text{ continuous} \end{cases}$$

The rationale for the definition is as follows. Suppose X and Y have a strong positive relationship to one another, by which we mean that large values of X tend to occur with large values of Y and small values of X with small values of Y. Then most of the probability mass or density will be associated with $(x - \mu_X)$ and $(y - \mu_Y)$ either both positive (both X and Y above their respective means) or both negative, so the product $(x - \mu_X)(y - \mu_Y)$ will tend to be positive. Thus, for a strong positive relationship, $\text{Cov}(X, Y)$ should be quite positive. For a strong negative relationship, the signs of $(x - \mu_X)$ and $(y - \mu_Y)$ will tend to be opposite, yielding a negative product. Thus, for a strong negative relationship, Cov(X, Y) should be quite negative. If X and Y are not strongly related, positive and negative products will tend to cancel one another, yielding a covariance near 0. Figure 5.4 illustrates the different possibilities. The covariance depends on *both* the set of possible pairs and the probabilities. In Figure 5.4, the probabilities could be changed without altering the set of possible pairs, and this could drastically change the value of Cov(X, Y).

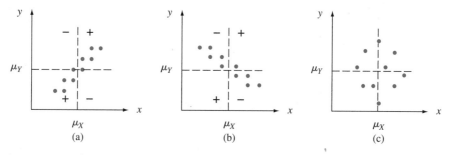

Figure 5.4 $p(x, y) = 1/10$ for each of ten pairs corresponding to indicated points; (a) positive covariance; (b) negative covariance; (c) covariance near zero

Example 5.15 The joint and marginal pmf's for $X =$ automobile policy deductible amount and $Y =$ homeowner policy deductible amount in Example 5.1 were

$p(x, y)$		y 0	100	200
x	100	.20	.10	.20
	250	.05	.15	.30

x	100	250
$p_X(x)$	.5	.5

y	0	100	200
$p_Y(y)$	.25	.25	.5

from which $\mu_X = \Sigma x p_X(x) = 175$ and $\mu_Y = 125$. Therefore,

$$\text{Cov}(X, Y) = \sum_{(x, y)} \sum (x - 175)(y - 125)p(x, y)$$

$$= (100 - 175)(0 - 125)(.20) + \cdots$$
$$+ (250 - 175)(200 - 125)(.30)$$

$$= 1875 \qquad \blacksquare$$

The following shortcut formula for Cov(X, Y) simplifies the computations.

PROPOSITION

$$\text{Cov}(X, Y) = E(XY) - \mu_X \cdot \mu_Y$$

According to this formula, no intermediate subtractions are necessary; only at the end of the computation is $\mu_X \cdot \mu_Y$ subtracted from $E(XY)$. The proof involves expanding $(X - \mu_X)(Y - \mu_Y)$ and then taking the expected value of each term separately. Note that $\text{Cov}(X, X) = E(X^2) - \mu_X^2 = V(X)$.

Example 5.16
(Example 5.5
continued)

The joint and marginal pdf's of X = amount of almonds and Y = amount of cashews were

$$f(x, y) = \begin{cases} 24xy & 0 \le x \le 1, 0 \le y \le 1, x + y \le 1 \\ 0 & \text{otherwise} \end{cases}$$

$$f_X(x) = \begin{cases} 12x(1 - x)^2 & 0 \le x \le 1 \\ 0 & \text{otherwise} \end{cases}$$

with $f_Y(y)$ obtained by replacing x by y in $f_X(x)$. It is easily verified that $\mu_X = \mu_Y = \frac{2}{5}$, and

$$E(XY) = \int_{-\infty}^{\infty}\int_{-\infty}^{\infty} xy\, f(x, y)dx\, dy = \int_0^1\int_0^{1-x} xy \cdot 24xy\, dy\, dx$$

$$= 8\int_0^1 x^2(1 - x)^3 dx = \frac{2}{15}$$

Thus $\text{Cov}(X, Y) = \frac{2}{15} - \left(\frac{2}{5}\right)\left(\frac{2}{5}\right) = \frac{2}{15} - \frac{4}{25} = -\frac{2}{75}$. A negative covariance is reasonable here because more almonds in the can implies fewer cashews. ∎

It would appear that the relationship in the insurance example is quite strong since $\text{Cov}(X, Y) = 1875$, whereas $\text{Cov}(X, Y) = -\frac{2}{75}$ in the nut example would seem to imply quite a weak relationship. Unfortunately, the covariance has a serious defect that makes it impossible to interpret a computed value of the covariance. In the insurance example, suppose we had expressed deductible amount in cents rather than in dollars. Then $100X$ would replace X, $100Y$ would replace Y, and the resulting covariance would be $\text{Cov}(100X, 100Y) = (100)(100)\text{Cov}(X, Y) = 18,750,000$. If, on the other hand, the deductible amount had been expressed in hundreds of dollars, the computed covariance would have been $(.01)(.01)(1875) = .1875$. *The defect of covariance is that its computed value depends critically on the units of measurement.* Ideally the choice of units should have no effect on a measure of strength of relationship. This is achieved by scaling the covariance.

Correlation

DEFINITION

The **correlation coefficient** of X and Y, denoted by $\text{Corr}(X, Y)$, $\rho_{X,Y}$, or just ρ, is defined by

$$\rho_{X,Y} = \frac{\text{Cov}(X, Y)}{\sigma_X \cdot \sigma_Y}$$

Example 5.17
It is easily verified that in the insurance problem of Example 5.15, $E(X^2) = 36,250$, $\sigma_X^2 = 36,250 - (175)^2 = 5625$, $\sigma_X = 75$, $E(Y^2) = 22,500$, $\sigma_Y^2 = 6875$, and $\sigma_Y = 82.92$. This gives

$$\rho = \frac{1875}{(75)(82.92)} = .301$$

∎

The following proposition shows that ρ remedies the defect of $\text{Cov}(X, Y)$ and also suggests how to recognize the existence of a strong (linear) relationship.

PROPOSITION

> **1.** If a and c are either both positive or both negative
>
> $$\text{Corr}(aX + b, cY + d) = \text{Corr}(X, Y)$$
>
> **2.** For any two rv's X and Y, $-1 \le \text{Corr}(X, Y) \le 1$.

Statement 1 says precisely that the correlation coefficient is not affected by a linear change in the units of measurement (if, say, X = temperature in °C, then $9X/5 + 32$ = temperature in °F). According to Statement 2, the strongest possible positive relationship is evidenced by $\rho = +1$, whereas the strongest possible negative relationship corresponds to $\rho = -1$. The proof of the first statement is sketched in Exercise 35, and that of the second appears in Supplementary Exercise 87 at the end of the chapter. For descriptive purposes, the relationship will be described as strong if $|\rho| \ge .8$, moderate if $.5 < |\rho| < .8$, and weak if $|\rho| \le .5$.

If we think of $p(x, y)$ or $f(x, y)$ as prescribing a mathematical model for how the two numerical variables X and Y are distributed in some population (height and weight, verbal SAT score and quantitative SAT score, etc.), then ρ is a population characteristic or parameter that measures how strongly X and Y are related in the population. In Chapter 12, we will consider taking a sample of pairs $(x_1, y_1), \ldots, (x_n, y_n)$ from the population. The sample correlation coefficient r will then be defined and used to make inferences about ρ.

The correlation coefficient ρ is actually not a completely general measure of the strength of a relationship.

PROPOSITION

> **1.** If X and Y are independent, then $\rho = 0$, but $\rho = 0$ does not imply independence.
>
> **2.** $\rho = 1$ or -1 iff $Y = aX + b$ for some numbers a and b with $a \ne 0$.

This proposition says that ρ is a measure of the degree of **linear** relationship between X and Y, and only when the two variables are perfectly related in a linear manner will ρ be as positive or negative as it can be. A ρ less than 1 in absolute value indicates only that the relationship is not completely linear, but there may still be a very strong nonlinear relation. Also, $\rho = 0$ does not imply that X and Y are independent, but only that there is complete absence of a linear relationship. When $\rho = 0$, X and Y are said to be **uncorrelated.** Two variables could be uncorrelated yet highly dependent because there is a strong nonlinear relationship, so be careful not to conclude too much from knowing that $\rho = 0$.

Example 5.18 Let X and Y be discrete rv's with joint pmf

$$p(x, y) = \begin{cases} \dfrac{1}{4} & (x, y) = (-4, 1), (4, -1), (2, 2), (-2, -2) \\ 0 & \text{otherwise} \end{cases}$$

The points that receive positive probability mass are identified on the (x, y) coordinate system in Figure 5.5. It is evident from the figure that the value of X is completely determined by the value of Y and vice versa, so the two variables are completely dependent. However, by symmetry $\mu_X = \mu_Y = 0$ and $E(XY) = (-4)\frac{1}{4} + (-4)\frac{1}{4} + (4)\frac{1}{4} + (4)\frac{1}{4} = 0$, so $\text{Cov}(X, Y) = E(XY) - \mu_X \cdot \mu_Y = 0$ and thus $\rho_{X,Y} = 0$. Although there is perfect dependence, there is also complete absence of any linear relationship!

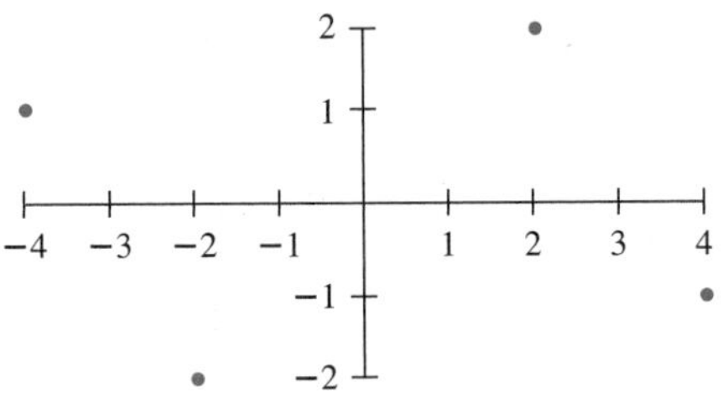

Figure 5.5 The population of pairs for Example 5.18 ■

A value of ρ near 1 does not necessarily imply that increasing the value of X *causes* Y to increase. It implies only that large X values are *associated* with large Y values. For example, in the population of children, vocabulary size and number of cavities are quite positively correlated, but it is certainly not true that cavities cause vocabulary to grow. Instead, the values of both these variables tend to increase as the value of age, a third variable, increases. For children of a fixed age, there is probably a very low correlation between number of cavities and vocabulary size. In summary, association (a high correlation) is not the same as causation.

Exercises | Section 5.2 (22–36)

22. An instructor has given a short quiz consisting of two parts. For a randomly selected student, let X = the number of points earned on the first part and Y = the number of points earned on the second part. Suppose that the joint pmf of X and Y is given in the accompanying table.

$p(x, y)$		y: 0	5	10	15
x	0	.02	.06	.02	.10
	5	.04	.15	.20	.10
	10	.01	.15	.14	.01

a. If the score recorded in the grade book is the total number of points earned on the two parts, what is the expected recorded score $E(X + Y)$?

b. If the maximum of the two scores is recorded, what is the expected recorded score?

23. The difference between the number of customers in line at the express checkout and the number in line at the superexpress checkout in Exercise 3 is $X_1 - X_2$. Calculate the expected difference.

24. Six individuals, including A and B, take seats around a circular table in a completely random fashion. Suppose the seats are numbered 1, . . . , 6. Let X = A's seat number and Y = B's seat number. If A sends a written message around the table to B in the direction in which they are closest, how many individuals (including A and B) would you expect to handle the message?

25. A surveyor wishes to lay out a square region with each side having length L. However, because of measurement error, he instead lays out a rectangle in which the north–south sides both have length X and the east–west sides both have length Y. Suppose that X and Y are independent and that each is uniformly distributed on the interval $[L - A, L + A]$ (where $0 < A < L$). What is the expected area of the resulting rectangle?

26. Consider a small ferry that can accommodate cars and buses. The toll for cars is $3, and the toll for buses is $10. Let X and Y denote the number of cars and buses, respectively, carried on a single trip. Suppose the joint distribution of X and Y is as given in the table of Exercise 7. Compute the expected revenue from a single trip.

27. Annie and Alvie have agreed to meet for lunch between noon (0:00 P.M.) and 1:00 P.M. Denote Annie's arrival time by X, Alvie's by Y, and suppose X and Y are independent with pdf's

$$f_X(x) = \begin{cases} 3x^2 & 0 \le x \le 1 \\ 0 & \text{otherwise} \end{cases}$$

$$f_Y(y) = \begin{cases} 2y & 0 \le y \le 1 \\ 0 & \text{otherwise} \end{cases}$$

What is the expected amount of time that the one who arrives first must wait for the other person? [*Hint:* $h(X, Y) = |X - Y|$.]

28. Show that if X and Y are independent rv's, then $E(XY) = E(X) \cdot E(Y)$. Then apply this in Exercise 25. [*Hint:* Consider the continuous case with $f(x, y) = f_X(x) \cdot f_Y(y)$.]

29. Compute the correlation coefficient ρ for X and Y of Example 5.16 (the covariance has already been computed).

30. **a.** Compute the covariance for X and Y in Exercise 22.
    b. Compute ρ for X and Y in the same exercise.

31. **a.** Compute the covariance between X and Y in Exercise 9.
    b. Compute the correlation coefficient ρ for this X and Y.

32. Reconsider the minicomputer component lifetimes X and Y as described in Exercise 12. Determine $E(XY)$. What can be said about Cov(X, Y) and ρ?

33. Use the result of Exercise 28 to show that when X and Y are independent, Cov(X, Y) = Corr(X, Y) = 0.

34. **a.** Recalling the definition of σ^2 for a single rv X, write a formula that would be appropriate for computing the variance of a function $h(X, Y)$ of two random variables. [*Hint:* Remember that variance is just a special expected value.]
    b. Use this formula to compute the variance of the recorded score $h(X, Y)$ [$= \max(X, Y)$] in part (b) of Exercise 22.

35. **a.** Use the rules of expected value to show that Cov($aX + b, cY + d$) = ac Cov(X, Y).
    b. Use part (a) along with the rules of variance and standard deviation to show that Corr($aX + b, cY + d$) = Corr(X, Y) when a and c have the same sign.
    c. What happens if a and c have opposite signs?

36. Show that if $Y = aX + b$ ($a \ne 0$), then Corr(X, Y) = $+1$ or -1. Under what conditions will $\rho = +1$?

5.3 | **Statistics and Their Distributions**

The observations in a single sample were denoted in Chapter 1 by $x_1, x_2, \ldots, x_n$. Consider selecting two different samples of size n from the same population distribution. The x_i's in the second sample will virtually always differ at least a bit from those in the first sample. For example, a first sample of $n = 3$ cars of a particular type might result in fuel efficiencies $x_1 = 30.7$, $x_2 = 29.4$, $x_3 = 31.1$, whereas a second sample may give $x_1 = 28.8$, $x_2 = 30.0$, and $x_3 = 31.1$. Before we obtain data, there is uncertainty about the value of each x_i. Because of this uncertainty, *before* the data becomes available we view each observation as a random variable and denote the sample by $X_1, X_2, \ldots, X_n$ (uppercase letters for random variables).

This variation in observed values in turn implies that the value of any function of the sample observations, such as the sample mean, sample standard deviation, or sample fourth spread, also varies from sample to sample. That is, prior to obtaining $x_1, \ldots, x_n$, there is uncertainty as to the value of $\bar{x}$, the value of s, and so on.

Example 5.19   Suppose that material strength for a randomly selected specimen of a particular type has a Weibull distribution with parameter values $\alpha = 2$ (shape) and $\beta = 5$ (scale). The corresponding density curve is shown in Figure 5.6. Formulas from Section 4.5 give

$$\mu = E(X) = 4.4311 \qquad \tilde{\mu} = 4.1628 \qquad \sigma^2 = V(X) = 5.365 \qquad \sigma = 2.316$$

The mean exceeds the median because of the distribution's positive skew.

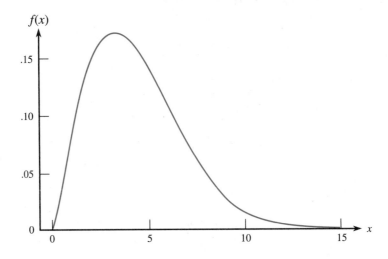

Figure 5.6   The Weibull density curve for Example 5.19

We asked MINITAB to generate six different samples, each with $n = 10$, from this distribution (material strengths for six different groups of ten specimens each). The results appear in Table 5.1, followed by the values of the sample mean, sample median, and sample standard deviation for each sample. Notice first that the ten observations in any particular sample are all different from those in any other sample. Second, the six values of the sample mean are all different from one another, as are the six values of the sample median and the six values of the sample standard deviation. The same is true of the sample 10% trimmed means, sample fourth spreads, and so on.

Table 5.1   Samples from the Weibull distribution of Example 5.19

Sample	1	2	3	4	5	6
1	6.1171	5.07611	3.46710	1.55601	3.12372	8.93795
2	4.1600	6.79279	2.71938	4.56941	6.09685	3.92487
3	3.1950	4.43259	5.88129	4.79870	3.41181	8.76202
4	0.6694	8.55752	5.14915	2.49759	1.65409	7.05569
5	1.8552	6.82487	4.99635	2.33267	2.29512	2.30932
6	5.2316	7.39958	5.86887	4.01295	2.12583	5.94195
7	2.7609	2.14755	6.05918	9.08845	3.20938	6.74166
8	10.2185	8.50628	1.80119	3.25728	3.23209	1.75468
9	5.2438	5.49510	4.21994	3.70132	6.84426	4.91827
10	4.5590	4.04525	2.12934	5.50134	4.20694	7.26081

Sample	1	2	3	4	5	6
$\bar{x}$	4.401	5.928	4.229	4.132	3.620	5.761
$\tilde{x}$	4.360	6.144	4.608	3.857	3.221	6.342
s	2.642	2.062	1.611	2.124	1.678	2.496

Furthermore, the value of the sample mean from any particular sample can be regarded as a *point estimate* ("point" because it is a single number, corresponding to a single point on the number line) of the population mean μ, whose value is known to be 4.4311. None of the estimates from these six samples is identical to what is being estimated. The estimates from the second and sixth samples are much too large, whereas the fifth sample gives a substantial underestimate. Similarly, the sample standard deviation gives a point estimate of the population standard deviation. All six of the resulting estimates are in error by at least a small amount.

In summary, the values of the individual sample observations vary from sample to sample, so in general does the value of any quantity computed from sample data, and the value of a sample characteristic used as an estimate of the corresponding population characteristic will virtually never coincide with what is being estimated. ■

DEFINITION

> A **statistic** is any quantity whose value can be calculated from sample data. Prior to obtaining data, there is uncertainty as to what value of any particular statistic will result. Therefore, a statistic is a random variable and will be denoted by an uppercase letter; a lowercase letter is used to represent the calculated or observed value of the statistic.

Thus, the sample mean, regarded as a statistic (before a sample has been selected or an experiment carried out), is denoted by $\bar{X}$; the calculated value of this statistic is $\bar{x}$. Similarly, S represents the sample standard deviation thought of as a statistic, and its computed value is s. If samples of two different types of bricks are selected and the individual compressive strengths are denoted by $X_1, \ldots, X_m$ and $Y_1, \ldots, Y_n$, respectively, then the statistic $\bar{X} - \bar{Y}$, the difference between the two sample mean compressive strengths, is often of great interest.

Any statistic, being a random variable, has a probability distribution. In particular, the sample mean $\bar{X}$ has a probability distribution. Suppose, for example, that $n = 2$ components are randomly selected and the number of breakdowns while under warranty is determined for each one. Possible values for the sample mean number of breakdowns $\bar{X}$ are 0 (if $X_1 = X_2 = 0$), .5 (if either $X_1 = 0$ and $X_2 = 1$ or $X_1 = 1$ and $X_2 = 0$), 1, 1.5, ... The probability distribution of $\bar{X}$ specifies $P(\bar{X} = 0)$, $P(\bar{X} = .5)$, and so on, from which other probabilities such as $P(1 \leq \bar{X} \leq 3)$ and $P(\bar{X} \geq 2.5)$ can be calculated. Similarly, if for a sample of size $n = 2$, the only possible values of the sample variance are 0, 12.5, and 50 (which is the case if X_1 and X_2 can each take on only the values 40, 45, and 50), then the probability distribution of S^2 gives $P(S^2 = 0)$, $P(S^2 = 12.5)$, and $P(S^2 = 50)$. The probability distribution of a statistic is sometimes referred to as its **sampling distribution** to emphasize that it describes how the statistic varies in value across all samples that might be selected.

Random Samples

The probability distribution of any particular statistic depends not only on the population distribution (normal, uniform, etc.) and the sample size n but also on the method of sampling. Consider selecting a sample of size $n = 2$ from a population consisting of just the three values 1, 5, and 10, and suppose that the statistic of interest is the sample variance. If sampling is done "with replacement," then $S^2 = 0$ will result if $X_1 = X_2$. However, S^2 cannot equal 0 if sampling is "without replacement." So $P(S^2 = 0) = 0$ for one sampling method, and this probability is positive for the other method. Our next definition describes a sampling method often encountered (at least approximately) in practice.

DEFINITION

> The rv's $X_1, X_2, \ldots, X_n$ are said to form a (simple) **random sample** of size n if
>
> **1.** The X_i's are independent rv's.
>
> **2.** Every X_i has the same probability distribution.

Conditions 1 and 2 can be paraphrased by saying that the X_i's are *independent and identically distributed* (iid). If sampling is either with replacement or from an infinite (conceptual) population, Conditions 1 and 2 are satisfied exactly. These conditions will be approximately satisfied if sampling is without replacement, yet the sample size n is much smaller than the population size N. In practice, if $n/N \le .05$ (at most 5% of the population is sampled), we can pretend that the X_i's form a random sample. The virtue of this sampling method is that the probability distribution of any statistic can be more easily obtained than for any other sampling method.

There are two general methods for obtaining information about a statistic's sampling distribution. One method involves calculations based on probability rules, and the other involves carrying out a simulation experiment.

Deriving the Sampling Distribution of a Statistic

Probability rules can be used to obtain the distribution of a statistic provided that it is a "fairly simple" function of the X_i's and either there are relatively few different X values in the population or else the population distribution has a "nice" form. Our next two examples illustrate such situations.

Example 5.20
A large automobile service center charges $40, $45, and $50 for a tune-up of four-, six-, and eight-cylinder cars, respectively. If 20% of its tune-ups are done on four-cylinder cars, 30% on six-cylinder cars, and 50% on eight-cylinder cars, then the probability distribution of revenue from a single randomly selected tune-up is given by

x	40	45	50	
$p(x)$	.2	.3	.5	with $\mu = 46.5$, $\sigma^2 = 15.25$

(5.2)

Suppose on a particular day only two servicing jobs involve tune-ups. Let $X_1 = $ the revenue from the first tune-up and $X_2 = $ the revenue from the second. Suppose that X_1 and X_2 are independent, each with the probability distribution shown in (5.2) [so that X_1 and X_2 constitute a random sample from the distribution (5.2)]. Table 5.2 lists possible (x_1, x_2) pairs, the probability of each (computed using (5.2) and the assumption of independence), and the resulting $\bar{x}$ and s^2 values. Now to obtain the probability distribution of $\bar{X}$, the sample average revenue per tune-up, we must consider each possible value $\bar{x}$ and compute its probability. For example, $\bar{x} = 45$ occurs three times in the table with probabilities .10, .09, and .10, so

$$p_{\bar{X}}(45) = P(\bar{X} = 45) = .10 + .09 + .10 = .29$$

Table 5.2   Outcomes, probabilities, and values of $\bar{x}$ and s^2 for Example 5.20

x_1	x_2	$p(x_1, x_2)$	$\bar{x}$	s^2
40	40	.04	40	0
40	45	.06	42.5	12.5
40	50	.10	45	50
45	40	.06	42.5	12.5
45	45	.09	45	0
45	50	.15	47.5	12.5
50	40	.10	45	50
50	45	.15	47.5	12.5
50	50	.25	50	0

Similarly,

$$p_{S^2}(50) = P(S^2 = 50) = P(X_1 = 40, X_2 = 50 \text{ or } X_1 = 50, X_2 = 40)$$
$$= .10 + .10 = .20$$

The complete sampling distributions of $\bar{X}$ and S^2 appear in (5.3) and (5.4).

$\bar{x}$		40	42.5	45	47.5	50	
$p_{\bar{X}}(\bar{x})$		.04	.12	.29	.30	.25	(5.3)

s^2		0	12.5	50	
$p_{S^2}(s^2)$		.38	.42	.20	(5.4)

Figure 5.7 pictures a probability histogram for both the original distribution (5.2) and the $\bar{X}$ distribution (5.3). The figure suggests first that the mean (expected value) of the $\bar{X}$ distribution is equal to the mean 46.5 of the original distribution, since both histograms appear to be centered at the same place.

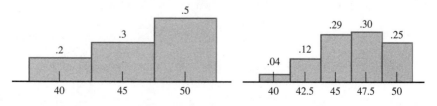

Figure 5.7   Probability histograms for the underlying distribution and $\bar{X}$ distribution in Example 5.20

From (5.3),

$$\mu_{\bar{X}} = E(\bar{X}) = \Sigma \bar{x} p_{\bar{X}}(\bar{x}) = (40)(.04) + \cdots + (50)(.25) = 46.5 = \mu$$

Second, it appears that the $\bar{X}$ distribution has smaller spread (variability) than the original distribution, since probability mass has moved in toward the mean. Again from (5.3),

$$\sigma_{\bar{X}}^2 = V(\bar{X}) = \Sigma \bar{x}^2 \cdot p_{\bar{X}}(\bar{x}) - \mu_{\bar{X}}^2$$
$$= (40)^2(.04) + \cdots + (50)^2(.25) - (46.5)^2$$
$$= 7.625 = \frac{15.25}{2} = \frac{\sigma^2}{2}$$

The variance of $\bar{X}$ is precisely half that of the original variance (because $n = 2$).
The mean value of S^2 is

$$\mu_{S^2} = E(S^2) = \Sigma s^2 \cdot p_{S^2}(s^2)$$
$$= (0)(.38) + (12.5)(.42) + (50)(.20) = 15.25 = \sigma^2$$

That is, the $\bar{X}$ sampling distribution is centered at the population mean μ, and the S^2 sampling distribution is centered at the population variance σ^2.

If four tune-ups had been done on the day of interest, the sample average revenue $\bar{X}$ would be based on a random sample of four X_i's, each having the distribution (5.2). More calculation eventually yields the pmf of $\bar{X}$ for $n = 4$ as

$\bar{x}$	40	41.25	42.5	43.75	45	46.25	47.5	48.75	50
$p_{\bar{X}}(\bar{x})$	.0016	.0096	.0376	.0936	.1761	.2340	.2350	.1500	.0625

From this, for $n = 4$, $\mu_{\bar{X}} = 46.50 = \mu$ and $\sigma_{\bar{X}}^2 = 3.8125 = \sigma^2/4$. Figure 5.8 is a probability histogram of this pmf.

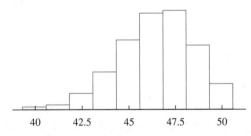

Figure 5.8   Probability histogram for $\bar{X}$ based on $n = 4$ in Example 5.20   ■

Example 5.20 should suggest first of all that the computation of $p_{\bar{X}}(\bar{x})$ and $p_{S^2}(s^2)$ can be tedious. If the original distribution (5.2) had allowed for more than three possible values 40, 45, and 50, then even for $n = 2$ the computations would have been more involved. The example should also suggest, however, that there are some general relationships between $E(\bar{X})$, $V(\bar{X})$, $E(S^2)$, and the mean μ and variance σ^2 of the original distribution. These are stated in the next section. Now consider an example in which the random sample is drawn from a continuous distribution.

Example 5.21   The time that it takes to serve a customer at the cash register in a minimarket is a random variable having an exponential distribution with parameter λ. Suppose X_1 and X_2

are service times for two different customers, assumed independent of each other. Consider the total service time $T_o = X_1 + X_2$ for the two customers, also a statistic. The cdf of T_o is, for $t \geq 0$,

$$F_{T_o}(t) = P(X_1 + X_2 \leq t) = \iint\limits_{\{(x_1, x_2):x_1+x_2\leq t\}} f(x_1, x_2)dx_1\,dx_2$$

$$= \int_0^t \int_0^{t-x_1} \lambda e^{-\lambda x_1} \cdot \lambda e^{-\lambda x_2}\, dx_2\, dx_1 = \int_0^t [\lambda e^{-\lambda x_1} - \lambda e^{-\lambda t}]\, dx_1$$

$$= 1 - e^{-\lambda t} - \lambda t e^{-\lambda t}$$

The region of integration is pictured in Figure 5.9.

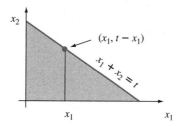

Figure 5.9 Region of integration to obtain cdf of T_o in Example 5.21

The pdf of T_o is obtained by differentiating $F_{T_o}(t)$:

$$f_{T_o}(t) = \begin{cases} \lambda^2 t e^{-\lambda t} & t \geq 0 \\ 0 & t < 0 \end{cases} \tag{5.5}$$

This is a gamma pdf ($\alpha = 2$ and $\beta = 1/\lambda$). The pdf of $\overline{X} = T_o/2$ is obtained from the relation $\{\overline{X} \leq \overline{x}\}$ iff $\{T_o \leq 2\overline{x}\}$ as

$$f_{\overline{X}}(\overline{x}) = \begin{cases} 4\lambda^2 \overline{x} e^{-2\lambda \overline{x}} & \overline{x} \geq 0 \\ 0 & \overline{x} < 0 \end{cases} \tag{5.6}$$

The mean and variance of the underlying exponential distribution are $\mu = 1/\lambda$ and $\sigma^2 = 1/\lambda^2$. From Expressions (5.5) and (5.6), it can be verified that $E(\overline{X}) = 1/\lambda$, $V(\overline{X}) = 1/(2\lambda^2)$, $E(T_o) = 2/\lambda$, and $V(T_o) = 2/\lambda^2$. These results again suggest some general relationships between means and variances of $\overline{X}$, T_o, and the underlying distribution. ∎

Simulation Experiments

The second method of obtaining information about a statistic's sampling distribution is to perform a simulation experiment. This method is usually used when a derivation via probability rules is too difficult or complicated to be carried out. Such an experiment is virtually always done with the aid of a computer. The following characteristics of an experiment must be specified:

1. The statistic of interest ($\overline{X}$, S, a particular trimmed mean, etc.)

2. The population distribution (normal with $\mu = 100$ and $\sigma = 15$, uniform with lower limit $A = 5$ and upper limit $B = 10$, etc.)

3. The sample size n (e.g., $n = 10$ or $n = 50$)

4. The number of replications k (e.g., $k = 500$)

Then use a computer to obtain k different random samples, each of size n, from the designated population distribution. For each such sample, calculate the value of the statistic and construct a histogram of the k calculated values. This histogram gives the *approximate* sampling distribution of the statistic. The larger the value of k, the better the approximation will tend to be (the actual sampling distribution emerges as $k \to \infty$). In practice, $k = 500$ or 1000 is usually enough if the statistic is "fairly simple."

Example 5.22   The population distribution for our first simulation study is normal with $\mu = 8.25$ and $\sigma = .75$, as pictured in Figure 5.10. [The article "Platelet Size in Myocardial Infarction" (*British Med. J.,* 1983: 449–451) suggests this distribution for platelet volume in individuals with no history of serious heart problems.]

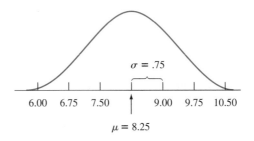

$\sigma = .75$

6.00   6.75   7.50   9.00   9.75   10.50

$\mu = 8.25$

Figure 5.10   Normal distribution, with $\mu = 8.25$ and $\sigma = .75$

We actually performed four different experiments, with 500 replications for each one. In the first experiment, 500 samples of $n = 5$ observations each were generated using MINITAB, and the sample sizes for the other three were $n = 10$, $n = 20$, and $n = 30$, respectively. The sample mean was calculated for each sample, and the resulting histograms of $\bar{x}$ values appear in Figure 5.11.

The first thing to notice about the histograms is their shape. To a reasonable approximation, each of the four looks like a normal curve. The resemblance would be even more striking if each histogram had been based on many more than 500 $\bar{x}$ values. Second, each histogram is centered approximately at 8.25, the mean of the population being sampled. Had the histograms been based on an unending sequence of $\bar{x}$ values, their centers would have been exactly the population mean, 8.25.

The final aspect of the histograms to note is their spread relative to one another. The smaller the value of n, the greater the extent to which the sampling distribution spreads out about the mean value. This is why the histograms for $n = 20$ and $n = 30$ are based on narrower class intervals than those for the two smaller sample sizes. For the larger sample sizes, most of the $\bar{x}$ values are quite close to 8.25. This is the effect of averaging. When n is small, a single unusual x value can result in an $\bar{x}$ value far from the center. With a larger sample size, any unusual x values, when averaged in with the other sample values, still tend to yield an $\bar{x}$ value close to μ. Combining these insights yields a result that should appeal to your intuition: $\overline{X}$ **based on a large n tends to be closer to μ than does $\overline{X}$ based on a small n.**

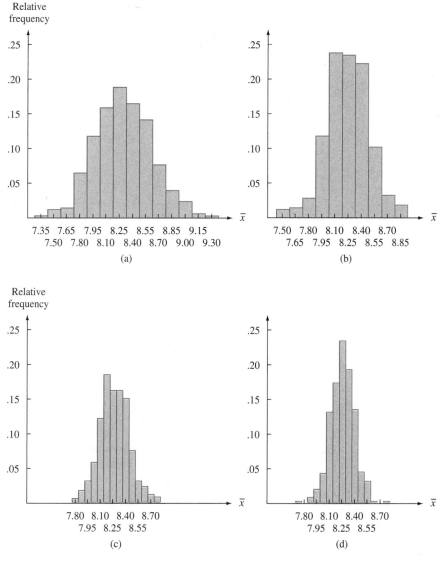

Figure 5.11 Sample histograms for $\bar{x}$ based on 500 samples, each consisting of n observations: (a) $n = 5$; (b) $n = 10$; (c) $n = 20$; (d) $n = 30$

Example 5.23 Consider a simulation experiment in which the population distribution is quite skewed. Figure 5.12 (page 230) shows the density curve for lifetimes of a certain type of electronic control (this is actually a lognormal distribution with $E(\ln(X)) = 3$ and $V(\ln(X)) = .4$). Again the statistic of interest is the sample mean $\bar{X}$. The experiment utilized 500 replications and considered the same four sample sizes as in Example 5.22. The resulting histograms along with a normal probability plot from MINITAB for the 500 $\bar{x}$ values based on $n = 30$ are shown in Figure 5.13 (page 231).

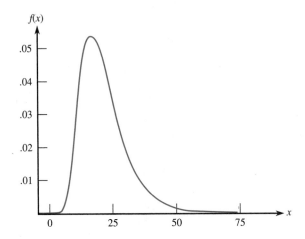

Figure 5.12   Density curve for the simulation experiment of Example 5.23
$[E(X) = \mu = 21.7584, V(X) = \sigma^2 = 82.1449]$

Unlike the normal case, these histograms all differ in shape. In particular, they become progressively less skewed as the sample size n increases. The average of the 500 $\bar{x}$ values for the four different sample sizes are all quite close to the mean value of the population distribution. If each histogram had been based on an unending sequence of $\bar{x}$ values rather than just 500, all four would have been centered at exactly 21.7584. Thus different values of n change the shape but not the center of the sampling distribution of $\bar{X}$. Comparison of the four histograms in Figure 5.13 also shows that as n increases, the spread of the histograms decreases. Increasing n results in a greater degree of concentration about the population mean value and makes the histogram look more like a normal curve. The histogram of Figure 5.13(d) and the normal probability plot in Figure 5.13(e) provide convincing evidence that a sample size of $n = 30$ is sufficient to overcome the skewness of the population distribution and give an approximately normal $\bar{X}$ sampling distribution. ∎

Exercises | Section 5.3 (37–45)

37. A particular brand of dishwasher soap is sold in three sizes: 25 oz, 40 oz, and 65 oz. Twenty percent of all purchasers select a 25 oz box, fifty percent select a 40 oz box, and the remaining thirty percent choose a 65 oz box. Let X_1 and X_2 denote the package sizes selected by two independently selected purchasers.
  a. Determine the sampling distribution of $\bar{X}$, calculate $E(\bar{X})$, and compare to μ.
  b. Determine the sampling distribution of the sample variance S^2, calculate $E(S^2)$, and compare to σ^2.

38. There are two traffic lights on my way to work. Let X_1 be the number of lights at which I must stop and suppose that the distribution of X_1 is as follows:

x_1	0	1	2	$\mu = 1.1, \sigma^2 = .49$
$p(x_1)$	.2	.5	.3	

Let X_2 be the number of lights at which I must stop on the way home; X_2 is independent of X_1. Assume that X_2 has the same distribution as X_1, so that X_1, X_2 is a random sample of size $n = 2$.

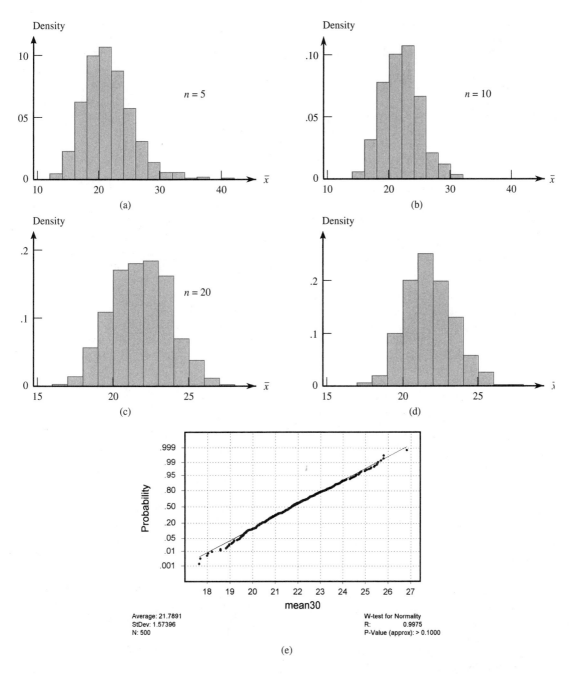

Figure 5.13 Results of the simulation experiment of Example 5.23: (a) $\bar{x}$ histogram for $n = 5$; (b) $\bar{x}$ histogram for $n = 10$; (c) $\bar{x}$ histogram for $n = 20$; (d) $\bar{x}$ histogram for $n = 30$; (e) normal probability plot for $n = 30$ (from MINITAB)

a. Let $T_o = X_1 + X_2$ and determine the probability distribution of T_o.

b. Calculate μ_{T_o}. How does it relate to μ, the population mean?

c. Calculate $\sigma_{T_o}^2$. How does it relate to σ^2, the population variance?

39. It is known that 80% of all brand A zip drives work in a satisfactory manner throughout the warranty period (are "successes"). Suppose that $n = 10$ drives are randomly selected. Let X = the number of successes in the sample. The statistic X/n is the sample proportion (fraction) of successes. Obtain the sampling distribution of this statistic. [*Hint:* One possible value of X/n is .3, corresponding to $X = 3$. What is the probability of this value (what kind of random variable is X)?]

40. A box contains ten sealed envelopes numbered $1, \ldots, 10$. The first five contain no money, the next three each contain $5, and there is a $10 bill in each of the last two. A sample of size 3 is selected *with* replacement (so we have a random sample), and you get the largest amount in any of the envelopes selected. If X_1, X_2, and X_3 denote the amounts in the selected envelopes, the statistic of interest is M = the maximum of X_1, X_2, and X_3.

a. Obtain the probability distribution of this statistic.

b. Describe how you would carry out a simulation experiment to compare the distributions of M for various sample sizes. How would you guess the distribution would change as n increases?

41. Let X be the number of packages being mailed by a randomly selected customer at a certain shipping facility. Suppose the distribution of X is as follows:

x	1	2	3	4
$p(x)$	.4	.3	.2	.1

a. Consider a random sample of size $n = 2$ (two customers), and let $\overline{X}$ be the sample mean number of packages shipped. Obtain the probability distribution of $\overline{X}$.

b. Refer to part (a) and calculate $P(\overline{X} \leq 2.5)$.

c. Again consider a random sample of size $n = 2$, but now focus on the statistic R = the sample range (difference between the largest and smallest values in the sample). Obtain the distribution of R. [*Hint:* Calculate the value of R for each outcome and use the probabilities from part (a).]

d. If a random sample of size $n = 4$ is selected, what is $P(\overline{X} \leq 1.5)$? (*Hint:* You should not have to list all possible outcomes, only those for which $\overline{x} \leq 1.5$.)

42. A company maintains three offices in a certain region, each staffed by two employees. Information concerning yearly salaries (1000's of dollars) is as follows:

Office	1	1	2	2	3	3
Employee	1	2	3	4	5	6
Salary	19.7	23.6	20.2	23.6	15.8	19.7

a. Suppose two of these employees are randomly selected from among the six (without replacement). Determine the sampling distribution of the sample mean salary $\overline{X}$.

b. Suppose one of the three offices is randomly selected. Let X_1 and X_2 denote the salaries of the two employees. Determine the sampling distribution of $\overline{X}$.

c. How does $E(\overline{X})$ from parts (a) and (b) compare to the population mean salary μ?

43. Suppose the amount of liquid dispensed by a certain machine is uniformly distributed with lower limit $A = 8$ oz and upper limit $B = 10$ oz. Describe how you would carry out simulation experiments to compare the sampling distribution of the (sample) fourth spread for sample sizes $n = 5, 10, 20$, and 30.

44. Carry out a simulation experiment using a statistical computer package or other software to study the sampling distribution of $\overline{X}$ when the population distribution is Weibull with $\alpha = 2$ and $\beta = 5$ as in Example 5.19. Consider the four sample sizes $n = 5, 10, 20$, and 30, and in each case use 500 replications. For which of these sample sizes does the $\overline{X}$ sampling distribution appear to be approximately normal?

45. Carry out a simulation experiment using a statistical computer package or other software to study the sampling distribution of $\overline{X}$ when the population distribution is lognormal with $E(\ln(X)) = 3$ and $V(\ln(X)) = 1$. Consider the four sample sizes $n = 10, 20, 30$, and 50, and in each case use 500 replications. For which of these sample sizes does the $\overline{X}$ sampling distribution appear to be approximately normal?

5.4 | **The Distribution of the Sample Mean**

The importance of the sample mean $\overline{X}$ springs from its use in drawing conclusions about the population mean μ. Some of the most frequently used inferential procedures are based on properties of the sampling distribution of $\overline{X}$. A preview of these properties appeared in the calculations and simulation experiments of the previous section, where we noted relationships between $E(\overline{X})$ and μ and also among $V(\overline{X})$, σ^2, and n.

PROPOSITION

Let $X_1, X_2, \ldots, X_n$ be a random sample from a distribution with mean value μ and standard deviation σ. Then

1. $E(\overline{X}) = \mu_{\overline{x}} = \mu$

2. $V(\overline{X}) = \sigma_{\overline{x}}^2 = \sigma^2/n$ and $\sigma_{\overline{x}} = \sigma/\sqrt{n}$

In addition, with $T_o = X_1 + \cdots + X_n$ (the sample total), $E(T_o) = n\mu$, $V(T_o) = n\sigma^2$, and $\sigma_{T_o} = \sqrt{n}\sigma$.

Proofs of these results are given in the next section. According to Result 1, the sampling (i.e., probability) distribution of $\overline{X}$ is centered precisely at the mean of the population from which the sample has been selected. Result 2 shows that the $\overline{X}$ distribution becomes more concentrated about μ as the sample size n increases. In marked contrast, the distribution of T_o becomes more spread out as n increases. Averaging moves probability in toward the middle, whereas totaling spreads probability out over a wider and wider range of values.

Example 5.24   In a notched tensile fatigue test on a titanium specimen, the expected number of cycles to first acoustic emission (used to indicate crack initiation) is $\mu = 28{,}000$, and the standard deviation of the number of cycles is $\sigma = 5000$. Let $X_1, X_2, \ldots, X_{25}$ be a random sample of size 25, where each X_i is the number of cycles on a different randomly selected specimen. Then the expected value of the sample mean number of cycles until first emission is $E(\overline{X}) = \mu = 28{,}000$, and the expected total number of cycles for the 25 specimens is $E(T_o) = n\mu = 25(28{,}000) = 700{,}000$. The standard deviations of $\overline{X}$ and T_o are

$$\sigma_{\overline{x}} = \sigma/\sqrt{n} = \frac{5000}{\sqrt{25}} = 1000$$

$$\sigma_{T_o} = \sqrt{n}\sigma = \sqrt{25}(5000) = 25{,}000$$

If the sample size increases to $n = 100$, $E(\overline{X})$ is unchanged, but $\sigma_{\overline{x}} = 500$, half of its previous value (the sample size must be quadrupled to halve the standard deviation of $\overline{X}$). ∎

The Case of a Normal Population Distribution

Looking back to the simulation experiment of Example 5.22, we see that when the population distribution is normal, each histogram of $\overline{x}$ values is well approximated by a normal curve. The precise result follows.

PROPOSITION

> Let $X_1, X_2, \ldots, X_n$ be a random sample from a *normal* distribution with mean μ and standard deviation σ. Then for *any n*, $\overline{X}$ is normally distributed (with mean μ and standard deviation $\sigma/\sqrt{n}$), as is T_o (with mean $n\mu$ and standard deviation $\sqrt{n}\sigma$).*

We know everything there is to know about the $\overline{X}$ and T_o distributions when the population distribution is normal. In particular, probabilities such as $P(a \le \overline{X} \le b)$ and $P(c \le T_o \le d)$ can be obtained simply by standardizing. Figure 5.14 illustrates the proposition.

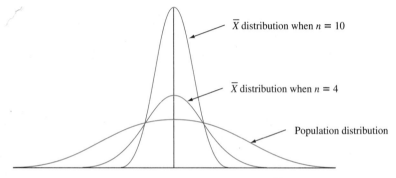

Figure 5.14   A normal population distribution and $\overline{X}$ sampling distributions

Example 5.25 The time that it takes a randomly selected rat of a certain subspecies to find its way through a maze is a normally distributed rv with $\mu = 1.5$ min and $\sigma = .35$ min. Suppose five rats are selected. Let $X_1, \ldots, X_5$ denote their times in the maze. Assuming the X_i's to be a random sample from this normal distribution, what is the probability that the total time $T_o = X_1 + \cdots + X_5$ for the five is between 6 and 8 min? By the proposition, T_o has a normal distribution with $\mu_{T_o} = n\mu = 5(1.5) = 7.5$ and variance $\sigma_{T_o}^2 = n\sigma^2 = 5(.1225) = .6125$, so $\sigma_{T_o} = .783$. To standardize T_o, subtract μ_{T_o} and divide by σ_{T_o}:

$$P(6 \le T_o \le 8) = P\left(\frac{6 - 7.5}{.783} \le Z \le \frac{8 - 7.5}{.783}\right)$$

$$= P(-1.92 \le Z \le .64) = \Phi(.64) - \Phi(-1.92) = .7115$$

Determination of the probability that the sample average time $\overline{X}$ (a normally distributed variable) is at most 2.0 min requires $\mu_{\overline{X}} = \mu = 1.5$ and $\sigma_{\overline{X}} = \sigma/\sqrt{n} = .35/\sqrt{5} = .1565$. Then

$$P(\overline{X} \le 2.0) = P\left(Z \le \frac{2.0 - 1.5}{.1565}\right) = P(Z \le 3.19) = \Phi(3.19) = .9993 \qquad ∎$$

---

*A proof of the result for T_o when $n = 2$ is possible using the method in Example 5.21, but the details are messy. The general result is usually proved using a theoretical tool called a *moment generating function*. One of the chapter references can be consulted for more information.

The Central Limit Theorem

When the X_i's are normally distributed, so is $\overline{X}$ for every sample size n. The simulation experiment of Example 5.23 suggests that even when the population distribution is highly nonnormal, averaging produces a distribution more bell-shaped than the one being sampled. A reasonable conjecture is that if n is large, a suitable normal curve will approximate the actual distribution of $\overline{X}$. The formal statement of this result is the most important theorem of probability.

THEOREM

The Central Limit Theorem (CLT)

Let $X_1, X_2, \ldots, X_n$ be a random sample from a distribution with mean μ and variance σ^2. Then if n is sufficiently large, $\overline{X}$ has approximately a normal distribution with $\mu_{\overline{X}} = \mu$ and $\sigma_{\overline{X}}^2 = \sigma^2/n$, and T_o also has approximately a normal distribution with $\mu_{T_o} = n\mu$, $\sigma_{T_o}^2 = n\sigma^2$. The larger the value of n, the better the approximation.

Figure 5.15 illustrates the Central Limit Theorem. According to the CLT, when n is large and we wish to calculate a probability such as $P(a \leq \overline{X} \leq b)$, we need only "pretend" that $\overline{X}$ is normal, standardize it, and use the normal table. The resulting answer will be approximately correct. The exact answer could be obtained only by first finding the distribution of $\overline{X}$, so the CLT provides a truly impressive shortcut. The proof of the theorem involves much advanced mathematics.

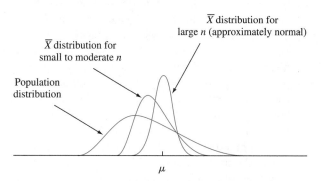

Figure 5.15 The Central Limit Theorem illustrated

Example 5.26   When a batch of a certain chemical product is prepared, the amount of a particular impurity in the batch is a random variable with mean value 4.0 g and standard deviation 1.5 g. If 50 batches are independently prepared, what is the (approximate) probability that the sample average amount of impurity $\overline{X}$ is between 3.5 and 3.8 g? According to the rule of thumb to be stated shortly, $n = 50$ is large enough for the CLT to be applicable. $\overline{X}$ then has approximately a normal distribution with mean value $\mu_{\overline{X}} = 4.0$ and $\sigma_{\overline{X}} = 1.5/\sqrt{50} = .2121$, so

$$P(3.5 \leq \overline{X} \leq 3.8) \approx P\left(\frac{3.5 - 4.0}{.2121} \leq Z \leq \frac{3.8 - 4.0}{.2121} \right)$$
$$= \Phi(-.94) - \Phi(-2.36) = .1645 \qquad \blacksquare$$

Example 5.27   A certain consumer organization customarily reports the number of major defects for each new automobile that it tests. Suppose the number of such defects for a certain model is a random variable with mean value 3.2 and standard deviation 2.4. Among 100 randomly selected cars of this model, how likely is it that the sample average number of major defects exceeds 4? Let X_i denote the number of major defects for the ith car in the random sample. Notice that X_i is a discrete rv, but the CLT is applicable whether the variable of interest is discrete or continuous. Also, although the fact that the standard deviation of this nonnegative variable is quite large relative to the mean value suggests that its distribution is positively skewed, the large sample size implies that $\overline{X}$ does have approximately a normal distribution. Using $\mu_{\overline{X}} = 3.2$ and $\sigma_{\overline{X}} = .24$,

$$P(\overline{X} > 4) \approx P\left(Z > \frac{4 - 3.2}{.24}\right) = 1 - \Phi(3.33) = .0004 \qquad \blacksquare$$

The CLT provides insight into why many random variables have probability distributions that are approximately normal. For example, the measurement error in a scientific experiment can be thought of as the sum of a number of underlying perturbations and errors of small magnitude.

Although the usefulness of the CLT for inference will soon be apparent, the intuitive content of the result gives many beginning students difficulty. Again looking back to Figure 5.7, the probability histogram on the left is a picture of the distribution being sampled. It is discrete and quite skewed, so does not look at all like a normal distribution. The distribution of $\overline{X}$ for $n = 2$ starts to exhibit some symmetry, and this is even more pronounced for $n = 4$ in Figure 5.8. Figure 5.16 contains the probability distribution of $\overline{X}$ for $n = 8$, as well as a probability histogram for this distribution. With $\mu_{\overline{X}} = \mu = 46.5$ and $\sigma_{\overline{X}} = \sigma/\sqrt{n} = 3.905/\sqrt{8} = 1.38$, if we fit a normal curve with this mean and standard deviation through the histogram of $\overline{X}$, the areas of rectangles in the probability histogram are reasonably well approximated by the normal curve areas, at least in the central part of the distribution. The picture for T_o is similar except that the horizontal scale is much more spread out, with T_o ranging from 320 ($\overline{x} = 40$) to 400 ($\overline{x} = 50$).

A practical difficulty in applying the CLT is in knowing when n is sufficiently large. The problem is that the accuracy of the approximation for a particular n depends on the shape of the original underlying distribution being sampled. If the underlying distribution is close to bell-shaped, then the approximation will be good even for a small n, whereas if it is far from bell-shaped then a large n will be required. We will use the following rule of thumb, which is frequently somewhat conservative.

RULE OF THUMB | If $n > 30$, the Central Limit Theorem can be used.

There are population distributions for which even an n of 40 or 50 does not suffice, but such distributions are rarely encountered in practice.

$\bar{x}$	40	40.625	41.25	41.875	42.5	43.125
$p(\bar{x})$	.0000	.0000	.0003	.0012	.0038	.0112
$\bar{x}$	43.75	44.375	45	45.625	46.25	46.875
$p(\bar{x})$	.0274	.0556	.0954	.1378	.1704	.1746
$\bar{x}$	47.5	48.125	48.75	49.375	50	
$p(\bar{x})$	.1474	.0998	.0519	.0188	.0039	

(a)

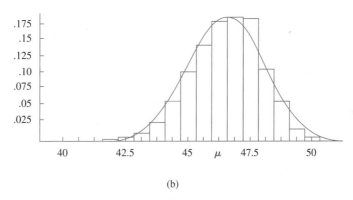

(b)

Figure 5.16 (a) Probability distribution of $\bar{X}$ for $n = 8$; (b) probability histogram and normal approximation to the distribution of $\bar{X}$, when the original distribution is as in Example 5.20

Other Applications of the Central Limit Theorem

The CLT can be used to justify the normal approximation to the binomial distribution discussed in Chapter 4. Recall that a binomial variable X is the number of successes in a binomial experiment consisting of n independent success/failure trials with $p = P(S)$ for any particular trial. Define new rv's $X_1, X_2, \ldots, X_n$ by

$$X_i = \begin{cases} 1 & \text{if the } i\text{th trial results in a success} \\ 0 & \text{if the } i\text{th trial results in a failure} \end{cases} \quad (i = 1, \ldots, n)$$

Because the trials are independent and $P(S)$ is constant from trial to trial, the X_i's are iid (a random sample from a Bernoulli distribution). The CLT then implies that if n is sufficiently large, both the sum and the average of the X_i's have approximately normal distributions. When the X_i's are summed, a 1 is added for every S that occurs and a 0 for every F, so $X_1 + \cdots + X_n = X$. The sample mean of the X_i's is X/n, the sample proportion of successes. That is, both X and X/n are approximately normal when n is large. The necessary sample size for this approximation depends on the value of p: When p is close to .5, the distribution of each X_i is reasonably symmetric (Figure 5.17 on page 238), whereas the distribution is quite skewed when p is near 0 or 1. Using the approximation only if both $np \geq 10$ and $n(1 - p) \geq 10$ ensures that n is large enough to overcome any skewness in the underlying Bernoulli distribution.

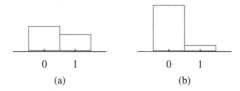

Figure 5.17   Two Bernoulli distributions: (a) $p = .4$ (reasonably symmetric); (b) $p = .1$ (very skewed)

Recall from Section 4.5 that X has a lognormal distribution if $\ln(X)$ has a normal distribution.

PROPOSITION

> Let $X_1, X_2, \ldots, X_n$ be a random sample from a distribution for which only positive values are possible [$P(X_i > 0) = 1$]. Then if n is sufficiently large, the product $Y = X_1 X_2 \cdot \cdots \cdot X_n$ has approximately a lognormal distribution.

To verify this, note that

$$\ln(Y) = \ln(X_1) + \ln(X_2) + \cdots + \ln(X_n)$$

Since $\ln(Y)$ is a sum of independent and identically distributed rv's [the $\ln(X_i)$'s], it is approximately normal when n is large, so Y itself has approximately a lognormal distribution. As an example of the applicability of this result, Bury (*Statistical Models in Applied Science,* Wiley, p. 590) argues that the damage process in plastic flow and crack propagation is a multiplicative process, so that variables such as percentage elongation and rupture strength have approximately lognormal distributions.

Exercises | Section 5.4 (46–57)

46. The inside diameter of a randomly selected piston ring is a random variable with mean value 12 cm and standard deviation .04 cm.
   a. If $\overline{X}$ is the sample mean diameter for a random sample of $n = 16$ rings, where is the sampling distribution of $\overline{X}$ centered, and what is the standard deviation of the $\overline{X}$ distribution?
   b. Answer the questions posed in part (a) for a sample size of $n = 64$ rings.
   c. For which of the two random samples, the one of part (a) or the one of part (b), is $\overline{X}$ more likely to be within .01 cm of 12 cm? Explain your reasoning.

47. Refer to Exercise 46. Suppose the distribution of diameter is normal.

   a. Calculate $P(11.99 \le \overline{X} \le 12.01)$ when $n = 16$.
   b. How likely is it that the sample mean diameter exceeds 12.01 when $n = 25$?

48. Let $X_1, X_2, \ldots, X_{100}$ denote the actual net weights of 100 randomly selected 50-lb bags of fertilizer.
   a. If the expected weight of each bag is 50 and the variance is 1, calculate $P(49.75 \le \overline{X} \le 50.25)$ (approximately) using the CLT.
   b. If the expected weight is 49.8 lb rather than 50 lb so that on average bags are underfilled, calculate $P(49.75 \le \overline{X} \le 50.25)$.

49. There are 40 students in an elementary statistics class. On the basis of years of experience, the instructor knows that the time needed to grade a randomly chosen first examination paper is a random variable with an expected value of 6 min and a standard deviation of 6 min.

a. If grading times are independent and the instructor begins grading at 6:50 P.M. and grades continuously, what is the (approximate) probability that he is through grading before the 11:00 P.M. TV news begins?

b. If the sports report begins at 11:10, what is the probability that he misses part of the report if he waits until grading is done before turning on the TV?

50. The breaking strength of a rivet has a mean value of 10,000 psi and a standard deviation of 500 psi.

a. What is the probability that the sample mean breaking strength for a random sample of 40 rivets is between 9900 and 10,200?

b. If the sample size had been 15 rather than 40, could the probability requested in part (a) be calculated from the given information?

51. The time taken by a randomly selected applicant for a mortgage to fill out a certain form has a normal distribution with mean value 10 min and standard deviation 2 min. If five individuals fill out a form on one day and six on another, what is the probability that the sample average amount of time taken on each day is at most 11 min?

52. The lifetime of a certain type of battery is normally distributed with mean value 10 hours and standard deviation 1 hour. There are four batteries in a package. What lifetime value is such that the total lifetime of all batteries in a package exceeds that value for only 5% of all packages?

53. Rockwell hardness of pins of a certain type is known to have a mean value of 50 and a standard deviation of 1.2.

a. If the distribution is normal, what is the probability that the sample mean hardness for a random sample of 9 pins is at least 51?

b. What is the (approximate) probability that the sample mean hardness for a random sample of 40 pins is at least 51?

54. Suppose the sediment density (g/cm) of a randomly selected specimen from a certain region is normally distributed with mean 2.65 and standard deviation .85 (suggested in "Modeling Sediment and Water Column Interactions for Hydrophobic Pollutants," *Water Research*, 1984: 1169–1174).

a. If a random sample of 25 specimens is selected, what is the probability that the sample average sediment density is at most 3.00? Between 2.65 and 3.00?

b. How large a sample size would be required to ensure that the first probability in part (a) is at least .99?

55. The first assignment in an introductory computer science class involves running a short program. If past experience indicates that 40% of all beginning students will make no programming errors, compute the (approximate) probability that in a class of 50 students

a. At least 25 will make no errors. (*Hint:* Normal approximation to the binomial)

b. Between 15 and 25 (inclusive) will make no errors.

56. The number of parking tickets issued in a certain city on any given weekday has a Poisson distribution with parameter $\lambda = 50$. What is the approximate probability that

a. Between 35 and 70 tickets are given out on a particular day? (*Hint:* When λ is large, a Poisson rv has approximately a normal distribution.)

b. The total number of tickets given out during a 5-day week is between 225 and 275?

57. Suppose the distribution of the time X (in hours) spent by students at a certain university on a particular project is gamma with parameters $\alpha = 50$ and $\beta = 2$. Because α is large, it can be shown that X has approximately a normal distribution. Use this fact to compute the probability that a randomly selected student spends at most 125 hours on the project.

5.5 | **The Distribution of a Linear Combination**

The sample mean $\overline{X}$ and sample total T_o are special cases of a type of random variable that arises very frequently in statistical applications.

DEFINITION

Given a collection of n random variables $X_1, \ldots, X_n$ and n numerical constants $a_1, \ldots, a_n$, the rv

$$Y = a_1 X_1 + \cdots + a_n X_n = \sum_{i=1}^{n} a_i X_i \tag{5.7}$$

is called a **linear combination** of the X_i's.

Taking $a_1 = a_2 = \cdots = a_n = 1$ gives $Y = X_1 + \cdots + X_n = T_o$, and $a_1 = a_2 = \cdots = a_n = \frac{1}{n}$ yields $Y = \frac{1}{n} X_1 + \cdots + \frac{1}{n} X_n = \frac{1}{n}(X_1 + \cdots + X_n) = \frac{1}{n} T_o = \overline{X}$. Notice that we are not requiring the X_i's to be independent or identically distributed. All the X_i's could have different distributions and therefore different mean values and variances. We first consider the expected value and variance of a linear combination.

PROPOSITION

Let $X_1, X_2, \ldots, X_n$ have mean values $\mu_1, \ldots, \mu_n$, respectively, and variances of $\sigma_1^2, \ldots, \sigma_n^2$, respectively.

1. Whether or not the X_i's are independent,

$$E(a_1 X_1 + a_2 X_2 + \cdots + a_n X_n) = a_1 E(X_1) + a_2 E(X_2) + \cdots + a_n E(X_n)$$
$$= a_1 \mu_1 + \cdots + a_n \mu_n \tag{5.8}$$

2. If $X_1, \ldots, X_n$ are independent,

$$V(a_1 X_1 + a_2 X_2 + \cdots + a_n X_n) = a_1^2 V(X_1) + a_2^2 V(X_2) + \cdots + a_n^2 V(X_n)$$
$$= a_1^2 \sigma_1^2 + \cdots + a_n^2 \sigma_n^2 \tag{5.9}$$

and

$$\sigma_{a_1 X_1 + \cdots + a_n X_n} = \sqrt{a_1^2 \sigma_1^2 + \cdots + a_n^2 \sigma_n^2} \tag{5.10}$$

3. For any $X_1, \ldots, X_n$,

$$V(a_1 X_1 + \cdots + a_n X_n) = \sum_{i=1}^{n} \sum_{j=1}^{n} a_i a_j \text{Cov}(X_i, X_j) \tag{5.11}$$

Proofs are sketched out at the end of the section. A paraphrase of (5.8) is that the expected value of a linear combination is the same linear combination of the expected values—for example, $E(2X_1 + 5X_2) = 2\mu_1 + 5\mu_2$. The result (5.9) in Statement 2 is a special case of (5.11) in Statement 3; when the X_i's are independent, $\text{Cov}(X_i, X_j) = 0$ for $i \neq j$ and $= V(X_i)$ for $i = j$ (this simplification actually occurs when the X_i's are uncorrelated, a weaker condition than independence). Specializing to the case of a random sample (X_i's iid) with $a_i = 1/n$ for every i gives $E(\overline{X}) = \mu$ and $V(\overline{X}) = \sigma^2/n$, as discussed in Section 5.4. A similar comment applies to the rules for T_o.

Example 5.28   A gas station sells three grades of gasoline: regular unleaded, extra unleaded, and super unleaded. These are priced at \$1.20, \$1.35, and \$1.50 per gallon, respectively. Let X_1, X_2, and X_3 denote the amounts of these grades purchased (gallons) on a particular day. Suppose the X_i's are independent with $\mu_1 = 1000$, $\mu_2 = 500$, $\mu_3 = 300$, $\sigma_1 = 100$, $\sigma_2 = 80$, and $\sigma_3 = 50$. The revenue from sales is $Y = 1.2X_1 + 1.35X_2 + 1.5X_3$, and

$$E(Y) = 1.2\mu_1 + 1.35\mu_2 + 1.5\mu_3 = \$2325$$
$$V(Y) = (1.2)^2\sigma_1^2 + (1.35)^2\sigma_2^2 + (1.5)^2\sigma_3^2 = 31{,}689$$
$$\sigma_Y = \sqrt{31{,}689} = \$178.01 \qquad \blacksquare$$

The Difference Between Two Random Variables

An important special case of a linear combination results from taking $n = 2$, $a_1 = 1$, and $a_2 = -1$:

$$Y = a_1X_1 + a_2X_2 = X_1 - X_2$$

We then have the following corollary to the proposition.

COROLLARY

> $E(X_1 - X_2) = E(X_1) - E(X_2)$ and, if X_1 and X_2 are independent, $V(X_1 - X_2) = V(X_1) + V(X_2)$.

The expected value of a difference is the difference of the two expected values, but the variance of a difference between two independent variables is the *sum, not* the difference, of the two variances. There is just as much variability in $X_1 - X_2$ as in $X_1 + X_2$ [writing $X_1 - X_2 = X_1 + (-1)X_2$, $(-1)X_2$ has the same amount of variability as X_2 itself].

Example 5.29 A certain automobile manufacturer equips a particular model with either a six-cylinder engine or a four-cylinder engine. Let X_1 and X_2 be fuel efficiencies for independently and randomly selected six-cylinder and four-cylinder cars, respectively. With $\mu_1 = 22$, $\mu_2 = 26$, $\sigma_1 = 1.2$, and $\sigma_2 = 1.5$,

$$E(X_1 - X_2) = \mu_1 - \mu_2 = 22 - 26 = -4$$
$$V(X_1 - X_2) = \sigma_1^2 + \sigma_2^2 = (1.2)^2 + (1.5)^2 = 3.69$$
$$\sigma_{X_1 - X_2} = \sqrt{3.69} = 1.92$$

If we relabel so that X_1 refers to the four-cylinder car, then $E(X_1 - X_2) = 4$, but the variance of the difference is still 3.69. $\blacksquare$

The Case of Normal Random Variables

When the X_i's form a random sample from a normal distribution, $\overline{X}$ and T_o are both normally distributed. Here is a more general result concerning linear combinations.

PROPOSITION

> If $X_1, X_2, \ldots, X_n$ are independent, normally distributed rv's (with possibly different means and/or variances), then any linear combination of the X_i's also has a normal distribution. In particular, the difference $X_1 - X_2$ between two independent, normally distributed variables is itself normally distributed.

Example 5.30
(Example 5.28 continued)

The total revenue from the sale of the three grades of gasoline on a particular day was $Y = 1.2X_1 + 1.35X_2 + 1.5X_3$, and we calculated $\mu_Y = 2325$ and (assuming independence) $\sigma_Y = 178.01$. If the X_i's are normally distributed, the probability that revenue exceeds 2500 is

$$P(Y > 2500) = P\left(Z > \frac{2500 - 2325}{178.01}\right)$$
$$= P(Z > .98) = 1 - \Phi(.98) = .1635 \qquad \blacksquare$$

The CLT can also be generalized so it applies to certain linear combinations. Roughly speaking, if n is large and no individual term is likely to contribute too much to the overall value, then Y has approximately a normal distribution.

Proofs for the case $n = 2$

For the result concerning expected values, suppose that X_1 and X_2 are continuous with joint pdf $f(x_1, x_2)$. Then

$$E(a_1X_1 + a_2X_2) = \int_{-\infty}^{\infty}\int_{-\infty}^{\infty}(a_1x_1 + a_2x_2)f(x_1, x_2)dx_1\,dx_2$$

$$= a_1\int_{-\infty}^{\infty}\int_{-\infty}^{\infty}x_1f(x_1, x_2)dx_2\,dx_1$$

$$+ a_2\int_{-\infty}^{\infty}\int_{-\infty}^{\infty}x_2f(x_1, x_2)dx_1\,dx_2$$

$$= a_1\int_{-\infty}^{\infty}x_1f_{X_1}(x_1)dx_1 + a_2\int_{-\infty}^{\infty}x_2f_{X_2}(x_2)dx_2$$

$$= a_1E(X_1) + a_2E(X_2)$$

Summation replaces integration in the discrete case. The argument for the variance result does not require specifying whether either variable is discrete or continuous. Recalling that $V(Y) = E[(Y - \mu_Y)^2]$,

$$V(a_1X_1 + a_2X_2) = E\{[a_1X_1 + a_2X_2 - (a_1\mu_1 + a_2\mu_2)]^2\}$$
$$= E\{a_1^2(X_1 - \mu_1)^2 + a_2^2(X_2 - \mu_2)^2 + 2a_1a_2(X_1 - \mu_1)(X_2 - \mu_2)\}$$

The expression inside the braces is a linear combination of the variables $Y_1 = (X_1 - \mu_1)^2$, $Y_2 = (X_2 - \mu_2)^2$, and $Y_3 = (X_1 - \mu_1)(X_2 - \mu_2)$, so carrying the E operation through to the three terms gives $a_1^2V(X_1) + a_2^2V(X_2) + 2a_1a_2\text{Cov}(X_1, X_2)$ as required. $\blacksquare$

Exercises | Section 5.5 (58–74)

58. A shipping company handles containers in three different sizes: (1) 27 ft³ ($3 \times 3 \times 3$), (2) 125 ft³, and (3) 512 ft³. Let X_i ($i = 1, 2, 3$) denote the number of type i containers shipped during a given week. With $\mu_i = E(X_i)$ and $\sigma_i^2 = V(X_i)$, suppose that the mean values and standard deviations are as follows:

$$\mu_1 = 200 \qquad \mu_2 = 250 \qquad \mu_3 = 100$$
$$\sigma_1 = 10 \qquad \sigma_2 = 12 \qquad \sigma_3 = 8$$

a. Assuming that X_1, X_2, X_3 are independent, calculate the expected value and variance of the total volume shipped. [*Hint:* Volume $= 27X_1 + 125X_2 + 512X_3$.]

b. Would your calculations necessarily be correct if the X_i's were not independent? Explain.

59. Let $X_1, X_2,$ and X_3 represent the times necessary to perform three successive repair tasks at a certain service facility. Suppose they are independent normal rv's with expected values $\mu_1, \mu_2,$ and μ_3 and variances $\sigma_1^2, \sigma_2^2,$ and σ_3^2, respectively.

a. If $\mu_1 = \mu_2 = \mu_3 = 60$ and $\sigma_1^2 = \sigma_2^2 = \sigma_3^2 = 15$, calculate $P(X_1 + X_2 + X_3 \le 200)$. What is $P(150 \le X_1 + X_2 + X_3 \le 200)$?

b. Using the μ_i's and σ_i's given in part (a), calculate $P(55 \le \bar{X})$ and $P(58 \le \bar{X} \le 62)$.

c. Using the μ_i's and σ_i's given in part (a), calculate $P(-10 \le X_1 - .5X_2 - .5X_3 \le 5)$.

d. If $\mu_1 = 40, \mu_2 = 50, \mu_3 = 60, \sigma_1^2 = 10, \sigma_2^2 = 12,$ and $\sigma_3^2 = 14$, calculate $P(X_1 + X_2 + X_3 \le 160)$ and $P(X_1 + X_2 \ge 2 X_3)$.

60. Five automobiles of the same type are to be driven on a 300-mile trip. The first two will use an economy brand of gasoline, and the other three will use a name brand. Let $X_1, X_2, X_3, X_4,$ and X_5 be the observed numbers of miles per gallon for the five cars. Suppose these variables are independent and normally distributed with $\mu_1 = \mu_2 = 20, \mu_3 = \mu_4 = \mu_5 = 21,$ and $\sigma^2 = 4$ for the economy brand and 3.5 for the name brand. Define an rv Y by

$$Y = \frac{X_1 + X_2}{2} - \frac{X_3 + X_4 + X_5}{3}$$

so that Y is a measure of the difference in efficiency between economy gas and name-brand gas. Compute $P(0 \le Y)$ and $P(-1 \le Y \le 1)$. [*Hint:* $Y = a_1X_1 + \cdots + a_5X_5$, with $a_1 = \frac{1}{2}, \ldots, a_5 = -\frac{1}{3}$.]

61. Exercise 26 introduced random variables X and Y, the number of cars and buses, respectively, carried by a ferry on a single trip. The joint pmf of X and Y is given in the table in Exercise 7. It is readily verified that X and Y are independent.

a. Compute the expected value, variance, and standard deviation of the total number of vehicles on a single trip.

b. If each car is charged $3 and each bus $10, compute the expected value, variance, and standard deviation of the revenue resulting from a single trip.

62. Manufacture of a certain component requires three different machining operations. Machining time for each operation has a normal distribution, and the three times are independent of one another. The mean values are 15, 30, and 20 min, respectively, and the standard deviations are 1, 2, and 1.5 min, respectively. What is the probability that it takes at most 1 hour of machining time to produce a randomly selected component?

63. Refer to Exercise 3.

a. Calculate the covariance between $X_1 =$ the number of customers in the express checkout and $X_2 =$ the number of customers in the super-express checkout.

b. Calculate $V(X_1 + X_2)$. How does this compare to $V(X_1) + V(X_2)$?

64. Suppose your waiting time for a bus in the morning is uniformly distributed on [0, 8], whereas waiting time in the evening is uniformly distributed on [0, 10] independent of morning waiting time.

a. If you take the bus each morning and evening for a week, what is your total expected waiting time? (*Hint:* Define rv's $X_1, \ldots, X_{10}$ and use a rule of expected value.)

b. What is the variance of your total waiting time?

c. What are the expected value and variance of the difference between morning and evening waiting times on a given day?

d. What are the expected value and variance of the difference between total morning waiting time and total evening waiting time for a particular week?

65. Suppose that when the pH of a certain chemical compound is 5.00, the pH measured by a randomly selected beginning chemistry student is a random variable with mean 5.00 and standard deviation .2. A large batch of the compound is subdivided and a sample given to each student in a morning lab and each student in an afternoon lab. Let $\bar{X} =$ the average pH as determined by the morning students and $\bar{Y} =$ the average pH as determined by the afternoon students.

a. If pH is a normal variable and there are 25 students in each lab, compute $P(-.1 \le \bar{X} - \bar{Y} \le .1)$. (*Hint:* $\bar{X} - \bar{Y}$ is a linear combination of normal variables, so is normally distributed. Compute $\mu_{\bar{X} - \bar{Y}}$ and $\sigma_{\bar{X} - \bar{Y}}$.)

b. If there are 36 students in each lab, but pH determinations are not assumed normal, calculate (approximately) $P(-.1 \le \bar{X} - \bar{Y} \le .1)$.

66. If two loads are applied to a cantilever beam as shown in the accompanying drawing, the bending moment at 0 due to the loads is $a_1X_1 + a_2X_2$.

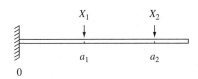

a. Suppose that X_1 and X_2 are independent rv's with means 2 and 4 kips, respectively, and standard deviations .5 and 1.0 kip, respectively. If $a_1 = 5$ ft and $a_2 = 10$ ft, what is the expected bending moment and what is the standard deviation of the bending moment?

b. If X_1 and X_2 are normally distributed, what is the probability that the bending moment will exceed 75 kip-ft?

c. Suppose the positions of the two loads are random variables. Denoting them by A_1 and A_2, assume that these variables have means of 5 and 10 ft, respectively, that each has a standard deviation of .5, and that all A_i's and X_i's are independent of one another. What is the expected moment now?

d. For the situation of part (c), what is the variance of the bending moment?

e. If the situation is as described in part (a) except that $\text{Corr}(X_1, X_2) = .5$ (so that the two loads are not independent), what is the variance of the bending moment?

67. One piece of PVC pipe is to be inserted inside another piece. The length of the first piece is normally distributed with mean value 20 in. and standard deviation .5 in. The length of the second piece is a normal rv with mean and standard deviation 15 in. and .4 in., respectively. The amount of overlap is normally distributed with mean value 1 in. and standard deviation .1 in. Assuming that the lengths and amount of overlap are independent of one another, what is the probability that the total length after insertion is between 34.5 in. and 35 in.?

68. Two airplanes are flying in the same direction in adjacent parallel corridors. At time $t = 0$, the first airplane is 10 km ahead of the second one. Suppose the speed of the first plane (km/hr) is normally distributed with mean 520 and standard deviation 10 and the second plane's speed is also normally distributed with mean and standard deviation 500 and 10, respectively.

a. What is the probability that after 2 hours of flying, the second plane has not caught up to the first plane?

b. Determine the probability that the planes are separated by at most 10 km after 2 hours.

69. Three different roads feed into a particular freeway entrance. Suppose that during a fixed time period, the number of cars coming from each road onto the freeway is a random variable, with expected value and standard deviation as given in the table.

	Road 1	Road 2	Road 3
Expected value	800	1000	600
Standard deviation	16	25	18

a. What is the expected total number of cars entering the freeway at this point during the period? (*Hint:* Let $X_i =$ the number from road i.)

b. What is the variance of the total number of entering cars? Have you made any assumptions about the relationship between the numbers of cars on the different roads?

c. With X_i denoting the number of cars entering from road i during the period, suppose that $\text{Cov}(X_1, X_2) = 80$, $\text{Cov}(X_1, X_3) = 90$, and $\text{Cov}(X_2, X_3) = 100$ (so that the three streams of traffic are not independent). Compute the expected total number of entering cars and the standard deviation of the total.

70. Suppose we take a random sample of size n from a continuous distribution having median 0, so that the probability of any one observation being positive is .5. We now disregard the signs of the observations, rank them from smallest to largest in absolute value, and then let $W =$ the sum of the ranks of the observations having positive signs. For example, if the observations are $-.3$, $+.7$, $+2.1$, and -2.5, then the ranks of positive observations are 2 and 3, so $W = 5$. In Chapter 15, W will be called *Wilcoxon's signed-rank statistic*. W can be represented as follows:

$$W = 1 \cdot Y_1 + 2 \cdot Y_2 + 3 \cdot Y_3 + \cdots + n \cdot Y_n$$

$$= \sum_{i=1}^{n} i \cdot Y_i$$

where the Y_i's are independent Bernoulli rv's, each with $p = .5$ ($Y_i = 1$ corresponds to the observation with rank i being positive). Compute the following:

a. $E(Y_i)$ and then $E(W)$ using the equation for W [*Hint:* The first n positive integers sum to $n(n + 1)/2$.]

b. $V(Y_i)$ and then $V(W)$ [*Hint:* The sum of the squares of the first n positive integers is $n(n + 1)(2n + 1)/6$.]

71. In Exercise 66, the weight of the beam itself contributes to the bending moment. Assume that the beam is of uniform thickness and density, so that the resulting load is uniformly distributed on the beam. If the weight of the beam is random, the resulting load from the weight is also random; denote this load by W (kip-ft).

a. If the beam is 12 ft long, W has mean 1.5 and standard deviation .25, and the fixed loads are as described in part (a) of Exercise 66, what are the expected value and variance of the bending moment? (*Hint:* If the load due to the beam were w kip-ft, the contribution to the bending moment would be $w \int_0^{12} x \, dx$.)

b. If all three variables (X_1, X_2, and W) are normally distributed, what is the probability that the bending moment will be at most 200 kip-ft?

72. I have three errands to take care of in the Administration Building. Let X_i = the time that it takes for the ith errand ($i = 1, 2, 3$) and let X_4 = the total time in minutes that I spend walking to and from the building and between each errand. Suppose the X_i's are independent, normally distributed, with the following means and standard deviations: $\mu_1 = 15$, $\sigma_1 = 4$, $\mu_2 = 5$, $\sigma_2 = 1$, $\mu_3 = 8$, $\sigma_3 = 2$, $\mu_4 = 12$, $\sigma_4 = 3$. I

plan to leave my office at precisely 10:00 A.M. and wish to post a note on my door that reads, "I will return by t A.M." What time t should I write down if I want the probability of my arriving after t to be .01?

73. Suppose the expected tensile strength of type-A steel is 105 ksi and the standard deviation of tensile strength is 8 ksi. For type-B steel, suppose the expected tensile strength and standard deviation of tensile strength are 100 ksi and 6 ksi, respectively. Let $\overline{X}$ = the sample average tensile strength of a random sample of 40 type-A specimens, and let $\overline{Y}$ = the sample average tensile strength of a random sample of 35 type-B specimens.

a. What is the approximate distribution of $\overline{X}$? Of $\overline{Y}$?

b. What is the approximate distribution of $\overline{X} - \overline{Y}$? Justify your answer.

c. Calculate (approximately) $P(-1 \le \overline{X} - \overline{Y} \le 1)$.

d. Calculate $P(\overline{X} - \overline{Y} \ge 10)$. If you actually observed $\overline{X} - \overline{Y} \ge 10$, would you doubt that $\mu_1 - \mu_2 = 5$?

74. In an area having sandy soil, 50 small trees of a certain type were planted, and another 50 trees were planted in an area having clay soil. Let X = the number of trees planted in sandy soil that survive 1 year and Y = the number of trees planted in clay soil that survive 1 year. If the probability that a tree planted in sandy soil will survive 1 year is .7 and the probability of 1-year survival in clay soil is .6, compute an approximation to $P(-5 \le X - Y \le 5)$ (do not bother with the continuity correction).

Supplementary Exercises (75–94)

75. A restaurant serves three fixed-price dinners costing $7, $9, and $10. For a randomly selected couple dining at this restaurant, let X = the cost of the man's dinner and Y = the cost of the woman's dinner. The joint pmf of X and Y is given in the following table:

$p(x, y)$		y		
		7	9	10
	7	.05	.05	.10
x	9	.05	.10	.35
	10	0	.20	.10

a. Compute the marginal pmf's of X and Y.

b. What is the probability that the man's and the woman's dinner cost at most $9 each?

c. Are X and Y independent? Justify your answer.

d. What is the expected total cost of the dinner for the two people?

e. Suppose that when a couple opens fortune cookies at the conclusion of the meal, they find the message "You will receive as a refund the difference between the cost of the more expensive and the less expensive meal that you have chosen." How much does the restaurant expect to refund?

76. In cost estimation, the total cost of a project is the sum of component task costs. Each of these costs is a random variable with a probability distribution. It is customary to obtain information about the total cost distribution by adding together characteristics of the individual component cost distributions—this is called the "roll-up" procedure. For example, $E(X_1 + \cdots + X_n) = E(X_1) + \cdots + E(X_n)$, so the roll-up procedure is valid for mean cost. Suppose that there are two component tasks and that X_1 and X_2 are independent normally distributed random variables. Is the roll-up procedure valid for the 75th percentile? That is, is the 75th percentile of the distribution of $X_1 + X_2$ the same as the sum of the 75th percentiles of the two individual distributions? If not, what is the relationship between the percentile of the sum and the sum of percentiles? For what percentiles is the roll-up procedure valid in this case?

77. A health-food store stocks two different brands of a certain type of grain. Let X = the amount (lb) of brand A on hand, and Y = the amount of brand B on hand. Suppose the joint pdf of X and Y is

$$f(x, y) = \begin{cases} kxy & x \geq 0, y \geq 0, 20 \leq x + y \leq 30 \\ 0 & \text{otherwise} \end{cases}$$

   a. Draw the region of positive density and determine the value of k.

   b. Are X and Y independent? Answer by first deriving the marginal pdf of each variable.

   c. Compute $P(X + Y \leq 25)$.

   d. What is the expected total amount of this grain on hand?

   e. Compute $\text{Cov}(X, Y)$ and $\text{Corr}(X, Y)$.

   f. What is the variance of the total amount of grain on hand?

78. Let $X_1, X_2, \ldots, X_n$ be random variables denoting n independent bids for an item that is for sale. Suppose each X_i is uniformly distributed on the interval $[100, 200]$. If the seller sells to the highest bidder, how much can he expect to earn on the sale? [*Hint:* Let $Y = \max(X_1, X_2, \ldots, X_n)$. First find $F_Y(y)$ by noting that $Y \leq y$ iff each X_i is $\leq y$. Then obtain the pdf and $E(Y)$.]

79. Suppose that for a certain individual, calorie intake at breakfast is a random variable with expected value 500 and standard deviation 50, calorie intake at lunch is random with expected value 900 and standard deviation 100, and calorie intake at dinner

is a random variable with expected value 2000 and standard deviation 180. Assuming that intakes at different meals are independent of one another, what is the probability that average calorie intake per day over the next (365-day) year is at most 3500? [*Hint:* Let X_i, Y_i, and Z_i denote the three calorie intakes on day i. Then total intake is given by $\Sigma(X_i + Y_i + Z_i)$.]

80. The mean weight of luggage checked by a randomly selected tourist-class passenger flying between two cities on a certain airline is 40 lb, and the standard deviation is 10 lb. The mean and standard deviation for a business-class passenger are 30 lb and 6 lb, respectively.

   a. If there are 12 business-class passengers and 50 tourist-class passengers on a particular flight, what are the expected value of total luggage weight and the standard deviation of total luggage weight?

   b. If individual luggage weights are independent normally distributed rv's, what is the probability that total luggage weight is at most 2500 lb?

81. We have seen that if $E(X_1) = E(X_2) = \cdots = E(X_n) = \mu$, then $E(X_1 + \cdots + X_n) = n\mu$. In some applications, the number of X_i's under consideration is not a fixed number n but instead is an rv N. For example, let N = the number of components that are brought into a repair shop on a particular day, and let X_i denote the repair shop time for the ith component. Then the total repair time is $X_1 + X_2 + \cdots + X_N$, the sum of a *random* number of random variables. When N is independent of the X_i's, it can be shown that

$$E(X_1 + \cdots + X_N) = E(N) \cdot \mu$$

   a. If the expected number of components brought in on a particularly day is 10 and expected repair time for a randomly submitted component is 40 min, what is the expected total repair time for components submitted on any particular day?

   b. Suppose components of a certain type come in for repair according to a Poisson process with a rate of 5 per hour. The expected number of defects per component is 3.5. What is the expected value of the total number of defects on components submitted for repair during a 4-hour period? Be sure to indicate how your answer follows from the general result just given.

82. Suppose the proportion of rural voters in a certain state who favor a particular gubernatorial candidate is .45 and the proportion of suburban and urban voters favoring the candidate is .60. If a sample of 200 rural voters and 300 urban and suburban voters is obtained, what is the approximate probability that at least 250 of these voters favor this candidate?

83. Let μ denote the true pH of a chemical compound. A sequence of n independent sample pH determinations will be made. Suppose each sample pH is a random variable with expected value μ and standard deviation .1. How many determinations are required if we wish the probability that the sample average is within .02 of the true pH to be at least .95? What theorem justifies your probability calculation?

84. If the amount of soft drink that I consume on any given day is independent of consumption on any other day and is normally distributed with $\mu = 13$ oz and $\sigma = 2$ and if I currently have two six-packs of 16-oz bottles, what is the probability that I still have some soft drink left at the end of 2 weeks (14 days)?

85. Refer to Exercise 58, and suppose that the X_i's are independent with each one having a normal distribution. What is the probability that the total volume shipped is at most 100,000 ft³?

86. A student has a class that is supposed to end at 9:00 A.M. and another that is supposed to begin at 9:10 A.M. Suppose the actual ending time of the 9 A.M. class is a normally distributed rv X_1 with mean 9:02 and standard deviation 1.5 min and that the starting time of the next class is also a normally distributed rv X_2 with mean 9:10 and standard deviation 1 min. Suppose also that the time necessary to get from one classroom to the other is a normally distributed rv X_3 with mean 6 min and standard deviation 1 min. What is the probability that the student makes it to the second class before the lecture starts? (Assume independence of X_1, X_2, and X_3, which is reasonable if the student pays no attention to the finishing time of the first class.)

87. a. Use the general formula for the variance of a linear combination to write an expression for $V(aX + Y)$. Then let $a = \sigma_Y/\sigma_X$ and show that $\rho \geq -1$. [*Hint:* Variance is always ≥ 0, and $\text{Cov}(X, Y) = \sigma_X \cdot \sigma_Y \cdot \rho$.]
b. By considering $V(aX - Y)$, conclude that $\rho \leq 1$.
c. Use the fact that $V(W) = 0$ only if W is a constant to show that $\rho = 1$ only if $Y = aX + b$.

88. Suppose a randomly chosen individual's verbal score X and quantitative score Y on a nationally administered aptitude examination have joint pdf

$$f(x, y) = \begin{cases} \dfrac{2}{5}(2x + 3y) & 0 \leq x \leq 1, 0 \leq y \leq 1 \\ 0 & \text{otherwise} \end{cases}$$

You are asked to provide a prediction t of the individual's total score $X + Y$. The error of prediction is the mean squared error $E[(X + Y - t)^2]$. What value of t minimizes the error of prediction?

89. a. Let X_1 have a chi-squared distribution with parameter ν_1 (see Section 4.4), and let X_2 be independent of X_1 and have a chi-squared distribution with parameter ν_2. Use the technique of Example 5.21 to show that $X_1 + X_2$ has a chi-squared distribution with parameter $\nu_1 + \nu_2$.
b. In Exercise 65 of Chapter 4, you were asked to show that if Z is a standard normal rv, then Z^2 has a chi-squared distribution with $\nu = 1$. Let $Z_1, Z_2, \ldots, Z_n$ be n independent standard normal rv's. What is the distribution of $Z_1^2 + \cdots + Z_n^2$? Justify your answer.
c. Let $X_1, \ldots, X_n$ be a random sample from a normal distribution with mean μ and variance σ^2. What is the distribution of the sum $Y = \sum_{i=1}^{n} [(X_i - \mu)/\sigma]^2$? Justify your answer.

90. a. Show that $\text{Cov}(X, Y + Z) = \text{Cov}(X, Y) + \text{Cov}(X, Z)$.
b. Let X_1 and X_2 be quantitative and verbal scores on one aptitude exam and let Y_1 and Y_2 be corresponding scores on another exam. If $\text{Cov}(X_1, Y_1) = 5$, $\text{Cov}(X_1, Y_2) = 1$, $\text{Cov}(X_2, Y_1) = 2$, and $\text{Cov}(X_2, Y_2) = 8$, what is the covariance between the two total scores $X_1 + X_2$ and $Y_1 + Y_2$?

91. A rock specimen from a particular area is randomly selected and weighed two different times. Let W denote the actual weight and X_1 and X_2 the two measured weights. Then $X_1 = W + E_1$ and $X_2 = W + E_2$, where E_1 and E_2 are the two measurement errors. Suppose that the E_i's are independent of one another and of W and that $V(E_1) = V(E_2) = \sigma_E^2$.
a. Express ρ, the correlation coefficient between the two measured weights X_1 and X_2, in terms of σ_W^2, the variance of actual weight, and σ_X^2, the variance of measured weight.
b. Compute ρ when $\sigma_W = 1$ kg and $\sigma_E = .01$ kg.

92. Let A denote the percentage of one constituent in a randomly selected rock specimen, and let B denote the percentage of a second constituent in that same specimen. Suppose D and E are measurement errors in determining the values of A and B, so that measured values are $X = A + D$ and $Y = B + E$, respectively. Assume that measurement errors are independent of one another and of actual values.

a. Show that

$$\text{Corr}(X, Y)$$
$$= \text{Corr}(A, B) \cdot \sqrt{\text{Corr}(X_1, X_2)} \cdot \sqrt{\text{Corr}(Y_1, Y_2)}$$

where X_1 and X_2 are replicate measurements on the value of A, and Y_1 and Y_2 are defined analogously with respect to B. What effect does the presence of measurement error have on the correlation?

b. What is the maximum value of $\text{Corr}(X, Y)$ when $\text{Corr}(X_1, X_2) = .8100$ and $\text{Corr}(Y_1, Y_2) = .9025$? Is this disturbing?

93. Let $X_1, \ldots, X_n$ be independent rv's with mean values $\mu_1, \ldots, \mu_n$ and variances $\sigma_1^2, \ldots, \sigma_n^2$. Consider a function $h(x_1, \ldots, x_n)$, and use it to define a new rv $Y = h(X_1, \ldots, X_n)$. Under rather general conditions on the h function, if the σ_i's are all small relative to the corresponding μ_i's, it can be shown that $E(Y) \approx h(\mu_1, \ldots, \mu_n)$ and

$$V(Y) \approx \left(\frac{\partial h}{\partial x_1} \right)^2 \cdot \sigma_1^2 + \cdots + \left(\frac{\partial h}{\partial x_n} \right)^2 \cdot \sigma_n^2$$

where each partial derivative is evaluated at $(x_1, \ldots, x_n) = (\mu_1, \ldots, \mu_n)$. Suppose three resistors with resistances X_1, X_2, X_3 are connected in parallel across a battery with voltage X_4. Then by Ohm's law, the current is

$$Y = X_4 \left[\frac{1}{X_1} + \frac{1}{X_2} + \frac{1}{X_3} \right]$$

Let $\mu_1 = 10$ ohms, $\sigma_1 = 1.0$ ohm, $\mu_2 = 15$ ohms, $\sigma_2 = 1.0$ ohm, $\mu_3 = 20$ ohms, $\sigma_3 = 1.5$ ohms, $\mu_4 = 120$ V, $\sigma_4 = 4.0$ V. Calculate the approximate expected value and standard deviation of the current (suggested by "Random Samplings," *CHEMTECH*, 1984: 696–697).

94. A more accurate approximation to $E[h(X_1, \ldots, X_n)]$ in Exercise 93 is

$$h(\mu_1, \ldots, \mu_n) + \frac{1}{2} \sigma_1^2 \left(\frac{\partial^2 h}{\partial x_1^2} \right) + \cdots + \frac{1}{2} \sigma_n^2 \left(\frac{\partial^2 h}{\partial x_n^2} \right)$$

Compute this for $Y = h(X_1, X_2, X_3, X_4)$ given in Exercise 93, and compare it to the leading term $h(\mu_1, \ldots, \mu_n)$.

Bibliography

Larsen, Richard, and Morris Marx, *Introduction to Mathematical Statistics* (2nd ed.), Prentice Hall, Englewood Cliffs, NJ, 1986. More limited coverage than in the book by Olkin et al., but well written and readable.

Olkin, Ingram, Cyrus Derman, and Leon Gleser, *Probability Models and Applications* (2nd ed.), Macmillan, New York, 1994. Contains a careful and comprehensive exposition of joint distributions, rules of expectation, and limit theorems.

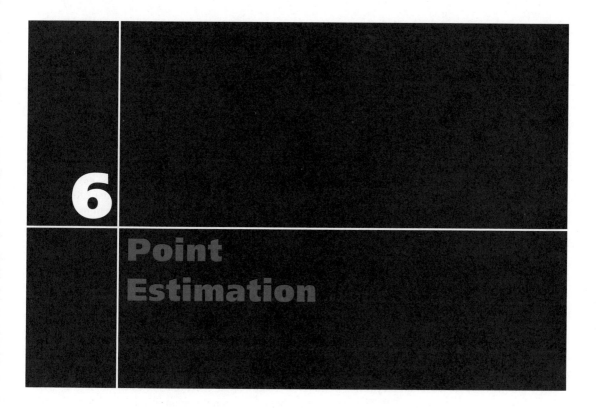

6

Point Estimation

Introduction

Given a parameter of interest, such as a population mean μ or population proportion p, the objective of point estimation is to use a sample to compute a number that represents in some sense a good guess for the true value of the parameter. The resulting number is called a *point estimate*. In Section 6.1, we present some general concepts of point estimation. In Section 6.2, we describe and illustrate two important methods for obtaining point estimates: the method of moments and the method of maximum likelihood.

6.1 | Some General Concepts of Point Estimation

Statistical inference is almost always directed toward drawing some type of conclusion about one or more parameters (population characteristics). To do so requires that an investigator obtain sample data from each of the populations under study. Conclusions can then be based on the computed values of various sample quantities. For example, let μ (a parameter) denote the true average breaking strength of wire connections used in bonding semiconductor wafers. A random sample of $n = 10$ connections might be made, and the breaking strength of each one determined, resulting in observed strengths $x_1, x_2, \ldots, x_{10}$. The sample mean breaking strength $\bar{x}$ could then be used to draw a conclusion about the value of μ. Similarly, if σ^2 is the variance of the breaking strength distribution (population variance, another parameter), the value of the sample variance s^2 can be used to infer something about σ^2.

When discussing general concepts and methods of inference, it is convenient to have a generic symbol for the parameter of interest. We will use the Greek letter θ for this purpose. The objective of point estimation is to select a single number, based on sample data, that represents a sensible value for θ. Suppose, for example, that the parameter of interest is μ, the true average lifetime of calculator batteries of a certain type. A random sample of $n = 3$ batteries might yield observed lifetimes (hours) $x_1 = 5.0$, $x_2 = 6.4$, $x_3 = 5.9$. The computed value of the sample mean lifetime is $\bar{x} = 5.77$, and it is reasonable to regard 5.77 as a very plausible value of μ—our "best guess" for the value of μ based on the available sample information.

Suppose we want to estimate a parameter of a single population (e.g., μ or σ) based on a random sample of size n. Recall from the previous chapter that before data is available, the sample observations must be considered random variables (rv's) $X_1, X_2, \ldots, X_n$. It follows that any function of the X_i's—that is, any statistic—such as the sample mean $\bar{X}$ or sample standard deviation S is also a random variable. The same is true if available data consists of more than one sample. For example, we can represent tensile strengths of m type 1 specimens and n type 2 specimens by $X_1, \ldots, X_m$ and $Y_1, \ldots, Y_n$, respectively. The difference between the two sample mean strengths is $\bar{X} - \bar{Y}$, the natural statistic for making inferences about $\mu_1 - \mu_2$, the difference between the population mean strengths.

DEFINITION

> A **point estimate** of a parameter θ is a single number that can be regarded as a sensible value for θ. A point estimate is obtained by selecting a suitable statistic and computing its value from the given sample data. The selected statistic is called the **point estimator** of θ.

In the calculator battery example just given, the estimator used to obtain the point estimate of μ was $\bar{X}$, and the point estimate of μ was 5.77. If the three observed lifetimes had instead been $x_1 = 5.6$, $x_2 = 4.5$, and $x_3 = 6.1$, use of the estimator $\bar{X}$ would have resulted in the estimate $\bar{x} = (5.6 + 4.5 + 6.1)/3 = 5.40$. The symbol $\hat{\theta}$ ("theta hat") is customarily used to denote both the estimator of θ and the point estimate resulting from a given sample.* Thus $\hat{\mu} = \overline{X}$ is read as "the point estimator of μ is the sample mean $\overline{X}$." The statement "the point estimate of μ is 5.77" can be written concisely as $\hat{\mu} = 5.77$. Notice that in writing $\hat{\theta} = 72.5$, there is no indication of how this point estimate was obtained (what statistic was used). It is recommended that both the estimator and the resulting estimate be reported.

Example 6.1 An automobile manufacturer has developed a new type of bumper, which is supposed to absorb impacts with less damage than previous bumpers. The manufacturer has used this bumper in a sequence of 25 controlled crashes against a wall, each at 10 mph, using one of its compact car models. Let X = the number of crashes that result in no visible damage to the automobile. The parameter to be estimated is p = the proportion of all such crashes that result in no damage [alternatively, $p = P(\text{no damage in a single crash})$]. If X is observed to be $x = 15$, the most reasonable estimator and estimate are

$$\text{estimator } \hat{p} = \frac{X}{n} \qquad \text{estimate} = \frac{x}{n} = \frac{15}{25} = .60 \qquad \blacksquare$$

If for each parameter of interest there were only one reasonable point estimator, there would not be much to point estimation. In most problems, though, there will be more than one reasonable estimator.

Example 6.2 Reconsider the accompanying 20 observations on dielectric breakdown voltage for pieces of epoxy resin first introduced in Example 4.29 (Section 4.6).

| 24.46 | 25.61 | 26.25 | 26.42 | 26.66 | 27.15 | 27.31 | 27.54 | 27.74 | 27.94 |
| 27.98 | 28.04 | 28.28 | 28.49 | 28.50 | 28.87 | 29.11 | 29.13 | 29.50 | 30.88 |

The pattern in the normal probability plot given there is quite straight, so we now assume that the distribution of breakdown voltage is normal with mean value μ. Because normal distributions are symmetric, μ is also the median lifetime of the distribution. The given observations are then assumed to be the result of a random sample $X_1, X_2, \ldots, X_{20}$ from this normal distribution. Consider the following estimators and resulting estimates for μ:

a. Estimator $= \overline{X}$, estimate $= \overline{x} = \Sigma x_i / n = 555.86/20 = 27.793$

b. Estimator $= \tilde{X}$, estimate $= \tilde{x} = (27.94 + 27.98)/2 = 27.960$

c. Estimator $= [\min(X_i) + \max(X_i)]/2$ = the average of the two extreme lifetimes, estimate $= [\min(x_i) + \max(x_i)]/2 = (24.46 + 30.88)/2 = 27.670$

d. Estimator $= \overline{X}_{\text{tr}(10)}$, the 10% trimmed mean (discard the smallest and largest 10% of the sample and then average),

$$\text{estimate} = \overline{x}_{\text{tr}(10)}$$
$$= \frac{555.86 - 24.46 - 25.61 - 29.50 - 30.88}{16}$$
$$= 27.838$$

---

*Following earlier notation, we could use $\hat{\Theta}$ (an uppercase theta) for the estimator, but this is cumbersome to write.

Each one of the estimators (a)–(d) uses a different measure of the center of the sample to estimate μ. Which of the estimates is closest to the true value? We cannot answer this without knowing the true value. A question that can be answered is, "Which estimator, when used on other samples of X_i's, will tend to produce estimates closest to the true value?" We will shortly consider this type of question.   ■

Example 6.3   In the near future there will be increasing interest in developing low-cost Mg-based alloys for various casting processes. It is therefore important to have practical ways of determining various mechanical properties of such alloys. The article "On the Development of a New Approach for the Determination of Yield Strength in Mg-based Alloys" (*Light Metal Age,* Oct. 1998: 50–53) proposed an ultrasonic method for this purpose. Consider the following sample of observations on elastic modulus (GPa) of AZ91D alloy specimens from a die-casting process:

$$44.2 \quad 43.9 \quad 44.7 \quad 44.2 \quad 44.0 \quad 43.8 \quad 44.6 \quad 43.1$$

Assume that these observations are the result of a random sample $X_1, \ldots, X_8$ from the population distribution of elastic modulus under such circumstances. We want to estimate the population variance σ^2. A natural estimator is the sample variance:

$$\hat{\sigma}^2 = S^2 = \frac{\Sigma(X_i - \overline{X})^2}{n-1} = \frac{\Sigma X_i^2 - (\Sigma X_i)^2/n}{n-1}$$

The corresponding estimate is

$$\hat{\sigma}^2 = s^2 = \frac{\Sigma x_i^2 - (\Sigma x_i)^2/8}{7} = \frac{15{,}533.79 - (352.5)^2/8}{7}$$

$$= .25125 \approx .251$$

The estimate of σ would then be $\hat{\sigma} = s = \sqrt{.25125} = .501$.

An alternative estimator would result from using divisor n instead of $n - 1$ (i.e., the average squared deviation):

$$\hat{\sigma}^2 = \frac{\Sigma(X_i - \overline{X})^2}{n} \qquad \text{estimate} = \frac{1.75875}{8} = .220$$

We will shortly indicate why many statisticians prefer S^2 to the estimator with divisor n.   ■

In the best of all possible worlds, we could find an estimator $\hat{\theta}$ for which $\hat{\theta} = \theta$ always. However, $\hat{\theta}$ is a function of the sample X_i's, so it is a random variable. For some samples, $\hat{\theta}$ will yield a value larger than θ, whereas for other samples $\hat{\theta}$ will underestimate θ. If we write

$$\hat{\theta} = \theta + \text{error of estimation}$$

then an accurate estimator would be one resulting in small estimation errors, so that estimated values will be near the true value. An estimator that has the properties of unbiasedness and minimum variance will often be accurate in this sense.

Unbiased Estimators

Suppose we have two measuring instruments; one instrument has been accurately calibrated, but the other systematically gives readings smaller than the true value being measured. When each instrument is used repeatedly on the same object, because of measurement error, the observed measurements will not be identical. However, the measurements produced by the first instrument will be distributed about the true value in such a way that on average this instrument measures what it purports to measure, so it is called an unbiased instrument. The second instrument yields observations that have a systematic error component or bias.

<table>
<tr><td>DEFINITION</td><td>A point estimator $\hat{\theta}$ is said to be an **unbiased estimator** of θ if $E(\hat{\theta}) = \theta$ for every possible value of θ. If $\hat{\theta}$ is not unbiased, the difference $E(\hat{\theta}) - \theta$ is called the **bias** of $\hat{\theta}$.</td></tr>
</table>

Thus, $\hat{\theta}$ is unbiased if its probability (i.e., sampling) distribution is always "centered" at the true value of the parameter. Figure 6.1 pictures the distributions of several biased and unbiased estimators. Note that "centered" here means that the expected value, not the median, of the distribution of $\hat{\theta}$ is equal to θ.

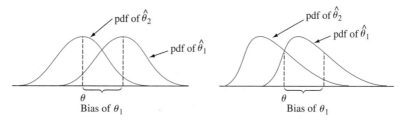

Figure 6.1 The pdf's of a biased estimator $\hat{\theta}_1$ and an unbiased estimator $\hat{\theta}_2$ for a parameter θ

It may seem as though it is necessary to know the value of θ (in which case estimation is unnecessary) to see whether $\hat{\theta}$ is unbiased. This is not usually the case because often a general expected value argument can be used to verify unbiasedness.

In Example 6.1, the sample proportion X/n was used as an estimator of p, where X, the number of sample successes, had a binomial distribution with parameters n and p. Thus,

$$E(\hat{p}) = E\left(\frac{X}{n}\right) = \frac{1}{n} E(X) = \frac{1}{n} (np) = p$$

<table>
<tr><td>PROPOSITION</td><td>When X is a binomial rv with parameters n and p, the sample proportion $\hat{p} = X/n$ is an unbiased estimator of p.</td></tr>
</table>

No matter what the true value of p is, the distribution of the estimator $\hat{p}$ will be centered at the true value.

Example 6.4 Suppose that X, the reaction time to a certain stimulus, has a uniform distribution on the interval from 0 to an unknown upper limit θ (so the density function of X is rectangular in shape with height $1/\theta$ for $0 \le x \le \theta$). It is desired to estimate θ on the basis of a random sample $X_1, X_2, \ldots, X_n$ of reaction times. Since θ is the largest possible time in the entire population of reaction times, consider as a first estimator the largest sample reaction time: $\hat{\theta}_1 = \max(X_1, \ldots, X_n)$. If $n = 5$ and $x_1 = 4.2$, $x_2 = 1.7$, $x_3 = 2.4$, $x_4 = 3.9$, $x_5 = 1.3$, the point estimate of θ is $\hat{\theta}_1 = \max(4.2, 1.7, 2.4, 3.9, 1.3) = 4.2$.

Unbiasedness implies that some samples will yield estimates that exceed θ and other samples will yield estimates smaller than θ—otherwise θ could not possibly be the center (balance point) of $\hat{\theta}_1$'s distribution. However, our proposed estimator will never overestimate θ (the largest sample value cannot exceed the largest population value) and will underestimate θ unless the largest sample value equals θ. This intuitive argument shows that $\hat{\theta}_1$ is a biased estimator. More precisely, it can be shown (see Exercise 32) that

$$E(\hat{\theta}_1) = \frac{n}{n+1} \cdot \theta < \theta \qquad \left(\text{since } \frac{n}{n+1} < 1\right)$$

The bias of $\hat{\theta}_1$ is given by $n\theta/(n+1) - \theta = -\theta/(n+1)$, which approaches 0 as n gets large.

It is easy to modify $\hat{\theta}_1$ to obtain an unbiased estimator of θ. Consider the estimator

$$\hat{\theta}_2 = \frac{n+1}{n} \cdot \max(X_1, \ldots, X_n)$$

Using this estimator on the data gives the estimate $(6/5)(4.2) = 5.04$. The fact that $(n+1)/n > 1$ implies that $\hat{\theta}_2$ will overestimate θ for some samples and underestimate it for others. The mean value of this estimator is

$$E(\hat{\theta}_2) = E\left[\frac{n+1}{n} \max(X_1, \ldots, X_n)\right] = \frac{n+1}{n} \cdot E[\max(X_1, \ldots, X_n)]$$

$$= \frac{n+1}{n} \cdot \frac{n}{n+1} \, \theta = \theta$$

If $\hat{\theta}_2$ is used repeatedly on different samples to estimate θ, some estimates will be too large and others will be too small, but in the long run there will be no systematic tendency to underestimate or overestimate θ. ∎

Principle of Unbiased Estimation

When choosing among several different estimators of θ, select one that is unbiased.

According to this principle, the unbiased estimator $\hat{\theta}_2$ in Example 6.4 should be preferred to the biased estimator $\hat{\theta}_1$. Consider now the problem of estimating σ^2.

PROPOSITION | Let $X_1, X_2, \ldots, X_n$ be a random sample from a distribution with mean μ and variance σ^2. Then the estimator
$$\hat{\sigma}^2 = S^2 = \frac{\sum(X_i - \overline{X})^2}{n - 1}$$
is an unbiased estimator of σ^2.

Proof For any rv Y, $V(Y) = E(Y^2) - [E(Y)]^2$, so $E(Y^2) = V(Y) + [E(Y)]^2$. Applying this to
$$S^2 = \frac{1}{n - 1}\left[\sum X_i^2 - \frac{(\sum X_i)^2}{n}\right]$$
gives
$$E(S^2) = \frac{1}{n - 1}\left\{\sum E(X_i^2) - \frac{1}{n} E[(\sum X_i)^2]\right\}$$
$$= \frac{1}{n - 1}\left\{\sum(\sigma^2 + \mu^2) - \frac{1}{n}\{V(\sum X_i) + [E(\sum X_i)]^2\}\right\}$$
$$= \frac{1}{n - 1}\left\{n\sigma^2 + n\mu^2 - \frac{1}{n} n\sigma^2 - \frac{1}{n}(n\mu)^2\right\}$$
$$= \frac{1}{n - 1}\{n\sigma^2 - \sigma^2\} = \sigma^2 \qquad \text{(as desired)} \qquad \blacksquare$$

The estimator that uses divisor n can be expressed as $(n - 1)S^2/n$, so
$$E\left[\frac{(n - 1)S^2}{n}\right] = \frac{n - 1}{n} E(S^2) = \frac{n - 1}{n}\sigma^2$$

This estimator is therefore not unbiased. The bias is $(n - 1)\sigma^2/n - \sigma^2 = -\sigma^2/n$. Because the bias is negative, the estimator with divisor n tends to underestimate σ^2, and this is why the divisor $n - 1$ is preferred by many statisticians (though when n is large, the bias is small and there is little difference between the two).

Although S^2 is unbiased for σ^2, S is a biased estimator of σ (its bias is small unless n is quite small). However, there are other good reasons to use S as an estimator, especially when the population distribution is normal. These will become more apparent when we discuss confidence intervals and hypothesis testing in the next several chapters.

In Example 6.2, we proposed several different estimators for the mean μ of a normal distribution. If there were a unique unbiased estimator for μ, the estimation problem would be resolved by using that estimator. Unfortunately, this is not the case.

PROPOSITION | If $X_1, X_2, \ldots, X_n$ is a random sample from a distribution with mean μ, then $\overline{X}$ is an unbiased estimator of μ. If in addition the distribution is continuous and symmetric, then $\tilde{X}$ and any trimmed mean are also unbiased estimators of μ.

The fact that $\overline{X}$ is unbiased is just a restatement of one of our rules of expected value: $E(\overline{X}) = \mu$ for every possible value of μ (for discrete as well as continuous distributions). The unbiasedness of the other estimators is more difficult to verify.

According to this proposition, the principle of unbiasedness by itself does not always allow us to select a single estimator. When the underlying population is normal, even the third estimator in Example 6.2 is unbiased, and there are many other unbiased estimators. What we now need is a way of selecting among unbiased estimators.

Estimators with Minimum Variance

Suppose $\hat{\theta}_1$ and $\hat{\theta}_2$ are two estimators of θ that are both unbiased. Then, although the distribution of each estimator is centered at the true value of θ, the spreads of the distributions about the true value may be different.

Principle of Minimum Variance Unbiased Estimation

Among all estimators of θ that are unbiased, choose the one that has minimum variance. The resulting $\hat{\theta}$ is called the **minimum variance unbiased estimator (MVUE)** of θ.

Figure 6.2 pictures the pdf's of two unbiased estimators, with $\hat{\theta}_1$ having smaller variance than $\hat{\theta}_2$. Then $\hat{\theta}_1$ is more likely than $\hat{\theta}_2$ to produce an estimate close to the true θ. The MVUE is, in a certain sense, the most likely among all unbiased estimators to produce an estimate close to the true θ.

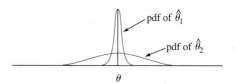

pdf of $\hat{\theta}_1$

pdf of $\hat{\theta}_2$

θ

Figure 6.2   Graphs of the pdf's of two different unbiased estimators

Example 6.5   We argued in Example 6.4 that when $X_1, \ldots, X_n$ is a random sample from a uniform distribution on $[0, \theta]$, the estimator

$$\hat{\theta}_1 = \frac{n+1}{n} \cdot \max(X_1, \ldots, X_n)$$

is unbiased for θ (we previously denoted this estimator by $\hat{\theta}_2$). This is not the only unbiased estimator of θ. The expected value of a uniformly distributed rv is just the midpoint of the interval of positive density, so $E(X_i) = \theta/2$. This implies that $E(\overline{X}) = \theta/2$, from which $E(2\overline{X}) = \theta$. That is, the estimator $\hat{\theta}_2 = 2\overline{X}$ is unbiased for θ.

If X is uniformly distributed on the interval $[A, B]$, then $V(X) = \sigma^2 = (B - A)^2/12$. Thus, in our situation, $V(X_i) = \theta^2/12$, $V(\overline{X}) = \sigma^2/n = \theta^2/(12n)$, and $V(\hat{\theta}_2) = V(2\overline{X}) = 4V(\overline{X}) = \theta^2/(3n)$. The results of Exercise 32 can be used to show that $V(\hat{\theta}_1) = \theta^2/[n(n+2)]$. The estimator $\hat{\theta}_1$ has smaller variance than does $\hat{\theta}_2$ if $3n < n(n+2)$—that is, if $0 < n^2 - n = n(n-1)$. As long as $n > 1$, $V(\hat{\theta}_1) < V(\hat{\theta}_2)$, so $\hat{\theta}_1$ is a better estimator than $\hat{\theta}_2$. More advanced methods can be used to show that $\hat{\theta}_1$ is the MVUE of θ—every other unbiased estimator of θ has variance that exceeds $\theta^2/[n(n + 2)]$. ∎

One of the triumphs of mathematical statistics has been the development of methodology for identifying the MVUE in a wide variety of situations. The most important result of this type for our purposes concerns estimating the mean μ of a normal distribution.

THEOREM

> Let $X_1, \ldots, X_n$ be a random sample from a normal distribution with parameters μ and σ. Then the estimator $\hat{\mu} = \overline{X}$ is the MVUE for μ.

Whenever we are convinced that the population being sampled is normal, the result says that $\overline{X}$ should be used to estimate μ. In Example 6.2, then, our estimate would be $\overline{x} = 27.793$.

In some situations, it is possible to obtain an estimator with small bias that would be preferred to the best unbiased estimator. This is illustrated in Figure 6.3. However, MVUEs are often easier to obtain than the type of biased estimator whose distribution is pictured.

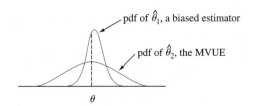

Figure 6.3   A biased estimator that is preferable to the MVUE

Some Complications

The last theorem does not say that in estimating a population mean μ, the estimator $\overline{X}$ should be used irrespective of the distribution being sampled.

Example 6.6

Suppose we wish to estimate the thermal conductivity μ of a certain material. Using standard measurement techniques, we will obtain a random sample $X_1, \ldots, X_n$ of n thermal conductivity measurements. Let's assume that the population distribution is a member of one of the following three families:

$$f(x) = \frac{1}{\sqrt{2\pi\sigma^2}} e^{-(x-\mu)^2/(2\sigma^2)} \qquad -\infty < x < \infty \qquad (6.1)$$

$$f(x) = \frac{1}{\pi[1 + (x - \mu)^2]} \qquad -\infty < x < \infty \qquad (6.2)$$

$$f(x) = \begin{cases} \dfrac{1}{2c} & -c \leq x - \mu \leq c \\ 0 & \text{otherwise} \end{cases} \qquad (6.3)$$

The pdf (6.1) is the normal distribution, (6.2) is called the Cauchy distribution, and (6.3) is a uniform distribution. All three distributions are symmetric about μ, and in fact the Cauchy distribution is bell-shaped but with much heavier tails (more probability farther out) than the normal curve. The uniform distribution has no tails. The four estimators for μ considered earlier are $\overline{X}$, $\tilde{X}$, $\overline{X}_e$ (the average of the two extreme observations), and $\overline{X}_{\text{tr}(10)}$, a trimmed mean.

The very important moral here is that the best estimator for μ depends crucially on which distribution is being sampled. In particular,

1. If the random sample comes from a normal distribution, then $\overline{X}$ is the best of the four estimators, since it has minimum variance among all unbiased estimators.

2. If the random sample comes from a Cauchy distribution, then $\overline{X}$ and $\overline{X}_e$ are terrible estimators for μ, whereas $\tilde{X}$ is quite good (the MVUE is not known); $\overline{X}$ is bad because it is very sensitive to outlying observations, and the heavy tails of the Cauchy distribution make a few such observations likely to appear in any sample.

3. If the underlying distribution is uniform, the best estimator is $\overline{X}_e$; this estimator is greatly influenced by outlying observations, but the lack of tails makes such observations impossible.

4. *The trimmed mean is best in none of these three situations, but works reasonably well in all three.* That is, $\overline{X}_{\text{tr}(10)}$ does not suffer too much in comparison with the best procedure in any of the three situations. ■

More generally, recent research in statistics has established that when estimating a point of symmetry μ of a continuous probability distribution, a trimmed mean with trimming proportion 10% or 20% (from each end of the sample) produces reasonably behaved estimates over a very wide range of possible models. For this reason, a trimmed mean with small trimming percentage is said to be a **robust estimator.**

In some situations, the choice is not between two different estimators constructed from the same sample, but instead between estimators based on two different experiments.

Example 6.7   Suppose a certain type of component has a lifetime distribution that is exponential with parameter λ, so that expected lifetime is $\mu = 1/\lambda$. A sample of n such components is selected, and each is put into operation. If the experiment is continued until all n lifetimes, $X_1, \ldots, X_n$, have been observed, then $\overline{X}$ is an unbiased estimator of μ.

In some experiments, though, the components are left in operation only until the time of the rth failure, where $r < n$. This procedure is referred to as **censoring.** Let Y_1 denote the time of the first failure (the minimum lifetime among the n components), Y_2 denote the time at which the second failure occurs (the second smallest lifetime), and so on. Since the experiment terminates at time Y_r, the total accumulated lifetime at termination is

$$T_r = \sum_{i=1}^{r} Y_i + (n - r)Y_r$$

We now demonstrate that $\hat{\mu} = T_r/r$ is an unbiased estimator for μ. To do so, we need two properties of exponential variables:

1. The memoryless property (see Section 4.4), which says that at any time point, remaining lifetime has the same exponential distribution as original lifetime.

2. If $X_1, \ldots, X_k$ are independent, each exponential with parameter λ, then $\min(X_1, \ldots, X_k)$ is exponential with parameter $k\lambda$ and has expected value $1/(k\lambda)$.

Since all n components last until Y_1, $n - 1$ last an additional $Y_2 - Y_1$, $n - 2$ an additional $Y_3 - Y_2$ amount of time, and so on, another expression for T_r is

$$T_r = nY_1 + (n - 1)(Y_2 - Y_1) + (n - 2)(Y_3 - Y_2) + \cdots$$
$$+ (n - r + 1)(Y_r - Y_{r-1})$$

But Y_1 is the minimum of n exponential variables, so $E(Y_1) = 1/(n\lambda)$. Similarly, $Y_2 - Y_1$ is the smallest of the $n - 1$ remaining lifetimes, each exponential with parameter λ (by the memoryless property), so $E(Y_2 - Y_1) = 1/[(n - 1)\lambda]$. Continuing, $E(Y_{i+1} - Y_i) = 1/[(n - i)\lambda]$, so

$$E(T_r) = nE(Y_1) + (n - 1)E(Y_2 - Y_1) + \cdots + (n - r + 1)E(Y_r - Y_{r-1})$$

$$= n \cdot \frac{1}{n\lambda} + (n - 1) \cdot \frac{1}{(n - 1)\lambda} + \cdots + (n - r + 1) \cdot \frac{1}{(n - r + 1)\lambda}$$

$$= \frac{r}{\lambda}$$

Therefore, $E(T_r/r) = (1/r)E(T_r) = (1/r) \cdot (r/\lambda) = 1/\lambda = \mu$ as claimed.

As an example, suppose 20 components are put on test and $r = 10$. Then if the first ten failure times are 11, 15, 29, 33, 35, 40, 47, 55, 58, and 72, the estimate of μ is

$$\hat{\mu} = \frac{11 + 15 + \cdots + 72 + (10)(72)}{10} = 111.5$$

The advantage of the experiment with censoring is that it terminates more quickly than the uncensored experiment. However, it can be shown that $V(T_r/r) = 1/(\lambda^2 r)$, which is larger than $1/(\lambda^2 n)$, the variance of $\overline{X}$ in the uncensored experiment. ∎

Reporting a Point Estimate: The Standard Error

Besides reporting the value of a point estimate, some indication of its precision should be given. The usual measure of precision is the standard error of the estimator used.

DEFINITION

> The **standard error** of an estimator $\hat{\theta}$ is its standard deviation $\sigma_{\hat{\theta}} = \sqrt{V(\hat{\theta})}$. If the standard error itself involves unknown parameters whose values can be estimated, substitution of these estimates into $\sigma_{\hat{\theta}}$ yields the **estimated standard error** (estimated standard deviation) of the estimator. The estimated standard error can be denoted either by $\hat{\sigma}_{\hat{\theta}}$ (the ^ over σ emphasizes that $\sigma_{\hat{\theta}}$ is being estimated) or by $s_{\hat{\theta}}$.

Example 6.8

(Example 6.2 continued)

Assuming that breakdown voltage is normally distributed, $\hat{\mu} = \overline{X}$ is the best estimator of μ. If the value of σ is known to be 1.5, the standard error of $\overline{X}$ is $\sigma_{\overline{X}} = \sigma/\sqrt{n} = 1.5/\sqrt{20} = .335$. If, as is usually the case, the value of σ is unknown, the estimate $\hat{\sigma} = s = 1.462$ is substituted into $\sigma_{\overline{X}}$ to obtain the estimated standard error $\hat{\sigma}_{\overline{X}} = s_{\overline{X}} = s/\sqrt{n} = 1.462/\sqrt{20} = .327$. ∎

Example 6.9

(Example 6.1 continued)

The standard error of $\hat{p} = X/n$ is

$$\sigma_{\hat{p}} = \sqrt{V(X/n)} = \sqrt{\frac{V(X)}{n^2}} = \sqrt{\frac{npq}{n^2}} = \sqrt{\frac{pq}{n}}$$

Since p and $q = 1 - p$ are unknown (else why estimate?), we substitute $\hat{p} = x/n$ and $\hat{q} = 1 - x/n$ into $\sigma_{\hat{p}}$, yielding the estimated standard error $\hat{\sigma}_{\hat{p}} = \sqrt{\hat{p}\hat{q}/n} = \sqrt{(.6)(.4)/25} = .098$. Alternatively, since the largest value of pq is attained when $p = q = .5$, an upper bound on the standard error is $\sqrt{1/(4n)} = .10$. ∎

When the point estimator $\hat{\theta}$ has approximately a normal distribution, which will often be the case when n is large, then we can be reasonably confident that the true value of θ lies within approximately 2 standard errors (standard deviations) of $\hat{\theta}$. Thus if a sample of $n = 36$ component lifetimes gives $\hat{\mu} = \bar{x} = 28.50$ and $s = 3.60$, then $s/\sqrt{n} = .60$, so within 2 estimated standard errors of $\hat{\mu}$ translates to the interval $28.50 \pm (2)(.60) = (27.30, 29.70)$.

If $\hat{\theta}$ is not necessarily approximately normal but is unbiased, then it can be shown that the estimate will deviate from θ by as much as four standard errors at most 6% of the time. We would then expect the true value to lie within 4 standard errors of $\hat{\theta}$ (and this is a very conservative statement, since it applies to *any* unbiased $\hat{\theta}$). Summarizing, the standard error tells us roughly within what distance of $\hat{\theta}$ we can expect the true value of θ to lie.

The form of the estimator $\hat{\theta}$ may be sufficiently complicated so that standard statistical theory cannot be applied to obtain an expression for $\sigma_{\hat{\theta}}$. This is true, for example, in the case $\theta = \sigma$, $\hat{\theta} = S$; the standard deviation of the statistic S, σ_S, cannot in general be determined. In recent years, a new computer-intensive method called the **bootstrap** has been introduced to address this problem. Suppose that the population pdf is $f(x; \theta)$, a member of a particular parametric family, and that data $x_1, x_2, \ldots, x_n$ gives $\hat{\theta} = 21.7$. We now use the computer to obtain "bootstrap samples" from the pdf $f(x; 21.7)$, and for each sample we calculate a "bootstrap estimate" $\hat{\theta}*$:

First bootstrap sample: $x_1^*, x_2^*, \ldots, x_n^*$; estimate $= \hat{\theta}_1^*$

Second bootstrap sample: $x_1^*, x_2^*, \ldots, x_n^*$; estimate $= \hat{\theta}_2^*$

$$\vdots$$

Bth bootstrap sample: $x_1^*, x_2^*, \ldots, x_n^*$; estimate $= \hat{\theta}_B^*$

$B = 100$ or 200 is often used. Now let $\bar{\theta}* = \Sigma\hat{\theta}_i^*/B$, the sample mean of the bootstrap estimates. The **bootstrap estimate** of $\hat{\theta}$'s standard error is now just the sample standard deviation of the $\hat{\theta}_i^*$'s:

$$S_{\hat{\theta}} = \sqrt{\frac{1}{B-1}\Sigma(\hat{\theta}_i^* - \bar{\theta}*)^2}$$

(In the bootstrap literature, B is often used in place of $B - 1$; for typical values of B, there is usually little difference between the resulting estimates.)

Example 6.10 A theoretical model suggests that X, the time to breakdown of an insulating fluid between electrodes at a particular voltage, has $f(x; \lambda) = \lambda e^{-\lambda x}$, an exponential distribution. A random sample of $n = 10$ breakdown times (min) gives the following data:

$$41.53 \quad 18.73 \quad 2.99 \quad 30.34 \quad 12.33 \quad 117.52 \quad 73.02 \quad 223.63 \quad 4.00 \quad 26.78$$

Since $E(X) = 1/\lambda$, $E(\bar{X}) = 1/\lambda$, so a reasonable estimate of λ is $\hat{\lambda} = 1/\bar{x} = 1/55.087 = .018153$. We then used a statistical computer package to obtain $B = 100$ bootstrap samples, each of size 10, from $f(x; .018153)$. The first such sample was 41.00, 109.70, 16.78, 6.31, 6.76, 5.62, 60.96, 78.81, 192.25, 27.61, from which $\sum x_i^* = 545.8$ and $\hat{\lambda}_1^* = 1/54.58 = .01832$. The average of the 100 bootstrap estimates is $\bar{\lambda}^* = .02153$, and the sample standard deviation of these 100 estimates is $s_{\hat{\lambda}} = .0091$, the bootstrap estimate of $\hat{\lambda}$'s standard error. A histogram of the $100 \, \hat{\lambda}_i^*$'s was somewhat positively skewed, suggesting that the sampling distribution of $\hat{\lambda}$ also has this property. ∎

Sometimes an investigator wishes to estimate a population characteristic without assuming that the population distribution belongs to a particular parametric family. An instance of this occurred in Example 6.6, where a 10% trimmed mean was proposed for estimating a symmetric population distribution's center θ. The data of Example 6.2 gave $\hat{\theta} = \bar{x}_{tr(10)} = 27.838$, but now there is no assumed $f(x; \theta)$, so how can we obtain a bootstrap sample? The answer is to regard the sample itself as constituting the population (the $n = 10$ observations in Example 6.2) and take B different samples, each of size n, *with* replacement from this population. The book by Bradley Efron and Robert Tibshirani or the one by John Rice listed in the chapter bibliography provides more information.

Exercises | Section 6.1 (1–19)

1. The accompanying data on flexural strength (MPa) for concrete beams of a certain type was introduced in Example 1.2.

5.9	7.2	7.3	6.3	8.1	6.8	7.0
7.6	6.8	6.5	7.0	6.3	7.9	9.0
8.2	8.7	7.8	9.7	7.4	7.7	9.7
7.8	7.7	11.6	11.3	11.8	10.7	

a. Calculate a point estimate of the mean value of strength for the conceptual population of all beams manufactured in this fashion, and state which estimator you used. *Hint:* $\sum x_i = 219.8$.

b. Calculate a point estimate of the strength value that separates the weakest 50% of all such beams from the strongest 50%, and state which estimator you used.

c. Calculate and interpret a point estimate of the population standard deviation σ. Which estimator did you use? *Hint:* $\sum x_i^2 = 1860.94$.

d. Calculate a point estimate of the proportion of all such beams whose flexural strength exceeds 10 MPa. *Hint:* Think of an observation as a "success" if it exceeds 10.

e. Calculate a point estimate of the population coefficient of variation σ/μ, and state which estimator you used.

2. A sample of 20 students who had recently taken elementary statistics yielded the following information on brand of calculator owned (T = Texas Instruments, H = Hewlett Packard, C = Casio, S = Sharp):

T T H T C T T S C H
S S T H C T T T H T

a. Estimate the true proportion of all such students who own a Texas Instruments calculator.

b. Some calculators made by Hewlett Packard utilize reverse Polish logic (no other manufacturers produce such calculators). Three out of the four HP calculators in the sample were of this type. Estimate the proportion of all such students who own a calculator that does not use reverse Polish logic.

3. Consider the following sample of observations on coating thickness for low-viscosity paint ("Achieving a Target Value for a Manufacturing Process: A Case Study," *J. of Quality Technology,* 1992: 22–26):

.83 .88 .88 1.04 1.09 1.12 1.29 1.31

1.48 1.49 1.59 1.62 1.65 1.71 1.76 1.83

Assume that the distribution of coating thickness is normal (a normal probability plot strongly supports this assumption).

a. Calculate a point estimate of the mean value of coating thickness, and state which estimator you used.

b. Calculate a point estimate of the median of the coating thickness distribution, and state which estimator you used.

c. Calculate a point estimate of the value that separates the largest 10% of all values in the thickness distribution from the remaining 90%, and state which estimator you used. *Hint:* Express what you are trying to estimate in terms of μ and σ.

d. Estimate $P(X < 1.5)$, i.e., the proportion of all thickness values less than 1.5. *Hint:* If you knew the values of μ and σ, you could calculate this probability. These values are not available, but they can be estimated.

e. What is the estimated standard error of the estimator that you used in part (b)?

4. The article from which the data of Exercise 1 was extracted also gave the accompanying strength observations for cylinders:

6.1 5.8 7.8 7.1 7.2 9.2 6.6 8.3 7.0 8.3

7.8 8.1 7.4 8.5 8.9 9.8 9.7 14.1 12.6 11.2

Prior to obtaining data, denote the beam strengths by $X_1, \ldots, X_m$ and the cylinder strengths by $Y_1, \ldots, Y_n$. Suppose that the X_i's constitute a random sample from a distribution with mean μ_1 and standard deviation σ_1, and that the Y_i's form a random sample (independent of the X_i's) from another distribution with mean μ_2 and standard deviation σ_2.

a. Use rules of expected value to show that $\overline{X} - \overline{Y}$ is an unbiased estimator of $\mu_1 - \mu_2$. Calculate the estimate for the given data.

b. Use rules of variance from Chapter 5 to obtain an expression for the variance and standard deviation (standard error) of the estimator in part (a), and then compute the estimated standard error.

c. Calculate a point estimate of the ratio σ_1/σ_2 of the two standard deviations.

d. Suppose a single beam and a single cylinder are randomly selected. Calculate a point estimate of the variance of the difference $X - Y$ between beam strength and cylinder strength.

5. As an example of a situation in which several different statistics could reasonably be used to calculate a point estimate, consider a population of N invoices. Associated with each invoice is its "book value," the recorded amount of that invoice. Let T denote the total book value, a known amount. Some of these book values are erroneous. An audit will be carried out by randomly selecting n invoices and determining the audited (correct) value for each one. Suppose that the sample gives the following results (in dollars).

	Invoice				
	1	**2**	**3**	**4**	**5**
Book value	300	720	526	200	127
Audited value	300	520	526	200	157
Error	0	200	0	0	-30

Let

$\overline{Y}$ = sample mean book value

$\overline{X}$ = sample mean audited value

$\overline{D}$ = sample mean error

Several different statistics for estimating the total audited (correct) value have been proposed (see "Statistical Models and Analysis in Auditing," *Statistical Science,* 1989: 2–33). These include

Mean per unit statistic = $N\overline{X}$

Difference statistic = $T - N\overline{D}$

Ratio statistic = $T \cdot (\overline{X}/\overline{Y})$

If $N = 5000$ and $T = 1,761,300$, calculate the three corresponding point estimates. (The cited article discusses properties of these estimators.)

6. Consider the accompanying observations on stream flow (1000's of acre-feet) recorded at a station in Colorado for the period April 1–August 31 over a 31-year span (from an article in the 1974 volume of *Water Resources Research*).

127.96	210.07	203.24	108.91	178.21
285.37	100.85	89.59	185.36	126.94
200.19	66.24	247.11	299.87	109.64
125.86	114.79	109.11	330.33	85.54
117.64	302.74	280.55	145.11	95.36
204.91	311.13	150.58	262.09	477.08
94.33				

An appropriate probability plot supports the use of the lognormal distribution (see Section 4.5) as a reasonable model for stream flow.
   a. Estimate the parameters of the distribution. [*Hint:* Remember that X has a lognormal distribution with parameters μ and σ^2 if $\ln(X)$ is normally distributed with mean μ and variance σ^2.]
   b. Use the estimates of part (a) to calculate an estimate of the expected value of stream flow. [*Hint:* What is $E(X)$?]

7. a. A random sample of 10 houses in a particular area, each of which is heated with natural gas, is selected and the amount of gas (therms) used during the month of January is determined for each house. The resulting observations are 103, 156, 118, 89, 125, 147, 122, 109, 138, 99. Let μ denote the average gas usage during January by all houses in this area. Compute a point estimate of μ.
   b. Suppose there are 10,000 houses in this area that use natural gas for heating. Let τ denote the total amount of gas used by all of these houses during January. Estimate τ using the data of part (a). What estimator did you use in computing your estimate?
   c. Use the data in part (a) to estimate p, the proportion of all houses that used at least 100 therms.
   d. Give a point estimate of the population median usage (the middle value in the population of all houses) based on the sample of part (a). What estimator did you use?

8. In a random sample of 80 components of a certain type, 12 are found to be defective.
   a. Give a point estimate of the proportion of all such components that are *not* defective.

b. A system is to be constructed by randomly selecting two of these components and connecting them in series, as shown here.

The series connection implies that the system will function if and only if neither component is defective (i.e., both components work properly). Estimate the proportion of all such systems that work properly. [*Hint:* If p denotes the probability that a component works properly, how can P(system works) be expressed in terms of p?]

9. Each of 150 newly manufactured items is examined and the number of scratches per item is recorded (the items are supposed to be free of scratches), yielding the following data:

Number of scratches per item	0	1	2	3	4	5	6	7
Observed frequency	18	37	42	30	13	7	2	1

Let X = the number of scratches on a randomly chosen item and assume that X has a Poisson distribution with parameter λ.
   a. Find an unbiased estimator of λ and compute the estimate for the data. [*Hint:* $E(X) = \lambda$ for X Poisson, so $E(\bar{X}) = ?$]
   b. What is the standard deviation (standard error) of your estimator? Compute the estimated standard error. (*Hint:* $\sigma_X^2 = \lambda$ for X Poisson.)

10. Using a long rod that has length μ, you are going to lay out a square plot in which the length of each side is μ. Thus, the area of the plot will be μ^2. However, you do not know the value of μ, so you decide to make n independent measurements $X_1, X_2, \ldots X_n$ of the length. Assume that each X_i has mean μ (unbiased measurements) and variance σ^2.
   a. Show that $\bar{X}^2$ is not an unbiased estimator for μ^2. [*Hint:* For any rv Y, $E(Y^2) = V(Y) + [E(Y)]^2$. Apply this with $Y = \bar{X}$.]
   b. For what value of k is the estimator $\bar{X}^2 - kS^2$ unbiased for μ^2? [*Hint:* Compute $E(\bar{X}^2 - kS^2)$.]

11. Of n_1 randomly selected male smokers, X_1 smoked filter cigarettes, whereas of n_2 randomly selected female smokers, X_2 smoked filter cigarettes. Let p_1 and p_2 denote the probabilities that a randomly selected male and female, respectively, smoke filter cigarettes.

a. Show that $(X_1/n_1) - (X_2/n_2)$ is an unbiased estimator for $p_1 - p_2$. [*Hint:* $E(X_i) = n_i p_i$ for $i = 1, 2$.]

b. What is the standard error of the estimator in part (a)?

c. How would you use the observed values x_1 and x_2 to estimate the standard error of your estimator?

d. If $n_1 = n_2 = 200$, $x_1 = 127$, and $x_2 = 176$, use the estimator of part (a) to obtain an estimate of $p_1 - p_2$.

e. Use the result of part (c) and the data of part (d) to estimate the standard error of the estimator.

12. Suppose a certain type of fertilizer has an expected yield per acre of μ_1 with variance σ^2, whereas the expected yield for a second type of fertilizer is μ_2 with the same variance σ^2. Let S_1^2 and S_2^2 denote the sample variances of yields based on sample sizes n_1 and n_2, respectively, of the two fertilizers. Show that the pooled (combined) estimator

$$\hat{\sigma}^2 = \frac{(n_1 - 1)S_1^2 + (n_2 - 1)S_2^2}{n_1 + n_2 - 2}$$

is an unbiased estimator of σ^2.

13. Consider a random sample $X_1, \ldots, X_n$ from the pdf

$$f(x; \theta) = .5(1 + \theta x) \qquad -1 \le x \le 1$$

where $-1 \le \theta \le 1$ (this distribution arises in particle physics). Show that $\hat{\theta} = 3\overline{X}$ is an unbiased estimator of θ. [*Hint:* First determine $\mu = E(X) = E(\overline{X})$.]

14. A sample of n captured Pandemonium jet fighters results in serial numbers $x_1, x_2, x_3, \ldots, x_n$. The CIA knows that the aircraft were numbered consecutively at the factory starting with α and ending with β, so that the total number of planes manufactured is $\beta - \alpha + 1$ (e.g., if $\alpha = 17$ and $\beta = 29$, then $29 - 17 + 1 = 13$ planes having serial numbers 17, 18, 19, . . . , 28, 29 were manufactured). However, the CIA does not know the values of α or β. A CIA statistician suggests using the estimator $\max(X_i) - \min(X_i) + 1$ to estimate the total number of planes manufactured.

a. If $n = 5$, $x_1 = 237$, $x_2 = 375$, $x_3 = 202$, $x_4 = 525$, and $x_5 = 418$, what is the corresponding estimate?

b. Under what conditions on the sample will the value of the estimate be exactly equal to the true total number of planes? Will the estimate ever

be larger than the true total? Do you think the estimator is unbiased for estimating $\beta - \alpha + 1$? Explain in one or two sentences.

15. Let $X_1, X_2, \ldots, X_n$ represent a random sample from a Rayleigh distribution with pdf

$$f(x; \theta) = \frac{x}{\theta} e^{-x^2/(2\theta)} \qquad x > 0$$

a. It can be shown that $E(X^2) = 2\theta$. Use this fact to construct an unbiased estimator of θ based on $\sum X_i^2$ (and use rules of expected value to show that it is unbiased).

b. Estimate θ from the following $n = 10$ observations on vibratory stress of a turbine blade under specified conditions:

| 16.88 | 10.23 | 4.59 | 6.66 | 13.68 |
| 14.23 | 19.87 | 9.40 | 6.51 | 10.95 |

16. Suppose the true average growth μ of one type of plant during a 1-year period is identical to that of a second type, but the variance of growth for the first type is σ^2, whereas for the second type, the variance is $4\sigma^2$. Let $X_1, \ldots, X_m$ be m independent growth observations on the first type [so $E(X_i) = \mu$, $V(X_i) = \sigma^2$], and let $Y_1, \ldots, Y_n$ be n independent growth observations on the second type [$E(Y_i) = \mu$, $V(Y_i) = 4\sigma^2$].

a. Show that for any δ between 0 and 1, the estimator $\hat{\mu} = \delta\overline{X} + (1 - \delta)\overline{Y}$ is unbiased for μ.

b. For fixed m and n, compute $V(\hat{\mu})$ and then find the value of δ that minimizes $V(\hat{\mu})$. [*Hint:* Differentiate $V(\hat{\mu})$ with respect to δ.]

17. In Chapter 3, we defined a negative binomial rv as the number of failures that occur before the rth success in a sequence of independent and identical success/failure trials. The probability mass function (pmf) of X, is

$$nb(x; r, p) =$$
$$\begin{cases} \dbinom{x + r - 1}{x} p^r (1 - p)^x & x = 0, 1, 2, \ldots \\ 0 & \text{otherwise} \end{cases}$$

a. Suppose that $r \ge 2$. Show that

$$\hat{p} = (r - 1)/(X + r - 1)$$

is an unbiased estimator for p. [*Hint:* Write out $E(\hat{p})$ and cancel $x + r - 1$ inside the sum.]

b. A reporter wishes to interview five individuals who support a certain candidate, so begins asking people whether (S) or not (F) they support the candidate. If the sequence of responses is *SFFSFFFSSS*, estimate p = the true proportion who support the candidate.

18. Let $X_1, X_2, \ldots, X_n$ be a random sample from a pdf $f(x)$ that is symmetric about μ, so that $\tilde{X}$ is an unbiased estimator of μ. If n is large, it can be shown that $V(\tilde{X}) \approx 1/(4n[f(\mu)]^2)$.

    a. Compare $V(\tilde{X})$ to $V(\overline{X})$ when the underlying distribution is normal.

    b. When the underlying pdf is Cauchy (see Example 6.6), $V(\overline{X}) = \infty$, so $\overline{X}$ is a terrible estimator. What is $V(\tilde{X})$ in this case when n is large?

19. An investigator wishes to estimate the proportion of students at a certain university who have violated the honor code. Having obtained a random sample of n students, she realizes that asking each, "Have you violated the honor code?" will probably result in some untruthful responses. Consider the following scheme, called a **randomized response** technique. The investigator makes up a deck of 100 cards, of which 50 are of type I and 50 are of type II.

Type I: Have you violated the honor code (yes or no)?

Type II: Is the last digit of your telephone number a 0, 1, or 2 (yes or no)?

Each student in the random sample is asked to mix the deck, draw a card, and answer the resulting question truthfully. Because of the irrelevant question on type II cards, a yes response no longer stigmatizes the respondent, so we assume that responses are truthful. Let p denote the proportion of honor-code violators (i.e., the probability of a randomly selected student being a violator), and let $\lambda = P(\text{yes response})$. Then λ and p are related by $\lambda = .5p + (.5)(.3)$.

    a. Let Y denote the number of yes responses, so $Y \sim \text{Bin}(n, \lambda)$. Thus, Y/n is an unbiased estimator of λ. Derive an estimator for p based on Y. If $n = 80$ and $y = 20$, what is your estimate? (*Hint:* Solve $\lambda = .5p + .15$ for p and then substitute Y/n for λ.)

    b. Use the fact that $E(Y/n) = \lambda$ to show that your estimator $\hat{p}$ is unbiased.

    c. If there were 70 type I and 30 type II cards, what would be your estimator for p?

6.2 | Methods of Point Estimation

The definition of unbiasedness does not in general indicate how unbiased estimators can be derived. We now discuss two "constructive" methods for obtaining point estimators: the method of moments and the method of maximum likelihood. By constructive we mean that the general definition of each type of estimator suggests explicitly how to obtain the estimator in any specific problem. Although maximum likelihood estimators are generally preferable to moment estimators because of certain efficiency properties, they often require significantly more computation than do moment estimators. It is sometimes the case that these methods yield unbiased estimators.

The Method of Moments

The basic idea of this method is to equate certain sample characteristics, such as the mean, to the corresponding population expected values. Then solving these equations for unknown parameter values yields the estimators.

DEFINITION

Let $X_1, \ldots, X_n$ be a random sample from a pmf or pdf $f(x)$. For $k = 1, 2, 3, \ldots$, the **kth population moment**, or **kth moment of the distribution $f(x)$**, is $E(X^k)$. The **kth sample moment** is $(1/n)\sum_{i=1}^{n}X_i^k$.

Thus, the first population moment is $E(X) = \mu$ and the first sample moment is $\sum X_i/n = \overline{X}$. The second population and sample moments are $E(X^2)$ and $\sum X_i^2/n$, respectively. The population moments will be functions of any unknown parameters $\theta_1, \theta_2, \ldots$.

DEFINITION

> Let $X_1, X_2, \ldots, X_n$ be a random sample from a distribution with pmf or pdf $f(x; \theta_1, \ldots, \theta_m)$, where $\theta_1, \ldots, \theta_m$ are parameters whose values are unknown. Then the **moment estimators** $\hat{\theta}_1, \ldots, \hat{\theta}_m$ are obtained by equating the first m sample moments to the corresponding first m population moments and solving for $\theta_1, \ldots, \theta_m$.

If, for example, $m = 2$, $E(X)$ and $E(X^2)$ will be functions of θ_1 and θ_2. Setting $E(X) = (1/n) \sum X_i (= \overline{X})$ and $E(X^2) = (1/n) \sum X_i^2$ gives two equations in θ_1 and θ_2. The solution then defines the estimators. For estimating a population mean μ, the method gives $\mu = \overline{X}$, so the estimator is the sample mean.

Example 6.11 Let $X_1, X_2, \ldots, X_n$ represent a random sample of service times of n customers at a certain facility, where the underlying distribution is assumed exponential with parameter λ. Since there is only one parameter to be estimated, the estimator is obtained by equating $E(X)$ to $\overline{X}$. Since $E(X) = 1/\lambda$ for an exponential distribution, this gives $1/\lambda = \overline{X}$ or $\lambda = 1/\overline{X}$. The moment estimator of λ is then $\hat{\lambda} = 1/\overline{X}$. ∎

Example 6.12 Let $X_1, \ldots, X_n$ be a random sample from a gamma distribution with parameters α and β. From Section 4.4, $E(X) = \alpha\beta$ and $E(X^2) = \beta^2\Gamma(\alpha + 2)/\Gamma(\alpha) = \beta^2(\alpha + 1)\alpha$. The moment estimators of α and β are obtained by solving

$$\overline{X} = \alpha\beta \qquad \frac{1}{n}\sum X_i^2 = \alpha(\alpha + 1)\beta^2$$

Since $\alpha(\alpha + 1)\beta^2 = \alpha^2\beta^2 + \alpha\beta^2$ and the first equation implies $\alpha^2\beta^2 = \overline{X}^2$, the second equation becomes

$$\frac{1}{n}\sum X_i^2 = \overline{X}^2 + \alpha\beta^2$$

Now dividing each side of this second equation by the corresponding side of the first equation and substituting back gives the estimators

$$\hat{\alpha} = \frac{\overline{X}^2}{(1/n)\sum X_i^2 - \overline{X}^2} \qquad \hat{\beta} = \frac{(1/n)\sum X_i^2 - \overline{X}^2}{\overline{X}}$$

To illustrate, the survival time data mentioned in Example 4.21 is

152	115	109	94	88	137	152	77	160	165
125	40	128	123	136	101	62	153	83	69

with $\overline{x} = 113.5$ and $(1/20)\sum x_i^2 = 14{,}087.8$. The estimates are

$$\hat{\alpha} = \frac{(113.5)^2}{14{,}087.8 - (113.5)^2} = 10.7 \qquad \hat{\beta} = \frac{14{,}087.8 - (113.5)^2}{113.5} = 10.6$$

These estimates of α and β differ from the values suggested by Gross and Clark because they used a different estimation technique.   ∎

Example 6.13   Let $X_1, \ldots, X_n$ be a random sample from a generalized negative binomial distribution with parameters r and p (Section 3.5). Since $E(X) = r(1 - p)/p$ and $V(X) = r(1 - p)/p^2$, $E(X^2) = V(X) + [E(X)]^2 = r(1 - p)(r - rp + 1)/p^2$. Equating $E(X)$ to $\overline{X}$ and $E(X^2)$ to $(1/n)\sum X_i^2$ eventually gives

$$\hat{p} = \frac{\overline{X}}{(1/n)\sum X_i^2 - \overline{X}^2} \qquad \hat{r} = \frac{\overline{X}^2}{(1/n)\sum X_i^2 - \overline{X}^2 - \overline{X}}$$

As an illustration, Reep, Pollard, and Benjamin ("Skill and Chance in Ball Games," *J. Royal Stat. Soc.,* 1971: 623–629) consider the negative binomial distribution as a model for the number of goals per game scored by National Hockey League teams. The data for 1966–1967 follows (420 games):

Goals	0	1	2	3	4	5	6	7	8	9	10
Frequency	29	71	82	89	65	45	24	7	4	1	3

Then,

$$\overline{x} = \sum x_i/420 = [(0)(29) + (1)(71) + \cdots + (10)(3)]/420 = 2.98$$

and

$$\sum x_i^2/420 = [(0)^2(29) + (1)^2(71) + \cdots + (10)^2(3)]/420 = 12.40$$

Thus,

$$\hat{p} = \frac{2.98}{12.40 - (2.98)^2} = .85 \qquad \hat{r} = \frac{(2.98)^2}{12.40 - (2.98)^2 - 2.98} = 16.5$$

Although r by definition must be positive, the denominator of $\hat{r}$ could be negative, indicating that the negative binomial distribution is not appropriate (or that the moment estimator is flawed).   ∎

Maximum Likelihood Estimation

The method of maximum likelihood was first introduced by R. A. Fisher, a geneticist and statistician, in the 1920s. Most statisticians recommend this method, at least when the sample size is large, since the resulting estimators have certain desirable efficiency properties (see the second proposition on page 271).

Example 6.14   A sample of ten new bike helmets manufactured by a certain company is obtained. Upon testing, it is found that the first, third, and tenth helmets are flawed, whereas the others are not. Let $p = P(\text{flawed helmet})$ and define $X_1, \ldots, X_{10}$ by $X_i = 1$ if the ith helmet is flawed and zero otherwise. Then the observed x_i's are 1, 0, 1, 0, 0, 0, 0, 0, 0, 1, so the joint pmf of the sample is

$$f(x_1, x_2, \ldots, x_{10}; p) = p(1 - p)p \cdot \cdots \cdot p = p^3(1 - p)^7 \qquad (6.4)$$

We now ask "For what value of p is the observed sample most likely to have occurred?" That is, we wish to find the value of p that maximizes the pmf (6.4), or, equivalently, maximizes the natural log of (6.4).* Since

$$\ln[f(x_1, \ldots, x_{10}; p)] = 3 \ln(p) + 7 \ln(1 - p) \tag{6.5}$$

which is a differentiable function of p, equating the derivative of (6.5) to zero gives the maximizing value†

$$\frac{d}{dp} \ln[f(x_1, \ldots, x_{10}; p)] = \frac{3}{p} - \frac{7}{1 - p} = 0 \Rightarrow p = \frac{3}{10} = \frac{x}{n}$$

where x is the observed number of successes (flawed helmets). The estimate of p is now $\hat{p} = \frac{3}{10}$. It is called the maximum likelihood estimate because for fixed $x_1, \ldots, x_{10}$, it is the parameter value that maximizes the likelihood (joint pmf) of the observed sample.

Note that if we had been told only that among the ten helmets there were three that were flawed, Equation (6.4) would be replaced by the binomial pmf $\binom{10}{3}p^3(1 - p)^7$, which is also maximized for $\hat{p} = \frac{3}{10}$. ∎

DEFINITION

Let $X_1, X_2, \ldots, X_n$ have joint pmf or pdf

$$f(x_1, x_2, \ldots, x_n; \theta_1, \ldots, \theta_m) \tag{6.6}$$

where the parameters $\theta_1, \ldots, \theta_m$ have unknown values. When $x_1, \ldots, x_n$ are the observed sample values and (6.6) is regarded as a function of $\theta_1, \ldots, \theta_m$, it is called the **likelihood function.** The maximum likelihood estimates (mle's) $\hat{\theta}_1, \ldots, \hat{\theta}_m$ are those values of the θ_i's that maximize the likelihood function, so that

$$f(x_1, \ldots, x_n; \hat{\theta}_1, \ldots, \hat{\theta}_m) \geq f(x_1, \ldots, x_n; \theta_1, \ldots, \theta_m) \text{ for all } \theta_1, \ldots, \theta_m$$

When the X_i's are substituted in place of the x_i's, the **maximum likelihood estimators** result.

The likelihood function tells us how likely the observed sample is as a function of the possible parameter values. Maximizing the likelihood gives the parameter values for which the observed sample is most likely to have been generated—that is, the parameter values that "agree most closely" with the observed data.

Example 6.15 Suppose $X_1, X_2, \ldots, X_n$ is a random sample from an exponential distribution with parameter λ. Because of independence, the likelihood function is a product of the individual pdf's:

$$f(x_1, \ldots, x_n; \lambda) = (\lambda e^{-\lambda x_1}) \cdot \cdots \cdot (\lambda e^{-\lambda x_n}) = \lambda^n e^{-\lambda \Sigma x_i}$$

---

*Since $\ln[g(x)]$ is a monotonic function of $g(x)$, finding x to maximize $\ln[g(x)]$ is equivalent to maximizing $g(x)$ itself. In statistics, taking the logarithm frequently changes a product to a sum, which is easier to work with.

†This conclusion requires checking the second derivative, but the details are omitted.

The ln(likelihood) is

$$\ln[f(x_1, \ldots, x_n; \lambda)] = n\ln(\lambda) - \lambda\Sigma x_i$$

Equating $(d/d\lambda)[\ln(\text{likelihood})]$ to zero results in $n/\lambda - \Sigma x_i = 0$, or $\lambda = n/\Sigma x_i = 1/\bar{x}$. Thus, the mle is $\hat{\lambda} = 1/\bar{X}$; it is identical to the method of moments estimator [but it is not an unbiased estimator, since $E(1/\bar{X}) \neq 1/E(\bar{X})$]. ∎

Example 6.16 Let $X_1, \ldots, X_n$ be a random sample from a normal distribution. The likelihood function is

$$f(x_1, \ldots, x_n; \mu, \sigma^2) = \frac{1}{\sqrt{2\pi\sigma^2}} e^{-(x_1-\mu)^2/(2\sigma^2)} \cdot \cdots \cdot \frac{1}{\sqrt{2\pi\sigma^2}} e^{-(x_n-\mu)^2/(2\sigma^2)}$$

$$= \left(\frac{1}{2\pi\sigma^2}\right)^{n/2} e^{-\Sigma(x_i-\mu)^2/(2\sigma^2)}$$

so

$$\ln[f(x_1, \ldots, x_n; \mu, \sigma^2)] = -\frac{n}{2}\ln(2\pi\sigma^2) - \frac{1}{2\sigma^2}\Sigma(x_i - \mu)^2$$

To find the maximizing values of μ and σ^2, we must take the partial derivatives of $\ln(f)$ with respect to μ and σ^2, equate them to zero, and solve the resulting two equations. Omitting the details, the resulting mle's are

$$\hat{\mu} = \bar{X} \qquad \hat{\sigma}^2 = \frac{\Sigma(X_i - \bar{X})^2}{n}$$

The mle of σ^2 is not the unbiased estimator, so two different principles of estimation (unbiasedness and maximum likelihood) yield two different estimators. ∎

Example 6.17 In Chapter 3, we discussed the use of the Poisson distribution for modeling the number of "events" that occur in a two-dimensional region. Assume that when the region R being sampled has area $a(R)$, the number X of events occurring in R has a Poisson distribution with parameter $\lambda a(R)$ (where λ is the expected number of events per unit area), and that nonoverlapping regions yield independent X's.

Suppose an ecologist selects n nonoverlapping regions $R_1, \ldots, R_n$ and counts the number of plants of a certain species found in each region. The joint pmf (likelihood) is then

$$p(x_1, \ldots, x_n; \lambda) = \frac{[\lambda \cdot a(R_1)]^{x_1} e^{-\lambda \cdot a(R_1)}}{x_1!} \cdot \cdots \cdot \frac{[\lambda \cdot a(R_n)]^{x_n} e^{-\lambda \cdot a(R_n)}}{x_n!}$$

$$= \frac{[a(R_1)]^{x_1} \cdot \cdots \cdot [a(R_n)]^{x_n} \cdot \lambda^{\Sigma x_i} \cdot e^{-\lambda\Sigma a(R_i)}}{x_1! \cdot \cdots \cdot x_n!}$$

The ln(likelihood) is

$$\ln[p(x_1, \ldots, x_n; \lambda)] = \Sigma x_i \cdot \ln[a(R_i)] + \ln(\lambda) \cdot \Sigma x_i - \lambda\Sigma a(R_i) - \Sigma\ln(x_i!)$$

Taking $d/d\lambda \ln(p)$ and equating it to zero yields

$$\frac{\Sigma x_i}{\lambda} - \Sigma a(R_i) = 0$$

so

$$\lambda = \frac{\sum x_i}{\sum a(R_i)}$$

The mle is then $\hat{\lambda} = \sum X_i/\sum a(R_i)$. This is intuitively reasonable because λ is the true density (plants per unit area), whereas $\hat{\lambda}$ is the sample density since $\sum a(R_i)$ is just the total area sampled. Because $E(X_i) = \lambda \cdot a(R_i)$, the estimator is unbiased.

Sometimes an alternative sampling procedure is used. Instead of fixing regions to be sampled, the ecologist will select n points in the entire region of interest and let $y_i =$ the distance from the ith point to the nearest plant. The cumulative distribution function (cdf) of $Y =$ distance to the nearest plant is

$$F_Y(y) = P(Y \le y) = 1 - P(Y > y) = 1 - P\begin{pmatrix} \text{no plants in a} \\ \text{circle of radius } y \end{pmatrix}$$

$$= 1 - \frac{e^{-\lambda \pi y^2}(\lambda \pi y^2)^0}{0!} = 1 - e^{-\lambda \cdot \pi y^2}$$

Taking the derivative of $F_Y(y)$ with respect to y yields

$$f_Y(y; \lambda) = \begin{cases} 2\pi\lambda y e^{-\lambda \pi y^2} & y \ge 0 \\ 0 & \text{otherwise} \end{cases}$$

If we now form the likelihood $f_Y(y_1; \lambda) \cdot \dots \cdot f_Y(y_n; \lambda)$, differentiate ln(likelihood), and so on, the resulting mle is

$$\hat{\lambda} = \frac{n}{\pi\sum Y_i^2} = \frac{\text{number of plants observed}}{\text{total area sampled}}$$

which is also a sample density. It can be shown that in a sparse environment (small λ), the distance method is in a certain sense better, whereas in a dense environment the first sampling method is better. ∎

Example 6.18 Let $X_1, \dots, X_n$ be a random sample from a Weibull pdf

$$f(x; \alpha, \beta) = \begin{cases} \dfrac{\alpha}{\beta^\alpha} \cdot x^{\alpha-1} \cdot e^{-(x/\beta)^\alpha} & x \ge 0 \\ 0 & \text{otherwise} \end{cases}$$

Writing the likelihood and ln(likelihood), then setting both $(\partial/\partial\alpha)[\ln(f)] = 0$ and $(\partial/\partial\beta)[\ln(f)] = 0$ yields the equations

$$\alpha = \left[\frac{\sum x_i^\alpha \cdot \ln(x_i)}{\sum x_i^\alpha} - \frac{\sum \ln(x_i)}{n}\right]^{-1} \qquad \beta = \left(\frac{\sum x_i^\alpha}{n}\right)^{1/\alpha}$$

These two equations cannot be solved explicitly to give general formulas for the mle's $\hat{\alpha}$ and $\hat{\beta}$. Instead, for each sample $x_1, \dots, x_n$, the equations must be solved using an iterative numerical procedure. Even moment estimators of α and β are somewhat complicated (see Exercise 21). ∎

Estimating Functions of Parameters

In Example 6.16, we obtained the mle of σ^2 when the underlying distribution is normal. The mle of $\sigma = \sqrt{\sigma^2}$, as well as many other mle's, can be easily derived using the following proposition.

PROPOSITION

The Invariance Principle

Let $\hat{\theta}_1, \hat{\theta}_2, \ldots, \hat{\theta}_m$ be the mle's of the parameters $\theta_1, \theta_2, \ldots, \theta_m$. Then the mle of any function $h(\theta_1, \theta_2, \ldots, \theta_m)$ of these parameters is the function $h(\hat{\theta}_1, \hat{\theta}_2, \ldots, \hat{\theta}_m)$ of the mle's.

Example 6.19
(Example 6.16 continued)

In the normal case, the mle's of μ and σ^2 are $\hat{\mu} = \overline{X}$ and $\hat{\sigma}^2 = \sum(X_i - \overline{X})^2/n$. To obtain the mle of the function $h(\mu, \sigma^2) = \sqrt{\sigma^2} = \sigma$, substitute the mle's into the function:

$$\hat{\sigma} = \sqrt{\hat{\sigma}^2} = \left[\frac{1}{n}\sum(X_i - \overline{X})^2\right]^{1/2}$$

The mle of σ is not the sample standard deviation S, though they are close unless n is quite small. ∎

Example 6.20
(Example 6.18 continued)

The mean value of an rv X that has a Weibull distribution is

$$\mu = \beta \cdot \Gamma(1 + 1/\alpha)$$

The mle of μ is therefore $\hat{\mu} = \hat{\beta}\Gamma(1 + 1/\hat{\alpha})$, where $\hat{\alpha}$ and $\hat{\beta}$ are the mle's of α and β. In particular, $\overline{X}$ is not the mle of μ, though it is an unbiased estimator. At least for large n, $\hat{\mu}$ is a better estimator than $\overline{X}$. ∎

A Desirable Property of the Maximum Likelihood Estimate

Although the principle of maximum likelihood estimation has considerable intuitive appeal, the following proposition provides additional rationale for the use of mle's.

PROPOSITION

Under very general conditions on the joint distribution of the sample, when the sample size n is large, the maximum likelihood estimator of any parameter θ is approximately unbiased $[E(\hat{\theta}) \approx \theta]$ and has variance that is nearly as small as can be achieved by any estimator. Stated another way, the mle $\hat{\theta}$ is approximately the MVUE of θ.

Because of this result and the fact that calculus-based techniques can usually be used to derive the mle's (though often numerical methods, such as Newton's method, are necessary), maximum likelihood estimation is the most widely used estimation technique among statisticians. Many of the estimators used in the remainder of the book are mle's. Obtaining an mle, however, does require that the underlying distribution be specified.

Some Complications

Sometimes calculus cannot be used to obtain mle's.

Example 6.21   Suppose my waiting time for a bus is uniformly distributed on $[0, \theta]$ and the results $x_1, \ldots, x_n$ of a random sample from this distribution have been observed. Since $f(x; \theta) = 1/\theta$ for $0 \leq x \leq \theta$ and 0 otherwise,

$$f(x_1, \ldots, x_n; \theta) = \begin{cases} \dfrac{1}{\theta^n} & 0 \leq x_1 \leq \theta, \ldots, 0 \leq x_n \leq \theta \\ 0 & \text{otherwise} \end{cases}$$

As long as $\max(x_i) \leq \theta$, the likelihood is $1/\theta^n$, which is positive, but as soon as $\theta < \max(x_i)$, the likelihood drops to 0. This is illustrated in Figure 6.4. Calculus will not work because the maximum of the likelihood occurs at a point of discontinuity, but the figure shows that $\hat{\theta} = \max(X_i)$. Thus, if my waiting times are 2.3, 3.7, 1.5, .4, and 3.2, then the mle is $\hat{\theta} = 3.7$.

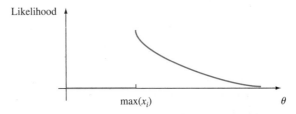

Figure 6.4   The likelihood function for Example 6.21   ■

Example 6.22   A method that is often used to estimate the size of a wildlife population involves performing a capture/recapture experiment. In this experiment, an initial sample of M animals is captured, each of these animals is tagged, and the animals are then returned to the population. After allowing enough time for the tagged individuals to mix into the population, another sample of size n is captured. With X = the number of tagged animals in the second sample, the objective is to use the observed x to estimate the population size N.

The parameter of interest is $\theta = N$, which can assume only integer values, so even after determining the likelihood function (pmf of X here), using calculus to obtain N would present difficulties. If we think of a success as a previously tagged animal being recaptured, then sampling is without replacement from a population containing M successes and $N - M$ failures, so that X is a hypergeometric rv and the likelihood function is

$$p(x; N) = h(x; n, M, N) = \frac{\binom{M}{x} \cdot \binom{N - M}{n - x}}{\binom{N}{n}}$$

The integer-valued nature of N notwithstanding, it would be difficult to take the derivative of $p(x; N)$. However, if we consider the ratio of $p(x; N)$ to $p(x; N - 1)$, we have

$$\frac{p(x; N)}{p(x; N - 1)} = \frac{(N - M) \cdot (N - n)}{N(N - M - n + x)}$$

This ratio is larger than 1 if and only if (iff) $N < Mn/x$. The value of N for which $p(x; N)$ is maximized is therefore the largest integer less than Mn/x. If we use standard mathematical notation $[r]$ for the largest integer less than or equal to r, the mle of N is $\hat{N} = [Mn/x]$. As an illustration, if $M = 200$ fish are taken from a lake and tagged, subsequently $n = 100$ fish are recaptured, and among the 100 there are $x = 11$ tagged fish, then $\hat{N} = [(200)(100)/11] = [1818.18] = 1818$. The estimate is actually rather intuitive; x/n is the proportion of the recaptured sample that is tagged, whereas M/N is the proportion of the entire population that is tagged. The estimate is obtained by equating these two proportions (estimating a population proportion by a sample proportion). ∎

Suppose $X_1, X_2, \ldots, X_n$ is a random sample from a pdf $f(x; \theta)$ that is symmetric about θ, but that the investigator is unsure of the form of the f function. It is then desirable to use an estimator $\hat{\theta}$ that is *robust*—that is, one that performs well for a wide variety of underlying pdf's. One such estimator is a trimmed mean. In recent years, statisticians have proposed another type of estimator, called an *M-estimator,* based on a generalization of maximum likelihood estimation. Instead of maximizing the log likelihood $\sum \ln[f(x; \theta)]$ for a specified f, one maximizes $\sum \rho(x_i; \theta)$. The "objective function" ρ is selected to yield an estimator with good robustness properties. The book by David Hoaglin et al. (see the bibliography) contains a good exposition on this subject.

Exercises | Section 6.2 (20–30)

20. A random sample of n bike helmets manufactured by a certain company is selected. Let $X =$ the number among the n that are flawed and let $p = P(\text{flawed})$. Assume that only X is observed, rather than the sequence of S's and F's.
  a. Derive the maximum likelihood estimator of p. If $n = 20$ and $x = 3$, what is the estimate?
  b. Is the estimator of part (a) unbiased?
  c. If $n = 20$ and $x = 3$, what is the mle of the probability $(1 - p)^5$ that none of the next five helmets examined is flawed?

21. Let X have a Weibull distribution with parameters α and β, so

$$E(X) = \beta \cdot \Gamma(1 + 1/\alpha)$$
$$V(X) = \beta^2 \{ \Gamma(1 + 2/\alpha) - [\Gamma(1 + 1/\alpha)]^2 \}$$

  a. Based on a random sample $X_1, \ldots, X_n$, write equations for the method of moments estimators of β and α. Show that, once the estimate of α has been obtained, the estimate of β can be found from a table of the gamma function and that the estimate of α is the solution to a complicated equation involving the gamma function.
  b. If $n = 20$, $\bar{x} = 28.0$, and $\sum x_i^2 = 16,500$, compute the estimates. [*Hint:* $[\Gamma(1.2)]^2/\Gamma(1.4) = .95$.]

22. Let X denote the proportion of allotted time that a randomly selected student spends working on a certain aptitude test. Suppose the pdf of X is

$$f(x; \theta) = \begin{cases} (\theta + 1)x^\theta & 0 \le x \le 1 \\ 0 & \text{otherwise} \end{cases}$$

where $-1 < \theta$. A random sample of ten students yields data $x_1 = .92$, $x_2 = .79$, $x_3 = .90$, $x_4 = .65$, $x_5 = .86$, $x_6 = .47$, $x_7 = .73$, $x_8 = .97$, $x_9 = .94$, $x_{10} = .77$.
  a. Use the method of moments to obtain an estimator of θ and then compute the estimate for this data.
  b. Obtain the maximum likelihood estimator of θ and then compute the estimate for the given data.

23. Two different computer systems are monitored for a total of n weeks. Let X_i denote the number of breakdowns of the first system during the ith week and suppose the X_i's are independent and drawn from a Poisson distribution with parameter λ_1. Similarly, let Y_i denote the number of breakdowns of the second system during the ith week and assume independence with each Y_i Poisson with parameter λ_2. Derive the mle's of λ_1, λ_2, and $\lambda_1 - \lambda_2$. [*Hint:*

Using independence, write the joint pmf (likelihood) of the X_i's and Y_i's together.]

24. Refer to Exercise 20. Instead of selecting $n = 20$ helmets to examine, suppose I examine helmets in succession until I have found $r = 3$ flawed ones. If the 20th helmet is the third flawed one (so that the number of helmets examined that were not flawed is $x = 17$), what is the mle of p? Is this the same as the estimate in Exercise 20? Why or why not? Is it the same as the estimate computed from the unbiased estimator of Exercise 17?

25. The shear strength of each of ten test spot welds is determined, yielding the following data (psi):

392   376   401   367   389   362   409   415   358   375

   a. Assuming that shear strength is normally distributed, estimate the true average shear strength and standard deviation of shear strength using the method of maximum likelihood.

   b. Again assuming a normal distribution, estimate the strength value below which 95% of all welds will have their strengths. (*Hint:* What is the 95th percentile in terms of μ and σ? Now use the invariance principle.)

26. Refer to Exercise 25. Suppose we decide to examine another test spot weld. Let $X =$ shear strength of the weld. Use the given data to obtain the mle of $P(X \leq 400)$. [*Hint:* $P(X \leq 400) = \Phi((400 - \mu)/\sigma)$.]

27. Let $X_1, \ldots, X_n$ be a random sample from a gamma distribution with parameters α and β.

   a. Derive the equations whose solution yields the maximum likelihood estimators of α and β. Do you think they can be solved explicitly?

   b. Show that the mle of $\mu = \alpha\beta$ is $\hat{\mu} = \bar{X}$.

28. Let $X_1, X_2, \ldots, X_n$ represent a random sample from the Rayleigh distribution with density function given in Exercise 15. Determine

   a. The maximum likelihood estimator of θ and then calculate the estimate for the vibratory stress data given in that exercise. Is this estimator the same as the unbiased estimator suggested in Exercise 15?

   b. The mle of the median of the vibratory stress distribution. (*Hint:* First express the median in terms of θ.)

29. Consider a random sample $X_1, X_2, \ldots, X_n$ from the shifted exponential pdf

$$f(x; \lambda, \theta) = \begin{cases} \lambda e^{-\lambda(x - \theta)} & x \geq \theta \\ 0 & \text{otherwise} \end{cases}$$

Taking $\theta = 0$ gives the pdf of the exponential distribution considered previously (with positive density to the right of zero). An example of the shifted exponential distribution appeared in Example 4.4, in which the variable of interest was time headway in traffic flow and $\theta = .5$ was the minimum possible time headway.

   a. Obtain the maximum likelihood estimators of θ and λ.

   b. If $n = 10$ time headway observations are made, resulting in the values 3.11, .64, 2.55, 2.20, 5.44, 3.42, 10.39, 8.93, 17.82, and 1.30, calculate the estimates of θ and λ.

30. At time $t = 0$, 20 identical components are put on test. The lifetime distribution of each is exponential with parameter λ. The experimenter then leaves the test facility unmonitored. On his return 24 hours later, the experimenter immediately terminates the test after noticing that $y = 15$ of the 20 components are still in operation (so 5 have failed). Derive the mle of λ. [*Hint:* Let $Y =$ the number that survive 24 hours. Then $Y \sim \text{Bin}(n, p)$. What is the mle of p? Now notice that $p = P(X_i \geq 24)$ where X_i is exponentially distributed. This relates λ to p, so the former can be estimated once the latter has been.]

Supplementary Exercises   (31–38)

31. An estimator $\hat{\theta}$ is said to be **consistent** if for any $\epsilon > 0$, $P(|\hat{\theta} - \theta| \geq \epsilon) \to 0$ as $n \to \infty$. That is, $\hat{\theta}$ is consistent if, as the sample size gets larger, it is less and less likely that $\hat{\theta}$ will be further than ϵ from the true value of θ. Show that $\bar{X}$ is a consistent estimator of μ when $\sigma^2 < \infty$ by using Chebyshev's in-

equality from Exercise 43 of Chapter 3. (*Hint:* The inequality can be rewritten in the form

$$P(|Y - \mu_Y| \geq \epsilon) \leq \sigma_Y^2/\epsilon$$

Now identify Y with $\bar{X}$.)

32. a. Let $X_1, \ldots, X_n$ be a random sample from a uniform distribution on $[0, \theta]$. Then the mle of θ is $\hat{\theta} = Y = \max(X_i)$. Use the fact that $Y \le y$ iff each $X_i \le y$ to derive the cdf of Y. Then show that the pdf of $Y = \max(X_i)$ is

$$f_Y(y) = \begin{cases} \dfrac{ny^{n-1}}{\theta^n} & 0 \le y \le \theta \\ 0 & \text{otherwise} \end{cases}$$

b. Use the result of part (a) to show that the mle is biased but that $(n + 1)\max(X_i)/n$ is unbiased.

33. At time $t = 0$, there is one individual alive in a certain population. A **pure birth process** then unfolds as follows. The time until the first birth is exponentially distributed with parameter λ. After the first birth, there are two individuals alive. The time until the first gives birth again is exponential with parameter λ, and similarly for the second individual. Therefore, the time until the next birth is the minimum of two exponential (λ) variables, which is exponential with parameter 2λ. Similarly, once the second birth has occurred, there are three individuals alive, so the time until the next birth is an exponential rv with parameter 3λ, and so on (the memoryless property of the exponential distribution is being used here). Suppose the process is observed until the sixth birth has occurred and the successive birth times are 25.2, 41.7, 51.2, 55.5, 59.5, 61.8 (from which you should calculate the times between successive births). Derive the mle of λ. (*Hint:* The likelihood is a product of exponential terms.)

34. The **mean squared error** of an estimator $\hat{\theta}$ is $\text{MSE}(\hat{\theta}) = E(\hat{\theta} - \theta)^2$. If $\hat{\theta}$ is unbiased, then $\text{MSE}(\hat{\theta}) = V(\hat{\theta})$, but in general $\text{MSE}(\hat{\theta}) = V(\hat{\theta}) + (\text{bias})^2$. Consider the estimator $\hat{\sigma}^2 = KS^2$, where $S^2 =$ sample variance. What value of K minimizes the mean squared error of this estimator when the population distribution is normal? [*Hint:* It can be shown that

$$E[(S^2)^2] = (n + 1)\sigma^4/(n - 1)$$

In general, it is difficult to find $\hat{\theta}$ to minimize $\text{MSE}(\hat{\theta})$, which is why we look only at unbiased estimators and minimize $V(\hat{\theta})$.]

35. Let $X_1, \ldots, X_n$ be a random sample from a pdf that is symmetric about μ. An estimator for μ that has been found to perform well for a variety of underlying distributions is the *Hodges–Lehmann estimator*. To define it, first compute for each $i \le j$ and each $j = 1, 2, \ldots, n$ the pairwise average $\bar{X}_{i,j} = (X_i + X_j)/2$. Then the estimator is $\hat{\mu} =$ the median of the $\bar{X}_{i,j}$'s. Compute the value of this estimate using the data of Exercise 44 of Chapter 1. (*Hint:* Construct a square table with the x_i's listed on the left margin and on top. Then compute averages on and above the diagonal.)

36. When the population distribution is normal, the statistic median$\{|X_1 - \tilde{X}|, \ldots, |X_n - \tilde{X}|\}/.6745$ can be used to estimate σ. This estimator is more resistant to the effects of outliers (observations far from the bulk of the data) than is the sample standard deviation. Compute both the corresponding point estimate and s for the data of Example 6.2.

37. When the sample standard deviation S is based on a random sample from a normal population distribution, it can be shown that

$$E(S) = \sqrt{2/(n - 1)}\,\Gamma(n/2)\sigma/\Gamma((n - 1)/2)$$

Use this to obtain an unbiased estimator for σ of the form cS. What is c when $n = 20$?

38. Each of n specimens is to be weighed twice on the same scale. Let X_i and Y_i denote the two observed weights for the ith specimen. Suppose X_i and Y_i are independent of one another, each normally distributed with mean value μ_i (the true weight of specimen i) and variance σ^2.

a. Show that the maximum likelihood estimator of σ^2 is $\hat{\sigma}^2 = \Sigma(X_i - Y_i)^2/(4n)$. [*Hint:* If $\bar{z} = (z_1 + z_2)/2$, then $\Sigma(z_i - \bar{z})^2 = (z_1 - z_2)^2/2$.]

b. Is the mle $\hat{\sigma}^2$ an unbiased estimator of σ^2? Find an unbiased estimator of σ^2. [*Hint:* For any rv Z, $E(Z^2) = V(Z) + [E(Z)]^2$. Apply this to $Z = X_i - Y_i$.]

Bibliography

DeGroot, Morris, *Probability and Statistics* (2nd ed.), Addison-Wesley, Reading, MA, 1986. Includes an

excellent discussion of both general properties and methods of point estimation; of particular interest are examples showing how general principles and methods can yield unsatisfactory estimators in particular situations.

Efron, Bradley, and Robert Tibshirani, *An Introduction to the Bootstrap,* Chapman and Hall, New York, 1993. The bible of the bootstrap.

Hoaglin, David, Frederick Mosteller, and John Tukey, *Understanding Robust and Exploratory Data Analysis,* Wiley, New York, 1983. Contains several good chapters on robust point estimation, including one on *M*-estimation.

Hogg, Robert, and Allen Craig, *Introduction to Mathematical Statistics* (5th ed.), Prentice Hall, Englewood Cliffs, NJ, 1995. A good discussion of unbiasedness.

Larsen, Richard, and Morris Marx, *Introduction to Mathematical Statistics* (2nd ed.), Prentice Hall, Englewood Cliffs, NJ, 1985. A very good discussion of point estimation from a slightly more mathematical perspective than the present text.

Rice, John, *Mathematical Statistics and Data Analysis* (2nd ed.), Duxbury Press, Belmont, CA, 1994. A nice blending of statistical theory and data.

7

Statistical Intervals Based on a Single Sample

Introduction

A point estimate, because it is a single number, by itself provides no information about the precision and reliability of estimation. Consider, for example, using the statistic $\overline{X}$ to calculate a point estimate for the true average breaking strength (g) of paper towels of a certain brand, and suppose that $\overline{x} = 9322.7$. Because of sampling variability, it is virtually never the case that $\overline{x} = \mu$. The point estimate says nothing about how close it might be to μ. An alternative to reporting a single sensible value for the parameter being estimated is to calculate and report an entire interval of plausible values—an *interval estimate* or *confidence interval* (CI). A confidence interval is always calculated by first selecting a *confidence level,* which is a measure of the degree of reliability of the interval. A confidence interval with a 95% confidence level for the true average breaking strength might have a lower limit of 9162.5 and an upper limit of 9482.9. Then at the 95% confidence level, any value of μ between 9162.5 and 9482.9 is plausible. A confidence level of 95% implies that 95% of all samples would give an interval that includes μ, or whatever other parameter is being estimated, and only 5% of all samples would yield an erroneous interval. The most frequently used confidence levels are 95%,

99%, and 90%. The higher the confidence level, the more strongly we believe that the value of the parameter being estimated lies within the interval (an interpretation of any particular confidence level will be given shortly).

Information about the precision of an interval estimate is conveyed by the width of the interval. If the confidence level is high and the resulting interval is quite narrow, our knowledge of the value of the parameter is reasonably precise. A very wide confidence interval, however, gives the message that there is a great deal of uncertainty concerning the value of what we are estimating. Figure 7.1 shows 95% confidence intervals for true average breaking strengths of two different brands of paper towels. One of these intervals suggests precise knowledge about μ, whereas the other suggests a very wide range of plausible values.

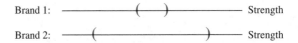

Figure 7.1   Confidence intervals indicating precise (brand 1) and imprecise (brand 2) information about μ

7.1 | Basic Properties of Confidence Intervals

The basic concepts and properties of confidence intervals (CIs) are most easily introduced by first focusing on a simple, albeit somewhat unrealistic, problem situation. Suppose that the parameter of interest is a population mean μ and that

1. The population distribution is normal.

2. The value of the population standard deviation σ is known.

Normality of the population distribution is often a reasonable assumption. However, if the value of μ is unknown, it is implausible that the value of σ would be available (knowledge of a population's center typically precedes information concerning spread). In later sections, we will develop methods based on less restrictive assumptions.

Example 7.1   Industrial engineers who specialize in ergonomics are concerned with designing workspace and devices operated by workers so as to achieve high productivity and comfort. The article "Studies on Ergonomically Designed Alphanumeric Keyboards" (*Human Factors*, 1985: 175–187) reports on a study of preferred height for an experimental keyboard with large forearm–wrist support. A sample of $n = 31$ trained typists was selected, and the preferred keyboard height was determined for each typist. The resulting sample average preferred height was $\bar{x} = 80.0$ cm. Assuming that preferred height is normally distributed with $\sigma = 2.0$ cm (a value suggested by data in the article), obtain a CI for μ, the true average preferred height for the population of all experienced typists. ∎

The actual sample observations $x_1, x_2, \ldots, x_n$ are assumed to be the result of a random sample $X_1, \ldots, X_n$ from a normal distribution with mean value μ and standard deviation σ. The results of Chapter 5 then imply that irrespective of the sample size n, the sample mean $\overline{X}$ is normally distributed with expected value μ and standard deviation $\sigma/\sqrt{n}$. Standardizing $\overline{X}$ by first subtracting its expected value and then dividing by its standard deviation yields the variable

$$Z = \frac{\overline{X} - \mu}{\sigma/\sqrt{n}} \qquad (7.1)$$

which has a standard normal distribution. Because the area under the standard normal curve between -1.96 and 1.96 is .95,

$$P\left(-1.96 < \frac{\overline{X} - \mu}{\sigma/\sqrt{n}} < 1.96\right) = .95 \qquad (7.2)$$

The next step in the development is to manipulate the inequalities inside the parentheses in (7.2) so that they appear in the equivalent form $l < \mu < u$, where the endpoints l and u involve $\overline{X}$ and $\sigma/\sqrt{n}$. This is achieved through the following sequence of operations, each one yielding inequalities equivalent to those we started with:

1. Multiply through by $\sigma/\sqrt{n}$ to obtain

$$-1.96 \cdot \frac{\sigma}{\sqrt{n}} < \overline{X} - \mu < 1.96 \cdot \frac{\sigma}{\sqrt{n}}$$

2. Subtract $\overline{X}$ from each term to obtain

$$-\overline{X} - 1.96 \cdot \frac{\sigma}{\sqrt{n}} < -\mu < -\overline{X} + 1.96 \cdot \frac{\sigma}{\sqrt{n}}$$

3. Multiply through by -1 to eliminate the minus sign in front of μ (which reverses the direction of each inequality) to obtain

$$\overline{X} + 1.96 \cdot \frac{\sigma}{\sqrt{n}} > \mu > \overline{X} - 1.96 \cdot \frac{\sigma}{\sqrt{n}}$$

that is,

$$\overline{X} - 1.96 \cdot \frac{\sigma}{\sqrt{n}} < \mu < \overline{X} + 1.96 \cdot \frac{\sigma}{\sqrt{n}}$$

Because each set of inequalities in the sequence is equivalent to the original one, the probability associated with each is .95. In particular,

$$P\left(\overline{X} - 1.96\frac{\sigma}{\sqrt{n}} < \mu < \overline{X} + 1.96\frac{\sigma}{\sqrt{n}}\right) = .95 \qquad (7.3)$$

The event inside the parentheses in (7.3) has a somewhat unfamiliar appearance; always before, the random quantity has appeared in the middle with constants on both ends, as in $a \leq Y \leq b$. In (7.3) the random quantity appears on the two ends, whereas the unknown constant μ appears in the middle. To interpret (7.3), think of a **random interval** having left endpoint $\overline{X} - 1.96 \cdot \sigma/\sqrt{n}$ and right endpoint $\overline{X} + 1.96 \cdot \sigma/\sqrt{n}$, which in interval notation is

$$\left(\overline{X} - 1.96 \cdot \frac{\sigma}{\sqrt{n}}, \overline{X} + 1.96 \cdot \frac{\sigma}{\sqrt{n}}\right) \qquad (7.4)$$

The interval (7.4) is random because the two endpoints of the interval involve a random variable (rv). Note that the interval is centered at the sample mean $\overline{X}$ and extends $1.96\sigma/\sqrt{n}$ to each side of $\overline{X}$. Thus, the interval's width is $2 \cdot (1.96) \cdot \sigma/\sqrt{n}$, which is not random; only the location of the interval (its midpoint $\overline{X}$) is random (Figure 7.2). Now (7.3) can be paraphrased as "*the probability is .95 that the random interval (7.4) includes or covers the true value of μ.*" Before any experiment is performed and any data is gathered, it is quite likely (probability .95) that μ will lie inside the interval in Expression (7.4).

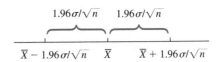

Figure 7.2   The random interval (7.4) centered at $\overline{X}$

DEFINITION

If after observing $X_1 = x_1$, $X_2 = x_2$, . . . , $X_n = x_n$, we compute the observed sample mean $\overline{x}$ and then substitute $\overline{x}$ into (7.4) in place of $\overline{X}$, the resulting fixed interval is called a **95% confidence interval for μ.** This CI can be expressed either as

$$\left(\overline{x} - 1.96 \cdot \frac{\sigma}{\sqrt{n}}, \overline{x} + 1.96 \cdot \frac{\sigma}{\sqrt{n}}\right) \quad \text{is a 95\% CI for } \mu$$

or as

$$\overline{x} - 1.96 \cdot \frac{\sigma}{\sqrt{n}} < \mu < \overline{x} + 1.96 \cdot \frac{\sigma}{\sqrt{n}} \quad \text{with 95\% confidence}$$

A concise expression for the interval is $\overline{x} \pm 1.96 \cdot \sigma/\sqrt{n}$, where $-$ gives the left endpoint (lower limit) and $+$ gives the right endpoint (upper limit).

Example 7.2
(Example 7.1 continued)

The quantities needed for computation of the 95% CI for true average preferred height are $\sigma = 2.0$, $n = 31$, and $\overline{x} = 80.0$. The resulting interval is

$$\overline{x} \pm 1.96 \cdot \frac{\sigma}{\sqrt{n}} = 80.0 \pm (1.96)\frac{2.0}{\sqrt{31}} = 80.0 \pm .7 = (79.3, 80.7)$$

That is, we can be highly confident that $79.3 < \mu < 80.7$. This interval is relatively narrow, indicating that μ has been rather precisely estimated. ■

Interpreting a Confidence Interval

The confidence level 95% for the interval just defined was inherited from the probability .95 for the random interval (7.4). Intervals having other levels of confidence will be introduced shortly. For now, though, consider how 95% confidence can be interpreted.

Because we started with an event whose probability was .95—that the random interval (7.4) would capture the true value of μ—and then used the data in Example 7.1 to compute the fixed interval (79.3, 80.7), it is tempting to conclude that μ is within this fixed interval with probability .95. But by substituting $\bar{x} = 80.0$ for $\bar{X}$, all randomness disappears; the interval (79.3, 80.7) is not a random interval, and μ is a constant (unfortunately unknown to us), so it is *incorrect* to write the statement $P(\mu$ lies in $(79.3, 80.7)) = .95$.

A correct interpretation of "95% confidence" relies on the long-run frequency interpretation of probability: To say that an event A has probability .95 is to say that if the experiment on which A is defined is performed over and over again, in the long run A will occur 95% of the time. Suppose we obtain another sample of typists' preferred heights and compute another 95% interval. Then suppose we consider repeating this for a third sample, a fourth sample, and so on. Let A be the event that $\bar{X} - 1.96 \cdot \sigma/\sqrt{n} < \mu < \bar{X} + 1.96 \cdot \sigma/\sqrt{n}$. Since $P(A) = .95$, in the long run 95% of our computed CIs will contain μ. This is illustrated in Figure 7.3, where the vertical line cuts the measurement axis at the true (but unknown) value of μ. Notice that of the 11 intervals pictured, only intervals 3 and 11 fail to contain μ. In the long run, only 5% of the intervals so constructed would fail to contain μ.

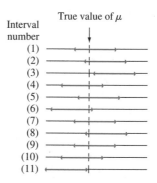

Figure 7.3 Repeated construction of 95% CIs

According to this interpretation, the confidence level 95% is not so much a statement about any particular interval such as (79.3, 80.7) but pertains to what would happen if a very large number of like intervals were to be constructed. Although this may seem unsatisfactory, the root of the difficulty lies with our interpretation of probability—it applies to a long sequence of replications of an experiment rather than just a single replication. There is another approach to the construction and interpretation of CIs that uses the notion of subjective probability and Bayes' theorem, but the technical details are beyond the scope of this text; the book by Winkler (see the Chapter 2 bibliography) is a good source. The interval presented here (as well as each interval presented subsequently) is called a "classical" CI because its interpretation rests on the classical notion of probability (though the main ideas were developed as recently as the 1930s).

Other Levels of Confidence

Suppose we want a 99% CI rather than a 95% interval. Rather than starting with a probability of .95, we must begin with a probability of .99. Since the area under the standard normal curve between −2.58 and 2.58 equals .99, replacing 1.96 with 2.58 in the definition yields a 99% interval.

This suggests that any desired level of confidence can be achieved by replacing 1.96 or 2.58 with the appropriate standard normal critical value. As Figure 7.4 shows, a probability of $1 - \alpha$ is achieved by using $z_{\alpha/2}$ in place of 1.96.

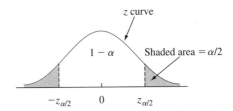

Figure 7.4 $P(-z_{\alpha/2} \leq Z < z_{\alpha/2}) = 1 - \alpha$

DEFINITION

A **100(1 − α)% confidence interval** for the mean μ of a normal population when the value of σ is known is given by

$$\left(\bar{x} - z_{\alpha/2} \cdot \frac{\sigma}{\sqrt{n}}, \bar{x} + z_{\alpha/2} \cdot \frac{\sigma}{\sqrt{n}}\right) \qquad (7.5)$$

or, equivalently, by $\bar{x} \pm z_{\alpha/2} \cdot \sigma/\sqrt{n}$.

Example 7.3

The production process for engine control housing units of a particular type has recently been modified. Prior to this modification, historical data had suggested that the distribution of hole diameters for bushings on the housings was normal with a standard deviation of .100 mm. It is believed that the modification has not affected the shape of the distribution or the standard deviation, but that the value of the mean diameter may have changed. A sample of 40 housing units is selected and hole diameter is determined for each one, resulting in a sample mean diameter of 5.426 mm. Let's calculate a confidence interval for true average hole diameter using a confidence level of 90%. This requires that 100(1 − α) = 90, from which α = .10 and $z_{\alpha/2} = z_{.05} = 1.645$ (corresponding to a cumulative z curve area of .9500). The desired interval is then

$$5.426 \pm (1.645)\frac{.100}{\sqrt{40}} = 5.426 \pm .026 = (5.400, 5.452)$$

With a reasonably high degree of confidence, we can say that $5.400 < \mu < 5.452$. This interval is rather narrow because of the small amount of variability in hole diameter ($\sigma = .100$). ∎

Confidence Level, Precision, and Choice of Sample Size

Why settle for a confidence level of 95% when a level of 99% is achievable? Because the price paid for the higher confidence level is a wider interval. Because the 95% interval extends $1.96 \cdot \sigma/\sqrt{n}$ to each side of $\bar{x}$, the width of the interval is $2(1.96) \cdot \sigma/\sqrt{n} = 3.92 \cdot \sigma/\sqrt{n}$. Similarly, the width of the 99% interval is $2(2.58) \cdot \sigma/\sqrt{n} = 5.16 \cdot \sigma/\sqrt{n}$. That is, we have more confidence in the 99% interval precisely because it is wider. The higher the desired degree of confidence, the wider the resulting interval. In fact, the only 100% CI for μ is $(-\infty, \infty)$, which is not terribly informative because, even before sampling, we knew that this interval covers μ.

If we think of the width of the interval as specifying its precision or accuracy, then the confidence level (or reliability) of the interval is inversely related to its precision. A highly reliable interval estimate may be imprecise in that the endpoints of the interval may be far apart, whereas a precise interval may entail relatively low reliability. Thus, it cannot be said unequivocally that a 99% interval is to be preferred to a 95% interval; the gain in reliability entails a loss in precision.

An appealing strategy is to specify both the desired confidence level and interval width and then determine the necessary sample size.

Example 7.4 Extensive monitoring of a computer time-sharing system has suggested that response time to a particular editing command is normally distributed with standard deviation 25 millisec. A new operating system has been installed, and we wish to estimate the true average response time μ for the new environment. Assuming that response times are still normally distributed with $\sigma = 25$, what sample size is necessary to ensure that the resulting 95% CI has a width of (at most) 10? The sample size n must satisfy

$$10 = 2 \cdot (1.96)(25/\sqrt{n})$$

Rearranging this equation gives

$$\sqrt{n} = 2 \cdot (1.96)(25)/10 = 9.80$$

so

$$n = (9.80)^2 = 96.04$$

Since n must be an integer, a sample size of 97 is required. ∎

The general formula for the sample size n necessary to ensure an interval width w is obtained from $w = 2 \cdot z_{\alpha/2} \cdot \sigma/\sqrt{n}$ as

$$n = \left(2z_{\alpha/2} \cdot \frac{\sigma}{w}\right)^2$$

The smaller the desired width w, the larger n must be. In addition, n is an increasing function of σ (more population variability necessitates a larger sample size) and of the confidence level $100(1 - \alpha)$ (as α decreases, $z_{\alpha/2}$ increases).

The half-width $1.96\sigma/\sqrt{n}$ of the 95% CI is sometimes called the **bound on the error of estimation** associated with a 95% confidence level; that is, with 95% confidence, the point estimate $\bar{x}$ will be no farther than this from μ. Before obtaining data, an investigator may wish to determine a sample size for which a particular value of the bound is achieved. For example, with μ representing the average fuel efficiency (mpg) for all cars of a certain type, the objective of an investigation may be to estimate μ to within 1 mpg with 95% confidence. More generally, if we wish to estimate μ to within an amount B (the specified bound on the error of estimation) with $100(1 - \alpha)$% confidence, the necessary sample size results from replacing $2/w$ by $1/B$ in the formula in the preceding box.

Deriving a Confidence Interval

Let $X_1, X_2, \ldots, X_n$ denote the sample on which the CI for a parameter θ is to be based. Suppose a random variable satisfying the following two properties can be found:

1. The variable depends functionally on both $X_1, \ldots, X_n$ and θ.

2. The probability distribution of the variable does not depend on θ or on any other unknown parameters.

Let $h(X_1, X_2, \ldots, X_n; \theta)$ denote this random variable. For example, if the population distribution is normal with known σ and $\theta = \mu$, the variable $h(X_1, \ldots, X_n; \mu) = (\bar{X} - \mu)/(\sigma/\sqrt{n})$ satisfies both properties; it clearly depends functionally on μ, yet has the standard normal probability distribution, which does not depend on μ. In general, the form of the h function is usually suggested by examining the distribution of an appropriate estimator $\hat{\theta}$.

For any α between 0 and 1, constants a and b can be found to satisfy

$$P(a < h(X_1, \ldots, X_n; \theta) < b) = 1 - \alpha \tag{7.6}$$

Because of the second property, a and b do not depend on θ. In the normal example, $a = -z_{\alpha/2}$ and $b = z_{\alpha/2}$. Now suppose that the inequalities in (7.6) can be manipulated to isolate θ, giving the equivalent probability statement

$$P(l(X_1, X_2, \ldots, X_n) < \theta < u(X_1, X_2, \ldots, X_n)) = 1 - \alpha$$

Then $l(x_1, x_2, \ldots, x_n)$ and $u(x_1, \ldots, x_n)$ are the lower and upper confidence limits, respectively, for a $100(1 - \alpha)$% CI. In the normal example, we saw that $l(X_1, \ldots, X_n) = \bar{X} - z_{\alpha/2} \cdot \sigma/\sqrt{n}$ and $u(X_1, \ldots, X_n) = \bar{X} + z_{\alpha/2} \cdot \sigma/\sqrt{n}$.

Example 7.5   A theoretical model suggests that the time to breakdown of an insulating fluid between electrodes at a particular voltage has an exponential distribution with parameter λ (see Section 4.4). A random sample of $n = 10$ breakdown times yields the following sample data (in min): $x_1 = 41.53$, $x_2 = 18.73$, $x_3 = 2.99$, $x_4 = 30.34$, $x_5 = 12.33$, $x_6 = 117.52$, $x_7 = 73.02$, $x_8 = 223.63$, $x_9 = 4.00$, $x_{10} = 26.78$. A 95% CI for λ and for the true average breakdown time are desired.

Let $h(X_1, X_2, \ldots, X_n; \lambda) = 2\lambda\sum X_i$. It can be shown that this random variable has a probability distribution called a chi-squared distribution with $2n$ degrees of freedom (df) ($\nu = 2n$, where ν is the parameter of a chi-squared distribution as discussed in Section 4.4). Appendix Table A.7 pictures a typical chi-squared density curve and tabulates critical values that capture specified tail areas. The relevant number of degrees of freedom here is $2(10) = 20$. The $\nu = 20$ row of the table shows that 34.170 captures upper-tail area .025 and 9.591 captures lower-tail area .025 (upper-tail area .975). Thus, for $n = 10$,

$$P(9.591 < 2\lambda\sum X_i < 34.170) = .95$$

Division by $2\sum X_i$ isolates λ, yielding

$$P(9.591/(2\sum X_i) < \lambda < (34.170/(2\sum X_i)) = .95$$

The lower limit of the 95% CI for λ is $9.591/(2\sum x_i)$, and the upper limit is $34.170/(2\sum x_i)$. For the given data, $\sum x_i = 550.87$, giving the interval (.00871, .03101). The expected value of an exponential rv is $\mu = 1/\lambda$. Since

$$P(2\sum X_i/34.170 < 1/\lambda < 2\sum X_i/9.591) = .95$$

the 95% CI for true average breakdown time is $(2\sum x_i/34.170, 2\sum x_i/9.591) = (32.24, 114.87)$. This interval is obviously quite wide, reflecting substantial variability in breakdown times and a small sample size. ∎

In general, the upper and lower confidence limits result from replacing each $<$ in (7.6) by $=$ and solving for θ. In the insulating fluid example just considered, $2\lambda\sum x_i = 34.170$ gives $\lambda = 34.170/(2\sum x_i)$ as the upper confidence limit, and the lower limit is obtained from the other equation. Notice that the two interval limits are not equidistant from the point estimate, since the interval is not of the form $\hat{\theta} \pm c$.

Bootstrap Confidence Intervals

The bootstrap technique was introduced in Chapter 6 as a way of estimating $\sigma_{\hat{\theta}}$. It can also be applied to obtain a CI for θ. Consider again estimating the mean μ of a normal distribution when σ is known. Let's replace μ by θ and use $\hat{\theta} = \overline{X}$ as the point estimator. Notice that $1.96\sigma/\sqrt{n}$ is the 97.5th percentile of the distribution of $\hat{\theta} - \theta$ [that is, $P(\overline{X} - \mu < 1.96\sigma/\sqrt{n}) = P(Z < 1.96) = .9750$]. Similarly, $-1.96\sigma/\sqrt{n}$ is the 2.5th percentile, so

$$.95 = P(\text{2.5th percentile} < \hat{\theta} - \theta < \text{97.5th percentile})$$
$$= P(\hat{\theta} - \text{2.5th percentile} > \theta > \hat{\theta} - \text{97.5th percentile})$$

That is, with

$$l = \hat{\theta} - \text{97.5th percentile of } \hat{\theta} - \theta$$
$$u = \hat{\theta} - \text{2.5th percentile of } \hat{\theta} - \theta$$

(7.7)

the CI for θ is (l, u). In many cases, the percentiles in (7.7) cannot be calculated, but they *can* be estimated from bootstrap samples. Suppose we obtain $B = 1000$ bootstrap samples and calculate $\hat{\theta}_1^*, \ldots, \hat{\theta}_{1000}^*$, and $\overline{\theta}^*$ and then the 1000 differences $\hat{\theta}_1^* - \overline{\theta}^*, \ldots,$ $\hat{\theta}_{1000}^* - \overline{\theta}^*$. The 25th largest and 25th smallest of these differences are estimates of the unknown percentiles in (7.7). Consult the Rice or Efron books cited in Chapter 6 for more information.

Exercises | Section 7.1 (1–11)

1. Consider a normal population distribution with the value of σ known.
   a. What is the confidence level for the interval $\overline{x} \pm 2.81\sigma/\sqrt{n}$?
   b. What is the confidence level for the interval $\overline{x} \pm 1.44\sigma/\sqrt{n}$?
   c. What value of $z_{\alpha/2}$ in the CI formula (7.5) results in a confidence level of 99.7%?
   d. Answer the question posed in part (c) for a confidence level of 75%.

2. Each of the following is a confidence interval for $\mu =$ true average (i.e., population mean) resonance frequency (Hz) for all tennis rackets of a certain type:

   (114.4, 115.6)   (114.1, 115.9)

   a. What is the value of the sample mean resonance frequency?
   b. Both intervals were calculated from the same sample data. The confidence level for one of these intervals is 90% and for the other is 99%. Which of the intervals has the 90% confidence level, and why?

3. Suppose that a random sample of 50 bottles of a particular brand of cough syrup is selected, and the alcohol content of each bottle is determined. Let μ denote the average alcohol content for the population of all bottles of the brand under study. Suppose that the resulting 95% confidence interval is (7.8, 9.4).
   a. Would a 90% confidence interval calculated from this same sample have been narrower or wider than the given interval? Explain your reasoning.
   b. Consider the following statement: There is a 95% chance that μ is between 7.8 and 9.4. Is this statement correct? Why or why not?
   c. Consider the following statement: We can be highly confident that 95% of all bottles of this type of cough syrup have an alcohol content that

is between 7.8 and 9.4. Is this statement correct? Why or why not?
   d. Consider the following statement: If the process of selecting a sample of size 50 and then computing the corresponding 95% interval is repeated 100 times, 95 of the resulting intervals will include μ. Is this statement correct? Why or why not?

4. A CI is desired for the true average stray-load loss μ (watts) for a certain type of induction motor when the line current is held at 10 amps for a speed of 1500 rpm. Assume that stray-load loss is normally distributed with $\sigma = 3.0$.
   a. Compute a 95% CI for μ when $n = 25$ and $\overline{x} = 58.3$.
   b. Compute a 95% CI for μ when $n = 100$ and $\overline{x} = 58.3$.
   c. Compute a 99% CI for μ when $n = 100$ and $\overline{x} = 58.3$.
   d. Compute an 82% CI for μ when $n = 100$ and $\overline{x} = 58.3$.
   e. How large must n be if the width of the 99% interval for μ is to be 1.0?

5. Assume that the helium porosity (in percentage) of coal samples taken from any particular seam is normally distributed with true standard deviation .75.
   a. Compute a 95% CI for the true average porosity of a certain seam if the average porosity for 20 specimens from the seam was 4.85.
   b. Compute a 98% CI for true average porosity of another seam based on 16 specimens with a sample average porosity of 4.56.
   c. How large a sample size is necessary if the width of the 95% interval is to be .40?
   d. What sample size is necessary to estimate true average porosity to within .2 with 99% confidence?

6. On the basis of extensive tests, the yield point of a particular type of mild steel-reinforcing bar is known to be normally distributed with $\sigma = 100$.

The composition of the bar has been slightly modified, but the modification is not believed to have affected either the normality or the value of σ.

a. Assuming this to be the case, if a sample of 25 modified bars resulted in a sample average yield point of 8439 lb, compute a 90% CI for the true average yield point of the modified bar.

b. How would you modify the interval in part (a) to obtain a confidence level of 92%?

7. By how much must the sample size n be increased if the width of the CI (7.5) is to be halved? If the sample size is increased by a factor of 25, what effect will this have on the width of the interval? Justify your assertions.

8. Let $\alpha_1 > 0$, $\alpha_2 > 0$, with $\alpha_1 + \alpha_2 = \alpha$. Then

$$P\left(-z_{\alpha_1} < \frac{\overline{X} - \mu}{\sigma/\sqrt{n}} < z_{\alpha_2}\right) = 1 - \alpha$$

a. Use this equation to derive a more general expression for a $100(1 - \alpha)\%$ CI for μ of which the interval (7.5) is a special case.

b. Let $\alpha = .05$ and $\alpha_1 = \alpha/4$, $\alpha_2 = 3\alpha/4$. Does this result in a narrower or wider interval than the interval (7.5)?

9. a. Under the same conditions as those leading to the interval (7.5), $P[(\overline{X} - \mu)/(\sigma/\sqrt{n}) < 1.645] = .95$. Use this to derive a one-sided interval for μ that has infinite width and provides a lower confidence bound on μ. What is this interval for the data in Exercise 5(a)?

b. Generalize the result of part (a) to obtain a lower bound with confidence level $100(1 - \alpha)\%$.

c. What is an analogous interval to that of part (b) that provides an upper bound on μ? Compute this 99% interval for the data of Exercise 4(a).

10. A random sample of $n = 15$ heat pumps of a certain type yielded the following observations on lifetime (in years):

2.0	1.3	6.0	1.9	5.1	.4	1.0	5.3
15.7	.7	4.8	.9	12.2	5.3	.6	

a. Assume that the lifetime distribution is exponential and use an argument parallel to that of Example 7.5 to obtain a 95% CI for expected (true average) lifetime.

b. How should the interval of part (a) be altered to achieve a confidence level of 99%?

c. What is a 95% CI for the standard deviation of the lifetime distribution? (*Hint:* What is the standard deviation of an exponential random variable?)

11. Consider the next 1000 95% CIs for μ that a statistical consultant will obtain for various clients. Suppose the data sets on which the intervals are based are selected independently of one another. How many of these 1000 intervals do you expect to capture the corresponding value of μ? What is the probability that between 940 and 960 of these intervals contain the corresponding value of μ? (*Hint:* Let $Y =$ the number among the 1000 intervals that contain μ. What kind of random variable is Y?)

7.2 Large-Sample Confidence Intervals for a Population Mean and Proportion

The CI for μ given in the previous section assumed that the population distribution is normal and that the value of σ is known. We now present a large-sample CI whose validity does not require these assumptions. After showing how the argument leading to this interval generalizes to yield other large-sample intervals, we focus on an interval for a population proportion p.

A Large-Sample Interval for μ

Let $X_1, X_2, \ldots, X_n$ be a random sample from a population having a mean μ and standard deviation σ. Provided that n is large, the Central Limit Theorem (CLT) implies that $\overline{X}$ has approximately a normal distribution whatever the nature of the population distribution. It then follows that $Z = (\overline{X} - \mu)/(\sigma/\sqrt{n})$ has approximately a standard normal distribution, so that

$$P\left(-z_{\alpha/2} < \frac{\overline{X} - \mu}{\sigma/\sqrt{n}} < z_{\alpha/2}\right) \approx 1 - \alpha$$

An argument parallel to that given in Section 7.1 yields $\bar{x} \pm z_{\alpha/2} \cdot \sigma/\sqrt{n}$ as a large-sample CI for μ with a confidence level of *approximately* $100(1 - \alpha)\%$. That is, when n is large, the CI for μ given previously remains valid whatever the population distribution, provided that the qualifier "approximately" is inserted in front of the confidence level.

One practical difficulty with this development is that computation of the interval requires the value of σ, which will almost never be known. Consider the standardized variable

$$Z = \frac{\overline{X} - \mu}{S/\sqrt{n}}$$

in which the sample standard deviation S replaces σ. Previously there was randomness only in the numerator of Z (by virtue of $\overline{X}$). Now there is randomness in both the numerator and the denominator—the values of both $\overline{X}$ and S vary from sample to sample. However, when n is large, the use of S rather than σ adds very little extra variability to Z. More specifically, in this case the new Z also has approximately a standard normal distribution. Manipulation of the inequalities in a probability statement involving this new Z yields a general large-sample interval for μ.

PROPOSITION

If n is sufficiently large, the standardized variable

$$Z = \frac{\overline{X} - \mu}{S/\sqrt{n}}$$

has approximately a standard normal distribution. This implies that

$$\bar{x} \pm z_{\alpha/2} \cdot \frac{s}{\sqrt{n}} \qquad (7.8)$$

is a **large-sample confidence interval for μ** with confidence level approximately $100(1 - \alpha)\%$. This formula is valid regardless of the shape of the population distribution.

Generally speaking, $n > 40$ will be sufficient to justify the use of this interval. This is somewhat more conservative than the rule of thumb for the CLT because of the additional variability introduced by using s in place of σ.

Example 7.6   The alternating-current (AC) breakdown voltage of an insulating liquid indicates its dielectric strength. The article "Testing Practices for the AC Breakdown Voltage Testing of Insulation Liquids," *IEEE Electrical Insulation Magazine,* 1995: 21–26) gave the accompanying sample observations on breakdown voltage (kV) of a particular circuit under certain conditions.

62	50	53	57	41	53	55	61	59	64	50	53	64	62	50	68
54	55	57	50	55	50	56	55	46	55	53	54	52	47	47	55
57	48	63	57	57	55	53	59	53	52	50	55	60	50	56	58

A boxplot of the data (Figure 7.5) shows a high concentration in the middle half of the data (narrow box width). There is a single outlier at the upper end, but this value is actually a bit closer to the median (55) than is the smallest sample observation.

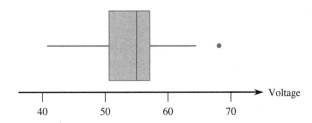

Figure 7.5 A boxplot for the breakdown voltage data from Example 7.6

Summary quantities include $n = 48$, $\Sigma x_i = 2626$, and $\Sigma x_i^2 = 144{,}950$, from which $\bar{x} = 54.7$ and $s = 5.23$. The 95% confidence interval is then

$$54.7 \pm 1.96 \frac{5.23}{\sqrt{48}} = 54.7 \pm 1.5 = (53.2, 56.2)$$

That is,

$$53.2 < \mu < 56.2$$

with a confidence level of approximately 95%. The interval is reasonably narrow, indicating that we have precisely estimated μ. ■

Unfortunately, the choice of sample size to yield a desired interval width is not as straightforward here as it was for the case of known σ. This is because the width of (7.8) is $2z_{\alpha/2}s/\sqrt{n}$. Since the value of s is not available before the data has been gathered, the width of the interval cannot be determined solely by the choice of n. The only option for an investigator who wishes to specify a desired width is to make an educated guess as to what the value of s might be. By being conservative and guessing a larger value of s, an n larger than necessary will be chosen. The investigator may be able to specify a reasonably accurate value of the population range (the difference between the largest and smallest values). Then if the population distribution is not too skewed, dividing the range by 4 gives a ballpark value of what s might be.

Example 7.7 Refer to Example 7.6 on breakdown voltage. Suppose the investigator believes that virtually all values in the population are between 40 and 70. Then $(70 - 40)/4 = 7.5$ gives a reasonable value for s. The appropriate sample size for estimating true average break-down voltage to within 1 kV with confidence level 95%—that is, for the 95% CI to have a width of 2 kV—is

$$n = [(1.96)(7.5)/1]^2 \approx 217$$ ∎

A General Large-Sample Confidence Interval

The large-sample intervals $\bar{x} \pm z_{\alpha/2} \cdot \sigma/\sqrt{n}$ and $\bar{x} \pm z_{\alpha/2} \cdot s/\sqrt{n}$ are special cases of a general large-sample CI for a parameter θ. Suppose that $\hat{\theta}$ is an estimator satisfying the following properties: (1) It has approximately a normal distribution; (2) it is (at least approximately) unbiased; and (3) an expression for $\sigma_{\hat{\theta}}$, the standard deviation of $\hat{\theta}$, is available. For example, in the case $\theta = \mu$, $\hat{\mu} = \bar{X}$ is an unbiased estimator whose distribution is approximately normal when n is large and $\sigma_{\hat{\mu}} = \sigma_{\bar{x}} = \sigma/\sqrt{n}$. Standardizing $\hat{\theta}$ yields the rv $Z = (\hat{\theta} - \theta)/\sigma_{\hat{\theta}}$, which has approximately a standard normal distribution. This justifies the probability statement

$$P\left(-z_{\alpha/2} < \frac{\hat{\theta} - \theta}{\sigma_{\hat{\theta}}} < z_{\alpha/2}\right) \approx 1 - \alpha \qquad (7.9)$$

Suppose, first, that $\sigma_{\hat{\theta}}$ does not involve any unknown parameters (e.g., known σ in the case $\theta = \mu$). Then replacing each $<$ in (7.9) by $=$ results in $\theta = \hat{\theta} \pm z_{\alpha/2} \cdot \sigma_{\hat{\theta}}$, so the lower and upper confidence limits are $\hat{\theta} - z_{\alpha/2} \cdot \sigma_{\hat{\theta}}$ and $\hat{\theta} + z_{\alpha/2} \cdot \sigma_{\hat{\theta}}$, respectively. Now suppose that $\sigma_{\hat{\theta}}$ does not involve θ but does involve at least one other unknown parameter. Let $s_{\hat{\theta}}$ be the estimate of $\sigma_{\hat{\theta}}$ obtained by using estimates in place of the unknown parameters (e.g., $s/\sqrt{n}$ estimates $\sigma/\sqrt{n}$). Under general conditions (essentially that $s_{\hat{\theta}}$ be close to $\sigma_{\hat{\theta}}$ for most samples), a valid CI is $\hat{\theta} \pm z_{\alpha/2} \cdot s_{\hat{\theta}}$. The interval $\bar{x} \pm z_{\alpha/2} \cdot s/\sqrt{n}$ is an example.

Finally, suppose that $\sigma_{\hat{\theta}}$ does involve the unknown θ. This is the case, for example, when $\theta = p$, a population proportion. Then $(\hat{\theta} - \theta)/\sigma_{\hat{\theta}} = z_{\alpha/2}$ can be difficult to solve. An approximate solution can often be obtained by replacing θ in $\sigma_{\hat{\theta}}$ by its estimate $\hat{\theta}$. This results in an estimated standard deviation $s_{\hat{\theta}}$, and the corresponding interval is again $\hat{\theta} \pm z_{\alpha/2} \cdot s_{\hat{\theta}}$.

A Large-Sample Confidence Interval for a Population Proportion

Let p denote the proportion of "successes" in a population, where *success* identifies an individual or object that has a specified property. A random sample of n individuals is to be selected, and X is the number of successes in the sample. Provided that n is small compared to the population size, X can be regarded as a binomial rv with $E(X) = np$ and $\sigma_X = \sqrt{np(1 - p)}$. Furthermore, if n is large ($np \geq 10$ and $nq \geq 10$), X has approximately a normal distribution.

The natural estimator of p is $\hat{p} = X/n$, the sample fraction of successes. Since $\hat{p}$ is just X multiplied by the constant $1/n$, $\hat{p}$ also has approximately a normal distribution. As shown in Section 6.1, $E(\hat{p}) = p$ (unbiasedness) and $\sigma_{\hat{p}} = \sqrt{p(1 - p)/n}$. The standard deviation $\sigma_{\hat{p}}$ involves the unknown parameter p. Standardizing $\hat{p}$ by subtracting p and dividing by $\sigma_{\hat{p}}$ then implies that

$$P\left(-z_{\alpha/2} < \frac{\hat{p} - p}{\sqrt{p(1 - p)/n}} < z_{\alpha/2}\right) \approx 1 - \alpha$$

Proceeding as suggested in the subsection Deriving a Confidence Interval (Section 7.1), the confidence limits result from replacing each $<$ by $=$ and solving the resulting quadratic equation for p. This gives the two roots

$$p = \frac{\hat{p} + \frac{z_{\alpha/2}^2}{2n} \pm z_{\alpha/2} \sqrt{\frac{\hat{p}\hat{q}}{n} + \frac{z_{\alpha/2}^2}{4n^2}}}{1 + (z_{\alpha/2}^2)/n}$$

PROPOSITION

A **confidence interval for a population proportion p** with confidence level approximately $100(1 - \alpha)\%$ has

$$\text{lower confidence limit} = \frac{\hat{p} + \frac{z_{\alpha/2}^2}{2n} - z_{\alpha/2} \sqrt{\frac{\hat{p}\hat{q}}{n} + \frac{z_{\alpha/2}^2}{4n^2}}}{1 + (z_{\alpha/2}^2)/n}$$

and (7.10)

$$\text{upper confidence limit} = \frac{\hat{p} + \frac{z_{\alpha/2}^2}{2n} + z_{\alpha/2} \sqrt{\frac{\hat{p}\hat{q}}{n} + \frac{z_{\alpha/2}^2}{4n^2}}}{1 + (z_{\alpha/2}^2)/n}$$

If the sample size is large, $z^2/(2n)$ is negligible compared to $\hat{p}$, $z^2/(4n^2)$ under the square root is negligible compared to $\hat{p}\hat{q}/n$, and z^2/n is negligible compared to 1. Discarding these negligible terms gives approximate confidence limits

$$\hat{p} \pm z_{\alpha/2} \sqrt{\hat{p}\hat{q}/n} \tag{7.11}$$

This is of the general form $\hat{\theta} \pm z_{\alpha/2}\, \hat{\sigma}_{\hat{\theta}}$ of a large-sample interval suggested in the last subsection. For decades this latter interval has been recommended as long as the normal approximation for $\hat{p}$ is justified. However, recent research has shown that the somewhat more complicated interval given in the proposition has an actual confidence level that tends to be closer to the nominal level than does the traditional interval (Agresti, Alan, and Brent Coull, "Approximate Is Better Than 'Exact' for Interval Estimation of a Binomial Proportion," *The American Statistician,* 1998: 119–126). That is, if $z_{\alpha/2} = 1.96$ is used, the confidence level for the "new" interval tends to be closer to 95% for almost all values of p than is the case for the traditional interval; this is also true for other confidence levels. In addition, Agresti and Coull state that the interval "can be recommended for use with nearly all sample sizes and parameter values," so the conditions $n\hat{p} \geq 10$ and $n\hat{q} \geq 10$ need not be checked.

Example 7.8

The article "Repeatability and Reproducibility for Pass/Fail Data" (*J. of Testing and Eval.,* 1997: 151–153) reported that in $n = 48$ trials in a particular laboratory, 16 resulted in ignition of a particular type of substrate by a lighted cigarette. Let p denote the long-run proportion of all such trials that would result in ignition. A point estimate for p is $\hat{p} = 16/48 = .333$. A confidence interval for p with a confidence level of approximately 95% is

$$\frac{.333 + (1.96)^2/96 \pm 1.96\sqrt{(.333)(.667)/48 + (1.96)^2/9216}}{1 + (1.96)^2/48}$$

$$= \frac{.373 \pm .139}{1.08} = (.217, .474)$$

The traditional interval is

$$.333 \pm 1.96\sqrt{(.333)(.667)/48} = .333 \pm .133 = (.200, .466)$$

These two intervals would be in much closer agreement were the sample size substantially larger. ∎

Equating the width of the CI for p to a prespecified width w gives a quadratic equation for the sample size n necessary to give an interval with a desired degree of precision. Suppressing the subscript in $z_{\alpha/2}$, the solution is

$$n = \frac{2z^2\hat{p}\hat{q} - z^2w^2 \pm \sqrt{4z^4\hat{p}\hat{q}(\hat{p}\hat{q} - w^2) + w^2z^4}}{w^2} \tag{7.12}$$

Neglecting the terms in the numerator involving w^2 gives

$$n \approx \frac{4z^2\hat{p}\hat{q}}{w^2}$$

This latter expression is what results from equating the width of the traditional interval to w.

These formulas unfortunately involve the unknown $\hat{p}$. The most conservative approach is to take advantage of the fact that $\hat{p}\hat{q} \; [= \hat{p}(1 - \hat{p})]$ is a maximum when $\hat{p} = .5$. Thus if $\hat{p} = \hat{q} = .5$ is used in (7.12), the width will be at most w regardless of what value of $\hat{p}$ results from the sample. Alternatively, if the investigator believes strongly, based on prior information, that $p \leq p_0 \leq .5$, then p_0 can be used in place of $\hat{p}$. A similar comment applies when $p \geq p_0 \geq .5$.

Example 7.9 The width of the 95% CI in Example 7.8 is .257. The value of n necessary to ensure a width of .10 irrespective of the value of $\hat{p}$ is

$$n = \frac{2(1.96)^2(.25) - (1.96)^2(.01) \pm \sqrt{4(1.96)^4(.25)(.25 - .01) + (.01)(1.96)^4}}{.01} = 380.3$$

Thus a sample size of 381 should be used. The expression for n based on the traditional CI gives a slightly larger value of 385. ∎

One-Sided Confidence Intervals

The confidence intervals discussed thus far give both a lower confidence bound *and* an upper confidence bound for the parameter being estimated. In some circumstances, an investigator will want only one of these two types of bounds. For example, a psychologist may wish to calculate a 95% upper confidence bound for true average reaction time to a particular stimulus, or a reliability engineer may want only a lower confidence bound for true average lifetime of components of a certain type. Because the cumulative area under the curve to the left of 1.645 is .95,

$$P\left(\frac{\overline{X} - \mu}{S/\sqrt{n}} < 1.645\right) \approx .95$$

Manipulating the inequality inside the parentheses to isolate μ on one side and replacing rv's by calculated values gives the inequality $\mu > \bar{x} - 1.645s/\sqrt{n}$; the expression on the right is the desired lower confidence bound. Starting with $P(-1.645 < z) \approx .95$ and manipulating the inequality results in the upper confidence bound. A similar argument gives a one-sided bound associated with any other confidence level.

PROPOSITION

> A **large-sample upper confidence bound for μ** is
>
> $$\mu < \bar{x} + z_\alpha \cdot \frac{s}{\sqrt{n}}$$
>
> and a **large-sample lower confidence bound for μ** is
>
> $$\mu > \bar{x} - z_\alpha \cdot \frac{s}{\sqrt{n}}$$
>
> A **one-sided confidence bound for p** results from replacing $z_{\alpha/2}$ by z_α and $\pm$ by either $+$ or $-$ in the CI formula (7.10) for p.

Example 7.10 The slant shear test is the most widely accepted procedure for assessing the quality of a bond between a repair material and its concrete substrate. The article "Testing the Bond Between Repair Materials and Concrete Substrate" (*ACI Materials J.,* 1996: 553–558) reported that in one particular investigation, a sample of 48 shear strength observations gave a sample mean strength of 17.17 N/mm² and a sample standard deviation of 3.28 N/mm². A lower confidence bound for true average shear strength μ with confidence level 95% is

$$17.17 - (1.645)\frac{(3.28)}{\sqrt{48}} = 17.17 - .78 = 16.39$$

That is, with a confidence level of 95%, the value of μ lies in the interval $(16.39, \infty)$. ■

Exercises | Section 7.2 (12–27)

12. A random sample of 110 lightning flashes in a certain region resulted in a sample average radar echo duration of .81 sec and a sample standard deviation of .34 sec ("Lightning Strikes to an Airplane in a Thunderstorm," *J. of Aircraft,* 1984: 607–611). Calculate a 99% (two-sided) confidence interval for the true average echo duration μ, and interpret the resulting interval.

13. The article "Extravisual Damage Detection? Defining the Standard Normal Tree" (*Photogrammetric Engr. and Remote Sensing,* 1981: 515–522) discusses the use of color infrared photography in identification of normal trees in Douglas fir stands. Among data reported were summary statistics for green-filter analytic optical densitometric measurements on samples of both healthy and diseased trees. For a sample of 69 healthy trees, the sample mean dye-layer density was 1.028, and the sample standard deviation was .163.

  a. Calculate a 95% (two-sided) CI for the true average dye-layer density for all such trees.

  b. Suppose the investigators had made a rough guess of .16 for the value of s before collecting data. What sample size would be necessary to obtain an interval width of .05 for a confidence level of 95%?

14. The article "Evaluating Tunnel Kiln Performance" (*Amer. Ceramic Soc. Bull.,* Aug. 1997: 59–63) gave the following summary information for fracture strengths (Mpa) of $n = 169$ ceramic bars fired in a particular kiln: $\bar{x} = 89.10$, $s = 3.73$.
  a. Calculate a (two-sided) confidence interval for true average fracture strength using a confidence level of 95%. Does it appear that true average fracture strength has been precisely estimated?
  b. Suppose the investigators had believed a priori that the population standard deviation was about 4 MPa. Based on this supposition, how large a sample would have been required to estimate μ to within .5 MPa with 95% confidence?

15. Determine the confidence level for each of the following large-sample one-sided confidence bounds:
  a. Upper bound: $\bar{x} + .84s/\sqrt{n}$
  b. Lower bound: $\bar{x} - 2.05s/\sqrt{n}$
  c. Upper bound: $\bar{x} + .67s/\sqrt{n}$

16. The charge-to-tap time (min) for a carbon steel in one type of open hearth furnace was determined for each heat in a sample of size 46, resulting in a sample mean time of 382.1 and a sample standard deviation of 31.5. Calculate a 95% upper confidence bound for true average charge-to-tap time.

17. A Brinell hardness test involves measuring the diameter of the indentation made when a hardened steel ball is pressed into material under a standard test load. Suppose that the Brinell hardness is determined for each specimen in a sample of size 50, resulting in a sample mean hardness of 64.3 and a sample standard deviation of 6.0. Calculate a 99% lower confidence bound for true average Brinell hardness for material specimens of this type.

18. The article "Ultimate Load Capacities of Expansion Anchor Bolts" (*J. of Energy Engr.,* 1993: 139–158) gave the following summary data on shear strength (kip) for a sample of 3/8 in. anchor bolts: $n = 78$, $\bar{x} = 4.25$, $s = 1.30$. Calculate a lower confidence bound using a confidence level of 90% for true average shear strength.

19. The article "Limited Yield Estimation for Visual Defect Sources" (*IEEE Trans. on Semiconductor Manuf.,* 1997: 17–23) reported that, in a study of a particular wafer inspection process, 356 dies were examined by an inspection probe and 201 of these passed the probe. Assuming a stable process, calculate a 95% (two-sided) confidence interval for the proportion of all dies that pass the probe.

20. The Associated Press (Dec. 16, 1991) reported that, in a sample of 507 adult Americans, only 142 correctly described the Bill of Rights as the first ten amendments to the U.S. Constitution. Calculate a (two-sided) confidence interval using a 99% confidence level for the proportion of all U.S. adults that could give a correct description of the Bill of Rights.

21. A random sample of 539 households from a certain midwestern city was selected, and it was determined that 133 of these households owned at least one firearm ("The Social Determinants of Gun Ownership: Self-Protection in an Urban Environment," *Criminology,* 1997: 629–640). Using a 95% confidence level, calculate a lower confidence bound for the proportion of all households in this city that own at least one firearm.

22. A random sample of 487 nonsmoking women of normal weight (body mass index between 19.8 and 26.0) who had given birth at a large metropolitan medical center was selected ("The Effects of Cigarette Smoking and Gestational Weight Change on Birth Outcomes in Obese and Normal-Weight Women," *Amer. J. of Public Health,* 1997: 591–596). It was determined that 7.2% of these births resulted in children of low birth weight (less than 2500 g). Calculate an upper confidence bound using a confidence level of 99% for the proportion of all such births that result in children of low birth weight.

23. The article "An Evaluation of Football Helmets Under Impact Conditions" (*Amer. J. Sports Medicine,* 1984: 233–237) reports that when each football helmet in a random sample of 37 suspension-type helmets was subjected to a certain impact test, 24 showed damage. Let p denote the proportion of all helmets of this type that would show damage when tested in the prescribed manner.
  a. Calculate a 99% CI for p.
  b. What sample size would be required for the width of a 99% CI to be at most .10, irrespective of $\hat{p}$?

24. A sample of 56 research cotton samples resulted in a sample average percentage elongation of 8.17 and a sample standard deviation of 1.42 ("An Apparent Relation Between the Spiral Angle ϕ, the Percent Elongation E_1, and the Dimensions of the Cotton Fiber," *Textile Research J.,* 1978: 407–410). Calculate a 95% large-sample CI for the true average percentage elongation μ. What assumptions are you making about the distribution of percentage elongation?

25. A state legislator wishes to survey residents of her district to see what proportion of the electorate is aware of her position on using state funds to pay for abortions.

a. What sample size is necessary if the 95% CI for p is to have width at most .10 irrespective of p?

b. If the legislator has strong reason to believe that at least $\frac{2}{3}$ of the electorate know of her position, how large a sample size would you recommend?

26. The superintendent of a large school district, having once had a course in probability and statistics, believes that the number of teachers absent on any given day has a Poisson distribution with parameter λ. Use the accompanying data on absences for 50 days to derive a large-sample CI for λ. [*Hint:* The mean and variance of a Poisson variable both equal λ, so

$$Z = \frac{\bar{X} - \lambda}{\sqrt{\lambda/n}}$$

has approximately a standard normal distribution. Now proceed as in the derivation of the interval for p by making a probability statement (with probability $1 - \alpha$) and solving the resulting inequalities for λ (see the argument just after (7.10)).]

Number of absences	0	1	2	3	4	5	6	7	8	9	10
Frequency	1	4	8	10	8	7	5	3	2	1	1

27. Reconsider the CI (7.10) for p, and focus on a confidence level of 95%. Show that the confidence limits agree quite well with those of the traditional interval (7.11) once two successes and two failures have been appended to the sample (i.e., (7.11) based on $x + 2$ S's in $n + 4$ trials). *Hint:* $1.96 \approx 2$. *Note:* Agresti and Coull showed that this adjustment of the traditional interval also has actual confidence level close to the nominal level.

7.3 | Intervals Based on a Normal Population Distribution

The CI for μ presented in Section 7.2 is valid provided that n is large. The resulting interval can be used whatever the nature of the population distribution. The CLT cannot be invoked, however, when n is small. In this case, one way to proceed is to make a specific assumption about the form of the population distribution and then derive a CI tailored to that assumption. For example, we could develop a CI for μ when the population is described by a gamma distribution, another interval for the case of a Weibull population, and so on. Statisticians have indeed carried out this program for a number of different distributional families. Because the normal distribution is more frequently appropriate as a population model than is any other type of distribution, we will focus here on a CI for this situation.

ASSUMPTION

> The population of interest is normal, so that $X_1, \ldots, X_n$ constitutes a random sample from a normal distribution with both μ and σ unknown.

The key result underlying the interval in Section 7.2 was that for large n, the rv $Z = (\bar{X} - \mu)/(S/\sqrt{n})$ has approximately a standard normal distribution. When n is small, S is no longer likely to be close to σ, so the variability in the distribution of Z arises from randomness in both the numerator and the denominator. This implies that the probability distribution of $(\bar{X} - \mu)/(S/\sqrt{n})$ will be more spread out than the standard normal distribution. The result on which inferences are based introduces a new family of probability distributions called the family of t distributions.

THEOREM

> When $\overline{X}$ is the mean of a random sample of size n from a normal distribution with mean μ, the rv
>
> $$T = \frac{\overline{X} - \mu}{S/\sqrt{n}} \qquad (7.13)$$
>
> has a probability distribution called a t distribution with $n - 1$ degrees of freedom (df).

Properties of t Distributions

Before applying this theorem, we must first discuss properties of t distributions. Although the variable of interest is still $(\overline{X} - \mu)/(S/\sqrt{n})$, we now denote it by T to emphasize that it does not have a standard normal distribution when n is small. Recall that a normal distribution is governed by two parameters, the mean μ and the standard deviation σ. A t distribution is governed by only one parameter, called the **number of degrees of freedom** of the distribution, abbreviated by df. We denote this parameter by the Greek letter ν. Possible values of ν are the positive integers 1, 2, 3, ... Each different value of ν corresponds to a different t distribution.

For any fixed value of the parameter ν, the density function that specifies the associated t curve has an even more complicated appearance than the normal density function. Fortunately, we need concern ourselves only with several of the more important features of these curves.

Properties of t Distributions

Let t_ν denote the density function curve for ν df.

1. Each t_ν curve is bell-shaped and centered at 0.

2. Each t_ν curve is more spread out than the standard normal (z) curve.

3. As ν increases, the spread of the corresponding t_ν curve decreases.

4. As $\nu \rightarrow \infty$, the sequence of t_ν curves approaches the standard normal curve (so the z curve is often called the t curve with df $= \infty$).

Figure 7.6 illustrates several of these properties for selected values of ν.

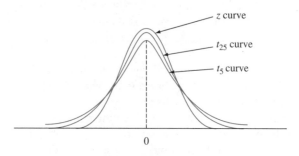

Figure 7.6 t_ν and z curves

The number of df for T in (7.13) is $n - 1$ because, although S is based on the n deviations $X_1 - \overline{X}, \ldots, X_n - \overline{X}, \sum(X_i - \overline{X}) = 0$ implies that only $n - 1$ of these are "freely determined." The number of df for a t variable is the number of freely determined deviations on which the estimated standard deviation in the denominator of T is based.

Since we want to use T to obtain a CI in the same way that Z was previously used, it is necessary to establish notation analogous to z_α for the t distribution.

Notation

Let $t_{\alpha,\nu}$ = the number on the measurement axis for which the area under the t curve with ν df to the right of $t_{\alpha,\nu}$ is α; $t_{\alpha,\nu}$ is called a **t critical value.**

This notation is illustrated in Figure 7.7. Appendix Table A.5 gives $t_{\alpha,\nu}$ for selected values of α and ν. This table also appears inside the back cover. The columns of the table correspond to different values of α. To obtain $t_{.05,15}$, go to the $\alpha = .05$ column, look down to the $\nu = 15$ row, and read $t_{.05,15} = 1.753$. Similarly, $t_{.05,22} = 1.717$ (.05 column, $\nu = 22$ row), and $t_{.01,22} = 2.508$.

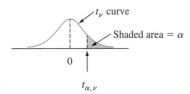

Figure 7.7 A pictorial definition of $t_{\alpha,\nu}$

The values of $t_{\alpha,\nu}$ exhibit regular behavior as we move across a row or down a column. For fixed ν, $t_{\alpha,\nu}$ increases as α decreases, since we must move farther to the right of zero to capture area α in the tail. For fixed α, as ν is increased (i.e., as we look down any particular column of the t table) the value of $t_{\alpha,\nu}$ decreases. This is because a larger value of ν implies a t distribution with smaller spread, so it is not necessary to go so far from zero to capture tail area α. Furthermore, $t_{\alpha,\nu}$ decreases more slowly as ν increases. Consequently, the table values are shown in increments of 2 between 30 df and 40 df, and then jump to $\nu = 50, 60, 120$, and finally ∞. Because t_∞ is the standard normal curve, the familiar z_α values appear in the last row of the table. The rule of thumb suggested earlier for use of the large-sample CI (if $n > 40$) comes from the approximate equality of the standard normal and t distributions for $\nu \geq 40$.

The One-Sample t Confidence Interval

The standardized variable T has a t distribution with $n - 1$ df, and the area under the corresponding t density curve between $-t_{\alpha/2,n-1}$ and $t_{\alpha/2,n-1}$ is $1 - \alpha$ (area $\alpha/2$ lies in each tail), so

$$P(-t_{\alpha/2,n-1} < T < t_{\alpha/2,n-1}) = 1 - \alpha \tag{7.14}$$

Expression (7.14) differs from expressions in previous sections in that T and $t_{\alpha/2,n-1}$ are used in place of Z and $z_{\alpha/2}$, but it can be manipulated in the same manner to obtain a confidence interval for μ.

PROPOSITION

Let $\bar{x}$ and s be the sample mean and sample standard deviation computed from the results of a random sample from a normal population with mean μ. Then a **$100(1-\alpha)\%$ confidence interval for μ** is

$$\left(\bar{x} - t_{\alpha/2,n-1} \cdot \frac{s}{\sqrt{n}}, \bar{x} + t_{\alpha/2,n-1} \cdot \frac{s}{\sqrt{n}} \right) \qquad (7.15)$$

or, more compactly, $\bar{x} \pm t_{\alpha/2,n-1} \cdot s/\sqrt{n}$.

An **upper confidence bound for μ** is

$$\bar{x} + t_{\alpha,n-1} \cdot \frac{s}{\sqrt{n}}$$

and replacing $+$ by $-$ in this latter expression gives a **lower confidence bound for μ,** both with confidence level $100(1-\alpha)\%$.

Example 7.11   As part of a larger project to study the behavior of stressed-skin panels, a structural component being used extensively in North America, the article "Time-Dependent Bending Properties of Lumber" (*J. of Testing and Eval.*, 1996: 187–193) reported on various mechanical properties of Scotch pine lumber specimens. Consider the following observations on modulus of elasticity (MPa) obtained 1 minute after loading in a certain configuration:

10,490	16,620	17,300	15,480	12,970	17,260	13,400	13,900
13,630	13,260	14,370	11,700	15,470	17,840	14,070	14,760

Figure 7.8 shows a normal probability plot obtained from MINITAB. The straightness of the pattern in the plot provides strong support for assuming that the population distribution of modulus of elasticity is at least approximately normal.

Hand calculation of the sample mean and standard deviation is simplified by subtracting 10,000 from each observation: $y_i = x_i - 10,000$. It is easily verified that $\sum y_i = 72,520$ and $\sum y_i^2 = 392,083,800$, from which $\bar{y} = 4532.5$ and $s_y = 2055.67$. Thus $\bar{x} = 14,532.5$ and $s_x = 2055.67$ (adding or subtracting the same constant from each observation does not affect variability). The sample size is 16, so a confidence interval for population mean modulus of elasticity is based on 15 df. A confidence level of 95% for a two-sided interval requires the t critical value of 2.131. The resulting interval is

$$\bar{x} \pm t_{.025,15} \cdot \frac{s}{\sqrt{n}} = 14,532.5 \pm (2.131)\frac{2055.67}{\sqrt{16}}$$

$$= 14,532.5 \pm 1095.2 = (13,437.3, 15,627.7)$$

This interval is quite wide both because of the small sample size and because of the large amount of variability in the sample. A 95% lower confidence bound is obtained by using $-$ and 1.753 in place of $\pm$ and 2.131, respectively.

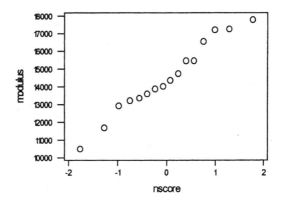

Figure 7.8 A normal probability plot of the modulus of elasticity data ■

Unfortunately, it is not easy to select n to control the width of the t interval. This is because the width involves the unknown (before the data is collected) s and because n enters not only through $1/\sqrt{n}$ but also through $t_{\alpha/2,n-1}$. As a result, an appropriate n can be obtained only by trial and error.

In Chapter 15, we will discuss a small-sample CI for μ that is valid provided only that the population distribution is symmetric, a weaker assumption than normality. However, when the population distribution is normal, the t interval tends to be shorter than would be *any* other interval with the same confidence level.

A Prediction Interval for a Single Future Value

In many applications, an investigator wishes to *predict* a single value of a variable to be observed at some future time, rather than to *estimate* the mean value of that variable.

Example 7.12 Consider the following sample of fat content (in percentage) of $n = 10$ randomly selected hot dogs ("Sensory and Mechanical Assessment of the Quality of Frankfurters," *J. Texture Studies,* 1990: 395–409):

$$25.2 \quad 21.3 \quad 22.8 \quad 17.0 \quad 29.8 \quad 21.0 \quad 25.5 \quad 16.0 \quad 20.9 \quad 19.5$$

Assuming that these were selected from a normal population distribution, a 95% CI for (interval estimate of) the population mean fat content is

$$\bar{x} \pm t_{.025,9} \cdot \frac{s}{\sqrt{n}} = 21.90 \pm 2.262 \cdot \frac{4.134}{\sqrt{10}} = 21.90 \pm 2.96$$

$$= (18.94, 24.86)$$

Suppose, however, you are going to eat a single hot dog of this type and want a *prediction* for the resulting fat content. A *point* prediction, analogous to a *point* estimate, is just $\bar{x} = 21.90$. This prediction unfortunately gives no information about reliability or precision. ■

The general setup is as follows: We will have available a random sample X_1, $X_2, \ldots, X_n$ from a normal population distribution, and we wish to predict the value of

X_{n+1}, a single future observation. A point predictor is $\overline{X}$, and the resulting prediction error is $\overline{X} - X_{n+1}$. The expected value of the prediction error is

$$E(\overline{X} - X_{n+1}) = E(\overline{X}) - E(X_{n+1}) = \mu - \mu = 0$$

Since X_{n+1} is independent of $X_1, \ldots, X_n$, it is independent of $\overline{X}$, so the variance of the prediction error is

$$V(\overline{X} - X_{n+1}) = V(\overline{X}) + V(X_{n+1}) = \frac{\sigma^2}{n} + \sigma^2 = \sigma^2\left(1 + \frac{1}{n}\right)$$

The prediction error is a linear combination of independent normally distributed rv's, so itself is normally distributed. Thus,

$$Z = \frac{(\overline{X} - X_{n+1}) - 0}{\sqrt{\sigma^2\left(1 + \frac{1}{n}\right)}} = \frac{\overline{X} - X_{n+1}}{\sqrt{\sigma^2\left(1 + \frac{1}{n}\right)}}$$

has a standard normal distribution. It can be shown that replacing σ by the sample standard deviation S (of $X_1, \ldots, X_n$) results in

$$T = \frac{\overline{X} - X_{n+1}}{S\sqrt{1 + \frac{1}{n}}} \sim t \text{ distribution with } n - 1 \text{ df}$$

Manipulating this T variable as $T = (\overline{X} - \mu)/(S/\sqrt{n})$ was manipulated in the development of a CI gives the following result.

PROPOSITION

> A **prediction interval** (PI) for a single observation to be selected from a normal population distribution is
>
> $$\overline{x} \pm t_{\alpha/2, n-1} \cdot s\sqrt{1 + \frac{1}{n}} \qquad (7.16)$$
>
> The *prediction level* is $100(1 - \alpha)\%$.

The interpretation of a 95% prediction level is similar to that of a 95% confidence level; if the interval (7.16) is calculated for sample after sample, in the long run 95% of these intervals will include the corresponding future values of X.

Example 7.13
(Example 7.12
continued)

With $n = 10$, $\overline{x} = 21.90$, $s = 4.134$, and $t_{.025,9} = 2.262$, a 95% PI for the fat content of a single hot dog is

$$21.90 \pm (2.262)(4.134)\sqrt{1 + \frac{1}{10}} = 21.90 \pm 9.81$$

$$= (12.09, 31.71)$$

This interval is quite wide, indicating substantial uncertainty about fat content. Notice that the width of the PI is more than three times that of the CI. ∎

The error of prediction is $\overline{X} - X_{n+1}$, a difference between two random variables, whereas the estimation error is $\overline{X} - \mu$, the difference between a random variable and a fixed (but unknown) value. The PI is wider than the CI because there is more variability in the prediction error (due to X_{n+1}) than in the estimation error. In fact, as n gets arbitrarily large, the CI shrinks to the single value μ, and the PI approaches $\mu \pm z_{\alpha/2} \cdot \sigma$. There is uncertainty about a single X value even when there is no need to estimate.

Tolerance Intervals

Consider a population of automobiles of a certain type, and suppose that under specified conditions, fuel efficiency (mpg) has a normal distribution with $\mu = 30$ and $\sigma = 2$. Then since the interval from -1.645 to 1.645 captures 90% of the area under the z curve, 90% of all these automobiles will have fuel efficiency values between $\mu - 1.645\sigma = 26.71$ and $\mu + 1.645\sigma = 33.29$. But what if the values of μ and σ are not known? We can take a sample of size n, determine the fuel efficiencies, $\overline{x}$, and s, and form the interval whose lower limit is $\overline{x} - 1.645s$ and whose upper limit is $\overline{x} + 1.645s$. However, because of sampling variability in the estimates of μ and σ, there is a good chance that the resulting interval will include less than 90% of the population values. Intuitively, to have an a priori 95% chance of the resulting interval including at least 90% of the population values, when $\overline{x}$ and s are used in place of μ and σ, we should also replace 1.645 by some larger number. For example, when $n = 20$, the value 2.310 is such that we can be 95% confident that the interval $\overline{x} \pm 2.310s$ will include at least 90% of the fuel efficiency values in the population.

> Let k be a number between 0 and 100. A **tolerance interval** for capturing at least $k\%$ of the values in a normal population distribution with a confidence level 95% has the form
>
> $$\overline{x} \pm (\text{tolerance critical value}) \cdot s$$
>
> Tolerance critical values for $k = 90, 95$, and 99 in combination with various sample sizes are given in Appendix Table A.6. This table also includes critical values for a confidence level of 99% (these values are larger than the corresponding 95% values). Replacing $\pm$ by $+$ gives an upper tolerance bound, and using $-$ in place of $\pm$ results in a lower tolerance bound. Critical values for obtaining these one-sided bounds also appear in Appendix Table A.6.

Example 7.14 Let's return to the modulus of elasticity data discussed in Example 7.11, where $n = 16$, $\overline{x} = 14,532.5$, $s = 2055.67$, and a normal probability plot of the data indicated that population normality was quite plausible. For a confidence level of 95%, a two-sided tolerance interval for capturing at least 95% of the modulus of elasticity values for specimens of lumber in the population sampled uses the tolerance critical value of 2.903. The resulting interval is

$$14,532.5 \pm (2.903)(2055.67) = 14,532.5 \pm 5967.6 = (8,564.9, 20,500.1)$$

We can be highly confident that at least 95% of all lumber specimens have modulus of elasticity values between 8,564.9 and 20,500.1.

The 95% CI for μ was (13,437.3, 15,627.7), and the 95% prediction interval for the modulus of elasticity of a single lumber specimen is (10,017.0, 19,048.0). Both the prediction interval and the tolerance interval are substantially wider than the confidence interval. ∎

Intervals Based on Nonnormal Population Distributions

The one-sample t CI for μ is robust to small or even moderate departures from normality unless n is quite small. By this we mean that if a critical value for 95% confidence, for example, is used in calculating the interval, the actual confidence level will be reasonably close to the nominal 95% level. If, however, n is small and the population distribution is highly nonnormal, then the actual confidence level may be considerably different from the one you think you are using when you obtain a particular critical value from the t table. It would certainly be distressing to believe that your confidence level is about 95% when in fact it was really more like 88%! The bootstrap technique, introduced in Section 7.1, has been found to be quite successful at estimating parameters in a wide variety of nonnormal situations.

In contrast to the confidence interval, the validity of the prediction and tolerance intervals described in this section is closely tied to the normality assumption. These latter intervals should not be used in the absence of compelling evidence for normality. The excellent reference *Statistical Intervals,* cited in the bibliography at the end of this chapter, discusses alternative procedures of this sort for various other situations.

Exercises | Section 7.3 (28–39)

28. Determine the values of the following quantities:
   a. $t_{.1,15}$ **b.** $t_{.05,15}$ **c.** $t_{.05,25}$ **d.** $t_{.05,40}$ **e.** $t_{.005,40}$

29. Determine the t critical value that will capture the desired t curve area in each of the following cases.
   a. Central area = .95, df = 10
   b. Central area = .95, df = 20
   c. Central area = .99, df = 20
   d. Central area = .99, df = 50
   e. Upper-tail area = .01, df = 25
   f. Lower-tail area = .025, df = 5

30. Determine the t critical value for a two-sided confidence interval in each of the following situations.
   a. Confidence level = 95%, df = 10
   b. Confidence level = 95%, df = 15
   c. Confidence level = 99%, df = 15
   d. Confidence level = 99%, n = 5
   e. Confidence level = 98%, df = 24
   f. Confidence level = 99%, n = 38

31. Determine the t critical value for a lower or an upper confidence bound for each of the situations described in Exercise 30.

32. A random sample of $n = 8$ E-glass fiber test specimens of a certain type yielded a sample mean interfacial shear yield stress of 30.2 and a sample standard deviation of 3.1 ("On Interfacial Failure in Notched Unidirectional Glass/Epoxy Composites," *J. of Composite Materials,* 1985: 276–286). Assuming that interfacial shear yield stress is normally distributed, compute a 95% CI for true average stress (as did the authors of the cited article).

33. The article "Measuring and Understanding the Aging of Kraft Insulating Paper in Power Transformers" (*IEEE Electrical Insul. Mag.,* 1996: 28–34) contained the following observations on degree of polymerization for paper specimens for which viscosity times concentration fell in a certain middle range.

418	421	421	422	425	427	431
434	437	439	446	447	448	453
454	463	465

   a. Construct a boxplot of the data and comment on any interesting features.

b. Is it plausible that the given sample observations were selected from a normal distribution?

c. Calculate a two-sided 95% confidence interval for true average degree of polymerization (as did the authors of the article). Does the interval suggest that 440 is a plausible value for true average degree of polymerization? What about 450?

34. A sample of 14 joint specimens of a particular type gave a sample mean proportional limit stress of 8.48 MPa and a sample standard deviation of .79 MPa ("Characterization of Bearing Strength Factors in Pegged Timber Connections," *J. of Structural Engr.*, 1997: 326–332).

a. Calculate and interpret a 95% lower confidence bound for the true average proportional limit stress of all such joints. What, if any, assumptions did you make about the distribution of proportional limit stress?

b. Calculate and interpret a 95% lower prediction bound for the proportional limit stress of a single joint of this type.

35. Exercise 46 in Chapter 1 introduced the following sample observations on stabilized viscosity of asphalt specimens: 2781, 2900, 3013, 2856, 2888. A normal probability plot supports the assumption that viscosity is at least approximately normally distributed.

a. Estimate true average viscosity in a way that conveys information about precision and reliability.

b. Predict the viscosity for a single asphalt specimen in a way that conveys information about precision and reliability. How does the prediction compare to the estimate calculated in part (a)?

36. The $n = 26$ observations on escape time given in Exercise 36 of Chapter 1 give a sample mean and sample standard deviation of 370.69 and 24.36, respectively.

a. Calculate an upper confidence bound for population mean escape time using a confidence level of 95%.

b. Calculate an upper prediction bound for the escape time of a single additional worker using a prediction level of 95%. How does this bound compare with the confidence bound of part (a)?

c. Suppose that two additional workers will be chosen to participate in the simulated escape exercise. Denote their escape times by X_{27} and X_{28}, and let $\overline{X}_{new}$ denote the average of these two values. Modify the formula for a PI for a single x value to obtain a PI for $\overline{X}_{new}$, and calculate a 95% two-sided interval based on the given escape data.

37. A study of the ability of individuals to walk in a straight line ("Can We Really Walk Straight?" *Amer. J. of Physical Anthro.*, 1992: 19–27) reported the accompanying data on cadence (strides per second) for a sample of $n = 20$ randomly selected healthy men:

.95 .85 .92   .95 .93   .86 1.00 .92 .85 .81

.78 .93 .93 1.05 .93 1.06 1.06 .96 .81 .96

A normal probability plot gives substantial support to the assumption that the population distribution of cadence is approximately normal. A descriptive summary of the data from MINITAB follows.

Variable	N	Mean	Median	TrMean	StDev	SEMean
cadence	20	0.9255	0.9300	0.9261	0.0809	0.0181

Variable	Min	Max	Q1	Q3
cadence	0.7800	1.0600	0.8525	0.9600

a. Calculate and interpret a 95% confidence interval for population mean cadence.

b. Calculate and interpret a 95% prediction interval for the cadence of a single individual randomly selected from this population.

c. Calculate an interval that includes at least 99% of the cadences in the population distribution using a confidence level of 95%.

38. A sample of 25 pieces of laminate used in the manufacture of circuit boards was selected and the amount of warpage (in.) under particular conditions was determined for each piece, resulting in a sample mean warpage of .0635 and a sample standard deviation of .0065.

a. Calculate a prediction for the amount of warpage of a single piece of laminate in a way that provides information about precision and reliability.

b. Calculate an interval for which you can have a high degree of confidence that at least 95% of all pieces of laminate result in amounts of warpage that are between the two limits of the interval.

39. A more extensive tabulation of t critical values than what appears in this book shows that for the t distribution with 20 df, the areas to the right of the values .687, .860, and 1.064 are .25, .20, and .15, respectively. What is the confidence level for each of the following three confidence intervals for the mean μ of a normal population distribution? Which of the three intervals would you recommend be used, and why?

a. $(\bar{x} - .687s/\sqrt{21}, \bar{x} + 1.725s/\sqrt{21})$

b. $(\bar{x} - .860s/\sqrt{21}, \bar{x} + 1.325s/\sqrt{21})$

c. $(\bar{x} - 1.064s/\sqrt{21}, \bar{x} + 1.064s/\sqrt{21})$

7.4 | Confidence Intervals for the Variance and Standard Deviation of a Normal Population

Although inferences concerning a population variance σ^2 or standard deviation σ are usually of less interest than those about a mean or proportion, there are occasions when such procedures are needed. In the case of a normal population distribution, inferences are based on the following result concerning the sample variance S^2.

THEOREM

Let $X_1, X_2, \ldots, X_n$ be a random sample from a normal distribution with parameters μ and σ^2. Then the rv

$$\frac{(n-1)S^2}{\sigma^2} = \frac{\Sigma(X_i - \overline{X})^2}{\sigma^2}$$

has a chi-squared (χ^2) probability distribution with $n-1$ df.

As discussed in Sections 4.4 and 7.1, the chi-squared distribution is a continuous probability distribution with a single parameter ν, called the number of degrees of freedom, with possible values 1, 2, 3, ... The graphs of several χ^2 probability distribution functions (pdf's) are illustrated in Figure 7.9. Each pdf $f(x; \nu)$ is positive only for $x > 0$, and each has a positive skew (long upper tail), though the distribution moves rightward and becomes more symmetric as ν increases. To specify inferential procedures that use the chi-squared distribution, we need notation analogous to that for a t critical value $t_{\alpha,\nu}$.

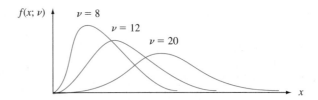

Figure 7.9 Graphs of chi-squared density functions

Notation

Let $\chi^2_{\alpha,\nu}$, called a **chi-squared critical value,** denote the number on the measurement axis such that α of the area under the chi-squared curve with ν df lies to the right of $\chi^2_{\alpha,\nu}$.

Because the t distribution is symmetric, it was necessary to tabulate only upper-tail critical values ($t_{\alpha,\nu}$ for small values of α). The chi-squared distribution is not symmetric, so Appendix Table A.7 contains values of $\chi^2_{\alpha,\nu}$ both for α near 0 and near 1, as illustrated in Figure 7.10(b). For example, $\chi^2_{.025,14} = 26.119$ and $\chi^2_{.95,20}$ (the 5th percentile) = 10.851.

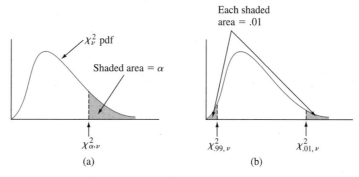

Figure 7.10 $\chi^2_{\alpha,\nu}$ notation illustrated

The rv $(n - 1)S^2/\sigma^2$ satisfies the two properties on which the general method for obtaining a CI is based: It is a function of the parameter of interest σ^2, yet its probability distribution (chi-squared) does not depend on this parameter. The area under a chi-squared curve with ν df to the right of $\chi^2_{\alpha/2,\nu}$ is $\alpha/2$, as is the area to the left of $\chi^2_{1-\alpha/2,\nu}$. Thus, the area captured between these two critical values is $1 - \alpha$. As a consequence of this and the theorem just stated,

$$P\left(\chi^2_{1-\alpha/2,n-1} < \frac{(n-1)S^2}{\sigma^2} < \chi^2_{\alpha/2,n-1}\right) = 1 - \alpha \qquad (7.17)$$

The inequalities in (7.17) are equivalent to

$$\frac{(n-1)S^2}{\chi^2_{\alpha/2,n-1}} < \sigma^2 < \frac{(n-1)S^2}{\chi^2_{1-\alpha/2,n-1}}$$

Substituting the computed value s^2 into the limits gives a CI for σ^2, and taking square roots gives an interval for σ.

A **100($1 - \alpha$)% confidence interval for the variance σ^2 of a normal population** has lower limit

$$(n - 1)s^2/\chi^2_{\alpha/2,n-1}$$

and upper limit

$$(n - 1)s^2/\chi^2_{1-\alpha/2,n-1}$$

A **confidence interval for σ** has lower and upper limits that are the square roots of the corresponding limits in the interval for σ^2.

Example 7.15 The accompanying data on breakdown voltage of electrically stressed circuits was read from a normal probability plot that appeared in the article "Damage of Flexible Printed

Wiring Boards Associated with Lightning-Induced Voltage Surges" (*IEEE Transactions on Components, Hybrids, and Manuf. Tech.*, 1985: 214–220). The straightness of the plot gave strong support to the assumption that breakdown voltage is approximately normally distributed.

1470	1510	1690	1740	1900	2000	2030	2100	2190
2200	2290	2380	2390	2480	2500	2580	2700	

Let σ^2 denote the variance of the breakdown voltage distribution. The computed value of the sample variance is $s^2 = 137{,}324.3$, the point estimate of σ^2. With df $= n - 1 = 16$, a 95% CI requires $\chi^2_{.975,16} = 6.908$ and $\chi^2_{.025,16} = 28.845$. The interval is

$$\left(\frac{16(137{,}324.3)}{28.845}, \frac{16(137{,}324.3)}{6.908} \right) = (76{,}172.3, 318{,}064.4)$$

Taking the square root of each endpoint yields (276.0, 564.0) as the 95% CI for σ. These intervals are quite wide, reflecting substantial variability in breakdown voltage in combination with a small sample size. ∎

CIs for σ^2 and σ when the population distribution is not normal can be difficult to obtain, even when the sample size is large. For such cases, consult a knowledgeable statistician.

Exercises Section 7.4 (40–44)

40. Determine the values of the following quantities:
   a. $\chi^2_{.1,15}$ **b.** $\chi^2_{.1,25}$
   c. $\chi^2_{.01,25}$ **d.** $\chi^2_{.005,25}$
   e. $\chi^2_{.99,25}$ **f.** $\chi^2_{.995,25}$

41. Determine the following:
   a. The 95th percentile of the chi-squared distribution with $\nu = 10$
   b. The 5th percentile of the chi-squared distribution with $\nu = 10$
   c. $P(10.98 \leq \chi^2 \leq 36.78)$ where χ^2 is a chi-squared rv with $\nu = 22$
   d. $P(\chi^2 < 14.611$ or $\chi^2 > 37.652)$ where χ^2 is a chi-squared rv with $\nu = 25$

42. The amount of lateral expansion (mils) was determined for a sample of $n = 9$ pulsed-power gas metal arc welds used in LNG ship containment tanks. The resulting sample standard deviation was $s = 2.81$ mils. Assuming normality, derive a 95% CI for σ^2 and for σ.

43. The following observations were made on fracture toughness of a base plate of 18% nickel maraging steel ["Fracture Testing of Weldments," *ASTM Special Publ. No. 381*, 1965: 328–356 (in ksi $\sqrt{\text{in.}}$, given in increasing order)]:

69.5	71.9	72.6	73.1	73.3	73.5	75.5	75.7
75.8	76.1	76.2	76.2	77.0	77.9	78.1	79.6
79.7	79.9	80.1	82.2	83.7	93.7		

Calculate a 99% CI for the standard deviation of the fracture toughness distribution. Is this interval valid whatever the nature of the distribution? Explain.

44. The results of a Wagner turbidity test performed on 15 samples of standard Ottawa testing sand were (in microamperes)

26.7	25.8	24.0	24.9	26.4	25.9	24.4	21.7
24.1	25.9	27.3	26.9	27.3	24.8	23.6	

   a. Is it plausible that this sample was selected from a normal population distribution?
   b. Calculate an upper confidence bound with confidence level 95% for the population standard deviation of turbidity.

Supplementary Exercises (45–60)

45. Example 1.11 introduced the accompanying observations on bond strength:

11.5	12.1	9.9	9.3	7.8	6.2	6.6	7.0
13.4	17.1	9.3	5.6	5.7	5.4	5.2	5.1
4.9	10.7	15.2	8.5	4.2	4.0	3.9	3.8
3.6	3.4	20.6	25.5	13.8	12.6	13.1	8.9
8.2	10.7	14.2	7.6	5.2	5.5	5.1	5.0
5.2	4.8	4.1	3.8	3.7	3.6	3.6	3.6

  a. Estimate true average bond strength in a way that conveys information about precision and reliability. *Hint:* $\sum x_i = 387.8$ and $\sum x_i^2 = 4247.08$.
  b. Calculate a 95% CI for the proportion of all such bonds whose strength values would exceed 10.

46. A triathlon consisting of swimming, cycling, and running is one of the more strenuous amateur sporting events. The article "Cardiovascular and Thermal Response of Triathlon Performance" (*Medicine and Science in Sports and Exercise,* 1988: 385–389) reports on a research study involving nine male triathletes. Maximum heart rate (beats/min) was recorded during performance of each of the three events. For swimming, the sample mean and sample standard deviation were 188.0 and 7.2, respectively. Assuming that the heart rate distribution is (approximately) normal, construct a 98% CI for true mean heart rate of triathletes while swimming.

47. For each of 18 preserved cores from oil-wet carbonate reservoirs, the amount of residual gas saturation after a solvent injection was measured at water flood-out. Observations, in percentage of pore volume, were

23.5	31.5	34.0	46.7	45.6	32.5
41.4	37.2	42.5	46.9	51.5	36.4
44.5	35.7	33.5	39.3	22.0	51.2

(See "Relative Permeability Studies of Gas-Water Flow Following Solvent Injection in Carbonate Rocks," *Soc. Petroleum Engineers J.,* 1976: 23–30.)
  a. Construct a boxplot of this data and comment on any interesting features.
  b. Is it plausible that the sample was selected from a normal population distribution?

  c. Calculate a 98% CI for the true average amount of residual gas saturation.

48. A journal article reports that a sample of size 5 was used as a basis for calculating a 95% CI for the true average natural frequency (Hz) of delaminated beams of a certain type. The resulting interval was (229.764, 233.504). You decide that a confidence level of 99% is more appropriate than the 95% level used. What are the limits of the 99% interval? *Hint:* Use the center of the interval and its width to determine $\bar{x}$ and s.

49. The financial manager of a large department-store chain selected a random sample of 200 of its credit card customers and found that 136 had incurred an interest charge during the previous year because of an unpaid balance.
  a. Compute a 90% CI for the true proportion of credit card customers who incurred an interest charge during the previous year.
  b. If the desired width of the 90% interval is .05, what sample size is necessary to ensure this?
  c. Does the upper limit of the interval in part (a) specify a 90% upper confidence bound for the proportion being estimated? Explain.

50. The reaction time (RT) to a stimulus is the interval of time commencing with stimulus presentation and ending with the first discernible movement of a certain type. The article "Relationship of Reaction Time and Movement Time in a Gross Motor Skill" (*Perceptual and Motor Skills,* 1973: 453–454) reports that the sample average RT for 16 experienced swimmers to a pistol start was .214 sec and the sample standard deviation was .036 sec.
  a. Making any necessary assumptions, derive a 90% CI for true average RT for all experienced swimmers.
  b. Calculate a 90% upper confidence bound for the standard deviation of the reaction time distribution.
  c. Predict RT for another such individual in a way that conveys information about precision and reliability.

51. Aphid infestation of fruit trees can be controlled either by spraying with pesticide or by inundation with ladybugs. In a particular area, four different groves of fruit trees are selected for experimentation. The first three groves are sprayed with pesticides 1, 2, and 3, respectively, and the fourth is treated with ladybugs, with the following results on yield:

Treatment	n_i = number of trees	$\bar{x}_i$ (bushels/tree)	s_i
1	100	10.5	1.5
2	90	10.0	1.3
3	100	10.1	1.8
4	120	10.7	1.6

Let μ_i = the true average yield (bushels/tree) after receiving the ith treatment. Then

$$\theta = \frac{1}{3}(\mu_1 + \mu_2 + \mu_3) - \mu_4$$

measures the difference in true average yields between treatment with pesticides and treatment with ladybugs. When n_1, n_2, n_3, and n_4 are all large, the estimator $\hat{\theta}$ obtained by replacing each μ_i by $\bar{X}_i$ is approximately normal. Use this to derive a large-sample $100(1 - \alpha)\%$ CI for θ, and compute the 95% interval for the given data.

52. It is important that face masks used by firefighters be able to withstand high temperatures because firefighters commonly work in temperatures of 200–500°F. In a test of one type of mask, 11 of 55 masks had lenses pop out at 250°. Construct a 90% CI for the true proportion of masks of this type whose lenses would pop out at 250°.

53. A manufacturer of college textbooks is interested in estimating the strength of the bindings produced by a particular binding machine. Strength can be measured by recording the force required to pull the pages from the binding. If this force is measured in pounds, how many books should be tested to estimate the average force required to break the binding to within .1 lb with 95% confidence? Assume that σ is known to be .8.

54. Chronic exposure to asbestos fiber is a well-known health hazard. The article "The Acute Effects of Chrysotile Asbestos Exposure on Lung Function" (*Environ. Research*, 1978: 360–372) reports results of a study based on a sample of construction workers who had been exposed to asbestos over a prolonged period. Among the data given in the article were the following (ordered) values of pulmonary compliance (cm³/cm H_2O) for each of 16 subjects 8 months after the exposure period (pulmonary

compliance is a measure of lung elasticity, or how effectively the lungs are able to inhale and exhale):

167.9	180.8	184.8	189.8	194.8	200.2
201.9	206.9	207.2	208.4	226.3	227.7
228.5	232.4	239.8	258.6		

a. Is it plausible that the population distribution is normal?

b. Compute a 95% CI for the true average pulmonary compliance after such exposure.

c. Calculate an interval that, with a confidence level of 95%, includes at least 95% of the pulmonary compliance values in the population distribution.

55. In Example 6.7, we introduced the concept of a censored experiment in which n components are put on test and the experiment terminates as soon as r of the components have failed. Suppose component lifetimes are independent, each having an exponential distribution with parameter λ. Let Y_1 denote the time at which the first failure occurs, Y_2 the time at which the second failure occurs, and so on, so that $T_r = Y_1 + \cdots + Y_r + (n - r)Y_r$ is the total accumulated lifetime at termination. Then it can be shown that $2\lambda T_r$ has a chi-squared distribution with $2r$ df. Use this fact to develop a $100(1 - \alpha)\%$ CI formula for true average lifetime $1/\lambda$. Compute a 95% CI from the data in Example 6.7.

56. Let $X_1, X_2, \ldots, X_n$ be a random sample from a continuous probability distribution having median $\tilde{\mu}$ (so that $P(X_i \le \tilde{\mu}) = P(X_i \ge \tilde{\mu}) = .5$).

a. Show that

$$P(\min(X_i) < \tilde{\mu} < \max(X_i)) = 1 - \left(\frac{1}{2}\right)^{n-1}$$

so that $(\min(x_i), \max(x_i))$ is a $100(1 - \alpha)\%$ confidence interval for $\tilde{\mu}$ with $\alpha = \left(\frac{1}{2}\right)^{n-1}$. [*Hint:* The complement of the event $\{\min(X_i) < \tilde{\mu} < \max(X_i)\}$ is $\{\max(X_i) \le \tilde{\mu}\} \cup \{\min(X_i) \ge \tilde{\mu}\}$. But $\max(X_i) \le \tilde{\mu}$ iff $X_i \le \tilde{\mu}$ for all i.]

b. For each of six normal male infants, the amount of the amino acid alanine (mg/100 mL) was determined while the infants were on an isoleucine-free diet, resulting in the following data:

2.84   3.54   2.80   1.44   2.94   2.70

Compute a 97% CI for the true median amount of alanine for infants on such a diet ("The Essential Amino Acid Requirements of Infants," *Amer. J. Nutrition*, 1964: 322–330).

c. Let $x_{(2)}$ denote the second smallest of the x_i's and $x_{(n-1)}$ denote the second largest of the x_i's. What is the confidence coefficient of the interval $(x_{(2)},$ $x_{(n-1)})$ for $\tilde{\mu}$?

57. Let $X_1, X_2, \ldots, X_n$ be a random sample from a uniform distribution on the interval $[0, \theta]$, so that

$$f(x) = \begin{cases} \dfrac{1}{\theta} & 0 \le x \le \theta \\ 0 & \text{otherwise} \end{cases}$$

Then if $Y = \max(X_i)$, it can be shown that the rv $U = Y/\theta$ has density function

$$f_U(u) = \begin{cases} nu^{n-1} & 0 \le u \le 1 \\ 0 & \text{otherwise} \end{cases}$$

a. Use $f_U(u)$ to verify that

$$P\left((\alpha/2)^{1/n} < \frac{Y}{\theta} \le (1 - \alpha/2)^{1/n}\right) = 1 - \alpha$$

and use this to derive a $100(1 - \alpha)\%$ CI for θ.

b. Verify that $P(\alpha^{1/n} \le Y/\theta \le 1) = 1 - \alpha$ and derive a $100(1 - \alpha)\%$ CI for θ based on this probability statement.

c. Which of the two intervals derived previously is shorter? If my waiting time for a morning bus is uniformly distributed and observed waiting times are $x_1 = 4.2$, $x_2 = 3.5$, $x_3 = 1.7$, $x_4 = 1.2$, and $x_5 = 2.4$, derive a 95% CI for θ by using the shorter of the two intervals.

58. Let $0 \le \gamma \le \alpha$. Then a $100(1 - \alpha)\%$ CI for μ when n is large is

$$\left(\bar{x} - z_\gamma \cdot \frac{s}{\sqrt{n}}, \bar{x} + z_{\alpha-\gamma} \cdot \frac{s}{\sqrt{n}}\right)$$

The choice $\gamma = \alpha/2$ yields the usual interval derived in Section 7.2; if $\gamma \ne \alpha/2$, this interval is not symmetric about $\bar{x}$. The width of this interval is $w = s(z_\gamma + z_{\alpha-\gamma})/\sqrt{n}$. Show that w is minimized for the choice $\gamma = \alpha/2$, so that the symmetric interval is the shortest. [*Hints:* (a) By definition of z_α, $\Phi(z_\alpha) = 1 - \alpha$, so that $z_\alpha = \Phi^{-1}(1 - \alpha)$; (b) the relationship between the derivative of a function $y = f(x)$ and the inverse function $x = f^{-1}(y)$ is $(d/dy)f^{-1}(y) = 1/f'(x)$.]

59. Suppose $x_1, x_2, \ldots, x_n$ are observed values resulting from a random sample from a symmetric but possibly heavy-tailed distribution. Let $\tilde{x}$ and f_s denote the sample median and fourth spread, respectively. Chapter 11 of *Understanding Robust and Exploratory Data Analysis* (see the bibliography in Chapter 6) suggests the following robust 95% CI for the population mean (point of symmetry):

$$\bar{x} \pm \left(\frac{\text{conservative } t \text{ critical value}}{1.075}\right) \cdot \frac{f_s}{\sqrt{n}}$$

The value of the quantity in parentheses is 2.10 for $n = 10$, 1.94 for $n = 20$, and 1.91 for $n = 30$. Compute this CI for the data of Exercise 43 and compare to the t CI appropriate for a normal population distribution.

60. a. Use the results of Example 7.5 to obtain a 95% lower confidence bound for the parameter λ of an exponential distribution and calculate the bound based on the data given in the example.

b. If lifetime X has an exponential distribution, the probability that lifetime exceeds t is $P(X > t) = e^{-\lambda t}$. Use the result of part (a) to obtain a 95% lower confidence bound for the probability that breakdown time exceeds 100 min.

Bibliography

DeGroot, Morris, *Probability and Statistics* (2nd ed.), Addison-Wesley, Reading, MA, 1986. A very good exposition of the general principles of statistical inference.

Hahn, Gerald, and William Meeker, *Statistical Intervals,* Wiley, New York, 1991. Everything you ever wanted to know about statistical intervals (confidence, prediction, tolerance, and others).

Larsen, Richard, and Morris Marx, *Introduction to Mathematical Statistics* (2nd ed.), Prentice Hall, Englewood Cliffs, NJ, 1986. Similar to DeGroot's presentation, but slightly less mathematical.

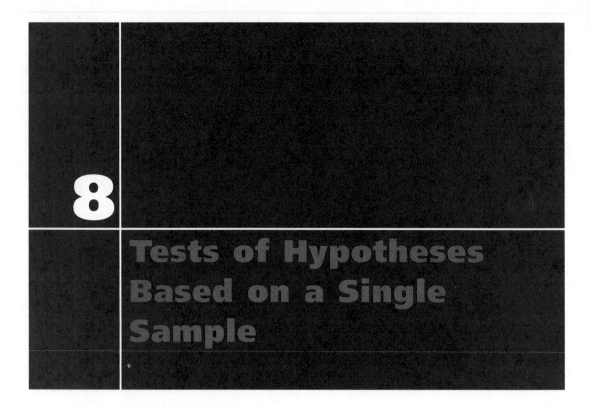

8

Tests of Hypotheses Based on a Single Sample

Introduction

A parameter can be estimated from sample data either by a single number (a point estimate) or an entire interval of plausible values (a confidence interval). Frequently, however, the objective of an investigation is not to estimate a parameter but to decide which of two contradictory claims about the parameter is correct. Methods for accomplishing this comprise the part of statistical inference called *hypothesis testing*. In this chapter, we first discuss some of the basic concepts and terminology in hypothesis testing and then develop decision procedures for the most frequently encountered testing problems based on a sample from a single population.

8.1 | Hypotheses and Test Procedures

A **statistical hypothesis,** or hypothesis, is a claim either about the value of a single population characteristic or about the values of several population characteristics. One example of a hypothesis is the claim $\mu = .75$, where μ is the true average inside diameter of a certain type of PVC pipe. Another example is the statement $p < .10$, where p is the proportion of defective circuit boards among all circuit boards produced by a certain manufacturer. If μ_1 and μ_2 denote the true average breaking strengths of two different types of twine, one hypothesis is the assertion that $\mu_1 - \mu_2 = 0$, and another is the statement $\mu_1 - \mu_2 > 5$.

In any hypothesis-testing problem, there are two contradictory hypotheses under consideration. One hypothesis might be the claim $\mu = .75$ and the other $\mu \neq .75$, or the two contradictory statements might be $p \geq .10$ and $p < .10$. The objective is to decide, based on sample information, which of the two hypotheses is correct. There is a familiar analogy to this in a criminal trial. One claim is the assertion that the accused individual is innocent. In the U.S. judicial system, this is the claim that is initially believed to be true. Only in the face of strong evidence to the contrary should the jury reject this claim in favor of the alternative assertion that the accused is guilty. In this sense, the claim of innocence is the favored or protected hypothesis, and the burden of proof is placed on those who believe in the alternative claim.

Similarly, in testing statistical hypotheses, the problem will be formulated so that one of the claims is initially favored. This initially favored claim will not be rejected in favor of the alternative claim unless sample evidence contradicts it and provides strong support for the alternative assertion.

DEFINITION

> The **null hypothesis,** denoted by H_0, is the claim about one or more population characteristics that is initially assumed to be true (the "prior belief" claim). The **alternative hypothesis,** denoted by H_a, is the assertion that is contradictory to H_0.
>
> The null hypothesis will be rejected in favor of the alternative hypothesis only if sample evidence suggests that H_0 is false. If the sample does not strongly contradict H_0, we will continue to believe in the truth of the null hypothesis. The two possible conclusions from a hypothesis-testing analysis are then *reject H_0* or *fail to reject H_0*.

A **test of hypotheses** is a method for using sample data to decide whether the null hypothesis should be rejected. Thus, we might test H_0: $\mu = .75$ against the alternative H_a: $\mu \neq .75$. Only if sample data strongly suggests that μ is something other than .75 should the null hypothesis be rejected. In the absence of such evidence, H_0 should not be rejected, since it is still quite plausible.

Sometimes an investigator does not want to accept a particular assertion unless and until data can provide strong support for the assertion. As an example, suppose a company is considering putting a new type of coating on bearings that it produces. The true average wear life with the current coating is known to be 1000 hours. With μ denoting the true average life for the new coating, the company would not want to make a change unless evidence strongly suggested that μ exceeds 1000. An appropriate problem formulation would involve testing H_0: $\mu = 1000$ against H_a: $\mu > 1000$. The conclusion that a change is justified is identified with H_a, and it would take conclusive evidence to justify rejecting H_0 and switching to the new coating.

Scientific research often involves trying to decide whether a current theory should be replaced by a more plausible and satisfactory explanation of the phenomenon under investigation. A conservative approach is to identify the current theory with H_0 and the researcher's alternative explanation with H_a. Rejection of the current theory will then occur only when evidence is much more consistent with the new theory. In many situations, H_a is referred to as the "researcher's hypothesis," since it is the claim that the researcher would really like to validate. The word *null* means "of no value, effect, or consequence," which suggests that H_0 should be identified with the hypothesis of no change (from current opinion), no difference, no improvement, and so on. Suppose, for example, that 10% of all circuit boards produced by a certain manufacturer during a recent period were defective. An engineer has suggested a change in the production process in the belief that it will result in a reduced defective rate. Let p denote the true proportion of defective boards resulting from the changed process. Then the research hypothesis, on which the burden of proof is placed, is the assertion that $p < .10$. Thus the alternative hypothesis is H_a: $p < .10$.

In our treatment of hypothesis testing, H_0 will always be stated as an equality claim. If θ denotes the parameter of interest, the null hypothesis will have the form H_0: $\theta = \theta_0$, where θ_0 is a specified number called the *null value* of the parameter (value claimed for θ by the null hypothesis). As an example, consider the circuit board situation just discussed. The suggested alternative hypothesis was H_a: $p < .10$, the claim that the defective rate is reduced by the process modification. A natural choice of H_0 in this situation is the claim that $p \geq .10$, according to which the new process is either no better *or* worse than the one currently used. We will instead consider H_0: $p = .10$ versus H_a: $p < .10$. The rationale for using this simplified null hypothesis is that any reasonable decision procedure for deciding between H_0: $p = .10$ and H_a: $p < .10$ will also be reasonable for deciding between the claim that $p \geq .10$ and H_a. The use of a simplified H_0 is preferred because it has certain technical benefits, which will be apparent shortly.

The alternative to the null hypothesis H_0: $\theta = \theta_0$ will look like one of the following three assertions: (1) H_a: $\theta > \theta_0$ (in which case the implicit null hypothesis is $\theta \leq \theta_0$), (2) H_a: $\theta < \theta_0$ (so the implicit null hypothesis states that $\theta \geq \theta_0$), or (3) H_a: $\theta \neq \theta_0$. For example, let σ denote the standard deviation of the distribution of inside diameters (inches) for a certain type of metal sleeve. If the decision was made to use the sleeve unless sample evidence conclusively demonstrated that $\sigma > .001$, the appropriate hypotheses would be H_0: $\sigma = .001$ versus H_a: $\sigma > .001$.

Test Procedures

A test procedure is a rule, based on sample data, for deciding whether to reject H_0. A test of H_0: $p = .10$ versus H_a: $p < .10$ in the circuit board problem might be based on examining a random sample of $n = 200$ boards. Let X denote the number of defective boards in the sample, a binomial random variable; x represents the observed value of X. If H_0 is true, $E(X) = np = 200(.10) = 20$, whereas we can expect fewer than 20 defective boards if H_a is true. A value x just a bit below 20 does not strongly contradict H_0, so it is reasonable to reject H_0 only if x is substantially less than 20. One such test procedure is to reject H_0 if $x \leq 15$ and not reject H_0 otherwise. This procedure has two constituents: (1) a *test statistic* or function of the sample data used to make a decision and (2) a *rejection region* consisting of those x values for which H_0 will be rejected in favor of H_a. For the rule just suggested, the rejection region consists of $x = 0, 1, 2, \ldots$, and 15. H_0 will not be rejected if $x = 16, 17, \ldots, 199$, or 200.

A test procedure is specified by the following:

1. A **test statistic,** a function of the sample data on which the decision (reject H_0 or do not reject H_0) is to be based

2. A **rejection region,** the set of all test statistic values for which H_0 will be rejected

The null hypothesis will then be rejected if and only if the observed or computed test statistic value falls in the rejection region.

As another example, suppose a cigarette manufacturer claims that the average nicotine content μ of brand B cigarettes is (at most) 1.5 mg. It would be unwise to reject the manufacturer's claim without strong contradictory evidence, so an appropriate problem formulation is to test H_0: $\mu = 1.5$ versus H_a: $\mu > 1.5$. Consider a decision rule based on analyzing a random sample of 32 cigarettes. Let $\overline{X}$ denote the sample average nicotine content. If H_0 is true, $E(\overline{X}) = \mu = 1.5$, whereas if H_0 is false, we expect $\overline{X}$ to exceed 1.5. Strong evidence against H_0 is provided by a value $\overline{x}$ that considerably exceeds 1.5. Thus, we might use $\overline{X}$ as a test statistic along with the rejection region $\overline{x} \geq 1.6$.

In both the circuit board and nicotine examples, the choice of test statistic and form of the rejection region make sense intuitively. However, the choice of cutoff value used to specify the rejection region is somewhat arbitrary. Instead of rejecting H_0: $p = .10$ in favor of H_a: $p < .10$ when $x \leq 15$, we could use the rejection region $x \leq 14$. For this region, H_0 would not be rejected if 15 defective boards are observed, whereas this occurrence would lead to rejection of H_0 if the initially suggested region is employed. Similarly, the rejection region $\overline{x} \geq 1.55$ might be used in the nicotine problem in place of the region $\overline{x} \geq 1.60$.

Errors in Hypothesis Testing

The basis for choosing a particular rejection region lies in an understanding of the errors that one might be faced with in drawing a conclusion. Consider the rejection region $x \leq 15$ in the circuit board problem. Even when H_0: $p = .10$ is true, it might happen that an unusual sample results in $x = 13$, so that H_0 is erroneously rejected. On the other hand, even when H_a: $p < .10$ is true, an unusual sample might yield $x = 20$, in which case H_0 would not be rejected, again an incorrect conclusion. Thus, it is possible that H_0 may be rejected when it is true or that H_0 may not be rejected when it is false. These possible errors are not consequences of a foolishly chosen rejection region. Either one of these two errors might result when the region $x \leq 14$ is employed, or indeed when any other region is used.

DEFINITION

> A **type I error** consists of rejecting the null hypothesis H_0 when it is true.
> A **type II error** involves not rejecting H_0 when H_0 is false.

In the nicotine problem, a type I error consists of rejecting the manufacturer's claim that $\mu = 1.5$ when it is actually true. If the rejection region $\bar{x} \geq 1.6$ is employed, it might happen that $\bar{x} = 1.63$ even when $\mu = 1.5$, resulting in a type I error. Alternatively, it may be that H_0 is false and yet $\bar{x} = 1.52$ is observed, leading to H_0 not being rejected (a type II error).

In the best of all possible worlds, test procedures for which neither type of error is possible could be developed. However, this ideal can be achieved only by basing a decision on an examination of the entire population, which is almost always impractical. The difficulty with using a procedure based on sample data is that because of sampling variability, an unrepresentative sample may result. Even though $E(\bar{X}) = \mu$, the observed value $\bar{x}$ may differ substantially from μ (at least if n is small). Thus, when $\mu = 1.5$ in the nicotine situation, $\bar{x}$ may be much larger than 1.5, resulting in erroneous rejection of H_0. Alternatively, it may be that $\mu = 1.6$ yet an $\bar{x}$ much smaller than this is observed, leading to a type II error.

Instead of demanding error-free procedures, we must look for procedures for which either type of error is unlikely to occur. That is, a good procedure is one for which the probability of making either type of error is small. The choice of a particular rejection region cutoff value fixes the probabilities of type I and type II errors. These error probabilities are traditionally denoted by α and β, respectively. Because H_0 specifies a unique value of the parameter, there is a single value of α. However, there is a different value of β for each value of the parameter consistent with H_a.

Example 8.1

A certain type of automobile is known to sustain no visible damage 25% of the time in 10-mph crash tests. A modified bumper design has been proposed in an effort to increase this percentage. Let p denote the proportion of all 10-mph crashes with this new bumper that result in no visible damage. The hypotheses to be tested are $H_0: p = .25$ (no improvement) versus $H_a: p > .25$. The test will be based on an experiment involving $n = 20$ independent crashes with prototypes of the new design. Intuitively, H_0 should be rejected if a substantial number of the crashes show no damage. Consider the following test procedure:

Test statistic: X = the number of crashes with no visible damage

Rejection region: $R_8 = \{8, 9, 10, \ldots, 19, 20\}$; that is, reject H_0 if $x \geq 8$, where x is the observed value of the test statistic.

This rejection region is called *upper-tailed* because it consists only of large values of the test statistic.

When H_0 is true, X has a binomial probability distribution with $n = 20$ and $p = .25$. Thus,

$$\alpha = P(\text{type I error}) = P(H_0 \text{ is rejected when it is true})$$
$$= P(X \geq 8 \text{ when } X \sim \text{Bin}(20, .25)) = 1 - B(7; 20, .25)$$
$$= 1 - .898 = .102$$

That is, when H_0 is actually true, roughly 10% of all experiments consisting of 20 crashes would result in H_0 being incorrectly rejected (a type I error).

In contrast to α, there is not a single β. Instead, there is a different β for each different p that exceeds .25. Thus there is a value of β for $p = .3$ (in which case $X \sim$ Bin(20, .3)), another value of β for $p = .5$, and so on. For example,

$$\beta(.3) = P(\text{type II error when } p = .3)$$
$$= P(H_0 \text{ is not rejected when it is false because } p = .3)$$
$$= P(X \leq 7 \text{ when } X \sim \text{Bin}(20, .3)) = B(7; 20, .3) = .772$$

When p is actually .3 rather than .25 (a "small" departure from H_0), roughly 77% of all experiments of this type would result in H_0 being incorrectly not rejected!

The accompanying table displays β for selected values of p (each calculated for the rejection region R_8). Clearly, β decreases as the value of p moves farther to the right of the null value .25. Intuitively, the greater the departure from H_0, the less likely it is that such a departure will not be detected.

p	.3	.4	.5	.6	.7	.8
$\beta(p)$	.772	.416	.132	.021	.001	.000

The proposed test procedure is still reasonable for testing the more realistic null hypothesis that $p \leq .25$. In this case, there is no longer a single α, but instead there is an α for each p that is at most .25: $\alpha(.25)$, $\alpha(.23)$, $\alpha(.20)$, $\alpha(.15)$, and so on. It is easily verified, though, that $\alpha(p) < \alpha(.25) = .102$ if $p < .25$. That is, the largest value of α occurs for the boundary value .25 between H_0 and H_a. Thus, if α is small for the simplified null hypothesis, it will also be as small as or smaller for the more realistic H_0. ∎

Example 8.2 The drying time of a certain type of paint under specified test conditions is known to be normally distributed with mean value 75 min and standard deviation 9 min. Chemists have proposed a new additive designed to decrease average drying time. It is believed that drying times with this additive will remain normally distributed with $\sigma = 9$. Because of the expense associated with the additive, evidence should strongly suggest an improvement in average drying time before such a conclusion is adopted. Let μ denote the true average drying time when the additive is used. The appropriate hypotheses are H_0: $\mu = 75$ versus H_a: $\mu < 75$. Only if H_0 can be rejected will the additive be declared successful and used.

Experimental data is to consist of drying times from $n = 25$ test specimens. Let $X_1, \ldots, X_{25}$ denote the 25 drying times—a random sample of size 25 from a normal distribution with mean value μ and standard deviation $\sigma = 9$. The sample mean drying time $\overline{X}$ then has a normal distribution with expected value $\mu_{\overline{X}} = \mu$ and standard deviation $\sigma_{\overline{X}} = \sigma/\sqrt{n} = 9/\sqrt{25} = 1.80$. When H_0 is true, $\mu_{\overline{X}} = 75$, so an $\overline{x}$ value somewhat less than 75 would not strongly contradict H_0. A reasonable rejection region has the form $\overline{X} \leq c$, where the cutoff value c is suitably chosen. Consider the choice $c = 70.8$, so that the test procedure consists of test statistic $\overline{X}$ and rejection region $\overline{x} \leq 70.8$. Because the rejection region consists only of small values of the test statistic, the test is said to be *lower-tailed*. Calculation of α and β now involves a routine standardization of $\overline{X}$ followed by reference to the standard normal probabilities of Appendix Table A.3:

$$\alpha = P(\text{type I error}) = P(H_0 \text{ is rejected when it is true})$$
$$= P(\overline{X} \leq 70.8 \text{ when } \overline{X} \sim \text{normal with } \mu_{\overline{X}} = 75, \sigma_{\overline{X}} = 1.8)$$
$$= \Phi\left(\frac{70.8 - 75}{1.8}\right) = \Phi(-2.33) = .01$$

$$\beta(72) = P(\text{type II error when } \mu = 72)$$
$$= P(H_0 \text{ is not rejected when it is false because } \mu = 72)$$
$$= P(\overline{X} > 70.8 \text{ when } \overline{X} \sim \text{normal with } \mu_{\overline{X}} = 72 \text{ and } \sigma_{\overline{X}} = 1.8)$$
$$= 1 - \Phi\left(\frac{70.8 - 72}{1.8}\right) = 1 - \Phi(-.67) = 1 - .2514 = .7486$$
$$\beta(70) = 1 - \Phi\left(\frac{70.8 - 70}{1.8}\right) = .3300 \qquad \beta(67) = .0174$$

For the specified test procedure, only 1% of all experiments carried out as described will result in H_0 being rejected when it is actually true. However, the chance of a type II error is very large when $\mu = 72$ (only a small departure from H_0), somewhat less when $\mu = 70$, and quite small when $\mu = 67$ (a very substantial departure from H_0). These error probabilities are illustrated in Figure 8.1. Notice that α is computed using the probability distribution of the test statistic when H_0 is true, whereas determination of β requires knowing the test statistic's distribution when H_0 is false.

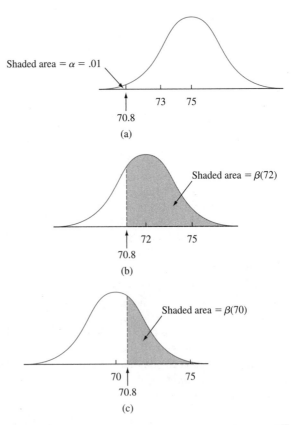

Figure 8.1 α and β illustrated for Example 8.2: (a) the distribution of $\overline{X}$ when $\mu = 75$ (H_0 true); (b) the distribution of $\overline{X}$ when $\mu = 72$ (H_0 false); (c) the distribution of $\overline{X}$ when $\mu = 70$ (H_0 false)

As in Example 8.1, if the more realistic null hypothesis $\mu \geq 75$ is considered, there is an α for each parameter value for which H_0 is true: $\alpha(75)$, $\alpha(75.8)$, $\alpha(76.5)$, and so on. It is easily verified, though, that $\alpha(75)$ is the largest of all these type I error probabilities. Focusing on the boundary value amounts to working explicitly with the "worst case." ■

The specification of a cutoff value for the rejection region in the examples just considered was somewhat arbitrary. Use of the rejection region $R_8 = \{8, 9, \ldots, 20\}$ in Example 8.1 resulted in $\alpha = .102$, $\beta(.3) = .772$, and $\beta(.5) = .132$. Many would think these error probabilities intolerably large. Perhaps they can be decreased by changing the cutoff value.

Example 8.3
(Example 8.1
continued)

Let us use the same experiment and test statistic X as previously described in the automobile bumper problem, but now consider the rejection region $R_9 = \{9, 10, \ldots, 20\}$. Since X still has a binomial distribution with parameters $n = 20$ and p,

$$\alpha = P(H_0 \text{ is rejected when } p = .25)$$
$$= P(X \geq 9 \text{ when } X \sim \text{Bin}(20, .25)) = 1 - B(8; 20, .25) = .041$$

The type I error probability has been decreased by using the new rejection region. However, a price has been paid for this decrease:

$$\beta(.3) = P(H_0 \text{ is not rejected when } p = .3)$$
$$= P(X \leq 8 \text{ when } X \sim \text{Bin}(20, .3)) = B(8; 20, .3) = .887$$
$$\beta(.5) = B(8; 20, .5) = .252$$

Both these β's are larger than the corresponding error probabilities .772 and .132 for the region R_8. In retrospect, this is not surprising; α is computed by summing over probabilities of test statistic values *in the rejection region*, whereas β is the probability that X falls *in the complement* of the rejection region. Making the rejection region smaller must therefore decrease α while increasing β for any fixed alternative value of the parameter. ■

Example 8.4
(Example 8.2
continued)

The use of cutoff value $c = 70.8$ in the paint-drying example resulted in a very small value of α (.01) but rather large β's. Consider the same experiment and test statistic $\overline{X}$ with the new rejection region $\bar{x} \leq 72$. Because $\overline{X}$ is still normally distributed with mean value $\mu_{\overline{X}} = \mu$ and $\sigma_{\overline{X}} = 1.8$,

$$\alpha = P(H_0 \text{ is rejected when it is true})$$
$$= P(\overline{X} \leq 72 \text{ when } \overline{X} \sim N(75, 1.8^2))$$
$$= \Phi\left(\frac{72 - 75}{1.8}\right) = \Phi(-1.67) = .0475 \approx .05$$

$$\beta(72) = P(H_0 \text{ is not rejected when } \mu = 72)$$
$$= P(\overline{X} > 72 \text{ when } \overline{X} \text{ is a normal rv with mean 72 and standard deviation 1.8})$$
$$= 1 - \Phi\left(\frac{72 - 72}{1.8}\right) = 1 - \Phi(0) = .5$$

$$\beta(70) = 1 - \Phi\left(\frac{72 - 70}{1.8}\right) = .1335 \qquad \beta(67) = .0027$$

The change in cutoff value has made the rejection region larger (it includes more $\bar{x}$ values), resulting in a decrease in β for each fixed μ less than 75. However, α for this new region has increased from the previous value .01 to approximately .05. If a type I error probability this large can be tolerated, though, the second region ($c = 72$) is preferable to the first ($c = 70.8$) because of the smaller β's. ■

The results of these examples can be generalized in the following manner.

PROPOSITION

> Suppose an experiment and a sample size are fixed, and a test statistic is chosen. Then decreasing the size of the rejection region to obtain a smaller value of α results in a larger value of β for any particular parameter value consistent with H_a.

This proposition says that once the test statistic and n are fixed, there is no rejection region that will simultaneously make both α and all β's small. A region must be chosen to effect a compromise between α and β.

Because of the suggested guidelines for specifying H_0 and H_a, a type I error is usually more serious than a type II error (this can always be achieved by proper choice of the hypotheses). The approach adhered to by most statistical practitioners is then to specify the largest value of α that can be tolerated and find a rejection region having that value of α rather than anything smaller. This makes β as small as possible subject to the bound on α. The resulting value of α is often referred to as the **significance level** of the test. Traditional levels of significance are .10, .05, and .01, though the level in any particular problem will depend on the seriousness of a type I error—the more serious this error, the smaller should be the significance level. The corresponding test procedure is called a **level α test** (e.g., a level .05 test or a level .01 test). A test with significance level α is one for which the type I error probability is controlled at the specified level.

Example 8.5   Consider the situation mentioned previously in which μ was the true average nicotine content of brand B cigarettes. The objective is to test H_0: $\mu = 1.5$ versus H_a: $\mu > 1.5$ based on a random sample $X_1, X_2, \ldots, X_{32}$ of nicotine contents. Suppose the distribution of nicotine content is known to be normal with $\sigma = .20$. Then $\bar{X}$ is normally distributed with mean value $\mu_{\bar{X}} = \mu$ and standard deviation $\sigma_{\bar{X}} = .20/\sqrt{32} = .0354$.

Rather than use $\bar{X}$ itself as the test statistic, let's standardize $\bar{X}$ assuming that H_0 is true.

Test statistic: $$Z = \frac{\bar{X} - 1.5}{\sigma/\sqrt{n}} = \frac{\bar{X} - 1.5}{.0354}$$

Z expresses the distance between $\bar{X}$ and its expected value when H_0 is true as some number of standard deviations. For example, $z = 3$ results from an $\bar{x}$ that is 3 standard deviations larger than we would have expected it to be were H_0 true.

Rejecting H_0 when $\bar{x}$ "considerably" exceeds 1.5 is equivalent to rejecting H_0 when z "considerably" exceeds 0. That is, the form of the rejection region is $z \geq c$. Let's now determine c so that $\alpha = .05$. When H_0 is true, Z has a standard normal distribution. Thus,

$$\alpha = P(\text{type I error}) = P(\text{rejecting } H_0 \text{ when } H_0 \text{ is true})$$
$$= P(Z \geq c \text{ when } Z \sim N(0, 1))$$

The value c must capture upper-tail area .05 under the z curve. Either from Section 4.3 or directly from Appendix Table A.3, $c = z_{.05} = 1.645$.

Notice that $z \geq 1.645$ is equivalent to $\bar{x} - 1.5 \geq (.0354)(1.645)$, that is, $\bar{x} \geq 1.56$. Then β is the probability that $\bar{X} < 1.56$ and can be calculated for any μ greater than 1.5. ∎

Exercises | Section 8.1 (1–14)

1. For each of the following assertions, state whether it is a legitimate statistical hypothesis and why.
  a. $H: \sigma > 100$ **b.** $H: \bar{x} = 45$
  c. $H: s \leq .20$ **d.** $H: \sigma_1/\sigma_2 < 1$
  e. $H: \bar{X} - \bar{Y} = 5$
  f. $H: \lambda \leq .01$, where λ is the parameter of an exponential distribution used to model component lifetime

2. For the following pairs of assertions, indicate which do not comply with our rules for setting up hypotheses and why (the subscripts 1 and 2 differentiate between quantities for two different populations or samples).
  a. $H_0: \mu = 100, H_a: \mu > 100$
  b. $H_0: \sigma = 20, H_a: \sigma \leq 20$
  c. $H_0: p \neq .25, H_a: p = .25$
  d. $H_0: \mu_1 - \mu_2 = 25, H_a: \mu_1 - \mu_2 > 100$
  e. $H_0: S_1^2 = S_2^2, H_a: S_1^2 \neq S_2^2$
  f. $H_0: \mu = 120, H_a: \mu = 150$
  g. $H_0: \sigma_1/\sigma_2 = 1, H_a: \sigma_1/\sigma_2 \neq 1$
  h. $H_0: p_1 - p_2 = -.1, H_a: p_1 - p_2 < -.1$

3. To determine whether the pipe welds in a nuclear power plant meet specifications, a random sample of welds is selected, and tests are conducted on each weld in the sample. Weld strength is measured as the force required to break the weld. Suppose the specifications state that mean strength of welds should exceed 100 lb/in.^{2}; the inspection team decides to test $H_0: \mu = 100$ versus $H_a: \mu > 100$. Explain why it might be preferable to use this H_a rather than $\mu < 100$.

4. Let μ denote the true average radioactivity level (picocuries per liter). The value 5 pCi/L is considered the dividing line between safe and unsafe water. Would you recommend testing $H_0: \mu = 5$ versus $H_a: \mu > 5$ or $H_0: \mu = 5$ versus $H_a: \mu < 5$? Explain your reasoning. (*Hint:* Think about the consequences of a type I and type II error for each possibility.)

5. Before agreeing to purchase a large order of polyethylene sheaths for a particular type of high-pressure oil-filled submarine power cable, a company wants to see conclusive evidence that the true standard deviation of sheath thickness is less than .05 mm. What hypotheses should be tested, and why? In this context, what are the type I and type II errors?

6. Many older homes have electrical systems that use fuses rather than circuit breakers. A manufacturer of 40-amp fuses wants to make sure that the mean amperage at which its fuses burn out is in fact 40. If the mean amperage is lower than 40, customers will complain because the fuses require replacement too often. If the mean amperage is higher than 40, the manufacturer might be liable for damage to an electrical system due to fuse malfunction. To verify the amperage of the fuses, a sample of fuses is to be selected and inspected. If a hypothesis test were to be performed on the resulting data, what null and alternative hypotheses would be of interest to the manufacturer? Describe type I and type II errors in the context of this problem situation.

7. Water samples are taken from water used for cooling as it is being discharged from a power plant into a river. It has been determined that as long as the mean temperature of the discharged water is at most 150°F, there will be no negative effects on the river's ecosystem. To investigate whether the plant is in compliance with regulations that prohibit a mean discharge-water temperature above 150°, 50 water samples will be taken at randomly selected times, and the temperature of each sample recorded. The resulting data will be used to test the hypotheses $H_0: \mu = 150°$ versus $H_a: \mu > 150°$. In the context of this situation, describe type I and type II errors. Which type of error would you consider more serious? Explain.

8. A regular type of laminate is currently being used by a manufacturer of circuit boards. A special laminate has been developed to reduce warpage. The regular laminate will be used on one sample of specimens and the special laminate on another sample, and the amount of warpage will then be determined for each specimen. The manufacturer will then switch to the special laminate only if it can be demonstrated that the true average amount of warpage for that laminate is less than for the regular laminate. State the relevant hypotheses, and describe the type I and type II errors in the context of this situation.

9. Two different companies have applied to provide cable television service in a certain region. Let p denote the proportion of all potential subscribers who favor the first company over the second. Consider testing $H_0: p = .5$ versus $H_a: p \neq .5$ based on a random sample of 25 individuals. Let X denote the number in the sample who favor the first company and x represent the observed value of X.
   a. Which of the following rejection regions is most appropriate and why?

   $R_1 = \{x: x \leq 7 \text{ or } x \geq 18\}, R_2 = \{x: x \leq 8\},$
   $R_3 = \{x: x \geq 17\}$

   b. In the context of this problem situation, describe what type I and type II errors are.
   c. What is the probability distribution of the test statistic X when H_0 is true? Use it to compute the probability of a type I error.
   d. Compute the probability of a type II error for the selected region when $p = .3$, again when $p = .4$, and also for both $p = .6$ and $p = .7$.
   e. Using the selected region, what would you conclude if 6 of the 25 queried favored company 1?

10. A mixture of pulverized fuel ash and Portland cement to be used for grouting should have a compressive strength of more than 1300 KN/m². The mixture will not be used unless experimental evidence indicates conclusively that the strength specification has been met. Suppose compressive strength for specimens of this mixture is normally distributed with $\sigma = 60$. Let μ denote the true average compressive strength.
   a. What are the appropriate null and alternative hypotheses?
   b. Let $\overline{X}$ denote the sample average compressive strength for $n = 20$ randomly selected specimens. Consider the test procedure with test statistic $\overline{X}$ and rejection region $\overline{x} \geq 1331.26$. What is the probability distribution of the test statistic when H_0 is true? What is the probability of a type I error for the test procedure?
   c. What is the probability distribution of the test statistic when $\mu = 1350$? Using the test procedure of part (b), what is the probability that the mixture will be judged unsatisfactory when in fact $\mu = 1350$ (a type II error)?
   d. How would you change the test procedure of part (b) to obtain a test with significance level .05? What impact would this change have on the error probability of part (c)?
   e. Consider the standardized test statistic $Z = (\overline{X} - 1300)/(\sigma/\sqrt{n}) = (\overline{X} - 1300)/13.42$. What are the values of Z corresponding to the rejection region of part (b)?

11. The calibration of a scale is to be checked by weighing a 10-kg test specimen 25 times. Suppose that the results of different weighings are independent of one another and that the weight on each trial is normally distributed with $\sigma = .200$ kg. Let μ denote the true average weight reading on the scale.
   a. What hypotheses should be tested?
   b. Suppose the scale is to be recalibrated if either $\overline{x} \geq 10.1032$ or $\overline{x} \leq 9.8968$. What is the probability that recalibration is carried out when it is actually unnecessary?
   c. What is the probability that recalibration is judged unnecessary when in fact $\mu = 10.1$? When $\mu = 9.8$?
   d. Let $z = (\overline{x} - 10)/(\sigma/\sqrt{n})$. For what value c is the rejection region of part (b) equivalent to the "two-tailed" region $either\ z \geq c\ or\ z \leq -c$?
   e. If the sample size were only 10 rather than 25, how should the procedure of part (d) be altered so that $\alpha = .05$?

f. Using the test of part (e), what would you conclude from the following sample data:

9.981 10.006   9.857 10.107 9.888

9.728 10.439 10.214 10.190 9.793

g. Reexpress the test procedure of part (b) in terms of the standardized test statistic $Z = (\overline{X} - 10)/(\sigma/\sqrt{n})$.

12. A new design for the braking system on a certain type of car has been proposed. For the current system, the true average braking distance at 40 mph under specified conditions is known to be 120 ft. It is proposed that the new design be implemented only if sample data strongly indicates a reduction in true average braking distance for the new design.
a. Define the parameter of interest and state the relevant hypotheses.
b. Suppose braking distance for the new system is normally distributed with $\sigma = 10$. Let $\overline{X}$ denote the sample average braking distance for a random sample of 36 observations. Which of the following rejection regions is appropriate: $R_1 = \{\overline{x}: \overline{x} \geq 124.80\}$, $R_2 = \{\overline{x}: \overline{x} \leq 115.20\}$, $R_3 = \{\overline{x}: \text{either } \overline{x} \geq 125.13 \text{ or } \overline{x} \leq 114.87\}$?
c. What is the significance level for the appropriate region of part (b)? How would you change the region to obtain a test with $\alpha = .001$?

d. What is the probability that the new design is not implemented when its true average braking distance is actually 115 ft and the appropriate region from part (b) is used?
e. Let $Z = (\overline{X} - 120)/(\sigma/\sqrt{n})$. What is the significance level for the rejection region $\{z: z \leq -2.33\}$? For the region $\{z: z \leq -2.88\}$?

13. Let $X_1, \ldots, X_n$ denote a random sample from a normal population distribution with a known value of σ.
a. For testing the hypotheses $H_0: \mu = \mu_0$ versus $H_a: \mu > \mu_0$ (where μ_0 is a fixed number), show that the test with test statistic $\overline{X}$ and rejection region $\overline{x} \geq \mu_0 + 2.33\sigma/\sqrt{n}$ has significance level .01.
b. Suppose the procedure of part (a) is used to test $H_0: \mu \leq \mu_0$ versus $H_a: \mu > \mu_0$. If $\mu_0 = 100$, $n = 25$, and $\sigma = 5$, what is the probability of committing a type I error when $\mu = 99$? When $\mu = 98$? In general, what can be said about the probability of a type I error when the actual value of μ is less than μ_0? Verify your assertion.

14. Reconsider the situation of Exercise 11 and suppose the rejection region is $\{\overline{x}: \overline{x} \geq 10.1004 \text{ or } \overline{x} \leq 9.8940\} = \{z: z \geq 2.51 \text{ or } z \leq -2.65\}$.
a. What is α for this procedure?
b. What is β when $\mu = 10.1$? When $\mu = 9.9$? Is this desirable?

8.2 | Tests About a Population Mean

The general discussion in Chapter 7 of confidence intervals for a population mean μ focused on three different cases. We now develop test procedures for these same three cases.

Case I: A Normal Population with Known σ

Although the assumption that the value of σ is known is rarely met in practice, this case provides a good starting point because of the ease with which general procedures and their properties can be developed. The null hypothesis in all three cases will state that μ has a particular numerical value, the *null value,* which we will denote by μ_0. Let $X_1, \ldots, X_n$ represent a random sample of size n from the normal population. Then the sample mean $\overline{X}$ has a normal distribution with expected value $\mu_{\overline{X}} = \mu$ and standard deviation $\sigma_{\overline{X}} = \sigma/\sqrt{n}$. When H_0 is true, $\mu_{\overline{X}} = \mu_0$. Consider now the statistic Z obtained by standardizing $\overline{X}$ under the assumption that H_0 is true:

$$Z = \frac{\overline{X} - \mu_0}{\sigma/\sqrt{n}}$$

Substitution of the computed sample mean $\bar{x}$ gives z, the distance between $\bar{x}$ and μ_0 expressed in "standard deviation units." For example, if the null hypothesis is H_0: $\mu = 100$, $\sigma_{\bar{x}} = \sigma/\sqrt{n} = 10/\sqrt{25} = 2.0$ and $\bar{x} = 103$, then the test statistic value is $z = (103 - 100)/2.0 = 1.5$. That is, the observed value of $\bar{x}$ is 1.5 standard deviations (of $\bar{X}$) above what we expect it to be when H_0 is true. The statistic Z is a natural measure of the distance between $\bar{X}$, the estimator of μ, and its expected value when H_0 is true. If this distance is too great in a direction consistent with H_a, the null hypothesis should be rejected.

Suppose first that the alternative hypothesis has the form H_a: $\mu > \mu_0$. Then an $\bar{x}$ value less than μ_0 certainly does not provide support for H_a. Such an $\bar{x}$ corresponds to a negative value of z (since $\bar{x} - \mu_0$ is negative and the divisor $\sigma/\sqrt{n}$ is positive). Similarly, an $\bar{x}$ value that exceeds μ_0 by only a small amount (corresponding to z which is positive but small) does not suggest that H_0 should be rejected in favor of H_a. The rejection of H_0 is appropriate only when $\bar{x}$ considerably exceeds μ_0—that is, when the z value is positive and large. In summary, the appropriate rejection region, based on the test statistic Z rather than $\bar{X}$, has the form $z \geq c$.

As discussed in Section 8.1, the cutoff value c should be chosen to control the probability of a type I error at the desired level α. This is easily accomplished because the distribution of the test statistic Z when H_0 is true is the standard normal distribution (that's why μ_0 was subtracted in standardizing). The required cutoff c is the z critical value that captures upper-tail area α under the standard normal curve. As an example, let $c = 1.645$, the value that captures tail area .05 ($z_{.05} = 1.645$). Then,

$$\alpha = P(\text{type I error}) = P(H_0 \text{ is rejected when } H_0 \text{ is true})$$
$$= P(Z \geq 1.645 \text{ when } Z \sim N(0, 1)) = 1 - \Phi(1.645) = .05$$

More generally, the rejection region $z \geq z_\alpha$ has type I error probability α. The test procedure is *upper-tailed* because the rejection region consists only of large values of the test statistic.

Analogous reasoning for the alternative hypothesis H_a: $\mu < \mu_0$ suggests a rejection region of the form $z \leq c$, where c is a suitably chosen negative number ($\bar{x}$ is far below μ_0 if and only if z is quite negative). Because Z has a standard normal distribution when H_0 is true, taking $c = -z_\alpha$ yields $P(\text{type I error}) = \alpha$. This is a *lower-tailed* test. For example, $z_{.10} = 1.28$ implies that the rejection region $z \leq -1.28$ specifies a test with significance level .10.

Finally, when the alternative hypothesis is H_a: $\mu \neq \mu_0$, H_0 should be rejected if $\bar{x}$ is too far to either side of μ_0. This is equivalent to rejecting H_0 either if $z \geq c$ or if $z \leq -c$. Suppose we desire $\alpha = .05$. Then,

$$.05 = P(Z \geq c \text{ or } Z \leq -c \text{ when } Z \text{ has a standard normal distribution})$$
$$= \Phi(-c) + 1 - \Phi(c) = 2[1 - \Phi(c)]$$

Thus, c is such that $1 - \Phi(c)$, the area under the standard normal curve to the right of c, is .025 (and not .05!). From Section 4.3 or Appendix Table A.3, $c = 1.96$, and the rejection region is $z \geq 1.96$ *or* $z \leq -1.96$. For any α, the *two-tailed* rejection region $z \geq z_{\alpha/2}$ *or* $z \leq -z_{\alpha/2}$ has type I error probability α (since area $\alpha/2$ is captured under each of the two tails of the z curve). Again, the key reason for using the standardized test statistic Z is that, because Z has a known distribution when H_0 is true (standard normal), a rejection region with desired type I error probability is easily obtained by using an appropriate critical value.

The test procedure for case I is summarized in the accompanying box, and the corresponding rejection regions are illustrated in Figure 8.2.

Null hypothesis: $H_0:\ \mu = \mu_0$

Test statistic value: $z = \dfrac{\bar{x} - \mu_0}{\sigma/\sqrt{n}}$

Alternative Hypothesis   **Rejection Region for Level α Test**

$H_a:\ \mu > \mu_0$ $z \geq z_\alpha$ (upper-tailed test)

$H_a:\ \mu < \mu_0$ $z \leq -z_\alpha$ (lower-tailed test)

$H_a:\ \mu \neq \mu_0$ either $z \geq z_{\alpha/2}$ or $z \leq -z_{\alpha/2}$ (two-tailed test)

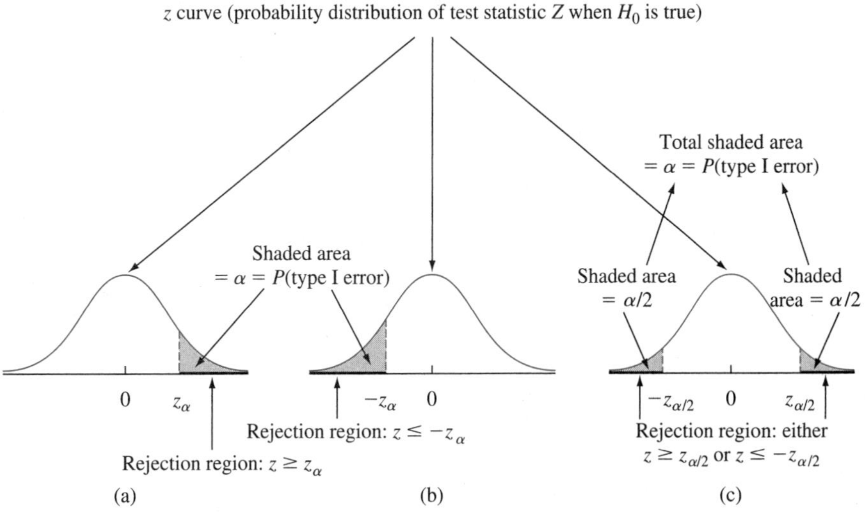

Figure 8.2 Rejection regions for z tests: (a) upper-tailed test; (b) lower-tailed test; (c) two-tailed test

Use of the following sequence of steps is recommended in any hypothesis-testing analysis.

1. Identify the parameter of interest and describe it in the context of the problem situation.

2. Determine the null value and state the null hypothesis.

3. State the appropriate alternative hypothesis.

4. Give the formula for the computed value of the test statistic (substituting the null value and the known values of any other parameters, but *not* those of any sample-based quantities).

5. State the rejection region for the selected significance level α.

6. Compute any necessary sample quantities, substitute into the formula for the test statistic value, and compute that value.

7. Decide whether H_0 should be rejected and state this conclusion in the problem context.

The formulation of hypotheses (Steps 2 and 3) should be done before examining the data.

Example 8.6   A manufacturer of sprinkler systems used for fire protection in office buildings claims that the true average system-activation temperature is 130°. A sample of $n = 9$ systems, when tested, yields a sample average activation temperature of 131.08°F. If the distribution of activation times is normal with standard deviation 1.5°F, does the data contradict the manufacturer's claim at significance level $\alpha = .01$?

1. Parameter of interest: μ = true average activation temperature

2. Null hypothesis: H_0: $\mu = 130$ (null value $= \mu_0 = 130$)

3. Alternative hypothesis: H_a: $\mu \neq 130$ (a departure from the claimed value in *either* direction is of concern)

4. Test statistic value:

$$z = \frac{\bar{x} - \mu_0}{\sigma/\sqrt{n}} = \frac{\bar{x} - 130}{1.5/\sqrt{n}}$$

5. Rejection region: The form of H_a implies use of a two-tailed test with rejection region *either* $z \geq z_{.005}$ *or* $z \leq -z_{.005}$. From Section 4.3 or Appendix Table A.3, $z_{.005} = 2.58$, so we reject H_0 if either $z \geq 2.58$ or $z \leq -2.58$.

6. Substituting $n = 9$ and $\bar{x} = 131.08$,

$$z = \frac{131.08 - 130}{1.5/\sqrt{9}} = \frac{1.08}{.5} = 2.16$$

That is, the observed sample mean is a bit more than 2 standard deviations above what would have been expected were H_0 true.

7. The computed value $z = 2.16$ does not fall in the rejection region ($-2.58 < 2.16 < 2.58$), so H_0 cannot be rejected at significance level .01. The data does not give strong support to the claim that the true average differs from the design value of 130. ■

β and Sample Size Determination

The z tests for case I are among the few in statistics for which there are simple formulas available for β, the probability of a type II error. Consider first the upper-tailed test with rejection region $z \geq z_\alpha$. This is equivalent to $\bar{x} \geq \mu_0 + z_\alpha \cdot \sigma/\sqrt{n}$, so H_0 will not be rejected if $\bar{x} < \mu_0 + z_\alpha \cdot \sigma/\sqrt{n}$. Now let μ' denote a particular value of μ that exceeds the null value μ_0. Then

$$\beta(\mu') = P(H_0 \text{ is not rejected when } \mu = \mu')$$
$$= P(\overline{X} < \mu_0 + z_\alpha \cdot \sigma/\sqrt{n} \text{ when } \mu = \mu')$$
$$= P\left(\frac{\overline{X} - \mu'}{\sigma/\sqrt{n}} < z_\alpha + \frac{\mu_0 - \mu'}{\sigma/\sqrt{n}} \text{ when } \mu = \mu'\right)$$
$$= \Phi\left(z_\alpha + \frac{\mu_0 - \mu'}{\sigma/\sqrt{n}}\right)$$

As μ' increases, $\mu_0 - \mu'$ becomes more negative, so $\beta(\mu')$ will be small when μ' greatly exceeds μ_0 (because the value at which Φ is evaluated will then be quite negative). Error probabilities for the lower-tailed and two-tailed tests are derived in an analogous manner.

If σ is large, the probability of a type II error can be large at an alternative value μ' that is of particular concern to an investigator. Suppose we fix α and also specify β for such an alternative value. In the sprinkler example, company officials might view $\mu' = 132$ as a very substantial departure from H_0: $\mu = 130$, and therefore wish $\beta(132) = .10$ in addition to $\alpha = .01$. More generally, consider the two restrictions $P(\text{type I error}) = \alpha$ and $\beta(\mu') = \beta$ for specified α, μ', and β. Then for an upper-tailed test, the sample size n should be chosen to satisfy

$$\Phi\left(z_\alpha + \frac{\mu_0 - \mu'}{\sigma/\sqrt{n}}\right) = \beta$$

This implies that

$$-z_\beta = \begin{array}{c} z \text{ critical value that} \\ \text{captures lower-tail area } \beta \end{array} = z_\alpha + \frac{\mu_0 - \mu'}{\sigma/\sqrt{n}}$$

It is easy to solve this equation for the desired n. A parallel argument yields the necessary sample size for lower- and two-tailed tests as summarized in the next box.

Alternative Hypothesis	Type II Error Probability $\beta(\mu')$ for a Level α Test
H_a: $\mu > \mu_0$	$\Phi\left(z_\alpha + \dfrac{\mu_0 - \mu'}{\sigma/\sqrt{n}}\right)$
H_a: $\mu < \mu_0$	$1 - \Phi\left(-z_\alpha + \dfrac{\mu_0 - \mu'}{\sigma/\sqrt{n}}\right)$
H_a: $\mu \neq \mu_0$	$\Phi\left(z_{\alpha/2} + \dfrac{\mu_0 - \mu'}{\sigma/\sqrt{n}}\right) - \Phi\left(-z_{\alpha/2} + \dfrac{\mu_0 - \mu'}{\sigma/\sqrt{n}}\right)$

where $\Phi(z)$ = the standard normal cdf.

The sample size n for which a level α test also has $\beta(\mu') = \beta$ at the alternative value μ' is

$$n = \begin{cases} \left[\dfrac{\sigma(z_\alpha + z_\beta)}{\mu_0 - \mu'}\right]^2 & \begin{array}{l} \text{for a one-tailed} \\ \text{(upper or lower) test} \end{array} \\[2em] \left[\dfrac{\sigma(z_{\alpha/2} + z_\beta)}{\mu_0 - \mu'}\right]^2 & \begin{array}{l} \text{for a two-tailed test} \\ \text{(an approximate solution)} \end{array} \end{cases}$$

Example 8.7 Let μ denote the true average tread life of a certain type of tire. Consider testing H_0: $\mu = 20{,}000$ versus H_a: $\mu > 20{,}000$ based on a sample of size $n = 16$ from a normal population distribution with $\sigma = 1500$. A test with $\alpha = .01$ requires $z_\alpha = z_{.01} = 2.33$. The probability of making a type II error when $\mu = 21{,}000$ is

$$\beta(21{,}000) = \Phi\left(2.33 + \frac{20{,}000 - 21{,}000}{1500/\sqrt{16}}\right) = \Phi(-.34) = .3669$$

Since $z_{.1} = 1.28$, the requirement that the level .01 test also have $\beta(21{,}000) = .1$ necessitates

$$n = \left[\frac{1500(2.33 + 1.28)}{20{,}000 - 21{,}000}\right]^2 = (-5.42)^2 = 29.32$$

The sample size must be an integer, so $n = 30$ tires should be used. ∎

Case II: Large-Sample Tests

When the sample size is large, the z tests for case I are easily modified to yield valid test procedures without requiring either a normal population distribution or known σ. The key result was used in Chapter 7 to justify large-sample confidence intervals: A large n implies that the sample standard deviation s will be close to σ for most samples, so that the standardized variable

$$Z = \frac{\overline{X} - \mu}{S/\sqrt{n}}$$

has *approximately* a standard normal distribution. Substitution of the null value μ_0 in place of μ yields the test statistic

$$Z = \frac{\overline{X} - \mu_0}{S/\sqrt{n}}$$

which has approximately a standard normal distribution when H_0 is true. The use of rejection regions given previously for case I (e.g., $z \geq z_\alpha$ when the alternative hypothesis is H_a: $\mu > \mu_0$) then results in test procedures for which the significance level is approximately (rather than exactly) α. The rule of thumb $n > 40$ will again be used to characterize a large sample size.

Example 8.8 A dynamic cone penetrometer (DCP) is used for measuring material resistance to penetration (mm/blow) as a cone is driven into pavement or subgrade. Suppose that for a particular application, it is required that the true average DCP value for a certain type of pavement be less than 30. The pavement will not be used unless there is conclusive evidence that the specification has been met. Let's state and test the appropriate hypotheses using the following data ("Probabilistic Model for the Analysis of Dynamic Cone Penetrometer Test Values in Pavement Structure Evaluation," *J. of Testing and Evaluation,* 1999: 7–14):

14.1	14.5	15.5	16.0	16.0	16.7	16.9	17.1	17.5	17.8
17.8	18.1	18.2	18.3	18.3	19.0	19.2	19.4	20.0	20.0
20.8	20.8	21.0	21.5	23.5	27.5	27.5	28.0	28.3	30.0
30.0	31.6	31.7	31.7	32.5	33.5	33.9	35.0	35.0	35.0
36.7	40.0	40.0	41.3	41.7	47.5	50.0	51.0	51.8	54.4
55.0	57.0								

Figure 8.3 shows a descriptive summary obtained from MINITAB. The sample mean DCP is less than 30. However, there is a substantial amount of variation in the data (sample coefficient of variation $= s/\bar{x} = .4265$), so the fact that the mean is less than the design specification cutoff may be a consequence just of sampling variability. Notice that the histogram does not resemble at all a normal curve (and a normal probability plot does not exhibit a linear pattern), but the large-sample z tests do not require a normal population distribution.

Descriptive Statistics

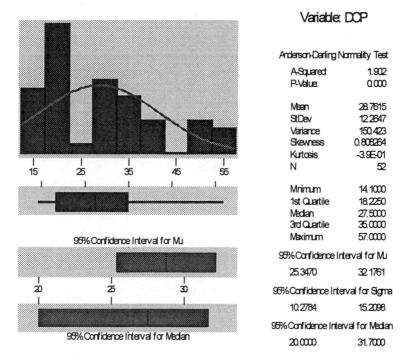

Variable: DCP

Anderson-Darling Normality Test

A-Squared	1.902
P-Value:	0.000
Mean	28.7615
StDev	12.2647
Variance	150.423
Skewness	0.808264
Kurtosis	-3.9E-01
N	52
Minimum	14.1000
1st Quartile	18.2250
Median	27.5000
3rd Quartile	35.0000
Maximum	57.0000

95% Confidence Interval for Mu

25.3470	32.1761

95% Confidence Interval for Sigma

10.2784	15.2098

95% Confidence Interval for Median

20.0000	31.7000

Figure 8.3 MINITAB descriptive summary for the DCP data of Example 8.8

1. μ = true average DCP value
2. H_0: $\mu = 30$
3. H_a: $\mu < 30$ (so the pavement will not be used unless the null hypothesis is rejected)
4. $z = \dfrac{\bar{x} - 30}{s/\sqrt{n}}$
5. Using a test with significance level .05, H_0 will be rejected if $z \leq -1.645$ (a lower-tailed test).
6. With $n = 52$, $\bar{x} = 28.76$, and $s = 12.2647$,

$$z = \frac{28.76 - 30}{12.2647/\sqrt{52}} = \frac{-1.24}{1.701} = -.73$$

7. Since $-.73 > -1.645$, H_0 cannot be rejected. We do not have compelling evidence for concluding that $\mu < 30$, so use of the pavement is not justified. ■

Determination of β and the necessary sample size for these large-sample tests can be based either on specifying a plausible value of σ and using the case I formulas (even though s is used in the test) or on using the curves to be introduced shortly in connection with case III.

Case III: A Normal Population Distribution

When n is small, the Central Limit Theorem (CLT) can no longer be invoked to justify the use of a large-sample test. We faced this same difficulty in obtaining a small-sample confidence interval (CI) for μ in Chapter 7. Our approach here will be the same one used there: We will assume that the population distribution is at least approximately normal and describe test procedures whose validity rests on this assumption. If an investigator has good reason to believe that the population distribution is quite nonnormal, a distribution-free test from Chapter 15 can be used. Alternatively, a statistician can be consulted regarding procedures valid for specific families of population distributions other than the normal family.

The key result on which tests for a normal population mean are based was used in Chapter 7 to derive the t CI: If $X_1, X_2, \ldots, X_n$ is a random sample from a normal distribution, the standardized variable

$$T = \frac{\overline{X} - \mu}{S/\sqrt{n}}$$

has a t distribution with $n - 1$ degrees of freedom (df). Consider testing H_0: $\mu = \mu_0$ against H_a: $\mu > \mu_0$ by using the test statistic $T = (\overline{X} - \mu_0)/(S/\sqrt{n})$. That is, the test statistic results from standardizing $\overline{X}$ under the assumption that H_0 is true (using $S/\sqrt{n}$, the estimated standard deviation of $\overline{X}$, rather than $\sigma/\sqrt{n}$). When H_0 is true, the test statistic has a t distribution with $n - 1$ df. Knowledge of the test statistic's distribution when H_0 is true (the "null distribution") allows us to construct a rejection region for which the type I error probability is controlled at the desired level. In particular, use of the upper-tail t critical value $t_{\alpha,n-1}$ to specify the rejection region $t \geq t_{\alpha,n-1}$ implies that

$$P(\text{type I error}) = P(H_0 \text{ is rejected when it is true})$$
$$= P(T \geq t_{\alpha,n-1} \text{ when } T \text{ has a } t \text{ distribution with } n - 1 \text{ df})$$
$$= \alpha$$

The test statistic is really the same here as in the large-sample case but is labeled T to emphasize that its null distribution is a t distribution with $n - 1$ df rather than the standard normal (z) distribution. The rejection region for the t test differs from that for the z test only in that a t critical value $t_{\alpha,n-1}$ replaces the z critical value z_α. Similar comments apply to alternatives for which a lower-tailed or two-tailed test is appropriate.

THE ONE-SAMPLE t TEST

Null hypothesis: H_0: $\mu = \mu_0$

Test statistic value: $t = \dfrac{\overline{x} - \mu_0}{s/\sqrt{n}}$

Alternative Hypothesis	Rejection Region for a Level α Test
H_a: $\mu > \mu_0$	$t \geq t_{\alpha, n-1}$ (upper-tailed)
H_a: $\mu < \mu_0$	$t \leq -t_{\alpha, n-1}$ (lower-tailed)
H_a: $\mu \neq \mu_0$	either $t \geq t_{\alpha/2, n-1}$ or $t \leq -t_{\alpha/2, n-1}$ (two-tailed)

Example 8.9 A well-designed and safe workplace can contribute greatly to increased productivity. It is especially important that workers not be asked to perform tasks, such as lifting, that exceed their capabilities. The accompanying data on maximum weight of lift (MAWL, in kg) for a frequency of four lifts/min was reported in the article "The Effects of Speed, Frequency, and Load on Measured Hand Forces for a Floor-to-Knuckle Lifting Task" (*Ergonomics,* 1992: 833–843); subjects were randomly selected from the population of healthy males age 18–30. Assuming that MAWL is normally distributed, does the following data suggest that the population mean MAWL exceeds 25?

$$25.8 \quad 36.6 \quad 26.3 \quad 21.8 \quad 27.2$$

Let's carry out a test using a significance level of .05.

1. μ = population mean MAWL

2. H_0: $\mu = 25$

3. H_a: $\mu > 25$

4. $t = \dfrac{\bar{x} - 25}{s/\sqrt{n}}$

5. Reject H_0 if $t \geq t_{\alpha, n-1} = t_{.05,4} = 2.132$.

6. $\Sigma x_i = 137.7$ and $\Sigma x_i^2 = 3911.97$, from which $\bar{x} = 27.54$, $s = 5.47$, and

$$t = \frac{27.54 - 25}{5.47/\sqrt{5}} = \frac{2.54}{2.45} = 1.04$$

The accompanying MINITAB output from a request for a one-sample t test has the same calculated values (the P-value is discussed in Section 8.4).

```
Test of mu = 25.00 vs mu > 25.00
Variable   N   Mean   StDev   SE Mean     T   P-Value
mawl       5   27.54   5.47      2.45   1.04    0.18
```

7. Since 1.04 does not fall in the rejection region ($1.04 < 2.132$), H_0 cannot be rejected at significance level .05. It is still plausible that μ is (at most) 25. ∎

β and Sample Size Determination

The calculation of β at the alternative value μ' in case I was carried out by expressing the rejection region in terms of $\bar{x}$ (e.g., $\bar{x} \geq \mu_0 + z_\alpha \cdot \sigma/\sqrt{n}$) and then subtracting μ' to standardize correctly. An equivalent approach involves noting that when $\mu = \mu'$, the test statistic $Z = (\bar{X} - \mu_0)/(\sigma/\sqrt{n})$ still has a normal distribution with variance 1, but now the mean value is given by $(\mu' - \mu_0)/(\sigma/\sqrt{n})$. That is, when $\mu = \mu'$, the test statistic still has a normal distribution though not the standard normal distribution. Because of this, $\beta(\mu')$ is an area under the normal curve corresponding to mean value $(\mu' - \mu_0)(\sigma/\sqrt{n})$ and variance 1. Both α and β involve working with normally distributed variables.

The calculation of $\beta(\mu')$ for the t test is much less straightforward. This is because the distribution of the test statistic $T = (\overline{X} - \mu_0)/(S/\sqrt{n})$ is quite complicated when H_0 is false and H_a is true. Thus, for an upper-tailed test, determining

$$\beta(\mu') = P(T < t_{\alpha,n-1} \text{ when } \mu = \mu' \text{ rather than } \mu_0)$$

involves integrating a very unpleasant density function. This must be done numerically, but fortunately it has been done by research statisticians for both one- and two-tailed t tests. The results are summarized in graphs of β that appear in Appendix Table A.17. There are four sets of graphs, corresponding to one-tailed tests at level .05 and level .01 and two-tailed tests at the same levels.

To understand how these graphs are used, note first that both β and the necessary sample size n in case I are functions not just of the absolute difference $|\mu_0 - \mu'|$ but of $d = |\mu_0 - \mu'|/\sigma$. Suppose, for example, that $|\mu_0 - \mu'| = 10$. This departure from H_0 will be much easier to detect (smaller β) when $\sigma = 2$, in which case μ_0 and μ' are 5 population standard deviations apart, than when $\sigma = 10$. The fact that β for the t test depends on d rather than just $|\mu_0 - \mu'|$ is unfortunate, since to use the graphs one must have some idea of the true value of σ. A conservative (large) guess for σ will yield a conservative (large) value of $\beta(\mu')$ and a conservative estimate of the sample size necessary for prescribed α and $\beta(\mu')$.

Once the alternative μ' and value of σ are selected, d is calculated and its value located on the horizontal axis of the relevant set of curves. The value of β is the height of the $n - 1$ df curve above the value of d (visual interpolation is necessary if $n - 1$ is not a value for which the corresponding curve appears), as illustrated in Figure 8.4.

Rather than fixing n (i.e., $n - 1$, and thus the particular curve from which β is read), one might prescribe both α (.05 or .01 here) and a value of β for the chosen μ' and σ. After computing d, the point (d, β) is located on the relevant set of graphs. The curve below and closest to this point gives $n - 1$ and thus n (again, interpolation is often necessary).

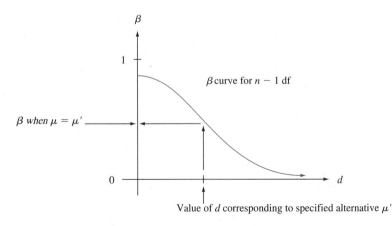

Figure 8.4 A typical β curve for the t test

Example 8.10   The true average voltage drop from collector to emitter of insulated gate bipolar transistors of a certain type is supposed to be at most 2.5 volts. An investigator selects a sample of $n = 10$ such transistors and uses the resulting voltages as a basis for testing H_0: $\mu = 2.5$ versus H_a: $\mu > 2.5$ using a t test with significance level $\alpha = .05$. If the standard deviation of the voltage distribution is $\sigma = .100$, how likely is it that H_0 will not be rejected when in fact $\mu = 2.6$? With $d = |2.5 - 2.6|/.100 = 1.0$, the point on the β curve at 9 df for a one-tailed test with $\alpha = .05$ above 1.0 has height approximately .1, so $\beta \approx .1$. The investigator might think that this is too large a value of β for such a substantial departure from H_0 and may wish to have $\beta = .05$ for this alternative value of μ. Since $d = 1.0$, the point $(d, \beta) = (1.0, .05)$ must be located. This point is very close to the 14 df curve, so using $n = 15$ will give both $\alpha = .05$ and $\beta = .05$ when the value of μ is 2.6 and $\sigma = .10$. A larger value of σ would give a larger β for this alternative, and an alternative value of μ closer to 2.5 would also result in an increased value of β. ∎

Most of the widely used statistical computer packages will also calculate type II error probabilities and determine necessary sample sizes. As an example, we asked MINITAB to do the calculations from Example 8.10. Its computations are based on **power,** which is simply $1 - \beta$. We want β to be small, which is equivalent to asking that the power of the test be large. For example, $\beta = .05$ corresponds to a value of .95 for power. Here is the resulting MINITAB output.

```
Power and Sample Size

Testing mean = null (versus > null)
Calculating power for mean = null + 0.1
Alpha = 0.05  Sigma = 0.1

Sample
Size      Power
10        0.8975
```

```
Power and Sample Size

1-Sample t Test

Testing mean = null (versus > null)
Calculating power for mean = null + 0.1
Alpha = 0.05  Sigma = 0.1

Sample   Target   Actual
Size     Power    Power
13       0.9500   0.9597
```

Notice from the second part of the output that the sample size necessary to obtain a power of .95 ($\beta = .05$) for an upper-tailed test with $\alpha = .05$ when $\sigma = .1$ and μ' is .1 larger than μ_0 is only $n = 13$, whereas eyeballing our β curves gave 15. When available, this type of software is more trustworthy than the curves.

Exercises | Section 8.2 (15–34)

15. Let the test statistic Z have a standard normal distribution when H_0 is true. Give the significance level for each of the following situations.

a. H_a: $\mu > \mu_0$, rejection region $z \geq 1.88$
b. H_a: $\mu < \mu_0$, rejection region $z \leq -2.75$
c. H_a: $\mu \neq \mu_0$, rejection region $z \geq 2.88$ or $z \leq -2.88$

16. Let the test statistic T have a t distribution when H_0 is true. Give the significance level for each of the following situations:
    a. H_a: $\mu > \mu_0$, df = 15, rejection region $t \geq 3.733$
    b. H_a: $\mu < \mu_0$, $n = 24$, rejection region $t \leq -2.500$
    c. H_a: $\mu \neq \mu_0$, $n = 31$, rejection region $t \geq 1.697$ or $t \leq -1.697$

17. Answer the following questions for the tire problem in Example 8.7.
    a. If $\bar{x} = 20{,}960$ and a level $\alpha = .01$ test is used, what is the decision?
    b. If a level .01 test is used, what is $\beta(20{,}500)$?
    c. If a level .01 test is used and it is also required that $\beta(20{,}500) = .05$, what sample size n is necessary?
    d. If $\bar{x} = 20{,}960$, what is the smallest α at which H_0 can be rejected (based on $n = 16$)?

18. Reconsider the paint-drying situation of Example 8.2, in which drying time for a test specimen is normally distributed with $\sigma = 9$. The hypotheses H_0: $\mu = 75$ versus H_a: $\mu < 75$ are to be tested using a random sample of $n = 25$ observations.
    a. How many standard deviations (of $\bar{X}$) below the null value is $\bar{x} = 72.3$?
    b. If $\bar{x} = 72.3$, what is the conclusion using $\alpha = .01$?
    c. What is α for the test procedure that rejects H_0 when $z \leq -2.88$?
    d. For the test procedure of part (c), what is $\beta(70)$?
    e. If the test procedure of part (c) is used, what n is necessary to ensure that $\beta(70) = .01$?
    f. If a level .01 test is used with $n = 100$, what is the probability of a type I error when $\mu = 76$?

19. The melting point of each of 16 samples of a certain brand of hydrogenated vegetable oil was determined, resulting in $\bar{x} = 94.32$. Assume that the distribution of melting point is normal with $\sigma = 1.20$.
    a. Test H_0: $\mu = 95$ versus H_a: $\mu \neq 95$ using a two-tailed level .01 test.
    b. If a level .01 test is used, what is $\beta(94)$, the probability of a type II error when $\mu = 94$?
    c. What value of n is necessary to ensure that $\beta(94) = .1$ when $\alpha = .01$?

20. Lightbulbs of a certain type are advertised as having an average lifetime of 750 hours. The price of these bulbs is very favorable, so a potential customer has decided to go ahead with a purchase arrangement unless it can be conclusively demonstrated that the true average lifetime is smaller than what is advertised. A random sample of 50 bulbs was selected, the lifetime of each bulb determined, and the appropriate hypotheses were tested using MINITAB, resulting in the accompanying output.

```
Variable  N   Mean StDev SEMean      Z P-Value
lifetime 50 738.44 38.20   5.40 -2.14   0.016
```

What conclusion would be appropriate for a significance level of .05? A significance level of .01? What significance level and conclusion would you recommend?

21. The true average diameter of ball bearings of a certain type is supposed to be .5 in. A one-sample t test will be carried out to see whether this is the case. What conclusion is appropriate in each of the following situations?
    a. $n = 13$, $t = 1.6$, $\alpha = .05$
    b. $n = 13$, $t = -1.6$, $\alpha = .05$
    c. $n = 25$, $t = -2.6$, $\alpha = .01$
    d. $n = 25$, $t = -3.9$

22. The article "The Foreman's View of Quality Control" (*Quality Engr.,* 1990: 257–280) described an investigation into the coating weights for large pipes resulting from a galvanized coating process. Production standards call for a true average weight of 200 lb per pipe. The accompanying descriptive summary and boxplot are from MINITAB.

```
Variable  N   Mean Median TrMean StDev SEMean
ctg wt   30 206.73 206.00 206.81  6.35   1.16

Variable     Min    Max      Q1      Q3
ctg wt    193.00 218.00  202.75  212.00
```

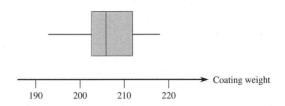

a. What does the boxplot suggest about the status of the specification for true average coating weight?
b. A normal probability plot of the data was quite straight. Use the descriptive output to test the appropriate hypotheses.

23. Exercise 36 in Chapter 1 gave $n = 26$ observations on escape time (sec) for oil workers in a simulated exercise, from which the sample mean and sample standard deviation are 370.69 and 24.36, respec-

tively. Suppose the investigators had believed a priori that true average escape time would be at most 6 minutes. Does the data contradict this prior belief? Assuming normality, test the appropriate hypotheses using a significance level of .05.

24. Reconsider the sample observations on stabilized viscosity of asphalt specimens introduced in Exercise 46 in Chapter 1 (2781, 2900, 3013, 2856, and 2888). Suppose that for a particular application, it is required that true average viscosity be 3000. Does this requirement appear to have been satisfied? State and test the appropriate hypotheses.

25. The desired percentage of SiO_2 in a certain type of aluminous cement is 5.5. To test whether the true average percentage is 5.5 for a particular production facility, 16 independently obtained samples are analyzed. Suppose that the percentage of SiO_2 in a sample is normally distributed with $\sigma = .3$ and that $\bar{x} = 5.25$.

    a. Does this indicate conclusively that the true average percentage differs from 5.5? Carry out the analysis using the sequence of steps suggested in the text.
    b. If the true average percentage is $\mu = 5.6$ and a level $\alpha = .01$ test based on $n = 16$ is used, what is the probability of detecting this departure from H_0?
    c. What value of n is required to satisfy $\alpha = .01$ and $\beta(5.6) = .01$?

26. To obtain information on the corrosion-resistance properties of a certain type of steel conduit, 45 specimens are buried in soil for a 2-year period. The maximum penetration (in mils) for each specimen is then measured, yielding a sample average penetration of $\bar{x} = 52.7$ and a sample standard deviation of $s = 4.8$. The conduits were manufactured with the specification that true average penetration be at most 50 mils. They will be used unless it can be demonstrated conclusively that the specification has not been met. What would you conclude?

27. The Charpy V-notch impact test is the basis for studying many material toughness criteria. This test was applied to 32 samples of a particular alloy at 110°F. The sample average amount of transverse lateral expansion was computed to be 73.1 mils, and the sample standard deviation was $s = 5.9$ mils. To be suitable for a particular application, the true average amount of expansion should be less than 75 mils. The alloy will not be

used unless the sample provides strong evidence that this criterion has been met. Test the relevant hypotheses using $\alpha = .01$ to decide whether the alloy is suitable.

28. Minor surgery on horses under field conditions requires a reliable short-term anesthetic producing good muscle relaxation, minimal cardiovascular and respiratory changes, and a quick, smooth recovery with minimal aftereffects so that horses can be left unattended. The article "A Field Trial of Ketamine Anesthesia in the Horse" (*Equine Vet. J.,* 1984: 176–179) reports that for a sample of $n = 73$ horses to which ketamine was administered under certain conditions, the sample average lateral recumbency (lying-down) time was 18.86 min and the standard deviation was 8.6 min. Does this data suggest that true average lateral recumbency time under these conditions is less than 20 min? Test the appropriate hypotheses at level of significance .10.

29. The amount of shaft wear (.0001 in.) after a fixed mileage was determined for each of $n = 8$ internal combustion engines having copper lead as a bearing material, resulting in $\bar{x} = 3.72$ and $s = 1.25$.

    a. Assuming that the distribution of shaft wear is normal with mean μ, use the t test at level .05 to test $H_0: \mu = 3.50$ versus $H_a: \mu > 3.50$.
    b. Using $\sigma = 1.25$, what is the type II error probability $\beta(\mu')$ of the test for the alternative $\mu' = 4.00$?

30. The recommended daily dietary allowance for zinc among males older than age 50 years is 15 mg/day. The article "Nutrient Intakes and Dietary Patterns of Older Americans: A National Study" (*J. Gerontology,* 1992: M145–150) reports the following summary data on intake for a sample of males age 65–74 years: $n = 115$, $\bar{x} = 11.3$, and $s = 6.43$. Does this data indicate that average daily zinc intake in the population of all males age 65–74 falls below the recommended allowance?

31. In an experiment designed to measure the time necessary for an inspector's eyes to become used to the reduced amount of light necessary for penetrant inspection, the sample average time for $n = 9$ inspectors was 6.32 sec and the sample standard deviation was 1.65 sec. It has previously been assumed that the average adaptation time was at least 7 sec. Assuming adaptation time to be normally distributed, does the data contradict prior belief? Use the t test with $\alpha = .1$.

32. A sample of 12 radon detectors of a certain type was selected, and each was exposed to 100 pCi/L of radon. The resulting readings were as follows:

105.6	90.9	91.2	96.9	96.5	91.3
100.1	105.0	99.6	107.7	103.3	92.4

a. Does this data suggest that the population mean reading under these conditions differs from 100? State and test the appropriate hypotheses using $\alpha = .05$.

b. Suppose that prior to the experiment, a value of $\sigma = 7.5$ had been assumed. How many determinations would then have been appropriate to obtain $\beta = .10$ for the alternative $\mu = 95$?

33. Show that for any $\Delta > 0$, when the population distribution is normal and σ is known, the two-tailed test satisfies $\beta(\mu_0 - \Delta) = \beta(\mu_0 + \Delta)$, so that $\beta(\mu')$ is symmetric about μ_0.

34. For a fixed alternative value μ', show that $\beta(\mu') \to 0$ as $n \to \infty$ for either a one-tailed or a two-tailed z test in the case of a normal population distribution with known σ.

8.3 | Tests Concerning a Population Proportion

Let p denote the proportion of individuals or objects in a population who possess a specified property (e.g., cars with manual transmissions or smokers who smoke a filter cigarette). If an individual or object with the property is labeled a success (S), then p is the population proportion of successes. Tests concerning p will be based on a random sample of size n from the population. Provided that n is small relative to the population size, X (the number of S's in the sample) has (approximately) a binomial distribution. Furthermore, if n itself is large, both X and the estimator $\hat{p} = X/n$ are approximately normally distributed. We first consider large-sample tests based on this latter fact and then turn to the small-sample case that directly uses the binomial distribution.

Large-Sample Tests

Large-sample tests concerning p are a special case of the more general large-sample procedures for a parameter θ. Let $\hat{\theta}$ be an estimator of θ that is (at least approximately) unbiased and has approximately a normal distribution. The null hypothesis has the form $H_0: \theta = \theta_0$, where θ_0 denotes a number (the null value) appropriate to the problem context. Suppose that when H_0 is true, the standard deviation of $\hat{\theta}$, $\sigma_{\hat{\theta}}$, involves no unknown parameters. For example, if $\theta = \mu$ and $\hat{\theta} = \overline{X}$, $\sigma_{\hat{\theta}} = \sigma_{\overline{X}} = \sigma/\sqrt{n}$, which involves no unknown parameters only if the value of σ is known. A large-sample test statistic results from standardizing $\hat{\theta}$ under the assumption that H_0 is true (so that $E(\hat{\theta}) = \theta_0$):

$$\text{Test statistic:} \quad Z = \frac{\hat{\theta} - \theta_0}{\sigma_{\hat{\theta}}}$$

If the alternative hypothesis is $H_a: \theta > \theta_0$, an upper-tailed test whose significance level is approximately α is specified by the rejection region $z \geq z_\alpha$. The other two alternatives, $H_a: \theta < \theta_0$ and $H_a: \theta \neq \theta_0$, are tested using a lower-tailed z test and a two-tailed z test, respectively.

In the case $\theta = p$, $\sigma_{\hat{\theta}}$ will not involve any unknown parameters when H_0 is true, but this is atypical. When $\sigma_{\hat{\theta}}$ does involve unknown parameters, it is often possible to use an estimated standard deviation $S_{\hat{\theta}}$ in place of $\sigma_{\hat{\theta}}$ and still have Z approximately normally distributed when H_0 is true (because when n is large, $s_{\hat{\theta}} \approx \sigma_{\hat{\theta}}$ for most samples). The large-sample test of the previous section furnishes an example of this: Because σ is usually unknown, we use $s_{\hat{\theta}} = s_{\bar{X}} = s/\sqrt{n}$ in place of $\sigma/\sqrt{n}$ in the denominator of z.

The estimator $\hat{p} = X/n$ is unbiased ($E(\hat{p}) = p$), has approximately a normal distribution, and its standard deviation is $\sigma_{\hat{p}} = \sqrt{p(1-p)/n}$. These facts were used in Section 7.2 to obtain a large-sample confidence interval for p. When H_0 is true, $E(\hat{p}) = p_0$ and $\sigma_{\hat{p}} = \sqrt{p_0(1-p_0)/n}$, so $\sigma_{\hat{p}}$ does not involve any unknown parameters. It then follows that when n is large and H_0 is true, the test statistic

$$Z = \frac{\hat{p} - p_0}{\sqrt{p_0(1-p_0)/n}}$$

has approximately a standard normal distribution. If the alternative hypothesis is H_a: $p > p_0$ and the upper-tailed rejection region $z \geq z_\alpha$ is used, then

$$P(\text{type I error}) = P(H_0 \text{ is rejected when it is true})$$
$$= P(Z \geq z_\alpha \text{ when } Z \text{ has approximately a standard normal distribution}) \approx \alpha$$

Thus, the desired level of significance α is attained by using the critical value that captures area α in the upper tail of the z curve. Rejection regions for the other two alternative hypotheses, lower-tailed for H_a: $p < p_0$ and two-tailed for H_a: $p \neq p_0$, are justified in an analogous manner.

Null hypothesis: H_0: $p = p_0$

Test statistic value: $z = \dfrac{\hat{p} - p_0}{\sqrt{p_0(1-p_0)/n}}$

Alternative Hypothesis	**Rejection Region**
H_a: $p > p_0$	$z \geq z_\alpha$ (upper-tailed)
H_a: $p < p_0$	$z \leq -z_\alpha$ (lower-tailed)
H_a: $p \neq p_0$	either $z \geq z_{\alpha/2}$ or $z \leq -z_{\alpha/2}$ (two-tailed)

These test procedures are valid provided that $np_0 \geq 10$ and $n(1 - p_0) \geq 10$.

Example 8.11 Many consumers are turning to generics as a way of reducing the cost of prescription medications. The article "Commercial Information on Drugs: Confusing to the Physician?" (*J. of Drug Issues,* 1988: 245–257) gives the results of a survey of 102 doctors. Only 47 of those surveyed knew the generic name for the drug methadone. Does this provide strong evidence for concluding that fewer than half of all physicians know the generic name for methadone? Let's carry out a test of hypotheses using a significance level of .01.

1. p = the proportion of all physicians who know the generic name for methadone

2. H_0: $p = .5$

3. H_a: $p < .5$

4. Since $102(.5) \geq 10$, the z test can be used. The test statistic value is

$$z = (\hat{p} - .5)/\sqrt{(.5)(.5)/n}$$

5. The form of H_a implies that a lower-tailed test is appropriate: Reject H_0 if $z \leq -z_{.01} = -2.33$.

6. $\hat{p} = 47/102 = .461$, so $z = (.461 - .5)/\sqrt{(.5)(.5)/102} = -.039/.0495 = -.79$.

7. Since $-.79$ is not in the rejection region, H_0 cannot be rejected at level .01. There is not conclusive evidence that fewer than 50% of all physicians know the generic name for methadone. ∎

β and Sample Size Determination

When H_0 is true, the test statistic Z has approximately a standard normal distribution. Now suppose that H_0 is *not* true and that $p = p'$. Then Z still has approximately a normal distribution (because it is a linear function of $\hat{p}$), but its mean value and variance are no longer 0 and 1, respectively. Instead,

$$E(Z) = \frac{p' - p_0}{\sqrt{p_0(1 - p_0)/n}} \qquad V(Z) = \frac{p'(1 - p')/n}{p_0(1 - p_0)/n}$$

The probability of a type II error for an upper-tailed test is $\beta(p') = P(Z < z_\alpha \text{ when } p = p')$. This can be computed by using the given mean and variance to standardize and then referring to the standard normal cdf. In addition, if it is desired that the level α test also have $\beta(p') = \beta$ for a specified value of β, this equation can be solved for the necessary n as in Section 8.2. General expressions for $\beta(p')$ and n are given in the accompanying box.

Alternative Hypothesis	$\beta(p')$
$H_a: \ p > p_0$	$\Phi\left[\dfrac{p_0 - p' + z_\alpha \sqrt{p_0(1 - p_0)/n}}{\sqrt{p'(1 - p')/n}} \right]$
$H_a: \ p < p_0$	$1 - \Phi\left[\dfrac{p_0 - p' - z_\alpha \sqrt{p_0(1 - p_0)/n}}{\sqrt{p'(1 - p')/n}} \right]$
$H_a: \ p \neq p_0$	$\Phi\left[\dfrac{p_0 - p' + z_{\alpha/2} \sqrt{p_0(1 - p_0)/n}}{\sqrt{p'(1 - p')/n}} \right]$
	$\quad - \Phi\left[\dfrac{p_0 - p' - z_{\alpha/2} \sqrt{p_0(1 - p_0)/n}}{\sqrt{p'(1 - p')/n}} \right]$

The sample size n for which the level α test also satisfies $\beta(p') = \beta$ is

$$n = \begin{cases} \left[\dfrac{z_\alpha \sqrt{p_0(1 - p_0)} + z_\beta \sqrt{p'(1 - p')}}{p' - p_0} \right]^2 & \text{one-tailed test} \\[4mm] \left[\dfrac{z_{\alpha/2} \sqrt{p_0(1 - p_0)} + z_\beta \sqrt{p'(1 - p')}}{p' - p_0} \right]^2 & \begin{array}{l}\text{two-tailed test (an}\\ \text{approximate solution)}\end{array} \end{cases}$$

Example 8.12 A package-delivery service advertises that at least 90% of all packages brought to its office by 9 A.M. for delivery in the same city are delivered by noon that day. Let p denote the true proportion of such packages that are delivered as advertised and consider the hypotheses $H_0: p = .9$ versus $H_a: p < .9$. If only 80% of the packages are delivered as advertised, how likely is it that a level .01 test based on $n = 225$ packages will detect such a departure from H_0? What should the sample size be to ensure that $\beta(.8) = .01$? With $\alpha = .01$, $p_0 = .9$, $p' = .8$, and $n = 225$,

$$\beta(.8) = 1 - \Phi\left(\frac{.9 - .8 - 2.33\sqrt{(.9)(.1)/225}}{\sqrt{(.8)(.2)/225}} \right)$$

$$= 1 - \Phi(2.00) = .0228$$

Thus, the probability that H_0 will be rejected using the test when $p = .8$ is .9772— roughly 98% of all samples will result in correct rejection of H_0.

Using $z_\alpha = z_\beta = 2.33$ in the sample size formula yields

$$n = \left[\frac{2.33\sqrt{(.9)(.1)} + 2.33\sqrt{(.8)(.2)}}{.8 - .9} \right]^2 \approx 266 \qquad \blacksquare$$

Small-Sample Tests

Test procedures when the sample size n is small are based directly on the binomial distribution rather than the normal approximation. Consider the alternative hypothesis $H_a: p > p_0$ and again let X be the number of successes in the sample. Then X is the test statistic, and the upper-tailed rejection region has the form $x \geq c$. When H_0 is true, X has a binomial distribution with parameters n and p_0, so

$$P(\text{type I error}) = P(H_0 \text{ is rejected when it is true})$$
$$= P(X \geq c \text{ when } X \sim \text{Bin}(n, p_0))$$
$$= 1 - P(X \leq c - 1 \text{ when } X \sim \text{Bin}(n, p_0))$$
$$= 1 - B(c - 1; n, p_0)$$

As the critical value c decreases, more x values are included in the rejection region and $P(\text{type I error})$ increases. Because X has a discrete probability distribution, it is usually not possible to find a value of c for which $P(\text{type I error})$ is exactly the desired significance level α (e.g., .05 or .01). Instead, the largest rejection region of the form $\{c, c + 1, \ldots, n\}$ satisfying $1 - B(c - 1; n, p_0) \leq \alpha$ is used.

Let p' denote an alternative value of p ($p' > p_0$). When $p = p', X \sim \text{Bin}(n, p')$, so

$$\beta(p') = P(\text{type II error when } p = p')$$
$$= P(X < c \text{ when } X \sim \text{Bin}(n, p')) = B(c - 1; n, p')$$

That is, $\beta(p')$ is the result of a straightforward binomial probability calculation. The sample size n necessary to ensure that a level α test also has specified β at a particular alternative value p' must be determined by trial and error using the binomial cdf.

Test procedures for $H_a: p < p_0$ and for $H_a: p \neq p_0$ are constructed in a similar manner. In the former case, the appropriate rejection region has the form $x \leq c$ (a lower-tailed test). The critical value c is the largest number satisfying $B(c; n, p_0) \leq \alpha$. The rejection region when the alternative hypothesis is $H_a: p \neq p_0$ consists of both large and small x values.

Example 8.13 A plastics manufacturer has developed a new type of plastic trash can and proposes to sell them with an unconditional 6-year warranty. To see whether this is economically feasible, 20 prototype cans are subjected to an accelerated life test to simulate 6 years of use. The proposed warranty will be modified only if the sample data strongly suggests that fewer than 90% of such cans would survive the 6-year period. Let p denote the proportion of all cans that survive the accelerated test. The relevant hypotheses are H_0: $p = .9$ versus H_a: $p < .9$. A decision will be based on the test statistic X, the number among the 20 that survive. If the desired significance level is $\alpha = .05$, c must satisfy $B(c; 20, .9) \leq .05$. From Appendix Table A.1, $B(15; 20, .9) = .043$, whereas $B(16; 20, .9) = .133$. The appropriate rejection region is therefore $x \leq 15$. If the accelerated test results in $x = 14$, H_0 would be rejected in favor of H_a, necessitating a modification of the proposed warranty. The probability of a type II error for the alternative value $p' = .8$ is

$$\beta(.8) = P(H_0 \text{ is not rejected when } X \sim \text{Bin}(20, .8))$$
$$= P(X \geq 16 \text{ when } X \sim \text{Bin}(20, .8))$$
$$= 1 - B(15; 20, .8) = 1 - .370 = .630$$

That is, when $p = .8$, 63% of all samples consisting of $n = 20$ cans would result in H_0 being incorrectly not rejected. This error probability is high because 20 is a small sample size and $p' = .8$ is close to the null value $p_0 = .9$. ∎

Exercises | Section 8.3 (35–43)

35. State DMV records indicate that of all vehicles undergoing emissions testing during the previous year, 70% passed on the first try. A random sample of 200 cars tested in a particular county during the current year yields 156 that passed on the initial test. Does this suggest that the true proportion for this county during the current year differs from the previous statewide proportion? Test the relevant hypotheses using $\alpha = .05$.

36. A manufacturer of nickel–hydrogen batteries randomly selects 100 nickel plates for test cells, cycles them a specified number of times, and determines that 14 of the plates have blistered.
   a. Does this provide compelling evidence for concluding that more than 10% of all plates blister under such circumstances? State and test the appropriate hypotheses using a significance level of .05. In reaching your conclusion, what type of error might you have committed?
   b. If it is really the case that 15% of all plates blister under these circumstances and a sample size of 100 is used, how likely is it that the null hypothesis of part (a) will not be rejected by the

level .05 test? Answer this question for a sample size of 200.
   c. How many plates would have to be tested to have $\beta(.15) = .10$ for the test of part (a)?

37. A random sample of 150 recent donations at a certain blood bank reveals that 92 were type A blood. Does this suggest that the actual percentage of type A donations differs from 40%, the percentage of the population having type A blood? Carry out a test of the appropriate hypotheses using a significance level of .01. Would your conclusion have been different if a significance level of .05 had been used?

38. A university library ordinarily has a complete shelf inventory done once every year. Because of new shelving rules instituted the previous year, the head librarian believes it may be possible to save money by postponing the inventory. The librarian decides to select at random 1000 books from the library's collection and have them searched in a preliminary manner. If evidence indicates strongly that the true proportion of misshelved or unlocatable books is less than .02, then the inventory will be postponed.

a. Among the 1000 books searched, 15 were mis-shelved or unlocatable. Test the relevant hypotheses and advise the librarian what to do (use $\alpha = .05$).

b. If the true proportion of misshelved and lost books is actually .01, what is the probability that the inventory will be (unnecessarily) taken?

c. If the true proportion is .05, what is the probability that the inventory will be postponed?

39. The article "Statistical Evidence of Discrimination" (*J. Amer. Stat. Assoc.*, 1982: 773–783) discusses the court case *Swain v. Alabama* (1965), in which it was alleged that there was discrimination against blacks in grand jury selection. Census data suggested that 25% of those eligible for grand jury service were black, yet a random sample of 1050 called to appear for possible duty yielded only 177 blacks. Using a level .01 test, does this data argue strongly for a conclusion of discrimination?

40. A plan for an executive traveler's club has been developed by an airline on the premise that 5% of its current customers would qualify for membership. A random sample of 500 customers yielded 40 who would qualify.

a. Using this data, test at level .01 the null hypothesis that the company's premise is correct against the alternative that it is not correct.

b. What is the probability that when the test of part (a) is used, the company's premise will be judged correct when in fact 10% of all current customers qualify?

41. Each of a group of 20 intermediate tennis players is given two rackets, one having nylon strings and the other synthetic gut strings. After several weeks of playing with the two rackets, each player will be asked to state a preference for one of the two types of strings. Let p denote the proportion of all such players who would prefer gut to nylon, and let X be the number of players in the sample who prefer gut. Because gut strings are more expensive, consider the null hypothesis that at most 50% of all such players prefer gut. We simplify this to H_0: $p = .5$,

planning to reject H_0 only if sample evidence strongly favors gut strings.

a. Which of the rejection regions $\{15, 16, 17, 18, 19, 20\}$, $\{0, 1, 2, 3, 4, 5\}$, or $\{0, 1, 2, 3, 17, 18, 19, 20\}$ is most appropriate, and why are the other two not appropriate?

b. What is the probability of a type I error for the chosen region of part (a)? Does the region specify a level .05 test? Is it the best level .05 test?

c. If 60% of all enthusiasts prefer gut, calculate the probability of a type II error using the appropriate region from part (a). Repeat if 80% of all enthusiasts prefer gut.

d. If 13 out of the 20 players prefer gut, should H_0 be rejected using a significance level of .10?

42. A manufacturer of plumbing fixtures has developed a new type of washerless faucet. Let $p = P(\text{a randomly selected faucet of this type will develop a leak within 2 years under normal use})$. The manufacturer has decided to proceed with production unless it can be determined that p is too large; the borderline acceptable value of p is specified as .10. The manufacturer decides to subject n of these faucets to accelerated testing (approximating 2 years of normal use). With $X =$ the number among the n faucets that leak before the test concludes, production will commence unless the observed X is too large. It is decided that if $p = .10$, the probability of not proceeding should be at most .10, whereas if $p = .30$ the probability of proceeding should be at most .10. Can $n = 10$ be used? $n = 20$? $n = 25$? What is the appropriate rejection region for the chosen n, and what are the actual error probabilities when this region is used?

43. Scientists think that robots will play a crucial role in factories in the next 20 years. Suppose that in an experiment to determine whether the use of robots to weave computer cables is feasible, a robot was used to assemble 500 cables. The cables were examined and there were 14 defectives. If human assemblers have a defect rate of .03 (3%), does this data support the hypothesis that the proportion of defectives is lower for robots than humans? Use a .01 significance level.

8.4 | *P*-Values

One way to report the result of a hypothesis-testing analysis is to simply say whether the null hypothesis was rejected at a specified level of significance. Thus, an investigator might state that H_0 was rejected at level of significance .05 or that use of a level .01 test resulted in not rejecting H_0. This type of statement is somewhat inadequate because it says nothing about whether the computed value of the test statistic just barely fell into the rejection region or whether it exceeded the critical value by a large amount. A related difficulty is that such a report imposes the specified significance level on other decision makers. In many decision situations, individuals may have different views concerning the consequences of a type I or type II error. Each individual would then want to select his or her own significance level—some selecting $\alpha = .05$, others $.01$, and so on—and reach a conclusion accordingly. This could result in some individuals rejecting H_0 while others conclude that the data does not show a strong enough contradiction of H_0 to justify its rejection.

Example 8.14 The true average time to initial relief of pain for a best-selling pain reliever is known to be 10 min. Let μ denote the true average time to relief for a company's newly developed reliever. The company wishes to produce and market this reliever only if it provides quicker relief than the best-seller, so it wishes to test H_0: $\mu = 10$ versus H_a: $\mu < 10$. Only if experimental evidence leads to rejection of H_0 will the new reliever be introduced. After weighing the relative seriousness of each type of error, a single level of significance must be agreed on and a decision—to reject H_0 and introduce the reliever or not to do so—made at that level.

Suppose the new reliever has been introduced. The company supports its claim of quicker relief by stating that, based on an analysis of experimental data, H_0: $\mu = 10$ was rejected in favor of H_a: $\mu < 10$ using level of significance $\alpha = .10$. Any individuals contemplating a switch to this new reliever would naturally want to reach their own conclusions concerning the validity of the claim. Individuals who are satisfied with the best-seller would view a type I error (concluding that the new product provides quicker relief when it actually does not) as serious, so might wish to use $\alpha = .05, .01$, or even smaller levels. Unfortunately, the nature of the company's statement prevents an individual decision maker from reaching a conclusion at such a level. The company has imposed its own choice of significance level on others. The report could have been done in a manner that allowed each individual flexibility in drawing a conclusion at a personally selected α. ∎

A *P-value* conveys much information about the strength of evidence against H_0 and allows an individual decision maker to draw a conclusion at any specified level α. Before we give a general definition, consider how the conclusion in a hypothesis-testing problem depends on the selected level α.

Example 8.15 The nicotine content problem discussed in Example 8.5 involved testing H_0: $\mu = 1.5$ versus H_a: $\mu > 1.5$. Because of the inequality in H_a, the rejection region is upper-tailed, with H_0 rejected if $z \geq z_\alpha$. Suppose $z = 2.10$. The accompanying table displays the rejection region for each of four different α's along with the resulting conclusion.

Level of Significance α	Rejection Region	Conclusion
.05	$z \geq 1.645$	Reject H_0
.025	$z \geq 1.96$	Reject H_0
.01	$z \geq 2.33$	Do not reject H_0
.005	$z \geq 2.58$	Do not reject H_0

For α relatively large, the z critical value z_α is not very far out in the upper tail; 2.10 exceeds the critical value, and so H_0 is rejected. However, as α decreases, the critical value increases. For small α, the z critical value is large, 2.10 is less than z_α, and H_0 is not rejected.

Recall that for an upper-tailed z test, α is just the area under the z curve to the right of the critical value z_α. That is, once α is specified, the critical value is chosen to capture upper-tail area α. Appendix Table A.3 shows that the area to the right of 2.10 is .0179. Using an α larger than .0179 corresponds to $z_\alpha < 2.10$. An α less than .0179 necessitates using a z critical value that exceeds 2.10. The decision at a particular level α thus depends on how the selected α compares to the tail area captured by the computed z. This is illustrated in Figure 8.5. Notice in particular that .0179, the captured tail area, is the smallest level α at which H_0 would be rejected, because using any smaller α results in a z critical value that exceeds 2.10, so that 2.10 is not in the rejection region.

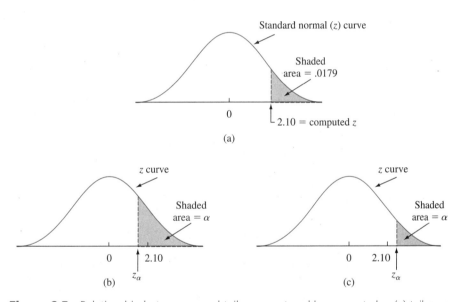

Figure 8.5 Relationship between α and tail area captured by computed z; (a) tail area captured by computed z; (b) when $\alpha > .0179$, $z_\alpha < 2.10$ and H_0 is rejected; (c) when $\alpha < .0179$, $z_\alpha > 2.10$ and H_0 is not rejected

In general, suppose the probability distribution of a test statistic when H_0 is true has been determined. Then, for specified α, the rejection region is determined by finding a critical value or values that capture tail area α (upper, lower, or two-tailed, whichever is appropriate) under the probability distribution curve. The smallest α for which H_0 would be rejected is the tail area captured by the computed value of the test statistic. This smallest α is the P-value.

DEFINITION

The **P-value** (or *observed significance level*) is the smallest level of significance at which H_0 would be rejected when a specified test procedure is used on a given data set. Once the P-value has been determined, the conclusion at any particular level α results from comparing the P-value to α:

1. P-value $\leq \alpha \Rightarrow$ reject H_0 at level α.

2. P-value $> \alpha \Rightarrow$ do not reject H_0 at level α.

It is customary to call the data *significant* when H_0 is rejected and *not significant* otherwise. The P-value is then the smallest level at which the data is significant. An easy way to visualize the comparison of the P-value with the chosen α is to draw a picture like that of Figure 8.6. The calculation of the P-value depends on whether the test is upper-, lower-, or two-tailed. However, once it has been calculated, the comparison with α does not depend on which type of test was used.

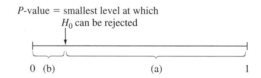

Figure 8.6   Comparing α and the P-value; (a) reject H_0 when α lies here; (b) do not reject H_0 when α lies here

Example 8.16
(Example 8.14 continued)

Suppose that when data from an experiment involving the new pain reliever was analyzed, the P-value for testing H_0: $\mu = 10$ versus H_a: $\mu < 10$ was calculated as .0384. Since $\alpha = .05$ is larger than the P-value (.05 lies in the interval (a) of Figure 8.6), H_0 would be rejected by anyone carrying out the test at level .05. However, at level .01, H_0 would not be rejected because .01 is smaller than the smallest level (.0384) at which H_0 can be rejected.   ∎

The most widely used statistical computer packages automatically calculate and print out a P-value when a hypothesis-testing analysis is performed. A conclusion can then be drawn directly from the output, without reference to a table of critical values.

A useful alternative definition equivalent to the one just given is as follows:

DEFINITION

The **P-value** is the probability, calculated assuming H_0 is true, of obtaining a test statistic value at least as contradictory to H_0 as the value that actually resulted. The smaller the P-value, the more contradictory is the data to H_0.

Thus, if $z = 2.10$ for an upper-tailed z test, P-value $= P(Z \geq 2.10,$ when H_0 is true) $= 1 - \Phi(2.10) = .0179$ as before.

The *P*-Value for a *z* Test

The P-value for a z test (one based on a test statistic whose distribution when H_0 is true is at least approximately standard normal) is easily determined from the information in Appendix Table A.3. Consider an upper-tailed test and let z denote the computed value of the test statistic Z. The null hypothesis is rejected if $z \geq z_\alpha$, and the *P*-value is the smallest α for which this is the case. Since z_α increases as α decreases, the *P*-value is the value of α for which $z = z_\alpha$. That is, the *P*-value is just the area captured by the computed value z in the upper tail of the standard normal curve. The corresponding cumulative area is $\Phi(z)$, so in this case *P*-value $= 1 - \Phi(z)$.

An analogous argument for a lower-tailed test shows that the *P*-value is the area captured by the computed value z in the lower tail of the standard normal curve. More care must be exercised in the case of a two-tailed test. Suppose first that z is positive. Then the *P*-value is the value of α satisfying $z = z_{\alpha/2}$ (i.e., computed $z =$ upper-tail critical value). This says that the area captured in the upper tail is half the *P*-value, so that *P*-value $= 2[1 - \Phi(z)]$. If z is negative, the *P*-value is the α for which $z = -z_{\alpha/2}$, or, equivalently, $-z = z_{\alpha/2}$, so *P*-value $= 2[1 - \Phi(-z)]$. Since $-z = |z|$ when z is negative, *P*-value $= 2[1 - \Phi(|z|)]$ for either positive or negative z.

$$
\text{\textit{P}-value:} \quad P = \begin{cases} 1 - \Phi(z) & \text{for an upper-tailed test} \\ \Phi(z) & \text{for a lower-tailed test} \\ 2[1 - \Phi(|z|)] & \text{for a two-tailed test} \end{cases}
$$

Each of these is the probability of getting a value at least as extreme as what was obtained (assuming H_0 true). The three cases are illustrated in Figure 8.7.

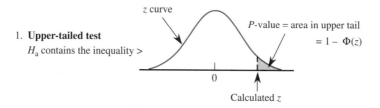

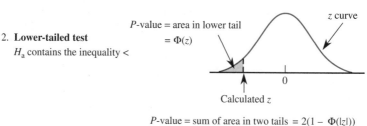

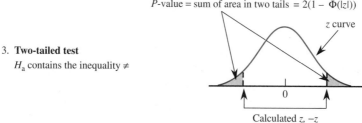

Figure 8.7 Determination of the *P*-value for a *z* test

The next example illustrates the use of the *P*-value approach to hypothesis testing by means of a sequence of steps modified from our previously recommended sequence.

Example 8.17 The target thickness for silicon wafers used in a certain type of integrated circuit is 245 μm. A sample of 50 wafers is obtained and the thickness of each one is determined, resulting in a sample mean thickness of 246.18 μm and a sample standard deviation of 3.60 μm. Does this data suggest that true average wafer thickness is something other than the target value?

1. Parameter of interest: μ = true average wafer thickness

2. Null hypothesis: H_0: $\mu = 245$

3. Alternative hypothesis: H_a: $\mu \neq 245$

4. Formula for test statistic value: $z = \dfrac{\bar{x} - 245}{s/\sqrt{n}}$

5. Calculation of test statistic value: $z = \dfrac{246.18 - 245}{3.60/\sqrt{50}} = 2.32$

6. Determination of *P*-value: Because the test is two-tailed,

$$P\text{-value} = 2(1 - \Phi(2.32)) = .0204$$

7. Conclusion: Using a significance level of .01, H_0 would not be rejected since .0204 > .01. At this significance level, there is insufficient evidence to conclude that true average thickness differs from the target value. ∎

P-Values for *t* Tests

Just as the *P*-value for a *z* test is a *z* curve area, the *P*-value for a *t* test will be a *t* curve area. Figure 8.8 illustrates the three different cases. The number of df for the one-sample *t* test is $n - 1$.

The table of *t* critical values used previously for confidence and prediction intervals doesn't contain enough information about any particular *t* distribution to allow for accurate determination of desired areas. So we have included another *t* table in Appendix Table A.8, one that contains a tabulation of upper-tail *t* curve areas. Each different column of the table is for a different number of df, and the rows are for calculated values of the test statistic *t* ranging from 0.0 to 4.0 in increments of .1. For example, the number .074 appears at the intersection of the 1.6 row and the 8 df column, so the area under the 8 df curve to the right of 1.6 (an upper-tail area) is .074. Because *t* curves are symmetric, .074 is also the area under the 8 df curve to the left of −1.6 (a lower-tail area).

Suppose, for example, that a test of H_0: $\mu = 100$ versus H_a: $\mu > 100$ is based on the 8 df *t* distribution. If the calculated value of the test statistic is $t = 1.6$, then the *P*-value for this upper-tailed test is .074. Because .074 exceeds .05, we would not be able to reject H_0 at a significance level of .05. If the alternative hypothesis is H_a: $\mu < 100$ and a test based on 20 df yields $t = -3.2$, then Appendix Table A.8 shows that the *P*-value is the captured lower-tail area .002. The null hypothesis can be rejected at either level .05 or .01. Consider testing H_0: $\mu_1 - \mu_2 = 0$ versus H_a: $\mu_1 - \mu_2 \neq 0$; the null

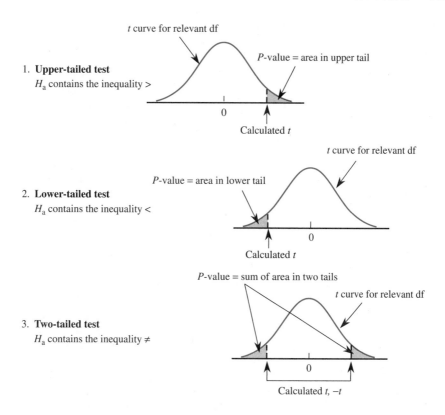

1. **Upper-tailed test**
   H_a contains the inequality >

   t curve for relevant df

   P-value = area in upper tail

   0

   Calculated *t*

2. **Lower-tailed test**
   H_a contains the inequality <

   P-value = area in lower tail

   t curve for relevant df

   0

   Calculated *t*

3. **Two-tailed test**
   H_a contains the inequality ≠

   P-value = sum of area in two tails

   t curve for relevant df

   0

   Calculated *t*, −*t*

Figure 8.8 *P*-values for *t* tests

hypothesis states that the means of the two populations are identical, whereas the alternative hypothesis states that they are different without specifying a direction of departure from H_0. If a *t* test is based on 20 df and *t* = 3.2, then the *P*-value for this two-tailed test is 2(.002) = .004. This would also be the *P*-value for *t* = −3.2. The tail area is doubled because values both larger than 3.2 and smaller than −3.2 are more contradictory to H_0 than what was calculated (values farther out in *either* tail of the *t* curve).

Example 8.18   In Example 8.9, we carried out a test of H_0: $\mu = 25$ versus H_a: $\mu > 25$ based on 4 df. The calculated value of *t* was 1.04. Looking to the 4 df column of Appendix Table A.8 and down to the 1.0 row, we see that the entry is .187, so *P*-value ≈ .187. This *P*-value is clearly larger than any reasonable significance level α (.01, .05, and even .10), so there is no reason to reject the null hypothesis. The MINITAB output included in Example 8.9 has *P*-value = .18. *P*-values from software packages will be more accurate than what results from Appendix Table A.8 since values of *t* in our table are accurate only to the tenths digit.   ■

Exercises | Section 8.4 (44–56)

44. For which of the given *P*-values would the null hypothesis be rejected when performing a level .05 test?

   a. .001 **b.** .021 **c.** .078
   d. .047 **e.** .148

45. Pairs of *P*-values and significance levels, α, are given. For each pair, state whether the observed *P*-value would lead to rejection of H_0 at the given significance level.
   a. *P*-value = .084, α = .05
   b. *P*-value = .003, α = .001
   c. *P*-value = .498, α = .05
   d. *P*-value = .084, α = .10
   e. *P*-value = .039, α = .01
   f. *P*-value = .218, α = .10

46. Let μ denote the mean reaction time to a certain stimulus. For a large-sample *z* test of H_0: μ = 5 versus H_a: $\mu > 5$, find the *P*-value associated with each of the given values of the *z* test statistic.
   a. 1.42   **b.** .90   **c.** 1.96   **d.** 2.48   **e.** −.11

47. Newly purchased tires of a certain type are supposed to be filled to a pressure of 30 lb/in.^{2}. Let μ denote the true average pressure. Find the *P*-value associated with each given *z* statistic value for testing H_0: μ = 30 versus H_a: $\mu \neq 30$.
   a. 2.10   **b.** −1.75   **c.** −.55   **d.** 1.41   **e.** −5.3

48. Give as much information as you can about the *P*-value of a *t* test in each of the following situations:
   a. Upper-tailed test, df = 8, *t* = 2.0
   b. Lower-tailed test, df = 11, *t* = −2.4
   c. Two-tailed test, df = 15, *t* = −1.6
   d. Upper-tailed test, df = 19, *t* = −.4
   e. Upper-tailed test, df = 5, *t* = 5.0
   f. Two-tailed test, df = 40, *t* = −4.8

49. The paint used to make lines on roads must reflect enough light to be clearly visible at night. Let μ denote the true average reflectometer reading for a new type of paint under consideration. A test of H_0: μ = 20 versus H_a: $\mu > 20$ will be based on a random sample of size *n* from a normal population distribution. What conclusion is appropriate in each of the following situations?
   a. *n* = 15, *t* = 3.2, α = .05
   b. *n* = 9, *t* = 1.8, α = .01
   c. *n* = 24, *t* = −.2

50. Let μ denote true average serum receptor concentration for all pregnant women. The average for all women is known to be 5.63. The article "Serum Transferrin Receptor for the Detection of Iron Deficiency in Pregnancy" (*Amer. J. Clinical Nutr.*, 1991: 1077–1081) reports that *P*-value > .10 for a test of H_0: μ = 5.63 versus H_a: $\mu \neq 5.63$ based on *n* = 176 pregnant women. Using a significance level of .01, what would you conclude?

51. An aspirin manufacturer fills bottles by weight rather than by count. Since each bottle should contain 100 tablets, the average weight per tablet should be 5 grains. Each of 100 tablets taken from a very large lot is weighed, resulting in a sample average weight per tablet of 4.87 grains and a sample standard deviation of .35 grain. Does this information provide strong evidence for concluding that the company is not filling its bottles as advertised? Test the appropriate hypotheses using α = .01 by first computing the *P*-value and then comparing it to the specified significance level.

52. Because of variability in the manufacturing process, the actual yielding point of a sample of mild steel subjected to increasing stress will usually differ from the theoretical yielding point. Let *p* denote the true proportion of samples that yield before their theoretical yielding point. If on the basis of a sample it can be concluded that more than 20% of all specimens yield before the theoretical point, the production process will have to be modified.
   a. If 15 of 60 specimens yield before the theoretical point, what is the *P*-value when the appropriate test is used, and what would you advise the company to do?
   b. If the true percentage of "early yields" is actually 50% (so that the theoretical point is the median of the yield distribution) and a level .01 test is used, what is the probability that the company concludes a modification of the process is necessary?

53. The times of first sprinkler activation for a series of tests with fire prevention sprinkler systems using an aqueous film-forming foam were (in sec)

27 41 22 27 23 35 30 33 24 27 28 22 24

(see "Use of AFFF in Sprinkler Systems," *Fire Technology*, 1976: 5). The system has been designed so that true average activation time is at most 25 sec under such conditions. Does the data strongly contradict the validity of this design specification? Test the relevant hypotheses at significance level .05 using the *P*-value approach.

54. A certain pen has been designed so that true average writing lifetime under controlled conditions (involving the use of a writing machine) is at least 10 hours. A random sample of 18 pens is selected, the writing lifetime of each is determined, and a normal probability plot of the resulting data supports the use of a one-sample *t* test.

a. What hypotheses should be tested if the investigators believe a priori that the design specification has been satisfied?

b. What conclusion is appropriate if the hypotheses of part (a) are tested, $t = -2.3$, and $\alpha = .05$?

c. What conclusion is appropriate if the hypotheses of part (a) are tested, $t = -1.8$, and $\alpha = .01$?

d. What should be concluded if the hypotheses of part (a) are tested and $t = -3.6$?

55. A spectrophotometer used for measuring CO concentration [ppm (parts per million) by volume] is checked for accuracy by taking readings on a manufactured gas (called span gas) in which the CO concentration is very precisely controlled at 70 ppm. If the readings suggest that the spectrophotometer is not working properly, it will have to be recalibrated. Assume that if properly calibrated, measured concentration for span gas samples is normally distributed. On the basis of the six readings, 85, 77, 82, 68, 72, and 69, is recalibration nec-

essary? Carry out a test of the relevant hypotheses using the P-value approach with $\alpha = .05$.

56. The relative conductivity of a semiconductor device is determined by the amount of impurity "doped" into the device during its manufacture. A silicon diode to be used for a specific purpose requires an average cut-on voltage of .60 V, and, if this is not achieved, the amount of impurity must be adjusted. A sample of diodes was selected and the cut-on voltage was determined. The accompanying SAS output resulted from a request to test the appropriate hypotheses.

```
 N      Mean     Std Dev         T  Prob>|T|
15  0.0453333  0.0899100  1.9527887    0.0711
```

[*Note:* SAS explicitly tests H_0: $\mu = 0$, so to test H_0: $\mu = .60$, the null value .60 must be subtracted from each x_i; the reported mean is then the average of the $(x_i - .60)$ values. Also, SAS's P-value is always for a two-tailed test.] What would be concluded for a significance level of .01? .05? .10?

8.5 | Some Comments on Selecting a Test Procedure

Once the experimenter has decided on the question of interest and the method for gathering data (the design of the experiment), construction of an appropriate test procedure consists of three distinct steps:

1. Specify a test statistic (the function of the observed values that will serve as the decision maker).

2. Decide on the general form of the rejection region (typically reject H_0 for suitably large values of the test statistic, reject for suitably small values, or reject for either small or large values).

3. Select the specific numerical critical value or values that will separate the rejection region from the acceptance region (by obtaining the distribution of the test statistic when H_0 is true, and then selecting a level of significance).

In the examples thus far, both Steps 1 and 2 were carried out in an ad hoc manner through intuition. For example, when the underlying population was assumed normal with mean μ and known σ, we were led from $\overline{X}$ to the standardized test statistic

$$Z = \frac{\overline{X} - \mu_0}{\sigma/\sqrt{n}}$$

For testing H_0: $\mu = \mu_0$ versus H_a: $\mu > \mu_0$, intuition then suggested rejecting H_0 when z was large. Finally the critical value was determined by specifying the level of significance α and using the fact that Z has a standard normal distribution when H_0 is true.

The reliability of the test in reaching a correct decision can be assessed by studying type II error probabilities.

Issues to be considered in carrying out Steps 1–3 encompass the following questions:

1. What are the practical implications and consequences of choosing a particular level of significance once the other aspects of a test procedure have been determined?

2. Does there exist a general principle, not dependent just on intuition, that can be used to obtain best or good test procedures?

3. When there exist two or more tests that are appropriate in a given situation, how can the tests be compared to decide which should be used?

4. If a test is derived under specific assumptions about the distribution or population being sampled, how well will the test procedure work when the assumptions are violated?

Statistical Versus Practical Significance

Although the process of reaching a decision by using the methodology of classical hypothesis testing involves selecting a level of significance and then rejecting or not rejecting H_0 at that level α, simply reporting the α used and the decision reached conveys little of the information contained in the sample data. Especially when the results of an experiment are to be communicated to a large audience, rejection of H_0 at level .05 will be much more convincing if the observed value of the test statistic greatly exceeds the 5% critical value than if it barely exceeds that value. This is precisely what led to the notion of P-value as a way of reporting significance without imposing a particular α on others who might wish to draw their own conclusions.

Even if a P-value is included in a summary of results, however, there may be difficulty in interpreting this value and in making a decision. This is because a small P-value, which would ordinarily indicate **statistical significance** in that it would strongly suggest rejection of H_0 in favor of H_a, may be the result of a large sample size in combination with a departure from H_0 that has little **practical significance.** In many experimental situations, only departures from H_0 of large magnitude would be worthy of detection, whereas a small departure from H_0 would have little practical significance.

Consider as an example testing H_0: $\mu = 100$ versus H_a: $\mu > 100$ where μ is the mean of a normal population with $\sigma = 10$. Suppose a true value of $\mu = 101$ would not represent a serious departure from H_0 in the sense that not rejecting H_0 when $\mu = 101$ would be a relatively inexpensive error. For a reasonably large sample size n, this μ would lead to an $\bar{x}$ value near 101, so we would not want this sample evidence to argue strongly for rejection of H_0 when $\bar{x} = 101$ is observed. For various sample sizes, Table 8.1 records both the P-value when $\bar{x} = 101$ and also the probability of not rejecting H_0 at level .01 when $\mu = 101$.

The second column in Table 8.1 shows that even for moderately large sample sizes, the P-value of $\bar{x} = 101$ argues very strongly for rejection of H_0, whereas the observed $\bar{x}$ itself suggests that in practical terms the true value of μ differs little from the null value $\mu_0 = 100$. The third column points out that even when there is little practical difference between the true μ and the null value, for a fixed level of significance a large sample size will almost always lead to rejection of the null hypothesis at that level. To

Table 8.1 An illustration of the effect of sample size on P-values and β

n	P-Value When $\bar{x} = 101$	$\beta(101)$ for Level .01 Test
25	.3085	.9664
100	.1587	.9082
400	.0228	.6293
900	.0013	.2514
1600	.0000335	.0475
2500	.000000297	.0038
10,000	7.69×10^{-24}	.0000

summarize, *one must be especially careful in interpreting evidence when the sample size is large, since any small departure from H_0 will almost surely be detected by a test, yet such a departure may have little practical significance.*

The Likelihood Ratio Principle

Let $x_1, x_2, \ldots, x_n$ be the observations in a random sample of size n from a probability distribution $f(x; \theta)$. The joint distribution evaluated at these sample values is the product $f(x_1; \theta) \cdot f(x_2; \theta) \cdot \cdots \cdot f(x_n; \theta)$. As in the discussion of maximum likelihood estimation, the *likelihood function* is this joint distribution regarded as a function of θ. Consider testing H_0: θ is in Ω_0 versus H_a: θ is in Ω_a, where Ω_0 and Ω_a are disjoint (for example, H_0: $\theta \le 100$ versus H_a: $\theta > 100$). The **likelihood ratio principle** for test construction proceeds as follows:

1. Find the largest value of the likelihood for any θ in Ω_0 (by finding the maximum likelihood estimate within Ω_0 and substituting back into the likelihood function).

2. Find the largest value of the likelihood for any θ in Ω_a.

3. Form the ratio

$$\lambda(x_1, \ldots, x_n) = \frac{\text{maximum likelihood for } \theta \text{ in } \Omega_0}{\text{maximum likelihood for } \theta \text{ in } \Omega_a}$$

The ratio $\lambda(x_1, \ldots, x_n)$ is called the *likelihood ratio statistic value.* The test procedure consists of rejecting H_0 when this ratio is small. That is, a constant k is chosen, and H_0 is rejected if $\lambda(x_1, \ldots, x_n) \le k$. Thus, H_0 is rejected when the denominator of λ greatly exceeds the numerator, indicating that the data is much more consistent with H_a than with H_0.

The constant k is selected to yield the desired type I error probability. Often the inequality $\lambda \le k$ can be manipulated to yield a simpler equivalent condition. For example, for testing H_0: $\mu \le \mu_0$ versus H_a: $\mu > \mu_0$ in the case of normality, $\lambda \le k$ is equivalent to $t \ge c$. Thus, with $c = t_{\alpha,n-1}$, the likelihood ratio test is the one-sample t test.

The likelihood ratio principle can also be applied when the X_i's have different distributions and even when they are dependent, though the likelihood function can be complicated in such cases. Many of the test procedures to be presented in subsequent chapters are obtained from the likelihood ratio principle. These tests often turn out to minimize β among all tests that have the desired α, so are truly best tests. For more details and some worked examples, refer to the book by DeGroot or the one by Hogg and Craig listed in the Chapter 6 bibliography.

A practical limitation on the use of the likelihood ratio principle is that, to construct the likelihood ratio test statistic, the form of the probability distribution from which the sample comes must be specified. To derive the t test from the likelihood ratio principle, the investigator must assume a normal pdf. If an investigator is willing to assume that the distribution is symmetric but does not want to be specific about its exact form (such as normal, uniform, or Cauchy), then the principle fails because there is no way to write a joint pdf simultaneously valid for all symmetric distributions. In Chapter 15, we will present several **distribution-free** test procedures, so called because the probability of a type I error is controlled simultaneously for many different underlying distributions. These procedures are useful when the investigator has limited knowledge of the underlying distribution. We shall also say more about issues 3 and 4 listed at the outset of this section.

Exercises | Section 8.5 (57–58)

57. Reconsider the paint-drying problem discussed in Example 8.2. The hypotheses were H_0: $\mu = 75$ versus H_a: $\mu < 75$, with σ assumed to have value 9.0. Consider the alternative value $\mu = 74$, which in the context of the problem would presumably not be a practically significant departure from H_0.
  a. For a level .01 test, compute β at this alternative for sample sizes $n = 100, 900$, and 2500.
  b. If the observed value of $\overline{X}$ is $\bar{x} = 74$, what can you say about the resulting P-value when $n = 2500$? Is the data statistically significant at any of the standard values of α?

  c. Would you really want to use a sample size of 2500 along with a level .01 test (disregarding the cost of such an experiment)? Explain.

58. Consider the large-sample level .01 test in Section 8.3 for testing H_0: $p = .2$ against H_a: $p > .2$.
  a. For the alternative value $p = .21$, compute $\beta(.21)$ for sample sizes $n = 100, 2500, 10,000$, 40,000, and 90,000.
  b. For $\hat{p} = x/n = .21$, compute the P-value when $n = 100, 2500, 10,000$, and 40,000.
  c. In most situations, would it be reasonable to use a level .01 test in conjunction with a sample size of 40,000? Why or why not?

Supplementary Exercises (59–80)

59. A sample of 50 lenses used in eyeglasses yields a sample mean thickness of 3.05 mm and a sample standard deviation of .34 mm. The desired true average thickness of such lenses is 3.20 mm. Does the data strongly suggest that the true average thickness of such lenses is something other than what is desired? Test using $\alpha = .05$.

60. In Exercise 59, suppose the experimenter had believed before collecting the data that the value of σ was approximately .30. If the experimenter wished the probability of a type II error to be .05

when $\mu = 3.00$, was a sample size 50 unnecessarily large?

61. It is specified that a certain type of iron should contain .85 gm of silicon per 100 gm of iron (.85%). The silicon content of each of 25 randomly selected iron specimens was determined, and the accompanying MINITAB output resulted from a test of the appropriate hypotheses.

Variable	N	Mean	StDev	SE Mean	T	P
sil cont	25	0.8880	0.1807	0.0361	1.05	0.30

a. What hypotheses were tested?

b. What conclusion would be reached for a significance level of .05, and why? Answer the same question for a significance level of .10.

62. One method for straightening wire before coiling it to make a spring is called "roller straightening." The article "The Effect of Roller and Spinner Wire Straightening on Coiling Performance and Wire Properties" (*Springs*, 1987: 27–28) reports on the tensile properties of wire. Suppose a sample of 16 wires is selected and each is tested to determine tensile strength (N/mm^2). The resulting sample mean and standard deviation are 2160 and 30, respectively.

a. The mean tensile strength for springs made using spinner straightening is 2150 N/mm^2. What hypotheses should be tested to determine whether the mean tensile strength for the roller method exceeds 2150?

b. Assuming that the tensile strength distribution is approximately normal, what test statistic would you use to test the hypotheses in part (a)?

c. What is the value of the test statistic for this data?

d. What is the *P*-value for the value of the test statistic computed in part (c)?

e. For a level .05 test, what conclusion would you reach?

63. A new method for measuring phosphorus levels in soil is described in the article "A Rapid Method to Determine Total Phosphorus in Soils" (*Soil Sci. Amer. J.*, 1988: 1301–1304). Suppose a sample of 11 soil specimens, each with a true phosphorus content of 548 mg/kg, is analyzed using the new method. The resulting sample mean and standard deviation for phosphorus level are 587 and 10, respectively.

a. Is there evidence that the mean phosphorus level reported by the new method differs significantly from the true value of 548 mg/kg? Use $\alpha = .05$.

b. What assumptions must you make for the test in part (a) to be appropriate?

64. The article "Orchard Floor Management Utilizing Soil-Applied Coal Dust for Frost Protection" (*Agri. and Forest Meteorology*, 1988: 71–82) reports the following values for soil heat flux of eight plots covered with coal dust.

34.7   35.4   34.7   37.7   32.5   28.0   18.4   24.9

The mean soil heat flux for plots covered only with grass is 29.0. Assuming that the heat-flux distribution is approximately normal, does the data suggest that the coal dust is effective in increasing the mean heat flux over that for grass? Test the appropriate hypotheses using $\alpha = .05$.

65. The article "Caffeine Knowledge, Attitudes, and Consumption in Adult Women" (*J. of Nutrition Educ.*, 1992: 179–184) reports the following summary data on daily caffeine consumption for a sample of adult women: $n = 47$, $\bar{x} = 215$ mg, $s = 235$ mg, and range = 5–1176.

a. Does it appear plausible that the population distribution of daily caffeine consumption is normal? Is it necessary to assume a normal population distribution to test hypotheses about the value of the population mean consumption? Explain your reasoning.

b. Suppose it had previously been believed that mean consumption was at most 200 mg. Does the given data contradict this prior belief? Test the appropriate hypotheses at significance level .10 and include a *P*-value in your analysis.

66. The accompanying output resulted when MINITAB was used to test the appropriate hypotheses about true average activation time based on the data in Exercise 53. Use this information to reach a conclusion at significance level .05 and also at level .01.

```
TEST OF MU = 25.000 VS MU G.T.  25.000
       N    MEAN STDEV SE MEAN     T P VALUE
time 13 27.923 5.619   1.559 1.88   0.043
```

67. The true average breaking strength of ceramic insulators of a certain type is supposed to be at least 10 psi. They will be used for a particular application unless sample data indicates conclusively that this specification has not been met. A test of hypotheses using $\alpha = .01$ is to be based on a random sample of ten insulators. Assume that the breaking-strength distribution is normal with unknown standard deviation.

a. If the true standard deviation is .80, how likely is it that insulators will be judged satisfactory when true average breaking strength is actually only 9.5? Only 9.0?

b. What sample size would be necessary to have a 75% chance of detecting that true average breaking strength is 9.5 when the true standard deviation is .80?

68. The accompanying observations on residual flame time (sec) for strips of treated children's nightwear were given in the article "An Introduction to Some Precision and Accuracy of Measurement Problems"

(*J. of Testing and Eval.*, 1982: 132–140). Suppose a true average flame time of at most 9.75 had been mandated. Does the data suggest that this condition has not been met? Carry out an appropriate test after first investigating the plausibility of assumptions that underlie your method of inference.

9.85	9.93	9.75	9.77	9.67	9.87	9.67
9.94	9.85	9.75	9.83	9.92	9.74	9.99
9.88	9.95	9.95	9.93	9.92	9.89	

69. The incidence of a certain type of chromosome defect in the U.S. adult male population is believed to be 1 in 75. A random sample of 800 individuals in U.S. penal institutions reveals 16 who have such defects. Can it be concluded that the incidence rate of this defect among prisoners differs from the presumed rate for the entire adult male population?
   a. State and test the relevant hypotheses using $\alpha = .05$. What type of error might you have made in reaching a conclusion?
   b. What P-value is associated with this test? Based on this P-value, could H_0 be rejected at significance level .20?

70. In an investigation of the toxin produced by a certain poisonous snake, a researcher prepared 26 different vials, each containing 1 g of the toxin, and then determined the amount of antitoxin needed to neutralize the toxin. The sample average amount of antitoxin necessary was found to be 1.89 mg, and the sample standard deviation was .42. Previous research had indicated that the true average neutralizing amount was 1.75 mg/g of toxin. Does the new data contradict the value suggested by prior research? Test the relevant hypotheses using the P-value approach. Does the validity of your analysis depend on any assumptions about the population distribution of neutralizing amount? Explain.

71. The sample average unrestrained compressive strength for 45 specimens of a particular type of brick was computed to be 3107 psi, and the sample standard deviation was 188. The distribution of unrestrained compressive strength may be somewhat skewed. Does the data strongly indicate that the true average unrestrained compressive strength is less than the design value of 3200? Test using $\alpha = .001$.

72. To test the ability of auto mechanics to identify simple engine problems, an automobile with a single such problem was taken in turn to 72 different car repair facilities. Only 42 of the 72 mechanics who worked on the car correctly identified the problem. Does this strongly indicate that the true proportion of mechanics who could identify this problem is less than .75? Compute the P-value and reach a conclusion accordingly.

73. When $X_1, X_2, \ldots, X_n$ are independent Poisson variables, each with parameter λ, and n is large, the sample mean $\overline{X}$ has approximately a normal distribution with $\mu = E(\overline{X}) = \lambda$ and $\sigma^2 = V(\overline{X}) = \lambda/n$. This implies that

$$Z = \frac{\overline{X} - \lambda}{\sqrt{\lambda/n}}$$

has approximately a standard normal distribution. For testing $H_0: \lambda = \lambda_0$, we can replace λ by λ_0 in the equation for Z to obtain a test statistic. This statistic is actually preferred to the large-sample statistic with denominator $S/\sqrt{n}$ (when the X_i's are Poisson) because it is tailored explicitly to the Poisson assumption. If the number of requests for consulting received by a certain statistician during a 5-day work week has a Poisson distribution and the total number of consulting requests during a 36-week period is 160, does this suggest that the true average number of weekly requests exceeds 4.0? Test using $\alpha = .02$.

74. A hot-tub manufacturer advertises that with its heating equipment, a temperature of 100°F can be achieved in at most 15 min. A random sample of 32 tubs is selected, and the time necessary to achieve a 100°F temperature is determined for each tub. The sample average time and sample standard deviation are 17.5 min and 2.2 min, respectively. Does this data cast doubt on the company's claim? Compute the P-value and use it to reach a conclusion at level .05 (assume that the heating-time distribution is approximately normal).

75. Chapter 7 presented a CI for the variance σ^2 of a normal population distribution. The key result there was that the rv $\chi^2 = (n - 1)S^2/\sigma^2$ has a chi-squared distribution with $n - 1$ df. Consider the null hypothesis $H_0: \sigma^2 = \sigma_0^2$ (equivalently, $\sigma = \sigma_0$). Then when H_0 is true, the test statistic $\chi^2 = (n - 1)S^2/\sigma_0^2$ has a chi-squared distribution with $n - 1$ df. If the relevant alternative is $H_a: \sigma^2 > \sigma_0^2$, rejecting H_0 if $(n - 1)s^2/\sigma_0^2 \geq \chi^2_{\alpha,n-1}$ gives a test with significance level α. To ensure reasonably uniform characteristics for a particular application, it is desired that the true standard deviation of the softening point of a certain type of petroleum pitch be at most .50°C.

The softening points of ten different specimens were determined, yielding a sample standard deviation of .58°C. Does this strongly contradict the uniformity specification? Test the appropriate hypotheses using $\alpha = .01$.

76. Referring to Exercise 75, suppose an investigator wishes to test H_0: $\sigma^2 = .04$ versus H_a: $\sigma^2 < .04$ based on a sample of 21 observations. The computed value of $20s^2/.04$ is 8.58. Place bounds on the P-value and then reach a conclusion at level .01.

77. When the population distribution is normal and n is large, the sample standard deviation S has approximately a normal distribution with $E(S) \approx \sigma$ and $V(S) \approx \sigma^2/(2n)$. We already know that in this case, for any n, $\overline{X}$ is normal with $E(\overline{X}) = \mu$ and $V(\overline{X}) = \sigma^2/n$.
   a. Assuming that the underlying distribution is normal, what is an approximately unbiased estimator of the 99th percentile $\theta = \mu + 2.33\sigma$?
   b. When the X_i's are normal, it can be shown that $\overline{X}$ and S are independent rv's (one measures location whereas the other measures spread). Use this to compute $V(\hat{\theta})$ and $\sigma_{\hat{\theta}}$ for the estimator $\hat{\theta}$ of part (a). What is the estimated standard error $\hat{\sigma}_{\hat{\theta}}$?
   c. Write a test statistic for testing H_0: $\theta = \theta_0$ that has approximately a standard normal distribution when H_0 is true. If soil pH is normally distributed in a certain region and 64 soil samples yield $\overline{x} = 6.33$, $s = .16$, does this provide strong evidence for concluding that at most 99% of all possible samples would have a pH of less than 6.75? Test using $\alpha = .01$.

78. Let $X_1, X_2, \ldots, X_n$ be a random sample from an exponential distribution with parameter λ. Then it can be shown that $2\lambda\Sigma X_i$ has a chi-squared distribution with $\nu = 2n$ (by first showing that $2\lambda X_i$ has a chi-squared distribution with $\nu = 2$).
   a. Use this fact to obtain a test statistic and rejection region that together specify a level α test for H_0: $\mu = \mu_0$ versus each of the three commonly encountered alternatives. [*Hint*: $E(X_i) = \mu = 1/\lambda$, so $\mu = \mu_0$ is equivalent to $\lambda = 1/\mu_0$.]
   b. Suppose that ten identical components, each having exponentially distributed time until failure, are tested. The resulting failure times are

   95 16 11 3 42 71 225 64 87 123

   Use the test procedure of part (a) to decide whether the data strongly suggests that the true average lifetime is less than the previously claimed value of 75.

79. Suppose the population distribution is normal with known σ. Let γ be such that $0 < \gamma < \alpha$. For testing H_0: $\mu = \mu_0$ versus H_a: $\mu \neq \mu_0$, consider the test that rejects H_0 if either $z \geq z_\gamma$ or $z \leq -z_{\alpha-\gamma}$, where the test statistic is $Z = (\overline{X} - \mu_0)/(\sigma/\sqrt{n})$.
   a. Show that $P(\text{type I error}) = \alpha$.
   b. Derive an expression for $\beta(\mu')$. (*Hint*: Express the test in the form "reject H_0 if either $\overline{x} \geq c_1$ or $\leq c_2$.")
   c. Let $\Delta > 0$. For what values of γ (relative to α) will $\beta(\mu_0 + \Delta) < \beta(\mu_0 - \Delta)$?

80. After a period of apprenticeship, an organization gives an exam that must be passed to be eligible for membership. Let $p = P(\text{randomly chosen apprentice passes})$. The organization wishes an exam that most but not all should be able to pass, so it decides that $p = .90$ is desirable. For a particular exam, the relevant hypotheses are H_0: $p = .90$ versus the alternative H_a: $p \neq .90$. Suppose ten people take the exam, and let $X =$ the number who pass.
   a. Does the lower-tailed region $\{0, 1, \ldots, 5\}$ specify a level .01 test?
   b. Show that even though H_a is two-sided, no two-tailed test is a level .01 test.
   c. Sketch a graph of $\beta(p')$ as a function of p' for this test. Is this desirable?

Bibliography

See the bibliographies at the end of Chapter 6 and Chapter 7.

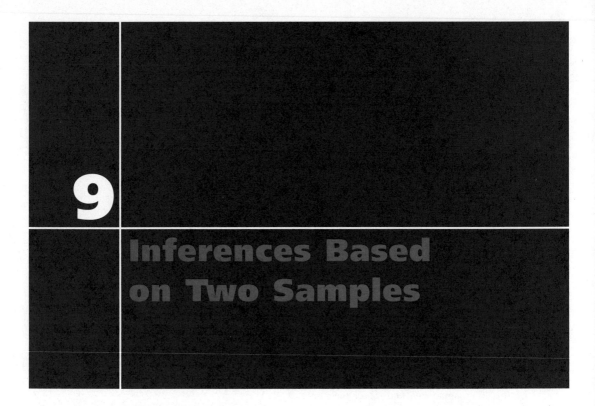

9

Inferences Based on Two Samples

Introduction

Chapters 7 and 8 presented confidence intervals (CIs) and hypothesis-testing procedures for a single population parameter (μ, p, and σ^2). Here we extend these methods to situations involving the means, proportions, and variances of two different population distributions.

9.1 | *z* Tests and Confidence Intervals for a Difference Between Two Population Means

The inferences discussed in this section concern a difference $\mu_1 - \mu_2$ between the means of two different population distributions. An investigator might, for example, wish to test hypotheses about the difference between true average breaking strengths of two different types of corrugated fiberboard. One such hypothesis would state that $\mu_1 - \mu_2 = 0$, that is, that $\mu_1 = \mu_2$. Alternatively, it may be appropriate to estimate $\mu_1 - \mu_2$ by computing a 95% CI. Such inferences are based on a sample of strength observations for each type of fiberboard.

Basic Assumptions

1. $X_1, X_2, \ldots, X_m$ is a random sample from a population with mean μ_1 and variance σ_1^2.

2. $Y_1, Y_2, \ldots, Y_n$ is a random sample from a population with mean μ_2 and variance σ_2^2.

3. The X and Y samples are independent of one another.

The natural estimator of $\mu_1 - \mu_2$ is $\overline{X} - \overline{Y}$, the difference between the corresponding sample means. The test statistic results from standardizing this estimator, so we need expressions for the expected value and standard deviation of $\overline{X} - \overline{Y}$.

PROPOSITION

The expected value of $\overline{X} - \overline{Y}$ is $\mu_1 - \mu_2$, so $\overline{X} - \overline{Y}$ is an unbiased estimator of $\mu_1 - \mu_2$. The standard deviation of $\overline{X} - \overline{Y}$ is

$$\sigma_{\overline{X}-\overline{Y}} = \sqrt{\frac{\sigma_1^2}{m} + \frac{\sigma_2^2}{n}}$$

Proof
sented in Chapter 5. Since the expected value of a difference is the difference of expected values,

$$E(\overline{X} - \overline{Y}) = E(\overline{X}) - E(\overline{Y}) = \mu_1 - \mu_2$$

Because the X and Y samples are independent, $\overline{X}$ and $\overline{Y}$ are independent quantities, so the variance of the difference is the *sum* of $V(\overline{X})$ and $V(\overline{Y})$:

$$V(\overline{X} - \overline{Y}) = V(\overline{X}) + V(\overline{Y}) = \frac{\sigma_1^2}{m} + \frac{\sigma_2^2}{n}$$

The standard deviation of $\overline{X} - \overline{Y}$ is the square root of this expression. ∎

If we think of $\mu_1 - \mu_2$ as a parameter θ, then its estimator is $\hat{\theta} = \overline{X} - \overline{Y}$ with standard deviation $\sigma_{\hat{\theta}}$ given by the proposition. When σ_1^2 and σ_2^2 both have known values, the test statistic will have the form $(\hat{\theta} - \text{null value})/\sigma_{\hat{\theta}}$; this form of a test statistic was used in several one-sample problems in the previous chapter. When σ_1^2 and σ_2^2 are unknown, the sample variances must be used to estimate $\sigma_{\hat{\theta}}$.

Test Procedures for Normal Populations with Known Variances

In Chapters 7 and 8, the first CI and test procedure for a population mean μ were based on the assumption that the population distribution was normal with the value of the population variance σ^2 known to the investigator. Similarly, we first assume here that *both* population distributions are normal and that the values of *both* σ_1^2 and σ_2^2 are known. Situations in which one or both of these assumptions can be dispensed with will be presented shortly.

Because the population distributions are normal, both $\overline{X}$ and $\overline{Y}$ have normal distributions. This implies that $\overline{X} - \overline{Y}$ is normally distributed, with expected value $\mu_1 - \mu_2$ and standard deviation $\sigma_{\overline{X}-\overline{Y}}$ given in the foregoing proposition. Standardizing $\overline{X} - \overline{Y}$ gives the standard normal variable

$$Z = \frac{\overline{X} - \overline{Y} - (\mu_1 - \mu_2)}{\sqrt{\dfrac{\sigma_1^2}{m} + \dfrac{\sigma_2^2}{n}}} \tag{9.1}$$

In a hypothesis-testing problem, the null hypothesis will state that $\mu_1 - \mu_2$ has a specified value. Denoting this null value by Δ_0, the null hypothesis becomes H_0: $\mu_1 - \mu_2 = \Delta_0$. Often $\Delta_0 = 0$, in which case H_0 says that $\mu_1 = \mu_2$. A test statistic results from replacing $\mu_1 - \mu_2$ in Expression (9.1) by the null value Δ_0. Because the test statistic Z is obtained by standardizing $\overline{X} - \overline{Y}$ under the assumption that H_0 is true, it has a standard normal distribution in this case. Consider the alternative hypothesis H_a: $\mu_1 - \mu_2 > \Delta_0$. A value $\overline{x} - \overline{y}$ that considerably exceeds Δ_0 (the expected value of $\overline{X} - \overline{Y}$ when H_0 is true) provides evidence against H_0 and for H_a. Such a value of $\overline{x} - \overline{y}$ corresponds to a positive and large value of z. Thus, H_0 should be rejected in favor of H_a if z is greater than or equal to an appropriately chosen critical value. Because the test statistic Z has a standard normal distribution when H_0 is true, the upper-tailed rejection region $z \geq z_\alpha$ gives a test with significance level (type I error probability) α. Rejection regions for H_a: $\mu_1 - \mu_2 < \Delta_0$ and H_a: $\mu_1 - \mu_2 \neq \Delta_0$ that yield tests with desired significance level α are lower-tailed and two-tailed, respectively.

Null hypothesis: H_0: $\mu_1 - \mu_2 = \Delta_0$

Test statistic value: $z = \dfrac{\overline{x} - \overline{y} - \Delta_0}{\sqrt{\dfrac{\sigma_1^2}{m} + \dfrac{\sigma_2^2}{n}}}$

Alternative Hypothesis	Rejection Region for Level α Test
H_a: $\mu_1 - \mu_2 > \Delta_0$	$z \geq z_\alpha$ (upper-tailed)
H_a: $\mu_1 - \mu_2 < \Delta_0$	$z \leq -z_\alpha$ (lower-tailed)
H_a: $\mu_1 - \mu_2 \neq \Delta_0$	either $z \geq z_{\alpha/2}$ or $z \leq -z_{\alpha/2}$ (two-tailed)

Because these are z tests, a P-value is computed as it was for the z tests in Chapter 8 (e.g., P-value $= 1 - \Phi(z)$ for an upper-tailed test).

Example 9.1 Analysis of a random sample consisting of $m = 20$ specimens of cold-rolled steel to determine yield strengths resulted in a sample average strength of $\bar{x} = 29.8$ ksi. A second random sample of $n = 25$ two-side galvanized steel specimens gave a sample average strength of $\bar{y} = 34.7$ ksi. Assuming that the two yield-strength distributions are normal with $\sigma_1 = 4.0$ and $\sigma_2 = 5.0$ (suggested by a graph in the article "Zinc-Coated Sheet Steel: An Overview," *Automotive Engr.*, Dec. 1984: 39–43), does the data indicate that the corresponding true average yield strengths μ_1 and μ_2 are different? Let's carry out a test at significance level $\alpha = .01$.

1. The parameter of interest is $\mu_1 - \mu_2$, the difference between the true average strengths for the two types of steel.

2. The null hypothesis is H_0: $\mu_1 - \mu_2 = 0$.

3. The alternative hypothesis is H_a: $\mu_1 - \mu_2 \neq 0$; if H_a is true, then μ_1 and μ_2 are different.

4. With $\Delta_0 = 0$, the test statistic value is

$$z = \frac{\bar{x} - \bar{y}}{\sqrt{\dfrac{\sigma_1^2}{m} + \dfrac{\sigma_2^2}{n}}}$$

5. The inequality in H_a implies that the test is two-tailed. For $\alpha = .01$, $\alpha/2 = .005$ and $z_{\alpha/2} = z_{.005} = 2.58$. H_0 will be rejected if $z \geq 2.58$ or if $z \leq -2.58$.

6. Substituting $m = 20$, $\bar{x} = 29.8$, $\sigma_1^2 = 16.0$, $n = 25$, $\bar{y} = 34.7$, and $\sigma_2^2 = 25.0$ into the formula for z yields

$$z = \frac{29.8 - 34.7}{\sqrt{\dfrac{16.0}{20} + \dfrac{25.0}{25}}} = \frac{-4.90}{1.34} = -3.66$$

That is, the observed value of $\bar{x} - \bar{y}$ is more than 3 standard deviations below what would be expected were H_0 true.

7. Since $-3.66 < -2.58$, the computed z does fall in the lower tail of the rejection region. H_0 is therefore rejected at level .01 in favor of the conclusion that $\mu_1 \neq \mu_2$. The sample data strongly suggests that the true average yield strength for cold-rolled

steel differs from that for galvanized steel. The *P*-value for this two-tailed test is $2(1 - \Phi(3.66)) \approx 2(1 - 1) = 0$, so H_0 should be rejected at *any* reasonable significance level.   ∎

Using a Comparison to Identify Causality

Investigators are often interested in comparing either the effects of two different treatments on a response or the response after treatment with the response after no treatment (treatment vs. control). If the individuals or objects to be used in the comparison are not assigned by the investigators to the two different conditions, the study is said to be **observational.** The difficulty with drawing conclusions based on an observational study is that, although statistical analysis may indicate a significant difference in response between the two groups, the difference may be due to some underlying factors that had not been controlled rather than to any difference in treatments.

Example 9.2   A letter in the *Journal of the American Medical Association* (May 19, 1978) reports that, of 215 male physicians who were Harvard graduates and died between November 1974 and October 1977, the 125 in full-time practice lived an average of 48.9 years beyond graduation, whereas the 90 with academic affiliations lived an average of 43.2 years beyond graduation. Does the data suggest that the mean lifetime after graduation for doctors in full-time practice exceeds the mean lifetime for those who have an academic affiliation (if so, those medical students who say that they are "dying to obtain an academic affiliation" may be closer to the truth than they realize; in other words, is "publish or perish" really "publish and perish")?

Let μ_1 denote the true average number of years lived beyond graduation for physicians in full-time practice, and let μ_2 denote the same quantity for physicians with academic affiliations. Assume the 125 and 90 physicians to be random samples from populations 1 and 2, respectively (which may not be reasonable if there is reason to believe that Harvard graduates have special characteristics that differentiate them from all other physicians—in this case inferences would be restricted just to the "Harvard populations"). The letter from which the data was taken gave no information about variances, so for illustration assume that $\sigma_1 = 14.6$ and $\sigma_2 = 14.4$. The hypotheses are H_0: $\mu_1 - \mu_2 = 0$ versus H_a: $\mu_1 - \mu_2 > 0$, so Δ_0 is zero. The computed value of the test statistic is

$$z = \frac{48.9 - 43.2}{\sqrt{\dfrac{(14.6)^2}{125} + \dfrac{(14.4)^2}{90}}} = \frac{5.70}{\sqrt{1.70 + 2.30}} = 2.85$$

The *P*-value for an upper-tailed test is $1 - \Phi(2.85) = .0022$. At significance level .01, H_0 is rejected (because $\alpha > P$-value) in favor of the conclusion that $\mu_1 - \mu_2 > 0$ ($\mu_1 > \mu_2$). This is consistent with the information reported in the letter.

This data resulted from a **retrospective** observational study; the investigator did not start out by selecting a sample of doctors and assigning some to the "academic affiliation" treatment and the others to the "full-time practice" treatment, but instead identified members of the two groups by looking backward in time (through obituaries!) to past records. Can the statistically significant result here really be attributed to a difference in the type of medical practice after graduation, or is there some other underlying factor (e.g., age at graduation, exercise regimens, etc.) that might also furnish a plausible explanation for the difference? Observational studies have been used to argue for a causal link between smoking and lung cancer. There are many studies that show that the incidence of lung cancer is significantly higher among smokers than among nonsmokers. However, individuals had decided whether to become smokers long before investigators arrived on the scene, and factors in making this decision may have played a causal role in the contraction of lung cancer.   ∎

A **randomized controlled experiment** results when investigators assign subjects to the two treatments in a random fashion. When statistical significance is observed in such an experiment, the investigator and other interested parties will have more confidence in the conclusion that the difference in response has been caused by a difference in treatments. A very famous example of this type of experiment and conclusion is the Salk polio vaccine experiment described in Section 9.4. These issues are discussed at greater length in the (nonmathematical) books by Moore and by Freedman et al., listed in the Chapter 1 references.

β and the Choice of Sample Size

The probability of a type II error is easily calculated when both population distributions are normal with known values of σ_1 and σ_2. Consider the case in which the alternative hypothesis is H_a: $\mu_1 - \mu_2 > \Delta_0$. Let Δ' denote a value of $\mu_1 - \mu_2$ that exceeds Δ_0 (a value for which H_0 is false). The upper-tailed rejection region $z \geq z_\alpha$ can be reexpressed in the form $\bar{x} - \bar{y} \geq \Delta_0 + z_\alpha \sigma_{\bar{X}-\bar{Y}}$. Thus, the probability of a type II error when $\mu_1 - \mu_2 = \Delta'$ is

$$\beta(\Delta') = P(\text{not rejecting } H_0 \text{ when } \mu_1 - \mu_2 = \Delta')$$
$$= P(\bar{X} - \bar{Y} < \Delta_0 + z_\alpha \sigma_{\bar{X}-\bar{Y}} \text{ when } \mu_1 - \mu_2 = \Delta')$$

When $\mu_1 - \mu_2 = \Delta'$, $\bar{X} - \bar{Y}$ is normally distributed with mean value Δ' and standard deviation $\sigma_{\bar{X}-\bar{Y}}$ (the same standard deviation as when H_0 is true); using these values to standardize the inequality in parentheses gives β.

Alternative Hypothesis	$\beta(\Delta') = P(\text{type II error when } \mu_1 - \mu_2 = \Delta')$
H_a: $\mu_1 - \mu_2 > \Delta_0$	$\Phi\left(z_\alpha - \dfrac{\Delta' - \Delta_0}{\sigma} \right)$
H_a: $\mu_1 - \mu_2 < \Delta_0$	$1 - \Phi\left(-z_\alpha - \dfrac{\Delta' - \Delta_0}{\sigma} \right)$
H_a: $\mu_1 - \mu_2 \neq \Delta_0$	$\Phi\left(z_{\alpha/2} - \dfrac{\Delta' - \Delta_0}{\sigma} \right) - \Phi\left(-z_{\alpha/2} - \dfrac{\Delta' - \Delta_0}{\sigma} \right)$

where $\sigma = \sigma_{\bar{X}-\bar{Y}} = \sqrt{(\sigma_1^2/m) + (\sigma_2^2/n)}$

Example 9.3
(Example 9.1
continued)

Suppose that when μ_1 and μ_2 (the true average yield strengths for the two types of steel) differ by as much as 5, the probability of detecting such a departure from H_0 should be .90. Does a level .01 test with sample sizes $m = 20$ and $n = 25$ satisfy this condition? The value of σ for these sample sizes (the denominator of z) was previously calculated as 1.34. The probability of a type II error for the two-tailed level .01 test when $\mu_1 - \mu_2 = \Delta' = 5$ is

$$\beta(5) = \Phi\left(2.58 - \frac{5 - 0}{1.34}\right) - \Phi\left(-2.58 - \frac{5 - 0}{1.34}\right)$$
$$= \Phi(-1.15) - \Phi(-6.31) = .1251$$

It is easy to verify that $\beta(-5) = .1251$ also (because the rejection region is symmetric). Thus, the probability of detecting such a departure is $1 - \beta(5) = .8749$. Because this is somewhat less than .9, slightly larger sample sizes should be used. ∎

As in Chapter 8, sample sizes m and n can be determined that will satisfy both $P(\text{type I error}) = a$ specified α and $P(\text{type II error when } \mu_1 - \mu_2 = \Delta') = a$ specified β. For an upper-tailed test, equating the previous expression for $\beta(\Delta')$ to the specified value of β gives

$$\frac{\sigma_1^2}{m} + \frac{\sigma_2^2}{n} = \frac{(\Delta' - \Delta_0)^2}{(z_\alpha + z_\beta)^2}$$

When the two sample sizes are equal, this equation yields

$$m = n = \frac{(\sigma_1^2 + \sigma_2^2)(z_\alpha + z_\beta)^2}{(\Delta' - \Delta_0)^2}$$

These expressions are also correct for a lower-tailed test, whereas α is replaced by $\alpha/2$ for a two-tailed test.

Large-Sample Tests

The assumptions of normal population distributions and known values of σ_1 and σ_2 are unnecessary when both sample sizes are large. In this case, the Central Limit Theorem guarantees that $\overline{X} - \overline{Y}$ has approximately a normal distribution regardless of the underlying population distributions. Furthermore, using S_1^2 and S_2^2 in place of σ_1^2 and σ_2^2 in Expression (9.1) gives a variable whose distribution is approximately standard normal:

$$Z = \frac{\overline{X} - \overline{Y} - (\mu_1 - \mu_2)}{\sqrt{\dfrac{S_1^2}{m} + \dfrac{S_2^2}{n}}}$$

A large-sample test statistic results from replacing $\mu_1 - \mu_2$ by Δ_0, the expected value of $\overline{X} - \overline{Y}$ when H_0 is true. This statistic Z then has approximately a standard normal distribution when H_0 is true, so level α tests are obtained by using z critical values exactly as before.

Use of the test statistic value

$$z = \frac{\bar{x} - \bar{y} - \Delta_0}{\sqrt{\dfrac{s_1^2}{m} + \dfrac{s_2^2}{n}}}$$

along with the previously stated upper-, lower-, and two-tailed rejection regions based on z critical values gives large-sample tests whose significance levels are approximately α. These tests are usually appropriate if both $m > 40$ and $n > 40$. A P-value is computed exactly as it was for our earlier z tests.

Example 9.4   In selecting a sulphur concrete for roadway construction in regions that experience heavy frost, it is important that the chosen concrete have a low value of thermal conductivity to minimize subsequent damage due to changing temperatures. Suppose two types of concrete, a graded aggregate and a no-fines aggregate, are being considered for a certain road. Table 9.1 summarizes data from an experiment carried out to compare the two types of concrete. Does this information suggest that true average conductivity for the graded concrete exceeds that for the no-fines concrete? Let's use a test with $\alpha = .01$.

Table 9.1   Data for Example 9.4

Type	Sample Size	Sample Average Conductivity	Sample SD
Graded	42	.486	.187
No-fines	42	.359	.158

Let μ_1 and μ_2 denote the true average thermal conductivity for the graded aggregate and no-fines aggregate concrete, respectively. The two hypotheses are $H_0: \mu_1 - \mu_2 = 0$ versus $H_a: \mu_1 - \mu_2 > 0$. H_0 will be rejected if $z \geq z_{.01} = 2.33$. We calculate

$$z = \frac{.486 - .359}{\sqrt{\dfrac{(.187)^2}{42} + \dfrac{(.158)^2}{42}}} = \frac{.127}{.0378} = 3.36$$

Since $3.36 \geq 2.33$, H_0 is rejected at significance level .01. Alternatively, the P-value for an upper-tailed z test is

$$P\text{-value} = 1 - \Phi(z) = 1 - \Phi(3.36) = .0004$$

H_0 should be rejected not only for a test with $\alpha = .01$ but also for $\alpha = .001$ or any other α exceeding .0004. The data argues strongly for the conclusion that true average thermal conductivity for the graded concrete does exceed that for the no-fines concrete. ∎

Confidence Intervals for $\mu_1 - \mu_2$

When both population distributions are normal, standardizing $\overline{X} - \overline{Y}$ gives a random variable Z with a standard normal distribution. Since the area under the z curve between $-z_{\alpha/2}$ and $z_{\alpha/2}$ is $1 - \alpha$, it follows that

$$P\left(-z_{\alpha/2} < \frac{\overline{X} - \overline{Y} - (\mu_1 - \mu_2)}{\sqrt{\dfrac{\sigma_1^2}{m} + \dfrac{\sigma_2^2}{n}}} < z_{\alpha/2}\right) = 1 - \alpha$$

Manipulation of the inequalities inside the parentheses to isolate $\mu_1 - \mu_2$ yields the equivalent probability statement

$$P\left(\overline{X} - \overline{Y} - z_{\alpha/2}\sqrt{\dfrac{\sigma_1^2}{m} + \dfrac{\sigma_2^2}{n}} < \mu_1 - \mu_2 < \overline{X} - \overline{Y} + z_{\alpha/2}\sqrt{\dfrac{\sigma_1^2}{m} + \dfrac{\sigma_2^2}{n}}\right) = 1 - \alpha$$

This implies that a $100(1 - \alpha)\%$ CI for $\mu_1 - \mu_2$ has lower limit $\overline{x} - \overline{y} - z_{\alpha/2} \cdot \sigma_{\overline{X} - \overline{Y}}$ and upper limit $\overline{x} - \overline{y} + z_{\alpha/2} \cdot \sigma_{\overline{X} - \overline{Y}}$, where $\sigma_{\overline{X} - \overline{Y}}$ is the square-root expression. This interval is a special case of the general formula $\hat{\theta} \pm z_{\alpha/2} \cdot \sigma_{\hat{\theta}}$.

   If both m and n are large, the CLT implies that this interval is valid even without the assumption of normal populations; in this case, the confidence level is *approximately* $100(1 - \alpha)\%$. Furthermore, use of the sample variances S_1^2 and S_2^2 in the standardized variable Z yields a valid interval in which s_1^2 and s_2^2 replace σ_1^2 and σ_2^2.

Provided that m and n are both large, a CI for $\mu_1 - \mu_2$ with a confidence level of approximately $100(1 - \alpha)\%$ is

$$\overline{x} - \overline{y} \pm z_{\alpha/2}\sqrt{\dfrac{s_1^2}{m} + \dfrac{s_2^2}{n}}$$

where $-$ gives the lower limit and $+$ the upper limit of the interval.

The standard rule of thumb for characterizing sample sizes as large is again $m > 40$ and $n > 40$.

Example 9.5 An experiment carried out to study various characteristics of anchor bolts resulted in 78 observations on shear strength (kip) of 3/8-in. diameter bolts and 88 observations on strength of 1/2-in. diameter bolts. Summary quantities from Minitab follow, and a comparative boxplot is presented in Figure 9.1. The sample sizes, sample means, and sample standard deviations agree with values given in the article "Ultimate Load Capacities of Expansion Anchor Bolts" (*J. Energy Engr.,* 1993: 139–158). The summaries suggest that the main difference between the two samples is in where they are centered.

Variable	N	Mean	Median	TrMean	StDev	SEMean
diam 3/8	78	4.250	4.230	4.238	1.300	0.147

Variable	Min	Max	Q1	Q3		
diam 3/8	1.634	7.327	3.389	5.075		

Variable	N	Mean	Median	TrMean	StDev	SEMean
diam 1/2	88	7.140	7.113	7.150	1.680	0.179

Variable	Min	Max	Q1	Q3		
diam 1/2	2.450	11.343	5.965	8.447		

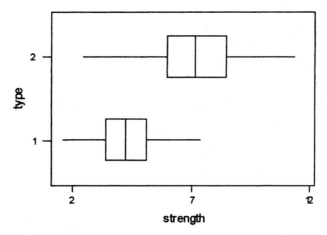

Figure 9.1 A comparative boxplot of the shear strength data

Let's now calculate a confidence interval for the difference between true average shear strength for 3/8-in. bolts (μ_1) and true average shear strength for 1/2-in. bolts (μ_2) using a confidence level of 95%:

$$4.25 - 7.14 \pm (1.96)\sqrt{\frac{(1.30)^2}{78} + \frac{(1.68)^2}{88}} = -2.89 \pm (1.96)(.2318)$$

$$= -2.89 \pm .45 = (-3.34, -2.44)$$

That is, with 95% confidence, $-3.34 < \mu_1 - \mu_2 < -2.44$. We can therefore be highly confident that the true average shear strength for the 1/2-in. bolts exceeds that for the 3/8-in. bolts by between 2.44 kip and 3.34 kip. Notice that if we relabel so that μ_1 refers to 1/2-in. bolts and μ_2 to 3/8-in. bolts, the confidence interval is now centered at +2.89 and the value .45 is still subtracted and added to obtain the confidence limits. The resulting interval is (2.44, 3.34), and the interpretation is identical to that for the interval previously calculated. ■

If the variances σ_1^2 and σ_2^2 are at least approximately known and the investigator uses equal sample sizes, then the sample size n for each sample that yields a $100(1 - \alpha)\%$ interval of width w is

$$n = \frac{4z_{\alpha/2}^2(\sigma_1^2 + \sigma_2^2)}{w^2}$$

which will generally have to be rounded up to an integer.

Exercises | Section 9.1 (1–16)

1. An article in the November 1983 *Consumer Reports* compared various types of batteries. The average lifetimes of Duracell Alkaline AA batteries and erase lifetimes of Duracell Alkaline AA batteries and Eveready Energizer Alkaline AA batteries were given as 4.1 hours and 4.5 hours, respectively. Suppose these are the population average lifetimes.

a. Let $\overline{X}$ be the sample average lifetime of 100 Duracell batteries and $\overline{Y}$ be the sample average lifetime of 100 Eveready batteries. What is the mean value of $\overline{X} - \overline{Y}$ (i.e., where is the distribution of $\overline{X} - \overline{Y}$ centered)? How does your answer depend on the specified sample sizes?

b. Suppose the population standard deviations of lifetime are 1.8 hours for Duracell batteries and 2.0 hours for Eveready batteries. With the sample sizes given in part (a), what is the variance of the statistic $\overline{X} - \overline{Y}$, and what is its standard deviation?

c. For the sample sizes given in part (a), draw a picture of the approximate distribution curve of $\overline{X} - \overline{Y}$ (include a measurement scale on the horizontal axis). Would the shape of the curve necessarily be the same for sample sizes of 10 batteries of each type? Explain.

2. Let μ_1 and μ_2 denote true average tread lives for two competing brands of size P205/65R15 radial tires. Test $H_0: \mu_1 - \mu_2 = 0$ versus $H_a: \mu_1 - \mu_2 \neq 0$ at level .05 using the following data: $m = 45$, $\overline{x} = 42{,}500$, $s_1 = 2200$, $n = 45$, $\overline{y} = 40{,}400$, and $s_2 = 1900$.

3. Let μ_1 denote true average tread life for a premium brand of P205/65R15 radial tire and let μ_2 denote the true average tread life for an economy brand of the same size. Test $H_0: \mu_1 - \mu_2 = 5000$ versus $H_a: \mu_1 - \mu_2 > 5000$ at level .01 using the following data: $m = 45$, $\overline{x} = 42{,}500$, $s_1 = 2200$, $n = 45$, $\overline{y} = 36{,}800$, and $s_2 = 1500$.

4. a. Use the data of Exercise 2 to compute a 95% CI for $\mu_1 - \mu_2$. Does the resulting interval suggest that $\mu_1 - \mu_2$ has been precisely estimated?

b. Use the data of Exercise 3 to compute a 95% upper confidence bound for $\mu_1 - \mu_2$.

5. Persons having Reynaud's syndrome are apt to suffer a sudden impairment of blood circulation in fingers and toes. In an experiment to study the extent of this impairment, each subject immersed a forefinger in water and the resulting heat output (cal/cm²/min) was measured. For $m = 10$ subjects with the syndrome, the average heat output was $\overline{x} = .64$, and for $n = 10$ nonsufferers, the average output was 2.05. Let μ_1 and μ_2 denote the true average heat outputs for the two types of subjects. Assume that the two distributions of heat output are normal with $\sigma_1 = .2$ and $\sigma_2 = .4$.

a. Test $H_0: \mu_1 - \mu_2 = -1.0$ versus $H_a: \mu_1 - \mu_2 < -1.0$ at level .01. (H_a says that the calorie output for sufferers is more than 1 cal/cm²/min below that for nonsufferers.)

b. Compute the P-value for the value of Z obtained in part (a).

c. What is the probability of a type II error when the actual difference between μ_1 and μ_2 is $\mu_1 - \mu_2 = -1.2$?

d. Assuming that $m = n$, what sample sizes are required to ensure that $\beta = .1$ when $\mu_1 - \mu_2 = -1.2$?

6. An experiment to compare the tension bond strength of polymer latex modified mortar (Portland cement mortar to which polymer latex emulsions have been added during mixing) to that of unmodified mortar resulted in $\overline{x} = 18.12$ kgf/cm² for the modified mortar ($m = 40$) and $\overline{y} = 16.87$ kgf/cm² for the unmodified mortar ($n = 32$). Let μ_1 and μ_2 be the true average tension bond strengths for the modified and unmodified mortars, respectively. Assume that the bond strength distributions are both normal.

a. Assuming that $\sigma_1 = 1.6$ and $\sigma_2 = 1.4$, test $H_0: \mu_1 - \mu_2 = 0$ versus $H_a: \mu_1 - \mu_2 > 0$ at level .01.

b. Compute the probability of a type II error for the test of part (a) when $\mu_1 - \mu_2 = 1$.

c. Suppose the investigator decided to use a level .05 test and wished $\beta = .10$ when $\mu_1 - \mu_2 = 1$. If $m = 40$, what value of n is necessary?

d. How would the analysis and conclusion of part (a) change if σ_1 and σ_2 were unknown but $s_1 = 1.6$ and $s_2 = 1.4$?

7. Are male college students more easily bored than their female counterparts? This question was examined in the article "Boredom in Young Adults—Gender and Cultural Comparisons" (*J. of Cross-Cultural Psych.*, 1991: 209–223). The authors administered a scale called the Boredom Proneness Scale to 97 male and 148 female U.S. college students. Does the accompanying data support the research hypothesis that the mean Boredom Proneness Rating is higher for men than for women? Test the appropriate hypotheses using a .05 significance level.

Gender	Sample Size	Sample Mean	Sample SD
Males	97	10.40	4.83
Females	148	9.26	4.68

8. Tensile strength tests were carried out on two different grades of wire rod ("Fluidized Bed Patenting of Wire Rods," *Wire J.*, June 1977: 56–61), resulting in the accompanying data:

Grade	Sample Size	Sample Mean (kg/mm²)	Sample SD
AISI 1064	$m = 129$	$\bar{x} = 107.6$	$s_1 = 1.3$
AISI 1078	$n = 129$	$\bar{y} = 123.6$	$s_2 = 2.0$

a. Does the data provide compelling evidence for concluding that true average strength for the 1078 grade exceeds that for the 1064 grade by more than 10 kg/mm²? Test the appropriate hypotheses using the *P*-value approach.

b. Estimate the difference between true average strengths for the two grades in a way that provides information about precision and reliability.

9. An experiment was performed to compare the fracture toughness of high-purity 18 Ni maraging steel with commercial-purity steel of the same type (*Corrosion Science*, 1971: 723–736). For $m = 32$ specimens, the sample average toughness was $\bar{x} = 65.6$ for the high-purity steel, whereas for $n = 38$ specimens of commercial steel $\bar{y} = 59.8$. Because the high-purity steel is more expensive, its use for a certain application can be justified only if its fracture toughness exceeds that of commercial-purity steel by more than 5. Suppose that both toughness distributions are normal.

a. Assuming that $\sigma_1 = 1.2$ and $\sigma_2 = 1.1$, test the relevant hypotheses using $\alpha = .001$.

b. Compute β for the test conducted in part (a) when $\mu_1 - \mu_2 = 6$.

10. The level of lead in the blood was determined for a sample of 152 male hazardous-waste workers age 20–30 and also for a sample of 86 female workers, resulting in a mean ± standard error of 5.5 ± 0.3 for the men and 3.8 ± 0.2 for the women ("Temporal Changes in Blood Lead Levels of Hazardous Waste Workers in New Jersey, 1984–1987," *Environ. Monitoring and Assessment*, 1993: 99–107). Calculate an estimate of the difference between true average blood lead levels for male and female workers in a way that provides information about reliability and precision.

11. The accompanying table gives summary data on cube compressive strength (N/mm²) for concrete specimens made with a pulverized fuel-ash mix ("A Study of Twenty-Five-Year-Old Pulverized Fuel Ash Concrete Used in Foundation Structures," *Proc. Inst. Civ. Engrs.*, Mar. 1985: 149–165):

Age (days)	Sample Size	Sample Mean	Sample SD
7	68	26.99	4.89
28	74	35.76	6.43

Calculate a 99% CI for the difference between true average 7-day strength and true average 28-day strength.

12. A mechanical engineer wishes to compare strength properties of steel beams with similar beams made with a particular alloy. The same number of beams, *n*, of each type will be tested. Each beam will be set in a horizontal position with a support on each end, a force of 2500 lb will be applied at the center, and the deflection will be measured. From past experience with such beams, the engineer is willing to assume that the true standard deviation of deflection for both types of beam is .05 in. Because the alloy is more expensive, the engineer wishes to test at level .01 whether it has smaller average deflection than the steel beam. What value of *n* is appropriate if the desired type II error probability is .05 when the difference in true average deflection favors the alloy by .04 in.?

13. The level of monoamine oxidase (MAO) activity in blood platelets (nm/mg protein/h) was determined for each individual in a sample of 43 chronic schizophrenics, resulting in $\bar{x} = 2.69$ and $s_1 = 2.30$, as well as for 45 normal subjects, resulting in $\bar{y} = 6.35$ and $s_2 = 4.03$. Does this data strongly suggest that true average MAO activity for normal subjects is more than twice the activity level for schizophrenics? Derive a test procedure and carry out the test using $\alpha = .01$. [*Hint:* H_0 and H_a here have a different form from the three standard cases. Let μ_1 and μ_2 refer to true average MAO activity for schizophrenics and normal subjects, respectively, and consider the parameter $\theta = 2\mu_1 - \mu_2$. Write H_0 and H_a in terms of θ, estimate θ, and derive $\hat{\sigma}_{\hat{\theta}}$ ("Reduced Monoamine Oxidase Activity in Blood Platelets from Schizophrenic Patients," *Nature*, July 28, 1972: 225–226).]

14. Show for the upper-tailed test with σ_1 and σ_2 known that as either *m* or *n* increases, β decreases when $\mu_1 - \mu_2 > \Delta_0$.

15. For the case of equal sample sizes ($m = n$) and fixed α, what happens to the necessary sample size *n* as β is decreased, where β is the desired type II error probability at a fixed alternative?

16. To decide whether two different types of steel have the same true average fracture toughness values, n specimens of each type are tested, yielding the following results:

Type	Sample Average	Sample SD
1	60.1	1.0
2	59.9	1.0

Calculate the P-value for the appropriate two-sample z test, assuming that the data was based on $n = 100$. Then repeat the calculation for $n = 400$. Is the small P-value for $n = 400$ indicative of a difference that has practical significance? Would you have been satisfied with just a report of the P-value? Comment briefly.

9.2 The Two-Sample t Test and Confidence Interval

In real problems, it is virtually always the case that the values of the population variances are unknown. In the previous section, we illustrated for large sample sizes the use of a test procedure and CI in which the sample variances were used in place of the population variances. In fact, for large samples, the CLT allows us to use these methods even when the two populations of interest are not normal.

There are many problems, though, in which at least one sample size is small and the population variances have unknown values. In the absence of the CLT, we proceed by making specific assumptions about the underlying population distributions. The use of inferential procedures that follow from these assumptions is then restricted to situations in which the assumptions are at least approximately satisfied.

ASSUMPTIONS

Both populations are normal, so that $X_1, X_2, \ldots, X_m$ is a random sample from a normal distribution and so is $Y_1, \ldots, Y_n$ (with the X's and Y's independent of one another). The plausibility of these assumptions can be judged by constructing a normal probability plot of the x_i's and another of the y_i's.

The test statistic and confidence interval formula are based on the same standardized variable developed in Section 9.1, but the relevant distribution is now t rather than z.

THEOREM

When the population distributions are both normal, the standardized variable

$$T = \frac{\overline{X} - \overline{Y} - (\mu_1 - \mu_2)}{\sqrt{\dfrac{S_1^2}{m} + \dfrac{S_2^2}{n}}} \tag{9.2}$$

has approximately a t distribution with df ν estimated from the data by

$$\nu = \frac{\left(\dfrac{s_1^2}{m} + \dfrac{s_2^2}{n}\right)^2}{\dfrac{(s_1^2/m)^2}{m-1} + \dfrac{(s_2^2/n)^2}{n-1}}$$

(round down to the nearest integer). The **two-sample *t* confidence interval for** $\mu_1 - \mu_2$ with confidence level $100(1 - \alpha)\%$ is then

$$\bar{x} - \bar{y} \pm t_{\alpha/2,\nu} \sqrt{\frac{s_1^2}{m} + \frac{s_2^2}{n}}$$

The **two-sample *t* test** for testing $H_0: \mu_1 - \mu_2 = \Delta_0$ is as follows:

$$\text{Test statistic value:} \quad t = \frac{\bar{x} - \bar{y} - \Delta_0}{\sqrt{\frac{s_1^2}{m} + \frac{s_2^2}{n}}}$$

Alternative Hypothesis	**Rejection Region for Approximate Level α Test**
$H_a: \mu_1 - \mu_2 > \Delta_0$	$t \geq t_{\alpha,\nu}$ (upper-tailed)
$H_a: \mu_1 - \mu_2 < \Delta_0$	$t \leq -t_{\alpha,\nu}$ (lower-tailed)
$H_a: \mu_1 - \mu_2 \neq \Delta_0$ either	$t \geq t_{\alpha/2,\nu}$ or $t \leq -t_{\alpha/2,\nu}$ (two-tailed)

A *P*-value can be computed as described in Section 8.4 for the one-sample *t* test.

Example 9.6 The void volume within a textile fabric affects comfort, flammability, and insulation properties. Permeability of a fabric refers to the accessibility of void space to the flow of a gas or liquid. The article "The Relationship Between Porosity and Air Permeability of Woven Textile Fabrics" (*J. of Testing and Eval.*, 1997: 108–114) gave summary information on air permeability ($cm^3/cm^2/sec$) for a number of different fabric types. Consider the following data on two different types of plain-weave fabric:

Fabric type	**Sample size**	**Sample mean**	**Sample standard deviation**
Cotton	10	51.71	.79
Triacetate	10	136.14	3.59

Assuming that the porosity distributions for both types of fabric are normal, let's calculate a confidence interval for the difference between true average porosity for the cotton fabric and that for the acetate fabric, using a 95% confidence level. Before the appropriate *t* critical value can be selected, df must be determined:

$$df = \frac{\left(\dfrac{.6241}{10} + \dfrac{12.8881}{10}\right)^2}{\dfrac{(.6241/10)^2}{9} + \dfrac{(12.8881/10)^2}{9}} = \frac{1.8258}{.1850} = 9.87$$

Thus we use $\nu = 9$; Appendix Table A.5 gives $t_{.025,9} = 2.262$. The resulting interval is

$$51.71 - 136.14 \pm (2.262) \sqrt{\frac{.6241}{10} + \frac{12.8881}{10}} = -84.43 \pm 2.63$$

$$= (-87.06, -81.80)$$

With a high degree of confidence, we can say that true average porosity for triacetate fabric specimens exceeds that for cotton specimens by between 81.80 and 87.06 cm³/cm²/sec. ∎

Example 9.7 · The deterioration of many municipal pipeline networks across the country is a growing concern. One technology proposed for pipeline rehabilitation uses a flexible liner threaded through existing pipe. The article "Effect of Welding on a High-Density Polyethylene Liner" (*J. of Materials in Civil Engr.*, 1996: 94–100) reported the following data on tensile strength (psi) of liner specimens both when a certain fusion process was used and when this process was not used.

No fusion	2748	2700	2655	2822	2511			
	3149	3257	3213	3220	2753			
	$m = 10$	$\bar{x} = 2902.8$		$s_1 = 277.3$				
Fused	3027	3356	3359	3297	3125	2910	2889	2902
	$n = 8$	$\bar{y} = 3108.1$		$s_2 = 205.9$				

Figure 9.2 shows normal probability plots from MINITAB. The linear pattern in each plot supports the assumption that the tensile strength distributions under the two conditions are both normal.

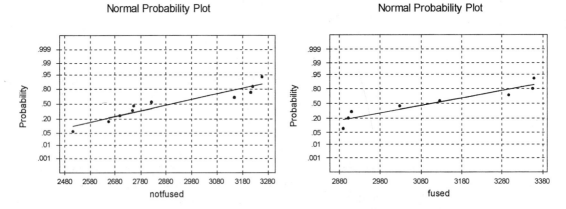

Figure 9.2 Normal probability plots from MINITAB for the tensile strength data

The authors of the article stated that the fusion process increased the average tensile strength. The message from the comparative boxplot of Figure 9.3 is not all that clear. Let's carry out a test of hypotheses to see whether the data supports this conclusion.

1. Let μ_1 be the true average tensile strength of specimens when the no-fusion treatment is used and μ_2 denote the true average tensile strength when the fusion treatment is used.

2. H_0: $\mu_1 - \mu_2 = 0$ (no difference in the true average tensile strengths for the two treatments)

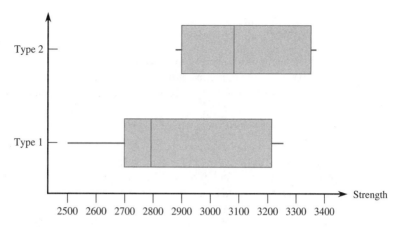

Figure 9.3   A comparative boxplot of the tensile strength data

3. H_a: $\mu_1 - \mu_2 < 0$ (true average tensile strength for the no-fusion treatment is less than that for the fusion treatment, so that the investigators' conclusion is correct)

4. The null value is $\Delta_0 = 0$, so the test statistic is

$$t = \frac{\bar{x} - \bar{y}}{\sqrt{\dfrac{s_1^2}{m} + \dfrac{s_2^2}{n}}}$$

5. We now compute both the test statistic value and the df for the test:

$$t = \frac{2902.8 - 3108.1}{\sqrt{\dfrac{(277.3)^2}{10} + \dfrac{(205.9)^2}{8}}} = \frac{-205.3}{113.97} = -1.8$$

Using $s_1^2/m = 7689.529$ and $s_2^2/n = 5299.351$,

$$\nu = \frac{(7689.529 + 5299.351)^2}{(7689.529)^2/9 + (5299.351)^2/7} = \frac{168{,}711{,}003.7}{10{,}581{,}747.35} = 15.94$$

so the test will be based on 15 df.

6. Appendix Table A.8 shows that the area under the 15 df *t* curve to the right of 1.8 is .046, so the *P*-value for a lower-tailed test is also .046. The following MINITAB output summarizes all the computations:

```
Twosample T for nofusion vs fused

             N      Mean    StDev    SE Mean
no fusion    10     2903    277      88
fused        8      3108    206      73

95% C.I. for mu nofusion-mu fused: (-488, 38)
T-Test mu nofusion = mu fused (vs <): T = -1.80  P = 0.046  DF = 15
```

7. Using a significance level of .05, we can barely reject the null hypothesis in favor of the alternative hypothesis, confirming the conclusion stated in the article. However,

someone demanding more compelling evidence might select $\alpha = .01$, a level for which H_0 cannot be rejected.

If the question posed had been whether fusing increased true average strength by more than 100 psi, then the relevant hypotheses would have been $H_0: \mu_1 - \mu_2 = -100$ versus $H_a: \mu_1 - \mu_2 < -100$; that is, the null value would have been $\Delta_0 = -100$. ∎

Pooled *t* Procedures

Alternatives to the two-sample t procedures just described result from assuming not only that the two population distributions are normal but also that they have equal variances ($\sigma_1^2 = \sigma_2^2$). That is, the two population distribution curves are assumed normal with equal spreads, the only possible difference between them being where they are centered. If we let σ^2 denote the common variance, this parameter can be estimated by combining ("pooling") information from the two samples (see Exercise 34). Standardizing $\overline{X} - \overline{Y}$ using this pooled estimator of σ^2 gives a t variable based on $m + n - 2$ df, from which a test procedure and confidence interval are easily obtained.

In the past, many statisticians recommended these pooled t procedures over the two-sample t procedures. The pooled t test, for example, can be derived from the likelihood ratio principle, whereas the two-sample t test is not a likelihood ratio test. Furthermore, the significance level for the pooled t test is exact, whereas it is only approximate for the two-sample t test. However, recent research has shown that, although the pooled t test does outperform the two-sample t test by a bit (smaller β's for the same α) when $\sigma_1^2 = \sigma_2^2$, the former test can easily lead to erroneous conclusions if applied when the variances are different. Analogous comments apply to the behavior of the two confidence intervals. That is, the pooled t procedures are not robust to violations of the equal variance assumption.

It has been suggested that one could carry out a preliminary test of $H_0: \sigma_1^2 = \sigma_2^2$ and use a pooled t procedure if this null hypothesis is rejected. Unfortunately, the usual "F test" of equal variances (Section 9.5) is quite sensitive to the assumption of normal population distributions, much more so than t procedures. We therefore recommend the conservative approach of using two-sample t procedures unless there is really compelling evidence for doing otherwise, particularly when the two sample sizes are different.

Type II Error Probabilities

Determining type II error probabilities (or equivalently, power $= 1 - \beta$) for the two-sample t test is complicated. There does not appear to be any simple way to use the β curves of Appendix Table A.17. The most recent version of MINITAB (Version 12) will calculate power for the pooled t test but not for the two-sample t test. However, the UCLA Statistics Department homepage (http://www.stat.ucla.edu) permits access to a power calculator that will do this. For example, we specified $m = 10$, $n = 8$, $\sigma_1 = 300$, $\sigma_2 = 225$ (these are the sample sizes for Example 9.7, whose sample standard deviations are somewhat smaller than these values of σ_1 and σ_2), and asked for the power of a two-tailed level .05 test of $H_0: \mu_1 - \mu_2 = 0$ when $\mu_1 - \mu_2 = 100$, 250, and 500. The resulting values of the power were .1089, .4609, and .9635 (corresponding to $\beta = .89$,

.54, and .04), respectively. In general, β will decrease as the sample sizes increase, as α increases, and as $\mu_1 - \mu_2$ moves farther from 0. The software will also calculate sample sizes necessary to obtain a specified value of power for a particular value of $\mu_1 - \mu_2$.

Exercises | Section 9.2 (17–35)

17. Determine the number of degrees of freedom for the two-sample *t* test or CI in each of the following situations:
a. $m = 10$, $n = 10$, $s_1 = 5.0$, $s_2 = 6.0$
b. $m = 10$, $n = 15$, $s_1 = 5.0$, $s_2 = 6.0$
c. $m = 10$, $n = 15$, $s_1 = 2.0$, $s_2 = 6.0$
d. $m = 12$, $n = 24$, $s_1 = 5.0$, $s_2 = 6.0$

18. Let μ_1 and μ_2 denote true average densities for two different types of brick. Assuming normality of the two density distributions, test H_0: $\mu_1 - \mu_2 = 0$ versus H_a: $\mu_1 - \mu_2 \neq 0$ using the following data: $m = 6$, $\bar{x} = 22.73$, $s_1 = .164$, $n = 5$, $\bar{y} = 21.95$, and $s_2 = .240$.

19. Suppose μ_1 and μ_2 are true mean stopping distances at 50 mph for cars of a certain type equipped with two different types of braking systems. Use the two-sample *t* test at significance level .01 to test H_0: $\mu_1 - \mu_2 = -10$ versus H_a: $\mu_1 - \mu_2 < -10$ for the following data: $m = 6$, $\bar{x} = 115.7$, $s_1 = 5.03$, $n = 6$, $\bar{y} = 129.3$, and $s_2 = 5.38$.

20. Use the data of Exercise 19 to calculate a 95% CI for the difference between true average stopping distance for cars equipped with system 1 and cars equipped with system 2. Does the interval suggest that precise information about the value of this difference is available?

21. Quantitative noninvasive techniques are needed for routinely assessing symptoms of peripheral neuropathies, such as carpal tunnel syndrome (CTS). The article "A Gap Detection Tactility Test for Sensory Deficits Associated with Carpal Tunnel Syndrome" (*Ergonomics*, 1995: 2588–2601) reported on a test that involved sensing a tiny gap in an otherwise smooth surface by probing with a finger; this functionally resembles many work-related tactile activities, such as detecting scratches or surface defects. When finger probing was not allowed, the sample average gap detection threshold for $m = 8$ normal subjects was 1.71 mm, and the sample standard deviation was .53; for $n = 10$ CTS subjects,

the sample mean and sample standard deviation were 2.53 and .87, respectively. Does this data suggest that true average gap detection threshold for CTS subjects exceeds that for normal subjects? State and test the relevant hypotheses using a significance level of .01.

22. The slant shear test is widely accepted for evaluating the bond of resinous repair materials to concrete; it utilizes cylinder specimens made of two identical halves bonded at 30°. The article "Testing the Bond Between Repair Materials and Concrete Substrate" (*ACI Materials J.*, 1996: 553–558) reported that for 12 specimens prepared using wire-brushing, the sample mean shear strength (N/mm²) and sample standard deviation were 19.20 and 1.58, respectively, whereas for 12 hand-chiseled specimens, the corresponding values were 23.13 and 4.01. Does the true average strength appear to be different for the two methods of surface preparation? State and test the relevant hypotheses using a significance level of .05. What are you assuming about the shear strength distributions?

23. Fusible interlinings are being used with increasing frequency to support outer fabrics and improve the shape and drape of various pieces of clothing. The article "Compatibility of Outer and Fusible Interlining Fabrics in Tailored Garments" (*Textile Res. J.*, 1997: 137–142) gave the accompanying data on extensibility (%) at 100 gm/cm for both high-quality fabric (H) and poor-quality fabric (P) specimens:

H	1.2	.9	.7	1.0	1.7	1.7	1.1	.9	1.7
	1.9	1.3	2.1	1.6	1.8	1.4	1.3	1.9	1.6
	.8	2.0	1.7	1.6	2.3	2.0			
P	1.6	1.5	1.1	2.1	1.5	1.3	1.0	2.6	

a. Construct normal probability plots to verify the plausibility of both samples having been selected from normal population distributions.
b. Construct a comparative boxplot. Does it suggest that there is a difference between true

average extensibility for high-quality fabric specimens and that for poor-quality specimens?

c. The sample mean and standard deviation for the high-quality sample are 1.508 and .444, respectively, and those for the poor-quality sample are 1.588 and .530. Use the two-sample *t* test to decide whether true average extensibility differs for the two types of fabrics.

24. The firmness of a piece of fruit is an important indicator of fruit ripeness. The Magness–Taylor firmness (N) was determined for one sample of 20 golden apples with a shelf life of zero days, resulting in a sample mean of 8.74 and a sample standard deviation of .66, and another sample of 20 apples with a shelf life of 20 days, with a sample mean and sample standard deviation of 4.96 and .39, respectively. Calculate a confidence interval for the difference between true average firmness for zero-day apples and true average firmness for 20-day apples using a confidence level of 95%, and interpret the interval.

25. Low-back pain (LBP) is a serious health problem in many industrial settings. The article "Isodynamic Evaluation of Trunk Muscles and Low-Back Pain Among Workers in a Steel Factory" (*Ergonomics,* 1995: 2107–2117) reported the accompanying summary data on lateral range of motion (degrees) for a sample of workers without a history of LBP and another sample with a history of this malady.

Condition	Sample size	Sample mean	Sample SD
No LBP	28	91.5	5.5
LBP	31	88.3	7.8

Calculate a 90% confidence interval for the difference between population mean extent of lateral motion for the two conditions. Does the interval suggest that population mean lateral motion differs for the two conditions? Is the message different if we use a confidence level of 95%?

26. The article "The Influence of Corrosion Inhibitor and Surface Abrasion on the Failure of Aluminum-Wired Twist-on Connections" (*IEEE Trans. on Components, Hybrids, and Manuf. Tech.,* 1984: 20–25) reported data on potential drop measurements for one sample of connectors wired with alloy aluminum and another sample wired with EC aluminum. Does the accompanying SAS output suggest that the true average potential drop for alloy connections (type 1) is higher than that for EC connections (type 1) as stated in the article)? Carry out the appropriate test using a significance level of .01. In reaching your conclusion, what type of error might you have committed? *Note:* SAS reports the *P*-value for a two-tailed test.

Type	N	Mean	Std Dev	Std Error
1	20	17.49900000	0.55012821	0.12301241
2	20	16.90000000	0.48998389	0.10956373

Variances	T	DF	Prob > |T|
Unequal	3.6362	37.5	0.0008
Equal	3.6362	38.0	0.0008

27. Tennis elbow is thought to be aggravated by the impact experienced when hitting the ball. The article "Forces on the Hand in the Tennis One-Handed Backhand" (*Intl. J. of Sport Biomechanics,* 1991: 282–292) reported the force (N) on the hand just after impact on a one-handed backhand drive for six advanced players and for eight intermediate players.

Type of player	Sample size	Sample mean	Sample SD
1. Advanced	6	40.3	11.3
2. Intermediate	8	21.4	8.3

In their analysis of the data, the authors assumed that both force distributions were normal. Calculate a 95% CI for the difference between true average force for advanced players (μ_1) and true average force for intermediate players (μ_2). Does your interval provide compelling evidence for concluding that the two μ's are different? Would you have reached the same conclusion by calculating a CI for $\mu_2 - \mu_1$ (i.e., by reversing the 1 and 2 labels on the two types of players)? Explain.

28. Refer to Exercise 27, and carry out a test of hypotheses at level .01 to decide whether true average force after impact is greater for advanced players than for intermediate players (use the *P*-value method).

29. The article "Effect of Internal Gas Pressure on the Compression Strength of Beverage Cans and Plastic Bottles" (*J. Testing and Evaluation,* 1993: 129–131) includes the accompanying data on compression strength (lb) for a sample of 12-oz alu-

minum cans filled with strawberry drink and another sample filled with cola. Does the data suggest that the extra carbonation of cola results in a higher average compression strength? Base your answer on a *P*-value. What assumptions are necessary for your analysis?

Beverage	Sample size	Sample mean	Sample SD
Strawberry drink	15	540	21
Cola	15	554	15

30. The article "Flexure of Concrete Beams Reinforced with Advanced Composite Orthogrids" (*J. of Aerospace Engr.*, 1997: 7–15) gave the accompanying data on ultimate load (kN) for two different types of beams.

Type	Sample size	Sample mean	Sample SD
Fiberglass grid	26	33.4	2.2
Commercial carbon grid	26	42.8	4.3

a. Assuming that the underlying distributions are normal, calculate and interpret a 99% CI for the difference between true average load for the fiberglass beams and that for the carbon beams.

b. Does the upper limit of the interval you calculated in part (a) give a 99% upper confidence bound for the difference between the two μ's? If not, calculate such a bound. Does it strongly suggest that true average load for the carbon beams is more than that for the fiberglass beams? Explain.

31. Refer to Exercise 33 in Section 7.3. The cited article also gave the following observations on degree of polymerization for specimens having viscosity times concentration in a higher range:

429   430   430   431   436   437

440   441   445   446   447

a. Construct a comparative boxplot for the two samples, and comment on any interesting features.

b. Calculate a 95% confidence interval for the difference between true average degree of polymerization for the middle range and that for the high range. Does the interval suggest that μ_1 and μ_2 may in fact be different? Explain your reasoning.

32. The article cited in Exercise 34 in Section 7.3 gave the following summary data on proportional stress limits for specimens constructed using two different types of wood.

Type of wood	Sample size	Sample mean	Sample SD
Red oak	14	8.48	.79
Douglas fir	10	6.65	1.28

Assuming that both samples were selected from normal distributions, carry out a test of hypotheses to decide whether the true average proportional stress limit for red oak joints exceeds that for Douglas fir joints by more than 1 MPa?

33. The accompanying table summarizes data on body weight gain (g) both for a sample of animals given a 1 mg/pellet dose of a certain soft steroid and for a sample of control animals ("The Soft Drug Approach," *CHEMTECH*, 1984: 28–38).

Treatment	Sample size	Sample mean	Sample SD
Steroid	8	32.8	2.6
Control	10	40.5	2.5

Does the data suggest that the true average weight gain in the control situation exceeds that for the steroid treatment by more than 5 g? State and test the appropriate hypotheses at a significance level of .01 using the *P*-value method.

34. Suppose not only that the two population or treatment response distributions are normal but also that they have equal variances. Let σ^2 denote the common variance. This variance can be estimated by a "pooled" (i.e., combined) sample variance as follows:

$$s_p^2 = \left(\frac{m-1}{m+n-2}\right)s_1^2 + \left(\frac{n-1}{m+n-2}\right)s_2^2$$

($m + n - 2$ is the sum of the df's contributed by the two samples). It can then be shown that the standardized variable

$$t = \frac{(\bar{X} - \bar{Y}) - (\mu_1 - \mu_2)}{S_p\sqrt{\frac{1}{m} + \frac{1}{n}}}$$

has a *t* distribution with $m + n - 2$ df.

a. Use this t variable to obtain a pooled t confidence interval formula for $\mu_1 - \mu_2$.

b. A sample of ultrasonic humidifiers of one particular brand was selected for which the observations on maximum output of moisture (oz) in a controlled chamber were 14.0, 14.3, 12.2, and 15.1. A sample of the second brand gave output values 12.1, 13.6, 11.9, and 11.2 ("Multiple Comparisons of Means Using Simultaneous Confidence Intervals," *J. of Quality Technology,* 1989: 232–241). Use the pooled t formula from part (a) to estimate the difference between true average outputs for the two brands with a 95% confidence interval.

c. Estimate the difference between the two μ's using the two-sample t interval discussed in this section, and compare it to the interval of part (b).

35. Refer to Exercise 34. Describe the pooled t test for testing $H_0: \mu_1 - \mu_2 = \Delta_0$ when both population distributions are normal with $\sigma_1 = \sigma_2$. Then use this test procedure to test the hypotheses suggested in Exercise 33.

9.3 | Analysis of Paired Data

In Sections 9.1 and 9.2, we considered testing for a difference between two means μ_1 and μ_2. This was done by utilizing the results of a random sample $X_1, X_2, \ldots, X_m$ from the distribution with mean μ_1 and a completely independent (of the X's) sample $Y_1, \ldots, Y_n$ from the distribution with mean μ_2. That is, either m individuals were selected from population 1 and n different individuals from population 2, or m individuals (or experimental objects) were given one treatment and another set of n individuals were given the other treatment. In contrast, there are a number of experimental situations in which there is only one set of n individuals or experimental objects, and two observations are made on each individual or object.

Example 9.8 Trace metals in drinking water affect the flavor, and unusually high concentrations can pose a health hazard. The article "Trace Metals of South Indian River" (*Envir. Studies,* 1982: 62–66) reports on a study in which six river locations were selected (six experimental objects) and the zinc concentration (mg/L) determined for both surface water and bottom water at each location. The six pairs of observations are displayed in the accompanying table. Does the data suggest that true average concentration in bottom water exceeds that of surface water?

	Location					
	1	2	3	4	5	6
Zinc concentration in bottom water (x)	.430	.266	.567	.531	.707	.716
Zinc concentration in surface water (y)	.415	.238	.390	.410	.605	.609
Difference	.015	.028	.177	.121	.102	.107

Figure 9.4(a) displays a plot of this data. At first glance, there appears to be little difference between the x and y samples. From location to location, there is a great deal of

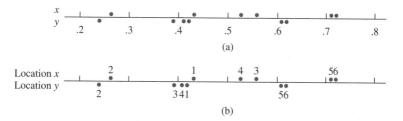

Figure 9.4 Plot of paired data from Example 9.8: (a) observations not identified by location; (b) observations identified by location ▪

variability in each sample, and it looks as though any differences between the samples can be attributed to this variability. However, when the observations are identified by location, as in Figure 9.4(b), a different view emerges. At each location, bottom concentration exceeds surface concentration. This is confirmed by the fact that all $x - y$ differences (bottom water concentration $-$ surface water concentration) displayed in the bottom row of the data table are positive. As we will see, a correct analysis of this data focuses on these differences.

ASSUMPTIONS

> The data consists of n independently selected pairs $(X_1, Y_1), (X_2, Y_2), \ldots, (X_n, Y_n)$, with $E(X_i) = \mu_1$ and $E(Y_i) = \mu_2$. Let $D_1 = X_1 - Y_1$, $D_2 = X_2 - Y_2, \ldots,$ $D_n = X_n - Y_n$, so the D_i's are the differences within pairs. Then the D_i's are assumed to be normally distributed with variance σ_D^2 (this is usually a consequence of the X_i's and Y_i's themselves being normally distributed).

We are again interested in testing hypotheses about the difference $\mu_1 - \mu_2$. The denominator of the two-sample t test was obtained by first applying the rule $V(\overline{X} - \overline{Y}) = V(\overline{X}) + V(\overline{Y})$. However, with paired data, the X and Y observations within each pair are often not independent, so that $\overline{X}$ and $\overline{Y}$ are not independent of one another, and the rule is not valid. We must therefore abandon the two-sample t test and look for an alternative method of analysis.

The Paired t Test

Because different pairs are independent, the D_i's are independent of one another. If we let $D = X - Y$, where X and Y are the first and second observations, respectively, within an arbitrary pair, then the expected difference is

$$\mu_D = E(X - Y) = E(X) - E(Y) = \mu_1 - \mu_2$$

(the rule of expected values used here is valid even when X and Y are dependent). Thus, any hypothesis about $\mu_1 - \mu_2$ can be phrased as a hypothesis about the mean difference μ_D. But since the D_i's constitute a normal random sample (of differences) with mean μ_D, hypotheses about μ_D can be tested using a one-sample t test. That is, *to test hypotheses about $\mu_1 - \mu_2$ when data is paired, form the differences $D_1, D_2, \ldots, D_n$ and carry out a one-sample t test (based on $n - 1$ df) on the differences.*

The Paired t Test

Null hypothesis: H_0: $\mu_D = \Delta_0$ (where $D = X - Y$ is the difference between the first and second observations within a pair ($\mu_D = \mu_1 - \mu_2$))

Test statistic value: $t = \dfrac{\bar{d} - \Delta_0}{s_D/\sqrt{n}}$ (where $\bar{d}$ and s_D are the sample mean and standard deviation, respectively, of the d_i's)

Alternative Hypothesis	Rejection Region for Level α Test
H_a: $\mu_D > \Delta_0$	$t \geq t_{\alpha, n-1}$
H_a: $\mu_D < \Delta_0$	$t \leq -t_{\alpha, n-1}$
H_a: $\mu_D \neq \Delta_0$	either $t \geq t_{\alpha/2, n-1}$ or $t \leq -t_{\alpha/2, n-1}$

Example 9.9 Musculoskeletal neck-and-shoulder disorders are all too common among office staff who perform repetitive tasks using visual display units. The article "Upper-Arm Elevation During Office Work" (*Ergonomics*, 1996: 1221–1230) reported on a study to determine whether more varied work conditions would have any impact on arm movement. The accompanying data was obtained from a sample of $n = 16$ subjects. Each observation is the amount of time, expressed as a proportion of total time observed, during which arm elevation was below 30°. The two measurements from each subject were obtained 18 months apart. During this period, work conditions were changed, and subjects were allowed to engage in a wider variety of work tasks. Does the data suggest that true average time during which elevation is below 30° differs after the change from what it was before the change?

Subject	1	2	3	4	5	6	7	8
Before	81	87	86	82	90	86	96	73
After	78	91	78	78	84	67	92	70
Difference	3	−4	8	4	6	19	4	3

Subject	9	10	11	12	13	14	15	16
Before	74	75	72	80	66	72	56	82
After	58	62	70	58	66	60	65	73
Difference	16	13	2	22	0	12	−9	9

Figure 9.5 shows a normal probability plot of the 16 differences; the pattern in the plot is quite straight, supporting the normality assumption. A boxplot of these differences appears in Figure 9.6; the boxplot is located considerably to the right of zero, suggesting that perhaps $\mu_D > 0$ (note also that 13 of the 16 differences are positive and only two are negative).

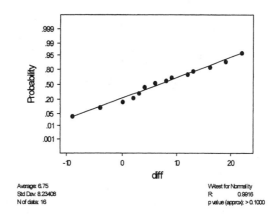

Average: 6.75
Std Dev: 8.23408
N of data: 16

W-test for Normality
R   0.9916
p value (approx): > 0.1000

Figure 9.5   A normal probability plot from MINITAB of the differences in Example 9.9

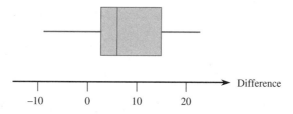

Figure 9.6   A boxplot of the differences in Example 9.9

Let's now use the recommended sequence of steps to test the appropriate hypotheses.

1. Let μ_D denote the true average difference between elevation time before the change in work conditions and time after the change.

2. H_0: $\mu_D = 0$ (there is no difference between true average time before the change and true average time after the change)

3. H_a: $\mu_D \neq 0$

4. $t = \dfrac{\bar{d} - 0}{s_D/\sqrt{n}} = \dfrac{\bar{d}}{s_D/\sqrt{n}}$

5. $n = 16$, $\sum d_i = 108$, $\sum d_i^2 = 1746$, from which $\bar{d} = 6.75$, $s_D = 8.234$, and

$$t = \frac{6.75}{8.234/\sqrt{16}} = 3.28 \approx 3.3$$

6. Appendix Table A.8 shows that the area to the right of 3.3 under the t curve with 15 df is .002. The inequality in H_a implies that a two-tailed test is appropriate, so the P-value is approximately $2(.002) = .004$ (MINITAB gives .0051).

7. Since $.004 < .01$, the null hypothesis can be rejected at either significance level .05 or .01. It does appear that the true average difference between times is something

other than zero, that is, true average time after the change is different from that before the change. ∎

When the number of pairs is large, the assumption of a normal difference distribution is not necessary. The CLT validates the resulting z test.

A Confidence Interval for μ_D

In the same way that the t CI for a single population mean μ is based on the t variable $T = (\bar{X} - \mu)/(S/\sqrt{n})$, a t confidence interval for μ_D $(= \mu_1 - \mu_2)$ is based on the fact that

$$T = \frac{\bar{D} - \mu_D}{S_D/\sqrt{n}}$$

has a t distribution with $n - 1$ df. Manipulation of this t variable, as in previous derivations of CIs, yields the following $100(1 - \alpha)\%$ CI:

The **paired t CI for μ_D** is

$$\bar{d} \pm t_{\alpha/2, n-1} \cdot s_D/\sqrt{n}$$

When n is small, the validity of this interval requires that the distribution of differences be at least approximately normal. For large n, the CLT ensures that the resulting z interval is valid without any restrictions on the distribution of differences.

Example 9.10 Example 7.11 gave data on the modulus of elasticity obtained 1 minute after loading in a certain configuration. The cited article also gave the values of modulus of elasticity obtained 4 weeks after loading for the same lumber specimens. The data is presented here.

Observation	1 min	4 weeks	Difference
1	10,490	9,110	1380
2	16,620	13,250	3370
3	17,300	14,720	2580
4	15,480	12,740	2740
5	12,970	10,120	2850
6	17,260	14,570	2690
7	13,400	11,220	2180
8	13,900	11,100	2800
9	13,630	11,420	2210
10	13,260	10,910	2350
11	14,370	12,110	2260
12	11,700	8,620	3080
13	15,470	12,590	2880
14	17,840	15,090	2750
15	14,070	10,550	3520
16	14,760	12,230	2530

The normal probability plot of the differences, shown in Figure 9.7, appears to be reasonably straight, though the point on the far left deviates somewhat from a line determined by the other points. (Use of a formal inferential procedure presented in Chapter 14 indicates that it is reasonable to assume that the population distribution of the differences is approximately normal.)

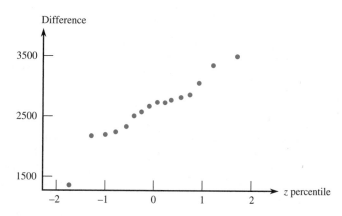

Figure 9.7   Normal probability plot of the differences in Example 9.10

The sample consists of 16 pairs, so a 99% confidence interval based on 15 df requires the t critical value 2.947. With $\bar{d} = 2635.6$ and $s_D = 508.64$, the interval is

$$2635.6 \pm (2.947)\frac{508.64}{\sqrt{16}} = 2635.6 \pm 374.7 = (2260.9, 3010.3)$$

We can be highly confident, at the 99% confidence level, that the true average modulus of elasticity after 1 minute exceeds that after 4 weeks by between roughly 2261 MPa and 3010 MPa. This interval is rather wide, partly because of the high confidence level and partly because there is a reasonable amount of variability in the sample differences. ■

Paired Data and Two-Sample t Procedures

Consider using the two-sample t test on paired data. The numerators of the paired t and two-sample t test statistics are identical, since $\bar{d} = \Sigma d_i/n = [\Sigma(x_i - y_i)]/n = (\Sigma x_i)/n - (\Sigma y_i)/n = \bar{x} - \bar{y}$. The difference between the two statistics is due entirely to the denominators. Each test statistic is obtained by standardizing $\bar{X} - \bar{Y}\,(= \bar{D})$, but in the presence of dependence the two-sample t standardization is incorrect. To see this, recall from Section 5.5 that

$$V(X \pm Y) = V(X) + V(Y) \pm 2\,\text{Cov}(X, Y)$$

Since the correlation between X and Y is

$$\rho = \text{Corr}(X, Y) = \text{Cov}(X, Y)/[\sqrt{V(X)} \cdot \sqrt{V(Y)}]$$

it follows that

$$V(X - Y) = \sigma_1^2 + \sigma_2^2 - 2\rho\sigma_1\sigma_2$$

Applying this to $\overline{X} - \overline{Y}$ yields

$$V(\overline{X} - \overline{Y}) = V(\overline{D}) = V\left(\frac{1}{n}\sum D_i\right) = \frac{V(D_i)}{n} = \frac{\sigma_1^2 + \sigma_2^2 - 2\rho\sigma_1\sigma_2}{n}$$

The two-sample t test is based on the assumption of independence, in which case $\rho = 0$. But in many paired experiments, there will be a strong *positive* dependence between X and Y (large X associated with large Y), so that ρ will be positive and the variance of $\overline{X} - \overline{Y}$ will be smaller than $\sigma_1^2/n + \sigma_2^2/n$. Thus, *whenever there is positive dependence within pairs, the denominator for the paired t statistic should be smaller than for t of the independent-samples test.* Often two-sample t will be much closer to zero than paired t, considerably understating the significance of the data.

Similarly, when data is paired, the paired t CI will usually be narrower than the (incorrect) two-sample t CI. Use of the two-sample t interval for the data in Example 9.10 gives the 99% CI (705, 4566). The width of this interval is 3861, whereas that of the paired t interval is roughly 749. The reason for this is that there is much less variability in the differences than there is in either the 1-min observations or the 4-week observations ($s_D = 509$, $s_1 = 2056$, and $s_2 = 1902$).

Paired Versus Unpaired Experiments

In our examples, paired data resulted from two observations on the same individual (Example 9.9) or experimental object (lumber specimen in Example 9.10). Even when this cannot be done, paired data with dependence within pairs can be obtained by matching individuals or objects on one or more characteristics thought to influence responses. For example, in a medical experiment to compare the efficacy of two drugs for lowering blood pressure, the experimenter's budget might allow for the treatment of 20 patients. If 10 patients are randomly selected for treatment with the first drug and another 10 independently selected for treatment with the second drug, an independent-samples experiment results.

However, the experimenter, knowing that blood pressure is influenced by age and weight, might decide to pair off patients so that within each of the resulting 10 pairs, age and weight were approximately equal (though there might be sizable differences between pairs). Then each drug would be given to a different patient within each pair for a total of 10 observations on each drug.

Without this matching (or "blocking"), one drug might appear to outperform the other just because patients in one sample were lighter and younger and thus more susceptible to a decrease in blood pressure than the heavier and older patients in the second sample. However, there is a price to be paid for pairing—a smaller number of degrees of freedom for the paired analysis—so we must ask when one type of experiment should be preferred to the other.

There is no straightforward and precise answer to this question, but there are some useful guidelines. If we have a choice between two t tests that are both valid (and carried out at the same level of significance α), we should prefer the test that has the larger number of degrees of freedom. The reason for this is that a larger number of degrees of freedom means smaller β for any fixed alternative value of the parameter or parameters. That is, for a fixed type I error probability, the probability of a type II error is decreased by increasing degrees of freedom.

However, if the experimental units are quite heterogeneous in their responses, it will be difficult to detect small but significant differences between two treatments. This is essentially what happened in the data set in Example 9.8; for both "treatments" (bottom water and surface water), there is great between-location variability, which tends to mask differences in treatments within locations. If there is a high positive correlation within experimental units or subjects, the variance of $\overline{D} = \overline{X} - \overline{Y}$ will be much smaller than the unpaired variance. Because of this reduced variance, it will be easier to detect a difference with paired samples than with independent samples. The pros and cons of pairing can now be summarized as follows.

1. If there is great heterogeneity between experimental units and a large correlation within experimental units (large positive ρ), then the loss in degrees of freedom will be compensated for by the increased precision associated with pairing, so a paired experiment is preferable to an independent-samples experiment.
2. If the experimental units are relatively homogeneous and the correlation within pairs is not large, the gain in precision due to pairing will be outweighed by the decrease in degrees of freedom, so an independent-samples experiment should be used.

Of course, values of σ_1^2, σ_2^2, and ρ will not usually be known very precisely, so an investigator will be required to make a seat-of-the-pants judgment as to whether Situation 1 or 2 obtains. In general, if the number of observations that can be obtained is large, then a loss in degrees of freedom (e.g., from 40 to 20) will not be serious; but if the number is small, then the loss (say, from 16 to 8) because of pairing may be serious if not compensated for by increased precision. Similar considerations apply when choosing between the two types of experiments to estimate $\mu_1 - \mu_2$ with a confidence interval.

Exercises | Section 9.3 (36–46)

36. Consider the accompanying data on breaking load (kg/25 mm width) for various fabrics in both an unabraded condition and an abraded condition ("The Effect of Wet Abrasive Wear on the Tensile Properties of Cotton and Polyester-Cotton Fabrics," *J. Testing and Evaluation*, 1993: 84–93). Use the paired t test, as did the authors of the cited article, to test $H_0: \mu_D = 0$ versus $H_a: \mu_D > 0$ at significance level .01.

				Fabric				
	1	2	3	4	5	6	7	8
U	36.4	55.0	51.5	38.7	43.2	48.8	25.6	49.8
A	28.5	20.0	46.0	34.5	36.5	52.5	26.5	46.5

37. Hexavalent chromium has been identified as an inhalation carcinogen and an air toxin of concern in a

number of different locales. The article "Airborne Hexavalent Chromium in Southwestern Ontario" (*J. of Air and Waste Mgmnt. Assoc.,* 1997: 905–910) gave the accompanying data on both indoor and outdoor concentration (nanograms/m³) for a sample of houses selected from a certain region.

House	1	2	3	4	5	6	7	8	9
Indoor	.07	.08	.09	.12	.12	.12	.13	.14	.15
Outdoor	.29	.68	.47	.54	.97	.35	.49	.84	.86

House	10	11	12	13	14	15	16	17
Indoor	.15	.17	.17	.18	.18	.18	.18	.19
Outdoor	.28	.32	.32	1.55	.66	.29	.21	1.02

House	18	19	20	21	22	23	24	25
Indoor	.20	.22	.22	.23	.23	.25	.26	.28
Outdoor	1.59	.90	.52	.12	.54	.88	.49	1.24

House	26	27	28	29	30	31	32	33
Indoor	.28	.29	.34	.39	.40	.45	.54	.62
Outdoor	.48	.27	.37	1.26	.70	.76	.99	.36

a. Calculate a confidence interval for the population mean difference between indoor and outdoor concentrations using a confidence level of 95%, and interpret the resulting interval.

b. If a 34th house were to be randomly selected from the population, between what values would you predict the difference in concentrations to lie?

38. Concrete specimens with varying height-to-diameter ratios cut from various positions on the original cylinder were obtained both from a normal-strength concrete mix and from a high-strength mix. The peak stress (MPa) was determined for each mix, resulting in the following data ("Effect of Length on Compressive Strain Softening of Concrete," *J. of Engr. Mechanics,* 1997: 25–35).

Test condition	1	2	3	4	5
Normal	42.8	55.6	49.0	48.7	44.1
High	90.9	93.1	86.3	90.3	88.5

Test condition	6	7	8	9	10
Normal	55.4	50.1	45.7	51.4	43.1
High	88.1	93.2	90.8	90.1	92.6

Test condition	11	12	13	14	15
Normal	46.8	46.7	47.7	45.8	45.4
High	88.2	88.6	91.0	90.0	90.1

a. Construct a comparative boxplot of peak stresses for the two types of concrete, and comment on any interesting features.

b. Estimate the difference between true average peak stresses for the two types of concrete in a

way that conveys information about precision and reliability. Be sure to check the plausibility of any assumptions needed in your analysis. Does it appear plausible that the true average peak stresses for the two types of concrete are identical? Why or why not?

39. Scientists and engineers frequently wish to compare two different techniques for measuring or determining the value of a variable. In such situations, interest centers on testing whether the mean difference in measurements is zero. The article "Evaluation of the Deuterium Dilution Technique Against the Test Weighing Procedure for the Determination of Breast Milk Intake" (*Amer. J. Clinical Nutr.,* 1983: 996–1003) reports the accompanying data on amount of milk ingested by each of 14 randomly selected infants.

Infant	1	2	3	4
Isotopic method	1509	1418	1561	1556
Test-weighing method	1498	1254	1336	1565
Difference	11	164	225	−9

Infant	5	6	7	8
Isotopic method	2169	1760	1098	1198
Test-weighing method	2000	1318	1410	1129
Difference	169	442	−312	69

Infant	9	10	11
Isotopic method	1479	1281	1414
Test-weighing method	1342	1124	1468
Difference	137	157	−54

Infant	12	13	14
Isotopic method	1954	2174	2058
Test-weighing method	1604	1722	1518
Difference	350	452	540

a. Is it plausible that the population distribution of differences is normal?

b. Does it appear that the true average difference between intake values measured by the two methods is something other than zero? Determine the *P*-value of the test and use it to reach a conclusion at significance level .05.

40. Two types of fish attractors, one made from vitrified clay pipes and the other from cement blocks and brush, were used during 16 different time periods spanning 4 years at Lake Tohopekaliga, Florida ("Two Types of Fish Attractors Compared in Lake

Tohopekaliga, Florida," *Trans. Amer. Fisheries Soc.,* 1978: 689–695). The following observations are of fish caught per fishing day.

	Period							
	1	2	3	4	5	6	7	8
Pipe	6.64	7.89	1.83	.42	.85	.29	.57	.63
Brush	9.73	8.21	2.17	.75	1.61	.75	.83	.56

	Period							
	9	10	11	12	13	14	15	16
Pipe	.32	.37	.00	.11	4.86	1.80	.23	.58
Brush	.76	.32	.48	.52	5.38	2.33	.91	.79

Does one attractor appear to be more effective on average than the other?
a. Use the paired *t* test with $\alpha = .01$ to test H_0: $\mu_D = 0$ versus H_a: $\mu_D \neq 0$.
b. What happens if the two-sample *t* test is used ($s_1 = 2.48$ and $s_2 = 2.91$)?

41. In an experiment designed to study the effects of illumination level on task performance ("Performance of Complex Tasks Under Different Levels of Illumination," *J. Illuminating Eng.,* 1976: 235–242), subjects were required to insert a fine-tipped probe into the eyeholes of ten needles in rapid succession both for a low light level with a black background and a higher level with a white background. Each data value is the time (sec) required to complete the task.

	Subject				
	1	2	3	4	5
Black	25.85	28.84	32.05	25.74	20.89
White	18.23	20.84	22.96	19.68	19.50

	Subject			
	6	7	8	9
Black	41.05	25.01	24.96	27.47
White	24.98	16.61	16.07	24.59

Does the data indicate that the higher level of illumination yields a decrease of more than 5 sec in true average task completion time? Test the appropriate hypotheses using the *P*-value approach.

42. It has been estimated that between 1945 and 1971, as many as 2 million children were born to mothers treated with diethylstilbestrol (DES), a nonsteroidal estrogen recommended for pregnancy maintenance. The FDA banned this drug in 1971 because research indicated a link with the incidence of cervical cancer. The article "Effects of Prenatal Exposure to Diethylstilbestrol (DES) on Hemispheric Laterality and Spatial Ability in Human Males" (*Hormones and Behavior,* 1992: 62–75) discussed a study in which 10 males exposed to DES and their unexposed brothers underwent various tests. This is the summary data on the results of a spatial ability test: $\bar{x} = 12.6$ (exposed), $\bar{y} = 13.7$, and standard error of mean difference = .5. Test at level .05 to see whether exposure is associated with reduced spatial ability by obtaining the *P*-value.

43. Cushing's disease is characterized by muscular weakness due to adrenal or pituitary dysfunction. To provide effective treatment, it is important to detect childhood Cushing's disease as early as possible. Age at onset of symptoms and age at diagnosis for 15 children suffering from the disease were given in the article "Treatment of Cushing's Disease in Childhood and Adolescence by Transphenoidal Microadenomectomy" (*New Engl. J. of Med.,* 1984: 889). Here are the values of the differences between age at onset of symptoms and age at diagnosis:

−24	−12	−55	−15	−30	−60	−14	−21
−48	−12	−25	−53	−61	−69	−80	

a. Does the accompanying normal probability plot cast strong doubt on the approximate normality of the population distribution of differences?

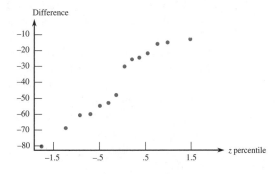

b. Calculate a lower 95% confidence bound for the population mean difference, and interpret the resulting bound.

c. Suppose the (age at diagnosis) − (age at onset) differences had been calculated. What would be a 95% upper confidence bound for the corresponding population mean difference?

44. The article "Agronomic Performance of Winter Versus Spring Wheat" (*Agronomy J.,* 1991: 527–531) presented the results of an experiment to compare the yield (kg/ha) of Sundance winter wheat and Manitou spring wheat. Data from nine test plots is given in the accompanying table. Is there sufficient evidence to conclude that true average yield for the winter wheat is more than 500 kg/ha higher than for spring wheat?

	6	**7**	**8**	**9**
Sundance	2860	3470	2042	3689
Manitou	2074	2308	1525	2779

a. Check the plausibility of any assumptions needed to carry out an appropriate test of hypotheses.

b. State and test the relevant hypotheses.

45. Refer to Exercise 41. Compute an interval estimate for the difference between true average task time under the high illumination level and true average time under the low level.

46. Construct a paired data set for which $t = \infty$, so that the data is highly significant when the correct analysis is used, yet t for the two-sample t test is quite near zero, so the incorrect analysis yields an insignificant result.

Location

	1	**2**	**3**	**4**	**5**
Sundance	3201	3095	3297	3644	3604
Manitou	2386	2011	2616	3094	3069

(continued)

9.4 | Inferences Concerning a Difference Between Population Proportions

Having presented methods for comparing the means of two different populations, we now turn to the comparison of two population proportions. The notation for this problem is an extension of the notation used in the corresponding one-population problem. We let p_1 and p_2 denote the proportions of individuals in populations 1 and 2, respectively, who possess a particular characteristic. Alternatively, if we use the label S for an individual who possesses the characteristic of interest (does favor a particular proposition, has read at least one book within the last month, etc.), then p_1 and p_2 represent the probabilities of seeing the label S on a randomly chosen individual from populations 1 and 2, respectively.

We will assume the availability of a sample of m individuals from the first population and n from the second. The variables X and Y will represent the number of individuals in each sample possessing the characteristic that defines p_1 and p_2. Provided the population sizes are much larger than the sample sizes, the distribution of X can be taken to be binomial with parameters m and p_1, and similarly, Y is taken to be a binomial variable with parameters n and p_2. Furthermore, the samples are assumed to be independent of one another, so that X and Y are independent rv's.

The obvious estimator for $p_1 - p_2$, the difference in population proportions, is the corresponding difference in sample proportions $X/m - Y/n$. With $\hat{p}_1 = X/m$ and $\hat{p}_2 = Y/n$, the estimator of $p_1 - p_2$ can be expressed as $\hat{p}_1 - \hat{p}_2$.

PROPOSITION

Let $X \sim \mathrm{Bin}(m, p_1)$ and $Y \sim \mathrm{Bin}(n, p_2)$ with X and Y independent variables. Then,

$$E(\hat{p}_1 - \hat{p}_2) = p_1 - p_2$$

so $\hat{p}_1 - \hat{p}_2$ is an unbiased estimator of $p_1 - p_2$, and

$$V(\hat{p}_1 - \hat{p}_2) = \frac{p_1 q_1}{m} + \frac{p_2 q_2}{n} \qquad \text{(where } q_i = 1 - p_i) \qquad (9.3)$$

Proof Since $E(X) = mp_1$ and $E(Y) = np_2$,

$$E\left(\frac{X}{m} - \frac{Y}{n}\right) = \frac{1}{m}E(X) - \frac{1}{n}E(Y) = \frac{1}{m}mp_1 - \frac{1}{n}np_2 = p_1 - p_2$$

Since $V(X) = mp_1 q_1$, $V(Y) = np_2 q_2$, and X and Y are independent,

$$V\left(\frac{X}{m} - \frac{Y}{n}\right) = V\left(\frac{X}{m}\right) + V\left(\frac{Y}{n}\right) = \frac{1}{m^2}V(X) + \frac{1}{n^2}V(Y) = \frac{p_1 q_1}{m} + \frac{p_2 q_2}{n} \qquad \blacksquare$$

We will focus first on situations in which both m and n are large. Then because $\hat{p}_1$ and $\hat{p}_2$ individually have approximately normal distributions, the estimator $\hat{p}_1 - \hat{p}_2$ also has approximately a normal distribution. Using the mean value and variance from this proposition to standardize $\hat{p}_1 - \hat{p}_2$ yields a variable Z whose distribution is approximately standard normal:

$$Z = \frac{\hat{p}_1 - \hat{p}_2 - (p_1 - p_2)}{\sqrt{\dfrac{p_1 q_1}{m} + \dfrac{p_2 q_2}{n}}}$$

A Large-Sample Test Procedure

Analogously to the hypotheses for $\mu_1 - \mu_2$, the most general null hypothesis an investigator might consider would be of the form $H_0: p_1 - p_2 = \Delta_0$, where Δ_0 is again a specified number. Although for population means the case $\Delta_0 \neq 0$ presented no difficulties, for population proportions the cases $\Delta_0 = 0$ and $\Delta_0 \neq 0$ must be considered separately. Since the vast majority of actual problems of this sort involve $\Delta_0 = 0$ (i.e., the null hypothesis $p_1 = p_2$), we will concentrate on this case. When $H_0: p_1 - p_2 = 0$ is true, let p denote the common value of p_1 and p_2 (and similarly for q). Then the standardized variable

$$Z = \frac{\hat{p}_1 - \hat{p}_2 - 0}{\sqrt{pq\left(\dfrac{1}{m} + \dfrac{1}{n}\right)}} \qquad (9.4)$$

has approximately a standard normal distribution when H_0 is true. However, this Z cannot serve as a test statistic because the value of p is unknown—H_0 asserts only that there is a common value of p, but does not say what that value is. To obtain a test statistic having approximately a standard normal distribution when H_0 is true (so that use of an appropriate z critical value specifies a level α test), p must be estimated from the sample data.

Assuming then that $p_1 = p_2 = p$, instead of separate samples of size m and n from two different populations (two different binomial distributions), we really have a single sample of size $m + n$ from one population with proportion p. Since the total number of individuals in this combined sample having the characteristic of interest is $X + Y$, the estimator of p is

$$\hat{p} = \frac{X + Y}{m + n} = \frac{m}{m + n}\hat{p}_1 + \frac{n}{m + n}\hat{p}_2 \tag{9.5}$$

The second expression for $\hat{p}$ shows that it is actually a weighted average of estimators $\hat{p}_1$ and $\hat{p}_2$ obtained from the two samples. If we take (9.5) (with $\hat{q} = 1 - \hat{p}$) and substitute back into (9.4), the resulting statistic has approximately a standard normal distribution when H_0 is true.

Null hypothesis: H_0: $p_1 - p_2 = 0$

Test statistic value (large samples): $z = \dfrac{\hat{p}_1 - \hat{p}_2}{\sqrt{\hat{p}\hat{q}(1/m + 1/n)}}$

Alternative Hypothesis	Rejection Region for Approximate Level α Test
H_a: $p_1 - p_2 > 0$	$z \geq z_\alpha$
H_a: $p_1 - p_2 < 0$	$z \leq -z_\alpha$
H_a: $p_1 - p_2 \neq 0$	either $z \geq z_{\alpha/2}$ or $z \leq -z_{\alpha/2}$

A P-value is calculated in the same way as for previous z tests.

Example 9.11 Some defendants in criminal proceedings plead guilty and are sentenced without a trial, whereas others who plead innocent are subsequently found guilty and then are sentenced. In recent years, legal scholars have speculated as to whether sentences of those who plead guilty differ in severity from sentences for those who plead innocent and are subsequently judged guilty. Consider the accompanying data on defendants from San Francisco County accused of robbery, all of whom had previous prison records ("Does It Pay to Plead Guilty? Differential Sentencing and the Functioning of Criminal Courts," *Law and Society Rev.,* 1981–1982: 45–69). Does this data suggest that the proportion of all defendants in these circumstances who plead guilty and are sent to prison differs from the proportion who are sent to prison after pleading innocent and being found guilty?

	Plea	
	Guilty	**Not Guilty**
Number judged guilty	$m = 191$	$n = 64$
Number sentenced to prison	$x = 101$	$y = 56$
Sample proportion	$\hat{p}_1 = .529$	$\hat{p}_2 = .875$

Let p_1 and p_2 denote the two population proportions. The hypotheses of interest are H_0: $p_1 - p_2 = 0$ versus H_a: $p_1 - p_2 \neq 0$. At level .01, H_0 should be rejected if either

$z \geq z_{.005} = 2.58$ or if $z \leq -2.58$. The combined estimate of the common success proportion is $\hat{p} = (101 + 56)/(191 + 64) = .616$. The value of the test statistic is then

$$z = \frac{.529 - .875}{\sqrt{(.616)(.384)(1/191 + 1/64)}} = \frac{-.346}{.070} = -4.94$$

Since $-4.94 \leq -2.58$, H_0 must be rejected.

The P-value for a two-tailed z test is

$$P\text{-value} = 2[1 - \Phi(|z|)] = 2[1 - \Phi(4.94)] < 2[1 - \Phi(3.49)] = .0004$$

A more extensive standard normal table yields P-value $\approx .0000006$. This P-value is so minuscule that at any reasonable level α, H_0 should be rejected. The data very strongly suggests that $p_1 \neq p_2$ and, in particular, that initially pleading guilty may be a good strategy as far as avoiding prison is concerned.

The cited article also reported data on defendants in several other counties. The authors broke down the data by type of crime (burglary or robbery) and by nature of prior record (none, some but no prison, and prison). In every case, the conclusion was the same: Among defendants judged guilty, those who pleaded that way were less likely to receive prison sentences. ∎

Type II Error Probabilities and Sample Sizes

Here the determination of β is a bit more cumbersome than it was for other large-sample tests. The reason is that the denominator of Z is an estimate of the standard deviation of $\hat{p}_1 - \hat{p}_2$ assuming that $p_1 = p_2 = p$. When H_0 is false, $\hat{p}_1 - \hat{p}_2$ must be restandardized using

$$\sigma_{\hat{p}_1 - \hat{p}_2} = \sqrt{\frac{p_1 q_1}{m} + \frac{p_2 q_2}{n}} \tag{9.6}$$

The form of σ implies that β is not a function of just $p_1 - p_2$, so we denote it by $\beta(p_1, p_2)$.

Alternative Hypothesis	$\beta(p_1, p_2)$
H_a: $p_1 - p_2 > 0$	$\Phi\left[\dfrac{z_\alpha \sqrt{\bar{p}\bar{q}(1/m + 1/n)} - (p_1 - p_2)}{\sigma}\right]$
H_a: $p_1 - p_2 < 0$	$1 - \Phi\left[\dfrac{-z_\alpha \sqrt{\bar{p}\bar{q}(1/m + 1/n)} - (p_1 - p_2)}{\sigma}\right]$
H_a: $p_1 - p_2 \neq 0$	$\Phi\left[\dfrac{z_{\alpha/2} \sqrt{\bar{p}\bar{q}(1/m + 1/n)} - (p_1 - p_2)}{\sigma}\right]$ $-\Phi\left[\dfrac{-z_{\alpha/2} \sqrt{\bar{p}\bar{q}(1/m + 1/n)} - (p_1 - p_2)}{\sigma}\right]$

where $\bar{p} = (mp_1 + np_2)/(m + n)$, $\bar{q} = (mq_1 + nq_2)/(m + n)$, and σ is given by (9.6).

Proof For the upper-tailed test ($H_a: p_1 - p_2 > 0$),

$$\beta(p_1, p_2) = P[\hat{p}_1 - \hat{p}_2 < z_\alpha \sqrt{\hat{p}\hat{q}(1/m + 1/n)}]$$

$$= P\left[\frac{(\hat{p}_1 - \hat{p}_2 - (p_1 - p_2)}{\sigma} < \frac{z_\alpha \sqrt{\hat{p}\hat{q}(1/m + 1/n)} - (p_1 - p_2)}{\sigma} \right]$$

When m and n are both large,

$$\hat{p} = (m\hat{p}_1 + n\hat{p}_2)/(m + n) \approx (mp_1 + np_2)/(m + n) = \bar{p}$$

and $\hat{q} \approx \bar{q}$, which yields the previous (approximate) expression for $\beta(p_1, p_2)$. ■

Alternatively, for specified p_1, p_2 with $p_1 - p_2 = d$, the sample sizes necessary to achieve $\beta(p_1, p_2) = \beta$ can be determined. For example, for the upper-tailed test, we equate $-z_\beta$ to the argument of $\Phi(\cdot)$ (i.e., what's inside the parentheses) in the foregoing box. If $m = n$, there is a simple expression for the common value.

For the case $m = n$, the level α test has type II error probability β at the alternative values p_1, p_2 with $p_1 - p_2 = d$ when

$$n = \frac{\left[z_\alpha \sqrt{(p_1 + p_2)(q_1 + q_2)/2} + z_\beta \sqrt{p_1 q_1 + p_2 q_2} \right]^2}{d^2} \quad (9.7)$$

for an upper- or lower-tailed test, with $\alpha/2$ replacing α for a two-tailed test.

Example 9.12 One of the truly impressive applications of statistics occurred in connection with the design of the 1954 Salk polio vaccine experiment and analysis of the resulting data. Part of the experiment focused on the efficacy of the vaccine in combating paralytic polio. Because it was thought that without a control group of children, there would be no sound basis for assessment of the vaccine, it was decided to administer the vaccine to one group and a placebo injection (visually indistinguishable from the vaccine but known to have no effect) to a control group. For ethical reasons and also because it was thought that the knowledge of vaccine administration might have an effect on treatment and diagnosis, the experiment was conducted in a **double-blind** manner. That is, neither the individuals receiving injections nor those administering them actually knew who was receiving vaccine and who was receiving the placebo (samples were numerically coded)—remember, at that point it was not at all clear whether the vaccine was beneficial.

Let p_1 and p_2 be the probabilities of a child getting paralytic polio for the control and treatment condition, respectively. The objective was to test $H_0: p_1 - p_2 = 0$ versus $H_a: p_1 - p_2 > 0$ (the alternative states that a vaccinated child is less likely to contract polio than an unvaccinated child). Supposing the true value of p_1 is .0003 (an incidence rate of 30 per 100,000), the vaccine would be a significant improvement if the incidence rate was halved—that is, $p_2 = .00015$. Using a level $\alpha = .05$ test, it would then be reasonable to ask for sample sizes for which $\beta = .1$ when

$p_1 = .0003$ and $p_2 = .00015$. Assuming equal sample sizes, the required n is obtained from (9.7) as

$$n = \frac{\left[1.645\sqrt{(.5)(.00045)(1.99955)} + 1.28\sqrt{(.00015)(.99985) + (.0003)(.9997)}\right]^2}{(.0003 - .00015)^2}$$

$$= [(.0349 + .0271)/.00015]^2 \approx 171,000$$

The actual data for this experiment follows. Sample sizes of approximately 200,000 were used. The reader can easily verify that $z = 6.43$, a highly significant value. The vaccine was judged a resounding success! An excellent expository article describing the experiment and results appears in *Statistics: A Guide to the Unknown*, edited by Judith Tanur et al. (see Chapter 1 bibliography).

Placebo: $m = 201,229$, $x =$ number of cases of paralytic polio $= 110$

Vaccine: $n = 200,745$, $y = 33$ ∎

A Large-Sample Confidence Interval for $p_1 - p_2$

As with means, many two-sample problems involve the objective of comparison through hypothesis testing, but sometimes an interval estimate for $p_1 - p_2$ is appropriate. Both $\hat{p}_1 = X/m$ and $\hat{p}_2 = Y/n$ have approximate normal distributions when m and n are both large. If we identify θ with $p_1 - p_2$, then $\hat{\theta} = \hat{p}_1 - \hat{p}_2$ satisfies the conditions necessary for obtaining a large-sample CI. In particular, the estimated standard deviation of $\hat{\theta}$ is $\sqrt{(\hat{p}_1\hat{q}_1/m) + (\hat{p}_2\hat{q}_2/n)}$. The $100(1 - \alpha)\%$ interval $\hat{\theta} \pm z_{\alpha/2} \cdot \hat{\sigma}_{\theta}$ then becomes

$$\hat{p}_1 - \hat{p}_2 \pm z_{\alpha/2}\sqrt{\frac{\hat{p}_1\hat{q}_1}{m} + \frac{\hat{p}_2\hat{q}_2}{n}}$$

Notice that the estimated standard deviation of $\hat{p}_1 - \hat{p}_2$ (the square-root expression) is different here from what it was for hypothesis testing when $\Delta_0 = 0$.

Example 9.13 The article "The Association of Marijuana Use with Outcome of Pregnancy" (*Amer. J. Public Health*, 1983: 1161–1164) reports the accompanying data on incidence of major malfunctions among newborns both for mothers who were marijuana users and for mothers who did not use marijuana.

	User	Nonuser
Sample size	1246	11,178
Number of major malfunctions	42	294
$\hat{p}$	.0337	.0263

Let p_1 denote the proportion of births that involve a major malfunction among all mothers who are marijuana users, and define p_2 similarly for nonusers. A 99% CI for $p_1 - p_2$ requires

$$2.58 \sqrt{\frac{\hat{p}_1 \hat{q}_1}{m} + \frac{\hat{p}_2 \hat{q}_2}{n}} = 2.58 \sqrt{\frac{(.0337)(.9663)}{1246} + \frac{(.0263)(.9737)}{11,178}}$$
$$= (2.58)(.00533) = .0138$$

The confidence interval is then $.0337 - .0263 \pm .0138 = .0074 \pm .0138 = (-.0064, .0212)$. This interval is quite narrow, which suggests that $p_1 - p_2$ has been precisely estimated. Notice that the interval contains zero, so this is one plausible value for $p_1 - p_2$.   ∎

Small-Sample Inferences

On occasion an inference concerning $p_1 - p_2$ may have to be based on samples for which at least one sample size is small. Appropriate methods for such situations are not as straightforward as those for large samples, and there is more controversy among statisticians as to recommended procedures. One frequently used test, called the Fisher–Irwin test, is based on the hypergeometric distribution. A statistician can be consulted for more information.

Exercises | Section 9.4 (47–56)

47. Is someone who switches brands because of a financial inducement less likely to remain loyal than someone who switches without inducement? Let p_1 and p_2 denote the true proportions of switchers to a certain brand with and without inducement, respectively, who subsequently make a repeat purchase. Test H_0: $p_1 - p_2 = 0$ versus H_a: $p_1 - p_2 < 0$ using $\alpha = .01$ and the following data:

$m = 200$ number of successes $= 30$

$n = 600$ number of successes $= 180$

(Similar data is given in "Impact of Deals and Deal Retraction on Brand Switching," *J. Marketing,* 1980: 62–70.)

48. A sample of 300 urban adult residents of a particular state revealed 63 who favored increasing the highway speed limit from 55 to 65 mph, whereas a sample of 180 rural residents yielded 75 who favored the increase. Does this data indicate that the sentiment for increasing the speed limit is different for the two groups of residents?

a. Test H_0: $p_1 = p_2$ versus H_a: $p_1 \neq p_2$ using $\alpha = .05$, where p_1 refers to the urban population.
b. If the true proportions favoring the increase are actually $p_1 = .20$ (urban) and $p_2 = .40$ (rural), what is the probability that H_0 will be rejected using a level .05 test with $m = 300$, $n = 180$?

49. It is thought that the front cover and the nature of the first question on mail surveys influence the response rate. The article "The Impact of Cover Design and First Questions on Response Rates for a Mail Survey of Skydivers" (*Leisure Sciences,* 1991: 67–76) tested this theory by experimenting with different cover designs. One cover was plain; the other used a picture of a skydiver. The researchers speculated that the return rate would be lower for the plain cover.

Cover	Number Sent	Number Returned
Plain	207	104
Skydiver	213	109

Does this data support the researchers' hypothesis? Test the relevant hypotheses using $\alpha = .10$ by first calculating a *P*-value.

50. Do teachers find their work rewarding and satisfying? The article "Work-Related Attitudes" (*Psychological Reports*, 1991: 443–450) reports the results of a survey of 395 elementary school teachers and 266 high school teachers. Of the elementary school teachers, 224 said they were very satisfied with their jobs, whereas 126 of the high school teachers were very satisfied with their work. Estimate the difference between the proportion of all elementary school teachers who are satisfied and all high school teachers who are satisfied by calculating a CI.

51. A random sample of 5726 telephone numbers from a certain region taken in March 1992 yielded 1105 that were unlisted, and 1 year later a sample of 5384 yielded 980 unlisted numbers.
   a. Test at level .10 to see whether there is a difference in true proportions of unlisted numbers between the two years.
   b. If $p_1 = .20$ and $p_2 = .18$, what sample sizes ($m = n$) would be necessary to detect such a difference with probability .90?

52. Ionizing radiation is being given increasing attention as a method for preserving horticultural products. The article "The Influence of Gamma-Irradiation on the Storage Life of Red Variety Garlic" (*J. of Food Processing and Preservation*, 1983: 179–183) reports that 153 of 180 irradiated garlic bulbs were marketable (no external sprouting, rotting, or softening) 240 days after treatment, whereas only 119 of 180 untreated bulbs were marketable after this length of time. Does this data suggest that ionizing radiation is beneficial as far as marketability is concerned?

53. In medical investigations, the ratio $\theta = p_1/p_2$ is often of more interest than the difference $p_1 - p_2$ (e.g., individuals given treatment 1 are how many times as likely to recover as those given treatment 2?). Let $\hat{\theta} = \hat{p}_1/\hat{p}_2$. When *m* and *n* are both large, the statistic $\ln(\hat{\theta})$ has approximately a normal distribution with approximate mean value $\ln(\theta)$ and approximate standard deviation $[(m - x)/(mx) + (n - y)/(ny)]^{1/2}$.
   a. Use these facts to obtain a large-sample 95% CI formula for estimating $\ln(\theta)$, and then a CI for θ itself.
   b. Return to the heart attack data of Example 1.3, and calculate an interval of plausible values for θ at the 95% confidence level. What does this in-

terval suggest about the efficacy of the aspirin treatment?

54. Sometimes experiments involving success or failure responses are run in a paired or before/after manner. Suppose that before a major policy speech by a political candidate, *n* individuals are selected and asked whether (*S*) or not (*F*) they favor the candidate. Then after the speech the same *n* people are asked the same question. The responses can be entered in a table as follows:

		After	
		S	*F*
Before	*S*	X_1	X_2
	F	X_3	X_4

where $X_1 + X_2 + X_3 + X_4 = n$. Let p_1, p_2, p_3, and p_4 denote the four cell probabilities, so that $p_1 = P(S$ before and S after), and so on. We wish to test the hypothesis that the true proportion of supporters (*S*) after the speech has not increased against the alternative that it has increased.
   a. State the two hypotheses of interest in terms of p_1, p_2, p_3, and p_4.
   b. Construct an estimator for the after/before difference in success probabilities.
   c. When *n* is large, it can be shown that the rv $(X_i - X_j)/n$ has approximately a normal distribution with variance given by $[p_i + p_j - (p_i - p_j)^2]/n$. Use this to construct a test statistic with approximately a standard normal distribution when H_0 is true (the result is called McNemar's test).
   d. If $x_1 = 350$, $x_2 = 150$, $x_3 = 200$, and $x_4 = 300$, what do you conclude?

55. Two different types of alloy, A and B, have been used to manufacture experimental specimens of a small tension link to be used in a certain engineering application. The ultimate strength (ksi) of each specimen was determined, and the results are summarized in the accompanying frequency distribution.

	A	B
26 – < 30	6	4
30 – < 34	12	9
34 – < 38	15	19
38 – < 42	7	10
	$m = 40$	$n = 42$

Compute a 95% CI for the difference between the true proportions of all specimens of alloys A and B that have an ultimate strength of at least 34 ksi.

56. Suppose a 95% CI for $p_1 - p_2$ is to be constructed based on equal sample sizes from the two populations. For what value of n $(= m)$ will the resulting interval have width at most .1 irrespective of the results of the sampling?

9.5 Inferences Concerning Two Population Variances

Methods for comparing two population variances (or standard deviations) are occasionally needed, though such problems arise much less frequently than those involving means or proportions. For the case in which the populations under investigation are normal, the procedures are based on a new family of probability distributions.

The *F* Distribution

The *F* probability distribution has two parameters, denoted by ν_1 and ν_2. The parameter ν_1 is called the *number of numerator degrees of freedom,* and ν_2 is the *number of denominator degrees of freedom;* here ν_1 and ν_2 are positive integers. A random variable that has an *F* distribution cannot assume a negative value. Since the density function is complicated and will not be used explicitly, we omit the formula. There is an important connection between an *F* variable and chi-squared variables. If X_1 and X_2 are independent chi-squared rv's with ν_1 and ν_2 df, respectively, then the rv

$$F = \frac{X_1/\nu_1}{X_2/\nu_2} \tag{9.8}$$

the ratio of the two chi-squared variables divided by their respective degrees of freedom, can be shown to have an *F* distribution.

Figure 9.8 illustrates the graph of a typical *F* density function. Analogous to the notation $t_{\alpha,\nu}$ and $\chi^2_{\alpha,\nu}$, we use F_{α,ν_1,ν_2} for the point on the axis that captures α of the area under the *F* density curve with ν_1 and ν_2 df in the upper tail. The density curve is not symmetric, so it would seem that both upper- and lower-tail critical values must be tabulated. This is not necessary, though, because of the following property.

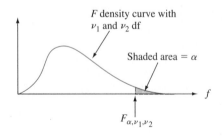

Figure 9.8 An *F* density curve and critical value

$$F_{1-\alpha,\nu_1,\nu_2} = 1/F_{\alpha,\nu_2,\nu_1} \qquad (9.9)$$

Appendix Table A.9 gives F_{α,ν_1,ν_2} for $\alpha = .10, .05, .01,$ and $.001,$ and various values of ν_1 (in different columns of the table) and ν_2 (in different groups of rows of the table). For example, $F_{.05,6,10} = 3.22$ and $F_{.05,10,6} = 4.06$. To obtain $F_{.95,6,10}$, the number which captures .95 of the area to its right (and thus .05 to the left) under the F curve with $\nu_1 = 6$ and $\nu_2 = 10$, we use (9.9): $F_{.95,6,10} = 1/F_{.05,10,6} = 1/4.06 = .246$.

Inferential Methods

A test procedure for hypotheses concerning the ratio σ_1^2/σ_2^2 as well as a CI for this ratio are based on the following result.

THEOREM

Let $X_1, \ldots, X_m$ be a random sample from a normal distribution with variance σ_1^2, let $Y_1, \ldots, Y_n$ be another random sample (independent of the X_i's) from a normal distribution with variance σ_2^2, and let S_1^2 and S_2^2 denote the two sample variances. Then the rv

$$F = \frac{S_1^2/\sigma_1^2}{S_2^2/\sigma_2^2} \qquad (9.10)$$

has an F distribution with $\nu_1 = m - 1$ and $\nu_2 = n - 1$.

This theorem results from combining (9.8) with the fact that the variables $(m - 1)S_1^2/\sigma_1^2$ and $(n - 1)S_2^2/\sigma_2^2$ each have a chi-squared distribution with $m - 1$ and $n - 1$ df, respectively (see Section 7.4). Because F involves a ratio rather than a difference, the test statistic is the ratio of sample variances; the claim that $\sigma_1^2 = \sigma_2^2$ is then rejected if the ratio differs by too much from 1.

Null hypothesis: $H_0: \sigma_1^2 = \sigma_2^2$

Test statistic value: $f = s_1^2/s_2^2$

Alternative Hypothesis	Rejection Region for a Level α Test
$H_a: \sigma_1^2 > \sigma_2^2$	$f \geq F_{\alpha,m-1,n-1}$
$H_a: \sigma_1^2 < \sigma_2^2$	$f \leq F_{1-\alpha,m-1,n-1}$
$H_a: \sigma_1^2 \neq \sigma_2^2$	either $f \geq F_{\alpha/2,m-1,n-1}$ or $f \leq F_{1-\alpha/2,m-1,n-1}$

Since critical values are tabled only for $\alpha = .10, .05, .01,$ and $.001,$ the two-tailed test can be performed only at levels .20, .10, .02, and .002. More extensive tabulations of F critical values are available elsewhere.

Example 9.14   On the basis of data reported in a 1979 article in the *Journal of Gerontology* ("Serum Ferritin in an Elderly Population," pp. 521–524), the authors concluded that the ferritin distribution in the elderly had a smaller variance than in the younger adults. (Serum ferritin is used in diagnosing iron deficiency.) For a sample of 28 elderly men, the sample standard deviation of serum ferritin (mg/L) was $s_1 = 52.6$; for 26 young men, the sample standard deviation was $s_2 = 84.2$. Does this data support the conclusion as applied to men?

Let σ_1^2 and σ_2^2 denote the variance of the serum ferritin distributions for elderly men and young men, respectively. We wish to test $H_0: \sigma_1^2 = \sigma_2^2$ versus $H_a: \sigma_1^2 < \sigma_2^2$. At level .01, H_0 will be rejected if $f \leq F_{.99,27,25}$. To obtain the critical value, we need $F_{.01,25,27}$. From Appendix Table A.9, $F_{.01,25,27} = 2.54$, so $F_{.99,27,25} = 1/2.54 = .394$. The computed value of F is $(52.6)^2/(84.2)^2 = .390$. Since $.390 \leq .394$, H_0 is rejected at level .01 in favor of H_a, so variability does appear to be greater in young men than in elderly men.   ∎

P-Values for F Tests

Recall that the *P*-value for an upper-tailed *t* test is the area under the relevant *t* curve (the one with appropriate df) to the right of the calculated *t*. In the same way, the *P*-value for an upper-tailed *F* test is the area under the *F* curve with appropriate numerator and denominator df to the right of the calculated *f*. Figure 9.9 illustrates this for a test based on $\nu_1 = 4$ and $\nu_2 = 6$.

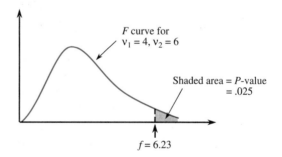

Figure 9.9   A *P*-value for an upper-tailed *F* test

Unfortunately, tabulation of *F* curve upper-tail areas is much more cumbersome than for *t* curves because two df's are involved. For each combination of ν_1 and ν_2, our *F* table gives only the four critical values that capture areas .10, .05, .01, and .001. Figure 9.10 shows what can be said about the *P*-value depending on where *f* falls relative to the four critical values.

For example, for a test with $\nu_1 = 4$ and $\nu_2 = 6$,

$$f = 5.70 \implies .01 < P\text{-value} < .05$$
$$f = 2.16 \implies P\text{-value} > .10$$
$$f = 25.03 \implies P\text{-value} < .001$$

Only if *f* equals a tabulated value do we obtain an exact *P*-value (e.g., if $f = 4.53$, then *P*-value $= .05$). Once we know that $.01 < P\text{-value} < .05$, H_0 would be rejected at a sig-

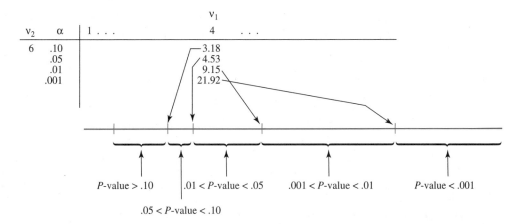

Figure 9.10 Obtaining *P*-value information from the *F* table for an upper-tailed *F* test

nificance level of .05 but not at a level of .01. When *P*-value $<$ .001, H_0 should be rejected at any reasonable significance level.

The *F* tests discussed in succeeding chapters will all be upper-tailed. If, however, a lower-tailed *F* test is appropriate, then (9.9) should be used to obtain lower-tailed critical values so that a bound or bounds on the *P*-value can be established. In the case of a two-tailed test, the bound or bounds from a one-tailed test should be multiplied by 2. For example, if $f = 5.82$ when $\nu_1 = 4$ and $\nu_2 = 6$, then since 5.82 falls between the .05 and .01 critical values, $2(.01) <$ *P*-value $< 2(.05)$, giving .02 $<$ *P*-value $< .10$. H_0 would then be rejected if $\alpha = .10$ but not if $\alpha = .01$. In this case, we cannot say from our table what conclusion is appropriate when $\alpha = .05$ (since we don't know whether the *P*-value is smaller or larger than this). However, statistical software shows that the area to the right of 5.82 under this *F* curve is .029, so the *P*-value is .058 and the null hypothesis should therefore not be rejected at level .05 (.058 is the smallest α for which H_0 can be rejected and our chosen α is smaller than this).

A Confidence Interval for σ_1/σ_2

The CI for σ_1^2/σ_2^2 is based on replacing *F* in the probability statement

$$P(F_{1-\alpha/2,\nu_1,\nu_2} < F < F_{\alpha/2,\nu_1,\nu_2}) = 1 - \alpha$$

by the *F* variable (9.10) and manipulating the inequalities to isolate σ_1^2/σ_2^2. An interval for σ_1/σ_2 results from taking the square root of each limit. The details are left for an exercise.

Exercises | Section 9.5 (57–64)

57. Obtain or compute the following quantities:
  a. $F_{.05,5,8}$ **b.** $F_{.05,8,5}$ **c.** $F_{.95,5,8}$ **d.** $F_{.95,8,5}$
  e. The 99th percentile of the *F* distribution with $\nu_1 = 10$, $\nu_2 = 12$

  f. The 1st percentile of the *F* distribution with $\nu_1 = 10$, $\nu_2 = 12$
  g. $P(F \le 6.16)$ for $\nu_1 = 6$, $\nu_2 = 4$
  h. $P(.177 \le F \le 4.74)$ for $\nu_1 = 10$, $\nu_2 = 5$

58. Give as much information as you can about the P-value of the F test in each of the following situations:

   a. $\nu_1 = 5$, $\nu_2 = 10$, upper-tailed test, $f = 4.75$
   b. $\nu_1 = 5$, $\nu_2 = 10$, upper-tailed test, $f = 2.00$
   c. $\nu_1 = 5$, $\nu_2 = 10$, two-tailed test, $f = 5.64$
   d. $\nu_1 = 5$, $\nu_2 = 10$, lower-tailed test, $f = .200$
   e. $\nu_1 = 35$, $\nu_2 = 20$, upper-tailed test, $f = 3.24$

59. The sample standard deviation of sodium concentration in whole blood (mEq/L) for $m = 20$ marine eels was found to be $s_1 = 40.5$, whereas the sample standard deviation of concentration for $n = 20$ freshwater eels was $s_2 = 32.1$ ("Ionic Composition of the Plasma and Whole Blood of Marine and Freshwater Eels," *Comp. Biochemistry and Physiology,* 1974: 541–544). Assuming normality of the two concentration distributions, test at level .10 to see whether the data suggests any difference between concentration variances for the two types of eels.

60. Refer to Example 9.7. Does the data suggest that the standard deviation of the strength distribution for fused specimens is smaller than that for not-fused specimens? Carry out a test at significance level .01 by obtaining as much information as you can about the P-value.

61. Toxaphene is an insecticide that has been identified as a pollutant in the Great Lakes ecosystem. To investigate the effect of toxaphene exposure on animals, groups of rats were given toxaphene in their diet. The article "Reproduction Study of Toxaphene in the Rat" (*J. of Environ. Sci. Health,* 1988: 101–126) reports weight gains (in grams) for rats given a low dose (4 ppm) and for control rats whose diet did not include the insecticide. The sample standard deviation for 23 female control rats was 32 g

and for 20 female low-dose rats was 54 g. Does this data suggest that there is more variability in low-dose weight gains than in control weight gains? Assuming normality, carry out a test of hypotheses at significance level .05.

62. In a study of copper deficiency in cattle, the copper values (μg Cu/100 mL blood) were determined both for cattle grazing in an area known to have well-defined molybdenum anomalies (metal values in excess of the normal range of regional variation) and for cattle grazing in a nonanomalous area ("An Investigation into Copper Deficiency in Cattle in the Southern Pennines," *J. Agricultural Soc. Cambridge,* 1972: 157–163), resulting in $s_1 = 21.5$ ($m = 48$) for the anomalous condition and $s_2 = 19.45$ ($n = 45$) for the nonanomalous condition. Test for the equality versus inequality of population variances at significance level .10 by using the P-value approach.

63. The article "Enhancement of Compressive Properties of Failed Concrete Cylinders with Polymer Impregnation" (*J. Testing and Evaluation,* 1977: 333–337) reports the following data on impregnated compressive modulus (psi $\times 10^6$) when two different polymers were used to repair cracks in failed concrete.

Epoxy	1.75	2.12	2.05	1.97
MMA prepolymer	1.77	1.59	1.70	1.69

Obtain a 90% CI for the ratio of variances by first using the method suggested in the text to obtain a general confidence interval formula.

64. Reconsider the data of Example 9.6, and calculate a 95% upper confidence bound for the ratio of the standard deviation of the triacetate porosity distribution to that of the cotton porosity distribution.

Supplementary Exercises (65–91)

65. The accompanying summary data on compression strength (lb) for $12 \times 10 \times 8$ in. boxes appeared in the article "Compression of Single-Wall Corrugated Shipping Containers Using Fixed and Floating Test Platens" (*J. Testing and Evaluation,* 1992: 318–320). The authors stated that "the difference between the compression strength using fixed and floating platen method was found to be small compared to normal variation in compression strength between identical boxes." Do you agree?

Method	Sample size	Sample mean	Sample SD
Fixed	10	807	27
Floating	10	757	41

66. The authors of the article "Dynamics of Canopy Structure and Light Interception in *Pinus elliotti*, North Florida" (*Ecological Monographs*, 1991: 33–51) planned an experiment to determine the effect of fertilizer on a measure of leaf area. A number of plots were available for the study, and half were selected at random to be fertilized. To ensure that the plots to receive the fertilizer and the control plots were similar, before beginning the experiment tree density (the number of trees per hectare) was recorded for eight plots to be fertilized and eight control plots, resulting in the given data. MINITAB output follows.

Fertilizer plots	1024	1216	1312	1280
	1216	1312	992	1120
Control plots	1104	1072	1088	1328
	1376	1280	1120	1200

```
Two sample T for fertilizer vs control

            N    Mean   StDev   SE Mean
fertilize   8    1184    126      44
control     8    1196    118      42

95% CI for mu fertilize - mu control:
(-144, 120)
```

a. Construct a comparative boxplot and comment on any interesting features.
b. Would you conclude that there is a significant difference in the mean tree density for fertilizer and control plots? Use $\alpha = .05$.
c. Interpret the given confidence interval.

67. Is the response rate for questionnaires affected by including some sort of incentive to respond along with the questionnaire? In one experiment, 110 questionnaires with no incentive resulted in 75 being returned, whereas 98 questionnaires that included a chance to win a lottery yielded 66 responses ("Charities, No; Lotteries, No; Cash, Yes," *Public Opinion Quarterly*, 1996: 542–562). Does this data suggest that including an incentive increases the likelihood of a response? State and test the relevant hypotheses at significance level .10 by using the *P*-value method.

68. The accompanying data was obtained in a study to evaluate the liquefaction potential at a proposed nuclear power station ("Cyclic Strengths Compared for Two Sampling Techniques," *J. Geotechnical Division, Am. Soc. Civil Engrs. Proceedings*, 1981: 563–576). Before cyclic strength testing, soil samples were gathered using both a pitcher tube method and a block method, resulting in the following observed values of dry density (lb/ft³).

Pitcher sampling	101.1	111.1	107.6	98.1
	99.5	98.7	103.3	108.9
	109.1	104.1	110.0	98.4
	105.1	104.5	105.7	103.3
	100.3	102.6	101.7	105.4
	99.6	103.3	102.1	104.3
Block sampling	107.1	105.0	98.0	97.9
	103.3	104.6	100.1	98.2
	97.9	103.2	96.9	

Calculate and interpret a 95% CI for the difference between true average dry densities for the two sampling methods.

69. The article "Quantitative MRI and Electrophysiology of Preoperative Carpal Tunnel Syndrome in a Female Population" (*Ergonomics*, 1997: 642–649) reported that (−473.3, 1691.9) was a large-sample 95% confidence interval for the difference between true average thenar muscle volume (mm³) for sufferers of carpal tunnel syndrome and true average volume for nonsufferers. Calculate a 90% confidence interval for this difference.

70. The following summary data on bending strength (lb-in./in.) of joints is taken from the article "Bending Strength of Corner Joints Constructed with Injection Molded Splines" (*Forest Products J.*, April, 1997: 89–92).

Type	Sample size	Sample mean	Sample SD
Without side coating	10	80.95	9.59
With side coating	10	63.23	5.96

a. Calculate a 95% lower confidence bound for true average strength of joints with a side coating.
b. Calculate a 95% lower prediction bound for the strength of a single joint with a side coating.
c. Calculate an interval that, with 95% confidence, includes the strength values for at least 95% of the population of all joints with side coatings.
d. Calculate a 95% confidence interval for the difference between true average strengths for the two types of joints.

71. An experiment was carried out to compare various properties of cotton/polyester spun yarn finished with softener only and yarn finished with softener plus 5% DP-resin ("Properties of a Fabric Made with Tandem Spun Yarns," *Textile Res. J.*, 1996:

607–611). One particularly important characteristic of fabric is its durability, that is, its ability to resist wear. For a sample of 40 softener-only specimens, the sample mean stoll-flex abrasion resistance (cycles) in the filling direction of the yarn was 3975.0, with a sample standard deviation of 245.1. Another sample of 40 softener-plus specimens gave a sample mean and sample standard deviation of 2795.0 and 293.7, respectively. Calculate a confidence interval with confidence level 99% for the difference between true average abrasion resistances for the two types of fabrics. Does your interval provide convincing evidence that true average resistances differ for the two types of fabrics? Why or why not?

72. The derailment of a freight train due to the catastrophic failure of a traction motor armature bearing provided the impetus for a study reported in the article "Locomotive Traction Motor Armature Bearing Life Study" (*Lubrication Engr.,* Aug. 1997: 12–19). A sample of 17 high-mileage traction motors was selected and the amount of cone penetration (mm/10) was determined both for the pinion bearing and for the commutator armature bearing, resulting in the following data.

Motor	1	2	3	4	5	6
Commutator	211	273	305	258	270	209
Pinion	226	278	259	244	273	236

Motor	7	8	9	10	11	12
Commutator	223	288	296	233	262	291
Pinion	290	287	315	242	288	242

Motor	13	14	15	16	17	
Commutator	278	275	210	272	264	
Pinion	278	208	281	274	268	

Calculate an estimate of the population mean difference between penetration for the commutator armature bearing and penetration for the pinion bearing, and do so in a way that conveys information about the reliability and precision of the estimate. (*Note:* A normal probability plot validates the necessary normality assumption.) Would you say that the population mean difference has been precisely estimated? Does it look as though population mean penetration differs for the two types of bearings? Explain.

73. Headability is the ability of a cylindrical piece of material to be shaped into the head of a bolt, screw, or other cold formed part without cracking. The article "New Methods for Assessing Cold Heading Quality" (*Wire J. Intl.,* Oct. 1996: 66–72) described the result of a headability impact test applied to 30 specimens

of aluminum killed steel and 30 specimens of silicon killed steel. The sample mean headability rating number for the steel specimens was 6.43 and the sample mean for aluminum specimens was 7.09. Suppose that the sample standard deviations were 1.08 and 1.19, respectively. Do you agree with the article's authors that the difference in headability ratings is significant at the 5% level (assuming that the two headability distributions are normal)?

74. The article "Two Parameters Limiting the Sensitivity of Laboratory Tests of Condoms as Viral Barriers" (*J. of Testing and Eval.,* 1996: 279–286) reported that, in brand A condoms, among 16 tears produced by a puncturing needle, the sample mean tear length was 74.0 μm, whereas for the 14 brand B tears, the sample mean length was 61.0 μm (determined using light microscopy and scanning electron micrographs). Suppose the sample standard deviations are 14.8 and 12.5, respectively (consistent with the sample ranges given in the article). The authors commented that the thicker brand B condom displayed a smaller mean tear length than the thinner brand A condom. Is this difference in fact statistically significant? State the appropriate hypotheses and test at $\alpha = .05$.

75. Information about hand posture and forces generated by the fingers during manipulation of various daily objects is needed for designing high-tech hand prosthetic devices. The article "Grip Posture and Forces During Holding Cylindrical Objects with Circular Grips" (*Ergonomics,* 1996: 1163–1176) reported that for a sample of 11 females, the sample mean four-finger pinch strength (N) was 98.1 and the sample standard deviation was 14.2. For a sample of 15 males, the sample mean and sample standard deviation were 129.2 and 39.1, respectively.
   a. A test carried out to see whether true average strengths for the two genders were different resulted in $t = 2.51$ and P-value $= .019$. Does the appropriate test procedure described in this chapter yield this value of t and the stated P-value?
   b. Is there substantial evidence for concluding that true average strength for males exceeds that for females by more than 25 N? State and test the relevant hypotheses.

76. The article "Pine Needles as Sensors of Atmospheric Pollution" (*Environ. Monitoring,* 1982: 273–286) reported on the use of neutron-activity analysis to determine pollutant concentration in pine needles. According to the article's authors, "These observations strongly indicated that for those elements which are

determined well by the analytical procedures, the distribution of concentration is lognormal. Accordingly, in tests of significance the logarithms of concentrations will be used." The given data refers to bromine concentration in needles taken from a site near an oil-fired steam plant and from a relatively clean site. The summary values are means and standard deviations of the log-transformed observations.

Site	Sample size	Mean log concentration	SD of log concentration
Steam plant	8	18.0	4.9
Clean	9	11.0	4.6

Let μ_1^* be the true average *log* concentration at the first site, and define μ_2^* analogously for the second site.

a. Use the pooled t test (based on assuming normality *and* equal standard deviations) to decide at significance level .05 whether the two concentration distribution means are equal.

b. If σ_1^* and σ_2^*, the standard deviations of the two log concentration distributions, are not equal, would μ_1 and μ_2, the means of the concentration distributions, be the same if $\mu_1^* = \mu_2^*$? Explain your reasoning.

77. Long-term exposure of textile workers to cotton dust released during processing can result in substantial health problems, so textile researchers have been investigating methods that will result in reduced risks while preserving important fabric properties. The accompanying data on roving cohesion strength (kN · m/kg) for specimens produced at five different twist multiples is from the article "Heat Treatment of Cotton: Effect on Endotoxin Content, Fiber and Yarn Properties, and Processability" (*Textile Research J.*, 1996: 727–738).

Twist multiple	1.054	1.141	1.245	1.370	1.481
Control strength	.45	.60	.61	.73	.69
Heated strength	.51	.59	.63	.73	.74

The authors of the cited article stated that strength for heated specimens appeared to be slightly higher on average than for the control specimens. Is the difference statistically significant? State and test the relevant hypotheses using $\alpha = .05$ by calculating the P-value.

78. The accompanying summary data on the ratio of strength to cross-sectional area for knee extensors

is taken from the article "Knee Extensor and Knee Flexor Strength: Cross-Sectional Area Ratios in Young and Elderly Men" (*J. of Gerontology*, 1992: M204–M210).

Group	Sample size	Sample mean	Standard error
Young	13	7.47	.22
Elderly men	12	6.71	.28

Does this data suggest that the true average ratio for young men exceeds that for elderly men? Carry out a test of appropriate hypotheses using $\alpha = .05$. Be sure to state any assumptions necessary for your analysis.

79. The accompanying data on response time appeared in the article "The Extinguishment of Fires Using Low-Flow Water Hose Streams—Part II" (*Fire Technology*, 1991: 291–320).

Good visibility:
.43 1.17 .37 .47 .68 .58 .50 2.75
Poor visibility:
1.47 .80 1.58 1.53 4.33 4.23 3.25 3.22

The authors analyzed the data with the pooled t test. Does the use of this test appear justified? (*Hint:* Check for normality. The normal scores for $n = 8$ are $-1.53, -.89, -.49, -.15, .15, .49, .89,$ and 1.53.)

80. The article "The Relationship Between Distress and Delight in Males' and Females' Reactions to Frightening Films" (*Human Communication Research*, 1991: 625–637) reports that investigators measured emotional responses of 50 men and 60 women after viewing a segment from a horror film. The article included the following statement: "Females were much more likely to express distress than were males. Although males did express higher levels of delight than females, the difference was not statistically significant." The accompanying summary data was also contained in the article.

	Distress Index		Delight Index	
Gender	Mean	SD	Mean	SD
Men	31.2	10.0	12.02	3.65
Women	40.4	9.1	9.09	5.55
	P-value < .001		Not significant (P-value > .05)	

Give a brief discussion of the relevant hypotheses and conclusions.

81. An experimenter wishes to obtain a CI for the difference between true average breaking strength for cables manufactured by company I and by company II. Suppose breaking strength is normally distributed for both types of cable with $\sigma_1 = 30$ psi and $\sigma_2 = 20$ psi.
    a. If costs dictate that the sample size for the type I cable should be three times the sample size for the type II cable, how many observations are required if the 99% CI is to be no wider than 20 psi?
    b. Suppose a total of 400 observations is to be made. How many of the observations should be made on type I cable samples if the width of the resulting interval is to be a minimum?

82. An experiment to determine the effects of temperature on the survival of insect eggs was described in the article "Development Rates and a Temperature-Dependent Model of Pales Weevil" (*Environ. Entomology*, 1987: 956–962). At 11°C, 73 of 91 eggs survived to the next stage of development. At 30°C, 102 of 110 eggs survived. Do the results of this experiment suggest that the survival rate (proportion surviving) differs for the two temperatures? Calculate the *P*-value and use it to test the appropriate hypotheses.

83. The insulin-binding capacity (pmol/mg protein) was measured for a sample of diabetic rats treated with a low dose of insulin and another sample of diabetic rats treated with a high dose, yielding the following data:

    Low dose: $m = 8, \bar{x} = 1.98, s_1 = .51$

    High dose: $n = 12, \bar{y} = 1.30, s_2 = .35$

    (*J. Clinical Investigation*, 1978: 552–560.)
    a. Does the data indicate that there is any difference in true average insulin-binding capacity due to the dosage level? Use a test with $\alpha = .001$.
    b. What can be said about the *P*-value?

84. In the article referenced in Exercise 83, there were actually four experimental treatments: control nondiabetic, untreated diabetic, low-dose diabetic, and high-dose diabetic. Denote the sample size for the *i*th treatment by n_i and the sample variance by S_i^2 ($i = 1, 2, 3, 4$). Assuming that the true variance for each treatment is σ^2, construct a pooled estimator of σ^2 that is unbiased, and verify using rules of expected value that it is indeed unbiased. What is your estimate for the following actual data? *Hint:* See Exercise 34.

	Treatment			
	1	**2**	**3**	**4**
Sample size	16	18	8	12
Sample SD	.64	.81	.51	.35

85. Suppose a level .05 test of H_0: $\mu_1 - \mu_2 = 0$ versus H_a: $\mu_1 - \mu_2 > 0$ is to be performed, assuming $\sigma_1 = \sigma_2 = 10$ and normality of both distributions, using equal sample sizes ($m = n$). Evaluate the probability of a type II error when $\mu_1 - \mu_2 = 1$ and $n = 25, 100, 2500,$ and $10,000$. Can you think of real problems in which the difference $\mu_1 - \mu_2 = 1$ has little practical significance? Would sample sizes of $n = 10,000$ be desirable in such problems?

86. The following data refers to airborne bacteria count (number of colonies/ft³) both for $m = 8$ carpeted hospital rooms and for $n = 8$ uncarpeted rooms ("Microbial Air Sampling in a Carpeted Hospital," *J. Environmental Health*, 1968: 405). Does there appear to be a difference in true average bacteria count between carpeted and uncarpeted rooms?

    Carpeted: 11.8 8.2 7.1 13.0 10.8 10.1 14.6 14.0

    Uncarpeted: 12.1 8.3 3.8   7.2 12.0 11.1 10.1 13.7

    Suppose you later learned that all carpeted rooms were in a veterans' hospital, whereas all uncarpeted rooms were in a children's hospital. Would you be able to assess the effect of carpeting? Comment.

87. In a study of the relationship between nightmare frequency and gender, each of 160 men and 192 women was asked whether nightmares occurred often (at least one per month) or seldom, with the following results:

    Men: $m = 160, x = 55$ (often = success)

    Women: $n = 192, y = 60$

    Does the data indicate at level .05 that there is any difference in the true proportions of men and women who often have nightmares? ("Personality Characteristics of Nightmare Sufferers," *J. Nervous and Mental Diseases*, 1971: 29–31.)

88. McNemar's test, developed in Exercise 54, can also be used when individuals are paired (matched) to yield n pairs and then one member of each pair is given treatment 1 and the other is given treatment 2. Then X_1 is the number of pairs in which both treat-

ments were successful, and similarly for X_2, X_3, and X_4. The test statistic for testing equal efficacy of the two treatments is given by $(X_2 - X_3)/\sqrt{(X_2 + X_3)}$, which has approximately a standard normal distribution when H_0 is true. Use this to test whether the drug ergotamine is effective in the treatment of migraine headaches.

Ergotamine

		S	F
Placebo	S	44	34
	F	46	30

The data is fictitious, but the conclusion agrees with that in the article "Controlled Clinical Trial of Ergotamine Tartrate" (*British Med. J.,* 1970: 325–327).

89. The article "Evaluating Variability in Filling Operations" (*Food Tech.,* 1984: 51–55) describes two different filling operations used in a ground-beef packing plant. Both filling operations were set to fill packages with 1400 g of ground beef. In a random sample of size 30 taken from each filling operation, the resulting means and standard deviations were 1402.24 g and 10.97 g for operation 1 and 1419.63 g and 9.96 g for operation 2.
   a. Using a .05 significance level, is there sufficient evidence to indicate that the true mean weight of the packages differs for the two operations?
   b. Does the data from operation 1 suggest that the true mean weight of packages produced by op-

eration 1 is higher than 1400 g? Use a .05 significance level.

90. Let $X_1, \ldots, X_m$ be a random sample from a Poisson distribution with parameter λ_1, and let $Y_1, \ldots, Y_n$ be a random sample from another Poisson distribution with parameter λ_2. We wish to test $H_0: \lambda_1 - \lambda_2 = 0$ against one of the three standard alternatives. Since $\mu = \lambda$ for a Poisson distribution, when m and n are large the large-sample z test of Section 9.1 can be used. However, the fact that $V(\overline{X}) = \lambda/n$ suggests that a different denominator should be used in standardizing $\overline{X} - \overline{Y}$. Develop a large-sample test procedure appropriate to this problem, and then apply it to the following data to test whether the plant densities for a particular species are equal in two different regions (where each observation is the number of plants found in a randomly located square sampling quadrat having area 1m², so for region 1, there were 40 quadrats in which one plant was observed, etc.).

Frequency

	0	1	2	3	4	5	6	7	
Region 1	28	40	28	17	8	2	1	1	$m = 125$
Region 2	14	25	30	18	49	2	1	1	$n = 140$

91. Referring to Exercise 90, develop a large-sample confidence interval formula for $\lambda_1 - \lambda_2$. Calculate the interval for the + data given there using a confidence level of 95%.

| **Bibliography**

See the bibliography at the end of Chapter 7.

10

The Analysis
of Variance

Introduction

In studying methods for the analysis of quantitative data, we first focused on problems involving a single sample of numbers and then turned to a comparative analysis of two different such samples. In one-sample problems, the data consisted of observations on or responses from individuals or experimental objects randomly selected from a single population. In two-sample problems, either the two samples were drawn from two different populations and the parameters of interest were the population means, or else two different treatments were applied to experimental units (individuals or objects) selected from a single population; in this latter case, the parameters of interest were referred to as true treatment means.

The **analysis of variance,** or more briefly **ANOVA,** refers broadly to a collection of experimental situations and statistical procedures for the analysis of quantitative responses from experimental units. The simplest ANOVA problem is referred to variously as a **single-factor, single-classification,** or **one-way ANOVA** and involves the analysis either of data sampled from more than two numerical populations (distributions) or of data from experiments in which more than two treatments have been used. The characteristic that differentiates the treatments or populations from one another is called the **factor** under study, and the different treatments or populations are referred to as the **levels** of the factor. Examples of such situations include the following:

1. An experiment to study the effects of five different brands of gasoline on automobile engine operating efficiency (mpg)
2. An experiment to study the effects of the presence of four different sugar solutions (glucose, sucrose, fructose, and a mixture of the three) on bacterial growth
3. An experiment to investigate whether hardwood concentration in pulp (%) has an effect on tensile strength of bags made from the pulp
4. An experiment to decide whether the color density of fabric specimens depends on the amount of dye used

In (1) the factor of interest is gasoline brand, and there are five different levels of the factor. In (2) the factor is sugar, with four levels (or five, if a control solution containing no sugar is used). In both (1) and (2), the factor is qualitative in nature, and the levels correspond to possible categories of the factor. In (3) and (4), the factors are concentration of hardwood and amount of dye, respectively; both these factors are quantitative in nature, so the levels identify different settings of the factor. When the factor of interest is quantitative, statistical techniques from regression analysis (discussed in Chapters 12 and 13) can also be used to analyze the data.

In this chapter, we focus on single-factor ANOVA. Section 10.1 presents the F test for testing the null hypothesis that the population or treatment means are identical. Section 10.2 considers further analysis of the data when H_0 has been rejected. Section 10.3 covers some other aspects of single-factor ANOVA. The next chapter introduces ANOVA experiments involving more than a single factor.

10.1 | Single-Factor ANOVA

Single-factor ANOVA focuses on a comparison of more than two population or treatment means. Let

I = the number of populations or treatments being compared

μ_1 = the mean of population 1 or the true average response when treatment 1 is applied

$\vdots$

μ_I = the mean of population I or the true average response when treatment I is applied

Then the hypotheses of interest are

$$H_0: \quad \mu_1 = \mu_2 = \cdots = \mu_I$$

versus

$$H_a: \quad \text{at least two of the } \mu_i\text{'s are different}$$

If $I = 4$, H_0 is true only if all four μ_i's are identical. H_a would be true, for example, if $\mu_1 = \mu_2 \neq \mu_3 = \mu_4$, if $\mu_1 = \mu_3 = \mu_4 \neq \mu_2$, or if all four μ_i's differ from one another.
   A test of these hypotheses requires that we have available a random sample from each population or treatment.

Example 10.1   The article "Compression of Single-Wall Corrugated Shipping Containers Using Fixed and Floating Test Platens" (*J. Testing and Evaluation,* 1992: 318–320) describes an experiment in which several different types of boxes were compared with respect to compression strength (lb). Table 10.1 presents the results of a single-factor ANOVA experiment involving $I = 4$ types of boxes (the sample means and standard deviations are in good agreement with values given in the article).

Table 10.1   The data and summary quantities for Example 10.1

Type of box	Compression strength (lb)	Sample mean	Sample SD
1	655.5 788.3 734.3 721.4 679.1 699.4	713.00	46.55
2	789.2 772.5 786.9 686.1 732.1 774.8	756.93	40.34
3	737.1 639.0 696.3 671.7 717.2 727.1	698.07	37.20
4	535.1 628.7 542.4 559.0 586.9 520.0	562.02	39.87
	Grand mean =	682.50	

With μ_i denoting the true average compression strength for boxes of type i ($i = 1, 2, 3, 4$), the null hypothesis is $H_0: \mu_1 = \mu_2 = \mu_3 = \mu_4$. Figure 10.1(a) shows a comparative boxplot for the four samples. There is a substantial amount of overlap among observations on the first three types of boxes, but compression strengths for the fourth type appear considerably smaller than for the other types. This suggests that H_0 is not true. The comparative boxplot in Figure 10.1(b) is based on adding 120 to each observation in the fourth sample (giving mean 682.02 and the same standard deviation) and leaving the other observations unaltered. It is no longer obvious whether H_0 is true or false. In situations such as this, we need a formal test procedure. ■

Notation and Assumptions

In two-sample problems, we used the letters X and Y to differentiate the observations in one sample from those in the other. Because this is cumbersome for three or more samples, it is customary to use a single letter with two subscripts. The first subscript identifies the sample number, corresponding to the population or treatment being

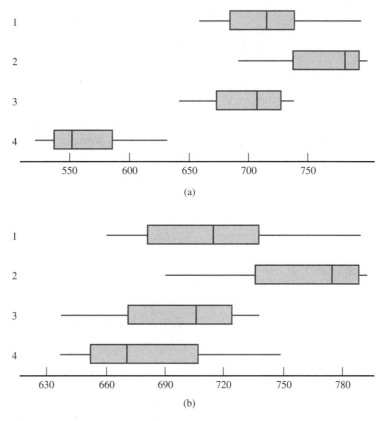

Figure 10.1   Boxplots for Example 10.1: (a) original data; (b) altered data

sampled, and the second subscript denotes the position of the observation within that sample. Let

$X_{i,j}$ = the random variable (rv) that denotes the jth measurement taken from the ith population, or the measurement taken on the jth experimental unit that receives the ith treatment

$x_{i,j}$ = the observed value of $X_{i,j}$ when the experiment is performed

The observed data is usually displayed in a rectangular table, such as Table 10.1. There samples from the different populations appear in different rows of the table, and $x_{i,j}$ is the jth number in the ith row. For example, $x_{2,3} = 786.9$ (the third observation from the second population), and $x_{4,1} = 535.1$. When there is no ambiguity, we will write x_{ij} rather than $x_{i,j}$ (e.g., if there were 15 observations on each of 12 treatments, x_{112} could mean $x_{1,12}$ or $x_{11,2}$). It is assumed that the X_{ij}'s within any particular sample are independent—a random sample from the ith population or treatment distribution—and that different samples are independent of one another.

In some experiments, different samples contain different numbers of observations. However, the concepts and methods of single-factor ANOVA are most easily developed for the case of equal sample sizes. Restricting ourselves for the moment to this case, let J denote the number of observations in each sample ($J = 6$ in Example 10.1). The data set consists of IJ observations. The individual sample means will be denoted by $\overline{X}_{1.}, \overline{X}_{2.}, \ldots, \overline{X}_{I.}$. That is,

$$\overline{X}_{i.} = \frac{\sum\limits_{j=1}^{J} X_{ij}}{J} \qquad i = 1, 2, \ldots, I$$

The dot in place of the second subscript signifies that we have added over all values of that subscript while holding the other subscript value fixed, and the horizontal bar indicates division by J to obtain an average. Similarly, the average of all IJ observations, called the **grand mean,** is

$$\overline{X}_{..} = \frac{\sum\limits_{i=1}^{I} \sum\limits_{j=1}^{J} X_{ij}}{IJ}$$

For the strength data in Table 10.1, $\bar{x}_{1.} = 713.00$, $\bar{x}_{2.} = 756.93$, $\bar{x}_{3.} = 698.07$, $\bar{x}_{4.} = 562.02$, and $\bar{x}_{..} = 682.50$. Additionally, let $S_1^2, S_2^2, \ldots, S_I^2$ represent the sample variances:

$$S_i^2 = \frac{\sum\limits_{j=1}^{J} (X_{ij} - \overline{X}_{i.})^2}{J - 1} \qquad i = 1, 2, \ldots, I$$

From Example 10.1, $s_1 = 46.55$, $s_1^2 = 2166.90$, and so on.

ASSUMPTIONS	The I population or treatment distributions are all normal with the same variance σ^2. That is, each X_{ij} is normally distributed with $$E(X_{ij}) = \mu_i \qquad V(X_{ij}) = \sigma^2$$

The I sample standard deviations will generally differ somewhat even when the corresponding σ's are identical. In Example 10.1, the largest among s_1, s_2, s_3, and s_4 is about 1.25 times the smallest. A rough rule of thumb is that if the largest s is not much more than two times the smallest, it is reasonable to assume equal σ^2's.

In previous chapters, a normal probability plot was suggested for checking normality. The individual sample sizes in ANOVA are typically too small for I separate plots to be informative. A single plot can be constructed by subtracting $\bar{x}_{1.}$ from each observation in the first sample, $\bar{x}_{2.}$ from each observation in the second, and so on, and then plotting these IJ deviations against the z percentiles. Figure 10.2 gives such a plot for the data of Example 10.1. The straightness of the pattern gives strong support to the normality assumption.

If either the normality assumption or the assumption of equal variances is judged implausible, a method of analysis other than the usual F test must be employed. Please seek expert advice in such situations (one possibility, a data transformation, is suggested in Section 10.3).

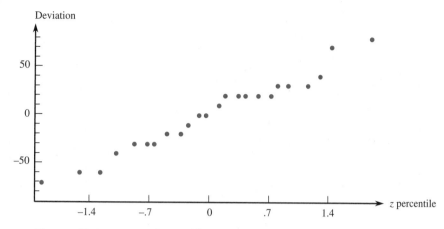

Figure 10.2   A normal probability plot based on the data of Example 10.1

The Test Statistic

If H_0 is true, the J observations in each sample come from a normal population distribution with the *same* mean value μ, in which case the sample means $\bar{x}_1., \ldots, \bar{x}_I.$ should be reasonably close. The test procedure is based on comparing a measure of differences among the $\bar{x}_i.$'s ("between-samples" variation) to a measure of variation calculated from *within* each of the samples.

DEFINITION

Mean square for treatments is given by

$$\text{MSTr} = \frac{J}{I-1}[(\bar{X}_1. - \bar{X}..)^2 + (\bar{X}_2. - \bar{X}..)^2 + \cdots + (\bar{X}_I. - \bar{X}..)^2]$$

$$= \frac{J}{I-1}\sum_i (\bar{X}_i. - \bar{X}..)^2$$

and **mean square for error** is

$$\text{MSE} = \frac{S_1^2 + S_2^2 + \cdots + S_I^2}{I}$$

The test statistic for single-factor ANOVA is $F = \text{MSTr}/\text{MSE}$.

The terminology "mean square" will be explained shortly. Notice that uppercase $\bar{X}$'s and S^2's are used, so MSTr and MSE are defined as statistics. We will follow tradition and also use MSTr and MSE (rather than mstr and mse) to denote the calculated values of these statistics. Each S_i^2 assesses variation within a particular sample, so MSE is a measure of within-samples variation.

What kind of value of F provides evidence for or against H_0? Notice first that if H_0 is true (all μ_i's are equal), the values of the individual sample means should be close to one another and therefore close to the grand mean, resulting in a relatively small value of MSTr. However, if the μ_i's are quite different, some $\bar{x}_i$'s should differ quite a bit from $\bar{x}..$. So the value of MSTr is affected by the status of H_0 (true or false). This is not the case with MSE, because the s_i^2's depend only on the underlying value of σ^2 and not on where the various distributions are centered. The following box presents an important property of $E(\text{MSTr})$ and $E(\text{MSE})$, the expected values of these two statistics.

PROPOSITION

> When H_0 is true,
>
> $$E(\text{MSTr}) = E(\text{MSE}) = \sigma^2$$
>
> whereas when H_0 is false,
>
> $$E(\text{MSTr}) > E(\text{MSE}) = \sigma^2$$
>
> That is, both statistics are unbiased for estimating the common population variance σ^2 when H_0 is true, but MSTr tends to overestimate σ^2 when H_0 is false.

The unbiasedness of MSE is a consequence of $E(S_i^2) = \sigma^2$ whether H_0 is true or false. When H_0 is true, each $\bar{X}_i$ has the same mean value μ and variance σ^2/J, so $\sum(\bar{X}_i - \bar{X}..)^2/(I-1)$, the "sample variance" of the $\bar{X}_i$'s, estimates σ^2/J unbiasedly; multiplying this by J gives MSTr as an unbiased estimator of σ^2 itself. The $\bar{X}_i$'s tend to spread out more when H_0 is false than when it is true, tending to inflate the value of MSTr in this case. Thus, a value of F that greatly exceeds 1, corresponding to MSTr much larger than MSE, casts considerable doubt on H_0. The form of the rejection region is $f \geq c$. The cutoff c should be chosen to give $P(F \geq c$ when H_0 is true$) = \alpha$, the desired significance level. We therefore need the distribution of F when H_0 is true.

F Distributions and the F Test

In Chapter 9, we introduced a family of probability distributions called F distributions. An F distribution arises in connection with a ratio in which there is one number of degrees of freedom (df) associated with the numerator and another number of degrees of freedom associated with the denominator. Let ν_1 and ν_2 denote the number of numerator and denominator degrees of freedom, respectively, for a variable with an F distribution. Both ν_1 and ν_2 are positive integers. Figure 10.3 pictures an F density curve and the corresponding upper-tail critical value F_{α,ν_1,ν_2}. Appendix Table A.9 gives these critical values for $\alpha = .10, .05, .01$, and $.001$. Values of ν_1 are identified with different columns of the table, and the rows are labeled with various values of ν_2. For example, the F critical value that captures upper-tail area .05 under the F curve with $\nu_1 = 4$ and $\nu_2 = 6$ is $F_{.05,4,6} = 4.53$, whereas $F_{.05,6,4} = 6.16$. Section 9.5 discussed how these F critical values could be used to obtain an upper and/or lower bound on the P-value of an F test. The standard statistical computer packages will include a P-value on single-factor ANOVA output. The key theoretical result is that the test statistic F has an F distribution when H_0 is true.

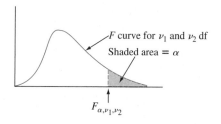

Figure 10.3   An F curve and critical value F_{α, ν_1, ν_2}

THEOREM

Let $F = \text{MSTr}/\text{MSE}$ be the test statistic in a single-factor ANOVA problem involving I populations or treatments with a random sample of J observations from each one. When H_0 is true and the basic assumptions of this section are satisfied, F has an F distribution with $\nu_1 = I - 1$ and $\nu_2 = I(J - 1)$. With f denoting the computed value of F, the rejection region $f \geq F_{\alpha, I-1, I(J-1)}$ then specifies a test with significance level α.

The rationale for $\nu_1 = I - 1$ is that, although MSTr is based on the I deviations $\overline{X}_{1.} - \overline{X}_{..}, \ldots, \overline{X}_{I.} - \overline{X}_{..}, \Sigma(\overline{X}_{i.} - \overline{X}_{..}) = 0$, so only $I - 1$ of these are freely determined. Because each sample contributes $J - 1$ df to MSE and these samples are independent, $\nu_2 = (J - 1) + \cdots + (J - 1) = I(J - 1)$.

Example 10.2
(Example 10.1
continued)

The values of I and J for the strength data are 4 and 6, respectively, so numerator df $= I - 1 = 3$ and denominator df $= I(J - 1) = 20$. At significance level .05, H_0: $\mu_1 = \mu_2 = \mu_3 = \mu_4$ will be rejected in favor of the conclusion that at least two μ_i's are different if $f \geq F_{.05,3,20} = 3.10$. The grand mean is $\bar{x}_{..} = \Sigma\Sigma x_{ij}/(IJ) = 682.50$

$$\text{MSTr} = \frac{6}{4 - 1}[(713.00 - 682.50)^2 + (756.93 - 682.50)^2$$

$$+ (698.07 - 682.50)^2 + (562.02 - 682.50)^2] = 42{,}455.86$$

$$\text{MSE} = \frac{1}{4}[(46.55)^2 + (40.34)^2 + (37.20)^2 + (39.87)^2] = 1691.92$$

$$f = \text{MSTr}/\text{MSE} = 42{,}455.86/1691.92 = 25.09$$

Since $25.09 \geq 3.10$, H_0 is resoundingly rejected at significance level .05. True average compression strength does appear to depend on box type. In fact, P-value $=$ area under F curve to right of $25.09 = .000$, so H_0 would be rejected at any reasonable significance level. ∎

Computational Formulas for ANOVA

The calculations leading to f can be done efficiently by using formulas similar to the computing formula for a single sample variance s^2. Let $x_{i.}$ represent the *sum* (not the average, since there is no bar) of the x_{ij}'s for i fixed (sum of the numbers in the ith row of the table) and $x_{..}$ denote the sum of *all* the x_{ij}'s (the **grand total**).

DEFINITION

The **total sum of squares (SST), treatment sum of squares (SSTr),** and **error sum of squares (SSE)** are given by

$$\text{SST} = \sum_{i=1}^{I} \sum_{j=1}^{J} (x_{ij} - \bar{x}_{..})^2 = \sum_{i=1}^{I} \sum_{j=1}^{J} x_{ij}^2 - \frac{1}{IJ} x_{..}^2$$

$$\text{SSTr} = \sum_{i=1}^{I} \sum_{j=1}^{J} (\bar{x}_{i.} - \bar{x}_{..})^2 = \frac{1}{J} \sum_{i=1}^{I} x_{i.}^2 - \frac{1}{IJ} x_{..}^2$$

$$\text{SSE} = \sum_{i=1}^{I} \sum_{j=1}^{J} (x_{ij} - \bar{x}_{i.})^2 \qquad \text{where } x_{i.} = \sum_{j=1}^{J} x_{ij} \quad x_{..} = \sum_{i=1}^{I} \sum_{j=1}^{J} x_{ij}$$

The sum of squares SSTr appears in the numerator of F, and SSE appears in the denominator of F; the reason for defining SST will be apparent shortly.

The expressions on the far right-hand side of SST and SSTr are the computing formulas for these sums of squares. Both SST and SSTr involve $x_{..}^2/(IJ)$ (the square of the grand total divided by IJ), which is usually called the **correction factor for the mean** (CF). After the correction factor is computed, SST is obtained by squaring each number in the data table, adding these squares together, and subtracting the correction factor. SSTr results from squaring each row total, summing them, dividing by J, and subtracting the correction factor.

A computing formula for SSE is a consequence of a simple relationship among the three sums of squares.

Fundamental Identity

$$\text{SST} = \text{SSTr} + \text{SSE} \tag{10.1}$$

Thus, if any two of the sums of squares are computed, the third can be obtained through (10.1); SST and SSTr are easiest to compute, and then SSE = SST − SSTr. The proof follows from squaring both sides of the relationship

$$x_{ij} - \bar{x}_{..} = (x_{ij} - \bar{x}_{i.}) + (\bar{x}_{i.} - \bar{x}_{..}) \tag{10.2}$$

and summing over all i and j. This gives SST on the left, and SSTr and SSE as the two extreme terms on the right. The cross-product term is easily seen to be zero.

The interpretation of the fundamental identity is an important aid to an understanding of ANOVA. SST is a measure of the total variation in the data—the sum of all squared deviations about the grand mean. The identity says that this total variation can be partitioned into two pieces. SSE measures variation that would be present (within rows) even if H_0 were true and is thus the part of total variation that is *unexplained* by the truth or falsity of H_0. SSTr is the amount of variation (between rows) that *can be explained* by possible differences in the μ_i's. If explained variation is large relative to unexplained variation, then H_0 is rejected in favor of H_a.

Once SSTr and SSE are computed, each is divided by its associated df to obtain a mean square (mean in the sense of average). Then F is the ratio of the two mean squares.

$$\text{MSTr} = \frac{\text{SSTr}}{I - 1} \qquad \text{MSE} = \frac{\text{SSE}}{I(J - 1)} \qquad F = \frac{\text{MSTr}}{\text{MSE}} \qquad (10.3)$$

The computations are often summarized in a tabular format, called an **ANOVA table**, as displayed in Table 10.2.

Table 10.2   An ANOVA table

Source of Variation	df	Sum of Squares	Mean Square	f
Treatments	$I - 1$	SSTr	MSTr = SSTr/$(I - 1)$	MSTr/MSE
Error	$I(J - 1)$	SSE	MSE = SSE/$[I(J - 1)]$	
Total	$IJ - 1$	SST		

Example 10.3   The accompanying data resulted from an experiment comparing the degree of soiling for fabric copolymerized with three different mixtures of methacrylic acid (similar data appeared in the article "Chemical Factors Affecting Soiling and Soil Release from Cotton DP Fabric," *American Dyestuff Reporter,* 1983: 25–30).

						$x_{i\cdot}$	$\bar{x}_{i\cdot}$
Mixture 1	.56	1.12	.90	1.07	.94	4.59	.918
Mixture 2	.72	.69	.87	.78	.91	3.97	.794
Mixture 3	.62	1.08	1.07	.99	.93	4.69	.938

$$x_{\cdot\cdot} = 13.25$$

Let μ_i denote the true average degree of soiling when mixture i is used ($i = 1, 2, 3$). The null hypothesis $H_0: \mu_1 = \mu_2 = \mu_3$ states that the true average degree of soiling is identical for the three mixtures. We will carry out a test at significance level .01 to see whether H_0 should be rejected in favor of the assertion that true average degree of soiling is not the same for all mixtures. Since $I - 1 = 2$ and $I(J - 1) = 12$, the F critical value for the rejection region is $F_{.01,2,12} = 6.93$. Squaring each of the 15 observations and summing gives $\sum\sum x_{ij}^2 = (.56)^2 + (1.12)^2 + \cdots + (.93)^2 = 12.1351$. The values of the three sums of squares are

$$SST = 12.1351 - (13.25)^2/15 = 12.1351 - 11.7042 = .4309$$

$$SSTr = \frac{1}{5}[(4.59)^2 + (3.97)^2 + (4.69)^2] - 11.7042$$

$$= 11.7650 - 11.7042 = .0608$$

$$SSE = .4309 - .0608 = .3701$$

The remaining computations are summarized in the accompanying ANOVA table. Because $f = .99$ is not at least $F_{.01,2,12} = 6.93$, H_0 is not rejected at significance level .01. The mixtures appear to be indistinguishable with respect to degree of soiling ($F_{.10,2,12} = 2.81 \Rightarrow P\text{-value} > .10$).

Source of Variation	df	Sum of Squares	Mean Square	f
Treatments	2	.0608	.0304	.99
Error	12	.3701	.0308	
Total	14	.4309		

When the F test causes H_0 to be rejected, the experimenter will often be interested in further analysis to decide which μ_i's differ from which others. Procedures for doing this are called multiple comparison procedures, and several are described in the next two sections.

Exercises | Section 10.1 (1–10)

1. In an experiment to compare the tensile strengths of $I = 5$ different types of copper wire, $J = 4$ samples of each type were used. The between-samples and within-samples estimates of σ^2 were computed as $MSTr = 2573.3$ and $MSE = 1394.2$, respectively. Use the F test at level .05 to test $H_0: \mu_1 = \mu_2 = \mu_3 = \mu_4 = \mu_5$ versus H_a: at least two μ_i's are unequal.

2. Suppose that the compression strength observations on the fourth type of box in Example 10.1 had been 655.1, 748.7, 662.4, 679.0, 706.9, and 640.0 (obtained by adding 120 to each previous x_{4j}). Assuming no change in the remaining observations, carry out an F test with $\alpha = .05$.

3. The lumen output was determined for each of $I = 3$ different brands of 60-watt soft-white lightbulbs,

with $J = 8$ bulbs of each brand tested. The sums of squares were computed as $SSE = 4773.3$ and $SSTr = 591.2$. State the hypotheses of interest (including word definitions of parameters), and use the F test of ANOVA ($\alpha = .05$) to decide whether there are any differences in true average lumen outputs among the three brands for this type of bulb by obtaining as much information as possible about the P-value.

4. In a study to assess the effects of malaria infection on mosquito hosts ("Plasmodium Cynomolgi: Effects of Malaria Infection on Laboratory Flight Performance of Anopheles Stephensi Mosquitos," *Experimental Parasitology*, 1977: 397–404), mosquitos were fed on either infective or noninfective rhesus monkeys. Subsequently the distance they flew during a 24-hour period was measured using a flight mill. The mosquitos were divided into four groups of eight mosquitos each: infective rhesus and sporozites present (IRS), infective rhesus and oocysts present (IRD), infective rhesus and no infection developed (IRN), and noninfective (C). The summary data values are $\bar{x}_{1.} = 4.39$ (IRS), $\bar{x}_{2.} = 4.52$ (IRD), $\bar{x}_{3.} = 5.49$ (IRN), $\bar{x}_{4.} = 6.36$ (C), $\bar{x}_{..} = 5.19$, and $\sum\sum x_{ij}^2 = 911.91$. Use the ANOVA F test at level .05 to decide whether there are any differences between true average flight times for the four treatments.

5. Consider the following summary data on the modulus of elasticity ($\times 10^6$ psi) for lumber of three different grades (in close agreement with values in the article "Bending Strength and Stiffness of Second-Growth Douglas-Fir Dimension Lumber" (*Forest Products J.,* 1991: 35–43), except that the sample sizes there were larger).

Grade	J	$\bar{x}_{i.}$	s_i
1	10	1.63	.27
2	10	1.56	.24
3	10	1.42	.26

Use this data and a significance level of .01 to test the null hypothesis of no difference in mean modulus of elasticity for the three grades.

6. The article "Origin of Precambrian Iron Formations" (*Econ. Geology,* 1964: 1025-1057) reports the following data on total Fe for four types of iron formation (1 = carbonate, 2 = silicate, 3 = magnetite, 4 = hematite).

1: 20.5 28.1 27.8 27.0 28.0
25.2 25.3 27.1 20.5 31.3

2: 26.3 24.0 26.2 20.2 23.7
34.0 17.1 26.8 23.7 24.9

3: 29.5 34.0 27.5 29.4 27.9
26.2 29.9 29.5 30.0 35.6

4: 36.5 44.2 34.1 30.3 31.4
33.1 34.1 32.9 36.3 25.5

Carry out an analysis of variance F test at significance level .01, and summarize the results in an ANOVA table.

7. In an experiment to investigate the performance of four different brands of spark plugs intended for use on a 125-cc two-stroke motorcycle, five plugs

of each brand were tested and the number of miles (at a constant speed) until failure was observed. The partial ANOVA table for the data is given here. Fill in the missing entries, state the relevant hypotheses, and carry out a test by obtaining as much information as you can about the P-value.

Source	df	Sum of Squares	Mean Square	f
Brand				
Error			14,713.69	
Total		310,500.76		

8. A study of the properties of metal plate-connected trusses used for roof support ("Modeling Joints Made with Light-Gauge Metal Connector Plates," *Forest Products J.,* 1979: 39-44) yielded the following observations on axial stiffness index (kips/in.) for plate lengths 4, 6, 8, 10, and 12 in.

4: 309.2 409.5 311.0 326.5 316.8
349.8 309.7

6: 402.1 347.2 361.0 404.5 331.0
348.9 381.7

8: 392.4 366.2 351.0 357.1 409.9
367.3 382.0

10: 346.7 452.9 461.4 433.1 410.6
384.2 362.6

12: 407.4 441.8 419.9 410.7 473.4
441.2 465.8

Does variation in plate length have any effect on true average axial stiffness? State and test the relevant hypotheses using analysis of variance with $\alpha = .01$. Display your results in an ANOVA table. (*Hint:* $\sum\sum x_{ij}^2 = 5,241,420.79$.)

9. Six samples of each of four types of cereal grain grown in a certain region were analyzed to determine thiamin content, resulting in the following data (μg/g):

Wheat	5.2	4.5	6.0	6.1	6.7	5.8
Barley	6.5	8.0	6.1	7.5	5.9	5.6
Maize	5.8	4.7	6.4	4.9	6.0	5.2
Oats	8.3	6.1	7.8	7.0	5.5	7.2

Does this data suggest that at least two of the grains differ with respect to true average thiamin content? Use a level $\alpha = .05$ test based on the P-value method.

10. In single-factor ANOVA with I treatments and J observations per treatment, let $\mu = (1/I)\sum\mu_i$.

  a. Express $E(\overline{X}_{..})$ in terms of μ. [*Hint:* $\overline{X}_{..} = (1/I)\sum\overline{X}_{i\cdot}$.]

  b. Compute $E(\overline{X}_{i\cdot}^2)$. [*Hint:* For any rv Y, $E(Y^2) = V(Y) + [E(Y)]^2$.]

  c. Compute $E(\overline{X}_{..}^2)$.

  d. Compute $E(\text{SSTr})$ and then show that

$$E(\text{MSTr}) = \sigma^2 + \frac{J}{I-1}\sum(\mu_i - \mu)^2$$

  e. Using the result of part (d), what is $E(\text{MSTr})$ when H_0 is true? When H_0 is false, how does $E(\text{MSTr})$ compare to σ^2?

10.2 | Multiple Comparisons in ANOVA

When the computed value of the F statistic in single-factor ANOVA is not significant, the analysis is terminated because no differences among the μ_i's have been identified. But when H_0 is rejected, the investigator will usually want to know which of the μ_i's are different from one another. A method for carrying out this further analysis is called a **multiple comparisons procedure.** There are a number of such procedures in the statistics literature. Here we present one that many statisticians recommend for deciding for each i and j whether it is plausible that $\mu_i = \mu_j$.

Tukey's Procedure (the *T* Method)

Tukey's procedure involves the use of another probability distribution called the **Studentized range distribution.** The distribution depends on two parameters: a numerator df m and a denominator df ν. Let $Q_{\alpha,m,\nu}$ denote the upper-tail α critical value of the Studentized range distribution with m numerator df and ν denominator df (analogous to F_{α,ν_1,ν_2}). Values of $Q_{\alpha,m,\nu}$ are given in Appendix Table A.10. Then $Q_{\alpha,I,I(J-1)}$ can be used to obtain simultaneous confidence intervals for all pairwise differences $\mu_i - \mu_j$. Notice that numerator df for the appropriate Q critical value is I, the number of treatments or populations, and not $I - 1$, as it was for F.

PROPOSITION

With probability $1 - \alpha$,

$$\overline{X}_{i\cdot} - \overline{X}_{j\cdot} - Q_{\alpha,I,I(J-1)}\sqrt{\text{MSE}/J} \le \mu_i - \mu_j$$
$$\le \overline{X}_{i\cdot} - \overline{X}_{j\cdot} + Q_{\alpha,I,I(J-1)}\sqrt{\text{MSE}/J} \quad (10.4)$$

for *every* i and j ($i = 1, \ldots, I$ and $j = 1, \ldots, I$) with $i \ne j$.

When the computed $\overline{x}_{i\cdot}$, $\overline{x}_{j\cdot}$, and MSE are substituted into (10.4), the result is a collection of **simultaneous confidence statements** about the true values of all differences $\mu_i - \mu_j$ between true treatment means. *Each interval from (10.4) that does not include zero yields the conclusion that μ_i and μ_j differ significantly at level α.*

The computation of all intervals in (10.4) looks complicated, but a straightforward sequence of steps culminates in a pictorial summary of the conclusions:

The *T* Method for Identifying Significantly Different μ_i's

1. Select α and find $Q_{\alpha,I,I(J-1)}$ from Appendix Table A.10.
2. Determine $w = Q_{\alpha,I,I(J-1)} \cdot \sqrt{MSE/J}$.
3. List the sample means in increasing order and underline those pairs that differ by less than w. Any pair of sample means not underscored by the same line corresponds to a pair of population or treatment means that are judged significantly different.

It is easily verified that a particular interval of the form (10.4) will contain zero if the corresponding pair of sample means is underscored by the same line.

Example 10.4 An experiment was carried out to compare five different brands of automobile oil filters with respect to their ability to capture foreign material. Let μ_i denote the true average amount of material captured by brand i filters ($i = 1, \ldots, 5$) under controlled conditions. A sample of nine filters of each brand was used, resulting in the following sample mean amounts: $\bar{x}_{1.} = 14.5, \bar{x}_{2.} = 13.8, \bar{x}_{3.} = 13.3, \bar{x}_{4.} = 14.3$, and $\bar{x}_{5.} = 13.1$. Table 10.3 is the ANOVA table summarizing the first part of the analysis.

Table 10.3 ANOVA table for Example 10.4

Source of Variation	df	Sum of Squares	Mean Square	f
Treatments (brands)	4	13.32	3.33	37.84
Error	40	3.53	.088	
Total	44	16.85		

Since $F_{.05,4,40} = 2.61$, H_0 is rejected (decisively) at level .05. We now use Tukey's procedure to look for significant differences among the μ_i's. From Appendix Table A.10, $Q_{.05,5,40} = 4.04$ (the second subscript on Q is I and not $I - 1$ as in F), so $w = 4.04\sqrt{.088/9} = .4$. The five sample means are arranged in increasing order, and every pair differing by less than .4 is underscored:

$$\begin{array}{ccccc} \bar{x}_{5.} & \bar{x}_{3.} & \bar{x}_{2.} & \bar{x}_{4.} & \bar{x}_{1.} \\ 13.1 & 13.3 & 13.8 & 14.3 & 14.5 \end{array}$$

Thus, brands 1 and 4 are not significantly different from one another, but are significantly higher than the other three brands in their true average contents. Brand 2 is significantly better than 3 and 5 but worse than 1 and 4, and brands 3 and 5 do not differ significantly. ∎

In Example 10.4, if $\bar{x}_{2.} = 14.15$ rather than 13.8 with the same computed w, then the configuration of underscored means would be

$\bar{x}_{5.}$	$\bar{x}_{3.}$	$\bar{x}_{2.}$	$\bar{x}_{4.}$	$\bar{x}_{1.}$
13.1	13.3	14.15	14.3	14.5

When there are no significant differences among a group of more than two means, it is customary to draw a continuous line underneath the entire group.

Example 10.5   A biologist wished to study the effects of ethanol on sleep time. A sample of 20 rats, matched for age and other characteristics, was selected, and each rat was given an oral injection having a particular concentration of ethanol per body weight. The rapid eye movement (REM) sleep time for each rat was then recorded for a 24-hour period, with the following results:

Treatment (concentration of ethanol)					$x_{i.}$	$\bar{x}_{i.}$
0 (control)	88.6	73.2	91.4	68.0	75.2	396.4	79.28
1 g/kg	63.0	53.9	69.2	50.1	71.5	307.7	61.54
2 g/kg	44.9	59.5	40.2	56.3	38.7	239.6	47.92
4 g/kg	31.0	39.6	45.3	25.2	22.7	163.8	32.76

$x_{..} = 1107.5$ $\bar{x}_{..} = 55.375$

Does the data indicate that the true average REM sleep time depends on the concentration of ethanol? (This example is based on an experiment reported in "Relationship of Ethanol Blood Level to REM and Non-REM Sleep Time and Distribution in the Rat," *Life Sciences,* 1978: 839–846.)

The $\bar{x}_{i.}$'s differ rather substantially from one another, but there is also a great deal of variability within each sample, so to answer the question precisely we must carry out the ANOVA. With $\sum\sum x_{ij}^2 = 68{,}697.6$ and correction factor $x_{..}^2/(IJ) = (1107.5)^2/20 = 61{,}327.8$, the computing formulas yield

$$\text{SST} = 68{,}697.6 - 61{,}327.8 = 7369.8$$

$$\text{SSTr} = \frac{1}{5}[(396.40)^2 + (307.70)^2 + (239.60)^2 + (163.80)^2] - 61{,}327.8$$

$$= 67{,}210.2 - 61{,}327.8 = 5882.4$$

and

$$\text{SSE} = 7369.8 - 5882.4 = 1487.4$$

Table 10.4 is a SAS ANOVA table. The last column gives the P-value, which is .0001. Using a significance level of .05, we reject the null hypothesis $H_0: \mu_1 = \mu_2 = \mu_3 = \mu_4$, since P-value $= .0001 < .05 = \alpha$. True average REM sleep time does appear to depend on concentration level.

Table 10.4 SAS ANOVA Table

```
                        Analysis of Variance Procedure
Dependent Variable:  TIME
                             Sum of              Mean
Source            DF         Squares            Square      F Value    Pr > F
Model              3       5882.35750        1960.78583      21.09     0.0001
Error             16       1487.40000          92.96250
Corrected
Total             19       7369.75750
```

There are $I = 4$ treatments and 16 df for error, so $Q_{.05,4,16} = 4.05$ and $w = 4.05\sqrt{93.0/5} = 17.47$. Ordering the means and underscoring yields

$\bar{x}_{4.}$	$\bar{x}_{3.}$	$\bar{x}_{2.}$	$\bar{x}_{1.}$
32.76	47.92	61.54	79.28

The interpretation of this underscoring must be done with care, since we seem to have concluded that treatments 2 and 3 do not differ, 3 and 4 do not differ, yet 2 and 4 do differ. The suggested way of expressing this is to say that, although evidence allows us to conclude that treatments 2 and 4 differ from one another, neither has been shown to be significantly different from 3. Treatment 1 has a significantly higher true average REM sleep time than any of the other treatments.

Figure 10.4 shows SAS output from the application of Tukey's procedure.

```
            Alpha = 0.05  df = 16  MSE = 92.9625
        Critical Value of Studentized Range = 4.046
            Minimum Significant Difference = 17.446

Means with the same letter are not significantly different.

Tukey Grouping                  Mean        N    TREATMENT
                  A           79.280        5    0 (control)

                  B           61.540        5    1 gm/kg
                  B
        C         B           47.920        5    2 gm/kg
        C
        C                     32.760        5    4 gm/kg
```

Figure 10.4 Tukey's method using SAS

The Interpretation of α in Multiple Comparison

Tukey's method involves the simultaneous construction of confidence intervals for all differences $\mu_i - \mu_j$ of pairs of treatment means. In the construction of confidence intervals (CIs) in the previous chapters, $\alpha = .05$ or 95% confidence referred to the error rate or confidence level for the individual statement about the parameter of interest. If an ANOVA experiment involves comparison of four treatments, then Tukey's procedure obtains simultaneously six different intervals (since there are six different pairs of treatment means). The error rate $\alpha = .05$ no longer refers to a particular interval, but instead refers to the experiment as a whole, so is called an **experiment-wise error rate.** If Tukey's method were used on a great many different ANOVA data sets, then in approximately 95% of these experiments no erroneous claim would be made about any of the $(\mu_i - \mu_j)$'s, whereas in only 5% would at least one incorrect claim be made. That is, in 95% of all experiments, every CI constructed from (10.4) would include the true value of $\mu_i - \mu_j$, and only 5% of the time would at least one interval fail to cover the true value of a $\mu_i - \mu_j$. Because the confidence level for the entire set of comparisons of means is 95%, the confidence level for any particular comparison is larger than 95% and increases as the number of comparisons (i.e., treatment means) increases. This distinction between the experimentwise error rate or confidence level and a "per-comparison" error rate or confidence level is important and should be kept in mind when interpreting the results of any multiple comparisons analysis.

Confidence Intervals for Other Parametric Functions

In some situations, a CI is desired for a function of the μ_i's more complicated than a difference $\mu_i - \mu_j$. Let $\theta = \sum c_i \mu_i$, where the c_i's are constants. One such function is $\frac{1}{2}(\mu_1 + \mu_2) - \frac{1}{3}(\mu_3 + \mu_4 + \mu_5)$, which in the context of Example 10.4 measures the difference between the group consisting of the first two brands and that of the last three brands. Because the X_{ij}'s are normally distributed with $E(X_{ij}) = \mu_i$ and $V(X_{ij}) = \sigma^2$, $\hat{\theta} = \sum_i c_i \overline{X}_{i\cdot}$ is normally distributed, unbiased for θ, and

$$V(\hat{\theta}) = V\left(\sum_i c_i \overline{X}_{i\cdot}\right) = \sum_i c_i^2 V(\overline{X}_{i\cdot}) = \frac{\sigma^2}{J} \sum_i c_i^2$$

Estimating σ^2 by MSE and forming $\hat{\sigma}_{\hat{\theta}}$ results in a t variable $(\hat{\theta} - \theta)/\hat{\sigma}_{\hat{\theta}}$, which can be manipulated to obtain the following $100(1 - \alpha)\%$ confidence interval for $\sum c_i \mu_i$:

$$\sum c_i \overline{x}_{i\cdot} \pm t_{\alpha/2, I(J-1)} \sqrt{\frac{\text{MSE} \sum c_i^2}{J}} \tag{10.5}$$

Example 10.6
(Example 10.4 continued)

The parametric function for comparing the first two (store) brands of oil filter with the last three (national) brands is $\theta = \frac{1}{2}(\mu_1 + \mu_2) - \frac{1}{3}(\mu_3 + \mu_4 + \mu_5)$, from which

$$\sum c_i^2 = \left(\frac{1}{2}\right)^2 + \left(\frac{1}{2}\right)^2 + \left(-\frac{1}{3}\right)^2 + \left(-\frac{1}{3}\right)^2 + \left(-\frac{1}{3}\right)^2 = \frac{5}{6}$$

With $\hat{\theta} = \frac{1}{2}(\overline{x}_{1\cdot} + \overline{x}_{2\cdot}) - \frac{1}{3}(\overline{x}_{3\cdot} + \overline{x}_{4\cdot} + \overline{x}_{5\cdot}) = .583$ and MSE $= .088$, a 95% interval is

$$.583 \pm 2.021\sqrt{5(.088)/[(6)(9)]} = .583 \pm .182 = (.401, .765) \qquad \blacksquare$$

Sometimes an experiment is carried out to compare each of several "new" treatments to a control treatment. In such situations, a multiple comparisons technique called Dunnett's method is appropriate.

Exercises | Section 10.2 (11–21)

11. An experiment to compare the spreading rates of five different brands of yellow interior latex paint available in a particular area used four gallons ($J = 4$) of each paint. The sample average spreading rates (ft²/gal) for the five brands were $\bar{x}_{1.} = 462.0$, $\bar{x}_{2.} = 512.8$, $\bar{x}_{3.} = 437.5$, $\bar{x}_{4.} = 469.3$, and $\bar{x}_{5.} = 532.1$. The computed value of F was found to be significant at level $\alpha = .05$. With MSE = 272.8, use Tukey's procedure to investigate significant differences in the true average spreading rates between brands.

12. In Exercise 11, suppose $\bar{x}_{3.} = 427.5$. Now which true average spreading rates differ significantly from one another? Be sure to use the method of underscoring to illustrate your conclusions, and write a paragraph summarizing your results.

13. Repeat Exercise 12 supposing that $\bar{x}_{2.} = 502.8$ in addition to $\bar{x}_{3.} = 427.5$.

14. Use Tukey's procedure on the data in Exercise 4 to identify differences in true average flight times among the four types of mosquitos.

15. Use Tukey's procedure on the data of Exercise 6 to identify differences in true average total Fe among the four types of formations (use MSE = 15.64).

16. Reconsider the axial stiffness data given in Exercise 8. ANOVA output from MINITAB follows.

```
Analysis of Variance for stiffness
Source   DF      SS      MS       F       P
length    4   43993   10998   10.48   0.000
Error    30   31475    1049
Total    34   75468
```

```
Level    N     Mean    StDev
4        7   333.21    36.59
6        7   368.06    28.57
8        7   375.13    20.83
10       7   407.36    44.51
12       7   437.17    26.00
```

```
Pooled StDev = 32.39
```

```
Tukey's pairwise comparisons

    Family error rate = 0.0500
Individual error rate = 0.00693

Critical value = 4.10
```

```
Intervals for (column level mean) − (row
level mean)
             4         6         8        10
 6       -85.0
          15.4
 8       -92.1     -57.3
           8.3      43.1
10      -124.3     -89.5     -82.4
         -23.9      10.9      18.0
12      -154.2    -119.3    -112.2    -80.0
         -53.8     -18.9     -11.8     20.4
```

a. Is it plausible that the variances of the five axial stiffness index distributions are identical? Explain.

b. Use the output (without reference to our F table) to test the relevant hypotheses.

c. Use the Tukey intervals given in the output to determine which means differ, and construct the corresponding underscoring pattern.

17. Refer to Exercise 5. Compute a 95% t CI for $\theta = \frac{1}{2}(\mu_1 + \mu_2) - \mu_3$.

18. Consider the accompanying data on plant growth after the application of different types of growth hormone.

	1	13	17	7	14
	2	21	13	20	17
Hormone	3	18	15	20	17
	4	7	11	18	10
	5	6	11	15	8

a. Perform an F test at level $\alpha = .05$.

b. What happens when Tukey's procedure is applied?

19. Consider a single-factor ANOVA experiment in which $I = 3$, $J = 5$, $\bar{x}_{1.} = 10$, $\bar{x}_{2.} = 12$, and $\bar{x}_{3.} = 20$. Find a value of SSE for which $f > F_{.05,2,12}$, so that $H_0: \mu_1 = \mu_2 = \mu_3$ is rejected, yet when Tukey's procedure is applied none of the μ_i's can be said to differ significantly from one another.

20. Refer to Exercise 19 and suppose $\bar{x}_{1.} = 10$, $\bar{x}_{2.} = 15$, and $\bar{x}_{3.} = 20$. Can you now find a value of SSE that produces such a contradiction between the F test and Tukey's procedure?

21. The article "The Effect of Enzyme Inducing Agents on the Survival Times of Rats Exposed to Lethal Levels of Nitrogen Dioxide" (*Toxicology and Applied Pharmacology,* 1978: 169–174) reports the following data on survival times for rats exposed to nitrogen dioxide (70 ppm) via different injection regimens. There were $J = 14$ rats in each group.

Regimen	$\bar{x}_{i.}$(min)	s_i
1. Control	166	32
2. 3-Methylcholanthrene	303	53
3. Allylisopropylacetamide	266	54
4. Phenobarbital	212	35
5. Chlorpromazine	202	34
6. *p*-Aminobenzoic Acid	184	31

a. Test the null hypothesis that true average survival time does not depend on injection regimen against the alternative that there is some dependence on injection regimen using $\alpha = .01$.

b. Suppose that $100(1 - \alpha)\%$ CIs for k different parametric functions are computed from the same ANOVA data set. Then it is easily verified that the simultaneous confidence level is at least $100(1 - k\alpha)\%$. Compute CIs with simultaneous confidence level at least 98% for $\mu_1 - \frac{1}{5}(\mu_2 + \mu_3 + \mu_4 + \mu_5 + \mu_6)$ and $\frac{1}{4}(\mu_2 + \mu_3 + \mu_4 + \mu_5) - \mu_6$.

10.3 | More on Single-Factor ANOVA

In this section, we briefly consider some issues relating to single-factor ANOVA that were not dealt with in the first two sections. These include an alternative description of the model parameters, β for the F test, the relationship of the test to procedures previously considered, data transformation, a random effects model, and formulas for the case of unequal sample sizes.

An Alternative Description of the ANOVA Model

The assumptions of single-factor ANOVA can be described succinctly by means of the "model equation"

$$X_{ij} = \mu_i + \epsilon_{ij}$$

where ϵ_{ij} represents a random deviation from the population or true treatment mean μ_i. The ϵ_{ij}'s are assumed to be independent, normally distributed rv's (implying that the X_{ij}'s are also) with $E(\epsilon_{ij}) = 0$ (so that $E(X_{ij}) = \mu_i$) and $V(\epsilon_{ij}) = \sigma^2$ (from which $V(X_{ij}) = \sigma^2$ for every i and j). An alternative description of single-factor ANOVA will give added insight and suggest appropriate generalizations to models involving more than one factor. Define a parameter μ by

$$\mu = \frac{1}{I} \sum_{i=1}^{I} \mu_i$$

and the parameters $\alpha_1, \ldots, \alpha_I$ by

$$\alpha_i = \mu_i - \mu \qquad (i = 1, \ldots, I)$$

Then the treatment mean μ_i can be written as $\mu + \alpha_i$, where μ represents the true average overall response in the experiment, and α_i is the effect, measured as a departure from μ, due to the ith treatment. Whereas we initially had I parameters, we now have

$I + 1$ ($\mu, \alpha_1, \ldots, \alpha_I$). However, because $\sum \alpha_i = 0$ (the average departure from the overall mean response is zero), only I of these new parameters are independently determined, so there are as many independent parameters as there were before. In terms of μ and the α_i's, the model becomes

$$X_{ij} = \mu + \alpha_i + \epsilon_{ij} \qquad (i = 1, \ldots, I, \quad j = 1, \ldots, J)$$

In Chapter 11, we will develop analogous models for multifactor ANOVA. The claim that the μ_i's are identical is equivalent to the equality of the α_i's, and because $\sum \alpha_i = 0$, the null hypothesis becomes

$$H_0: \alpha_1 = \alpha_2 = \cdots = \alpha_I = 0$$

In Section 10.1, it was stated that MSTr is an unbiased estimator of σ^2 when H_0 is true but otherwise tends to overestimate σ^2. More precisely,

$$E(\text{MSTr}) = \sigma^2 + \frac{J}{I - 1} \sum \alpha_i^2$$

When H_0 is true, $\sum \alpha_i^2 = 0$ so $E(\text{MSTr}) = \sigma^2$ (MSE is unbiased whether or not H_0 is true). If $\sum \alpha_i^2$ is used as a measure of the extent to which H_0 is false, then a larger value of $\sum \alpha_i^2$ will result in a greater tendency for MSTr to overestimate σ^2. In the next chapter, formulas for expected mean squares for multifactor models will be used to suggest how to form F ratios to test various hypotheses.

Proof of the Formula for $E(\text{MSTr})$ For any rv Y, $E(Y^2) = V(Y) + [E(Y)]^2$, so

$$E(\text{SSTr}) = E\left(\frac{1}{J}\sum_i X_{i\cdot}^2 - \frac{1}{IJ}X_{\cdot\cdot}^2\right) = \frac{1}{J}\sum_i E(X_{i\cdot}^2) - \frac{1}{IJ}E(X_{\cdot\cdot}^2)$$

$$= \frac{1}{J}\sum_i \{V(X_{i\cdot}) + [E(X_{i\cdot})]^2\} - \frac{1}{IJ}\{V(X_{\cdot\cdot}) + [E(X_{\cdot\cdot})]^2\}$$

$$= \frac{1}{J}\sum_i \{J\sigma^2 + [J(\mu + \alpha_i)]^2\} - \frac{1}{IJ}[IJ\sigma^2 + (IJ\mu)^2]$$

$$= I\sigma^2 + IJ\mu^2 + 2\mu J\sum_i \alpha_i + J\sum_i \alpha_i^2 - \sigma^2 - IJ\mu^2$$

$$= (I - 1)\sigma^2 + J\sum_i \alpha_i^2 \qquad (\text{since } \sum \alpha_i = 0)$$

The result then follows from the relationship MSTr = SSTr/($I - 1$).   ∎

β for the F Test

Consider a set of parameter values $\alpha_1, \alpha_2, \ldots, \alpha_I$ for which H_0 is not true. The probability of a type II error, β, is the probability that H_0 is not rejected when that set is the set of true values. One might think that β would have to be determined separately for each different configuration of α_i's. Fortunately, since β for the F test depends on the α_i's and σ^2 only through $\sum \alpha_i^2/\sigma^2$, it can be simultaneously evaluated for many different alternatives. For example, $\sum \alpha_i^2 = 4$ for each of the following sets of α_i's for which H_0 is false, so β is identical for all three alternatives:

1. $\alpha_1 = -1$, $\alpha_2 = -1$, $\alpha_3 = 1$, $\alpha_4 = 1$

2. $\alpha_1 = -\sqrt{2}$, $\alpha_2 = \sqrt{2}$, $\alpha_3 = 0$, $\alpha_4 = 0$

3. $\alpha_1 = -\sqrt{3}$, $\alpha_2 = \sqrt{1/3}$, $\alpha_3 = \sqrt{1/3}$, $\alpha_4 = \sqrt{1/3}$

The quantity $J\sum \alpha_i^2 / \sigma^2$ is called the **noncentrality parameter** for one-way ANOVA (because when H_0 is false the test statistic has a *noncentral F* distribution with this as one of its parameters), and β is a decreasing function of the value of this parameter. Thus, for fixed values of σ^2 and J, the null hypothesis is more likely to be rejected for alternatives far from H_0 (large $\sum \alpha_i^2$) than for alternatives close to H_0. For a fixed value of $\sum \alpha_i^2$, β decreases as the sample size J on each treatment increases, and it increases as the variance σ^2 increases (since greater underlying variability makes it more difficult to detect any given departure from H_0).

Because hand computation of β and sample size determination for the F test are quite difficult (as in the case of t tests), statisticians have constructed sets of curves from which β can be read. These are called *power curves*. Sets of curves for numerator df $\nu_1 = 3$ and $\nu_1 = 4$ are displayed in Figure 10.5* and Figure 10.6*, respectively. After the values of σ^2 and the α_i's for which β is desired are specified, these are used to compute the value of ϕ, where $\phi^2 = (J/I)\sum \alpha_i^2 / \sigma^2$. We then enter the appropriate set of curves at the value of ϕ on the horizontal axis, move up to the curve associated with error df ν_2, and move over to the value of power on the vertical axis. Finally, $\beta = 1 -$ power.

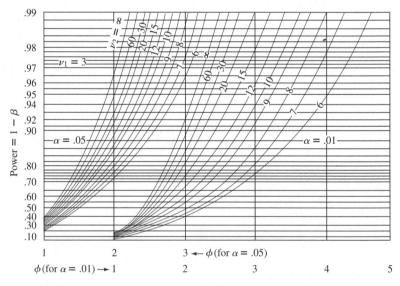

Figure 10.5 Power curves for the ANOVA F test ($\nu_1 = 3$)

*Reproduced with permission from E. S. Pearson and H. O. Hartley, "Charts of the Power Function for Analysis of Variance Tests, Derived from the Non-central F Distribution," *Biometrika*, vol. 38, 1951: 112.

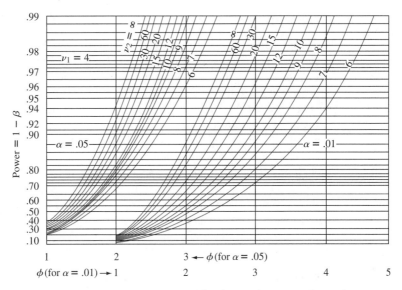

Figure 10.6 Power curves for the ANOVA F test ($\nu_1 = 4$)

Example 10.7 The effects of four different heat treatments on yield point (tons/in.^{2}) of steel ingots are to be investigated. A total of eight ingots will be cast using each treatment. Suppose the true standard deviation of yield point for any of the four treatments is $\sigma = 1$. How likely is it that H_0 will not be rejected at level .05 if three of the treatments have the same expected yield point and the other treatment has an expected yield point that is 1 ton/in.^{2} greater than the common value of the other three (i.e., the fourth yield is on average 1 standard deviation above those for the first three treatments)?

Suppose that $\mu_1 = \mu_2 = \mu_3$ and $\mu_4 = \mu_1 + 1$, $\mu = (\sum \mu_i)/4 = \mu_1 + \frac{1}{4}$. Then $\alpha_1 = \mu_1 - \mu = -\frac{1}{4}$, $\alpha_2 = -\frac{1}{4}$, $\alpha_3 = -\frac{1}{4}$, $\alpha_4 = \frac{3}{4}$ so

$$\phi^2 = \frac{8}{4}\left[\left(\frac{1}{4}\right)^2 + \left(-\frac{1}{4}\right)^2 + \left(-\frac{1}{4}\right)^2 + \left(\frac{3}{4}\right)^2\right] = \frac{3}{2}$$

and $\phi = 1.22$. The degrees of freedom are $\nu_1 = I - 1 = 3$ and $\nu_2 = I(J - 1) = 28$, so interpolating visually between $\nu_2 = 20$ and $\nu_2 = 30$ gives power $\approx .47$ and $\beta \approx .53$. This β is rather large, so we might decide to increase the value of J. How many ingots of each type would be required to yield $\beta \approx .05$ for the alternative under consideration? By trying different values of J, we can verify that $J = 24$ will meet the requirement, but any smaller J will not. ∎

As an alternative to the use of power curves, the SAS package has a function that calculates the cumulative area under a noncentral F curve (inputs F_α, numerator df, denominator df, and ϕ^2), and this area is β. Version 12 of MINITAB will calculate the power when sample sizes are equal and will determine the necessary common sample size to achieve a specified power.

Relationship of the *F* Test to the *t* Test

When the number of treatments or populations is $I = 2$, all formulas and results connected with the *F* test still make sense, so ANOVA can be used to test $H_0: \mu_1 = \mu_2$ versus $H_a: \mu_1 \neq \mu_2$. In this case, a two-tailed, two-sample *t* test can also be used. In Section 9.3, we mentioned the pooled *t* test, which requires equal variances, as an alternative to the two-sample *t* procedure. It can be shown that the single-factor ANOVA *F* test and the two-tailed pooled *t* test are equivalent; for any given data set, the *P*-values for the two tests will be identical, so the same conclusion will be reached by either test.

The two-sample *t* test is more flexible than the *F* test when $I = 2$ for two reasons. First, it is valid without the assumption that $\sigma_1 = \sigma_2$; second, it can be used to test $H_a: \mu_1 > \mu_2$ (an upper-tailed *t* test) or $H_a: \mu_1 < \mu_2$ as well as $H_a: \mu_1 \neq \mu_2$. In the case of $I \geq 3$, there is unfortunately no general test procedure known to have good properties without assuming equal variances.

Single-Factor ANOVA When Sample Sizes Are Unequal

When the sample sizes from each population or treatment are not equal, let $J_1, J_2, \ldots, J_I$ denote the I sample sizes and let $n = \sum_i J_i$ denote the total number of observations. The accompanying box gives ANOVA formulas and the test procedure.

$$\text{SST} = \sum_{i=1}^{I} \sum_{j=1}^{J_i} (X_{ij} - \overline{X}_{..})^2 = \sum_{i=1}^{I} \sum_{j=1}^{J_i} X_{ij}^2 - \frac{1}{n} X_{..}^2 \qquad \text{df} = n - 1$$

$$\text{SSTr} = \sum_{i=1}^{I} \sum_{j=1}^{J_i} (\overline{X}_{i.} - \overline{X}_{..})^2 = \sum_{i=1}^{I} \frac{1}{J_i} X_{i.}^2 - \frac{1}{n} X_{..}^2 \qquad \text{df} = I - 1$$

$$\text{SSE} = \sum_{i=1}^{I} \sum_{j=1}^{J_i} (X_{ij} - \overline{X}_{i.})^2 = \text{SST} - \text{SSTr} \qquad \text{df} = \sum(J_i - 1) = n - I$$

Test statistic value:

$$f = \frac{\text{MSTr}}{\text{MSE}} \qquad \text{where MSTr} = \frac{\text{SSTr}}{I - 1} \quad \text{MSE} = \frac{\text{SSE}}{n - I}$$

Rejection region: $f \geq F_{\alpha, I-1, n-I}$

Example 10.8 The article "On the Development of a New Approach for the Determination of Yield Strength in Mg-based Alloys" (*Light Metal Age,* Oct., 1998: 51–53) presented the following data on elastic modulus (GPa) obtained by a new ultrasonic method for specimens of a certain alloy produced using three different casting processes.

									J_i	$x_{i.}$	$\overline{x}_{i.}$
Permanent molding	45.5	45.3	45.4	44.4	44.6	43.9	44.6	44.0	8	357.7	44.71
Die casting	44.2	43.9	44.7	44.2	44.0	43.8	44.6	43.1	8	352.5	44.06
Plaster molding	46.0	45.9	44.8	46.2	45.1	45.5			6	273.5	45.58
									22	983.7	

Let μ_1, μ_2, and μ_3 denote the true average elastic moduli for the three different processes under the given circumstances. The relevant hypotheses are $H_0: \mu_1 = \mu_2 = \mu_3$ versus H_a: at least two of the μ_i's are different. The test statistic is of course $F = $ MSTr/MSE, based on $I - 1 = 2$ numerator df and $n - I = 22 - 3 = 19$ denominator df. Relevant quantities include

$$\Sigma\Sigma x_{ij}^2 = 43{,}998.73 \qquad CF = \frac{983.7^2}{22} = 43{,}984.80$$

$$SST = 43{,}998.73 - 43{,}984.80 = 13.93$$

$$SSTr = \frac{357.7^2}{8} + \frac{352.5^2}{8} + \frac{273.5^2}{6} - 43{,}984.80 = 7.93$$

$$SSE = 13.93 - 7.93 = 6.00$$

The remaining computations are displayed in the accompanying ANOVA table. Since $F_{.001,2,19} = 10.16 < 12.56 = f$, the P-value is smaller than .001. Thus the null hypothesis should be rejected at any reasonable significance level; there is compelling evidence for concluding that true average elastic modulus somehow depends on which casting process is used.

Source of Variation	df	Sum of Squares	Mean Square	f
Treatments	2	7.93	3.965	12.56
Error	19	6.00	.3158	
Total	21	13.93		

Multiple Comparisons When Sample Sizes Are Unequal

There is more controversy among statisticians regarding which multiple comparisons procedure to use when sample sizes are unequal than there is in the case of equal sample sizes. The procedure that we present here is recommended in the excellent book *Beyond ANOVA: Basics of Applied Statistics* (see the chapter bibliography) for use when the I sample sizes $J_1, J_2, \ldots, J_I$ are reasonably close to one another ("mild imbalance"). It modifies Tukey's method by using averages of pairs of $1/J_i$'s in place of $1/J$.

Let

$$w_{ij} = Q_{\alpha,I,n-I} \cdot \sqrt{\frac{MSE}{2}\left(\frac{1}{J_i} + \frac{1}{J_j}\right)}$$

Then the probability is *approximately* $1 - \alpha$ that

$$\overline{X}_{i.} - \overline{X}_{j.} - w_{ij} \leq \mu_i - \mu_j \leq \overline{X}_{i.} - \overline{X}_{j.} + w_{ij}$$

for every i and j ($i = 1, \ldots, I$ and $j = 1, \ldots, I$) with $i \neq j$.

The simultaneous confidence level $100(1 - \alpha)\%$ is only approximate rather than exact as it is with equal sample sizes. The underscoring method can still be used, but now the w_{ij} factor used to decide whether $\bar{x}_{i\cdot}$ and $\bar{x}_{j\cdot}$ can be connected will depend on J_i and J_j.

Example 10.9
(Example 10.8 continued)

The sample sizes for the elastic modulus data were $J_1 = 8$, $J_2 = 8$, $J_3 = 6$, and $I = 3$, $n - I = 19$, MSE $= .316$. A simultaneous confidence level of approximately 95% requires $Q_{.05,3,19} = 3.59$, from which

$$w_{12} = 3.59 \sqrt{\frac{.316}{2}\left(\frac{1}{8} + \frac{1}{8}\right)} = .713, \qquad w_{13} = .771 \qquad w_{23} = .771$$

Since $\bar{x}_{1\cdot} - \bar{x}_{2\cdot} = 44.71 - 44.06 = .65 < w_{12}$, μ_1 and μ_2 are judged not significantly different. The accompanying underscoring scheme shows that μ_1 and μ_3 appear to differ significantly, as do μ_2 and μ_3:

	2. Die	1. Permanent	3. Plaster
	44.06	44.71	45.58

■

Data Transformation

The use of ANOVA methods can be invalidated by substantial differences in the variances $\sigma_1^2, \ldots, \sigma_I^2$ (which until now have been assumed equal with common value σ^2). It sometimes happens that $V(X_{ij}) = \sigma_i^2 = g(\mu_i)$, a known function of μ_i (so that when H_0 is false, the variances are not equal). For example, if X_{ij} has a Poisson distribution with parameter λ_i (approximately normal if $\lambda_i \geq 10$), then $\mu_i = \lambda_i$ and $\sigma_i^2 = \lambda_i$ so $g(\mu_i) = \mu_i$ is the known function. In such cases, one can often transform the X_{ij}'s to $h(X_{ij})$'s so that they will have approximately equal variances (while leaving the transformed variables approximately normal), and then the F test can be used on the transformed observations. The key idea in choosing a transformation $h(\cdot)$ is that often $V[h(X_{ij})] \approx V(X_{ij}) \cdot [h'(\mu_i)]^2 = g(\mu_i) \cdot [h'(\mu_i)]^2$. We wish to find the function $h(\cdot)$ for which $g(\mu_i) \cdot [h'(\mu_i)]^2 = c$ (a constant) for every i.

PROPOSITION

> If $V(X_{ij}) = g(\mu_i)$, a known function of μ_i, then a transformation $h(X_{ij})$ that "stabilizes the variance" so that $V[h(X_{ij})]$ is approximately the same for each i is given by $h(x) \propto \int [g(x)]^{-1/2}dx$.

In the Poisson case, $g(x) = x$, so $h(x)$ should be proportional to $\int x^{-1/2}dx = 2x^{1/2}$. Thus, Poisson data should be transformed to $h(x_{ij}) = \sqrt{x_{ij}}$ before the analysis.

A Random Effects Model

The single-factor problems considered so far have all been assumed to be examples of a **fixed effects** ANOVA model. By this we mean that the chosen levels of the factor under study are the only ones considered relevant by the experimenter. The single-factor fixed effects model is

$$X_{ij} = \mu + \alpha_i + \epsilon_{ij} \qquad \sum \alpha_i = 0 \tag{10.6}$$

where the ϵ_{ij}'s are random and both μ and the α_i's are fixed parameters whose values are unknown.

In some single-factor problems, the particular levels studied by the experimenter are chosen, either by design or through sampling, from a large population of levels. For example, to study the effects on task performance time of using different operators on a particular machine, a sample of five operators might be chosen from a large pool of operators. Similarly, the effect of soil pH on the yield of maize plants might be studied by using soils with four specific pH values chosen from among the many possible pH levels. When the levels used are selected at random from a larger population of possible levels, the factor is said to be random rather than fixed, and the fixed effects model (10.6) is no longer appropriate. An analogous **random effects** model is obtained by replacing the fixed α_i's in (10.6) by random variables. The resulting model description is

$$X_{ij} = \mu + A_i + \epsilon_{ij} \quad \text{with } E(A_i) = E(\epsilon_{ij}) = 0$$

$$V(\epsilon_{ij}) = \sigma^2 \qquad V(A_i) = \sigma_A^2 \tag{10.7}$$

all A_i's and ϵ_{ij}'s normally distributed and independent of one another.

The condition $E(A_i) = 0$ in (10.7) is similar to the condition $\sum \alpha_i = 0$ in (10.6); it states that the expected or average effect of the ith level measured as a departure from μ is zero.

For the random effects model (10.7), the hypothesis of no effects due to different levels is $H_0: \sigma_A^2 = 0$, which says that different levels of the factor contribute nothing to variability of the response. *Although the hypotheses in the single-factor fixed and random effects models are different, they are tested in exactly the same way,* by forming $F = \text{MSTr/MSE}$ and rejecting H_0 if $f \geq F_{\alpha,I-1,n-I}$. This can be justified intuitively by noting that $E(\text{MSE}) = \sigma^2$ (as for fixed effects), whereas

$$E(\text{MSTr}) = \sigma^2 + \frac{1}{I-1}\left(n - \frac{\sum J_i^2}{n}\right)\sigma_A^2 \tag{10.8}$$

where $J_1, J_2, \ldots, J_I$ are the sample sizes and $n = \sum J_i$. The factor in parentheses on the right side of (10.8) is nonnegative, so again $E(\text{MSTr}) = \sigma^2$ if H_0 is true and $E(\text{MSTr}) > \sigma^2$ if H_0 is false.

Example 10.10 The study of nondestructive forces and stresses in materials furnishes important information for efficient engineering design. The article "Zero-Force Travel-Time Parameters for Ultrasonic Head-Waves in Railroad Rail" (*Materials Evaluation*, 1985: 854–858) reports on a study of travel time for a certain type of wave that results from longitudinal stress of rails used for railroad track. Three measurements were made on each of six rails randomly selected from a population of rails. The investigators used random effects ANOVA to decide whether some variation in travel time could be attributed to "between-rail variability." The data is given in the accompanying table (each value, in nanoseconds, resulted from subtracting 36.1 μs from the original observation) along with the derived ANOVA table. The value of the F ratio is highly significant, so

$H_0: \sigma_A^2 = 0$ is rejected in favor of the conclusion that differences between rails is a source of travel-time variability.

				$x_i.$
1:	55	53	54	162
2:	26	37	32	95
3:	78	91	85	254
4:	92	100	96	288
5:	49	51	50	150
6:	80	85	83	248

$$x.. = 1197$$

Source of variation	df	Sum of squares	Mean square	f
Treatments	5	9310.5	1862.1	115.2
Error	12	194.0	16.17	
Total	17	9504.5		

■

Exercises | Section 10.3 (22–34)

22. The following data refers to yield of tomatoes (kg/plot) for four different levels of salinity; salinity level here refers to electrical conductivity (EC), where the chosen levels were EC = 1.6, 3.8, 6.0, and 10.2 nmhos/cm:

    1.6 59.5 53.3 56.8 63.1 58.7

    3.8 55.2 59.1 52.8 54.5

    6.0 51.7 48.8 53.9 49.0

    10.2 44.6 48.5 41.0 47.3 46.1

    Use the F test at level $\alpha = .05$ to test for any differences in true average yield due to the different salinity levels.

23. Apply the modified Tukey's method to the data in Exercise 22 to identify significant differences among the μ_i's.

24. The following partial ANOVA table is taken from the article "Perception of Spatial Incongruity" (*J. Nervous and Mental Disease,* 1961: 222) in which the abilities of three different groups to identify a

perceptual incongruity were assessed and compared. All individuals in the experiment had been hospitalized to undergo psychiatric treatment. There were 21 individuals in the depressive group, 32 individuals in the functional "other" group, and 21 individuals in the brain-damaged group. Complete the ANOVA table and carry out the F test at level $\alpha = .01$.

Source	df	Sum of squares	Mean square	f
Groups			76.09	
Error				
Total		1123.14		

25. An article in the *Canadian Entomologist* ("Influence of Natural Diets and Larval Density on Gypsy Moth, Lymantria Dispor, Egg Mass Characteristics," 1977: 1313–1318) reports the following data on egg mass diameters for moths reared on five different diets.

Diet	J_i	$\bar{x}_{i\cdot}$ (mm)	s_i
Red maple '74	13	1.134	.0252
Red oak/red maple	10	1.148	.0253
Red maple '75	20	1.159	.0179
Red oak	16	1.191	.0200
Red oak/white pine	16	1.217	.0160

a. Compute the grand mean $\bar{x}_{\cdot\cdot}$, SSTr, and MSTr. (*Hint:* $x_{\cdot\cdot} = \sum J_i \bar{x}_{i\cdot}$.)

b. Compute SSE and MSE. [*Hint:* SSE $= \sum (J_i - 1)s_i^2$.]

c. Does the data suggest that there are any differences between true average egg diameters for the different diets? Test using $\alpha = .05$ by first obtaining an upper and/or lower bound on the P-value.

d. Which means appear to differ significantly from one another?

26. Samples of six different brands of diet/imitation margarine were analyzed to determine the level of physiologically active polyunsaturated fatty acids (PAPFUA, in percentages), resulting in the following data:

Imperial	14.1 13.6 14.4 14.3
Parkay	12.8 12.5 13.4 13.0 12.3
Blue Bonnet	13.5 13.4 14.1 14.3
Chiffon	13.2 12.7 12.6 13.9
Mazola	16.8 17.2 16.4 17.3 18.0
Fleischmann's	18.1 17.2 18.7 18.4

(The preceding numbers are fictitious, but the sample means agree with data reported in the January 1975 issue of *Consumer Reports.*)

a. Use ANOVA to test for differences among the true average PAPFUA percentages for the different brands.

b. Compute CIs for all $(\mu_i - \mu_j)$'s.

c. Mazola and Fleischmann's are corn-based, whereas the others are soybean-based. Compute a CI for

$$\frac{(\mu_1 + \mu_2 + \mu_3 + \mu_4)}{4} - \frac{(\mu_5 + \mu_6)}{2}$$

[*Hint:* Modify the expression for $V(\hat{\theta})$ that led to (10.5) in the previous section.]

27. Although tea is the world's most widely consumed beverage after water, little is known about its nutritional value. Folacin is the only B vitamin present in any significant amount in tea, and recent advances in assay methods have made accurate determination of folacin content feasible. Consider the accompanying data on folacin content for randomly selected specimens of the four leading brands of green tea.

Brand	Observations
1	7.9 6.2 6.6 8.6 8.9 10.1 9.6
2	5.7 7.5 9.8 6.1 8.4
3	6.8 7.5 5.0 7.4 5.3 6.1
4	6.4 7.1 7.9 4.5 5.0 4.0

(Data is based on "Folacin Content of Tea," *J. Amer. Dietetic Assoc.*, 1983: 627–632.) Does this data suggest that true average folacin content is the same for all brands?

a. Carry out a test using $\alpha = .05$ via the P-value method.

b. Assess the plausibility of any assumptions required for your analysis in part (a).

c. Perform a multiple comparisons analysis to identify significant differences among brands.

28. For a single-factor ANOVA with sample sizes J_i $(i = 1, 2, \ldots, I)$, show that SSTr $= \sum J_i(\bar{X}_{i\cdot} - \bar{X}_{\cdot\cdot})^2 = \sum_i J_i \bar{X}_{i\cdot}^2 - n\bar{X}_{\cdot\cdot}^2$, where $n = \sum J_i$.

29. When sample sizes are equal ($J_i = J$), the parameters $\alpha_1, \alpha_2, \ldots, \alpha_I$ of the alternative parameterization are restricted by $\sum \alpha_i = 0$. For unequal sample sizes, the most natural restriction is $\sum J_i \alpha_i = 0$. Use this to show that

$$E(\text{MSTr}) = \sigma^2 + \frac{1}{I-1}\sum J_i \alpha_i^2$$

What is $E(\text{MSTr})$ when H_0 is true? [This expectation is correct if $\sum J_i \alpha_i = 0$ is replaced by the restriction $\sum \alpha_i = 0$ (or any other single linear restriction on the α_i's used to reduce the model to I independent parameters), but $\sum J_i \alpha_i = 0$ simplifies the algebra and yields natural estimates for the model parameters (in particular, $\hat{\alpha}_i = \bar{x}_{i\cdot} - \bar{x}_{\cdot\cdot}$).]

30. Reconsider Example 10.7 involving an investigation of the effects of different heat treatments on the yield point of steel ingots.

a. If $J = 8$ and $\sigma = 1$, what is β for a level .05 F test when $\mu_1 = \mu_2$, $\mu_3 = \mu_1 - 1$, and $\mu_4 = \mu_1 + 1$?

b. For the alternative of part (a), what value of J is necessary to obtain $\beta = .05$?

c. If there are $I = 5$ heat treatments, $J = 10$, and $\sigma = 1$, what is β for the level .05 F test when four of the μ_i's are equal and the fifth differs by 1 from the other four?

31. When sample sizes are not equal, the noncentrality parameter is $\sum J_i \alpha_i^2 / \sigma^2$ and $\phi^2 = (1/I) \sum J_i \alpha_i^2 / \sigma^2$. Referring to Exercise 22, what is the power of the test when $\mu_2 = \mu_3$, $\mu_1 = \mu_2 - \sigma$, and $\mu_4 = \mu_2 + \sigma$?

32. In an experiment to compare the quality of four different brands of reel-to-reel recording tape, five 2400-ft reels of each brand (A–D) were selected and the number of flaws in each reel was determined.

    A: 10   5 12 14   8

    B: 14 12 17   9   8

    C: 13 18 10 15 18

    D: 17 16 12 22 14

It is believed that the number of flaws has approximately a Poisson distribution for each brand. Analyze the data at level .01 to see whether the expected number of flaws per reel is the same for each brand.

33. Suppose that X_{ij} is a binomial variable with parameters n and p_i (so approximately normal when $np_i \geq 5$ and $nq_i \geq 5$). Then since $\mu_i = np_i$, $V(X_{ij}) = \sigma_i^2 = np_i(1 - p_i) = \mu_i(1 - \mu_i/n)$. How should the X_{ij}'s be transformed so as to stabilize the variance? [*Hint:* $g(\mu_i) = \mu_i(1 - \mu_i/n)$.]

34. Simplify $E(\text{MSTr})$ for the random effects model when $J_1 = J_2 = \cdots = J_I = J$.

Supplementary Exercises (35–46)

35. An experiment was carried out to compare flow rates for four different types of nozzles.
    a. Sample sizes were 5, 6, 7, and 6, respectively, and calculations gave $f = 3.68$. State and test the relevant hypotheses using $\alpha = .01$
    b. Analysis of the data using a statistical computer package yielded *P*-value = .029. At level .01, what would you conclude, and why?

36. The article "Computer-Assisted Instruction Augmented with Planned Teacher/Student Contacts" (*J. Exp. Educ.*, Winter, 1980–1981: 120–126) compared five different methods for teaching descriptive statistics. The five methods were traditional lecture and discussion (L/D), programmed textbook instruction (R), programmed text with lectures (R/L), computer instruction (C), and computer instruction with lectures (C/L). Forty-five students were randomly assigned, 9 to each method. After completing the course, the students took a 1-hour exam. In addition, a 10-minute retention test was administered 6 weeks later. Summary quantities are given.

Method	Exam $\bar{x}_{i.}$	s_i	Retention Test $\bar{x}_{i.}$	s_i
L/D	29.3	4.99	30.20	3.82
R	28.0	5.33	28.80	5.26
R/L	30.2	3.33	26.20	4.66
C	32.4	2.94	31.10	4.91
C/L	34.2	2.74	30.20	3.53

The grand mean for the exam was 30.82, and the grand mean for the retention test was 29.30.
    a. Does the data suggest that there is a difference among the five teaching methods with respect to true mean exam score? Use $\alpha = .05$.
    b. Using a .05 significance level, test the null hypothesis of no difference among the true mean retention test scores for the five different teaching methods.

37. Numerous factors contribute to the smooth running of an electric motor ("Increasing Market Share Through Improved Product and Process Design: An Experimental Approach," *Quality Engineering*, 1991: 361–369). In particular, it is desirable to keep motor noise and vibration to a minimum. To study the effect that the brand of bearing has on motor vibration, five different motor bearing brands were examined by installing each type of bearing on different random samples of six motors. The amount of motor vibration (measured in microns) was recorded when each of the 30 motors was running. The data for this study follows. State and test the relevant hypotheses at significance level .05, and then carry out a multiple comparisons analysis if appropriate.

							Mean
Brand 1	13.1	15.0	14.0	14.4	14.0	11.6	13.68
Brand 2	16.3	15.7	17.2	14.9	14.4	17.2	15.95
Brand 3	13.7	13.9	12.4	13.8	14.9	13.3	13.67
Brand 4	15.7	13.7	14.4	16.0	13.9	14.7	14.73
Brand 5	13.5	13.4	13.2	12.7	13.4	12.3	13.08

38. An article in the British scientific journal *Nature* ("Sucrose Induction of Hepatic Hyperplasia in the Rat," August 25, 1972: 461) reports on an experiment in which each of five groups consisting of six rats was put on a diet with a different carbohydrate. At the conclusion of the experiment, the DNA content of the liver of each rat was determined (mg/g liver), with the following results:

Carbohydrate	$\bar{x}_{i\cdot}$
Starch	2.58
Sucrose	2.63
Fructose	2.13
Glucose	2.41
Maltose	2.49

Assuming also that $\sum\sum x_{ij}^2 = 183.4$, does the data indicate that true average DNA content is affected by the type of carbohydrate in the diet? Construct an ANOVA table and use a .05 level of significance.

39. Referring to Exercise 38, construct a *t* CI for

$$\theta = \mu_1 - (\mu_2 + \mu_3 + \mu_4 + \mu_5)/4$$

which measures the difference between the average DNA content for the starch diet and the combined average for the four other diets. Does the resulting interval include zero?

40. Refer to Exercise 38. What is β for the test when true average DNA content is identical for three of the diets and falls below this common value by 1 standard deviation (σ) for the other two diets?

41. Four laboratories (1–4) are randomly selected from a large population, and each is asked to make three determinations of the percentage of methyl alcohol in specimens of a compound taken from a single batch. Based on the accompanying data, are differences among laboratories a source of variation in the percentage of methyl alcohol? State and test the relevant hypotheses using significance level .05.

1: 85.06 85.25 84.87

2: 84.99 84.28 84.88

3: 84.48 84.72 85.10

4: 84.10 84.55 84.05

42. The critical flicker frequency (cff) is the highest frequency (in cycles/sec) at which a person can detect the flicker in a flickering light source. At frequencies above the cff, the light source appears to be continuous even though it is actually flickering. An investigation carried out to see whether true average cff depends on iris color yielded the following data (based on the article "The Effects of Iris Color on Critical Flicker Frequency," *J. General Psych.*, 1973: 91–95).

Iris color

	1. Brown	2. Green	3. Blue
	26.8	26.4	25.7
	27.9	24.2	27.2
	23.7	28.0	29.9
	25.0	26.9	28.5
	26.3	29.1	29.4
	24.8		28.3
	25.7		
	24.5		
J_i	8	5	6
$x_{i\cdot}$	204.7	134.6	169.0
$\bar{x}_{i\cdot}$	25.59	26.92	28.17

$n = 19$ $x_{\cdot\cdot} = 508.3$

a. State and test the relevant hypotheses at significance level .05 by using the *F* table to obtain an upper and/or lower bound on the *P*-value. *Hint:* $\sum\sum x_{ij}^2 = 13{,}659.67$ and CF $= 13{,}598.36$.

b. Investigate differences between iris colors with respect to mean cff.

43. Let $c_1, c_2, \ldots, c_I$ be numbers satisfying $\sum c_i = 0$. Then $\sum c_i\mu_i = c_1\mu_1 + \cdots + c_I\mu_I$ is called a *contrast* in the μ_i's. Notice that with $c_1 = 1$, $c_2 = -1$, $c_3 = \cdots = c_I = 0$, $\sum c_i\mu_i = \mu_1 - \mu_2$, which implies that every pairwise difference between μ_i's is a contrast (so is, e.g., $\mu_1 - .5\mu_2 - .5\mu_3$). A method attributed to Scheffé gives simultaneous CIs with simultaneous confidence level $100(1 - \alpha)\%$ for *all* possible contrasts (an infinite number of them!). The interval for $\sum c_i\mu_i$ is

$$\sum c_i\bar{x}_{i\cdot} \pm (\sum c_i^2/J_i)^{1/2} \cdot [(I - 1) \cdot \text{MSE} \cdot F_{\alpha, I-1, n-I}]^{1/2}$$

Using the critical flicker frequency data of Exercise 42, calculate the Scheffé intervals for the contrasts $\mu_1 - \mu_2$, $\mu_1 - \mu_3$, $\mu_2 - \mu_3$, and $.5\mu_1 + .5\mu_2 - \mu_3$ (this last contrast compares blue to the average of brown and green). Which contrasts appear to differ significantly from 0, and why?

44. Four types of mortars—ordinary cement mortar (OCM), polymer impregnated mortar (PIM), resin

mortar (RM), and polymer cement mortar (PCM)—
were subjected to a compression test to measure
strength (MPa). Three strength observations for
each mortar type are given in the article. "Polymer
Mortar Composite Matrices for Maintenance-Free
Highly Durable Ferrocement" (*J. Ferrocement*,
1984: 337–345) and are reproduced here. Construct
an ANOVA table. Using a .05 significance level, determine whether the data suggests that the true
mean strength is not the same for all four mortar
types. If you determine that the true mean strengths
are not all equal, use Tukey's method to identify the
significant differences.

OCM	32.15	35.53	34.20
PIM	126.32	126.80	134.79
RM	117.91	115.02	114.58
PCM	29.09	30.87	29.80

45. Suppose the x_{ij}'s are "coded" by $y_{ij} = cx_{ij} + d$. How
does the value of the F statistic computed from the
y_{ij}'s compare to the value computed from the x_{ij}'s?
Justify your assertion.

46. In Example 10.10, subtract $\bar{x}_i$ from each observation in the ith sample ($i = 1, \ldots, 6$) to obtain a set
of 18 residuals. Then construct a normal probability plot and comment on the plausibility of the normality assumption.

Bibliography

Miller, Rupert, *Beyond ANOVA: The Basics of Applied
Statistics,* Wiley, New York, 1986. An excellent
source of information about assumption checking
and alternative methods of analysis.

Montgomery, Douglas, *Design and Analysis of Experiments* (4th ed.), Wiley, New York, 1997. A very up-
to-date presentation of ANOVA models and methodology.

Neter, John, William Wasserman, and Michael Kutner,
Applied Linear Statistical Models (4th ed.), Irwin,
Homewood, IL, 1996. The second half of this book

contains a very well-presented survey of ANOVA;
the level is comparable to that of the present text, but
the discussion is more comprehensive, making the
book an excellent reference.

Ott, Lyman, *An Introduction to Statistical Methods and
Data Analysis* (4th ed.), Duxbury Press, Belmont,
CA, 1993. Includes several chapters on ANOVA
methodology that can profitably be read by students
desiring a very nonmathematical exposition; there is
a good chapter on various multiple comparison
methods.

11

Multifactor Analysis of Variance

Introduction

In the previous chapter, we used the analysis of variance (ANOVA) to test for equality of either I different population means or the true average responses associated with I different levels of a single factor (alternatively referred to as I different treatments). In many experimental situations, there are two or more factors that are of simultaneous interest. This chapter extends the methods of Chapter 10 to investigate such multifactor situations.

In the first two sections, we concentrate on the case of two factors of interest. We will use I to denote the number of levels of the first factor (A) and J to denote the number of levels of the second factor (B). Then there are IJ possible combinations consisting of one level of factor A and one of factor B; each such combination is called a treatment, so there are IJ different treatments. The number of observations made on treatment (i, j) will be denoted by K_{ij}. In Section 11.1 we present the model and analysis when $K_{ij} = 1$. An important special case of this type is a randomized block design, in which a single factor A is of primary interest but another factor, "blocks," is created to control for extraneous variability in experimental units or subjects. In Section 11.2, we focus on the case $K_{ij} = K > 1$, and mention briefly the difficulties associated with unequal K_{ij}'s.

Section 11.3 considers experiments involving more than two factors, including a Latin square design, which controls for the effects of two extraneous factors thought to influence the response variable. When the number of factors is large, an experiment consisting of at least one observation for each treatment would be expensive and time-consuming. One important special case, which we discuss in Section 11.4, is that in which there are p factors, each of which has two levels. There are then 2^p different treatments. We consider both the case in which observations are made on all these treatments (a complete design) and the case in which observations are made for only a selected subset of treatments (an incomplete design).

11.1 Two-Factor ANOVA with $K_{ij} = 1$

When factor A consists of I levels and factor B consists of J levels, there are IJ different combinations (pairs) of levels of the two factors, each called a treatment. With $K_{ij} =$ the number of observations on the treatment consisting of factor A at level i and factor B at level j, we focus in this section on the case $K_{ij} = 1$, so that the data consists of IJ observations. We will first discuss the fixed effects model, in which the only levels of interest for the two factors are those actually represented in the experiment. The case in which one or both factors are random is discussed briefly at the end of the section.

Example 11.1 Is it really as easy to remove marks on fabrics from erasable pens as the word *erasable* might imply? Consider the following data from an experiment to compare three different brands of pens and four different wash treatments with respect to their ability to remove marks on a particular type of fabric (based on "An Assessment of the Effects of Treatment, Time, and Heat on the Removal of Erasable Pen Marks from Cotton and Cotton/Polyester Blend Fabrics," *J. of Testing and Evaluation,* 1991: 394–397). The response variable is a quantitative indicator of overall specimen color change; the lower this value, the more marks were removed.

		Washing Treatment				
		1	**2**	**3**	**4**	**Total**
	1	.97	.48	.48	.46	2.39
Brand of Pen	**2**	.77	.14	.22	.25	1.38
	3	.67	.39	.57	.19	1.82
	Total	2.41	1.01	1.27	.90	5.59

Is there any difference in the true average amount of color change due either to the different brands of pen or to the different washing treatments?   ■

As in single-factor ANOVA, double subscripts are used to identify random variables and observed values. Let

X_{ij} = the random variable (rv) denoting the measurement when factor A is held
at level i and factor B is held at level j

x_{ij} = the observed value of X_{ij}

The x_{ij}'s are usually presented in a two-way table in which the ith row contains the observed values when factor A is held at level i and the jth column contains the observed values when factor B is held at level j. In the erasable-pen experiment of Example 11.1, the number of levels of factor A is $I = 3$, the number of levels of factor B is $J = 4$, $x_{13} = .48$, $x_{22} = .14$, and so on.

Whereas in single-factor ANOVA we were interested only in row means and the grand mean, here we are interested also in column means. Let

$$\overline{X}_{i.} = \begin{array}{c} \text{the average of measurements obtained} \\ \text{when factor } A \text{ is held at level } i \end{array} = \frac{\displaystyle\sum_{j=1}^{J} X_{ij}}{J}$$

$$\overline{X}_{.j} = \begin{array}{c} \text{the average of measurements obtained} \\ \text{when factor } B \text{ is held at level } j \end{array} = \frac{\displaystyle\sum_{i=1}^{I} X_{ij}}{I}$$

$$\overline{X}_{..} = \text{the grand mean} = \frac{\displaystyle\sum_{i=1}^{I}\sum_{j=1}^{J} X_{ij}}{IJ}$$

with observed values $\overline{x}_{i.}$, $\overline{x}_{.j}$, and $\overline{x}_{..}$. Totals rather than averages are denoted by omitting the horizontal bar (so $x_{.j} = \sum_i x_{ij}$, etc.). Intuitively, to see whether there is any effect due to the levels of factor A, we should compare the observed $\overline{x}_{i.}$'s with one another, and information about the different levels of factor B should come from the $\overline{x}_{.j}$'s.

The Model

Proceeding by analogy to single-factor ANOVA, one's first inclination in specifying a model is to let μ_{ij} = the true average response when factor A is at level i and factor B at level j, giving IJ mean parameters. Then let

$$X_{ij} = \mu_{ij} + \epsilon_{ij}$$

where ϵ_{ij} is the random amount by which the observed value differs from its expectation and the ϵ_{ij}'s are assumed normal and independent with common variance σ^2. Unfortunately, there is no valid test procedure for this choice of parameters. The reason is that under the alternative hypothesis of interest, the μ_{ij}'s are free to take on any values whatsoever, whereas σ^2 can be any value greater than zero, so that there are $IJ + 1$ freely varying parameters. But there are only IJ observations, so after using each x_{ij} as an estimate of μ_{ij}, there is no way to estimate σ^2.

To rectify this problem of a model having more parameters than observed values, we must specify a model that is realistic yet involves relatively few parameters.

Assume the existence of I parameters $\alpha_1, \alpha_2, \ldots, \alpha_I$ and J parameters $\beta_1, \beta_2, \ldots,$ β_J such that

$$X_{ij} = \alpha_i + \beta_j + \epsilon_{ij} \qquad (i = 1, \ldots, I, \quad j = 1, \ldots, J) \qquad (11.1)$$

so that

$$\mu_{ij} = \alpha_i + \beta_j \qquad (11.2)$$

Including σ^2, there are now $I + J + 1$ model parameters, so if $I \geq 3$ and $J \geq 3$, then there will be fewer parameters than observations (in fact, we will shortly modify (11.2) so that even $I = 2$ and/or $J = 2$ will be accommodated).

The model specified in (11.1) and (11.2) is called an **additive model** because each mean response μ_{ij} is the sum of an effect due to factor A at level i (α_i) and an effect due to factor B at level j (β_j). The difference between mean responses for factor A at level i and level i' when B is held at level j is $\mu_{ij} - \mu_{i'j}$. When the model is additive,

$$\mu_{ij} - \mu_{i'j} = (\alpha_i + \beta_j) - (\alpha_{i'} + \beta_j) = \alpha_i - \alpha_{i'}$$

which is independent of the level j of the second factor. A similar result holds for $\mu_{ij} - \mu_{ij'}$. Thus, additivity means that the difference in mean responses for two levels of one of the factors is the same for all levels of the other factor. Figure 11.1(a) shows a set of mean responses that satisfy the condition of additivity, and Figure 11.1(b) shows a non-additive configuration of mean responses.

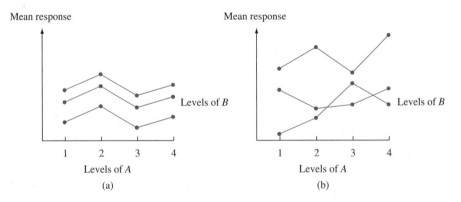

Figure 11.1 Mean responses for two types of model: (a) additive; (b) nonadditive

Example 11.2
(Example 11.1 continued)

When we plot the observed x_{ij}'s in a manner analogous to that of Figure 11.1, we get the result shown in Figure 11.2. Although there is some "crossing over" in the observed x_{ij}'s, the configuration is reasonably representative of what would be expected under additivity with just one observation per treatment.

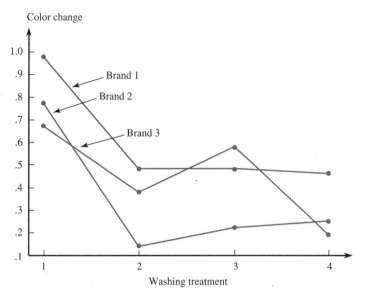

Figure 11.2 Plot of data from Example 11.1   ■

Expression (11.2) is not quite the final model description because the α_i's and β_j's are not uniquely determined. Following are two different configurations of the α_i's and β_j's that yield the same additive μ_{ij}'s.

	$\beta_1 = 1$	$\beta_2 = 4$
$\alpha_1 = 1$	$\mu_{11} = 2$	$\mu_{12} = 5$
$\alpha_2 = 2$	$\mu_{21} = 3$	$\mu_{22} = 6$

	$\beta_1 = 2$	$\beta_2 = 5$
$\alpha_1 = 0$	$\mu_{11} = 2$	$\mu_{12} = 5$
$\alpha_2 = 1$	$\mu_{21} = 3$	$\mu_{22} = 6$

By subtracting any constant c from all α_i's and adding c to all β_j's, other configurations corresponding to the same additive model are obtained. This nonuniqueness is eliminated by use of the following model.

$$X_{ij} = \mu + \alpha_i + \beta_j + \epsilon_{ij} \tag{11.3}$$

where $\displaystyle\sum_{i=1}^{I} \alpha_i = 0$, $\displaystyle\sum_{j=1}^{J} \beta_j = 0$, and the ϵ_{ij}'s are assumed independent, normally distributed, with mean 0 and common variance σ^2.

This is analogous to the alternative choice of parameters for single-factor ANOVA discussed in Section 10.3. It is not difficult to verify that (11.3) is an additive model in which the parameters are uniquely determined (for example, for the μ_{ij}'s mentioned previously, $\mu = 4$, $\alpha_1 = -.5$, $\alpha_2 = .5$, $\beta_1 = -1.5$, and $\beta_2 = 1.5$). Notice that there are only $I - 1$ independently determined α_i's and $J - 1$ independently determined β_j's, so (including μ) (11.3) specifies $I + J - 1$ mean parameters.

The interpretation of the parameters of (11.3) is straightforward: μ is the true grand mean (mean response averaged over all levels of both factors), α_i is the effect of factor A at level i (measured as a deviation from μ), and β_j is the effect of factor B at level j. Unbiased (and maximum likelihood) estimators for these parameters are

$$\hat{\mu} = \overline{X}_{..} \qquad \hat{\alpha}_i = \overline{X}_{i.} - \overline{X}_{..} \qquad \hat{\beta}_j = \overline{X}_{.j} - \overline{X}_{..}$$

There are two different hypotheses of interest in a two-factor experiment with $K_{ij} = 1$. The first, denoted by H_{0A}, states that the different levels of factor A have no effect on true average response. The second, denoted by H_{0B}, asserts that there is no factor B effect.

$$H_{0A}: \; \alpha_1 = \alpha_2 = \cdots = \alpha_I = 0$$
$$\text{versus } H_{aA}: \text{ at least one } \alpha_i \neq 0$$
$$H_{0B}: \; \beta_1 = \beta_2 = \cdots = \beta_J = 0 \tag{11.4}$$
$$\text{versus } H_{aB}: \text{ at least one } \beta_j \neq 0$$

(No factor A effect implies that all α_i's are equal, so they must all be 0 since they sum to 0, and similarly for the β_j's.)

The Test Procedures

The description and analysis now follow closely that for single-factor ANOVA. The relevant sums of squares and their computing forms are given by

$$\text{SST} = \sum_{i=1}^{I}\sum_{j=1}^{J} (X_{ij} - \overline{X}_{..})^2 = \sum_{i=1}^{I}\sum_{j=1}^{J} X_{ij}^2 - \frac{1}{IJ}X_{..}^2 \qquad \text{df} = IJ - 1$$

$$\text{SSA} = \sum_{i=1}^{I}\sum_{j=1}^{J} (\overline{X}_{i.} - \overline{X}_{..})^2 = \frac{1}{J}\sum_{i=1}^{I} X_{i.}^2 - \frac{1}{IJ}X_{..}^2 \qquad \text{df} = I - 1$$

$$\text{SSB} = \sum_{i=1}^{I}\sum_{j=1}^{J} (\overline{X}_{.j} - \overline{X}_{..})^2 = \frac{1}{I}\sum_{j=1}^{J} X_{.j}^2 - \frac{1}{IJ}X_{..}^2 \qquad \text{df} = J - 1 \tag{11.5}$$

$$\text{SSE} = \sum_{i=1}^{I}\sum_{j=1}^{J} (X_{ij} - \overline{X}_{i.} - \overline{X}_{.j} + \overline{X}_{..})^2 \qquad \text{df} = (I-1)(J-1)$$

and the fundamental identity

$$\text{SST} = \text{SSA} + \text{SSB} + \text{SSE} \tag{11.6}$$

allows SSE to be determined by subtraction.

The expression for SSE results from replacing μ, α_i, and β_j in $\sum[X_{ij} - (\mu + \alpha_i + \beta_j)]^2$ by their estimators. Error df is $IJ - $ number of mean parameters estimated $= IJ - [1 +$

$(I - 1) + (J - 1)] = (I - 1)(J - 1)$. As in single-factor ANOVA, total variation is split into a part (SSE) that is not explained by either the truth or the falsity of H_{0A} or H_{0B}, and two parts that can be explained by possible falsity of the two null hypotheses.

Statistical theory now says that if we form F ratios as in single-factor ANOVA, when H_{0A} (H_{0B}) is true, the corresponding F ratio has an F distribution with numerator df $= I - 1$ ($J - 1$) and denominator df $= (I - 1)(J - 1)$.

Hypotheses	Test Statistic Value	Rejection Region
H_{0A} versus H_{aA}	$f_A = \dfrac{\text{MSA}}{\text{MSE}}$	$f_A \geq F_{\alpha, I-1, (I-1)(J-1)}$
H_{0B} versus H_{aB}	$f_B = \dfrac{\text{MSB}}{\text{MSE}}$	$f_B \geq F_{\alpha, J-1, (I-1)(J-1)}$

Example 11.3
(Example 11.2 continued)

The $x_{i\cdot}$'s (row totals) and $x_{\cdot j}$'s (column totals) for the color change data are displayed along the right and bottom margins of the data table given previously. In addition, $\sum\sum x_{ij}^2 = 3.2987$ and the correction factor is $x_{\cdot\cdot}^2/(IJ) = (5.59)^2/12 = 2.6040$. The sums of squares are then

$$\text{SST} = 3.2987 - 2.6040 = .6947$$

$$\text{SSA} = \frac{1}{4}[(2.39)^2 + (1.38)^2 + (1.82)^2] - 2.6040 = .1282$$

$$\text{SSB} = \frac{1}{3}[(2.41)^2 + (1.01)^2 + (1.27)^2 + (.90)^2] - 2.6040 = .4797$$

$$\text{SSE} = .6947 - (.1282 + .4797) = .0868$$

The accompanying ANOVA table (Table 11.1) summarizes further calculations.

Table 11.1 ANOVA table for Example 11.3

Source of Variation	df	Sum of Squares	Mean Square	f
Factor A (brand)	$I - 1 = 2$	SSA = .1282	MSA = .0641	$f_A = 4.43$
Factor B (wash treatment)	$J - 1 = 3$	SSB = .4797	MSB = .1599	$f_B = 11.05$
Error	$(I - 1)(J - 1) = 6$	SSE = .0868	MSE = .01447	
Total	$IJ - 1 = 11$	SST = .6947		

The critical value for testing H_{0A} at level of significance .05 is $F_{.05, 2, 6} = 5.14$. Since $4.43 < 5.14$, H_{0A} cannot be rejected at significance level .05. True average color change does not appear to depend on brand of pen. Because $F_{.05, 3, 6} = 4.76$ and $11.05 \geq 4.76$, H_{0B} is rejected at significance level .05 in favor of the assertion that color change varies with washing treatment. A statistical computer package gives P-values of .066 and .007 for these two tests. ∎

Expected Mean Squares

The plausibility of using the F tests just described is demonstrated by computing the expected mean squares. After some tedious algebra,

$$E(\text{MSE}) = \sigma^2 \quad \text{(when the model is additive)}$$

$$E(\text{MSA}) = \sigma^2 + \frac{J}{I-1} \sum_{i=1}^{I} \alpha_i^2$$

$$E(\text{MSB}) = \sigma^2 + \frac{I}{J-1} \sum_{j=1}^{J} \beta_j^2$$

When H_{0A} is true, MSA is an unbiased estimator of σ^2, so F is a ratio of two unbiased estimators of σ^2. When H_{0A} is false, MSA tends to overestimate σ^2, so H_{0A} should be rejected when the ratio F_A is too large. Similar comments apply to MSB and H_{0B}.

Multiple Comparisons in Two-Factor ANOVA

When either H_{0A} or H_{0B} has been rejected, Tukey's procedure can be used to identify significant differences between the levels of the factor under investigation. The steps in the analysis are identical to those for a single-factor ANOVA:

1. For comparing levels of factor A, obtain $Q_{\alpha,I,(I-1)(J-1)}$.
For comparing levels of factor B, obtain $Q_{\alpha,J,(I-1)(J-1)}$.

2. Compute

$$w = Q \cdot \text{(estimated standard deviation of the sample means being compared)}$$

$$= \begin{cases} Q_{\alpha,I,(I-1)(J-1)} \cdot \sqrt{\text{MSE}/J} & \text{for factor } A \text{ comparisons} \\ Q_{\alpha,J,(I-1)(J-1)} \cdot \sqrt{\text{MSE}/I} & \text{for factor } B \text{ comparisons} \end{cases}$$

(because, e.g., the standard deviation of $\overline{X}_{i\cdot}$ is $\sigma/\sqrt{J}$).

3. Arrange the sample means in increasing order, underscore those pairs differing by less than w, and identify pairs not underscored by the same line as corresponding to significantly different levels of the given factor.

Example 11.4
(Example 11.3 continued)

Identification of significant differences among the four washing treatments requires $Q_{.05,4,6} = 4.90$ and $w = 4.90\sqrt{(.01447)/3} = .340$. The four factor B sample means (column averages) are now listed in increasing order, and any pair differing by less than .340 is underscored by a line segment:

$$\begin{array}{cccc} \overline{x}_{4\cdot} & \overline{x}_{2\cdot} & \overline{x}_{3\cdot} & \overline{x}_{1\cdot} \\ .300 & .337 & .423 & .803 \end{array}$$

Washing treatment 1 appears to differ significantly from the other three treatments, but no other significant differences are identified. In particular, it is not apparent which among treatments 2, 3, and 4 is best at removing marks. ∎

Randomized Block Experiments

In using single-factor ANOVA to test for the presence of effects due to the I different treatments under study, once the IJ subjects or experimental units have been chosen, treatments should be allocated in a completely random fashion. That is, J subjects should be chosen at random for the first treatment, then another sample of J chosen at random from the remaining $IJ - J$ subjects for the second treatment, and so on.

It frequently happens, though, that subjects or experimental units exhibit heterogeneity with respect to other variables that may affect the observed responses. When this is the case, the presence or absence of a significant F value may be due to this extraneous variation rather than to the presence or absence of factor effects. This was the reason for introducing a paired experiment in Chapter 9. The analogy to a paired experiment when $I > 2$ is called a **randomized block** experiment. An extraneous factor, "blocks," is constructed by dividing the IJ units into J groups with I units in each group. This grouping or blocking is done in such a way that within each block, the I units are homogeneous with respect to other factors thought to affect the responses. Then within each homogeneous block, the I treatments are randomly assigned to the I units or subjects in the block.

Example 11.5 A consumer product-testing organization wished to compare the annual power consumption for five different brands of dehumidifier. Because power consumption depends on the prevailing humidity level, it was decided to monitor each brand at four different levels ranging from moderate to heavy humidity (thus blocking on humidity level). Within each level, brands were randomly assigned to the five selected locations. The resulting amount of power consumption (annual kwh) appears in Table 11.2.

Table 11.2 Power consumption data for Example 11.5

Treatments (brands)	Blocks (humidity level) 1	2	3	4	$x_{i.}$	$\bar{x}_{i.}$
1	685	792	838	875	3190	797.50
2	722	806	893	953	3374	843.50
3	733	802	880	941	3356	839.00
4	811	888	952	1005	3656	914.00
5	828	920	978	1023	3749	937.25
$x_{.j}$	3779	4208	4541	4797	17,325	

Since $\sum\sum x_{ij}^2 = 15{,}178{,}901.00$ and $x_{..}^2/(IJ) = 15{,}007{,}781.25$,

$$\text{SST} = 15{,}178{,}901.00 - 15{,}007{,}781.25 = 171{,}119.75$$

$$\text{SSA} = \frac{1}{4}[60{,}244{,}049] - 15{,}007{,}781.25 = 53{,}231.00$$

$$\text{SSB} = \frac{1}{5}[75{,}619{,}995] - 15{,}007{,}781.25 = 116{,}217.75$$

and

$$\text{SSE} = 171{,}119.75 - 53{,}231.00 - 116{,}217.75 = 1671.00$$

The ANOVA calculations are summarized in Table 11.3 on page 442.

Table 11.3 ANOVA table for Example 11.5

Source of Variation	df	Sum of Squares	Mean Square	f
Treatments (brands)	4	53,231.00	13,307.75	$f_A = 95.57$
Blocks	3	116,217.75	38,739.25	$f_B = 278.20$
Error	12	1671.00	139.25	
Total	19	171,119.75		

Since $F_{.05,4,12} = 3.26$ and $f_A = 95.57 \geq 3.26$, H_0 is rejected in favor of H_a, and we conclude that power consumption does depend on the brand of humidifier. To identify significantly different brands, we use Tukey's procedure. $Q_{.05,5,12} = 4.51$ and $w = 4.51\sqrt{139.25/4} = 26.6$.

$\bar{x}_{1\cdot}$	$\bar{x}_{3\cdot}$	$\bar{x}_{2\cdot}$	$\bar{x}_{4\cdot}$	$\bar{x}_{5\cdot}$
797.50	839.00	843.50	914.00	937.25

The underscoring indicates that the brands can be divided into three groups with respect to power consumption.

Because the block factor is of secondary interest, $F_{.05,3,12}$ is not needed, though the computed value of F_B is clearly highly significant. Figure 11.3 shows SAS output for this data. Notice that in the first part of the ANOVA table, the sums of squares (SS's) for treatments (brands) and blocks (humidity levels) are combined into a single "model" SS.

```
                    Analysis of Variance Procedure
Dependent Variable: POWERUSE
                                Sum of              Mean
Source                  DF      Squares             Square      F Value      Pr > F
Model                    7     169448.750         24206.964     173.84       0.0001
Error                   12       1671.000           139.250
Corrected Total         19     171119.750

              R-Square          C.V.          Root MSE      POWERUSE Mean
              0.990235        1.362242          11.8004        866.25000

Source                  DF      Anova SS        Mean Square      F Value      PR > F
BRAND                    4     53231.000        13307.750         95.57       0.0001
HUMIDITY                 3    116217.750        38739.250        278.20       0.0001

         Alpha = 0.05  df = 12  MSE = 139.25
     Critical Value of Studentized Range = 4.508
         Minimum Significant Difference = 26.597

Means with the same letter are not significantly different.
      Tukey Grouping          Mean        N        BRAND
                    A       937.250        4          5
                    A
                    A       914.000        4          4
                    B       843.500        4          2
                    B
                    B       839.000        4          3
                    C       797.500        4          1
```

Figure 11.3 SAS output for power consumption data

In many experimental situations in which treatments are to be applied to subjects, a single subject can receive all I of the treatments. Blocking is then often done on the subjects themselves to control for variability between subjects; each subject is then said to act as its own control. Social scientists sometimes refer to such experiments as repeated-measures designs. The "units" within a block are then the different "instances" of treatment application. Similarly, blocks are often taken as different time periods, locations, or observers.

Example 11.6 The data in Table 11.4 is from the article "Compounding of Discriminative Stimuli from the Same and Different Sensory Modalities" (*J. Experimental Analysis Behavior,* 1971: 337–342). Rat response was maintained by fixed interval schedules of reinforcement in the presence of a tone or two separate lights. The lights were either of moderate (L1) or low intensity (L2). Observations are given as the mean number of responses emitted by each subject during single and compound stimuli presentations over a 4-day period.

Table 11.4 Response data for Example 11.6

Stimulus	Subject 1	2	3	4	$x_{i.}$	$\bar{x}_{i.}$
L1	8.0	17.3	52.0	22.0	99.3	24.8
L2	6.9	19.3	63.7	21.6	111.5	27.9
Tone (T)	9.3	18.8	60.0	28.3	116.4	29.1
L1 + L2	9.2	24.9	82.4	44.9	161.4	40.3
L1 + T	12.0	31.7	83.8	37.4	164.9	41.2
L2 + T	9.4	33.6	96.6	40.6	180.2	45.1
$x_{.j}$	54.8	145.6	438.5	194.8	833.7	

With $\sum\sum x_{ij}^2 = 44{,}614.21$, SST $= 15{,}653.56$, SSA $= 1428.28$, SSB $= 13{,}444.63$, and SSE $= 780.65$. Further calculations are summarized in Table 11.5.

Table 11.5 ANOVA table for Example 11.6

Source of Variation	df	Sum of Squares	Mean Square	f
Stimuli (A)	5	1428.28	285.66	$f_A = 5.49$
Subjects (B)	3	13,444.63	4481.54	$f_B = 86.12$
Error	15	780.65	52.04	
Total	23	15,653.56		

Since $F_{.05,5,15} = 2.90$ and $5.49 \geq 2.90$, we conclude that there are differences in the true average responses associated with the different stimuli. For Tukey's procedure, $w = 4.59\sqrt{52.04/4} = 16.56$.

$\bar{x}_{1.}$	$\bar{x}_{2.}$	$\bar{x}_{3.}$	$\bar{x}_{4.}$	$\bar{x}_{5.}$	$\bar{x}_{6.}$
24.8	27.9	29.1	40.3	41.2	45.1

Thus, both L1 and L2 are significantly different from L2 + T, and there are no other significant differences among the stimuli. ∎

In most randomized block experiments in which subjects serve as blocks, the subjects actually participating in the experiment are selected from a large population. The subjects then contribute random rather than fixed effects. This does not affect the procedure for comparing treatments when $K_{ij} = 1$ (one observation per "cell," as in this section), but the procedure is altered if $K_{ij} = K > 1$. We will shortly consider two-factor models in which effects are random.

More on Blocking

When $I = 2$, either the F test or the paired differences t test can be used to analyze the data. The resulting conclusion will not depend on which procedure is used, since $T^2 = F$ and $t_{\alpha/2,\nu}^2 = F_{\alpha,1,\nu}$.

Just as with pairing, blocking entails both a potential gain and a potential loss in precision. If there is a great deal of heterogeneity in experimental units, the value of the variance parameter σ^2 in the one-way model will be large. The effect of blocking is to filter out the variation represented by σ^2 in the two-way model appropriate for a randomized block experiment. Other things being equal, a smaller value of σ^2 results in a test that is more likely to detect departures from H_0 (i.e., a test with greater power).

However, other things are not equal here, since the single-factor F test is based on $I(J - 1)$ degrees of freedom (df) for error, whereas the two-factor F test is based on $(I - 1)(J - 1)$ df for error. Fewer degrees of freedom for error results in a decrease in power, essentially because the denominator estimator of σ^2 is not as precise. This loss in degrees of freedom can be especially serious if the experimenter can afford only a small number of observations. Nevertheless, if it appears that blocking will significantly reduce variability, it is probably worth the loss in degrees of freedom.

Models for Random Effects

In many experiments, the actual levels of a factor used in the experiment, rather than being the only ones of interest to the experimenter, have been selected from a much larger population of possible levels of the factor. In a two-factor situation, when this is the case for both factors, a **random effects model** is appropriate. The case in which the levels of one factor are the only ones of interest and the levels of the other factor are selected from a population of levels leads to a **mixed effects model.** The two-factor random effects model when $K_{ij} = 1$ is

$$X_{ij} = \mu + A_i + B_j + \epsilon_{ij} \qquad (i = 1, \ldots, I, \quad j = 1, \ldots, J)$$

where the A_i's, B_j's, and ϵ_{ij}'s are all independent, normally distributed rv's with mean 0 and variances σ_A^2, σ_B^2, and σ^2, respectively. The hypotheses of interest are then H_{0A}: $\sigma_A^2 = 0$ (level of factor A does not contribute to variation in the response) versus H_{aA}:

$\sigma_A^2 > 0$ and H_{0B}: $\sigma_B^2 = 0$ versus H_{aB}: $\sigma_B^2 > 0$. Whereas $E(MSE) = \sigma^2$ as before, the expected mean squares for factors A and B are now

$$E(MSA) = \sigma^2 + J\sigma_A^2 \qquad E(MSB) = \sigma^2 + I\sigma_B^2$$

Thus, when H_{0A} (H_{0B}) is true, F_A (F_B) is still a ratio of two unbiased estimators of σ^2. It can be shown that a level α test for H_{0A} versus H_{aA} still rejects H_{0A} if $f_A \geq F_{\alpha, I-1, (I-1)(J-1)}$, and similarly the same procedure as before is used to decide between H_{0B} and H_{aB}.

For the case in which factor A is fixed and factor B is random, the mixed model is

$$X_{ij} = \mu + \alpha_i + B_j + \epsilon_{ij} \qquad (i = 1, \ldots, I, \quad j = 1, \ldots, J)$$

where $\sum \alpha_i = 0$ and the B_j's and ϵ_{ij}'s are normally distributed with mean 0 and variances σ_B^2 and σ^2, respectively. Now the two null hypotheses are

$$H_{0A}: \alpha_1 = \cdots = \alpha_I = 0 \qquad \text{and} \qquad H_{0B}: \sigma_B^2 = 0$$

with expected mean squares

$$E(MSE) = \sigma^2 \qquad E(MSA) = \sigma^2 + \frac{J}{I-1}\sum \alpha_i^2 \qquad E(MSB) = \sigma^2 + I\sigma_B^2$$

The test procedures for H_{0A} versus H_{aA} and H_{0B} versus H_{aB} are exactly as before. For example, in the analysis of the color change data in Example 11.1, if the four wash treatments were randomly selected, then because $f_B = 11.05$ and $F_{.05,3,6} = 4.76$, H_{0B}: $\sigma_B^2 = 0$ is rejected in favor of H_{aB}: $\sigma_B^2 > 0$. An estimate of the "variance component" σ_B^2 is then given by $(MSB - MSE)/I = .0485$.

Summarizing, when $K_{ij} = 1$, although the hypotheses and expected mean squares differ from the case of both effects fixed, the test procedures are identical.

Exercises | Section 11.1 (1–15)

1. The number of miles of useful tread wear (in 1000's) was determined for tires of each of five different makes of subcompact car (factor A, with $I = 5$) in combination with each of four different brands of radial tires (factor B, with $J = 4$), resulting in $IJ = 20$ observations. The values SSA = 30.6, SSB = 44.1, and SSE = 59.2 were then computed. Assume that an additive model is appropriate.

   a. Test H_0: $\alpha_1 = \alpha_2 = \alpha_3 = \alpha_4 = \alpha_5 = 0$ (no differences in true average tire lifetime due to makes of cars) versus H_a: at least one $\alpha_i \neq 0$ using a level .05 test.

   b. H_0: $\beta_1 = \beta_2 = \beta_3 = \beta_4 = 0$ (no differences in true average tire lifetime due to brands of tires) versus H_a: at least one $\beta_j \neq 0$ using a level .05 test.

2. Four different coatings are being considered for corrosion protection of metal pipe. The pipe will be buried in three different types of soil. To investigate whether the amount of corrosion depends either on the coating or on the type of soil, 12 pieces of pipe are selected. Each piece is coated with one of the four coatings and buried in one of the three types of soil for a fixed time, after which the amount of corrosion (depth of maximum pits, in .0001 in.) is determined. The data appears in the accompanying table:

		Soil Type (B)		
		1	**2**	**3**
	1	64	49	50
Coating (A)	2	53	51	48
	3	47	45	50
	4	51	43	52

a. Assuming the validity of the additive model, carry out the ANOVA analysis using an ANOVA table to see whether the amount of corrosion depends on either the type of coating used or the type of soil. Use $\alpha = .05$.

b. Compute $\hat{\mu}, \hat{\alpha}_1, \hat{\alpha}_2, \hat{\alpha}_3, \hat{\alpha}_4, \hat{\beta}_1, \hat{\beta}_2,$ and $\hat{\beta}_3$.

3. The article "Adiabatic Humidification of Air with Water in a Packed Tower" (*Chem. Eng. Prog.*, 1952: 362–370) reports data on gas film heat transfer coefficient (Btu/hr ft^2 on °F) as a function of gas rate (factor A) and liquid rate (factor B).

			B		
		1(190)	**2(250)**	**3(300)**	**4(400)**
	1(200)	200	226	240	261
A	**2(400)**	278	312	330	381
	3(700)	369	416	462	517
	4(1100)	500	575	645	733

a. After constructing an ANOVA table, test at level .01 both the hypothesis of no gas-rate effect against the appropriate alternative and the hypothesis of no liquid-rate effect against the appropriate alternative.

b. Use Tukey's procedure to investigate differences in expected heat transfer coefficient due to different gas rates.

c. Repeat part (b) for liquid rates.

4. In an experiment to see whether the amount of coverage of light-blue interior latex paint depends either on the brand of paint or on the brand of roller used, 1 gallon of each of four brands of paint was applied using each of three brands of roller, resulting in the following data (number of square feet covered).

		Roller Brand		
		1	**2**	**3**
	1	454	446	451
Paint	2	446	444	447
Brand	3	439	442	444
	4	444	437	443

a. Construct the ANOVA table. [*Hint:* The computations can be expedited by subtracting 400 (or any other convenient number) from each observation. This does not affect the final results.]

b. State and test hypotheses appropriate for deciding whether paint brand has any effect on coverage. Use $\alpha = .05$.

c. Repeat part (b) for brand of roller.

d. Use Tukey's method to identify significant differences among brands. Is there one brand that seems clearly preferable to the others?

5. In an experiment to assess the effect of the angle of pull on the force required to cause separation in electrical connectors, four different angles (factor A) were used and each of a sample of five connectors (factor B) was pulled once at each angle ("A Mixed Model Factorial Experiment in Testing Electrical Connectors," *Industrial Quality Control*, 1960: 12–16). The data appears in the accompanying table:

				B		
		1	**2**	**3**	**4**	**5**
	0°	45.3	42.2	39.6	36.8	45.8
	2°	44.1	44.1	38.4	38.0	47.2
A	**4°**	42.7	42.7	42.6	42.2	48.9
	6°	43.5	45.8	47.9	37.9	56.4

Does the data suggest that true average separation force is affected by the angle of pull? State and test the appropriate hypotheses at level .01 by first constructing an ANOVA table (SST = 396.13, SSA = 58.16, and SSB = 246.97).

6. A particular county employs three assessors who are responsible for determining the value of residential property in the county. To see whether these assessors differ systematically in their assessments, five houses are selected, and each assessor is asked to determine the market value of each house. With factor A denoting assessors ($I = 3$) and factor B denoting houses ($J = 5$), suppose SSA = 11.7, SSB = 113.5, and SSE = 25.6.

a. Test $H_0: \alpha_1 = \alpha_2 = \alpha_3 = 0$ at level .05. (H_0 states that there are no systematic differences among assessors.)

b. Explain why a randomized block experiment with only 5 houses was used rather than a one-way ANOVA experiment involving a total of 15 different houses with each assessor asked to assess 5 different houses (a different group of 5 for each assessor).

7. The article "Rate of Stuttering Adaptation Under Two Electro-Shock Conditions" (*Behavior Research Therapy,* 1967: 49–54) gives adaptation scores for three different treatments: (1) no shock, (2) shock following each stuttered word, and (3) shock during each moment of stuttering. These treatments were used on each of 18 stutterers.

    a. Summary statistics include $x_{1.} = 905$, $x_{2.} = 913$, $x_{3.} = 936$, $x_{..} = 2754$, $\sum_j x_{.j}^2 = 430,295$, and $\sum\sum x_{ij}^2 = 143,930$. Construct the ANOVA table and test at level .05 to see whether true average adaptation score depends on the treatment given.

    b. Judging from the F ratio for subjects (factor B), do you think that blocking on subjects was effective in this experiment? Explain.

8. The accompanying table gives plasma epinephrine concentration for ten experimental subjects during (1) isoflurane, (2) halothane, and (3) cyclopropane anesthesia ("Sympathoadrenal and Hemodynamic Effects of Isoflurane, Halothane, and Cyclopropane in Dogs," *Anesthesiology,* 1974: 465–470).

    a. Does the choice of anesthetic affect true average concentration? Test H_0: $\alpha_1 = \alpha_2 = \alpha_3 = 0$ at level .05 after constructing the ANOVA table.

		Subject (B)			
	1	**2**	**3**	**4**	**5**
Anesthetic 1	.28	.51	1.00	.39	.29
(A) 2	.30	.39	.63	.38	.21
3	1.07	1.35	.69	.28	1.24

	6	**7**	**8**	**9**	**10**
Anesthetic 1	.36	.32	.69	.17	.33
(A) 2	.88	.39	.51	.32	.42
3	1.53	.49	.56	1.02	.30

$$\sum\sum x_{ij}^2 = 13.7980$$

    b. Use Tukey's procedure to investigate significant differences among the anesthetics.

9. The article "The Effects of a Pneumatic Stool and a One-Legged Stool on Lower Limb Joint Load and Muscular Activity During Sitting and Rising" (*Ergonomics,* 1993: 519–535) gives the accompanying data on the effort required of a subject to arise from four different types of stools (Borg scale). Perform an analysis of variance using $\alpha = .05$ and follow this with a multiple comparisons analysis if appropriate.

		Subject									
		1	**2**	**3**	**4**	**5**	**6**	**7**	**8**	**9**	$\bar{x}_{i.}$
Type of Stool	**1**	12	10	7	7	8	9	8	7	9	8.56
	2	15	14	14	11	11	11	12	11	13	12.44
	3	12	13	13	10	8	11	12	8	10	10.78
	4	10	12	9	9	7	10	11	7	8	9.22

10. The strength of concrete used in commercial construction tends to vary from one batch to another. Consequently, small test cylinders of concrete sampled from a batch are "cured" for periods up to about 28 days in temperature- and moisture-controlled environments before strength measurements are made. Concrete is then "bought and sold on the basis of strength test cylinders" (ASTM C 31 Standard Test Method for Making and Curing Concrete Test Specimens in the Field). The accompanying data resulted from an experiment carried out to compare three different curing methods with respect to compressive strength (MPa). Analyze this data.

Batch	Method A	Method B	Method C
1	30.7	33.7	30.5
2	29.1	30.6	32.6
3	30.0	32.2	30.5
4	31.9	34.6	33.5
5	30.5	33.0	32.4
6	26.9	29.3	27.8
7	28.2	28.4	30.7
8	32.4	32.4	33.6
9	26.6	29.5	29.2
10	28.6	29.4	33.2

11. The "residuals" from a two-factor ANOVA with $K_{ij} = 1$ are the quantities $x_{ij} - (\hat{\mu} + \hat{\alpha}_i + \hat{\beta}_j)$. A normal probability plot of these residuals can be used as a plausibility check of the normality assumption. Construct such a plot for the data of Example 11.1 and comment.

12. Suppose that in the experiment described in Exercise 6 the five houses had actually been selected at random from among those of a certain age and size, so that factor B is random rather than fixed. Test H_0: $\sigma_B^2 = 0$ versus H_a: $\sigma_B^2 > 0$ using a level .01 test.

13. a. Show that a constant d can be added to (or subtracted from) each x_{ij} without affecting any of the ANOVA sums of squares.

    b. Suppose that each x_{ij} is multiplied by a nonzero constant c. How does this affect the ANOVA sums of squares? How does this affect the values of the F statistics F_A and F_B? What effect does "coding" the data by $y_{ij} = cx_{ij} + d$ have on the conclusions resulting from the ANOVA procedures?

14. Use the fact that $E(X_{ij}) = \mu + \alpha_i + \beta_j$ with $\Sigma\alpha_i = \Sigma\beta_j = 0$ to show that $E(\overline{X}_{i.} - \overline{X}_{..}) = \alpha_i$, so that $\hat{\alpha}_i = \overline{X}_{i.} - \overline{X}_{..}$ is an unbiased estimator for α_i.

15. The power curves of Figures 10.5 and 10.6 can be used to obtain $\beta = P(\text{type II error})$ for the F test in two-factor ANOVA. For fixed values of $\alpha_1, \alpha_2, \ldots, \alpha_I$, the quantity $\phi^2 = (J/I)\Sigma\alpha_i^2/\sigma^2$ is computed.

Then the figure corresponding to $\nu_1 = I - 1$ is entered on the horizontal axis at the value ϕ, the power is read on the vertical axis from the curve labeled $\nu_2 = (I - 1)(J - 1)$, and $\beta = 1 - \text{power}$.

a. For the corrosion experiment described in Exercise 2, find β when $\alpha_1 = 4$, $\alpha_2 = 0$, $\alpha_3 = \alpha_4 = -2$, and $\sigma = 4$. Repeat for $\alpha_1 = 6$, $\alpha_2 = 0$, $\alpha_3 = \alpha_4 = -3$, and $\sigma = 4$.

b. By symmetry, what is β for the test of H_{0B} versus H_{aB} in Example 11.1 when $\beta_1 = .3$, $\beta_2 = \beta_3 = \beta_4 = -.1$, and $\sigma = .3$?

11.2 Two-Factor ANOVA with $K_{ij} > 1$

In Section 11.1, we analyzed data from a two-factor experiment in which there was one observation for each of the IJ combinations of levels of the two factors. To obtain valid test procedures, the μ_{ij}'s were assumed to have an additive structure with $\mu_{ij} = \mu + \alpha_i + \beta_j$, $\Sigma\alpha_i = \Sigma\beta_j = 0$. Additivity means that the difference in true average responses for any two levels of the factors is the same for each level of the other factor. For example, $\mu_{ij} - \mu_{i'j} = (\mu + \alpha_i + \beta_j) - (\mu + \alpha_{i'} + \beta_j) = \alpha_i - \alpha_{i'}$ independent of the level j of the second factor. This is shown in Figure 11.1(a), in which the lines connecting true average responses are parallel.

Figure 11.1(b) depicts a set of true average responses that does not have additive structure. The lines connecting these μ_{ij}'s are not parallel, which means that the difference in true average responses for different levels of one factor does depend on the level of the other factor. When additivity does not hold, we say that there is **interaction** between the different levels of the factors. The assumption of additivity allowed us in Section 11.1 to obtain an estimator of the random error variance σ^2 (MSE) that was unbiased whether or not either null hypothesis of interest was true. When $K_{ij} > 1$ for at least one (i, j) pair, a valid estimator of σ^2 can be obtained without assuming additivity. In specifying the appropriate model and deriving test procedures, we will focus on the case $K_{ij} = K > 1$, so the number of observations per "cell" (for each combination of levels) is constant.

Parameters for the Fixed Effects Model with Interaction

Rather than use the μ_{ij}'s themselves as model parameters, it is usual to use an equivalent set that reveals more clearly the role of interaction. Let

$$\mu = \frac{1}{IJ}\sum_i\sum_j\mu_{ij} \qquad \mu_{i.} = \frac{1}{J}\sum_j\mu_{ij.} \qquad \mu_{.j} = \frac{1}{I}\sum_i\mu_{ij} \qquad (11.7)$$

Thus μ is the expected response averaged over all levels of both factors (the true grand mean), $\mu_{i.}$ is the expected response averaged over levels of the second factor when the first factor A is held at level i, and similarly for $\mu_{.j}$. Now define

$$\alpha_i = \mu_{i.} - \mu = \text{the effect of factor } A \text{ at level } i$$

$$\beta_j = \mu_{.j} - \mu = \text{the effect of factor } B \text{ at level } j \qquad (11.8)$$

$$\gamma_{ij} = \mu_{ij} - (\mu + \alpha_i + \beta_j)$$

from which

$$\mu_{ij} = \mu + \alpha_i + \beta_j + \gamma_{ij} \qquad (11.9)$$

The model is additive if and only if all γ_{ij}'s $= 0$. The γ_{ij}'s are referred to as the **inter-action parameters.** The α_i's are called the **main effects for factor A,** whereas the β_j's are the **main effects for factor B.** Although there are I α_i's, J β_j's, and IJ γ_{ij}'s in addition to μ, the conditions $\sum \alpha_i = 0$, $\sum \beta_j = 0$, $\sum_j \gamma_{ij} = 0$ for any i, and $\sum_i \gamma_{ij} = 0$ for any j [all by virtue of (11.7) and (11.8)] imply that only IJ of these new parameters are independently determined: μ, $I - 1$ of the α_i's, $J - 1$ of the β_j's, and $(I - 1)(J - 1)$ of the γ_{ij}'s.

There are now three sets of hypotheses that will be considered:

H_{0AB}: $\gamma_{ij} = 0$ for all i, j versus H_{aAB}: at least one $\gamma_{ij} \neq 0$

H_{0A}: $\alpha_1 = \cdots = \alpha_I = 0$ versus H_{aA}: at least one $\alpha_i \neq 0$

H_{0B}: $\beta_1 = \cdots = \beta_J = 0$ versus H_{aB}: at least one $\beta_j \neq 0$

The no-interaction hypothesis H_{0AB} is usually tested first. If H_{0AB} is not rejected, then the other two hypotheses can be tested to see whether the main effects are significant. If H_{0AB} is rejected and H_{0A} is then tested and not rejected, the resulting model $\mu_{ij} = \mu + \beta_j + \gamma_{ij}$ does not lend itself to straightforward interpretation. In such a case, it is best to construct a picture similar to that of Figure 11.1(b) to try to visualize the way in which the factors interact.

Notation, Model, and Analysis

We now use triple subscripts for both random variables and observed values, with X_{ijk} and x_{ijk} referring to the kth observation (replication) when factor A is at level i and factor B is at level j. The model is then

$$X_{ijk} = \mu + \alpha_i + \beta_j + \gamma_{ij} + \epsilon_{ijk} \qquad (11.10)$$

$$i = 1, \ldots, I, \quad j = 1, \ldots, J, \quad k = 1, \ldots, K$$

where the ϵ_{ij}'s are independent and normally distributed, each with mean 0 and variance σ^2.

Again a dot in place of a subscript means that we have summed over all values of that subscript, whereas a horizontal bar denotes averaging. Thus, $X_{ij.}$ is the total of all

K observations made for factor A at level i and factor B at level j [all observations in the (i, j)th cell], and $\overline{X}_{ij\cdot}$, is the average of these K observations.

Example 11.7 Three different varieties of tomato (Harvester, Pusa Early Dwarf, and Ife No. 1) and four different plant densities (10, 20, 30, and 40 thousand plants per hectare) are being considered for planting in a particular region. To see whether either variety or plant density affects yield, each combination of variety and plant density is used in three different plots, resulting in the data on yields in Table 11.6 (based on the article "Effects of Plant Density on Tomato Yields in Western Nigeria," *Experimental Agriculture,* 1976: 43–47):

Table 11.6 Yield data for Example 11.7

| Variety | Planting Density | | | | $x_{i\cdot\cdot}$ | $\overline{x}_{i\cdot\cdot}$ |
	10,000	20,000	30,000	40,000		
H	10.5 9.2 7.9	12.8 11.2 13.3	12.1 12.6 14.0	10.8 9.1 12.5	136.0	11.33
Ife	8.1 8.6 10.1	12.7 13.7 11.5	14.4 15.4 13.7	11.3 12.5 14.5	146.5	12.21
P	16.1 15.3 17.5	16.6 19.2 18.5	20.8 18.0 21.0	18.4 18.9 17.2	217.5	18.13
$x_{\cdot j\cdot}$	103.3	129.5	142.0	125.2	500.00	
$\overline{x}_{\cdot j\cdot}$	11.48	14.39	15.78	13.91		13.89

Here $I = 3$, $J = 4$, and $K = 3$, for a total of $IJK = 36$ observations. ∎

To test the hypotheses of interest, we again define sums of squares and present computing formulas:

$$\text{SST} = \sum_i \sum_j \sum_k (X_{ijk} - \overline{X}_{\cdots})^2 = \sum_i \sum_j \sum_k X_{ijk}^2 - \frac{1}{IJK} X_{\cdots}^2 \qquad \text{df} = IJK - 1$$

$$\text{SSE} = \sum_i \sum_j \sum_k (X_{ijk} - \overline{X}_{ij\cdot})^2$$

$$= \sum_i \sum_j \sum_k X_{ijk}^2 - \frac{1}{K} \sum_i \sum_j X_{ij\cdot}^2 \qquad \text{df} = IJ(K - 1)$$

$$\text{SSA} = \sum_i \sum_j \sum_k (\overline{X}_{i\cdots} - \overline{X}_{\cdots})^2 = \frac{1}{JK} \sum_i X_{i\cdots}^2 - \frac{1}{IJK} X_{\cdots}^2 \qquad \text{df} = I - 1$$

$$\text{SSB} = \sum_i \sum_j \sum_k (\overline{X}_{\cdot j\cdot} - \overline{X}_{\cdots})^2 = \frac{1}{IK} \sum_j X_{\cdot j\cdot}^2 - \frac{1}{IJK} X_{\cdots}^2 \qquad \text{df} = J - 1$$

$$\text{SSAB} = \sum_i \sum_j \sum_k (\overline{X}_{ij\cdot} - \overline{X}_{i\cdots} - \overline{X}_{\cdot j\cdot} + \overline{X}_{\cdots})^2 \qquad \text{df} = (I - 1)(J - 1)$$

The fundamental identity

$$\text{SST} = \text{SSA} + \text{SSB} + \text{SSAB} + \text{SSE}$$

implies that **interaction sum of squares** SSAB can be obtained by subtraction.

Total variation is thus partitioned into four pieces: unexplained (SSE—which would be present whether or not any of the three null hypotheses was true) and three pieces that may be explained by the truth or falsity of the three H_0's. Each of four mean squares is defined by MS = SS/df. The expected mean squares suggest that each set of hypotheses should be tested using the appropriate ratio of mean squares with MSE in the denominator:

$$E(\text{MSE}) = \sigma^2$$

$$E(\text{MSA}) = \sigma^2 + \frac{JK}{I-1} \sum_{i=1}^{I} \alpha_i^2$$

$$E(\text{MSB}) = \sigma^2 + \frac{IK}{J-1} \sum_{j=1}^{J} \beta_j^2$$

$$E(\text{MSAB}) = \sigma^2 + \frac{K}{(I-1)(J-1)} \sum_{i=1}^{I} \sum_{j=1}^{J} \gamma_{ij}^2$$

Each of the three mean square ratios can be shown to have an F distribution when the associated H_0 is true, which yields the following level α test procedures.

Hypotheses	Test Statistic Value	Rejection Region
H_{0A} versus H_{aA}	$f_A = \dfrac{\text{MSA}}{\text{MSE}}$	$f_A \geq F_{\alpha, I-1, IJ(K-1)}$
H_{0B} versus H_{aB}	$f_B = \dfrac{\text{MSB}}{\text{MSE}}$	$f_B \geq F_{\alpha, J-1, IJ(K-1)}$
H_{0AB} versus H_{aAB}	$f_{AB} = \dfrac{\text{MSAB}}{\text{MSE}}$	$f_{AB} \geq F_{\alpha, (I-1)(J-1), IJ(K-1)}$

As before, the results of the analysis are summarized in an ANOVA table.

Example 11.8
(Example 11.7 continued)

From the given data, $x_{\cdot\cdot\cdot}^2 = (500)^2 = 250{,}000$,

$$\sum_i \sum_j \sum_k x_{ijk}^2 = (10.5)^2 + (9.2)^2 + \cdots + (18.9)^2 + (17.2)^2 = 7404.80$$

$$\sum_i x_{i\cdot\cdot}^2 = (136.0)^2 + (146.5)^2 + (217.5)^2 = 87{,}264.50$$

and

$$\sum_j x_{\cdot j\cdot}^2 = 63{,}280.18$$

The cell totals ($x_{ij\cdot}$'s) are

	10,000	20,000	30,000	40,000
H	27.6	37.3	38.7	32.4
Ife	26.8	37.9	43.5	38.3
P	48.9	54.3	59.8	54.5

from which $\sum_i \sum_j x_{ij.}^2 = (27.6)^2 + \cdots + (54.5)^2 = 22{,}100.28$. Then

$$\text{SST} = 7404.80 - \frac{1}{36}(250{,}000) = 7404.80 - 6944.44 = 460.36$$

$$\text{SSA} = \frac{1}{12}(87{,}264.50) - 6944.44 = 327.60$$

$$\text{SSB} = \frac{1}{9}(63{,}280.18) - 6944.44 = 86.69$$

$$\text{SSE} = 7404.80 - \frac{1}{3}(22{,}100.28) = 38.04$$

and

$$\text{SSAB} = 460.36 - 327.60 - 86.69 - 38.04 = 8.03$$

Table 11.7 summarizes the computations.

Table 11.7 ANOVA Table for Example 11.8

Source of Variation	df	Sum of Squares	Mean Square	f
Varieties	2	327.60	163.8	$f_A = 103.02$
Density	3	86.69	28.9	$f_B = 18.18$
Interaction	6	8.03	1.34	$f_{AB} = .84$
Error	24	38.04	1.59	
Total	35	460.36		

Since $F_{.01,6,24} = 3.67$ and $f_{AB} = .84$ is not ≥ 3.67, H_{0AB} cannot be rejected at level .01, so we conclude that the interaction effects are not significant. Now the presence or absence of main effects can be investigated. Since $F_{.01,2,24} = 5.61$ and $f_A = 103.02 \geq 5.61$, H_{0A} is rejected at level .01 in favor of the conclusion that different varieties do affect the true average yields. Similarly, $f_B = 18.18 \geq 4.72 = F_{.01,3,24}$, so we conclude that true average yield also depends on plant density. ∎

Multiple Comparisons

When the no-interaction hypothesis H_{0AB} is not rejected and at least one of the two main effect null hypotheses is rejected, Tukey's method can be used to identify significant differences in levels. For identifying differences among the α_i's when H_{0A} is rejected,

1. Obtain $Q_{\alpha,I,IJ(K-1)}$, where the second subscript I identifies the number of levels being compared and the third subscript refers to the number of degrees of freedom for error.

2. Compute $w = Q\sqrt{\text{MSE}/(JK)}$, where JK is the number of observations averaged to obtain each of the $\bar{x}_{i..}$'s compared in Step 3.

3. Order the $\bar{x}_{i..}$'s from smallest to largest and, as before, underscore all pairs that differ by less than w. Pairs not underscored correspond to significantly different levels of factor A.

To identify different levels of factor B when H_{0B} is rejected, replace the second subscript in Q by J, replace JK by IK in w, and replace $\bar{x}_{i..}$ by $\bar{x}_{.j.}$.

Example 11.9
(Example 11.8 continued)

For factor A (varieties), $I = 3$, so with $\alpha = .01$ and $IJ(K - 1) = 24$, $Q_{.01,3,24} = 4.55$. Then $w = 4.55\sqrt{1.59/12} = 1.66$, so ordering and underscoring gives

$\bar{x}_{1..}$	$\bar{x}_{2..}$	$\bar{x}_{3..}$
11.33	12.21	18.13

The Harvester and Ife varieties do not appear to differ significantly from one another in effect on true average yield, but both differ from the Pusa variety.

For factor B (density), $J = 4$ so $Q_{.01,4,24} = 4.91$ and $w = 4.91\sqrt{1.59/9} = 2.06$.

$\bar{x}_{.1.}$	$\bar{x}_{.4.}$	$\bar{x}_{.2.}$	$\bar{x}_{.3.}$
11.48	13.91	14.39	15.78

Thus, with experimentwise error rate .01, which is quite conservative, only the lowest density appears to differ significantly from all others. Even with $\alpha = .05$ (so that $w = 1.64$), densities 2 and 3 cannot be judged significantly different from one another in their effect on yield. ∎

Models with Mixed and Random Effects

In some problems, the levels of either factor may have been chosen from a large population of possible levels, so that the effects contributed by the factor are random rather than fixed. As in Section 11.1, if both factors contribute random effects, the model is referred to as a random effects model, whereas if one factor is fixed and the other is random, a mixed effects model results. We will now consider the analysis for a mixed effects model in which factor A (rows) is the fixed factor and factor B (columns) is the random factor. The case in which both factors are random is dealt with in Exercise 26.

The mixed effects model in this situation is

$$X_{ijk} = \mu + \alpha_i + B_j + G_{ij} + \epsilon_{ijk}$$
$$i = 1, \ldots, I, \quad j = 1, \ldots, J, \quad k = 1, \ldots, K$$

Here μ and α_i's are constants with $\Sigma\alpha_i = 0$, and the B_j's, G_{ij}'s, and ϵ_{ijk}'s are independent, normally distributed random variables with expected value 0 and variances σ_B^2, σ_G^2, and σ^2, respectively.*

---

*This is referred to as an "unrestricted" model. An alternative "restricted" model requires that $\sum_i G_{ij} = 0$ for each j (so the G_{ij}'s are no longer independent). Expected mean squares and F ratios appropriate for testing certain hypotheses depend on the choice of model. MINITAB's default option gives output for the unrestricted model.

$$H_{0A}: \alpha_1 = \alpha_2 = \cdots = \alpha_I = 0 \quad \text{versus} \quad H_{aA}: \text{at least one } \alpha_i \neq 0$$

$$H_{0B}: \sigma_B^2 = 0 \qquad\qquad\qquad \text{versus} \quad H_{aB}: \sigma_B^2 > 0$$

$$H_{0G}: \sigma_G^2 = 0 \qquad\qquad\qquad \text{versus} \quad H_{aG}: \sigma_G^2 > 0$$

It is customary to test H_{0A} and H_{0B} only if the no-interaction hypothesis H_{0G} cannot be rejected.

The relevant sums of squares and mean squares needed for the test procedures are defined and computed exactly as in the fixed effects case. The expected mean squares for the mixed model are

$$E(MSE) = \sigma^2$$

$$E(MSA) = \sigma^2 + K\sigma_G^2 + \frac{JK}{I-1}\Sigma\alpha_i^2$$

$$E(MSB) = \sigma^2 + K\sigma_G^2 + IK\sigma_B^2$$

and

$$E(MSAB) = \sigma^2 + K\sigma_G^2$$

Thus, to test the no-interaction hypothesis, the ratio $f_{AB} = MSAB/MSE$ is again appropriate, with H_{0G} rejected if $f_{AB} \geq F_{\alpha,(I-1)(J-1),IJ(K-1)}$. However, for testing H_{0A} versus H_{aA}, the expected mean squares suggest that, although the numerator of the F ratio should still be MSA, the denominator should be MSAB rather than MSE. MSAB is also the denominator of the F ratio for testing H_{0B}.

> For testing H_{0A} versus H_{aA} (factors A fixed, B random), the test statistic value is $f_A = MSA/MSAB$, and the rejection region is $f_A \geq F_{\alpha,I-1,(I-1)(J-1)}$. The test of H_{0B} versus H_{aB} utilizes $f_B = MSB/MSAB$, and the rejection region is $f_B \geq F_{\alpha,J-1,(I-1)(J-1)}$.

Example 11.10 A process engineer has identified two potential causes of electric motor vibration, the material used for the motor casing (factor A) and the supply source of bearings used in the motor (factor B). The accompanying data on the amount of vibration (microns) resulted from an experiment in which motors with casings made of steel, aluminum, and plastic were constructed using bearings supplied by five randomly selected sources.

		Supply source				
		1	**2**	**3**	**4**	**5**
	Steel	13.1 13.2	16.3 15.8	13.7 14.3	15.7 15.8	13.5 12.5
Material	**Aluminum**	15.0 14.8	15.7 16.4	13.9 14.3	13.7 14.2	13.4 13.8
	Plastic	14.0 14.3	17.2 16.7	12.4 12.3	14.4 13.9	13.2 13.1

Only the three casing materials used in the experiment are under consideration for use in production, so factor A is fixed. However, the five supply sources were randomly selected from a much larger population, so factor B is random. The relevant null hypotheses are

$$H_{0A}: \alpha_1 = \alpha_2 = \alpha_3 = 0 \qquad H_{0B}: \sigma_B^2 = 0 \qquad H_{0AB}: \sigma_G^2 = 0$$

MINITAB output appears in Figure 11.4. The P-value column in the ANOVA table indicates that the latter two null hypotheses should be rejected at significance level .05. Different casing materials by themselves do not appear to affect vibration, but interaction between material and supplier is a significant source of variation in vibration.

```
Factor          Type      Levels   Values
casmater        fixed          3   1    2    3
source          random         5   1    2    3    4    5


Source              DF          SS          MS         F          P
casmater             2      0.7047      0.3523      0.24      0.790
source               4     36.6747      9.1687      6.32      0.013
casmater*source      8     11.6053      1.4507     13.03      0.000
Error               15      1.6700      0.1113
Total               29     50.6547


Source              Variance   Error    Expected Mean Square for Each Term
                    component  term     (using unrestricted model)
1 casmater                        3     (4) + 2(3) + Q[1]
2 source             1.2863       3     (4) + 2(3) + 6(2)
3 casmater*source    0.6697       4     (4) + 2(3)
4 Error              0.1113             (4)
```

Figure 11.4 Output from MINITAB's balanced ANOVA option for the data of Example 11.10

When at least two of the K_{ij}'s are unequal, the ANOVA computations are much more complex than for the case $K_{ij} = K$, and there are no nice formulas for the appropriate test statistics. One of the chapter references can be consulted for more information.

Exercises | Section 11.2 (16–26)

16. In an experiment to assess the effects of curing time (factor A) and type of mix (factor B) on the compressive strength of hardened cement cubes, three different curing times were used in combination with four different mixes, with three observations obtained for each of the 12 curing time–mix combinations. The resulting sums of squares were computed to be SSA = 30,763.0, SSB = 34,185.6, SSE = 97,436.8, and SST = 205,966.6.

a. Construct an ANOVA table.

b. Test at level .05 the null hypothesis H_{0AB}: all γ_{ij}'s $= 0$ (no interaction of factors) against H_{aAB}: at least one $\gamma_{ij} \neq 0$.

c. Test at level .05 the null hypothesis H_{0A}: $\alpha_1 = \alpha_2 = \alpha_3 = 0$ (factor A main effects are absent) against H_{aA}: at least one $\alpha_i \neq 0$.

d. Test H_{0B}: $\beta_1 = \beta_2 = \beta_3 = \beta_4 = 0$ versus H_{aB}: at least one $\beta_j \neq 0$ using a level .05 test.

e. The values of the $\bar{x}_{i..}$'s were $\bar{x}_{1..} = 4010.88$, $\bar{x}_{2..} = 4029.10$, and $\bar{x}_{3..} = 3960.02$. Use Tukey's procedure to investigate significant differences among the three curing times.

17. The article "Towards Improving the Properties of Plaster Moulds and Castings" (*J. Engr. Manuf.*, 1991: 265–269) describes several ANOVAs carried out to study how the amount of carbon fiber and sand additions affect various characteristics of the molding process. Here we give data on casting hardness and on wet-mold strength.

Sand Addition (%)	Carbon Fiber Addition (%)	Casting Hardness	Wet-Mold Strength
0	0	61.0	34.0
0	0	63.0	16.0
15	0	67.0	36.0
15	0	69.0	19.0
30	0	65.0	28.0
30	0	74.0	17.0
0	.25	69.0	49.0
0	.25	69.0	48.0
15	.25	69.0	43.0
15	.25	74.0	29.0
30	.25	74.0	31.0
30	.25	72.0	24.0
0	.50	67.0	55.0
0	.50	69.0	60.0
15	.50	69.0	45.0
15	.50	74.0	43.0
30	.50	74.0	22.0
30	.50	74.0	48.0

a. An ANOVA for wet-mold strength gives SS-Sand = 705, SSFiber = 1278, SSE = 843, and SST = 3105. Test for the presence of any effects using $\alpha = .05$.

b. Carry out an ANOVA on the casting hardness observations using $\alpha = .05$.

c. Plot sample mean hardness against sand percentage for different levels of carbon fiber. Is the plot consistent with your analysis in part (b)?

18. The accompanying data resulted from an experiment to investigate whether yield from a certain chemical process depended either on the formulation of a particular input or on mixer speed.

		Speed		
		60	**70**	**80**
		189.7	185.1	189.0
	1	188.6	179.4	193.0
		190.1	177.3	191.1
Formulation				
		165.1	161.7	163.3
	2	165.9	159.8	166.6
		167.6	161.6	170.3

A statistical computer package gave SS(Form) = 2253.44, SS(Speed) = 230.81, SS(Form*Speed) = 18.58, and SSE = 71.87.

a. Does there appear to be interaction between the factors?

b. Does yield appear to depend on either formulation or speed?

c. Calculate estimates of the main effects.

d. The *fitted values* are $\hat{x}_{ijk} = \hat{\mu} + \hat{\alpha}_i + \hat{\beta}_j + \hat{\gamma}_{ij}$, and the *residuals* are $x_{ijk} - \hat{x}_{ijk}$. Verify that the residuals are .23, −.87, .63, 4.50, −1.20, −3.30, −2.03, 1.97, .07, −1.10, −.30, 1.40, .67, −1.23, .57, −3.43, −.13, and 3.57.

e. Construct a normal probability plot from the residuals given in part (d). Do the ϵ_{ijk}'s appear to be normally distributed?

19. The accompanying data table gives observations on total acidity of coal samples of three different types, with determinations made using three different concentrations of ethanolic NaOH ("Chemistry of Brown Coals," *Australian J. Applied Science,* 1958: 375–379).

		Type of Coal		
		Morwell	**Yallourn**	**Maddingley**
NaOH Conc.	**.404N**	8.27, 8.17	8.66, 8.61	8.14, 7.96
	.626N	8.03, 8.21	8.42, 8.58	8.02, 7.89
	.786N	8.60, 8.20	8.61, 8.76	8.13, 8.07

Additionally, $\sum_i \sum_j \sum_k x_{ijk}^2 = 1240.1525$ and $\sum_i \sum_j x_{ij\cdot}^2 = 2479.9991$.

a. Assuming both effects to be fixed, construct an ANOVA table, test for the presence of interaction, and then test for the presence of main effects for each factor (all using level .01).

b. Use Tukey's procedure to identify significant differences among the types of coal.

20. The current (in μA) necessary to produce a certain level of brightness of a television tube was measured for two different types of glass and three different types of phosphor, resulting in the accompanying data ("Fundamentals of Analysis of Variance," *Industrial Quality Control,* 1956: 5–8):

	Phosphor Type		
	1	**2**	**3**
Glass 1	280, 290, 285	300, 310, 295	270, 285, 290
Type 2	230, 235, 240	260, 240, 235	220, 225, 230

Assuming that both factors are fixed, test H_{0AB} versus H_{aAB} at level .01. Then if H_{0AB} cannot be rejected, test the two sets of main effect hypotheses.

21. In an experiment to investigate the effect of "cement factor" (number of sacks of cement per cubic yard) on flexural strength of the resulting concrete ("Studies of Flexural Strength of Concrete. Part 3: Effects of Variation in Testing Procedure," *Proceedings ASTM*, 1957: 1127–1139), $I = 3$ different factor values were used, $J = 5$ different batches of cement were selected, and $K = 2$ beams were cast from each cement factor/batch combination. Summary values include $\sum\sum\sum x_{ijk}^2 = 12{,}280{,}103$, $\sum\sum x_{ij.}^2 = 24{,}529{,}699$, $\sum x_{i..}^2 = 122{,}380{,}901$, $\sum x_{.j.}^2 = 73{,}427{,}483$, and $x_{...} = 19{,}143$.
    a. Construct the ANOVA table.
    b. Assuming a mixed model with cement factor (A) fixed and batches (B) random, test the three pairs of hypotheses of interest at level .05.

22. A study was carried out to compare the writing lifetimes of four premium brands of pens. It was thought that the writing surface might affect lifetime, so three different surfaces were randomly selected. A writing machine was used to ensure that conditions were otherwise homogeneous (e.g., constant pressure and a fixed angle). The accompanying table shows the two lifetimes (min) obtained for each brand–surface combination. In addition, $\sum\sum\sum x_{ijk}^2 = 11{,}499{,}492$ and $\sum\sum x_{ij.}^2 = 22{,}982{,}552$.

Writing Surface

		1	2	3	$x_{i..}$
	1	709, 659	713, 726	660, 645	4112
Brand	2	668, 685	722, 740	692, 720	4227
of Pen	3	659, 685	666, 684	678, 750	4122
	4	698, 650	704, 666	686, 733	4137
$x_{.j.}$		5413	5621	5564	16,598

Carry out an appropriate ANOVA, and state your conclusions.

23. The accompanying data was obtained in an experiment to investigate whether compressive strength of concrete cylinders depends on the type of capping material used or variability in different batches ("The Effect of Type of Capping Material on the Compressive Strength of Concrete Cylinders," *Proceedings ASTM*, 1958: 1166–1186). Each number is a cell total ($x_{ij.}$) based on $K = 3$ observations.

Batch

		1	2	3	4	5
	1	1847	1942	1935	1891	1795
Capping Material	2	1779	1850	1795	1785	1626
	3	1806	1892	1889	1891	1756

In addition, $\sum\sum\sum x_{ijk}^2 = 16{,}815{,}853$ and $\sum\sum x_{ij.}^2 = 50{,}443{,}409$. Obtain the ANOVA table and then test at level .01 the hypotheses H_{0G} versus H_{aG}, H_{0A} versus H_{aA}, and H_{0B} versus H_{aB}, assuming that capping is a fixed effect and batches is a random effect.

24. a. Show that $E(\overline{X}_{i..} - \overline{X}_{...}) = \alpha_i$, so that $\overline{X}_{i..} - \overline{X}_{...}$ is an unbiased estimator for α_i (in the fixed effects model).
    b. With $\hat{\gamma}_{ij} = \overline{X}_{ij.} - \overline{X}_{i..} - \overline{X}_{.j.} + \overline{X}_{...}$, show that $\hat{\gamma}_{ij}$ is an unbiased estimator for γ_{ij} (in the fixed effects model).

25. Show how a $100(1 - \alpha)\%$ t CI for $\alpha_i - \alpha_{i'}$ can be obtained. Then compute a 95% interval for $\alpha_2 - \alpha_3$ using the data from Exercise 19. [*Hint:* With $\theta = \alpha_2 - \alpha_3$, the result of Exercise 24a indicates how to obtain $\hat{\theta}$. Then compute $V(\hat{\theta})$ and $\sigma_{\hat{\theta}}$, and obtain an estimate of $\sigma_{\hat{\theta}}$ by using $\sqrt{\text{MSE}}$ to estimate σ (which identifies the appropriate number of df).]

26. When both factors are random in a two-way ANOVA experiment with K replications per combination of factor levels, the expected mean squares are $E(\text{MSE}) = \sigma^2$, $E(\text{MSA}) = \sigma^2 + K\sigma_G^2 + JK\sigma_A^2$, $E(\text{MSB}) = \sigma^2 + K\sigma_G^2 + IK\sigma_B^2$, and $E(\text{MSAB}) = \sigma^2 + K\sigma_G^2$.
    a. What F ratio is appropriate for testing $H_{0G}: \sigma_G^2 = 0$ versus $H_{aG}: \sigma_G^2 > 0$?
    b. Answer part (a) for testing $H_{0A}: \sigma_A^2 = 0$ versus $H_{aA}: \sigma_A^2 > 0$ and $H_{0B}: \sigma_B^2 = 0$ versus $H_{aB}: \sigma_B^2 > 0$.

11.3 Three-Factor ANOVA

To indicate the nature of models and analyses when ANOVA experiments involve more than two factors, we will focus here on the case of three fixed factors—A, B, and C. The numbers of levels of the three factors will be denoted by I, J, and K, respectively, and

L_{ijk} = the number of observations made with factor A at level i, factor B at level j, and factor C at level k. As with two-factor ANOVA, the analysis is quite complicated when the L_{ijk}'s are not all equal, so we further specialize to $L_{ijk} = L$. Then X_{ijkl} and x_{ijkl} denote the observed value, before and after the experiment is performed, of the lth replication ($l = 1, 2, \ldots, L$) when the three factors are fixed at levels $i, j,$ and k.

To understand the parameters that will appear in the three-factor ANOVA model, first recall that in two-factor ANOVA with replications, $E(X_{ijk}) = \mu_{ij} = \mu + \alpha_i + \beta_j + \gamma_{ij}$, where the restrictions $\sum_i \alpha_i = \sum_j \beta_j = 0$, $\sum_i \gamma_{ij} = 0$ for every j, and $\sum_j \gamma_{ij} = 0$ for every i were necessary to obtain a unique set of parameters. If we use dot subscripts on the μ_{ij}'s to denote averaging (rather than summation), then

$$\mu_{i\cdot} - \mu_{\cdot\cdot} = \frac{1}{J}\sum_j \mu_{ij} - \frac{1}{IJ}\sum_i\sum_j \mu_{ij} = \alpha_i$$

is the effect of factor A at level i averaged over levels of factor B, whereas

$$\mu_{ij} - \mu_{\cdot j} = \mu_{ij} - \frac{1}{I}\sum_i \mu_{ij} = \alpha_i + \gamma_{ij}$$

is the effect of factor A at level i specific to factor B at level j. If the effect of A at level i depends on the level of B, then there is interaction between the factors, and the γ_{ij}'s are not all zero. In particular,

$$\mu_{ij} - \mu_{\cdot j} - \mu_{i\cdot} + \mu_{\cdot\cdot} = \gamma_{ij} \tag{11.11}$$

The Three-Factor Fixed Effects Model

The model for three-factor ANOVA with $L_{ijk} = L$ is

$$X_{ijkl} = \mu_{ijk} + \epsilon_{ijkl} \qquad i = 1, \ldots, I, \quad j = 1, \ldots, J \tag{11.12}$$
$$k = 1, \ldots, K, \quad l = 1, \ldots, L$$

where the ϵ_{ijkl}'s are normally distributed with mean 0 and variance σ^2, and

$$\mu_{ijk} = \mu + \alpha_i + \beta_j + \delta_k + \gamma_{ij}^{AB} + \gamma_{ik}^{AC} + \gamma_{jk}^{BC} + \gamma_{ijk} \tag{11.13}$$

The restrictions necessary to obtain uniquely defined parameters are that the sum over any subscript of any parameter on the right-hand side of (11.13) equal 0.

The parameters γ_{ij}^{AB}, γ_{ik}^{AC}, and γ_{jk}^{BC} are called two-factor interactions, and γ_{ijk} is called a three-factor interaction; the α_i's, β_j's, and δ_k's are the main effects parameters. For any fixed level k of the third factor, analogous to (11.11),

$$\mu_{ijk} - \mu_{i\cdot k} - \mu_{\cdot jk} + \mu_{\cdot\cdot k} = \gamma_{ij}^{AB} + \gamma_{ijk}$$

is the interaction of the ith level of A with the jth level of B specific to the kth level of C, whereas

$$\mu_{ij\cdot} - \mu_{i\cdot\cdot} - \mu_{\cdot j\cdot} + \mu_{\cdot\cdot\cdot} = \gamma_{ij}^{AB}$$

is the interaction between A at level i and B at level j averaged over levels of C. If the interaction of A at level i and B at level j does not depend on k, then all γ_{ijk}'s equal 0. Thus nonzero γ_{ijk}'s represent nonadditivity of the two-factor γ_{ij}^{AB}'s over the various levels of the third factor C. If the experiment included more than three factors, there would be corresponding higher-order interaction terms with analogous interpretations. Note that in the previous argument, if we had considered fixing the level of either A or B (rather than C, as was done) and examining the γ_{ijk}'s, their interpretation would be the same—if any of the interactions of two factors depend on the level of the third factor, then there are nonzero γ_{ijk}'s.

The Analysis of a Three-Factor Experiment

When $L > 1$, there is a sum of squares for each main effect, each two-factor interaction, and the three-factor interaction. To write these in a way that indicates how sums of squares are defined when there are more than three factors, note that any of the model parameters in (11.13) can be estimated unbiasedly by averaging X_{ijkl} over appropriate subscripts and taking differences. Thus,

$$\hat{\mu} = \overline{X}_{....} \qquad \hat{\alpha}_i = \overline{X}_{i...} - \overline{X}_{....} \qquad \hat{\gamma}_{ij}^{AB} = \overline{X}_{ij..} - \overline{X}_{i...} - \overline{X}_{.j..} + \overline{X}_{....}$$

$$\hat{\gamma}_{ijk} = \overline{X}_{ijk.} - \overline{X}_{ij..} - \overline{X}_{i\cdot k\cdot} - \overline{X}_{\cdot jk\cdot} + \overline{X}_{i...} + \overline{X}_{.j..} + \overline{X}_{..k\cdot} - \overline{X}_{....}$$

with other main effects and interaction estimators obtained by symmetry. Then sums of squares are

$$\text{SST} = \sum_i \sum_j \sum_k \sum_l (X_{ijkl} - \overline{X}_{....})^2 \qquad df = IJKL - 1$$

$$\text{SSA} = \sum_i \sum_j \sum_k \sum_l \hat{\alpha}_i^2 = JKL \sum_i (\overline{X}_{i...} - \overline{X}_{....})^2 \qquad df = I - 1$$

$$\text{SSAB} = \sum_i \sum_j \sum_k \sum_l (\hat{\gamma}_{ij}^{AB})^2 = KL \sum_i \sum_j (\overline{X}_{ij..} - \overline{X}_{i...} - \overline{X}_{.j..} + \overline{X}_{....})^2 \qquad df = (I-1)(J-1)$$

$$\text{SSABC} = \sum_i \sum_j \sum_k \sum_l \hat{\gamma}_{ijk}^2 = L \sum_i \sum_j \sum_k \hat{\gamma}_{ijk}^2 \qquad df = (I-1)(J-1)(K-1)$$

$$\text{SSE} = \sum_i \sum_j \sum_k \sum_l (X_{ijkl} - \overline{X}_{ijk.})^2 \qquad df = IJK(L-1)$$

with the other main effect and two-factor interaction sums of squares obtained by symmetry. SST is the sum of the other eight SS's.

Even the computational formulas for these SS's are quite tedious to use, so we eschew them in favor of output from a statistical computer package. The current version of MINITAB, for example, will fit a three-factor model with fixed, mixed, or random effects.

Each sum of squares (excepting SST) when divided by its df gives a mean square, with

$$E(\text{MSE}) = \sigma^2$$

$$E(\text{MSA}) = \sigma^2 + \frac{JKL}{I-1} \sum_i \alpha_i^2$$

$$E(\text{MSAB}) = \sigma^2 + \frac{KL}{(I-1)(J-1)} \sum_i \sum_j (\gamma_{ij}^{AB})^2$$

$$E(\text{MSABC}) = \sigma^2 + \frac{L}{(I-1)(J-1)(K-1)} \sum_i \sum_j \sum_k (\gamma_{ijk})^2$$

and similar expressions for the other expected mean squares. Main effect and interaction hypotheses are tested by forming F ratios with MSE in each denominator.

Null Hypothesis	Test Statistic Value	Rejection Region
H_{0A}: all α_i's $= 0$	$f_A = \dfrac{\text{MSA}}{\text{MSE}}$	$f_A \geq F_{\alpha, I-1, IJK(L-1)}$
H_{0AB}: all γ_{ij}^{AB}'s $= 0$	$f_{AB} = \dfrac{\text{MSAB}}{\text{MSE}}$	$f_{AB} \geq F_{\alpha,(I-1)(J-1), IJK(L-1)}$
H_{0ABC}: all γ_{ijk}'s $= 0$	$f_{ABC} = \dfrac{\text{MSABC}}{\text{MSE}}$	$f_{ABC} \geq F_{\alpha,(I-1)(J-1)(K-1), IJK(L-1)}$

Usually the main effect hypotheses are tested only if all interactions are judged not significant.

This analysis assumes that $L_{ijk} = L > 1$. If $L = 1$, then as in the two-factor case, the highest-order interactions must be assumed to equal 0 to obtain an MSE that estimates σ^2. Setting $L = 1$ and disregarding the fourth subscript summation over l, the foregoing formulas for sums of squares are still valid, and $\text{SSE} = \sum_i \sum_j \sum_k \hat{\gamma}_{ijk}^2$ with $\overline{X}_{ijk.} = X_{ijk}$ in the expression for $\hat{\gamma}_{ijk}$.

Example 11.11 The following observations (body temperature $-$ 100°F) were reported in an experiment to study heat tolerance of cattle ("The Significance of the Coat in Heat Tolerance of Cattle," *Australian J. Agriculture Research,* 1959: 744–748). Measurements were made at four different periods (factor A, with $I = 4$) on two different strains of cattle (factor B, with $J = 2$) having four different types of coat (factor C, with $K = 4$); $L = 3$ observations were made for each of the $4 \times 2 \times 4 = 32$ combinations of levels of the three factors.

	B_1				B_2			
	C_1	C_2	C_3	C_4	C_1	C_2	C_3	C_4
A_1	3.6	3.4	2.9	2.5	4.2	4.4	3.6	3.0
	3.8	3.7	2.8	2.4	4.0	3.9	3.7	2.8
	3.9	3.9	2.7	2.2	3.9	4.2	3.4	2.9
A_2	3.8	3.8	2.9	2.4	4.4	4.2	3.8	2.0
	3.6	3.9	2.9	2.2	4.4	4.3	3.7	2.9
	4.0	3.9	2.8	2.2	4.6	4.7	3.4	2.8
A_3	3.7	3.8	2.9	2.1	4.2	4.0	4.0	2.0
	3.9	4.0	2.7	2.0	4.4	4.6	3.8	2.4
	4.2	3.9	2.8	1.8	4.5	4.5	3.3	2.0
A_4	3.6	3.6	2.6	2.0	4.0	4.0	3.8	2.0
	3.5	3.7	2.9	2.0	4.1	4.4	3.7	2.2
	3.8	3.9	2.9	1.9	4.2	4.2	3.5	2.3

The table of cell totals ($x_{ijk.}$'s) for all combinations of the three factors is

$x_{ijk.}$	B_1				B_2			
	C_1	C_2	C_3	C_4	C_1	C_2	C_3	C_4
A_1	11.3	11.0	8.4	7.1	12.1	12.5	10.7	8.7
A_2	11.4	11.6	8.6	6.8	13.4	13.2	10.9	7.7
A_3	11.8	11.7	8.4	5.9	13.1	13.1	11.1	6.4
A_4	10.9	11.2	8.4	5.9	12.3	12.6	11.0	6.5

Figure 11.5 displays plots of the corresponding cell means $\bar{x}_{ijk.} = x_{ijk.}/3$. We will return to these plots after considering tests of various hypotheses. The basis for these tests is the ANOVA table given in Table 11.8.

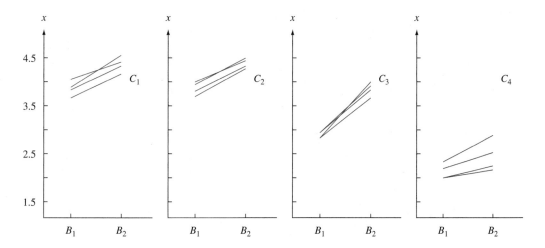

Figure 11.5   Plots of $x_{ijk.}$ for Example 11.11

Table 11.8   ANOVA table for Example 11.11

Source	df	Sum of Squares	Mean Square	f
A	$I - 1 = 3$	.49	.163	4.13
B	$J - 1 = 1$	6.45	6.45	163.29
C	$K - 1 = 3$	48.93	16.31	412.91
AB	$(I - 1)(J - 1) = 3$	.02	.0067	.170
AC	$(I - 1)(K - 1) = 9$	1.61	.179	4.53
BC	$(J - 1)(K - 1) = 3$	.88	.293	7.42
ABC	$(I - 1)(J - 1)(K - 1) = 9$	.25	.0278	.704
Error	$IJK(L - 1) = 64$	2.53	.0395	
Total	$IJKL - 1 = 95$	61.16		

Since $F_{.01,9,64} \approx 2.70$ and $f_{ABC} = MSABC/MSE = .704$ does not exceed 2.70, we conclude that three-factor interactions are not significant. However, although the AB interactions are also not significant, both AC and BC interactions as well as all main effects seem to be necessary in the model. When there are no ABC or AB interactions, a plot of the $\bar{x}_{ijk}$'s ($= \hat{\mu}_{ijk}$) separately for each level of C should reveal no substantial interactions (if only the ABC interactions are zero, plots are more difficult to interpret; see the article "Two-Dimensional Plots for Interpreting Interactions in the Three-Factor Analysis of Variance Model," *Amer. Statistician,* May 1979: 63–69). ■

Tukey's procedure can be used in three-factor (or more) ANOVA. The second subscript on Q is the number of sample means being compared, and the third is degrees of freedom for error.

Models with random and mixed effects can also be analyzed. Sums of squares and degrees of freedom are identical to the fixed effects case, but expected mean squares are of course different for the random main effects or interactions. A good reference is the book by Douglas Montgomery listed in the chapter bibliography.

Latin Square Designs

When several factors are to be studied simultaneously, an experiment in which there is at least one observation for every possible combination of levels is referred to as a **complete layout.** If the factors are A, B, and C with I, J, and K levels, respectively, a complete layout requires at least IJK observations. Frequently an experiment of this size is either impracticable because of cost, time, or space constraints, or literally impossible. For example, if the response variable is sales of a certain product and the factors are different display configurations, different stores, and different time periods, then only one display configuration can realistically be used in a given store during a given time period.

A three-factor experiment in which fewer than IJK observations are made is called an **incomplete layout.** There are some incomplete layouts in which the pattern of combinations of factors is such that the analysis is straightforward. One such three-factor design is called a **Latin square.** It is appropriate when $I = J = K$ (e.g., four display configurations, four stores, and four time periods) and all two- and three-factor interaction effects are assumed absent. If the levels of factor A are identified with the rows of a two-way table and the levels of B with the columns of the table, then the defining characteristic of a Latin square design is that *every level of factor C appears exactly once in each row and exactly once in each column.* Pictured in Figure 11.6 are examples of 3×3, 4×4, and 5×5 Latin squares. There are 12 different 3×3 Latin squares, and the number of different $N \times N$ Latin squares increases rapidly with N (e.g., every permutation of rows of a given Latin square yields a Latin square, and similarly for column permutations). It is recommended that the square actually used in a particular experiment be chosen at random from the set of all possible squares of the desired dimension; for further details, consult one of the chapter references.

The Model and Analysis for Latin Squares

The letter N will be used to denote the common value of I, J, and K. Then a complete layout with one observation per combination would require N^3 observations, whereas a Latin square requires only N^2 observations. Once a particular square has been chosen,

Figure 11.6 Examples of Latin squares

the value of k (the level of factor C) is completely determined by the values of i and j. To emphasize this, we use $x_{ij(k)}$ to denote the observed value when the three factors are at levels i, j, and k, respectively, with k taking on only one value for each i, j pair. The model is then

$$X_{ij(k)} = \mu + \alpha_i + \beta_j + \delta_k + \epsilon_{ij(k)} \qquad i, j, k = 1, \ldots, N$$

where $\sum \alpha_i = \sum \beta_j = \sum \delta_k = 0$ and the $\epsilon_{ij(k)}$'s are independent and normally distributed with mean 0 and variance σ^2.

We employ the following notation for totals and averages:

$$X_{i\cdot\cdot} = \sum_j X_{ij(k)} \qquad X_{\cdot j\cdot} = \sum_i X_{ij(k)} \qquad X_{\cdot\cdot k} = \sum_{ij} X_{ij(k)} \qquad X_{\cdots} = \sum_i \sum_j X_{ij(k)}$$

$$\overline{X}_{i\cdot\cdot} = \frac{X_{i\cdot\cdot}}{N} \qquad \overline{X}_{\cdot j\cdot} = \frac{X_{\cdot j\cdot}}{N} \qquad \overline{X}_{\cdot\cdot k} = \frac{X_{\cdot\cdot k}}{N} \qquad \overline{X}_{\cdots} = \frac{X_{\cdots}}{N^2}$$

Note that although $X_{i\cdot\cdot}$ previously suggested a double summation, now it corresponds to a single sum over all j (and the associated values of k). The sums of squares used in the analysis follow.

$$\text{SST} = \sum_i \sum_j (X_{ij(k)} - \overline{X}_{\cdots})^2 = \sum_i \sum_j X_{ij(k)}^2 - \frac{X_{\cdots}^2}{N^2} \qquad \text{df} = N^2 - 1$$

$$\text{SSA} = \sum_i \sum_j (\overline{X}_{i\cdot\cdot} - \overline{X}_{\cdots})^2 = \frac{1}{N} \sum_i X_{i\cdot\cdot}^2 - \frac{X_{\cdots}^2}{N^2} \qquad \text{df} = N - 1$$

$$\text{SSB} = \sum_i \sum_j (\overline{X}_{\cdot j\cdot} - \overline{X}_{\cdots})^2 = \frac{1}{N} \sum_j X_{\cdot j\cdot}^2 - \frac{X_{\cdots}^2}{N^2} \qquad \text{df} = N - 1$$

$$\text{SSC} = \sum_i \sum_j (\overline{X}_{\cdot\cdot k} - \overline{X}_{\cdots})^2 = \frac{1}{N} \sum_k X_{\cdot\cdot k}^2 - \frac{X_{\cdots}^2}{N^2} \qquad \text{df} = N - 1$$

$$\text{SSE} = \sum_i \sum_j [X_{ij(k)} - (\hat{\mu} + \hat{\alpha}_i + \hat{\beta}_j + \hat{\delta}_k)]^2$$

$$= \sum_i \sum_j (X_{ij(k)} - \overline{X}_{i\cdot\cdot} - \overline{X}_{\cdot j\cdot} - \overline{X}_{\cdot\cdot k} + 2\overline{X}_{\cdots})^2 \qquad \text{df} = (N-1)(N-2)$$

$$\text{SST} = \text{SSA} + \text{SSB} + \text{SSC} + \text{SSE}$$

Each mean square is, of course, the ratio SS/df. For testing H_{0C}: $\delta_1 = \delta_2 = \cdots = \delta_N = 0$, the test statistic value is $f_C = \text{MSC/MSE}$, with H_{0C} rejected if $f_C \geq F_{\alpha, N-1, (N-1)(N-2)}$. The other two main effect null hypotheses are also rejected if the corresponding F ratio exceeds $F_{\alpha, N-1, (N-1)(N-2)}$.

If any of the null hypotheses is rejected, significant differences can be identified by using Tukey's procedure. After computing $w = Q_{\alpha, N, (N-1)(N-2)} \cdot \sqrt{\text{MSE}/N}$, pairs of sample means (the $\bar{x}_{i..}$'s, $\bar{x}_{.j.}$'s, or $\bar{x}_{..k}$'s) differing by more than w correspond to significant differences between associated factor effects (the α_i's, β_j's, or δ_k's).

The hypothesis H_{0C} is frequently the one of central interest. A Latin square design is used to control for extraneous variation in the A and B factors, as was done by a randomized block design for the case of a single extraneous factor. Thus, in the product sales example mentioned previously, variation due to both stores and time periods is controlled by a Latin square design, enabling an investigator to test for the presence of effects due to different product display configurations.

Example 11.12 In an experiment to investigate the effect of relative humidity on abrasion resistance of leather cut from a rectangular pattern ("The Abrasion of Leather," *J. Inter. Soc. Leather Trades' Chemists,* 1946: 287), a 6×6 Latin square was used to control for possible variability due to row and column position in the pattern. The six levels of relative humidity studied were $1 = 25\%$, $2 = 37\%$, $3 = 50\%$, $4 = 62\%$, $5 = 75\%$, and $6 = 87\%$, with the following results:

	B (columns)						
	1	**2**	**3**	**4**	**5**	**6**	$x_{i..}$
1	^{3}7.38	^{4}5.39	^{6}5.03	^{2}5.50	^{5}5.01	^{1}6.79	35.10
2	^{2}7.15	^{1}8.16	^{5}4.96	^{4}5.78	^{3}6.24	^{6}5.06	37.35
3	^{4}6.75	^{6}5.64	^{3}6.34	^{5}5.31	^{1}7.81	^{2}8.05	39.90
A (rows) 4	^{1}8.05	^{3}6.45	^{2}6.31	^{6}5.46	^{4}6.05	^{5}5.51	37.83
5	^{6}5.65	^{5}5.44	^{1}7.27	^{3}6.54	^{2}7.03	^{4}5.96	37.89
6	^{5}6.00	^{2}6.55	^{4}5.93	^{1}8.02	^{6}5.80	^{3}6.61	38.91
$x_{.j.}$	40.98	37.63	35.84	36.61	37.94	37.98	

Also, $x_{..1} = 46.10$, $x_{..2} = 40.59$, $x_{..3} = 39.56$, $x_{..4} = 35.86$, $x_{..5} = 32.23$, $x_{..6} = 32.64$, $x_{...} = 226.98$, and $\sum_i \sum_j x_{ij(k)}^2 = 1462.89$. Further computations are summarized in Table 11.9.

Table 11.9 ANOVA table for Example 11.12

Source of Variation	df	Sum of Squares	Mean Square	f
A (rows)	5	2.19	.438	2.50
B (columns)	5	2.57	.514	2.94
C (treatments)	5	23.53	4.706	26.89
Error	20	3.49	.175	
Total	35	31.78		

Since $F_{.05,5,20} = 2.71$ and $26.89 \geq 2.71$, H_{0C} is rejected in favor of the hypothesis that relative humidity does on average affect abrasion resistance.

To apply Tukey's procedure, $w = Q_{.05,6,20} \cdot \sqrt{MSE/6} = 4.45\sqrt{.175/6} = .76$. Ordering the $\bar{x}_{..k}$'s and underscoring yields

75%	87%	62%	50%	37%	25%
5.37	5.44	5.98	6.59	6.77	7.68

In particular, the lowest relative humidity appears to result in a true average abrasion resistance significantly higher than for any other relative humidity studied. ∎

Exercises | Section 11.3 (27–37)

27. The output of a continuous extruding machine that coats steel pipe with plastic was studied as a function of the thermostat temperature profile (*A*, at three levels), type of plastic (*B*, at three levels), and the speed of the rotating screw that forces the plastic through a tube-forming die (*C*, at three levels). There were two replications (*L* = 2) at each combination of levels of the factors, yielding a total of 54 observations on output. The sums of squares were SSA = 14,144.44, SSB = 5511.27, SSC = 244,696.39, SSAB = 1069.62, SSAC = 62.67, SSBC = 331.67, SSE = 3127.50, and SST = 270,024.33.

a. Construct the ANOVA table.

b. Use appropriate *F* tests to show that none of the *F* ratios for two- or three-factor interactions is significant at level .05.

c. Which main effects appear significant?

d. With $x_{..1.} = 8242$, $x_{..2.} = 9732$, and $x_{..3.} = 11,210$, use Tukey's procedure to identify significant differences among the levels of factor *C*.

28. To see whether thrust force in drilling is affected by drilling speed (*A*), feed rate (*B*), or material used (*C*), an experiment using four speeds, three rates, and two materials was performed, with two samples (*L* = 2) drilled at each combination of levels of the three factors. Sums of squares were calculated as follows: SSA = 19,149.73, SSB = 2,589,047.62, SSC = 157,437.52, SSAB = 53,238.21, SSAC = 9033.73, SSBC = 91,880.04, SSE = 56,819.50, and SST = 2,983,164.81. Construct the ANOVA table and identify significant interactions using $\alpha = .01$. Is there any single factor that appears to have no effect on thrust force? (In other words, does any factor appear nonsignificant in every effect in which it appears?)

29. The article "An Analysis of Variance Applied to Screw Machines" (*Industrial Quality Control*, 1956: 8–9) describes an experiment to investigate how the length of steel bars was affected by time of day (*A*), heat treatment applied (*B*), and screw machine used (*C*). The three times were 8:00 A.M., 11:00 A.M., and 3:00 P.M., and there were two treatments and four machines (a 3 × 2 × 4 factorial experiment), resulting in the accompanying data [coded as 1000(length − 4.380), which does not affect the analysis].

B_1	C_1	C_2	C_3	C_4
A_1	6, 9, 1, 3	7, 9, 5, 5	1, 2, 0, 4	6, 6, 7, 3
A_2	6, 3, 1, −1	8, 7, 4, 8	3, 2, 1, 0	7, 9, 11, 6
A_3	5, 4, 9, 6	10, 11, 6, 4	−1, 2, 6, 1	10, 5, 4, 8

B_2	C_1	C_2	C_3	C_4
A_1	4, 6, 0, 1	6, 5, 3, 4	−1, 0, 0, 1	4, 5, 5, 4
A_2	3, 1, 1, −2	6, 4, 1, 3	2, 0, −1, 1	9, 4, 6, 3
A_3	6, 0, 3, 7	8, 7, 10, 0	0, −2, 4, −4	4, 3, 7, 0

Sums of squares include SSAB = 1.646, SSAC = 71.021, SSBC = 1.542, SSE = 447.500, and SST = 1037.833.

a. Construct the ANOVA table for this data.

b. Test to see whether any of the interaction effects are significant at level .05.

c. Test to see whether any of the main effects are significant at level .05 (i.e., H_{0A} versus H_{aA}, etc.).

d. Use Tukey's procedure to investigate significant differences among the four machines.

30. The following summary quantities were computed from an experiment involving four levels of nitrogen (A), two times of planting (B), and two levels of potassium (C) ("Use and Misuse of Multiple Comparison Procedures," *Agronomy J.,* 1977: 205–208). Only one observation (N content, in percentage, of corn grain) was made for each of the 16 combinations of levels.

SSA = .22625 SSB = .000025 SSC = .0036
SSAB = .004325 SSAC = .00065
SSBC = .000625 SST = .2384.

a. Construct the ANOVA table.

b. Assume that there are no three-way interaction effects, so that MSABC is a valid estimate of σ^2, and test at level .05 for interaction and main effects.

c. The nitrogen averages are $\bar{x}_{1..} = 1.1200$, $\bar{x}_{2..} = 1.3025$, $\bar{x}_{3..} = 1.3875$, and $\bar{x}_{4..} = 1.4300$. Use Tukey's method to examine differences in percentage N among the nitrogen levels ($Q_{.05,4,3} = 6.82$).

31. The article "Kolbe–Schmitt Carbonation of 2-Naphthol" (*Industrial and Eng. Chemistry: Process and Design Development,* 1969: 165–173) presented the accompanying data on percentage yield of BON acid as a function of reaction time (1, 2, and 3 hours), temperature (30, 70, and 100°C), and pressure (30, 70, and 100 psi). Assuming that there is no three-factor interaction, so that SSE = SSABC provides an estimate of σ^2, MINITAB gave the accompanying ANOVA table. Carry out all appropriate tests.

| | B_1 | | |
	C_1	C_2	C_3
A_1	68.5	73.0	68.7
A_2	74.5	75.0	74.6
A_3	70.5	72.5	74.7

| | B_2 | | |
	C_1	C_2	C_3
A_1	72.8	80.1	72.0
A_2	72.0	81.5	76.0
A_3	69.5	84.5	76.0

| | B_3 | | |
	C_1	C_2	C_3
A_1	72.5	72.5	73.1
A_2	75.5	70.0	76.0
A_3	65.0	66.5	70.5

Analysis of Variance for Yield

Source	DF	SS	MS	F	P
time	2	42.112	21.056	8.76	0.010
temp	2	110.732	55.366	23.04	0.000
press	2	68.136	34.068	14.18	0.002
time*temp	4	67.761	16.940	7.05	0.010
time*press	4	35.184	8.796	3.66	0.056
temp*press	4	136.437	34.109	14.20	0.001
Error	8	19.223	2.403		
Total	26	479.585			

32. When factors A and B are fixed but factor C is random and the restricted model is used (see the footnote on page 453; there is a technical complication with the unrestricted model here).

$$E(MSE) = \sigma^2$$

$$E(MSA) = \sigma^2 + JL\sigma_{AC}^2 + \frac{JKL}{I-1}\Sigma\alpha_i^2$$

$$E(MSB) = \sigma^2 + IL\sigma_{BC}^2 + \frac{IKL}{J-1}\Sigma\beta_j^2$$

$$E(MSC) = \sigma^2 + IJL\sigma_C^2$$

$$E(MSAB) = \sigma^2 + L\sigma_{ABC}^2$$

$$+ \frac{KL}{(I-1)(J-1)}\Sigma_i\Sigma_j(\gamma_{ij}^{AB})^2$$

$$E(MSAC) = \sigma^2 + JL\sigma_{AC}^2$$

$$E(MSBC) = \sigma^2 + IL\sigma_{BC}^2$$

$$E(MSABC) = \sigma^2 + L\sigma_{ABC}^2$$

a. Based on these expected mean squares, what F ratios would you use to test H_0: $\sigma_{ABC}^2 = 0$; H_0: $\sigma_C^2 = 0$; H_0: $\gamma_{ij}^{AB} = 0$ for all i, j; and H_0: $\alpha_1 = \cdots = \alpha_I = 0$?

b. In an experiment to assess the effects of age, type of soil, and day of production on compressive strength of cement/soil mixtures, two ages (*A*), four types of soil (*B*), and 3 days (*C*, assumed random) were used, with $L = 2$ observations made for each combination of factor levels. The resulting sums of squares were SSA = 14,318.24, SSB = 9656.40, SSC = 2270.22, SSAB = 3408.93, SSAC = 1442.58, SSBC = 3096.21, SSABC = 2832.72, and SSE = 8655.60. Obtain the ANOVA table and carry out all tests using level .01.

33. Because of potential variability in aging due to different castings and segments on the castings, a Latin square design with $N = 7$ was used to investigate the effect of heat treatment on aging. With A = castings, B = segments, C = heat treatments, summary statistics include $x_{...} = 3815.8$, $\sum x_{i..}^2 = 297,216.90$, $\sum x_{.j.}^2 = 297,200.64$, $\sum x_{..k}^2 = 297,155.01$, and $\sum\sum x_{ij(k)}^2 = 297,317.65$. Obtain the ANOVA table and test at level .05 the hypothesis that heat treatment has no effect on aging.

34. The article "The Responsiveness of Food Sales to Shelf Space Requirements" (*J. Marketing Research,* 1964: 63–67) reports the use of a Latin square design to investigate the effect of shelf space on food sales. The experiment was carried out over a 6-week period using six different stores, resulting in the following data on sales of powdered coffee cream (with shelf space index in parentheses).

		Week		
		1	2	3
	1	27 (5)	14 (4)	18 (3)
	2	34 (6)	31 (5)	34 (4)
Store	3	39 (2)	67 (6)	31 (5)
	4	40 (3)	57 (1)	39 (2)
	5	15 (4)	15 (3)	11 (1)
	6	16 (1)	15 (2)	14 (6)

		Week		
		4	5	6
	1	35 (1)	28 (6)	22 (2)
	2	46 (3)	37 (2)	23 (1)
Store	3	49 (4)	38 (1)	48 (3)
	4	70 (6)	37 (4)	50 (5)
	5	9 (2)	18 (5)	17 (6)
	6	12 (5)	19 (3)	22 (4)

Construct the ANOVA table, and state and test at level .01 the hypothesis that shelf space does not affect sales against the appropriate alternative.

35. The article "Variation in Moisture and Ascorbic Acid Content from Leaf to Leaf and Plant to Plant in Turnip Greens" (*Southern Cooperative Services Bull.,* 1951: 13–17) uses a Latin square design in which factor *A* is plant, factor *B* is leaf size (smallest to largest), factor *C* (in parentheses) is time of weighing, and the response variable is moisture content.

		Leaf Size (*B*)		
		1	2	3
	1	6.67 (5)	7.15 (4)	8.29 (1)
	2	5.40 (2)	4.77 (5)	5.40 (4)
Plant (*A*)	3	7.32 (3)	8.53 (2)	8.50 (5)
	4	4.92 (1)	5.00 (3)	7.29 (2)
	5	4.88 (4)	6.16 (1)	7.83 (3)

		Leaf Size (*B*)	
		4	5
	1	8.95 (3)	9.62 (2)
	2	7.54 (1)	6.93 (3)
Plant (*A*)	3	9.99 (4)	9.68 (1)
	4	7.85 (5)	7.08 (4)
	5	5.83 (2)	8.51 (5)

When all three factors are random, the expected mean squares are $E(\text{MSA}) = \sigma^2 + N\sigma_A^2$, $E(\text{MSB}) = \sigma^2 + N\sigma_B^2$, $E(\text{MSC}) = \sigma^2 + N\sigma_C^2$, and $E(\text{MSE}) = \sigma^2$. This implies that the *F* ratios for testing H_{0A}: $\sigma_A^2 = 0$, H_{0B}: $\sigma_B^2 = 0$, and H_{0C}: $\sigma_C^2 = 0$ are identical to those for fixed effects. Obtain the ANOVA table and test at level .05 to see whether there is any variation in moisture content due to the factors.

36. The article "An Assessment of the Effects of Treatment, Time, and Heat on the Removal of Erasable Pen Marks from Cotton and Cotton/Polyester Blend Fabrics (*J. Testing and Eval.,* 1991: 394–397) reports the following sums of squares for the response variable *degree of removal of marks:* SSA = 39.171, SSB = .665, SSC = 21.508, SSAB = 1.432, SSAC = 15.953, SSBC = 1.382, SSABC = 9.016, and SSE = 115.820. Four different laundry treatments, three different types of pen, and six different fabrics were used in the experiment, and there were three observations for each treatment–pen–fabric combination. Perform an analysis of variance using $\alpha = .01$ for each test, and state your conclusions (assume fixed effects for all three factors).

37. A four-factor ANOVA experiment was carried out to investigate the effects of fabric (A), type of exposure (B), level of exposure (C), and fabric direction (D) on extent of color change in exposed fabric as measured by a spectrocolorimeter. Two observations were made for each of the three fabrics, two types, three levels, and two directions, resulting in MSA = 2207.329,

MSB = 47.255, MSC = 491.783, MSD = .044, MSAB = 15.303, MSAC = 275.446, MSAD = .470, MSBC = 2.141, MSBD = .273, MSCD = .247, MSABC = 3.714, MSABD = 4.072, MSACD = .767, MSBCD = .280, MSE = .977, and MST = 93.621 ("Accelerated Weathering of Marine Fabrics," *J. Testing and Eval.*, 1992: 139–143). Assuming fixed effects for all factors, carry out an analysis of variance using $\alpha = .01$ for all tests and summarize your conclusions.

11.4 2^p Factorial Experiments

If an experimenter wishes to study simultaneously the effect of p different factors on a response variable and the factors have $I_1, I_2, \ldots, I_p$ levels, respectively, then a complete experiment requires at least $I_1 \cdot I_2 \cdot \cdots \cdot I_p$ observations. In such situations, the experimenter can often perform a "screening experiment" with each factor at only two levels to obtain preliminary information about factor effects. An experiment in which there are p factors, each at two levels, is referred to as a **2^p factorial experiment.** The analysis of data from such an experiment is computationally simpler than for more general factorial experiments. In addition, a 2^p experiment provides a simple setting for introducing the important concepts of confounding and fractional replications.

2^3 Experiments

As in Section 11.3, we let X_{ijkl} and x_{ijkl} refer to the observation from the lth replication with factors *A, B,* and *C* at levels $i, j,$ and k, respectively. The model for this situation is

$$X_{ijkl} = \mu + \alpha_i + \beta_j + \delta_k + \gamma_{ij}^{AB} + \gamma_{ik}^{AC} + \gamma_{jk}^{BC} + \gamma_{ijk} + \epsilon_{ijkl} \quad (11.14)$$

for $i = 1, 2; j = 1, 2; k = 1, 2; l = 1, \ldots, n$. The ϵ_{ijkl}'s are assumed independent, normally distributed, with mean 0 and variance σ^2. Because there are only two levels of each factor, the side conditions on the parameters of (11.14) that uniquely specify the model are simply stated: $\alpha_1 + \alpha_2 = 0, \ldots, \gamma_{11}^{AB} + \gamma_{21}^{AB} = 0, \gamma_{12}^{AB} + \gamma_{22}^{AB} = 0, \gamma_{11}^{AB} + \gamma_{12}^{AB} = 0, \gamma_{21}^{AB} + \gamma_{22}^{AB} = 0$, and the like. These conditions imply that there is only one functionally independent parameter of each type (for each main effect and interaction). For example, $\alpha_2 = -\alpha_1$, whereas $\gamma_{21}^{AB} = -\gamma_{11}^{AB}, \gamma_{12}^{AB} = -\gamma_{11}^{AB}$, and $\gamma_{22}^{AB} = \gamma_{11}^{AB}$. Because of this, each sum of squares in the analysis will have 1 df.

The parameters of the model can be estimated by taking averages over various subscripts of the X_{ijkl}'s and then forming appropriate linear combinations of the averages. For example,

$$\hat{\alpha}_1 = \overline{X}_{1\cdots} - \overline{X}_{\cdots\cdots}$$

$$= \frac{(X_{111\cdot} + X_{121\cdot} + X_{112\cdot} + X_{122\cdot} - X_{211\cdot} - X_{212\cdot} - X_{221\cdot} - X_{222\cdot})}{8n}$$

and

$$\hat{\gamma}_{11}^{AB} = \overline{X}_{11\cdots} - \overline{X}_{1\cdots} - \overline{X}_{\cdot1\cdot\cdot} + \overline{X}_{\cdots\cdots}$$

$$= \frac{(X_{111\cdot} - X_{121\cdot} - X_{211\cdot} + X_{221\cdot} + X_{112\cdot} - X_{122\cdot} - X_{212\cdot} + X_{222\cdot})}{8n}$$

Each estimator is, except for the factor $1/(8n)$, a linear function of the cell totals ($X_{ijk\cdot}$'s) in which each coefficient is $+1$ or -1, with an equal number of each; such functions are called **contrasts** in the $X_{ijk\cdot}$'s. Furthermore, the estimators satisfy the same side conditions satisfied by the parameters themselves. For example,

$$\hat{\alpha}_1 + \hat{\alpha}_2 = \overline{X}_{1\cdots} - \overline{X}_{\cdots\cdots} + \overline{X}_{2\cdots} - \overline{X}_{\cdots\cdots} = \overline{X}_{1\cdots} + \overline{X}_{2\cdots} - 2\overline{X}_{\cdots\cdots}$$

$$= \frac{1}{4n}X_{1\cdots} + \frac{1}{4n}X_{2\cdots} - \frac{2}{8n}X_{\cdots\cdots} = \frac{1}{4n}X_{\cdots\cdots} - \frac{1}{4n}X_{\cdots\cdots} = 0$$

Example 11.13 In an experiment to investigate the compressive strength properties of cement–soil mixtures, two different aging periods were used in combination with two different aging temperatures and two different soils. Two replications were made for each combination of levels of the three factors, resulting in the following data:

Age	Temperature	Soil 1	Soil 2
1	1	471, 413	385, 434
	2	485, 552	530, 593
2	1	712, 637	770, 705
	2	712, 789	741, 806

The computed cell totals are $x_{111\cdot} = 884$, $x_{211\cdot} = 1349$, $x_{121\cdot} = 1037$, $x_{221\cdot} = 1501$, $x_{112\cdot} = 819$, $x_{212\cdot} = 1475$, $x_{122\cdot} = 1123$, and $x_{222\cdot} = 1547$, so $x_{\cdots\cdots} = 9735$. Then

$$\hat{\alpha}_1 = (884 - 1349 + 1037 - 1501 + 819 - 1475 + 1123 - 1547)/16$$
$$= -125.5625 = -\hat{\alpha}_2$$
$$\hat{\gamma}_{11}^{AB} = (884 - 1349 - 1037 + 1501 + 819 - 1475 - 1123 + 1547)/16$$
$$= -14.5625 = -\hat{\gamma}_{12}^{AB} = -\hat{\gamma}_{21}^{AB} = \hat{\gamma}_{22}^{AB}$$

The other parameter estimates can be computed in the same manner. ■

Sums of Squares and Analysis for a 2^3 Experiment

The reason for computing parameter estimates is that sums of squares for the various effects are easily obtained from the estimates. For example,

$$\text{SSA} = \sum_i \sum_j \sum_k \sum_l \hat{\alpha}_i^2 = 4n\sum_{i=1}^{2} \hat{\alpha}_i^2 = 4n[\hat{\alpha}_1^2 + (-\hat{\alpha}_1)^2] = 8n\hat{\alpha}_1^2$$

and

$$\text{SSAB} = \sum_i \sum_j \sum_k \sum_l (\hat{\gamma}_{ij}^{AB})^2$$

$$= 2n\sum_{i=1}^{2}\sum_{j=1}^{2}(\hat{\gamma}_{ij}^{AB})^2 = 2n[(\hat{\gamma}_{11}^{AB})^2 + (-\hat{\gamma}_{11}^{AB})^2 + (-\hat{\gamma}_{11}^{AB})^2 + (\hat{\gamma}_{11}^{AB})^2]$$

$$= 8n(\hat{\gamma}_{11}^{AB})^2$$

Since each estimate is a contrast in the cell totals multiplied by $1/(8n)$, each sum of squares has the form $(\text{contrast})^2/(8n)$. Thus, to compute the various sums of squares, we need to know the coefficients $(+1$ or $-1)$ of the appropriate contrasts. The signs $(+$ or $-)$ on each $x_{ijk\cdot}$ in each effect contrast are most conveniently displayed in a table. We will use the notation (1) for the experimental condition $i = 1, j = 1, k = 1$, a for $i = 2$, $j = 1, k = 1$, ab for $i = 2, j = 2, k = 1$, and so on. If level 1 is thought of as "low" and level 2 as "high," any letter that appears denotes a high level of the associated factor. In Table 11.10, each column gives the signs for a particular effect contrast in the $x_{ijk\cdot}$'s associated with the different experimental conditions.

Table 11.10 Signs for computing effect contrasts

Experimental Condition	Cell Total	A	B	C	AB	AC	BC	ABC
(1)	$x_{111\cdot}$	−	−	−	+	+	+	−
a	$x_{211\cdot}$	+	−	−	−	−	+	+
b	$x_{121\cdot}$	−	+	−	−	+	−	+
ab	$x_{221\cdot}$	+	+	−	+	−	−	−
c	$x_{112\cdot}$	−	−	+	+	−	−	+
ac	$x_{212\cdot}$	+	−	+	−	+	−	−
bc	$x_{122\cdot}$	−	+	+	−	−	+	−
abc	$x_{222\cdot}$	+	+	+	+	+	+	+

In each of the first three columns, the sign is $+$ if the corresponding factor is at the high level and $-$ if it is at the low level. Every sign in the AB column is then the "product" of the signs in the A and B columns, with $(+)(+) = (-)(-) = +$ and $(+)(-) = (-)(+) = -$, and similarly for the AC and BC columns. Finally, the signs in the ABC column are the products of AB with C (or B with AC or A with BC). Thus, for example,

$$AC \text{ contrast} = + x_{111\cdot} - x_{211\cdot} + x_{121\cdot} - x_{221\cdot} - x_{112\cdot} + x_{212\cdot} - x_{122\cdot} + x_{222\cdot}$$

Once the seven effect contrasts are computed,

$$\text{SS(effect)} = \frac{(\text{effect contrast})^2}{8n}$$

Even with a table of signs, calculation of the contrasts is tedious. An efficient computational technique, due to Yates, is as follows. Write in a column the eight cell totals in the **standard order** as given in the table of signs, and establish three additional columns. In each of these three columns, the first four entries are the sums of entries 1 and 2, 3 and 4, 5 and 6, 7 and 8 of the previous columns. The last four entries are the differences between entries 2 and 1, 4 and 3, 6 and 5, and 8 and 7 of the previous column. The last column then contains $x_{....}$ and the seven effect contrasts in standard order. Squaring each contrast and dividing by $8n$ then gives the seven sums of squares.

Example 11.14
(Example 11.13 continued)
Since $n = 2$, $8n = 16$. Yates's method is illustrated in Table 11.11.

Table 11.11   Yates's method of computation

Treatment Condition	$x_{ijk.}$	1	2	Effect Contrast	SS = (contrast)²/16
(1) = $x_{111.}$	884	2233	4771	9735	
$a = x_{211.}$	1349	2538	4964	2009	252,255.06
$b = x_{121.}$	1037	2294	929	681	28,985.06
$ab = x_{221.}$	1501	2670	1080	−233	3,393.06
$c = x_{112.}$	819	465	305	193	2,328.06
$ac = x_{212.}$	1475	464	376	151	1,425.06
$bc = x_{122.}$	1123	656	−1	71	315.06
$abc = x_{222.}$	1547	424	−232	−231	3,335.06
					292,036.42

From the original data, $\sum_i \sum_j \sum_k \sum_l x_{ijkl}^2 = 6{,}232{,}289$, and

$$\frac{x_{....}^2}{16} = 5{,}923{,}139.06$$

so

$$SST = 6{,}232{,}289 - 5{,}923{,}139.06 = 309{,}149.94$$
$$SSE = SST - [SSA + \cdots + SSABC] = 309{,}149.94 - 292{,}036.42$$
$$= 17{,}113.52$$

The ANOVA calculations are summarized in Table 11.12.

Table 11.12   ANOVA table for Example 11.14

Source of Variation	df	Sum of Squares	Mean Square	f
A	1	252,255.06	252,255.06	117.92
B	1	28,985.06	28,985.06	13.55
C	1	2,328.06	2,328.06	1.09
AB	1	3,393.06	3,393.06	1.59
AC	1	1,425.06	1,425.06	.67
BC	1	315.06	315.06	.15
ABC	1	3,335.06	3,335.06	1.56
Error	8	17,113.52	2,139.19	
Total	15	309,149.94		

Figure 11.7 shows SAS output for this example. Only the *P*-values for age (*A*) and temperature (*B*) are less than .01, so only these effects are judged significant.

```
Analysis of Variance Procedure
Dependent Variable: STRENGTH
                                 Sum of            Mean
Source               DF          Squares          Square       F Value      Pr > F
Model                7         292036.4375      41719.4911       19.50      0.0002
Error                8          17113.5000       2139.1875
Corrected Total     15         309149.9375

            R-Square              C.V.          Root MSE            POWERUSE Mean

            0.944643            7.601660         46.25135              608.437500

Source               DF         Anova SS       Mean Square      F Value      Pr > F

AGE                  1         252255.0625     252255.0625      117.92       0.0001
TEMP                 1          28985.0625      28985.0625       13.55       0.0062
AGE*TEMP             1           3393.0625       3393.0625        1.59       0.2434
SOIL                 1           2328.0625       2328.0625        1.09       0.3273
AGE*SOIL             1           1425.0625       1425.0625        0.67       0.4380
TEMP*SOIL            1            315.0625        315.0625        0.15       0.7111
AGE*TEMP*SOIL        1           3335.0625       3335.0625        1.56       0.2471
```

Figure 11.7 SAS output for strength data of Example 11.14 ■

2^p Experiments for $p > 3$

Although the computations when $p > 3$ are quite tedious, the analysis parallels that of the three-factor case. For example, if there are four factors *A, B, C,* and *D*, there are 16 different experimental conditions. The first 8 in standard order are exactly those already listed for a three-factor experiment. The second 8 are obtained by placing the letter *d* beside each condition in the first group. Yates's method is then initiated by computing totals across replications, listing these totals in standard order, and proceeding as before; with *p* factors, the *p*th column to the right of the treatment totals will give the effect contrasts.

For $p > 3$, there will often be no replications of the experiment (so only one complete replicate is available). One possible way to test hypotheses is to assume that certain higher-order effects are absent and then add the corresponding sums of squares to obtain an SSE. Such an assumption can, however, be misleading in the absence of prior knowledge (see the book by Montgomery listed in the chapter bibliography). An alternative approach involves working directly with the effect contrasts. Each contrast has a normal distribution with the same variance. When a particular effect is absent, the expected value of the corresponding contrast is 0, but this is not so when the effect is present. The suggested method of analysis is to construct a normal probability plot of the effect contrasts (or, equivalently, the effect parameter estimates, since estimate = contrast/2^p when $n = 1$). Points corresponding to absent effects will tend to fall close to a straight line, whereas points associated with substantial effects will typically be far from this line.

Example 11.15 The accompanying data is from the article "Quick and Easy Analysis of Unreplicated Factorials" (*Technometrics,* 1989: 469–473). The four factors are A = acid strength, B = time, C = amount of acid, and D = temperature, and the response variable is the yield of isatin. The observations, in standard order, are .08, .04, .53, .43, .31, .09, .12, .36, .79, .68, .73, .08, .77, .38, .49, and .23. Table 11.13 displays the effect estimates as given in the article (which used contrast/8 rather than contrast/16).

Table 11.13   Effect estimates for Example 11.15

Effect	A	B	AB	C	AC	BC	ABC	D
estimate	−.191	−.021	−.001	−.076	.034	−.066	.149	.274

Effect	AD	BD	ABD	CD	ACD	BCD	$ABCD$
estimate	−.161	−.251	−.101	−.026	−.066	.124	.019

Figure 11.8 is a normal probability plot of the effect estimates. All points in the plot fall close to the same straight line, suggesting the complete absence of any effects (we will shortly give an example in which this is not the case).

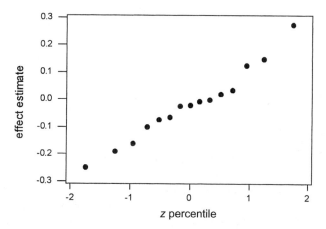

Figure 11.8   A normal probability plot of effect estimates from Example 11.15   ■

Visual judgments of deviation from straightness in a normal probability plot are rather subjective. The article cited in Example 11.15 describes a more objective technique for identifying significant effects in an unreplicated experiment.

Confounding

It is often not possible to carry out all 2^p experimental conditions of a 2^p factorial experiment in a homogeneous experimental environment. In such situations, it may be possible to separate the experimental conditions into 2^r homogeneous blocks ($r < p$), so that there are 2^{p-r} experimental conditions in each block. The blocks may, for example, correspond to different laboratories, different time periods, or different operators or work crews. In the simplest case, $p = 3$ and $r = 1$, so that there are two blocks with each block consisting of four of the eight experimental conditions.

As always, blocking is effective in reducing variation associated with extraneous sources. However, when the 2^p experimental conditions are placed in 2^r blocks, the price paid for this blocking is that $2^r - 1$ of the factor effects cannot be estimated. This is because $2^r - 1$ factor effects (main effects and/or interactions) are mixed up or **confounded** with the block effects. The allocation of experimental conditions to blocks is then usually done so that only higher-level interactions are confounded, whereas main effects and low-order interactions remain estimable and hypotheses can be tested.

To see how allocation to blocks is accomplished, consider first a 2^3 experiment with two blocks ($r = 1$) and four treatments per block. Suppose we select ABC as the effect to be confounded with blocks. Then any experimental condition having an odd number of letters in common with ABC, such as b (one letter) or abc (three letters), is placed in one block, whereas any condition having an even number of letters in common with ABC (where 0 is even) goes in the other block. Figure 11.9 shows this allocation of treatments to the two blocks.

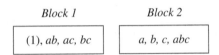

Block 1   *Block 2*

(1), *ab, ac, bc*   *a, b, c, abc*

Figure 11.9 Confounding ABC in a 2^3 experiment

In the absence of replications, the data from such an experiment would usually be analyzed by assuming that there were no two-factor interactions (additivity) and using SSE = SSAB + SSAC + SSBC with 3 df to test for the presence of main effects. Alternatively, a normal probability plot of effect contrasts or effect parameter estimates could be examined. Most frequently, though, there are replications when just three factors are being studied. Suppose there are u replicates, resulting in a total of $2^r \cdot u$ blocks in the experiment. Then after subtracting from SST all sums of squares associated with effects not confounded with blocks (computed using Yates's method), the block sum of squares is computed using the $2^r \cdot u$ block totals and then subtracted to yield SSE (so there are $2^r \cdot u - 1$ df for blocks).

Example 11.16 The article "Factorial Experiments in Pilot Plant Studies" (*Industrial and Eng. Chemistry,* 1951: 1300–1306) reports the results of an experiment to assess the effects of reactor temperature (A), gas throughput (B), and concentration of active constituent (C) on strength of the product solution (measured in arbitrary units) in a recirculation unit. Two blocks were used, with the ABC effect confounded with blocks, and there were two replications, resulting in the data in Figure 11.10. The four block × replication totals are 288, 212, 88, and 220, with a grand total of 808, so

$$\text{SSBl} = \frac{(288)^2 + (212)^2 + (88)^2 + (220)^2}{4} - \frac{(808)^2}{16} = 5204.00$$

Replication 1

Block 1		Block 2	
(1)	99	a	18
ab	52	b	51
ac	42	c	108
bc	95	abc	35

Replication 2

Block 1		Block 2	
(1)	46	a	18
ab	−47	b	62
ac	22	c	104
bc	67	abc	36

Figure 11.10 Data for Example 11.16

The other sums of squares are computed by Yates's method using the eight experimental condition totals, resulting in the ANOVA table given as Table 11.14. By comparison with $F_{.05,1,6} = 5.99$, we conclude that only the main effects for A and C differ significantly from zero.

Table 11.14 ANOVA table for Example 11.16

Source of Variation	df	Sum of Squares	Mean Square	f
A	1	12,996	12,996	39.82
B	1	702.25	702.25	2.15
C	1	2,756.25	2,756.25	8.45
AB	1	210.25	210.25	.64
AC	1	30.25	30.25	.093
BC	1	25	25	.077
Blocks	3	5,204	1,734.67	5.32
Error	6	1,958	326.33	
Total	15	23,882		

■

Confounding Using More Than Two Blocks

In the case $r = 2$ (four blocks), three effects are confounded with blocks. The experimenter first chooses two defining effects to be confounded. For example, in a five-factor experiment (A, B, C, D, and E), the two three-factor interactions BCD and CDE might be chosen for confounding. The third effect confounded is then the **generalized interaction** of the two, obtained by writing the two chosen effects side by side and then cancelling any letters common to both: $(BCD)(CDE) = BE$. Notice that if ABC and CDE are chosen for confounding, their generalized interaction is $(ABC)(CDE) = ABDE$, so that no main effects or two-factor interactions are confounded.

Once the two defining effects have been selected for confounding, one block consists of all treatment conditions having an even number of letters in common with both defining effects. The second block consists of all conditions having an even number of letters in common with the first defining contrast and an odd number of letters in common with the second contrast, and the third and fourth blocks consist of the "odd/even" and "odd/odd" contrasts. In a five-factor experiment with defining effects *ABC* and *CDE*, this results in the allocation to blocks as shown in Figure 11.11 (with the number of letters in common with each defining contrast appearing beside each experimental condition).

Block 1		Block 2		Block 3		Block 4	
(1)	(0, 0)	*d*	(0, 1)	*a*	(1, 0)	*c*	(1, 1)
ab	(2, 0)	*e*	(0, 1)	*b*	(1, 0)	*ad*	(1, 1)
de	(0, 2)	*ac*	(2, 1)	*cd*	(1, 2)	*ae*	(1, 1)
acd	(2, 2)	*bc*	(2, 1)	*ce*	(1, 2)	*bd*	(1, 1)
ace	(2, 2)	*abd*	(2, 1)	*ade*	(1, 2)	*be*	(1, 1)
bcd	(2, 2)	*abe*	(2, 1)	*bde*	(1, 2)	*abc*	(3, 1)
bce	(2, 2)	*acde*	(2, 3)	*abcd*	(3, 2)	*cde*	(1, 3)
abde	(2, 2)	*bcde*	(2, 3)	*abce*	(3, 2)	*abcde*	(3, 3)

Figure 11.11 Four blocks in a 2^5 factorial experiment with defining effects *ABC* and *CDE*

The block containing (1) is called the **principal block.** Once it has been constructed, a second block can be obtained by selecting any experimental condition not in the principal block and obtaining its generalized interaction with every condition in the principal block. The other blocks are then constructed in the same way by first selecting a condition not in a block already constructed and finding generalized interactions with the principal block.

For experimental situations with $p > 3$, there is often no replication, so sums of squares associated with nonconfounded higher-order interactions are usually pooled to obtain an error sum of squares that can be used in the denominators of the various F statistics. All computations can again be carried out using Yates's technique, with SSBl being the sum of sums of squares associated with confounded effects.

When $r > 2$, one first selects r defining effects to be confounded with blocks, making sure that no one of the effects chosen is the generalized interaction of any other two selected. The additional $2^r - r - 1$ effects confounded with the blocks are then the generalized interactions of all effects in the defining set (including not only generalized interactions of pairs of effects, but also of sets of three, four, and so on). Consult the book by Montgomery for details.

Fractional Replication

When the number p of factors is large, even a single replicate of a 2^p experiment can be expensive and time-consuming. For example, one replicate of a 2^6 factorial experiment involves an observation for each of the 64 different experimental conditions. An appealing strategy in such situations is to make observations for only a fraction of the 2^p conditions. Provided that care is exercised in the choice of conditions to be observed, much information about factor effects can still be obtained.

Suppose we decide to include only 2^{p-1} (half) of the 2^p possible conditions in our experiment; this is usually called a **half-replicate.** The price paid for this economy is twofold. First, information about a single effect (determined by the 2^{p-1} conditions selected for observation) is completely lost to the experimenter in the sense that no reasonable estimate of the effect is possible. Second, the remaining $2^p - 2$ main effects and interactions are paired up so that any one effect in a particular pair is confounded with the other effect in the same pair. For example, one such pair may be {A, BCD}, so that separate estimates of the A main effect and BCD interaction are not possible. It is desirable, then, to select a half-replicate for which main effects and low-order interactions are paired off (confounded) only with higher-order interactions rather than with one another.

The first step in selecting a half-replicate is to select a defining effect as the nonestimable effect. Suppose that in a five-factor experiment, $ABCDE$ is chosen as the defining effect. Now the $2^5 = 32$ possible treatment conditions are divided into two groups with 16 conditions each, one group consisting of all conditions having an odd number of letters in common with $ABCDE$ and the other containing an even number of letters in common with the defining contrast. Then either group of 16 conditions is used as the half-replicate. The "odd" group is

$a, b, c, d, e, abc, abd, abe, acd, ace, ade, bcd, bce, bde, cde, abcde$

Each main effect and interaction other than $ABCDE$ is then confounded with (**aliased** with) its generalized interaction with $ABCDE$. Thus $(AB)(ABCDE) = CDE$, so the AB interaction and CDE interaction are confounded with each other. The resulting **alias pairs** are

{A, $BCDE$}	{B, $ACDE$}	{C, $ABDE$}	{D, $ABCE$}	{E, $ABCD$}
{AB, CDE}	{AC, BDE}	{AD, BCE}	{AE, BCD}	{BC, ADE}
{BD, ACE}	{BE, ACD}	{CD, ABE}	{CE, ABD}	{DE, ABC}

Note in particular that every main effect is aliased with a four-factor interaction. Assuming these interactions to be negligible allows us to test for the presence of main effects.

To select a quarter-replicate of a 2^p factorial experiment (2^{p-2} of the 2^p possible treatment conditions), two defining effects must be selected. These two and their generalized interaction become the nonestimable effects. Instead of alias pairs as in the half-replicate, each remaining effect is now confounded with three other effects, each being its generalized interaction with one of the three nonestimable effects.

Example 11.17 The article "More on Planning Experiments to Increase Research Efficiency" (*Industrial and Eng. Chemistry,* 1970: 60–65) reports on the results of a quarter-replicate of a 2^5 experiment in which the five factors were A = condensation temperature, B = amount of material B, C = solvent volume, D = condensation time, and E = amount of material E. The response variable was the yield of the chemical process. The chosen defining contrasts were ACE and BDE, with generalized interaction $(ACE)(BDE)$ = $ABCD$. The remaining 28 main effects and interactions can now be partitioned into seven groups of four effects each such that the effects within a group cannot be assessed separately. For example, the generalized interactions of A with the nonestimable effects are $(A)(ACE) = CE$, $(A)(BDE) = ABDE$, and $(A)(ABCD) = BCD$, so one alias group is {A, CE, $ABDE$, BCD}. The complete set of alias groups is

$$\{A, CE, ABDE, BCD\} \qquad \{B, ABCE, DE, ACD\} \qquad \{C, AE, BCDE, ABD\}$$
$$\{D, ACDE, BE, ABC\} \qquad \{E, AC, BD, ABCDE\} \qquad \{AB, BCE, ADE, CD\}$$
$$\{AD, CDE, ABE, BC\} \qquad\qquad\qquad\qquad\qquad\qquad\qquad ■$$

Analysis of a Fractional Replicate

Once the defining contrasts have been chosen for a quarter-replicate, they are used as in the discussion of confounding to divide the 2^p treatment conditions into four groups of 2^{p-2} conditions each. Then any one of the four groups is selected as the set of conditions for which data will be collected. Similar comments apply to a $1/2^r$ replicate of a 2^p factorial experiment.

Having made observations for the selected treatment combinations, a table of signs similar to Table 11.10 is constructed. The table contains a row only for each of the treatment combinations actually observed rather than the full 2^p rows, and there is a single column for each alias group (since each effect in the group would have the same set of signs for the treatment conditions selected for observation). The signs in each column indicate as usual how contrasts for the various sums of squares are computed. Yates's method can also be used, but the rule for arranging observed conditions in standard order must be modified.

The difficult part of a fractional replication analysis typically involves deciding what to use for error sum of squares. Since there will usually be no replication (though one could observe, e.g., two replicates of a quarter-replicate), some effect sums of squares must be pooled to obtain an error sum of squares. In a half-replicate of a 2^8 experiment, for example, an alias structure can be chosen so that the eight main effects and 28 two-factor interactions are each confounded only with higher-order interactions and that there are an additional 27 alias groups involving only higher-order interactions. Assuming the absence of higher-order interaction effects, the resulting 27 sums of squares can then be added to yield an error sum of squares, allowing 1-df tests for all main effects and two-factor interactions. However, in many cases tests for main effects can be obtained only by pooling some or all of the sums of squares associated with alias groups involving two-factor interactions, and the corresponding two-factor interactions cannot be investigated.

Example 11.18
(Example 11.17 continued)

The set of treatment conditions chosen and resulting yields for the quarter-replicate of the 2^5 experiment were

e	ab	ad	bc	cd	ace	bde	$abcde$
23.2	15.5	16.9	16.2	23.8	23.4	16.8	18.1

The abbreviated table of signs is displayed in Table 11.15.

With SSA denoting the sum of squares for effects in the alias group $\{A, CE, ABDE, BCD\}$,

$$SSA = \frac{(-23.2 + 15.5 + 16.9 - 16.2 - 23.8 + 23.4 - 16.8 + 18.1)^2}{8} = 4.65$$

Similarly, SSB = 53.56, SSC = 10.35, SSD = .91, SSE′ = 10.35 (the ′ differentiates this quantity from error sum of squares SSE), SSAB = 6.66, and SSAD = 3.25, giving SST = 4.65 + 53.56 + · · · + 3.25 = 89.73. To test for main effects, we use SSE = SSAB + SSAD = 9.91 with 2 df. The ANOVA table is in Table 11.16.

Table 11.15 Table of signs for Example 11.18

	A	B	C	D	E	AB	AD
e	−	−	−	−	+	+	+
ab	+	+	−	−	−	+	−
ad	+	−	−	+	−	−	+
bc	−	+	+	−	−	−	+
cd	−	−	+	+	−	+	−
ace	+	−	+	−	+	−	−
bde	−	+	−	+	+	−	−
abcde	+	+	+	+	+	+	+

Table 11.16 ANOVA table for Example 11.18

Source	df	Sum of Squares	Mean Square	f
A	1	4.65	4.65	.94
B	1	53.56	53.56	10.80
C	1	10.35	10.35	2.09
D	1	.91	.91	.18
E	1	10.35	10.35	2.09
Error	2	9.91	4.96	
Total	7	89.73		

Since $F_{.05,1,2} = 18.51$, none of the five main effects can be judged significant. Of course, with only 2 df for error, the test is not very powerful (i.e., it is quite likely to fail to detect the presence of effects). The article from *Industrial and Engineering Chemistry* from which the data came actually had an independent estimate of the standard error of the treatment effects based on prior experience, so used a somewhat different analysis. Our analysis was done here only for illustrative purposes, since one would ordinarily want many more than 2 df for error. ■

As an alternative to F tests based on pooling sums of squares to obtain SSE, a normal probability plot of effect contrasts can be examined.

Example 11.19 An experiment was carried out to investigate shrinkage in the plastic casing material used for speedometer cables ("An Explanation and Critique of Taguchi's Contribution to Quality Engineering," *Quality and Reliability Engr. Intl.,* 1988: 123–131). The engineers started with 15 factors: liner outside diameter, liner die, liner material, liner line speed, wire braid type, braiding tension, wire diameter, liner tension, liner temperature, coating material, coating die type, melt temperature, screen pack, cooling method, and line speed. It was suspected that only a few of these factors were important, so a screening experiment in the form of a 2^{15-11} factorial (a $1/2^{11}$ fraction of a 2^{15} factorial experiment) was carried out. The resulting alias structure is quite complicated; in particular, every main effect is confounded with two-factor interactions. The response variable was the percentage shrinkage for a cable specimen produced at designated levels of the factors.

Figure 11.12 displays a normal probability plot of the effect contrasts. All but two of the points fall quite close to a straight line. The discrepant points correspond to effects E = wire braid type and G = wire diameter, suggesting that these two factors are the only ones that affect the amount of shrinkage.

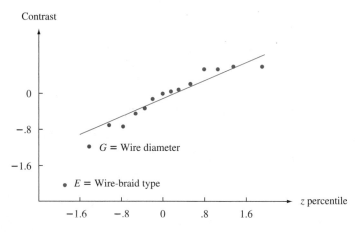

Figure 11.12 Normal probability plot of contrasts from Example 11.19 ■

The subjects of factorial experimentation, confounding, and fractional replication encompass many models and techniques we have not discussed. For more information, the chapter references should be consulted.

Exercises | Section 11.4 (38–49)

38. The accompanying data resulted from an experiment to study the nature of dependence of welding current on the three factors: welding voltage, wire feed speed, and tip-to-workpiece distance. There were two levels of each factor (a 2^3 experiment) with two replications per combination of levels (the averages across replications agree with values given in the article "A Study on Prediction of Welding Current in Gas Metal Arc Welding," *J. Engr. Manuf.,* 1991: 64–69). The first two given numbers are for the treatment (1), the next two for a, and so on in standard order: 200.0, 204.2, 215.5, 219.5, 272.7, 276.9, 299.5, 302.7, 166.6, 172.6, 186.4, 192.0, 232.6, 240.8, 253.4, 261.6.

a. Verify that the sums of squares are as given in the accompanying ANOVA table from MINITAB.

b. Which effects appear to be important, and why?

Analysis of Variance for current

Source	DF	SS	MS	F	P
Volt	1	1685.1	1685.1	102.38	0.000
Speed	1	21272.2	21272.2	1292.37	0.000
Dist	1	5076.6	5076.6	308.42	0.000
Volt*speed	1	36.6	36.6	2.22	0.174
Volt*dist	1	0.4	0.4	0.03	0.877
Speed*dist	1	109.2	109.2	6.63	0.033
Volt*speed*dist	1	23.5	23.5	1.43	0.266
Error	8	131.7	16.5		
Total	15	28335.3			

39. The accompanying data resulted from a 2^3 experiment with three replications per combination of treatments designed to study the effects of concentration of detergent (A), concentration of sodium carbonate (B), and concentration of sodium carboxymethyl cellulose (C) on cleaning ability of a solution in washing tests (a larger number indicates better cleaning ability than a smaller number).

Factor Levels

A	B	C	Condition	Observations
1	1	1	(1)	106, 93, 116
2	1	1	a	198, 200, 214
1	2	1	b	197, 202, 185
2	2	1	ab	329, 331, 307
1	1	2	c	149, 169, 135
2	1	2	ac	243, 247, 220
1	2	2	bc	255, 230, 252
2	2	2	abc	383, 360, 364

a. After obtaining cell totals $x_{ijk\cdot}$, compute estimates of β_1, γ^{AC}_{11}, and γ^{AC}_{21}.

b. Use the cell totals along with Yates's method to compute the effect contrasts and sums of squares. Then construct an ANOVA table and test all appropriate hypotheses using $\alpha = .05$.

40. In a study of processes used to remove impurities from cellulose goods ("Optimization of Rope-Range Bleaching of Cellulosic Fabrics," *Textile Research J.,* 1976: 493–496), the following data resulted from a 2^4 experiment involving the desizing process. The four factors were enzyme concentration (A), pH (B), temperature (C), and time (D).

Treatment	Enzyme (g/l)	pH	Temp. (°C)	Time (h)	Starch % by Weight 1st Repl.	Starch % by Weight 2nd Repl.
(1)	.50	6.0	60.0	6	9.72	13.50
a	.75	6.0	60.0	6	9.80	14.04
b	.50	7.0	60.0	6	10.13	11.27
ab	.75	7.0	60.0	6	11.80	11.30
c	.50	6.0	70.0	6	12.70	11.37
ac	.75	6.0	70.0	6	11.96	12.05
bc	.50	7.0	70.0	6	11.38	9.92
abc	.75	7.0	70.0	6	11.80	11.10
d	.50	6.0	60.0	8	13.15	13.00
ad	.75	6.0	60.0	8	10.60	12.37
bd	.50	7.0	60.0	8	10.37	12.00
abd	.75	7.0	60.0	8	11.30	11.64
cd	.50	6.0	70.0	8	13.05	14.55
acd	.75	6.0	70.0	8	11.15	15.00
bcd	.50	7.0	70.0	8	12.70	14.10
abcd	.75	7.0	70.0	8	13.20	16.12

a. Use Yates's algorithm to obtain sums of squares and the ANOVA table.

b. Do there appear to be any second-, third-, or fourth-order interaction effects present? Explain your reasoning. Which main effects appear to be significant?

41. In Exercise 39, suppose a low water temperature has been used to obtain the data. The entire experiment is then repeated with a higher water temperature to obtain the following data. Use Yates's algorithm on the entire set of 48 observations to obtain the sums of squares and ANOVA table, and then test appropriate hypotheses at level .05.

Condition	Observations
d	144, 154, 158
ad	239, 227, 244
bd	232, 242, 246
abd	364, 362, 346
cd	194, 162, 203
acd	284, 295, 291
bcd	291, 287, 297
abcd	411, 406, 395

42. The following data on power consumption in electric furnace heats (kW consumed per ton of melted product) resulted from a 2^4 factorial experiment with three replicates ("Studies on a 10-cwt Arc Furnace," *J. Iron and Steel Institute,* 1956: 22). The factors were nature of roof A (low, high), power setting B (low, high), scrap used C (tube, plate), and charge D (700 lb, 1000 lb).

Treatment	x_{ijklm}	Treatment	x_{ijklm}
(1)	866, 862, 800	d	988, 808, 650
a	946, 800, 840	ad	966, 976, 876
b	774, 834, 746	bd	702, 658, 650
ab	709, 789, 646	abd	784, 700, 596
c	1017, 990, 954	cd	922, 808, 868
ac	1028, 906, 977	acd	1056, 870, 908
bc	817, 783, 771	bcd	798, 726, 700
abc	829, 806, 691	abcd	752, 714, 714

Construct the ANOVA table and test all hypotheses of interest using $\alpha = .01$.

43. The article "Statistical Design and Analysis of Qualification Test Program for a Small Rocket Engine" (*Industrial Quality Control,* 1964: 14–18) presents data from an experiment to assess the effects of vibration (A), temperature cycling (B), altitude cycling (C), and temperature for altitude cycling and firing (D) on thrust duration. A subset of the data is given here. (In the article, there were four levels of D rather than just two.) Use the Yates method to obtain sums of squares and the ANOVA table. Then assume that three- and four-factor interactions are absent, pool the corresponding sums of squares to obtain an estimate of σ^2, and test all appropriate hypotheses at level .05.

		D_1		D_2	
		C_1	C_2	C_1	C_2
A_1	B_1	21.60	21.60	11.54	11.50
	B_2	21.09	22.17	11.14	11.32
A_2	B_1	21.60	21.86	11.75	9.82
	B_2	19.57	21.85	11.69	11.18

44. a. In a 2^4 experiment, suppose two blocks are to be used, and it is decided to confound the $ABCD$ interaction with the block effect. Which treatments should be carried out in the first block [the one containing the treatment (1)], and which treatments are allocated to the second block?

b. In an experiment to investigate niacin retention in vegetables as a function of cooking temperature (A), sieve size (B), type of processing (C), and cooking time (D), each factor was held at two levels. Two blocks were used, with the allocation of blocks as given in part (a) to confound only the $ABCD$ interaction with blocks. Use Yates's procedure to obtain the ANOVA table for the accompanying data.

Treatment	x_{ijkl}	Treatment	x_{ijkl}
(1)	91	d	72
a	85	ad	78
b	92	bd	68
ab	94	abd	79
c	86	cd	69
ac	83	acd	75
bc	85	bcd	72
abc	90	$abcd$	71

c. Assume that all three-way interaction effects are absent, so that the associated sums of squares can be combined to yield an estimate of σ^2, and carry out all appropriate tests at level .05.

45. a. An experiment was carried out to investigate the effects on audio sensitivity of varying resistance (A), two capacitances (B, C), and inductance of a coil (D) in part of a television circuit. If four blocks were used with four treatments per

block, and the defining effects for confounding were AB and CD, which treatments appeared in each block?

b. Suppose two replications of the experiment described in part (a) were performed, resulting in the accompanying data. Obtain the ANOVA table and test all relevant hypotheses at level .01.

Treatment	x_{ijkl1}	x_{ijkl2}	Treatment	x_{ijkl1}	x_{ijkl2}
(1)	618	598	d	598	585
a	583	560	ad	587	541
b	477	525	bd	480	508
ab	421	462	abd	462	449
c	601	595	cd	603	577
ac	550	589	acd	571	552
bc	505	484	bcd	502	508
abc	452	451	$abcd$	449	455

46. In an experiment involving four factors (A, B, C, and D) and four blocks, show that at least one main effect or two-factor interaction effect must be confounded with the block effect.

47. a. In a seven-factor experiment (A, ..., G), suppose a quarter-replicate is actually carried out. If the defining effects are $ABCDE$ and $CDEFG$, what is the third nonestimable effect and what treatments are in the group containing (1)? What are the alias groups of the seven main effects?

b. If the quarter-replicate is to be carried out using four blocks (with eight treatments per block), what are the blocks if the chosen confounding effects are ACF and BDG?

48. Suppose that in the rocket thrust problem of Exercise 43, enough resources had been available for only a half-replicate of the 2^4 experiment.

a. If the effect $ABCD$ is chosen as the defining effect for the replicate and the group of eight treatments for which data is obtained includes treatment (1), what other treatments are in the observed group and what are the alias pairs?

b. Suppose the results of carrying out the experiment as described in part (a) are as recorded here (given in standard order after deleting the half not observed). Assuming that two- and three-factor interactions are negligible, test at level .05 for the presence of main effects. Also construct a normal probability plot:

19.09   20.11   21.66   20.44
13.72   11.26   11.72   12.29

49. A half-replicate of a 2^5 experiment to investigate the effects of heating time (A), quenching time (B), drawing time (C), position of heating coils (D), and measurement position (E) on hardness of steel castings resulted in the accompanying data. Construct the ANOVA table and (assuming second- and higher-order interactions to be negligible) test at level .01 for the presence of main effects. Also construct a normal probability plot.

Treat-ment	Observation	Treat-ment	Observation
a	70.4	acd	66.6
b	72.1	ace	67.5
c	70.4	ade	64.0
d	67.4	bcd	66.8
e	68.0	bce	70.3
abc	73.8	bde	67.9
abd	67.0	cde	65.9
abe	67.8	$abcde$	68.0

Supplementary Exercises (50–61)

50. The results of a study on the effectiveness of line drying on the smoothness of fabric were summarized in the article "Line-Dried vs. Machine-Dried Fabrics: Comparison of Appearance, Hand, and Consumer Acceptance" (*Home Econ. Research J.,* 1984: 27–35). Smoothness scores were given for nine different types of fabric and five different drying methods: (1) machine dry, (2) line dry, (3) line dry followed by 15-min tumble, (4) line dry with softener, and (5) line dry with air movement. Regarding the different types of fabric as blocks, construct an ANOVA table. Using a .05 significance level, test to see whether there is a difference in the true mean smoothness score for the drying methods.

Drying Method

		1	2	3	4	5
	Crepe	3.3	2.5	2.8	2.5	1.9
	Double knit	3.6	2.0	3.6	2.4	2.3
	Twill	4.2	3.4	3.8	3.1	3.1
	Twill mix	3.4	2.4	2.9	1.6	1.7
Fabric	Terry	3.8	1.3	2.8	2.0	1.6
	Broadcloth	2.2	1.5	2.7	1.5	1.9
	Sheeting	3.5	2.1	2.8	2.1	2.2
	Corduroy	3.6	1.3	2.8	1.7	1.8
	Denim	2.6	1.4	2.4	1.3	1.6

51. The water absorption of two types of mortar used to repair damaged cement was discussed in the article "Polymer Mortar Composite Matrices for Maintenance-Free, Highly Durable Ferrocement" (*J. Ferrocement,* 1984: 337–345). Specimens of ordinary cement mortar (OCM) and polymer cement mortar (PCM) were submerged for varying lengths of time (5, 9, 24, or 48 hours) and water absorption (% by weight) was recorded. With mortar type as factor A (with two levels) and submersion period as factor B (with four levels), three observations were made for each factor level combination. Data included in the article was used to compute the sums of squares, which were SSA = 322.667, SSB = 35.623, SSAB = 8.557, and SST = 372.113. Use this information to construct an ANOVA table. Test the appropriate hypotheses at a .05 significance level.

52. Four plots were available for an experiment to compare clover accumulation for four different sowing rates ("Performance of Overdrilled Red Clover with Different Sowing Rates and Initial Grazing Managements," *N. Zeal. J. Exp. Ag.,* 1984: 71–81). Since the four plots had been grazed differently prior to the experiment and it was thought that this might affect clover accumulation, a randomized block experiment was used with all four sowing rates tried on a section of each plot. Use the given data to test the null hypothesis of no difference in true mean clover accumulation (kg DM/ha) for the different sowing rates.

Sowing Rate (kg/ha)

		3.6	6.6	10.2	13.5
	1	1155	2255	3505	4632
	2	123	406	564	416
Plot	3	68	416	662	379
	4	62	75	362	564

53. In an automated chemical coating process, the speed with which objects on a conveyor belt are passed through a chemical spray (belt speed), the amount of chemical sprayed (spray volume), and the brand of chemical used (brand) are factors that may affect the uniformity of the coating applied. A replicated 2^3 experiment was conducted in an effort to increase the coating uniformity. In the following table, higher values of the response variable are associated with higher surface uniformity:

| | | | | Surface Uniformity | |
| | | | | Repli- cation 1 | Repli- cation 2 |
Run	Spray Volume	Belt Speed	Brand		
1	−	−	−	40	36
2	+	−	−	25	28
3	−	+	−	30	32
4	+	+	−	50	48
5	−	−	+	45	43
6	+	−	+	25	30
7	−	+	+	30	29
8	+	+	+	52	49

Analyze this data and state your conclusions.

54. Coal-fired power plants used in the electrical industry have gained increased public attention because of the environmental problems associated with solid wastes generated by large-scale combustion ("Fly Ash Binders in Stabilization of FGD Wastes," *J. of Environmental Engineering,* 1998: 43–49). A study was conducted to analyze the influence of three factors, binder type (*A*), amount of water (*B*), and land disposal scenario (*C*), that affect certain leaching characteristics of solid wastes from combustion. Each factor was studied at two levels. An unreplicated 2^3 experiment was run, and a response value EC50 (the Effective Concentration, in mg/L, that decreases 50% of the light in a luminescence bioassay) was measured for each combination of factor levels. The experimental data is given in the following table:

| | | Factor | | Response |
Run	*A*	*B*	*C*	EC50
1	−1	−1	−1	23,100
2	1	−1	−1	43,000
3	−1	1	−1	71,400
4	1	1	−1	76,000
5	−1	−1	1	37,000
6	1	−1	1	33,200
7	−1	1	1	17,000
8	1	1	1	16,500

Carry out an appropriate ANOVA and state your conclusions.

55. Impurities in the form of iron oxides lower the economic value and usefulness of industrial minerals, such as kaolins, to ceramic and paper-processing industries. A 2^4 experiment was conducted to assess the effects of four factors on the percentage of iron removed from kaolin samples ("Factorial Experiments in the Development of a Kaolin Bleaching Process Using Thiourea in Sulphuric Acid Solutions," *Hydrometallurgy,* 1997: 181–197). The factors and their levels are listed in the following table:

Factor	Description	Units	Low Level	High Level
A	H_2SO_4	M	.10	.25
B	Thiourea	g/l	0.0	5.0
C	Temperature	°C	70	90
D	Time	min	30	150

The data from an unreplicated 2^4 experiment is listed in the next table.

Test Run	Iron Extraction (%)	Test Run	Iron Extraction (%)
(1)	7	*d*	28
a	11	*ad*	51
b	7	*bd*	33
ab	12	*abd*	57
c	21	*cd*	70
ac	41	*acd*	95
bc	27	*bcd*	77
abc	48	*abcd*	99

a. Calculate estimates of all main effects and two-factor interaction effects for this experiment.

b. Create a probability plot of the effects. Which effects appear to be important?

56. Factorial designs have been used in forestry to assess the effects of various factors on the growth behavior of trees. In one such experiment, researchers thought that healthy spruce seedlings should bud sooner than diseased spruce seedlings ("Practical Analysis of Factorial Experiments in Forestry," *Canadian J. of Forestry,* 1995: 446–461). In addition, before planting, seedlings were also exposed to three levels of pH to see whether this factor has an effect on virus uptake into the root system. The following table shows data from a 2 × 3 experiment to study both factors:

		pH		
		3	**5.5**	**7**
	Diseased	1.2, 1.4,	.8, .6,	1.0, 1.0,
		1.0, 1.2,	.8, 1.0,	1.2, 1.4,
		1.4	.8	1.2
Health	Healthy	1.4, 1.6,	1.0, 1.2,	1.2, 1.4,
		1.6, 1.6,	1.2, 1.4,	1.2, 1.2,
		1.4	1.4	1.4

The response variable is an average rating of five buds from a seedling. The ratings are 0 (bud not broken), 1 (bud partially expanded), and 2 (bud fully expanded). Analyze this data.

57. One property of automobile air bags that contributes to their ability to absorb energy is the permeability ($ft^3/ft^2/min$) of the woven material used to construct the air bags. Understanding how permeability is influenced by various factors is important for increasing the effectiveness of air bags. In one study, the effects of three factors, each at three levels, were studied ("Analysis of Fabrics used in Passive Restraint Systems—Airbags," *J. of the Textile Institute,* 1996: 554–571):

A (Temperature): 8°C, 50°C, 75°C
B (Fabric denier): 420-D, 630-D, 840-D
C (Air pressure): 17.2 kPa, 34.4 kPa, 103.4 kPa

Temperature 8°

	Pressure		
Denier	**17.2**	**34.4**	**103.4**
420-D	73	157	332
	80	155	322
630-D	35	91	288
	433	98	271
840-D	125	234	477
	111	233	464

Temperature 50°

	Pressure		
Denier	**17.2**	**34.4**	**103.4**
420-D	52	125	281
	51	118	264
630-D	16	72	169
	12	78	173
840-D	96	149	338
	100	155	350

(continued)

Temperature 75°

	Pressure		
Denier	**17.2**	**34.4**	**103.4**
420-D	37	95	276
	31	106	281
630-D	30	91	213
	41	100	211
840-D	102	170	307
	98	160	311

Analyze this data and state your conclusions (assume that all factors are fixed).

58. A chemical engineer has carried out an experiment to study the effects of the fixed factors vat pressure (A), cooking time of pulp (B), and hardwood concentration (C) on the strength of paper. The experiment involved two pressures, four cooking times, three concentrations, and two observations at each combination of these levels. Calculated sums of squares are SSA = 6.94, SSB = 5.61, SSC = 12.33, SSAB = 4.05, SSAC = 7.32, SSBC = 15.80, SSE = 14.40, and SST = 70.82. Construct the ANOVA table and carry out appropriate tests at significance level .05.

59. The bond strength when mounting an integrated circuit on a metalized glass substrate was studied as a function of factor A = adhesive type, factor B = cure time, and factor C = conductor material (copper and nickel). The data follows, along with an ANOVA table from MINITAB. What conclusions can you draw from the data?

Copper		Cure Time		
		1	**2**	**3**
	1	72.7	74.6	80.0
		80.0	77.5	82.7
Adhesive	2	77.8	78.5	84.6
		75.3	81.1	78.3
	3	77.3	80.9	83.9
		76.5	82.6	85.0

Nickel		**1**	**2**	**3**
	1	74.7	75.7	77.2
		77.4	78.2	74.6
Adhesive	2	79.3	78.8	83.0
		77.8	75.4	83.9
	3	77.2	84.5	89.4
		78.4	77.5	81.2

```
Analysis of Variance for strength
Source          DF        SS        MS      F      P
Adhesive         2   101.317    50.659   6.54  0.007
Curetime         2   151.317    75.659   9.76  0.001
Conmater         1     0.722     0.722   0.09  0.764
Adhes*curet      4    30.526     7.632   0.98  0.441
Adhes*conm       2     8.015     4.008   0.52  0.605
Curet*conm       2     5.952     2.976   0.38  0.687
Adh*curet*conm   4    33.298     8.325   1.07  0.398
Error           18   139.515     7.751
Total           35   470.663
```

60. The article "Food Consumption and Energy Requirements of Captive Bald Eagles" (*J. Wildlife Mgmt.*, 1982: 646–654) investigated mean gross daily energy intake (the response variable) for different diet types (factor A, with three levels) and temperature (factor B, with three levels). Summary quantities given in the article were used to generate data, resulting in SSA = 18,138, SSB = 5182, SSAB = 1737, SST = 36,348, and error df = 36. Construct an ANOVA table and test the relevant hypotheses.

61. Analogous to a Latin square, a Greco–Latin square design can be used when it is suspected that three extraneous factors may affect the response variable and all four factors (the three extraneous ones and the one of interest) have the same number of levels. In a Latin square, each level of the factor of interest (C) appears once in each row (with each level of A) and once in each column (with each level of B). In a Greco–Latin square, each level of factor D appears once in each row, in each column, and also with each level of the third extraneous factor C. Alternatively, the design can be used when the four factors are all of equal interest, the number of levels of each is N, and resources are available for only N^2 observations. A 5×5 square is pictured in (a), with (k, l) in each cell denoting the kth level of C and lth level of D. In (b) we present data on weight loss in silicon bars used for semiconductor material as a function

of volume of etch (A), color of nitric acid in the etch solution (B), size of bars (C), and time in the etch solution (D) (from "Applications of Analytic Techniques to the Semiconductor Industry," Fourteenth Midwest Quality Control Conference, 1959).

Let $X_{ij(kl)}$ denote the observed weight loss when factor A is at level i, B is at level j, C is at level k, and D is at level l. Assuming no interaction between factors, total sum of squares SST (with $N^2 - 1$ df) can be partitioned into SSA, SSB, SSC, SSD, and SSE. Give expressions for these sums of squares, including computing formulas, obtain the ANOVA table for the given data, and test each of the four main effect hypotheses using $\alpha = .05$.

		B				
(C, D)		**1**	**2**	**3**	**4**	**5**
	1	(1, 1)	(2, 3)	(3, 5)	(4, 2)	(5, 4)
	2	(2, 2)	(3, 4)	(4, 1)	(5, 3)	(1, 5)
A	**3**	(3, 3)	(4, 5)	(5, 2)	(1, 4)	(2, 1)
	4	(4, 4)	(5, 1)	(1, 3)	(2, 5)	(3, 2)
	5	(5, 5)	(1, 2)	(2, 4)	(3, 1)	(4, 3)

(a)

65	82	108	101	126
84	109	73	97	83
105	129	89	89	52
119	72	76	117	84
97	59	94	78	106

(b)

Bibliography

Box, George, William Hunter, and Stuart Hunter, *Statistics for Experimenters,* Wiley, New York, 1978. Contains a wealth of suggestions and insights on data analysis based on the authors' extensive consulting experience.

DeVor, R., T. Chang, and J. W. Sutherland, *Statistical Quality Design and Control,* Macmillan, New York, 1992. Includes a modern survey of factorial and fractional factorial experimentation with a minimum of mathematics.

Hocking, Ronald, *The Analysis of Linear Models,* Brooks/Cole, Pacific Grove, CA, 1985. A very general treatment of analysis of variance written by one of the foremost authorities in this field.

Kleinbaum, David, Lawrence Kupper, Keith Muller, and Azhar Nizam, *Applied Regression Analysis and Other Multivariable Methods* (3rd ed.), Duxbury Press, Pacific Grove, 1998. Contains an especially good discussion of problems associated with analysis of "unbalanced data"—that is, unequal K_{ij}'s.

Kuehl, Robert O., *Statistical Principles of Research Design and Analysis,* Wadsworth, Belmont, CA, 1994. An up-to-date and comprehensive treatment of designed experiments and analysis of the resulting data.

Montgomery, Douglas, *Design and Analysis of Experiments* (4th ed.), Wiley, New York, 1997. See the Chapter 10 bibliography.

Neter, John, William Wasserman, and Michael Kutner, *Applied Linear Statistical Models* (4th ed.), Irwin, Homewood, IL, 1996. See the Chapter 10 bibliography.

Vardeman, Stephen, *Statistics for Engineering Problem Solving,* PWS, Boston, 1994. A general introduction for engineers, with much descriptive and inferential methodology for data from designed experiments.

12

Simple Linear Regression and Correlation

Introduction

In the two-sample problems discussed in Chapter 9, we were interested in comparing values of parameters for the x distribution and the y distribution. Even when observations were paired, we did not try to use information about one of the variables in studying the other variable. This is precisely the objective of regression analysis: to exploit the relationship between two (or more) variables so that we can gain information about one of them through knowing values of the other(s).

Much of mathematics is devoted to studying variables that are *deterministically* related. Saying that x and y are related in this manner means that once we are told the value of x, the value of y is completely specified. For example, suppose we decide to rent a van for a day and that the rental cost is $25.00 plus $.30 per mile driven. If we let $x =$ the number of miles driven and $y =$ the rental charge, then $y = 25 + .3x$. If we drive the van 100 miles ($x = 100$), then $y = 25 + .3(100) = 55$. As another example, if the initial velocity of a particle is v_0 and it undergoes constant acceleration a, then distance traveled $= y = v_0 x + \frac{1}{2}ax^2$ where $x =$ time.

There are many variables x and y that would appear to be related to one another, but not in a deterministic fashion. A familiar example to many students is given by variables x = high school grade point average (GPA) and y = college GPA. The value of y cannot be determined just from knowledge of x, and two different students could have the same x value but have very different y values. Yet there is a tendency for those students who have high (low) high school GPAs also to have high (low) college GPAs. Knowledge of a student's high school GPA should be quite helpful in enabling us to predict how that person will do in college.

Other examples of variables related in a nondeterministic fashion include x = age of a child and y = size of that child's vocabulary, x = size of an engine in cubic centimeters and y = fuel efficiency for an automobile equipped with that engine, and x = applied tensile force and y = amount of elongation in a metal strip.

Regression analysis is the part of statistics that deals with investigation of the relationship between two or more variables related in a nondeterministic fashion. In this chapter, we generalize the linear relation $y = \beta_0 + \beta_1 x$ to a linear probabilistic relationship, develop procedures for making inferences about the parameters of the model, and obtain a quantitative measure (the correlation coefficient) of the extent to which the two variables are related. In the next chapter, we will consider techniques for validating a particular model and investigate nonlinear relationships and relationships involving more than two variables.

12.1 | The Simple Linear Regression Model

The simplest deterministic mathematical relationship between two variables x and y is a linear relationship $y = \beta_0 + \beta_1 x$. The set of pairs (x, y) for which $y = \beta_0 + \beta_1 x$ determines a straight line with slope β_1 and y-intercept β_0.* The objective of this section is to develop a linear probabilistic model.

If the two variables are not deterministically related, then for a fixed value of x, the value of the second variable is random. For example, if we are investigating the relationship between age of child and size of vocabulary and decide to select a child of age $x = 5.0$ years, then before the selection is made, vocabulary size is a random variable Y. After a particular 5-year-old child has been selected and tested, a vocabulary of

---

*The slope of a line is the change in y for a 1-unit increase in x. For example, if $y = -3x + 10$, then y decreases by 3 when x increases by 1, so the slope is -3. The y-intercept is the height at which the line crosses the vertical axis and is obtained by setting $x = 0$ in the equation.

2000 words may result. We would then say that the observed value of Y associated with fixing $x = 5.0$ was $y = 2000$.

More generally, the variable whose value is fixed by the experimenter will be denoted by x and will be called the **independent, predictor,** or **explanatory variable.** For fixed x, the second variable will be random; we denote this random variable and its observed value by Y and y, respectively, and refer to it as the **dependent** or **response variable.**

Usually observations will be made for a number of settings of the independent variable. Let $x_1, x_2, \ldots, x_n$ denote values of the independent variable for which observations are made, and let Y_i and y_i respectively denote the random variable and observed value associated with x_i. The available data then consists of the n pairs $(x_1, y_1), (x_2, y_2), \ldots, (x_n, y_n)$. A first step in regression analysis involving two variables is to construct a **scatter plot** of the observed data. In such a plot, each (x_i, y_i) is represented as a point plotted on a two-dimensional coordinate system.

Example 12.1   Visual and musculoskeletal problems associated with the use of visual display terminals (VDTs) have become rather common in recent years. Some researchers have focused on vertical gaze direction as a source of eye strain and irritation. This direction is known to be closely related to ocular surface area (OSA), so a method of measuring OSA is needed. The accompanying representative data on $y = $ OSA (cm²) and $x = $ width of the palprebal fissure (that is, the horizontal width of the eye opening, in cm) is from the article "Analysis of Ocular Surface Area for Comfortable VDT Workstation Layout" (*Ergonomics*, 1996: 877–884). The order in which observations were obtained was not given, so for convenience they are listed in increasing order of x values.

i	1	2	3	4	5	6	7	8	9	10	11	12	13	14	15
x_i	.40	.42	.48	.51	.57	.60	.70	.75	.75	.78	.84	.95	.99	1.03	1.12
y_i	1.02	1.21	.88	.98	1.52	1.83	1.50	1.80	1.74	1.63	2.00	2.80	2.48	2.47	3.05

i	16	17	18	19	20	21	22	23	24	25	26	27	28	29	30
x_i	1.15	1.20	1.25	1.25	1.28	1.30	1.34	1.37	1.40	1.43	1.46	1.49	1.55	1.58	1.60
y_i	3.18	3.76	3.68	3.82	3.21	4.27	3.12	3.99	3.75	4.10	4.18	3.77	4.34	4.21	4.92

Thus $(x_1, y_1) = (.40, 1.02)$, $(x_5, y_5) = (.57, 1.52)$, and so on. A MINITAB scatter plot is shown in Figure 12.1; we used an option that produced a dotplot of both the x values and y values individually along the right and top margins of the plot, which makes it easier to visualize the distributions of the individual variables (histograms or boxplots are alternative options). Here are some things to notice about the data and plot:

- Several observations have identical x values yet different y values (for example, $x_8 = x_9 = .75$, but $y_8 = 1.80$ and $y_9 = 1.74$). Thus the value of y is *not* determined solely by x but also by various other factors.

- There is a strong tendency for y to increase as x increases. That is, larger values of OSA tend to be associated with larger values of fissure width—a positive relationship between the variables.

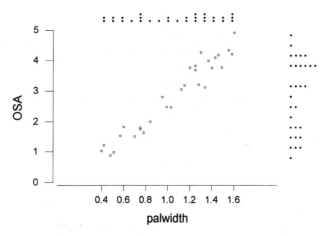

Figure 12.1 Scatter plot from MINITAB for the data from Example 12.1, along with dotplots of *x* and *y* values

• It appears that the value of *y* could be predicted from *x* by finding a line that is reasonably close to the points in the plot (the authors of the cited article superimposed such a line on their plot). In other words, there is evidence of a substantial (though not perfect) linear relationship between the two variables. ■

The horizontal and vertical axes in the scatter plot of Figure 12.1 intersect at the point (0, 0). In many data sets, the values of *x* or *y* or the values of both variables differ considerably from zero relative to the range(s) of the values. For example, a study of how air conditioner efficiency is related to maximum daily outdoor temperature might involve observations for temperatures ranging from 80°F to 100°F. When this is the case, a more informative plot would show the appropriately labeled axes intersecting at some point other than (0, 0).

Example 12.2 Forest growth and decline phenomena throughout the world have attracted considerable public and scientific interest. The article "Relationships Among Crown Condition, Growth, and Stand Nutrition in Seven Northern Vermont Sugarbushes" (*Canad. J. of Forest Res.,* 1995: 386–397) included a scatter plot of *y* = mean crown dieback (%), one indicator of growth retardation, and *x* = soil pH (higher pH corresponds to more acidic soil), from which the following observations were taken:

x	3.3	3.4	3.4	3.5	3.6	3.6	3.7	3.7	3.8	3.8
y	7.3	10.8	13.1	10.4	5.8	9.3	12.4	14.9	11.2	8.0

x	3.9	4.0	4.1	4.2	4.3	4.4	4.5	5.0	5.1	
y	6.6	10.0	9.2	12.4	2.3	4.3	3.0	1.6	1.0	

Figure 12.2 (page 492) shows two MINITAB scatter plots of this data. In Figure 12.2(a), MINITAB selected the scale for both axes. We obtained Figure 12.2(b) by specifying minimum and maximum values for *x* and *y*, so that the axes would intersect

roughly at the point (0, 0). The second plot is more crowded than the first one; such crowding can make it more difficult to ascertain the general nature of any relationship. For example, it can be more difficult to spot curvature in a crowded plot.

Large values of percentage dieback tend to be associated with low soil pH, a negative or inverse relationship. Furthermore, the two variables appear to be at least approximately linearly related, although the points would be spread out about any straight line drawn through the plot.

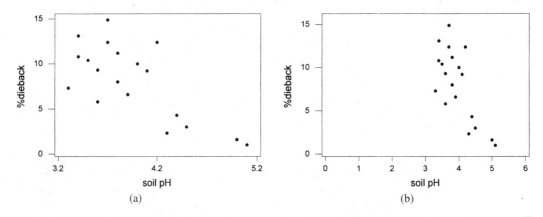

Figure 12.2 MINITAB scatter plots of data in Example 12.2

A Linear Probabilistic Model

For the deterministic model $y = \beta_0 + \beta_1 x$, the actual observed value of y is a linear function of x. The appropriate generalization of this to a probabilistic model assumes that *the expected value of Y is a linear function of x,* but that for fixed x, the variable Y differs from its expected value by a random amount.

The Simple Linear Regression Model

There exist parameters β_0, β_1, and σ^2 such that for any fixed value of the independent variable x, the dependent variable is related to x through the model equation

$$Y = \beta_0 + \beta_1 x + \epsilon \qquad (12.1)$$

The quantity ϵ in the model equation is a random variable, assumed to be normally distributed with $E(\epsilon) = 0$ and $V(\epsilon) = \sigma^2$.

The variable ϵ is usually referred to as the **random deviation** or **random error term** in the model. Without ϵ, any observed pair (x, y) would correspond to a point falling exactly on the line $y = \beta_0 + \beta_1 x$, called the **true** (or **population**) **regression line.** The inclusion of the random error term allows (x, y) to fall either above the true regression line (when $\epsilon > 0$) or below the line (when $\epsilon < 0$). The points $(x_1, y_1), \ldots,$

(x_n, y_n) resulting from n independent observations will then be scattered about the true regression line, as illustrated in Figure 12.3. On occasion, the appropriateness of the simple linear regression model may be suggested by theoretical considerations (e.g., there is an exact linear relationship between the two variables, with ϵ representing measurement error). Much more frequently, though, the reasonableness of the model is indicated by a scatter plot exhibiting a substantial linear pattern (as in Figure 12.1).

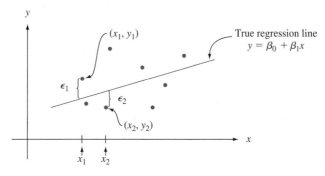

Figure 12.3 Points corresponding to observations from the simple linear regression model

Implications of the model equation (12.1) can best be understood with the aid of the following notation. Let x^* denote a particular value of the independent variable x and

$\mu_{Y \cdot x^*}$ = the expected (or mean) value of Y when $x = x^*$

$\sigma^2_{Y \cdot x^*}$ = the variance of Y when $x = x^*$

Alternative notation is $E(Y|x^*)$ and $V(Y|x^*)$. For example, if $x =$ applied stress (kg/mm^2) and $y =$ time to fracture (hr), then $\mu_{Y \cdot 20}$ would denote the expected value of time-to-fracture when applied stress is 20 kg/mm^2. If we think of an entire population of (x, y) pairs, then $\mu_{Y \cdot x^*}$ is the mean of all y values for which $x = x^*$, and $\sigma^2_{Y \cdot x^*}$ is a measure of how much these values of y spread out about the mean value. If, for example, $x =$ age of a child and $y =$ vocabulary size, then $\mu_{Y \cdot 5}$ is the average vocabulary size for all 5-year-old children in the population, and $\sigma^2_{Y \cdot 5}$ describes the amount of variability in vocabulary size for this part of the population. Once x is fixed, the only randomness on the right-hand side of the model equation (12.1) is in the random error ϵ, and its mean value and variance are 0 and σ^2, respectively, whatever the value of x. This implies that

$$\mu_{Y \cdot x^*} = E(\beta_0 + \beta_1 x^* + \epsilon) = \beta_0 + \beta_1 x^* + E(\epsilon) = \beta_0 + \beta_1 x^*$$
$$\sigma^2_{Y \cdot x^*} = V(\beta_0 + \beta_1 x^* + \epsilon) = V(\beta_0 + \beta_1 x^*) + V(\epsilon) = 0 + \sigma^2 = \sigma^2$$

Replacing x^* in $\mu_{Y \cdot x^*}$ by x gives the relation $\mu_{Y \cdot x} = \beta_0 + \beta_1 x$, which says that the *mean value* of Y, rather than Y itself, is a linear function of x. The true regression line $y = \beta_0 + \beta_1 x$ is thus the *line of mean values;* its height above any particular x value is the expected value of Y for that value of x. The slope β_1 of the true regression line is interpreted as the *expected* change in Y associated with a 1-unit increase in the value of x. The second relation states that the amount of variability in the distribution of

Y values is the same at each different value of x (homogeneity of variance). In the example involving age of a child and vocabulary size, the model implies that average vocabulary size changes linearly with age (hopefully β_1 is positive) and that the amount of variability in vocabulary size at any particular age is the same as at any other age. Finally, for fixed x, Y is the sum of a constant $\beta_0 + \beta_1 x$ and a normally distributed rv ϵ, so itself has a normal distribution. These properties are illustrated in Figure 12.4. The variance parameter σ^2 determines the extent to which each normal curve spreads out about its mean value (the height of the line). When σ^2 is small, an observed point (x, y) will almost always fall quite close to the true regression line, whereas observations may deviate considerably from their expected values (corresponding to points far from the line) when σ^2 is large.

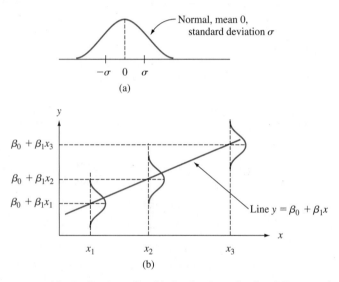

Figure 12.4 (a) Distribution of ϵ; (b) distribution of Y for different values of x

Example 12.3 Suppose the relationship between applied stress x and time-to-failure y is described by the simple linear regression model with true regression line $y = 65 - 1.2x$ and $\sigma = 8$. Then for any fixed value x^* of stress, time-to-failure has a normal distribution with mean value $65 - 1.2x^*$ and standard deviation 8. Roughly speaking, in the population consisting of all (x, y) points, the magnitude of a typical deviation from the true regression line is about 8. For $x = 20$, Y has mean value $\mu_{Y \cdot 20} = 65 - 1.2(20) = 41$, so

$$P(Y > 50 \text{ when } x = 20) = P\left(Z > \frac{50 - 41}{8}\right) = 1 - \Phi(1.13) = .1292$$

The probability that time-to-failure exceeds 50 when applied stress is 25 is, because $\mu_{Y \cdot 25} = 35$,

$$P(Y > 50 \text{ when } x = 25) = P\left(Z > \frac{50 - 35}{8}\right) = 1 - \Phi(1.88) = .0301$$

These probabilities are illustrated as the shaded areas in Figure 12.5.

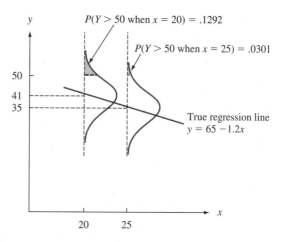

Figure 12.5 Probabilities based on the simple linear regression model

Suppose that Y_1 denotes an observation on time-to-failure made with $x = 25$ and Y_2 denotes an independent observation made with $x = 24$. Then $Y_1 - Y_2$ is normally distributed with mean value $E(Y_1 - Y_2) = \beta_1 = -1.2$, variance $V(Y_1 - Y_2) = \sigma^2 + \sigma^2 = 128$, and standard deviation $\sqrt{128} = 11.314$. The probability that Y_1 exceeds Y_2 is

$$P(Y_1 - Y_2 > 0) = P\left(Z > \frac{0 - (-1.2)}{11.314}\right) = P(Z > .11) = .4562$$

That is, even though we expected Y to decrease when x increases by 1 unit, it is not unlikely that the observed Y at $x + 1$ will be larger than the observed Y at x. ■

Exercises | Section 12.1 (1–11)

1. The efficiency ratio for a steel specimen immersed in a phosphating tank is the weight of the phosphate coating divided by the metal loss (both in mg/ft²). The article "Statistical Process Control of a Phosphate Coating Line" (*Wire J. Intl.*, May, 1997: 78–81) gave the accompanying data on tank temperature (x) and efficiency ratio (y).

Temp.	170	172	173	174	174	175	176
Ratio	.84	1.31	1.42	1.03	1.07	1.08	1.04

Temp.	177	180	180	180	180	180	181
Ratio	1.80	1.45	1.60	1.61	2.13	2.15	.84

Temp.	181	182	182	182	182	184	184
Ratio	1.43	.90	1.81	1.94	2.68	1.49	2.52

Temp.	185	186	188
Ratio	3.00	1.87	3.08

a. Construct stem-and-leaf displays of both temperature and efficiency ratio, and comment on interesting features.

b. Is the value of efficiency ratio completely and uniquely determined by tank temperature? Explain your reasoning.

c. Construct a scatter plot of the data. Does it appear that efficiency ratio could be very well predicted by the value of temperature? Explain your reasoning.

2. The article "Exhaust Emissions from Four-Stroke Lawn Mower Engines" (*J. of the Air and Water Mgmnt. Assoc.*, 1997: 945–952) reported data from a study in which both a baseline gasoline mixture and a reformulated gasoline were used. Consider the following observations on age (yr) and NO_x emissions (g/kW-h):

Engine	1	2	3	4	5
Age	0	0	2	11	7
Baseline	1.72	4.38	4.06	1.26	5.31
Reformulated	1.88	5.93	5.54	2.67	6.53

Engine	6	7	8	9	10
Age	16	9	0	12	4
Baseline	.57	3.37	3.44	.74	1.24
Reformulated	.74	4.94	4.89	.69	1.42

Construct scatter plots of NO_x emissions versus age. What appears to be the nature of the relationship between these two variables? (*Note:* The authors of the cited article commented on the relationship.)

3. Bivariate data often arises from the use of two different techniques to measure the same quantity. As an example, the accompanying observations on $x =$ hydrogen concentration (ppm) using a gas chromatography method and $y =$ concentration using a new sensor method were read from a graph in the article "A New Method to Measure the Diffusible Hydrogen Content in Steel Weldments Using a Polymer Electrolyte-Based Hydrogen Sensor" (*Welding Res.,* July 1997: 251s–256s).

x	47	62	65	70	70	78	95	100	114	118
y	38	62	53	67	84	79	93	106	117	116

x	124	127	140	140	140	150	152	164	198	221
y	127	114	134	139	142	170	149	154	200	215

Construct a scatter plot. Does there appear to be a very strong relationship between the two types of concentration measurements? Do the two methods appear to be measuring roughly the same quantity? Explain your reasoning.

4. A study to assess the capability of subsurface flow wetland systems to remove biochemical oxygen demand (BOD) and various other chemical constituents resulted in the accompanying data on $x =$ BOD mass loading (kg/ha/d) and $y =$ BOD mass removal (kg/ha/d) ("Subsurface Flow Wetlands— A Performance Evaluation," *Water Envir. Res.,* 1995: 244–247).

x	3	8	10	11	13	16	27	30	35	37	38	44	103	142
y	4	7	8	8	10	11	16	26	21	9	31	30	75	90

a. Construct boxplots of both mass loading and mass removal, and comment on any interesting features.

b. Construct a scatter plot of the data and comment on any interesting features.

5. The article "Objective Measurement of the Stretchability of Mozzarella Cheese" (*J. of Texture Studies,* 1992: 185–194) reported on an experiment to investigate how the behavior of mozzarella cheese varied with temperature. Consider the accompanying data on $x =$ temperature and $y =$ elongation (%) at failure of the cheese. (*Note:* The researchers were Italian and used *real* mozzarella cheese, not the poor cousin widely available in the United States.)

x	59	63	68	72	74	78	83
y	118	182	247	208	197	135	132

a. Construct a scatter plot in which the axes intersect at (0, 0). Mark 0, 20, 40, 60, 80, and 100 on the horizontal axis and 0, 50, 100, 150, 200, and 250 on the vertical axis.

b. Construct a scatter plot in which the axes intersect at (55, 100), as was done in the cited article. Does this plot seem preferable to the one in part (a)? Explain your reasoning.

c. What do the plots of parts (a) and (b) suggest about the nature of the relationship between the two variables?

6. One factor in the development of tennis elbow, a malady that strikes fear in the hearts of all serious tennis players, is the impact-induced vibration of the racket-and-arm system at ball contact. It is well known that the likelihood of getting tennis elbow depends on various properties of the racket used. Consider the scatter plot of $x =$ racket resonance frequency (Hz) and $y =$ sum of peak-to-peak acceleration (a characteristic of arm vibration, in m/sec/sec) for $n = 23$ different rackets ("Transfer of Tennis Racket Vibrations into the Human Forearm," *Medicine and Science in Sports and Exercise,* 1992: 1134–1140). Discuss interesting features of the data and scatter plot.

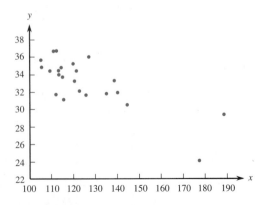

7. The article "Some Field Experience in the Use of an Accelerated Method in Estimating 28-Day Strength of Concrete" (*J. Amer. Concrete Institute,* 1969: 895) considered regressing y = 28-day standard-cured strength (psi) against x = accelerated strength (psi). Suppose the equation of the true regression line is $y = 1800 + 1.3x$.
   a. What is the expected value of 28-day strength when accelerated strength = 2500?
   b. By how much can we expect 28-day strength to change when accelerated strength increases by 1 psi?
   c. Answer part (b) for an increase of 100 psi.
   d. Answer part (b) for a decrease of 100 psi.

8. Referring to Exercise 7, suppose that the standard deviation of the random deviation ϵ is 350 psi.
   a. What is the probability that the observed value of 28-day strength will exceed 5000 psi when the value of accelerated strength is 2000?
   b. Repeat part (a) with 2500 in place of 2000.
   c. Consider making two independent observations on 28-day strength, the first for an accelerated strength of 2000 and the second for $x = 2500$. What is the probability that the second observation will exceed the first by more than 1000 psi?
   d. Let Y_1 and Y_2 denote observations on 28-day strength when $x = x_1$ and $x = x_2$, respectively. By how much would x_2 have to exceed x_1 in order that $P(Y_2 > Y_1) = .95$?

9. The flow rate y (m³/min) in a device used for air-quality measurement depends on the pressure drop x (in. of water) across the device's filter. Suppose that for x values between 5 and 20, the two variables are related according to the simple linear regression model with true regression line $y = -.12 + .095x$.
   a. What is the expected change in flow rate associated with a 1-in. increase in pressure drop? Explain.

   b. What change in flow rate can be expected when pressure drop decreases by 5 in.?
   c. What is the expected flow rate for a pressure drop of 10 in.? A drop of 15 in.?
   d. Suppose $\sigma = .025$ and consider a pressure drop of 10 in. What is the probability that the observed value of flow rate will exceed .835? That observed flow rate will exceed .840?
   e. What is the probability that an observation on flow rate when pressure drop is 10 in. will exceed an observation on flow rate made when pressure drop is 11 in.?

10. Suppose the expected cost of a production run is related to the size of the run by the equation $y = 4000 + 10x$. Let Y denote an observation on the cost of a run. If the variables *size* and *cost* are related according to the simple linear regression model, could it be the case that $P(Y > 5500$ when $x = 100) = .05$ and $P(Y > 6500$ when $x = 200) = .10$? Explain.

11. Suppose that in a certain chemical process the reaction time y (hour) is related to the temperature (°F) in the chamber in which the reaction takes place according to the simple linear regression model with equation $y = 5.00 - .01x$ and $\sigma = .075$.
    a. What is the expected change in reaction time for a 1°F increase in temperature? For a 10°F increase in temperature?
    b. What is the expected reaction time when temperature is 200°F? When temperature is 250°F?
    c. Suppose five observations are made independently on reaction time, each one for a temperature of 250°F. What is the probability that all five times are between 2.4 and 2.6 hours?
    d. What is the probability that two independently observed reaction times for temperatures 1° apart are such that the time at the higher temperature exceeds the time at the lower temperature?

12.2 | Estimating Model Parameters

We will assume in this and the next several sections that the variables x and y are related according to the simple linear regression model. The values of β_0, β_1, and σ^2 will almost never be known to an investigator. Instead, sample data consisting of n observed pairs $(x_1, y_1), \ldots, (x_n, y_n)$ will be available, from which the model parameters and the true regression line itself can be estimated. These observations are assumed to have been obtained independently of one another. That is, y_i is the observed value of an rv Y_i, where $Y_i = \beta_0 + \beta_1 x_i + \epsilon_i$ and the n deviations $\epsilon_1, \epsilon_2, \ldots, \epsilon_n$ are independent rv's. Independence of $Y_1, Y_2, \ldots, Y_n$ follows from the independence of the ϵ_i's.

According to the model, the observed points will be distributed about the true regression line in a random manner. Figure 12.6 shows a typical plot of observed pairs along with two candidates for the estimated regression line, $y = a_0 + a_1 x$ and $y = b_0 + b_1 x$. Intuitively, the line $y = a_0 + a_1 x$ is not a reasonable estimate of the true line $y = \beta_0 + \beta_1 x$ because, if $y = a_0 + a_1 x$ were the true line, the observed points would almost surely have been closer to this line. The line $y = b_0 + b_1 x$ is a more plausible estimate because the observed points are scattered rather closely about this line.

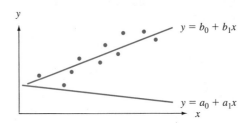

Figure 12.6 Two different estimates of the true regression line

Figure 12.6 and the foregoing discussion suggest that our estimate of $y = \beta_0 + \beta_1 x$ should be a line that provides in some sense a best fit to the observed data points. This is what motivates the principle of least squares, which can be traced back to the German mathematician Gauss (1777–1855). According to this principle, a line provides a good fit to the data if the vertical distances (deviations) from the observed points to the line are small (see Figure 12.7). The measure of the goodness of fit is the sum of the squares of these deviations. The best-fit line is then the one having the smallest possible sum of squared deviations.

Principle of Least Squares

The vertical deviation of the point (x_i, y_i) from the line $y = b_0 + b_1 x$ is

$$\text{height of point} - \text{height of line} = y_i - (b_0 + b_1 x_i)$$

The sum of squared vertical deviations from the points $(x_1, y_1), \ldots, (x_n, y_n)$ to the line is then

$$f(b_0, b_1) = \sum_{i=1}^{n} [y_i - (b_0 + b_1 x_i)]^2$$

The point estimates of β_0 and β_1, denoted by $\hat{\beta}_0$ and $\hat{\beta}_1$ and called the **least squares estimates,** are those values that minimize $f(b_0, b_1)$. That is, $\hat{\beta}_0$ and $\hat{\beta}_1$ are such that $f(\hat{\beta}_0, \hat{\beta}_1) \leq f(b_0, b_1)$ for any b_0 and b_1. The **estimated regression line** or **least squares line** is then the line whose equation is $y = \hat{\beta}_0 + \hat{\beta}_1 x$.

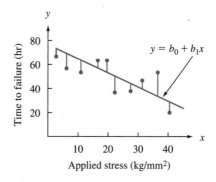

Figure 12.7   Deviations of observed data from line $y = b_0 + b_1x$

The minimizing values of b_0 and b_1 are found by taking partial derivatives of $f(b_0, b_1)$ with respect to both b_0 and b_1, equating them both to zero [analogously to $f'(b) = 0$ in univariate calculus], and solving the equations

$$\frac{\partial f(b_0, b_1)}{\partial b_0} = \Sigma 2(y_i - b_0 - b_1x_i)(-1) = 0$$

$$\frac{\partial f(b_0, b_1)}{\partial b_1} = \Sigma 2(y_i - b_0 - b_1x_i)(-x_i) = 0$$

Cancellation of the -2 factor and rearrangement gives the following system of equations, called the **normal equations:**

$$nb_0 + (\Sigma x_i)b_1 = \Sigma y_i$$
$$(\Sigma x_i)b_0 + (\Sigma x_i^2)b_1 = \Sigma x_i y_i$$

The normal equations are linear in the two unknowns b_0 and b_1. Provided that at least two of the x_i's are different, the least squares estimates are the unique solution to this system.

The least squares estimate of the slope coefficient β_1 of the true regression line is

$$b_1 = \hat{\beta}_1 = \frac{\Sigma(x_i - \bar{x})(y_i - \bar{y})}{\Sigma(x_i - \bar{x})^2} = \frac{S_{xy}}{S_{xx}} \tag{12.2}$$

Computing formulas for the numerator and denominator of $\hat{\beta}_1$ are

$$S_{xy} = \Sigma x_i y_i - (\Sigma x_i)(\Sigma y_i)/n \qquad S_{xx} = \Sigma x_i^2 - (\Sigma x_i)^2/n$$

The least squares estimate of the intercept β_0 of the true regression line is

$$b_0 = \hat{\beta}_0 = \frac{\Sigma y_i - \hat{\beta}_1 \Sigma x_i}{n} = \bar{y} - \hat{\beta}_1 \bar{x} \tag{12.3}$$

The computational formulas for S_{xy} and S_{xx} require only the summary statistics $\Sigma x_i, \Sigma y_i,$ $\Sigma x_i^2, \Sigma x_i y_i$ (Σy_i^2 will be needed shortly) and minimize the effects of rounding. In computing $\hat{\beta}_0$, use extra digits in $\hat{\beta}_1$ because, if $\bar{x}$ is large in magnitude, rounding will affect the final answer. We emphasize that *before $\hat{\beta}_1$ and $\hat{\beta}_0$ are computed, a scatter plot should be examined to see whether a linear probabilistic model is plausible.* If the points do not tend to cluster about a straight line with roughly the same degree of spread for all x, other models should be investigated. In practice, plots and regression calculations are usually done by using a statistical computer package.

Example 12.4 No-fines concrete, made from a uniformly graded coarse aggregate and a cement–water paste, is beneficial in areas prone to excessive rainfall because of its excellent drainage properties. The article "Pavement Thickness Design for No-Fines Concrete Parking Lots" (*J. of Transportation Engr.*, 1995: 476–484) employed a least squares analysis in studying how y = porosity (%) is related to x = unit weight (pcf) in concrete specimens. Consider the following representative data, displayed in a tabular format convenient for calculating the values of the summary statistics.

Obs.	x	y	x²	xy	y²
1	99.0	28.8	9801.00	2851.20	829.44
2	101.1	27.9	10221.21	2820.69	778.41
3	102.7	27.0	10547.29	2772.90	729.00
4	103.0	25.2	10609.00	2595.60	635.04
5	105.4	22.8	11109.16	2403.12	519.84
6	107.0	21.5	11449.00	2300.50	462.25
7	108.7	20.9	11815.69	2271.83	436.81
8	110.8	19.6	12276.64	2171.68	384.16
9	112.1	17.1	12566.41	1916.91	292.41
10	112.4	18.9	12633.76	2124.36	357.21
11	113.6	16.0	12904.96	1817.60	256.00
12	113.8	16.7	12950.44	1900.46	278.89
13	115.1	13.0	13248.01	1496.30	169.00
14	115.4	13.6	13317.16	1569.44	184.96
15	120.0	10.8	14400.00	1296.00	116.64
Sum	1640.1	299.8	179,849.73	32,308.59	6430.06

Thus $\bar{x} = 109.34$, $\bar{y} = 19.986667$, and

$$\hat{\beta}_1 = \frac{S_{xy}}{S_{xx}} = \frac{32,308.59 - (1640.1)(299.8)/15}{179849.73 - (1640.1)^2/15}$$

$$= \frac{-471.542}{521.196} = -.90473066 \approx -.905$$

$$\hat{\beta}_0 = 19.986667 - (-.90473066)(109.34) = 118.909917 \approx 118.91$$

We estimate that the expected change in porosity associated with a 1-pcf increase in unit weight is $-.905\%$ (a decrease of .905%). The equation of the estimated regression line (least squares line) is then $y = 118.91 - .905x$. Figure 12.8, generated by the statistical computer package S-Plus, shows that the least squares line provides an excellent summary of the relationship between the two variables.

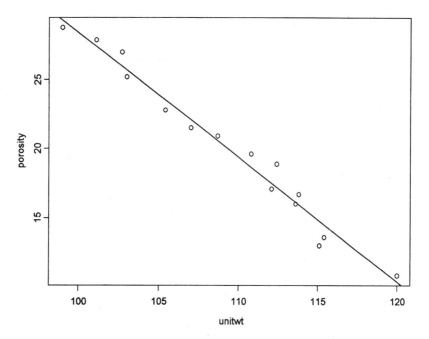

Figure 12.8   A scatter plot of the data in Example 12.4 with the least squares line superimposed, from S-Plus   ▪

The estimated regression line can immediately be used for two different purposes. For a fixed x value x^*, $\hat{\beta}_0 + \hat{\beta}_1 x^*$ (the height of the line above x^*) gives either (1) a point estimate of the expected value of Y when $x = x^*$ or (2) a point prediction of the Y value that will result from a single new observation made at $x = x^*$.

Example 12.5   Refer to the unit weight–porosity data in the previous example. A point estimate for true average porosity for all specimens whose unit weight is 110 is

$$\hat{\mu}_{Y \cdot 110} = \hat{\beta}_0 + \hat{\beta}_1(110) = 118.91 - .905(110) = 19.4\%$$

If a single specimen whose unit weight is 110 pcf is to be selected, 19.4% is also a point prediction for the porosity of this specimen.   ▪

The least squares line should not be used to make a prediction for an x value much beyond the range of the data, such as $x = 90$ or $x = 135$ in Example 12.4. The **danger of extrapolation** is that the fitted relationship (a line here) may not be valid for such x values. (In the foregoing example, $x = 135$ gives $\hat{y} = -3.3$, a patently ridiculous value of porosity, but extrapolation will not always result in such inconsistencies.)

Estimating σ^2

The parameter σ^2 determines the amount of variability inherent in the regression model. A large value of σ^2 will lead to observed (x_i, y_i)'s that are quite spread out about the true regression line, whereas when σ^2 is small the observed points will tend to fall very close

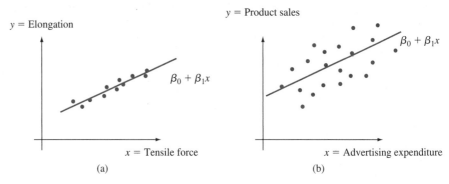

Figure 12.9 Typical sample for σ^2: (a) small; (b) large

to the true line (see Figure 12.9). An estimate of σ^2 will be used in confidence interval (CI) formulas and hypothesis-testing procedures presented in the next two sections. Because the equation of the true line is unknown, the estimate is based on the extent to which the sample observations deviate from the estimated line. Many large deviations (residuals) suggest a large value of σ^2, whereas deviations all of which are small in magnitude suggest that σ^2 is small.

DEFINITION

> The **fitted** (or **predicted**) **values** $\hat{y}_1, \hat{y}_2, \ldots, \hat{y}_n$ are obtained by successively substituting $x_1, \ldots, x_n$ into the equation of the estimated regression line: $\hat{y}_1 = \hat{\beta}_0 + \hat{\beta}_1 x_1$, $\hat{y}_2 = \hat{\beta}_0 + \hat{\beta}_1 x_2, \ldots, \hat{y}_n = \hat{\beta}_0 + \hat{\beta}_1 x_n$. The **residuals** are the vertical deviations $y_1 - \hat{y}_1, y_2 - \hat{y}_2, \ldots, y_n - \hat{y}_n$ from the estimated line.

In words, the predicted value $\hat{y}_i$ is the value of y that we would predict or expect when using the estimated regression line with $x = x_i$; $\hat{y}_i$ is the height of the estimated regression line above the value x_i for which the ith observation was made. The residual $y_i - \hat{y}_i$ is the difference between the observed y_i and the predicted $\hat{y}_i$. If the residuals are all small in magnitude, then much of the variability in observed y values appears to be due to the linear relationship between x and y, whereas many large residuals suggest quite a bit of inherent variability in y relative to the amount due to the linear relation. Assuming that the line in Figure 12.7 is the least squares line, the residuals are identified by the vertical line segments from the observed points to the line.

Example 12.6 An investigation of the relationship between traffic flow x (1000's of cars per 24 hours) and lead content y of bark on trees near the highway (μg/g dry wt) yielded the data in the x_i and y_i columns of the accompanying table. The summary statistics are $\sum x_i = 198.3$, $\sum x_i^2 = 4198.03$, $\sum y_i = 7034$, $\sum y_i^2 = 5{,}390{,}382$, and $\sum x_i y_i = 149{,}354.4$, so

$$\hat{\beta}_1 = \frac{(149{,}354.4) - (198.3)(7034)/11}{(4198.03) - (198.3)^2/11} = 36.18385 \qquad \hat{\beta}_0 = -12.84159$$

i	x_i	y_i	$\hat{y}_i$	$y_i - \hat{y}_i$
1	8.3	227	287.48	−60.48
2	8.3	312	287.48	24.52
3	12.1	362	424.98	−62.98
4	12.1	521	424.98	96.02
5	17.0	640	602.28	37.72
6	17.0	539	602.28	−63.28
7	17.0	728	602.28	125.72
8	24.3	945	866.43	78.57
9	24.3	738	866.43	−128.43
10	24.3	759	866.43	−107.43
11	33.6	1263	1202.94	60.06

The estimated regression line is $y = -12.84 + 36.18x$. For numerical accuracy, the fitted values $\hat{y}_i$ are calculated from $\hat{y}_i = -12.84159 + 36.18385x_i$. The residuals should sum to 0 (a consequence of the first normal equation), but the sum here is .01 because of rounding. A positive residual results from a point lying above the estimated regression line ($y_i > \hat{y}_i$) and a negative residual from a point lying below the line.   ■

In much the same way that the deviations from the mean in a one-sample situation were combined to obtain the estimate $s^2 = \Sigma(x_i - \bar{x})^2/(n - 1)$, the estimate of σ^2 in regression analysis is based on squaring and summing the residuals. We will continue to use the symbol s^2 for this estimated variance, so don't confuse it with our previous s^2.

DEFINITION

> The **error sum of squares,** (equivalently, residual sum of squares) denoted by SSE, is
>
> $$\text{SSE} = \Sigma(y_i - \hat{y}_i)^2 = \Sigma[y_i - (\hat{\beta}_0 + \hat{\beta}_1 x_i)]^2$$
>
> and the estimate of σ^2 is
>
> $$\hat{\sigma}^2 = s^2 = \frac{\text{SSE}}{n - 2} = \frac{\Sigma(y_i - \hat{y}_i)^2}{n - 2}$$

The divisor $n - 2$ in s^2 is the number of degrees of freedom (df) associated with the estimate (or, equivalently, with error sum of squares). This is because to obtain s^2, the two parameters β_0 and β_1 must first be estimated, which results in a loss of 2 df (just as μ had to be estimated in one-sample problems, resulting in an estimated variance based on $n - 1$ df). Replacing each y_i in the formula for s^2 by the rv Y_i gives the estimator S^2. It can be shown that S^2 is an unbiased estimator for σ^2 (though the estimator S is not unbiased for σ).

Example 12.7
(Example 12.6 continued)

The residuals for the traffic flow/lead-content data were calculated previously. The corresponding error sum of squares is

$$\text{SSE} = (-60.48)^2 + (24.52)^2 + \cdots + (60.06)^2 = 76{,}493.98$$

The estimate of σ^2 is then $\hat{\sigma}^2 = s^2 = 76,493.98/(11 - 2) = 8499.33$, and the estimated standard deviation is $\hat{\sigma} = s = \sqrt{8499.33} = 92.19$. Roughly speaking, 92.19 is the magnitude of a typical deviation from the estimated regression line. ∎

Computation of SSE from the defining formula involves much tedious arithmetic because both the predicted values and residuals must first be calculated. Use of the following computational formula does not require these quantities.

$$\text{SSE} = \Sigma y_i^2 - \hat{\beta}_0 \Sigma y_i - \hat{\beta}_1 \Sigma x_i y_i$$

This expression results from substituting $\hat{y}_i = \hat{\beta}_0 + \hat{\beta}_1 x_i$ into $\Sigma(y_i - \hat{y}_i)^2$, squaring the summand, carrying through the sum to the resulting three terms, and simplifying. This computational formula is especially sensitive to the effects of rounding in $\hat{\beta}_0$ and $\hat{\beta}_1$, so carrying as many digits as possible in intermediate computations will protect against round-off error.

Example 12.8 The article "Promising Quantitative Nondestructive Evaluation Techniques for Composite Materials" (*Materials Evaluation,* 1985: 561–565) reports on a study to investigate how the propagation of an ultrasonic stress wave through a substance depends on the properties of the substance. The accompanying data on fracture strength (x, as a percentage of ultimate tensile strength) and attenuation (y, in neper/cm, the decrease in amplitude of the stress wave) in fiberglass-reinforced polyester composites was read from a graph that appeared in the article. The simple linear regression model is suggested by the substantial linear pattern in the scatter plot.

x	12	30	36	40	45	57	62	67	71	78	93	94	100	105
y	3.3	3.2	3.4	3.0	2.8	2.9	2.7	2.6	2.5	2.6	2.2	2.0	2.3	2.1

The necessary summary quantities are $n = 14$, $\Sigma x_i = 890$, $\Sigma x_i^2 = 67,182$, $\Sigma y_i = 37.6$, $\Sigma y_i^2 = 103.54$, and $\Sigma x_i y_i = 2234.30$, from which $S_{xx} = 10,603.4285714$, $S_{xy} = -155.98571429$, $\hat{\beta}_1 = -.0147109$, and $\hat{\beta}_0 = 3.6209072$. The computational formula for SSE gives

$$\text{SSE} = 103.54 - (3.6209072)(37.6) - (-.0147109)(2234.30)$$
$$= .2624532$$

so $s^2 = .2624532/12 = .0218711$ and $s = .1479$. When $\hat{\beta}_0$ and $\hat{\beta}_1$ are rounded to three decimal places in the computational formula for SSE, the result is

$$\text{SSE} = 103.54 - (3.621)(37.6) - (-.015)(2234.30) = .905$$

which is more than three times the correct value. ∎

The Coefficient of Determination

Figure 12.10 shows three different scatter plots of bivariate data. In all three plots, the heights of the different points vary substantially, indicating that there is much

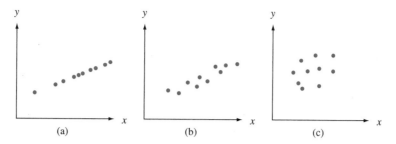

Figure 12.10   Using the model to explain y variation: (a) data for which all variation is explained; (b) data for which most variation is explained; (c) data for which little variation is explained

variability in observed y values. The points in the first plot all fall exactly on a straight line. In this case, all (100%) of the sample variation in y can be attributed to the fact that x and y are linearly related in combination with variation in x. The points in Figure 12.10(b) do not fall exactly on a line, but compared to overall y variability, the deviations from the least squares line are small. It is reasonable to conclude in this case that much of the observed y variation can be attributed to the approximate linear relationship between the variables postulated by the simple linear regression model. When the scatter plot looks like that of Figure 12.10(c), there is substantial variation about the least squares line relative to overall y variation, so the simple linear regression model fails to explain variation in y by relating y to x.

The error sum of squares SSE can be interpreted as a measure of how much variation in y is left unexplained by the model—that is, how much cannot be attributed to a linear relationship. In Figure 12.10(a), SSE = 0, and there is no unexplained variation, whereas unexplained variation is small for the data of Figure 12.10(b) and much larger in Figure 12.10(c). A quantitative measure of the total amount of variation in observed y values is given by the **total sum of squares**

$$\text{SST} = S_{yy} = \Sigma(y_i - \bar{y})^2 = \Sigma y_i^2 - (\Sigma y_i)^2/n$$

Total sum of squares is the sum of squared deviations about the sample mean of the observed y values. Thus, the same number $\bar{y}$ is subtracted from each y_i in SST, whereas SSE involves subtracting each different predicted value $\hat{y}_i$ from the corresponding observed y_i. Just as SSE is the sum of squared deviations about the least squares line $y = \hat{\beta}_0 + \hat{\beta}_1 x$, SST is the sum of squared deviations about the horizontal line at height $\bar{y}$ (since then vertical deviations are $y_i - \bar{y}$) as pictured in Figure 12.11 (page 506). Furthermore, because the sum of squared deviations about the least squares line is smaller than the sum of squared deviations about *any* other line, SSE < SST unless the horizontal line itself is the least squares line. The ratio SSE/SST is the proportion of total variation that cannot be explained by the simple linear regression model, and $1 -$ SSE/SST (a number between 0 and 1) is the proportion of observed y variation explained by the model.

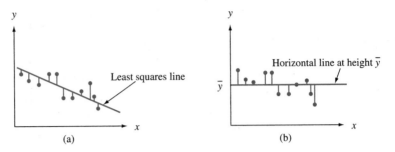

Figure 12.11   Sums of squares illustrated; (a) SSE = sum of squared deviations about the least squares line; (b) SST = sum of squared deviations about the horizontal line

---

DEFINITION

> The **coefficient of determination,** denoted by r^2, is given by
>
> $$r^2 = 1 - \frac{\text{SSE}}{\text{SST}}$$
>
> It is interpreted as the proportion of observed y variation that can be explained by the simple linear regression model (attributed to an approximate linear relationship between y and x).

The higher the value of r^2, the more successful is the simple linear regression model in explaining y variation. When regression analysis is done by a statistical computer package, either r^2 or $100r^2$ (the percentage of variation explained by the model) is a prominent part of the output. If r^2 is small, an analyst will usually want to search for an alternative model (either a nonlinear model or a multiple regression model that involves more than a single independent variable) that can more effectively explain y variation.

Example 12.9
(Example 12.4 continued)

The scatter plot of the no-fines concrete data in Figure 12.8 certainly portends a very high r^2 value. With

$$\hat{\beta}_0 = 118.909917 \qquad \hat{\beta}_1 = -.90473066 \qquad \Sigma y_i = 299.8$$
$$\Sigma x_i y_i = 32,308.59 \qquad \Sigma y_i^2 = 6430.06$$

we have

$$\text{SST} = 6430.06 - \frac{299.8^2}{15} = 438.057333 \approx 438.06$$

$$\text{SSE} = 6430.06 - (118.909917)(299.8) - (-.90473066)(32,308.59)$$
$$= 11.4388 \approx 11.44$$

The coefficient of determination is then

$$r^2 = 1 - \frac{11.44}{438.06} = 1 - .026 = .974$$

That is, 97.4% of the observed variation in porosity is attributable to (can be explained by) the approximate linear relationship between porosity and unit weight, a very impressive result. (Many social scientists would die for an r^2 value much above .5!)

Figure 12.12 shows partial MINITAB output for the porosity-unit weight data of Examples 12.4 and 12.9; the package will also provide the predicted values and residuals upon request, as well as other information. The formats used by other packages differ slightly from that of MINITAB, but the information content is very similar. Quantities such as the standard deviations, t-ratios, and the ANOVA table are discussed in Section 12.3.

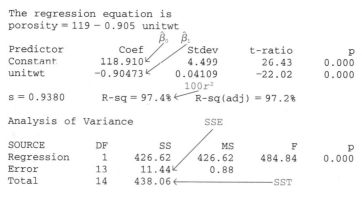

Figure 12.12 MINITAB output for the regression of Examples 12.4 and 12.9

The coefficient of determination can be written in a slightly different way by introducing a third sum of squares—**regression sum of squares,** SSR—given by SSR = SST − SSE. Regression sum of squares is interpreted as the amount of total variation that *is* explained by the model. Then we have

$$r^2 = 1 - \text{SSE/SST} = (\text{SST} - \text{SSE})/\text{SST} = \text{SSR/SST}$$

the ratio of explained variation to total variation. The ANOVA table in Figure 12.12 shows that SSR = 426.62, from which $r^2 = 426.62/438.06 = .974$.

Further Comments: Terminology and Scope of Regression Analysis

The term *regression analysis* was first used by Francis Galton in the late nineteenth century in connection with his work on the relationship between father's height x and son's height y. After collecting a number of pairs (x_i, y_i), Galton used the principle of least squares to obtain the equation of the estimated regression line with the objective of using it to predict son's height from father's height. In using the derived line, Galton found that if a father was above average in height, the son would also be expected to be above average in height, *but not by as much as the father was.* Similarly, the son of a shorter-than-average father would also be expected to be shorter than average, but not by as much as the father. Thus, the predicted height of a son was "pulled back in" toward the mean; because *regression* means a coming or going back, Galton adopted the terminology *regression line.* This phenomenon of being pulled back in toward the mean has been observed in many other situations (e.g., batting averages from year to year in baseball) and is called the **regression effect.**

Our discussion thus far has presumed that the independent variable is under the control of the investigator, so that only the dependent variable Y is random. This was not, however, the case with Galton's experiment; fathers' heights were not preselected, but instead both X and Y were random. Methods and conclusions of regression analysis can be applied both when the values of the independent variable are fixed in advance and when they are random, but because the derivations and interpretations are more straightforward in the former case, we will continue to work explicitly with it. For more commentary, see the excellent book by John Neter et al. listed in the chapter bibliography.

Exercises | Section 12.2 (12–29)

12. Exercise 4 gave data on $x = $ BOD mass loading and $y = $ BOD mass removal. Values of relevant summary quantities are

$n = 14 \qquad \Sigma x_i = 517$

$\Sigma y_i = 346 \qquad \Sigma x_i^2 = 39{,}095$

$\Sigma y_i^2 = 17{,}454 \qquad \Sigma x_i y_i = 25{,}825$

a. Obtain the equation of the least squares line.
b. Predict the value of BOD mass removal for a single observation made when BOD mass loading is 35, and calculate the value of the corresponding residual.
c. Calculate SSE and then a point estimate of σ.
d. What proportion of observed variation in removal can be explained by the approximate linear relationship between the two variables?
e. The last two x values, 103 and 142, are much larger than the others. How are the equation of the least squares line and the value of r^2 affected by deletion of the two corresponding observations from the sample? Adjust the given values of the summary quantities, and use the fact that the new value of SSE is 311.79.

13. The accompanying data on $x = $ current density (mA/cm^2) and $y = $ rate of deposition (μm/min) appeared in the article "Plating of 60/40 Tin/Lead Solder for Head Termination Metallurgy" (*Plating and Surface Finishing,* Jan. 1997: 38–40). Do you agree with the claim by the article's author that "a linear relationship was obtained from the tin-lead rate of deposition as a function of current density"? Explain your reasoning.

x	20	40	60	80
y	.24	1.20	1.71	2.22

14. Refer to the tank temperature–efficiency ratio data given in Exercise 1.
a. Determine the equation of the estimated regression line.
b. Calculate a point estimate for true average efficiency ratio when tank temperature is 182.
c. Calculate the values of the residuals from the least squares line for the four observations for which temperature is 182. Why do they not all have the same sign?
d. What proportion of the observed variation in efficiency ratio can be attributed to the simple linear regression relationship between the two variables?

15. Values of modulus of elasticity (MoE, the ratio of stress, i.e., force per unit area, to strain, i.e., deformation per unit length, in GPa) and flexural strength (a measure of the ability to resist failure in bending, in MPa) were determined for a sample of concrete beams of a certain type, resulting in the following data (read from a graph in the article "Effects of Aggregates and Microfillers on the Flexural Properties of Concrete," *Magazine of Concrete Research,* 1997: 81–98).

MoE	29.8	33.2	33.7	35.3	35.5	36.1	36.2
Strength	5.9	7.2	7.3	6.3	8.1	6.8	7.0

MoE	36.3	37.5	37.7	38.7	38.8	39.6	41.0
Strength	7.6	6.8	6.5	7.0	6.3	7.9	9.0

(*continued*)

MoE	42.8	42.8	43.5	45.6	46.0	46.9	48.0
Strength	8.2	8.7	7.8	9.7	7.4	7.7	9.7

MoE	49.3	51.7	62.6	69.8	79.5	80.0
Strength	7.8	7.7	11.6	11.3	11.8	10.7

a. Construct a stem-and-leaf display of the MoE values, and comment on any interesting features.

b. Is the value of strength completely and uniquely determined by the value of MoE? Explain.

c. Use the accompanying MINITAB output to obtain the equation of the least squares line for predicting strength from modulus of elasticity, and then predict strength for a beam whose modulus of elasticity is 40. Would you feel comfortable using the least squares line to predict strength when modulus of elasticity is 100? Explain.

```
Predictor      Coef     Stdev  t-ratio       p
Constant     3.2925    0.6008     5.48   0.000
mod elas    0.10748   0.01280     8.40   0.000
s = 0.8657   R-sq = 73.8%   R-sq (adj) = 72.8%

Analysis of Variance

SOURCE       DF      SS       MS       F       p
Regression    1  52.870   52.870   70.55   0.000
Error        25  18.736    0.749
Total        26  71.605
```

d. What are the values of SSE, SST, and the coefficient of determination? Do these values suggest that the simple linear regression model effectively describes the relationship between the two variables? Explain.

16. The article "Characterization of Highway Runoff in Austin, Texas, Area" (*J. of Envir. Engr.*, 1998: 131–137) gave a scatter plot, along with the least squares line, of x = rainfall volume (m³) and y = runoff volume (m³) for a particular location. The accompanying values were read from the plot.

x	5	12	14	17	23	30	40	47
y	4	10	13	15	15	25	27	46

x	55	67	72	81	96	112	127
y	38	46	53	70	82	99	100

a. Does a scatter plot of the data support the use of the simple linear regression model?

b. Calculate point estimates of the slope and intercept of the population regression line.

c. Calculate a point estimate of the true average runoff volume when rainfall volume is 50.

d. Calculate a point estimate of the standard deviation σ.

e. What proportion of the observed variation in runoff volume can be attributed to the simple linear regression relationship between runoff and rainfall?

17. A regression of y = calcium content (g/l) on x = dissolved material (mg/cm²) was reported in the article "Use of Fly Ash or Silica Fume to Increase the Resistance of Concrete to Feed Acids" (*Magazine of Concrete Research*, 1997: 337–344). The equation of the estimated regression line was $y = 3.678 + .144x$, with $r^2 = .860$, based on $n = 23$.

a. Interpret the estimated slope .144 and the coefficient of determination .860.

b. Calculate a point estimate of the true average calcium content when the amount of dissolved material is 50 mg/cm².

c. The value of total sum of squares was SST = 320.398. Calculate an estimate of the error standard deviation σ in the simple linear regression model.

18. The following summary statistics were obtained from a study that used regression analysis to investigate the relationship between pavement deflection and surface temperature of the pavement at various locations on a state highway. Here x = temperature (°F) and y = deflection adjustment factor ($y \geq 0$):

$$n = 15 \quad \Sigma x_i = 1425 \quad \Sigma y_i = 10.68$$
$$\Sigma x_i^2 = 139{,}037.25 \quad \Sigma x_i y_i = 987.645$$
$$\Sigma y_i^2 = 7.8518$$

(Many more than 15 observations were made in the study; the reference is "Flexible Pavement Evaluation and Rehabilitation," *Transportation Eng. J.*, 1977: 75–85.)

a. Compute $\hat{\beta}_1$, $\hat{\beta}_0$, and the equation of the estimated regression line. Graph the estimated line.

b. What is the estimate of expected change in the deflection adjustment factor when temperature is increased by 1°F?

c. Suppose temperature were measured in °C rather than in °F. What would be the estimated regression line? Answer part (b) for an increase of 1°C. (*Hint:* °F = (9/5)°C + 32; now substitute for the "old x" in terms of the "new x.")

d. If a 200°F surface temperature were within the realm of possibility, would you use the estimated line of part (a) to predict deflection factor for this temperature? Why or why not?

19. The following data is representative of that reported in the article "An Experimental Correlation of Oxides of Nitrogen Emissions from Power Boilers Based on Field Data" (*J. Eng. for Power,* July 1973: 165–170), with x = burner area liberation rate (MBtu/hr-ft^2) and y = NO$_X$ emission rate (ppm):

x	100	125	125	150	150	200	200
y	150	140	180	210	190	320	280

x	250	250	300	300	350	400	400
y	400	430	440	390	600	610	670

a. Assuming that the simple linear regression model is valid, obtain the least squares estimate of the true regression line.

b. What is the estimate of expected NO$_X$ emission rate when burner area liberation rate equals 225?

c. Estimate the amount by which you expect NO$_X$ emission rate to change when burner area liberation rate is decreased by 50.

d. Would you use the estimated regression line to predict emission rate for a liberation rate of 500? Why or why not?

20. A number of studies have shown lichens (certain plants composed of an alga and a fungus) to be excellent bioindicators of air pollution. The article "The Epiphytic Lichen Hypogymnia Physodes as a Biomonitor of Atmospheric Nitrogen and Sulphur Deposition in Norway" (*Envir. Monitoring and Assessment,* 1993: 27–47) gives the following data (read from a graph) on x = NO$_3^-$ wet deposition (g N/m^2) and y = lichen N (% dry weight):

x	.05	.10	.11	.12	.31	.37	.42
y	.48	.55	.48	.50	.58	.52	1.02

x	.58	.68	.68	.73	.85	.92
y	.86	.86	1.00	.88	1.04	1.70

The author used simple linear regression to analyze the data. Use the accompanying MINITAB output to answer the following questions:

a. What are the least squares estimates of β_0 and β_1?

b. Predict lichen N for an NO$_3^-$ deposition value of .5.

c. What is the estimate of σ?

d. What is the value of total variation, and how much of it can be explained by the model relationship?

```
The regression equation is lichen
N = 0.365 + 0.967 no3 depo

Predictor     Coef     Stdev    t-ratio       p
Constant   0.36510   0.09904       3.69   0.004
no3 depo   0.9668    0.1829        5.29   0.000

s = 0.1932   R-sq = 71.7%   R-sq (adj) = 69.2%

Analysis of Variance

SOURCE        DF       SS       MS       F       P
Regression     1   1.0427   1.0427   27.94   0.000
Error         11   0.4106   0.0373
Total         12   1.4533
```

21. The article "Effects of Bike Lanes on Driver and Bicyclist Behavior" (*ASCE Transportation Eng. J.,* 1977: 243–256) reports the results of a regression analysis with x = available travel space in feet (a convenient measure of roadway width, defined as the distance between a cyclist and the roadway center line) and separation distance y between a bike and a passing car (determined by photography). The data, for ten streets with bike lanes, follows:

x	12.8	12.9	12.9	13.6	14.5
y	5.5	6.2	6.3	7.0	7.8

x	14.6	15.1	17.5	19.5	20.8
y	8.3	7.1	10.0	10.8	11.0

a. Verify that $\Sigma x_i = 154.20$, $\Sigma y_i = 80$, $\Sigma x_i^2 = 2452.18$, $\Sigma x_i y_i = 1282.74$, and $\Sigma y_i^2 = 675.16$.

b. Derive the equation of the estimated regression line.

c. What separation distance would you predict for another street that has 15.0 as its available travel space value?

d. What would be the estimate of expected separation distance for all streets having available travel space value 15.0?

22. a. Use the summary quantities from Exercise 18 to estimate the standard deviation of the random deviation ϵ in the simple linear regression model.

b. Based on the summary quantities from Exercise 18, what proportion of variation in deflection adjustment factor can be explained by the simple linear regression relationship between adjustment factor and temperature?

23. a. Obtain SSE for the data in Exercise 19 from the defining formula [SSE $= \sum(y_i - \hat{y}_i)^2$], and compare to the value calculated from the computational formula.

b. Calculate the value of total sum of squares. Does the simple linear regression model appear to do an effective job of explaining variation in emission rate? Justify your assertion.

24. The accompanying data was read from a graph that appeared in the article "Reactions on Painted Steel Under the Influence of Sodium Chloride, and Combinations Thereof" (*Ind. Engr. Chem. Prod. Res. Dev.*, 1985: 375–378). The independent variable is SO_2 deposition rate (mg/m²/day) and the dependent variable is steel weight loss (g/m²).

x	14	18	40	43	45	112
y	280	350	470	500	560	1200

a. Construct a scatter plot. Does the simple linear regression model appear to be reasonable in this situation?

b. Calculate the equation of the estimated regression line.

c. What percentage of observed variation in steel weight loss can be attributed to the model relationship in combination with variation in deposition rate?

d. Because the largest x value in the sample greatly exceeds the others, this observation may have been very influential in determining the equation of the estimated line. Delete this observation and recalculate the equation. Does the new equation appear to differ substantially from the original one (you might consider predicted values)?

25. Show that b_1 and b_0 of expressions (12.2) and (12.3) satisfy the normal equations.

26. Show that the "point of averages" $(\bar{x}, \bar{y})$ lies on the estimated regression line.

27. Suppose an investigator has data on the amount of shelf space x devoted to display of a particular product and sales revenue y for that product. The investigator may wish to fit a model for which the true regression line passes through $(0, 0)$. The appropriate model is $Y = \beta_1 x + \epsilon$. Assume that $(x_1, y_1), \ldots, (x_n, y_n)$ are observed pairs generated from this model and derive the least squares estimator of β_1. (*Hint:* Write the sum of squared deviations as a function of b_1, a trial value, and use calculus to find the minimizing value of b_1.)

28. a. Consider the data in Exercise 20. Suppose that instead of the least squares line passing through the points $(x_1, y_1), \ldots, (x_n, y_n)$ we wish the least squares line passing through $(x_1 - \bar{x}, y_1), \ldots, (x_n - \bar{x}, y_n)$. Construct a scatter plot of the (x_i, y_i) points and then of the $(x_i - \bar{x}, y_i)$ points. Use the plots to explain intuitively how the two least squares lines are related to one another.

b. Suppose that instead of the model $Y_i = \beta_0 + \beta_1 x_i + \epsilon_i$ $(i = 1, \ldots, n)$, we wish to fit a model of the form $Y_i = \beta_0^* + \beta_1^*(x_i - \bar{x}) + \epsilon_i$ $(i = 1, \ldots, n)$. What are the least squares estimators of β_0^* and β_1^*, and how do they relate to $\hat{\beta}_0$ and $\hat{\beta}_1$?

29. Consider the following three data sets, in which the variables of interest are $x =$ commuting distance and $y =$ commuting time. Based on a scatter plot and the values of s and r^2, in which situation would simple linear regression be most (least) effective, and why?

Data Set	1		2		3	
	x	y	x	y	x	y
	15	42	5	16	5	8
	16	35	10	32	10	16
	17	45	15	44	15	22
	18	42	20	45	20	23
	19	49	25	63	25	31
	20	46	50	115	50	60
S_{xx}	17.50		1270.8333		1270.8333	
S_{xy}	29.50		2722.5		1431.6667	
$\hat{\beta}_1$	1.685714		2.142295		1.126557	
$\hat{\beta}_0$	13.666672		7.868852		3.196729	
SST	114.83		5897.5		1627.33	
SSE	65.10		65.10		14.48	

12.3 | Inferences About the Slope Parameter β_1

The slope β_1 of the population regression line is the true average change in the dependent variable y associated with a 1-unit increase in the independent variable x. The slope of the least squares line, $\hat{\beta}_1$, gives a point estimate of β_1. In general, just as the value of the sample mean or sample standard deviation will vary from sample to sample, so will the value of $\hat{\beta}_1$. For example, if the slope of the population line is 25.0, a first sample may result in the least squares line slope 24.3, a second may give $\hat{\beta}_1 = 26.8$, a third may yield a slope of 25.4, and so on. In the same way that a confidence interval for μ and procedures for testing hypotheses about μ were based on properties of the sampling distribution of $\bar{X}$, further inferences about β_1 are based on thinking of $\hat{\beta}_1$ as a statistic and investigating its sampling distribution.

The values of the x_i's are assumed to be chosen before the experiment is performed, so only the Y_i's are random. The estimators (statistics, and thus random variables) for β_0 and β_1 are obtained by replacing y_i by Y_i in (12.2) and (12.3):

$$\hat{\beta}_1 = \frac{\sum(x_i - \bar{x})(Y_i - \bar{Y})}{\sum(x_i - \bar{x})^2}$$

$$\hat{\beta}_0 = \frac{\sum Y_i - \hat{\beta}_1\sum x_i}{n}$$

Similarly, the estimator for σ^2 results from replacing each y_i in the formula for s^2 by the rv Y_i:

$$\hat{\sigma}^2 = S^2 = \frac{\sum Y_i^2 - \hat{\beta}_0\sum Y_i - \hat{\beta}_1\sum x_iY_i}{n - 2}$$

The denominator of $\hat{\beta}_1$, $S_{xx} = \sum(x_i - \bar{x})^2$, depends only on the x_i's and not on the Y_i's, so it is a constant. Then because $\sum(x_i - \bar{x})\bar{Y} = \bar{Y}\sum(x_i - \bar{x}) = \bar{Y} \cdot 0 = 0$, the slope estimator can be written as

$$\hat{\beta}_1 = \frac{\sum(x_i - \bar{x})Y_i}{S_{xx}} = \sum c_iY_i \qquad \text{where } c_i = (x_i - \bar{x})/S_{xx}$$

That is, $\hat{\beta}_1$ is a linear function of the independent rv's $Y_1, Y_2, \ldots, Y_n$, each of which is normally distributed. Invoking properties of a linear function of random variables discussed in Section 5.5 leads to the following results.

1. The mean value of $\hat{\beta}_1$ is $E(\hat{\beta}_1) = \mu_{\hat{\beta}_1} = \beta_1$, so $\hat{\beta}_1$ is an unbiased estimator of β_1 (the distribution of $\hat{\beta}_1$ is always centered at the value of β_1).

2. The variance and standard deviation of $\hat{\beta}_1$ are

$$V(\hat{\beta}_1) = \sigma_{\hat{\beta}_1}^2 = \frac{\sigma^2}{S_{xx}} \qquad \sigma_{\hat{\beta}_1} = \frac{\sigma}{\sqrt{S_{xx}}} \qquad (12.4)$$

where $S_{xx} = \sum(x_i - \bar{x})^2 = \sum x_i^2 - (\sum x_i)^2/n$. Replacing σ by its estimate s gives an estimate for $\sigma_{\hat{\beta}_1}$ (the estimated standard deviation, i.e., estimated standard error, of $\hat{\beta}_1$):

$$s_{\hat{\beta}_1} = \frac{s}{\sqrt{S_{xx}}}$$

(This estimate can also be denoted by $\hat{\sigma}_{\hat{\beta}_1}$.)

3. The estimator $\hat{\beta}_1$ has a normal distribution (because it is a linear function of independent normal rv's).

According to (12.4), the variance of $\hat{\beta}_1$ equals the variance σ^2 of the random error term—or, equivalently, of any Y_i—divided by $\sum(x_i - \bar{x})^2$. Because $\sum(x_i - \bar{x})^2$ is a measure of how spread out the x_i's are about $\bar{x}$, we conclude that making observations at x_i values that are quite spread out results in a more precise estimator of the slope parameter (smaller variance of $\hat{\beta}_1$), whereas values of x_i all close to one another imply a highly variable estimator. Of course, if the x_i's are spread out too far, a linear model may not be appropriate throughout the range of observation.

Many inferential procedures discussed previously were based on standardizing an estimator by first subtracting its mean value and then dividing by its estimated standard deviation. In particular, test procedures and a CI for the mean μ of a normal population utilized the fact that the standardized variable $(\bar{X} - \mu)/(S/\sqrt{n})$—that is, $(\bar{X} - \mu)/S_{\bar{x}}$—had a t distribution with $n - 1$ df. A similar result here provides the key to further inferences concerning β_1.

THEOREM

The assumptions of the simple linear regression model imply that the standardized variable

$$T = \frac{\hat{\beta}_1 - \beta_1}{S/\sqrt{S_{xx}}} = \frac{\hat{\beta}_1 - \beta_1}{S_{\hat{\beta}_1}}$$

has a t distribution with $n - 2$ df.

A Confidence Interval for β_1

As in the derivation of previous CIs, we begin with a probability statement:

$$P(-t_{\alpha/2,n-2} < \frac{\hat{\beta}_1 - \beta_1}{S_{\hat{\beta}_1}} < t_{\alpha/2,n-2}) = 1 - \alpha$$

Manipulation of the inequalities inside the parentheses to isolate β_1 and substitution of estimates in place of the estimators gives the CI formula.

A 100$(1 - \alpha)$% **CI for the slope β_1** of the true regression line is

$$\hat{\beta}_1 \pm t_{\alpha/2,n-2} \cdot s_{\hat{\beta}_1}$$

This interval has the same general form as did many of our previous intervals. It is centered at the point estimate of the parameter, and the amount it extends out to each side of the estimate depends on the desired confidence level (through the t critical value) and on the amount of variability in the estimator $\hat{\beta}_1$ (through $s_{\hat{\beta}_1}$, which will tend to be small when there is little variability in the distribution of $\hat{\beta}_1$ and large otherwise).

Example 12.10 Variations in clay brick masonry weight have implications not only for structural and acoustical design, but also for design of heating, ventilating, and air conditioning systems. The article "Clay Brick Masonry Weight Variation" (*J. of Architectural Engr.,* 1996: 135–137) gave a scatter plot of y = mortar dry density (lb/ft^3) versus x = mortar air content (%) for a sample of mortar specimens, from which the following representative data was read:

x	5.7	6.8	9.6	10.0	10.7	12.6	14.4	15.0	15.3
y	119.0	121.3	118.2	124.0	112.3	114.1	112.2	115.1	111.3

x	16.2	17.8	18.7	19.7	20.6	25.0
y	107.2	108.9	107.8	111.0	106.2	105.0

The scatter plot of this data in Figure 12.13 certainly suggests the appropriateness of the simple linear regression model; there appears to be a substantial negative linear relationship between air content and density, one in which density tends to decrease as air content increases.

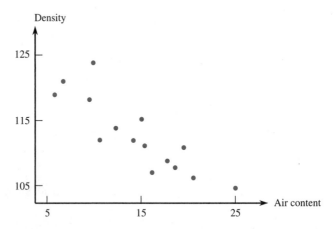

Figure 12.13 Scatter plot of the data from Example 12.10

The values of the summary statistics required for calculation of the least squares estimates are

$$\Sigma x_i = 218.1 \qquad \Sigma y_i = 1693.6 \qquad \Sigma x_i^2 = 3577.01$$
$$\Sigma x_i y_i = 24{,}252.54 \qquad \Sigma y_i^2 = 191{,}672.90$$

from which $S_{xy} = -372.404$, $S_{xx} = 405.836$, $\hat{\beta}_1 = -.917622$, $\hat{\beta}_0 = 126.248889$, SST $=$ 454.1693, SSE $= 112.4432$, and $r^2 = 1 - 112.4432/454.1693 = .752$. Roughly 75% of the observed variation in density can be attributed to the simple linear regression model relationship between density and air content. Error df is $15 - 2 = 13$, from which $s^2 = 112.4432/13 = 8.6495$ and $s = 2.941$.

The estimated standard deviation of $\hat{\beta}_1$ is

$$s_{\hat{\beta}_1} = \frac{s}{\sqrt{S_{xx}}} = \frac{2.941}{\sqrt{405.836}} = .1460$$

The t critical value for a confidence level of 95% is $t_{.025,13} = 2.160$. The confidence interval is

$$-.918 \pm (2.160)(.1460) = -.918 \pm .315 = (-1.233, -.603)$$

With a high degree of confidence, we estimate that an average decrease in density of between .603 lb/ft³ and 1.233 lb/ft³ is associated with a 1% increase in air content (at least for air content values between roughly 5% and 25%, corresponding to the x values in our sample). The interval is reasonably narrow, indicating that the slope of the population line has been precisely estimated. Notice that the interval includes only negative values, so we can be quite confident of the tendency for density to decrease as air content increases.

Looking at the SAS output of Figure 12.14, we find the value of $s_{\hat{\beta}_1}$ under Parameter Estimates as the second number in the Standard Error column. All of the widely used

```
Dependent Variable: DENSITY

                      Analysis of Variance

Source      DF     Sum of Squares    Mean Square    F Value    Prob > F
Model        1         341.72606       341.72606     39.508      0.0001
Error       13         112.44327         8.64948
C Total     14         454.16933

           Root MSE         2.94100         R-square        0.7524
           Dep Mean       112.90667         Adj R-sq        0.7334
           C.V.             2.60481

                       Parameter Estimates

                   Parameter      Standard      T for H0:
Variable    DF     Estimate         Error     Parameter = 0    Prob > |T|
INTERCEP     1    126.248889     2.25441683       56.001          0.0001
AIRCONT      1     -0.917622     0.14598888       -6.286          0.0001

                       Dep Var      Predict
                Obs    DENSITY        Value      Residual
                 1      119.0        121.0       -2.0184
                 2      121.3        120.0        1.2909
                 3      118.2        117.4        0.7603
                 4      124.0        117.1        6.9273
                 5      112.3        116.4       -4.1303
                 6      114.1        114.7       -0.5869
                 7      112.2        113.0       -0.8351
                 8      115.1        112.5        2.6154
                 9      111.3        112.2       -0.9093
                10      107.2        111.4       -4.1834
                11      108.9        109.9       -1.0152
                12      107.8        109.1       -1.2894
                13      111.0        108.2        2.8283
                14      106.2        107.3       -1.1459
                15      105.0        103.3        1.6917

                Sum of Residuals                        0
                Sum of Squared Residuals          112.4433
                Predicted Resid SS (Press)        146.4144
```

Figure 12.14   SAS output for the data of Example 12.10

statistical packages include this estimated standard error in output. There is also an estimated standard error for the statistic $\hat{\beta}_0$ from which a confidence interval for the intercept β_0 of the population regression line can be calculated. ■

Hypothesis-Testing Procedures

As before, the null hypothesis in a test about β_1 will be an equality statement. The null value (value of β_1 claimed true by the null hypothesis) will be denoted by β_{10} (read "beta one nought," *not* "beta ten"). The test statistic results from replacing β_1 in the standardized variable T by the null value β_{10}—that is, from standardizing the estimator of β_1 under the assumption that H_0 is true. The test statistic thus has a t distribution with $n - 2$ df when H_0 is true, so the type I error probability is controlled at the desired level α by using an appropriate t critical value.

The most commonly encountered pair of hypotheses about β_1 is $H_0: \beta_1 = 0$ versus $H_a: \beta_1 \neq 0$. When this null hypothesis is true, $\mu_{Y \cdot x} = \beta_0$ independent of x, so knowledge of x gives no information about the value of the dependent variable. A test of these two hypotheses is often referred to as the *model utility test* in simple linear regression. Unless n is quite small, H_0 will be rejected and the utility of the model confirmed precisely when r^2 is reasonably large. The simple linear regression model should not be used for further inferences (estimates of mean value or predictions of future values) unless the model utility test results in rejection of H_0 for a suitably small α.

Null hypothesis: H_0: $\beta_1 = \beta_{10}$

Test statistic value: $t = \dfrac{\hat{\beta}_1 - \beta_{10}}{s_{\hat{\beta}_1}}$

Alternative Hypothesis	Rejection Region for Level α Test
H_a: $\beta_1 > \beta_{10}$	$t \geq t_{\alpha, n-2}$
H_a: $\beta_1 < \beta_{10}$	$t \leq -t_{\alpha, n-2}$
H_a: $\beta_1 \neq \beta_{10}$	either $t \geq t_{\alpha/2, n-2}$ or $t \leq -t_{\alpha/2, n-2}$

A P-value based on $n - 2$ df can be calculated just as was done previously for t tests in Chapters 8 and 9.

The **model utility test** is the test of $H_0: \beta_1 = 0$ versus $H_a: \beta_1 \neq 0$, in which case the test statistic value is the t ratio $t = \hat{\beta}_1 / s_{\hat{\beta}_1}$.

Example 12.11 The cleanliness of molten aluminum metal or alloy prior to casting is determined mainly by the hydrogen and inclusion content of the melt. The article "Effect of Melt Cleanliness on the Properties of an A1-10 Wt Pct Si-10 Vol Pct SiC(p) Composite" (*Metallurgical Trans.*, 1993: 1631–1645) reports on a study in which various tensile properties were related to $x = $ volume fraction of oxides/inclusions (%). Here we present data (read from a graph) on $y = $ elongation (%) of test bars. The authors state that the scatter plot shows a linear relationship and give the equation of the least squares line. Let's use the MINITAB output of Figure 12.15 to carry out the model utility test at significance level $\alpha = .01$.

x	.10	.16	.31	.37	.37	.46	.50	.50	.60	.70
y	.96	1.10	.80	.84	.77	.87	.60	.87	.60	.61

x	.75	.80	.90	1.00	1.07	1.08	1.11	1.30	1.37	1.54
y	.70	.41	.40	.41	.45	.59	.25	.25	.08	.10

```
The regression equation is
elon = 1.07 − 0.649 volfrac
```

$$s_{\hat{\beta}_1} \qquad t = \dfrac{\hat{\beta}_1}{s_{\hat{\beta}_1}}$$

```
Predictor         Coef        Stdev      t-ratio        P       P-value
Constant       1.06930      0.04966       21.53      0.000    for model
volfrac       -0.64884      0.05840      -11.11      0.000    utility test

s = 0.1049        R-sq = 87.3%          R-sq(adj) = 86.6%

Analysis of Variance

SOURCE           DF          SS           MS          F           P
Regression        1        1.3583       1.3583     123.42       0.000
Error            18        0.1981       0.0110
Total            19        1.5564
```

Figure 12.15   MINITAB output for Example 12.11

The parameter of interest is β_1, the expected change in percentage elongation associated with a 1% increase in the volume fraction of oxides/inclusions. H_0: $\beta_1 = 0$ will be rejected in favor of H_a: $\beta_1 \neq 0$ if the t ratio $t = \hat{\beta}_1/s_{\hat{\beta}_1}$ satisfies either $t \geq t_{\alpha/2,n-2} = t_{.005,18} = 2.878$ or $t \leq -2.878$. From Figure 12.15, $\hat{\beta}_1 = -.64884$, $s_{\hat{\beta}_1} = .05840$, and

$$t = \frac{-.64884}{.05840} = -11.11 \qquad \text{(also on output)}$$

Clearly, $-11.11 \leq -2.878$, so H_0 is resoundingly rejected. Alternatively, the P-value is twice the area captured under the 18 df t curve to the left of -11.11. MINITAB gives P-value $= .000$, so H_0 should be rejected at any reasonable α. This confirmation of the utility of the simple linear regression model gives us license to calculate various estimates and predictions as described in Section 12.4.   ■

Regression and ANOVA

The splitting of the total sum of squares $\sum(y_i - \bar{y})^2$ into a part SSE, which measures unexplained variation, and a part SSR, which measures variation explained by the linear relationship, is strongly reminiscent of one-way ANOVA. In fact, the null hypothesis H_0: $\beta_1 = 0$ can be tested against H_a: $\beta_1 \neq 0$ by constructing an ANOVA table (Table 12.1, page 518) and rejecting H_0 if $f \geq F_{\alpha,1,n-2}$.

The F test gives exactly the same result as the model utility t test because $t^2 = f$ and $t^2_{\alpha/2,n-2} = F_{\alpha,1,n-2}$. Virtually all computer packages that have regression options include such an ANOVA table in the output. For example, Figure 12.14 shows SAS output for the mortar data of Example 12.10. The ANOVA table at the top of the output has $f = 39.508$ with a P-value of .0001 for the model utility test. The table of parameter estimates gives $t = -6.286$, again with $P = .0001$, and $(-6.286)^2 = 39.51$.

Table 12.1 ANOVA Table for Simple Linear Regression

Source of Variation	df	Sum of Squares	Mean Square	f
Regression	1	SSR	SSR	$\dfrac{SSR}{SSE/(n-2)}$
Error	$n-2$	SSE	$s^2 = \dfrac{SSE}{n-2}$	
Total	$n-1$	SST		

Exercises | Section 12.3 (30–43)

30. Reconsider the situation described in Exercise 7, in which x = accelerated strength of concrete and y = 28-day cured strength. Suppose the simple linear regression model is valid for x between 1000 and 4000 and that $\beta_1 = 1.25$ and $\sigma = 350$. Consider an experiment in which $n = 7$, and the x values at which observations are made are $x_1 = 1000$, $x_2 = 1500$, $x_3 = 2000$, $x_4 = 2500$, $x_5 = 3000$, $x_6 = 3500$, and $x_7 = 4000$.
   a. Calculate $\sigma_{\hat\beta_1}$, the standard deviation of $\hat\beta_1$.
   b. What is the probability that the estimated slope based on such observations will be between 1.00 and 1.50?
   c. Suppose it is also possible to make a single observation at each of the $n = 11$ values $x_1 = 2000$, $x_2 = 2100, \ldots, x_{11} = 3000$. If a major objective is to estimate β_1 as accurately as possible, would the experiment with $n = 11$ be preferable to the one with $n = 7$?

31. Reconsider the summary quantities given in Exercise 18 for the regression of y = deflection factor on x = temperature.
   a. Compute the estimated standard deviation $s_{\hat\beta_1}$.
   b. Calculate a 95% CI for β_1, the expected change in deflection factor associated with a 1°F increase in temperature.

32. Exercise 16 of Section 12.2 gave data on x = rainfall volume and y = runoff volume (both in m³). Use the accompanying MINITAB output to decide whether there is a useful linear relationship between rainfall and runoff, and then calculate a confidence interval for the true average change in runoff volume associated with a 1-m³ increase in rainfall volume.

```
The regression equation is
runoff = -1.13 + 0.827 rainfall

Predictor     Coef    Stdev   t-ratio      p
Constant    -1.128    2.368     -0.48   0.642
rainfall   0.82697  0.03652     22.64   0.000

s = 5.240    R-sq = 97.5%    R-sq(adj) = 97.3%
```

33. Exercise 15 of Section 12.2 included MINITAB output for a regression of flexural strength of concrete beams on modulus of elasticity.
   a. Use the output to calculate a confidence interval with a confidence level of 95% for the slope β_1 of the population regression line, and interpret the resulting interval.
   b. Suppose it had previously been believed that when modulus of elasticity increased by 1 GPa, the associated true average change in flexural strength would be at most .1 MPa. Does the sample data contradict this belief? State and test the relevant hypotheses.

34. Refer to the MINITAB output of Exercise 20, in which x = NO_3^- wet deposition and y = lichen N (%).
   a. Carry out the model utility test at level .01, using the rejection region approach.
   b. Repeat part (a) using the P-value approach.
   c. Suppose it had previously been believed that when NO_3^- wet deposition increases by .1 g N/m², the associated change in expected lichen N is at least .15%. Carry out a test of hypotheses at level .01 to decide whether the data contradicts this prior belief.

35. The article "Root Dentine Transparency: Age Determination of Human Teeth Using Computerized Densitometric Analysis" (*Amer. J. Phys. Anthro.*, 1991: 25–30) reports on an investigation of methods for age determination based on tooth characteristics. With x = percentage of root with transparent dentine and y = age (years), consider the following representative data for anterior teeth:

x	15	19	31	39	41	44	47	48	55	64
y	23	52	65	55	32	60	78	59	61	60

a. Calculate a 95% CI for the expected change in age associated with a 1% increase in transparent dentine content. What does the interval suggest about usefulness of the model?

b. Carry out a test of model utility based on the P-value. Would you use the least squares line to predict age from transparent dentine content? Explain.

36. An article in the *Journal of Public Health Engineering* reports the results of a regression analysis based on $n = 15$ observations in which x = filter application temperature (°C) and y = % efficiency of BOD removal. Calculated quantities include $\sum x_i = 402$, $\sum x_i^2 = 11,098$, $s = 3.725$, and $\hat{\beta}_1 = 1.7035$.

a. Test at level .01 $H_0: \beta_1 = 1$, which states that the expected increase in % BOD removal is 1 when filter application temperature increases by 1°C, against the alternative $H_a: \beta_1 > 1$.

b. Compute a 99% CI for β_1, the expected increase in % BOD removal for a 1°C increase in filter application temperature.

37. The article "Hydrogen, Oxygen, and Nitrogen in Cobalt Metal" (*Metallurgia,* 1969: 121–127) contains a plot of the following data pairs, where x = pressure of extracted gas (microns) and y = extraction time (min):

x	40	130	155	160	260	275	325	370	420	480
y	2.5	3.0	3.1	3.3	3.7	4.1	4.3	4.8	5.0	5.4

a. Estimate σ and the standard deviation of $\hat{\beta}_1$.

b. Suppose the investigators had believed prior to the experiment that on average there would be an increase of .006 min. in extraction time asso-

ciated with an increase of 1 micron in pressure. Use the P-value approach with a significance level of .10 to decide whether the data contradicts this prior belief.

38. Refer to the data on x = liberation rate and $y = \text{NO}_X$ emission rate given in Exercise 19.

a. Does the simple linear regression model specify a useful relationship between the two rates? Use the appropriate test procedure to obtain information about the P-value and then reach a conclusion at significance level .01.

b. Compute a 95% CI for the expected change in emission rate associated with a 10 MBtu/hr-ft² increase in liberation rate.

39. Carry out the model utility test using the ANOVA approach for the traffic flow/lead-content data of Example 12.6. Verify that it gives a result equivalent to that of the t test.

40. Use the rules of expected value to show that $\hat{\beta}_0$ is an unbiased estimator for β_0 (assuming that $\hat{\beta}_1$ is unbiased for β_1).

41. a. Verify that $E(\hat{\beta}_1) = \beta_1$ by using the rules of expected value from Chapter 5.

b. Use the rules of variance from Chapter 5 to verify the expression for $V(\hat{\beta}_1)$ given in this section.

42. Verify that if each x_i is multiplied by a positive constant c and each y_i is multiplied by another positive constant d, the t statistic for testing $H_0: \beta_1 = 0$ versus $H_a: \beta_1 \neq 0$ is unchanged in value (the value of $\hat{\beta}_1$ will change, which shows that the magnitude of $\hat{\beta}_1$ is not by itself indicative of model utility).

43. The probability of a type II error for the t test for $H_0: \beta_1 = \beta_{10}$ can be computed in the same manner as it was computed for the t tests of Chapter 8. If the alternative value of β_1 is denoted by β_1', the value of

$$d = \frac{|\beta_{10} - \beta_1'|}{\sigma\sqrt{\dfrac{n-1}{\sum x_i^2 - (\sum x_i)^2/n}}}$$

is first calculated, then the appropriate set of curves in Appendix Table A.17 is entered on the horizontal axis at the value of d, and β is read from the curve for $n - 2$ df. Use this to compute P(type II error) for the test of Exercise 36 when $\beta_1' = 2$ and $\sigma = 4$.

12.4 | Inferences Concerning $\mu_{Y \cdot x^*}$ and the Prediction of Future Y Values

Let x^* denote a specified value of the independent variable x. Then once the estimates $\hat{\beta}_0$ and $\hat{\beta}_1$ have been calculated, $\hat{\beta}_0 + \hat{\beta}_1 x^*$ can be regarded either as a point estimate of $\mu_{Y \cdot x^*}$ (the expected or true average value of Y when $x = x^*$) or as a prediction of the Y value that will result from a single observation made when $x = x^*$. The point estimate or prediction by itself gives no information concerning how precisely $\mu_{Y \cdot x^*}$ has been estimated or Y has been predicted. This can be remedied by developing a CI for $\mu_{Y \cdot x^*}$ and a prediction interval (PI) for a single Y value.

Before we obtain sample data, both $\hat{\beta}_0$ and $\hat{\beta}_1$ are subject to sampling variability—that is, they are both statistics whose values will vary from sample to sample. Suppose, for example, that $\beta_0 = 50$ and $\beta_1 = 2$. Then a first sample of (x, y) pairs might give $\hat{\beta}_0 = 52.35$, $\hat{\beta}_1 = 1.895$, a second sample might result in $\hat{\beta}_0 = 46.52$, $\hat{\beta}_1 = 2.056$, and so on. It follows that $\hat{Y} = \hat{\beta}_0 + \hat{\beta}_1 x^*$ itself varies in value from sample to sample, so it is a statistic. If the intercept and slope of the population line are the aforementioned values 50 and 2, respectively, and $x^* = 10$, then this statistic is trying to estimate the value $50 + 2(10) = 70$. The estimate from a first sample might be $52.35 + 1.895(10) = 71.30$, from second sample might be $46.52 + 2.056(10) = 67.08$, etc. In the same way that a confidence interval for β_1 was based on properties of the sampling distribution of $\hat{\beta}_1$, a confidence interval for a mean y value in regression is based on properties of the sampling distribution of the statistic $\hat{\beta}_0 + \hat{\beta}_1 x^*$.

Substitution of the expressions for $\hat{\beta}_0$ and $\hat{\beta}_1$ into $\hat{\beta}_0 + \hat{\beta}_1 x^*$ followed by some algebraic manipulation leads to the representation of $\hat{\beta}_0 + \hat{\beta}_1 x^*$ as a linear function of the Y_i's:

$$\hat{\beta}_0 + \hat{\beta}_1 x^* = \sum_{i=1}^{n} \left[\frac{1}{n} + \frac{(x^* - \bar{x})(x_i - \bar{x})}{\sum(x_i - \bar{x})^2} \right] Y_i = \sum_{i=1}^{n} d_i Y_i$$

The coefficients $d_1, d_2, \ldots, d_n$ in this linear function involve the x_i's and x^*, all of which are fixed. Application of the rules of Section 5.5 to this linear function gives the following properties.

Let $\hat{Y} = \hat{\beta}_0 + \hat{\beta}_1 x^*$, where x^* is some fixed value of x. Then

1. The mean value of $\hat{Y}$ is

$$E(\hat{Y}) = E(\hat{\beta}_0 + \hat{\beta}_1 x^*) = \mu_{\hat{\beta}_0 + \hat{\beta}_1 x^*} = \beta_0 + \beta_1 x^*$$

Thus, $\hat{\beta}_0 + \hat{\beta}_1 x^*$ is an unbiased estimator for $\beta_0 + \beta_1 x^*$ (i.e., for $\mu_{Y \cdot x^*}$).

2. The variance of $\hat{Y}$ is

$$V(\hat{Y}) = \sigma_{\hat{Y}}^2 = \sigma^2 \left[\frac{1}{n} + \frac{(x^* - \bar{x})^2}{\sum x_i^2 - (\sum x_i)^2/n} \right] = \sigma^2 \left[\frac{1}{n} + \frac{(x^* - \bar{x})^2}{S_{xx}} \right]$$

and the standard deviation $\sigma_{\hat{Y}}$ is the square root of this expression. The estimated standard deviation of $\hat{\beta}_0 + \hat{\beta}_1 x^*$, denoted by $s_{\hat{Y}}$ or $s_{\hat{\beta}_0 + \hat{\beta}_1 x^*}$, results from replacing σ by its estimate s:

$$s_{\hat{Y}} = s_{\hat{\beta}_0 + \hat{\beta}_1 x^*} = s \sqrt{\frac{1}{n} + \frac{(x^* - \bar{x})^2}{S_{xx}}}$$

3. $\hat{Y}$ has a normal distribution.

The variance of $\hat{\beta}_0 + \hat{\beta}_1 x^*$ is smallest when $x^* = \bar{x}$ and increases as x^* moves away from $\bar{x}$ in either direction. Thus, the estimator of $\mu_{Y \cdot x^*}$ is more precise when x^* is near the center of the x_i's than when it is far from the x values at which observations have been made. This will imply that both the CI and PI are narrower for an x^* near $\bar{x}$ than for an x^* far from $\bar{x}$. Most statistical computer packages will provide both $\hat{\beta}_0 + \hat{\beta}_1 x^*$ and $s_{\hat{\beta}_0 + \hat{\beta}_1 x^*}$ for any specified x^* upon request.

Inferences Concerning $\mu_{Y \cdot x^*}$

Just as inferential procedures for β_1 were based on the t variable obtained by standardizing β_1, a t variable obtained by standardizing $\hat{\beta}_0 + \hat{\beta}_1 x^*$ leads to a CI and test procedures here.

THEOREM

The variable

$$T = \frac{\hat{\beta}_0 + \hat{\beta}_1 x^* - (\beta_0 + \beta_1 x^*)}{S_{\hat{\beta}_0 + \hat{\beta}_1 x^*}} = \frac{\hat{Y} - (\beta_0 + \beta_1 x^*)}{S_{\hat{Y}}} \qquad (12.5)$$

has a t distribution with $n - 2$ df.

As for β_1 in the previous section, a probability statement involving this standardized variable can be manipulated to yield a confidence interval for $\mu_{Y \cdot x^*}$.

A $100(1 - \alpha)\%$ **CI for $\mu_{Y \cdot x^*}$**, the expected value of Y when $x = x^*$, is

$$\hat{\beta}_0 + \hat{\beta}_1 x^* \pm t_{\alpha/2, n-2} \cdot s_{\hat{\beta}_0 + \hat{\beta}_1 x^*} = \hat{y} \pm t_{\alpha/2, n-2} \cdot s_{\hat{Y}} \qquad (12.6)$$

This CI is centered at the point estimate for $\mu_{Y \cdot x^*}$ and extends out to each side by an amount that depends on the confidence level and on the extent of variability in the estimator on which the point estimate is based.

Example 12.12 Corrosion of steel reinforcing bars is the most important durability problem for reinforced concrete structures. Carbonation of concrete results from a chemical reaction that lowers the pH value by enough to initiate corrosion of the rebar. Representative data on x = carbonation depth (mm) and y = strength (MPa) for a sample of core specimens taken from a particular building follow (read from a plot in the article "The Carbonation of Concrete Structures in the Tropical Environment of Singapore," *Magazine of Concrete Res.*, 1996: 293–300).

x	8.0	15.0	16.5	20.0	20.0	27.5	30.0	30.0	35.0
y	22.8	27.2	23.7	17.1	21.5	18.6	16.1	23.4	13.4

x	38.0	40.0	45.0	50.0	50.0	55.0	55.0	59.0	65.0
y	19.5	12.4	13.2	11.4	10.3	14.1	9.7	12.0	6.8

A scatter plot of the data (see Figure 12.16) gives strong support to use of the simple linear regression model. Relevant quantities are as follows:

$$\Sigma x_i = 659.0 \qquad \Sigma x_i^2 = 28{,}967.50 \qquad \bar{x} = 36.6111 \qquad S_{xx} = 4840.7778$$
$$\Sigma y_i = 293.2 \qquad \Sigma x_i y_i = 9293.95 \qquad \Sigma y_i^2 = 5335.76$$
$$\hat{\beta}_1 = -.297561 \qquad \hat{\beta}_0 = 27.182936 \qquad SSE = 131.2402$$
$$r^2 = .766 \qquad s = 2.8640$$

Let's now calculate a confidence interval, using a 95% confidence level, for the mean strength for all core specimens having a carbonation depth of 45 mm—that is, a confidence interval for $\beta_0 + \beta_1(45)$. The interval is centered at

$$\hat{y} = \hat{\beta}_0 + \hat{\beta}(45) = 27.18 - .2976(45) = 13.79$$

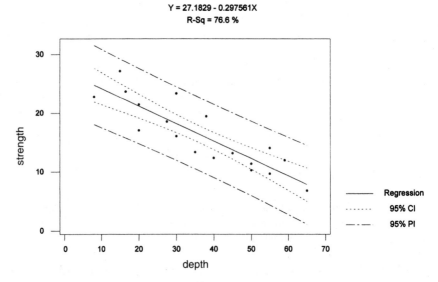

Figure 12.16 MINITAB scatter plot with confidence intervals and prediction intervals for the data of Example 12.12

The estimated standard deviation of the statistic $\hat{Y}$ is

$$s_{\hat{Y}} = 2.8640 \sqrt{\frac{1}{18} + \frac{(45 - 36.6111)^2}{4840.7778}} = .7582$$

The 16 df t critical value for a 95% confidence level is 2.120, from which we determine the desired interval to be

$$13.79 \pm (2.120)(.7582) = 13.79 \pm 1.61 = (12.18, 15.40)$$

The narrowness of this interval suggests that we have reasonably precise information about the mean value being estimated. Remember that if we recalculated this interval for sample after sample, in the long run about 95% of the calculated intervals would include $\beta_0 + \beta_1(45)$. We can only hope that this mean value lies in the single interval that we have calculated.

Figure 12.17 shows MINITAB output resulting from a request to fit the simple linear regression model and calculate confidence intervals for the mean value of strength at depths of 45 mm and 35 mm. The intervals are at the bottom of the output; note that the second interval is narrower than the first, because 35 is much closer to $\bar{x}$ than is 45. Figure 12.16 shows (1) curves corresponding to the confidence limits for each different x value and (2) prediction limits, to be discussed shortly. Notice how the curves get farther and farther apart as x moves away from $\bar{x}$.

```
The regression equation is strength = 27.2 − 0.298 depth
Predictor        Coef         Stdev       t-ratio        p

Constant        27.183        1.651        16.46       0.000
   depth       −0.29756      0.04116       −7.23       0.000

s = 2.864    R-sq = 76.6%     R-sq(adj) = 75.1%

Analysis of Variance

SOURCE           DF          SS           MS         F         p

Regression        1        428.62       428.62     52.25     0.000
    Error        16        131.24         8.20
    Total        17        559.86

   Fit      Stdev.Fit           95.0% C.I.              95.0% P.I.
13.793        0.758        (12.185, 15.401)        (7.510, 20.075)

   Fit      Stdev.Fit           95.0% C.I.              95.0% P.I.
16.768        0.678        (15.330, 18.207)        (10.527, 23.009)
```

Figure 12.17   MINITAB regression output for the data of Example 12.12   ■

In some situations, a CI is desired not just for a single x value but for two or more x values. Suppose an investigator wishes a CI both for $\mu_{Y \cdot v}$ and for $\mu_{Y \cdot w}$, where v and w are two different values of the independent variable. It is tempting to compute the interval (12.6) first for $x = v$ and then for $x = w$. Suppose we use $\alpha = .05$ in each computation to get two 95% intervals. Then if the variables involved in computing the two intervals were independent of one another, the joint confidence coefficient would be $(.95) \cdot (.95) \approx .90$.

However, the intervals are not independent because the same $\hat{\beta}_0$, $\hat{\beta}_1$, and S are used in each. We therefore cannot assert that the joint confidence coefficient for the two intervals is exactly 90%. It can be shown, though, that if the $100(1 - \alpha)\%$ CI (12.6) is computed both for $x = v$ and for $x = w$ to obtain joint CIs for $\mu_{Y \cdot v}$ and $\mu_{Y \cdot w}$, then *the joint confidence coefficient on the resulting pair of intervals is at least* $100(1 - 2\alpha)\%$. In particular, using $\alpha = .05$ results in a joint confidence coefficient of *at least* 90%, whereas using $\alpha = .01$ results in at least 98% confidence. For example, in Example 12.12 a 95% CI for $\mu_{Y \cdot 45}$ was (12.185, 15.401) and a 95% CI for $\mu_{Y \cdot 35}$ was (15.330, 18.207). The simultaneous or joint confidence level for the two statements $12.185 < \mu_{Y \cdot 45} < 15.401$ and $15.330 < \mu_{Y \cdot 35} < 18.207$ is at least 90%.

The validity of these joint or simultaneous CIs rests on a probability result called the **Bonferroni inequality,** so the joint CIs are referred to as **Bonferroni intervals.** The method is easily generalized to yield joint intervals for k different $\mu_{Y \cdot x}$'s. *Using the interval* (12.6) *separately first for* $x = x_1^*$, *then for* $x = x_2^*, \ldots$, *and finally for* $x = x_k^*$ *yields a set of* k *CIs for which the joint or simultaneous confidence level is guaranteed to be at least* $100(1 - k\alpha)\%$.

Tests of hypotheses about $\beta_0 + \beta_1 x^*$ are based on the test statistic T obtained by replacing $\beta_0 + \beta_1 x^*$ in the numerator of (12.5) by the null value μ_0. For example, H_0: $\beta_0 + \beta_1(45) = 15$ in Example 12.12 says that when carbonation depth is 45, expected (i.e., true average) strength is 15. The test statistic value is $t = [\hat{\beta}_0 + \hat{\beta}_1(45) - 15]/s_{\hat{\beta}_0 + \hat{\beta}_1(45)}$, and the test is upper-, lower-, or two-tailed according to the inequality in H_a.

A Prediction Interval for a Future Value of Y

Analogous to the CI (12.6) for $\mu_{Y \cdot x^*}$, one frequently wishes to obtain an interval of plausible values for the value of Y associated with some future observation when the independent variable has value x^*. For instance, in the example in which vocabulary size y is related to the age x of a child, for $x = 6$ years (12.6) would provide a CI for the true average vocabulary size of all 6-year-old children. Alternatively, we might wish an interval of plausible values for the vocabulary size of a particular 6-year-old child.

A CI refers to a parameter, or population characteristic, whose value is fixed but unknown to us. In contrast, a future value of Y is not a parameter but instead a random variable; for this reason we refer to an interval of plausible values for a future Y as a **prediction interval** rather than a confidence interval. The error of estimation is $\beta_0 + \beta_1 x^* - (\hat{\beta}_0 + \hat{\beta}_1 x^*)$, a difference between a fixed (but unknown) quantity and a random variable. The error of prediction is $Y - (\hat{\beta}_0 + \hat{\beta}_1 x^*)$, a difference between two random variables. There is thus more uncertainty in prediction than in estimation, so a PI will be wider than a CI. Because the future value Y is independent of the observed Y_i's,

$$
\begin{aligned}
V[Y - (\hat{\beta}_0 + \hat{\beta}_1 x^*)] &= \text{variance of prediction error} \\
&= V(Y) + V(\hat{\beta}_0 + \hat{\beta}_1 x^*) \\
&= \sigma^2 + \sigma^2 \left[\frac{1}{n} + \frac{(x^* - \bar{x})^2}{S_{xx}} \right] \\
&= \sigma^2 \left[1 + \frac{1}{n} + \frac{(x^* - \bar{x})^2}{S_{xx}} \right]
\end{aligned}
$$

Furthermore, because $E(Y) = \beta_0 + \beta_1 x^*$ and $E(\hat{\beta}_0 + \hat{\beta}_1 x^*) = \beta_0 + \beta_1 x^*$, the expected value of the prediction error is $E(Y - (\hat{\beta}_0 + \hat{\beta}_1 x^*)) = 0$. It can then be shown that the standardized variable

$$T = \frac{Y - (\hat{\beta}_0 + \hat{\beta}_1 x^*)}{S \sqrt{1 + \dfrac{1}{n} + \dfrac{(x^* - \bar{x})^2}{S_{xx}}}}$$

has a t distribution with $n - 2$ df. Substituting this T into the probability statement $P(-t_{\alpha/2, n-2} < T < t_{\alpha/2, n-2}) = 1 - \alpha$ and manipulating to isolate Y between the two inequalities yields the following interval.

A $100(1 - \alpha)\%$ **PI for a future Y observation to be made when $x = x^*$** is

$$\hat{\beta}_0 + \hat{\beta}_1 x^* \pm t_{\alpha/2, n-2} \cdot s \sqrt{1 + \frac{1}{n} + \frac{(x^* - \bar{x})^2}{S_{xx}}} \qquad (12.7)$$

$$= \hat{\beta}_0 + \hat{\beta}_1 x^* \pm t_{\alpha/2, n-2} \cdot \sqrt{s^2 + s^2_{\hat{\beta}_0 + \hat{\beta}_1 x^*}}$$

$$= \hat{y} \pm t_{\alpha/2, n-2} \cdot \sqrt{s^2 + s^2_{\hat{Y}}}$$

The interpretation of the prediction level $100(1 - \alpha)\%$ is identical to that of previous confidence levels—if (12.7) is used repeatedly, in the long run the resulting intervals will actually contain the observed y values $100(1 - \alpha)\%$ of the time. Notice that the 1 underneath the initial square root symbol makes the PI (12.7) wider than the CI (12.6), though the intervals are both centered at $\hat{\beta}_0 + \hat{\beta}_1 x^*$. Also, as $n \to \infty$, the width of the CI approaches 0, whereas the width of the PI does not (because even with perfect knowledge of β_0 and β_1, there will still be uncertainty in prediction).

Example 12.13 Let's return to the carbonation depth-strength data of Example 12.12 and calculate a 95% prediction interval for a strength value that would result from selecting a single core specimen whose carbonation depth is 45 mm. Relevant quantities from that example are

$$\hat{y} = 13.79 \qquad s_{\hat{Y}} = .7582 \qquad s = 2.8640$$

For a prediction level of 95% based on $n - 2 = 16$ df, the t critical value is 2.120, exactly what we previously used for a 95% confidence level. The prediction interval is then

$$13.79 \pm (2.120)\sqrt{(2.8640)^2 + (.7582)^2} = 13.79 \pm (2.120)(2.963)$$

$$= 13.79 \pm 6.28 = (7.51, 20.07)$$

Plausible values for a single observation on strength when depth is 45 mm are (at the 95% prediction level) between 7.51 MPa and 20.07 MPa. The 95% confidence interval for mean strength when depth is 45 was (12.18, 15.40). The prediction interval is much wider than this because of the extra $(2.8640)^2$ under the square root. Figure 12.16, the MINITAB output in Example 12.12, shows this interval as well as the confidence interval. ∎

The Bonferroni technique can be employed as in the case of confidence intervals. If a $100(1 - \alpha)\%$ PI is calculated for each of k different values of x, the simultaneous or joint prediction level for all k intervals is at least $100(1 - k\alpha)\%$.

Exercise Section 12.4 (44–56)

44. Fitting the simple linear regression model to the $n = 27$ observations on $x =$ modulus of elasticity and $y =$ flexural strength given in Exercise 15 of Section 12.2 resulted in $\hat{y} = 7.592$, $s_{\hat{Y}} = .179$ when $x = 40$ and $\hat{y} = 9.741$, $s_{\hat{Y}} = .253$ for $x = 60$.
   a. Explain why $s_{\hat{Y}}$ is larger when $x = 60$ than when $x = 40$.
   b. Calculate a confidence interval with a confidence level of 95% for the true average strength of all beams whose modulus of elasticity is 40.
   c. Calculate a prediction interval with a prediction level of 95% for the strength of a single beam whose modulus of elasticity is 40.
   d. If a 95% CI is calculated for true average strength when modulus of elasticity is 60, what will be the simultaneous confidence level for both this interval and the interval calculated in part (b)?

45. Reconsider the filter application temperature/% BOD removal experiment described in Exercise 36. In addition to information given there, $\hat{\beta}_0 = 8.2141$.
   a. Compute a 90% CI for $\beta_0 + 25\beta_1$, the expected % BOD removal when filter application temperature is 25°C.
   b. Test at level .01 the hypotheses H_0: $\beta_0 + 25\beta_1 = 50$ versus H_a: $\beta_0 + 25\beta_1 > 50$ (the alternative hypothesis states that expected % BOD removal exceeds 50 when filter application temperature is 25°C).

46. The article "The Incorporation of Uranium and Silver by Hydrothermally Synthesized Galena" (*Econ. Geology,* 1964: 1003–1024) reports on the determination of silver content of galena crystals grown in a closed hydrothermal system over a range of temperature. With $x =$ crystallization temperature in °C and $y =$ Ag$_2$S in mol %, the data follows:

x	398	292	352	575	568	450	550	408	484	350	503	600	600
y	.15	.05	.23	.43	.23	.40	.44	.44	.45	.09	.59	.63	.60

   from which $\sum x_i = 6130$, $\sum x_i^2 = 3,022,050$, $\sum y_i = 4.73$, $\sum y_i^2 = 2.1785$, $\sum x_i y_i = 2418.74$, $\hat{\beta}_1 = .00143$, $\hat{\beta}_0 = -.311$, and $s = .131$.

   a. Estimate true average silver content when temperature is 500°C using a 95% confidence interval.
   b. How would the width of a 95% CI for true average silver content when temperature is 400°C compare to the width of the interval in part (a)? Answer without computing this new interval.
   c. Calculate a 95% CI for the true average change in silver content associated with a 1°C increase in temperature.
   d. Suppose it had previously been believed that when crystallization temperature was 400°C, true average silver content would be .25. Carry out a test at significance level .05 to decide whether the sample data contradicts this prior belief.

47. The simple linear regression model provides a very good fit to the data on rainfall and runoff volume given in Exercise 16 of Section 12.2. The equation of the least squares line is $\hat{y} = -1.128 + .82697x$, $r^2 = .975$, and $s = 5.24$.
   a. Use the fact that $s_{\hat{Y}} = 1.44$ when rainfall volume is 40 m^3 to predict runoff in a way that conveys information about reliability and precision. Does the resulting interval suggest that precise information about the value of runoff for this future observation is available? Explain your reasoning.
   b. Calculate a PI for runoff when rainfall is 50 using the same prediction level as in part (a). What can be said about the simultaneous prediction level for the two intervals you have calculated?

48. A study reported in the article "The Effects of Water Vapor Concentration on the Rate of Combustion of an Artificial Graphite in Humid Air Flow" (*Combustion and Flame,* 1983: 107–118) gives data on $x =$ temperature of a nitrogen–oxygen mixture (1000°F) under specified conditions and $y =$ oxygen diffusivity. Summary quantities are $n = 9$, $\sum x_i = 12.6$, $\sum y_i = 27.68$, $\sum x_i^2 = 18.24$, $\sum x_i y_i = 40.968$, and $\sum y_i^2 = 93.3448$. Assume that the two variables are related according to the simple linear regression model.

a. Estimate true average oxygen diffusivity when temperature is 1500°F, and do so in a way that conveys information about reliability and precision.

b. Predict oxygen diffusivity for a single observation to be made when temperature is 1500°F, and do so in a way that conveys information about reliability and precision. How does your prediction calculated here compare to the estimate you calculated in part (a)?

c. Would a prediction interval for diffusivity when temperature is 1200°F using the same prediction level as in part (b) be wider or narrower than the interval calculated there? Answer without computing this second interval.

49. You are told that a 95% CI for expected lead content when traffic flow is 15, based on a sample of $n = 10$ observations, is (462.1, 597.7). Calculate a CI with confidence level 99% for expected lead content when traffic flow is 15.

50. An experiment to measure the macroscopic magnetic relaxation time in crystals (μsec) as a function of the strength of the external biasing magnetic field (KG) yielded the following data ("An Optical Faraday Rotation Technique for the Determination of Magnetic Relaxation Times," *IEEE Trans. Magnetics,* June 1968: 175–178, with data read from a graph that appeared in the article).

x	11.0	12.5	15.2	17.2	19.0	20.8
y	187	225	305	318	367	365

x	22.0	24.2	25.3	27.0	29.0
y	400	435	450	506	558

The summary statistics are $\sum x_i = 223.2$, $\sum y_i = 4116$, $\sum x_i^2 = 4877.50$, $\sum x_i y_i = 90{,}096.1$, and $\sum y_i^2 = 1{,}666{,}782$. Compute the following:

a. A 95% CI for expected relaxation time when field strength equals 18.

b. A 95% PI for future relaxation time when field strength equals 18.

c. Simultaneous CIs for expected relaxation time when field strength equals 15, 18, and 20; your joint confidence coefficient should be at least 97%.

51. Refer to Example 12.11 in which x = volume fraction of oxides/inclusions and y = % elongation.

a. MINITAB gave $s_{\hat{\beta}_0 + \hat{\beta}_1(.40)} = .0311$ and $s_{\hat{\beta}_0 + \hat{\beta}_1(1.20)} = .0352$. Why is the former estimated standard deviation smaller than the latter one?

b. Use the MINITAB output from the example to calculate a 95% CI for expected % elongation when volume fraction = .40.

c. Use the MINITAB output to calculate a 95% PI for a single value of % elongation to be observed when volume fraction = 1.20.

52. Plasma etching is essential to the fine-line pattern transfer in current semiconductor processes. The article "Ion Beam-Assisted Etching of Aluminum with Chlorine" (*J. Electrochem. Soc.,* 1985: 2010–2012) gives the accompanying data (read from a graph) on chlorine flow (x, in SCCM) through a nozzle used in the etching mechanism and etch rate (y, in 100A/min).

x	1.5	1.5	2.0	2.5	2.5	3.0	3.5	3.5	4.0
y	23.0	24.5	25.0	30.0	33.5	40.0	40.5	47.0	49.0

The summary statistics are $\sum x_i = 24.0$, $\sum y_i = 312.5$, $\sum x_i^2 = 70.50$, $\sum x_i y_i = 902.25$, $\sum y_i^2 = 11{,}626.75$, $\hat{\beta}_0 = 6.448718$, $\hat{\beta}_1 = 10.602564$.

a. Does the simple linear regression model specify a useful relationship between chlorine flow and etch rate?

b. Estimate the true average change in etch rate associated with a 1-SCCM increase in flow rate using a 95% confidence interval, and interpret the interval.

c. Calculate a 95% CI for $\mu_{Y \cdot 3.0}$, the true average etch rate when flow = 3.0. Has this average been precisely estimated?

d. Calculate a 95% PI for a single future observation on etch rate to be made when flow = 3.0. Is the prediction likely to be accurate?

e. Would the 95% CI and PI when flow = 2.5 be wider or narrower than the corresponding intervals of parts (c) and (d)? Answer without actually computing the intervals.

f. Would you recommend calculating a 95% PI for a flow of 6.0? Explain.

53. Consider the following four intervals based on the data of Example 12.4 (Section 12.2):

a. A 95% CI for mean porosity when unit weight is 110

b. A 95% PI for porosity when unit weight is 110

c. A 95% CI for mean porosity when unit weight is 115

d. A 95% PI for porosity when unit weight is 115

Without computing any of these intervals, what can be said about their widths relative to one another?

54. The decline of water supplies in certain areas of the United States has created the need for increased understanding of relationships between economic factors such as crop yield and hydrologic and soil factors. The article "Variability of Soil Water Properties and Crop Yield in a Sloped Watershed" (*Water Resources Bull.*, 1988: 281–288) gives data on grain sorghum yield (*y*, in g/m-row) and distance upslope (*x*, in m) on a sloping watershed. Selected observations are given in the accompanying table:

x	0	10	20	30	45	50	70
y	500	590	410	470	450	480	510

x	80	100	120	140	160	170	190
y	450	360	400	300	410	280	350

a. Construct a scatter plot. Does the simple linear regression model appear to be plausible?
b. Carry out a test of model utility.
c. Estimate true average yield when distance upslope is 75 by giving an interval of plausible values.

55. Infestation of crops by insects has long been of great concern to farmers and agricultural scientists. The article "Cotton Square Damage by the Plant Bug, *Lygus hesperus*, and Abscission Rates" (*J. of Econ. Entom.*, 1988: 1328–1337) reports data on *x* = age of a cotton plant (days) and *y* = % damaged squares. Consider the accompanying *n* = 12 observations (read from a scatter plot in the article):

x	9	12	12	15	18	18
y	11	12	23	30	29	52

x	21	21	27	30	30	33
y	41	65	60	72	84	93

a. Why is the relationship between *x* and *y* not deterministic?
b. Does a scatter plot suggest that the simple linear regression model will describe the relationship between the two variables?
c. The summary statistics are $\Sigma x_i = 246$, $\Sigma x_i^2 = 5742$, $\Sigma y_i = 572$, $\Sigma y_i^2 = 35{,}634$, and $\Sigma x_i y_i = 14{,}022$. Determine the equation of the least squares line.
d. Predict the percentage of damaged squares when the age is 20 days by giving an interval of plausible values.

56. Verify that $V(\hat{\beta}_0 + \hat{\beta}_1 x)$ is indeed given by the expression in the text. [*Hint*: $V(\Sigma d_i Y_i) = \Sigma d_i^2 \cdot V(Y_i)$.]

12.5 Correlation

There are many situations in which the objective in studying the joint behavior of two variables is to see whether they are related, rather than to use one to predict the value of the other. In this section, we first develop the sample correlation coefficient *r* as a measure of how strongly related two variables *x* and *y* are in a sample and then relate *r* to the correlation coefficient *ρ* defined in Chapter 5.

The Sample Correlation Coefficient *r*

Given *n* pairs of observations $(x_1, y_1), (x_2, y_2), \ldots, (x_n, y_n)$, it is natural to speak of *x* and *y* having a positive relationship if large *x*'s are paired with large *y*'s and small *x*'s with small *y*'s. Similarly, if large *x*'s are paired with small *y*'s and small *x*'s with large *y*'s, then a negative relationship between the variables is implied. Consider the quantity

$$S_{xy} = \sum_{i=1}^{n} (x_i - \bar{x})(y_i - \bar{y}) = \sum_{i=1}^{n} x_i y_i - \left(\sum_{i=1}^{n} x_i\right)\left(\sum_{i=1}^{n} y_i\right)\bigg/ n$$

Then if the relationship is strongly positive, an x_i above the mean $\bar{x}$ will tend to be paired with a y_i above the mean $\bar{y}$, so that $(x_i - \bar{x})(y_i - \bar{y}) > 0$, and this product will also be positive whenever both x_i and y_i are below their respective means. Thus, a positive relationship implies that S_{xy} will be positive. An analogous argument shows that when the relationship is negative, S_{xy} will be negative, since most of the products $(x_i - \bar{x})(y_i - \bar{y})$ will be negative. This is illustrated in Figure 12.18.

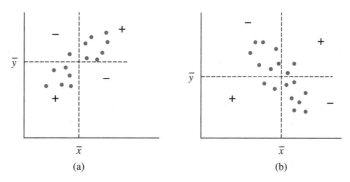

Figure 12.18 (a) Scatter plot with S_{xy} positive; (b) scatter plot with S_{xy} negative [+ means $(x_i - \bar{x})(y_i - \bar{y}) > 0$, and − means $(x_i - \bar{x})(y_i - \bar{y}) < 0$]

Although S_{xy} seems a plausible measure of the strength of a relationship, we do not yet have any idea of how positive or negative it can be. Unfortunately, S_{xy} has a serious defect: By changing the units of measurement of either x or y, S_{xy} can be made either arbitrarily large in magnitude or arbitrarily close to zero. For example, if $S_{xy} = 25$ when x is measured in meters, then $S_{xy} = 25,000$ when x is measured in millimeters and .025 when x is expressed in kilometers. A reasonable condition to impose on any measure of how strongly x and y are related is that the calculated measure should not depend on the particular units used to measure them. This condition is achieved by modifying S_{xy} to obtain the sample correlation coefficient.

DEFINITION

The **sample correlation coefficient** for the n pairs $(x_1, y_1), \ldots, (x_n, y_n)$ is

$$r = \frac{S_{xy}}{\sqrt{\sum(x_i - \bar{x})^2}\sqrt{\sum(y_i - \bar{y})^2}} = \frac{S_{xy}}{\sqrt{S_{xx}}\sqrt{S_{yy}}} \qquad (12.8)$$

Example 12.14 An accurate assessment of soil productivity is critical to rational land-use planning. Unfortunately, as the author of the article "Productivity Ratings Based on Soil Series" (*Prof. Geographer,* 1980: 158–163) argues, an acceptable soil productivity index is not so easy to come by. One difficulty is that productivity is determined partly by which crop is planted, and the relationship between yield of two different crops planted in the same soil may not be very strong. To illustrate, the article presents the accompanying data on corn yield x and peanut yield y (mT/Ha) for eight different types of soil:

x	2.4	3.4	4.6	3.7	2.2	3.3	4.0	2.1
y	1.33	2.12	1.80	1.65	2.00	1.76	2.11	1.63

With $\sum x_i = 25.7$, $\sum y_i = 14.40$, $\sum x_i^2 = 88.31$, $\sum x_i y_i = 46.856$, and $\sum y_i^2 = 26.4324$,

$$S_{xx} = 88.31 - \frac{(25.7)^2}{8} = 88.31 - 82.56 = 5.75$$

$$S_{yy} = 26.4324 - \frac{(14.40)^2}{8} = .5124$$

$$S_{xy} = 46.856 - \frac{(25.7)(14.40)}{8} = .5960$$

from which

$$r = \frac{.5960}{\sqrt{5.75}\sqrt{.5124}} = .347 \qquad \blacksquare$$

Properties of r

The most important properties of r are as follows:

1. The value of r does not depend on which of the two variables under study is labeled x and which is labeled y.

2. The value of r is independent of the units in which x and y are measured.

3. $-1 \leq r \leq 1$

4. $r = 1$ if and only if (iff) all (x_i, y_i) pairs lie on a straight line with positive slope, and $r = -1$ iff all (x_i, y_i) pairs lie on a straight line with negative slope.

5. The square of the sample correlation coefficient gives the value of the coefficient of determination that would result from fitting the simple linear regression model—in symbols, $(r)^2 = r^2$.

Property 1 stands in marked contrast to what happens in regression analysis, where virtually all quantities of interest (the estimated slope, estimated y-intercept, s^2, etc.) depend on which of the two variables is treated as the dependent variable. However, Property 5 shows that the proportion of variation in the dependent variable explained by fitting the simple linear regression model does not depend on which variable plays this role.

Property 2 is equivalent to saying that r is unchanged if each x_i is replaced by cx_i and if each y_i is replaced by dy_i (a change in the scale of measurement), as well as if each x_i is replaced by $x_i - a$ and y_i by $y_i - b$ (which changes the location of zero on the measurement axis). This implies, for example, that r is the same whether temperature is measured in °F or °C.

Property 3 tells us that the maximum value of r, corresponding to the largest possible degree of positive relationship, is $r = 1$, whereas the most negative relationship is identified with $r = -1$. According to Property 4, the largest positive and largest negative correlations are achieved only when all points lie along a straight line. Any other configuration of points, even if the configuration suggests a deterministic relationship between variables, will yield an r value less than 1 in absolute magnitude. Thus, *r measures the degree of linear relationship* among variables. A value of r near 0 is not evidence of the lack of a strong relationship, but only the absence of a linear relation, so that such a value of r must be interpreted with caution. Figure 12.19 illustrates several configurations of points associated with different values of r.

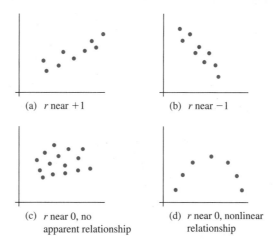

(a) r near $+1$ (b) r near -1

(c) r near 0, no   (d) r near 0, nonlinear
     apparent relationship   relationship

Figure 12.19   Data plots for different values of r

A frequently asked question is, "When can it be said that there is a strong correlation between the variables, and when is the correlation weak?" A reasonable rule of thumb is to say that the correlation is weak if $0 \le |r| \le .5$, strong if $.8 \le |r| \le 1$, and moderate otherwise. It may surprise you that $r = .5$ is considered weak, but $r^2 = .25$ implies that in a regression of y on x, only 25% of observed y variation would be explained by the model. In Example 12.14, the correlation between corn yield and peanut yield would be described as weak.

The Population Correlation Coefficient ρ and Inferences About Correlation

The correlation coefficient r is a measure of how strongly related x and y are in the observed sample. We can think of the pairs (x_i, y_i) as having been drawn from a bivariate population of pairs, with (X_i, Y_i) having joint probability distribution $f(x, y)$. In Chapter 5, we defined the correlation coefficient $\rho(X, Y)$ by

$$\rho = \rho(X, Y) = \frac{\text{Cov}(X, Y)}{\sigma_X \cdot \sigma_Y}$$

where

$$\text{Cov}(X, Y) = \begin{cases} \sum_x \sum_y (x - \mu_X)(y - \mu_Y)f(x, y) & (X, Y) \text{ discrete} \\ \int_{-\infty}^{\infty}\int_{-\infty}^{\infty} (x - \mu_X)(y - \mu_Y)f(x, y)dx\, dy & (X, Y) \text{ continuous} \end{cases}$$

If we think of $f(x, y)$ as describing the distribution of pairs of values within the entire population, ρ becomes a measure of how strongly related x and y are in that population. Properties of ρ analogous to those for r were given in Chapter 5.

The population correlation coefficient ρ is a parameter or population characteristic, just as μ_X, μ_Y, σ_X, and σ_Y are, so we can use the sample correlation coefficient to make various inferences about ρ. In particular, r is a point estimate for ρ, and the corresponding estimator is

$$\hat{\rho} = R = \frac{\sum(X_i - \bar{X})(Y_i - \bar{Y})}{\sqrt{\sum(X_i - \bar{X})^2}\sqrt{\sum(Y_i - \bar{Y})^2}}$$

Example 12.15 In some locations, there is a strong association between concentrations of two different pollutants. The article "The Carbon Component of the Los Angeles Aerosol: Source Apportionment and Contributions to the Visibility Budget" (*J. Air Pollution Control Fed.*, 1984: 643–650) reports the accompanying data on ozone concentration x (ppm) and secondary carbon concentration y ($\mu g/m^3$):

x	.066	.088	.120	.050	.162	.186	.057	.100
y	4.6	11.6	9.5	6.3	13.8	15.4	2.5	11.8

x	.112	.055	.154	.074	.111	.140	.071	.110
y	8.0	7.0	20.6	16.6	9.2	17.9	2.8	13.0

The summary quantities are $n = 16$, $\sum x_i = 1.656$, $\sum y_i = 170.6$, $\sum x_i^2 = .196912$, $\sum x_i y_i = 20.0397$, and $\sum y_i^2 = 2253.56$, from which

$$r = \frac{20.0397 - (1.656)(170.6)/16}{\sqrt{.196912 - (1.656)^2/16}\sqrt{2253.56 - (170.6)^2/16}}$$

$$= \frac{2.3826}{(.1597)(20.8456)} = .716$$

The point estimate of the population correlation coefficient ρ between ozone concentration and secondary carbon concentration is $\hat{\rho} = r = .716$. ∎

The small-sample intervals and test procedures presented in Chapters 7–9 were based on an assumption of population normality. To test hypotheses about ρ, we must make an analogous assumption about the distribution of pairs of (x, y) values in the population. We are now assuming that *both* X and Y are random, whereas much of our regression work focused on x fixed by the experimenter.

ASSUMPTION

The joint probability distribution of (X, Y) is specified by

$$f(x, y) = \frac{1}{2\pi \cdot \sigma_1\sigma_2\sqrt{1 - \rho^2}}e^{-[((x-\mu_1)/\sigma_1)^2 - 2\rho(x-\mu_1)(y-\mu_2)/\sigma_1\sigma_2 + ((y-\mu_2)/\sigma_2)^2]/[2(1-\rho^2)]}$$

$$-\infty < x < \infty$$

$$-\infty < y < \infty \qquad (12.9)$$

where μ_1 and σ_1 are the mean and standard deviation of X, and μ_2 and σ_2 are the mean and standard deviation of Y; $f(x, y)$ is called the **bivariate normal probability distribution.**

The bivariate normal distribution is obviously rather complicated, but for our purposes we need only a passing acquaintance with several of its properties. The surface determined by $f(x, y)$ lies entirely above the x–y plane [$f(x, y) \geq 0$] and has a three-dimensional bell- or mound-shaped appearance, as illustrated in Figure 12.20. If we slice through the surface with any plane perpendicular to the x–y plane and look at the shape of the curve sketched out on the "slicing plane," the result is a normal curve. More precisely, if $X = x$, it can be shown that the (conditional) distribution of Y is normal with mean $\mu_{Y \cdot x} = \mu_2 - \rho\mu_1\sigma_2/\sigma_1 + \rho\sigma_2 x/\sigma_1$ and variance $(1 - \rho^2)\sigma_2^2$. This is exactly the model used in simple linear regression with $\beta_0 = \mu_2 - \rho\mu_1\sigma_2/\sigma_1$, $\beta_1 = \rho\sigma_2/\sigma_1$, and $\sigma^2 = (1 - \rho^2)\sigma_2^2$ independent of x. The implication is that *if the observed pairs* (x_i, y_i) *are actually drawn from a bivariate normal distribution, then the simple linear regression model is an appropriate way of studying the behavior of Y for fixed x.* If $\rho = 0$, then $\mu_{Y \cdot x} = \mu_2$ independent of x; in fact, when $\rho = 0$ the joint probability density function $f(x, y)$ of (12.9) can be factored into a part involving x only and a part involving y only, which implies that X and Y are independent variables.

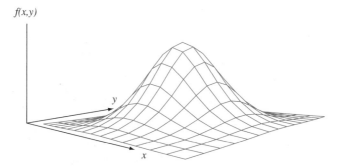

Figure 12.20   A graph of the bivariate normal pdf

Assuming that the pairs are drawn from a bivariate normal distribution allows us to test hypotheses about ρ and to construct a CI. There is no completely satisfactory way to check the plausibility of the bivariate normality assumption. A partial check involves constructing two separate normal probability plots, one for the sample x_i's and another for the sample y_i's, since bivariate normality implies that the marginal distributions of both X and Y are normal. If either plot deviates substantially from a straight-line pattern, the following inferential procedures should not be used when the sample size n is small.

Procedure for Testing H_0: $\rho = 0$

Test statistic: $\quad T = \dfrac{R\sqrt{n-2}}{\sqrt{1-R^2}}$

When H_0 is true, T has a t distribution with $n - 2$ df.

Alternative Hypothesis	Rejection Region for Level α Test
H_a: $\rho > 0$	$t \geq t_{\alpha, n-2}$
H_a: $\rho < 0$	$t \leq -t_{\alpha, n-2}$
H_a: $\rho \neq 0$	either $t \geq t_{\alpha/2, n-2}$ or $t \leq -t_{\alpha/2, n-2}$

A P-value based on $n - 2$ df can be calculated as described previously.

Example 12.16 Neurotoxic effects of manganese are well known and are usually caused by high occupational exposure over long periods of time. In the fields of occupational hygiene and environmental hygiene, the relationship between lipid peroxidation, which is responsible for deterioration of foods and damage to live tissue, and occupational exposure has not been previously reported. The article "Lipid Peroxidation in Workers Exposed to Manganese" (*Scand. J. Work and Environ. Health*, 1996: 381–386) gave data on $x =$ manganese concentration in blood (ppb) and $y =$ concentration (μmol/1) of malondialdehyde, which is a stable product of lipid peroxidation, both for a sample of 22 workers exposed to manganese and for a control sample of 45 individuals. The value of r for the control sample was .29, from which

$$t = \frac{(.29)\sqrt{45-2}}{\sqrt{1-(.29)^2}} \approx 2.0$$

The corresponding P-value for a two-tailed t test based on 43 df is roughly .052 (the cited article reported only that P-value $> .05$). We would not want to reject the assertion that $\rho = 0$ at either significance level .01 or .05. For the sample of exposed workers, $r = .83$ and $t \approx 6.7$, clear evidence that there is a linear relationship in the entire population of exposed workers from which the sample was selected. ∎

Because ρ measures the extent to which there is a linear relationship between the two variables in the population, the null hypothesis H_0: $\rho = 0$ states that there is no such population relationship. In Section 12.3, we used the t-ratio $\hat{\beta}_1/s_{\hat{\beta}_1}$ to test for a linear relationship between the two variables in the context of regression analysis. It turns out that the two test procedures are completely equivalent because $r\sqrt{n-2}/\sqrt{1-r^2} = \hat{\beta}_1/s_{\hat{\beta}_1}$. When interest lies only in assessing the strength of any linear relationship rather than in fitting a model and using it to estimate or predict, the test statistic formula just presented requires fewer computations than does the t-ratio.

Other Inferences Concerning ρ

The procedure for testing H_0: $\rho = \rho_0$ when $\rho_0 \neq 0$ is not equivalent to any procedure from regression analysis. The test statistic is based on a transformation of R called the Fisher transformation.

PROPOSITION

When $(X_1, Y_1), \ldots, (X_n, Y_n)$ is a sample from a bivariate normal distribution, the rv

$$V = \frac{1}{2} \ln\left(\frac{1 + R}{1 - R}\right) \qquad (12.10)$$

has approximately a normal distribution with mean and variance

$$\mu_V = \frac{1}{2} \ln\left(\frac{1 + \rho}{1 - \rho}\right) \qquad \sigma_V^2 = \frac{1}{n - 3}$$

The rationale for the transformation is to obtain a function of R that has a variance independent of ρ; this would not be the case with R itself. Also, the transformation should not be used if n is quite small, since the approximation will not be valid.

Test statistic: $Z = \dfrac{V - \dfrac{1}{2} \ln[(1 + \rho_0)/(1 - \rho_0)]}{1/\sqrt{n - 3}}$

Alternative Hypothesis	**Rejection Region for Level α Test**
H_a: $\rho > \rho_0$	$z \geq z_\alpha$
H_a: $\rho < \rho_0$	$z \leq -z_\alpha$
H_a: $\rho \neq \rho_0$	either $z \geq z_{\alpha/2}$ or $z \leq -z_{\alpha/2}$

Example 12.17 The article "A Study of a Partial Nutrient Removal System for Wastewater Treatment Plants" (*Water Research,* 1972: 1389–1397) reports on a method of nitrogen removal that involves the treatment of the supernatant from an aerobic digester. Both the influent total nitrogen x (mg/L) and the percentage y of nitrogen removed were recorded for 20 days, with resulting summary statistics $\sum x_i = 285.90$, $\sum x_i^2 = 4409.55$, $\sum y_i = 690.30$, $\sum y_i^2 = 29{,}040.29$, and $\sum x_i y_i = 10{,}818.56$. Does the data indicate that influent total nitrogen and percentage of nitrogen removed are at least moderately positively correlated?

Our previous interpretation of moderate positive correlation was $.5 < \rho < .8$, so we wish to test H_0: $\rho = .5$ versus H_a: $\rho > .5$. The computed value of r is .733, so

$$\frac{1}{2} \ln\left(\frac{1 + .733}{1 - .733}\right) = .935 \qquad \frac{1}{2} \ln\left(\frac{1 + .5}{1 - .5}\right) = .549$$

This gives $z = (.935 - .549)\sqrt{17} = 1.59$. Since $1.59 < 1.645$, at level $.05$ we cannot conclude that $\rho > .5$, so the relationship has not been shown to be even moderately strong (a somewhat surprising conclusion since $r = .73$, but when n is small a large r may result even when ρ is small). ∎

To obtain a CI for ρ, we first derive an interval for $\mu_V = \frac{1}{2} \ln[(1 + \rho)/(1 - \rho)]$. Standardizing V, writing a probability statement, and manipulating the resulting inequalities yields

$$\left(v - \frac{z_{\alpha/2}}{\sqrt{n - 3}}, v + \frac{z_{\alpha/2}}{\sqrt{n - 3}} \right) \tag{12.11}$$

as a $100(1 - \alpha)\%$ interval for μ_V, where $v = \frac{1}{2} \ln[(1 + r)/(1 - r)]$. This interval can then be manipulated to yield a CI for ρ.

The interval

$$\left(\frac{e^{2c_1} - 1}{e^{2c_1} + 1}, \frac{e^{2c_2} - 1}{e^{2c_2} + 1} \right)$$

is a $100(1 - \alpha)\%$ CI for ρ, where c_1 and c_2 are the left and right endpoints, respectively, of the interval (12.11).

Example 12.18 The sample correlation coefficient between influent nitrogen and percentage nitrogen
(Example 12.17 removed was $r = .733$, giving $v = .935$. With $n = 20$, a 95% confidence interval for μ_V is
continued) $(.935 - 1.96/\sqrt{17}, .935 + 1.96/\sqrt{17}) = (.460, 1.410) = (c_1, c_2)$. The 95% interval for ρ is

$$\left[\frac{e^{2(.46)} - 1}{e^{2(.46)} + 1}, \frac{e^{2(1.41)} - 1}{e^{2(1.41)} + 1} \right] = (.43, .89)$$ ∎

In Chapter 5, we cautioned that a large value of the correlation coefficient (near 1 or -1) implies only association and not causation. This applies to both ρ and r.

Exercises | Section 12.5 (57–67)

57. The article "Behavioural Effects of Mobile Telephone Use During Simulated Driving" (*Ergonomics*, 1995: 2536–2562) reported that for a sample of 20 experimental subjects, the sample correlation coefficient for x = age and y = time since the subject had acquired a driving license (yr) was .97. Why do you think the value of r is so close to 1? (The article's authors gave an explanation.)

58. The Turbine Oil Oxidation Test (TOST) and the Rotating Bomb Oxidation Test (RBOT) are two different procedures for evaluating the oxidation stability of steam turbine oils. The article "Dependence of Oxidation Stability of Steam Turbine Oil on Base Oil Composition" (*J. of the Society of Tribologists and Lubrication Engrs.*, Oct. 1997: 19–24) reported the accompanying observations on x = TOST time (hr) and y = RBOT time (min) for 12 oil specimens.

TOST	4200	3600	3750	3675	4050	2770
RBOT	370	340	375	310	350	200

TOST	4870	4500	3450	2700	3750	3300
RBOT	400	375	285	225	345	285

a. Calculate and interpret the value of the sample correlation coefficient (as did the article's authors).

b. How would the value of r be affected if we had let x = RBOT time and y = TOST time?

c. How would the value of r be affected if RBOT time were expressed in hours?

d. Construct normal probability plots and comment.

e. Carry out a test of hypotheses to decide whether RBOT time and TOST time are linearly related.

59. Toughness and fibrousness of asparagus are major determinants of quality. This was the focus of a study reported in "Post-Harvest Glyphosate Application Reduces Toughening, Fiber Content, and Lignification of Stored Asparagus Spears" (*J. of the Amer. Soc. of Horticultural Science*, 1988: 569–572). The article reported the accompanying data (read from a graph) on x = shear force (kg) and y = percent fiber dry weight.

x	46	48	55	57	60	72	81	85	94
y	2.18	2.10	2.13	2.28	2.34	2.53	2.28	2.62	2.63

x	109	121	132	137	148	149	184	185	187
y	2.50	2.66	2.79	2.80	3.01	2.98	3.34	3.49	3.26

$n = 18$ $\sum x_i = 1950$ $\sum x_i^2 = 251{,}970$

$\sum y_i = 47.92$ $\sum y_i^2 = 130.6074$ $\sum x_i y_i = 5530.92$

a. Calculate the value of the sample correlation coefficient. Based on this value, how would you describe the nature of the relationship between the two variables?

b. If a first specimen has a larger value of shear force than does a second specimen, what tends to be true of percent dry fiber weight for the two specimens?

c. If shear force is expressed in pounds, what happens to the value of r? Why?

d. If the simple linear regression model were fit to this data, what proportion of observed variation in percent fiber dry weight could be explained by the model relationship?

e. Carry out a test at significance level .01 to decide whether there is a positive linear association between the two variables.

60. The article "A Dual-Buffer Titration Method for Lime Requirement of Acid Mine-soils" (*J. of Envi-*

ron. Qual., 1988: 452–456) reports on the results of a study relating to revegetation of soil at mine reclamation sites. With x = KCl extractable aluminum and y = amount of lime required to bring soil pH to 7.0, data in the article resulted in the following summary statistics: $n = 24$, $\sum x = 48.15$, $\sum x^2 = 155.4685$, $\sum y = 263.5$, $\sum y^2 = 3750.53$, and $\sum xy = 658.455$. Carry out a test at significance level .01 to see whether the population correlation coefficient is something other than 0.

61. The following summary quantities for x = particulate pollution ($\mu g/m^3$) and y = luminance (.01 cd/m²) were calculated from a representative sample of data that appeared in the article "Luminance and Polarization of the Sky Light at Seville (Spain) Measured in White Light" (*Atmos. Environ.*, 1988: 595–599): $n = 15$, $\sum x = 860$, $\sum y = 348$, $\sum x^2 = 56{,}700$, $\sum y^2 = 8954$, and $\sum xy = 22{,}265$.

a. Test to see whether there is a positive correlation between particulate pollution and luminance in the population from which the data was selected.

b. What proportion of observed variation in luminance can be attributed to the approximate linear relationship between luminance and particulate pollution?

62. Hydrogen content is conjectured to be an important factor in porosity of aluminum alloy castings. The article "The Reduced Pressure Test as a Measuring Tool in the Evaluation of Porosity/Hydrogen Content in A1–7 Wt Pct Si-10 Vol Pct SiC(p) Metal Matrix Composite" (*Metallurgical Trans.*, 1993: 1857–1868) gives the accompanying data on x = content and y = gas porosity for one particular measurement technique:

x	.18	.20	.21	.21	.21	.22	.23
y	.46	.70	.41	.45	.55	.44	.24

x	.23	.24	.24	.25	.28	.30	.37
y	.47	.22	.80	.88	.70	.72	.75

MINITAB gives the following output in response to a CORRELATION command:

```
Correlation of Hydrcon and
Porosity = 0.449
```

a. Test at level .05 to see whether the population correlation coefficient differs from 0.

b. If a simple linear regression analysis had been carried out, what percentage of observed variation in porosity could be attributed to the model relationship?

63. Physical properties of six flame-retardant fabric samples were investigated in the article "Sensory and Physical Properties of Inherently Flame-Retardant Fabrics" (*Textile Research*, 1984: 61–68). Use the accompanying data and a .05 significance level to determine whether a linear relationship exists between stiffness x (mg-cm) and thickness y (mm). Is the result of the test surprising in light of the value of r?

x	7.98	24.52	12.47	6.92	24.11	35.71
y	.28	.65	.32	.27	.81	.57

64. The article "Increases in Steroid Binding Globulins Induced by Tamoxifen in Patients with Carcinoma of the Breast" (*J. Endocrinology*, 1978: 219–226) reports data on the effects of the drug tamoxifen on change in the level of cortisol-binding globulin (CBG) of patients during treatment. With age = x and ΔCBG = y, summary values are $n = 26$, $\Sigma x_i = 1613$, $\Sigma(x_i - \bar{x})^2 = 3756.96$, $\Sigma y_i = 281.9$, $\Sigma(y_i - \bar{y})^2 = 465.34$, and $\Sigma x_i y_i = 16{,}731$.

a. Compute a 90% CI for the true correlation coefficient ρ.

b. Test $H_0: \rho = -.5$ versus $H_a: \rho < -.5$ at level .05.

c. In a regression analysis of y on x, what proportion of variation in change of cortisol-binding globulin level could be explained by variation in patient age within the sample?

d. If you decide to perform a regression analysis with age as the dependent variable, what proportion of variation in age is explainable by variation in ΔCBG?

65. The article "Chronological Trend in Blood Lead Levels" (*N. Engl. J. Med.*, 1983: 1373–1377) gives the following data on y = average blood lead level of white children age 6 months to 5 years and x = amount of lead used in gasoline production (in 1000 tons) for ten 6-month periods:

x	48	59	79	80	95
y	9.3	11.0	12.8	14.1	13.6

x	95	97	102	102	107
y	13.8	14.6	14.6	16.0	18.2

a. Construct separate normal probability plots for x and y. Do you think it is reasonable to assume that the (x, y) pairs are from a bivariate normal population?

b. Does the data provide sufficient evidence to conclude that there is a linear relationship between blood lead level and the amount of lead used in gasoline production? Use $\alpha = .01$.

66. A sample of $n = 500$ (x, y) pairs was collected and a test of $H_0: \rho = 0$ versus $H_a: \rho \neq 0$ was carried out. The resulting *P*-value was computed to be .00032.

a. What conclusion would be appropriate at level of significance .001?

b. Does this small *P*-value indicate that there is a very strong linear relationship between x and y (a value of ρ that differs considerably from 0)? Explain.

67. A sample of $n = 10{,}000$ (x, y) pairs resulted in $r = .022$. Test $H_0: \rho = 0$ versus $H_a: \rho \neq 0$ at level .05. Is the result statistically significant? Comment on the practical significance of your analysis.

Supplementary Exercises (68–83)

68. The article "Refuse-Derived Fuel Evaluation in an Industrial Spreader-Stoker Boiler" (*J. Engr. for Gas Turbines and Power*, 1984: 782–788) reports the accompanying data on x = % refuse-derived fuel (RDF) heat input and y = % efficiency for a certain boiler:

x	37	30	48	29	27	16	0	20
y	78.0	77.2	74.4	77.7	76.9	79.0	82.1	76.5

a. Obtain the equation of the estimated regression line.

b. Does the simple linear regression model specify a useful relationship between % RDF heat input and % efficiency? State and test the appropriate hypotheses.

c. To obtain an accurate estimate of β_1, would it have been preferable to make four observations at $x = 0$ and four observations at $x = 50$ (assuming that the model is valid for x between 0 and 50)? What about three observations at $x = 0$ and three at $x = 50$? Explain.

d. Estimate true average % efficiency when % RDF heat input is 25 using a 95% CI. Does it appear that true average % efficiency has been precisely estimated? Explain.

69. The accompanying data on $x =$ diesel oil consumption rate measured by the drain–weigh method and $y =$ rate measured by the CI-trace method, both in g/hr, was read from a graph in the article "A New Measurement Method of Diesel Engine Oil Consumption Rate" (*J. Society Auto Engr.*, 1985: 28–33):

x	4	5	8	11	12	16	17	20	22	28	30	31	39
y	5	7	10	10	14	15	13	25	20	24	31	28	39

a. Assuming that x and y are related by the simple linear regression model, test $H_0: \beta_1 = 1$ versus $H_a: \beta_1 \neq 1$ using a significance level of .05.

b. Calculate the value of the sample correlation coefficient for this data.

70. The accompanying SAS output is based on data from the article "Evidence for and the Rate of Denitrification in the Arabian Sea" (*Deep Sea Research*, 1978: 431–435). The variables under study are $x =$ salinity level (%) and $y =$ nitrate level (μM/L).

a. What is the sample size n? [*Hint*: Look for degrees of freedom for SSE.]

b. Calculate a point estimate of expected nitrate level when salinity level is 35.5.

c. Does there appear to be a useful linear relationship between the two variables?

d. What is the value of the sample correlation coefficient?

e. Would you use the simple linear regression model to draw conclusions when the salinity level is 40?

71. The presence of hard alloy carbides in high chromium white iron alloys results in excellent abrasion resistance, making them suitable for materials handling in the mining and materials processing industries. The accompanying data on $x =$ retained austentite content (%) and $y =$ abrasive wear loss (mm^3) in pin wear tests with garnet as the abrasive was read from a plot in the article "Microstructure-Property Relationships in High Chromium White Iron Alloys" (*Intl. Materials Reviews*, 1996: 59–82).

x	4.6	17.0	17.4	18.0	18.5	22.4	26.5	30.0	34.0
y	.66	.92	1.45	1.03	.70	.73	1.20	.80	.91

x	38.8	48.2	63.5	65.8	73.9	77.2	79.8	84.0
y	1.19	1.15	1.12	1.37	1.45	1.50	1.36	1.29

Use the data and the SAS output on page 540 to answer the following questions.

SAS Output for Exercise 70

Dependent Variable: NITRLVL

Analysis of Variance

Source	DF	Sum of Squares	Mean Square	F Value	Prob > F
Model	1	64.49622	64.49622	63.309	0.0002
Error	6	6.11253	1.01875		
C Total	7	70.60875			

Root MSE	1.00933	R-square	0.9134	
Dep Mean	26.91250	Adj R-sq	0.8990	
C.V.	3.75043			

Parameter Estimates

Variable	DF	Parameter Estimate	Standard Error	T for H0: Parameter = 0	Prob > :T:
INTERCEP	1	326.976038	37.71380243	8.670	0.0001
SALINITY	1	− 8.403964	1.05621381	− 7.957	0.0002

SAS Output for Exercise 71

Dependent Variable: ABRLOSS
```
                               Analysis of Variance

Source            DF      Sum of Squares      Mean Square       F Value      Prob > F

Model              1            0.63690          0.63690         15.444        0.0013
Error             15            0.61860          0.04124
C Total           16            1.25551

           Root MSE       0.20308     R-square     0.5073
           Dep Mean       1.10765     Adj R-sq     0.4744
           C.V.          18.33410

                              Parameter Estimates

                     Parameter        Standard        T for H0:
Variable      DF      Estimate          Error       Parameter = 0      Prob > |T|

INTERCEP       1      0.787218       0.09525879          8.264           0.0001
AUSTCONT       1      0.007570       0.00192626          3.930           0.0013
```

a. What proportion of observed variation in wear loss can be attributed to the simple linear regression model relationship?

b. What is the value of the sample correlation coefficient?

c. Test the utility of the simple linear regression model using $\alpha = .01$.

d. Estimate the true average wear loss when content is 50% and do so in a way that conveys information about reliability and precision.

e. What value of wear loss would you predict when content is 30%, and what is the value of the corresponding residual?

72. The accompanying data was read from a scatter plot in the article "Urban Emissions Measured with Aircraft" (*J. of the Air and Waste Mgmt. Assoc.*, 1998: 16–25). The response variable is ΔNO_y and the explanatory variable is ΔCO.

ΔCO	50	60	95	108	135
ΔNO_y	2.3	4.5	4.0	3.7	8.2

ΔCO	210	214	315	720
ΔNO_y	5.4	7.2	13.8	32.1

a. Fit an appropriate model to the data and judge the utility of the model.

b. Predict the value of ΔNO_y that would result from making one more observation when ΔCO is 400, and do so in a way that conveys information about precision and reliability. Does it appear that ΔNO_y can be accurately predicted? Explain.

c. The largest value of ΔCO is much greater than the other values. Does this observation appear to have had a substantial impact on the fitted equation?

73. The accompanying data is a subset of the data that appeared in the article "Radial Tension Strength of Pipe and Other Curved Flexural Members" (*J. Amer. Concrete Inst.*, 1980: 33–39). The variables are age of a pipe specimen (x, in days) and load necessary to obtain a first crack (y, in 1000 lb/ft).

x	20	20	20	25	25
y	11.45	10.42	11.14	10.84	11.17

x	25	31	31	31
y	10.54	9.47	9.19	9.54

a. Calculate the equation of the estimated regression line.

b. Suppose a theoretical model suggests that the expected decrease in load associated with a 1-day increase in age is at most .10. Does the data contradict this assertion? State and test the appropriate hypotheses at significance level .05.

c. For purposes of estimating the slope of the true regression line as accurately as possible, would it have been preferable to make a single observation at each of the ages 20, 21, 22, . . . , 30, and 31? Explain.

d. Calculate an estimate of true average load to first crack when age is 28 days. Your estimate should convey information regarding precision of estimation.

74. An investigation was carried out to study the relationship between speed (ft/sec) and stride rate (number of steps taken/sec) among female marathon run-

ners. Resulting summary quantities included $n = 11$, $\Sigma(\text{speed}) = 205.4$, $\Sigma(\text{speed})^2 = 3880.08$, $\Sigma(\text{rate}) = 35.16$, $\Sigma(\text{rate})^2 = 112.681$, and $\Sigma(\text{speed})(\text{rate}) = 660.130$.

a. Calculate the equation of the least squares line that you would use to predict stride rate from speed.

b. Calculate the equation of the least squares line that you would use to predict speed from stride rate.

c. Calculate the coefficient of determination for the regression of stride rate on speed of part (a) and for the regression of speed on stride rate of part (b). How are these related?

75. The article "Photocharge Effects in Dye Sensitized Ag[Br,I] Emulsions at Millisecond Range Exposures" (*Photographic Sci. and Engr.*, 1981: 138–144) gives the accompanying data on $x = \%$ light absorption at 5800A and $y =$ peak photovoltage:

x	4.0	8.7	12.7	19.1	21.4
y	.12	.28	.55	.68	.85

x	24.6	28.9	29.8	30.5
y	1.02	1.15	1.34	1.29

a. Construct a scatter plot of this data. What does it suggest?

b. Assuming that the simple linear regression model is appropriate, obtain the equation of the estimated regression line.

c. What proportion of the observed variation in peak photovoltage can be explained by the model relationship?

d. Predict peak photovoltage when % absorption is 19.1, and compute the value of the corresponding residual.

e. The article's authors claim that there is a useful linear relationship between % absorption and peak photovoltage. Do you agree? Carry out a formal test.

f. Give an estimate of the change in expected peak photovoltage associated with a 1% increase in light absorption. Your estimate should convey information about the precision of estimation.

g. Repeat part (f) for the expected value of peak photovoltage when % light absorption is 20.

76. In Section 12.4, we presented a formula for $V(\hat{\beta}_0 + \hat{\beta}_1 x^*)$ and a CI for $\beta_0 + \beta_1 x^*$. Taking $x^* = 0$ gives

$\sigma_{\hat{\beta}_0}^2$ and a CI for β_0. Use the data of Example 12.10 to calculate the estimated standard deviation of $\hat{\beta}_0$ and a 95% CI for the y-intercept of the true regression line.

77. Show that $\text{SSE} = S_{yy} - \hat{\beta}_1 S_{xy}$, which gives an alternative computational formula for SSE.

78. Suppose that x and y are positive variables and that a sample of n pairs results in $r \approx 1$. If the sample correlation coefficient is computed for the (x, y^2) pairs, will the resulting value also be approximately 1? Explain.

79. Let s_x and s_y denote the sample standard deviations of the observed x's and y's, respectively [so $s_x^2 = \Sigma(x_i - \bar{x})^2/(n - 1)$ and similarly for s_y^2].

a. Show that an alternative expression for the estimated regression line $y = \hat{\beta}_0 + \hat{\beta}_1 x$ is

$$ y = \bar{y} + r \cdot \frac{s_y}{s_x}(x - \bar{x}) $$

b. This expression for the regression line can be interpreted as follows. Suppose $r = .5$. What then is the predicted y for an x that lies 1 SD (s_x units) above the mean of the x_i's? If r were 1, the prediction would be for y to lie 1 SD above its mean $\bar{y}$, but since $r = .5$, we predict a y that is only .5 SD ($.5s_y$ unit) above $\bar{y}$. Using the data in Exercise 64 for a patient whose age is 1 SD below the average age in the sample, by how many standard deviations is the patient's predicted ΔCBG above or below the average ΔCBG for the sample?

80. Verify that the t statistic for testing $H_0: \beta_1 = 0$ in Section 12.3 is identical to the t statistic in Section 12.5 for testing $H_0: \rho = 0$.

81. Use the formula for computing SSE to verify that $r^2 = 1 - \text{SSE}/\text{SST}$.

82. The article "Increased Oxygen Consumption During the Uptake of Water by the Eversible Vesicles of *Petrobius Brevistylis*" (*J. Insect Physiology*, 1977: 1285–1294) presents the results of a regression of $y =$ increased oxygen uptake (in μL) above the mean resting rate on $x =$ weight increase (mg) when dehydrated insects were allowed access to distilled water. A sample size of $n = 20$ was used, and the computed summary statistics were (approximately, based on numbers read from a graph) $\Sigma x_i = 63.5$, $\Sigma y_i = 17.26$, $\Sigma x_i^2 = 311.74$, $\Sigma x_i y_i = 71.51$, and $\Sigma y_i^2 = 19.9625$.

a. Compute the equation of the estimated regression line.

b. There was only one observation made for an x value larger than 7: for $x_{20} = 9.8$, $y_{20} = 1.9$. The investigator would like to know whether the exclusion of this point greatly alters the estimated regression relationship. Compute the estimated regression line based just on the 19 pairs with (9.8, 1.9) deleted from the sample. What y would you predict using this new line when $x = 9.8$? [*Hint:* First recompute the summary statistics; for example, new $\sum x_i =$ old $\sum x_i - 9.8$.]

83. Reconsider the situation of Exercise 68, in which $x = \%$ RDF heat input and $y = \%$ efficiency for a particular boiler were related via the simple linear regression model $Y = \beta_0 + \beta_1 x + \epsilon$. Suppose that for a second boiler, these variables are also related via the simple linear regression model $Y = \gamma_0 + \gamma_1 x + \epsilon$ and that $V(\epsilon) = \sigma^2$ for both boilers. If the data set consists of n_1 observations on the first boiler and n_2 on the second and if SSE_1 and SSE_2 denote the two error sums of squares, then a pooled estimate of σ^2 is $\hat{\sigma}^2 = (SSE_1 + SSE_2)/(n_1 + n_2 - 4)$. Let SS_{x1} and SS_{x2} denote $\sum (x_i - \bar{x})^2$ for the data on the first and second boilers, respectively. A test of H_0: $\beta_1 - \gamma_1 = 0$ (equal slopes) is based on the statistic

$$T = \frac{\hat{\beta}_1 - \hat{\gamma}_1}{\hat{\sigma}\sqrt{\dfrac{1}{SS_{x1}} + \dfrac{1}{SS_{x2}}}}$$

When H_0 is true, T has a t distribution with $n_1 + n_2 - 4$ df. Suppose the six observations on the second boiler are (0, 81.3), (10, 78.4), (20, 78.2), (25, 79.1), (30, 77.6), and (40, 77.4). Using this along with the data of Exercise 68, carry out a test at level .05 to see whether expected change in % efficiency associated with a 1% increase in RDF heat input is identical for the two boilers.

Bibliography

Draper, Norman, and Harry Smith, *Applied Regression Analysis* (3rd ed.), Wiley, New York, 1999. The most comprehensive and authoritative book on regression analysis currently in print.

Neter, John, Michael Kutner, Christopher Nachtsheim, and William Wasserman, *Applied Linear Statistical Models* (4th ed.), Irwin, Homewood, IL, 1996. The first 15 chapters constitute an extremely readable and informative survey of regression analysis.

13

Nonlinear and Multiple Regression

Introduction

The probabilistic model studied in Chapter 12 specified that the observed value of the dependent variable Y deviated from the linear regression function $\mu_{Y \cdot x} = \beta_0 + \beta_1 x$ by a random amount. Here we consider two ways of generalizing the simple linear regression model. The first way is to replace $\beta_0 + \beta_1 x$ by a nonlinear function of x, and the second is to use a regression function involving more than a single independent variable. After fitting a regression function of the chosen form to the given data, it is of course important to have methods available for making inferences about the parameters of the chosen model. Before these methods are used, though, the data analyst should first assess the validity of the chosen model. In Section 13.1, we discuss methods, based primarily on a graphical analysis of the residuals (observed minus predicted y's), for checking the aptness of a fitted model.

In Section 13.2, we consider nonlinear regression functions of a single independent variable x that are "intrinsically linear." By this we mean that it is possible to transform one or both of the variables so that the relationship between the new variables is linear. Another class of nonlinear relations is obtained by using polynomial regression functions of the form $\mu_{Y \cdot x} = \beta_0 + \beta_1 x + \beta_2 x^2 + \cdots + \beta_k x^k$; these polynomial models are the subject of Section 13.3. Multiple regression analysis involves building models for relating y to two or more independent variables. The focus in Section 13.4 is on interpretation of the parameters of various multiple regression models and on understanding and using the regression output from various statistical computer packages. The last section of the chapter surveys some extensions and pitfalls of multiple regression modeling.

13.1 Aptness of the Model and Model Checking

A plot of the observed pairs (x_i, y_i) is a necessary first step in deciding on the form of a mathematical relationship between x and y. It is possible to fit many functions other than a linear one ($y = b_0 + b_1 x$) to the data, using either the principle of least squares or another fitting method. Once a function of the chosen form has been fitted, it is important to check the fit of the model to see whether it is in fact appropriate. One way to study the fit is to superimpose a graph of the best-fit function on the scatter plot of the data. However, any tilt or curvature of the best-fit function may obscure some aspects of the fit that should be investigated. Furthermore, the scale on the vertical axis may make it difficult to assess the extent to which observed values deviate from the best-fit functions.

Residuals and Standardized Residuals

A more effective approach to assessment of model adequacy is to compute the fitted or predicted values $\hat{y}_i$ and the residuals $e_i = y_i - \hat{y}_i$ and then plot various functions of these computed quantities. We then examine the plots either to confirm our choice of model or for indications that the model is not appropriate. Suppose the simple linear regression model is correct, and let $y = \hat{\beta}_0 + \hat{\beta}_1 x$ be the equation of the estimated regression line. Then the ith residual is $e_i = y_i - (\hat{\beta}_0 + \hat{\beta}_1 x)$. To derive properties of the residuals, let $e_i = Y_i - \hat{Y}_i$ represent the ith residual as a random variable (rv) (before observations are actually made). Then

$$E(Y_i - \hat{Y}_i) = E(Y_i) - E(\hat{\beta}_0 + \hat{\beta}_1 x_i) = \beta_0 + \beta_1 x_i - (\beta_0 + \beta_1 x_i) = 0 \quad (13.1)$$

so each residual has expected value 0. Because $\hat{Y}_i \, (= \hat{\beta}_0 + \hat{\beta}_1 x_i)$ is a linear function of the Y_j's, so is $Y_i - \hat{Y}_i$ (where the coefficients depend on the x_j's). Thus, the normality of the Y_j's implies that each residual is normally distributed. It can also be shown that

$$V(Y_i - \hat{Y}_i) = \sigma^2 \cdot \left[1 - \frac{1}{n} - \frac{(x_i - \bar{x})^2}{\sum(x_j - \bar{x})^2} \right] \quad (13.2)$$

Replacing σ^2 by s^2 and taking the square root of Equation (13.2) gives the estimated standard deviation of a residual.

Let's now standardize each residual by subtracting the mean value (zero) and then dividing by the estimated standard deviation.

The standardized residuals are given by

$$e_i^* = \frac{y_i - \hat{y}_i}{s\sqrt{1 - \dfrac{1}{n} - \dfrac{(x_i - \bar{x})^2}{\sum_j (x_j - \bar{x})^2}}} \qquad i = 1, \ldots, n \tag{13.3}$$

If, for example, a particular standardized residual is 1.5, then the residual itself is 1.5 (estimated) standard deviations larger than what would be expected from fitting the correct model. Notice that the variances of the residuals differ from one another. If n is reasonably large, though, the bracketed term in (13.2) will be approximately 1, so some sources use e_i/s as the standardized residual. Computation of the e_i^*'s can be tedious, but the most widely used statistical computer packages automatically provide these values and (upon request) can construct various plots involving them.

Example 13.1 Exercise 19 in Chapter 12 presented data on $x =$ burner area liberation rate and $y =$ NO$_x$ emissions. Here we reproduce the data and give the fitted values, residuals, and standardized residuals. The estimated regression line is $y = -45.55 + 1.71x$, and $r^2 = .961$. Notice that the standardized residuals are not a constant multiple of the residuals (i.e., $e_i^* \neq e_i/s$):

x_i	y_i	$\hat{y}_i$	e_i	e_i^*
100	150	125.6	24.4	.75
125	140	168.4	−28.4	−.84
125	180	168.4	11.6,	.35
150	210	211.1	−1.1	−.03
150	190	211.1	−21.1	−.62
200	320	296.7	23.3	.66
200	280	296.7	−16.7	−.47
250	400	382.3	17.7	.50
250	430	382.3	47.7	1.35
300	440	467.9	−27.9	−.80
300	390	467.9	−77.9	−2.24
350	600	553.4	46.6	1.39
400	610	639.0	−29.0	−.92
400	670	639.0	31.0	.99

■

Diagnostic Plots

The basic plots that many statisticians recommend for an assessment of model validity and usefulness are the following:

1. e_i^* (or e_i) on the vertical axis versus x_i on the horizontal axis

2. e_i^* (or e_i) on the vertical axis versus $\hat{y}_i$ on the horizontal axis

3. $\hat{y}_i$ on the vertical axis versus y_i on the horizontal axis

4. A normal probability plot of the standardized residuals

Plots 1 and 2 are called **residual plots** (against the independent variable and fitted values, respectively), whereas Plot 3 is fitted against observed values.

    If Plot 3 yields points close to the 45° line [slope +1 through (0, 0)], then the estimated regression function gives accurate predictions of the values actually observed. Thus, Plot 3 provides a visual assessment of model effectiveness in making predictions. Provided that the model is correct, neither residual plot should exhibit distinct patterns. The residuals should be randomly distributed about 0 according to a normal distribution, so all but a very few standardized residuals should lie between −2 and +2 (i.e., all but a few residuals within 2 standard deviations of their expected value 0). The plot of standardized residuals versus $\hat{y}$ is really a combination of the two other plots, showing implicitly both how residuals vary with x and how fitted values compare with observed values. This latter plot is the single one most often recommended for multiple regression analysis. Plot 4 allows the analyst to assess the plausibility of the assumption that ϵ has a normal distribution.

Example 13.2
(Example 13.1 continued)

Figure 13.1 presents a scatter plot of the data and the four plots just recommended. The plot of $\hat{y}$ versus y confirms the impression given by r^2 that x is effective in predicting y and also indicates that there is no observed y for which the predicted value is terribly

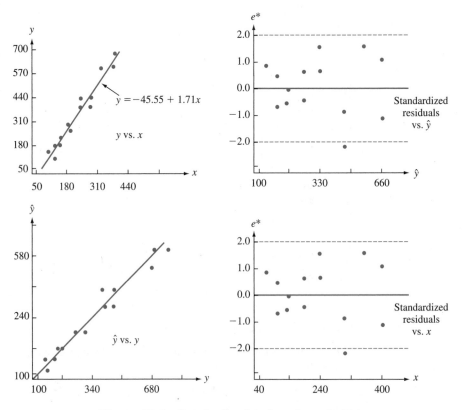

Figure 13.1 Plots for the data from Example 13.1

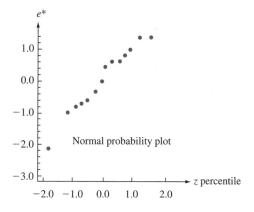

Figure 13.1 (cont'd)   Plots for the data from Example 13.1

far off the mark. Both residual plots show no unusual pattern or discrepant values. There is one standardized residual slightly outside the interval (−2, 2), but this is not surprising in a sample of size 14. The normal probability plot of the standardized residuals is reasonably straight. In summary, the plots leave us with no qualms about either the appropriateness of a simple linear relationship or the fit to the given data. ∎

Difficulties and Remedies

Although we hope that our analysis will yield plots like those of Figure 13.1, quite frequently the plots will suggest one or more of the following difficulties:

1. A nonlinear probabilistic relationship between x and y is appropriate.

2. The variance of ϵ (and of Y) is not a constant σ^2 but depends on x.

3. The selected model fits the data well except for a very few discrepant or outlying data values, which may have greatly influenced the choice of the best-fit function.

4. The error term ϵ does not have a normal distribution.

5. When the subscript i indicates the time order of the observations, the ϵ_i's exhibit dependence over time.

6. One or more relevant independent variables have been omitted from the model.

Figure 13.2 (page 548) presents residual plots corresponding to items 1–3, 5, and 6. In Chapter 4, we discussed patterns in normal probability plots that cast doubt on the assumption of an underlying normal distribution. Notice that the residuals from the data in Figure 13.2(d) with the circled point included would not by themselves necessarily suggest further analysis, yet when a new line is fit with that point deleted, the new line differs considerably from the original line. This type of behavior is more difficult to identify in multiple regression. It is most likely to arise when there is a single (or very few) data point(s) with independent variable value(s) far removed from the remainder of the data.

We now indicate briefly what remedies are available for the types of difficulties. For a more comprehensive discussion, one or more of the references on regression

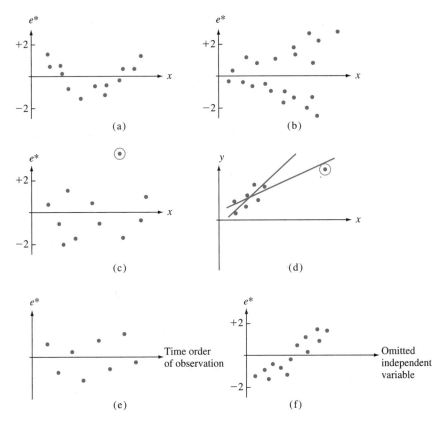

Figure 13.2   Plots that indicate abnormality in data: (a) nonlinear relationship; (b) non-constant variance; (c) discrepant observation; (d) observation with large influence; (e) dependence in errors; (f) variable omitted

analysis should be consulted. If the residual plot looks something like that of Figure 13.2(a), exhibiting a curved pattern, then a nonlinear function of x may be fit.

The residual plot of Figure 13.2(b) suggests that, although a straight-line relationship may be reasonable, the assumption that $V(Y_i) = \sigma^2$ for each i is of doubtful validity. When the assumptions of Chapter 12 are valid, it can be shown that among all unbiased estimators of β_0 and β_1, the ordinary least squares estimators have minimum variance. These estimators give equal weight to each (x_i, Y_i). If the variance of Y increases with x, then Y_i's for large x_i should be given less weight than those with small x_i. This suggests that β_0 and β_1 should be estimated by minimizing

$$f_w(b_0, b_1) = \Sigma w_i[y_i - (b_0 + b_1 x_i)]^2 \qquad (13.4)$$

where the w_i's are weights that decrease with increasing x_i. Minimization of Expression (13.4) yields **weighted least squares** estimates. For example, if the standard deviation of Y is proportional to x (for $x > 0$)—that is, $V(Y) = kx^2$—then it can be shown that the weights $w_i = 1/x_i^2$ yield best estimators of β_0 and β_1. The books by John Neter et al. and by S. Chatterjee and Bertram Price contain more detail (see the chapter bibliography). Weighted least squares is used quite frequently by econometricians (economists who use statistical methods) to estimate parameters.

When plots or other evidence suggest that the data set contains outliers or points having large influence on the resulting fit, one possible approach is to omit these outlying points and recompute the estimated regression equation. This would certainly be correct if it were found that the outliers resulted from errors in recording data values or experimental errors. If no assignable cause can be found for the outliers, it is still desirable to report the estimated equation both with and without outliers omitted. Yet another approach is to retain possible outliers but to use an estimation principle that puts relatively less weight on outlying values than does the principle of least squares. One such principle is MAD (minimize absolute deviations), which selects $\hat{\beta}_0$ and $\hat{\beta}_1$ to minimize $\sum |y_i - (b_0 + b_1 x_i)|$. Unlike the estimates of least squares, there are no nice formulas for the MAD estimates; their values must be found by using an iterative computational procedure. Such procedures are also used when it is suspected that the ϵ_i's have a distribution that is not normal but instead has "heavy tails" (making it much more likely than for the normal distribution that discrepant values will enter the sample); robust regression procedures are those that produce reliable estimates for a wide variety of underlying error distributions. Least squares estimators are not robust in the same way that the sample mean $\bar{X}$ is not a robust estimator for μ.

When a plot suggests time dependence in the error terms, an appropriate analysis may involve a transformation of the y's or else a model explicitly including a time variable. Lastly, a plot such as that of Figure 13.2(f), which shows a pattern in the residuals when plotted against an omitted variable, suggests that a multiple regression model that includes the previously omitted variable should be considered.

Exercises | Section 13.1 (1–14)

1. Suppose the variables x = commuting distance and y = commuting time are related according to the simple linear regression model with $\sigma = 10$.
  a. If $n = 5$ observations are made at the x values $x_1 = 5$, $x_2 = 10$, $x_3 = 15$, $x_4 = 20$, and $x_5 = 25$, calculate the standard deviations of the five corresponding residuals.
  b. Repeat part (a) for $x_1 = 5$, $x_2 = 10$, $x_3 = 15$, $x_4 = 20$, and $x_5 = 50$.
  c. What do the results of parts (a) and (b) imply about the deviation of the estimated line from the observation made at the largest sampled x value?

2. The x values and standardized residuals for the chlorine flow/etch rate data of Exercise 52 (Section 12.4) are displayed in the accompanying table. Construct a standardized residual plot and comment on its appearance.

x	1.50	1.50	2.00	2.50	2.50
$e*$	.31	1.02	−1.15	−1.23	.23

x	3.00	3.50	3.50	4.00
$e*$	.73	−1.36	1.53	.07

3. Example 12.6 presented the residuals from a simple linear regression of bark lead content y on traffic flow x.
  a. Plot the residuals against x. Does the resulting plot suggest that a straight-line regression function is a reasonable choice of model? Explain your reasoning.
  b. Using $s = 92.19$, compute the values of the standardized residuals. Is $e_i^* \approx e_i/s$ for $i = 1, \ldots, n$, or are the e_i^*'s not close to being proportional to the e_i's?
  c. Plot the standardized residuals against x. Does the plot differ significantly in general appearance from the plot of part (a)?

4. Wear resistance of certain nuclear reactor components made of Zircaloy-2 is partly determined by properties of the oxide layer. The following data appears in an article that proposed a new nondestructive testing method to monitor thickness of the layer ("Monitoring of Oxide Layer Thickness on Zircaloy-2 by the Eddy Current Test Method," *J. of Testing and Eval.*, 1987: 333–336). The variables are x = oxide-layer thickness (μm) and y = eddy-current response (arbitrary units).

x	0	7	17	114	133
y	20.3	19.8	19.5	15.9	15.1

x	142	190	218	237	285
y	14.7	11.9	11.5	8.3	6.6

The authors summarized the relationship by giving the equation of the least squares line as $y = 20.6 - .047x$. Calculate and plot the residuals against x and then comment on the appropriateness of the simple linear regression model.

5. Refer to Exercise 4 and use $s = .7921$ to calculate the standardized residuals from a simple linear regression. Construct a standardized residual plot and comment. Also construct a normal probability plot and comment.

6. The accompanying data on x = true density (kg/mm^3) and y = moisture content (% d.b.) was read from a plot in the article "Physical Properties of Cumin Seed" (*J. Agric. Engr. Res.*, 1996: 93–98).

x	7.0	9.3	13.2	16.3	19.1	22.0
y	1046	1065	1094	1117	1130	1135

The equation of the least squares line is $y = 1008.14 + 6.19268x$ (this differs very slightly from the equation given in the article); $s = 7.265$ and $r^2 = .968$.
   a. Carry out a test of model utility and comment.
   b. Compute the values of the residuals and plot the residuals against x. Does the plot suggest that a linear regression function is inappropriate?
   c. Compute the values of the standardized residuals and plot them against x. Are there any unusually large (positive or negative) standardized residuals? Does this plot give the same message as the plot of part (b) regarding the appropriateness of a linear regression function?

7. The article "Effects of Gamma Radiation on Juvenile and Mature Cuttings of Quaking Aspen" (*Forest Science*, 1967: 240–245) reports the following data on exposure time to radiation (x, in kr/16 hr) and dry weight of roots (y, in mg $\times$ 10^{-1}).

x	0	2	4	6	8
y	110	123	119	86	62

   a. Construct a scatter plot. Does the plot suggest that a linear probabilistic relationship is appropriate?
   b. A linear regression results in the least squares line $y = 127 - 6.65x$, with $s = 16.94$. Compute the residuals and standardized residuals and then construct residual plots. What do these plots suggest? What type of function should provide a better fit to the data than a straight line does?

8. The article "Pedunculate Oak Woodland in a Severe Environment" (*J. Ecology*, 1978: 707–740) reports the following data on x = age (years) and y = annual trunk diameter growth increment (mm) for a sample of trees in a certain region:

x	17	23	30	37.5	40
y	2.20	1.25	.85	1.30	1.70

x	46.5	50	54	55	93
y	.75	.75	.50	1.00	.70

The estimated regression line is $y = 1.78 - .0154x$, with $s = .434$. Does a scatter plot suggest that a particular observation has excessively influenced the choice of a best-fit line? Is this impression confirmed by the standardized residuals?

9. Consider the following four (x, y) data sets; the first three have the same x values, so these values are listed only once (from Frank Anscombe, "Graphs in Statistical Analysis," *Amer. Statistician*, 1973: 17–21):

Data Set	1–3	1	2	3	4	4
Variable	x	y	y	y	x	y
	10.0	8.04	9.14	7.46	8.0	6.58
	8.0	6.95	8.14	6.77	8.0	5.76
	13.0	7.58	8.74	12.74	8.0	7.71
	9.0	8.81	8.77	7.11	8.0	8.84
	11.0	8.33	9.26	7.81	8.0	8.47
	14.0	9.96	8.10	8.84	8.0	7.04
	6.0	7.24	6.13	6.08	8.0	5.25
	4.0	4.26	3.10	5.39	19.0	12.50
	12.0	10.84	9.13	8.15	8.0	5.56
	7.0	4.82	7.26	6.42	8.0	7.91
	5.0	5.68	4.74	5.73	8.0	6.89

For each of these four data sets, the values of the summary statistics $\sum x_i, \sum x_i^2, \sum y_i, \sum y_i^2,$ and $\sum x_i y_i$ are virtually identical, so all quantities computed from these five will be essentially identical for the four sets—the least squares line ($y = 3 + .5x$), SSE, s^2, r^2, t intervals, t statistics, and so on. The summary statistics provide no way of distinguishing among the four data sets. Based on a scatter plot and a residual plot for each set, comment on the appropriateness or inappropriateness of fitting a straight-line model; include in your comments any specific suggestions for how a "straight-line analysis" might be modified or qualified.

10. a. Show that $\sum_{i=1}^{n} e_i = 0$ when the e_i's are the residuals from a simple linear regression.

b. Are the residuals from a simple linear regression independent of one another, positively correlated, or negatively correlated? Explain.

c. Show that $\sum_{i=1}^{n} x_i e_i = 0$ for the residuals from a simple linear regression. [This result along with part (a) shows that there are two linear restrictions on the e_i's, resulting in a loss of 2 df when the squared residuals are used to estimate σ^2.]

d. Is it true that $\sum_{i=1}^{n} e_i^* = 0$? Give a proof or a counterexample.

11. a. Express the ith residual $Y_i - \hat{Y}_i$ (where $\hat{Y}_i = \hat{\beta}_0 + \hat{\beta}_1 x_i$) in the form $\sum c_j Y_j$, a linear function of the Y_j's. Then use rules of variance to verify that $V(Y_i - \hat{Y}_i)$ is given by Expression (13.2).

b. It can be shown that $\hat{Y}_i$ and $Y_i - \hat{Y}_i$ (the ith predicted value and residual) are independent of one another. Use this fact, the relation $Y_i = \hat{Y}_i + (Y_i - \hat{Y}_i)$, and the expression for $V(\hat{Y})$ from Section 12.4 to again verify Expression (13.2).

c. As x_i moves farther away from $\bar{x}$, what happens to $V(\hat{Y}_i)$ and to $V(Y_i - \hat{Y}_i)$?

12. a. Could a linear regression result in residuals 23, −27, 5, 17, −8, 9, and 15? Why or why not?

b. Could a linear regression result in residuals 23, −27, 5, 17, −8, −12, and 2 corresponding to x values 3, −4, 8, 12, −14, −20, and 25? Why or why not? (*Hint:* See Exercise 10.)

13. Recall that $\hat{\beta}_0 + \hat{\beta}_1 x$ has a normal distribution with expected value $\beta_0 + \beta_1 x$ and variance

$$\sigma^2 \left\{ \frac{1}{n} + \frac{(x - \bar{x})^2}{\sum(x_i - \bar{x})^2} \right\}$$

so that

$$Z = \frac{\hat{\beta}_0 + \hat{\beta}_1 x - (\beta_0 + \beta_1 x)}{\sigma \left(\frac{1}{n} + \frac{(x - \bar{x})^2}{\sum(x_i - \bar{x})^2} \right)^{1/2}}$$

has a standard normal distribution. If $S = \sqrt{SSE/(n - 2)}$ is substituted for σ, the resulting variable has a t distribution with $n - 2$ df. By analogy, what is the distribution of any particular standardized residual? If $n = 25$, what is the probability that a particular standardized residual falls outside the interval (−2.50, 2.50)?

14. If there is at least one x value at which more than one observation has been made, there is a formal test procedure for testing

$$H_0: \quad \mu_{Y \cdot x} = \beta_0 + \beta_1 x \text{ for some values } \beta_0, \beta_1 \text{ (the true regression function is linear)}$$

versus

$$H_a: \quad H_0 \text{ is not true (the true regression function is not linear)}$$

Suppose observations are made at $x_1, x_2, \ldots, x_c$. Let $Y_{11}, Y_{12}, \ldots, Y_{1n_1}$ denote the n_1 observations when $x = x_1; \ldots; Y_{c1}, Y_{c2}, \ldots, Y_{cn_c}$ denote the n_c observations when $x = x_c$. With $n = \sum n_i$ (the total number of observations), SSE has $n - 2$ df. We break SSE into two pieces, SSPE (pure error) and SSLF (lack of fit), as follows:

$$SSPE = \sum_i \sum_j (Y_{ij} - \bar{Y}_{i \cdot})^2$$
$$= \sum \sum Y_{ij}^2 - \sum n_i \bar{Y}_{i \cdot}^2$$
$$SSLF = SSE - SSPE$$

The n_i observations at x_i contribute $n_i - 1$ df to SSPE, so the number of degrees of freedom for SSPE is $\sum_i (n_i - 1) = n - c$ and the degrees of freedom for SSLF is $n - 2 - (n - c) = c - 2$. Let MSPE $= SSPE/(n - c)$ and MSLF $=$ SSLF$/(c - 2)$. Then it can be shown that whereas $E(\text{MSPE}) = \sigma^2$ whether or not H_0 is true, $E(\text{MSLF}) = \sigma^2$ if H_0 is true and $E(\text{MSLF}) > \sigma^2$ if H_0 is false.

Test statistic: $\quad F = \dfrac{\text{MSLF}}{\text{MSPE}}$

Rejection region: $\quad f \geq F_{\alpha, c-2, n-c}$

The following data comes from the article "Changes in Growth Hormone Status Related to Body Weight of Growing Cattle" (*Growth*, 1977: 241–247), with $x =$ body weight and $y =$ metabolic clearance rate/body weight.

x	110	110	110	230	230	230	360
y	235	198	173	174	149	124	115

(continued)

x	360	360	360	505	505	505	505
y	130	102	95	122	112	98	96

(So $c = 4$, $n_1 = n_2 = 3$, $n_3 = n_4 = 4$.)

a. Test H_0 versus H_a at level .05 using the lack-of-fit test just described.

b. Does a scatter plot of the data suggest that the relationship between x and y is linear? How does this compare with the result of part (a)? (A nonlinear regression function was used in the article.)

13.2 Regression with Transformed Variables

The necessity for an alternative model to the linear probabilistic model $Y = \beta_0 + \beta_1 x + \epsilon$ may be suggested either by a theoretical argument or else by examining diagnostic plots from a linear regression analysis. In either case, settling on a model whose parameters can be easily estimated is desirable. An important class of such models is specified by means of functions that are "intrinsically linear."

DEFINITION

> A function relating y to x is **intrinsically linear** if by means of a transformation on x and/or y, the function can be expressed as $y' = \beta_0 + \beta_1 x'$, where $x' =$ the transformed independent variable and $y' =$ the transformed dependent variable.

Four of the most useful intrinsically linear functions are given in Table 13.1. In each case, the appropriate transformation is either a log transformation—either base 10 or natural logarithm (base e)—or a reciprocal transformation. Representative graphs of the four functions appear in Figure 13.3.

Table 13.1 Useful intrinsically linear functions*

Function	Transformation(s) to Linearize	Linear Form
a. Exponential: $y = \alpha e^{\beta x}$	$y' = \ln(y)$	$y' = \ln(\alpha) + \beta x$
b. Power: $y = \alpha x^\beta$	$y' = \log(y)$, $x' = \log(x)$	$y' = \log(\alpha) + \beta x'$
c. $y = \alpha + \beta \cdot \log(x)$	$x' = \log(x)$	$y = \alpha + \beta x'$
d. Reciprocal: $y = \alpha + \beta \cdot \dfrac{1}{x}$	$x' = \dfrac{1}{x}$	$y = \alpha + \beta x'$

*When log($\cdot$) appears, either a base-10 or a base-e logarithm can be used.

Thus, for an exponential function relationship, only y is transformed to achieve linearity, whereas for a power function relationship, both x and y are transformed. Because the variable x is in the exponent in an exponential relationship, y increases (if $\beta > 0$) or decreases (if $\beta < 0$) much more rapidly as x increases than is the case for the power function, though over a short interval of x values it can be difficult to differentiate between the two functions. Examples of functions that are not intrinsically linear are $y = \alpha + \gamma e^{\beta x}$ and $y = \alpha + \gamma x^\beta$.

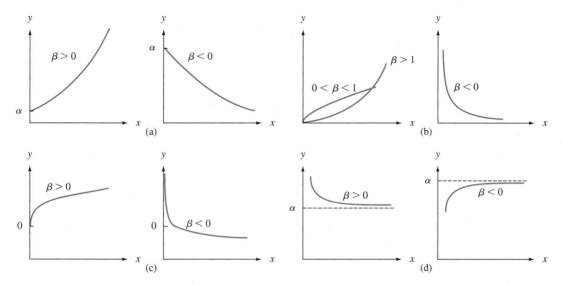

Figure 13.3 Graphs of the intrinsically linear functions given in Table 13.1

Intrinsically linear functions lead directly to probabilistic models which, though not linear in x as a function, have parameters whose values are easily estimated using ordinary least squares.

DEFINITION

A probabilistic model relating Y to x is **intrinsically linear** if, by means of a transformation on Y and/or x, it can be reduced to a linear probabilistic model $Y' = \beta_0 + \beta_1 x' + \epsilon'$.

The intrinsically linear probabilistic models that correspond to the four functions of Table 13.1 are as follows:

a. $Y = \alpha e^{\beta x} \cdot \epsilon$, a multiplicative exponential model, so that $\ln(Y) = Y' = \beta_0 + \beta_1 x' + \epsilon'$ with $x' = x$, $\beta_0 = \ln(\alpha)$, $\beta_1 = \beta$, and $\epsilon' = \ln(\epsilon)$.

b. $Y = \alpha x^{\beta} \cdot \epsilon$, a multiplicative power model, so that $\log(Y) = Y' = \beta_0 + \beta_1 x' + \epsilon'$ with $x' = \log(x)$, $\beta_0 = \log(\alpha)$, $\beta_1 = \beta$, and $\epsilon' = \log(\epsilon)$.

c. $Y = \alpha + \beta \log(x) + \epsilon$, so that $x' = \log(x)$ immediately linearizes the model.

d. $Y = \alpha + \beta \cdot 1/x + \epsilon$, so that $x' = 1/x$ yields a linear model.

The additive exponential and power models, $Y = \alpha e^{\beta x} + \epsilon$ and $Y = \alpha x^{\beta} + \epsilon$, are not intrinsically linear. Notice that both (a) and (b) require a transformation on Y and, as a result, a transformation on the error variable ϵ. In fact, if ϵ has a lognormal distribution (see Chapter 4) with $E(\epsilon) = e^{\sigma^2/2}$ and $V(\epsilon) = \tau^2$ independent of x, then the transformed models for both (a) and (b) will satisfy all the assumptions of Chapter 12 regarding the linear probabilistic model; this in turn implies that all inferences for the parameters of the transformed model based on these assumptions will be valid. If σ^2 is small, $\mu_{Y \cdot x} \approx \alpha e^{\beta x}$ in (a) or αx^β in (b).

The major advantage of an intrinsically linear model is that the parameters β_0 and β_1 of the transformed model can be immediately estimated using the principle of least squares simply by substituting x' and y' into the estimating formulas:

$$\hat{\beta}_1 = \frac{\sum x_i' y_i' - \sum x_i' \sum y_i'/n}{\sum (x_i')^2 - (\sum x_i')^2/n}$$

$$\hat{\beta}_0 = \frac{\sum y_i' - \hat{\beta}_1 \sum x_i'}{n} = \bar{y}' - \hat{\beta}_1 \bar{x}'$$

(13.5)

Parameters of the original nonlinear model can then be estimated by transforming back $\hat{\beta}_0$ and/or $\hat{\beta}_1$ if necessary. In cases (a) and (b), when σ^2 is small, an approximate confidence interval for $\mu_{Y \cdot x}$ results from taking antilogs in the interval for $\beta_0 + \beta_1 x$.*

Example 13.3 Taylor's equation for tool life y as a function of cutting time x states that $xy^c = k$ or, equivalently, that $y = \alpha x^\beta$. The article "The Effect of Experimental Error on the Determination of Optimum Metal Cutting Conditions" (*J. Eng. for Industry,* 1967: 315–322) observes that the relationship is not exact (deterministic) and that the parameters α and β must be estimated from data. Thus, an appropriate model is the multiplicative power model $Y = \alpha \cdot x^\beta \cdot \epsilon$, which the author fit to the accompanying data consisting of 12 carbide tool life observations (Table 13.2). In addition to the x, y, x', and y' values, the predicted transformed values ($\hat{y}'$) and the predicted values on the original scale ($\hat{y}$, after transforming back) are given.

Table 13.2 Data for Example 13.3

	x	y	$x' = \ln(x)$	$y' = \ln(y)$	$\hat{y}'$	$\hat{y} = e^{\hat{y}'}$
1	600.	2.3500	6.39693	.85442	1.12754	3.0881
2	600.	2.6500	6.39693	.97456	1.12754	3.0881
3	600.	3.0000	6.39693	1.09861	1.12754	3.0881
4	600.	3.6000	6.39693	1.28093	1.12754	3.0881
5	500.	6.4000	6.21461	1.85630	2.11203	8.2650
6	500.	7.8000	6.21461	2.05412	2.11203	8.2650
7	500.	9.8000	6.21461	2.28238	2.11203	8.2650
8	500.	16.5000	6.21461	2.80336	2.11203	8.2650
9	400.	21.5000	5.99146	3.06805	3.31694	27.5760
10	400.	24.5000	5.99146	3.19867	3.31694	27.5760
11	400.	26.0000	5.99146	3.25810	3.31694	27.5760
12	400.	33.0000	5.99146	3.49651	3.31694	27.5760

---

*Strictly speaking, taking antilogs gives a confidence interval for the median of the Y distribution for the fixed x value—that is, for $\tilde{\mu}_{Y \cdot x}$. Because the lognormal distribution is positively skewed, $\mu_{Y \cdot x} > \tilde{\mu}_{Y \cdot x}$; the two are approximately equal if σ^2 is close to 0.

The summary statistics for fitting a straight line to the transformed data are $\sum x_i' = 74.41200$, $\sum y_i' = 26.22601$, $\sum x_i'^2 = 461.75874$, $\sum y_i'^2 = 67.74609$, and $\sum x_i'y_i' = 160.84601$, so

$$\hat{\beta}_1 = \frac{(160.84601) - (74.41200)(26.22601)/12}{461.75874 - (74.41200)^2/12} = -5.3996$$

$$\hat{\beta}_0 = \frac{26.22601 - (-5.3996)(74.41200)}{12} = 35.6684$$

The estimated values of α and β, the parameters of the power function model, are $\hat{\beta} = \hat{\beta}_1 = -5.3996$ and $\hat{\alpha} = e^{\hat{\beta}_0} = 3.094491530 \cdot 10^{15}$. Thus, the estimated regression function is $\hat{\mu}_{Y \cdot x} \approx 3.094491530 \cdot 10^{15} \cdot x^{-5.3996}$. To recapture Taylor's (estimated) equation, set $y = 3.094491530 \cdot 10^{15} \cdot x^{-5.3996}$, whence, $xy^{.185} = 740$.

Figure 13.4(a) gives a plot of the standardized residuals from the linear regression using transformed variables (for which $r^2 = .922$); there is no apparent pattern in the plot, though one standardized residual is a bit large, and the residuals look as they should for a simple linear regression. Figure 13.4(b) pictures a plot of $\hat{y}$ versus y, which indicates satisfactory predictions on the original scale.

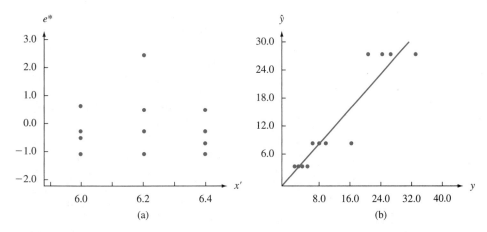

Figure 13.4   (a) Standardized residuals versus x' from Example 13.3; (b) $\hat{y}$ versus y from Example 13.3

To obtain a confidence interval for median tool life when cutting time is 500, we transform $x = 500$ to $x' = 6.21461$. Then $\hat{\beta}_0 + \hat{\beta}_1 x' = 2.1120$, and a 95% CI for $\beta_0 + \beta_1(6.21461)$ is (from Section 12.4) $2.1120 \pm (2.228)(.0824) = (1.928, 2.296)$. The 95% CI for $\tilde{\mu}_{Y \cdot 500}$ is then obtained by taking antilogs: $(e^{1.928}, e^{2.296}) = (6.876, 9.930)$. It is easily checked that for the transformed data $s^2 = \hat{\sigma}^2 \approx .081$. Because this is quite small, $(6.876, 9.930)$ is an approximate interval for $\mu_{Y \cdot 500}$. ∎

Example 13.4 In the article "Ethylene Synthesis in Lettuce Seeds: Its Physiological Significance" (*Plant Physiology*, 1972: 719–722), ethylene content of lettuce seeds (y, in nL/g dry wt) was studied as a function of exposure time (x, in min) to an ethylene absorbent. Figure 13.5 (page 556) presents both a scatter plot of the data and a plot of

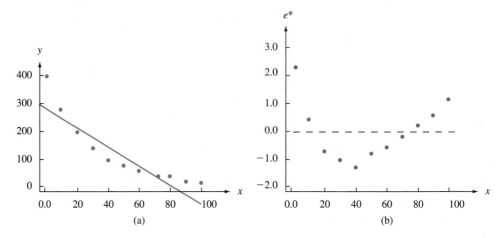

Figure 13.5 (a) Scatter plot; (b) residual plot from linear regression for the data in Example 13.4

the residuals generated from a linear regression of y on x. Both plots show a strong curved pattern, suggesting that a transformation to achieve linearity is appropriate. In addition, a linear regression gives negative predictions for $x = 90$ and $x = 100$.

The author did not give any argument for a theoretical model, but his plot of $y' = \ln(y)$ versus x shows a strong linear relationship, suggesting that an exponential function will provide a good fit to the data. Table 13.3 shows the data values and other information from a linear regression of y' on x. The estimates of parameters of the linear model are $\hat{\beta}_1 = -.0323$ and $\hat{\beta}_0 = 5.941$, with $r^2 = .995$. The estimated regression function for the exponential model is $\hat{\mu}_{Y \cdot x} \approx e^{\hat{\beta}_0} \cdot e^{\hat{\beta}_1 x} = 380.32e^{-.0323x}$. The predicted values $\hat{y}_i$ can then be obtained by substitution of x_i ($i = 1, \ldots, n$) into $\hat{\mu}_{Y \cdot x}$ or else by computing $\hat{y}_i = e^{\hat{y}_i'}$ where the $\hat{y}_i'$'s are the predictions from the transformed straight-line model. Figure 13.6 presents both a plot of $e'*$ versus x (the standardized residuals from a linear regression) and a plot of $\hat{y}$ versus y. These plots support the choice of an exponential model.

Table 13.3 Data for Example 13.4

x	y	$y' = \ln(y)$	$\hat{y}'$	$\hat{y} = e^{\hat{y}'}$
2	408	6.01	5.876	353.32
10	274	5.61	5.617	275.12
20	196	5.28	5.294	199.12
30	137	4.92	4.971	144.18
40	90	4.50	4.647	104.31
50	78	4.36	4.324	75.50
60	51	3.93	4.001	54.64
70	40	3.69	3.677	39.55
80	30	3.40	3.354	28.62
90	22	3.09	3.031	20.72
100	15	2.71	2.708	15.00

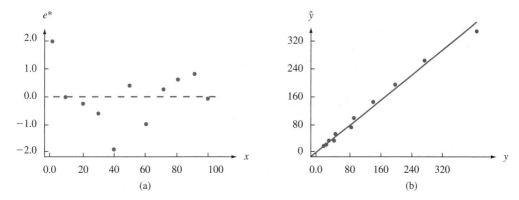

Figure 13.6 Plot of (a) standardized residuals (after transforming) versus x; (b) $\hat{y}$ versus y for data in Example 13.4 ∎

In analyzing transformed data, one should keep in mind the following points:

1. Estimating β_1 and β_0 as in (13.5) and then transforming back to obtain estimates of the original parameters is not equivalent to using the principle of least squares directly on the original model. Thus, for the exponential model, we could estimate α and β by minimizing $\sum(y_i - \alpha e^{\beta x_i})^2$. The resulting estimates would not be equal: $\hat{\alpha} \neq e^{\hat{\beta}_0}$ and $\hat{\beta} \neq \hat{\beta}_1$.

2. If the chosen model is not intrinsically linear, the approach summarized in (13.5) cannot be used. Instead, least squares (or some other fitting procedure) would have to be applied to the untransformed model. Thus, for the additive exponential model $Y = \alpha e^{\beta x} + \epsilon$, least squares would involve minimizing $\sum(y_i - \alpha e^{\beta x_i})^2$. Taking partial derivatives with respect to α and β results in two nonlinear normal equations in α and β; these equations must then be solved using an iterative procedure.

3. When the transformed linear model satisfies all the assumptions listed in Chapter 12, the method of least squares yields best estimates of the transformed parameters. However, estimates of the original parameters may not be best in any sense, though they will be reasonable. For example, in the exponential model, the estimator $\hat{\alpha} = e^{\hat{\beta}_0}$ will not be unbiased, though it will be the maximum likelihood estimator of α if the error variable ϵ' is normally distributed. It is conceivable that using least squares directly (without transforming) could yield better estimates, though the computations would be quite burdensome.

4. If a transformation on y has been made and one wishes to use the standard formulas to test hypotheses or construct CIs, ϵ' should be at least approximately normally distributed. To check this, the residuals from the transformed regression should be examined.

5. When y is transformed, the r^2 value from the resulting regression refers to variation in the y_i''s explained by the transformed regression model. Although a high value of r^2 here indicates a good fit of the estimated original nonlinear model to the observed y_i's, r^2 does not refer to these original observations. Perhaps the best way to assess the quality of the fit is to compute the predicted values $\hat{y}_i'$ using the transformed model, transform them back to the original y scale to obtain $\hat{y}_i$, and then plot $\hat{y}$ versus y. A good fit is then evidenced by points close to the 45° line. One could compute SSE $= \sum(y_i - \hat{y}_i)^2$

as a numerical measure of the goodness of fit. When the model was linear, we compared this to $SST = \sum(y_i - \bar{y})^2$, the total variation about the horizontal line at height $\bar{y}$; this led to r^2. In the nonlinear case, though, it is not necessarily informative to measure total variation in this way, so an r^2 value is not as useful as in the linear case.

Logistic Regression

The simple linear regression model is appropriate for relating a quantitative response variable y to a quantitative predictor x. Suppose that y is a dichotomous variable with possible values 1 and 0 corresponding to success and failure. Let $p = P(S) = P(y = 1)$. Frequently, the value of p will depend on the value of some quantitative variable x. For example, the probability that a car needs warranty service of a certain kind might well depend on the car's mileage, or the probability of avoiding an infection of a certain type might depend on the dosage in an inoculation. Instead of using just the symbol p for the success probability, we now use $p(x)$ to emphasize the dependence of this probability on the value of x. The simple linear regression equation $Y = \beta_0 + \beta_1 x + \epsilon$ is no longer appropriate, for taking the mean value on each side of the equation gives

$$\mu_{Y \cdot x} = 1 \cdot p(x) + 0 \cdot (1 - p(x)) = p(x) = \beta_0 + \beta_1 x$$

Whereas $p(x)$ is a probability and therefore must be between 0 and 1, $\beta_0 + \beta_1 x$ need not be in this range.

Instead of letting the mean value of y be a linear function of x, we now consider a model in which some function of the mean value of y is a linear function of x. In other words, we allow $p(x)$ to be a function of $\beta_0 + \beta_1 x$ rather than $\beta_0 + \beta_1 x$ itself. A function that has been found quite useful in many applications is the **logit function**

$$p(x) = \frac{e^{\beta_0 + \beta_1 x}}{1 + e^{\beta_0 + \beta_1 x}}$$

Figure 13.7 shows a graph of $p(x)$ for particular values of β_0 and β_1 with $\beta_1 > 0$. As x increases, the probability of success increases. For β_1 negative, the success probability would be a decreasing function of x.

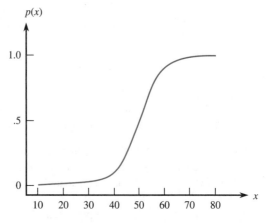

Figure 13.7 A graph of a logit function

Logistic regression means assuming that $p(x)$ is related to x by the logit function. Straightforward algebra shows that

$$\frac{p(x)}{1 - p(x)} = e^{\beta_0 + \beta_1 x}$$

The expression on the left-hand side is called the *odds ratio*. If, for example, $\dfrac{p(60)}{1 - p(60)} = 3$, then when $x = 60$ a success is three times as likely as a failure. We now see that the logarithm of the odds ratio is a linear function of the predictor. In particular, the slope parameter β_1 is the change in the log odds associated with a 1-unit increase in x. This implies that the odds ratio itself changes by the multiplicative factor e^{β_1} when x increases by 1 unit.

Fitting the logistic regression to sample data requires that the parameters β_0 and β_1 be estimated. This is usually done using the maximum likelihood technique described in Chapter 6. The details are quite involved, but fortunately the most popular statistical computer packages will do this on request and provide quantitative and pictorial indications of how well the model fits.

Example 13.5 Here is data on launch temperature and the incidence of failure for O-rings in 24 space shuttle launches prior to the *Challenger* disaster of January 1986.

Temperature	Failure	Temperature	Failure	Temperature	Failure
53	Y	68	N	75	N
56	Y	69	N	75	Y
57	Y	70	N	76	N
63	N	70	Y	76	N
66	N	70	Y	78	N
67	N	70	Y	79	N
67	N	72	N	80	N
67	N	73	N	81	N

Figure 13.8 (page 560) shows JMP output for a logistic regression analysis. We have chosen to let p denote the probability of failure. Failures tended to occur at lower temperatures and successes at higher temperatures, so the graph of $\hat{p}$ decreases as temperature increases. The estimate of β_1 is $\hat{\beta}_1 = -.1713$, and the estimated standard deviation of $\hat{\beta}_1$ is $s_{\hat{\beta}_1} = .08344$. Provided that n is large enough, and we assume it is in this case, $\hat{\beta}_1$ has approximately a normal distribution. If $\beta_1 = 0$ (temperature does not affect the likelihood of O-ring failure), $z = \hat{\beta}_1 / s_{\hat{\beta}_1}$ has approximately a standard normal distribution. The value of this z-ratio is -2.05, and the P-value for a two-tailed test is .0404 (twice the area captured under the z curve to the left of -2.05). JMP reports the value of a chi-squared statistic, which is just z^2, and the chi-squared P-value differs from that for z only because of rounding. For each 1-degree increase in temperature, the odds of failure decreases by a factor of roughly .84. The launch temperature for the *Challenger* mission was only 31°F. Because this value is much smaller than any temperature in our sample, it is dangerous to extrapolate the estimated relationship. Nevertheless, it appears that for a temperature this small, O-ring failure is almost a sure thing.

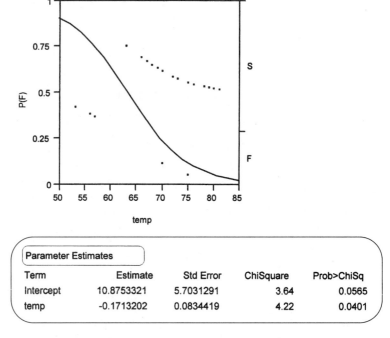

Figure 13.8 Logistic regression output from JMP ■

Exercises | Section 13.2 (15–25)

15. No tortilla chip aficionado likes soggy chips, so it is important to find characteristics of the production process that produce chips with an appealing texture. The following data on x = frying time (sec) and y = moisture content (%) appeared in the article "Thermal and Physical Properties of Tortilla Chips as a Function of Frying Time" (*J. of Food Processing and Preservation*, 1995: 175–189).

x	5	10	15	20	25	30	45	60
y	16.3	9.7	8.1	4.2	3.4	2.9	1.9	1.3

a. Construct a scatter plot of y versus x and comment.

b. Construct a scatter plot of the $(\ln(x), \ln(y))$ pairs and comment.

c. What probabilistic relationship between x and y is suggested by the linear pattern in the plot of part (b)?

d. Predict the value of moisture content when frying time is 20 in a way that conveys information about reliability and precision.

e. Analyze the residuals from fitting the simple linear regression model to the transformed data and comment.

16. The article "The Luminosity–Spectral Index Relationship for Radio Galaxies" (*Nature*, 1972: 88–89) suggested that for class S galaxies, $\ln(L_{178})$ is linearly related to spectral index, where L_{178} denotes luminosity at 178 MHz. Representative data appears here:

Spectral Index	.59	.67	.72	.80	.85	.90
$\ln(L_{178})$	23.6	25.6	26.4	25.7	26.8	26.7

Spectral Index	.94	.66	1.00	.86	1.03	.70
$\ln(L_{178})$	27.0	24.9	27.1	27.2	26.9	25.2

a. Estimate the parameters of the exponential model implied by the linear relationship between $\ln(L_{178})$ and spectral index.

b. What value of L_{178} would you predict for a spectral index of .75?

c. Compute a 95% prediction interval for luminosity when the spectral index is .95.

17. The following data on mass rate of burning x and flame length y is representative of that which appeared in the article "Some Burning Characteristics of Filter Paper" (*Combustion Science and Technology*, 1971: 103–120):

x	1.7	2.2	2.3	2.6	2.7	3.0	3.2
y	1.3	1.8	1.6	2.0	2.1	2.2	3.0

x	3.3	4.1	4.3	4.6	5.7	6.1
y	2.6	4.1	3.7	5.0	5.8	5.3

a. Estimate the parameters of a power function model.
b. Construct diagnostic plots to check whether a power function is an appropriate model choice.
c. Test H_0: $\beta = \frac{4}{3}$ versus H_a: $\beta < \frac{4}{3}$, using a level .05 test.
d. Test the null hypothesis that states that the median flame length when burning rate is 5.0 is twice the median flame length when burning rate is 2.5 against the alternative that this is not the case.

18. An investigation of the influence of sodium benzoate concentration on the critical minimum pH necessary for the inhibition of Fe ("Mechanism of the Corrosion Inhibition of Fe by Sodium Benzoate," *Corrosion Science*, 1971: 675–682) yielded the accompanying data, which suggests that expected critical minimum pH is linearly related to the natural logarithm of concentration:

Concentration	.01	.025	.1	.95
pH	5.1	5.5	6.1	7.3

a. What is the implied probabilistic model, and what are the estimates of the model parameters?
b. What critical minimum pH would you predict for a concentration of 1.0? Obtain a 95% PI for critical minimum pH when concentration is 1.0.

19. Thermal endurance tests were performed to study the relationship between temperature and lifetime of polyester enameled wire ("Thermal Endurance of Polyester Enameled Wires Using Twisted Wire Specimens," *IEEE Trans. Insulation*, 1965: 38–44), resulting in the following data.

Temp.	200	200	200	200	200	200
Lifetime	5933	5404	4947	4963	3358	3878

Temp.	220	220	220	220	220	220
Lifetime	1561	1494	747	768	609	777

Temp.	240	240	240	240	240	240
Lifetime	258	299	209	144	180	184

a. Does a scatter plot of the data suggest a linear probabilistic relationship between lifetime and temperature?
b. What model is implied by a linear relationship between expected ln(lifetime) and 1/temperature? Does a scatter plot of the transformed data appear consistent with this relationship?
c. Estimate the parameters of the model suggested in part (b). What lifetime would you predict for a temperature of 220?
d. Because there are multiple observations at each x value, the method in Exercise 14 can be used to test the null hypothesis that states the model suggested in part (b) is correct. Carry out the test at level .01.

20. Exercise 14 presented data on body weight x and metabolic clearance rate/body weight y. Consider the following intrinsically linear functions for specifying the relationship between the two variables: (a) $\ln(y)$ versus x, (b) $\ln(y)$ versus $\ln(x)$, (c) y versus $\ln(x)$, (d) y versus $1/x$, and (e) $\ln(y)$ versus $1/x$. Use any appropriate diagnostic plots and analyses to decide which of these functions you would select to specify a probabilistic model. Explain your reasoning.

21. A plot in the article "Thermal Conductivity of Polyethylene: The Effects of Crystal Size, Density, and Orientation on the Thermal Conductivity" (*Polymer Eng. and Science*, 1972: 204–208) suggests that the expected value of thermal conductivity y is a linear function of $10^4 \cdot 1/x$ where x is lamellar thickness.

x	240	410	460	490	520	590	745	8300
y	12.0	14.7	14.7	15.2	15.2	15.6	16.0	18.1

a. Estimate the parameters of the regression function and the regression function itself.
b. Predict the value of thermal conductivity when lamellar thickness is 500 angstroms.

22. In each of the following cases, decide whether the given function is intrinsically linear. If so, identify x' and y' and then explain how a random error term ϵ can be introduced to yield an intrinsically linear probabilistic model.

a. $y = 1/(\alpha + \beta x)$
b. $y = 1/(1 + e^{\alpha + \beta x})$
c. $y = e^{e^{\alpha + \beta x}}$ (a Gompertz curve)
d. $y = \alpha + \beta e^{\lambda x}$

23. Suppose x and y are related according to a probabilistic exponential model $Y = \alpha e^{\beta x} \cdot \epsilon$ with $V(\epsilon)$ a constant independent of x (as was the case in the simple linear model $Y = \beta_0 + \beta_1 x + \epsilon$). Is $V(Y)$ a constant independent of x [as was the case for $Y = \beta_0 + \beta_1 x + \epsilon$, where $V(Y) = \sigma^2$]? Explain your reasoning. Draw a picture of a prototype scatter plot resulting from this model. Answer the same questions for the power model $Y = \alpha x^\beta \cdot \epsilon$.

24. Kyphosis refers to severe forward flexion of the spine following corrective spinal surgery. A study carried out to determine risk factors for kyphosis reported the accompanying ages (months) for 40 subjects at the time of the operation; the first 18 subjects did have kyphosis and the remaining 22 did not.

Kyphosis	12	15	42	52	59	73
	82	91	96	105	114	120
	121	128	130	139	139	157

No kyphosis	1	1	2	8	11	18
	22	31	37	61	72	81
	97	112	118	127	131	140
	151	159	177	206		

Use the accompanying MINITAB logistic regression output to decide whether age appears to have a significant impact on the presence of kyphosis.

25. The following data resulted from a study commissioned by a large management consulting company to investigate the relationship between amount of job experience (months) for a junior consultant and the likelihood of the consultant being able to perform a certain complex task.

Success	8	13	14	18	20	21	21	22	25
	26	28	29	30	32				
Failure	4	5	6	6	7	9	10	11	11
	13	15	18	19	20	23	27		

Interpret the accompanying MINITAB logistic regression output, and sketch a graph of the estimated probability of task performance as a function of experience.

Logistic Regression Table for Exercise 24

Predictor	Coef	StDev	Z	P	Odds Ratio	95% Lower	CI Upper
Constant	-0.5727	0.6024	-0.95	0.342			
age	0.004296	0.005849	0.73	0.463	1.00	0.99	1.02

Logistic Regression Table for Exercise 25

Predictor	Coef	StDev	Z	P	Odds Ratio	95% Lower	CI Upper
Constant	-3.211	1.235	-2.60	0.009			
age	0.17772	0.06573	2.70	0.007	1.19	1.05	1.36

13.3 Polynomial Regression

The nonlinear yet intrinsically linear models of Section 13.2 involved functions of the independent variable x that were either strictly increasing or strictly decreasing. In many situations, either theoretical reasoning or else a scatter plot of the data suggests that the true regression function $\mu_{Y \cdot x}$ has one or more peaks or valleys—that is, at least one relative minimum or maximum. In such cases, a polynomial function $y = \beta_0 + \beta_1 x + \cdots + \beta_k x^k$ may provide a satisfactory approximation to the true regression function.

DEFINITION

The **kth-degree polynomial regression model equation** is

$$Y = \beta_0 + \beta_1 x + \beta_2 x^2 + \cdots + \beta_k x^k + \epsilon \tag{13.6}$$

where ϵ is a normally distributed random variable with

$$\mu_\epsilon = 0 \qquad \sigma_\epsilon^2 = \sigma^2 \tag{13.7}$$

From (13.6) and (13.7), it follows immediately that

$$\mu_{Y \cdot x} = \beta_0 + \beta_1 x + \cdots + \beta_k x^k \qquad \sigma_{Y \cdot x}^2 = \sigma^2 \qquad (13.8)$$

In words, the expected value of Y is a kth-degree polynomial function of x, whereas the variance of Y, which controls the spread of observed values about the regression function, is the same for each value of x. The observed pairs $(x_1, y_1), \ldots, (x_n, y_n)$ are assumed to have been generated independently from the model (13.6). Figure 13.9 illustrates both a quadratic and cubic model.

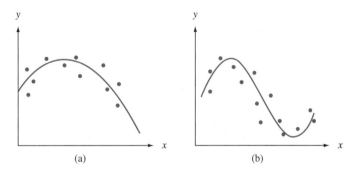

Figure 13.9 (a) Quadratic regression model; (b) cubic regression model

Estimation of Parameters Using Least Squares

To estimate $\beta_0, \beta_1, \ldots, \beta_k$, consider a trial regression function $y = b_0 + b_1 x + \cdots + b_k x^k$. Then the goodness of fit of this function to the observed data can be assessed by computing the sum of squared deviations

$$f(b_0, b_1, \ldots, b_k) = \sum_{i=1}^{n} [y_i - (b_0 + b_1 x_i + b_2 x_i^2 + \cdots + b_k x_i^k)]^2 \qquad (13.9)$$

According to the principle of least squares, the estimates $\hat{\beta}_0, \hat{\beta}_1, \ldots, \hat{\beta}_k$ are those values of $b_0, b_1, \ldots, b_k$ that minimize Expression (13.9). It should be noted that when $x_1, x_2, \ldots, x_n$ are all different, there is a polynomial of degree $n - 1$ that fits the data perfectly, so that the minimizing value of (13.9) is zero when $k = n - 1$. However, in virtually all applications, the polynomial model (13.6) with k large is quite unrealistic, and in most applications $k = 2$ (quadratic) or $k = 3$ (cubic) is appropriate.

To find the minimizing values in (13.9), we take the $k + 1$ partial derivatives $\partial f / \partial b_0, \partial f / \partial b_1, \ldots, \partial f / \partial b_k$ and equate them to 0, resulting in the system of normal equations for the estimates. Because the trial function $b_0 + b_1 x + \cdots + b_k x^k$ is linear in $b_0, \ldots, b_k$ (though not in x), the $k + 1$ normal equations are linear in these unknowns:

$$
\begin{aligned}
b_0 n + b_1 \Sigma x_i + b_2 \Sigma x_i^2 + \cdots + b_k \Sigma x_i^k &= \Sigma y_i \\
b_0 \Sigma x_i + b_1 \Sigma x_i^2 + b_2 \Sigma x_i^3 + \cdots + b_k \Sigma x_i^{k+1} &= \Sigma x_i y_i \\
\vdots \qquad \vdots \qquad \vdots \qquad \qquad \vdots \qquad & \\
b_0 \Sigma x_i^k + b_1 \Sigma x_i^{k+1} + \cdots + b_k \Sigma x_i^{2k} &= \Sigma x_i^k y_i
\end{aligned}
\qquad (13.10)
$$

All standard statistical computer packages will automatically solve the equations in (13.10) and provide the estimates as well as much other information.*

Example 13.6 The article "Determination of Biological Maturity and Effect of Harvesting and Drying Conditions on Milling Quality of Paddy" (*J. Agricultural Eng. Research,* 1975: 353–361) reports the following data on date x of harvesting (number of days after flowering) and yield y (kg/ha) of paddy, a grain farmed in India:

x	16	18	20	22	24	26	28	30
y	2508	2518	3304	3423	3057	3190	3500	3883

x	32	34	36	38	40	42	44	46
y	3823	3646	3708	3333	3517	3241	3103	2776

The scatter plot in Figure 13.10(a) suggests that the author's choice of a quadratic model is reasonable. Figure 13.10(b) also gives MINITAB output from a quadratic regression, from which

$$\hat{\beta}_0 = -1070.4 \qquad \hat{\beta}_1 = 293.48 \qquad \hat{\beta}_2 = -4.5358$$

and the estimated regression function is

$$y = -1070.4 + 293.48x - 4.5358x^2$$

The estimated regression functions can be used to compute predicted values (substitute $x_1 = 16$, then $x_2 = 18$, etc., to obtain $\hat{y}_1, \hat{y}_2, \ldots$) and then residuals $(y_i - \hat{y}_i)$. A residual plot indicates that a quadratic regression function is appropriate, though there is somewhat more spread in the residuals for small x than for large x. Weighted least squares can be used to obtain a more satisfactory fit, but the authors did not consider this.

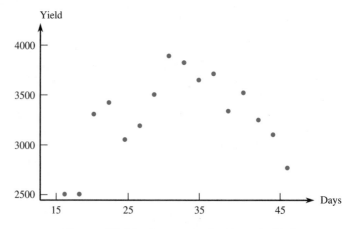

Figure 13.10 Scatter plot for Example 13.6

---

*We will see in Section 13.4 that polynomial regression is a special case of multiple regression, so a command appropriate for this latter task is generally used.

```
THE REGRESSION EQUATION is YIELD = -1070 + 293 DAYS - 4.54 DAYSSQD

Predictor       Coef      Stdev     t-ratio        p
Constant      -1070.4      617.3      -1.73     0.107
DAYS           293.48      42.18       6.96     0.000
DAYSSQD       -4.5358     0.6744      -6.73     0.000

s = 203.9    R-sq = 79.4%    R-sq(adj) = 76.2%

Analysis of Variance

SOURCE          DF         SS         MS        F         p
Regression       2    2084779    1042389    25.08     0.000
Error           13     540388      41568
Total           15    2625167

Obs.   DAYS     YIELD       Fit    Stdev.Fit     Residual    St.Resid
  1    16.0    2508.0    2464.2      135.6         43.8        0.29
  2    18.0    2518.0    2742.7      104.8       -224.7       -1.28
  3    20.0    3304.0    2984.9       83.0        319.1        1.71
  4    22.0    3423.0    3190.9       71.3        232.1        1.22
  5    24.0    3057.0    3360.6       68.4       -303.6       -1.58
  6    26.0    3190.0    3494.0       70.7       -304.0       -1.59
  7    28.0    3500.0    3591.1       74.2        -91.1       -0.48
  8    30.0    3883.0    3651.9       76.4        231.1        1.22
  9    32.0    3823.0    3676.4       76.4        146.6        0.78
 10    34.0    3646.0    3664.6       74.2        -18.6       -0.10
 11    36.0    3708.0    3616.6       70.7         91.4        0.48
 12    38.0    3333.0    3532.3       68.4       -199.3       -1.04
 13    40.0    3517.0    3411.6       71.3        105.4        0.55
 14    42.0    3241.0    3254.7       83.0        -13.7       -0.07
 15    44.0    3103.0    3061.5      104.8         41.5        0.24
 16    46.0    2776.0    2832.1      135.6        -56.1       -0.37

   Fit    Stdev.Fit        95% C.I.              95% P.I.
 3431.8        69.2     (3282.2, 3581.4)    (2966.5, 3897.1)
```

Figure 13.10(b)   MINITAB output for Example 13.6   ∎

$\hat{\sigma}^2$ and R^2

To make further inferences about the parameters of the regression function, the error variance σ^2 must be estimated. With $\hat{y}_i = \hat{\beta}_0 + \hat{\beta}_1 x_i + \cdots + \hat{\beta}_k x_i^k$ the ith residual is $y_i - \hat{y}_i$, and the sum of squared residuals (error sum of squares) is SSE $= \Sigma(y_i - \hat{y}_i)^2$. The estimate of σ^2 is then

$$\hat{\sigma}^2 = s^2 = \frac{\text{SSE}}{n - (k + 1)} = \text{MSE} \qquad (13.11)$$

where the denominator $n - (k + 1)$ is used because $k + 1$ df are lost in estimating β_0, $\beta_1, \ldots, \beta_k$.

If we again let SST $= \Sigma(y_i - \bar{y})^2$, then SSE/SST is the proportion of the total variation in the observed y_i's that is not explained by the polynomial model. The quantity $1 - \text{SSE/SST}$, the proportion of variation explained by the model, is called the **coefficient of multiple determination** and is denoted by R^2.

Suppose we consider fitting a cubic model to the data in Example 13.6. Because the cubic model includes the quadratic as a special case, the fit to a cubic will be at least as good as the fit to a quadratic. More generally, with SSE_k = the error sum of squares from a kth-degree polynomial, $\text{SSE}_{k'} \le \text{SSE}_k$ and $R_{k'}^2 \ge R_k^2$ whenever $k' > k$. Because the objective of regression analysis is to find a model that is both simple (relatively few parameters) and provides a good fit to the data, a higher-degree polynomial may not specify a better model than a lower-degree model despite its higher R^2 value. To balance the cost of using more parameters against the gain in R^2, many statisticians use the **adjusted coefficient of multiple determination**

$$\text{adjusted } R^2 = 1 - \frac{n-1}{n-(k+1)} \cdot \frac{\text{SSE}}{\text{SST}} = \frac{(n-1)R^2 - k}{n-1-k} \qquad (13.12)$$

Adjusted R^2 adjusts the proportion of unexplained variation upward [since $(n-1)/(n-k-1) > 1$], which results in adjusted $R^2 < R^2$. Thus if $R_2^2 = .66$, $R_3^2 = .70$, and $n = 10$, then

$$\text{adjusted } R_2^2 = \frac{9(.66) - 2}{10 - 3} = .563 \qquad \text{adjusted } R_3^2 = \frac{9(.70) - 3}{10 - 4} = .550$$

so the small gain in R^2 in going from a quadratic to a cubic model is not enough to offset the cost of adding an extra parameter to the model.

Example 13.7
(Example 13.6 continued)

SSE and SST are typically found on computer output in an ANOVA table. Figure 13.10 gives SSE = 540,388 and SST = 2,625,167 for the paddy yield data, from which

$$R^2 = 1 - \frac{\text{SSE}}{\text{SST}} = 1 - \frac{540,388}{2,625,167} = 1 - .206 = .794$$

Thus, 79.4% of the observed variation in yield can be attributed to the model relationship. Adjusted $R^2 = .762$, only a small downward change in R^2. The estimates of σ^2 and σ are

$$s^2 = \frac{\text{SSE}}{n-(k+1)} = \frac{540,388}{16-(2+1)} = 41,568.3$$

$$s = \sqrt{s^2} = 203.9 \qquad \blacksquare$$

Besides computing R^2 and adjusted R^2, one should examine the usual diagnostic plots to determine whether model assumptions are valid or whether modification may be appropriate.

Confidence Intervals and Test Procedures

Because the y_i's appear in the normal equations (13.10) only on the right-hand side and in a linear fashion, the resulting estimates $\hat{\beta}_0, \ldots, \hat{\beta}_k$ are themselves linear functions of the y_i's. Thus, the estimators are linear functions of the Y_i's, so each $\hat{\beta}_i$ has a normal distribution. It can also be shown that each $\hat{\beta}_i$ is an unbiased estimator of β_i.

Let $\sigma_{\hat{\beta}_i}$ denote the standard deviation of the estimator $\hat{\beta}_i$. This standard deviation has the form

$$\sigma_{\hat{\beta}_i} = \sigma \cdot \left\{ \begin{array}{l} \text{a complicated expression involving } \textit{all} \\ x_j\text{'s, } x_j^2\text{'s, } \ldots, \text{ and } x_j^k\text{'s} \end{array} \right\}$$

Fortunately, the expression in braces has been programmed into all of the most frequently used statistical computer packages. The estimated standard deviation of $\hat{\beta}_i$, $s_{\hat{\beta}_i}$, results from substituting s in place of σ in the expression for $\sigma_{\hat{\beta}_i}$. These estimated standard deviations $s_{\hat{\beta}_0}$, $s_{\hat{\beta}_1}$, . . . , and $s_{\hat{\beta}_k}$ appear on output from all the aforementioned statistical packages. Let $S_{\hat{\beta}_i}$ denote the estimator of $\sigma_{\hat{\beta}_i}$—that is, the random variable whose observed value is $s_{\hat{\beta}_i}$. Then it can be shown that the standardized variable

$$T = \frac{\hat{\beta}_i - \beta_i}{S_{\hat{\beta}_i}} \qquad (13.13)$$

has a t distribution based on $n - (k + 1)$ df. This leads to the following inferential procedures.

A $100(1 - \alpha)\%$ CI for β_i, the coefficient of x^i in the polynomial regression function, is

$$\hat{\beta}_i \pm t_{\alpha/2, n-(k+1)} \cdot s_{\hat{\beta}_i}$$

A test of H_0: $\beta_i = \beta_{i0}$ is based on the t statistic value

$$t = \frac{\hat{\beta}_i - \beta_{i0}}{s_{\hat{\beta}_i}}$$

The test is based on $n - (k + 1)$ df and is upper-, lower-, or two-tailed according to whether the inequality in H_a is $>$, $<$, or $\neq$.

A point estimate of $\mu_{Y \cdot x}$—that is, of $\beta_0 + \beta_1 x + \cdots + \beta_k x^k$—is $\hat{\mu}_{Y \cdot x} = \hat{\beta}_0 + \hat{\beta}_1 x + \cdots + \hat{\beta}_k x^k$. The estimated standard deviation of the corresponding estimator is rather complicated. Many computer packages will give this estimated standard deviation for *any* x value when requested to do so by a user. This, along with an appropriate standardized t variable, can be used to justify the following procedures.

Let x^* denote a specified value of x. A $100(1 - \alpha)\%$ CI for $\mu_{Y \cdot x^*}$ is

$$\hat{\mu}_{Y \cdot x^*} \pm t_{\alpha/2, n-(k+1)} \cdot \left\{ \begin{array}{c} \text{estimated SD of} \\ \hat{\mu}_{Y \cdot x^*} \end{array} \right\}$$

With $\hat{Y} = \hat{\beta}_0 + \hat{\beta}_1 x^* + \cdots + \hat{\beta}_k (x^*)^k$, $\hat{y}$ denoting the calculated value of $\hat{Y}$ for the given data, and $s_{\hat{y}}$ denoting the estimated standard deviation of the statistic $\hat{Y}$, the formula for the CI is much like the one in the case of simple linear regression:

$$\hat{y} \pm t_{\alpha/2, n-(k+1)} \cdot s_{\hat{y}}$$

A $100(1 - \alpha)\%$ PI for a future y value to be observed when $x = x^*$ is

$$\hat{\mu}_{Y \cdot x^*} \pm t_{\alpha/2, n-(k+1)} \cdot \left\{ s^2 + \left(\begin{array}{c} \text{estimated SD} \\ \text{of } \hat{\mu}_{Y \cdot x^*} \end{array} \right)^2 \right\}^{1/2} = \hat{y} \pm t_{\alpha/2, n-(k+1)} \cdot \sqrt{s^2 + s_{\hat{Y}}^2}$$

Example 13.8 Figure 13.10(b) shows that $\hat{\beta}_2 = -4.5358$ and $s_{\hat{\beta}_2} = .6744$ (from the Stdev column at the top of the output). For testing $H_0: \beta_2 = 0$ versus $H_a: \beta_2 \neq 0$ (H_0 here says that the quadratic term in the model is unnecessary), the test statistic is $T = \hat{\beta}_2/S_{\hat{\beta}_2}$, with computed value $-4.5358/.6744 = -6.73$. The test is based on $n - (k + 1) = 16 - 3 = 13$ df; with $t_{.025,13} = 2.160$, H_0 is rejected at level .05 if either $t \geq 2.160$ or $t \leq -2.160$. Since $-6.73 \leq -2.160$, H_0 is rejected at level .05, validating the inclusion of the quadratic term; the corresponding P-value is zero.

The last line of output in Figure 13.10(b) contains estimation and prediction information for $x = 25$:

$$\hat{y} = \hat{\beta}_0 + \hat{\beta}_1(25) + \hat{\beta}_2(25)^2 = \text{Fit} = 3431.8$$

$$\left(\begin{array}{c} \text{estimated SD of} \\ \hat{\beta}_0 + \hat{\beta}_1(25) + \hat{\beta}_2(25)^2 \end{array}\right) = s_{\hat{y}} = \text{Stdev.Fit} = 69.2$$

Thus, a 95% CI for mean yield when date of harvesting $= 25$ is

$$3431.8 \pm (2.160)(69.2) = 3431.8 \pm 149.5 = (3282.3, 3581.3)$$

whereas a 95% PI for yield when $x = 25$ is

$$3431.8 \pm (2.160)[41{,}568.3 + (69.2)^2]^{1/2} = 3431.8 \pm 465.1$$
$$= (2966.7, 3896.9)$$

The slight discrepancies between these interval limits and those appearing on the output are attributable to rounding. Notice again that the PI is much wider than the corresponding CI. This is because s^2 is quite large compared to the Stdev.Fit. ∎

Centering x Values

For the quadratic model with regression function $\mu_{Y \cdot x} = \beta_0 + \beta_1 x + \beta_2 x^2$, the parameters β_0, β_1, and β_2 characterize the behavior of the function near $x = 0$. For example, β_0 is the height at which the regression function crosses the vertical axis $x = 0$, whereas β_1 is the first derivative of the function at $x = 0$ (instantaneous rate of change of $\mu_{Y \cdot x}$ at $x = 0$). If the x_i's all lie far from 0, we may not have precise information about the values of these parameters. Let $\bar{x} =$ the average of the x_i's for which observations are to be taken and consider the model

$$Y = \beta_0^* + \beta_1^*(x - \bar{x}) + \beta_2^*(x - \bar{x})^2 + \epsilon \tag{13.14}$$

In the model (13.14), $\mu_{Y \cdot x} = \beta_0^* + \beta_1^*(x - \bar{x}) + \beta_2^*(x - \bar{x})^2$, and the parameters now describe the behavior of the regression function near the center $\bar{x}$ of the data.

To estimate the parameters of (13.14), we simply subtract $\bar{x}$ from each x_i to obtain $x_i' = x_i - \bar{x}$, and then use the x_i''s in place of the x_i's. An important benefit of this is that the coefficients of $b_0, \ldots, b_k$ in the normal equations (13.10) will be of much smaller magnitude than would be the case were the original x_i's used. When the system is solved by computer, this centering protects against any round-off error that may result.

Example 13.9 The article "A Method for Improving the Accuracy of Polynomial Regression Analysis" (*J. Quality Technology*, 1971: 149–155) reports the following data on $x =$ cure temperature (°F) and y = ultimate shear strength of a rubber compound (psi), with $\bar{x} = 297.13$:

x	280	284	292	295	298	305	308	315
x'	−17.13	−13.13	−5.13	−2.13	.87	7.87	10.87	17.87
y	770	800	840	810	735	640	590	560

A computer analysis yielded the results shown in Table 13.4.

Table 13.4 Estimated coefficients and standard deviations for Example 13.9

Parameter	Estimate	Estimated SD	Parameter	Estimate	Estimated SD
β_0	−26,219.64	11,912.78	β_0^*	759.36	23.20
β_1	189.21	80.25	β_1^*	−7.61	1.43
β_2	−.3312	.1350	β_2^*	−.3312	.1350

The estimated regression function using the original model is $y = -26{,}219.64 + 189.21x - .3312x^2$, whereas for the centered model the function is $y = 759.36 - 7.61(x - 297.13) - .3312(x - 297.13)^2$. These estimated functions are identical; the only difference is that different parameters have been estimated for the two models. The estimated standard deviations indicate clearly that β_0^* and β_1^* have been more accurately estimated than β_0 and β_1. The quadratic parameters are identical ($\beta_2 = \beta_2^*$), as can be seen by comparing the x^2 term in (13.14) with the original model. We emphasize again that a major benefit of centering is the gain in computational accuracy, not only in quadratic but also in higher-degree models. ∎

The book by Neter et al., listed in the chapter bibliography, is a good source for more information about polynomial regression.

Exercises | Section 13.3 (26–35)

26. In addition to a linear regression of true density on moisture content, the article cited in Exercise 6 considered a quadratic regression of bulk density versus moisture content. Data from a graph in the article follows, along with a MINITAB output from the quadratic fit.

The regression equation is
bulkdens = 403 + 16.2 moiscont − 0.706 contsqd

Predictor	Coef	StDev	T	P
Constant	403.24	36.45	11.06	0.002
moiscont	16.164	5.451	2.97	0.059
contsqd	−0.7063	0.1852	−3.81	0.032

S = 10.15   R-Sq = 93.8%   R-Sq(adj) = 89.6%

Analysis of Variance

Source	DF	SS	MS	F	P
Regression	2	4637.7	2318.9	22.51	0.016
Residual Error	3	309.1	103.0		
Total	5	4946.8			

Obs	moiscont	bulkdens	Fit	StDev Fit	Residual	St Resid
1	7.0	479.00	481.78	9.35	−2.78	−0.70
2	10.3	503.00	494.79	5.78	8.21	0.98
3	13.7	487.00	492.12	6.49	−5.12	−0.66
4	16.6	470.00	476.93	6.10	−6.93	−0.85
5	19.8	458.00	446.39	5.69	11.61	1.38
6	22.0	412.00	416.99	8.75	−4.99	−0.97

Fit	StDev Fit	95.0% CI	95.0% PI
491.10	6.52	(470.36, 511.83)	(452.71, 529.48)

a. Does a scatter plot of the data appear consistent with the quadratic regression model?

b. What proportion of observed variation in density can be attributed to the model relationship?

c. Does the quadratic model appear to be useful? Carry out a test at significance level .05.

d. The last line of output is from a request for estimation and prediction information when moisture content is 14. Calculate a 99% PI for density when moisture content is 14.

e. Does the quadratic predictor appear to provide useful information? Test the appropriate hypotheses at significance level .05.

27. The following data on $y = $ glucose concentration (g/L) and $x = $ fermentation time (days) for a particular blend of malt liquor was read from a scatter plot in the article "Improving Fermentation Productivity with Reverse Osmosis" (*Food Tech.*, 1984: 92–96):

x	1	2	3	4	5	6	7	8
y	74	54	52	51	52	53	58	71

a. Verify that a scatter plot of the data is consistent with the choice of a quadratic regression model.

b. The estimated quadratic regression equation is $y = 84.482 - 15.875x + 1.7679x^2$. Predict the value of glucose concentration for a fermentation time of 6 days and compute the corresponding residual.

c. Using SSE = 61.77, what proportion of observed variation can be attributed to the quadratic regression relationship?

d. The $n = 8$ standardized residuals based on the quadratic model are 1.91, −1.95, −.25, .58, .90, .04, −.66, and .20. Construct a plot of the standardized residuals versus x and a normal probability plot. Do the plots exhibit any troublesome features?

e. The estimated standard deviation of $\hat{\mu}_{Y \cdot 6}$—that is, $\hat{\beta}_0 + \hat{\beta}_1(6) + \hat{\beta}_2(36)$—is 1.69. Compute a 95% CI for $\mu_{Y \cdot 6}$.

f. Compute a 95% PI for a glucose concentration observation made after 6 days of fermentation time.

28. The viscosity (y) of an oil was measured by a cone and plate viscometer at six different cone speeds (x). It was assumed that a quadratic regression model was appropriate, and the estimated regression function resulting from the $n = 6$ observations was

$$y = -113.0937 + 3.3684x - .01780x^2$$

a. Estimate $\mu_{Y \cdot 75}$, the expected viscosity when speed is 75 rpm.

b. What viscosity would you predict for a cone speed of 60 rpm?

c. If $\sum y_i^2 = 8386.43$, $\sum y_i = 210.70$, $\sum x_i y_i = 17,002.00$, and $\sum x_i^2 y_i = 1,419,780$, compute SSE $[= \sum y_i^2 - \hat{\beta}_0 \sum y_i - \hat{\beta}_1 \sum x_i y_i - \hat{\beta}_2 \sum x_i^2 y_i]$, s^2, and s.

d. From part (c), SST $= 8386.43 - (210.70)^2/6 = 987.35$. Using SSE computed in part (c), what is the computed value of R^2?

e. If the estimated standard deviation of $\hat{\beta}_2$ is $s_{\hat{\beta}_2} = .00226$, test $H_0: \beta_2 = 0$ versus $H_a: \beta_2 \neq 0$ at level .01.

29. Exercise 7 presented the following data on exposure time to radiation x and dry weight of roots y:

x	0	2	4	6	8
y	110	123	119	86	62

The estimated quadratic regression function is $y = 111.89 + 8.06x - 1.84x^2$.

a. Compute the predicted values and residuals. Then compute SSE and s^2. [See Exercise 28(c).]

b. Compute the coefficient of multiple determination R^2.

c. The estimated standard deviation of $\hat{\beta}_2$, the estimator of the quadratic coefficient β_2, is $s_{\hat{\beta}_2} = .480$. Does the quadratic term belong in the model? State and test the appropriate hypotheses at level .05.

d. The estimated standard deviation of $\hat{\beta}_1$ is $s_{\hat{\beta}_1} = 4.01$. Use this and the information in part (c) to obtain joint CIs for β_1 and β_2 with joint confidence level (at least) 95%.

e. The estimated standard deviation of $\hat{\mu}_{Y \cdot 4}$ ($= \hat{\beta}_0 + 4\hat{\beta}_1 + 16\hat{\beta}_2$) is 5.01. Compute a 90% CI for $\mu_{Y \cdot 4}$.

f. Estimate the exposure time that maximizes expected dry weight of roots.

30. The article "A Simulation-Based Evaluation of Three Cropping Systems on Cracking-Clay Soils in a Summer Rainfall Environment" (*Agricultural Meteorology*, 1976: 211–229) proposes a quadratic model for the relationship between water supply index (x) and farm wheat yield (y). Representative data and the resulting MINITAB output follow:

x	1.2	1.3	1.5	1.8	2.1	2.3	2.5
y	790	950	740	1230	1000	1465	1370

(continued)

x	2.9	3.1	3.2	3.3	3.9	4.0	4.3
y	1420	1625	1600	1720	1500	1550	1560

```
The regression equation is
yield = -252 + 1000 index - 135 indexsqd

Predictor    Coef    Stdev    t-ratio       p
Constant   -251.6    285.1      -0.88   0.396
Index      1000.1    229.5       4.36   0.001
Indexsqd  -135.44    41.97      -3.23   0.008

s = 135.6   R-sq = 85.3%   R-sq(adj) = 82.6%

Analysis of Variance

SOURCE       DF       SS      MS      F       p
Regression    2  1170208  585104  31.83   0.000
Error        11   202228   18384
Total        13  1372435
```

a. Interpret the value of the coefficient of multiple determination.

b. Calculate a 95% CI for the coefficient of the quadratic predictor.

c. The estimated standard deviation of $\hat{\beta}_0 + \hat{\beta}_1 x + \hat{\beta}_2 x^2$ when $x = 2.5$ is 53.5. Test $H_0: \mu_{Y\cdot 2.5} = 1500$ versus $H_a: \mu_{Y\cdot 2.5} < 1500$ using $\alpha = .01$.

d. Obtain a 95% PI for wheat yield when the water supply index is 2.5 by using the information given in part (c).

31. The accompanying data was obtained from a study of a certain method for preparing pure alcohol from refinery streams ("Direct Hydration of Olefins," *Industrial and Eng. Chemistry,* 1961: 209–211). The independent variable x is volume hourly space velocity, and the dependent variable y is the amount of conversion of iso-butylene.

x	1	1	2	4	4	4	6
y	23.0	24.5	28.0	30.9	32.0	33.6	20.0

a. Assuming that a quadratic probabilistic model is appropriate, estimate the regression function.

b. Determine the predicted values and residuals and construct a residual plot. Does the plot look roughly as expected when the quadratic model is correct? Does the plot indicate that any observation has had a great influence on the fit? Does a scatter plot identify a point having large influence? If so, which point?

c. Obtain s^2 and R^2. Does the quadratic model provide a good fit to the data?

d. In Exercise 11, it was noted that the predicted value $\hat{Y}_j$ and the residual $Y_j - \hat{Y}_j$ are indepen-

dent of one another, so that $\sigma^2 = V(Y_j) = V(\hat{Y}_j) + V(Y_j - \hat{Y}_j)$. A computer printout gives the estimated standard deviations of the predicted values as .955, .955, .712, .777, .777, .777, and 1.407. Use these values along with s^2 to compute the estimated standard deviation of each residual. Then compute the standardized residuals and plot them against x. Does the plot look much like the plot of part (b)? Suppose you had standardized the residuals using just s in the denominator. Would the resulting values differ much from the correct values?

e. Using information given in part (d), compute a 90% PI for isobutylene conversion when volume hourly space velocity is 4.

32. The following data is a subset of data obtained in an experiment to study the relationship between soil pH x and $y =$ A1. Concentration/EC ("Root Responses of Three *Gramineae* Species to Soil Acidity in an Oxisol and an Ultisol," *Soil Science,* 1973: 295–302):

x	4.01	4.07	4.08	4.10	4.18	
y	1.20	.78	.83	.98	.65	

x	4.20	4.23	4.27	4.30	4.41	
y	.76	.40	.45	.39	.30	

x	4.45	4.50	4.58	4.68	4.70	4.77
y	.20	.24	.10	.13	.07	.04

A cubic model was proposed in the article, but the version of MINITAB used by the author of the present text refused to include the x^3 term in the model, stating that "x^3 is highly correlated with other predictor variables." To remedy this, $\bar{x} = 4.3456$ was subtracted from each x value to yield $x' = x - \bar{x}$. A cubic regression was then requested to fit the model having regression function:

$$y = \beta_0^* + \beta_1^* x' + \beta_2^* (x')^2 + \beta_3^* (x')^3$$

The following computer output resulted:

Parameter	Estimate	Estimated SD
β_0^*	.3463	.0366
β_1^*	-1.2933	.2535
β_2^*	2.3964	.5699
β_3^*	-2.3968	2.4590

a. What is the estimated regression function for the "centered" model?

b. What is the estimated value of the coefficient β_3 in the "uncentered" model with regression function $y = \beta_0 + \beta_1 x + \beta_2 x^2 + \beta_3 x^3$? What is the estimate of β_2?

c. Using the cubic model, what value of y would you predict when soil pH is 4.5?

d. Carry out a test to decide whether the cubic term should be retained in the model.

33. In many polynomial regression problems, rather than fitting a "centered" regression function using $x' = x - \bar{x}$, computational accuracy can be improved by using a function of the standardized independent variable $x' = (x - \bar{x})/s_x$, where s_x is the standard deviation of the x_i's. Consider fitting the cubic regression function $y = \beta_0^* + \beta_1^* x' + \beta_2^*(x')^2 + \beta_3^*(x')^3$ to the following data resulting from a study of the relation between thrust efficiency y of supersonic propelling rockets and the half-divergence angle x of the rocket nozzle ("More on Correlating Data," *CHEMTECH*, 1976: 266–270):

x	5	10	15	20	25	30	35
y	.985	.996	.988	.962	.940	.915	.878

Parameter	Estimate	Estimated SD
β_0^*	.9671	.0026
β_1^*	−.0502	.0051
β_2^*	−.0176	.0023
β_3^*	.0062	.0031

a. What value of y would you predict when the half-divergence angle is 20? When $x = 25$?

b. What is the estimated regression function $\hat{\beta}_0 + \hat{\beta}_1 x + \hat{\beta}_2 x^2 + \hat{\beta}_3 x^3$ for the "unstandardized" model?

c. Use a level .05 test to decide whether the cubic term should be deleted from the model.

d. What can you say about the relationship between SSE's and R^2's for the standardized and unstandardized models? Explain.

e. SSE for the cubic model is .00006300, whereas for a quadratic model SSE is .00014367. Compute R^2 for each model. Does the difference between the two suggest that the cubic term can be deleted?

34. The following data resulted from an experiment to assess the potential of unburnt colliery spoil as a medium for plant growth. The variables are $x =$ acid extractable cations and $y =$ exchangeable acidity/total cation exchange capacity ("Exchangeable Acidity in Unburnt Colliery Spoil," *Nature*, 1969: 161):

x	−23	−5	16	26	30	38	52
y	1.50	1.46	1.32	1.17	.96	.78	.77

x	58	67	81	96	100	113
y	.91	.78	.69	.52	.48	.55

Standardizing the independent variable x to obtain $x' = (x - \bar{x})/s_x$ and fitting the regression function $y = \beta_0^* + \beta_1^* x' + \beta_2^*(x')^2$ yielded the accompanying computer output:

Parameter	Estimate	Estimated SD
β_0^*	.8733	.0421
β_1^*	−.3255	.0316
β_2^*	.0448	.0319

a. Estimate $\mu_{Y \cdot 50}$.

b. Compute the value of the coefficient of multiple determination. [See Exercise 28(c).]

c. What is the estimated regression function $\hat{\beta}_0 + \hat{\beta}_1 x + \hat{\beta}_2 x^2$ using the unstandardized variable x?

d. What is the estimated standard deviation of $\hat{\beta}_2$ computed in part (c)?

e. Carry out a test using the standardized estimates to decide whether the quadratic term should be retained in the model. Repeat using the unstandardized estimates. Do your conclusions differ?

35. The article "The Respiration in Air and in Water of the Limpets *Patella caerulea* and *Patella lusitanica*" (*Comp. Biochemistry and Physiology*, 1975: 407–411) proposed a simple power model for the relationship between respiration rate y and temperature x for *P. caerulea* in air. However, a plot of $\ln(y)$ versus x exhibits a curved pattern. Fit the quadratic power model $Y = \alpha e^{\beta x + \gamma x^2} \cdot \epsilon$ to the accompanying data:

x	10	15	20	25	30
y	37.1	70.1	109.7	177.2	222.6

13.4 | Multiple Regression Analysis

In multiple regression, the objective is to build a probabilistic model that relates a dependent variable y to more than one independent or predictor variable. Let k represent the number of predictor variables ($k \geq 2$) and denote these predictors by $x_1, x_2, \ldots, x_k$. For example, in attempting to predict the selling price of a house, we might have $k = 3$ with $x_1 = $ size (ft^2), $x_2 = $ age (years), and $x_3 = $ number of rooms.

DEFINITION

The **general additive multiple regression model equation** is

$$Y = \beta_0 + \beta_1 x_1 + \beta_2 x_2 + \cdots + \beta_k x_k + \epsilon \qquad (13.15)$$

where $E(\epsilon) = 0$ and $V(\epsilon) = \sigma^2$. In addition, for purposes of testing hypotheses and calculating CIs or PIs, it is assumed that ϵ is normally distributed.

Let $x_1^*, x_2^*, \ldots, x_k^*$ be particular values of $x_1, \ldots, x_k$. Then (13.15) implies that

$$\mu_{Y \cdot x_1^*, \ldots, x_k^*} = \beta_0 + \beta_1 x_1^* + \cdots + \beta_k x_k^* \qquad (13.16)$$

Thus, just as $\beta_0 + \beta_1 x$ describes the mean Y value as a function of x in simple linear regression, the **true** (or **population**) **regression function** $\beta_0 + \beta_1 x_1 + \cdots + \beta_k x_k$ gives the expected value of Y as a function of $x_1, \ldots, x_k$. The β_i's are the **true** (or **population**) **regression coefficients.** The regression coefficient β_1 is interpreted as the expected change in Y associated with a 1-unit increase in x_1 *while $x_2, \ldots, x_k$ are held fixed.* Analogous interpretations hold for $\beta_2, \ldots, \beta_k$.

Models with Interaction and Quadratic Predictors

If an investigator has obtained observations on y, x_1, and x_2, one possible model is $Y = \beta_0 + \beta_1 x_1 + \beta_2 x_2 + \epsilon$. However, it is possible to construct other models by forming predictors that are mathematical functions of x_1 and/or x_2. For example, with $x_3 = x_1^2$ and $x_4 = x_1 x_2$, the model

$$Y = \beta_0 + \beta_1 x_1 + \beta_2 x_2 + \beta_3 x_3 + \beta_4 x_4 + \epsilon$$

has the general form of (13.15). In general, it is not only permissible for some predictors to be mathematical functions of others but also often highly desirable in the sense that the resulting model may be much more successful in explaining variation in y than any model without such predictors. This discussion also shows that polynomial regression is indeed a special case of multiple regression. For example, the quadratic model $Y = \beta_0 + \beta_1 x + \beta_2 x^2 + \epsilon$ has the form of (13.15) with $k = 2$, $x_1 = x$, and $x_2 = x^2$.

For the case of two independent variables, x_1 and x_2, there are four useful multiple regression models:

1. The first-order model:

$$Y = \beta_0 + \beta_1 x_1 + \beta_2 x_2 + \epsilon$$

2. The second-order no-interaction model:

$$Y = \beta_0 + \beta_1 x_1 + \beta_2 x_2 + \beta_3 x_1^2 + \beta_4 x_2^2 + \epsilon$$

3. The model with first-order predictors and interaction:

$$Y = \beta_0 + \beta_1 x_1 + \beta_2 x_2 + \beta_3 x_1 x_2 + \epsilon$$

4. The complete second-order or full quadratic model:

$$Y = \beta_0 + \beta_1 x_1 + \beta_2 x_2 + \beta_3 x_1^2 + \beta_4 x_2^2 + \beta_5 x_1 x_2 + \epsilon$$

Understanding the differences among these models is an important first step in building realistic regression models from the independent variables under study.

The first-order model is the most straightforward generalization of simple linear regression. It states that, for a fixed value of either variable, the expected value of Y is a linear function of the other variable and that the expected change in Y for a unit increase in x_1 (x_2) is β_1 (β_2) independent of the level of x_2 (x_1). Thus, if we graph the regression function as a function of x_1 for several different values of x_2, we obtain as contours of the regression function a collection of parallel lines, as pictured in Figure 13.11(a). The function $y = \beta_0 + \beta_1 x_1 + \beta_2 x_2$ specifies a plane in three-dimensional space; the first-order model says that each observed value of the dependent variable corresponds to a point which deviates from this plane by a random amount ϵ.

According to the second-order no-interaction model, if x_2 is fixed, the expected change in Y for a 1-unit increase in x_1 is

$$\beta_0 + \beta_1(x_1 + 1) + \beta_2 x_2 + \beta_3(x_1 + 1)^2 + \beta_4 x_2^2$$
$$- (\beta_0 + \beta_1 x_1 + \beta_2 x_2 + \beta_3 x_1^2 + \beta_4 x_2^2) = \beta_1 + \beta_3 + 2\beta_3 x_1$$

Because this expected change does not depend on x_2, the contours of the regression function for different values of x_2 are still parallel to one another. However, the dependence of the expected change on the value of x_1 means that the contours are now curves rather than straight lines. This is pictured in Figure 13.11(b). In this case, the regression surface is no longer a plane in three-dimensional space but is instead a curved surface.

The contours of the regression function for the first-order interaction model are nonparallel straight lines. This is because the expected change in Y when x_1 is increased by 1 is

$$\beta_0 + \beta_1(x_1 + 1) + \beta_2 x_2 + \beta_3(x_1 + 1)x_2 - (\beta_0 + \beta_1 x_1 + \beta_2 x_2 + \beta_3 x_1 x_2) = \beta_1 + \beta_3 x_2$$

This expected change depends on the value of x_2, so each contour line must have a different slope as in Figure 13.11(c). The word *interaction* reflects the fact that an expected change in Y when one variable increases in value depends on the value of the other variable.

Finally, for the complete second-order model, the expected change in Y when x_2 is held fixed while x_1 is increased by 1 unit is $\beta_1 + \beta_3 + 2\beta_3 x_1 + \beta_5 x_2$, which is a function of both x_1 and x_2. This implies that the contours of the regression function are both curved and not parallel to one another, as illustrated in Figure 13.11(d).

Similar considerations apply to models constructed from more than two independent variables. In general, the presence of interaction terms in the model implies that the expected change in Y depends not only on the variable being increased or decreased but

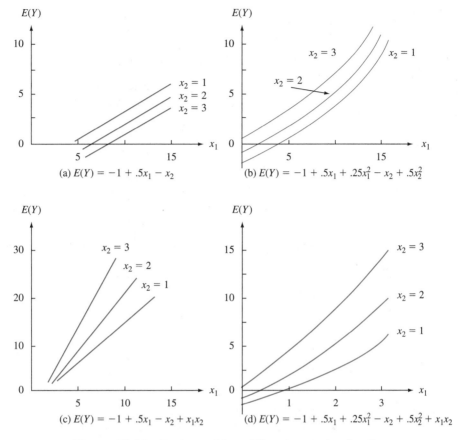

Figure 13.11 Contours of four different regression functions

also on the values of some of the fixed variables. As in ANOVA, it is possible to have higher-way interaction terms (e.g., $x_1 x_2 x_3$), making model interpretation more difficult.

Note that if the model contains interaction or quadratic predictors, the generic interpretation of a β_i given previously will not usually apply. This is because it is not then possible to increase x_i by 1 unit and hold the values of all other predictors fixed.

Models with Predictors for Categorical Variables

Thus far we have explicitly considered the inclusion of only quantitative (numerical) predictor variables in a multiple regression model. Using simple numerical coding, qualitative (categorical) variables, such as bearing material (aluminum or copper/lead) or type of wood (pine, oak, or walnut), can also be incorporated into a model. Let's first focus on the case of a dichotomous variable, one with just two possible categories— male or female, U.S. or foreign manufacture, and so on. With any such variable, we associate a **dummy** or **indicator variable** x whose possible values 0 and 1 indicate which category is relevant for any particular observation.

Example 13.10 The article "Estimating Urban Travel Times: A Comparative Study" (*Trans. Res.,* 1980: 173–175) described a study relating the dependent variable y = travel time between locations in a certain city and the independent variable x_2 = distance between locations. Two types of vehicles, passenger cars and trucks, were used in the study. Let

$$x_1 = \begin{cases} 1 & \text{if the vehicle is a truck} \\ 0 & \text{if the vehicle is a passenger car} \end{cases}$$

One possible multiple regression model is

$$Y = \beta_0 + \beta_1 x_1 + \beta_2 x_2 + \epsilon$$

The mean value of travel time depends on whether a vehicle is a car or a truck:

$$\text{mean time} = \beta_0 + \beta_2 x_2 \qquad \text{when } x_1 = 0 \quad \text{(cars)}$$
$$\text{mean time} = \beta_0 + \beta_1 + \beta_2 x_2 \qquad \text{when } x_1 = 1 \quad \text{(trucks)}$$

The coefficient β_1 is the difference in mean times between trucks and cars with distance held fixed; if $\beta_1 > 0$, on average it will take trucks longer to traverse any particular distance than it will for cars.

A second possibility is a model with an interaction predictor:

$$Y = \beta_0 + \beta_1 x_1 + \beta_2 x_2 + \beta_3 x_1 x_2 + \epsilon$$

Now the mean times for the two types of vehicles are

$$\text{mean time} = \beta_0 + \beta_2 x_2 \qquad\qquad \text{when } x_1 = 0$$
$$\text{mean time} = \beta_0 + \beta_1 + (\beta_2 + \beta_3) x_2 \qquad \text{when } x_1 = 1$$

For each model, the graph of the mean time versus distance is a straight line for either type of vehicle, as illustrated in Figure 13.12. The two lines are parallel for the first (no-interaction) model, but in general they will have different slopes when the second model is correct. For this latter model, the change in mean travel time associated with a 1-mile increase in distance depends on which type of vehicle is involved—the two variables

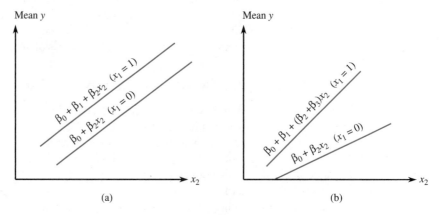

Figure 13.12 Regression functions for models with one dummy variable (x_1) and one quantitative variable x_2: (a) no interaction; (b) interaction

"vehicle type" and "travel time" interact. Indeed, data collected by the authors of the cited article suggested the presence of interaction. ■

You might think that the way to handle a three-category situation is to define a single numerical variable with coded values such as 0, 1, and 2 corresponding to the three categories. This is incorrect, because it imposes an ordering on the categories that is not necessarily implied by the problem context. The correct approach to incorporating three categories is to define *two* different dummy variables. Suppose, for example, that y is the lifetime of a certain cutting tool, x_1 is cutting speed, and that there are three brands of tool being investigated. Then let

$$x_2 = \begin{cases} 1 & \text{if a brand A tool is used} \\ 0 & \text{otherwise} \end{cases} \qquad x_3 = \begin{cases} 1 & \text{if a brand B tool is used} \\ 0 & \text{otherwise} \end{cases}$$

When an observation on a brand A tool is made, $x_2 = 1$ and $x_3 = 0$, whereas for a brand B tool, $x_2 = 0$ and $x_3 = 1$. An observation made on a brand C tool has $x_2 = x_3 = 0$, and it is not possible that $x_2 = x_3 = 1$ because a tool cannot simultaneously be both brand A and brand B. The no-interaction model would have only the predictors x_1, x_2, and x_3. The following interaction model allows the mean change in lifetime associated with a 1-unit increase in speed to depend on the brand of tool:

$$Y = \beta_0 + \beta_1 x_1 + \beta_2 x_2 + \beta_3 x_3 + \beta_4 x_1 x_2 + \beta_5 x_1 x_3 + \epsilon$$

Construction of a picture like Figure 13.12 with a graph for each of the three possible (x_2, x_3) pairs gives three nonparallel lines (unless $\beta_4 = \beta_5 = 0$).

More generally, incorporating a categorical variable with c possible categories into a multiple regression model requires the use of $c - 1$ indicator variables (e.g., five brands of tools would necessitate using four indicator variables). Thus even one categorical variable can add many predictors to a model.

Estimating Parameters

The data in simple linear regression consists of n pairs $(x_1, y_1), \ldots, (x_n, y_n)$. Suppose that a multiple regression model contains two predictor variables x_1 and x_2. Then the data set will consist of n triples $(x_{11}, x_{21}, y_1), (x_{12}, x_{22}, y_2), \ldots, (x_{1n}, x_{2n}, y_n)$. Here the first subscript on x refers to the predictor and the second to the observation number. More generally, with k predictors, the data consists of n $(k + 1)$-tuples $(x_{11}, x_{21}, \ldots, x_{k1}, y_1), (x_{12}, x_{22}, \ldots, x_{k2}, y_2), \ldots, (x_{1n}, x_{2n}, \ldots, x_{kn}, y_n)$, where x_{ij} is the value of the ith predictor x_i associated with the observed value y_j. The y_j's are assumed to have been observed independently of one another according to the model (13.15). To estimate the parameters $\beta_0, \beta_1, \ldots, \beta_k$ using the principle of least squares, form the sum of squared deviations of the observed y_j's from a trial function $y = b_0 + b_1 x_1 + \cdots + b_k x_k$:

$$f(b_0, b_1, \ldots, b_k) = \sum_j [y_j - (b_0 + b_1 x_{1j} + b_2 x_{2j} + \cdots + b_k x_{kj})]^2 \qquad (13.17)$$

The least squares estimates are those values of $b_0, b_1, \ldots, b_k$ that minimize $f(b_0, \ldots, b_k)$. Taking the partial derivative of f with respect to each b_i ($i = 0, 1, \ldots, k$) and equating all partials to zero yields the following system of **normal equations:**

$$b_0 n + b_1 \sum x_{1j} + b_2 \sum x_{2j} + \cdots + b_k \sum x_{kj} = \sum y_j$$
$$b_0 \sum x_{1j} + b_1 \sum x_{1j}^2 + b_2 \sum x_{1j} x_{2j} + \cdots + b_k \sum x_{1j} x_{kj} = \sum x_{1j} y_j$$

$$b_0 \sum x_{kj} + b_1 \sum x_{1j} x_{kj} + \cdots + b_{k-1} \sum x_{k-1,j} x_{kj} + b_k \sum x_{kj}^2 = \sum x_{kj} y_j$$

(13.18)

That these equations are linear in the unknowns $b_0, b_1, \ldots, b_k$ is a consequence of the regression function being linear in the parameters. Solving Equations (13.18) yields the least squares estimates $\hat{\beta}_0, \hat{\beta}_1, \ldots, \hat{\beta}_k$. In general, the system (13.18) can be solved by hand by first computing all coefficients of the b_i's and then using a technique such as Gaussian elimination. This is quite tedious, but fortunately any of the standard statistical regression packages will automatically solve for $\hat{\beta}_0, \ldots, \hat{\beta}_k$.

Example 13.11 The article "How to Optimize and Control the Wire Bonding Process: Part II" (*Solid State Technology,* Jan. 1991: 67–72) described an experiment carried out to assess the impact of the variables x_1 = force (gm), x_2 = power (mw), x_3 = temperature (°C), and x_4 = time (ms) on y = ball bond shear strength (gm). The following data* was generated to be consistent with the information given in the article:

Observation	Force	Power	Temperature	Time	Strength
1	30	60	175	15	26.2
2	40	60	175	15	26.3
3	30	90	175	15	39.8
4	40	90	175	15	39.7
5	30	60	225	15	38.6
6	40	60	225	15	35.5
7	30	90	225	15	48.8
8	40	90	225	15	37.8
9	30	60	175	25	26.6
10	40	60	175	25	23.4
11	30	90	175	25	38.6
12	40	90	175	25	52.1
13	30	60	225	25	39.5
14	40	60	225	25	32.3
15	30	90	225	25	43.0
16	40	90	225	25	56.0
17	25	75	200	20	35.2
18	45	75	200	20	46.9

(continued)

*From the book *Statistics Engineering Problem Solving* by Stephen Vardeman, an excellent exposition of the territory covered by our book, albeit at a somewhat higher level.

Observation	Force	Power	Temperature	Time	Strength
19	35	45	200	20	22.7
20	35	105	200	20	58.7
21	35	75	150	20	34.5
22	35	75	250	20	44.0
23	35	75	200	10	35.7
24	35	75	200	30	41.8
25	35	75	200	20	36.5
26	35	75	200	20	37.6
27	35	75	200	20	40.3
28	35	75	200	20	46.0
29	35	75	200	20	27.8
30	35	75	200	20	40.3

A statistical computer package gave the following least squares estimates:

$$\hat{\beta}_0 = -37.48 \qquad \hat{\beta}_1 = .2117 \qquad \hat{\beta}_2 = .4983 \qquad \hat{\beta}_3 = .1297 \qquad \hat{\beta}_4 = .2583$$

Thus we estimate that .1297 gm is the average change in strength associated with a 1-degree increase in temperature when the other three predictors are held fixed; the other estimated coefficients are interpreted in a similar manner.

The estimated regression equation is

$$y = -37.48 + .2117x_1 + .4983x_2 + .1297x_3 + .2583x_4$$

A point prediction of strength resulting from a force of 35 gm, power of 75 mw, temperature of 200 degrees, and time of 20 ms is

$$\hat{y} = -37.48 + (.2117)(35) + (.4983)(75) + (.1297)(200) + (.2583)(20)$$
$$= 38.41 \text{ gm}$$

This is also a point estimate of the mean value of strength for the specified values of force, power, temperature, and time. ∎

σ^2 and R^2

Substituting the values of the predictors from the successive observations into the equation for the estimated regression gives the **predicted** or **fitted values** $\hat{y}_1, \hat{y}_2, \ldots, \hat{y}_n$. For example, since the values of the four predictors for the last observation in Example 13.11 are 35, 75, 200, and 20, respectively, the corresponding predicted value is $\hat{y}_{30} = 38.41$. The **residuals** are the differences $y_1 - \hat{y}_1, \ldots, y_n - \hat{y}_n$. The last residual in Example 13.11 is $40.3 - 38.41 = 1.89$. The closer the residuals are to zero, the better the job our estimated equation is doing in predicting the y values corresponding to values of the predictors in our sample.

As for both simple linear regression and polynomial regression, the estimate of σ^2 is based on the sum of squared residuals

$$\text{SSE} = \Sigma(y_j - \hat{y}_j)^2 = \Sigma[y_j - (\hat{\beta}_0 + \hat{\beta}_1 x_{1j} + \cdots + \hat{\beta}_k x_{kj})]^2$$

An efficient calculation formula for SSE is employed by most statistical computer packages. Because $k + 1$ parameters $(\beta_0, \beta_1, \ldots, \beta_k)$ have been estimated, $k + 1$ df are lost, so $n - (k + 1)$ df is associated with SSE, and

$$\hat{\sigma}^2 = s^2 = \frac{\text{SSE}}{n - (k + 1)} = \text{MSE}$$

With $\text{SST} = \sum(y_i - \bar{y})^2$, the proportion of total variation explained by the multiple regression model is $R^2 = 1 - \text{SSE/SST}$, the **coefficient of multiple determination.** As in polynomial regression, R^2 is often adjusted for the number of parameters in the model by the formula

$$R_a^2 = [(n - 1)R^2 - k]/[n - (k + 1)]$$

The positive square root of the coefficient of multiple determination is called the multiple correlation coefficient R. It can be shown that R is the sample correlation coefficient r between the observed y_j's and the predicted $\hat{y}_j$'s (i.e., using $x_j = \hat{y}_j$ in the formula for r results in $r = R$).

Example 13.12 Investigators carried out a study to see how various characteristics of concrete are influenced by x_1 = % limestone powder and x_2 = water-cement ratio, resulting in the accompanying data ("Durability of Concrete with Addition of Limestone Powder," *Magazine of Concrete Research*, 1996: 131–137):

x_1	x_2	x_1x_2	28-day comp str. (MPa)	Adsorbability (%)
21	.65	13.65	33.55	8.42
21	.55	11.55	47.55	6.26
7	.65	4.55	35.00	6.74
7	.55	3.85	35.90	6.59
28	.60	16.80	40.90	7.28
0	.60	0.00	39.10	6.90
14	.70	9.80	31.55	10.80
14	.50	7.00	48.00	5.63
14	.60	8.40	42.30	7.43
			$\bar{y} = 39.317$, SST $= 278.52$	$\bar{y} = 7.339$, SST $= 18.356$

Consider first compressive strength as the dependent variable y. Fitting $\beta_0 + \beta_1x_1 + \beta_2x_2$ results in

$$y = 84.82 + .1643x_1 - 79.67x_2 \qquad \text{SSE} = 72.25 \ (df = 6) \qquad R^2 = .741 \qquad R_a^2 = .654$$

whereas including an interaction predictor gives

$$y = 6.22 + 5.779x_1 + 51.33x_2 - 9.357x_1x_2$$

$$\text{SSE} = 29.35 \ (df = 5) \qquad R^2 = .895 \quad R_a^2 = .831$$

Based on this latter fit, a prediction for compressive strength when % limestone = 14 and water-cement ratio = .60 is

$$y = 6.22 + 5.779(14) + 51.33(.60) - 9.357(8.4) = 39.32$$

Fitting the full quadratic relationship results in virtually no change in the R^2 value. However, when the dependent variable is adsorbability, the following results are obtained: $R^2 = .747$ when just two predictors are used, .802 when the interaction predictor is added, and .889 when the five predictors for the full quadratic relationship are used. ∎

A Model Utility Test

A scatter plot of bivariate data [(x, y) pairs] will almost always indicate whether the simple linear regression model or a higher-degree polynomial will specify a useful relationship between the two variables. Unfortunately, with multivariate data, no analogous picture suggests whether a particular multiple regression model will be judged useful. The value of R^2 certainly communicates a preliminary message, but this value is sometimes deceptive because it can be greatly inflated by using a large number of predictors (large k) relative to the sample size n (this is the rationale behind adjusting R^2).

The model utility test in simple linear regression involved the null hypothesis H_0: $\beta_1 = 0$, according to which there is no useful relation between y and the single predictor x. Here we consider the assertion that $\beta_1 = 0, \beta_2 = 0, \ldots, \beta_k = 0$, which says that there is no useful relationship between y and *any* of the k predictors. If at least one of these β's is not 0, the corresponding predictor(s) is (are) useful. The test is based on a statistic that has a particular F distribution when H_0 is true.

Null hypothesis: H_0: $\beta_1 = \beta_2 = \cdots = \beta_k = 0$

Alternative hypothesis: H_a: at least one $\beta_i \neq 0$ $(i = 1, \ldots, k)$

Test statistic value*: $f = \dfrac{R^2/k}{(1 - R^2)/[n - (k + 1)]}$ (13.19)

Rejection region for a level α test: $f \geq F_{\alpha, k, n-(k+1)}$

Except for a constant multiple, the test statistic here is $R^2/(1 - R^2)$, the ratio of explained to unexplained variation. If the proportion of explained variation is high relative to unexplained, we would naturally want to reject H_0 and confirm the utility of the model. However, if k is large relative to n, the factor $[(n - (k + 1))/k]$ will decrease f considerably.

Example 13.13   Returning to the bond shear strength data of Example 13.11, a model with $k = 4$ predictors was fit, so the relevant hypotheses are

$$H_0: \beta_1 = \beta_2 = \beta_3 = \beta_4 = 0$$

H_a: at least one of these four β's is not zero

---

*f can also be written as MSR/MSE, where MS = SS/df, SSR = SST − SSE, and df for SSR = k.

Figure 13.13 shows output from the JMP statistical package. The values of s (Root Mean Square Error), R^2, and adjusted R^2 certainly suggest a useful model. The value of the model utility F ratio is

$$f = \frac{R^2/k}{(1 - R^2)/[n - (k + 1)]} = \frac{.713959/4}{.286041/(30 - 5)} = 15.60$$

This value also appears in the F Ratio column of the ANOVA table in Figure 13.13. The largest F critical value for 4 numerator and 25 denominator df in Appendix Table A.9 is 6.49, which captures an upper-tail area of .001. Thus P-value $< .001$. The ANOVA table in the JMP output shows that P-value $< .0001$. This is a highly significant result. The null hypothesis should be rejected at any reasonable significance level. We conclude that there *is* a useful linear relationship between y and *at least one* of the four predictors in the model. This does not mean that all four predictors are useful; we will say more about this subsequently.

Response: strength

Summary of Fit

RSquare	0.713959
RSquare Adj	0.668193
Root Mean Square Error	5.157979
Mean of Response	38.40667
Observations (or Sum Wgts)	30

Parameter Estimates

Term	Estimate	Std Error	t Ratio	Prob>|t|
Intercept	-37.47667	13.09964	-2.86	0.0084
force	0.2116667	0.210574	1.01	0.3244
power	0.4983333	0.070191	7.10	<.0001
temp	0.1296667	0.042115	3.08	0.0050
time	0.2583333	0.210574	1.23	0.2313

Whole-Model Test

Analysis of Variance

Source	DF	Sum of Squares	Mean Square	F Ratio
Model	4	1660.1400	415.035	15.6000
Error	25	665.1187	26.605	Prob>F
C Total	29	2325.2587		<.0001

Figure 13.13 Multiple regression output from JMP for the data of Example 13.13 ■

Inferences About Model Parameters

Before testing hypotheses, constructing CIs, and making predictions, one should first examine diagnostic plots to see whether the model needs modification or whether there are outliers in the data. The recommended plots are (standardized) residuals versus each independent variable, residuals versus $\hat{y}$, $\hat{y}$ versus y, and a normal probability plot of the standardized residuals. Potential problems are suggested by the same patterns discussed in Section 13.1. Of particular importance is the identification of observations that have a large influence on the fit. In the next section, we describe several diagnostic tools suitable for this task.

Because each $\hat{\beta}_i$ is a linear function of the y_i's, the standard deviation of each $\hat{\beta}_i$ is the product of σ and a function of the x_{ij}'s, so an estimate $s_{\hat{\beta}_i}$ is obtained by substituting s for σ. Unfortunately, the function of the x_{ij}'s is quite complicated, but all standard regression computer packages compute and show the $s_{\hat{\beta}_i}$'s. Inferences concerning a single β_i are based on the standardized variable

$$T = \frac{\hat{\beta}_i - \beta_i}{S_{\hat{\beta}_i}}$$

which has a t distribution with $n - (k + 1)$ df.

The point estimate of $\mu_{Y \cdot x_1^*, \ldots, x_k^*}$, the expected value of Y when $x_1 = x_1^*, \ldots, x_k = x_k^*$, is $\hat{\mu}_{Y \cdot x_1^*, \ldots, x_k^*} = \hat{\beta}_0 + \hat{\beta}_1 x_1^* + \cdots + \hat{\beta}_k x_k^*$. The estimated standard deviation of the corresponding estimator is again a complicated expression involving the sample x_{ij}'s. However, the better statistical computer packages will calculate it on request. Inferences about $\mu_{Y \cdot x_1^*, \ldots, x_k^*}$ are based on standardizing its estimator to obtain a t variable having $n - (k + 1)$ df.

Inferences Based on the Model $Y = \beta_0 + \beta_1 x_1 + \cdots + \beta_k x_k + \epsilon$

1. A $100(1 - \alpha)\%$ CI for β_i, the coefficient of x_i in the regression function, is

$$\hat{\beta}_i \pm t_{\alpha/2, n-(k+1)} \cdot s_{\hat{\beta}_i}$$

Simultaneous confidence intervals for several β_i's for which the simultaneous confidence level is controlled can be obtained by applying the Bonferroni technique.

2. A test for H_0: $\beta_i = \beta_{i0}$ uses the t statistic value $t = (\hat{\beta}_i - \beta_{i0})/s_{\hat{\beta}_i}$ based on $n - (k + 1)$ df. The test is upper-, lower-, or two-tailed according to whether H_a contains the inequality $>$, $<$, or $\neq$.

3. A $100(1 - \alpha)\%$ CI for $\mu_{Y \cdot x_1^*, \ldots, x_k^*}$ is

$$\hat{\mu}_{Y \cdot x_1^*, \ldots, x_k^*} \pm t_{\alpha/2, n-(k+1)} \cdot \{\text{estimated SD of } \hat{\mu}_{Y \cdot x_1^*, \ldots, x_k^*}\} = \hat{y} \pm t_{\alpha/2, n-(k+1)} \cdot s_{\hat{Y}}$$

where $\hat{Y}$ is the statistic $\hat{\beta}_0 + \hat{\beta}_1 x_1^* + \cdots + \hat{\beta}_k x_k^*$ and $\hat{y}$ is the calculated value of $\hat{Y}$.

4. A $100(1 - \alpha)\%$ PI for a future y value is

$$\hat{\mu}_{Y \cdot x_1^*, \ldots, x_k^*} \pm t_{\alpha/2, n-(k+1)} \cdot \{s^2 + (\text{estimated SD of } \hat{\mu}_{Y \cdot x_1^*, \ldots, x_k^*})^2\}^{1/2}$$

$$= \hat{y} \pm t_{\alpha/2, n-(k+1)} \cdot \sqrt{s^2 + s_{\hat{Y}}^2}$$

Example 13.14 Soil and sediment adsorption, the extent to which chemicals collect in a condensed form on the surface, is an important characteristic influencing the effectiveness of pesticides and various agricultural chemicals. The article "Adsorption of Phosphate, Arsenate, Methanearsonate, and Cacodylate by Lake and Stream Sediments: Comparisons with Soils" (*J. of Environ. Qual.*, 1984: 499–504) gives the accompanying data (Table 13.5) on y = phosphate adsorption index, x_1 = amount of extractable iron, and x_2 = amount of extractable aluminum.

Table 13.5 Data for Example 13.14

Observation	x_1 = Extractable Iron	x_2 = Extractable Aluminum	y = Adsorption Index
1	61	13	4
2	175	21	18
3	111	24	14
4	124	23	18
5	130	64	26
6	173	38	26
7	169	33	21
8	169	61	30
9	160	39	28
10	244	71	36
11	257	112	65
12	333	88	62
13	199	54	40

The article proposed the model

$$Y = \beta_0 + \beta_1 x_1 + \beta_2 x_2 + \epsilon$$

A computer analysis yielded the following information:

Parameter β_i	Estimate $\hat{\beta}_i$	Estimated SD $s_{\hat{\beta}_i}$
β_0	−7.351	3.485
β_1	.11273	.02969
β_2	.34900	.07131

$R^2 = .948$ adjusted $R^2 = .938$ $s = 4.379$

$\hat{\mu}_{Y \cdot 160,39} = \hat{y} = -7.351 + (.11273)(160) + (.34900)(39) = 24.30$

estimated SD of $\hat{\mu}_{Y \cdot 160,39} = s_{\hat{y}} = 1.30$

A 99% CI for β_1, the change in expected adsorption associated with a 1-unit increase in extractable iron while extractable aluminum is held fixed, requires $t_{.005,13-(2+1)} = t_{.005,10} = 3.169$. The CI is

$$.11273 \pm (3.169)(.02969) = .11273 \pm .09409 \approx (.019, .207)$$

Similarly, a 99% interval for β_2 is

$$.34900 \pm (3.169)(.07131) = .34900 \pm .22598 \approx (.123, .575)$$

The Bonferroni technique implies that the simultaneous confidence level for both intervals is at least 98%.

A 95% CI for $\mu_{Y \cdot 160, 39}$, expected adsorption when extractable iron $= 160$ and extractable aluminum $= 39$, is

$$24.30 \pm (2.228)(1.30) = 24.30 \pm 2.90 = (21.40, 27.20)$$

A 95% PI for a future value of adsorption to be observed when $x_1 = 160$ and $x_2 = 39$ is

$$24.30 \pm (2.228)\{(4.379)^2 + (1.30)^2\}^{1/2} = 24.30 \pm 10.18 = (14.12, 34.48) \quad \blacksquare$$

Frequently, the hypothesis of interest has the form $H_0: \beta_i = 0$ for a particular i. For example, after fitting the four-predictor model in Example 13.11, the investigator might wish to test $H_0: \beta_4 = 0$. According to H_0, as long as the predictors x_1, x_2, and x_3 remain in the model, x_4 contains no useful information about y. The test statistic value is the *t*-**ratio** $\hat{\beta}_i / s_{\hat{\beta}_i}$. Many statistical computer packages report the *t*-ratio and corresponding *P*-value for each predictor included in the model. For example, Figure 13.13 shows that as long as power, temperature, and time are retained in the model, the predictor $x_1 = $ force can be deleted.

An *F* Test for a Group of Predictors

The model utility F test was appropriate for testing whether there is useful information about the dependent variable in *any* of the k predictors (i.e., whether $\beta_1 = \cdots = \beta_k = 0$). In many situations, one first builds a model containing k predictors and then wishes to know whether any of the predictors in a particular subset provide useful information about Y. For example, a model to be used to predict students' test scores might include a group of background variables such as family income and education levels and also some school characteristic variables such as class size and spending per pupil. One interesting hypothesis is that the school characteristic predictors can be dropped from the model.

Let's label the predictors as $x_1, x_2, \ldots, x_l, x_{l+1}, \ldots, x_k$, so that it is the last $k - l$ that we are considering deleting from the model. We then wish to test

---

$H_0: \beta_{l+1} = \beta_{l+2} = \cdots = \beta_k = 0$
    (so the "reduced" model $Y = \beta_0 + \beta_1 x_1 + \cdots + \beta_l x_l + \epsilon$ is correct)

versus

H_a: at least one among $\beta_{l+1}, \ldots, \beta_k$ is not 0
    (so in the "full" model $Y = \beta_0 + \beta_1 x_1 + \cdots + \beta_k x_k + \epsilon$, at least one of the last $k - l$ predictors provides useful information)

The test is carried out by fitting both the full and reduced models. Because the full model contains not only the predictors of the reduced model but also some extra predictors, it should fit the data at least as well as the reduced model. That is, if we let SSE_k be the sum of squared residuals for the full model and SSE_l be the corresponding sum for the reduced model, then $SSE_k \leq SSE_l$.* Intuitively, if SSE_k is a great deal smaller than SSE_l, the full model provides a much better fit than the reduced model; the appropriate test statistic should then depend on the reduction $SSE_l - SSE_k$ in unexplained variation. The formal procedure is

SSE_k = unexplained variation for the full model

SSE_l = unexplained variation for the reduced model

Test statistic value: $f = \dfrac{(SSE_l - SSE_k)/(k - l)}{SSE_k/[n - (k + 1)]}$ (13.20)

Rejection region: $f \geq F_{\alpha, k-l, n-(k+1)}$

Example 13.15 The data in Table 13.6 was taken from the article "Applying Stepwise Multiple Regression Analysis to the Reaction of Formaldehyde with Cotton Cellulose" (*Textile Research J.*, 1984: 157–165). The dependent variable y is durable press rating, a quantitative measure of wrinkle resistance. The four independent variables used in the model building process are x_1 = HCHO (formaldehyde) concentration, x_2 = catalyst ratio, x_3 = curing temperature, and x_4 = curing time.

Table 13.6 Data for Example 13.15

Observation	x_1	x_2	x_3	x_4	y	Observation	x_1	x_2	x_3	x_4	y
1	8	4	100	1	1.4	16	4	10	160	5	4.6
2	2	4	180	7	2.2	17	4	13	100	7	4.3
3	7	4	180	1	4.6	18	10	10	120	7	4.9
4	10	7	120	5	4.9	19	5	4	100	1	1.7
5	7	4	180	5	4.6	20	8	13	140	1	4.6
6	7	7	180	1	4.7	21	10	1	180	1	2.6
7	7	13	140	1	4.6	22	2	13	140	1	3.1
8	5	4	160	7	4.5	23	6	13	180	7	4.7
9	4	7	140	3	4.8	24	7	1	120	7	2.5
10	5	1	100	7	1.4	25	5	13	140	1	4.5
11	8	10	140	3	4.7	26	8	1	160	7	2.1
12	2	4	100	3	1.6	27	4	1	180	7	1.8
13	4	10	180	3	4.5	28	6	1	160	1	1.5
14	6	7	120	7	4.7	29	4	1	100	1	1.3
15	10	13	180	3	4.8	30	7	10	100	7	4.6

---

*The estimates $\hat{\beta}_0, \hat{\beta}_1, \ldots, \hat{\beta}_l$ will in general be different for the full and reduced models, so in general two different multiple regressions must be run to obtain SSE_l and SSE_k. If the variables are listed in the suggested order, though, most computer packages provide an ANOVA table for the full model that can be used to avoid fitting the reduced model.

Consider the full model consisting of $k = 14$ predictors: $x_1, x_2, x_3, x_4, x_5 = x_1^2, \ldots,$ $x_8 = x_4^2, x_9 = x_1 x_2, \ldots, x_{14} = x_3 x_4$ (all first- and second-order predictors). Is the inclusion of the second-order predictors justified? That is, should the reduced model consisting of just the predictors x_1, x_2, x_3, and x_4 ($l = 4$) be used? Output resulting from fitting the two models follows:

Parameter	Estimate for Reduced Model	Estimate for Full Model
β_0	−.9122	−8.807
β_1	.16073	.1768
β_2	.21978	.7580
β_3	.011226	.10400
β_4	.10197	.5052
β_5	—	−.04393
β_6	—	−.035887
β_7	—	−.00003271
β_8	—	−.01646
β_9	—	.00588
β_{10}	—	.002702
β_{11}	—	.01178
β_{12}	—	−.0006547
β_{13}	—	.00242
β_{14}	—	.002526
R^2	.692	.921
SSE	17.4951	4.4782

The hypotheses to be tested are

$$H_0: \ \beta_5 = \beta_6 = \cdots = \beta_{14} = 0$$

versus

$$H_a: \ \text{at least one among } \beta_5, \ldots, \beta_{14} \text{ is not } 0$$

With $k = 14$ and $l = 4$, the F critical value for a test with $\alpha = .01$ is $F_{.01,10,15} = 3.80$. The test statistic value is

$$f = \frac{(17.4951 - 4.4782)/10}{4.4782/15} = \frac{1.3017}{.2985} = 4.36$$

Since $4.36 \geq 3.80$, H_0 is rejected. We conclude that the appropriate model should include at least one of the second-order predictors. ∎

Assessing Model Adequacy

The standardized residuals in multiple regression result from dividing each residual by its estimated standard deviation; the formula for these standard deviations is substantially more complicated than in the case of simple linear regression. We recommend a normal probability plot of the standardized residuals as a basis for validating the normality assumption. Plots of the standardized residuals versus each predictor and versus $\hat{y}$ should show no discernible pattern.

Example 13.16 Figure 13.14 shows a normal probability plot of the standardized residuals for the adsorption data and fitted model given in Example 13.14. The straightness of the plot casts little doubt on the assumption that the random deviation ϵ is normally distributed.

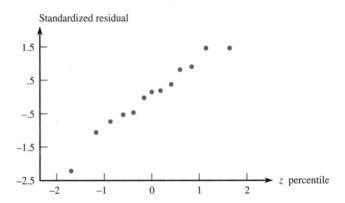

Figure 13.14 A normal probability plot of the standardized residuals for the data and model of Example 13.14

Figure 13.15 shows the other suggested plots for the adsorption data. Given that there are only 13 observations in the data set, there is not much evidence of a pattern in any of the first three plots other than randomness. The point at the bottom of each of these three plots corresponds to the observation with the large residual. We will say more about such observations subsequently. For the moment, there is no compelling reason for remedial action.

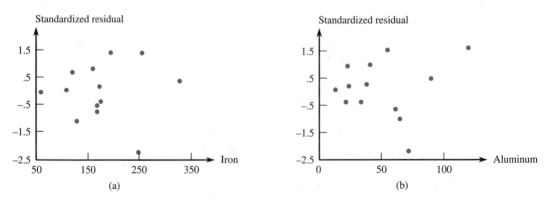

Figure 13.15 Diagnostic plots for the adsorption data: (a) standardized residual versus x_1; (b) standardized residual versus x_2

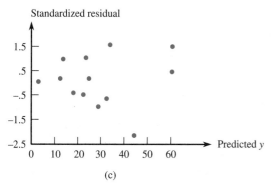

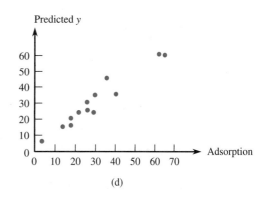

Figure 13.15 (cont'd)   Diagnostic plots for the adsorption data: (c) standardized residual versus $\hat{y}$; (d) $\hat{y}$ versus y ■

Exercises | Section 13.4 (36–52)

36. Cardiorespiratory fitness is widely recognized as a major component of overall physical well-being. Direct measurement of maximal oxygen uptake (VO_2max) is the single best measure of such fitness, but direct measurement is time-consuming and expensive. It is therefore desirable to have a prediction equation for VO_2max in terms of easily obtained quantities. Consider the variables

$y = VO_2$max (L/min) $x_1 = $ weight (kg)
$x_2 = $ age (yr)
$x_3 = $ time necessary to walk 1 mile (min)
$x_4 = $ heart rate at the end of the walk (beats/min)

Here is one possible model, for male students, consistent with the information given in the article "Validation of the Rockport Fitness Walking Test in College Males and Females" (*Research Quarterly for Exercise and Sport,* 1994: 152–158):

$Y = 5.0 + .01x_1 - .05x_2 - .13x_3 - .01x_4 + \epsilon$
$\sigma = .4$

a. Interpret β_1 and β_3.
b. What is the expected value of VO_2max when weight is 76 kg, age is 20 yr, walk time is 12 min, and heart rate is 140 b/m?
c. What is the probability that VO_2max will be between 1.00 and 2.60 for a single observation made when the values of the predictors are as stated in part (b)?

37. A trucking company considered a multiple regression model for relating the dependent variable $y = $ total daily travel time for one of its drivers (hours) to the predictors $x_1 = $ distance traveled (miles) and $x_2 = $ the number of deliveries made. Suppose that the model equation is

$$Y = -.800 + .060x_1 + .900x_2 + \epsilon$$

a. What is the mean value of travel time when distance traveled is 50 miles and three deliveries are made?
b. How would you interpret $\beta_1 = .060$, the coefficient of the predictor x_1? What is the interpretation of $\beta_2 = .900$?
c. If $\sigma = .5$ hour, what is the probability that travel time will be at most 6 hours when three deliveries are made and the distance traveled is 50 miles?

38. Let $y = $ wear life of a bearing, $x_1 = $ oil viscosity, and $x_2 = $ load. Suppose that the multiple regression model relating life to viscosity and load is

$$Y = 125.0 + .775x_1 + .0950x_2 - .0090x_1x_2 + \epsilon$$

a. What is the mean value of life when viscosity is 40 and load is 1100?
b. When viscosity is 30, what is the change in mean life associated with an increase of 1 in load? When viscosity is 40, what is the change in mean life associated with an increase of 1 in load?

39. Let $y = $ sales at a fast food outlet (1000's of $), $x_1 = $ number of competing outlets within a 1-mile radius,

x_2 = population within a 1-mile radius (1000's of people), and x_3 be an indicator variable that equals 1 if the outlet has a drive-up window and 0 otherwise. Suppose that the true regression model is

$$Y = 10.00 - 1.2x_1 + 6.8x_2 + 15.3x_3 + \epsilon$$

a. What is the mean value of sales when the number of competing outlets is 2, there are 8000 people within a 1-mile radius, and the outlet has a drive-up window?
b. What is the mean value of sales for an outlet without a drive-up window that has three competing outlets and 5000 people within a 1-mile radius?
c. Interpret β_3.

40. The article "Readability of Liquid Crystal Displays: A Response Surface" (*Human Factors*, 1983: 185–190) used a multiple regression model with four independent variables to study accuracy in reading liquid crystal displays. The variables were

y = error percentage for subjects reading a four-digit liquid crystal display

x_1 = level of backlight (ranging from 0 to 122 cd/m^2)

x_2 = character subtense (ranging from .025° to 1.34°)

x_3 = viewing angle (ranging from 0° to 60°)

x_4 = level of ambient light (ranging from 20 to 1500 lux)

The model fit to data was $Y = \beta_0 + \beta_1 x_1 + \beta_2 x_2 + \beta_3 x_3 + \beta_4 x_4 + \epsilon$. The resulting estimated coefficients were $\hat{\beta}_0 = 1.52$, $\hat{\beta}_1 = .02$, $\hat{\beta}_2 = -1.40$, $\hat{\beta}_3 = .02$, and $\hat{\beta}_4 = -.0006$.

a. Calculate an estimate of expected error percentage when $x_1 = 10$, $x_2 = .5$, $x_3 = 50$, and $x_4 = 100$.
b. Estimate the mean error percentage associated with a backlight level of 20, character subtense of .5, viewing angle of 10, and ambient light level of 30.
c. What is the estimated expected change in error percentage when the level of ambient light is increased by 1 unit while all other variables are fixed at the values given in part (a)? Answer for a 100-unit increase in ambient light level.
d. Explain why the answers in part (c) do not depend on the fixed values of x_1, x_2, and x_3. Under what conditions would there be such a dependence?

e. The estimated model was based on $n = 30$ observations, with SST = 39.2 and SSE = 20.0. Calculate and interpret the coefficient of multiple determination, and then carry out the model utility test using $\alpha = .05$.

41. The ability of ecologists to identify regions of greatest species richness could have an impact on the preservation of genetic diversity, a major objective of the World Conservation Strategy. The article "Prediction of Rarities from Habitat Variables: Coastal Plain Plants on Nova Scotian Lakeshores" (*Ecology*, 1992: 1852–1859) used a sample of $n = 37$ lakes to obtain the estimated regression equation

$$y = 3.89 + .033x_1 + .024x_2 + .023x_3 - .0080x_4 - .13x_5 - .72x_6$$

where y = species richness, x_1 = watershed area, x_2 = shore width, x_3 = poor drainage (%), x_4 = water color (total color units), x_5 = sand (%), and x_6 = alkalinity. The coefficient of multiple determination was reported as $R^2 = .83$. Carry out a test of model utility.

42. An investigation of a die casting process resulted in the accompanying data on x_1 = furnace temperature, x_2 = die close time, and y = temperature difference on the die surface ("A Multiple-Objective Decision-Making Approach for Assessing Simultaneous Improvement in Die Life and Casting Quality in a Die Casting Process," *Quality Engineering*, 1994: 371–383).

x_1	1250	1300	1350	1250	1300
x_2	6	7	6	7	6
y	80	95	101	85	92

x_1	1250	1300	1350	1350
x_2	8	8	7	8
y	87	96	106	108

MINITAB output from fitting the multiple regression model with predictors x_1 and x_2 is given here.

```
The regression equation is
tempdiff = -200 + 0.210 furntemp
              + 3.00 clostime

Predictor    Coef     Stdev   t-ratio      p
Constant  -199.56     11.64    -17.14  0.000
furntemp  0.210000  0.008642    24.30  0.000
clostime    3.0000    0.4321     6.94  0.000
```

(continued)

s = 1.058   R-sq = 99.1%   R-sq(adj) = 98.8%

Analysis of Variance

SOURCE	DF	SS	MS	F	p
Regression	2	715.50	357.75	319.31	0.000
Error	6	6.72	1.12		
Total	8	722.22			

a. Carry out the model utility test.

b. Calculate and interpret a 95% confidence interval for β_2, the population regression coefficient of x_2.

c. When $x_1 = 1300$ and $x_2 = 7$, the estimated standard deviation of $\hat{y}$ is $s_{\hat{y}} = .353$. Calculate a 95% confidence interval for true average temperature difference when furnace temperature is 1300 and die close time is 7.

d. Calculate a 95% prediction interval for the temperature difference resulting from a single experimental run with a furnace temperature of 1300 and a die close time of 7.

43. An experiment carried out to study the effect of the mole contents of cobalt (x_1) and the calcination temperature (x_2) on the surface area of an iron-cobalt hydroxide catalyst (y) resulted in the accompanying data ("Structural Changes and Surface Properties of $Co_xFe_{3-x}O_4$ Spinels," *J. of Chemical Tech. and Biotech.*, 1994: 161–170). A request to the SAS package to fit $\beta_0 + \beta_1 x_1 + \beta_2 x_2 + \beta_3 x_3$, where $x_3 = x_1 x_2$ (an interaction predictor) yielded the following output.

x_1	.6	.6	.6	.6	.6	1.0	1.0
x_2	200	250	400	500	600	200	250
y	90.6	82.7	58.7	43.2	25.0	127.1	112.3

(continued at top of next column)

x_1	1.0	1.0	1.0	2.6	2.6	2.6	2.6
x_2	400	500	600	200	250	400	500
y	19.6	17.8	9.1	53.1	52.0	43.4	42.4

x_1	2.6	2.8	2.8	2.8	2.8	2.8
x_2	600	200	250	400	500	600
y	31.6	40.9	37.9	27.5	27.3	19.0

a. Predict the value of surface area when cobalt content is 2.6 and temperature is 250, and calculate the value of the corresponding residual.

b. Since $\hat{\beta}_1 = -46.0$, is it legitimate to conclude that if cobalt content increases by 1 unit while the values of the other predictors remain fixed, surface area can be expected to decrease by roughly 46 units? Explain your reasoning.

c. Does there appear to be a useful linear relationship between y and the predictors?

d. Given that mole contents and calcination temperature remain in the model, does the interaction predictor x_3 provide useful information about y? State and test the appropriate hypotheses using a significance level of .01.

e. The estimated standard deviation of $\hat{Y}$ when mole contents is 2.0 and calcination temperature is 500 is $s_{\hat{y}} = 4.69$. Calculate a 95% confidence interval for the mean value of surface area under these circumstances.

Dependent Variable: SURFAREA

Analysis of Variance

Source	DF	Sum of Squares	Mean Square	F Value	Prob>F
Model	3	15223.52829	5074.50943	18.924	0.0001
Error	16	4290.53971	268.15873		
C Total	19	19514.06800			

Root MSE	16.37555	R-square	0.7801	
Dep Mean	48.06000	Adj R-sq	0.7389	
C.V.	34.07314			

Parameter Estimates

Variable	DF	Parameter Estimate	Standard Error	T for H0: Parameter=0	Prob >|T|
INTERCEP	1	185.485740	21.19747682	8.750	0.0001
COBCON	1	-45.969466	10.61201173	-4.332	0.0005
TEMP	1	-0.301503	0.05074421	-5.942	0.0001
CONTEMP	1	0.088801	0.02540388	3.496	0.0030

44. The authors of the article "An Ultracentrifuge Flour Absorption Method" (*Cereal Chemistry*, 1978: 96–101) studied the relationship between water absorption for wheat flour and various characteristics of the flour. In particular, the authors used a first-order multiple linear regression model to relate absorption y (%) to flour protein x_1 (%) and starch damage x_2 (Farrand units). The data and accompanying SPSS output follow.

x_1	x_2	y	x_1	x_2	y
8.5	2	30.9	12.9	24	47.0
8.9	3	32.7	12.0	25	46.8
10.6	3	36.7	12.9	28	45.9
10.2	20	41.9	13.1	28	48.8
9.8	22	40.9	11.4	32	46.2
10.8	20	42.9	13.2	28	47.8
11.6	31	46.3	11.6	35	49.2
12.0	32	47.6	12.1	34	48.3
12.5	31	47.2	11.3	35	48.6
10.9	28	44.0	11.1	40	50.2
12.2	36	47.7	11.5	45	49.6
11.9	28	43.9	11.6	50	53.2
11.3	30	46.8	11.7	55	54.3
13.0	27	46.2	11.7	57	55.8

a. Interpret $\hat{\beta}_1$ and $\hat{\beta}_2$.
b. What proportion of observed variation in absorption can be explained by the model relationship?
c. Does the chosen model appear to specify a useful linear relationship between absorption and at least one of the two predictors?

d. If flour protein remains in the model, is elimination of the predictor starch damage justified?
e. When $x_1 = 10$ and $x_2 = 25$, $\hat{y} = 42.253$ and $s_{\hat{y}} = .350$. Calculate and interpret a confidence interval and a prediction interval.
f. Including an interaction predictor x_3 gives $\hat{\beta}_3 = -.04304$ and $s_{\hat{\beta}_3} = .01773$. At significance level .01, should this predictor be retained?

45. The article "Analysis of the Modeling Methodologies for Predicting the Strength of Air-Jet Spun Yarns" (*Textile Res. J.*, 1997: 39–44) reported on a study carried out to relate yarn tenacity (y, in g/tex) to yarn count (x_1, in tex), percentage polyester (x_2), first nozzle pressure (x_3, in kg/cm^2), and second nozzle pressure (x_4, in kg/cm^2). The estimate of the constant term in the corresponding multiple regression equation was 6.121. The estimated coefficients for the four predictors were $-.082$, .113, .256, and $-.219$, respectively, and the coefficient of multiple determination was .946.

a. Assuming that the sample size was $n = 25$, state and test the appropriate hypotheses to decide whether the fitted model specifies a useful linear relationship between the dependent variable and at least one of the four model predictors.
b. Again using $n = 25$, calculate the value of adjusted R^2.
c. Calculate a 99% confidence interval for true mean yarn tenacity when yarn count is 16.5, yarn contains 50% polyester, first nozzle pressure is 3, and second nozzle pressure is 5 if the estimated standard deviation of predicted tenacity under these circumstances is .350.

SPSS output for Exercise 44

```
Multiple R            .98207
R Square              .96447
Adjusted R Square     .96163
Standard Error       1.09412


Analysis of Variance
                DF    Sum of Squares    Mean Square
Regression       2         812.37959      406.18980
Residual        25          29.92755        1.19710

F = 339.31092     Signif F = .0000

                Variables in the equation

Variable            B        SE B    95% Confdnce    Intrvl B
STARCH         .33563      .01814         .29828      .37298
FLOUR         1.44228      .20764        1.01465     1.86991
(Constant)   19.43976     2.18829       14.93290    23.94662
```

46. A regression analysis carried out to relate $y =$ repair time for a water filtration system (hr) to $x_1 =$ elapsed time since the previous service (months) and $x_2 =$ type of repair (1 if electrical and 0 if mechanical) yielded the following model based on $n = 12$ observations: $y = .950 + .400x_1 + 1.250x_2$. In addition, SST $= 12.72$, SSE $= 2.09$, and $s_{\hat{\beta}_2} = .312$.

  a. Does there appear to be a useful linear relationship between repair time and the two model predictors? Carry out a test of the appropriate hypotheses using a significance level of .05.

  b. Given that elapsed time since the last service remains in the model, does type of repair provide useful information about repair time? State and test the appropriate hypotheses using a significance level of .01.

  c. Calculate and interpret a 95% CI for β_2.

  d. The estimated standard deviation of a prediction for repair time when elapsed time is 6 months and the repair is electrical is .192. Predict repair time under these circumstances by calculating a 99% prediction interval. Does the interval suggest that the estimated model will give an accurate prediction? Why or why not?

47. Efficient design of certain types of municipal waste incinerators requires that information about energy content of the waste be available. The authors of the article "Modeling the Energy Content of Municipal Solid Waste Using Multiple Regression Analysis" (*J. of the Air and Waste Mgmt. Assoc.,* 1996: 650–656) kindly provided us with the accompanying data on $y =$ energy content (kcal/kg), the three physical composition variables $x_1 =$ % plastics by weight, $x_2 =$ % paper by weight, and $x_3 =$ % garbage by weight, and the proximate analysis variable $x_4 =$ % moisture by weight for waste specimens obtained from a certain region.

Obs.	Plastics	Paper	Garbage	Water	Energy content
1	18.69	15.65	45.01	58.21	947
2	19.43	23.51	39.69	46.31	1407
3	19.24	24.23	43.16	46.63	1452
4	22.64	22.20	35.76	45.85	1553
5	16.54	23.56	41.20	55.14	989
6	21.44	23.65	35.56	54.24	1162
7	19.53	24.45	40.18	47.20	1466
8	23.97	19.39	44.11	43.82	1656
9	21.45	23.84	35.41	51.01	1254
10	20.34	26.50	34.21	49.06	1336
11	17.03	23.46	32.45	53.23	1097
12	21.03	26.99	38.19	51.78	1266
13	20.49	19.87	41.35	46.69	1401
14	20.45	23.03	43.59	53.57	1223
15	18.81	22.62	42.20	52.98	1216
16	18.28	21.87	41.50	47.44	1334
17	21.41	20.47	41.20	54.68	1155
18	25.11	22.59	37.02	48.74	1453
19	21.04	26.27	38.66	53.22	1278
20	17.99	28.22	44.18	53.37	1153
21	18.73	29.39	34.77	51.06	1225
22	18.49	26.58	37.55	50.66	1237
23	22.08	24.88	37.07	50.72	1327
24	14.28	26.27	35.80	48.24	1229
25	17.74	23.61	37.36	49.92	1205
26	20.54	26.58	35.40	53.58	1221
27	18.25	13.77	51.32	51.38	1138
28	19.09	25.62	39.54	50.13	1295
29	21.25	20.63	40.72	48.67	1391
30	21.62	22.71	36.22	48.19	1372

Using MINITAB to fit a multiple regression model with the four aforementioned variables as predictors of energy content resulted in the following output.

```
The regression equation is
enercont = 2245 + 28.9 plastics
             + 7.64 paper + 4.30 garbage
             − 37.4 water

Predictor      Coef     StDev        T        p
Constant     2244.9     177.9    12.62    0.000
plastics     28.925     2.824    10.24    0.000
paper         7.644     2.314     3.30    0.003
garbage       4.297     1.916     2.24    0.034
water       −37.354     1.834   −20.36    0.000

s = 31.48     R-Sq = 96.4%      R-Sq(adj) = 95.8%
```

(continued)

Analysis of Variance

Source	DF	SS	MS	F	p
Regression	4	664931	166233	167.71	0.000
Error	25	24779	991		
Total	29	689710			

a. Interpret the values of the estimated regression coefficients $\hat{\beta}_1$ and $\hat{\beta}_4$.

b. State and test the appropriate hypotheses to decide whether the model fit to the data specifies a useful linear relationship between energy content and at least one of the four predictors.

c. Given that % plastics, % paper, and % water remain in the model, does % garbage provide useful information about energy content? State and test the appropriate hypotheses using a significance level of .05.

d. Use the fact that $s_{\hat{y}} = 12.47$ when $x_1 = 20$, $x_2 = 25$, $x_3 = 40$, and $x_4 = 45$ to calculate a 95% confidence interval for true average energy content under these circumstances. Does the resulting interval suggest that mean energy content has been precisely estimated?

e. Use the information given in part (d) to predict energy content for a waste sample having the specified characteristics in a way that conveys information about precision and reliability.

48. An experiment to investigate the effects of a new technique for degumming of silk yarn was described in the article "Some Studies in Degumming of Silk with Organic Acids" (*J. Society of Dyers and Colourists,* 1992: 79–86). One response variable of interest was y = weight loss (%). The experimenters made observations on weight loss for various values of three independent variables: x_1 = temperature (°C) = 90, 100, 110; x_2 = time of treatment (min) = 30, 75, 120; x_3 = tartaric acid concentration (g/L) = 0, 8, 16. In the regression analyses, the three values of each variable were coded as −1, 0, and 1, respectively, giving the accompanying data (the value y_8 = 19.3 was reported, but our value y_8 = 20.3 results in regression output identical to that appearing in the article).

Obs.	1	2	3	4	5	6	7	8
x_1	−1	−1	1	1	−1	−1	1	1
x_2	−1	1	−1	1	0	0	0	0
x_3	0	0	0	0	−1	1	−1	1
y	18.3	22.2	23.0	23.0	3.3	19.3	19.3	20.3

Obs.	9	10	11	12	13	14	15
x_1	0	0	0	0	0	0	0
x_2	−1	−1	1	1	0	0	0
x_3	−1	1	−1	1	0	0	0
y	13.1	23.0	20.9	21.5	22.0	21.3	22.6

A multiple regression model with $k = 9$ predictors—x_1, x_2, x_3, $x_4 = x_1^2$, $x_5 = x_2^2$, $x_6 = x_3^2$, $x_7 = x_1 x_2$, $x_8 = x_1 x_3$, and $x_9 = x_2 x_3$—was fit to the data, resulting in $\hat{\beta}_0 = 21.967$, $\hat{\beta}_1 = 2.8125$, $\hat{\beta}_2 = 1.2750$, $\hat{\beta}_3 = 3.4375$, $\hat{\beta}_4 = -2.208$, $\hat{\beta}_5 = 1.867$, $\hat{\beta}_6 = -4.208$, $\hat{\beta}_7 = -.975$, $\hat{\beta}_8 = -3.750$, $\hat{\beta}_9 = -2.325$, SSE $= 23.379$, and $R^2 = .938$.

a. Does this model specify a useful relationship? State and test the appropriate hypotheses using a significance level of .01.

b. The estimated standard deviation of $\hat{\mu}_Y$ when $x_1 = \cdots = x_9 = 0$ (i.e., when temperature = 100, time = 75, and concentration = 8) is 1.248. Calculate a 95% CI for expected weight loss when temperature, time, and concentration have the specified values.

c. Calculate a 95% PI for a single weight-loss value to be observed when temperature, time, and concentration have values 100, 75, and 8, respectively.

d. Fitting the model with only x_1, x_2, and x_3 as predictors gave $R^2 = .456$ and SSE $= 203.82$. Is there a useful relationship between weight loss and at least one of the second-order predictors x_4, x_5, $\ldots$, x_9? State and test the appropriate hypotheses.

49. The article "The Influence of Temperature and Sunshine on the Alpha-Acid Contents of Hops (*Agricultural Meteorology,* 1974: 375–382) reports the following data on yield (y), mean temperature over the period between date of coming into hops and date of picking (x_1), and mean percentage of sunshine during the same period (x_2), for the fuggle variety of hop.

x_1	16.7	17.4	18.4	16.8	18.9	17.1
x_2	30	42	47	47	43	41
y	210	110	103	103	91	76

(continued)

x_1	17.3	18.2	21.3	21.2	20.7	18.5
x_2	48	44	43	50	56	60
y	73	70	68	53	45	31

Here is partial MINITAB output from fitting the first-order model $Y = \beta_0 + \beta_1 x_1 + \beta_2 x_2 + \epsilon$ used in the article:

```
Predictor    Coef    Stdev    t-ratio       p
Constant    415.11   82.52      5.03    0.000
Temp         -6.593   4.859     -1.36    0.208
Sunshine     -4.504   1.071     -4.20    0.002

s = 24.45    R-sq = 76.8%    R-sq(adj) = 71.6%
```

a. What is $\hat{\mu}_{Y \cdot 18.9,43}$, and what is the corresponding residual?

b. Test $H_0: \beta_1 = \beta_2 = 0$ versus H_a: either β_1 or $\beta_2 \neq 0$ at level .05.

c. The estimated standard deviation of $\hat{\beta}_0 + \hat{\beta}_1 x_1 + \hat{\beta}_2 x_2$ when $x_1 = 18.9$ and $x_2 = 43$ is 8.20. Use this to obtain a 95% CI for $\mu_{Y \cdot 18.9,43}$.

d. Use the information in part (c) to obtain a 95% PI for yield in a future experiment when $x_1 = 18.9$ and $x_2 = 43$.

e. MINITAB reported that a 95% PI for yield when $x_1 = 18$ and $x_2 = 45$ is (35.94, 151.63). What is a 90% PI in this situation?

f. Given that x_2 is in the model, would you retain x_1?

g. When the model $Y = \beta_0 + \beta_2 x_2 + \epsilon$ is fit, the resulting value of R^2 is .721. Verify that the F statistic for testing $H_0: Y = \beta_0 + \beta_2 x_2 + \epsilon$ versus H_a: $Y = \beta_0 + \beta_1 x_1 + \beta_2 x_2 + \epsilon$ satisfies $t^2 = f$, where t is the value of the t statistic from part (f).

50. a. When the model $Y = \beta_0 + \beta_1 x_1 + \beta_2 x_2 + \beta_3 x_1^2 + \beta_4 x_2^2 + \beta_5 x_1 x_2 + \epsilon$ is fit to the hops data of Exercise 49, the estimate of β_5 is $\hat{\beta}_5 = .557$ with estimated standard deviation $s_{\hat{\beta}_5} = .94$. Test $H_0: \beta_5 = 0$ versus $H_a: \beta_5 \neq 0$.

b. Each t ratio $\hat{\beta}_i / s_{\hat{\beta}_i}$ ($i = 1, 2, 3, 4, 5$) for the model of part (a) is less than 2 in absolute value, yet $R^2 = .861$ for this model. Would it be correct to drop each term from the model because of its small t ratio? Explain.

c. Using $R^2 = .861$ for the model of part (a), test $H_0: \beta_3 = \beta_4 = \beta_5 = 0$ (which says that all second-order terms can be deleted).

51. The article "The Undrained Strength of Some Thawed Permafrost Soils" (*Canadian Geotechnical J.*, 1979: 420–427) contains the following data on undrained shear strength of sandy soil (y, in kPa), depth (x_1, in m), and water content (x_2, in %).

	y	x_1	x_2	$\hat{y}$	$y - \hat{y}$	e^*
1	14.7	8.9	31.5	23.35	-8.65	-1.50
2	48.0	36.6	27.0	46.38	1.62	.54
3	25.6	36.8	25.9	27.13	-1.53	-.53
4	10.0	6.1	39.1	10.99	-.99	-.17
5	16.0	6.9	39.2	14.10	1.90	.33
6	16.8	6.9	38.3	16.54	.26	.04
7	20.7	7.3	33.9	23.34	-2.64	-.42
8	38.8	8.4	33.8	25.43	13.37	2.17
9	16.9	6.5	27.9	15.63	1.27	.23
10	27.0	8.0	33.1	24.29	2.71	.44
11	16.0	4.5	26.3	15.36	.64	.20
12	24.9	9.9	37.8	29.61	-4.71	-.91
13	7.3	2.9	34.6	15.38	-8.08	-1.53
14	12.8	2.0	36.4	7.96	4.84	1.02

The predicted values and residuals were computed by fitting a full quadratic model, which resulted in the estimated regression function

$$y = -151.36 - 16.22x_1 + 13.48x_2 + .094x_1^2 - .253x_2^2 + .492x_1 x_2$$

a. Do plots of e^* versus x_1, e^* versus x_2, and e^* versus $\hat{y}$ suggest that the full quadratic model should be modified? Explain your answer.

b. The value of R^2 for the full quadratic model is .759. Test at level .05 the null hypothesis stating that there is no linear relationship between the dependent variable and any of the five predictors.

c. It can be shown that $V(Y) = \sigma^2 = V(\hat{Y}) + V(Y - \hat{Y})$. The estimate of σ is $\hat{\sigma} = s = 6.99$ (from the full quadratic model). First obtain the estimated standard deviation of $Y - \hat{Y}$ and then estimate the standard deviation of $\hat{Y}$ (i.e., $\hat{\beta}_0 + \hat{\beta}_1 x_1 + \hat{\beta}_2 x_2 + \hat{\beta}_3 x_1^2 + \hat{\beta}_4 x_2^2 + \hat{\beta}_5 x_1 x_2$) when $x_1 = 8.0$ and $x_2 = 33.1$. Finally, compute a 95% CI for mean strength. [*Hint:* What is $(y - \hat{y})/e^*$?]

d. Fitting the first-order model with regression function $\mu_{Y \cdot x_1 \cdot x_2} = \beta_0 + \beta_1 x_1 + \beta_2 x_2$ results in SSE = 894.95. Test at level .05 the null hypothesis that states that all quadratic terms can be deleted from the model.

52. The accompanying data resulted from a study of the relationship between brightness of finished paper (y) and the variables $H_2O_2\%$ by weight (x_1), NaOH% by weight (x_2), silicate % by weight (x_3), and process temperature (x_4) ("Advantages of

CEHDP Bleaching for High Brightness Kraft Pulp Production," *TAPPI*, 1964: 170A–173A). Each independent variable was allowed to assume five different values, and these values were coded for regression analysis as −2, −1, 0, 1, and 2.

Test No.	H$_2$O$_2$ (x_1)	NaOH Conc. (x_2)	Silicate Conc. (x_3)	Temp. (x_4)	Bright. (y)
1	−1	−1	−1	−1	83.9
2	+1	−1	−1	−1	84.9
3	−1	+1	−1	−1	83.4
4	+1	+1	−1	−1	84.2
5	−1	−1	+1	−1	83.8
6	+1	−1	+1	−1	84.7
7	−1	+1	+1	−1	84.0
8	+1	+1	+1	−1	84.8
9	−1	−1	−1	+1	84.5
10	+1	−1	−1	+1	86.0
11	−1	+1	−1	+1	82.6
12	+1	+1	−1	+1	85.1
13	−1	−1	+1	+1	84.5
14	+1	−1	+1	+1	86.0
15	−1	+1	+1	+1	84.0
16	+1	+1	+1	+1	85.4
17	−2	0	0	0	82.9
18	+2	0	0	0	85.5
19	0	−2	0	0	85.2
20	0	+2	0	0	84.5
21	0	0	−2	0	84.7
22	0	0	+2	0	85.0
23	0	0	0	−2	84.9
24	0	0	0	+2	84.0
25	0	0	0	0	84.5
26	0	0	0	0	84.7
27	0	0	0	0	84.6
28	0	0	0	0	84.9
29	0	0	0	0	84.9
30	0	0	0	0	84.5
31	0	0	0	0	84.6

Variables	−2	−1	0	+1	+2
x_1 Hydrogen peroxide (100%), %wt	.1	.2	.3	.4	.5
x_2 NaOH, %wt	.1	.2	.3	.4	.5
x_3 Silicate (41°Bé), %wt	5	1.5	2.5	3.5	4.5
x_4 Process temp., °F	130	145	160	175	190

a. When a (coded) model involving all linear terms, all quadratic terms, and all cross-product terms was fit, the estimated regression function was

$$y = 84.67 + .650x_1 - .258x_2 + .133x_3$$
$$+ .108x_4 - .135x_1^2 + .028x_2^2 + .028x_3^2$$
$$- .072x_4^2 + .038x_1x_2 - .075x_1x_3$$
$$+ .213x_1x_4 + .200x_2x_3 - .188x_2x_4$$
$$+ .050x_3x_4$$

Use this estimated model to predict brightness when H$_2$O$_2$ is .4%, NaOH is .4%, silicate is 3.5%, and temperature is 175. What are the values of the residuals for these values of the variables?

b. Express the estimated regression function in uncoded form.

c. SST = 17.2567, and R^2 for the model of part (a) is .885. When a model that includes only the four linear terms is fit, the resulting value of R^2 is .721. State and test at level .05 the null hypothesis that specifies that the coefficients of all quadratic and cross-product terms in the regression function are zero.

d. The estimated (coded) regression function when only linear terms are included is $\hat{\mu}_{Y \cdot x_1, x_2, x_3, x_4} = 85.5548 + .6500x_1 - .2583x_2 + .1333x_3 + .1083x_4$. When $x_1 = x_2 = x_3 = x_4 = 0$, the estimated standard deviation of $\hat{\mu}_{Y \cdot 0,0,0,0}$ is .0772. Suppose it had been believed that expected brightness for these values of the x_i's was at least 85.0. Does the given information contradict this belief? State and test the appropriate hypotheses.

13.5 | **Other Issues in Multiple Regression**

In this section, we touch upon a number of issues that may arise when a multiple regression analysis is carried out. Consult the chapter references for a more extensive treatment of any particular topic.

Transformations in Multiple Regression

Often theoretical considerations suggest a nonlinear relation between a dependent variable and two or more independent variables, whereas on other occasions diagnostic plots indicate that some type of nonlinear function should be used. Frequently a transformation will linearize the model.

Example 13.17 An article in *Lubrication Eng.* ("Accelerated Testing of Solid Film Lubricants," 1972: 365–372) reports on an investigation of wear life for solid film lubricant. Three sets of journal bearing tests were run on a Mil-L-8937-type film at each combination of three loads (3000, 6000, and 10,000 psi) and three speeds (20, 60, and 100 rpm), and the wear life (hours) was recorded for each run, as shown in Table 13.7.

Table 13.7   Wear-life data for Example 13.17

s	l(1000's)	w	s	l(1000's)	w
20	3	300.2	60	6	65.9
20	3	310.8	60	10	10.7
20	3	333.0	60	10	34.1
20	6	99.6	60	10	39.1
20	6	136.2	100	3	26.5
20	6	142.4	100	3	22.3
20	10	20.2	100	3	34.8
20	10	28.2	100	6	32.8
20	10	102.7	100	6	25.6
60	3	67.3	100	6	32.7
60	3	77.9	100	10	2.3
60	3	93.9	100	10	4.4
60	6	43.0	100	10	5.8
60	6	44.5			

The article contains the comment that a lognormal distribution is appropriate for W, since $\ln(W)$ is known to follow a normal law (recall from Chapter 4 that this is what defines a lognormal distribution). The model that appears is $W = (c/s^a l^b) \cdot \epsilon$, from which $\ln(W) = \ln(c) - a \ln(s) - b \ln(l) + \ln(\epsilon)$; so with $Y = \ln(W)$, $x_1 = \ln(s)$, $x_2 = \ln(l)$, $\beta_0 = \ln(c)$, $\beta_1 = -a$, and $\beta_2 = -b$, we have a multiple linear regression model. After computing $\ln(w_i)$, $\ln(s_i)$, and $\ln(l_i)$ for the data, a first-order model in the transformed variables yielded the results shown in Table 13.8 (page 598).

Table 13.8 Estimated coefficients and *t*-ratios for Example 13.17

Parameter β_i	Estimate $\hat{\beta}_i$	Estimated SD $s_{\hat{\beta}_i}$	$t = \hat{\beta}_i / s_{\hat{\beta}_i}$
β_0	10.8719	.7871	13.81
β_1	−1.2054	.1710	−7.05
β_2	−1.3979	.2327	−6.01

The coefficient of multiple determination (for the transformed observations) has value $R^2 = .781$. The estimated regression function for the transformed variables is

$$\ln(w) = 10.87 - 1.21 \ln(s) - 1.40 \ln(l)$$

so that the original regression function is estimated as

$$w = e^{10.87} \cdot s^{-1.21} \cdot l^{-1.40}$$

The Bonferroni approach can be used to obtain simultaneous CIs for β_1 and β_2, and because $\beta_1 = -a$ and $\beta_2 = -b$, intervals for a and b are then immediately available. ■

In Section 13.2, the logistic regression model was introduced to relate a dichotomous variable y to a single predictor. This model can be extended in an obvious way to incorporate more than one predictor.

Standardizing Variables

In Section 13.3, we considered transforming x to $x' = x - \bar{x}$ before fitting a polynomial. For multiple regression, especially when values of variables are large in magnitude, it is advantageous to carry this coding one step further. Let $\bar{x}_i$ and s_i be the sample average and sample standard deviation of the x_{ij}'s ($j = 1, \ldots, n$). We now code each variable x_i by $x'_i = (x_i - \bar{x}_i)/s_i$. The coded variable x'_i simply reexpresses any x_i value in units of standard deviation above or below the mean. Thus, if $\bar{x}_i = 100$ and $s_i = 20$, $x_i = 130$ becomes $x'_i = 1.5$ because 130 is 1.5 standard deviations above the mean of the values of x_i. For example, the coded full second-order model with two independent variables has regression function

$$E(Y) = \beta_0 + \beta_1\left(\frac{x_1 - \bar{x}_1}{s_1}\right) + \beta_2\left(\frac{x_2 - \bar{x}_2}{s_2}\right) + \beta_3\left(\frac{x_1 - \bar{x}_1}{s_1}\right)^2$$
$$+ \beta_4\left(\frac{x_2 - \bar{x}_2}{s_2}\right)^2 + \beta_5\left(\frac{x_1 - \bar{x}_1}{s_1}\right)\left(\frac{x_2 - \bar{x}_2}{s_2}\right)$$
$$= \beta_0 + \beta_1 x'_1 + \beta_2 x'_2 + \beta_3 x'_3 + \beta_4 x'_4 + \beta_5 x'_5$$

The benefits of coding are (1) increased numerical accuracy in all computations (through less computer round-off error) and (2) more accurate estimation than for the parameters of the uncoded model because the individual parameters of the coded model characterize the behavior of the regression function near the center of the data rather than near the origin.

Example 13.18 The article "The Value and the Limitations of High-Speed Turbo-Exhausters for the Removal of Tar-Fog from Carburetted Water-Gas" (*Soc. Chemical Industry J.*, 1946: 166–168) presents the data (in Table 13.9) on y = tar content (grains/100 ft³) of a gas stream as a function of x_1 = rotor speed (rpm) and x_2 = gas inlet temperature (°F). The data is also considered in the article "Some Aspects of Nonorthogonal Data Analysis" (*J. Quality Technology*, 1973: 67–79), which suggests using the coded model described previously.

Table 13.9 Data for Example 13.18

Run	y	x_1	x_2	x_1'	x_2'
1	60.0	2400	54.5	−1.52428	−.57145
2	61.0	2450	56.0	−1.39535	−.35543
3	65.0	2450	58.5	−1.39535	.00461
4	30.5	2500	43.0	−1.26642	−2.22763
5	63.5	2500	58.0	−1.26642	−.06740
6	65.0	2500	59.0	−1.26642	.07662
7	44.0	2700	52.5	−.75070	−.85948
8	52.0	2700	65.5	−.75070	1.01272
9	54.5	2700	68.0	−.75070	1.37276
10	30.0	2750	45.0	−.62177	−1.93960
11	26.0	2775	45.5	−.55731	−1.86759
12	23.0	2800	48.0	−.49284	−1.50755
13	54.0	2800	63.0	−.49284	.65268
14	36.0	2900	58.5	−.23499	.00461
15	53.5	2900	64.5	−.23499	.86870
16	57.0	3000	66.0	.02287	1.08472
17	33.5	3075	57.0	.21627	−.21141
18	34.0	3100	57.5	.28073	−.13941
19	44.0	3150	64.0	.40966	.79669
20	33.0	3200	57.0	.53859	−.21141
21	39.0	3200	64.0	.53859	.79669
22	53.0	3200	69.0	.53859	1.51677
23	38.5	3225	68.0	.60305	1.37276
24	39.5	3250	62.0	.66752	.50866
25	36.0	3250	64.5	.66752	.86870
26	8.5	3250	48.0	.66752	−1.50755
27	30.0	3500	60.0	1.31216	.22063
28	29.0	3500	59.0	1.31216	.07662
29	26.5	3500	58.0	1.31216	−.06740
30	24.5	3600	58.0	1.57002	−.06740
31	26.5	3900	61.0	2.34360	.36465

The means and standard deviations are $\bar{x}_1 = 2991.13$, $s_1 = 387.81$, $\bar{x}_2 = 58.468$, and $s_2 = 6.944$, so $x_1' = (x_1 - 2991.13)/387.81$ and $x_2' = (x_2 - 58.468)/6.944$. With $x_3' = (x_1')^2$, $x_4' = (x_2')^2$, $x_5' = x_1' \cdot x_2'$, fitting the full second-order model requires solving the system of six normal equations in six unknowns. A computer analysis yielded $\hat{\beta}_0 = 40.2660$, $\hat{\beta}_1 = -13.4041$, $\hat{\beta}_2 = 10.2553$, $\hat{\beta}_3 = 2.3313$, $\hat{\beta}_4 = -2.3405$, and $\hat{\beta}_5 = 2.5978$. The estimated regression equation is then

$$\hat{y} = 40.27 - 13.40x_1' + 10.26x_2' + 2.33x_3' - 2.34x_4' + 2.60x_5'$$

Thus, if $x_1 = 3200$ and $x_2 = 57.0$, $x_1' = .539$, $x_2' = -.211$, $x_3' = (.539)^2 = .2901$, $x_4' = (-.211)^2 = .0447$, and $x_5' = (.539)(-.211) = -.1139$, so

$$\hat{y} = 40.27 - (13.40)(.539) + (10.26)(-.211) + (2.33)(.2901)$$
$$- (2.34)(.0447) + (2.60)(-.1139) = 31.16 \qquad \blacksquare$$

Variable Selection

Often an experimenter will have data on a large number of predictors and then wish to build a regression model involving a subset of the predictors. The use of the subset will make the resulting model more manageable, especially if more data is to be subsequently collected, and also result in a model that is easier to interpret and understand than one with many more predictors. Two fundamental questions in connection with variable selection are the following:

1. If we can examine regressions involving all possible subsets of the predictors for which data is available, what criteria should be used to select a model?

2. If the number of predictors is too large to permit all regressions to be examined, is there a way of examining a reduced number of subsets among which a good model (or models) will be found?

To address Question (1) first, if the number of predictors is small (≤ 5, say), then it would not be too tedious to examine all possible regressions using any one of the readily available statistical computer packages. If data on at least six predictors is available, all possible regressions involve at least 64 ($= 2^6$) different models. Several packages will, for any specified m between 1 and 10, give the m best one-predictor models, the m best two-predictor models, and so on ("best" here means smallest SSE or, equivalently, largest R^2). MINITAB will do this for as many as 20 predictors, whereas BMDP will handle up to 27. The corresponding SSEs (or functions of them) can then be compared according to any of the criteria described next. The reason for specifying an m greater than 1 is to see whether the best models have similar SSE or R^2 values.

Criteria for Variable Selection   Again, we use a subscript k to denote a quantity (say, SSE_k) computed from a model with k predictors (and thus $k + 1$ β_i's, because β_0 will always be included). For a fixed value of k, it is reasonable to identify the best model as the one having minimum SSE_k. The more difficult issue concerns comparison of SSE_k's for different values of k. Three different criteria, each one a simple function of SSE_k, are widely used.

1. R_k^2, the coefficient of multiple determination for a k-predictor model. Because R_k^2 will virtually always increase as k does (and can never decrease), we are not interested in the k that maximizes R_k^2. Instead we wish to identify a small k for which R_k^2 is nearly as large as R^2 for all predictors in the model.

2. $MSE_k = SSE/(n - k - 1)$, the mean squared error for a k-predictor model. This is often used in place of R_k^2, because although R_k^2 never decreases with increasing k, a small decrease in SSE_k obtained with one extra predictor can be more than offset by a decrease of one in the denominator of MSE_k. The objective is then to find the

model having minimum MSE_k. Since adjusted $R_k^2 = 1 - MSE_k/MST$ where $MST = SST/(n - 1)$ is constant in k, examination of adjusted R_k^2 is equivalent to consideration of MSE_k.

3. The rationale for the third criterion, C_k, is more difficult to understand, but the criterion is gaining increasing acceptance among data analysts. Suppose the true regression model is specified by m predictors—that is,

$$Y = \beta_0 + \beta_1 x_1 + \cdots + \beta_m x_m + \epsilon \qquad V(\epsilon) = \sigma^2$$

so that

$$E(Y) = \beta_0 + \beta_1 x_1 + \cdots + \beta_m x_m$$

Consider fitting a model by using a subset of k of these m predictors; for simplicity of notation, suppose we use $x_1, x_2, \ldots, x_k$. Then by solving the system of normal equations, estimates $\hat{\beta}_0, \hat{\beta}_1, \ldots, \hat{\beta}_k$ are obtained (but not, of course, estimates of any β's corresponding to predictors not in the fitted model). The true expected value $E(Y)$ can then be estimated by $\hat{Y} = \hat{\beta}_0 + \hat{\beta}_1 x_1 + \cdots + \hat{\beta}_k x_k$. Now consider the **normalized expected total error of estimation**

$$\Gamma_k = \frac{E\left(\sum_{i=1}^{n} [\hat{Y}_i - E(Y_i)]^2\right)}{\sigma^2} = \frac{E(SSE_k)}{\sigma^2} + 2(k + 1) - n \qquad (13.21)$$

The second equality in (13.21) must be taken on faith because it requires a tricky expected-value argument. A particular subset is then appealing if its Γ_k value is small. Unfortunately, though, $E(SSE_k)$ and σ^2 are not known. To remedy this, let s^2 denote the estimate of σ^2 based on the model that includes all predictors for which data is available and define

$$C_k = \frac{SSE_k}{s^2} + 2(k + 1) - n$$

A desirable model is then specified by a subset of predictors for which C_k is small.

Example 13.19 The review article by Ron Hocking listed in the chapter bibliography reports on an analysis of data taken from the 1974 issues of *Motor Trend* magazine. The dependent variable y was gas mileage, there were $n = 32$ observations, and the predictors for which data was obtained were x_1 = engine shape (1 = straight and 0 = V), x_2 = number of cylinders, x_3 = transmission type (1 = manual and 0 = auto), x_4 = number of transmission speeds, x_5 = engine size, x_6 = horsepower, x_7 = number of carburetor barrels, x_8 = final drive ratio, x_9 = weight, and x_{10} = quarter-mile time. In Table 13.10 (page 602), we present summary information from the analysis. The table describes for each k the subset having minimum SSE_k; reading down the variables column indicates which variable is added in going from k to $k + 1$ (in going from $k = 2$ to $k = 3$, both x_3 and x_{10} are added, and x_2 is deleted). Figure 13.16 contains plots of R_k^2, adjusted R_k^2, and C_k against k; these plots are an important visual aid in selecting a subset. The estimate of σ^2 is $s^2 = 6.24$, which is MSE_{10}. A simple model that rates highly according to all criteria is the one containing predictors x_3, x_9, and x_{10}.

Table 13.10 Best subsets for gas mileage data of Example 13.19

k = Number of Predictors	Variables	SSE_k	R_k^2	Adjusted R_k^2	C_k
1	9	247.2	.756	.748	11.6
2	2	169.7	.833	.821	1.2
3	3, 10, −2	150.4	.852	.836	.1
4	6	142.3	.860	.839	.8
5	5	136.2	.866	.840	1.8
6	8	133.3	.869	.837	3.4
7	4	132.0	.870	.832	5.2
8	7	131.3	.871	.826	7.1
9	1	131.1	.871	.818	9.0
10	2	131.0	.871	.809	11.0

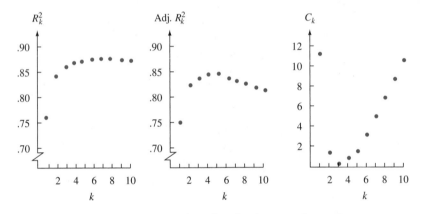

Figure 13.16 R_k^2 and C_k plots for the gas mileage data ■

Generally speaking, when a subset of k predictors ($k < m$) is used to fit a model, the estimators $\hat{\beta}_0, \hat{\beta}_1, \ldots, \hat{\beta}_k$ will be biased for $\beta_0, \beta_1, \ldots, \beta_k$ and $\hat{Y}$ will also be a biased estimator for the true $E(Y)$ (all this because $m - k$ predictors are missing from the fitted model). However, as measured by the total normalized expected error Γ_k, estimates based on a subset can provide more precision than would be obtained using all possible predictors; essentially this greater precision is obtained at the price of introducing a bias in the estimators. A value of k for which $C_k \approx k + 1$ indicates that the bias associated with this k-predictor model would be small.

Example 13.20 The bond shear strength data introduced in Example 13.11 contains values of four different independent variables x_1–x_4. We found that the model with only these four variables as predictors was useful, and there is no compelling reason to consider the inclusion of second-order predictors. Figure 13.17 is the MINITAB output that results from a request to identify the two best models of each given size.

The best two-predictor model, with predictors power and temperature, seems to be a very good choice on all counts: R^2 is significantly higher than for models with fewer predictors yet almost as large as for any larger models, adjusted R^2 is almost at its maximum for this data, and C_k is small and close to $2 + 1 = 3$.

```
Response is strength                                    f     p
                                                        o     o    t    t
                                                        r     w    e    i
                        Adj.                            c     e    m    m
Vars    R-sq    R-sq        C-p         s               e     r    p    e
  1     57.7    56.2       11.0      5.9289                        X
  1     10.8     7.7       51.9      8.6045                   X
  2     68.5    66.2        3.5      5.2070             X     X
  2     59.4    56.4       11.5      5.9136             X               X
  3     70.2    66.8        4.0      5.1590             X     X         X
  3     69.7    66.2        4.5      5.2078       X     X     X
  4     71.4    66.8        5.0      5.1580       X     X     X         X
```

Figure 13.17   Output from MINITAB's Best Subsets option ■

Stepwise Regression   When the number of predictors is too large to allow for explicit or implicit examination of all possible subsets, several alternative selection procedures generally will identify good models. The simplest such procedure is the **backward elimination** (BE) method. This method starts with the model in which all predictors under consideration are used. Let the set of all such predictors be $x_1, \ldots, x_m$. Then each t-ratio $\hat{\beta}_i/s_{\hat{\beta}_i}$ ($i = 1, \ldots, m$) appropriate for testing H_0: $\beta_i = 0$ versus H_a: $\beta_i \neq 0$ is examined. If the t-ratio with the smallest absolute value is less than a prespecified constant t_{out}, that is, if

$$\min_{i=1,\ldots,m} \left| \frac{\hat{\beta}_i}{s_{\hat{\beta}_i}} \right| < t_{out}$$

then the predictor corresponding to the smallest ratio is eliminated from the model. The reduced model is now fit, the $m - 1$ t-ratios are again examined, and another predictor is eliminated if it corresponds to the smallest absolute t-ratio smaller than t_{out}. In this way, the algorithm continues until at some stage, all absolute t-ratios are at least t_{out}. The model used is the one containing all predictors that were not eliminated. The value $t_{out} = 2$ is often recommended since most $t_{.05}$ values are near 2.

Example 13.21
(Example 13.18 continued)   For the coded full quadratic model in which $y = $ tar content, the five potential predictors are x_1', x_2', $x_3' = x_1'^2$, $x_4' = x_2'^2$, and $x_5' = x_1'x_2'$ (so $m = 5$). Without specifying t_{out}, the predictor with the smallest absolute t-ratio (asterisked) was eliminated at each stage, resulting in the sequence of models shown in Table 13.11.

Table 13.11   Backward elimination results for the data of Example 13.18

		$\mid t$-ratio $\mid$				
Step	**Predictors**	**1**	**2**	**3**	**4**	**5**
1	1, 2, 3, 4, 5	16.0	10.8	2.9	2.8	1.8*
2	1, 2, 3, 4	15.4	10.2	3.7	2.0*	—
3	1, 2, 3	14.5	12.2	4.3*	—	—
4	1, 2	10.9	9.1*	—	—	—
5	1	4.4*	—	—	—	—

Using $t_{out} = 2$, the resulting model would be based on x_1', x_2', and x_3', since at Step 3 no predictor could be eliminated. It can be verified that each subset is actually the best subset of its size, though this is by no means always the case. ■

An alternative to the BE procedure is **forward selection** (FS). FS starts with no predictors in the model and considers fitting in turn the model with only x_1, only x_2, . . . , and finally only x_m. The variable that, when fit, yields the largest absolute t-ratio enters the model provided that the ratio exceeds the specified constant t_{in}. Suppose x_1 enters the model. Then models with (x_1, x_2), (x_1, x_3), . . . , (x_1, x_m) are considered in turn. The largest $|\hat{\beta}_j/s_{\hat{\beta}_j}|$ $(j = 2, \ldots, m)$ then specifies the entering predictor provided that this maximum also exceeds t_{in}. This continues until at some step no absolute t-ratios exceed t_{in}. The entered predictors then specify the model. The value $t_{in} = 2$ is often used for the same reason that $t_{out} = 2$ is used in BE. For the tar-content data, FS resulted in the sequence of models given in Steps 5, 4, . . . , 1 in Table 13.11, and thus is in agreement with BE. This will not always be the case.

The stepwise procedure most widely used is a combination of FS and BE, denoted by FB. This procedure starts as does forward selection, by adding variables to the model, but after each addition it examines those variables previously entered to see whether any is a candidate for elimination. For example, if there are eight predictors under consideration and the current set consists of x_2, x_3, x_5, and x_6 with x_5 having just been added, the t-ratios $\hat{\beta}_2/s_{\hat{\beta}_2}$, $\hat{\beta}_3/s_{\hat{\beta}_3}$, and $\hat{\beta}_6/s_{\hat{\beta}_6}$ are examined. If the smallest absolute ratio is less than t_{out}, then the corresponding variable is eliminated from the model. The idea behind FB is that with forward selection, a single variable may be more strongly related to y than either of two or more other variables individually, but the combination of these variables may make the single variable subsequently redundant. This actually happened with the gas mileage data discussed in Example 13.19, with x_2 entering and subsequently leaving the model.

The FB procedure is part of several standard computer packages. The BMDP package specifies $t_{in} = 2$ and $t_{out} = \sqrt{3.9}$ (most packages actually use $f = t^2$ rather than t itself).

Although in most situations these automatic selection procedures will identify a good model, there is no guarantee that the best or even a nearly best model will result. Close scrutiny should be given to data sets for which there appear to be strong relationships among some of the potential predictors; we will say more about this shortly.

Identification of Influential Observations

In simple linear regression, it is easy to spot an observation whose x value is much larger or much smaller than other x values in the sample. Such an observation may have a great impact on the estimated regression equation (whether it actually does depends on how consistent the corresponding y value is to the remainder of the data). In multiple regression, it is also desirable to know whether the values of the predictors for a particular observation are such that it has the potential for exerting great influence on the estimated equation. One method for identifying potentially influential observations relies on the fact that because each $\hat{\beta}_i$ is a linear function of $y_1, y_2, \ldots, y_n$, each predicted y value of the form $\hat{y} = \hat{\beta}_0 + \hat{\beta}_1 x_1 + \cdots + \hat{\beta}_k x_k$ is also a linear function of the y_j's. In particular, the predicted values corresponding to sample observations can be written as follows:

$$\hat{y}_1 = h_{11}y_1 + h_{12}y_2 + \cdots + h_{1n}y_n$$
$$\hat{y}_2 = h_{21}y_1 + h_{22}y_2 + \cdots + h_{2n}y_n$$
$$\vdots \qquad \vdots \qquad \vdots \qquad \vdots$$
$$\hat{y}_n = h_{n1}y_1 + h_{n2}y_2 + \cdots + h_{nn}y_n$$

Each coefficient h_{ij} is a function only of the x_{ij}'s in the sample and not of the y_j's. It can be shown that $h_{ij} = h_{ji}$ and that $0 \leq h_{ij} \leq 1$.

Let's focus on the "diagonal" coefficients $h_{11}, h_{22}, \ldots, h_{nn}$. The coefficient h_{jj} is the weight given to y_j in computing the corresponding predicted value $\hat{y}_j$. This quantity can also be expressed as a measure of the distance between the point $(x_{1j}, \ldots, x_{kj})$ in k-dimensional space and the center of the data $(\bar{x}_1, \ldots, \bar{x}_k)$. It is therefore natural to characterize an observation whose h_{jj} is relatively large as one that has potentially large influence. Unless there is a perfect linear relationship among the k predictors, $\sum_{j=1}^{n} h_{jj} = k + 1$, so the average of the h_{jj}'s is $(k + 1)/n$. Some statisticians suggest that if $h_{jj} > 2(k + 1)/n$, the jth observation be cited as being potentially influential; others use $3(k + 1)/n$ as the dividing line.

Example 13.22 The accompanying data appeared in the article "Testing for the Inclusion of Variables in Linear Regression by a Randomization Technique" (*Technometrics*, 1966: 695–699) and was reanalyzed in Hoaglin and Welsch, "The Hat Matrix in Regression and ANOVA" (*Amer. Statistician*, 1978: 17–23). The h_{ij}'s (with elements below the diagonal omitted by symmetry) follow the data.

Beam Number	Specific Gravity (x_1)	Moisture Content (x_2)	Strength (y)
1	.499	11.1	11.14
2	.558	8.9	12.74
3	.604	8.8	13.13
4	.441	8.9	11.51
5	.550	8.8	12.38
6	.528	9.9	12.60
7	.418	10.7	11.13
8	.480	10.5	11.70
9	.406	10.5	11.02
10	.467	10.7	11.41

	1	2	3	4	5	6	7	8	9	10
1	.418	−.002	.079	−.274	−.046	.181	.128	.222	.050	.242
2		.242	.292	.136	.243	.128	−.041	.033	−.035	.004
3			.417	−.019	.273	.187	−.126	.044	−.153	.004
4				.604	.197	−.038	.168	−.022	.275	−.028
5					.252	.111	−.030	.019	−.010	−.010
6						.148	.042	.117	.012	.111
7							.262	.145	.277	.174
8								.154	.120	.168
9									.315	.148
10										.187

Here $k = 2$ so $(k + 1)/n = 3/10 = .3$; since $h_{44} = .604 > 2(.3)$, the fourth data point is identified as potentially influential. ∎

Another technique for assessing the influence of the jth observation that takes into account y_j as well as the predictor values involves deleting the jth observation from the data set and performing a regression based on the remaining observations. If the estimated coefficients from the "deleted observation" regression differ greatly from the estimates based on the full data, the jth observation has clearly had a substantial impact on the fit. One way to judge whether estimated coefficients change greatly is to express each change relative to the estimated standard deviation of the coefficient:

$$\frac{(\hat{\beta}_i \text{ before deletion}) - (\hat{\beta}_i \text{ after deletion})}{s_{\hat{\beta}_i}} = \frac{\text{change in } \hat{\beta}_i}{s_{\hat{\beta}_i}}$$

There exist efficient computational formulas that allow all this information to be obtained from the "no-deletion" regression, so that the additional n regressions are unnecessary.

Example 13.23 Consider separately deleting observations 1 and 6, whose residuals are the largest, and observation 4, where h_{jj} is large. Table 13.12 contains the relevant information.
(Example 13.22 continued)

Table 13.12 Changes in estimated coefficients for Example 13.23

Parameter	No-deletions Estimates	Estimated SD	Change When Point j Is Deleted		
			$j = 1$	$j = 4$	$j = 6$
β_0	10.302	1.896	2.710	−2.109	−.642
β_1	8.495	1.784	−1.772	1.695	.748
β_2	.2663	.1273	−.1932	.1242	.0329
		e_j:	−3.25	−.96	2.20
		h_{jj}:	.418	.604	.148

For deletion of both point 1 and point 4, the change in each estimate is in the range 1–1.5 standard deviations, which is reasonably substantial (this does not tell us what would happen if both points were simultaneously omitted). For point 6, however, the change is roughly .25 standard deviation. Thus, points 1 and 4, but not 6, might well be omitted in calculating a regression equation. ■

Multicollinearity

In many multiple regression data sets, the predictors $x_1, x_2, \ldots, x_k$ are highly interdependent. Suppose we consider the usual model

$$Y = \beta_0 + \beta_1 x_1 + \cdots + \beta_k x_k + \epsilon$$

with data $(x_{1j}, \ldots, x_{kj}, y_j)$ $(j = 1, \ldots, n)$ available for fitting. If we use the principle of least squares to regress x_i on the other predictors $x_1, \ldots, x_{i-1}, x_{i+1}, \ldots, x_k$, obtaining

$$\hat{x}_i = a_0 + a_1 x_1 + \cdots + a_{i-1} x_{i-1} + a_{i+1} x_{i+1} + \cdots + a_k x_k$$

it can be shown that

$$V(\hat{\beta}_i) = \frac{\sigma^2}{\sum_{j=1}^{n} (x_{ij} - \hat{x}_{ij})^2} \tag{13.22}$$

When the sample x_i values can be predicted very well from the other predictor values, the denominator of (13.22) will be small, so $V(\hat{\beta}_i)$ will be quite large. If this is the case for at least one predictor, the data is said to exhibit **multicollinearity.** Multicollinearity is often suggested by a regression computer output in which R^2 is large but some of the t-ratios $\hat{\beta}_i/s_{\hat{\beta}_i}$ are small for predictors that, based on prior information and intuition, seem important. Another clue to the presence of multicollinearity lies in a $\hat{\beta}_i$ value that has the opposite sign from that which intuition would suggest, indicating that another predictor or collection of predictors is serving as a "proxy" for x_i.

An assessment of the extent of multicollinearity can be obtained by regressing each predictor in turn on the remaining $k - 1$ predictors. Let R_i^2 denote the value of R^2 in the regression with dependent variable x_i and predictors $x_1, \ldots, x_{i-1}, x_{i+1}, \ldots, x_k$. It has been suggested that severe multicollinearity is present if $R_i^2 > .9$ for any i. MINITAB will refuse to include a predictor in the model when its R_i^2 value is quite close to 1.

There is unfortunately no consensus among statisticians as to what remedies are appropriate when severe multicollinearity is present. One possibility involves continuing to use a model that includes all the predictors but estimating parameters by using something other than least squares. Consult a chapter reference for more details.

Exercises | Section 13.5 (53–61)

53. The article "Bank Full Discharge of Rivers" (*Water Resources J.,* 1978: 1141–1154) reports data on discharge amount (q, in m³/sec), flow area (a, in m²), and slope of the water surface (b, in m/m) obtained at a number of floodplain stations. A subset of the data follows. The article proposed a multiplicative power model $Q = \alpha a^\beta b^\gamma \epsilon$.

q	17.6	23.8	5.7	3.0	7.5
a	8.4	31.6	5.7	1.0	3.3
b	.0048	.0073	.0037	.0412	.0416

q	89.2	60.9	27.5	13.2	12.2
a	41.1	26.2	16.4	6.7	9.7
b	.0063	.0061	.0036	.0039	.0025

a. Use an appropriate transformation to make the model linear and then estimate the regression parameters for the transformed model. Finally, estimate α, β, and γ (the parameters of the original model). What would be your prediction of discharge amount when flow area is 10 and slope is .01?

b. Without actually doing any analysis, how would you fit a multiplicative exponential model $Q = \alpha e^{\beta a} e^{\gamma b} \epsilon$?

c. After the transformation to linearity in part (a), a 95% CI for the value of the transformed regression function when $a = 3.3$ and $b = .0046$ was obtained from computer output as (.217, 1.755). Obtain a 95% CI for $\alpha a^\beta b^\gamma$ when $a = 3.3$ and $b = .0046$.

54. In an experiment to study factors influencing wood specific gravity ("Anatomical Factors Influencing Wood Specific Gravity of Slash Pines and the Implications for the Development of a High-Quality Pulpwood," *TAPPI,* 1964: 401–404), a sample of 20 mature wood samples was obtained, and measurements were taken on number of fibers/mm² in springwood (x_1), number of fibers/mm² in summerwood (x_2), % springwood (x_3), light absorption in springwood (x_4), and light absorption in summerwood (x_5).

a. Fitting the regression function $\mu_{Y \cdot x_1, x_2, x_3, x_4, x_5} = \beta_0 + \beta_1 x_1 + \cdots + \beta_5 x_5$ resulted in $R^2 = .769$. Does the data indicate that there is a linear relationship between specific gravity and at least one of the predictors? Test using $\alpha = .01$.

b. When x_2 is dropped from the model, the value of R^2 remains at .769. Compute adjusted R^2 for both the full model and the model with x_2 deleted.

c. When x_1, x_2, and x_4 are all deleted, the resulting value of R^2 is .654. The total sum of squares is SST = .0196610. Does the data suggest that all of x_1, x_2, and x_4 have zero coefficients in the true regression model? Test the relevant hypotheses at level .05.

d. The mean and standard deviation of x_3 were 52.540 and 5.4447, respectively, whereas those of x_5 were 89.195 and 3.6660, respectively. When the model involving these two standardized variables was fit, the estimated regression equation was $y = .5255 - .0236x_3' + .0097x_5'$. What value of specific gravity would you predict for a wood sample with % springwood = 50 and % light absorption in summerwood = 90?

e. The estimated standard deviation of the estimated coefficient $\hat{\beta}_3$ of x_3' (i.e., for $\hat{\beta}_3$ of the standardized model) was .0046. Obtain a 95% CI for β_3.

f. Using the information in parts (d) and (e), what is the estimated coefficient of x_3 in the unstandardized model (using only predictors x_3 and x_5), and what is the estimated standard deviation of the coefficient estimator (i.e., $s_{\hat{\beta}_3}$ for $\hat{\beta}_3$ in the unstandardized model)?

g. The estimate of σ for the two-predictor model is $s = .02001$, whereas the estimated standard deviation of $\hat{\beta}_0 + \hat{\beta}_3 x_3' + \hat{\beta}_5 x_5'$ when $x_3' = -.3747$ and $x_5' = -.2769$ (i.e., when $x_3 = 50.5$ and $x_5 = 88.9$) is .00482. Compute a 95% PI for specific gravity when % springwood = 50.5 and % light absorption in summerwood = 88.9.

55. In the accompanying table, we give the smallest SSE for each number of predictors k ($k = 1, 2, 3, 4$) for a regression problem in which y = cumulative heat of hardening in cement, x_1 = % tricalcium aluminate, x_2 = % tricalcium silicate, x_3 = % aluminum ferrate, and x_4 = % dicalcium silicate.

Number of Predictors k	Predictor(s)	SSE
1	x_4	880.85
2	x_1, x_2	58.01
3	x_1, x_2, x_3	49.20
4	x_1, x_2, x_3, x_4	47.86

In addition, $n = 13$, and SST = 2715.76.

a. Use the criteria discussed in the text to recommend the use of a particular regression model.

b. Would forward selection result in the best two-predictor model? Explain.

56. The article "Creep and Fatigue Characteristics of Ferrocement Slabs" (*J. Ferrocement*, 1984: 309–322) reported data on y = tensile strength (MPa), x_1 = slab thickness (cm), x_2 = load (kg), x_3 = age at loading (days), and x_4 = time under test (days) resulting from stress tests of $n = 9$ reinforced concrete slabs. The results of applying the BE elimination method of variable selection are summarized in the accompanying tabular format. Explain what occurred at each step of the procedure.

Step	1	2	3
Constant	8.496	12.670	12.989
x_1	−.29	−.42	−.49
T-RATIO	−1.33	−2.89	−3.14
x_2	.0104	.0110	.0116
T-RATIO	6.30	7.40	7.33
x_3	.0059		
T-RATIO	.83		
x_4	−.023	−.023	
T-RATIO	−1.48	−1.53	
S	.533	.516	.570
R-SQ	95.81	95.10	92.82

57. MINITAB's Best Regression option was used on the wood specific gravity data referred to in Exercise 54, resulting in the accompanying output. Which model(s) would you recommend investigating in more detail?

```
Response is spgrav
                                              s     %        s
                                              p  s  s  s     u
                                              r  u  p  p     m
                                              n  m  r  l     l
                                              g  r  w  t     t
                                              f  f  o  a     a
                              R-Sq            i  i  o  b     b
      Vars  R-Sq  (adj)   C-p        s        b  b  d  s     s
         1  56.4   53.9  10.6  0.021832                X
         1  10.6    5.7  38.5  0.031245                            X
         1   5.3    0.1  41.7  0.032155          X
         2  65.5   61.4   7.0  0.019975                X           X
         2  62.1   57.6   9.1  0.020950          X     X
         2  60.3   55.6  10.2  0.021439    X           X
         3  72.3   67.1   4.9  0.018461    X           X           X
         3  71.2   65.8   5.6  0.018807                X  X  X
         3  71.1   65.7   5.6  0.018846          X     X           X
         4  77.0   70.9   4.0  0.017353    X           X  X  X
         4  74.8   68.1   5.4  0.018179          X     X  X  X     X
         4  72.7   65.4   6.7  0.018919    X     X     X           X
         5  77.0   68.9   6.0  0.017953    X     X     X  X  X     X
```

58. The accompanying MINITAB output resulted from applying both the backward elimination method and the forward selection method to the wood specific gravity data referred to in Exercise 54. For each method, explain what occurred at every iteration of the algorithm.

```
Response is spgrav on 5 predictors,
with N = 20

Step            1       2       3       4
Constant     0.4421  0.4384  0.4381  0.5179

sprngfib     0.00011 0.00011 0.00012
T-Value         1.17    1.95    1.98

sumrfib      0.00001
T-Value         0.12

%sprwood    -0.00531 -0.00526 -0.00498 -0.00438
T-Value        -5.70    -6.56    -5.96    -5.20

spltabs      -0.0018 -0.0019
T-Value        -1.63   -1.76

sumltabs      0.0044  0.0044  0.0031  0.0027
T-Value         3.01    3.31    2.63    2.12

S             0.0180  0.0174  0.0185  0.0200
R-Sq           77.05   77.03   72.27   65.50
```

(continued at top of next column)

```
Response is spgrav on 5 predictors,
with N = 20

Step             1       2
Constant      0.7585  0.5179

%sprwood     -0.00444 -0.00438
T-Value         -4.82    -5.20

sumltabs               0.0027
T-Value                  2.12

S             0.0218  0.0200
R-Sq           56.36   65.50
```

59. Reconsider the wood specific gravity data referred to in Exercise 54. The following R^2 values resulted from regressing each predictor on the other four predictors (in the first regression, the dependent variable was x_1 and the predictors were x_2–x_5, etc.): .628, .711, .341, .403, and .403. Does multicollinearity appear to be a substantial problem? Explain.

60. A study carried out to investigate the relationship between a response variable relating to pressure drops in a screen-plate bubble column and the predictors x_1 = superficial fluid velocity, x_2 = liquid viscosity, and x_3 = opening mesh size resulted in the accompanying data ("A Correlation of Two-Phase Pressure Drops in Screen-Plate Bubble Column," *Canad. J. of Chem. Engr.*, 1993: 460–463). The standardized residuals and h_{ii} values resulted from the model with just x_1, x_2, and x_3 as predictors. Are there any unusual observations?

Observation	Velocity	Viscosity	Mesh size	Response	Standardized residual	h_{ii}
1	2.14	10.00	.34	28.9	2.01721	.202242
2	4.14	10.00	.34	26.1	1.34706	.066929
3	8.15	10.00	.34	22.8	.96537	.274393
4	2.14	2.63	.34	24.2	1.29177	.224518
5	4.14	2.63	.34	15.7	−.68311	.079651
6	8.15	2.63	.34	18.3	.23785	.267959
7	5.60	1.25	.34	18.1	.06456	.076001
8	4.30	2.63	.34	19.1	.13131	.074927
9	4.30	2.63	.34	15.4	−.74091	.074927
10	5.60	10.10	.25	12.0	−1.38857	.152317
11	5.60	10.10	.34	19.8	−.03585	.068468
12	4.30	10.10	.34	18.6	−.40699	.062849
13	2.40	10.10	.34	13.2	−1.92274	.175421
14	5.60	10.10	.55	22.8	−1.07990	.712933
15	2.14	112.00	.34	41.8	−1.19311	.516298
16	4.14	112.00	.34	48.6	1.21302	.513214
17	5.60	10.10	.25	19.2	.38451	.152317
18	5.60	10.10	.25	18.4	.18750	.152317
19	5.60	10.10	.25	15.0	−.64979	.152317

61. Refer to the water-discharge data given in Exercise 53 and let $y = \ln(q)$, $x_1 = \ln(a)$, and $x_2 = \ln(b)$. Consider fitting the model $Y = \beta_0 + \beta_1 x_1 + \beta_2 x_2 + \epsilon$.

a. The resulting h_{ii}'s are .138, .302, .266, .604, .464, .360, .215, .153, .214, and .284. Does any observation appear to be influential?

b. The estimated coefficients are $\hat{\beta}_0 = 1.5652$, $\hat{\beta}_1 = .9450$, $\hat{\beta}_2 = .1815$, and the corresponding estimated standard deviations are $s_{\hat{\beta}_0} = .7328$,

$s_{\hat{\beta}_1} = .1528$, and $s_{\hat{\beta}_2} = .1752$. The second standardized residual is $e_2^* = 2.19$. When the second observation is omitted from the data set, the resulting estimated coefficients are $\hat{\beta}_0 = 1.8982$, $\hat{\beta}_1 = 1.025$, and $\hat{\beta}_2 = .3085$. Do any of these changes indicate that the second observation is influential?

c. Deletion of the fourth observation (why?) yields $\hat{\beta}_0 = 1.4592$, $\hat{\beta}_1 = .9850$, and $\hat{\beta}_2 = .1515$. Is this observation influential?

Supplementary Exercises (62–78)

62. The authors of the article "Long-Term Effects of Cathodic Protection on Prestressed Concrete Structures" (*Corrosion*, 1997: 891–908) presented a scatter plot of $y =$ steady-state permeation flux (μA/cm^2) versus $x =$ inverse foil thickness (cm^{-1}); the substantial linear pattern was used as a basis for an important conclusion about material behavior. This is the MINITAB output from fitting the simple linear regression model to the data.

The regression equation is
flux = −0.398 + 0.260 invthick

Predictor	Coef	Stdev	t-ratio	p
Constant	−0.3982	0.5051	−0.79	0.460
invthick	0.26042	0.01502	17.34	0.000

s = 0.4506   R-sq = 98.0%   R-sq(adj) = 97.7%

Analysis of Variance

Source	DF	SS	MS	F	p
Regression	1	61.050	61.050	300.64	0.000
Error	6	1.218	0.203		
Total	7	62.269			

Obs.	inv-thick	flux	Fit	Stdev. Fit	Residual	St. Resid
1	19.8	4.3	4.758	0.242	−0.458	−1.20
2	20.6	5.6	4.966	0.233	0.634	1.64
3	23.5	6.1	5.722	0.203	0.378	0.94
4	26.1	6.2	6.399	0.182	−0.199	−0.48
5	30.3	6.9	7.493	0.161	−0.593	−1.41
6	43.5	11.2	10.930	0.236	0.270	0.70
7	45.0	11.3	11.321	0.253	−0.021	−0.06
8	46.5	11.7	11.711	0.271	−0.011	−0.03

a. Interpret the estimated slope and the coefficient of determination.

b. Calculate a point estimate of true average flux when inverse foil thickness is 23.5.

c. Does the model appear to be useful?

d. Predict flux when inverse thickness is 45 in a way that conveys information about precision and reliability.

e. Investigate model adequacy.

63. The article "Validation of the Rockport Fitness Walking Test in College Males and Females" (*Research Quarterly for Exercise and Sport*, 1994: 152–158) recommended the following estimated regression equation for relating $y =$ VO$_2$max (L/min, a measure of cardiorespiratory fitness) to the predictors $x_1 =$ gender (female = 0, male = 1), $x_2 =$ weight (lb), $x_3 =$ 1-mile walk time (min), and $x_4 =$ heart rate at the end of the walk (beats/min):

$$y = 3.5959 + .6566x_1 + .0096x_2 - .0096x_3 - .0080x_4$$

a. How would you interpret the estimated coefficient $\hat{\beta}_3 = -.0096$?

b. How would you interpret the estimated coefficient $\hat{\beta}_1 = .6566$?

c. Suppose that an observation made on a male whose weight was 170 lb, walk time was 11 min, and heart rate was 140 beats/min resulted in VO$_2$max = 3.15. What would you have predicted for VO$_2$max in this situation, and what is the value of the corresponding residual?

d. Using SSE = 30.1033 and SST = 102.3922, what proportion of observed variation in VO$_2$max can be attributed to the model relationship?

e. Assuming a sample size of $n = 20$, carry out a test of hypotheses to decide whether the chosen model specifies a useful relationship between VO$_2$max and at least one of the predictors.

64. Feature recognition from surface models of complicated parts is becoming increasingly important in the development of efficient computer-aided design (CAD) systems. The article "A Computationally Efficient Approach to Feature Abstraction in Design-Manufacturing Integration" (*J. of Engr. for Industry*, 1995: 16–27) contained a graph of $\log_{10}$(total recognition time), with time in sec, versus $\log_{10}$(number of edges of a part), from which the following representative values were read.

Log(edges)	1.1	1.5	1.7	1.9	2.0	2.1
Log(time)	.30	.50	.55	.52	.85	.98

Log(edges)	2.2	2.3	2.7	2.8	3.0	3.3
Log(time)	1.10	1.00	1.18	1.45	1.65	1.84

Log(edges)	3.5	3.8	4.2	4.3
Log(time)	2.05	2.46	2.50	2.76

a. Does a scatter plot of log(time) versus log(edges) suggest an approximate linear relationship between these two variables?

b. What probabilistic model for relating $y =$ recognition time to $x =$ number of edges is implied by the simple linear regression relationship between the transformed variables?

c. Summary quantities calculated from the data are

$$n = 16 \qquad \sum x_i' = 42.4 \qquad \sum y_i' = 21.69$$

$$\sum (x_i')^2 = 126.34 \qquad \sum (y_i')^2 = 38.5305$$

$$\sum x_i' y_i' = 68.640$$

Calculate estimates of the parameters for the model in part (b), and then obtain a point prediction of time when the number of edges is 300.

65. Air pressure (psi) and temperature (°F) were measured for a compression process in a certain piston-cylinder device, resulting in the following data (from *Introduction to Engineering Experimentation*, Prentice-Hall, Inc., 1996, p. 153):

Pressure	20.0	40.4	60.8	80.2	100.4
Temperature	44.9	102.4	142.3	164.8	192.2

Pressure	120.3	141.1	161.4	181.9	201.4
Temperature	221.4	228.4	249.5	269.4	270.8

Pressure	220.8	241.8	261.1	280.4	300.1
Temperature	291.5	287.3	313.3	322.3	325.8

(continued)

Pressure	320.6	341.1	360.8
Temperature	337.0	332.6	342.9

a. Would you fit the simple linear regression model to the data and use it as a basis for predicting temperature from pressure? Why or why not?

b. Find a suitable probabilistic model and use it as a basis for predicting the value of temperature that would result from a pressure of 200 in the most informative way possible.

66. ·An aeronautical engineering student carried out an experiment to study how $y =$ lift/drag ratio related to the variables $x_1 =$ position of a certain forward lifting surface relative to the main wing and $x_2 =$ tail placement relative to the main wing, obtaining the following data (*Statistics for Engineering Problem Solving*, p. 133):

x_1 (in.)	x_2 (in.)	y
−1.2	−1.2	.858
−1.2	0	3.156
−1.2	1.2	3.644
0	−1.2	4.281
0	0	3.481
0	1.2	3.918
1.2	−1.2	4.136
1.2	0	3.364
1.2	1.2	4.018

$$\bar{y} = 3.428, \; SST = 8.55$$

a. Fitting the first-order model gives SSE = 5.18, whereas including $x_3 = x_1 x_2$ as a predictor results in SSE = 3.07. Calculate and interpret the coefficient of multiple determination for each model.

b. Carry out a test of model utility using $\alpha = .05$ for each of the models described in part (a). Does either result surprise you?

67. An ammonia bath is the one most widely used for depositing Pd-Ni alloy coatings. The article "Modelling of Palladium and Nickel in an Ammonia Bath in a Rotary Device" (*Plating and Surface Finishing*, 1997: 102–104) reported on an investigation into how bath composition characteristics affect coating properties. Consider the following data on $x_1 =$ Pd concentration (g/dm³), $x_2 =$ Ni concentration (g/dm³), $x_3 =$ pH, $x_4 =$ temperature (°C), $x_5 =$ cathode current density (A/dm²), and $y =$ palladium content (%) of the coating.

Obs	pdconc	niconc	pH	temp	currdens	pallcont
1	16	24	9.0	35	5	61.5
2	8	24	9.0	35	3	51.0
3	16	16	9.0	35	3	81.0
4	8	16	9.0	35	5	50.9
5	16	24	8.0	35	3	66.7
6	8	24	8.0	35	5	48.8
7	16	16	8.0	35	5	71.3
8	8	16	8.0	35	3	62.8
9	16	24	9.0	25	3	64.0
10	8	24	9.0	25	5	37.7
11	16	16	9.0	25	5	68.7
12	8	16	9.0	25	3	54.1
13	16	24	8.0	25	5	61.6
14	8	24	8.0	25	3	48.0
15	16	16	8.0	25	3	73.2
16	8	16	8.0	25	5	43.3
17	4	20	8.5	30	4	35.0
18	20	20	8.5	30	4	69.6
19	12	12	8.5	30	4	70.0
20	12	28	8.5	30	4	48.2
21	12	20	7.5	30	4	56.0
22	12	20	9.5	30	4	77.6
23	12	20	8.5	20	4	55.0
24	12	20	8.5	40	4	60.6
25	12	20	8.5	30	2	54.9
26	12	20	8.5	30	6	49.8
27	12	20	8.5	30	4	54.1
28	12	20	8.5	30	4	61.2
29	12	20	8.5	30	4	52.5
30	12	20	8.5	30	4	57.1
31	12	20	8.5	30	4	52.5
32	12	20	8.5	30	4	56.6

a. Fit the first-order model with five predictors and assess its utility. Do all the predictors appear to be important?

b. Fit the complete second-order model and assess its utility.

c. Does the group of second-order predictors (interaction and quadratic) appear to provide more useful information about y than is contributed by the first-order predictors? Carry out an appropriate test of hypotheses.

d. The authors of the cited article recommended the use of all five first-order predictors plus the additional predictor $x_6 = (pH)^2$. Fit this model. Do all six predictors appear to be important?

68. The article "An Experimental Study of Resistance Spot Welding in 1 mm Thick Sheet of Low Carbon Steel" (*J. of Engr. Manufacture,* 1996: 341–348) discussed a statistical analysis whose basic aim was to establish a relationship that could explain the variation in weld strength (y) by relating strength to the process characteristics weld current (wc), weld time (wt), and electrode force (ef).

a. SST = 16.18555, and fitting the complete second-order model gave SSE = .80017. Calculate and interpret the coefficient of multiple determination.

b. Assuming that $n = 37$, carry out a test of model utility (the ANOVA table in the article states that $n - (k + 1) = 1$, but other information given contradicts this and is consistent with the sample size we suggest).

c. The given F ratio for the current-time interaction was 2.32. If all other predictors are retained in the model, can this interaction predictor be eliminated? *Hint:* As in simple linear regression, an F ratio for a coefficient is the square of its t-ratio.

d. The authors proposed eliminating two interaction predictors and a quadratic predictor, and recommended the estimated equation $y = 3.352 + .098wc + .222wt + .297ef - .0102(wt)^2 - .037(ef)^2 + .0128(wc)(wt)$. Consider a weld current of 10 kA, a weld time of 12 ac cycles, and an electrode force of 6 kN. Supposing that the estimated standard deviation of the predicted strength in this situation is .0750, calculate a 95% PI for strength. Does the interval suggest that the value of strength can be accurately predicted?

69. The accompanying data on x = frequency (MHz) and y = output power (W) for a certain laser configuration was read from a graph in the article "Frequency Dependence in RF Discharge Excited Waveguide CO_2 Lasers" (*IEEE J. Quantum Electronics,* 1984: 509–514):

x	60	63	77	100	125	157	186	222
y	16	17	19	21	22	20	15	5

A computer analysis yielded the following information for a quadratic regression model: $\hat{\beta}_0 = -1.5127$, $\hat{\beta}_1 = .391901$, $\hat{\beta}_2 = -.00163141$, $s_{\beta_2} = .00003391$, SSE = .29, SST = 202.88, and $s_{\hat{Y}} = .1141$ when $x = 100$.

a. Does the quadratic model appear to be suitable for explaining observed variation in output power by relating it to frequency?

b. Would the simple linear regression model be nearly as satisfactory as the quadratic model?

c. Do you think it would be worth considering a cubic model?

d. Compute a 95% CI for expected power output when frequency is 100.

e. Use a 95% PI to predict the power from a single experimental run when frequency is 100.

70. Conductivity is one important characteristic of glass. The article "Structure and Properties of

Rapidly Quenched Li_2O-Al_2O-Nb_2O_5 Glasses" (*J. Amer. Ceramic Soc.*, 1983: 890–892) reports the accompanying data on $x = Li_2O$ content of a certain type of glass and $y =$ conductivity at 500 K:

x	19	20	24	27	29	30
y	$10^{-8.0}$	$10^{-7.1}$	$10^{-7.2}$	$10^{-6.7}$	$10^{-6.2}$	$10^{-6.8}$

x	31	39	40	43	45	50
y	$10^{-5.8}$	$10^{-5.3}$	$10^{-6.0}$	$10^{-4.7}$	$10^{-5.4}$	$10^{-5.1}$

(This is a subset of the data that appeared in the article.) Propose a suitable model for relating y to x, estimate the model parameters, and predict conductivity when Li_2O content is 35.

71. The effect of manganese (Mn) on wheat growth is examined in the article "Manganese Deficiency and Toxicity Effects on Growth, Development and Nutrient Composition in Wheat" (*Agronomy J.*, 1984: 213–217). A quadratic regression model was used to relate $y =$ plant height (cm) to $x = \log_{10}($added Mn$)$, with μM as the units for added Mn. The accompanying data was read from a scatter diagram appearing in the article:

x	−1.0	−.4	0	.2	1.0
y	32	37	44	45	46

x	2.0	2.8	3.2	3.4	4.0
y	42	42	40	37	30

In addition, $\hat{\beta}_0 = 41.7422$, $\hat{\beta}_1 = 6.581$, $\hat{\beta}_2 = -2.3621$, $s_{\hat{\beta}_0} = .8522$, $s_{\hat{\beta}_1} = 1.002$, $s_{\hat{\beta}_2} = .3073$, and SSE = 26.98.

a. Is the quadratic model useful for describing the relationship between x and y? [*Hint:* Quadratic regression is a special case of multiple regression with $k = 2$, $x_1 = x$, and $x_2 = x^2$.] Apply an appropriate procedure.

b. Should the quadratic predictor be eliminated?

c. Estimate expected height for wheat treated with 10 μM of Mn using a 90% CI. [*Hint:* The estimated standard deviation of $\hat{\beta}_0 + \hat{\beta}_1 + \hat{\beta}_2$ is 1.031.]

72. The article "Chemithermomechanical Pulp from Mixed High Density Hardwoods" (*TAPPI*, July 1988: 145–146) reports on a study in which the accompanying data was obtained to relate $y =$

specific surface area (cm³/g) to $x_1 = \%$ NaOH used as a pretreatment chemical and $x_2 =$ treatment time (min) for a batch of pulp:

x_1	x_2	y
3	30	5.95
3	60	5.60
3	90	5.44
9	30	6.22
9	60	5.85
9	90	5.61
15	30	8.36
15	60	7.30
15	90	6.43

The accompanying MINITAB output resulted from a request to fit the model $Y = \beta_0 + \beta_1 x_1 + \beta_2 x_2 + \epsilon$.

```
The regression equation is
AREA = 6.05 + 0.142 NAOH - 0.0169 TIME

Predictor      Coef     Stdev   t-ratio      p
Constant     6.0483    0.5208     11.61  0.000
NAOH        0.14167   0.03301      4.29  0.005
TIME      -0.016944  0.006601     -2.57  0.043

s = 0.4851    R-sq = 80.7%    R-sq (adj) 74.2%

Analysis of Variance

SOURCE      DF       SS       MS      F      p
Regression   2   5.8854   2.9427  12.51  0.007
Error        6   1.4118   0.2353
Total        8   7.2972
```

a. What proportion of observed variation in specific surface area can be explained by the model relationship?

b. Does the chosen model appear to specify a useful relationship between the dependent variable and the predictors?

c. Provided that % NaOH remains in the model, would you suggest that the predictor *treatment time* be eliminated?

d. Calculate a 95% CI for the expected change in specific surface area associated with an increase of 1% in NaOH when treatment time is held fixed.

e. MINITAB reported that the estimated standard deviation of $\hat{\beta}_0 + \hat{\beta}_1(9) + \hat{\beta}_2(60)$ is .162. Calculate a prediction interval for the value of specific surface area to be observed when % NaOH = 9 and treatment time = 60.

73. A multiple regression analysis was carried out to relate $y =$ tensile strength of a synthetic-fiber specimen to the variables $x_1 =$ percent cotton and $x_2 =$ drying time. The data set consisted of $n = 12$ observations.

a. The estimated coefficients were $\hat{\beta}_0 = 180.00$, $\hat{\beta}_1 = 1.000$, and $\hat{\beta}_2 = 10.500$. Calculate a point estimate of the expected tensile strength when percent cotton = 15 and drying time = 3.5.

b. Sums of squares were SST = 1210.30 and SSE = 117.40. What proportion of observed variation in tensile strength can be attributed to the model relationship?

c. Use the information in part (b) to decide whether the model with variables x_1 and x_2 specifies a useful relationship.

d. The estimated standard deviation of $\hat{\beta}_0 + \hat{\beta}_1 x_1 + \hat{\beta}_2 x_2$ when $x_1 = 18$ and $x_2 = 3.0$ was 1.20. Calculate a 95% PI for tensile strength of a fabric specimen for which $x_1 = 18$ and $x_2 = 3.0$.

74. A study was carried out to relate time to failure (y) of a certain machine component to the variables operating voltage (x_1), motor speed (x_2), and operating temperature (x_3). The resulting data set consisted of $n = 20$ observations. When the model with the three variables x_1, x_2, and x_3 was fit to the data, the value of error sum of squares was 8212.5. Fitting the second-order interaction model (with quadratic predictors and all products of pairs of variables) gave an error sum of squares of 5027.1. Should at least one of the quadratic or interaction predictors be retained in the model? State and test the relevant hypotheses.

75. The article "A Statistical Analysis of the Notch Toughness of 9% Nickel Steels Obtained from Production Heats" (*J. of Testing and Eval.*, 1987: 355–363) reports on the results of a multiple regression analysis relating Charpy v-notch toughness y (Joules) to the following variables: x_1 = plate thickness (mm), x_2 = carbon content (%), x_3 = manganese content (%), x_4 = phosphorus content (%), x_5 = sulphur content (%), x_6 = silicon content (%), x_7 = nickel content (%), x_8 = yield strength (Pa), and x_9 = tensile strength (Pa).

a. The best possible subsets involved adding variables in the order $x_5, x_8, x_6, x_3, x_2, x_7, x_9, x_1$, and x_4. The values of R_k^2, MSE_k, and C_k are as follows:

No. of Predictors	1	2	3	4
R_k^2	.354	.453	.511	.550
MSE_k	2295	1948	1742	1607
C_k	314	173	89.6	35.7

No. of Predictors	5	6	7	8	9
R_k^2	.562	.570	.572	.575	.575
MSE_k	1566	1541	1535	1530	1532
C_k	19.9	11.0	9.4	8.2	10.0

Which model would you recommend? Explain the rationale for your choice.

b. The authors also considered second-order models involving predictors x_i^2 and $x_i x_j$. Information on the best such models starting with the variables x_2, x_3, x_5, x_6, x_7, and x_8 is as follows (in going from the best four-predictor model to the best five-predictor model, x_8 was deleted and both $x_2 x_6$ and $x_7 x_8$ were entered; x_8 reentered at a later stage):

No. of Predictors	1	2	3	4	5
R_k^2	.415	.541	.600	.629	.650
MSE_k	2079	1636	1427	1324	1251
C_k	433	109	104	52.4	16.5

No. of Predictors	6	7	8	9	10
R_k^2	.652	.655	.658	.659	.659
MSE_k	1246	1237	1229	1229	1230
C_k	14.9	11.2	8.5	9.2	11.0

Which of these models would you recommend, and why? (*Note:* Models based on eight of the original variables did not yield marked improvement on those under consideration here.)

76. A sample of $n = 20$ companies was selected, and the values of y = stock price and $k = 15$ predictor variables (such as quarterly dividend, previous year's earnings, and debt ratio) were determined. When the multiple regression model using these 15 predictors was fit to the data, $R^2 = .90$ resulted.

a. Does the model appear to specify a useful relationship between y and the predictor variables? Carry out a test using significance level .05. [*Hint:* The F critical value for 15 numerator and 4 denominator df is 5.86.]

b. Based on the result of part (a), does a high R^2 value by itself imply that a model is useful? Under what circumstances might you be suspicious of a model with a high R^2 value?

c. With n and k as given previously, how large would R^2 have to be for the model to be judged useful at the .05 level of significance?

77. Does exposure to air pollution result in decreased life expectancy? This question was examined in the article "Does Air Pollution Shorten Lives?" (*Statistics and Public Policy,* Reading, Mass., Addison-Wesley, 1977). Data on

y = total mortality rate (deaths per 10,000)

x_1 = mean suspended particle reading ($\mu g/m^3$)

x_2 = smallest sulfate reading ($[\mu g/m^3] \times 10$)

x_3 = population density (people/mi^2)

x_4 = (percent nonwhite) $\times$ 10

x_5 = (percent over 65) $\times$ 10

for the year 1960 was recorded for $n = 117$ randomly selected standard metropolitan statistical areas. The estimated regression equation was

$$y = 19.607 + .041x_1 + .071x_2 \\ + .001x_3 + .041x_4 + .687x_5$$

a. For this model, $R^2 = .827$. Using a .05 significance level, perform a model utility test.

b. The estimated standard deviation of $\hat{\beta}_1$ was .016. Calculate and interpret a 90% CI for β_1.

c. Given that the estimated standard deviation of $\hat{\beta}_4$ is .007, determine whether percent nonwhite is an important variable in the model. Use a .01 significance level.

d. In 1960 the values of x_1, x_2, x_3, x_4, and x_5 for Pittsburgh were 166, 60, 788, 68, and 95, respectively. Use the given regression equation to predict Pittsburgh's mortality rate. How does your prediction compare with the actual 1960 value of 103 deaths per 10,000?

78. Given that $R^2 = .723$ for the model containing predictors x_1, x_4, x_5, and x_8 and $R^2 = .689$ for the model with predictors x_1, x_3, x_5, and x_6, what can you say about R^2 for the model containing predictors

a. x_1, x_3, x_4, x_5, x_6, and x_8? Explain.

b. x_1 and x_4? Explain.

Bibliography

Chatterjee, S., and Bertram Price, *Regression Analysis by Example* (2nd ed.), Wiley, New York, 1990. A brief but informative discussion of selected topics, especially multicollinearity and the use of biased estimation methods.

Daniel, Cuthbert, and Fred Wood, *Fitting Equations to Data* (2nd ed.), Wiley, New York, 1980. Contains many insights and methods that evolved from the authors' extensive consulting experience.

Draper, Norman, and Harry Smith, *Applied Regression Analysis* (3rd ed.), Wiley, New York, 1999. See Chapter 12 bibliography.

Hoaglin, David, and Roy Welsch, "The Hat Matrix in Regression and ANOVA," *American Statistician,* 1978: 17–23. Describes methods for detecting influential observations in a regression data set.

Hocking, Ron, "The Analysis and Selection of Variables in Linear Regression," *Biometrics,* 1976: 1–49. An excellent survey of this topic.

Neter, John, Michael Kutner, Christopher Nachtsheim, and William Wasserman, *Applied Linear Statistical Models* (4th ed.), Irwin, Homewood, IL, 1996. See Chapter 12 bibliography.

14

The Analysis of Categorical Data

Introduction

In the simplest type of situation considered in this chapter, each observation in a sample is classified as belonging to one of a finite number of categories (e.g., blood type could be one of the four categories O, A, B, or AB). With p_i denoting the probability that any particular observation belongs in category i (or the proportion of the population belonging to category i), we wish to test a null hypothesis that completely specifies the values of all the p_i's (such as H_0: $p_1 = .45$, $p_2 = .35$, $p_3 = .15$, $p_4 = .05$, when there are four categories). The test statistic will be a measure of the discrepancy between the observed numbers in the categories and the expected numbers when H_0 is true. Because a decision will be reached by comparing the computed value of the test statistic to a critical value of the chi-squared distribution, the procedure is called a chi-squared goodness-of-fit test.

Sometimes the null hypothesis specifies that the p_i's depend on some smaller number of parameters without specifying the values of these parameters. For example, with three categories the null hypothesis might state that $p_1 = \theta^2$, $p_2 = 2\theta(1 - \theta)$, and $p_3 = (1 - \theta)^2$. For a chi-squared test to be performed, the values of any unspecified parameters must be estimated from the sample data. These problems are discussed in Section 14.2. The methods are then applied to test a null hypothesis that states that the sample comes from a particular family of distributions, such as the Poisson family (with λ estimated from the sample) or the normal family (with μ and σ estimated).

Chi-squared tests for two different situations are presented in Section 14.3. In the first, the null hypothesis states that the p_i's are the same for several different populations. The second type of situation involves taking a sample from a single population and classifying each individual with respect to two different categorical factors (such as religious preference and political party registration). The null hypothesis in this situation is that the two factors are independent within the population.

14.1 | Goodness-of-Fit Tests When Category Probabilities Are Completely Specified

A binomial experiment consists of a sequence of independent trials in which each trial can result in one of two possible outcomes (the same two possibilities for each trial). The two possibilities are labeled S (for success) and F (for failure). The probability of success, denoted by p, is assumed to be constant from trial to trial, and the number n of trials is fixed at the outset of the experiment. In Chapter 8, we presented a large-sample z test for testing H_0: $p = p_0$. Notice that this null hypothesis specifies both $P(S)$ and $P(F)$, since if $P(S) = p_0$, then $P(F) = 1 - p_0$. Denoting $P(F)$ by q and $1 - p_0$ by q_0, the null hypothesis can alternatively be written as H_0: $p = p_0$, $q = q_0$. The z test is two-tailed when the alternative of interest is $p \neq p_0$.

A **multinomial experiment** generalizes a binomial experiment by allowing each trial to result in one of k possible outcomes, where k is an integer greater than 2. For example, suppose a store accepts three different types of credit cards. A multinomial experiment would result from observing the type of credit card used—type 1, type 2, or type 3—by each of the next n customers who pay with a credit card. In general, we will refer to the k possible outcomes on any given trial as categories, and p_i will denote the probability that a trial results in category i. If the experiment consists of selecting n individuals or objects from a population and categorizing each one, then p_i is the proportion of the population falling in the ith category (such an experiment will be approximately multinomial provided that n is much smaller than the population size).

The null hypothesis of interest will specify the value of each p_i. For example, in the case $k = 3$, we might have H_0: $p_1 = .5$, $p_2 = .3$, $p_3 = .2$. The alternative hypothesis will state that H_0 is not true—that is, that at least one of the p_i's has a value different from that asserted by H_0 (in which case at least two must be different, since they sum to 1). The symbol p_{i0} will represent the value of p_i claimed by the null hypothesis. In the example just given, $p_{10} = .5$, $p_{20} = .3$, and $p_{30} = .2$.

Before the multinomial experiment is performed, the number of trials that will result in category i ($i = 1, 2, \ldots$, or k) is a random variable—just as the number of successes and the number of failures in a binomial experiment are random variables. This random variable will be denoted by N_i and its observed value by n_i. Since each trial results in exactly one of the k categories, $\sum N_i = n$, and the same is true of the n_i's. As an example, an experiment with $n = 100$ and $k = 3$ might yield $N_1 = 46$, $N_2 = 35$, and $N_3 = 19$.

The expected number of successes and expected number of failures in a binomial experiment are np and nq, respectively. When H_0: $p = p_0$, $q = q_0$ is true, the expected numbers of successes and failures are np_0 and nq_0, respectively. Similarly, in a multinomial experiment the expected number of trials resulting in category i is $E(N_i) = np_i$ ($i = 1, \ldots, k$). When H_0: $p_1 = p_{10}, \ldots, p_k = p_{k0}$ is true, these expected values become $E(N_1) = np_{10}$, $E(N_2) = np_{20}, \ldots$, $E(N_k) = np_{k0}$. For the case $k = 3$, H_0: $p_1 = .5$, $p_2 = .3$, $p_3 = .2$, and $n = 100$, $E(N_1) = 100(.5) = 50$, $E(N_2) = 30$, and $E(N_3) = 20$ when H_0 is true. The n_i's are often displayed in a tabular format consisting of a row of k cells, one for each category, as illustrated in Table 14.1. The expected values when H_0 is true are displayed just below the observed values. The N_i's and n_i's are usually referred to as *observed cell counts* (or *observed cell frequencies*), and $np_{10}, np_{20}, \ldots, np_{k0}$ are the corresponding *expected cell counts* under H_0.

Table 14.1   Observed and expected cell counts

Category	$i = 1$	$i = 2$	$\ldots$	$i = k$	Row total
Observed	n_1	n_2	$\ldots$	n_k	n
Expected	np_{10}	np_{20}	$\ldots$	np_{k0}	n

The n_i's should all be reasonably close to the corresponding np_{i0}'s when H_0 is true. On the other hand, several of the observed counts should differ substantially from these expected counts when the actual values of the p_i's differ markedly from what the null hypothesis asserts. The test procedure involves measuring the discrepancy between the n_i's and the np_{i0}'s, with H_0 being rejected when the measured discrepancy is sufficiently large. It is natural to base a measure of discrepancy on the squared deviations $(n_1 - np_{10})^2$, $(n_2 - np_{20})^2, \ldots$, $(n_k - np_{k0})^2$. An obvious way to combine these into an overall measure is to add them together to obtain $\sum(n_i - np_{i0})^2$. However, suppose $np_{10} = 100$ and $np_{20} = 10$. Then if $n_1 = 95$ and $n_2 = 5$, the two categories contribute the same squared deviations to the proposed measure. Yet n_1 is only 5% less than what would be expected when H_0 is true, whereas n_2 is 50% less. To take relative magnitudes of the deviations into account, we will divide each squared deviation by the corresponding expected count and then combine.

Before giving a more detailed description, we must discuss a type of probability distribution called the *chi-squared distribution*. This distribution was first introduced in Section 4.4 and was used in Chapter 7 to obtain a confidence interval for the variance σ^2 of a normal population. The chi-squared distribution has a single parameter ν, called the number of degrees of freedom (df) of the distribution, with possible values 1, 2, 3, ... Analogous to the critical value $t_{\alpha,\nu}$ for the t distribution, $\chi^2_{\alpha,\nu}$ is the value such that

α of the area under the χ^2 curve with ν df lies to the right of $\chi^2_{\alpha,\nu}$ (see Figure 14.1). Selected values of $\chi^2_{\alpha,\nu}$ are given in Appendix Table A.7.

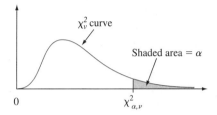

Figure 14.1   A critical value for a chi-squared distribution

THEOREM

Provided that $np_i \geq 5$ for every i ($i = 1, 2, \ldots, k$), the random variable

$$\chi^2 = \sum_{i=1}^{k} \frac{(N_i - np_i)^2}{np_i} = \sum_{\text{all cells}} \frac{(\text{observed} - \text{expected})^2}{\text{expected}}$$

has approximately a chi-squared distribution with $k - 1$ df.

The fact that df $= k - 1$ is a consequence of the restriction $\sum N_i = n$. Although there are k observed cell counts, once any $k - 1$ are known, the remaining one is uniquely determined. That is, there are only $k - 1$ "freely determined" cell counts, and thus $k - 1$ df.

If np_{i0} is substituted for np_i in χ^2, the resulting test statistic has a chi-squared distribution when H_0 is true. Rejection of H_0 is appropriate when $\chi^2 \geq c$ (because large discrepancies between observed and expected counts lead to a large value of χ^2), and the choice $c = \chi^2_{\alpha,k-1}$ yields a test with significance level α.

Null hypothesis: H_0: $p_1 = p_{10}, p_2 = p_{20}, \ldots, p_k = p_{k0}$

Alternative hypothesis: H_a: at least one p_i does not equal p_{i0}

Test statistic value: $\chi^2 = \displaystyle\sum_{\text{all cells}} \frac{(\text{observed} - \text{expected})^2}{\text{expected}} = \sum_{i=1}^{k} \frac{(n_i - np_{i0})^2}{np_{i0}}$

Rejection region: $\chi^2 \geq \chi^2_{\alpha,k-1}$

Example 14.1   If we focus on two different characteristics of an organism, each controlled by a single gene, and cross a pure strain having genotype AABB with a pure strain having genotype aabb (capital letters denoting dominant alleles and small letters recessive alleles), the resulting genotype will be AaBb. If these first-generation organisms are then crossed among themselves (a dihybrid cross), there will be four phenotypes depending on whether a dominant allele of either type is present. Mendel's laws of inheritance imply that these four phenotypes should have probabilities $\frac{9}{16}, \frac{3}{16}, \frac{3}{16}$, and $\frac{1}{16}$ of arising in any given dihybrid cross.

The article "Linkage Studies of the Tomato" (*Trans. Royal Canadian Institute,* 1931: 1–19) reports the following data on phenotypes from a dihybrid cross of tall cut-leaf tomatoes with dwarf potato-leaf tomatoes. There are $k = 4$ categories corresponding to the four possible phenotypes, with the null hypothesis being

$$H_0: p_1 = \frac{9}{16}, p_2 = \frac{3}{16}, p_3 = \frac{3}{16}, p_4 = \frac{1}{16}$$

The expected cell counts are $9n/16$, $3n/16$, $3n/16$, and $n/16$, and the test is based on $k - 1 = 3$ df. The total sample size was $n = 1611$. Observed and expected counts are given in Table 14.2.

Table 14.2 Observed and expected cell counts for Example 14.1

	$i = 1$ Tall, cut-leaf	$i = 2$ Tall, potato-leaf	$i = 3$ Dwarf, cut-leaf	$i = 4$ Dwarf, potato-leaf
n_i	926	288	293	104
np_{i0}	906.2	302.1	302.1	100.7

The contribution to χ^2 from the first cell is

$$\frac{(n_1 - np_{10})^2}{np_{10}} = \frac{(926 - 906.2)^2}{906.2} = .433$$

Cells 2, 3, and 4 contribute .658, .274, and .108, respectively, so $\chi^2 = .433 + .658 + .274 + .108 = 1.473$. A test with significance level .10 requires $\chi^2_{.10,3}$, the number in the 3 df row and .10 column of Appendix Table A.7. This critical value is 6.251. Since 1.473 is not at least 6.251, H_0 cannot be rejected even at this rather large level of significance. The data is quite consistent with Mendel's laws. ■

Although we have developed the chi-squared test for situations in which $k > 2$, it can also be used when $k = 2$. The null hypothesis in this case can be stated as H_0: $p_1 = p_{10}$, since the relations $p_2 = 1 - p_1$ and $p_{20} = 1 - p_{10}$ make the inclusion of $p_2 = p_{20}$ in H_0 redundant. The alternative hypothesis is H_a: $p_1 \neq p_{10}$. These hypotheses can also be tested using a two-tailed z test with test statistic

$$Z = \frac{(N_1/n) - p_{10}}{\sqrt{\dfrac{p_{10}(1 - p_{10})}{n}}} = \frac{\hat{p}_1 - p_{10}}{\sqrt{\dfrac{p_{10}p_{20}}{n}}}$$

Surprisingly, the two test procedures are completely equivalent. This is because it can be shown that $Z^2 = \chi^2$ and $(z_{\alpha/2})^2 = \chi^2_{1,\alpha}$, so that $\chi^2 \geq \chi^2_{1,\alpha}$ if and only if (iff) $|Z| \geq z_{\alpha/2}$.*

---

*The fact that $(z_{\alpha/2})^2 = \chi^2_{1,\alpha}$ is a consequence of the relationship between the standard normal distribution and the chi-squared distribution with 1 df; if $Z \sim N(0, 1)$, then Z^2 has a chi-squared distribution with $\nu = 1$.

If the alternative hypothesis is either H_a: $p_1 > p_{10}$ or H_a: $p_1 < p_{10}$, the chi-squared test cannot be used. One must then revert to an upper- or lower-tailed z test.

As is the case with all test procedures, one must be careful not to confuse statistical significance with practical significance. A computed χ^2 that exceeds $\chi^2_{\alpha,k-1}$ may be a result of a very large sample size rather than any practical differences between the hypothesized p_{i0}'s and true p_i's. Thus, if $p_{10} = p_{20} = p_{30} = \frac{1}{3}$, but the true p_i's have values .330, .340, and .330, a large value of χ^2 is sure to arise with a sufficiently large n. Before rejecting H_0, the $\hat{p}_i$'s should be examined to see whether they suggest a model different from that of H_0 from a practical point of view.

P-Values for Chi-Squared Tests

The chi-squared tests in this chapter are all upper-tailed, so we focus on this case. Just as the P-value for an upper-tailed t test is the area under the t_ν curve to the right of the calculated t, the P-value for an upper-tailed chi-squared test is the area under the χ^2_ν curve to the right of the calculated χ^2. Appendix Table A.7 provides limited P-value information because only five upper-tail critical values are tabulated for each different ν. We have therefore included another appendix table, analogous to Table A.8, that facilitates making more precise P-value statements.

The fact that t curves were all centered at zero allowed us to tabulate t curve tail areas in a relatively compact way, with the left margin giving values ranging from 0.0 to 4.0 on the horizontal t scale and various columns displaying corresponding upper-tail areas for various df's. The rightward movement of chi-squared curves as df increases necessitates a somewhat different type of tabulation. The left margin of Appendix Table A.11 displays various upper-tail areas: .100, .095, .090, . . . , .005, and .001. Each column of the table is for a different value of df, and the entries are values on the horizontal chi-squared axis that capture these corresponding tail areas. For example, moving down to tail area .085 and across to the 2 df column, we see that the area to the right of 4.93 under the 2 df chi-squared curve is .085 (see Figure 14.2).

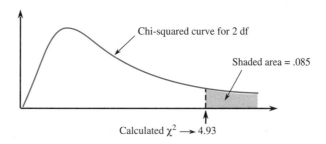

Figure 14.2 A P-value for an upper-tailed chi-squared test

To capture this same upper-tail area under the 10 df curve, we must go out to 16.54. In the 2 df column, the top row shows that if the calculated value of the chi-squared variable is smaller than 4.60, the captured tail area (the P-value) exceeds .10. Similarly, the bottom row in this column indicates that if the calculated value exceeds 13.81, the tail area is smaller than .001 (P-value $< .001$).

χ^2 When the p_i's Are Functions of Other Parameters

Frequently the p_i's are hypothesized to depend on a smaller number of parameters $\theta_1, \ldots, \theta_m$ $(m < k)$. Then a specific hypothesis involving the θ_i's yields specific p_{i0}'s, which are then used in the χ^2 test.

Example 14.2 In a well-known genetics article ("The Progeny in Generations F_{12} to F_{17} of a Cross Between a Yellow-Wrinkled and a Green-Round Seeded Pea," *J. Genetics,* 1923: 255–331), the early statistician G. U. Yule analyzed data resulting from crossing garden peas. The dominant alleles in the experiment were Y = yellow color and R = round shape, resulting in the double dominant YR. Yule examined 269 four-seed pods resulting from a dihybrid cross and counted the number of YR seeds in each pod. Letting X denote the number of YR's in a randomly selected pod, possible X values are 0, 1, 2, 3, 4, which we identify with cells 1, 2, 3, 4, and 5 of a rectangular table (so, e.g., a pod with $X = 4$ yields an observed count in cell 5).

The hypothesis that the Mendelian laws are operative and that genotypes of individual seeds within a pod are independent of one another implies that X has a binomial distribution with $n = 4$ and $\theta = \frac{9}{16}$. We thus wish to test $H_0: p_1 = p_{10}, \ldots, p_5 = p_{50}$, where

$$p_{i0} = P(i - 1 \text{ YR's among 4 seeds when } H_0 \text{ is true})$$

$$= \binom{4}{i - 1} \theta^{i-1}(1 - \theta)^{4-(i-1)} \qquad i = 1, 2, 3, 4, 5; \ \theta = \frac{9}{16}$$

Yule's data and the computations are in Table 14.3 with expected cell counts $np_{i0} = 269p_{i0}$.

Table 14.3 Observed and expected cell counts for Example 14.2

Cell i YR peas/pods	1 0	2 1	3 2	4 3	5 4
Observed	16	45	100	82	26
Expected	9.86	50.68	97.75	83.78	26.93
$\dfrac{(\text{observed} - \text{expected})^2}{\text{expected}}$	3.823	.637	.052	.038	.032

Thus, $\chi^2 = 3.823 + \cdots + .032 = 4.582$. Since $\chi^2_{.01,k-1} = \chi^2_{.01,4} = 13.277$, H_0 is not rejected at level .01. Appendix Table A.11 shows that because $4.582 < 7.77$, the P-value for the test exceeds .10. H_0 should not be rejected at any reasonable significance level. ∎

χ^2 When the Underlying Distribution Is Continuous

We have so far assumed that the k categories are naturally defined in the context of the experiment under consideration. The χ^2 test can also be used to test whether a sample comes from a specific underlying continuous distribution. Let X denote the variable being sampled and suppose the hypothesized pdf of X is $f_0(x)$. As in the construction of a frequency distribution in Chapter 1, subdivide the measurement scale of X into k intervals $[a_0, a_1), [a_1, a_2), \ldots, [a_{k-1}, a_k)$, where the interval $[a_{i-1}, a_i)$ includes the value a_{i-1} but not a_i. The cell probabilities specified by H_0 are then

$$p_{i0} = P(a_{i-1} \le X < a_i) = \int_{a_{i-1}}^{a_i} f_0(x)dx$$

The cells should be chosen so that $np_{i0} \ge 5$ for $i = 1, \ldots, k$. Often they are selected so that the np_{i0}'s are equal.

Example 14.3 To see whether the time of onset of labor among expectant mothers is uniformly distributed throughout a 24-hour day, we can divide a day into k periods, each of length $24/k$. The null hypothesis states that $f(x)$ is the uniform pdf on the interval $[0, 24]$, so that $p_{i0} = 1/k$. The article "The Hour of Birth" (*British J. Preventive and Social Medicine,* 1953: 43–59) reports on 1186 onset times, which were categorized into $k = 24$ 1-hour intervals beginning at midnight, resulting in cell counts of 52, 73, 89, 88, 68, 47, 58, 47, 48, 53, 47, 34, 21, 31, 40, 24, 37, 31, 47, 34, 36, 44, 78, and 59. Each expected cell count is $1186 \cdot \frac{1}{24} = 49.42$, and the resulting value of χ^2 is 162.77. Since $\chi^2_{.01,23} = 41.637$, the computed value is highly significant, and the null hypothesis is resoundingly rejected. Generally speaking, it appears that labor is much more likely to commence very late at night than during normal waking hours. ■

For testing whether a sample comes from a specific normal distribution, the fundamental parameters are $\theta_1 = \mu$ and $\theta_2 = \sigma$, and each p_{i0} will be a function of these parameters.

Example 14.4 At a certain university, final exams are supposed to last 2 hours. The psychology department constructed a departmental final for an elementary course that was believed to satisfy the following criteria: (1) actual time taken to complete the exam is normally distributed, (2) $\mu = 100$ min, and (3) exactly 90% of all students will finish within the 2-hour period. To see whether this is actually the case, 120 students were randomly selected, and their completion times recorded. It was decided that $k = 8$ intervals should be used. The criteria imply that the 90th percentile of the completion time distribution is $\mu + 1.28\sigma = 120$. Since $\mu = 100$, this implies that $\sigma = 15.63$.

The eight intervals that divide the standard normal scale into eight equally likely segments are $[0, .32), [.32, .675), [.675, 1.15), [1.15, \infty)$, and their four counterparts on the other side of 0. For $\mu = 100$ and $\sigma = 15.63$, these intervals become $[100, 105)$, $[105, 110.55), [110.55, 117.97)$, and $[117.97, \infty)$. Thus, $p_{i0} = \frac{1}{8} = .125$ $(i = 1, \ldots, 8)$, so each expected cell count is $np_{i0} = 120(.125) = 15$. The observed cell counts were 21, 17, 12, 16, 10, 15, 19, and 10, resulting in a χ^2 of 7.73. Since $\chi^2_{.10,7} = 12.017$ and 7.73 is not ≥ 12.017, there is no evidence for concluding that the criteria have not been met. ■

Exercises | Section 14.1 (1–11)

1. What conclusion would be appropriate for an upper-tailed chi-squared test in each of the following situations?
   a. $\alpha = .05$, df $= 4$, $\chi^2 = 12.25$
   b. $\alpha = .01$, df $= 3$, $\chi^2 = 8.54$
   c. $\alpha = .10$, df $= 2$, $\chi^2 = 4.36$
   d. $\alpha = .01$, $k = 6$, $\chi^2 = 10.20$

2. Say as much as you can about the P-value for an upper-tailed chi-squared test in each of the following situations:
   a. $\chi^2 = 7.5$, df $= 2$ b. $\chi^2 = 13.0$, df $= 6$
   c. $\chi^2 = 18.0$, df $= 9$ d. $\chi^2 = 21.3$, $k = 5$
   e. $\chi^2 = 5.0$, $k = 4$

3. A statistics department at a large university maintains a tutoring service for students in its introductory service courses. The service has been staffed with the expectation that 40% of its clients would be from the business statistics course, 30% from engineering statistics, 20% from the statistics course for social science students, and the other 10% from the course for agriculture students. A random sample of $n = 120$ clients revealed 52, 38, 21, and 9 from the four courses. Does this data suggest that the percentages on which staffing was based are not correct? State and test the relevant hypotheses using $\alpha = .05$.

4. It is hypothesized that when homing pigeons are disoriented in a certain manner, they will exhibit no preference for any direction of flight after takeoff (so that the direction X should be uniformly distributed on the interval from 0° to 360°). To test this, 120 pigeons are disoriented, let loose, and the direction of flight of each is recorded; the resulting data follows. Use the chi-squared test at level .10 to see whether the data supports the hypothesis.

Direction	0–<45°	45–<90°	90–<135°
Frequency	12	16	17
Direction	135–<180°	180–<225°	225–<270°
Frequency	15	13	20
Direction	270–<315°	315–<360°	
Frequency	17	10	

5. An information retrieval system has ten storage locations. Information has been stored with the expectation that the long-run proportion of requests for location i is given by $p_i = (5.5 - |i - 5.5|)/30$. A sample of 200 retrieval requests gave the following frequencies for locations 1–10, respectively: 4, 15, 23, 25, 38, 31, 32, 14, 10, and 8. Use a chi-squared test at significance level .10 to decide whether the data is consistent with the a priori proportions (use the P-value approach).

6. Sorghum is an important cereal crop whose quality and appearance could be affected by the presence of pigments in the pericarp (the walls of the plant ovary). The article "A Genetic and Biochemical Study on Pericarp Pigments in a Cross Between Two Cultivars of Grain Sorghum, Sorghum Bicolor" (*Heredity*, 1976: 413–416) reports on an experiment that involved an initial cross between CK60 sorghum (an American variety with white seeds) and Abu Taima (an Ethiopian variety with yellow seeds) to produce plants with red seeds and then a self-cross of the red-seeded plants. According to genetic theory, this F_2 cross should produce plants with red, yellow, or white seeds in the ratio $9 : 3 : 4$. The data from the experiment follow; does the data confirm or contradict the genetic theory? Test at level .05 using the P-value approach.

Seed Color	Red	Yellow	White
Observed Frequency	195	73	100

7. Criminologists have long debated whether there is a relationship between weather conditions and the incidence of violent crime. The author of the article "Is There a Season for Homicide?" (*Criminology*, 1988: 287–296) classified 1361 homicides according to season, resulting in the accompanying data. Test the null hypothesis of equal proportions using $\alpha = .01$ by using the chi-squared table to say as much as possible about the P-value.

Winter	Spring	Summer	Fall
328	334	372	327

8. The article "Psychiatric and Alcoholic Admissions Do Not Occur Disproportionately Close to Patients' Birthdays" (*Psychological Reports*, 1992: 944–946) focuses on the existence of any relationship between date of patient admission for treatment of alcoholism and patient's birthday. Assuming a 365-day year (i.e., excluding leap year), in the absence of any relation, a

patient's admission date is equally likely to be any one of the 365 possible days. The investigators established four different admission categories: (1) within 7 days of birthday, (2) between 8 and 30 days, inclusive, from the birthday, (3) between 31 and 90 days, inclusive, from the birthday, and (4) more than 90 days from the birthday. A sample of 200 patients gave observed frequencies of 11, 24, 69, and 96 for categories 1, 2, 3, and 4, respectively. State and test the relevant hypotheses using a significance level of .01.

9. The response time of a computer system to a request for a certain type of information is hypothesized to have an exponential distribution with parameter $\lambda = 1$ sec (so if $X =$ response time, the pdf of X under H_0 is $f_0(x) = e^{-x}$ for $x \geq 0$).
   a. If you had observed $X_1, X_2, \ldots, X_n$ and wanted to use the chi-squared test with five class intervals having equal probability under H_0, what would be the resulting class intervals?
   b. Carry out the chi-squared test using the following data resulting from a random sample of 40 response times:

.10	.99	1.14	1.26	3.24	.12	.26	.80
.79	1.16	1.76	.41	.59	.27	2.22	.66
.71	2.21	.68	.43	.11	.46	.69	.38
.91	.55	.81	2.51	2.77	.16	1.11	.02
2.13	.19	1.21	1.13	2.93	2.14	.34	.44

10. a. Show that another expression for the chi-squared statistic is

$$\chi^2 = \sum_{i=1}^{k} \frac{N_i^2}{np_{i0}} - n$$

Why is it more efficient to compute χ^2 using this formula?
   b. When the null hypothesis is $H_0: p_1 = p_2 = \cdots = p_k = 1/k$ (i.e., $p_{i0} = 1/k$ for all i), how does the formula of part (a) simplify? Use the simplified expression to calculate χ^2 for the pigeon/direction data in Exercise 4.

11. a. Having obtained a random sample from a population, you wish to use a chi-squared test to decide whether the population distribution is standard normal. If you base the test on six class intervals having equal probability under H_0, what should be the class intervals?
   b. If you wish to use a chi-squared test to test H_0: the population distribution is normal with $\mu = .5$, $\sigma = .002$ and the test is to be based on six equiprobable (under H_0) class intervals, what should be these intervals?
   c. Use the chi-squared test with the intervals of part (b) to decide, based on the following 45 bolt diameters, whether bolt diameter is a normally distributed variable with $\mu = .5$ in., $\sigma = .002$ in.

.4974	.4976	.4991	.5014	.5008	.4993
.4994	.5010	.4997	.4993	.5013	.5000
.5017	.4984	.4967	.5028	.4975	.5013
.4972	.5047	.5069	.4977	.4961	.4987
.4990	.4974	.5008	.5000	.4967	.4977
.4992	.5007	.4975	.4998	.5000	.5008
.5021	.4959	.5015	.5012	.5056	.4991
.5006	.4987	.4968			

14.2 Goodness of Fit for Composite Hypotheses

In the previous section, we presented a goodness-of-fit test based on a χ^2 statistic for deciding between $H_0: p_1 = p_{10}, \ldots, p_k = p_{k0}$ and the alternative H_a stating that H_0 is not true. The null hypothesis was a **simple hypothesis** in the sense that each p_{i0} was a specified number, so that the expected cell counts when H_0 was true were uniquely determined numbers.

In many situations, there are k naturally occurring categories, but H_0 states only that the p_i's are functions of other parameters $\theta_1, \ldots, \theta_m$ without specifying the values of these θ's. For example, a population may be in equilibrium with respect to proportions of the three genotypes AA, Aa, and aa. With $p_1, p_2,$ and p_3 denoting these proportions (probabilities), one may wish to test

$$H_0: \quad p_1 = \theta^2, p_2 = 2\theta(1 - \theta), p_3 = (1 - \theta)^2 \tag{14.1}$$

where θ represents the proportion of gene A in the population. This hypothesis is **composite** because knowing that H_0 is true does not uniquely determine the cell probabilities and expected cell counts but only their general form. To carry out a χ^2 test, the unknown θ_i's must first be estimated.

Similarly, we may be interested in testing to see whether a sample came from a particular family of distributions without specifying any particular member of the family. To use the χ^2 test to see whether the distribution is Poisson, for example, the parameter λ must be estimated. In addition, because there are actually an infinite number of possible values of a Poisson variable, these values must be grouped so that there are a finite number of cells. If H_0 states that the underlying distribution is normal, use of a χ^2 test must be preceded by a choice of cells and estimation of μ and σ.

χ^2 When Parameters Are Estimated

As before, k will denote the number of categories or cells and p_i will denote the probability of an observation falling in the ith cell. The null hypothesis now states that each p_i is a function of a small number of parameters $\theta_1, \ldots, \theta_m$ with the θ_i's otherwise unspecified:

$$H_0: p_1 = \pi_1(\boldsymbol{\theta}), \ldots, p_k = \pi_k(\boldsymbol{\theta}) \qquad \text{where } \boldsymbol{\theta} = (\theta_1, \ldots, \theta_m)$$
$$H_a: \text{the hypothesis } H_0 \text{ is not true} \tag{14.2}$$

For example, for H_0 of (14.1), $m = 1$ (there is only one θ), $\pi_1(\theta) = \theta^2$, $\pi_2(\theta) = 2\theta(1 - \theta)$, and $\pi_3(\theta) = (1 - \theta)^2$.

In the case $k = 2$, there is really only a single rv, N_1 (since $N_1 + N_2 = n$), which has a binomial distribution. The joint probability that $N_1 = n_1$ and $N_2 = n_2$ is then

$$P(N_1 = n_1, N_2 = n_2) = \binom{n}{n_1} p_1^{n_1} \cdot p_2^{n_2} \propto p_1^{n_1} \cdot p_2^{n_2}$$

where $p_1 + p_2 = 1$ and $n_1 + n_2 = n$. For general k, the joint distribution of $N_1, \ldots, N_k$ is the multinomial distribution (Section 5.1) with

$$P(N_1 = n_1, \ldots, N_k = n_k) \propto p_1^{n_1} \cdot p_2^{n_2} \cdot \cdots \cdot p_k^{n_k} \tag{14.3}$$

When H_0 is true, (14.3) becomes

$$P(N_1 = n_1, \ldots, N_k = n_k) \propto [\pi_1(\boldsymbol{\theta})]^{n_1} \cdot \cdots \cdot [\pi_k(\boldsymbol{\theta})]^{n_k} \tag{14.4}$$

To apply a chi-squared test, $\boldsymbol{\theta} = (\theta_1, \ldots, \theta_m)$ must be estimated.

METHOD OF ESTIMATION | Let $n_1, n_2, \ldots, n_k$ denote the observed values of $N_1, \ldots, N_k$. Then $\hat{\theta}_1, \ldots, \hat{\theta}_m$ are those values of the θ_i's that maximize (14.4).

The resulting estimators $\hat{\theta}_1, \ldots, \hat{\theta}_m$ are the **maximum likelihood estimators** of $\theta_1, \ldots, \theta_m$; this principle of estimation was discussed in Section 6.2.

Example 14.5 In humans there is a blood group, the MN group, that is composed of individuals having one of the three blood types M, MN, and N. Type is determined by two alleles, and there is no dominance, so the three possible genotypes give rise to three phenotypes. A population consisting of individuals in the MN group is in equilibrium if

$$P(M) = p_1 = \theta^2$$
$$P(MN) = p_2 = 2\theta(1 - \theta)$$
$$P(N) = p_3 = (1 - \theta)^2$$

for some θ. Suppose a sample from such a population yielded the results shown in Table 14.4.

Table 14.4 Observed counts for Example 14.5

Type	M	MN	N	
Observed	125	225	150	$n = 500$

Then

$$[\pi_1(\theta)]^{n_1}[\pi_2(\theta)]^{n_2}[\pi_3(\theta)]^{n_3} = [(\theta^2)]^{n_1}[2\theta(1 - \theta)]^{n_2}[(1 - \theta)^2]^{n_3}$$
$$= 2^{n_2} \cdot \theta^{2n_1+n_2} \cdot (1 - \theta)^{n_2+2n_3}$$

Maximizing this with respect to θ (or, equivalently, maximizing the natural logarithm of this quantity, which is easier to differentiate) yields

$$\hat{\theta} = \frac{2n_1 + n_2}{[(2n_1 + n_2) + (n_2 + 2n_3)]} = \frac{2n_1 + n_2}{2n}$$

With $n_1 = 125$ and $n_2 = 225$, $\hat{\theta} = 475/1000 = .475$. ∎

Once $\boldsymbol{\theta} = (\theta_1, \ldots, \theta_m)$ has been estimated by $\hat{\boldsymbol{\theta}} = (\hat{\theta}_1, \ldots, \hat{\theta}_m)$, the estimated expected cell counts are the $n\pi_i(\hat{\boldsymbol{\theta}})$'s. These are now used in place of the np_{i0}'s of Section 14.1 to specify a χ^2 statistic.

THEOREM

Under general "regularity" conditions on $\theta_1, \ldots, \theta_m$ and the $\pi_i(\boldsymbol{\theta})$'s, if $\theta_1, \ldots, \theta_m$ are estimated by the method of maximum likelihood as described previously and n is large,

$$\chi^2 = \sum_{\text{all cells}} \frac{(\text{observed} - \text{estimated expected})^2}{\text{estimated expected}} = \sum_{i=1}^{k} \frac{[N_i - n\pi_i(\hat{\boldsymbol{\theta}})]^2}{n\pi_i(\hat{\boldsymbol{\theta}})}$$

has approximately a chi-squared distribution with $k - 1 - m$ df when H_0 of (14.2) is true. An approximately level α test of H_0 versus H_a is then to reject H_0 if $\chi^2 \geq \chi^2_{\alpha,k-1-m}$. In practice, the test can be used if $n\pi_i(\hat{\boldsymbol{\theta}}) \geq 5$ for every i.

Notice that *the number of degrees of freedom is reduced by the number of θ_i's estimated.*

Example 14.6
(Example 14.5
continued)

With $\hat{\theta} = .475$ and $n = 500$, the estimated expected cell counts are $n\pi_1(\hat{\theta}) = 500(\hat{\theta})^2 = 112.81, n\pi_2(\hat{\theta}) = (500)(2)(.475)(1 - .475) = 249.38,$ and $n\pi_3(\hat{\theta}) = 500 - 112.81 - 249.38 = 137.81$. Then

$$\chi^2 = \frac{(125 - 112.81)^2}{112.81} + \frac{(225 - 249.38)^2}{249.38} + \frac{(150 - 137.81)^2}{137.81} = 4.78$$

Since $\chi^2_{.05,k-1-m} = \chi^2_{.05,3-1-1} = \chi^2_{.05,1} = 3.843$ and $4.78 \geq 3.843$, H_0 is rejected. Appendix Table A.11 shows that P-value $\approx .029$. ∎

Example 14.7

Consider a series of games between two teams, I and II, that terminates as soon as one team has won four games (with no possibility of a tie). A simple probability model for such a series assumes that outcomes of successive games are independent and that the probability of team I winning any particular game is a constant θ. We arbitrarily designate I the better team, so that $\theta \geq .5$. Any particular series can then terminate after 4, 5, 6, or 7 games. Let $\pi_1(\theta), \pi_2(\theta), \pi_3(\theta), \pi_4(\theta)$ denote the probability of termination in 4, 5, 6, and 7 games, respectively. Then

$$\pi_1(\theta) = P(\text{I wins in 4 games}) + P(\text{II wins in 4 games})$$
$$= \theta^4 + (1 - \theta)^4$$
$$\pi_2(\theta) = P(\text{I wins 3 of the first 4 and the fifth})$$
$$\quad + P(\text{I loses 3 of the first 4 and the fifth})$$
$$= \binom{4}{3} \theta^3(1 - \theta) \cdot \theta + \binom{4}{1} \theta(1 - \theta)^3 \cdot (1 - \theta)$$
$$= 4\theta(1 - \theta)[\theta^3 + (1 - \theta)^3]$$
$$\pi_3(\theta) = 10\theta^2(1 - \theta)^2[\theta^2 + (1 - \theta)^2]$$
$$\pi_4(\theta) = 20\theta^3(1 - \theta)^3$$

The article "Seven-Game Series in Sports" by Groeneveld and Meeden (*Mathematics Magazine*, 1975: 187–192) tested the fit of this model to results of National Hockey League playoffs during the period 1943–1967 (when league membership was stable). The data appears in Table 14.5.

Table 14.5 Observed and expected counts for the simple model

Cell Number of games played	1 4	2 5	3 6	4 7	
Observed frequency	15	26	24	18	$n = 83$
Estimated expected frequency	16.351	24.153	23.240	19.256	

The estimated expected cell counts are $83\pi_i(\hat{\theta})$, where $\hat{\theta}$ is the value of θ that maximizes

$$\{\theta^4 + (1 - \theta)^4\}^{15} \cdot \{4\theta(1 - \theta)[\theta^3 + (1 - \theta)^3]\}^{26}$$
$$\cdot \{10\theta^2(1 - \theta)^2[\theta^2 + (1 - \theta)^2]\}^{24} \cdot \{20\theta^3(1 - \theta)^3\}^{18} \quad (14.5)$$

Standard calculus methods fail to yield a nice formula for the maximizing value $\hat{\theta}$, so it must be computed using numerical methods. The result is $\hat{\theta} = .654$, from which $\pi_i(\hat{\theta})$ and the estimated expected cell counts are computed. The computed value of χ^2 is .360, and (since $k - 1 - m = 4 - 1 - 1 = 2$) $\chi^2_{.10,2} = 4.605$. There is thus no reason to reject the simple model as applied to NHL playoff series.

The cited article also considered World Series data for the period 1903–1973. For the simple model, $\chi^2 = 5.97$, so the model does not seem appropriate. The suggested reason for this is that for the simple model

$$P(\text{series lasts six games} \mid \text{series lasts at least six games}) \geq .5 \qquad (14.6)$$

whereas of the 38 series that actually lasted at least six games, only 13 lasted exactly six. The following alternative model is then introduced:

$$\pi_1(\theta_1, \theta_2) = \theta_1^4 + (1 - \theta_1)^4$$
$$\pi_2(\theta_1, \theta_2) = 4\theta_1(1 - \theta_1)[\theta_1^3 + (1 - \theta_1)^3]$$
$$\pi_3(\theta_1, \theta_2) = 10\theta_1^2(1 - \theta_1)^2\theta_2$$
$$\pi_4(\theta_1, \theta_2) = 10\theta_1^2(1 - \theta_1)^2(1 - \theta_2)$$

The first two π_i's are identical to the simple model, whereas θ_2 is the conditional probability of (14.6) (which can now be any number between 0 and 1). The values of $\hat{\theta}_1$ and $\hat{\theta}_2$ that maximize the expression analogous to expression (14.5) are determined numerically as $\hat{\theta}_1 = .614$, $\hat{\theta}_2 = .342$. A summary appears in Table 14.6, and $\chi^2 = .384$. Since two parameters are estimated, df $= k - 1 - m = 1$ with $\chi^2_{.10,1} = 2.706$, indicating a good fit of the data to this new model.

Table 14.6 Observed and expected counts for the more complex model

Number of games played	4	5	6	7
Observed frequency	12	16	13	25
Estimated expected frequency	10.85	18.08	12.68	24.39

One of the regularity conditions on the θ_i's in the theorem is that they be functionally independent of one another. That is, no single θ_i can be determined from the values of other θ_i's, so that m is the number of functionally independent parameters estimated. A general rule of thumb for degrees of freedom in a chi-squared test is the following.

$$\chi^2 \text{ df} = \left(\begin{array}{c}\text{number of freely} \\ \text{determined cell counts}\end{array}\right) - \left(\begin{array}{c}\text{number of independent} \\ \text{parameters estimated}\end{array}\right)$$

This rule will be used in connection with several different chi-squared tests in the next section.

Goodness of Fit for Discrete Distributions

Many experiments involve observing a random sample $X_1, X_2, \ldots, X_n$ from some discrete distribution. One may then wish to investigate whether the underlying distribution is a member of a particular family, such as the Poisson or negative binomial family. In the case of both a Poisson and a negative binomial distribution, the set of possible values is infinite, so the values must be grouped into k subsets before a chi-squared test can be used. The groupings should be done so that the expected frequency in each cell (group) is at least five. The last cell will then correspond to X values of $c, c + 1, c + 2, \ldots$ for some value c.

This grouping can considerably complicate the computation of the $\hat{\theta}_i$'s and estimated expected cell counts. This is because the theorem requires that the $\hat{\theta}_i$'s be obtained from the cell counts $N_1, \ldots, N_k$ rather than the sample values $X_1, \ldots, X_n$.

Example 14.8 Table 14.7 presents count data on the number of *Larrea divaricata* plants found in each of 48 sampling quadrats, as reported in the article "Some Sampling Characteristics of Plants and Arthropods of the Arizona Desert" (*Ecology*, 1962: 567–571).

Table 14.7 Observed counts for Example 14.8

Cell	1	2	3	4	5
Number of plants	0	1	2	3	≥ 4
Frequency	9	9	10	14	6

The author fit a Poisson distribution to the data. Let λ denote the Poisson parameter and suppose for the moment that the six counts in cell 5 were actually 4, 4, 5, 5, 6, 6. Then denoting sample values by $x_1, \ldots, x_{48}$, nine of the x_i's were 0, nine were 1, and so on. The likelihood of the observed sample is

$$\frac{e^{-\lambda}\lambda^{x_1}}{x_1!} \cdots \cdots \frac{e^{-\lambda}\lambda^{x_{48}}}{x_{48}!} = \frac{e^{-48\lambda}\lambda^{\Sigma x_i}}{x_1! \cdots \cdots x_{48}!} = \frac{e^{-48\lambda}\lambda^{101}}{x_1! \cdots \cdots x_{48}!}$$

The value of λ for which this is maximized is $\hat{\lambda} = \Sigma x_i / n = 101/48 = 2.10$ (the value reported in the article).

However, the $\hat{\lambda}$ required for χ^2 is obtained by maximizing Expression (14.4) rather than the likelihood of the full sample. The cell probabilities are

$$\pi_i(\lambda) = \frac{e^{-\lambda}\lambda^{i-1}}{(i-1)!} \qquad i = 1, 2, 3, 4$$

$$\pi_5(\lambda) = 1 - \sum_{i=0}^{3} \frac{e^{-\lambda}\lambda^i}{i!}$$

so the right-hand side of (14.4) becomes

$$\left[\frac{e^{-\lambda}\lambda^0}{0!} \right]^9 \left[\frac{e^{-\lambda}\lambda^1}{1!} \right]^9 \left[\frac{e^{-\lambda}\lambda^2}{2!} \right]^{10} \left[\frac{e^{-\lambda}\lambda^3}{3!} \right]^{14} \left[1 - \sum_{i=0}^{3} \frac{e^{-\lambda}\lambda^i}{i!} \right]^6$$

There is no nice formula for $\hat{\lambda}$, the maximizing value of λ, in this latter expression, so it must be obtained numerically. ∎

Because the parameter estimates are usually much more difficult to compute from the grouped data than from the full sample, they are virtually always computed using this latter method. When these "full" estimators are used in the chi-squared statistic, the distribution of the statistic is altered and a level α test is no longer specified by the critical value $\chi^2_{\alpha,k-1-m}$.

THEOREM

> Let $\hat{\theta}_1, \ldots, \hat{\theta}_m$ be the maximum likelihood estimators of $\theta_1, \ldots, \theta_m$ based on the full sample $X_1, \ldots, X_n$ and let χ^2 denote the statistic based on these estimators. Then the critical value c_α that specifies a level α upper-tailed test satisfies
>
> $$\chi^2_{\alpha,k-1-m} \leq c_\alpha \leq \chi^2_{\alpha,k-1} \tag{14.7}$$

The test procedure implied by this theorem is the following:

> If $\chi^2 \geq \chi^2_{\alpha,k-1}$, reject H_0.
> If $\chi^2 \leq \chi^2_{\alpha,k-1-m}$, do not reject H_0.
> If $\chi^2_{\alpha,k-1-m} < \chi^2 < \chi^2_{\alpha,k-1}$, withhold judgment. $\tag{14.8}$

Example 14.9
(Example 14.8
continued)

Using $\hat{\lambda} = 2.10$, the estimated expected cell counts are computed from $n\pi_i(\hat{\lambda})$ where $n = 48$. For example,

$$n\pi_1(\hat{\lambda}) = 48 \cdot \frac{e^{-2.1}(2.1)^0}{0!} = (48)(e^{-2.1}) = 5.88$$

Similarly, $n\pi_2(\hat{\lambda}) = 12.34$, $n\pi_3(\hat{\lambda}) = 12.96$, $n\pi_4(\hat{\lambda}) = 9.07$, and $n\pi_5(\hat{\lambda}) = 48 - 5.88 - \cdots - 9.07 = 7.75$. Then

$$\chi^2 = \frac{(9 - 5.88)^2}{5.88} + \cdots + \frac{(6 - 7.75)^2}{7.75} = 6.31$$

Since $m = 1$ and $k = 5$, at level .05 we need $\chi^2_{.05,3} = 7.815$ and $\chi^2_{.05,4} = 9.488$. Because $6.31 \leq 7.815$, we do not reject H_0; at the 5% level, the Poisson distribution provides a reasonable fit to the data. Notice that $\chi^2_{.10,3} = 6.251$ and $\chi^2_{.10,4} = 7.779$, so at level .10 we would have to withhold judgment on whether the Poisson distribution was appropriate. ∎

Sometimes even the maximum likelihood estimates based on the full sample are quite difficult to compute. This is the case, for example, for the two-parameter (generalized) negative binomial distribution. In such situations, method-of-moments estimates are often used, and the resulting χ^2 compared to $\chi^2_{\alpha,k-1-m}$, though it is not known to what extent the use of moments estimators affects the true critical value.

Goodness of Fit for Continuous Distributions

The chi-squared test can also be used to test whether the sample comes from a specified family of continuous distributions, such as the exponential family or the normal family. The choice of cells (class intervals) is even more arbitrary in the continuous case than in the discrete case. To ensure that the chi-squared test is valid, the cells should be chosen independently of the sample observations. Once the cells are chosen, it is almost always quite difficult to estimate unspecified parameters (such as μ and σ in the normal case) from the observed cell counts, so instead mle's based on the full sample are computed. The critical value c_α again satisfies (14.7), and the test procedure is given by (14.8).

Example 14.10 The Institute of Nutrition of Central America and Panama (INCAP) has carried out extensive dietary studies and research projects in Central America. In one study reported in the November 1964 issue of the *American Journal of Clinical Nutrition* ("The Blood Viscosity of Various Socioeconomic Groups in Guatemala"), serum total cholesterol measurements for a sample of 49 low-income rural Indians were reported as follows (in mg/L):

204 108 140 152 158 129 175 146 157 174 192 194 144 152 135 223 145

231 115 131 129 142 114 173 226 155 166 220 180 172 143 148 171 143

124 158 144 108 189 136 136 197 131 95 139 181 165 142 162

Is it plausible that serum cholesterol level is normally distributed for this population? Suppose that prior to sampling, it was believed that plausible values for μ and σ were 150 and 30, respectively. The seven equiprobable class intervals for the standard normal distribution are $(-\infty, -1.07)$, $(-1.07, -.57)$, $(-.57, -.18)$, $(-.18, .18)$, $(.18, .57)$, $(.57, 1.07)$, and $(1.07, \infty)$, with each endpoint also giving the distance in standard deviations from the mean for any other normal distribution. For $\mu = 150$ and $\sigma = 30$, these intervals become $(-\infty, 117.9)$, $(117.9, 132.9)$, $(132.9, 144.6)$, $(144.6, 155.4)$, $(155.4, 167.1)$, $(167.1, 182.1)$, and $(182.1, \infty)$.

To obtain the estimated cell probabilities $\pi_1(\hat\mu, \hat\sigma), \ldots, \pi_7(\hat\mu, \hat\sigma)$, we first need the mle's $\hat\mu$ and $\hat\sigma$. In Chapter 6, the mle of σ was shown to be $\left[\sum(x_i - \bar x)^2/n\right]^{1/2}$ (rather than s), so with $s = 31.75$

$$\hat\mu = \bar x = 157.02 \qquad \hat\sigma = \left[\frac{\sum(x_i - \bar x)^2}{n}\right]^{1/2} = \left[\frac{(n-1)s^2}{n}\right]^{1/2} = 31.42$$

Each $\pi_i(\hat\mu, \hat\sigma)$ is then the probability that a normal rv X with mean 157.02 and standard deviation 31.42 falls in the ith class interval. For example,

$$\pi_2(\hat\mu, \hat\sigma) = P(117.9 \leq X \leq 132.9) = P(-1.25 \leq Z \leq -.77) = .1150$$

so $n\pi_2(\hat\mu, \hat\sigma) = 49(.1150) = 5.64$. Observed and estimated expected cell counts are shown in Table 14.8.

The computed χ^2 is 4.60. With $k = 7$ cells and $m = 2$ parameters estimated, $\chi^2_{.05,k-1} = \chi^2_{.05,6} = 12.592$ and $\chi^2_{.05,k-1-m} = \chi^2_{.05,4} = 9.488$. Since $4.60 \leq 9.488$, a normal distribution provides quite a good fit to the data.

Table 14.8 Observed and expected counts for Example 14.10

Cell	$(-\infty, 117.9)$	$(117.9, 132.9)$	$(132.9, 144.6)$	$(144.6, 155.4)$
Observed	5	5	11	6
Estimated expected	5.17	5.64	6.08	6.64

Cell	$(155.4, 167.1)$	$(167.1, 182.1)$	$(182.1, \infty)$
Observed	6	7	9
Estimated expected	7.12	7.97	10.38

∎

Example 14.11 The article "Some Studies on Tuft Weight Distribution in the Opening Room" (*Textile Research J.*, 1976: 567–573) reports the accompanying data on the distribution of output tuft weight X (mg) of cotton fibers for the input weight $x_0 = 70$.

Interval	0–8	8–16	16–24	24–32	32–40	40–48	48–56	56–64	64–70
Observed frequency	20	8	7	1	2	1	0	1	0
Expected frequency	18.0	9.9	5.5	3.0	1.8	.9	.5	.3	.1

The authors postulated a truncated exponential distribution:

$$H_0: f(x) = \frac{\lambda e^{-\lambda x}}{1 - e^{-\lambda x_0}} \qquad 0 \le x \le x_0$$

The mean of this distribution is

$$\mu = \int_0^{x_0} x f(x)\,dx = \frac{1}{\lambda} - \frac{x_0 e^{-\lambda x_0}}{1 - e^{-\lambda x_0}}$$

The parameter λ was estimated by replacing μ by $\bar{x} = 13.086$ and solving the resulting equation to obtain $\hat{\lambda} = .0742$ (so $\hat{\lambda}$ is a method-of-moments estimate and not an mle). Then with $\hat{\lambda}$ replacing λ in $f(x)$, the estimated expected cell frequencies as displayed previously are computed as

$$40\pi_i(\hat{\lambda}) = 40P(a_{i-1} \le X < a_i) = 40 \int_{a_{i-1}}^{a_i} f(x)\,dx = \frac{40(e^{-\hat{\lambda}a_{i-1}} - e^{-\hat{\lambda}a_i})}{1 - e^{-\hat{\lambda}x_0}}$$

where $[a_{i-1}, a_i]$ is the ith class interval. To obtain expected cell counts of at least 5, the last six cells are combined to yield observed counts of 20, 8, 7, 5 and expected counts of 18.0, 9.9, 5.5, 6.6. The computed value of chi-squared is then $\chi^2 = 1.34$.

Because $\chi^2_{.05,2} = 5.992$, H_0 is not rejected, so the truncated exponential model provides a good fit. ∎

A Special Test for Normality

Probability plots were introduced in Section 4.6 as an informal method for assessing the plausibility of any specified population distribution as the one from which the given sample was selected. The straighter the probability plot, the more plausible is the distribution on which the plot is based. A normal probability plot is used for checking whether *any* member of the normal distribution family is plausible. Let's denote the sample x_i's when ordered from smallest to largest by $x_{(1)}, x_{(2)}, \ldots, x_{(n)}$. Then the plot suggested for checking normality was a plot of the points $(x_{(i)}, y_i)$, where $y_i = \Phi^{-1}((i - .5)/n)$.

A quantitative measure of the extent to which points cluster about a straight line is the sample correlation coefficient r introduced in Chapter 12. Consider calculating r for the n pairs $(x_{(1)}, y_1), \ldots, (x_{(n)}, y_n)$. The y_i's here are not observed values in a random sample from a y population, so properties of this r are quite different from those described in Section 12.5. However, it is true that the more r deviates from 1, the less the probability plot resembles a straight line (remember that a probability plot must slope upward). This idea can be extended to yield a formal test procedure: Reject the hypothesis of population normality if $r \leq c_\alpha$, where c_α is a critical value chosen to yield the desired significance level α. That is, the critical value is chosen so that when the population distribution is actually normal, the probability of obtaining an r value that is at most c_α (and thus incorrectly rejecting H_0) is the desired α. The developers of the MINITAB statistical computer package give critical values for $\alpha = .10$, .05, and .01 in combination with different sample sizes. These critical values are based on a slightly different definition of the y_i's than that given previously.

MINITAB will also construct a normal probability plot based on these y_i's. The plot will be almost identical in appearance to that based on the previous y_i's. When there are several tied $x_{(i)}$'s, MINITAB computes r by using the average of the corresponding y_i's as the second number in each pair.

Let $y_i = \Phi^{-1}[(i - .375)/(n + .25)]$ and compute the sample correlation coefficient r for the n pairs $(x_{(1)}, y_1), \ldots, (x_{(n)}, y_n)$. The Ryan–Joiner test of

H_0: the population distribution is normal

versus

H_a: the population distribution is not normal

consists of rejecting H_0 when $r \leq c_\alpha$. Critical values c_α are given in Appendix Table A.12 for various significance levels α and sample sizes n.

Example 14.12 The following sample of $n = 20$ observations on dielectric breakdown voltage of a piece of epoxy resin first appeared in Example 4.29.

y_i	−1.871	−1.404	−1.127	−.917	−.742	−.587	−.446	−.313	−.186	−.062
$x_{(i)}$	24.46	25.61	26.25	26.42	26.66	27.15	27.31	27.54	27.74	27.94

y_i	.062	.186	.313	.446	.587	.742	.917	1.127	1.404	1.871
$x_{(i)}$	27.98	28.04	28.28	28.49	28.50	28.87	29.11	29.13	29.50	30.88

We asked MINITAB to carry out the Ryan–Joiner test, and the result appears in Figure 14.3. The test statistic value is $r = .9881$, and Appendix Table A.12 gives .9600 as the critical value that captures lower-tail area .10 under the r sampling distribution curve when $n = 20$ and the underlying distribution is actually normal. Since $.9881 > .9600$, the null hypothesis of normality cannot be rejected even for a significance level as large as .10.

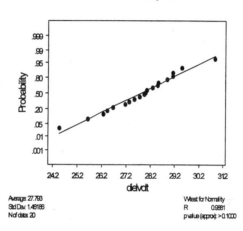

Figure 14.3 MINITAB output from the Ryan–Joiner test for the data of Example 14.12

■

Exercises | Section 14.2 (12–23)

12. Consider a large population of families in which each family has exactly three children. If the genders of the three children in any family are independent of one another, the number of male children in a randomly selected family will have a binomial distribution based on three trials.

a. Suppose a random sample of 160 families yields the following results. Test the relevant hypotheses by proceeding as in Example 14.5.

Number of male children	0	1	2	3
Frequency	14	66	64	16

b. Suppose a random sample of families in a non-human population resulted in observed frequencies of 15, 20, 12, and 3, respectively. Would the chi-squared test be based on the same number of degrees of freedom as the test in part (a)? Explain.

13. A study of sterility in the fruit fly ("Hybrid Dysgenesis in *Drosophila melanogaster:* The Biology of Female and Male Sterility," *Genetics,* 1979: 161–174) reports the following data on the number of ovaries developed for each female fly in a sample of size 1388. One model for unilateral sterility states that each ovary develops with some probability p independently of the other ovary. Test the fit of this model using χ^2.

x = Number of ovaries developed	0	1	2
Observed count	1212	118	58

14. The article "Feeding Ecology of the Red-Eyed Vireo and Associated Foliage-Gleaning Birds" (*Ecological Monographs*, 1971: 129–152) presents the accompanying data on the variable X = the number of hops before the first flight and preceded by a flight. The author then proposed and fit a geometric probability distribution [$p(x) = P(X = x) = p^{x-1} \cdot q$ for $x = 1, 2, \ldots$ where $q = 1 - p$] to the data. The total sample size was $n = 130$.

x	1	2	3	4	5	6	7	8	9	10	11	12
Number of times x observed	48	31	20	9	6	5	4	2	1	1	2	1

   a. The likelihood is $(p^{x_1-1} \cdot q) \cdot \cdots \cdot (p^{x_n-1} \cdot q) = p^{\Sigma x_i - n} \cdot q^n$. Show that the mle of p is given by $\hat{p} = (\Sigma x_i - n)/\Sigma x_i$ and compute $\hat{p}$ for the given data.

   b. Estimate the expected cell counts using $\hat{p}$ of part (a) [expected cell counts $= n \cdot (\hat{p})^{x-1} \cdot \hat{q}$ for $x = 1, 2, \ldots$] and test the fit of the model using a χ^2 test by combining the counts for $x = 7, 8, \ldots$, and 12 into one cell ($x \geq 7$).

15. A certain type of flashlight is sold with the four batteries included. A random sample of 150 flashlights is obtained, and the number of defective batteries in each is determined, resulting in the following data:

Number defective	0	1	2	3	4
Frequency	26	51	47	16	10

Let X be the number of defective batteries in a randomly selected flashlight. Test the null hypothesis that the distribution of X is Bin(4, θ). That is, with $p_i = P(i$ defectives), test

$$H_0: \quad p_i = \binom{4}{i} \theta^i (1 - \theta)^{4-i} \quad i = 0, 1, 2, 3, 4$$

[*Hint:* To obtain the mle of θ, write the likelihood (the function to be maximized) as $\theta^u (1 - \theta)^v$ where the exponents u and v are linear functions of the cell counts. Then take the natural log, differentiate with respect to θ, equate the result to 0, and solve for $\hat{\theta}$.]

16. In a genetics experiment, investigators looked at 300 chromosomes of a particular type and counted the number of sister-chromatid exchanges on each ("On the Nature of Sister-Chromatid Exchanges in 5-Bromodeoxyuridine-Substituted Chromosomes," *Genetics*, 1979: 1251–1264). A Poisson model was hypothesized for the distribution of the number of exchanges. Test the fit of a Poisson distribution to the data by first estimating λ and then combining the counts for $x = 8$ and $x = 9$ into one cell.

X = Number of exchanges	0	1	2	3	4	5	6	7	8	9
Observed counts	6	24	42	59	62	44	41	14	6	2

17. An article in *Annals of Mathematical Statistics* reports the following data on the number of borers in each of 120 groups of borers. Does the Poisson pmf provide a plausible model for the distribution of the number of borers in a group? (*Hint:* Add the frequencies for 7, 8, ..., 12 to establish a single category " ≥ 7.")

Number of borers	0	1	2	3	4	5	6	7	8	9	10	11	12
Frequency	24	16	16	18	15	9	6	5	3	4	3	0	1

18. The article "A Probabilistic Analysis of Dissolved Oxygen–Biochemical Oxygen Demand Relationship in Streams" (*J. Water Resources Control Fed.*, 1969: 73–90) reports data on the rate of oxygenation in streams at 20°C in a certain region. The sample mean and standard deviation were computed as $\bar{x} = .173$ and $s = .066$, respectively. Based on the accompanying frequency distribution, can it be concluded that oxygenation rate is a normally distributed variable? Use the chi-squared test with $\alpha = .05$.

Rate (per day)	Frequency
Below .100	12
.100–below .150	20
.150–below .200	23
.200–below .250	15
.250 or more	13

19. Each headlight on an automobile undergoing an annual vehicle inspection can be focused either too high (*H*), too low (*L*), or properly (*N*). Checking the two headlights simultaneously (and not distinguishing between left and right) results in the six possible outcomes *HH, LL, NN, HL, HN*, and *LN*. If the

probabilities (population proportions) for the single headlight focus direction are $P(H) = \theta_1$, $P(L) = \theta_2$, and $P(N) = 1 - \theta_1 - \theta_2$ and the two headlights are focused independently of one another, the probabilities of the six outcomes for a randomly selected car are the following:

$$p_1 = \theta_1^2 \qquad p_2 = \theta_2^2 \qquad p_3 = (1 - \theta_1 - \theta_2)^2$$
$$p_4 = 2\theta_1\theta_2 \qquad p_5 = 2\theta_1(1 - \theta_1 - \theta_2)$$
$$p_6 = 2\theta_2(1 - \theta_1 - \theta_2)$$

Use the accompanying data to test the null hypothesis

$$H_0: \quad p_1 = \pi_1(\theta_1, \theta_2), \ldots, p_6 = \pi_6(\theta_1, \theta_2)$$

where the $\pi_1(\theta_1, \theta_2)$'s are given previously.

Outcome	HH	LL	NN	HL	HN	LN
Frequency	49	26	14	20	53	38

(*Hint:* Write the likelihood as a function of θ_1 and θ_2, take the natural log, then compute $\partial/\partial\theta_1$ and $\partial/\partial\theta_2$, equate them to 0, and solve for $\hat{\theta}_1$, $\hat{\theta}_2$.)

20. The article "Compatibility of Outer and Fusible Interlining Fabrics in Tailored Garments (*Textile Res. J.,* 1997: 137–142) gave the following observations on bending rigidity (μN · m) for medium-quality fabric specimens, from which the accompanying MINITAB output was obtained.

24.6 12.7 14.4   30.6 16.1   9.5 31.5 17.2
46.9 68.3 30.8 116.7 39.5 73.8 80.6 20.3
25.8 30.9 39.2   36.8 46.6 15.6 32.3

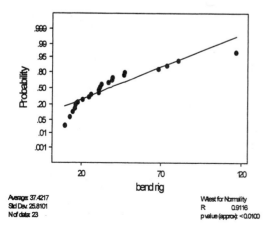

Normal Probability Plot

Average: 37.4217
Std Dev: 25.8101
N of data: 23

W-test for Normality
R   0.9116
p value (approx): < 0.0100

Would you use a one-sample t confidence interval to estimate true average bending rigidity? Explain your reasoning.

21. The article from which the data in Exercise 20 was obtained also gave the accompanying data on the composite mass/outer fabric mass ratio for high-quality fabric specimens:

1.15 1.40 1.34 1.29 1.36 1.26 1.22 1.40
1.29 1.41 1.32 1.34 1.26 1.36 1.36 1.30
1.28 1.45 1.29 1.28 1.38 1.55 1.46 1.32

MINITAB gave $r = .9852$ as the value of the Ryan–Joiner test statistic and reported that P-value $> .10$. Would you use the one-sample t test to test hypotheses about the value of the true average ratio? Why or why not?

22. The following data set consists of 25 observations on fracture toughness of base plate of 18% nickel maraging steel (from "Fracture Testing of Weldments," *ASTM Special Publ. No. 381,* 1965: 328–356). Use the normal probability plot correlation coefficient test to decide whether a normal distribution provides a plausible model for fracture toughness.

69.5 71.9 72.6 73.1 73.3 73.5 74.1 74.2 75.3
75.5 75.7 75.8 76.1 76.2 76.9 77.0 77.9 78.1
79.6 79.7 79.9 80.1 82.2 83.7 93.7

23. The article "Nonbloated Burned Clay Aggregate Concrete" (*J. Materials,* 1972: 555–563) reports the following data on 7-day flexural strength of nonbloated burned clay aggregate concrete samples (psi):

257 327 317 300 340 340 343 374 377 386
383 393 407 407 434 427 440 407 450 440
456 460 456 476 480 490 497 526 546 700

Test at level .10 to decide whether flexural strength is a normally distributed variable.

14.3 | Two-Way Contingency Tables

In the previous two sections, we discussed inferential problems in which the count data was displayed in a rectangular table of cells. Each table consisted of one row and a specified number of columns, where the columns corresponded to categories into which the population had been divided. We now study problems in which the data also consists of counts or frequencies, but the data table will now have I rows ($I \geq 2$) and J columns, so IJ cells. There are two commonly encountered situations in which such data arises:

1. There are I populations of interest, each corresponding to a different row of the table, and each population is divided into the same J categories. A sample is taken from the ith population ($i = 1, \ldots, I$), and the counts are entered in the cells in the ith row of the table.

2. There is a single population of interest, with each individual in the population categorized with respect to two different factors. There are I categories associated with the first factor, and J categories associated with the second factor. A single sample is taken, and the number of individuals belonging in both category i of factor 1 and category j of factor 2 is entered in the cell in row i, column j ($i = 1, \ldots, I; j = 1, \ldots, J$).

Let n_{ij} denote the number of individuals in the sample(s) falling in the (i, j)th cell (row i, column j) of the table—that is, the (i, j)th cell count. The table displaying the n_{ij}'s is called a **two-way contingency table;** a prototype is shown in Table 14.9.

Table 14.9 A two-way contingency table

	1	**2**	**...**	**j**	**...**	**J**
1	n_{11}	n_{12}	...	n_{1j}	...	n_{1J}
2	n_{21}					$\vdots$
$\vdots$	$\vdots$					
i	n_{i1}	...		n_{ij}	...	
$\vdots$	$\vdots$					
I	n_{I1}	...				n_{IJ}

In situations of type 1, we want to investigate whether the proportions in the different categories are the same for all populations. The null hypothesis states that the populations are **homogeneous** with respect to these categories. In type 2 situations, we investigate whether the categories of the two factors occur independently of one another in the population.

Testing for Homogeneity

We assume that each individual in every one of the I populations belongs in exactly one of J categories. A sample of n_i individuals is taken from the ith population; let $n = \sum n_i$ and

$n_{ij} = $ the number of individuals in the ith sample who fall into category j

$$n_{\cdot j} = \sum_{i=1}^{I} n_{ij} = \begin{array}{l} \text{the total number of individuals among} \\ \text{the } n \text{ sampled who fall into category } j \end{array}$$

The n_{ij}'s are recorded in a two-way contingency table with I rows and J columns. The sum of the n_{ij}'s in the ith row is n_i, whereas the sum of entries in the jth column is $n_{\cdot j}$.

Let

$$p_{ij} = \begin{array}{l} \text{the proportion of the individuals in} \\ \text{population } i \text{ who fall into category } j \end{array}$$

Thus, for population 1, the J proportions are $p_{11}, p_{12}, \ldots, p_{1J}$ (which sum to 1) and similarly for the other populations. The **null hypothesis of homogeneity** states that the proportion of individuals in category j is the same for each population, and that this is true for every category; that is, for every j, $p_{1j} = p_{2j} = \cdots = p_{Ij}$.

When H_0 is true, we can use $p_1, p_2, \ldots, p_J$ to denote the population proportions in the J different categories; these proportions are common to all I populations. The expected number of individuals in the ith sample who fall in the jth category when H_0 is true is then $E(N_{ij}) = n_i \cdot p_j$. To estimate $E(N_{ij})$, we must first estimate p_j, the proportion in category j. Among the total sample of n individuals, $N_{\cdot j}$ fall into category j, so we use $\hat{p}_j = N_{\cdot j}/n$ as the estimator (this can be shown to be the maximum likelihood estimator of p_j). Substitution of the estimate $\hat{p}_j$ for p_j in $n_i p_j$ yields a simple formula for estimated expected counts under H_0:

$$\hat{e}_{ij} = \text{estimated expected count in cell } (i, j) = n_i \cdot \frac{n_{\cdot j}}{n}$$

$$= \frac{(i\text{th row total})(j\text{th column total})}{n} \qquad (14.9)$$

The test statistic also has the same form as in previous problem situations. The number of degrees of freedom comes from the general rule of thumb. In each row of Table 14.9 there are $J - 1$ freely determined cell counts (each sample size n_i is fixed), so there are a total of $I(J - 1)$ freely determined cells. Parameters $p_1, \ldots, p_J$ are estimated, but because $\sum p_i = 1$, only $J - 1$ of these are independent. Thus df $= I(J - 1) - (J - 1) = (J - 1)(I - 1)$.

Null hypothesis: H_0: $p_{1j} = p_{2j} = \cdots = p_{Ij}$ $\quad j = 1, 2, \ldots, J$

Alternative hypothesis: H_a: H_0 is not true

Test statistic value:

$$\chi^2 = \sum_{\text{all cells}} \frac{(\text{observed} - \text{estimated expected})^2}{\text{estimated expected}} = \sum_{i=1}^{I} \sum_{j=1}^{J} \frac{(n_{ij} - \hat{e}_{ij})^2}{\hat{e}_{ij}}$$

Rejection region: $\chi^2 \geq \chi^2_{\alpha,(I-1)(J-1)}$

The test can safely be applied as long as $\hat{e}_{ij} \geq 5$ for all cells.

Example 14.13 A company packages a particular product in cans of three different sizes, each one using a different production line. Most cans conform to specifications, but a quality control engineer has identified the following reasons for nonconformance:

1. Blemish on can

2. Crack in can

3. Improper pull tab location

4. Pull tab missing

5. Other

A sample of nonconforming units is selected from each of the three lines, and each unit is categorized according to reason for nonconformity, resulting in the following contingency table data:

		Reason for nonconformity					**Sample Size**
		Blemish	**Crack**	**Location**	**Missing**	**Other**	
Production Line	1	34	65	17	21	13	150
	2	23	52	25	19	6	125
	3	32	28	16	14	10	100
	Total	89	145	58	54	29	375

Does the data suggest that the proportions falling in the various nonconformance categories are not the same for the three lines? The parameters of interest are the various proportions, and the relevant hypotheses are

H_0: the production lines are homogeneous with respect to the five nonconformance categories

H_a: the production lines are not homogeneous with respect to the categories

To calculate χ^2, we must first compute the estimated expected frequencies (assuming homogeneity). Consider the first nonconformance category for the first production line. When the lines are homogeneous,

estimated expected number among the 150 selected units that are blemished

$$= \frac{\text{(first row total)(first column total)}}{\text{total of sample sizes}} = \frac{(150)(89)}{375} = 35.60$$

The contribution of the cell in the upper-left corner to χ^2 is then

$$\frac{\text{(observed} - \text{estimated expected)}^2}{\text{estimated expected}} = \frac{(34 - 35.60)^2}{35.60} = .072$$

The other contributions are calculated in a similar manner. Figure 14.4 shows MINITAB output for the chi-squared test. The observed count is the top number in each cell, and directly below it is the estimated expected count. The contribution of each cell to χ^2 appears below the counts, and the test statistic value is $\chi^2 = 14.159$. All estimated expected counts are at least 5, so combining categories is unnecessary. The test is based on $(3 - 1)(5 - 1) = 8$ df. Appendix Table A.11 shows that the values that capture upper-tail areas of .08 and .075 under the 8 df curve are 14.06 and 14.26, respectively. Thus the P-value is between .075 and .08; MINITAB gives P-value $= .079$. The null hypothesis of homogeneity should not be rejected at the usual significance levels of .05 or .01, but it would be rejected for the higher α of .10.

```
Expected counts are printed below observed counts
          blem      crack       loc    missing      other      Total
1           34         65         17         21         13        150
         35.60      58.00      23.20      21.60      11.60

2           23         52         25         19          6        125
         29.67      48.33      19.33      18.00       9.67

3           32         28         16         14         10        100
         23.73      38.67      15.47      14.40       7.73

Total       89        145         58         54         29        375

Chisq = 0.072 + 0.845 + 1.657 + 0.017 + 0.169 + 1.498 + 0.278 +
        1.661 + 0.056 + 1.391 + 2.879 + 2.943 + 0.018 + 0.011 +
        0.664 = 14.159

df = 8, p = 0.079
```

Figure 14.4 MINITAB output for the chi-squared test of Example 14.13 ■

Testing for Independence

We focus now on the relationship between two different factors in a single population. The number of categories of the first factor will be denoted by I and the number of categories of the second factor by J. Each individual in the population is assumed to belong in exactly one of the I categories associated with the first factor and exactly one of the J categories associated with the second factor. For example, the population of interest might consist of all individuals who regularly watch the national news on television, with the first factor being preferred network (ABC, CBS,

NBC, or PBS, so $I = 4$) and the second factor political philosophy (liberal, moderate, or conservative, giving $J = 3$).

For a sample of n individuals taken from the population, let n_{ij} denote the number among the n who fall both in category i of the first factor and category j of the second factor. The n_{ij}'s can be displayed in a two-way contingency table with I rows and J columns. In the case of homogeneity for I populations, the row totals were fixed in advance, and only the J column totals were random. Now only the total sample size is fixed, and both the $n_{i\cdot}$'s and $n_{\cdot j}$'s are observed values of random variables. To state the hypotheses of interest, let

p_{ij} = the proportion of individuals in the population who belong in category i of factor 1 and category j of factor 2

= P(a randomly selected individual falls in both category i of factor 1 and category j of factor 2)

Then

$$p_{i\cdot} = \sum_j p_{ij} = P(\text{a randomly selected individual falls in category } i \text{ of factor 1})$$

$$p_{\cdot j} = \sum_i p_{ij} = P(\text{a randomly selected individual falls in category } j \text{ of factor 2})$$

Recall that two events A and B are independent if $P(A \cap B) = P(A) \cdot P(B)$. *The null hypothesis here says that an individual's category with respect to factor 1 is independent of the category with respect to factor 2.* In symbols, this becomes $p_{ij} = p_{i\cdot} \cdot p_{\cdot j}$ for every pair (i, j).

The expected count in cell (i, j) is $n \cdot p_{ij}$, so when the null hypothesis is true, $E(N_{ij}) = n \cdot p_{i\cdot} \cdot p_{\cdot j}$. To obtain a chi-squared statistic, we must therefore estimate the $p_{i\cdot}$'s $(i = 1, \ldots, I)$ and $p_{\cdot j}$'s $(j = 1, \ldots, J)$. The (maximum likelihood) estimates are

$$\hat{p}_{i\cdot} = \frac{n_{i\cdot}}{n} = \text{sample proportion for category } i \text{ of factor 1}$$

and

$$\hat{p}_{\cdot j} = \frac{n_{\cdot j}}{n} = \text{sample proportion for category } j \text{ of factor 2}$$

This gives estimated expected cell counts identical to those in the case of homogeneity.

$$\hat{e}_{ij} = n \cdot \hat{p}_{i\cdot} \cdot \hat{p}_{\cdot j} = n \cdot \frac{n_{i\cdot}}{n} \cdot \frac{n_{\cdot j}}{n} = \frac{n_{i\cdot} \cdot n_{\cdot j}}{n}$$

$$= \frac{(i\text{th row total})(j\text{th column total})}{n}$$

The test statistic is also identical to that used in testing for homogeneity, as is the number of degrees of freedom. This is because the number of freely determined cell counts is $IJ - 1$, since only the total n is fixed in advance. There are I estimated $p_{i\cdot}$'s, but only

$I - 1$ are independently estimated since $\Sigma p_{i.} = 1$, and similarly $J - 1$ $p_{.j}$'s are independently estimated, so $I + J - 2$ parameters are independently estimated. The rule of thumb now yields df $= IJ - 1 - (I + J - 2) = IJ - I - J + 1 = (I - 1) \cdot (J - 1)$.

Null hypothesis: H_0: $p_{ij} = p_{i.} \cdot p_{.j}$ $i = 1, \ldots, I; j = 1, \ldots, J$

Alternative hypothesis: H_a: H_0 is not true

Test statistic value:

$$\chi^2 = \sum_{\text{all cells}} \frac{(\text{observed} - \text{estimated expected})^2}{\text{estimated expected}} = \sum_{i=1}^{I} \sum_{j=1}^{J} \frac{(n_{ij} - \hat{e}_{ij})^2}{\hat{e}_{ij}}$$

Rejection region: $\chi^2 \geq \chi^2_{\alpha,(I-1)(J-1)}$

The test can safely be applied as long as $\hat{e}_{ij} \geq 5$ for all cells.

Example 14.14 A study of the relationship between facility conditions at gasoline stations and aggressiveness in the pricing of gasoline ("An Analysis of Price Aggressiveness in Gasoline Marketing," *J. Marketing Research,* 1970: 36–42) reports the accompanying data based on a sample of $n = 441$ stations. At level .01, does the data suggest that facility conditions and pricing policy are independent of one another? Observed and estimated expected counts are given in Table 14.10.

Table 14.10   Observed and estimated expected counts for Example 14.14

		Observed pricing policy				**Expected pricing policy**			
		Aggressive	**Neutral**	**Nonaggressive**	$n_{i.}$				
	Substandard	24	15	17	56	17.02	22.10	16.89	56
Condition	**Standard**	52	73	80	205	62.29	80.88	61.83	205
	Modern	58	86	36	180	54.69	71.02	54.29	180
	$n_{.j}$	134	174	133	441	134	174	133	441

Thus,

$$\chi^2 = \frac{(24 - 17.02)^2}{17.02} + \cdots + \frac{(36 - 54.29)^2}{54.29} = 22.47$$

and because $\chi^2_{.01,4} = 13.277$, the hypothesis of independence is rejected.

We conclude that knowledge of a station's pricing policy does give information about the condition of facilities at the station. In particular, stations with an aggressive pricing policy appear more likely to have substandard facilities than stations with a neutral or nonaggressive policy. ∎

Models and methods for analyzing data in which each individual is categorized with respect to three or more factors (multidimensional contingency tables) are discussed in several of the chapter references.

Exercises | Section 14.3 (24–36)

24. The accompanying two-way table was constructed using data in the article "Television Viewing and Physical Fitness in Adults" (*Research Quarterly for Exercise and Sport*, 1990: 315–320). The author hoped to determine whether time spent watching television is associated with cardiovascular fitness. Subjects were asked about their television-viewing habits and were classified as physically fit if they scored in the excellent or very good category on a step test. We include MINITAB output from a chi-squared analysis. The four TV groups corresponded to different amounts of time per day spent watching TV (0, 1–2, 3–4, or 5 or more hours). The 168 individuals represented in the first column were those judged physically fit. Expected counts appear below observed counts, and MINITAB displays the contribution to χ^2 from each cell. State and test the appropriate hypotheses using $\alpha = .05$.

```
                1          2      Total
      1        35        147        182
           25.48     156.52

      2       101        629        730
          102.20     627.80

      3        28        222        250
           35.00     215.00

      4         4         34         38
            5.32      32.68

  Total       168       1032       1200

ChiSq = 3.557 + 0.579 +
        0.014 + 0.002 +
        1.400 + 0.228 +
        0.328 + 0.053 = 6.161

df = 3
```

25. The accompanying data refers to leaf marks found on white clover samples selected from both long-grass areas and short-grass areas ("The Biology of the Leaf Mark Polymorphism in *Trifolium repens* L.," *Heredity*, 1976: 306–325). Use a χ^2 test to decide whether the true proportions of different marks are identical for the two types of regions.

| | | Type of mark | | | | Sample |
	L	LL	Y + YL	O	Others	size
Long-grass areas	409	11	22	7	277	726
Short-grass areas	512	4	14	11	220	761

26. The following data resulted from an experiment to study the effects of leaf removal on the ability of fruit of a certain type to mature ("Fruit Set, Herbivory, Fruit Reproduction, and the Fruiting Strategy of *Catalpa speciosa*," *Ecology*, 1980: 57–64):

Treatment	Number of Fruits Matured	Number of Fruits Aborted
Control	141	206
Two leaves removed	28	69
Four leaves removed	25	73
Six leaves removed	24	78
Eight leaves removed	20	82

Does the data suggest that the chance of a fruit maturing is affected by the number of leaves removed? State and test the appropriate hypotheses at level .01.

27. The article "Human Lateralization from Head to Foot: Sex-Related Factors" (*Science*, 1978: 1291–1292) reports for both a sample of right-handed men and a sample of right-handed women the number of individuals whose feet were the same size, had a bigger left than right foot (a difference of half a shoe size or more), or had a bigger right than left foot.

	L > R	L = R	L < R	Sample size
Men	2	10	28	40
Women	55	18	14	87

Does the data indicate that gender has a strong effect on the development of foot asymmetry? State the appropriate null and alternative hypotheses, compute the value of χ^2, and obtain information about the P-value.

28. The article "Susceptibility of Mice to Audiogenic Seizure Is Increased by Handling Their Dams During Gestation" (*Science,* 1976: 427–428) reports on research into the effect of different injection treatments on the frequencies of audiogenic seizures.

Treatment	No response	Wild running	Clonic seizure	Tonic seizure
Thienylalanine	21	7	24	44
Solvent	15	14	20	54
Sham	23	10	23	48
Unhandled	47	13	28	32

Does the data suggest that the true percentages in the different response categories depend on the nature of the injection treatment? State and test the appropriate hypotheses using $\alpha = .005$.

29. The accompanying data on sex combinations of two recombinants resulting from six different male genotypes appears in the article "A New Method for Distinguishing Between Meiotic and Premeiotic Recombinational Events in *Drosophila melanogaster*" (*Genetics,* 1979: 543–554). Does the data support the hypothesis that the frequency distribution among the three sex combinations is homogeneous with respect to the different genotypes? Define the parameters of interest, state the appropriate H_0 and H_a, and perform the analysis.

Sex combination

		M/M	M/F	F/F
	1	35	80	39
	2	41	84	45
Male	3	33	87	31
genotype	4	8	26	8
	5	5	11	6
	6	30	65	20

30. Three different design configurations are being considered for a particular component. There are four possible failure modes for the component. An engineer obtained the following data on number of failures in each mode for each of the three configurations. Does the configuration appear to have an effect on type of failure?

Failure Mode

		1	2	3	4
	1	20	44	17	9
Configuration	2	4	17	7	12
	3	10	31	14	5

31. A random sample of individuals who drive alone to work in a large metropolitan area was obtained, and each individual was categorized with respect to both size of car and commuting distance. Does the accompanying data suggest that commuting distance and size of car are related in the population sampled? State the appropriate hypotheses and use a level .05 chi-squared test.

		Commuting distance		
		0–<10	10–<20	≥20
	Subcompact	6	27	19
Size	Compact	8	36	17
of	Midsize	21	45	33
car	Full-size	14	18	6

32. Each individual in a random sample of high school and college students was cross-classified with respect to both political views and marijuana usage, resulting in the data displayed in the accompanying two-way table ("Attitudes About Marijuana and Political Views," *Psychological Reports,* 1973: 1051–1054). Does the data support the hypothesis that political views and marijuana usage level are independent within the population? Test the appropriate hypotheses using level of significance .01.

		Usage level		
		Never	Rarely	Frequently
	Liberal	479	173	119
Political views	Conservative	214	47	15
	Other	172	45	85

33. Show that the chi-squared statistic for the test of independence can be written in the form

$$\chi^2 = \sum_{i=1}^{I} \sum_{j=1}^{J} \left(\frac{N_{ij}^2}{\hat{E}_{ij}} \right) - n$$

Why is this formula more efficient computationally than the defining formula for χ^2?

34. Suppose that in Exercise 32 each student had been categorized with respect to political views, marijuana usage, and religious preference, with the categories of this latter factor being Protestant, Catholic, and other. The data could be displayed in three different two-way tables, one corresponding to each category of the third factor. With $p_{ijk} = P($political category i, marijuana category j, and religious category $k)$, the null hypothesis of independence of all three factors states that $p_{ijk} = p_{i\cdot\cdot} \cdot p_{\cdot j\cdot} \cdot p_{\cdot\cdot k}$. Let n_{ijk} denote the observed frequency in cell (i, j, k). Show how to estimate the expected cell counts assuming that H_0 is true ($\hat{e}_{ijk} = n\hat{p}_{ijk}$, so the $\hat{p}_{ijk}$'s must be determined). Then use the general rule of thumb to determine the number of degrees of freedom for the chi-squared statistic.

35. Suppose that in a particular state consisting of four distinct regions, a random sample of n_k voters is obtained from the kth region for $k = 1, 2, 3, 4$. Each voter is then classified according to which candidate (1, 2, or 3) he or she prefers and according to voter registration (1 = Dem., 2 = Rep., 3 =

Indep.). Let p_{ijk} denote the proportion of voters in region k who belong in candidate category i and registration category j. The null hypothesis of homogeneous regions is H_0: $p_{ij1} = p_{ij2} = p_{ij3} = p_{ij4}$ for all i, j (i.e., the proportion within each candidate/registration combination is the same for all four regions). Assuming that H_0 is true, determine $\hat{p}_{ijk}$ and $\hat{e}_{ijk}$ as functions of the observed n_{ijk}'s and use the general rule of thumb to obtain the number of degrees of freedom for the chi-squared test.

36. Consider the accompanying 2×3 table displaying the sample proportions that fell in the various combinations of categories (e.g., 13% of those in the sample were in the first category of both factors).

	1	2	3
1	.13	.19	.28
2	.07	.11	.22

a. Suppose the sample consisted of $n = 100$ people. Use the chi-squared test for independence with significance level .10.
b. Repeat part (a) assuming that the sample size was $n = 1000$.
c. What is the smallest sample size n for which these observed proportions would result in rejection of the independence hypothesis?

Supplementary Exercises (37–45)

37. The article "Birth Order and Political Success" (*Psych. Reports,* 1971: 1239–1242) reports that among 31 randomly selected candidates for political office who came from families with four children, 12 were firstborn, 11 were middleborn, and 8 were lastborn. Use this data to test the null hypothesis that a political candidate from such a family is equally likely to be in any one of the four ordinal positions.

38. The results of an experiment to assess the effect of crude oil on fish parasites are described in the article "Effects of Crude Oils on the Gastrointestinal Parasites of Two Species of Marine Fish" (*J. Wildlife Diseases,* 1983: 253–258). Three treatments (corresponding to populations in the proce-

dure described) were compared: (1) no contamination, (2) contamination by 1-year-old weathered oil, and (3) contamination by new oil. For each treatment condition, a sample of fish was taken, and then each fish was classified as either parasitized or not parasitized. Data compatible with that in the article is given. Does the data indicate that the three treatments differ with respect to the true proportion of parasitized and nonparasitized fish? Test using $\alpha = .01$.

Treatment	Parasitized	Nonparasitized
Control	30	3
Old oil	16	8
New oil	16	16

39. Qualifications of male and female head and assistant college athletic coaches were compared in the article "Sex Bias and the Validity of Believed Differences Between Male and Female Interscholastic Athletic Coaches" (*Research Quarterly for Exercise and Sport,* 1990: 259–267). Each person in random samples of 2225 male coaches and 1141 female coaches was classified according to number of years of coaching experience to obtain the accompanying two-way table. Is there enough evidence to conclude that the proportions falling into the experience categories are different for men and women? Use $\alpha = .01$.

Years of Experience

Gender	1–3	4–6	7–9	10–12	13+
Male	202	369	482	361	811
Female	230	251	238	164	258

40. The authors of the article "Predicting Professional Sports Game Outcomes from Intermediate Game Scores" (*Chance,* 1992: 18–22) used a chi-squared test to determine whether there was any merit to the idea that basketball games are not settled until the last quarter, whereas baseball games are over by the seventh inning. They also considered football and hockey. Data was collected for 189 basketball games, 92 baseball games, 80 hockey games, and 93 football games. The games analyzed were sampled randomly from all games played during the 1990 season for baseball and football and for the 1990–1991 season for basketball and hockey. For each game, the late-game leader was determined, and then it was noted whether the late-game leader actually ended up winning the game. The resulting data is summarized in the accompanying table.

Sport	Late-game Leader Wins	Late-game Leader Loses
Basketball	150	39
Baseball	86	6
Hockey	65	15
Football	72	21

The authors state that "*Late-game leader* is defined as the team that is ahead after three quarters in basketball and football, two periods in hockey, and seven innings in baseball. The chi-square value on three degrees of freedom is 10.52 ($P < .015$)."

a. State the relevant hypotheses and reach a conclusion using $\alpha = .05$.

b. Do you think that your conclusion in part (a) can be attributed to a single sport being an anomaly?

41. The accompanying two-way frequency table appears in the article "Marijuana Use in College" (*Youth and Society,* 1979: 323–334). Each of 445 college students was classified according to both frequency of marijuana use and parental use of alcohol and psychoactive drugs. Does the data suggest that parental usage and student usage are independent in the population from which the sample was drawn? Use the *P*-value method to reach a conclusion.

		Standard level of marijuana use		
		Never	Occasional	Regular
Parental use of alcohol and drugs	Neither	141	54	40
	One	68	44	51
	Both	17	11	19

42. In a study of 2989 cancer deaths, the location of death (home, acute-care hospital, or chronic-care facility) and age at death were recorded, resulting in the given two-way frequency table ("Where Cancer Patients Die," *Public Health Reports,* 1983: 173). Using a .01 significance level, test the null hypothesis that age at death and location of death are independent.

		Location		
		Home	Acute-care	Chronic-care
Age	15–54	94	418	23
	55–64	116	524	34
	65–74	156	581	109
	Over 74	138	558	238

43. In a study to investigate the extent to which individuals are aware of industrial odors in a certain region ("Annoyance and Health Reactions to Odor from Refineries and Other Industries in Carson, California," *Environmental Research,* 1978: 119–132), a sample of individuals was obtained from each of three different areas near industrial facilities. Each individual was asked whether he or she noticed odors (1) every day, (2) at least once/week, (3) at least once/month, (4) less often than once/month, or (5) not at all, resulting in the data and SPSS output at the top of page 648. State and test the appropriate hypotheses.

```
Crosstabulation: AREA By CATEGORY
          Count
          Exp  Val
CATEGORY ─► Row  Pct                                           Row
AREA      Col  Pct │ 1.00 │ 2.00 │ 3.00 │ 4.00 │ 5.00 │      Total
          1.00     │  20  │  28  │  23  │  14  │  12  │   97
                   │ 12.7 │ 24.7 │ 18.0 │ 16.0 │ 25.7 │  33.3%
                   │ 20.6%│ 28.9%│ 23.7%│ 14.4%│ 12.4%│
                   │ 52.6%│ 37.8%│ 42.6%│ 29.2%│ 15.6%│
          2.00     │  14  │  34  │  21  │  14  │  12  │   95
                   │ 12.4 │ 24.2 │ 17.6 │ 15.7 │ 25.1 │  32.6%
                   │ 14.7%│ 35.8%│ 22.1%│ 14.7%│ 12.6%│
                   │ 36.8%│ 45.9%│ 38.9%│ 29.2%│ 15.6%│
          3.00     │   4  │  12  │  10  │  20  │  53  │   99
                   │ 12.9 │ 25.2 │ 18.4 │ 16.3 │ 26.2 │  34.0%
                   │  4.0%│ 12.1%│ 10.1%│ 20.2%│ 53.5%│
                   │ 10.5%│ 16.2%│ 18.5%│ 41.7%│ 68.8%│
          Column      38     74     54     48     77     291
          Total     13.1%  25.4%  18.6%  16.5%  26.5%  100.0%

Chi-Square   D.F.    Significance    Min E.F.   Cells with E.F. < 5
----------   ----    ------------    --------   -------------------
 70.64156     8         .0000         12.405    None
```

44. Many shoppers have expressed unhappiness because grocery stores have stopped putting prices on individual grocery items. The article "The Impact of Item Price Removal on Grocery Shopping Behavior" (*J. Marketing,* 1980: 73–93) reports on a study in which each shopper in a sample was classified by age and by whether he or she felt the need for item pricing. Based on the accompanying data, does the need for item pricing appear to be independent of age?

	Age				
	<30	**30–39**	**40–49**	**50–59**	**≥60**
Number in sample	150	141	82	63	49
Number who want item pricing	127	118	77	61	41

45. Let p_1 denote the proportion of successes in a particular population. The test statistic value in Chapter 8 for testing H_0: $p_1 = p_{10}$ was $z = (\hat{p}_1 - p_{10})/\sqrt{p_{10}p_{20}/n}$, where $p_{20} = 1 - p_{10}$. Show that for the case $k = 2$, the chi-squared test statistic value of Section 14.1 satisfies $\chi^2 = z^2$. [*Hint:* First show that $(n_1 - np_{10})^2 = (n_2 - np_{20})^2$.]

Bibliography

Agvesti, Alan, *An Introduction to Categorical Data Analysis,* Wiley, New York, 1996. An excellent treatment of various aspects of categorical data analysis by one of the most prominent researchers in this area.

Everitt, B. S., *The Analysis of Contingency Tables,* 2nd ed., Halsted Press, New York, 1992. A compact but informative survey of methods for analyzing categorical data, exposited with a minimum of mathematics.

Mosteller, Frederick, and Richard Rourke, *Sturdy Statistics,* Addison-Wesley, Reading, MA, 1973. Contains several very readable chapters on the varied uses of chi-square.

Distribution-Free Procedures

Introduction

When the underlying population or populations are nonnormal, the t and F tests and t confidence intervals of Chapters 7–13 will in general have actual levels of significance or confidence levels that differ from the nominal levels (those prescribed by the experimenter through the choice of, say, $t_{.025}$, $F_{.01}$, etc.) α and $100(1 - \alpha)\%$, although the difference between actual and nominal levels may not be large when the departure from normality is not too severe. Because the t and F procedures require the distributional assumption of normality, they are not "distribution-free" procedures—alternatively, because they are based on a particular parametric family of distributions (normal), they are not "nonparametric" procedures.

In this chapter, we describe procedures that are valid [actual level α or confidence level $100(1 - \alpha)\%$] simultaneously for many different types of underlying distributions. Such procedures are called **distribution-free** or **nonparametric.** In Section 15.1, we discuss a test procedure for analyzing a single sample of data; Section 15.2 presents a test procedure for use in two-sample problems. In Section 15.3, we develop distribution-free confidence intervals for μ and $\mu_1 - \mu_2$.

Section 15.4 describes distribution-free ANOVA procedures. These procedures are all competitors of the parametric (*t* and *F*) procedures described in previous chapters, so it is important to compare the performance of the two types of procedures under both normal and nonnormal population models. Generally speaking, the distribution-free procedures perform almost as well as their *t* and *F* counterparts on the "home ground" of the normal distribution and will often yield a considerable improvement under nonnormal conditions.

15.1 | The Wilcoxon Signed-Rank Test

A research chemist performed a particular chemical experiment a total of ten times under identical conditions, obtaining the following ordered values of reaction temperature:

$$-.57 \quad -.19 \quad -.05 \quad .76 \quad 1.30 \quad 2.02 \quad 2.17 \quad 2.46 \quad 2.68 \quad 3.02$$

The distribution of reaction temperature is of course continuous. Suppose the investigator is willing to assume that the reaction temperature distribution is symmetric, that is, there is a point of symmetry such that the density curve to the left of that point is the mirror image of the density curve to its right. This point of symmetry is the median of the distribution, as illustrated in Figure 15.1 (and is also the mean value μ provided that the mean is finite). The assumption of symmetry may at first thought seem quite bold, but remember that any normal distribution is symmetric, so symmetry is actually a weaker assumption than normality.

Figure 15.1 Several symmetric distributions

Let's now consider testing the null hypothesis that the median of the reaction temperature distribution is zero, that is, H_0: $\tilde{\mu} = 0$. This amounts to saying that a temperature of any particular magnitude, e.g., 1.50, is no more likely to be positive (+1.50) than it is to be negative (−1.50). A glance at the data suggests that this hypothesis is not very tenable; for example, the sample median is 1.66, which is far larger than the magnitude of any of the three negative observations.

Figure 15.2 shows two different symmetric pdf's, one for which H_0 is true and one for which H_a is true. When H_0 is true, we expect the magnitudes of the negative observations in the sample to be comparable to the magnitudes of the positive observations. If, however, H_0 is "grossly" untrue as in Figure 15.2(b), then observations of large absolute magnitude will tend to be positive rather than negative.

Figure 15.2 (a) $\tilde{\mu} = 0$; (b) $\tilde{\mu} > 0$

For the ten observations in our sample, suppose we proceed as follows:

1. Disregarding the signs of the observations, rank the collection of ten numbers in order of increasing magnitude.

2. Use as a test statistic S_+ = sum of the ranks associated with the positive observations.

Absolute magnitude	.05	.19	.57	.76	1.30	2.02	2.17	2.46	2.68	3.02
Rank	1	2	3	4	5	6	7	8	9	10
Signed rank	−1	−2	−3	4	5	6	7	8	9	10

The observed value of S_+ is $s_+ = 4 + 5 + 6 + 7 + 8 + 9 + 10 = 49$. H_0 is now rejected in favor of H_a: $\tilde{\mu} > 0$ when s_+ is too large because a large value of s_+ indicates that most of the observations with large absolute magnitude are positive, which in turn indicates a median greater than 0. If, on the other hand, observations with negative signs are intermingled with positive observations when signs are disregarded, then the observed s_+ will not be very large.

The Distribution of S_+

To decide between H_0 and H_a, we must know whether 49 is a "surprisingly large" value of S_+ when the null hypothesis is true. This entails obtaining the probability distribution of S_+ when H_0 is true and selecting as the rejection region a set of large s_+ values with small probability under H_0 (small α). If our observed s_+ is in the rejection region, we can then reject H_0 at the desired level α, and otherwise not reject H_0.

The distribution of S_+ is most easily obtained when n is small, so let's consider the case $n = 5$. The key observation is that when H_0 is true, *any* collection of signed ranks is *just as likely* as any other collection. Worded differently, the smallest observation in absolute magnitude is just as likely to be positive as negative, the second smallest observation in absolute magnitude is just as likely to be positive as negative, and similarly for the other observations. Thus, for $n = 5$, the sequence $-1, +2, +3, -4, +5$ of signed ranks is just as likely to be observed (when H_0 is true) as is the sequence $-1, -2, -3, -4, +5$, or any other sequence. Table 15.1 (page 652) lists the possible signed rank sequences for $n = 5$ and the s_+ value associated with each sequence. Since there are 32 such sequences (two possible signs for the smallest observation, two for the second smallest, and so on, or $2 \cdot 2 \cdot 2 \cdot 2 \cdot 2 = 32$ sequences), each sequence has probability

Table 15.1   Possible signed-rank sequences for $n = 5$

Sequence					s_+	Sequence					s_+
−1	−2	−3	−4	−5	0	−1	−2	−3	+4	−5	4
+1	−2	−3	−4	−5	1	+1	−2	−3	+4	−5	5
−1	+2	−3	−4	−5	2	−1	+2	−3	+4	−5	6
−1	−2	+3	−4	−5	3	−1	−2	+3	+4	−5	7
+1	+2	−3	−4	−5	3	+1	+2	−3	+4	−5	7
+1	−2	+3	−4	−5	4	+1	−2	+3	+4	−5	8
−1	+2	+3	−4	−5	5	−1	+2	+3	+4	−5	9
+1	+2	+3	−4	−5	6	+1	+2	+3	+4	−5	10
−1	−2	−3	−4	+5	5	−1	−2	−3	+4	+5	9
+1	−2	−3	−4	+5	6	+1	−2	−3	+4	+5	10
−1	+2	−3	−4	+5	7	−1	+2	−3	+4	+5	11
−1	−2	+3	−4	+5	8	−1	−2	+3	+4	+5	12
+1	+2	−3	−4	+5	8	+1	+2	−3	+4	+5	12
+1	−2	+3	−4	+5	9	+1	−2	+3	+4	+5	13
−1	+2	+3	−4	+5	10	−1	+2	+3	+4	+5	14
+1	+2	+3	−4	+5	11	+1	+2	+3	+4	+5	15

1/32 of being observed when H_0 is true. From this the pmf of S_+ under H_0 can immediately be obtained as in Table 15.2. Note that for $n = 10$ observations there are $2^{10} = 1024$ possible signed-rank sequences, and to list these sequences would be very tedious. Each sequence, though, would receive probability 1/1024 under H_0, from which the null distribution (distribution when H_0 is true) of S_+ can be easily obtained.

Table 15.2   Null distribution of S_+ when $n = 5$

s_+	0	1	2	3	4	5	6	7
$p(s_+)$	1/32	1/32	1/32	2/32	2/32	3/32	3/32	3/32

s_+	8	9	10	11	12	13	14	15
$p(s_+)$	3/32	3/32	3/32	2/32	2/32	1/32	1/32	1/32

As an example of how the probabilities in Table 15.2 were obtained, focus on the S_+ value 7; there are three signed-rank sequences for which $S_+ = 7$, and each receives probability 1/32 under H_0, so that $P(S_+ = 7$ when H_0 is true$) = 3/32$.

Table 15.2 can now be used to specify a rejection region for H_0: $\tilde{\mu} = 0$ versus H_a: $\tilde{\mu} > 0$ for which α can be calculated. Consider the rejection region $R = \{13, 14, 15\}$. Then

$$\alpha = P(\text{reject } H_0 \text{ when } H_0 \text{ is true})$$
$$= P(S_+ = 13, 14, \text{ or } 15 \text{ when } H_0 \text{ is true})$$
$$= 1/32 + 1/32 + 1/32 = 3/32$$
$$= .094$$

so that $R = \{13, 14, 15\}$ specifies a test with approximate level .1. For the rejection region $\{14, 15\}$, $\alpha = 2/32 = .063$. For the sample $x_1 = .58$, $x_2 = 2.50$, $x_3 = -.21$, $x_4 = 1.23$, $x_5 = .97$, the signed rank sequence is $-1, +2, +3, +4, +5$, so $s_+ = 14$ and at level .063 H_0 would be rejected.

A General Description of the Wilcoxon Signed-Rank Test

Because the underlying distribution is assumed symmetric, $\mu = \tilde{\mu}$, so we will state the hypotheses of interest in terms of μ rather than $\tilde{\mu}$.*

ASSUMPTION

> $X_1, X_2, \ldots, X_n$ is a random sample from a continuous and symmetric probability distribution with mean (and median) μ.

When the hypothesized value of μ is μ_0, the absolute differences $|x_1 - \mu_0|, \ldots, |x_n - \mu_0|$ must be ranked from smallest to largest.

> Null hypothesis: H_0: $\mu = \mu_0$
>
> Test statistic value: s_+ = the sum of the ranks associated with positive $(x_i - \mu_0)$'s
>
> | Alternative Hypothesis | Rejection Region for Level α Test |
> |---|---|
> | H_a: $\mu > \mu_0$ | $s_+ \geq c_1$ |
> | H_a: $\mu < \mu_0$ | $s_+ \leq c_2$ [where $c_2 = n(n + 1)/2 - c_1$] |
> | H_a: $\mu \neq \mu_0$ | either $s_+ \geq c$ or $s_+ \leq n(n + 1)/2 - c$ |
>
> where the critical values c_1 and c obtained from Appendix Table A.13 satisfy $P(S_+ \geq c_1) \approx \alpha$ and $P(S_+ \geq c) \approx \alpha/2$ when H_0 is true.

Example 15.1   A manufacturer of electric irons, wishing to test the accuracy of the thermostat control at the 500°F setting, instructs a test engineer to obtain actual temperatures at that setting for 15 irons using a thermocouple. The resulting measurements are as follows:

494.6 510.8 487.5 493.2 502.6 485.0 495.9 498.2

501.6 497.3 492.0 504.3 499.2 493.5 505.8

The engineer believes it is reasonable to assume that a temperature deviation from 500° of any particular magnitude is just as likely to be positive as negative (the assumption

---

*If the tails of the distribution are "too heavy," as was the case with the Cauchy distribution mentioned in Chapter 6, then μ will not exist. In such cases, the Wilcoxon test will still be valid for tests concerning $\tilde{\mu}$.

of symmetry) but wants to protect against possible nonnormality of the actual temperature distribution, so she decides to use the Wilcoxon signed-rank test to see whether the data strongly suggests incorrect calibration of the iron.

The hypotheses are H_0: $\mu = 500$ versus H_a: $\mu \neq 500$, where $\mu =$ the true average actual temperature at the 500°F setting. Subtracting 500 from each x_i gives

$$-5.6 \quad 10.8 \quad -12.5 \quad -6.8 \quad 2.6 \quad -15.0 \quad -4.1 \quad -1.8 \quad 1.6 \quad -2.7$$
$$-8.0 \quad 4.3 \quad -.8 \quad -6.5 \quad 5.8$$

The ranks are obtained by ordering these from smallest to largest without regard to sign:

Absolute magnitude	.8	1.6	1.8	2.6	2.7	4.1	4.3	5.6	5.8	6.5	6.8	8.0	10.8	12.5	15.0
Rank	1	2	3	4	5	6	7	8	9	10	11	12	13	14	15
Sign	−	+	−	+	−	−	+	−	+	−	−	−	+	−	−

Thus, $s_+ = 2 + 4 + 7 + 9 + 13 = 35$. From Appendix Table A.13, $P(S_+ \geq 95) = P(S_+ \leq 25) = .024$ when H_0 is true, so the two-tailed test with approximate level .05 rejects H_0 when either $s_+ \geq 95$ or ≤ 25 [the exact α is $2(.024) = .048$]. Since $s_+ = 35$ is not in the rejection region, it cannot be concluded at level .05 that μ is anything other than 500. Even at level .094 (approximately .1), H_0 is not rejected, since $P(S_+ \leq 30) = .047$ implies that s_+ values between 30 and 90 are not significant at that level. The P-value of the data is thus greater than .1. ∎

Paired Observations

When the data consisted of pairs $(X_1, Y_1), \ldots, (X_n, Y_n)$ and the differences $D_1 = X_1 - Y_1, \ldots, D_n = X_n - Y_n$ were normally distributed, in Chapter 9 we used a paired t test to test hypotheses about the expected difference μ_D. If normality is not assumed, hypotheses about μ_D can be tested by using the Wilcoxon signed-rank test on the D_i's provided that the distribution of the differences is continuous and symmetric. *If X_i and Y_i both have continuous distributions that differ only with respect to their means* (so the Y distribution is the X distribution shifted by $\mu_1 - \mu_2 = \mu_D$), *then D_i will have a continuous symmetric distribution* (it is *not* necessary for the X and Y distributions to be symmetric individually). The null hypothesis is H_0: $\mu_D = \Delta_0$, and the test statistic S_+ is the sum of the ranks associated with the positive $(D_i - \Delta_0)$'s.

Example 15.2 An experiment to compare the abilities of two different solvents to extract creosote impregnated in test logs involved the use of eight different logs. After dividing each log into two segments, one segment was randomly selected for application of the first solvent with the second solvent used on the other segment, yielding the following data:

Log	1	2	3	4	5	6	7	8
Solvent 1	3.92	3.79	3.70	4.08	3.87	3.95	3.55	3.76
Solvent 2	4.25	4.20	4.41	3.89	4.39	3.75	4.20	3.90
Difference	−.33	−.41	−.71	.19	−.52	.20	−.65	−.14
Signed rank	−4	−5	−8	2	−6	3	−7	−1

The first solvent is currently used, and the second is a new formulation designed to result in improved extraction capability. Does this data suggest that the true average amount extracted by the second solvent exceeds that for the first solvent? The relevant hypotheses are H_0: $\mu_D = 0$ versus H_a: $\mu_D < 0$. Appendix Table A.13 shows that for a test with significance level approximately .05, the null hypothesis should be rejected if $s_+ \leq (8)(9)/2 - 30 = 6$. The test statistic value is $2 + 3 = 5$, which falls in the rejection region. We therefore reject H_0 at significance level .05 in favor of the conclusion that the new solvent does outperform the one currently used. The accompanying MINITAB output gives the test statistic value and also the corresponding P-value, which is $P(S_+ \leq 5$ when H_0 is true).

```
Test of median = 0.000000 versus median < 0.000000


              N for    Wilcoxon              Estimated
         N    Test    Statistic      P        Median
diff     8      8          5.0    0.040      -0.3025
```

Ties in the Wilcoxon Signed-Rank Test

Although a theoretical implication of the continuity of the underlying distribution is that ties will not occur, in practice they often do because of the discreteness of measuring instruments. If there are several data values with the same absolute magnitude, then they would be assigned the average of the ranks they would receive if they differed very slightly from one another. For example, if in Example 15.1 $x_8 = 498.2$ is changed to 498.4, then two different values of $(x_i - 500)$ would have absolute magnitude 1.6. The ranks to be averaged would be 2 and 3, so each would be assigned rank 2.5.

A Large-Sample Approximation

Appendix Table A.13 provides critical values for level α tests only when $n \leq 20$. For $n > 20$, it can be shown that S_+ has approximately a normal distribution with

$$\mu_{S_+} = \frac{n(n+1)}{4} \qquad \sigma^2_{S_+} = \frac{n(n+1)(2n+1)}{24}$$

when H_0 is true.

The mean and variance result from noting that when H_0 is true (the symmetric distribution is centered at μ_0), then the rank i is just as likely to receive a $+$ sign as it is to receive a $-$ sign. Thus,

$$S_+ = W_1 + W_2 + W_3 + \cdots + W_n$$

where

$$W_1 = \begin{cases} 1 & \text{with probability .5} \\ 0 & \text{with probability .5} \end{cases} \quad \cdots \quad W_n = \begin{cases} n & \text{with probability .5} \\ 0 & \text{with probability .5} \end{cases}$$

($W_i = 0$ is equivalent to rank i being associated with a $-$, so i does not contribute to S_+.)

S_+ is then a sum of random variables, and when H_0 is true, these W_i's can be shown to be independent. Application of the rules of expected value and variance gives the mean and variance of S_+. Because the W_i's are not identically distributed, our version of the Central Limit Theorem cannot be applied, but there is a more general version of the theorem that can be used to justify the normality conclusion.

The large-sample test statistic is now given by

$$Z = \frac{S_+ - n(n+1)/4}{\sqrt{n(n+1)(2n+1)/24}} \tag{15.1}$$

For the three standard alternatives, the critical values for level α tests are the usual standard normal values z_α, $-z_\alpha$, and $\pm z_{\alpha/2}$.

Example 15.3 A particular type of steel beam has been designed to have a compressive strength (lb/in.^{2}) of at least 50,000. For each beam in a sample of 25 beams, the compressive strength was determined and is given in Table 15.3. Assuming that actual compressive strength is distributed symmetrically about the true average value, use the Wilcoxon test to decide whether the true average compressive strength is less than the specified value. That is, test H_0: $\mu = 50,000$ versus H_a: $\mu < 50,000$ (favoring the claim that average compressive strength is at least 50,000).

Table 15.3 Data for Example 15.3

$x_i - 50,000$	Signed Rank	$x_i - 50,000$	Signed Rank	$x_i - 50,000$	Signed Rank
-10	-1	-99	-10	165	$+18$
-27	-2	113	$+11$	-178	-19
36	$+3$	-127	-12	-183	-20
-55	-4	-129	-13	-192	-21
73	$+5$	136	$+14$	-199	-22
-77	-6	-150	-15	-212	-23
-81	-7	-155	-16	-217	-24
90	$+8$	-159	-17	-229	-25
-95	-9				

The sum of the positively signed ranks is $3 + 5 + 8 + 11 + 14 + 18 = 59$, $n(n+1)/4 = 162.5$, and $n(n+1)(2n+1)/24 = 1381.25$, so

$$z = \frac{59 - 162.5}{\sqrt{1381.25}} = -2.78$$

The lower-tailed level .01 test rejects H_0 if $z \le -2.33$. Since $-2.78 \le -2.33$, H_0 is rejected in favor of the conclusion that true average compressive strength is less than 50,000. ∎

When there are ties in the absolute magnitudes, so that average ranks must be used, it is still correct to standardize S_+ by subtracting $n(n + 1)/4$, but the following corrected formula for variance should be used:

$$\sigma_{S_+}^2 = \frac{1}{24}n(n + 1)(2n + 1) - \frac{1}{48}\Sigma(\tau_i - 1)(\tau_i)(\tau_i + 1) \qquad (15.2)$$

where τ_i is the number of ties in the ith set of tied values and the sum is over all sets of tied values. If, for example, $n = 10$ and the signed ranks are 1, 2, −4, −4, 4, 6, 7, 8.5, 8.5, and 10, then there are two tied sets with $\tau_1 = 3$ and $\tau_2 = 2$, so the summation is $(2)(3)(4) + (1)(2)(3) = 30$ and $\sigma_{S_+}^2 = 96.25 - 30/48 = 95.62$. The denominator in (15.1) should be replaced by the square root of (15.2), though as this example shows, the correction is usually insignificant.

Efficiency of the Wilcoxon Signed-Rank Test

When the underlying distribution being sampled is normal, either the t test or the signed-rank test can be used to test a hypothesis about μ. The t test is the best test in such a situation because among all level α tests it is the one having minimum β. Since it is generally agreed that there are many experimental situations in which normality can be reasonably assumed, as well as some in which it should not be, there are two questions that must be addressed in an attempt to compare the two tests:

1. When the underlying distribution is normal (the "home ground" of the t test), how much is lost by using the signed-rank test?

2. When the underlying distribution is not normal, can a significant improvement be achieved by using the signed-rank test?

If the Wilcoxon test does not suffer much with respect to the t test on the "home ground" of the latter, and performs significantly better than the t test for a large number of other distributions, then there will be a strong case for using the Wilcoxon test.

Unfortunately, there are no simple answers to the two questions. Upon reflection, it is not surprising that the t test can perform poorly when the underlying distribution has "heavy tails" (i.e., when observed values lying far from μ are relatively more likely than they are when the distribution is normal). This is because the behavior of the t test depends on the sample mean, which can be very unstable in the presence of heavy tails. The difficulty in producing answers to the two questions is that β for the Wilcoxon test is very difficult to obtain and study for *any* underlying distribution, and the same can be said for the t test when the distribution is not normal. Even if β were easily obtained, any measure of efficiency would clearly depend on which underlying distribution was postulated. A number of different efficiency measures have been proposed by statisticians; one that many statisticians regard as credible is called **asymptotic relative efficiency** (ARE). The ARE of one test with respect to another is essentially the limiting ratio of sample sizes necessary to obtain identical error probabilities for the two tests. Thus, if the ARE of one test with respect to a second equals .5, then when sample sizes are large, twice as large a sample size will be required of the first test to perform as well as the second test. Although the ARE does not characterize test performance for small sample sizes, the following results can be shown to hold:

1. When the underlying distribution is normal, the ARE of the Wilcoxon test with respect to the t test is approximately .95.

2. For any distribution, the ARE will be at least .86 and for many distributions will be much greater than 1.

We can summarize these results by saying that, in large-sample problems, the Wilcoxon test is never very much less efficient than the t test and may be much more efficient if the underlying distribution is far from normal. Though the issue is far from resolved in the case of sample sizes obtained in most practical problems, studies have shown that the Wilcoxon test performs reasonably and is thus a viable alternative to the t test.

Exercises | Section 15.1 (1–9)

1. Reconsider the situation described in Exercise 32 of Section 8.2, and use the Wilcoxon test with $\alpha = .05$ to test the relevant hypotheses.

2. Use the Wilcoxon test to analyze the data given in Example 8.9.

3. The accompanying data is a subset of the data reported in the article "Synovial Fluid pH, Lactate, Oxygen and Carbon Dioxide Partial Pressure in Various Joint Diseases" (*Arthritis and Rheumatism,* 1971: 476–477). The observations are pH values of synovial fluid (which lubricates joints ánd tendons) taken from the knees of individuals suffering from arthritis. Assuming that true average pH for nonarthritic individuals is 7.39, test at level .05 to see whether the data indicates a difference between average pH values for arthritic and nonarthritic individuals.

   7.02 7.35 7.34 7.17 7.28 7.77 7.09

   7.22 7.45 6.95 7.40 7.10 7.32 7.14

4. A random sample of 15 automobile mechanics certified to work on a certain type of car was selected, and the time (in minutes) necessary for each one to diagnose a particular problem was determined, resulting in the following data:

   30.6 30.1 15.6 26.7 27.1 25.4 35.0 30.8

   31.9 53.2 12.5 23.2 8.8 24.9 30.2

   Use the Wilcoxon test at significance level .10 to decide whether the data suggests that true average diagnostic time is less than 30 minutes.

5. Both a gravimetric and a spectrophotometric method are under consideration for determining phosphate content of a particular material. Twelve samples of the material are obtained, each is split in half, and a determination is made on each half using one of the two methods, resulting in the following data.

Sample	1	2	3	4
Gravimetric	54.7	58.5	66.8	46.1
Spectrophotometric	55.0	55.7	62.9	45.5

Sample	5	6	7	8
Gravimetric	52.3	74.3	92.5	40.2
Spectrophotometric	51.1	75.4	89.6	38.4

Sample	9	10	11	12
Gravimetric	87.3	74.8	63.2	68.5
Spectrophotometric	86.8	72.5	62.3	66.0

Use the Wilcoxon test to decide whether one technique gives on average a different value than the other technique for this type of material.

6. Reconsider the situation described in Exercise 41 of Section 9.3, and use the Wilcoxon test to test the appropriate hypotheses.

7. Use the large-sample version of the Wilcoxon test at significance level .05 on the data of Exercise 37 in Section 9.3 to decide whether the true mean

difference between outdoor and indoor concentrations exceeds .20.

8. The accompanying 25 observations on fracture toughness of base plate of 18% nickel maraging steel were reported in the article "Fracture Testing of Weldments" (*ASTM Special Publ. No. 381,* 1965: 328–356). Suppose a company will agree to purchase this steel for a particular application only if it can be strongly demonstrated from experimental evidence that true average toughness exceeds 75. Assuming that the fracture toughness distribution is symmetric, state and test the appropriate hypotheses at level .05 and compute a *P*-value.

69.5 71.9 72.6 73.1 73.3 73.5 74.1 74.2 75.3

75.5 75.7 75.8 76.1 76.2 76.2 76.9 77.0 77.9

78.1 79.6 79.7 80.1 82.2 83.7 93.7

9. Suppose that observations $X_1, X_2, \ldots, X_n$ are made on a process at times $1, 2, \ldots, n$. On the basis of this data, we wish to test

H_0: the X_i's constitute an independent and identically distributed sequence

versus

H_a: X_{i+1} tends to be larger than X_i for $i = 1, \ldots, n$ (an increasing trend)

Suppose the X_i's are ranked from 1 to n. Then when H_a is true, larger ranks tend to occur later in the sequence, whereas if H_0 is true, large and small ranks tend to be mixed together. Let R_i be the rank of X_i and consider the test statistic $D = \sum_{i=1}^{n} (R_i - i)^2$. Then small values of D give support to H_a (e.g., the smallest value is 0 for $R_1 = 1, R_2 = 2, \ldots, R_n = n$), so H_0 should be rejected in favor of H_a if $d \le c$. When H_0 is true, any sequence of ranks has probability $1/n!$. Use this to find c for which the test has level as close to .10 as possible in the case $n = 4$. [*Hint:* List the 4! rank sequences, compute d for each one, and then obtain the null distribution of D. See the Lehmann book (in the chapter bibliography), p. 290, for more information.]

15.2 | The Wilcoxon Rank-Sum Test

When at least one of the sample sizes in a two-sample problem is small, the *t* test requires the assumption of normality (at least approximately). There are situations, though, in which an investigator would want to use a test that is valid even if the underlying distributions are quite nonnormal. We now describe such a test, called the Wilcoxon rank-sum test. An alternative name for the procedure is the Mann–Whitney test, though the Mann–Whitney test statistic is sometimes expressed in a slightly different form from that of the Wilcoxon test. The Wilcoxon test procedure is distribution-free because it will have the desired level of significance for a very large class of underlying distributions.

ASSUMPTIONS

> $X_1, \ldots, X_m$ and $Y_1, \ldots, Y_n$ are two independent random samples from continuous distributions with means μ_1 and μ_2, respectively. The X and Y distributions have the same shape and spread, the only possible difference between the two being in the values of μ_1 and μ_2.

When H_0: $\mu_1 - \mu_2 = \Delta_0$ is true, the X distribution is shifted by the amount Δ_0 to the right of the Y distribution; whereas when H_0 is false, the shift is by an amount other than Δ_0.

Development of the Test When $m = 3$, $n = 4$

Consider first testing H_0: $\mu_1 - \mu_2 = 0$. If μ_1 is actually much larger than μ_2, then most of the observed x's will fall to the right of the observed y's. However, if H_0 is true, then the observed values from the two samples should be intermingled. The test statistic will provide a quantification of how much intermingling there is in the two samples.

Consider the case $m = 3$, $n = 4$. Then if all three observed x's were to the right of all four observed y's, this would provide strong evidence for rejecting H_0 in favor of H_a: $\mu_1 - \mu_2 \neq 0$, with a similar conclusion being appropriate if all three x's fall below all four of the y's. Suppose we pool the X's and Y's into a combined sample of size $m + n = 7$ and rank these observations from smallest to largest, with the smallest receiving rank 1 and the largest, rank 7. If either most of the largest ranks or most of the smallest ranks were associated with X observations, we would begin to doubt H_0. This suggests the test statistic

$$W = \text{the sum of the ranks in the combined sample} \atop \text{associated with } X \text{ observations} \tag{15.3}$$

For the values of m and n under consideration, the smallest possible value of W is $w = 1 + 2 + 3 = 6$ (if all three x's are smaller than all four y's), and the largest possible value is $w = 5 + 6 + 7 = 18$ (if all three x's are larger than all four y's).

As an example, suppose $x_1 = -3.10$, $x_2 = 1.67$, $x_3 = 2.01$, $y_1 = 5.27$, $y_2 = 1.89$, $y_3 = 3.86$, and $y_4 = .19$. Then the pooled ordered sample is $-3.10, .19, 1.67, 1.89, 2.01, 3.86$, and 5.27. The X ranks for this sample are 1 (for -3.10), 3 (for 1.67), and 5 (for 2.01), so the computed value of W is $w = 1 + 3 + 5 = 9$.

The test procedure based on the statistic (15.3) is to reject H_0 if the computed value w is "too extreme"—that is, $\geq c$ for an upper-tailed test, $\leq c$ for a lower-tailed test, and either $\geq c_1$ or $\leq c_2$ for a two-tailed test. The critical constant(s) c (c_1, c_2) should be chosen so that the test has the desired level of significance α. To see how this should be done, recall that when H_0 is true, all seven observations come from the same population. This means that under H_0, any possible triple of ranks associated with the three x's—such as (1, 4, 5), (3, 5, 6), or (5, 6, 7)—has the same probability as any other possible rank triple. Since there are $\binom{7}{3} = 35$ possible rank triples, under H_0 each rank triple has probability $\frac{1}{35}$. From a list of all 35 rank triples and the w value associated with each, the probability distribution of W can immediately be determined. For example, there are four rank triples that have w value 11—(1, 3, 7), (1, 4, 6), (2, 3, 6), and (2, 4, 5)—so $P(W = 11) = \frac{4}{35}$. The summary of the listing and computations appears in Table 15.4.

Table 15.4 Probability distribution of W ($m = 3$, $n = 4$) when H_0 is true

w	6	7	8	9	10	11	12	13	14	15	16	17	18
$P(W = w)$	$\frac{1}{35}$	$\frac{1}{35}$	$\frac{2}{35}$	$\frac{3}{35}$	$\frac{4}{35}$	$\frac{4}{35}$	$\frac{5}{35}$	$\frac{4}{35}$	$\frac{4}{35}$	$\frac{3}{35}$	$\frac{2}{35}$	$\frac{1}{35}$	$\frac{1}{35}$

The distribution of Table 15.4 is symmetric about the value $w = (6 + 18)/2 = 12$, which is the middle value in the ordered list of possible W values. This is because the two rank triples (r, s, t) (with $r < s < t$) and $(8 - t, 8 - s, 8 - r)$ have values of w symmetric about 12, so for each triple with w value below 12, there is a triple with w value above 12 by the same amount.

If the alternative hypothesis is H_a: $\mu_1 - \mu_2 > 0$, then H_0 should be rejected in favor of H_a for large W values. Choosing as the rejection region the set of W values $\{17, 18\}$, $\alpha = P(\text{type I error}) = P(\text{reject } H_0 \text{ when } H_0 \text{ is true}) = P(W = 17 \text{ or } 18 \text{ when } H_0 \text{ is true}) = \frac{1}{35} + \frac{1}{35} = \frac{2}{35} = .057$; the region $\{17, 18\}$ therefore specifies a test with level of significance approximately .05. Similarly, the region $\{6, 7\}$, which is appropriate for H_a: $\mu_1 - \mu_2 < 0$, has $\alpha = .057 \approx .05$. The region $\{6, 7, 17, 18\}$, which is appropriate for the two-sided alternative, has $\alpha = \frac{4}{35} = .114$. The W value for the data given several paragraphs previously was $w = 9$, which is rather close to the middle value 12, so H_0 would not be rejected at any reasonable level α for any one of the three H_a's.

General Description of the Wilcoxon Rank-Sum Test

The null hypothesis H_0: $\mu_1 - \mu_2 = \Delta_0$ is handled by subtracting Δ_0 from each X_i and using the $(X_i - \Delta_0)$'s as the X_i's were previously used. Recalling that for any positive integer K, the sum of the first K integers is $K(K + 1)/2$, the smallest possible value of the statistic W is $m(m + 1)/2$, which occurs when the $(X_i - \Delta_0)$'s are all to the left of the Y sample. The largest possible value of W occurs when the $(X_i - \Delta_0)$'s lie entirely to the right of the Y's; in this case, $W = (n + 1) + \cdots + (m + n) = (\text{sum of first } m + n \text{ integers}) - (\text{sum of first } n \text{ integers})$, which gives $m(m + 2n + 1)/2$. As with the special case $m = 3$, $n = 4$, the distribution of W is symmetric about the value that is halfway between the smallest and largest values; this middle value is $m(m + n + 1)/2$. Because of this symmetry, probabilities involving lower-tail critical values can be obtained from corresponding upper-tail values.

Null hypothesis: H_0: $\mu_1 - \mu_2 = \Delta_0$

Test statistic value: $w = \sum\limits_{i=1}^{m} r_i$ where $r_1 = $ rank of $(x_i - \Delta_0)$ in the combined sample of $m + n$ $(x - \Delta_0)$'s and y's

Alternative Hypothesis	**Rejection Region**
H_a: $\mu_1 - \mu_2 > \Delta_0$	$w \geq c_1$
H_a: $\mu_1 - \mu_2 < \Delta_0$	$w \leq m(m + n + 1) - c_1$
H_a: $\mu_1 - \mu_2 \neq \Delta_0$	either $w \geq c$ or $w \leq m(m + n + 1) - c$

where $P(W \geq c_1 \text{ when } H_0 \text{ is true}) = \alpha$, $P(W \geq c \text{ when } H_0 \text{ is true}) = \alpha/2$.

Because W has a discrete probability distribution, there will not always exist a critical value corresponding exactly to one of the usual levels of significance. Appendix Table A.14 gives upper-tail critical values for probabilities closest to .05, .025, .01, and .005, from which level .05 or .01 one- and two-tailed tests can be obtained. The table gives information only for $m = 3, 4, \ldots, 8$ and $n = m, m + 1, \ldots, 8$ (that is,

$3 \le m \le n \le 8$). For values of m and n that exceed 8, a normal approximation can be used. To use the table for small m and n, though, *the X and Y samples should be labeled so that $m \le n$.*

Example 15.4 The urinary fluoride concentration (parts per million) was measured both for a sample of livestock grazing in an area previously exposed to fluoride pollution and for a similar sample grazing in an unpolluted region:

Polluted	21.3	18.7	23.0	17.1	16.8	20.9	19.7
Unpolluted	14.2	18.3	17.2	18.4	20.0		

Does the data indicate strongly that the true average fluoride concentration for livestock grazing in the polluted region is larger than for the unpolluted region? Use the Wilcoxon rank-sum test at level $\alpha = .01$.

The sample sizes here are 7 and 5. To obtain $m \le n$, label the unpolluted observations as the x's ($x_1 = 14.2, \ldots, x_5 = 20.0$) and the polluted observations as the y's. Thus, μ_1 is the true average fluoride concentration without pollution, and μ_2 is the true average concentration with pollution. The alternative hypothesis is H_a: $\mu_1 - \mu_2 < 0$ (pollution causes an increase in concentration), so a lower-tailed test is appropriate. From Appendix Table A.14 with $m = 5$ and $n = 7$, $P(W \ge 47$ when H_0 is true) $\approx .01$. The critical value for the lower-tailed test is therefore $m(m + n + 1) - 47 = 5(13) - 47 = 18$; H_0 will now be rejected if $w \le 18$. The pooled ordered sample follows; the computed W is $w = r_1 + r_2 + \cdots + r_5$ (where r_i is the rank of x_i) $= 1 + 5 + 4 + 6 + 9 = 25$. Since 25 is not ≤ 18, H_0 is not rejected at (approximately) level .01.

x	y	y	x	x	x	y	y	x	y	y	y
14.2	16.8	17.1	17.2	18.3	18.4	18.7	19.7	20.0	20.9	21.3	23.0
1	2	3	4	5	6	7	8	9	10	11	12

■

A Normal Approximation for *W*

When both m and n exceed 8, the distribution of W can be approximated by an appropriate normal curve, and this approximation can be used in place of Appendix Table A.14. To obtain the approximation, we need μ_W and σ_W^2 when H_0 is true. In this case, the rank R_i of $X_i - \Delta_0$ is equally likely to be any one of the possible values $1, 2, 3, \ldots, m + n$ (R_i has a discrete uniform distribution on the first $m + n$ positive integers), so $\mu_{R_i} = (m + n + 1)/2$. This gives, since $W = \sum R_i$,

$$\mu_W = \mu_{R_1} + \mu_{R_2} + \cdots + \mu_{R_m} = \frac{m(m + n + 1)}{2} \qquad (15.4)$$

The variance of R_i is also easily computed to be $(m + n + 1)(m + n - 1)/12$. However, because the R_i's are not independent variables, $V(W) \neq mV(R_i)$. Using the fact that, for any two distinct integers a and b between 1 and $m + n$ inclusive, $P(R_i = a, R_j = b) = 1/[(m + n)(m + n - 1)]$ (two integers are being sampled without replacement), $\text{Cov}(R_i, R_j) = -(m + n + 1)/12$, which yields

$$\sigma_W^2 = \sum_{i=1}^{m} V(R_i) + \sum\sum_{i \neq j} \text{Cov}(R_i, R_j) = \frac{mn(m + n + 1)}{12} \qquad (15.5)$$

A central limit theorem can then be used to conclude that when H_0 is true, the *test statistic*

$$Z = \frac{W - m(m + n + 1)/2}{\sqrt{mn(m + n + 1)/12}}$$

has approximately a standard normal distribution. This statistic is used in conjunction with the critical values z_α, $-z_\alpha$, and $\pm z_{\alpha/2}$ for upper-, lower-, and two-tailed tests, respectively.

Example 15.5 An article in the *Journal of Applied Physiology* ("Histamine Content in Sputum from Allergic and Non-Allergic Individuals," 1969: 535–539) reports the following data on sputum histamine level ($\mu g/g$ dry weight of sputum) for a sample of 9 individuals classified as allergics and another sample of 13 individuals classified as nonallergics.

Allergics	67.6	39.6	1651.0	100.0	65.9	1112.0	31.0	102.4	64.7				
Nonallergics	34.3	27.3	35.4	48.1	5.2	29.1	4.7	41.7	48.0	6.6	18.9	32.4	45.5

Does the data indicate that there is a difference in true average sputum histamine level between allergics and nonallergics?

Since both sample sizes exceed 8, we use the normal approximation. The null hypothesis is H_0: $\mu_1 - \mu_2 = 0$, and observed ranks of the x_i's are $r_1 = 18$, $r_2 = 11$, $r_3 = 22$, $r_4 = 19$, $r_5 = 17$, $r_6 = 21$, $r_7 = 7$, $r_8 = 20$, and $r_9 = 16$, so $w = \Sigma r_i = 151$. The mean and variance of W are given by $\mu_W = 9(23)/2 = 103.5$ and $\sigma_W^2 = 9(13)(23)/12 = 224.25$. Thus,

$$z = \frac{151 - 103.5}{\sqrt{224.25}} = 3.17$$

The alternative hypothesis is H_a: $\mu_1 - \mu_2 \neq 0$, so at level .01 H_0 is rejected if either $z \geq 2.58$ or $z \leq -2.58$. Because $3.17 \geq 2.58$, H_0 is rejected, and we conclude that there is a difference in true average sputum histamine levels (the article also used the Wilcoxon test). ∎

Ties in the Wilcoxon Rank-Sum Test

Theoretically, the assumption of continuity of the two distributions ensures that all $m + n$ observed x's and y's will have different values. In practice, though, there will often be ties in the observed values. As with the Wilcoxon signed-rank test, the common practice in dealing with ties is to assign each of the tied observations in a particular set of ties the average of the ranks they would receive if they differed very slightly from one another.

If the sample sizes both exceed 8, the numerator of Z is still appropriate, but the denominator should be replaced by the square root of the adjusted variance

$$\sigma_W^2 = \frac{mn(m + n + 1)}{12}$$

$$- \frac{mn}{12(m + n)(m + n - 1)} \Sigma(\tau_i - 1)(\tau_i)(\tau_i + 1) \tag{15.6}$$

where τ_i is the number of tied observations in the ith set of ties and the sum is over all sets of ties. Unless there are a great many ties, there is little difference between Equations (15.6) and (15.5).

Efficiency of the Wilcoxon Rank-Sum Test

When the distributions being sampled are both normal with $\sigma_1 = \sigma_2$, and therefore have the same shapes and spreads, either the pooled t test or the Wilcoxon test can be used (the two-sample t test assumes normality but not equal variances, so assumptions underlying its use are more restrictive in one sense and less in another than those for Wilcoxon's test). In this situation, the pooled t test is best among all possible tests in the sense of minimizing β for any fixed α. However, an investigator can never be absolutely certain that underlying assumptions are satisfied. It is therefore relevant to ask (1) how much is lost by using Wilcoxon's test rather than the pooled t test when the distributions are normal with equal variances and (2) how W compares to T in nonnormal situations.

The notion of test efficiency was discussed in the previous section in connection with the one-sample t test and Wilcoxon signed-rank test. The results for the two-sample tests are the same as those for the one-sample tests. When normality and equal variances both hold, the rank-sum test is approximately 95% as efficient as the pooled t test in large samples. That is, the t test will give the same error probabilities as the Wilcoxon test using slightly smaller sample sizes. On the other hand, the Wilcoxon test will always be at least 86% as efficient as the pooled t test and may be much more efficient if the underlying distributions are very nonnormal, especially with heavy tails. The comparison of the Wilcoxon test with the two-sample (unpooled) t test is less clearcut. The t test is not known to be the best test in any sense, so it seems safe to conclude that as long as the population distributions have similar shapes and spreads, the behavior of the Wilcoxon test should compare quite favorably to the two-sample t test.

Lastly, we note that β calculations for the Wilcoxon test are quite difficult. This is because the distribution of W when H_0 is false depends not only on $\mu_1 - \mu_2$ but also on the shapes of the two distributions. For most underlying distributions, the nonnull distribution of W is virtually intractable. This is why statisticians have developed large-sample (asymptotic relative) efficiency as a means of comparing tests. With the capabilities of modern-day computer software, another approach to calculation of β is to carry out a simulation experiment.

Exercises | Section 15.2 (10–16)

10. In an experiment to compare the bond strength of two different adhesives, each adhesive was used in five bondings of two surfaces, and the force necessary to separate the surfaces was determined for each bonding. For adhesive 1, the resulting values were 229, 286, 245, 299, and 250, whereas the adhesive 2 observations were 213, 179, 163, 247, and 225. Let μ_i denote the true average bond strength of adhesive type i. Use the Wilcoxon rank-sum test at level .05 to test H_0: $\mu_1 = \mu_2$ versus H_a: $\mu_1 > \mu_2$.

11. The article "A Study of Wood Stove Particulate Emissions" (*J. Air Pollution Control Assn.*, 1979: 724–728) reports the following data on burn time (hours) for samples of oak and pine. Test at level .05 to see whether there is any difference in true average burn time for the two types of wood.

    Oak 1.72 .67 1.55 1.56 1.42 1.23 1.77 .48
    Pine .98 1.40 1.33 1.52 .73 1.20

12. A modification has been made to the process for producing a certain type of "time-zero" film (film that begins to develop as soon as a picture is taken). Because the modification involves extra cost, it will be incorporated only if sample data strongly indicates that the modification has decreased true average developing time by more than 1 second. Assuming that the developing-time distributions differ only with respect to location if at all, use the Wilcoxon rank-sum test at level .05 on the accompanying data to test the appropriate hypotheses.

    Original process 8.6 5.1 4.5 5.4 6.3 6.6 5.7 8.5
    Modified process 5.5 4.0 3.8 6.0 5.8 4.9 7.0 5.7

13. The accompanying data resulted from an experiment to compare the effects of vitamin C in orange juice and in synthetic ascorbic acid on the length of odontoblasts in guinea pigs over a 6-week period ("The Growth of the Odontoblasts of the Incisor Tooth as a Criterion of the Vitamin C Intake of the Guinea Pig," *J. Nutrition*, 1947: 491–504). Use the Wilcoxon rank-sum test at level .01 to decide whether true average length differs for the two types of vitamin C intake. Compute also an approximate *P*-value.

    Orange juice 8.2 9.4 9.6 9.7 10.0 14.5 15.2 16.1 17.6 21.5
    Ascorbic acid 4.2 5.2 5.8 6.4 7.0 7.3 10.1 11.2 11.3 11.5

14. Test the hypotheses suggested in Exercise 13 using the following data:

    Orange juice 8.2 9.5 9.5 9.7 10.0 14.5 15.2 16.1 17.6 21.5
    Ascorbic acid 4.2 5.2 5.8 6.4 7.0 7.3 9.5 10.0 11.5 11.5

15. The article "Measuring the Exposure of Infants to Tobacco Smoke" (*N. Engl. J. Med.*, 1984: 1075–1078) reports on a study in which various measurements were taken both from a random sample of infants who had been exposed to household smoke and from a sample of unexposed infants. The accompanying data consists of observations on urinary concentration of cotanine, a major metabolite of nicotine (the values constitute a subset of the original data and were read from a plot that appeared in the article). Does the data suggest that true average cotanine level is higher in exposed infants than in unexposed infants by more than 25? Carry out a test at significance level .05.

    Unexposed 8 11 12 14 20 43 111
    Exposed 35 56 83 92 128 150 176 208

16. Reconsider the situation described in Exercise 79 of Chapter 9 and the accompanying MINITAB output (the Greek letter eta is used to denote a median).

    ```
    Mann-Whitney Confidence Interval and Test
    good              N = 8           Median = 0.540
    poor              N = 8           Median = 2.400
    Point estimate for ETA1-ETA2 is         -1.155
    95.9 Percent CI for ETA1-ETA2 is (-3.160, -0.409)
    W = 41.0
    Test of ETA1 = ETA2 vs ETA1 < ETA2 is significant
    at 0.0027
    ```

    a. Verify that the value of MINTAB's test statistic is correct.

    b. Carry out an appropriate test of hypotheses using a significance level of .01.

15.3 | Distribution-Free Confidence Intervals

The method we have used so far to construct a confidence interval (CI) can be described as follows: Start with a random variable (Z, T, χ^2, F, or the like) that depends on the parameter of interest and a probability statement involving the variable, manipulate the inequalities of the statement to isolate the parameter between random endpoints, and finally substitute computed values for random variables. Another general method for obtaining CIs takes advantage of a relationship between test procedures and CIs; a $100(1 - \alpha)\%$ CI for a parameter θ can be obtained from a level α test for $H_0: \theta = \theta_0$ versus $H_a: \theta \neq \theta_0$. This method will be used to derive intervals associated with the Wilcoxon signed-rank test and the Wilcoxon rank-sum test.

Before using the method to derive new intervals, reconsider the t test and the t interval. Suppose a random sample of $n = 25$ observations from a normal population yields summary statistics $\bar{x} = 100$, $s = 20$. Then a 90% CI for μ is

$$\left(\bar{x} - t_{.05,24} \cdot \frac{s}{\sqrt{25}}, \quad \bar{x} + t_{.05,24} \cdot \frac{s}{\sqrt{25}} \right) = (93.16, \ 106.84) \qquad (15.7)$$

Suppose that instead of a CI, we had wished to test a hypothesis about μ. For $H_0: \mu = \mu_0$ versus $H_a: \mu \neq \mu_0$, the t test at level .10 specifies that H_0 should be rejected if t is either ≥ 1.711 or ≤ -1.711, where

$$t = \frac{\bar{x} - \mu_0}{s/\sqrt{25}} = \frac{100 - \mu_0}{20/\sqrt{25}} = \frac{100 - \mu_0}{4} \qquad (15.8)$$

Consider now the null value $\mu_0 = 95$. Then $t = 1.25$ so H_0 is not rejected. Similarly, if $\mu_0 = 104$, then $t = -1$, so again H_0 is not rejected. However, if $\mu_0 = 90$, then $t = 2.5$, so H_0 is rejected, and if $\mu_0 = 108$, then $t = -2$, so H_0 is again rejected. By considering other values of μ_0 and the decision resulting from each one, the following general fact emerges: *Every number inside the interval* (15.7) *specifies a value of* μ_0 *for which* t *of* (15.8) *leads to nonrejection of* H_0, *whereas every number outside interval* (15.7) *corresponds to a* t *for which* H_0 *is rejected.* That is, for the fixed values of n, $\bar{x}$, and s, the set of all μ_0 values for which testing $H_0: \mu = \mu_0$ versus $H_a: \mu \neq \mu_0$ results in nonrejection of H_0 is precisely the interval (15.7).

PROPOSITION

> Suppose we have a level α test procedure for testing $H_0: \theta = \theta_0$ versus $H_a: \theta \neq \theta_0$. For fixed sample values, let A denote the set of all values θ_0 for which H_0 is not rejected. Then A is a $100(1 - \alpha)\%$ CI for θ.

There are actually pathological examples in which the set A defined in the proposition is not an interval of θ values, but instead the complement of an interval or something even stranger. To be more precise, we should really replace the notion of a CI with that of a confidence set. In the cases of interest here, the set A does turn out to be an interval.

The Wilcoxon Signed-Rank Interval

To test H_0: $\mu = \mu_0$ versus H_a: $\mu \neq \mu_0$ using the Wilcoxon signed-rank test, where μ is the mean of a continuous symmetric distribution, the absolute values $|x_1 - \mu_0|, \ldots,$ $|x_n - \mu_0|$ are ordered from smallest to largest, with the smallest receiving rank 1 and the largest, rank n. Each rank is then given the sign of its associated $x_i - \mu_0$, and the test statistic is the sum of the positively signed ranks. The two-tailed test rejects H_0 if s_+ is either $\geq c$ or $\leq n(n + 1)/2 - c$, where c is obtained from Appendix Table A.13 once the desired level of significance α is specified. For fixed $x_1, \ldots, x_n$, the $100(1 - \alpha)\%$ signed-rank interval will consist of all μ_0 for which H_0: $\mu = \mu_0$ is not rejected at level α. To identify this interval, it is convenient to express the test statistic S_+ in another form:

$$S_+ = \text{the number of pairwise averages } (X_i + X_j)/2 \text{ with } i \leq j \text{ that} \\ \text{are} \geq \mu_0 \tag{15.9}$$

That is, if we average each x_j in the list with each x_i to its left, including $(x_j + x_j)/2$ (which is just x_j), and count the number of these averages that are $\geq \mu_0$, s_+ results. In moving from left to right in the list of sample values, we are simply averaging every pair of observations in the sample [again including $(x_j + x_j)/2$] exactly once, so the order in which the observations are listed before averaging is not important. The equivalence of the two methods for computing s_+ is not difficult to verify. The number of pairwise averages is $\binom{n}{2} + n$ (the first term due to averaging of different observations and the second due to averaging each x_i with itself), which equals $n(n + 1)/2$. If either too many or too few of these pairwise averages are $\geq \mu_0$, H_0 is rejected.

Example 15.6 The following observations are values of cerebral metabolic rate for rhesus monkeys: $x_1 = 4.51$, $x_2 = 4.59$, $x_3 = 4.90$, $x_4 = 4.93$, $x_5 = 6.80$, $x_6 = 5.08$, $x_7 = 5.67$. The 28 pairwise averages are, in increasing order,

4.51	4.55	4.59	4.705	4.72	4.745	4.76	4.795	4.835	4.90
4.915	4.93	4.99	5.005	5.08	5.09	5.13	5.285	5.30	5.375
5.655	5.67	5.695	5.85	5.865	5.94	6.235	6.80		

The first few and the last few of these are pictured on a measurement axis in Figure 15.3.

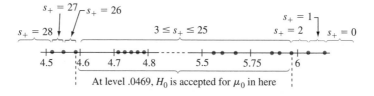

Figure 15.3 Plot of the data for Example 15.6

Because of the discreteness of the distribution of S_+, $\alpha = .05$ cannot be obtained exactly. The rejection region $\{0, 1, 2, 26, 27, 28\}$ has $\alpha = .046$, which is as close as possible to .05, so the level is approximately .05. Thus, if the number of pairwise averages $\geq \mu_0$ is between 3 and 25, inclusive, H_0 is not rejected. From Figure 15.3 the (approximate) 95% CI for μ is (4.59, 5.94). ■

In general, once the pairwise averages are ordered from smallest to largest, the endpoints of the Wilcoxon interval are two of the "extreme" averages. To express this precisely, let the smallest pairwise average be denoted by $\bar{x}_{(1)}$, the next smallest by $\bar{x}_{(2)}, \ldots$, and the largest by $\bar{x}_{(n(n+1)/2)}$.

PROPOSITION

> If the level α Wilcoxon signed-rank test for H_0: $\mu = \mu_0$ versus H_a: $\mu \neq \mu_0$ is to reject H_0 if either $s_+ \geq c$ or $s_+ \leq n(n + 1)/2 - c$, then a $100(1 - \alpha)$% CI for μ is
>
> $$(\bar{x}_{(n(n+1)/2-c+1)}, \bar{x}_{(c)}) \tag{15.10}$$

In words, the interval extends from the dth smallest pairwise average to the dth largest average, where $d = n(n + 1)/2 - c + 1$. Appendix Table A.15 gives the values of c that correspond to the usual confidence levels for $n = 5, 6, \ldots, 25$.

Example 15.7
(Example 15.6 continued)

For $n = 7$, an 89.1% interval (approximately 90%) is obtained by using $c = 24$ (since the rejection region $\{0, 1, 2, 3, 4, 24, 25, 26, 27, 28\}$ has $\alpha = .109$). The interval is $(\bar{x}_{(28-24+1)}, \bar{x}_{(24)}) = (\bar{x}_{(5)}, \bar{x}_{(24)}) = (4.72, 5.85)$, which extends from the fifth smallest to the fifth largest pairwise average. ■

Other Uses and Properties of the Wilcoxon Signed-Rank Interval

The derivation of the interval depended on having a single sample from a continuous symmetric distribution with mean (median) μ. When the data is paired, the interval constructed from the differences $d_1, d_2, \ldots, d_n$ is a CI for the mean (median) difference μ_D. In this case, the symmetry of X and Y distributions need not be assumed; as long as the X and Y distributions have the same shape, the $X - Y$ distribution will be symmetric, so only continuity is required.

For $n > 20$, the large-sample approximation to the Wilcoxon test based on standardizing S_+ gives an approximation to c in (15.10). The result [for a $100(1 - \alpha)$% interval] is

$$c \approx \frac{n(n + 1)}{4} + z_{\alpha/2}\sqrt{\frac{n(n + 1)(2n + 1)}{24}}$$

The efficiency of the Wilcoxon interval relative to the t interval is roughly the same as that for the Wilcoxon test relative to the t test. In particular, for large samples when the underlying population is normal, the Wilcoxon interval will tend to be slightly longer than the t interval, but if the population is quite nonnormal (symmetric but with heavy tails), then the Wilcoxon interval will tend to be much shorter than the t interval.

The Wilcoxon Rank-Sum Interval

The Wilcoxon rank-sum test for testing H_0: $\mu_1 - \mu_2 = \Delta_0$ is carried out by first combining the $(X_i - \Delta_0)$'s and Y_j's into one sample of size $m + n$ and ranking them from smallest (rank 1) to largest (rank $m + n$). The test statistic W is then the sum of the ranks of the $(X_i - \Delta_0)$'s. For the two-sided alternative, H_0 is rejected if w is either too small or too large.

To obtain the associated CI for fixed x_i's and y_j's, we must determine the set of all Δ_0 values for which H_0 is not rejected. This is easiest to do if we first express the test statistic in a slightly different form. The smallest possible value of W is $m(m + 1)/2$, corresponding to every $(X_i - \Delta_0)$ less than every Y_j, and there are mn differences of the form $(X_i - \Delta_0) - Y_j$. A bit of manipulation gives

$$W = [\text{number of } (X_i - Y_j - \Delta_0)\text{'s} \geq 0] + \frac{m(m + 1)}{2}$$

$$= [\text{number of } (X_i - Y_j)\text{'s} \geq \Delta_0] + \frac{m(m + 1)}{2}$$

(15.11)

Thus, rejecting H_0 if the number of $(x_i - y_j)$'s $\geq \Delta_0$ is either too small or too large is equivalent to rejecting H_0 for small or large w.

Expression (15.11) suggests that we compute $x_i - y_j$ for each i and j and order these mn differences from smallest to largest. Then if the null value Δ_0 is neither smaller than most of the differences nor larger than most, H_0: $\mu_1 - \mu_2 = \Delta_0$ is not rejected. Varying Δ_0 now shows that a CI for $\mu_1 - \mu_2$ will have as its lower endpoint one of the ordered $(x_i - y_j)$'s, and similarly for the upper endpoint.

PROPOSITION

Let $x_1, \ldots, x_m$ and $y_1, \ldots, y_n$ be the observed values in two independent samples from continuous distributions that differ only in location (and not in shape). With $d_{ij} = x_i - y_j$ and the ordered differences denoted by $d_{ij(1)}, d_{ij(2)}, \ldots, d_{ij(mn)}$, the general form of a $100(1 - \alpha)\%$ CI for $\mu_1 - \mu_2$ is

$$(d_{ij(mn-c+1)}, d_{ij(c)})$$

(15.12)

where c is the critical constant for the two-tailed level α Wilcoxon rank-sum test.

Notice that the form of the Wilcoxon rank-sum interval (15.12) is very similar to the Wilcoxon signed-rank interval (15.10); (15.10) uses pairwise averages from a single sample, whereas (15.12) uses pairwise differences from two samples. Appendix Table A.16 gives values of c for selected values of m and n.

Example 15.8 The article "Some Mechanical Properties of Impregnated Bark Board" (*Forest Products J.*, 1977: 31–38) reports the following data on maximum crushing strength (psi) for a sample of epoxy-impregnated bark board and for a sample of bark board impregnated with another polymer:

Epoxy (x's)	10,860	11,120	11,340	12,130	14,380	13,070
Other (y's)	4,590	4,850	6,510	5,640	6,390	

Obtain a 95% CI for the true average difference in crushing strength between the epoxy-impregnated board and the other type of board.

From Appendix Table A.16, since the smaller sample size is 5 and the larger sample size is 6, $c = 26$ for a confidence level of approximately 95%. The d_{ij}'s appear in Table 15.5. The five smallest d_{ij}'s $[d_{ij(1)}, \ldots, d_{ij(5)}]$ are 4350, 4470, 4610, 4730, and 4830, and the five largest d_{ij}'s are (in descending order) 9790, 9530, 8740, 8480, and 8220. Thus, the CI is $(d_{ij(5)}, d_{ij(26)}) = (4830, 8220)$.

Table 15.5   Differences for the rank-sum interval in Example 15.8

	d_{ij}	4590	4850	y_j 5640	6390	6510
	10,860	6270	6010	5220	4470	4350
	11,120	6530	6270	5480	4730	4610
x_i	11,340	6750	6490	5700	4950	4830
	12,130	7540	7280	6490	5740	5620
	13,070	8480	8220	7430	6680	6560
	14,380	9790	9530	8740	7990	7870

∎

When m and n are both large, the Wilcoxon test statistic has approximately a normal distribution. This can be used to derive a large-sample approximation for the value c in interval (15.12). The result is

$$c \approx \frac{mn}{2} + z_{\alpha/2} \sqrt{\frac{mn(m + n + 1)}{12}} \tag{15.13}$$

As with the signed-rank interval, the rank-sum interval (15.12) is quite efficient with respect to the t interval; in large samples, (15.12) will tend to be only a bit longer than the t interval when the underlying populations are normal and may be considerably shorter than the t interval if the underlying populations have heavier tails than do normal populations.

Exercises | Section 15.3 (17–22)

17. The article "The Lead Content and Acidity of Christchurch Precipitation" (*New Zealand J. Science,* 1980: 311–312) reports the accompanying data on lead concentration (μg/L) in samples gathered during eight different summer rainfalls: 17.0, 21.4, 30.6, 5.0, 12.2, 11.8, 17.3, and 18.8. Assuming that the lead-content distribution is symmetric, use the Wilcoxon signed-rank interval to obtain a 95% CI for μ.

18. Compute the 99% signed-rank interval for true average pH μ (assuming symmetry) using the data in Exercise 3. [*Hint:* Try to compute only those pair-

wise averages having relatively small or large values (rather than all 105 averages).]

19. Compute a CI for μ_D of Example 15.2 using the data given there; your confidence level should be roughly 95%.

20. The following observations are amounts of hydrocarbon emissions resulting from road wear of bias-belted tires under a 522-kg load inflated at 228 kPa and driven at 64 km/hr for 6 hours ("Characterization of Tire Emissions Using an

Indoor Test Facility," *Rubber Chemistry and Technology,* 1978: 7–25): .045, .117, .062, and .072. What confidence levels are achievable for this sample size using the signed-rank interval? Select an appropriate confidence level and compute the interval.

21. Compute the 90% rank-sum CI for $\mu_1 - \mu_2$ using the data in Exercise 10.

22. Compute a 99% CI for $\mu_1 - \mu_2$ using the data in Exercise 11.

15.4 | Distribution-Free ANOVA

The single-factor ANOVA model of Chapter 10 for comparing I population or treatment means assumed that for $i = 1, 2, \ldots, I$, a random sample of size J_i was drawn from a normal population with mean μ_i and variance σ^2. This can be written as

$$X_{ij} = \mu_i + \epsilon_{ij} \qquad j = 1, \ldots, J_i; i = 1, \ldots, I \qquad (15.14)$$

where the ϵ_{ij}'s are independent and normally distributed with mean zero and variance σ^2. Although the normality assumption was required for the validity of the F test described in Chapter 10, the validity of the Kruskal–Wallis test for testing equality of the μ_i's depends only on the ϵ_{ij}'s having the same continuous distribution.

The Kruskal–Wallis Test

Let $N = \Sigma J_i$, the total number of observations in the data set, and suppose we rank all N observations from 1 (the smallest X_{ij}) to N (the largest X_{ij}). When H_0: $\mu_1 = \mu_2 = \cdots = \mu_I$ is true, the N observations all come from the same distribution, in which case all possible assignments of the ranks $1, 2, \ldots, N$ to the I samples are equally likely and we expect ranks to be intermingled in these samples. If, however, H_0 is false, then some samples will consist mostly of observations having small ranks in the combined sample whereas others will consist mostly of observations having large ranks. More specifically, if R_{ij} denotes the rank of X_{ij} among the N observations, and $R_{i\cdot}$ and $\bar{R}_{i\cdot}$ denote, respectively, the total and average of the ranks in the ith sample, then when H_0 is true

$$E(R_{ij}) = \frac{N+1}{2} \qquad E(\bar{R}_{i\cdot}) = \frac{1}{J_i}\sum_j E(R_{ij}) = \frac{N+1}{2}$$

The Kruskal–Wallis test statistic is a measure of the extent to which the $\bar{R}_{i\cdot}$'s deviate from their common expected value $(N+1)/2$, and H_0 is rejected if the computed value of the statistic indicates too great a discrepancy between observed and expected rank averages.

TEST STATISTIC

$$K = \frac{12}{N(N+1)} \sum_{j=1}^{I} J_i\left(\bar{R}_{i\cdot} - \frac{N+1}{2}\right)^2$$

$$= \frac{12}{N(N+1)} \sum_{i=1}^{I} \frac{R_{i\cdot}^2}{J_i} - 3(N+1)$$

$\qquad (15.15)$

The second expression for K is the computational formula; it involves the rank totals (R_i.'s) rather than the averages and requires only one subtraction.

If H_0 is rejected when $k \geq c$, then c should be chosen so that the test has level α. That is, c should be the upper-tail critical value of the distribution of K when H_0 is true. Under H_0, each possible assignment of the ranks to the I samples is equally likely, so in theory all such assignments can be enumerated, the value of K determined for each one, and the null distribution obtained by counting the number of times each value of K occurs. Clearly this computation is tedious, so even though there are tables of the exact null distribution and critical values for small values of the J_i's, we will use the following "large-sample" approximation.

PROPOSITION

When H_0 is true and either

$$I = 3 \qquad J_i \geq 6 \qquad (i = 1, 2, 3)$$

or

$$I > 3 \qquad J_i \geq 5 \qquad (i = 1, \ldots, I)$$

then K has approximately a chi-squared distribution with $I - 1$ df. This implies that a test with approximate significance level α rejects H_0 if $k \geq \chi^2_{\alpha, I-1}$.

Example 15.9 The accompanying observations (Table 15.6) on axial stiffness index resulted from a study of metal plate–connected trusses in which five different plate lengths—4 in., 6 in., 8 in., 10 in., and 12 in.—were used ("Modeling Joints Made with Light-Gauge Metal Connector Plates," *Forest Products J.,* 1979: 39–44).

Table 15.6 Data and Ranks for Example 15.9

								r_i.	$\bar{r}_i$.
$i = 1$ (4"):	309.2	309.7	311.0	316.8	326.5	349.8	409.5		
$i = 2$ (6"):	331.0	347.2	348.9	361.0	381.7	402.1	404.5		
$i = 3$ (8"):	351.0	357.1	366.2	367.3	382.0	392.4	409.9		
$i = 4$ (10"):	346.7	362.6	384.2	410.6	433.1	452.9	461.4		
$i = 5$ (12"):	407.4	410.7	419.9	441.2	441.8	465.8	473.4		

									r_i.	$\bar{r}_i$.
	$i = 1$:	1	2	3	4	5	10	24	49	7.00
	$i = 2$:	6	8	9	13	17	21	22	96	13.71
Ranks	$i = 3$:	11	12	15	16	18	20	25	117	16.71
	$i = 4$:	7	14	19	26	29	32	33	160	22.86
	$i = 5$:	23	27	28	30	31	34	35	208	29.71

The computed value of K is

$$k = \frac{12}{35(36)} \left[\frac{(49)^2}{7} + \frac{(96)^2}{7} + \frac{(117)^2}{7} + \frac{(160)^2}{7} + \frac{(208)^2}{7} \right] - 3(36)$$

$$= 20.12$$

At level .01, $\chi^2_{.01,4} = 13.277$, and since $20.12 \geq 13.277$, H_0 is rejected and we conclude that expected axial stiffness does depend on plate length. ∎

Friedman's Test for a Randomized Block Experiment

Suppose $X_{ij} = \mu + \alpha_i + \beta_j + \epsilon_{ij}$, where α_i is the ith treatment effect, β_j is the jth block effect, and the ϵ_{ij}'s are drawn independently from the same continuous (but not necessarily normal) distribution. Then to test H_0: $\alpha_1 = \alpha_2 = \cdots = \alpha_I = 0$, the null hypothesis of no treatment effects, the observations are first ranked separately from 1 to I within each block, and then the rank average $\bar{r}_{i\cdot}$ is computed for each of the I treatments. When H_0 is true, the $\bar{r}_{i\cdot}$'s should be close to one another, since within each block all $I!$ assignments of ranks to treatments are equally likely. Friedman's test statistic measures the discrepancy between the expected value $(I + 1)/2$ of each rank average and the $\bar{r}_{i\cdot}$'s.

TEST STATISTIC

$$F_r = \frac{12J}{I(I + 1)} \sum_{i=1}^{I} \left(\bar{R}_{i\cdot} - \frac{I + 1}{2} \right)^2 = \frac{12}{IJ(I + 1)} \sum R_{i\cdot}^2 - 3J(I + 1)$$

As with the Kruskal–Wallis test, Friedman's test rejects H_0 when the computed value of the test statistic is too large. For the cases $I = 3$, $J = 2, \ldots, 15$ and $I = 4$, $J = 2, \ldots, 8$, Lehmann's book (see the chapter bibliography) gives the upper-tail critical values for the test. Alternatively, for even moderate values of J, the test statistic F_r has approximately a chi-squared distribution with $I - 1$ df when H_0 is true, so H_0 can be rejected if $f_r \geq \chi^2_{\alpha, I-1}$.

Example 15.10 The article "Physiological Effects During Hypnotically Requested Emotions" (*Psychosomatic Med.*, 1963: 334–343) reports the following data (Table 15.7) on skin potential (mV) when the emotions of fear, happiness, depression, and calmness were requested from each of eight subjects.

Table 15.7 Data and ranks for Example 15.10

x_{ij}	1	2	3	4	5	6	7	8		
				Blocks (Subjects)						
Fear	23.1	57.6	10.5	23.6	11.9	54.6	21.0	20.3		
Happiness	22.7	53.2	9.7	19.6	13.8	47.1	13.6	23.6		
Depression	22.5	53.7	10.8	21.1	13.7	39.2	13.7	16.3		
Calmness	22.6	53.1	8.3	21.6	13.3	37.0	14.8	14.8		
Ranks	1	2	3	4	5	6	7	8	$r_{i\cdot}$	$r_{i\cdot}^2$
Fear	4	4	3	4	1	4	4	3	27	729
Happiness	3	2	2	1	4	3	1	4	20	400
Depression	1	3	4	2	3	2	2	2	19	361
Calmness	2	1	1	3	2	1	3	1	14	196
										1686

Thus,

$$f_r = \frac{12}{4(8)(5)}(1686) - 3(8)(5) = 6.45$$

At level .05, $\chi^2_{.05,3} = 7.815$, and because 6.45 is not ≥ 7.815, H_0 is not rejected. There is no evidence that average skin potential depends on which emotion is requested. ∎

The book by Myles Hollander and Douglas Wolfe (see chapter bibliography) discusses multiple comparisons procedures associated with the Kruskal–Wallis and Friedman tests, as well as other aspects of distribution-free ANOVA.

Exercises | Section 15.4 (23–27)

23. The accompanying data refers to concentration of the radioactive isotope strontium-90 in milk samples obtained from five randomly selected dairies in each of four different regions.

	1	6.4	5.8	6.5	7.7	6.1
Region	2	7.1	9.9	11.2	10.5	8.8
	3	5.7	5.9	8.2	6.6	5.1
	4	9.5	12.1	10.3	12.4	11.7

Test at level .10 to see whether true average strontium-90 concentration differs for at least two of the regions.

24. The article "Production of Gaseous Nitrogen in Human Steady-State Conditions" (*J. Applied Physiology*, 1972: 155–159) reports the following observations on the amount of nitrogen expired (in liters) under four dietary regimens: fasting (1), 23% protein (2), 32% protein (3), and 67% protein (4). Use the Kruskal–Wallis test at level .05 to test equality of the corresponding μ_i's.

1. 4.079 4.859 3.540 5.047 3.298
2. 4.368 5.668 3.752 5.848 3.802
3. 4.169 5.709 4.416 5.666 4.123
4. 4.928 5.608 4.940 5.291 4.674

1. 4.679 2.870 4.648 3.847
2. 4.844 3.578 5.393 4.374
3. 5.059 4.403 4.496 4.688
4. 5.038 4.905 5.208 4.806

25. The accompanying data on cortisol level was reported in the article "Cortisol, Cortisone, and 11-Deoxycortisol Levels in Human Umbilical and Maternal Plasma in Relation to the Onset of Labor" (*J. Obstetric Gynaecology British Commonwealth*, 1974: 737–745). Experimental subjects were pregnant women whose babies were delivered between 38 and 42 weeks gestation. Group 1 individuals elected to deliver by Caesarean section before labor onset, group 2 delivered by emergency Caesarean during induced labor, and group 3 individuals experienced spontaneous labor. Use the Kruskal–Wallis test at level .05 to test for equality of the three population means.

Group 1	262 307 211	323 454 339
	304 154 287	356
Group 2	465 501 455	355 468 362
Group 3	343 772 207	1048 838 687

26. In a test to determine whether soil pretreated with small amounts of Basic-H makes the soil more permeable to water, soil samples were divided into blocks, and each block received each of the four treatments under study. The treatments were (A) water with .001% Basic-H flooded on control soil, (B) water without Basic-H on control soil, (C) water with Basic-H flooded on soil pretreated with Basic-H, and (D) water without Basic-H on soil pretreated with Basic-H. Test at level .01 to see whether there are any effects due to the different treatments.

			Blocks		
	1	**2**	**3**	**4**	**5**
A	37.1	31.8	28.0	25.9	25.5
B	33.2	25.3	20.2	20.3	18.3
C	58.9	54.2	49.2	47.9	38.2
D	56.7	49.6	46.4	40.9	39.4
	6	**7**	**8**	**9**	**10**
A	25.3	23.7	24.4	21.7	26.2
B	19.3	17.3	17.0	16.7	18.3
C	48.8	47.8	40.2	44.0	46.4
D	37.1	37.5	39.6	35.1	36.5

27. In an experiment to study the way in which different anesthetics affect plasma epinephrine concentration, ten dogs were selected and concentration was measured while they were under the influence of the anesthetics isoflurane, halothane, and cyclopropane ("Sympathoadrenal and Hemodynamic Effects of Isoflurane, Halothane, and Cyclopropane in Dogs," *Anesthesiology,* 1974: 465–470). Test at level .05 to see whether there is an anesthetic effect on concentration.

			Dog		
	1	**2**	**3**	**4**	**5**
Isoflurane	.28	.51	1.00	.39	.29
Halothane	.30	.39	.63	.38	.21
Cyclopropane	1.07	1.35	.69	.28	1.24
	6	**7**	**8**	**9**	**10**
Isoflurane	.36	.32	.69	.17	.33
Halothane	.88	.39	.51	.32	.42
Cyclopropane	1.53	.49	.56	1.02	.30

Supplementary Exercises (28–36)

28. The article "Effects of a Rice-Rich Versus Potato-Rich Diet on Glucose, Lipoprotein, and Cholesterol Metabolism in Noninsulin-Dependent Diabetics" (*Amer. J. Clinical Nutr.,* 1984: 598–606) gives the accompanying data on cholesterol-synthesis rate for eight diabetic subjects. Subjects were fed a standardized diet with potato or rice as the major carbohydrate source. Participants received both diets for specified periods of time, with cholesterol-synthesis rate (mmol/day) measured at the end of each dietary period. The analysis presented in this article used a distribution-free test. Use such a test with significance level .05 to determine whether the true mean cholesterol-synthesis rate differs significantly for the two sources of carbohydrates.

Cholesterol-Synthesis Rate

Subject	1	2	3	4	5	6	7	8
Potato	1.88	2.60	1.38	4.41	1.87	2.89	3.96	2.31
Rice	1.70	3.84	1.13	4.97	.86	1.93	3.36	2.15

29. High-pressure sales tactics or door-to-door salespeople can be quite offensive. Many people succumb to such tactics, sign a purchase agreement, and later regret their actions. In the mid-1970s, the Federal Trade Commission implemented regulations clarifying and extending rights of purchasers to cancel such agreements. The accompanying data is a subset of that given in the article "Evaluating the FTC Cooling-Off Rule" (*J. Consumer Affairs,* 1977: 101–106). Individual observations are cancellation rates for each of nine salespeople during each of 4 years. Use an appropriate test at level .05 to see whether true average cancellation rate depends on the year.

				Salesperson					
	1	**2**	**3**	**4**	**5**	**6**	**7**	**8**	**9**
1973	2.8	5.9	3.3	4.4	1.7	3.8	6.6	3.1	0.0
1974	3.6	1.7	5.1	2.2	2.1	4.1	4.7	2.7	1.3
1975	1.4	.9	1.1	3.2	.8	1.5	2.8	1.4	.5
1976	2.0	2.2	.9	1.1	.5	1.2	1.4	3.5	1.2

30. The given data on phosphorus concentration in topsoil for four different soil treatments appeared in the article "Fertilisers for Lotus and Clover Establishment on a Sequence of Acid Soils on the East Otago Uplands" (*N. Zeal. J. Exper. Ag.,* 1984: 119–129). Use a distribution-free procedure to test the null hypothesis of no difference in true mean phosphorus concentration (mg/g) for the four soil treatments.

	I	8.1	5.9	7.0	8.0	9.0
Treatment	II	11.5	10.9	12.1	10.3	11.9
	III	15.3	17.4	16.4	15.8	16.0
	IV	23.0	33.0	28.4	24.6	27.7

31. Refer to the data of Exercise 30 and compute a 95% CI for the difference between true average concentrations for treatments II and III.

32. The study reported in "Gait Patterns During Free Choice Ladder Ascents" (*Human Movement Sci.,* 1983: 187–195) was motivated by publicity concerning the increased accident rate for individuals climbing ladders. A number of different gait patterns were used by subjects climbing a portable straight ladder according to specified instructions. The ascent times for seven subjects who used a lateral gait and six subjects who used a four-beat diagonal gait are given.

Lateral .86 1.31 1.64 1.51 1.53 1.39 1.09
Diagonal 1.27 1.82 1.66 .85 1.45 1.24

a. Carry out a test using $\alpha = .05$ to see whether the data suggests any difference in the true average ascent times for the two gaits.
b. Compute a 95% CI for the difference between the true average gait times.

33. The **sign test** is a very simple procedure for testing hypotheses about a population median assuming only that the underlying distribution is continuous. To illustrate, consider the following sample of 20 observations on component lifetime (hr):

1.7	3.3	5.1	6.9	12.6	14.4	16.4
24.6	26.0	26.5	32.1	37.4	40.1	40.5
41.5	72.4	80.1	86.4	87.5	100.2	

We wish to test H_0: $\tilde{\mu} = 25.0$ versus H_a: $\tilde{\mu} > 25.0$. The test statistic is $Y =$ the number of observations that exceed 20.
a. Consider rejecting H_0 if $Y \geq 15$. What is the value of α (the probability of a type I error) for this test? *Hint:* Think of a "success" as a lifetime that exceeds 25.0. Then Y is the number of successes in the sample. What kind of a distribution does Y have when $\tilde{\mu} = 25.0$?
b. What rejection region of the form $Y \geq c$ specifies a test with a significance level as close to .05 as possible? Use this region to carry out the test for the given data.

Note: The test statistic is the number of differences $X_i - 25.0$ that have positive signs, hence the name *sign test.*

34. Refer to Exercise 33, and consider a confidence interval associated with the sign test, the **sign interval.** The relevant hypotheses are now H_0: $\tilde{\mu} = \tilde{\mu}_0$ versus H_a: $\tilde{\mu} \neq \tilde{\mu}_0$. Let's use the following rejection region: either $Y \geq 15$ or $Y \leq 5$.
a. What is the significance level for this test?
b. The confidence interval will consist of all values $\tilde{\mu}_0$ for which H_0 is not rejected. Determine the CI for the given data, and state the confidence level.

35. Suppose we wish to test

H_0: the X and Y distributions are identical

versus

H_a: the X distribution is less spread out than the Y distribution

The accompanying figure pictures X and Y distributions for which H_a is true. The Wilcoxon rank-sum test is not appropriate in this situation because when H_a is true as pictured, the Y's will tend to be at the extreme ends of the combined sample (resulting in small and large Y ranks), so the sum of X ranks will result in a W value that is neither large nor small.

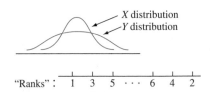

Consider modifying the procedure for assigning ranks as follows: After the combined sample of $m + n$ observations is ordered, the smallest observation is given rank 1, the largest observation is given rank 2, the second smallest is given rank 3, the second largest is given rank 4, and so on. Then if H_a is true as pictured, the X values will tend to be in the middle of the sample and thus receive large ranks. Let W′ denote the sum of the X ranks and consider rejecting H_0 in favor of H_a when w′ ≥ c. When H_0 is true, every possible set of X ranks has the same probability, so W′ *has the same distribu-*

tion as does W when H_0 is true. Thus, c can be chosen from Appendix Table A.14 to yield a level α test. The accompanying data refers to medial muscle thickness for arterioles from the lungs of children who died from sudden infant death syndrome (x's) and a control group of children (y's). Carry out the test of H_0 versus H_a at level .05.

SIDS 4.0 4.4 4.8 4.9
Control 3.7 4.1 4.3 5.1 5.6

Consult the Lehmann book (in the chapter bibliography) for more information on this test, called the Siegel–Tukey test.

36. The ranking procedure described in Exercise 35 is somewhat asymmetric, because the smallest observation receives rank 1 whereas the largest receives rank 2, and so on. Suppose both the smallest and the largest receive rank 1, the second smallest and second largest receive rank 2, and so on, and let W'' be the sum of the X ranks. The null distribution of W'' is not identical to the null distribution of W, so different tables are needed. Consider the case $m = 3$, $n = 4$. List all 35 possible orderings of the three X values among the seven observations (e.g., 1, 3, 7 or 4, 5, 6), assign ranks in the manner described, compute the value of W'' for each possibility, and then tabulate the null distribution of W''. For the test that rejects if $w'' \geq c$, what value of c prescribes approximately a level .10 test? This is the Ansari–Bradley test; for additional information, see the book by Hollander and Wolfe in the chapter bibliography.

Bibliography

Hollander, Myles, and Douglas Wolfe, *Nonparametric Statistical Methods* (2nd ed.), Wiley, New York, 1999. A very good reference on distribution-free methods with an excellent collection of tables.

Lehmann, Erich, *Nonparametrics: Statistical Methods Based on Ranks,* Holden-Day, San Francisco, 1975. An excellent discussion of the most important distribution-free methods, presented with a great deal of insightful commentary.

Quality Control Methods

Introduction

Quality characteristics of manufactured products have received much attention from design engineers and production personnel as well as those concerned with financial management. An article of faith over the years was that very high quality levels and economic well-being were incompatible goals. Recently, however, it has become increasingly apparent that raising quality levels can lead to decreased costs, a greater degree of consumer satisfaction, and thus increased profitability. This has resulted in renewed emphasis on statistical techniques for designing quality into products and for identifying quality problems at various stages of production and distribution.

Control charting is now used extensively in industry as a diagnostic technique for monitoring production processes to identify instability and unusual circumstances. After an introduction to basic ideas in Section 16.1, a number of different control charts are presented in the next four sections. The basis for most of these lies in our previous work concerning probability distributions of various statistics such as the sample mean $\overline{X}$ and sample proportion $\hat{p} = X/n$.

Another commonly encountered situation in industrial settings involves a decision by a customer as to whether a batch of items offered by a supplier is of acceptable quality. In the last section of the chapter, we briefly survey some acceptance sampling methods for deciding, based on sample data, on the disposition of a batch.

Besides control charts and acceptance sampling plans, which were first developed in the 1920s and 1930s, statisticians and engineers have recently introduced many new statistical methods for identifying types and levels of production inputs that will ensure high-quality output. Japanese investigators, and in particular the engineer/statistician G. Taguchi and his disciples, have been very influential in this respect, and there is now a large body of material known as "Taguchi methods." The ideas of experimental design, and in particular fractional factorial experiments, are key ingredients. There is still much controversy in the statistical community as to which designs and methods of analysis are best suited to the task at hand. A recent critique is contained in the expository article by George Box et al., cited in the chapter bibliography; the book by Thomas Ryan listed there is also a good source of information.

16.1 | General Comments on Control Charts

A central message throughout this book has been the pervasiveness of naturally occurring variation associated with any characteristic or attribute of different individuals or objects. In a manufacturing context, no matter how carefully machines are calibrated, environmental factors are controlled, materials and other inputs are monitored, and workers are trained, diameter will vary from bolt to bolt, some plastic sheets will be stronger than others, some fuses will be defective whereas others are not, and so on. We might think of such natural random variation as uncontrollable background noise.

There are, however, other sources of variation that may have a pernicious impact on the quality of items produced by some process. Such variation may be attributable to contaminated material, incorrect machine settings, unusual tool wear, and the like. These sources of variation have been termed *assignable causes* in the quality control literature. **Control charts** provide a mechanism for recognizing situations where assignable causes may be adversely affecting product quality. Once a chart indicates an out-of-control situation, an investigation can be launched to identify causes and take corrective action.

A basic element of control charting is that samples have been selected from the process of interest at a sequence of time points. Depending on the aspect of the process under investigation, some statistic, such as the sample mean or sample proportion of defective items, is chosen. The value of this statistic is then calculated for each sample in turn. A traditional control chart then results from plotting these calculated values over time, as illustrated in Figure 16.1 (page 680).

Notice that in addition to the plotted points themselves, the chart has a center line and two control limits. The basis for the choice of a center line is sometimes a target

Value of quality
statistic

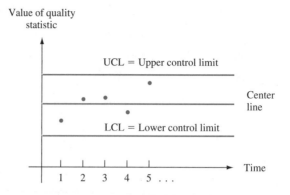

Figure 16.1 A prototypical control chart

value or design specification, for example, a desired value of the bearing diameter. In other cases, the height of the center line is estimated from the data. If the points on the chart all lie between the two control limits, the process is deemed to be in control. That is, the process is believed to be operating in a stable fashion reflecting only natural random variation. An out-of-control "signal" occurs whenever a plotted point falls outside the limits. This is assumed to be attributable to some assignable cause, and a search for such causes commences. The limits are designed so that an in-control process generates very few false alarms, whereas a process not in control quickly gives rise to a point outside the limits.

There is a strong analogy between the logic of control charting and our previous work in hypothesis testing. The null hypothesis here is that the process is in control. When an in-control process yields a point *outside* the control limits (an out-of-control signal), a type I error has occurred. On the other hand, a type II error results when an out-of-control process produces a point *inside* the control limits. Appropriate choice of sample size and control limits (the latter corresponding to specifying a rejection region in hypothesis testing) will make the associated error probabilities suitably small.

We emphasize that "in control" is not synonymous with "meets design specifications or tolerance." The extent of natural variation may be such that the percentage of items not conforming to specification is much higher than can be tolerated. In such cases, a major restructuring of the process will be necessary to improve process capability. An in-control process is simply one whose behavior with respect to variation is stable over time, showing no indications of unusual extraneous causes.

As a final introductory comment, software for control charting is now widely available. The journal *Quality Progress* contains many advertisements for statistical quality control computer packages. In addition, SAS and the newest version of MINITAB, among other general-purpose packages, have attractive quality control capabilities.

Exercises | Section 16.1 (1–3)

1. A control chart for thickness of rolled-steel sheets is based on an upper control limit of .0520 in. and a lower limit of .0475 in. The first ten values of the quality statistic (in this case $\overline{X}$, the sample mean thickness of $n = 5$ sample sheets) are .0506, .0493, .0502, .0501, .0512, .0498, .0485, .0500, .0505, and .0483. Construct the initial part of the quality control chart, and comment on its appearance.

2. Refer to Exercise 1 and suppose the ten most recent values of the quality statistic are .0493, .0485, .0490, .0503, .0492, .0486, .0495, .0494, .0493, and .0488. Construct the relevant portion of the corresponding control chart, and comment on its appearance.

3. Suppose a control chart is constructed so that the probability of a point falling outside the control limits when the process is actually in control is .002. What is the probability that ten successive points (based on independently selected samples) will be within the control limits? What is the probability that 25 successive points will all lie within the control limits? What is the smallest number of successive points plotted for which the probability of observing at least one outside the control limits exceeds .10?

16.2 | Control Charts for Process Location

Suppose the quality characteristic of interest is associated with a variable whose observed values result from making measurements. For example, the characteristic might be resistance of electrical wire (ohms), internal diameter of molded rubber expansion joints (cm), or hardness of a certain alloy (Brinell units). One important use of control charts is to see whether some measure of location of the variable's distribution remains stable over time. The most popular chart for this purpose is the $\overline{X}$ chart.

The $\overline{X}$ Chart Based on Known Parameter Values

Because there is uncertainty about the value of the variable for any particular item or specimen, we denote such a *random* variable (rv) by X. Assume that for an in-control process, X has a normal distribution with mean value μ and standard deviation σ. Then if $\overline{X}$ denotes the sample mean for a random sample of size n selected at a particular time point, we know that

1. $E(\overline{X}) = \mu$

2. $\sigma_{\overline{X}} = \sigma/\sqrt{n}$

3. $\overline{X}$ has a normal distribution.

It follows that

$$P(\mu - 3\sigma_{\overline{X}} \le \overline{X} \le \mu + 3\sigma_{\overline{X}}) = P(-3.00 \le Z \le 3.00) = .9974$$

where Z is a standard normal rv.* It is thus highly likely that for an in-control process, the sample mean will fall within 3 standard deviations ($3\sigma_{\overline{X}}$) of the process mean μ.

Consider first the case in which the values of both μ and σ are known. Suppose that at each of the time points 1, 2, 3, . . . , a random sample of size n is available. Let $\overline{x}_1, \overline{x}_2, \overline{x}_3, \ldots$ denote the calculated values of the corresponding sample means. An $\overline{X}$ chart results from plotting these $\overline{x}_i$'s over time—that is, plotting the points $(1, \overline{x}_1)$, $(2, \overline{x}_2)$, $(3, \overline{x}_3)$, and so on—and then drawing horizontal lines across the plot at

$$\text{LCL} = \text{lower control limit} = \mu - 3 \cdot \frac{\sigma}{\sqrt{n}}$$

$$\text{UCL} = \text{upper control limit} = \mu + 3 \cdot \frac{\sigma}{\sqrt{n}}$$

---

*The use of charts based on 3 SD limits is traditional, but tradition is certainly not inviolable.

Such a plot is often called a 3-sigma chart. Any point outside the control limits suggests that the process may have been out of control at that time, so a search for assignable causes should be initiated.

Example 16.1 Once each day, three specimens of motor oil are randomly selected from the production process, and each is analyzed to determine viscosity. The accompanying data (Table 16.1) is for a 25-day period. Extensive experience with this process suggests that when the process is in control, viscosity of a specimen is normally distributed with mean 10.5 and standard deviation .18. Thus, $\sigma_{\bar{x}} = \sigma/\sqrt{n} = .18/\sqrt{3} = .104$, so the 3 SD control limits are

$$LCL = \mu - 3 \cdot \frac{\sigma}{\sqrt{n}} = 10.5 - 3(.104) = 10.188$$

$$UCL = \mu + 3 \cdot \frac{\sigma}{\sqrt{n}} = 10.5 + 3(.104) = 10.812$$

All points on the control chart shown in Figure 16.2 are between the control limits, indicating stable behavior of the process mean over this time period (the standard deviation and range for each sample will be used in the next subsection).

Table 16.1 Viscosity data for Example 16.1

Day	Viscosity Observations			$\bar{x}$	s	Range
1	10.37	10.19	10.36	10.307	.101	.18
2	10.48	10.24	10.58	10.433	.175	.34
3	10.77	10.22	10.54	10.510	.276	.55
4	10.47	10.26	10.31	10.347	.110	.21
5	10.84	10.75	10.53	10.707	.159	.31
6	10.48	10.53	10.50	10.503	.025	.05
7	10.41	10.52	10.46	10.463	.055	.11
8	10.40	10.38	10.69	10.490	.173	.31
9	10.33	10.35	10.49	10.390	.087	.16
10	10.73	10.45	10.30	10.493	.218	.43
11	10.41	10.68	10.25	10.447	.217	.43
12	10.00	10.60	10.71	10.437	.382	.71
13	10.37	10.50	10.34	10.403	.085	.16
14	10.47	10.60	10.75	10.607	.140	.28
15	10.46	10.46	10.56	10.493	.058	.10
16	10.44	10.68	10.32	10.480	.183	.36
17	10.65	10.42	10.26	10.443	.196	.39
18	10.73	10.72	10.83	10.760	.061	.11
19	10.39	10.75	10.27	10.470	.250	.48
20	10.59	10.23	10.35	10.390	.183	.36
21	10.47	10.67	10.64	10.593	.108	.20
22	10.40	10.55	10.38	10.443	.093	.17
23	10.24	10.71	10.27	10.407	.263	.47
24	10.37	10.69	10.40	10.487	.177	.32
25	10.46	10.35	10.37	10.393	.059	.11

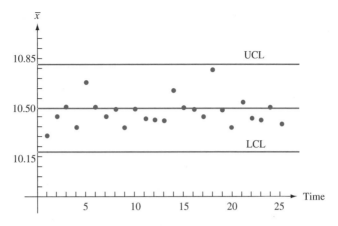

Figure 16.2 $\overline{X}$ chart for the viscosity data of Example 16.1   ■

$\overline{X}$ Charts Based on Estimated Parameters

In practice it frequently happens that values of μ and σ are unknown, so they must be estimated from sample data prior to determining the control limits. This is especially true when a process is first subjected to a quality control analysis. Again denote the number of observations in each sample by n and let k represent the number of samples available. Typical values of n are 3, 4, 5, or 6, and it is recommended that k be at least 20. We assume that the k samples were gathered during a period when the process was believed to be in control. More will be said about this assumption shortly.

With $\overline{x}_1, \overline{x}_2, \ldots, \overline{x}_k$ denoting the k calculated sample means, the usual estimate of μ is simply the average of these means:

$$\hat{\mu} = \overline{\overline{x}} = \frac{\sum_{i=1}^{k} \overline{x}_i}{k}$$

We present two different commonly used methods for estimating σ, one based on the k sample standard deviations and the other on the k sample ranges (recall that the sample range is the difference between the largest and smallest sample observations). Prior to the wide availability of good calculators and statistical computer software, ease of hand calculation was of paramount consideration, so the range method predominated. However, in the case of a normal population distribution, the unbiased estimator of σ based on S is known to have smaller variance than that based on the sample range. Statisticians say that the former estimator is more *efficient* than the latter. The loss in efficiency for the estimator is slight when n is very small but becomes important for $n > 4$.

Recall that the sample standard deviation is not an unbiased estimator for σ. When $X_1, \ldots, X_n$ is a random sample from a normal distribution, it can be shown (cf. Exercise 6.37) that

$$E(S) = a_n \cdot \sigma$$

where

$$a_n = \frac{\sqrt{2}\,\Gamma(n/2)}{\sqrt{n-1}\,\Gamma[(n-1)/2]}$$

and $\Gamma(\cdot)$ denotes the gamma function (see Section 4.4). A tabulation of a_n for selected n follows:

n	3	4	5	6	7	8
a_n	.886	.921	.940	.952	.959	.965

Let

$$\bar{S} = \frac{\sum_{i=1}^{k} S_i}{k}$$

where $S_1, S_2, \ldots, S_k$ are the sample standard deviations for the k samples. Then

$$E(\bar{S}) = \frac{1}{k} E\left(\sum_{i=1}^{k} S_i\right) = \frac{1}{k} \sum_{i=1}^{k} E(S_i) = \frac{1}{k} \sum_{i=1}^{k} a_n \cdot \sigma = a_n \cdot \sigma$$

Thus,

$$E\left(\frac{\bar{S}}{a_n}\right) = \frac{1}{a_n} E(\bar{S}) = \frac{1}{a_n} \cdot a_n \cdot \sigma = \sigma$$

so $\hat{\sigma} = \bar{S}/a_n$ is an unbiased estimator of σ.

CONTROL LIMITS BASED ON THE SAMPLE STANDARD DEVIATIONS	$$\text{LCL} = \bar{\bar{x}} - 3 \cdot \frac{\bar{s}}{a_n \sqrt{n}}$$ $$\text{UCL} = \bar{\bar{x}} + 3 \cdot \frac{\bar{s}}{a_n \sqrt{n}}$$ where $$\bar{\bar{x}} = \frac{\sum_{i=1}^{k} \bar{x}_i}{k} \qquad \bar{s} = \frac{\sum_{i=1}^{k} s_i}{k}$$

Example 16.2 Referring to the viscosity data of Example 16.1, we had $n = 3$ and $k = 25$. The values of $\bar{x}_i$ and s_i ($i = 1, \ldots, 25$) appear in Table 16.1, from which it follows that $\bar{\bar{x}} = 261.896/25 = 10.476$ and $\bar{s} = 3.834/25 = .153$. With $a_3 = .886$, we have

$$\text{LCL} = 10.476 - 3 \cdot \frac{.153}{.886\sqrt{3}} = 10.476 - .299 = 10.177$$

$$\text{UCL} = 10.476 + 3 \cdot \frac{.153}{.886\sqrt{3}} = 10.476 + .299 = 10.775$$

These limits differ a bit from previous limits based on $\mu = 10.5$ and $\sigma = .18$ because now $\hat{\mu} = 10.476$ and $\hat{\sigma} = \bar{s}/a_3 = .173$. Inspection of Table 16.1 shows that every $\bar{x}_i$ is between these new limits, so again no out-of-control situation is evident. ∎

To obtain an estimate of σ based on the sample range, note that if $X_1, \ldots, X_n$ form a random sample from a normal distribution, then

$$R = \text{range}(X_1, \ldots, X_n)$$
$$= \max(X_1, \ldots, X_n) - \min(X_1, \ldots, X_n)$$
$$= \max(X_1 - \mu, \ldots, X_n - \mu) - \min(X_1 - \mu, \ldots, X_n - \mu)$$
$$= \sigma\left\{\max\left(\frac{X_1 - \mu}{\sigma}, \ldots, \frac{X_n - \mu}{\sigma}\right) - \min\left(\frac{X_1 - \mu}{\sigma}, \ldots, \frac{X_n - \mu}{\sigma}\right)\right\}$$
$$= \sigma \cdot \{\max(Z_1, \ldots, Z_n) - \min(Z_1, \ldots, Z_n)\}$$

where $Z_1, \ldots, Z_n$ are independent standard normal rv's. Thus,

$$E(R) = \sigma \cdot E(\text{range of a standard normal sample})$$
$$= \sigma \cdot b_n$$

so that R/b_n is an unbiased estimator of σ.

Now denote the ranges for the k samples in the quality control data set by $r_1, r_2, \ldots, r_k$. The argument just given implies that the estimate

$$\hat{\sigma} = \frac{\dfrac{1}{k}\sum_{i=1}^{k} r_i}{b_n} = \frac{\bar{r}}{b_n}$$

comes from an unbiased estimator for σ. Selected values of b_n appear in the accompanying table [their computation is based on using statistical theory and numerical integration to determine $E(\min(Z_1, \ldots, Z_n))$ and $E(\max(Z_1, \ldots, Z_n))$].

n	3	4	5	6	7	8
b_n	1.693	2.058	2.325	2.536	2.706	2.844

CONTROL LIMITS BASED ON THE SAMPLE RANGES

$$\text{LCL} = \bar{\bar{x}} - 3 \cdot \frac{\bar{r}}{b_n\sqrt{n}}$$

$$\text{UCL} = \bar{\bar{x}} + 3 \cdot \frac{\bar{r}}{b_n\sqrt{n}}$$

where $\bar{r} = \sum_{i=1}^{k} r_i/k$ and $r_1, \ldots, r_k$ are the k individual sample ranges.

Example 16.3
(Example 16.2 continued)

It is easily verified from Table 16.1 that $\bar{r} = .292$, so $\hat{\sigma} = .292/b_3 = .292/1.693 = .172$ and

$$\text{LCL} = 10.476 - 3 \cdot \frac{.292}{1.693\sqrt{3}} = 10.476 - .299 = 10.177$$

$$\text{UCL} = 10.476 + 3 \cdot \frac{.292}{1.693\sqrt{3}} = 10.476 + .299 = 10.775$$

These limits are identical to those based on $\bar{s}$, and again every $\bar{x}_i$ lies between the limits. ∎

Recomputing Control Limits

We have assumed that the sample data used for estimating μ and σ was obtained from an in-control process. Suppose, though, that one of the points on the resulting control chart falls outside the control limits. Then if an assignable cause for this out-of-control situation can be found and verified, it is recommended that new control limits be calculated after deleting the corresponding sample from the data set. Similarly, if more than one point falls outside the original limits, new limits should be determined after eliminating any such point for which an assignable cause can be identified and dealt with. It may even happen that one or more points fall outside the new limits, in which case the deletion/recomputation process must be repeated.

Performance Characteristics of Control Charts

Generally speaking, a control chart will be effective if it gives very few out-of-control signals when the process is in control, but shows a point outside the control limits almost as soon as the process goes out of control. One assessment of a chart's effectiveness is based on the notion of "error probabilities." Suppose the variable of interest is normally distributed with known σ (the same value for an in-control or out-of-control process). In addition, consider a 3-sigma chart based on the target value μ_0, with $\mu = \mu_0$ when the process is in control. One error probability is

$$\alpha = P(\text{a single sample gives a point outside the control limits when } \mu = \mu_0)$$
$$= P(\overline{X} > \mu_0 + 3\sigma/\sqrt{n} \text{ or } \overline{X} < \mu_0 - 3\sigma/\sqrt{n} \text{ when } \mu = \mu_0)$$
$$= P\left(\frac{\overline{X} - \mu_0}{\sigma/\sqrt{n}} > 3 \text{ or } \frac{\overline{X} - \mu_0}{\sigma/\sqrt{n}} < -3 \text{ when } \mu = \mu_0\right)$$

The standardized variable $Z = (\overline{X} - \mu_0)/(\sigma/\sqrt{n})$ has a standard normal distribution when $\mu = \mu_0$, so

$$\alpha = P(Z > 3 \text{ or } Z < -3) = \Phi(-3.00) + 1 - \Phi(3.00) = .0026$$

If 3.09 rather than 3 had been used to determine the control limits (this is customary in Great Britain), then

$$\alpha = P(Z > 3.09 \text{ or } Z < -3.09) = .0020$$

The use of 3-sigma limits makes it highly unlikely that an out-of-control signal will result from an in-control process.

Now suppose the process goes out of control because μ has shifted to $\mu + \Delta\sigma$ (Δ might be positive or negative); Δ is the number of standard deviations by which μ has changed. A second error probability is

$$\beta = P\left(\begin{array}{c}\text{a single sample gives a point inside} \\ \text{the control limits when } \mu = \mu_0 + \Delta\sigma\end{array}\right)$$
$$= P(\mu_0 - 3\sigma/\sqrt{n} < \overline{X} < \mu_0 + 3\sigma/\sqrt{n} \text{ when } \mu = \mu_0 + \Delta\sigma)$$

We now standardize by first subtracting $\mu_0 + \Delta\sigma$ from each term inside the parentheses and then dividing by $\sigma/\sqrt{n}$. This gives

$$\beta = P(-3 - \sqrt{n}\Delta < \text{standard normal rv} < 3 - \sqrt{n}\Delta)$$
$$= \Phi(3 - \sqrt{n}\Delta) - \Phi(-3 - \sqrt{n}\Delta)$$

This error probability depends on Δ, which determines the size of the shift, and on the sample size n. In particular, for fixed Δ, β will decrease as n increases (the larger the sample size, the more likely it is that an out-of-control signal will result), and for fixed n, β decreases as $|\Delta|$ increases (the larger the magnitude of a shift, the more likely it is that an out-of-control signal will result). The accompanying table gives β for selected values of Δ when $n = 4$:

Δ	.25	.50	.75	1.00	1.50	2.00	2.50	3.00
β when $n = 4$	.9936	.9772	.9332	.8413	.5000	.1587	.0668	.0013

It is clear that a small shift is quite likely to go undetected in a single sample.

If 3 is replaced by 3.09 in the control limits, then α decreases from .0026 to .002, but for any fixed n and σ, β will increase. This is just a manifestation of the inverse relationship between the two types of error probabilities in hypothesis testing. For example, changing 3 to 2.5 will increase α and decrease β.

The error probabilities discussed thus far are computed under the assumption that the variable of interest is normally distributed. If the distribution is only slightly nonnormal, the Central Limit Theorem effect implies that $\overline{X}$ will have approximately a normal distribution even when n is small, in which case the stated error probabilities will be approximately correct. This is, of course, no longer the case when the variable's distribution deviates considerably from normality.

A second performance assessment involves expected or average run length needed to observe an out-of-control signal. When the process is in control, we should expect to observe many samples before seeing one whose $\overline{x}$ lies outside the control limits. On the other hand, if a process goes out of control, the expected number of samples necessary to detect this should be small.

Let p denote the probability that a single sample yields an $\overline{x}$ value outside the control limits, that is,

$$p = P(\overline{X} < \mu_0 - 3\sigma/\sqrt{n} \text{ or } \overline{X} > \mu_0 + 3\sigma/\sqrt{n})$$

Consider first an in-control process, so that $\overline{X}_1, \overline{X}_2, \overline{X}_3, \ldots$ are all normally distributed with mean value μ_0 and standard deviation $\sigma/\sqrt{n}$. Define an rv Y by

$$Y = \text{the first } i \text{ for which } \overline{X}_i \text{ falls outside the control limits}$$

If we think of each sample number as a trial and an out-of-control sample as a success, then Y is the number of (independent) trials necessary to observe a success. This Y has a geometric distribution, and we showed in Example 3.18 that $E(Y) = 1/p$. The acronym ARL (for *average run length*) is often used in place of $E(Y)$. Because $p = \alpha$ for an in-control process, we have

$$\text{ARL} = E(Y) = \frac{1}{p} = \frac{1}{\alpha} = \frac{1}{.0026} = 384.62$$

Replacing 3 in the control limits by 3.09 gives ARL $= 1/.002 = 500$.

Now suppose that, at a particular time point, the process mean shifts to $\mu = \mu_0 + \Delta\sigma$. If we define Y to be the first i subsequent to the shift for which a sample generates an out-of-control signal, it is again true that ARL $= E(Y) = 1/p$, but now $p = 1 - \beta$. The accompanying table gives selected ARLs for a 3-sigma chart when $n = 4$. These results again show the chart's effectiveness in detecting large shifts but also its inability to quickly identify small shifts. When sampling is done rather infrequently, a great many items are likely to be produced before a small shift in μ is detected. The CUSUM procedures discussed in Section 16.5 were developed to address this deficiency.

Δ	.25	.50	.75	1.00	1.50	2.00	2.50	3.00
ARL when $n = 4$	156.25	43.86	14.97	6.30	2.00	1.19	1.07	1.0013

Supplemental Rules for $\overline{X}$ Charts

The inability of $\overline{X}$ charts with 3-sigma limits to quickly detect small shifts in the process mean has prompted investigators to develop procedures that provide improved behavior in this respect. One approach involves introducing additional conditions that cause an out-of-control signal to be generated. The following conditions were recommended by Western Electric (then a subsidiary of AT&T). An intervention to take corrective action is appropriate whenever one of these conditions is satisfied:

1. Two out of three successive points fall outside 2-sigma limits on the same side of the center line.

2. Four out of five successive points fall outside 1-sigma limits on the same side of the center line.

3. Eight successive points fall on the same side of the center line.

A quality control text should be consulted for a discussion of these and other supplemental rules. In Section 16.5, we present a different type of procedure to overcome the $\overline{X}$ chart's deficiency.

Robust Control Charts

The presence of outliers in the sample data tends to reduce the sensitivity of control-charting procedures when parameters must be estimated. This is because the control limits are moved outward from the center line, making the identification of unusual points more difficult. We do *not* want the statistic whose values are plotted to be resistant to outliers because that would mask any out-of-control signal. For example, plotting sample medians would be less effective than plotting $\overline{x}_1, \overline{x}_2, \ldots$ as is done on an $\overline{X}$ chart.

The article "Robust Control Charts" by David M. Rocke (*Technometrics,* 1989: 173–184) presents a study of procedures for which control limits are based on statistics resistant to the effects of outliers. Rocke recommends control limits calculated from the *interquartile range* (IQR), which is very similar to the fourth spread introduced in Chapter 1. In particular,

$$\text{IQR} = \begin{cases} (\text{2nd largest } x_i) - (\text{2nd smallest } x_i) & n = 4, 5, 6, 7 \\ (\text{3rd largest } x_i) - (\text{3rd smallest } x_i) & n = 8, 9, 10, 11 \end{cases}$$

For a random sample from a normal distribution, $E(\text{IQR}) = k_n \sigma$; the values of k_n are given in the accompanying table:

n	4	5	6	7	8
k_n	.596	.990	1.282	1.512	.942

The suggested control limits are

$$LCL = \bar{\bar{x}} - 3 \cdot \frac{IQR}{k_n\sqrt{n}} \qquad UCL = \bar{\bar{x}} + 3 \cdot \frac{IQR}{k_n\sqrt{n}}$$

The values of $\bar{x}_1, \bar{x}_2, \bar{x}_3, \ldots$ are plotted. Simulations reported in the article indicated that the performance of the chart with these limits is superior to that of the traditional $\overline{X}$ chart.

Exercises | Section 16.2 (4–13)

4. In the case of known μ and σ, what control limits are necessary for the probability of a single point being outside the limits for an in-control process to be .005?

5. Consider a 3-sigma control chart with center line at μ_0 and based on $n = 5$. Assuming normality, calculate the probability that a single point will fall outside the control limits when the actual process mean is

   a. $\mu_0 + .5\sigma$
   b. $\mu_0 - \sigma$
   c. $\mu_0 + 2\sigma$

6. The accompanying table gives data on moisture content for specimens of a certain type of fabric. Determine control limits for a chart with center line at height 13.00 based on $\sigma = .600$, construct the control chart, and comment on its appearance.

Data for Exercise 6

Sample No.	Moisture-Content Observations					$\bar{x}$	s	Range
1	12.2	12.1	13.3	13.0	13.0	12.72	.536	1.2
2	12.4	13.3	12.8	12.6	12.9	12.80	.339	.9
3	12.9	12.7	14.2	12.5	12.9	13.04	.669	1.7
4	13.2	13.0	13.0	12.6	13.9	13.14	.477	1.3
5	12.8	12.3	12.2	13.3	12.0	12.52	.526	1.3
6	13.9	13.4	13.1	12.4	13.2	13.20	.543	1.5
7	12.2	14.4	12.4	12.4	12.5	12.78	.912	2.2
8	12.6	12.8	13.5	13.9	13.1	13.18	.526	1.3
9	14.6	13.4	12.2	13.7	12.5	13.28	.963	2.4
10	12.8	12.3	12.6	13.2	12.8	12.74	.329	.9
11	12.6	13.1	12.7	13.2	12.3	12.78	.370	.9
12	13.5	12.3	12.8	13.1	12.9	12.92	.438	1.2
13	13.4	13.3	12.0	12.9	13.1	12.94	.559	1.4
14	13.5	12.4	13.0	13.6	13.4	13.18	.492	1.2
15	12.3	12.8	13.0	12.8	13.5	12.88	.432	1.2
16	12.6	13.4	12.1	13.2	13.3	12.92	.554	1.3
17	12.1	12.7	13.4	13.0	13.9	13.02	.683	1.8
18	13.0	12.8	13.0	13.3	13.1	13.04	.182	.5
19	12.4	13.2	13.0	14.0	13.1	13.14	.573	1.6
20	12.7	12.4	12.4	13.9	12.8	12.84	.619	1.5
21	12.6	12.8	12.7	13.4	13.0	12.90	.316	.8
22	12.7	13.4	12.1	13.2	13.3	12.94	.541	1.3

7. Refer to the data given in Exercise 6, and construct a control chart with an estimated center line and limits based on using the sample standard deviations to estimate σ. Is there any evidence that the process is out of control?

8. Refer to Exercises 6 and 7, and now employ control limits based on using the sample ranges to estimate σ. Does the process appear to be in control?

9. The accompanying table gives sample means and standard deviations, each based on $n = 6$ observations of the refractive index of fiber-optic cable. Construct a control chart and comment on its appearance. [*Hint:* $\Sigma \bar{x}_i = 2317.07$ and $\Sigma s_i = 30.34$.]

Day	$\bar{x}$	s	Day	$\bar{x}$	s
1	95.47	1.30	9	96.63	1.48
2	97.38	.88	10	96.50	.80
3	96.85	1.43	11	97.22	1.42
4	96.64	1.59	12	96.55	1.65
5	96.87	1.52	13	97.02	1.28
6	96.52	1.27	14	95.55	1.14
7	96.08	1.16	15	96.29	1.37
8	96.48	.79	16	96.80	1.40

Day	$\bar{x}$	s	Day	$\bar{x}$	s
17	96.01	1.58	21	97.06	1.34
18	95.39	.98	22	98.34	1.60
19	96.58	1.21	23	96.42	1.22
20	96.43	.75	24	95.99	1.18

10. Refer to Exercise 9. An assignable cause was found for the unusually high sample average refractive index on day 22. Recompute control limits after deleting the data from this day. What do you conclude?

11. Consider the control chart based on control limits $\mu_0 \pm 2.81 \, \sigma/\sqrt{n}$.
    a. What is the ARL when the process is in control?
    b. What is the ARL when $n = 4$ and the process mean has shifted to $\mu = \mu_0 + \sigma$?
    c. How do the values of parts (a) and (b) compare to the corresponding values for a 3-sigma chart?

12. Apply the supplemental rules suggested in the text to the data of Exercise 6. Are there any out-of-control signals?

13. Calculate control limits for the data of Exercise 6, using the robust procedure presented in this section.

16.3 | Control Charts for Process Variation

The control charts discussed in the previous section were designed to control the location (equivalently, central tendency) of a process, with particular attention to the mean as a measure of location. It is equally important to ensure that a process is under control with respect to variation. In fact, most practitioners recommend that control be established on variation *prior to* constructing an $\overline{X}$ chart or any other chart for controlling location. In this section, we consider charts for variation based on the sample standard deviation S and also charts based on the sample range R. The former are generally preferred because the standard deviation gives a more efficient assessment of variation than does the range, but R charts were used first and tradition dies hard.

The S Chart

We again suppose that k independently selected samples are available, each one consisting of n observations on a normally distributed variable. Denote the sample standard deviations by $s_1, s_2, \ldots, s_k$, with $\bar{s} = \Sigma s_i/k$. The values $s_1, s_2, s_3, \ldots$ are plotted in sequence on an S chart. The center line of the chart will be at height $\bar{s}$, and the 3-sigma limits necessitate determining $3\sigma_S$ (just as 3-sigma limits of an $\overline{X}$ chart required $3\sigma_{\overline{x}} = 3\sigma/\sqrt{n}$, with σ then estimated from the data).

Recall that for any rv Y, $V(Y) = E(Y^2) - [E(Y)]^2$, and that a sample variance S^2 is an unbiased estimator of σ^2, that is, $E(S^2) = \sigma^2$. Thus,

$$V(S) = E(S^2) - [E(S)]^2 = \sigma^2 - (a_n\sigma)^2 = \sigma^2(1 - a_n^2)$$

where values of a_n for $n = 3, \ldots, 8$ are tabulated in the previous section. The standard deviation of S is then

$$\sigma_S = \sqrt{V(S)} = \sigma\sqrt{1 - a_n^2}$$

It is natural to estimate σ using $s_1, \ldots, s_k$ as was done in the previous section, namely, $\hat{\sigma} = \bar{s}/a_n$. Substituting $\hat{\sigma}$ for σ in the expression for σ_S gives the quantity used to calculate 3-sigma limits.

The 3-sigma control limits for an S control chart are

$$LCL = \bar{s} - 3\bar{s}\sqrt{1 - a_n^2}/a_n$$
$$UCL = \bar{s} + 3\bar{s}\sqrt{1 - a_n^2}/a_n$$

The expression for LCL will be negative if $n \leq 5$, in which case it is customary to use LCL $= 0$.

Example 16.4 Table 16.2 displays observations on stress resistance of plastic sheets (the force, in psi, necessary to crack a sheet). There are $k = 22$ samples, obtained at equally spaced time points, and $n = 4$ observations in each sample. It is easily verified that $\sum s_i = 51.10$ and $\bar{s} = 2.32$, so the center of the S chart will be at 2.32 (though because $n = 4$, LCL $= 0$ and the center line will not be midway between the control limits). From the previous section, $a_4 = .921$, from which the UCL is

$$UCL = 2.32 + 3(2.32)(\sqrt{1 - (.921)^2})/.921$$
$$= 2.32 + 2.94$$
$$= 5.26$$

Table 16.2 Stress-resistance data for Example 16.4

Sample No.	Observations				SD	Range
1	29.7	29.0	28.8	30.2	.64	1.4
2	32.2	29.3	32.2	32.9	1.60	3.6
3	35.9	29.1	32.1	31.3	2.83	6.8
4	28.8	27.2	28.5	35.7	3.83	8.5
5	30.9	32.6	28.3	28.3	2.11	4.3
6	30.6	34.3	34.8	26.3	3.94	8.5
7	32.3	27.7	30.9	27.8	2.30	4.6
8	32.0	27.9	31.0	30.8	1.76	4.1
9	24.2	27.5	28.5	31.1	2.85	6.9
10	33.7	24.4	34.3	31.0	4.53	9.9
11	35.3	33.2	31.4	28.0	3.09	7.3

(continued)

Sample No.	Observations				SD	Range
12	28.1	34.0	31.0	30.8	2.41	5.9
13	28.7	28.9	25.8	29.7	1.71	3.9
14	29.0	33.0	30.2	30.1	1.71	4.0
15	33.5	32.6	33.6	29.2	2.07	4.4
16	26.9	27.3	32.1	28.5	2.37	5.2
17	30.4	29.6	31.0	33.8	1.83	4.2
18	29.0	28.9	31.8	26.7	2.09	5.1
19	33.8	30.9	31.7	28.2	2.32	5.6
20	29.7	27.9	29.1	30.1	.96	2.2
21	27.9	27.7	30.2	32.9	2.43	5.2
22	30.0	31.4	27.7	28.1	1.72	3.7

The resulting control chart is shown in Figure 16.3. All plotted points are well within the control limits, suggesting stable process behavior with respect to variation.

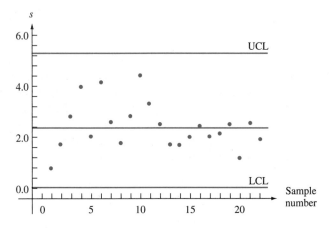

Figure 16.3 *S* chart for stress-resistance data for Example 16.4 ∎

The *R* Chart

Let $r_1, r_2, \ldots, r_k$ denote the k sample ranges and $\bar{r} = \sum r_i/k$. The center line of an R chart will be at height $\bar{r}$. Determination of the control limits requires σ_R, where R denotes the range (prior to making observations—as a random variable) of a random sample of size n from a normal distribution with mean value μ and standard deviation σ. Because

$$R = \max(X_1, \ldots, X_n) - \min(X_1, \ldots, X_n)$$
$$= \sigma\{\max(Z_1, \ldots, Z_n) - \min(Z_1, \ldots, Z_n)\}$$

where $Z_i = (X_i - \mu)/\sigma$, and the Z_i's are standard normal rv's, it follows that

$$\sigma_R = \sigma \cdot \left(\begin{array}{l}\text{standard deviation of the range of a random sample} \\ \text{of size } n \text{ from a standard normal distribution}\end{array}\right)$$

$$= \sigma \cdot c_n$$

The values of c_n for $n = 3, \ldots , 8$ appear in the accompanying table:

n	3	4	5	6	7	8
c_n	.888	.880	.864	.848	.833	.820

It is customary to estimate σ by $\hat{\sigma} = \bar{r}/b_n$ as discussed in the previous section. This gives $\hat{\sigma}_R = c_n\bar{r}/b_n$ as the estimated standard deviation of R.

The 3-sigma limits for an R chart are

$$\text{LCL} = \bar{r} - 3c_n\bar{r}/b_n$$
$$\text{UCL} = \bar{r} + 3c_n\bar{r}/b_n$$

The expression for LCL will be negative if $n \le 6$, in which case LCL = 0 should be used.

Example 16.5
(Example 16.4 continued)

Table 16.2 yields $\sum r_i = 115.3$ and $\bar{r} = 5.24$. Since $n = 4$, LCL = 0. With $b_4 = 2.058$ and $c_4 = .880$,

$$\text{UCL} = 5.24 + 3(.880)(5.24)/2.058 = 11.96$$

The R chart appears as Figure 16.4. As with the S chart, all points are between the limits, indicating an in-control process as far as variation is concerned.

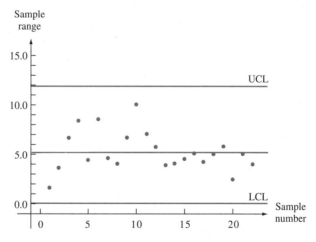

Figure 16.4 R chart for stress-resistance data of Example 16.5

Charts Based on Probability Limits

Consider an $\bar{X}$ chart based on the in-control (target) value μ_0 and known σ. When the variable of interest is normally distributed and the process is in control,

$$P(\bar{X}_i > \mu_0 + 3\sigma/\sqrt{n}) = .0013 = P(\bar{X}_i < \mu_0 - 3\sigma/\sqrt{n})$$

That is, the probability that a point on the chart falls above the UCL is .0013, as is the probability that the point falls below the LCL (using 3.09 in place of 3 gives .001 for each probability). When control limits are based on estimates of μ and σ, these probabilities will be approximately correct provided that n is not too small and k is at least 20.

By contrast, it is *not* the case for a 3-sigma S chart that $P(S_i > \text{UCL}) = P(S_i < \text{LCL}) = .0013$, nor is it true for a 3-sigma R chart that $P(R_i > \text{UCL}) = P(R_i < \text{LCL}) = .0013$. This is because neither the sample standard deviation S nor the sample range R has a normal distribution even when the population distribution is normal. Instead, both S and R have skewed distributions. The best that can be said for 3-sigma S and R charts is that an in-control process is quite unlikely to yield a point at any particular time that is outside the control limits. Some authors have advocated the use of control limits for which the "exceedance probability" for each limit is approximately .001. The book *Statistical Methods for Quality Improvement* (see the chapter bibliography) contains more information on this topic.

Exercises | Section 16.3 (14–18)

14. A manufacturer of dustless chalk instituted a quality control program to monitor chalk density. The sample standard deviations of densities for 24 different subgroups, each consisting of $n = 8$ chalk specimens, were as follows:

.204 .315 .096 .184 .230 .212 .322 .287

.145 .211 .053 .145 .272 .351 .159 .214

.388 .187 .150 .229 .276 .118 .091 .056

Calculate limits for an S chart, construct the chart, and check for out-of-control points. If there is an out-of-control point, delete it and repeat the process.

15. Subgroups of power supply units are selected once each hour from an assembly line, and the high-voltage output of each unit is determined.
  a. Suppose the sum of the resulting sample ranges for 30 subgroups, each consisting of four units, is 85.2. Calculate control limits for an R chart.
  b. Repeat part (a) if each subgroup consists of eight units and the sum is 106.2.

16. Calculate control limits for both an S chart and an R chart using the moisture content data from Exercise 6. Then check for the presence of any out-of-control signals.

17. Calculate control limits for an S chart from the refractive index data of Exercise 9. Does the process appear to be in control with respect to variability? Why or why not?

18. When S^2 is the sample variance of a normal random sample, $(n - 1)S^2/\sigma^2$ has a chi-squared distribution with $n - 1$ df, so

$$P\left(\chi^2_{.999,n-1} < \frac{(n-1)S^2}{\sigma^2} < \chi^2_{.001,n-1}\right) = .998$$

from which

$$P\left(\frac{\sigma^2\chi^2_{.999,n-1}}{n-1} < S^2 < \frac{\sigma^2\chi^2_{.001,n-1}}{n-1}\right) = .998$$

This suggests that an alternative chart for controlling process variation involves plotting the sample variances and using the control limits

$$\text{LCL} = \overline{s^2}\chi^2_{.999,n-1}/(n-1)$$
$$\text{UCL} = \overline{s^2}\chi^2_{.001,n-1}/(n-1)$$

Construct the corresponding chart for the data of Exercise 9. [*Hint:* The lower- and upper-tailed chi-squared critical values for 5 df are .210 and 20.515, respectively.]

16.4 | **Control Charts for Attributes**

The term *attribute data* is used in the quality control literature to describe two situations:

1. Each item produced is either defective or nondefective (conforms to specifications or does not).

2. A single item may have one or more defects, and the number of defects is determined.

In the former case, a control chart is based on the binomial distribution; in the latter case, the Poisson distribution is the basis for a chart.

The *p* Chart for Fraction Defective

Suppose that when a process is in control, the probability that any particular item is defective is p (equivalently, p is the long-run proportion of defective items for an in-control process) and that different items are independent of one another with respect to their conditions. Consider a sample of n items obtained at a particular time, and let X be the number of defectives and $\hat{p} = X/n$. Because X has a binomial distribution, $E(X) = np$ and $V(X) = np(1 - p)$, so

$$E(\hat{p}) = p \qquad V(\hat{p}) = \frac{p(1 - p)}{n}$$

Also, if $np \geq 10$ and $n(1 - p) \geq 10$, $\hat{p}$ has approximately a normal distribution.

In the case of known p (or a chart based on target value), the control limits are

$$LCL = p - 3\sqrt{\frac{p(1 - p)}{n}} \qquad UCL = p + 3\sqrt{\frac{p(1 - p)}{n}}$$

If each sample consists of n items, the number of defective items in the ith sample is x_i, and $\hat{p}_i = x_i/n$, then $\hat{p}_1, \hat{p}_2, \hat{p}_3, \ldots$ are plotted on the control chart.

Usually the value of p must be estimated from the data. Suppose that k samples from what is believed to be an in-control process are available, and let

$$\bar{p} = \frac{\sum_{i=1}^{k} \hat{p}_i}{k}$$

The estimate $\bar{p}$ is then used in place of p in the aforementioned control limits.

The p chart for the fraction of defective items has its center line at height $\bar{p}$ and control limits

$$LCL = \bar{p} - 3\sqrt{\frac{\bar{p}(1 - \bar{p})}{n}}$$

$$UCL = \bar{p} + 3\sqrt{\frac{\bar{p}(1 - \bar{p})}{n}}$$

If LCL is negative, it is replaced by 0.

Example 16.6 A sample of 100 cups from a particular dinnerware pattern was selected on each of 25 successive days, and each was examined for defects. The resulting numbers of unacceptable cups and corresponding sample proportions are as follows:

Day (i)	1	2	3	4	5	6	7	8	9	10	11	12	13
x_i	7	4	3	6	4	9	6	7	5	3	7	8	4
$\hat{p}_i$	.07	.04	.03	.06	.04	.09	.06	.07	.05	.03	.07	.08	.04

Day (i)	14	15	16	17	18	19	20	21	22	23	24	25
x_i	6	2	9	7	6	7	11	6	7	4	8	6
$\hat{p}_i$	.06	.02	.09	.07	.06	.07	.11	.06	.07	.04	.08	.06

Assuming that the process was in control during this period, let's establish control limits and construct a p chart. We have that $\sum \hat{p}_i = 1.52$, giving $\bar{p} = 1.52/25 = .0608$ and

$$\text{LCL} = .0608 - 3\sqrt{(.0608)(.9392)/100} = .0608 - .0717 = -.0109$$
$$\text{UCL} = .0608 + 3\sqrt{(.0608)(.9392)/100} = .0608 + .0717 = .1325$$

The LCL is therefore set at 0. The control chart pictured in Figure 16.5 shows that all points are within the control limits. This is consistent with the assumption of an in-control process.

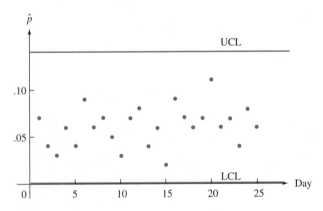

Figure 16.5 Control chart for fraction-defective data of Example 16.6 ■

The c Chart for Number of Defectives

We now consider situations in which the observation at each time point is the number of defects in a unit of some sort. The unit may consist of a single item (e.g., one automobile) or a group of items (e.g., blemishes on a set of four tires). In the second case, the group size is assumed to be the same at each time point.

The control chart for number of defectives is based on the Poisson probability distribution. Recall that if Y is a Poisson random variable with parameter θ, then

$$E(Y) = \theta \qquad V(Y) = \theta \qquad \sigma_Y = \sqrt{\theta}$$

Also, Y has approximately a normal distribution when θ is large ($\theta \geq 10$ will suffice for most purposes). Furthermore, if $Y_1, Y_2, \ldots, Y_n$ are independent Poisson variables with parameters $\theta_1, \theta_2, \ldots, \theta_n$, it can be shown that $Y_1 + \cdots + Y_n$ has a Poisson distribution with parameter $\theta_1 + \cdots + \theta_n$. In particular, if $\theta_1 = \cdots = \theta_n = \theta$ (the distribution of the number of defects per item is the same for each item), then the Poisson parameter is $\lambda = n\theta$.

Let λ denote the Poisson parameter for the number of defects in a unit (it is the expected number of defects per unit). In the case of known λ (or a chart based on a target value),

$$\text{LCL} = \lambda - 3\sqrt{\lambda} \qquad \text{UCL} = \lambda + 3\sqrt{\lambda}$$

With x_i denoting the total number of defects in the ith unit ($i = 1, 2, 3, \ldots$), then points at heights $x_1, x_2, x_3, \ldots$ are plotted on the chart.

Usually the value of λ must be estimated from the data. Since $E(X_i) = \lambda$, it is natural to use the estimate $\hat{\lambda} = \bar{x}$ (based on $x_1, x_2, \ldots, x_k$).

The c chart for the number of defectives in a unit has center line at height $\bar{x}$ and

$$\text{LCL} = \bar{x} - 3\sqrt{\bar{x}}$$
$$\text{UCL} = \bar{x} + 3\sqrt{\bar{x}}$$

If LCL is negative, it is replaced by 0.

Example 16.7 A company manufactures metal panels that are baked after first being coated with a slurry of powdered ceramic. Flaws sometimes appear in the finish of these panels, and the company wishes to establish a control chart for the number of flaws. The numbers of flaws in each of the 24 panels sampled at regular time intervals are as follows:

$$7 \ \ 10 \ \ \ 9 \ \ 12 \ \ 13 \ \ 6 \ \ 13 \ \ \ 7 \ \ 5 \ \ 11 \ \ \ 8 \ \ 10$$
$$13 \ \ \ 9 \ \ 21 \ \ 10 \ \ \ 6 \ \ 8 \ \ \ 3 \ \ 12 \ \ 7 \ \ 11 \ \ 14 \ \ 10$$

with $\Sigma x_i = 235$ and $\hat{\lambda} = \bar{x} = 235/24 = 9.79$. The control limits are

$$\text{LCL} = 9.79 - 3\sqrt{9.79} = .40 \qquad \text{UCL} = 9.79 + 3\sqrt{9.79} = 19.18$$

The control chart is in Figure 16.6 (page 698). The point corresponding to the fifteenth panel lies above the UCL. Upon investigation, the slurry used on that panel was discovered to be of unusually low viscosity (an assignable cause). Eliminating that observation from the data set gives $\bar{x} = 214/23 = 9.30$ and new control limits

$$\text{LCL} = 9.30 - 3\sqrt{9.30} = .15 \qquad \text{UCL} = 9.30 + 3\sqrt{9.30} = 18.45$$

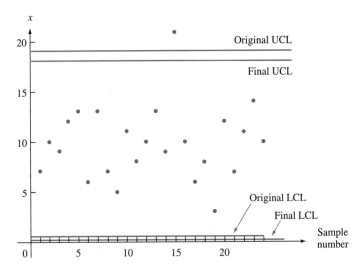

Figure 16.6 Control chart for number of flaws data of Example 16.7

The remaining 23 observations all lie between these limits, indicating an in-control process. ■

Control Charts Based on Transformed Data

The use of 3-sigma control limits is presumed to result in $P(\text{statistic} < \text{LCL}) \approx P(\text{statistic} > \text{UCL}) \approx .0013$ when the process is in control. However, when p is small, the normal approximation to the distribution of $\hat{p} = X/n$ will often not be very accurate in the extreme tails. Table 16.3 gives evidence of this behavior for selected values of p and n (the value of p is used to calculate the control limits). In many cases, the probability that a single point falls outside the control limits is very different from the nominal probability of .0026.

Table 16.3 In-control probabilities for a p chart

p	n	$P(\hat{p} < \text{LCL})$	$P(\hat{p} > \text{UCL})$	$P(\text{out-of-control point})$
.10	100	.00003	.00198	.00201
.10	200	.00048	.00299	.00347
.10	400	.00044	.00171	.00215
.05	200	.00004	.00266	.00270
.05	400	.00020	.00207	.00227
.05	600	.00031	.00189	.00220
.02	600	.00007	.00275	.00282
.02	800	.00036	.00374	.00410
.02	1000	.00023	.00243	.00266

This problem can be remedied by applying a transformation to the data. Let $h(X)$ denote a function applied to transform the binomial variable X. Then $h(\cdot)$ should be chosen so that $h(X)$ has approximately a normal distribution *and* this approximation is accurate in the tails. A recommended transformation is based on the arcsin (i.e., $\sin^{-1}$) function:

$$Y = h(X) = \sin^{-1}(\sqrt{X/n})$$

Then Y is approximately normal with mean value $\sin^{-1}(\sqrt{p})$ and variance $1/(4n)$; note that the variance is independent of p. Let $y_i = \sin^{-1}(\sqrt{x_i/n})$. Then points on the control chart are plotted at heights $y_1, y_2, \ldots$. For known n, the control limits are

$$\text{LCL} = \sin^{-1}(\sqrt{p}) - 3\sqrt{1/(4n)} \qquad \text{UCL} = \sin^{-1}(\sqrt{p}) + 3\sqrt{1/(4n)}$$

When p is not known, $\sin^{-1}(\sqrt{p})$ is replaced by $\bar{y}$.

Similar comments apply to the Poisson distribution when λ is small. The suggested transformation is $Y = h(X) = 2\sqrt{X}$, which has mean value $2\sqrt{\lambda}$ and variance 1. Resulting control limits are $2\sqrt{\lambda} \pm 3$ when λ is known and $\bar{y} \pm 3$ otherwise. The book *Statistical Methods for Quality Improvement* listed in the chapter bibliography discusses these issues in greater detail.

Exercises | Section 16.4 (19–26)

19. On each of the previous 25 days, 100 electronic devices of a certain type were randomly selected and subjected to a severe heat stress test. The total number of items that failed to pass the test was 578.
    a. Determine control limits for a 3-sigma p chart.
    b. The highest number of failed items on a given day was 39, and the lowest number was 13. Does either of these correspond to an out-of-control point? Explain.

20. A sample of 200 ROM computer chips was selected on each of 30 consecutive days, and the number of nonconforming chips on each day was as follows: 10, 18, 24, 17, 37, 19, 7, 25, 11, 24, 29, 15, 16, 21, 18, 17, 15, 22, 12, 20, 17, 18, 12, 24, 30, 16, 11, 20, 14, 28. Construct a p chart and examine it for any out-of-control points.

21. When $n = 150$, what is the smallest value of $\bar{p}$ for which the LCL in a p chart is positive?

22. Refer to the data of Exercise 20, and construct a control chart using the $\sin^{-1}$ transformation as suggested in the text.

23. The accompanying observations are numbers of defects in 25 1-square-yard specimens of woven fabric of a certain type: 3, 7, 5, 3, 4, 2, 8, 4, 3, 3, 6, 7, 2, 3, 2, 4, 7, 3, 2, 4, 4, 1, 5, 4, 6. Construct a c chart for the number of defects.

24. For what $\bar{x}$ values will the LCL in a c chart be negative?

25. In some situations, the sizes of sampled specimens vary, and larger specimens are expected to have more defects than smaller ones. For example, sizes of fabric samples inspected for flaws might vary over time. Alternatively, the number of items inspected might change with time. Let

$$u_i = \frac{\text{the number of defects observed at time } i}{\text{size of entity inspected at time } i}$$
$$= \frac{x_i}{g_i}$$

where "size" might refer to area, length, volume, or simply the number of items inspected. Then a **u chart** plots $u_1, u_2, \ldots$, has center line $\bar{u}$, and the control limits for the ith observations are $\bar{u} \pm 3\sqrt{\bar{u}/g_i}$.

Painted panels were examined in time sequence, and for each one, the number of blemishes in a specified sampling region was determined. The surface area (ft²) of the region examined varied from panel to panel. Results are given at the top of page 700. Construct a u chart.

Panel	Area Examined	No. of Blemishes		Panel	Area Examined	No. of Blemishes
1	.8	3		13	.8	5
2	.6	2		14	1.0	4
3	.8	3		15	1.0	6
4	.8	2		16	1.0	12
5	1.0	5		17	.8	3
6	1.0	5		18	.6	3
7	.8	10		19	.6	5
8	1.0	12		20	.6	1
9	.6	4				
10	.6	2				
11	.6	1		**26.** Construct a control chart for the data of Exercise 23		
12	.8	3		by using the transformation suggested in the text.		

16.5 | CUSUM Procedures

A defect of the traditional X chart is its inability to detect a relatively small change in a process mean. This is largely a consequence of the fact that whether a process is judged out of control at a particular time depends only on the sample at that time, and not on the past history of the process. **Cumulative sum (CUSUM)** control charts and procedures have been designed to remedy this defect.

There are two equivalent versions of a CUSUM procedure for a process mean, one graphical and the other computational. The computational version is used almost exclusively in practice, but the logic behind the procedure is most easily grasped by first considering the graphical form.

The V-Mask

Let μ_0 denote a target value or goal for the process mean and define *cumulative sums* by

$$S_1 = \bar{x}_1 - \mu_0$$
$$S_2 = (\bar{x}_1 - \mu_0) + (\bar{x}_2 - \mu_0) = \sum_{i=1}^{2}(\bar{x}_i - \mu_0)$$
$$\vdots$$
$$S_l = (\bar{x}_1 - \mu_0) + \cdots + (\bar{x}_l - \mu_0) = \sum_{i=1}^{l}(\bar{x}_i - \mu_0)$$

(in the absence of a target value, $\bar{\bar{x}}$ is used in place of μ_0). These cumulative sums are plotted over time. That is, at time l, we plot a point at height S_l. At the current time point r, the plotted points are $(1, S_1), (2, S_2), (3, S_3), \ldots, (r, S_r)$.

Now a V-shaped "mask" is superimposed on the plot, as shown in Figure 16.7. The point 0, which lies a distance d behind the point at which the two arms of the mask intersect, is positioned at the current CUSUM point (r, S_r). At time r, the process is judged out of control if any of the plotted points lies outside the V-mask—either above the upper arm or below the lower arm. When the process is in control, the $\bar{x}_i$'s will vary around the target

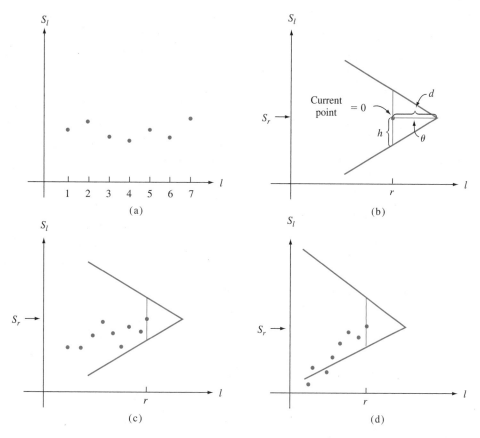

Figure 16.7 CUSUM plots: (a) successive points (l, S_l) in a CUSUM plot; (b) a V-mask with 0 = (r, S_r); (c) an in-control process; (d) an out-of-control process

value μ_0, so successive S_i's should vary around 0. Suppose, however, that at a certain time, the process mean shifts to a value larger than the target. From that point on, differences $\bar{x}_i - \mu_0$ will tend to be positive, so that successive S_l's will increase and plotted points will drift upward. If a shift has occurred prior to the current time point r, there is a good chance that (r, S_r) will be substantially higher than some other points in the plot, in which case these other points will be below the lower arm of the mask. Similarly, a shift to a value smaller than the target will subsequently result in points above the upper arm of the mask.

Any particular V-mask is determined by specifying the "lead distance" d and "half-angle" θ, or, equivalently, by specifying d and the length h of the vertical line segment from 0 to the lower (or to the upper) arm of the mask. One method for deciding which mask to use involves specifying the size of a shift in the process mean that is of particular concern to an investigator. Then the parameters of the mask are chosen to give desired values of α and β, the false-alarm probability and the probability of not detecting the specified shift, respectively. An alternative method involves selecting the mask that yields specified values of the ARL (average run length) both for an in-control process and for a process in which the mean has shifted by a designated amount. After developing the computational form of the CUSUM procedure, we will illustrate the second method of construction.

Example 16.8 A wood products company manufactures charcoal briquettes for barbecues. It packages these briquettes in bags of various sizes, the largest of which is supposed to contain 40 lb. Table 16.4 displays the weights of bags from 16 different samples, each of size $n = 4$. The first 10 of these were drawn from a normal distribution with $\mu = \mu_0 = 40$ and $\sigma = .5$. Starting with the eleventh sample, the mean has shifted upward to $\mu = 40.3$.

Table 16.4 Observations, $\bar{x}$'s, and cumulative sums for Example 16.8

Sample Number	Observations				$\bar{x}$	$\sum(\bar{x}_i - 40)$
1	40.77	39.95	40.86	39.21	40.20	.20
2	38.94	39.70	40.37	39.88	39.72	−.08
3	40.43	40.27	40.91	40.05	40.42	.34
4	39.55	40.10	39.39	40.89	39.98	.32
5	41.01	39.07	39.85	40.32	40.06	.38
6	39.06	39.90	39.84	40.22	39.76	.14
7	39.63	39.42	40.04	39.50	39.65	−.21
8	41.05	40.74	40.43	39.40	40.41	.20
9	40.28	40.89	39.61	40.48	40.32	.52
10	39.28	40.49	38.88	40.72	39.84	.36
11	40.57	40.04	40.85	40.51	40.49	.85
12	39.90	40.67	40.51	40.53	40.40	1.25
13	40.70	40.54	40.73	40.45	40.61	1.86
14	39.58	40.90	39.62	39.83	39.98	1.84
15	40.16	40.69	40.37	39.69	40.23	2.07
16	40.46	40.21	40.09	40.58	40.34	2.41

Figure 16.8 displays an $\bar{X}$ chart with center line at height 40 and control limits at

$$\mu_0 \pm 3\sigma_{\bar{x}} = 40 \pm 3 \cdot (.5/\sqrt{4}) = 40 \pm .75$$

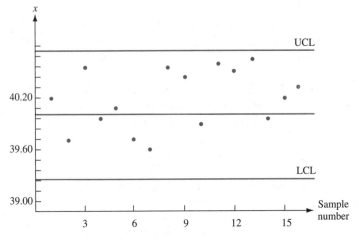

Figure 16.8 $\bar{X}$ control chart for the data of Example 16.8

No point on the chart lies outside the control limits. This chart suggests a stable process for which the mean has remained on target.

Figure 16.9 shows CUSUM plots with a particular V-mask superimposed. The plot in Figure 16.9(a) is for current time $r = 12$. All points in this plot lie inside the arms of the mask. However, the plot for $r = 13$ displayed in Figure 16.9(b) gives an out-of-control signal. The point falling below the lower arm of the mask suggests an increase in the value of the process mean. The mask at $r = 16$ is even more emphatic in its out-of-control message. This is in marked contrast to the ordinary $\overline{X}$ chart.

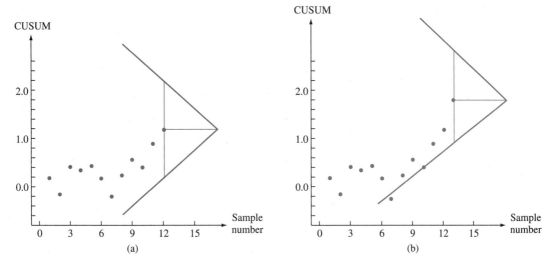

Figure 16.9 CUSUM plots and V-masks for data of Example 16.8: (a) V-mask at time $r = 12$; process in control; (b) V-mask at time $r = 13$; out-of-control signal ∎

Computational Form of the CUSUM Procedure

We first describe the computational version of the CUSUM procedure, then show its equivalence to the graphical form, and finally discuss designing a procedure to meet specified criteria.

Let $d_0 = e_0 = 0$, and calculate $d_1, d_2, d_3, \ldots$ and $e_1, e_2, e_3, \ldots$ recursively using the relationships

$$d_l = \max[0, d_{l-1} + (\overline{x}_l - (\mu_0 + k))]$$
$$e_l = \max[0, e_{l-1} - (\overline{x}_l - (\mu_0 - k))] \qquad (l = 1, 2, 3, \ldots)$$

Here the symbol k denotes the slope of the lower arm of the V-mask, and its value is customarily taken as $\Delta/2$ (where Δ is the size of a shift in μ on which attention is focused).

If at current time r, either $d_r > h$ or $e_r > h$, the process is judged to be out of control. The first inequality suggests the process mean has shifted to a value greater than the target, whereas $e_r > h$ indicates a shift to a smaller value.

Example 16.9   Reconsider the charcoal briquette data displayed in Table 16.4 of Example 16.8. The target value is $\mu_0 = 40$, and the size of a shift to be quickly detected is $\Delta = .3$. Thus,

$$k = \frac{\Delta}{2} = .15 \qquad \mu_0 + k = 40.15 \qquad \mu_0 - k = 39.85$$

so

$$d_l = \max[0, d_{l-1} + (\bar{x}_l - 40.15)]$$
$$e_l = \max[0, e_{l-1} - (\bar{x}_l - 39.85)]$$

Calculations of the first few d_l's proceeds as follows:

$$d_0 = 0$$
$$d_1 = \max[0, d_0 + (\bar{x}_1 - 40.15)]$$
$$= \max[0, 0 + (40.20 - 40.15)]$$
$$= .05$$
$$d_2 = \max[0, d_1 + (\bar{x}_2 - 40.15)]$$
$$= \max[0, .05 + (39.72 - 40.15)]$$
$$= 0$$
$$d_3 = \max[0, d_2 + (\bar{x}_3 - 40.15)]$$
$$= \max[0, 0 + (40.42 - 40.15)]$$
$$= .27$$

The remaining calculations are summarized in Table 16.5.

Table 16.5   CUSUM calculations for Example 16.9

Sample Number	$\bar{x}_l$	$\bar{x}_l - 40.15$	d_l	$\bar{x}_l - 39.85$	e_l
1	40.20	.05	.05	.35	0
2	39.72	−.43	0	−.13	.13
3	40.42	.27	.27	.57	0
4	39.98	−.17	.10	.13	0
5	40.06	−.09	.01	.21	0
6	39.76	−.39	0	−.09	.09
7	39.65	−.50	0	−.20	.29
8	40.41	.26	.26	.56	.0
9	40.32	.17	.43	.47	0
10	39.84	−.31	.12	−.01	.01
11	40.49	.34	.46	.64	0
12	40.40	.25	.71	.55	0
13	40.61	.46	1.17	.76	0
14	39.98	−.17	1.00	.13	0
15	40.23	.08	1.08	.38	0
16	40.34	.19	1.27	.49	0

The value $h = .95$ gives a CUSUM procedure with desirable properties—false alarms (incorrect out-of-control signals) rarely occur, yet a shift of $\Delta = .3$ will usually be detected rather quickly. With this value of h, the first out-of-control signal comes after the thirteenth sample is available. Since $d_{13} = 1.17 > .95$, it appears that the mean has shifted to a value larger than the target. This is the same message as the one given by the V-mask in Figure 16.9(b). ■

Equivalence of the V-Mask and Computational Form

Again let r denote the current time point, so that $\bar{x}_1, \bar{x}_2, \ldots, \bar{x}_r$ are available. Figure 16.10 displays a V-mask with the point labeled 0 at (r, S_r). The slope of the lower arm, which we denote by k, is h/d. Thus, the points on the lower arm above $r, r - 1, r - 2, \ldots$ are at heights $S_r - h, S_r - h - k, S_r - h - 2k$, and so on.

The process is in control if all points are on or between the arms of the mask. We wish to describe this condition algebraically. To do so, let

$$T_l = \sum_{i=1}^{l} [\bar{x}_i - (\mu_0 + k)] \qquad l = 1, 2, 3, \ldots, r$$

The conditions under which all points are on or above the lower arm are

$$S_r - h \le S_r \qquad \text{(trivially satisfied) i.e., } S_r \le S_r + h$$
$$S_r - h - k \le S_{r-1} \qquad\qquad\qquad \text{i.e., } S_r \le S_{r-1} + h + k$$
$$S_r - h - 2k \le S_{r-2} \qquad\qquad\qquad \text{i.e., } S_r \le S_{r-2} + h + 2k$$

$$\vdots \qquad\qquad\qquad\qquad\qquad \vdots \qquad \vdots$$

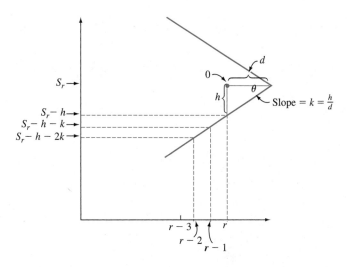

Figure 16.10   A V-mask with slope of lower arm $= k$

Now subtract rk from both sides of each inequality to obtain

$$S_r - rk \leq S_r - rk + h \qquad \text{i.e., } T_r \leq T_r + h$$
$$S_r - rk \leq S_{r-1} - (r-1)k + h \quad \text{i.e., } T_r \leq T_{r-1} + h$$
$$S_r - rk \leq S_{r-2} - (r-2)k + h \quad \text{i.e., } T_r \leq T_{r-2} + h$$
$$\cdot \qquad\qquad\qquad\qquad \cdot$$
$$\cdot \qquad\qquad\qquad\qquad \cdot$$
$$\cdot \qquad\qquad\qquad\qquad \cdot$$

Thus, all plotted points lie on or above the lower arm if and only if (iff) $T_r - T_r \leq h$, $T_r - T_{r-1} \leq h$, $T_r - T_{r-2} \leq h$, and so on. This is equivalent to

$$T_r - \min(T_1, T_2, \ldots, T_r) \leq h$$

In a similar manner, if we let

$$V_r = \sum_{i=1}^{r} [\bar{x}_i - (\mu_0 - k)] = S_r + rk$$

it can be shown that all points lie on or below the upper arm iff

$$\max(V_1, \ldots, V_r) - V_r \leq h$$

If we now let

$$d_r = T_r - \min(T_1, \ldots, T_r)$$
$$e_r = \max(V_1, \ldots, V_r) - V_r$$

it is easily seen that $d_1, d_2, \ldots$ and $e_1, e_2, \ldots$ can be calculated recursively as illustrated previously. For example, the expression for d_r follows from consideration of two cases:

1. $\min(T_1, \ldots, T_r) = T_r$, whence $d_r = 0$
2. $\min(T_1, \ldots, T_r) = \min(T_1, \ldots, T_{r-1})$, so that
$$d_r = T_r - \min(T_1, \ldots, T_{r-1})$$
$$= \bar{x}_r - (\mu_0 + k) + T_{r-1} - \min(T_1, \ldots, T_{r-1})$$
$$= \bar{x}_r - (\mu_0 + k) + d_{r-1}$$

Since d_r cannot be negative, it is the larger of these two quantities.

Designing a CUSUM Procedure

Let Δ denote the size of a shift in μ that is to be quickly detected using a CUSUM procedure.* It is common practice to let $k = \Delta/2$. Now suppose a quality control practitioner specifies desired values of two average run lengths:

1. ARL when the process is in control ($\mu = \mu_0$)
2. ARL when the process is out of control because the mean has shifted by Δ ($\mu = \mu_0 + \Delta$ or $\mu = \mu_0 - \Delta$)

A chart developed by Kenneth Kemp ("The Use of Cumulative Sums for Sampling Inspection Schemes," *Applied Statistics,* 1962: 23), called a *nomogram,* can then be used

---

*This contrasts with previous notation, where Δ represented the number of standard deviations by which μ changed.

to determine values of h and n that achieve the specified ARLs.* This chart is shown as Figure 16.11. The method for using the chart is described in the accompanying box. Either the value of σ must be known or an estimate is used in its place.

<table>
<tr>
<td>USING THE
KEMP
NOMOGRAM</td>
<td>

1. Locate the desired ARLs on the in-control and out-of-control scales. Connect these two points with a line.

2. Note where the line crosses the k' scale, and solve for n using the equation

$$k' = \frac{\Delta/2}{\sigma/\sqrt{n}}$$

Then round n up to the nearest integer.

3. Connect the point on the k' scale with the point on the in-control ARL scale using a second line, and note where this line crosses the h' scale. Then $h = (\sigma/\sqrt{n}) \cdot h'$.

</td>
</tr>
</table>

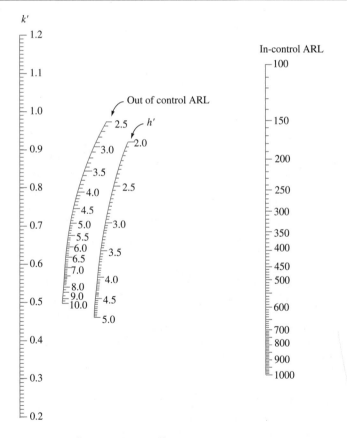

Figure 16.11   The Kemp nomogram†

---

*The word *nomogram* is not specific to this chart; nomograms are used for many other purposes.
†SOURCE: Kemp, Kenneth W., "The Use of Cumulative Sums for Sampling Inspection Schemes," *Applied Statistics,* Vol. XI, 1962: 23. With permission of the Royal Statistical Society.

The value $h = .95$ was used in Example 16.9. In that situation, it follows that the in-control ARL is 500 and the out-of-control ARL (for $\Delta = .3$) is 7.

Example 16.10 The target value for the diameter of the interior core of a hydraulic pump is 2.250 in. If the standard deviation of core diameter is $\sigma = .004$, what CUSUM procedure will yield an in-control ARL of 500 and an ARL of 5 when the mean core diameter shifts by the amount .003 in.?

Connecting the point 500 on the in-control ARL scale to the point 5 on the out-of-control ARL scale and extending the line to the k' scale on the far left in Figure 16.11 gives $k' = .74$. Thus,

$$k' = .74 = \frac{\Delta/2}{\sigma/\sqrt{n}} = \frac{.0015}{.004/\sqrt{n}} = .375\sqrt{n}$$

so

$$\sqrt{n} = \frac{.74}{.375} = 1.973 \qquad n = (1.973)^2 = 3.894$$

The CUSUM procedure should therefore be based on the sample size $n = 4$. Now connecting .74 on the k' scale to 500 on the in-control ARL scale gives $h' = 3.2$, from which

$$h = (\sigma/\sqrt{n}) \cdot (3.2) = (.004/\sqrt{4})(3.2) = .0064$$

An out-of-control signal results as soon as either $d_r > .0064$ or $e_r > .0064$. ∎

We have discussed CUSUM procedures for controlling process location. There are also CUSUM procedures for controlling process variation and for attribute data. The chapter references should be consulted for information on these procedures.

Exercises | Section 16.5 (27–30)

27. Containers of a certain treatment for septic tanks are supposed to contain 16 oz of liquid. A sample of five containers is selected from the production line once each hour and the sample average content is determined. Consider the following results: 15.992, 16.051, 16.066, 15.912, 16.030, 16.060, 15.982, 15.899, 16.038, 16.074, 16.029, 15.935, 16.032, 15.960, 16.055. Using $\Delta = .10$ and $h = .20$, employ the computational form of the CUSUM procedure to investigate the behavior of this process.

28. The target value for the diameter of a certain type of drive shaft is .75 in. The size of the shift in the average diameter considered important to detect is .002 in. Sample average diameters for successive groups of $n = 4$ shafts are as follows: .7507, .7504, .7492, .7501, .7503, .7510, .7490, .7497, .7488,

.7504, .7516, .7472, .7489, .7483, .7471, .7498, .7460, .7482, .7470, .7493, .7462, .7481. Use the computational form of the CUSUM procedure with $h = .003$ to see whether the process mean remained on target throughout the time of observation.

29. The standard deviation of a certain dimension on an aircraft part is .005 cm. What CUSUM procedure will give an in-control ARL of 600 and an out-of-control ARL of 4 when the mean value of the dimension shifts by .004 cm?

30. When the out-of-control ARL corresponds to a shift of 1 standard deviation in the process mean, what are the characteristics of the CUSUM procedure that has ARLs of 250 and 4.8 for the in-control and out-of-control conditions, respectively?

16.6 | **Acceptance Sampling**

Items coming from a production process are often sent in groups to another company or commercial establishment. A group might consist of all units from a particular production run or shift, in a shipping container of some sort, sent in response to a particular order, and so on. The group of items is usually called a *lot,* the sender is referred to as a *producer,* and the recipient of the lot is the *consumer.* Our focus will be on situations in which each item is either defective or nondefective, with p denoting the proportion of defective units in the lot. The consumer would naturally want to accept the lot only if the value of p is suitably small. Acceptance sampling is that part of applied statistics dealing with methods for deciding whether the consumer should accept or reject a lot.

Until quite recently, control chart procedures and acceptance sampling techniques were regarded by practitioners as equally important parts of quality control methodology. This is no longer the case. The reason is that the use of control charts and other recently developed strategies offers the opportunity to design quality into a product, whereas acceptance sampling deals with what has already been produced and thus does not provide for any direct control over process quality. This led the late American quality control expert W. E. Deming, a major force in persuading the Japanese to make substantial use of quality control methodology, to argue strongly against the use of acceptance sampling in many situations. In a similar vein, the recent book by Ryan (see the chapter bibliography) devotes several chapters to control charts and mentions acceptance sampling only in passing. As a reflection of this deemphasis, we content ourselves here with a brief introduction to basic concepts. More information can be found in several references of the chapter bibliography.

Single-Sampling Plans

The most straightforward type of acceptance sampling plan involves selecting a single random sample of size n and then rejecting the lot if the number of defectives in the sample exceeds a specified critical value c. Let the rv X denote the number of defective items in the lot and A denote the event that the lot is accepted. Then $P(A) = P(X \le c)$ is a function of p; the larger the value of p, the smaller will be the probability of accepting the lot.

If the sample size n is large relative to N, $P(A)$ is calculated using the hypergeometric distribution (the number of defectives in the lot is Np):

$$P(X \le c) = \sum_{x=0}^{c} h(x; n, Np, N) = \sum_{x=0}^{c} \frac{\binom{Np}{x} \cdot \binom{N(1-p)}{n-x}}{\binom{N}{n}}$$

When n is small relative to N (the rule of thumb suggested previously was $n \le .05N$, but some authors employ the less conservative rule $n \le .10N$), the binomial distribution can be used:

$$P(X \le c) = \sum_{x=0}^{c} b(x; n, p) = \sum_{x=0}^{c} \binom{n}{x} p^x (1-p)^{n-x}$$

Finally, if $P(A)$ is large only when p is small (this depends on the value of c), the Poisson approximation to the binomial distribution is justified:

$$P(X \le c) \approx \sum_{x=0}^{c} p(x; np) = \sum_{x=0}^{c} \frac{e^{-np}(np)^x}{x!}$$

The behavior of a sampling plan can be nicely summarized by graphing $P(A)$ as a function of p. Such a graph is called the **operating characteristic (OC) curve** for the plan.

Example 16.11 Consider the sampling plan with critical value $c = 2$ and sample size $n = 50$, and suppose that the lot size N exceeds 1000, so the binomial distribution can be used. This gives

$$P(A) = P(X \le 2) = (1 - p)^{50} + 50p(1 - p)^{49} + 1255p^2(1 - p)^{48}$$

The accompanying table shows $P(A)$ for selected values of p, and the corresponding operating characteristic (OC) curve is shown in Figure 16.12.

p	.01	.02	.03	.04	.05	.06	.07	.08	.09	.10	.12	.15
$P(A)$	.986	.922	.811	.677	.541	.416	.311	.226	.161	.112	.051	.014

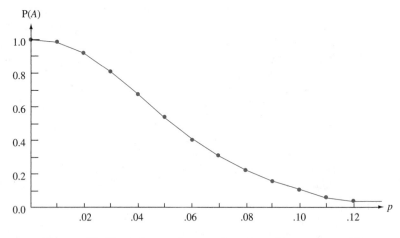

Figure 16.12   OC curve for sampling plan with $c = 2$, $n = 50$

The OC curve for the plan of Example 16.11 has $P(A)$ near 1 for p very close to 0. However, in many applications a defective rate of 8% [for which $P(A) = .226$] or even just 5% [$P(A) = .541$] would be considered excessive, in which case the acceptance probabilities are too high. Increasing the critical value c while holding n fixed gives a plan for which $P(A)$ increases at each p (except 0 and 1), so the new OC curve lies above the old one. This is desirable for p near 0 but not for larger values of p. Holding c constant while increasing n gives a lower OC curve, which is fine for larger p but not for p close to 0. We want an OC curve that is higher for very small p and lower for larger p. This can be achieved by increasing n and adjusting c.

Designing a Single-Sample Plan

An effective sampling plan is one with the following characteristics:

1. It has a specified high probability of accepting lots that the producer considers to be of good quality.

2. It has a specified low probability of accepting lots that the consumer considers to be of poor quality.

A plan of this sort can be developed by proceeding as follows. Let's designate two different values of p, one for which $P(A)$ is a specified value close to 1 and the other for which $P(A)$ is a specified value near 0. These two values of p—say, p_1 and p_2—are often called the **acceptable quality level (AQL)** and the **lot tolerance percent defective (LTPD).** That is, we require a plan for which

1. $P(A) = 1 - \alpha$ when $p = p_1 = $ AQL (α small)

2. $P(A) = \beta$ when $p = p_2 = $ LTPD (β small)

This is analogous to seeking a test procedure with specified type I error probability α and specified type II error probability β when testing hypotheses. For example, we might have

$$AQL = .01 \qquad \alpha = .05 \quad (P(A) = .95)$$
$$LTPD = .045 \qquad \beta = .10 \quad (P(A) = .10)$$

Because X is discrete, we must typically be content with values of n and c that approximately satisfy these conditions.

Table 16.6 gives information from which n and c can be determined in the case $\alpha = .05, \beta = .10$.

Table 16.6 Factors for determining n and c for a single-sample plan with $\alpha = .05, \beta = .10$

c	np_1	np_2	p_2/p_1	c	np_1	np_2	p_2/p_1
0	.051	2.30	45.10	8	4.695	12.99	2.77
1	.355	3.89	10.96	9	5.425	14.21	2.62
2	.818	5.32	6.50	10	6.169	15.41	2.50
3	1.366	6.68	4.89	11	6.924	16.60	2.40
4	1.970	7.99	4.06	12	7.690	17.78	2.31
5	2.613	9.28	3.55	13	8.464	18.86	2.24
6	3.285	10.53	3.21	14	9.246	20.13	2.18
7	3.981	11.77	2.96	15	10.040	21.29	2.12

Example 16.12 Let's determine a plan for which AQL $= p_1 = .01$ and LTPD $= p_2 = .045$. The ratio of p_2 to p_1 is

$$\frac{LTPD}{AQL} = \frac{p_2}{p_1} = \frac{.045}{.01} = 4.50$$

This value lies between the ratio 4.89 given in Table 16.6, for which $c = 3$, and 4.06, for which $c = 4$. Once one of these values of c is chosen, n can be determined either by dividing the np_1 value in Table 16.6 by p_1 or via np_2/p_2. Thus four different plans (two values of c, and for each two values of n) give approximately the specified value of α and β. Consider, for example, using $c = 3$ and

$$n = \frac{np_1}{p_1} = \frac{1.366}{.01} = 136.6 \approx 137$$

Then

$$\alpha = 1 - P(X \leq 3 \text{ when } p = p_1)$$

$$= 1 - \sum_{x=0}^{3} \binom{137}{x}(.01)^x(.99)^{137-x} = .050$$

(the Poisson approximation with $\lambda = 1.37$ also gives .050) and

$$\beta = P(X \leq 3 \text{ when } p = p_2) = .131$$

The plan with $c = 4$ and n determined from $np_2 = 7.99$ has $n = 178$, $\alpha = .034$, and $\beta = .094$. The larger sample size results in a plan with both α and β smaller than the corresponding specified values. ∎

The book by Douglas Montgomery cited in the chapter bibliography contains a chart from which c and n can be determined for *any* specified α and β.

It may happen that the number of defective items in the sample reaches $c + 1$ before all items have been examined. For example, in the case $c = 3$ and $n = 137$, it may be that the 125th item examined is the fourth defective item, so that the remaining 12 items need not be examined. However, it is generally recommended that all items be examined even when this does occur in order to provide a lot-by-lot quality history and estimates of p over time.

Double-Sampling Plans

In a double-sampling plan, the number of defective items x_1 in an initial sample of size n_1 is determined. There are then three possible courses of action: immediately accept the lot, immediately reject the lot, or take a second sample of n_2 items and reject or accept the lot depending on the total number $x_1 + x_2$ of defective items in the two samples. Besides the two sample sizes, a specific plan is characterized by three further numbers c_1, r_1, and c_2 as follows:

1. Reject the lot if $x_1 \geq r_1$.

2. Accept the lot if $x_1 \leq c_1$.

3. If $c_1 < x_1 < r_1$, take a second sample; then accept the lot if $x_1 + x_2 \leq c_2$ and reject it otherwise.

Example 16.13 Consider the double-sampling plan with $n_1 = 80$, $n_2 = 80$, $c_1 = 2$, $r_1 = 5$, and $c_2 = 6$. Thus, the lot will be accepted if (1) $x_1 = 0, 1,$ or 2; (2) $x_1 = 3$ and $x_2 = 0, 1, 2,$ or 3; or (3) $x_1 = 4$ and $x_2 = 0, 1,$ or 2.

Assuming that the lot size is large enough for the binomial approximation to apply, the probability $P(A)$ of accepting the lot is

$$P(A) = P(X_1 = 0, 1, \text{ or } 2) + P(X_1 = 3, X_2 = 0, 1, 2, \text{ or } 3)$$
$$+ P(X_1 = 4, X_2 = 0, 1, \text{ or } 2)$$

$$= \sum_{x_1=0}^{2} b(x_1; 80, p) + b(3; 80, p) \sum_{x_2=0}^{3} b(x_2; 80, p)$$

$$+ b(4; 80, p) \sum_{x_2=0}^{2} b(x_2; 80, p)$$

Again the graph of $P(A)$ versus p is the plan's OC curve. The OC curve for this plan appears in Figure 16.13.

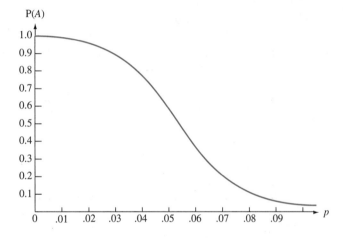

Figure 16.13   OC curve for the double-sampling plan of Example 16.13   ■

One standard method for designing a double-sampling plan involves proceeding as suggested earlier for single-sample plans. Specify values p_1 and p_2 along with corresponding acceptance probabilities $1 - \alpha$ and β. Then find a plan that satisfies these conditions. The book by Montgomery provides tables similar to Table 16.6 for this purpose in the cases $n_2 = n_1$ and $n_2 = 2n_1$ with $1 - \alpha = .95$, $\beta = .10$. Much more extensive tabulations of plans are available in other sources.

Analogous to standard practice with single-sample plans, it is recommended that all items in the first sample be examined even when the $(r_1 + 1)$st defective is discovered prior to inspection of the n_1th item. However, it is customary to terminate inspection of the second sample if the number of defectives is sufficient to justify rejection before all items have been examined. This is referred to as *curtailment* in the second sample. Under curtailment, it can be shown that the expected number of items inspected in a double-sampling plan is smaller than the number of items examined in a single-sampling plan when the OC curves of the two plans are close to being identical. This is the major virtue of double-sampling plans. For more on these matters as well as a discussion of multiple and sequential sampling plans (which involve selecting items for inspection one by one rather than in groups), a book on quality control should be consulted.

Rectifying Inspection and Other Design Criteria

In some situations, sampling inspection is carried out using *rectification*. For single-sample plans, this means that each defective item in the sample is replaced with a satisfactory one, and if the number of defectives in the sample exceeds the acceptance cutoff c, *all* items in the lot are examined and good items are substituted for any defectives. Let N denote the lot size. One important characteristic of a sampling plan with rectifying inspection is **average outgoing quality,** denoted by **AOQ.** This is the long-run proportion of defective items among those sent on after the sampling plan is employed. Now defectives will occur only among the $N - n$ items not inspected in a lot judged acceptable on the basis of a sample. Suppose, for example, that $P(A) = P(X \leq c) = .985$ when $p = .01$. Then in the long run, 98.5% of the $N - n$ items not in the sample will not be inspected, of which we expect 1% to be defective. This implies that the expected number of defectives in a randomly selected batch is $(N - n) \cdot P(A) \cdot p = .00985(N - n)$. Dividing this by the number of items in a lot gives average outgoing quality:

$$\text{AOQ} = \frac{(N - n) \cdot P(A) \cdot p}{N}$$

$$\approx P(A) \cdot p \quad \text{if } N \gg n$$

Because AOQ $= 0$ when either $p = 0$ or $p = 1$ [$P(A) = 0$ in the latter case], it follows that there is a value of p between 0 and 1 for which AOQ is a maximum. The maximum value of AOQ is called the **average outgoing quality limit, AOQL.** For example, for the plan with $n = 137$ and $c = 3$ discussed previously, AOQL $= .0142$, the value of AOQ at $p \approx .02$.

Proper choices of n and c will yield a sampling plan for which AOQL is a specified small number. Such a plan is not, however, unique, so another condition can be imposed. Frequently this second condition will involve the **average** (i.e., expected) **total number inspected,** denoted by **ATI.** The number of items inspected in a randomly chosen lot is a random variable that takes on the value n with probability $P(A)$ and N with probability $1 - P(A)$. Thus, the expected number of items inspected in a randomly selected lot is

$$\text{ATI} = n \cdot P(A) + N \cdot (1 - P(A))$$

It is common practice to select a sampling plan that has a specified AOQL and, in addition, minimum ATI at a particular quality level p.

Standard Sampling Plans

It may seem as though the determination of a sampling plan that simultaneously satisfies several criteria would be quite difficult. Fortunately, others have already laid the groundwork in the form of extensive tabulations of such plans. MIL STD 105D, developed by the military after World War II, is the most widely used set of plans. A civilian version, ANSI/ASQC Z1.4, is quite similar to the military version. A third set of plans that is quite popular was developed at Bell Laboratories prior to World War II by two applied statisticians named Dodge and Romig. The book by Montgomery (see the chapter bibliography) contains a readable introduction to the use of these plans.

Exercises | Section 16.6 (31–38)

31. Consider the single-sample plan with $c = 2$ and $n = 50$, as discussed in Example 16.11, but now suppose that the lot size is $N = 500$. Calculate $P(A)$, the probability of accepting the lot, for $p = .01, .02, \ldots, .10$ using the hypergeometric distribution. Does the binomial approximation give satisfactory results in this case?

32. A sample of 50 items is to be selected from a batch consisting of 5000 items. The batch will be accepted if the sample contains at most one defective item. Calculate the probability of lot acceptance for $p = .01, .02, \ldots, .10$, and sketch the OC curve.

33. Refer to Exercise 32 and consider the plan with $n = 100$ and $c = 2$. Calculate $P(A)$ for $p = .01, .02, \ldots, .05$, and sketch the two OC curves on the same set of axes. Which of the two plans is preferable (leaving aside the cost of sampling) and why?

34. Develop a single-sample plan for which AQL = .02 and LTPD = .07 in the case $\alpha = .05$, $\beta = .10$. Once values of n and c have been determined, calculate the achieved values of α and β for the plan.

35. Consider the double-sampling plan for which both sample sizes are 50. The lot is accepted after the first sample if the number of defectives is at most one, rejected if the number of defectives is at least four, and rejected after the second sample if the to-

tal number of defectives is six or more. Calculate the probability of accepting the lot when $p = .01$, .05, and .10.

36. Some sources advocate a somewhat more restrictive type of doubling-sampling plan in which $r_1 = c_2 + 1$, that is, the lot is rejected if at either stage the (total) number of defectives is at least r_1 (see the book by Montgomery). Consider this type of sampling plan with $n_1 = 50$, $n_2 = 100$, $c_1 = 1$, and $r_1 = 4$. Calculate the probability of lot acceptance when $p = .02$, .05, and .10.

37. Refer to Example 16.11, in which a single-sample plan with $n = 50$ and $c = 2$ was employed.
   a. Calculate AOQ for $p = .01, .02, \ldots, .10$. What does this suggest about the value of p for which AOQ is a maximum and the corresponding AOQL?
   b. Determine the value of p for which AOQ is a maximum and the corresponding value of AOQL. [*Hint:* Use calculus.]
   c. For $N = 2000$, calculate ATI for the values of p given in part (a).

38. Consider the single-sample plan that utilizes $n = 50$ and $c = 1$ when $N = 2000$. Determine the values of AOQ and ATI for selected values of p and graph each of these against p. Also determine the value of AOQL.

Supplementary Exercises (39–44)

39. Observations on shear strength for 26 subgroups of test spot welds, each consisting of six welds, yield $\sum \bar{x}_i = 10{,}980$, $\sum s_i = 402$, and $\sum r_i = 1074$. Calculate control limits for any relevant control charts.

40. The number of scratches on the surface of each of 24 rectangular metal plates is determined, yielding the following data: 8, 1, 7, 5, 2, 0, 2, 3, 4, 3, 1, 2, 5,

7, 3, 4, 6, 5, 2, 4, 0, 10, 2, 6. Construct an appropriate control chart and comment.

41. The following numbers are observations on tensile strength of synthetic fabric specimens selected from a production process at equally spaced time intervals. Construct appropriate control charts and comment (assume an assignable cause is identifiable for any out-of-control observations).

1. 51.3	51.7	49.5	**12.** 49.6 48.4 50.0
2. 51.0	50.0	49.3	**13.** 49.8 51.2 49.7
3. 50.8	51.1	49.0	**14.** 50.4 49.9 50.7
4. 50.6	51.1	49.0	**15.** 49.4 49.5 49.0
5. 49.6	50.5	50.9	**16.** 50.7 49.0 50.0
6. 51.3	52.0	50.3	**17.** 50.8 49.5 50.9
7. 49.7	50.5	50.3	**18.** 48.5 50.3 49.3
8. 51.8	50.3	50.0	**19.** 49.6 50.6 49.4
9. 48.6	50.5	50.7	**20.** 50.9 49.4 49.7
10. 49.6	49.8	50.5	**21.** 54.1 49.8 48.5
11. 49.9	50.7	49.8	**22.** 50.2 49.6 51.5

42. An alternative to the p chart for the fraction defective is the *np chart for number defective*. This chart has $\text{UCL} = n\bar{p} + 3\sqrt{n\bar{p}(1-\bar{p})}$, $\text{LCL} = n\bar{p} - 3\sqrt{n\bar{p}(1-\bar{p})}$, and the *number* of defectives from each sample is plotted on the chart. Construct such a chart for the data of Example 16.6. Will the use of an *np* chart always give the same message as the use of a *p* chart (that is, are the two charts equivalent)?

43. Resistance observations (ohms) for subgroups of a certain type of register gave the following summary quantities:

i	n_i	$\bar{x}_i$	s_i	i	n_i	$\bar{x}_i$	s_i
1	4	430.0	22.5	7	4	420.8	25.4
2	4	418.2	20.6	8	4	431.4	24.0
3	3	435.5	25.1	9	4	428.7	21.2
4	4	427.6	22.3	10	4	440.1	25.8
5	4	444.0	21.5	11	4	445.2	27.3
6	3	431.4	28.9	12	4	430.1	22.2

i	n_i	$\bar{x}_i$	s_i	i	n_i	$\bar{x}_i$	s_i
13	4	427.2	24.0	17	3	447.0	19.8
14	4	439.6	23.3	18	4	434.4	23.7
15	3	415.9	31.2	19	4	422.2	25.1
16	4	419.8	27.5	20	4	425.7	24.4

Construct appropriate control limits. [*Hint:* Use $\bar{x} = \sum n_i \bar{x}_i / \sum n_i$ and $s^2 = \sum (n_i - 1)s_i^2 / \sum (n_i - 1)$.]

44. Let α be a number between 0 and 1 and define a sequence $W_1, W_2, W_3, \ldots$ by $W_0 = \mu$ and $W_t = \alpha \bar{X}_t + (1 - \alpha)W_{t-1}$ for $t = 1, 2, \ldots$. Substituting for W_{t-1} its representation in terms of $\bar{X}_{t-1}$ and W_{t-2}, then substituting for W_{t-2}, etc., results in

$$W_t = \alpha \bar{X}_t + \alpha(1-\alpha)\bar{X}_{t-1} + \cdots + \alpha(1-\alpha)^{t-1}\bar{X}_1 + (1-\alpha)^t\mu$$

The fact that W_t depends not only on $\bar{X}_t$ but also on averages for past time points, albeit with (exponentially) decreasing weights, suggests that changes in the process mean will be more quickly reflected in the W_t's than in the individual $\bar{X}_t$'s.

a. Show that $E(W_t) = \mu$.

b. Let $\sigma_t^2 = V(W_t)$, and show that

$$\sigma_t^2 = \frac{\alpha[1 - (1-\alpha)^{2t}]}{2 - \alpha} \cdot \frac{\sigma^2}{n}$$

c. An *exponentially weighted moving-average control chart* plots the W_t's and uses control limits $\mu_0 \pm 3\sigma_t$ (or $\bar{\bar{x}}$ in place of μ_0). Construct such a chart for the data of Example 16.9, using $\mu_0 = 40$.

Bibliography

Box, George, Soren Bisgaard, and Conrad Fung, "An Explanation and Critique of Taguchi's Contributions to Quality Engineering," *Quality and Reliability Engineering International,* 1988: 123–131.

Montgomery, Douglas C., *Introduction to Statistical Quality Control* (3rd ed.), Wiley, New York, 1996. This is a comprehensive introduction to many aspects of quality control at roughly the same level as this book.

Ryan, Thomas P., *Statistical Methods for Quality Improvement,* Wiley, New York, 1989. Captures very nicely the modern flavor of quality control with minimal demands on the background of readers.

Wadsworth, Harrison, Kenneth Stephens, and A. Blanton Godfrey, *Modern Methods for Quality Control and Improvement,* Wiley, New York, 1986. Provides broad coverage of topics at a level in between the two aforementioned books.

Appendix Tables

Table A.1 Cumulative Binomial Probabilities $\qquad B(x; n, p) = \sum_{y=0}^{x} b(y; n, p)$

a. $n = 5$

		0.01	0.05	0.10	0.20	0.25	0.30	0.40	0.50	0.60	0.70	0.75	0.80	0.90	0.95	0.99
									p							
	0	.951	.774	.590	.328	.237	.168	.078	.031	.010	.002	.001	.000	.000	.000	.000
	1	.999	.977	.919	.737	.633	.528	.337	.188	.087	.031	.016	.007	.000	.000	.000
x	2	1.000	.999	.991	.942	.896	.837	.683	.500	.317	.163	.104	.058	.009	.001	.000
	3	1.000	1.000	1.000	.993	.984	.969	.913	.812	.663	.472	.367	.263	.081	.023	.001
	4	1.000	1.000	1.000	1.000	.999	.998	.990	.969	.922	.832	.763	.672	.410	.226	.049

b. $n = 10$

		0.01	0.05	0.10	0.20	0.25	0.30	0.40	0.50	0.60	0.70	0.75	0.80	0.90	0.95	0.99
									p							
	0	.904	.599	.349	.107	.056	.028	.006	.001	.000	.000	.000	.000	.000	.000	.000
	1	.996	.914	.736	.376	.244	.149	.046	.011	.002	.000	.000	.000	.000	.000	.000
	2	1.000	.988	.930	.678	.526	.383	.167	.055	.012	.002	.000	.000	.000	.000	.000
	3	1.000	.999	.987	.879	.776	.650	.382	.172	.055	.011	.004	.001	.000	.000	.000
x	4	1.000	1.000	.998	.967	.922	.850	.633	.377	.166	.047	.020	.006	.000	.000	.000
	5	1.000	1.000	1.000	.994	.980	.953	.834	.623	.367	.150	.078	.033	.002	.000	.000
	6	1.000	1.000	1.000	.999	.996	.989	.945	.828	.618	.350	.224	.121	.013	.001	.000
	7	1.000	1.000	1.000	1.000	1.000	.998	.988	.945	.833	.617	.474	.322	.070	.012	.000
	8	1.000	1.000	1.000	1.000	1.000	1.000	.998	.989	.954	.851	.756	.624	.264	.086	.004
	9	1.000	1.000	1.000	1.000	1.000	1.000	1.000	.999	.994	.972	.944	.893	.651	.401	.096

c. $n = 15$

		0.01	0.05	0.10	0.20	0.25	0.30	0.40	0.50	0.60	0.70	0.75	0.80	0.90	0.95	0.99
									p							
	0	.860	.463	.206	.035	.013	.005	.000	.000	.000	.000	.000	.000	.000	.000	.000
	1	.990	.829	.549	.167	.080	.035	.005	.000	.000	.000	.000	.000	.000	.000	.000
	2	1.000	.964	.816	.398	.236	.127	.027	.004	.000	.000	.000	.000	.000	.000	.000
	3	1.000	.995	.944	.648	.461	.297	.091	.018	.002	.000	.000	.000	.000	.000	.000
	4	1.000	.999	.987	.836	.686	.515	.217	.059	.009	.001	.000	.000	.000	.000	.000
	5	1.000	1.000	.998	.939	.852	.722	.402	.151	.034	.004	.001	.000	.000	.000	.000
	6	1.000	1.000	1.000	.982	.943	.869	.610	.304	.095	.015	.004	.001	.000	.000	.000
x	7	1.000	1.000	1.000	.996	.983	.950	.787	.500	.213	.050	.017	.004	.000	.000	.000
	8	1.000	1.000	1.000	.999	.996	.985	.905	.696	.390	.131	.057	.018	.000	.000	.000
	9	1.000	1.000	1.000	1.000	.999	.996	.966	.849	.597	.278	.148	.061	.002	.000	.000
	10	1.000	1.000	1.000	1.000	1.000	.999	.991	.941	.783	.485	.314	.164	.013	.001	.000
	11	1.000	1.000	1.000	1.000	1.000	1.000	.998	.982	.909	.703	.539	.352	.056	.005	.000
	12	1.000	1.000	1.000	1.000	1.000	1.000	1.000	.996	.973	.873	.764	.602	.184	.036	.000
	13	1.000	1.000	1.000	1.000	1.000	1.000	1.000	1.000	.995	.965	.920	.833	.451	.171	.010
	14	1.000	1.000	1.000	1.000	1.000	1.000	1.000	1.000	1.000	.995	.987	.965	.794	.537	.140

(*continued*)

Table A.1 Cumulative Binomial Probabilities *(cont.)*

$$B(x; n, p) = \sum_{y=0}^{x} b(y; n, p)$$

d. $n = 20$

									p							
		0.01	0.05	0.10	0.20	0.25	0.30	0.40	0.50	0.60	0.70	0.75	0.80	0.90	0.95	0.99
	0	.818	.358	.122	.012	.003	.001	.000	.000	.000	.000	.000	.000	.000	.000	.000
	1	.983	.736	.392	.069	.024	.008	.001	.000	.000	.000	.000	.000	.000	.000	.000
	2	.999	.925	.677	.206	.091	.035	.004	.000	.000	.000	.000	.000	.000	.000	.000
	3	1.000	.984	.867	.411	.225	.107	.016	.001	.000	.000	.000	.000	.000	.000	.000
	4	1.000	.997	.957	.630	.415	.238	.051	.006	.000	.000	.000	.000	.000	.000	.000
	5	1.000	1.000	.989	.804	.617	.416	.126	.021	.002	.000	.000	.000	.000	.000	.000
	6	1.000	1.000	.998	.913	.786	.608	.250	.058	.006	.000	.000	.000	.000	.000	.000
	7	1.000	1.000	1.000	.968	.898	.772	.416	.132	.021	.001	.000	.000	.000	.000	.000
	8	1.000	1.000	1.000	.990	.959	.887	.596	.252	.057	.005	.001	.000	.000	.000	.000
	9	1.000	1.000	1.000	.997	.986	.952	.755	.412	.128	.017	.004	.001	.000	.000	.000
x	10	1.000	1.000	1.000	.999	.996	.983	.872	.588	.245	.048	.014	.003	.000	.000	.000
	11	1.000	1.000	1.000	1.000	.999	.995	.943	.748	.404	.113	.041	.010	.000	.000	.000
	12	1.000	1.000	1.000	1.000	1.000	.999	.979	.868	.584	.228	.102	.032	.000	.000	.000
	13	1.000	1.000	1.000	1.000	1.000	1.000	.994	.942	.750	.392	.214	.087	.002	.000	.000
	14	1.000	1.000	1.000	1.000	1.000	1.000	.998	.979	.874	.584	.383	.196	.011	.000	.000
	15	1.000	1.000	1.000	1.000	1.000	1.000	1.000	.994	.949	.762	.585	.370	.043	.003	.000
	16	1.000	1.000	1.000	1.000	1.000	1.000	1.000	.999	.984	.893	.775	.589	.133	.016	.000
	17	1.000	1.000	1.000	1.000	1.000	1.000	1.000	1.000	.996	.965	.909	.794	.323	.075	.001
	18	1.000	1.000	1.000	1.000	1.000	1.000	1.000	1.000	.999	.992	.976	.931	.608	.264	.017
	19	1.000	1.000	1.000	1.000	1.000	1.000	1.000	1.000	1.000	.999	.997	.988	.878	.642	.182

(continued)

Table A.1 Cumulative Binomial Probabilities *(cont.)*

$$B(x; n, p) = \sum_{y=0}^{x} b(y; n, p)$$

e. $n = 25$

		p														
		0.01	**0.05**	**0.10**	**0.20**	**0.25**	**0.30**	**0.40**	**0.50**	**0.60**	**0.70**	**0.75**	**0.80**	**0.90**	**0.95**	**0.99**
	0	.778	.277	.072	.004	.001	.000	.000	.000	.000	.000	.000	.000	.000	.000	.000
	1	.974	.642	.271	.027	.007	.002	.000	.000	.000	.000	.000	.000	.000	.000	.000
	2	.998	.873	.537	.098	.032	.009	.000	.000	.000	.000	.000	.000	.000	.000	.000
	3	1.000	.966	.764	.234	.096	.033	.002	.000	.000	.000	.000	.000	.000	.000	.000
	4	1.000	.993	.902	.421	.214	.090	.009	.000	.000	.000	.000	.000	.000	.000	.000
	5	1.000	.999	.967	.617	.378	.193	.029	.002	.000	.000	.000	.000	.000	.000	.000
	6	1.000	1.000	.991	.780	.561	.341	.074	.007	.000	.000	.000	.000	.000	.000	.000
	7	1.000	1.000	.998	.891	.727	.512	.154	.022	.001	.000	.000	.000	.000	.000	.000
	8	1.000	1.000	1.000	.953	.851	.677	.274	.054	.004	.000	.000	.000	.000	.000	.000
	9	1.000	1.000	1.000	.983	.929	.811	.425	.115	.013	.000	.000	.000	.000	.000	.000
	10	1.000	1.000	1.000	.994	.970	.902	.586	.212	.034	.002	.000	.000	.000	.000	.000
	11	1.000	1.000	1.000	.998	.980	.956	.732	.345	.078	.006	.001	.000	.000	.000	.000
x	12	1.000	1.000	1.000	1.000	.997	.983	.846	.500	.154	.017	.003	.000	.000	.000	.000
	13	1.000	1.000	1.000	1.000	.999	.994	.922	.655	.268	.044	.020	.002	.000	.000	.000
	14	1.000	1.000	1.000	1.000	1.000	.998	.966	.788	.414	.098	.030	.006	.000	.000	.000
	15	1.000	1.000	1.000	1.000	1.000	1.000	.987	.885	.575	.189	.071	.017	.000	.000	.000
	16	1.000	1.000	1.000	1.000	1.000	1.000	.996	.946	.726	.323	.149	.047	.000	.000	.000
	17	1.000	1.000	1.000	1.000	1.000	1.000	.999	.978	.846	.488	.273	.109	.002	.000	.000
	18	1.000	1.000	1.000	1.000	1.000	1.000	1.000	.993	.926	.659	.439	.220	.009	.000	.000
	19	1.000	1.000	1.000	1.000	1.000	1.000	1.000	.998	.971	.807	.622	.383	.033	.001	.000
	20	1.000	1.000	1.000	1.000	1.000	1.000	1.000	1.000	.991	.910	.786	.579	.098	.007	.000
	21	1.000	1.000	1.000	1.000	1.000	1.000	1.000	1.000	.998	.967	.904	.766	.236	.034	.000
	22	1.000	1.000	1.000	1.000	1.000	1.000	1.000	1.000	1.000	.991	.968	.902	.463	.127	.002
	23	1.000	1.000	1.000	1.000	1.000	1.000	1.000	1.000	1.000	.998	.993	.973	.729	.358	.026
	24	1.000	1.000	1.000	1.000	1.000	1.000	1.000	1.000	1.000	1.000	.999	.996	.928	.723	.222

Table A.2 Cumulative Poisson Probabilities

$$F(x; \lambda) = \sum_{y=0}^{x} \frac{e^{-\lambda}\lambda^y}{y!}$$

		λ									
		.1	**.2**	**.3**	**.4**	**.5**	**.6**	**.7**	**.8**	**.9**	**1.0**
	0	.905	.819	.741	.670	.607	.549	.497	.449	.407	.368
	1	.995	.982	.963	.938	.910	.878	.844	.809	.772	.736
	2	1.000	.999	.996	.992	.986	.977	.966	.953	.937	.920
x	3		1.000	1.000	.999	.998	.997	.994	.991	.987	.981
	4				1.000	1.000	1.000	.999	.999	.998	.996
	5							1.000	1.000	1.000	.999
	6										1.000

(continued)

Table A.2 Cumulative Poisson Probabilities *(cont.)*
$$F(x; \lambda) = \sum_{y=0}^{x} \frac{e^{-\lambda}\lambda^y}{y!}$$

	λ										
	2.0	**3.0**	**4.0**	**5.0**	**6.0**	**7.0**	**8.0**	**9.0**	**10.0**	**15.0**	**20.0**
0	.135	.050	.018	.007	.002	.001	.000	.000	.000	.000	.000
1	.406	.199	.092	.040	.017	.007	.003	.001	.000	.000	.000
2	.677	.423	.238	.125	.062	.030	.014	.006	.003	.000	.000
3	.857	.647	.433	.265	.151	.082	.042	.021	.010	.000	.000
4	.947	.815	.629	.440	.285	.173	.100	.055	.029	.001	.000
5	.983	.916	.785	.616	.446	.301	.191	.116	.067	.003	.000
6	.995	.966	.889	.762	.606	.450	.313	.207	.130	.008	.000
7	.999	.988	.949	.867	.744	.599	.453	.324	.220	.018	.001
8	1.000	.996	.979	.932	.847	.729	.593	.456	.333	.037	.002
9		.999	.992	.968	.916	.830	.717	.587	.458	.070	.005
10		1.000	.997	.986	.957	.901	.816	.706	.583	.118	.011
11			.999	.995	.980	.947	.888	.803	.697	.185	.021
12			1.000	.998	.991	.973	.936	.876	.792	.268	.039
13				.999	.996	.987	.966	.926	.864	.363	.066
14				1.000	.999	.994	.983	.959	.917	.466	.105
15					.999	.998	.992	.978	.951	.568	.157
16					1.000	.999	.996	.989	.973	.664	.221
17						1.000	.998	.995	.986	.749	.297
18							.999	.998	.993	.819	.381
19							1.000	.999	.997	.875	.470
20								1.000	.998	.917	.559
21									.999	.947	.644
22									1.000	.967	.721
23										.981	.787
24										.989	.843
25										.994	.888
26										.997	.922
27										.998	.948
28										.999	.966
29										1.000	.978
30											.987
31											.992
32											.995
33											.997
34											.999
35											.999
36											1.000

x

Table A.3 Standard Normal Curve Areas

$\Phi(z) = P(Z \le z)$ $\Phi(z) = P(Z \le z)$

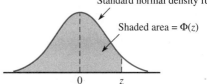

Standard normal density function

Shaded area = $\Phi(z)$

z	0.00	0.01	0.02	0.03	0.04	0.05	0.06	0.07	0.08	0.09
−3.4	0.0003	0.0003	0.0003	0.0003	0.0003	0.0003	0.0003	0.0003	0.0003	0.0002
−3.3	0.0005	0.0005	0.0005	0.0004	0.0004	0.0004	0.0004	0.0004	0.0004	0.0003
−3.2	0.0007	0.0007	0.0006	0.0006	0.0006	0.0006	0.0006	0.0005	0.0005	0.0005
−3.1	0.0010	0.0009	0.0009	0.0009	0.0008	0.0008	0.0008	0.0008	0.0007	0.0007
−3.0	0.0013	0.0013	0.0013	0.0012	0.0012	0.0011	0.0011	0.0011	0.0010	0.0010
−2.9	0.0019	0.0018	0.0017	0.0017	0.0016	0.0016	0.0015	0.0015	0.0014	0.0014
−2.8	0.0026	0.0025	0.0024	0.0023	0.0023	0.0022	0.0021	0.0021	0.0020	0.0019
−2.7	0.0035	0.0034	0.0033	0.0032	0.0031	0.0030	0.0029	0.0028	0.0027	0.0026
−2.6	0.0047	0.0045	0.0044	0.0043	0.0041	0.0040	0.0039	0.0038	0.0037	0.0036
−2.5	0.0062	0.0060	0.0059	0.0057	0.0055	0.0054	0.0052	0.0051	0.0049	0.0038
−2.4	0.0082	0.0080	0.0078	0.0075	0.0073	0.0071	0.0069	0.0068	0.0066	0.0064
−2.3	0.0107	0.0104	0.0102	0.0099	0.0096	0.0094	0.0091	0.0089	0.0087	0.0084
−2.2	0.0139	0.0136	0.0132	0.0129	0.0125	0.0122	0.0119	0.0116	0.0113	0.0110
−2.1	0.0179	0.0174	0.0170	0.0166	0.0162	0.0158	0.0154	0.0150	0.0146	0.0143
−2.0	0.0228	0.0222	0.0217	0.0212	0.0207	0.0202	0.0197	0.0192	0.0188	0.0183
−1.9	0.0287	0.0281	0.0274	0.0268	0.0262	0.0256	0.0250	0.0244	0.0239	0.0233
−1.8	0.0359	0.0352	0.0344	0.0336	0.0329	0.0322	0.0314	0.0307	0.0310	0.0294
−1.7	0.0446	0.0436	0.0427	0.0418	0.0409	0.0401	0.0392	0.0384	0.0375	0.0367
−1.6	0.0548	0.0537	0.0526	0.0516	0.0505	0.0495	0.0485	0.0475	0.0465	0.0455
−1.5	0.0668	0.0655	0.0643	0.0630	0.0618	0.0606	0.0594	0.0582	0.0571	0.0559
−1.4	0.0808	0.0793	0.0778	0.0764	0.0749	0.0735	0.0722	0.0708	0.0694	0.0681
−1.3	0.0968	0.0951	0.0934	0.0918	0.0901	0.0885	0.0869	0.0853	0.0838	0.0823
−1.2	0.1151	0.1131	0.1112	0.1093	0.1075	0.1056	0.1038	0.1020	0.1003	0.0985
−1.1	0.1357	0.1335	0.1314	0.1292	0.1271	0.1251	0.1230	0.1210	0.1190	0.1170
−1.0	0.1587	0.1562	0.1539	0.1515	0.1492	0.1469	0.1446	0.1423	0.1401	0.1379
−0.9	0.1841	0.1814	0.1788	0.1762	0.1736	0.1711	0.1685	0.1660	0.1635	0.1611
−0.8	0.2119	0.2090	0.2061	0.2033	0.2005	0.1977	0.1949	0.1922	0.1894	0.1867
−0.7	0.2420	0.2389	0.2358	0.2327	0.2296	0.2266	0.2236	0.2206	0.2177	0.2148
−0.6	0.2743	0.2709	0.2676	0.2643	0.2611	0.2578	0.2546	0.2514	0.2483	0.2451
−0.5	0.3085	0.3050	0.3015	0.2981	0.2946	0.2912	0.2877	0.2843	0.2810	0.2776
−0.4	0.3446	0.3409	0.3372	0.3336	0.3300	0.3264	0.3228	0.3192	0.3156	0.3121
−0.3	0.3821	0.3783	0.3745	0.3707	0.3669	0.3632	0.3594	0.3557	0.3520	0.3482
−0.2	0.4207	0.4168	0.4129	0.4090	0.4052	0.4013	0.3974	0.3936	0.3897	0.3859
−0.1	0.4602	0.4562	0.4522	0.4483	0.4443	0.4404	0.4364	0.4325	0.4286	0.4247
−0.0	0.5000	0.4960	0.4920	0.4880	0.4840	0.4801	0.4761	0.4721	0.4681	0.4641

(continued)

Table A.3 Standard Normal Curve Areas *(cont.)* $\Phi(z) = P(Z \le z)$

z	0.00	0.01	0.02	0.03	0.04	0.05	0.06	0.07	0.08	0.09
0.0	0.5000	0.5040	0.5080	0.5120	0.5160	0.5199	0.5239	0.5279	0.5319	0.5359
0.1	0.5398	0.5438	0.5478	0.5517	0.5557	0.5596	0.5636	0.5675	0.5714	0.5753
0.2	0.5793	0.5832	0.5871	0.5910	0.5948	0.5987	0.6026	0.6064	0.6103	0.6141
0.3	0.6179	0.6217	0.6255	0.6293	0.6331	0.6368	0.6406	0.6443	0.6480	0.6517
0.4	0.6554	0.6591	0.6628	0.6664	0.6700	0.6736	0.6772	0.6808	0.6844	0.6879
0.5	0.6915	0.6950	0.6985	0.7019	0.7054	0.7088	0.7123	0.7157	0.7190	0.7224
0.6	0.7257	0.7291	0.7324	0.7357	0.7389	0.7422	0.7454	0.7486	0.7517	0.7549
0.7	0.7580	0.7611	0.7642	0.7673	0.7704	0.7734	0.7764	0.7794	0.7823	0.7852
0.8	0.7881	0.7910	0.7939	0.7967	0.7995	0.8023	0.8051	0.8078	0.8106	0.8133
0.9	0.8159	0.8186	0.8212	0.8238	0.8264	0.8289	0.8315	0.8340	0.8365	0.8389
1.0	0.8413	0.8438	0.8461	0.8485	0.8508	0.8531	0.8554	0.8577	0.8599	0.8621
1.1	0.8643	0.8665	0.8686	0.8708	0.8729	0.8749	0.8770	0.8790	0.8810	0.8830
1.2	0.8849	0.8869	0.8888	0.8907	0.8925	0.8944	0.8962	0.8980	0.8997	0.9015
1.3	0.9032	0.9049	0.9066	0.9082	0.9099	0.9115	0.9131	0.9147	0.9162	0.9177
1.4	0.9192	0.9207	0.9222	0.9236	0.9251	0.9265	0.9278	0.9292	0.9306	0.9319
1.5	0.9332	0.9345	0.9357	0.9370	0.9382	0.9394	0.9406	0.9418	0.9429	0.9441
1.6	0.9452	0.9463	0.9474	0.9484	0.9495	0.9505	0.9515	0.9525	0.9535	0.9545
1.7	0.9554	0.9564	0.9573	0.9582	0.9591	0.9599	0.9608	0.9616	0.9625	0.9633
1.8	0.9641	0.9649	0.9656	0.9664	0.9671	0.9678	0.9686	0.9693	0.9699	0.9706
1.9	0.9713	0.9719	0.9726	0.9732	0.9738	0.9744	0.9750	0.9756	0.9761	0.9767
2.0	0.9772	0.9778	0.9783	0.9788	0.9793	0.9798	0.9803	0.9808	0.9812	0.9817
2.1	0.9821	0.9826	0.9830	0.9834	0.9838	0.9842	0.9846	0.9850	0.9854	0.9857
2.2	0.9861	0.9864	0.9868	0.9871	0.9875	0.9878	0.9881	0.9884	0.9887	0.9890
2.3	0.9893	0.9896	0.9898	0.9901	0.9904	0.9906	0.9909	0.9911	0.9913	0.9916
2.4	0.9918	0.9920	0.9922	0.9925	0.9927	0.9929	0.9931	0.9932	0.9934	0.9936
2.5	0.9938	0.9940	0.9941	0.9943	0.9945	0.9946	0.9948	0.9949	0.9951	0.9952
2.6	0.9953	0.9955	0.9956	0.9957	0.9959	0.9960	0.9961	0.9962	0.9963	0.9964
2.7	0.9965	0.9966	0.9967	0.9968	0.9969	0.9970	0.9971	0.9972	0.9973	0.9974
2.8	0.9974	0.9975	0.9976	0.9977	0.9977	0.9978	0.9979	0.9979	0.9980	0.9981
2.9	0.9981	0.9982	0.9982	0.9983	0.9984	0.9984	0.9985	0.9985	0.9986	0.9986
3.0	0.9987	0.9987	0.9987	0.9988	0.9988	0.9989	0.9989	0.9989	0.9990	0.9990
3.1	0.9990	0.9991	0.9991	0.9991	0.9992	0.9992	0.9992	0.9992	0.9993	0.9993
3.2	0.9993	0.9993	0.9994	0.9994	0.9994	0.9994	0.9994	0.9995	0.9995	0.9995
3.3	0.9995	0.9995	0.9995	0.9996	0.9996	0.9996	0.9996	0.9996	0.9996	0.9997
3.4	0.9997	0.9997	0.9997	0.9997	0.9997	0.9997	0.9997	0.9997	0.9997	0.9998

Table A.4 The Incomplete Gamma Function

$$F(x; \alpha) = \int_0^x \frac{1}{\Gamma(\alpha)} y^{\alpha-1} e^{-y} dy$$

x \ α	1	2	3	4	5	6	7	8	9	10
1	.632	.264	.080	.019	.004	.001	.000	.000	.000	.000
2	.865	.594	.323	.143	.053	.017	.005	.001	.000	.000
3	.950	.801	.577	.353	.185	.084	.034	.012	.004	.001
4	.982	.908	.762	.567	.371	.215	.111	.051	.021	.008
5	.993	.960	.875	.735	.560	.384	.238	.133	.068	.032
6	.998	.983	.938	.849	.715	.554	.394	.256	.153	.084
7	.999	.993	.970	.918	.827	.699	.550	.401	.271	.170
8	1.000	.997	.986	.958	.900	.809	.687	.547	.407	.283
9		.999	.994	.979	.945	.884	.793	.676	.544	.413
10		1.000	.997	.990	.971	.933	.870	.780	.667	.542
11			.999	.995	.985	.962	.921	.857	.768	.659
12			1.000	.998	.992	.980	.954	.911	.845	.758
13				.999	.996	.989	.974	.946	.900	.834
14				1.000	.998	.994	.986	.968	.938	.891
15					.999	.997	.992	.982	.963	.930

Table A.5   Critical Values for *t* Distributions

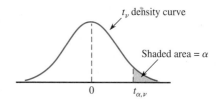

t_ν density curve

Shaded area = α

0    $t_{\alpha,\nu}$

	α						
ν	.10	.05	.025	.01	.005	.001	.0005
1	3.078	6.314	12.706	31.821	63.657	318.31	636.62
2	1.886	2.920	4.303	6.965	9.925	22.326	31.598
3	1.638	2.353	3.182	4.541	5.841	10.213	12.924
4	1.533	2.132	2.776	3.747	4.604	7.173	8.610
5	1.476	2.015	2.571	3.365	4.032	5.893	6.869
6	1.440	1.943	2.447	3.143	3.707	5.208	5.959
7	1.415	1.895	2.365	2.998	3.499	4.785	5.408
8	1.397	1.860	2.306	2.896	3.355	4.501	5.041
9	1.383	1.833	2.262	2.821	3.250	4.297	4.781
10	1.372	1.812	2.228	2.764	3.169	4.144	4.587
11	1.363	1.796	2.201	2.718	3.106	4.025	4.437
12	1.356	1.782	2.179	2.681	3.055	3.930	4.318
13	1.350	1.771	2.160	2.650	3.012	3.852	4.221
14	1.345	1.761	2.145	2.624	2.977	3.787	4.140
15	1.341	1.753	2.131	2.602	2.947	3.733	4.073
16	1.337	1.746	2.120	2.583	2.921	3.686	4.015
17	1.333	1.740	2.110	2.567	2.898	3.646	3.965
18	1.330	1.734	2.101	2.552	2.878	3.610	3.922
19	1.328	1.729	2.093	2.539	2.861	3.579	3.883
20	1.325	1.725	2.086	2.528	2.845	3.552	3.850
21	1.323	1.721	2.080	2.518	2.831	3.527	3.819
22	1.321	1.717	2.074	2.508	2.819	3.505	3.792
23	1.319	1.714	2.069	2.500	2.807	3.485	3.767
24	1.318	1.711	2.064	2.492	2.797	3.467	3.745
25	1.316	1.708	2.060	2.485	2.787	3.450	3.725
26	1.315	1.706	2.056	2.479	2.779	3.435	3.707
27	1.314	1.703	2.052	2.473	2.771	3.421	3.690
28	1.313	1.701	2.048	2.467	2.763	3.408	3.674
29	1.311	1.699	2.045	2.462	2.756	3.396	3.659
30	1.310	1.697	2.042	2.457	2.750	3.385	3.646
32	1.309	1.694	2.037	2.449	2.738	3.365	3.622
34	1.307	1.691	2.032	2.441	2.728	3.348	3.601
36	1.306	1.688	2.028	2.434	2.719	3.333	3.582
38	1.304	1.686	2.024	2.429	2.712	3.319	3.566
40	1.303	1.684	2.021	2.423	2.704	3.307	3.551
50	1.299	1.676	2.009	2.403	2.678	3.262	3.496
60	1.296	1.671	2.000	2.390	2.660	3.232	3.460
120	1.289	1.658	1.980	2.358	2.617	3.160	3.373
∞	1.282	1.645	1.960	2.326	2.576	3.090	3.291

Table A.6 Tolerance critical values for normal population distributions

	Two-sided intervals						One-sided intervals					
Confidence level	**95%**			**99%**			**95%**			**99%**		
% of population captured	$\ge 90\%$	$\ge 95\%$	$\ge 99\%$	$\ge 90\%$	$\ge 95\%$	$\ge 99\%$	$\ge 90\%$	$\ge 95\%$	$\ge 99\%$	$\ge 90\%$	$\ge 95\%$	$\ge 99\%$
2	32.019	37.674	48.430	160.193	188.491	242.300	20.581	26.260	37.094	103.029	131.426	185.617
3	8.380	9.916	12.861	18.930	22.401	29.055	6.156	7.656	10.553	13.995	17.370	23.896
4	5.369	6.370	8.299	9.398	11.150	14.527	4.162	5.144	7.042	7.380	9.083	12.387
5	4.275	5.079	6.634	6.612	7.855	10.260	3.407	4.203	5.741	5.362	6.578	8.939
6	3.712	4.414	5.775	5.337	6.345	8.301	3.006	3.708	5.062	4.411	5.406	7.335
7	3.369	4.007	5.248	4.613	5.488	7.187	2.756	3.400	4.642	3.859	4.728	6.412
8	3.136	3.732	4.891	4.147	4.936	6.468	2.582	3.187	4.354	3.497	4.285	5.812
9	2.967	3.532	4.631	3.822	4.550	5.966	2.454	3.031	4.143	3.241	3.972	5.389
10	2.839	3.379	4.433	3.582	4.265	5.594	2.355	2.911	3.981	3.048	3.738	5.074
11	2.737	3.259	4.277	3.397	4.045	5.308	2.275	2.815	3.852	2.898	3.556	4.829
12	2.655	3.162	4.150	3.250	3.870	5.079	2.210	2.736	3.747	2.777	3.410	4.633
13	2.587	3.081	4.044	3.130	3.727	4.893	2.155	2.671	3.659	2.677	3.290	4.472
14	2.529	3.012	3.955	3.029	3.608	4.737	2.109	2.615	3.585	2.593	3.189	4.337
15	2.480	2.954	3.878	2.945	3.507	4.605	2.068	2.566	3.520	2.522	3.102	4.222
16	2.437	2.903	3.812	2.872	3.421	4.492	2.033	2.524	3.464	2.460	3.028	4.133
17	2.400	2.858	3.754	2.808	3.345	4.393	2.002	2.486	3.414	2.405	2.963	4.037
18	2.366	2.819	3.702	2.753	3.279	4.307	1.974	2.453	3.370	2.357	2.905	3.960
19	2.337	2.784	3.656	2.703	3.221	4.230	1.949	2.423	3.331	2.314	2.854	3.892
20	2.310	2.752	3.615	2.659	3.168	4.161	1.926	2.396	3.295	2.276	2.808	3.832
25	2.208	2.631	3.457	2.494	2.972	3.904	1.838	2.292	3.158	2.129	2.633	3.601
30	2.140	2.549	3.350	2.385	2.841	3.733	1.777	2.220	3.064	2.030	2.516	3.447
35	2.090	2.490	3.272	2.306	2.748	3.611	1.732	2.167	2.995	1.957	2.430	3.334
40	2.052	2.445	3.213	2.247	2.677	3.518	1.697	2.126	2.941	1.902	2.364	3.249
45	2.021	2.408	3.165	2.200	2.621	3.444	1.669	2.092	2.898	1.857	2.312	3.180
50	1.996	2.379	3.126	2.162	2.576	3.385	1.646	2.065	2.863	1.821	2.269	3.125
60	1.958	2.333	3.066	2.103	2.506	3.293	1.609	2.022	2.807	1.764	2.202	3.038
70	1.929	2.299	3.021	2.060	2.454	3.225	1.581	1.990	2.765	1.722	2.153	2.974
80	1.907	2.272	2.986	2.026	2.414	3.173	1.559	1.965	2.733	1.688	2.114	2.924
90	1.889	2.251	2.958	1.999	2.382	3.130	1.542	1.944	2.706	1.661	2.082	2.883
100	1.874	2.233	2.934	1.977	2.355	3.096	1.527	1.927	2.684	1.639	2.056	2.850
150	1.825	2.175	2.859	1.905	2.270	2.983	1.478	1.870	2.611	1.566	1.971	2.741
200	1.798	2.143	2.816	1.865	2.222	2.921	1.450	1.837	2.570	1.524	1.923	2.679
250	1.780	2.121	2.788	1.839	2.191	2.880	1.431	1.815	2.542	1.496	1.891	2.638
300	1.767	2.106	2.767	1.820	2.169	2.850	1.417	1.800	2.522	1.476	1.868	2.608
∞	1.645	1.960	2.576	1.645	1.960	2.576	1.282	1.645	2.326	1.282	1.645	2.326

Sample size n

Table A.7 Critical Values for Chi-Squared Distributions

χ_ν^2 density curve

Shaded area $= \alpha$

$\chi_{\alpha,\nu}^2$

ν	.995	.99	.975	.95	.90	.10	.05	.025	.01	.005
1	0.000	0.000	0.001	0.004	0.016	2.706	3.843	5.025	6.637	7.882
2	0.010	0.020	0.051	0.103	0.211	4.605	5.992	7.378	9.210	10.597
3	0.072	0.115	0.216	0.352	0.584	6.251	7.815	9.348	11.344	12.837
4	0.207	0.297	0.484	0.711	1.064	7.779	9.488	11.143	13.277	14.860
5	0.412	0.554	0.831	1.145	1.610	9.236	11.070	12.832	15.085	16.748
6	0.676	0.872	1.237	1.635	2.204	10.645	12.592	14.440	16.812	18.548
7	0.989	1.239	1.690	2.167	2.833	12.017	14.067	16.012	18.474	20.276
8	1.344	1.646	2.180	2.733	3.490	13.362	15.507	17.534	20.090	21.954
9	1.735	2.088	2.700	3.325	4.168	14.684	16.919	19.022	21.665	23.587
10	2.156	2.558	3.247	3.940	4.865	15.987	18.307	20.483	23.209	25.188
11	2.603	3.053	3.816	4.575	5.578	17.275	19.675	21.920	24.724	26.755
12	3.074	3.571	4.404	5.226	6.304	18.549	21.026	23.337	26.217	28.300
13	3.565	4.107	5.009	5.892	7.041	19.812	22.362	24.735	27.687	29.817
14	4.075	4.660	5.629	6.571	7.790	21.064	23.685	26.119	29.141	31.319
15	4.600	5.229	6.262	7.261	8.547	22.307	24.996	27.488	30.577	32.799
16	5.142	5.812	6.908	7.962	9.312	23.542	26.296	28.845	32.000	34.267
17	5.697	6.407	7.564	8.682	10.085	24.769	27.587	30.190	33.408	35.716
18	6.265	7.015	8.231	9.390	10.865	25.989	28.869	31.526	34.805	37.156
19	6.843	7.632	8.906	10.117	11.651	27.203	30.143	32.852	36.190	38.580
20	7.434	8.260	9.591	10.851	12.443	28.412	31.410	34.170	37.566	39.997
21	8.033	8.897	10.283	11.591	13.240	29.615	32.670	35.478	38.930	41.399
22	8.643	9.542	10.982	12.338	14.042	30.813	33.924	36.781	40.289	42.796
23	9.260	10.195	11.688	13.090	14.848	32.007	35.172	38.075	41.637	44.179
24	9.886	10.856	12.401	13.848	15.659	33.196	36.415	39.364	42.980	45.558
25	10.519	11.523	13.120	14.611	16.473	34.381	37.652	40.646	44.313	46.925
26	11.160	12.198	13.844	15.379	17.292	35.563	38.885	41.923	45.642	48.290
27	11.807	12.878	14.573	16.151	18.114	36.741	40.113	43.194	46.962	49.642
28	12.461	13.565	15.308	16.928	18.939	37.916	41.337	44.461	48.278	50.993
29	13.120	14.256	16.147	17.708	19.768	39.087	42.557	45.772	49.586	52.333
30	13.787	14.954	16.791	18.493	20.599	40.256	43.773	46.979	50.892	53.672
31	14.457	15.655	17.538	19.280	21.433	41.422	44.985	48.231	52.190	55.000
32	15.134	16.362	18.291	20.072	22.271	42.585	46.194	49.480	53.486	56.328
33	15.814	17.073	19.046	20.866	23.110	43.745	47.400	50.724	54.774	57.646
34	16.501	17.789	19.806	21.664	23.952	44.903	48.602	51.966	56.061	58.964
35	17.191	18.508	20.569	22.465	24.796	46.059	49.802	53.203	57.340	60.272
36	17.887	19.233	21.336	23.269	25.643	47.212	50.998	54.437	58.619	61.581
37	18.584	19.960	22.105	24.075	26.492	48.363	52.192	55.667	59.891	62.880
38	19.289	20.691	22.878	24.884	27.343	49.513	53.384	56.896	61.162	64.181
39	19.994	21.425	23.654	25.695	28.196	50.660	54.572	58.119	62.426	65.473
40	20.706	22.164	24.433	26.509	29.050	51.805	55.758	59.342	63.691	66.766

For $\nu > 40$, $\chi_{\alpha,\nu}^2 \approx \nu\left(1 - \dfrac{2}{9\nu} + z_\alpha \sqrt{\dfrac{2}{9\nu}}\right)^3$

Table A.8 t Curve Tail Areas

t curve
Area to the right of t
0
t

t \ ν	1	2	3	4	5	6	7	8	9	10	11	12	13	14	15	16	17	18
0.0	.500	.500	.500	.500	.500	.500	.500	.500	.500	.500	.500	.500	.500	.500	.500	.500	.500	.500
0.1	.468	.465	.463	.463	.462.	.462	.462	.461	.461	.461	.461	.461	.461	.461	.461	.461	.461	.461
0.2	.437	.430	.427	.426	.425	.424	.424	.423	.423	.423	.423	.422	.422	.422	.422	.422	.422	.422
0.3	.407	.396	.392	.390	.388	.387	.386	.386	.386	.385	.385	.385	.384	.384	.384	.384	.384	.384
0.4	.379	.364	.358	.355	.353	.352	.351	.350	.349	.349	.348	.348	.348	.347	.347	.347	.347	.347
0.5	.352	.333	.326	.322	.319	.317	.316	.315	.315	.314	.313	.313	.313	.312	.312	.312	.312	.312
0.6	.328	.305	.295	.290	.287	.285	.284	.283	.282	.281	.280	.280	.279	.279	.279	.278	.278	.278
0.7	.306	.278	.267	.261	.258	.255	.253	.252	.251	.250	.249	.249	.248	.247	.247	.247	.247	.246
0.8	.285	.254	.241	.234	.230	.227	.225	.223	.222	.221	.220	.220	.219	.218	.218	.218	.217	.217
0.9	.267	.232	.217	.210	.205	.201	.199	.197	.196	.195	.194	.193	.192	.191	.191	.191	.190	.190
1.0	.250	.211	.196	.187	.182	.178	.175	.173	.172	.170	.169	.169	.168	.167	.167	.166	.166	.165
1.1	.235	.193	.176	.167	.162	.157	.154	.152	.150	.149	.147	.146	.146	.144	.144	.144	.143	.143
1.2	.221	.177	.158	.148	.142	.138	.135	.132	.130	.129	.128	.127	.126	.124	.124	.124	.123	.123
1.3	.209	.162	.142	.132	.125	.121	.117	.115	.113	.111	.110	.109	.108	.107	.107	.106	.105	.105
1.4	.197	.148	.128	.117	.110	.106	.102	.100	.098	.096	.095	.093	.092	.091	.091	.090	.090	.089
1.5	.187	.136	.115	.104	.097	.092	.089	.086	.084	.082	.081	.080	.079	.077	.077	.077	.076	.075
1.6	.178	.125	.104	.092	.085	.080	.077	.074	.072	.070	.069	.068	.067	.065	.065	.065	.064	.064
1.7	.169	.116	.094	.082	.075	.070	.065	.064	.062	.060	.059	.057	.056	.055	.055	.054	.054	.053
1.8	.161	.107	.085	.073	.066	.061	.057	.055	.053	.051	.050	.049	.048	.046	.046	.045	.045	.044
1.9	.154	.099	.077	.065	.058	.053	.050	.047	.045	.043	.042	.041	.040	.038	.038	.038	.037	.037
2.0	.148	.092	.070	.058	.051	.046	.043	.040	.038	.037	.035	.034	.033	.032	.032	.031	.031	.030
2.1	.141	.085	.063	.052	.045	.040	.037	.034	.033	.031	.030	.029	.028	.027	.027	.026	.025	.025
2.2	.136	.079	.058	.046	.040	.035	.032	.029	.028	.026	.025	.024	.023	.022	.022	.021	.021	.021
2.3	.131	.074	.052	.041	.035	.031	.027	.025	.023	.022	.021	.020	.019	.018	.018	.018	.017	.017
2.4	.126	.069	.048	.037	.031	.027	.024	.022	.020	.019	.018	.017	.016	.015	.015	.014	.014	.014
2.5	.121	.065	.044	.033	.027	.023	.020	.018	.017	.016	.015	.014	.013	.012	.012	.012	.011	.011
2.6	.117	.061	.040	.030	.024	.020	.018	.016	.014	.013	.012	.012	.011	.010	.010	.010	.009	.009
2.7	.113	.057	.037	.027	.021	.018	.015	.014	.012	.011	.010	.010	.009	.008	.008	.008	.008	.007
2.8	.109	.054	.034	.024	.019	.016	.013	.012	.010	.009	.009	.008	.008	.007	.007	.006	.006	.006
2.9	.106	.051	.031	.022	.017	.014	.011	.010	.009	.008	.007	.007	.006	.005	.005	.005	.005	.005
3.0	.102	.048	.029	.020	.015	.012	.010	.009	.007	.007	.006	.006	.005	.004	.004	.004	.004	.004
3.1	.099	.045	.027	.018	.013	.011	.009	.007	.006	.006	.005	.005	.004	.004	.004	.003	.003	.003
3.2	.096	.043	.025	.016	.012	.009	.008	.006	.005	.005	.004	.004	.003	.003	.003	.003	.003	.002
3.3	.094	.040	.023	.015	.011	.008	.007	.005	.005	.004	.004	.003	.003	.002	.002	.002	.002	.002
3.4	.091	.038	.021	.014	.010	.007	.006	.005	.004	.003	.003	.003	.002	.002	.002	.002	.002	.002
3.5	.089	.036	.020	.012	.009	.006	.005	.004	.003	.003	.002	.002	.002	.002	.002	.001	.001	.001
3.6	.086	.035	.018	.011	.008	.006	.004	.004	.003	.002	.002	.002	.002	.001	.001	.001	.001	.001
3.7	.084	.033	.017	.010	.007	.005	.004	.003	.002	.002	.002	.002	.001	.001	.001	.001	.001	.001
3.8	.082	.031	.016	.010	.006	.004	.003	.003	.002	.002	.001	.001	.001	.001	.001	.001	.001	.001
3.9	.080	.030	.015	.009	.006	.004	.003	.002	.002	.001	.001	.001	.001	.001	.001	.001	.001	.001
4.0	.078	.029	.014	.008	.005	.004	.003	.002	.002	.001	.001	.001	.001	.001	.001	.001	.000	.000

(continued)

Table A.8   *t* Curve Tail Areas *(cont.)*

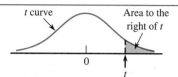

t curve

Area to the right of *t*

0

t

t \ *ν*	19	20	21	22	23	24	25	26	27	28	29	30	35	40	60	120	∞(= *z*)
0.0	.500	.500	.500	.500	.500	.500	.500	.500	.500	.500	.500	.500	.500	.500	.500	.500	.500
0.1	.461	.461	.461	.461	.461	.461	.461	.461	.461	.461	.461	.461	.460	.460	.460	.460	.460
0.2	.422	.422	.422	.422	.422	.422	.422	.422	.421	.421	.421	.421	.421	.421	.421	.421	.421
0.3	.384	.384	.384	.383	.383	.383	.383	.383	.383	.383	.383	.383	.383	.383	.383	.382	.382
0.4	.347	.347	.347	.347	.346	.346	.346	.346	.346	.346	.346	.346	.346	.346	.345	.345	.345
0.5	.311	.311	.311	.311	.311	.311	.311	.311	.311	.310	.310	.310	.310	.310	.309	.309	.309
0.6	.278	.278	.278	.277	.277	.277	.277	.277	.277	.277	.277	.277	.276	.276	.275	.275	.274
0.7	.246	.246	.246	.246	.245	.245	.245	.245	.245	.245	.245	.245	.244	.244	.243	.243	.242
0.8	.217	.217	.216	.216	.216	.216	.216	.215	.215	.215	.215	.215	.215	.214	.213	.213	.212
0.9	.190	.189	.189	.189	.189	.189	.188	.188	.188	.188	.188	.188	.187	.187	.186	.185	.184
1.0	.165	.165	.164	.164	.164	.164	.163	.163	.163	.163	.163	.163	.162	.162	.161	.160	.159
1.1	.143	.142	.142	.142	.141	.141	.141	.141	.141	.140	.140	.140	.139	.139	.138	.137	.136
1.2	.122	.122	.122	.121	.121	.121	.121	.120	.120	.120	.120	.120	.119	.119	.117	.116	.115
1.3	.105	.104	.104	.104	.103	.103	.103	.103	.102	.102	.102	.102	.101	.101	.099	.098	.097
1.4	.089	.089	.088	.088	.087	.087	.087	.087	.086	.086	.086	.086	.085	.085	.083	.082	.081
1.5	.075	.075	.074	.074	.074	.073	.073	.073	.073	.072	.072	.072	.071	.071	.069	.068	.067
1.6	.063	.063	.062	.062	.062	.061	.061	.061	.061	.060	.060	.060	.059	.059	.057	.056	.055
1.7	.053	.052	.052	.052	.051	.051	.051	.051	.050	.050	.050	.050	.049	.048	.047	.046	.045
1.8	.044	.043	.043	.043	.042	.042	.042	.042	.042	.041	.041	.041	.040	.040	.038	.037	.036
1.9	.036	.036	.036	.035	.035	.035	.035	.034	.034	.034	.034	.034	.033	.032	.031	.030	.029
2.0	.030	.030	.029	.029	.029	.028	.028	.028	.028	.028	.027	.027	.027	.026	.025	.024	.023
2.1	.025	.024	.024	.024	.023	.023	.023	.023	.023	.022	.022	.022	.022	.021	.020	.019	.018
2.2	.020	.020	.020	.019	.019	.019	.019	.018	.018	.018	.018	.018	.017	.017	.016	.015	.014
2.3	.016	.016	.016	.016	.015	.015	.015	.015	.015	.015	.014	.014	.014	.013	.012	.012	.011
2.4	.013	.013	.013	.013	.012	.012	.012	.012	.012	.012	.012	.011	.011	.011	.010	.009	.008
2.5	.011	.011	.010	.010	.010	.010	.010	.010	.009	.009	.009	.009	.009	.008	.008	.007	.006
2.6	.009	.009	.008	.008	.008	.008	.008	.008	.007	.007	.007	.007	.007	.007	.006	.005	.005
2.7	.007	.007	.007	.007	.006	.006	.006	.006	.006	.006	.006	.006	.005	.005	.004	.004	.003
2.8	.006	.006	.005	.005	.005	.005	.005	.005	.005	.005	.005	.004	.004	.004	.003	.003	.003
2.9	.005	.004	.004	.004	.004	.004	.004	.004	.004	.004	.004	.003	.003	.003	.003	.002	.002
3.0	.004	.004	.003	.003	.003	.003	.003	.003	.003	.003	.003	.003	.002	.002	.002	.002	.001
3.1	.003	.003	.003	.003	.003	.002	.002	.002	.002	.002	.002	.002	.002	.002	.001	.001	.001
3.2	.002	.002	.002	.002	.002	.002	.002	.002	.002	.002	.002	.002	.001	.001	.001	.001	.001
3.3	.002	.002	.002	.002	.002	.001	.001	.001	.001	.001	.001	.001	.001	.001	.001	.001	.000
3.4	.002	.001	.001	.001	.001	.001	.001	.001	.001	.001	.001	.001	.001	.001	.001	.000	.000
3.5	.001	.001	.001	.001	.001	.001	.001	.001	.001	.001	.001	.001	.001	.001	.000	.000	.000
3.6	.001	.001	.001	.001	.001	.001	.001	.001	.001	.001	.001	.001	.000	.000	.000	.000	.000
3.7	.001	.001	.001	.001	.001	.001	.001	.001	.000	.000	.000	.000	.000	.000	.000	.000	.000
3.8	.001	.001	.001	.000	.000	.000	.000	.000	.000	.000	.000	.000	.000	.000	.000	.000	.000
3.9	.000	.000	.000	.000	.000	.000	.000	.000	.000	.000	.000	.000	.000	.000	.000	.000	.000
4.0	.000	.000	.000	.000	.000	.000	.000	.000	.000	.000	.000	.000	.000	.000	.000	.000	.000

Table A.9 Critical Values for F Distributions

	α	\multicolumn{9}{c}{ν_1 = numerator df}

				ν_1 = numerator df							
ν_2 = denominator df	α	1	2	3	4	5	6	7	8	9
1	.100	39.86	49.50	53.59	55.83	57.24	58.20	58.91	59.44	59.86
	.050	161.45	199.50	215.71	224.58	230.16	233.99	236.77	238.88	240.54
	.010	4052.2	4999.5	5403.4	5624.6	5763.6	5859.0	5928.4	5981.1	6022.5
	.001	405284	500000	540379	562500	576405	585937	592873	598144	602284
2	.100	8.53	9.00	9.16	9.24	9.29	9.33	9.35	9.37	9.38
	.050	18.51	19.00	19.16	19.25	19.30	19.33	19.35	19.37	19.38
	.010	98.50	99.00	99.17	99.25	99.30	99.33	99.36	99.37	99.39
	.001	998.50	999.00	999.17	999.25	999.30	999.33	999.36	999.37	999.39
3	.100	5.54	5.46	5.39	5.34	5.31	5.28	5.27	5.25	5.24
	.050	10.13	9.55	9.28	9.12	9.01	8.94	8.89	8.85	8.81
	.010	34.12	30.82	29.46	28.71	28.24	27.91	27.67	27.49	27.35
	.001	167.03	148.50	141.11	137.10	134.58	132.85	131.58	130.62	129.86
4	.100	4.54	4.32	4.19	4.11	4.05	4.01	3.98	3.95	3.94
	.050	7.71	6.94	6.59	6.39	6.26	6.16	6.09	6.04	6.00
	.010	21.20	18.00	16.69	15.98	15.52	15.21	14.98	14.80	14.66
	.001	74.14	61.25	56.18	53.44	51.71	50.53	49.66	49.00	48.47
5	.100	4.06	3.78	3.62	3.52	3.45	3.40	3.37	3.34	3.32
	.050	6.61	5.79	5.41	5.19	5.05	4.95	4.88	4.82	4.77
	.010	16.26	13.27	12.06	11.39	10.97	10.67	10.46	10.29	10.16
	.001	47.18	37.12	33.20	31.09	29.75	28.83	28.16	27.65	27.24
6	.100	3.78	3.46	3.29	3.18	3.11	3.05	3.01	2.98	2.96
	.050	5.99	5.14	4.76	4.53	4.39	4.28	4.21	4.15	4.10
	.010	13.75	10.92	9.78	9.15	8.75	8.47	8.26	8.10	7.98
	.001	35.51	27.00	23.70	21.92	20.80	20.03	19.46	19.03	18.69
7	.100	3.59	3.26	3.07	2.96	2.88	2.83	2.78	2.75	2.72
	.050	5.59	4.74	4.35	4.12	3.97	3.87	3.79	3.73	3.68
	.010	12.25	9.55	8.45	7.85	7.46	7.19	6.99	6.84	6.72
	.001	29.25	21.69	18.77	17.20	16.21	15.52	15.02	14.63	14.33
8	.100	3.46	3.11	2.92	2.81	2.73	2.67	2.62	2.59	2.56
	.050	5.32	4.46	4.07	3.84	3.69	3.58	3.50	3.44	3.39
	.010	11.26	8.65	7.59	7.01	6.63	6.37	6.18	6.03	5.91
	.001	25.41	18.49	15.83	14.39	13.48	12.86	12.40	12.05	11.77
9	.100	3.36	3.01	2.81	2.69	2.61	2.55	2.51	2.47	2.44
	.050	5.12	4.26	3.86	3.63	3.48	3.37	3.29	3.23	3.18
	.010	10.56	8.02	6.99	6.42	6.06	5.80	5.61	5.47	5.35
	.001	22.86	16.39	13.90	12.56	11.71	11.13	10.70	10.37	10.11
10	.100	3.29	2.92	2.73	2.61	2.52	2.46	2.41	2.38	2.35
	.050	4.96	4.10	3.71	3.48	3.33	3.22	3.14	3.07	3.02
	.010	10.04	7.56	6.55	5.99	5.64	5.39	5.20	5.06	4.94
	.001	21.04	14.91	12.55	11.28	10.48	9.93	9.52	9.20	8.96
11	.100	3.23	2.86	2.66	2.54	2.45	2.39	2.34	2.30	2.27
	.050	4.84	3.98	3.59	3.36	3.20	3.09	3.01	2.95	2.90
	.010	9.65	7.21	6.22	5.67	5.32	5.07	4.89	4.74	4.63
	.001	19.69	13.81	11.56	10.35	9.58	9.05	8.66	8.35	8.12
12	.100	3.18	2.81	2.61	2.48	2.39	2.33	2.28	2.24	2.21
	.050	4.75	3.89	3.49	3.26	3.11	3.00	2.91	2.85	2.80
	.010	9.33	6.93	5.95	5.41	5.06	4.82	4.64	4.50	4.39
	.001	18.64	12.97	10.80	9.63	8.89	8.38	8.00	7.71	7.48

(continued)

Table A.9 Critical Values for *F* Distributions *(cont.)*

ν_1 = numerator df										
10	12	15	20	25	30	40	50	60	120	1000
60.19	60.71	61.22	61.74	62.05	62.26	62.53	62.69	62.79	63.06	63.30
241.88	243.91	245.95	248.01	249.26	250.10	251.14	251.77	252.20	253.25	254.19
6055.8	6106.3	6157.3	6208.7	6239.8	6260.6	6286.8	6302.5	6313.0	6339.4	6362.7
605621	610668	615764	620908	624017	626099	628712	630285	631337	633972	636301
9.39	9.41	9.42	9.44	9.45	9.46	9.47	9.47	9.47	9.48	9.49
19.40	19.41	19.43	19.45	19.46	19.46	19.47	19.48	19.48	19.49	19.49
99.40	99.42	99.43	99.45	99.46	99.47	99.47	99.48	99.48	99.49	99.50
999.40	999.42	999.43	999.45	999.46	999.47	999.47	999.48	999.48	999.49	999.50
5.23	5.22	5.20	5.18	5.17	5.17	5.16	5.15	5.15	5.14	5.13
8.79	8.74	8.70	8.66	8.63	8.62	8.59	8.58	8.57	8.55	8.53
27.23	27.05	26.87	26.69	26.58	26.50	26.41	26.35	26.32	26.22	26.14
129.25	128.32	127.37	126.42	125.84	125.45	124.96	124.66	124.47	123.97	123.53
3.92	3.90	3.87	3.84	3.83	3.82	3.80	3.80	3.79	3.78	3.76
5.96	5.91	5.86	5.80	5.77	5.75	5.72	5.70	5.69	5.66	5.63
14.55	14.37	14.20	14.02	13.91	13.84	13.75	13.69	13.65	13.56	13.47
48.05	47.41	46.76	46.10	45.70	45.43	45.09	44.88	44.75	44.40	44.09
3.30	3.27	3.24	3.21	3.19	3.17	3.16	3.15	3.14	3.12	3.11
4.74	4.68	4.62	4.56	4.52	4.50	4.46	4.44	4.43	4.40	4.37
10.05	9.89	9.72	9.55	9.45	9.38	9.29	9.24	9.20	9.11	9.03
26.92	26.42	25.91	25.39	25.08	24.87	24.60	24.44	24.33	24.06	23.82
2.94	2.90	2.87	2.84	2.81	2.80	2.78	2.77	2.76	2.74	2.72
4.06	4.00	3.94	3.87	3.83	3.81	3.77	3.75	3.74	3.70	3.67
7.87	7.72	7.56	7.40	7.30	7.23	7.14	7.09	7.06	6.97	6.89
18.41	17.99	17.56	17.12	16.85	16.67	16.44	16.31	16.21	15.98	15.77
2.70	2.67	2.63	2.59	2.57	2.56	2.54	2.52	2.51	2.49	2.47
3.64	3.57	3.51	3.44	3.40	3.38	3.34	3.32	3.30	3.27	3.23
6.62	6.47	6.31	6.16	6.06	5.99	5.91	5.86	5.82	5.74	5.66
14.08	13.71	13.32	12.93	12.69	12.53	12.33	12.20	12.12	11.91	11.72
2.54	2.50	2.46	2.42	2.40	2.38	2.36	2.35	2.34	2.32	2.30
3.35	3.28	3.22	3.15	3.11	3.08	3.04	3.02	3.01	2.97	2.93
5.81	5.67	5.52	5.36	5.26	5.20	5.12	5.07	5.03	4.95	4.87
11.54	11.19	10.84	10.48	10.26	10.11	9.92	9.80	9.73	9.53	9.36
2.42	2.38	2.34	2.30	2.27	2.25	2.23	2.22	2.21	2.18	2.16
3.14	3.07	3.01	2.94	2.89	2.86	2.83	2.80	2.79	2.75	2.71
5.26	5.11	4.96	4.81	4.71	4.65	4.57	4.52	4.48	4.40	4.32
9.89	9.57	9.24	8.90	8.69	8.55	8.37	8.26	8.19	8.00	7.84
2.32	2.28	2.24	2.20	2.17	2.16	2.13	2.12	2.11	2.08	2.06
2.98	2.91	2.85	2.77	2.73	2.70	2.66	2.64	2.62	2.58	2.54
4.85	4.71	4.56	4.41	4.31	4.25	4.17	4.12	4.08	4.00	3.92
8.75	8.45	8.13	7.80	7.60	7.47	7.30	7.19	7.12	6.94	6.78
2.25	2.21	2.17	2.12	2.10	2.08	2.05	2.04	2.03	2.00	1.98
2.85	2.79	2.72	2.65	2.60	2.57	2.53	2.51	2.49	2.45	2.41
4.54	4.40	4.25	4.10	4.01	3.94	3.86	3.81	3.78	3.69	3.61
7.92	7.63	7.32	7.01	6.81	6.68	6.52	6.42	6.35	6.18	6.02
2.19	2.15	2.10	2.06	2.03	2.01	1.99	1.97	1.96	1.93	1.91
2.75	2.69	2.62	2.54	2.50	2.47	2.43	2.40	2.38	2.34	2.30
4.30	4.16	4.01	3.86	3.76	3.70	3.62	3.57	3.54	3.45	3.37
7.29	7.00	6.71	6.40	6.22	6.09	5.93	5.83	5.76	5.59	5.44

(continued)

Table A.9 Critical Values for *F* Distributions *(cont.)*

		ν_1 = numerator df								
	α	1	2	3	4	5	6	7	8	9
13	.100	3.14	2.76	2.56	2.43	2.35	2.28	2.23	2.20	2.16
	.050	4.67	3.81	3.41	3.18	3.03	2.92	2.83	2.77	2.71
	.010	9.07	6.70	5.74	5.21	4.86	4.62	4.44	4.30	4.19
	.001	17.82	12.31	10.21	9.07	8.35	7.86	7.49	7.21	6.98
14	.100	3.10	2.73	2.52	2.39	2.31	2.24	2.19	2.15	2.12
	.050	4.60	3.74	3.34	3.11	2.96	2.85	2.76	2.70	2.65
	.010	8.86	6.51	5.56	5.04	4.69	4.46	4.28	4.14	4.03
	.001	17.14	11.78	9.73	8.62	7.92	7.44	7.08	6.80	6.58
15	.100	3.07	2.70	2.49	2.36	2.27	2.21	2.16	2.12	2.09
	.050	4.54	3.68	3.29	3.06	2.90	2.79	2.71	2.64	2.59
	.010	8.68	6.36	5.42	4.89	4.56	4.32	4.14	4.00	3.89
	.001	16.59	11.34	9.34	8.25	7.57	7.09	6.74	6.47	6.26
16	.100	3.05	2.67	2.46	2.33	2.24	2.18	2.13	2.09	2.06
	.050	4.49	3.63	3.24	3.01	2.85	2.74	2.66	2.59	2.54
	.010	8.53	6.23	5.29	4.77	4.44	4.20	4.03	3.89	3.78
	.001	16.12	10.97	9.01	7.94	7.27	6.80	6.46	6.19	5.98
17	.100	3.03	2.64	2.44	2.31	2.22	2.15	2.10	2.06	2.03
	.050	4.45	3.59	3.20	2.96	2.81	2.70	2.61	2.55	2.49
	.010	8.40	6.11	5.19	4.67	4.34	4.10	3.93	3.79	3.68
	.001	15.72	10.66	8.73	7.68	7.02	6.56	6.22	5.96	5.75
18	.100	3.01	2.62	2.42	2.29	2.20	2.13	2.08	2.04	2.00
	.050	4.41	3.55	3.16	2.93	2.77	2.66	2.58	2.51	2.46
	.010	8.29	6.01	5.09	4.58	4.25	4.01	3.84	3.71	3.60
	.001	15.38	10.39	8.49	7.46	6.81	6.35	6.02	5.76	5.56
19	.100	2.99	2.61	2.40	2.27	2.18	2.11	2.06	2.02	1.98
	.050	4.38	3.52	3.13	2.90	2.74	2.63	2.54	2.48	2.42
	.010	8.18	5.93	5.01	4.50	4.17	3.94	3.77	3.63	3.52
	.001	15.08	10.16	8.28	7.27	6.62	6.18	5.85	5.59	5.39
20	.100	2.97	2.59	2.38	2.25	2.16	2.09	2.04	2.00	1.96
	.050	4.35	3.49	3.10	2.87	2.71	2.60	2.51	2.45	2.39
	.010	8.10	5.85	4.94	4.43	4.10	3.87	3.70	3.56	3.46
	.001	14.82	9.95	8.10	7.10	6.46	6.02	5.69	5.44	5.24
21	.100	2.96	2.57	2.36	2.23	2.14	2.08	2.02	1.98	1.95
	.050	4.32	3.47	3.07	2.84	2.68	2.57	2.49	2.42	2.37
	.010	8.02	5.78	4.87	4.37	4.04	3.81	3.64	3.51	3.40
	.001	14.59	9.77	7.94	6.95	6.32	5.88	5.56	5.31	5.11
22	.100	2.95	2.56	2.35	2.22	2.13	2.06	2.01	1.97	1.93
	.050	4.30	3.44	3.05	2.82	2.66	2.55	2.46	2.40	2.34
	.010	7.95	5.72	4.82	4.31	3.99	3.76	3.59	3.45	3.35
	.001	14.38	9.61	7.80	6.81	6.19	5.76	5.44	5.19	4.99
23	.100	2.94	2.55	2.34	2.21	2.11	2.05	1.99	1.95	1.92
	.050	4.28	3.42	3.03	2.80	2.64	2.53	2.44	2.37	2.32
	.010	7.88	5.66	4.76	4.26	3.94	3.71	3.54	3.41	3.30
	.001	14.20	9.47	7.67	6.70	6.08	5.65	5.33	5.09	4.89
24	.100	2.93	2.54	2.33	2.19	2.10	2.04	1.98	1.94	1.91
	.050	4.26	3.40	3.01	2.78	2.62	2.51	2.42	2.36	2.30
	.010	7.82	5.61	4.72	4.22	3.90	3.67	3.50	3.36	3.26
	.001	14.03	9.34	7.55	6.59	5.98	5.55	5.23	4.99	4.80

ν_2 = denominator df

(continued)

Table A.9   Critical Values for *F* Distributions *(cont.)*

				ν_1 = numerator df						
10	**12**	**15**	**20**	**25**	**30**	**40**	**50**	**60**	**120**	**1000**
2.14	2.10	2.05	2.01	1.98	1.96	1.93	1.92	1.90	1.88	1.85
2.67	2.60	2.53	2.46	2.41	2.38	2.34	2.31	2.30	2.25	2.21
4.10	3.96	3.82	3.66	3.57	3.51	3.43	3.38	3.34	3.25	3.18
6.80	6.52	6.23	5.93	5.75	5.63	5.47	5.37	5.30	5.14	4.99
2.10	2.05	2.01	1.96	1.93	1.91	1.89	1.87	1.86	1.83	1.80
2.60	2.53	2.46	2.39	2.34	2.31	2.27	2.24	2.22	2.18	2.14
3.94	3.80	3.66	3.51	3.41	3.35	3.27	3.22	3.18	3.09	3.02
6.40	6.13	5.85	5.56	5.38	5.25	5.10	5.00	4.94	4.77	4.62
2.06	2.02	1.97	1.92	1.89	1.87	1.85	1.83	1.82	1.79	1.76
2.54	2.48	2.40	2.33	2.28	2.25	2.20	2.18	2.16	2.11	2.07
3.80	3.67	3.52	3.37	3.28	3.21	3.13	3.08	3.05	2.96	2.88
6.08	5.81	5.54	5.25	5.07	4.95	4.80	4.70	4.64	4.47	4.33
2.03	1.99	1.94	1.89	1.86	1.84	1.81	1.79	1.78	1.75	1.72
2.49	2.42	2.35	2.28	2.23	2.19	2.15	2.12	2.11	2.06	2.02
3.69	3.55	3.41	3.26	3.16	3.10	3.02	2.97	2.93	2.84	2.76
5.81	5.55	5.27	4.99	4.82	4.70	4.54	4.45	4.39	4.23	4.08
2.00	1.96	1.91	1.86	1.83	1.81	1.78	1.76	1.75	1.72	1.69
2.45	2.38	2.31	2.23	2.18	2.15	2.10	2.08	2.06	2.01	1.97
3.59	3.46	3.31	3.16	3.07	3.00	2.92	2.87	2.83	2.75	2.66
5.58	5.32	5.05	4.78	4.60	4.48	4.33	4.24	4.18	4.02	3.87
1.98	1.93	1.89	1.84	1.80	1.78	1.75	1.74	1.72	1.69	1.66
2.41	2.34	2.27	2.19	2.14	2.11	2.06	2.04	2.02	1.97	1.92
3.51	3.37	3.23	3.08	2.98	2.92	2.84	2.78	2.75	2.66	2.58
5.39	5.13	4.87	4.59	4.42	4.30	4.15	4.06	4.00	3.84	3.69
1.96	1.91	1.86	1.81	1.78	1.76	1.73	1.71	1.70	1.67	1.64
2.38	2.31	2.23	2.16	2.11	2.07	2.03	2.00	1.98	1.93	1.88
3.43	3.30	3.15	3.00	2.91	2.84	2.76	2.71	2.67	2.58	2.50
5.22	4.97	4.70	4.43	4.26	4.14	3.99	3.90	3.84	3.68	3.53
1.94	1.89	1.84	1.79	1.76	1.74	1.71	1.69	1.68	1.64	1.61
2.35	2.28	2.20	2.12	2.07	2.04	1.99	1.97	1.95	1.90	1.85
3.37	3.23	3.09	2.94	2.84	2.78	2.69	2.64	2.61	2.52	2.43
5.08	4.82	4.56	4.29	4.12	4.00	3.86	3.77	3.70	3.54	3.40
1.92	1.87	1.83	1.78	1.74	1.72	1.69	1.67	1.66	1.62	1.59
2.32	2.25	2.18	2.10	2.05	2.01	1.96	1.94	1.92	1.87	1.82
3.31	3.17	3.03	2.88	2.79	2.72	2.64	2.58	2.55	2.46	2.37
4.95	4.70	4.44	4.17	4.00	3.88	3.74	3.64	3.58	3.42	3.28
1.90	1.86	1.81	1.76	1.73	1.70	1.67	1.65	1.64	1.60	1.57
2.30	2.23	2.15	2.07	2.02	1.98	1.94	1.91	1.89	1.84	1.79
3.26	3.12	2.98	2.83	2.73	2.67	2.58	2.53	2.50	2.40	2.32
4.83	4.58	4.33	4.06	3.89	3.78	3.63	3.54	3.48	3.32	3.17
1.89	1.84	1.80	1.74	1.71	1.69	1.66	1.64	1.62	1.59	1.55
2.27	2.20	2.13	2.05	2.00	1.96	1.91	1.88	1.86	1.81	1.76
3.21	3.07	2.93	2.78	2.69	2.62	2.54	2.48	2.45	2.35	2.27
4.73	4.48	4.23	3.96	3.79	3.68	3.53	3.44	3.38	3.22	3.08
1.88	1.83	1.78	1.73	1.70	1.67	1.64	1.62	1.61	1.57	1.54
2.25	2.18	2.11	2.03	1.97	1.94	1.89	1.86	1.84	1.79	1.74
3.17	3.03	2.89	2.74	2.64	2.58	2.49	2.44	2.40	2.31	2.22
4.64	4.39	4.14	3.87	3.71	3.59	3.45	3.36	3.29	3.14	2.99

(continued)

Table A.9 Critical Values for *F* Distributions *(cont.)*

						v_1 = numerator df				
	α	1	2	3	4	5	6	7	8	9
25	.100	2.92	2.53	2.32	2.18	2.09	2.02	1.97	1.93	1.89
	.050	4.24	3.39	2.99	2.76	2.60	2.49	2.40	2.34	2.28
	.010	7.77	5.57	4.68	4.18	3.85	3.63	3.46	3.32	3.22
	.001	13.88	9.22	7.45	6.49	5.89	5.46	5.15	4.91	4.71
26	.100	2.91	2.52	2.31	2.17	2.08	2.01	1.96	1.92	1.88
	.050	4.23	3.37	2.98	2.74	2.59	2.47	2.39	2.32	2.27
	.010	7.72	5.53	4.64	4.14	3.82	3.59	3.42	3.29	3.18
	.001	13.74	9.12	7.36	6.41	5.80	5.38	5.07	4.83	4.64
27	.100	2.90	2.51	2.30	2.17	2.07	2.00	1.95	1.91	1.87
	.050	4.21	3.35	2.96	2.73	2.57	2.46	2.37	2.31	2.25
	.010	7.68	5.49	4.60	4.11	3.78	3.56	3.39	3.26	3.15
	.001	13.61	9.02	7.27	6.33	5.73	5.31	5.00	4.76	4.57
28	.100	2.89	2.50	2.29	2.16	2.06	2.00	1.94	1.90	1.87
	.050	4.20	3.34	2.95	2.71	2.56	2.45	2.36	2.29	2.24
	.010	7.64	5.45	4.57	4.07	3.75	3.53	3.36	3.23	3.12
	.001	13.50	8.93	7.19	6.25	5.66	5.24	4.93	4.69	4.50
29	.100	2.89	2.50	2.28	2.15	2.06	1.99	1.93	1.89	1.86
	.050	4.18	3.33	2.93	2.70	2.55	2.43	2.35	2.28	2.22
	.010	7.60	5.42	4.54	4.04	3.73	3.50	3.33	3.20	3.09
	.001	13.39	8.85	7.12	6.19	5.59	5.18	4.87	4.64	4.45
30	.100	2.88	2.49	2.28	2.14	2.05	1.98	1.93	1.88	1.85
	.050	4.17	3.32	2.92	2.69	2.53	2.42	2.33	2.27	2.21
	.010	7.56	5.39	4.51	4.02	3.70	3.47	3.30	3.17	3.07
	.001	13.29	8.77	7.05	6.12	5.53	5.12	4.82	4.58	4.39
40	.100	2.84	2.44	2.23	2.09	2.00	1.93	1.87	1.83	1.79
	.050	4.08	3.23	2.84	2.61	2.45	2.34	2.25	2.18	2.12
	.010	7.31	5.18	4.31	3.83	3.51	3.29	3.12	2.99	2.89
	.001	12.61	8.25	6.59	5.70	5.13	4.73	4.44	4.21	4.02
50	.100	2.81	2.41	2.20	2.06	1.97	1.90	1.84	1.80	1.76
	.050	4.03	3.18	2.79	2.56	2.40	2.29	2.20	2.13	2.07
	.010	7.17	5.06	4.20	3.72	3.41	3.19	3.02	2.89	2.78
	.001	12.22	7.96	6.34	5.46	4.90	4.51	4.22	4.00	3.82
60	.100	2.79	2.39	2.18	2.04	1.95	1.87	1.82	1.77	1.74
	.050	4.00	3.15	2.76	2.53	2.37	2.25	2.17	2.10	2.04
	.010	7.08	4.98	4.13	3.65	3.34	3.12	2.95	2.82	2.72
	.001	11.97	7.77	6.17	5.31	4.76	4.37	4.09	3.86	3.69
100	.100	2.76	2.36	2.14	2.00	1.91	1.83	1.78	1.73	1.69
	.050	3.94	3.09	2.70	2.46	2.31	2.19	2.10	2.03	1.97
	.010	6.90	4.82	3.98	3.51	3.21	2.99	2.82	2.69	2.59
	.001	11.50	7.41	5.86	5.02	4.48	4.11	3.83	3.61	3.44
200	.100	2.73	2.33	2.11	1.97	1.88	1.80	1.75	1.70	1.66
	.050	3.89	3.04	2.65	2.42	2.26	2.14	2.06	1.98	1.93
	.010	6.76	4.71	3.88	3.41	3.11	2.89	2.73	2.60	2.50
	.001	11.15	7.15	5.63	4.81	4.29	3.92	3.65	3.43	3.26
1000	.100	2.71	2.31	2.09	1.95	1.85	1.78	1.72	1.68	1.64
	.050	3.85	3.00	2.61	2.38	2.22	2.11	2.02	1.95	1.89
	.010	6.66	4.63	3.80	3.34	3.04	2.82	2.66	2.53	2.43
	.001	10.89	6.96	5.46	4.65	4.14	3.78	3.51	3.30	3.13

v_2 = denominator df

(continued)

Table A.9   Critical Values for *F* Distributions *(cont.)*

					v_1 = numerator df					
10	**12**	**15**	**20**	**25**	**30**	**40**	**50**	**60**	**120**	**1000**
1.87	1.82	1.77	1.72	1.68	1.66	1.63	1.61	1.59	1.56	1.52
2.24	2.16	2.09	2.01	1.96	1.92	1.87	1.84	1.82	1.77	1.72
3.13	2.99	2.85	2.70	2.60	2.54	2.45	2.40	2.36	2.27	2.18
4.56	4.31	4.06	3.79	3.63	3.52	3.37	3.28	3.22	3.06	2.91
1.86	1.81	1.76	1.71	1.67	1.65	1.61	1.59	1.58	1.54	1.51
2.22	2.15	2.07	1.99	1.94	1.90	1.85	1.82	1.80	1.75	1.70
3.09	2.96	2.81	2.66	2.57	2.50	2.42	2.36	2.33	2.23	2.14
4.48	4.24	3.99	3.72	3.56	3.44	3.30	3.21	3.15	2.99	2.84
1.85	1.80	1.75	1.70	1.66	1.64	1.60	1.58	1.57	1.53	1.50
2.20	2.13	2.06	1.97	1.92	1.88	1.84	1.81	1.79	1.73	1.68
3.06	2.93	2.78	2.63	2.54	2.47	2.38	2.33	2.29	2.20	2.11
4.41	4.17	3.92	3.66	3.49	3.38	3.23	3.14	3.08	2.92	2.78
1.84	1.79	1.74	1.69	1.65	1.63	1.59	1.57	1.56	1.52	1.48
2.19	2.12	2.04	1.96	1.91	1.87	1.82	1.79	1.77	1.71	1.66
3.03	2.90	2.75	2.60	2.51	2.44	2.35	2.30	2.26	2.17	2.08
4.35	4.11	3.86	3.60	3.43	3.32	3.18	3.09	3.02	2.86	2.72
1.83	1.78	1.73	1.68	1.64	1.62	1.58	1.56	1.55	1.51	1.47
2.18	2.10	2.03	1.94	1.89	1.85	1.81	1.77	1.75	1.70	1.65
3.00	2.87	2.73	2.57	2.48	2.41	2.33	2.27	2.23	2.14	2.05
4.29	4.05	3.80	3.54	3.38	3.27	3.12	3.03	2.97	2.81	2.66
1.82	1.77	1.72	1.67	1.63	1.61	1.57	1.55	1.54	1.50	1.46
2.16	2.09	2.01	1.93	1.88	1.84	1.79	1.76	1.74	1.68	1.63
2.98	2.84	2.70	2.55	2.45	2.39	2.30	2.25	2.21	2.11	2.02
4.24	4.00	3.75	3.49	3.33	3.22	3.07	2.98	2.92	2.76	2.61
1.76	1.71	1.66	1.61	1.57	1.54	1.51	1.48	1.47	1.42	1.38
2.08	2.00	1.92	1.84	1.78	1.74	1.69	1.66	1.64	1.58	1.52
2.80	2.66	2.52	2.37	2.27	2.20	2.11	2.06	2.02	1.92	1.82
3.87	3.64	3.40	3.14	2.98	2.87	2.73	2.64	2.57	2.41	2.25
1.73	1.68	1.63	1.57	1.53	1.50	1.46	1.44	1.42	1.38	1.33
2.03	1.95	1.87	1.78	1.73	1.69	1.63	1.60	1.58	1.51	1.45
2.70	2.56	2.42	2.27	2.17	2.10	2.01	1.95	1.91	1.80	1.70
3.67	3.44	3.20	2.95	2.79	2.68	2.53	2.44	2.38	2.21	2.05
1.71	1.66	1.60	1.54	1.50	1.48	1.44	1.41	1.40	1.35	1.30
1.99	1.92	1.84	1.75	1.69	1.65	1.59	1.56	1.53	1.47	1.40
2.63	2.50	2.35	2.20	2.10	2.03	1.94	1.88	1.84	1.73	1.62
3.54	3.32	3.08	2.83	2.67	2.55	2.41	2.32	2.25	2.08	1.92
1.66	1.61	1.56	1.49	1.45	1.42	1.38	1.35	1.34	1.28	1.22
1.93	1.85	1.77	1.68	1.62	1.57	1.52	1.48	1.45	1.38	1.30
2.50	2.37	2.22	2.07	1.97	1.89	1.80	1.74	1.69	1.57	1.45
3.30	3.07	2.84	2.59	2.43	2.32	2.17	2.08	2.01	1.83	1.64
1.63	1.58	1.52	1.46	1.41	1.38	1.34	1.31	1.29	1.23	1.16
1.88	1.80	1.72	1.62	1.56	1.52	1.46	1.41	1.39	1.30	1.21
2.41	2.27	2.13	1.97	1.87	1.79	1.69	1.63	1.58	1.45	1.30
3.12	2.90	2.67	2.42	2.26	2.15	2.00	1.90	1.83	1.64	1.43
1.61	1.55	1.49	1.43	1.38	1.35	1.30	1.27	1.25	1.18	1.08
1.84	1.76	1.68	1.58	1.52	1.47	1.41	1.36	1.33	1.24	1.11
2.34	2.20	2.06	1.90	1.79	1.72	1.61	1.54	1.50	1.35	1.16
2.99	2.77	2.54	2.30	2.14	2.02	1.87	1.77	1.69	1.49	1.22

Table A.10 Critical Values for Studentized Range Distributions

							m					
v	*α*	2	3	4	5	6	7	8	9	10	11	12
5	.05	3.64	4.60	5.22	5.67	6.03	6.33	6.58	6.80	6.99	7.17	7.32
	.01	5.70	6.98	7.80	8.42	8.91	9.32	9.67	9.97	10.24	10.48	10.70
6	.05	3.46	4.34	4.90	5.30	5.63	5.90	6.12	6.32	6.49	6.65	6.79
	.01	5.24	6.33	7.03	7.56	7.97	8.32	8.61	8.87	9.10	9.30	9.48
7	.05	3.34	4.16	4.68	5.06	5.36	5.61	5.82	6.00	6.16	6.30	6.43
	.01	4.95	5.92	6.54	7.01	7.37	7.68	7.94	8.17	8.37	8.55	8.71
8	.05	3.26	4.04	4.53	4.89	5.17	5.40	5.60	5.77	5.92	6.05	6.18
	.01	4.75	5.64	6.20	6.62	6.96	7.24	7.47	7.68	7.86	8.03	8.18
9	.05	3.20	3.95	4.41	4.76	5.02	5.24	5.43	5.59	5.74	5.87	5.98
	.01	4.60	5.43	5.96	6.35	6.66	6.91	7.13	7.33	7.49	7.65	7.78
10	.05	3.15	3.88	4.33	4.65	4.91	5.12	5.30	5.46	5.60	5.72	5.83
	.01	4.48	5.27	5.77	6.14	6.43	6.67	6.87	7.05	7.21	7.36	7.49
11	.05	3.11	3.82	4.26	4.57	4.82	5.03	5.20	5.35	5.49	5.61	5.71
	.01	4.39	5.15	5.62	5.97	6.25	6.48	6.67	6.84	6.99	7.13	7.25
12	.05	3.08	3.77	4.20	4.51	4.75	4.95	5.12	5.27	5.39	5.51	5.61
	.01	4.32	5.05	5.50	5.84	6.10	6.32	6.51	6.67	6.81	6.94	7.06
13	.05	3.06	3.73	4.15	4.45	4.69	4.88	5.05	5.19	5.32	5.43	5.53
	.01	4.26	4.96	5.40	5.73	5.98	6.19	6.37	6.53	6.67	6.79	6.90
14	.05	3.03	3.70	4.11	4.41	4.64	4.83	4.99	5.13	5.25	5.36	5.46
	.01	4.21	4.89	5.32	5.63	5.88	6.08	6.26	6.41	6.54	6.66	6.77
15	.05	3.01	3.67	4.08	4.37	4.59	4.78	4.94	5.08	5.20	5.31	5.40
	.01	4.17	4.84	5.25	5.56	5.80	5.99	6.16	6.31	6.44	6.55	6.66
16	.05	3.00	3.65	4.05	4.33	4.56	4.74	4.90	5.03	5.15	5.26	5.35
	.01	4.13	4.79	5.19	5.49	5.72	5.92	6.08	6.22	6.35	6.46	6.56
17	.05	2.98	3.63	4.02	4.30	4.52	4.70	4.86	4.99	5.11	5.21	5.31
	.01	4.10	4.74	5.14	5.43	5.66	5.85	6.01	6.15	6.27	6.38	6.48
18	.05	2.97	3.61	4.00	4.28	4.49	4.67	4.82	4.96	5.07	5.17	5.27
	.01	4.07	4.70	5.09	5.38	5.60	5.79	5.94	6.08	6.20	6.31	6.41
19	.05	2.96	3.59	3.98	4.25	4.47	4.65	4.79	4.92	5.04	5.14	5.23
	.01	4.05	4.67	5.05	5.33	5.55	5.73	5.89	6.02	6.14	6.25	6.34
20	.05	2.95	3.58	3.96	4.23	4.45	4.62	4.77	4.90	5.01	5.11	5.20
	.01	4.02	4.64	5.02	5.29	5.51	5.69	5.84	5.97	6.09	6.19	6.28
24	.05	2.92	3.53	3.90	4.17	4.37	4.54	4.68	4.81	4.92	5.01	5.10
	.01	3.96	4.55	4.91	5.17	5.37	5.54	5.69	5.81	5.92	6.02	6.11
30	.05	2.89	3.49	3.85	4.10	4.30	4.46	4.60	4.72	4.82	4.92	5.00
	.01	3.89	4.45	4.80	5.05	5.24	5.40	5.54	5.65	5.76	5.85	5.93
40	.05	2.86	3.44	3.79	4.04	4.23	4.39	4.52	4.63	4.73	4.82	4.90
	.01	3.82	4.37	4.70	4.93	5.11	5.26	5.39	5.50	5.60	5.69	5.76
60	.05	2.83	3.40	3.74	3.98	4.16	4.31	4.44	4.55	4.65	4.73	4.81
	.01	3.76	4.28	4.59	4.82	4.99	5.13	5.25	5.36	5.45	5.53	5.60
120	.05	2.80	3.36	3.68	3.92	4.10	4.24	4.36	4.47	4.56	4.64	4.71
	.01	3.70	4.20	4.50	4.71	4.87	5.01	5.12	5.21	5.30	5.37	5.44
∞	.05	2.77	3.31	3.63	3.86	4.03	4.17	4.29	4.39	4.47	4.55	4.62
	.01	3.64	4.12	4.40	4.60	4.76	4.88	4.99	5.08	5.16	5.23	5.29

Table A.11 Chi-Squared Curve Tail Areas

Upper-tail area	$\nu = 1$	$\nu = 2$	$\nu = 3$	$\nu = 4$	$\nu = 5$
> .100	< 2.70	< 4.60	< 6.25	< 7.77	< 9.23
.100	2.70	4.60	6.25	7.77	9.23
.095	2.78	4.70	6.36	7.90	9.37
.090	2.87	4.81	6.49	8.04	9.52
.085	2.96	4.93	6.62	8.18	9.67
.080	3.06	5.05	6.75	8.33	9.83
.075	3.17	5.18	6.90	8.49	10.00
.070	3.28	5.31	7.06	8.66	10.19
.065	3.40	5.46	7.22	8.84	10.38
.060	3.53	5.62	7.40	9.04	10.59
.055	3.68	5.80	7.60	9.25	10.82
.050	3.84	5.99	7.81	9.48	11.07
.045	4.01	6.20	8.04	9.74	11.34
.040	4.21	6.43	8.31	10.02	11.64
.035	4.44	6.70	8.60	10.34	11.98
.030	4.70	7.01	8.94	10.71	12.37
.025	5.02	7.37	9.34	11.14	12.83
.020	5.41	7.82	9.83	11.66	13.38
.015	5.91	8.39	10.46	12.33	14.09
.010	6.63	9.21	11.34	13.27	15.08
.005	7.87	10.59	12.83	14.86	16.74
.001	10.82	13.81	16.26	18.46	20.51
< .001	> 10.82	> 13.81	> 16.26	> 18.46	> 20.51

Upper-tail area	$\nu = 6$	$\nu = 7$	$\nu = 8$	$\nu = 9$	$\nu = 10$
> .100	< 10.64	< 12.01	< 13.36	< 14.68	< 15.98
.100	10.64	12.01	13.36	14.68	15.98
.095	10.79	12.17	13.52	14.85	16.16
.090	10.94	12.33	13.69	15.03	16.35
.085	11.11	12.50	13.87	15.22	16.54
.080	11.28	12.69	14.06	15.42	16.75
.075	11.46	12.88	14.26	15.63	16.97
.070	11.65	13.08	14.48	15.85	17.20
.065	11.86	13.30	14.71	16.09	17.44
.060	12.08	13.53	14.95	16.34	17.71
.055	12.33	13.79	15.22	16.62	17.99
.050	12.59	14.06	15.50	16.91	18.30
.045	12.87	14.36	15.82	17.24	18.64
.040	13.19	14.70	16.17	17.60	19.02
.035	13.55	15.07	16.56	18.01	19.44
.030	13.96	15.50	17.01	18.47	19.92
.025	14.44	16.01	17.53	19.02	20.48
.020	15.03	16.62	18.16	19.67	21.16
.015	15.77	17.39	18.97	20.51	22.02
.010	16.81	18.47	20.09	21.66	23.20
.005	18.54	20.27	21.95	23.58	25.18
.001	22.45	24.32	26.12	27.87	29.58
< .001	> 22.45	> 24.32	> 26.12	> 27.87	> 29.58

(continued)

Table A.11 Chi-Squared Curve Tail Areas *(cont.)*

Upper-tail area	$\nu = 11$	$\nu = 12$	$\nu = 13$	$\nu = 14$	$\nu = 15$
> .100	< 17.27	< 18.54	< 19.81	< 21.06	< 22.30
.100	17.27	18.54	19.81	21.06	22.30
.095	17.45	18.74	20.00	21.26	22.51
.090	17.65	18.93	20.21	21.47	22.73
.085	17.85	19.14	20.42	21.69	22.95
.080	18.06	19.36	20.65	21.93	23.19
.075	18.29	19.60	20.89	22.17	23.45
.070	18.53	19.84	21.15	22.44	23.72
.065	18.78	20.11	21.42	22.71	24.00
.060	19.06	20.39	21.71	23.01	24.31
.055	19.35	20.69	22.02	23.33	24.63
.050	19.67	21.02	22.36	23.68	24.99
.045	20.02	21.38	22.73	24.06	25.38
.040	20.41	21.78	23.14	24.48	25.81
.035	20.84	22.23	23.60	24.95	26.29
.030	21.34	22.74	24.12	25.49	26.84
.025	21.92	23.33	24.73	26.11	27.48
.020	22.61	24.05	25.47	26.87	28.25
.015	23.50	24.96	26.40	27.82	29.23
.010	24.72	26.21	27.68	29.14	30.57
.005	26.75	28.29	29.81	31.31	32.80
.001	31.26	32.90	34.52	36.12	37.69
< .001	> 31.26	> 32.90	> 34.52	> 36.12	> 37.69

Upper-tail area	$\nu = 16$	$\nu = 17$	$\nu = 18$	$\nu = 19$	$\nu = 20$
> .100	< 23.54	< 24.77	< 25.98	< 27.20	< 28.41
.100	23.54	24.76	25.98	27.20	28.41
.095	23.75	24.98	26.21	27.43	28.64
.090	23.97	25.21	26.44	27.66	28.88
.085	24.21	25.45	26.68	27.91	29.14
.080	24.45	25.70	26.94	28.18	29.40
.075	24.71	25.97	27.21	28.45	29.69
.070	24.99	26.25	27.50	28.75	29.99
.065	25.28	26.55	27.81	29.06	30.30
.060	25.59	26.87	28.13	29.39	30.64
.055	25.93	27.21	28.48	29.75	31.01
.050	26.29	27.58	28.86	30.14	31.41
.045	26.69	27.99	29.28	30.56	31.84
.040	27.13	28.44	29.74	31.03	32.32
.035	27.62	28.94	30.25	31.56	32.85
.030	28.19	29.52	30.84	32.15	33.46
.025	28.84	30.19	31.52	32.85	34.16
.020	29.63	30.99	32.34	33.68	35.01
.015	30.62	32.01	33.38	34.74	36.09
.010	32.00	33.40	34.80	36.19	37.56
.005	34.26	35.71	37.15	38.58	39.99
.001	39.25	40.78	42.31	43.81	45.31
< .001	> 39.25	> 40.78	> 42.31	> 43.81	> 45.31

Table A.12 Critical Values for the Ryan–Joiner Test of Normality

		α	
	.10	**.05**	**.01**
5	.9033	.8804	.8320
10	.9347	.9180	.8804
15	.9506	.9383	.9110
20	.9600	.9503	.9290
25	.9662	.9582	.9408
30	.9707	.9639	.9490
40	.9767	.9715	.9597
50	.9807	.9764	.9664
60	.9835	.9799	.9710
75	.9865	.9835	.9757

n

Table A.13 Critical Values for the Wilcoxon Signed-Rank Test

$$P_0(S_+ \geq c_1) = P(S_+ \geq c_1 \text{ when } H_0 \text{ is true})$$

n	c_1	$P_0(S_+ \geq c_1)$	n	c_1	$P_0(S_+ \geq c_1)$
3	6	.125		78	.011
4	9	.125		79	.009
	10	.062		81	.005
5	13	.094	14	73	.108
	14	.062		74	.097
	15	.031		79	.052
6	17	.109		84	.025
	19	.047		89	.010
	20	.031		92	.005
	21	.016	15	83	.104
7	22	.109		84	.094
	24	.055		89	.053
	26	.023		90	.047
	28	.008		95	.024
8	28	.098		100	.011
	30	.055		101	.009
	32	.027		104	.005
	34	.012	16	93	.106
	35	.008		94	.096
	36	.004		100	.052
9	34	.102		106	.025
	37	.049		112	.011
	39	.027		113	.009
	42	.010		116	.005
	44	.004	17	104	.103
10	41	.097		105	.095
	44	.053		112	.049
	47	.024		118	.025
	50	.010		125	.010
	52	.005		129	.005
11	48	.103	18	116	.098
	52	.051		124	.049
	55	.027		131	.024
	59	.009		138	.010
	61	.005		143	.005
12	56	.102	19	128	.098
	60	.055		136	.052
	61	.046		137	.048
	64	.026		144	.025
	68	.010		152	.010
	71	.005		157	.005
13	64	.108	20	140	.101
	65	.095		150	.049
	69	.055		158	.024
	70	.047		167	.010
	74	.024		172	.005

Table A.14 Critical Values for the Wilcoxon Rank-Sum Test

$$P_0(W \geq c) = P(W \geq c \text{ when } H_0 \text{ is true})$$

m	n	c	$P_0(W \geq c)$	m	n	c	$P_0(W \geq c)$
3	3	15	.05			40	.004
	4	17	.057		6	40	.041
		18	.029			41	.026
	5	20	.036			43	.009
		21	.018			44	.004
	6	22	.048		7	43	.053
		23	.024			45	.024
		24	.012			47	.009
	7	24	.058			48	.005
		26	.017		8	47	.047
		27	.008			49	.023
	8	27	.042			51	.009
		28	.024			52	.005
		29	.012	6	6	50	.047
		30	.006			52	.021
4	4	24	.057			54	.008
		25	.029			55	.004
		26	.014		7	54	.051
	5	27	.056			56	.026
		28	.032			58	.011
		29	.016			60	.004
		30	.008		8	58	.054
	6	30	.057			61	.021
		32	.019			63	.01
		33	.010			65	.004
		34	.005	7	7	66	.049
	7	33	.055			68	.027
		35	.021			71	.009
		36	.012			72	.006
		37	.006		8	71	.047
	8	36	.055			73	.027
		38	.024			76	.01
		40	.008			78	.005
		41	.004	8	8	84	.052
5	5	36	.048			87	.025
		37	.028			90	.01
		39	.008			92	.005

Table A.15 Critical Values for the Wilcoxon Signed-Rank Interval $(\bar{x}_{(n(n+1)/2-c+1)}, \bar{x}_{(c)})$

n	Confidence Level (%)	c	n	Confidence Level (%)	c	n	Confidence Level (%)	c
5	93.8	15	13	99.0	81	20	99.1	173
	87.5	14		95.2	74		95.2	158
6	96.9	21		90.6	70		90.3	150
	93.7	20	14	99.1	93	21	99.0	188
	90.6	19		95.1	84		95.0	172
7	98.4	28		89.6	79		89.7	163
	95.3	26	15	99.0	104	22	99.0	204
	89.1	24		95.2	95		95.0	187
8	99.2	36		90.5	90		90.2	178
	94.5	32	16	99.1	117	23	99.0	221
	89.1	30		94.9	106		95.2	203
9	99.2	44		89.5	100		90.2	193
	94.5	39	17	99.1	130	24	99.0	239
	90.2	37		94.9	118		95.1	219
10	99.0	52		90.2	112		89.9	208
	95.1	47	18	99.0	143	25	99.0	257
	89.5	44		95.2	131		95.2	236
11	99.0	61		90.1	124		89.9	224
	94.6	55	19	99.1	158			
	89.8	52		95.1	144			
12	99.1	71		90.4	137			
	94.8	64						
	90.8	61						

Table A.16   Critical Values for the Wilcoxon Rank-Sum Interval $(d_{ij(mn-c+1)}, d_{ij(c)})$

			Smaller Sample Size								
	5			**6**			**7**			**8**	
Larger Sample Size	**Confidence Level (%)**	**c**		**Confidence Level (%)**	**c**		**Confidence Level (%)**	**c**		**Confidence Level (%)**	**c**
5	99.2	25									
	94.4	22									
	90.5	21									
6	99.1	29		99.1	34						
	94.8	26		95.9	31						
	91.8	25		90.7	29						
7	99.0	33		99.2	39		98.9	44			
	95.2	30		94.9	35		94.7	40			
	89.4	28		89.9	33		90.3	38			
8	98.9	37		99.2	44		99.1	50		99.0	56
	95.5	34		95.7	40		94.6	45		95.0	51
	90.7	32		89.2	37		90.6	43		89.5	48
9	98.8	41		99.2	49		99.2	56		98.9	62
	95.8	38		95.0	44		94.5	50		95.4	57
	88.8	35		91.2	42		90.9	48		90.7	54
10	99.2	46		98.9	53		99.0	61		99.1	69
	94.5	41		94.4	48		94.5	55		94.5	62
	90.1	39		90.7	46		89.1	52		89.9	59
11	99.1	50		99.0	58		98.9	66		99.1	75
	94.8	45		95.2	53		95.6	61		94.9	68
	91.0	43		90.2	50		89.6	57		90.9	65
12	99.1	54		99.0	63		99.0	72		99.0	81
	95.2	49		94.7	57		95.5	66		95.3	74
	89.6	46		89.8	54		90.0	62		90.2	70

			Smaller Sample Size								
	9			**10**			**11**			**12**	
Larger Sample Size	**Confidence Level (%)**	**c**		**Confidence Level (%)**	**c**		**Confidence Level (%)**	**c**		**Confidence Level (%)**	**c**
9	98.9	69									
	95.0	63									
	90.6	60									
10	99.0	76		99.1	84						
	94.7	69		94.8	76						
	90.5	66		89.5	72						
11	99.0	83		99.0	91		98.9	99			
	95.4	76		94.9	83		95.3	91			
	90.5	72		90.1	79		89.9	86			
12	99.1	90		99.1	99		99.1	108		99.0	116
	95.1	82		95.0	90		94.9	98		94.8	106
	90.5	78		90.7	86		89.6	93		89.9	101

Table A.17 β Curves for t Tests

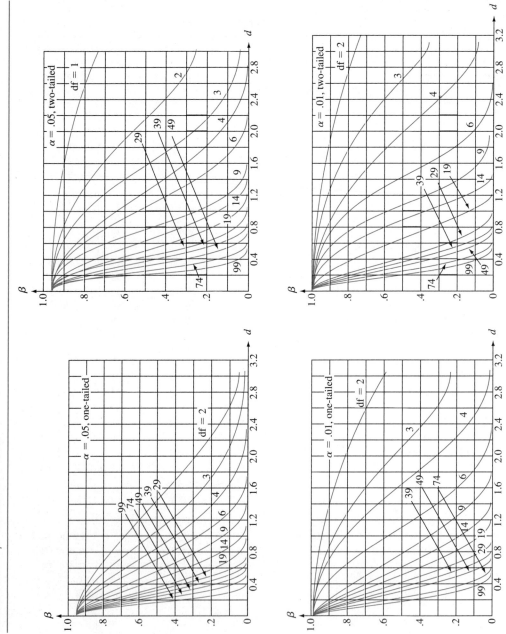

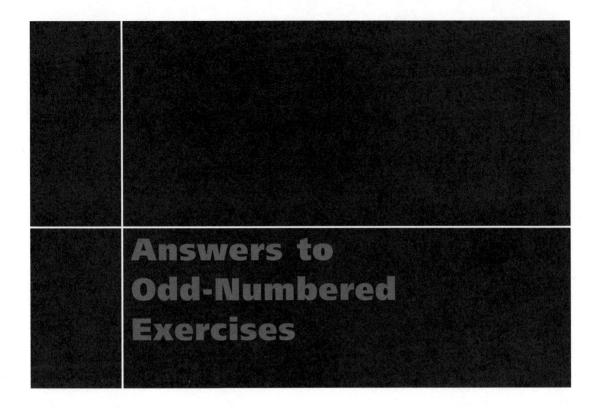

Answers to Odd-Numbered Exercises

Chapter 1

1. a. *Los Angeles Times, Oberlin Tribune, Gainesville Sun, Washington Post*
b. Duke Energy, Clorox, Seagate, Neiman Marcus
c. Vince Correa, Catherine Miller, Michael Cutler, Ken Lee
d. 2.97, 3.56, 2.20, 2.97

3. a. How likely is it that more than half of the sampled VCRs will need or have needed warranty service? What is the expected number among the 100 that need warranty service? How likely is it that the number needing warranty service will exceed the expected number by more than 10?
b. Suppose that 15 of the 100 sampled needed warranty service. How confident can we be that the proportion of *all* such VCRs needing warranty service is between .08 and .22? Does the sample provide compelling evidence for concluding that more than 10% of all such VCRs need warranty service?

5. a. No. All students taking a large statistics course who participate in an SI program of this sort.

b. Randomization protects against various biases and helps ensure that those in the SI group are as similar as possible to the students in the control group.
c. There would be no firm basis for assessing the effectiveness of SI (nothing to which the SI scores could reasonably be compared).

7. One could generate a simple random sample of all single-family homes in the city, or a stratified random sample by taking a simple random sample from each of the 10 district neighborhoods. From each of the selected homes, values of all desired variables would be determined. This would be an enumerative study because there exists a finite, identifiable population of objects from which to sample.

9. a. Possibly measurement error, recording error, differences in environmental conditions at the time of measurement, etc.

b. No. There is no sampling frame.

11.

83	4
84	3
85	3
86	77
87	758694
88	5632353679
89	9209863837
90	8319414643507
91	5100828611620
92	73767222
93	324307
94	7224
95	6
96	1
97	
98	8
99	
100	3

stem: tens and ones
leaf: tenths

The display is reasonably symmetric and in fact roughly bell-shaped. A representative value is in the low 90's. The extent of spread about the center is rather substantial. There is a single peak. The two largest observations are perhaps outliers.

13. Crunchy   Creamy

	2	2
644	3	0069
77220	4	00145
6320	5	003666
222	6	258
55	7	
0	8	

stem: tens
leaf: ones

Both sets of scores are rather spread out. There appear to be no outliers. The distribution of crunchy scores appears to be shifted to the right (toward larger values) of that for creamy scores by something on the order of 10.

15.

6L	430
6H	769689
7L	42014202
7H	
8L	011211410342
8H	9595578
9L	30
9H	58

The gap in the data—no scores in the high 70's.

17. a.

# Nonconforming	Frequency	Rel. freq.
0	7	.117
1	12	.200
2	13	.217
3	14	.233
4	6	.100
5	3	.050
6	3	.050
7	1	.017
8	1	.017
	60	1.001

b. .917, .867, $1 - .867 = .133$
c. The histogram has a substantial positive skew. It is centered somewhere between 2 and 3, and spreads out quite a bit about its center.

19. a. .99 (99%), .71 (71%)   **b.** .64 (64%), .44 (44%)
c. Strictly speaking, the histogram is not unimodal, but is close to being so with a moderate positive skew. A much larger sample size would likely give a smoother picture.

21. a.

y	Freq.	Rel. freq.
0	17	.362
1	22	.468
2	6	.128
3	1	.021
4	0	.000
5	1	.021
	47	1.000

.362, .638

b.

z	Freq.	Rel. freq.
0	13	.277
1	11	.234
2	3	.064
3	7	.149
4	5	.106
5	3	.064
6	3	.064
7	0	.000
8	2	.043
	47	1.001

.894, .830

23. a.

Class	Freq.	Rel. freq.
0–<100	21	.21
100–<200	32	.32
200–<300	26	.26
300–<400	12	.12
400–<500	4	.04
500–<600	3	.03
600–<700	1	.01
700–<800	0	.00
800–<900	1	.01
	100	1.00

b.

Class	Freq.	Rel. freq.	Density
0–<50	8	.08	.0016
50–<100	13	.13	.0026
100–<150	11	.11	.0022
150–<200	21	.21	.0042
200–<300	26	.26	.0026
300–<400	12	.12	.0012
400–<500	4	.04	.0004
500–<600	3	.03	.0003
600–<900	2	.02	.00007
	100	1.00	

c. .79

25.

Class	Freq.	Class	Freq.
10–<20	8	1.1–<1.2	2
20–<30	14	1.2–<1.3	6
30–<40	8	1.3–<1.4	7
40–<50	4	1.4–<1.5	9
50–<60	3	1.5–<1.6	6
60–<70	2	1.6–<1.7	4
70–<80	1	1.7–<1.8	5
	40	1.8–<1.9	1
			40

Original: positively skewed;
Transformed: much more symmetric, not far from bell-shaped.

27. a.

2	71
3	70 38 91 55 17 67 00 98
4	23 59 46 72 00 73 80 95
5	58 60 63 26 82 91
6	09 58 06 90 12 15 23 82 52 18 50 07
	93 98 49 63
7	60 17 45 53 14 88 71
8	32 71 stem: tens
9	13 46 leaf: ones and tenths

b. There are several observations that fall on class boundaries, and each must be placed in exactly one interval.

c.

Class	Freq.	Rel. freq.
20–<30	1	.02
30–<40	8	.16
40–<50	8	.16
50–<60	6	.12
60–<70	16	.32
70–<80	7	.14
80–<90	2	.04
90–<100	2	.04
	50	1.00

d. .34, .54

29.

Complaint	Freq.	Rel. freq.
J	10	.1667
F	9	.1500
B	7	.1167
M	4	.0667
C	3	.0500
N	6	.1000
O	21	.3500
	60	1.0001

31.

Class	Freq.	Cum. freq.	Cum. rel. freq.
0–<4	2	2	.050
4–<8	14	16	.400
8–<12	11	27	.675
12–<16	8	35	.875
16–<20	4	39	.975
20–<24	0	39	.975
24–<28	1	40	1.000

33. a. $\bar{x} = 192.57$, $\tilde{x} = 189.0$
b. New $\bar{x} = 189.71$; $\tilde{x}$ unchanged
c. 191.0, 7.14%   **d.** 122.6

35. a. $\bar{x} = 12.55$, $\tilde{x} = 12.50$, $\bar{x}_{tr(12.5)} = 12.40$. Deletion of the largest observation (18.0) causes $\tilde{x}$ and $\bar{x}_{tr}$ to be a bit smaller than $\bar{x}$.
b. By at most 4.0   **c.** No; multiply the values of $\bar{x}$ and $\tilde{x}$ by the conversion factor 1/2.2.

37. $\bar{x}_{tr(10)} = 11.46$

39. a. $\bar{x} = 1.0297$, $\tilde{x} = 1.009$ **b.** .383

41. a. .7   **b.** Also .7   **c.** 13

43. $\tilde{x} = 68.0$, $\bar{x}_{tr(20)} = 66.2$, $\bar{x}_{tr(30)} = 67.5$

45. a. $\bar{x} = 115.58$; the deviations are .82, .32, −.98, −.38, .22
b. .482, .694   **c.** .482   **d.** .482

47. $\bar{x} = 116.2$, $s = 25.75$. The magnitude of s indicates a substantial amount of variation about the center (a "representative" deviation of roughly 25).

49. a. .56.80, 197.8040   **b.** .5016, .708

51. a. 1264.766, 35.564   **b.** .351, .593

53. a. 2.74, 3.88   **b.** 1.14   **c.** Unchanged
d. At most .40   **e.** 1.19

55. a. 33   **b.** No
c. Slight positive skewness in the middle half, but rather symmetric overall. The extent of variability appears substantial.
d. At most 32

57. a. Yes. 125.8 is an extreme outlier and 250.2 is a mild outlier.

b. In addition to the presence of outliers, there is positive skewness both in the middle 50% of the data and, excepting the outliers, overall. Except for the two outliers, there appears to be a relatively small amount of variability in the data.

59. a. ED: .4, .10, 2.75, 2.65;

Non-Ed: 1.60, .30, 7.90, 7.60

b. ED: 8.9 and 9.2 are mild outliers, and 11.7 and 21.0 are extreme outliers.

There are not outliers in the non-ED sample.

c. Four outliers for ED, none for non-ED. Substantial positive skewness in both samples; less variability in ED (smaller f_s), and non-ED observations tend to be somewhat larger than ED observations.

61. Outliers, both mild and extreme, only at 6 A.M. Distributions at other times are quite symmetric. Variability increases somewhat until 2 P.M. and then decreases slightly, and the same is true of "typical" gasoline-vapor coefficient values.

63. Based on a comparative boxplot, the three samples appear to differ markedly with respect to center. There is somewhat less variability for the 160 flow rate than for the other two rates. No outliers are present. There is evidence of mild positive skewness in the middle 50% of each sample, but not for the 160 flow-rate sample or the 200 flow-rate sample as a whole.

65. a. 9.59, 59.41

b. $CV = .396$ for HC and $= .323$ for CO, so there is more relative variability in the HC data even though its SD is much smaller.

67. 10.65

69. a. $\bar{y} = a\bar{x} + b$, $s_y^2 = a^2 s_x^2$ **b.** 189.14, 1.87

71. $\bar{x} = .9255$, $s = .0809$, $\tilde{x} = .93$, small amount of variability, slight bit of skewness

73. a. The "five-number summaries" ($\tilde{x}$, the two fourths, and the smallest and largest observations) are identical and there are no outliers, so the three individual boxplots are identical.

b. Differences in variability, nature of gaps, and existence of clusters for the three samples.

c. No. Detail is lost.

75. a.

```
 0 | 2355566777888
 1 | 0000135555
 2 | 00257
 3 | 0033
 4 | 0057          stem: ones
 5 | 044           leaf: tenths
 6 |
 7 | 05
 8 | 8
 9 | 0
10 | 3
HI | 22.0, 24.5
```

b.

Class	Freq.	Rel. freq.	Density
0–<2	23	.500	.250
2–<4	9	.196	.098
4–<6	7	.152	.076
6–<10	4	.087	.022
10–<20	1	.022	.002
20–<30	2	.043	.004

77. a. $\bar{x}_{n+1} = (n\bar{x}_n + x_{n+1})/(n + 1)$

c. 12.53, .532

79. A substantial positive skew (assuming unimodality)

81. a. All points fall on a 45° line. Points fall below a 45° line.   **b.** Points fall well below a 45° line, indicating a substantial positive skew.

Chapter 2

1. a. $\mathscr{S} = \{1324, 3124, 1342, 3142, 1423, 1432, 4123,$ $4132, 2314, 2341, 3214, 3241, 2413, 2431, 4213,$ $4231\}$

b. $A = \{1324, 1342, 1423, 1432\}$

c. $B = \{2314, 2341, 3214, 3241, 2413, 2431, 4213,$ $4231\}$

d. $A \cup B = \{1324, 1342, 1423, 1432, 2314, 2341,$ $3214, 3241, 2413, 2431, 4213, 4231\}$,

$A \cap B$ contains no outcomes (A and B are disjoint), $A' = \{3124, 3142, 4123, 4132, 2314, 2341, 3214,$ $3241, 2413, 2431, 4213, 4231\}$

3. a. $A = \{SSF, SFS, FSS\}$

b. $B = \{SSF, SFS, FSS, SSS\}$

c. $C = \{SFS, SSF, SSS\}$

d. $C' = \{FFF, FSF, FFS, FSS, SFF\}$,

$A \cup C = \{SSF, SFS, FSS, SSS\}$,

$A \cap C = \{SSF, SFS\}$,

$B \cup C = \{SSF, SFS, FSS, SSS\} = B$,

$B \cap C = \{SSF, SFS, SSS\} = C$

5. a. $\mathscr{S} = \{(1, 1, 1), (1, 1, 2), (1, 1, 3), (1, 2, 1), (1, 2, 2),$ $(1, 2, 3), (1, 3, 1), (1, 3, 2), (1, 3, 3), (2, 1, 1), (2, 1, 2),$

(2, 1, 3), (2, 2, 1), (2, 2, 2), (2, 2, 3), (2, 3, 1), (2, 3, 2), (2, 3, 3), (3, 1, 1), (3, 1, 2), (3, 1, 3), (3, 2, 1), (3, 2, 2), (3, 2, 3), (3, 3, 1), (3, 3, 2), (3, 3, 3)}   **b.** {(1, 1, 1), (2, 2, 2), (3, 3, 3)}   **c.** {(1, 2, 3), (1, 3, 2), (2, 1, 3), (2, 3, 1), (3, 1, 2), (3, 2, 1)}   **d.** {(1, 1, 1), (1, 1, 3), (1, 3, 1), (1, 3, 3), (3, 1, 1), (3, 1, 3), (3, 3, 1), (3, 3, 3)}

7. a. There are 35 outcomes in $\mathcal{S}$.   **b.** {*AABABAB, AABAABB, AAABBAB, AAABABB, AAAABBB*}

11. a. .07   **b.** .30   **c.** .57

13. a. .36   **b.** .64   **c.** .53
d. .47   **e.** .17   **f.** .75

15. a. .913   **b.** .913

17. a. Other types of items (e.g., magazines) can be checked out.   **b.** .65   **c.** .85   **d.** .15

19. a. .8841   **b.** .0435

21. a. .10   **b.** .18, .19   **c.** .41   **d.** .59
e. .31   **f.** .69

23. a. .10   **b.** .30   **c.** .70   **d.** .60

25. a. .98   **b.** .02   **c.** .03   **d.** .24

27. a. .1   **b.** .7   **c.** .6

29. a. 20   **b.** 60   **c.** 10

31. a. 243   **b.** 3645 days (roughly 10 yr)

33. a. 15,504   **b.** 840   **c.** .0542   **d.** .0578

35. .20

37. .0456

39. a. .0839   **b.** .24975

41. a. .929   **b.** .0714   **c.** .99997520

43. .000394, .00394, .00001539

45. a. .45, .25, .10   **b.** .40, .222   **c.** .50, .429

47. a. .50   **b.** .50   **c.** .625
d. .375   **e.** .769

49. .217, .178

51. .436, .581

53. .083

55. a. .0111   **b.** .333   **c.** .2

59. a. .21   **b.** .455   **c.** .264, .274

61. a. .578, .278, .144   **b.** 0, .457, .543

63. b. .54   **c.** .68   **d.** .74   **e.** .7941

65. $P(\text{Mean} \mid S) = .3922, P(\text{Median} \mid S) = .2941$, so Mean and Median are most and least likely, respectively.

67. a. .126   **b.** .05   **c.** .1125   **d.** .2725
e. .5325   **f.** .2113

69. a. .7   **b.** .44   **c.** .318

73. .096, .222

75. a. .00601   **b.** .00421

77. .0059

79. a. .10, .20   **b.** 0

81. a. $p(2 - p)$ **b.** $1 - (1 - p)^n$ **c.** $(1 - p)^3$
d. $.9 + (1 - p)^3(.1)$
e. $.1(1 - p)^3/[.9 + .1(1 - p)^3] = .0137$ for $p = .5$

83. .8588, .9949

85. $[2\pi(1 - \pi)]/(1 - \pi^2)$

87. a. .30, .53   **b.** $.321 > .3$ **c.** .414

89. .0204, .0098

91. a. .0083   **b.** .2   **c.** .2

93. .1071, .5357

95. .905

97. a. .956   **b.** .994

99. .926

101. a. .018   **b.** .601

103. a. .883, .117   **b.** 23   **c.** .156

105. $1 - (1 - p_1)(1 - p_2) \cdots \cdots (1 - p_n)$

107. a. .0417   **b.** .375

109. $P(\text{hire \#1}) = 6/24$ for $s = 0$, $= 11/24$ for $s = 1$, $= 10/24$ for $s = 2$, and $= 6/24$ for $s = 3$, so $s = 1$ is best.

111. $1/4 = P(A_1 \cap A_2 \cap A_3)$
$\neq P(A_1) \cdot P(A_2) \cdot P(A_3) = 1/8$

Chapter 3

1. $x = 0$ for *FFF*; $x = 1$ for *SFF, FSF,* and *FFS*; $x = 2$ for *SSF, SFS,* and *FSS,* and $x = 3$ for *SSS*

3. $Z =$ average of the two numbers, with possible values $2/2, 3/2, \ldots, 12/2$; $W =$ absolute value of the difference, with possible values 0, 1, 2, 3, 4, 5

5. No. In Example 3.4, let $Y = 1$ if at most three batteries are examined and let $Y = 0$ otherwise. Then Y has only two values.

7. a. {0, 1, . . . , 12}; discrete   **c.** {1, 2, 3, . . . };
discrete   **e.** {0, c, $2c$, . . . , 10,000c}, where c is
the royalty per book; discrete   **g.** {x: $m \le x \le M$}
where m (M) is the minimum (maximum) possible
tension; continuous

9. a. {2, 4, 6, 8, . . . }, that is, {2(1), 2(2), 2(3),
2(4), . . . }, an infinite sequence; discrete
b. {2, 3, 4, 5, 6, . . . }, that is, {1 + 1, 1 + 2, 1 + 3,
1 + 4, . . . }, an infinite sequence; discrete

11. a. $p(4) = .45$, $p(6) = .40$, $p(8) = .15$, $p(x) = 0$ for
$x \ne 4, 6,$ or 8

13. a. .70   **b.** .45   **c.** .55
d. .71   **e.** .65   **f.** .45

15. a. (1, 2), (1, 3), (1, 4), (1, 5), (2, 3), (2, 4), (2, 5),
(3, 4), (3, 5), (4, 5)   **b.** $p(0) = .3$, $p(1) = .6$,
$p(2) = .1$ **c.** $F(x) = 0$ for $x < 0$, $= .3$ for $0 \le$
$x < 1$, $= .9$ for $1 \le x < 2$, and $= 1$ for $2 \le x$

17. a. .81   **b.** .162   **c.** It is A; $AUUUA$, $UAUUA$,
$UUAUA$, $UUUAA$; .00405

19. $p(1) = p(2) = p(3) = p(4) = .25$

21. $F(x) = 0$ for $x < 0$, $= .10$ for $0 \le x < 1$, $= .25$ for
$1 \le x < 2$, $= .45$ for $2 \le x < 3$, $= .70$ for $3 \le x <$
4, $= .90$ for $4 \le x < 5$, $= .96$ for $5 \le x < 6$, and $=$
1 for $6 \le x$

23. a. $p(1) = .30$, $p(3) = .10$, $p(4) = .05$, $p(6) = .15$,
$p(12) = .40$ **b.** .30, .60

25. a. $p(x) = (\frac{2}{3})^{x-1} \cdot (\frac{1}{3})$ for $x = 1, 2, 3, \ldots$
b. $p(y) = (\frac{2}{3})^{y-2} \cdot (\frac{1}{3})$ for $y = 2, 3, 4, 5, \ldots$
c. $p(z) = (\frac{25}{54}) \cdot (\frac{2}{3})^{2z-2}$ for $z = 1, 2, 3, \ldots$

29. a. .60   **b.** \$110

31. a. 16.38, 272.298, 3.9936   **b.** 401   **c.** 2496
d. 13.66

31. Yes, because $\sum_{x=1}^{\infty} 1/x^2 < \infty$.

35. \$700

37. $E(h(X)) = .408$, so gamble.

39. $V(-X) = V(X)$

41. a. 32.5   **b.** 7.5
c. $V(X) = E[X(X - 1)] + E(X) - [E(X)]^2$

43. a. .25, .11, .06, .04, .01   **b.** 2.64, 1.54, .04, conservative   **c.** They are equal.   **d.** $p(-1) =$
$.02 = p(1)$, $p(0) = .96$

45. a. .850   **b.** .200   **c.** .200   **d.** .701
e. .851   **f.** .000   **g.** .570

47. a. .354   **b.** .115   **c.** .918

49. a. .403   **b.** .213   **c.** .774

51. .1478

53. a. .017   **b.** .811, .425   **c.** .006, .902, .586

55. When $p = .9$, the probability is .99 for A and .9963
for B. If $p = .5$, these probabilities are .75 and
.6875, respectively.

57. The tabulation for $p > .5$ is unnecessary.

59. a. 20, 16   **b.** 70, 21

61. $P(|X - \mu| \ge 2\sigma) = .042$ when $p = .5$ and $= .065$
when $p = .75$, compared to the upper bound of .25.
Using $k = 3$ in place of $k = 2$, these probabilities are
.002 and .004, respectively, whereas the upper bound
is .11.

63. a. .1136   **b.** .039   **c.** .7955

65. a. $h(x; 15, 10, 20)$ for $x = 5, \ldots, 10$
b. .0325   **c.** .697

67. a. $h(x; 10, 10, 20)$ **b.** .033

69. a. $nb(x; 2, .5)$ **b.** .188   **c.** .688   **d.** 2, 4

71. $nb(x; 6, .5)$, 6

73. a. .932   **b.** .065   **c.** .068   **d.** .492
e. .251

75. a. .011   **b.** .441   **c.** .554, .459   **d.** .034

77. Poisson(5)   **a.** .492   **b.** .133

79. a. .091, .900, .283   **b.** 12, 3.464
c. .530, .011

81. a. .099   **b.** .135   **c.** 2

83. a. 4   **b.** .215   **c.** At least $-\ln(.1)/2 \approx 1.1513$
years

85. a. .221   **b.** 6,800,000   **c.** $p(x; 20.106)$

89. b. 3.114, .405, .636

91. a. $b(x; 15, .75)$ **b.** .314
c. .747   **d.** 11.25, 2.81   **e.** .310

93. .991

95. a. $p(x; 2.5)$ **b.** .067   **c.** .109

97. 1.813, 3.05

99. $p(2) = p^2$, $p(3) = (1 - p)p^2$, $p(4) = (1 - p)p^2$, $p(x) =$
$[1 - p(2) - \cdots - p(x - 3)](1 - p)p^2$ for $x = 5, 6,$
$7, \ldots$; .99950841

101. a. .0029   **b.** .0767, .9702

103. a. .135   **b.** .00144   **c.** $\sum_{x=0}^{\infty} [p(x; 2)]^5$

105. 3.590

107. a. No   **b.** .0273

109. b. $.6p(x; \lambda) + .4p(x; \mu)$ **c.** $(\lambda + \mu)/2$
d. $(\lambda - \mu)^2/4 + (\lambda + \mu)/2$

111. $\sum_{i=1}^{10} (p_{i+j+1} + p_{i-j-1})p_i$ where $p_k = 0$ if $k < 0$ or
$k > 10$.

Chapter 4

1. a. .25   **b.** .50   **c.** .4375

3. b. .5   **c.** .6875   **d.** .6328

5. a. .375   **b.** .125   **c.** .297   **d.** .578

7. a. $f(x) = .1$ for $25 \le x \le 35$ and 0 otherwise
   b. .20   **c.** .40   **d.** .20

9. a. .562   **b.** .438, .438   **c.** .071

11. a. .25   **b.** .1875   **c.** .9375   **d.** 1.4142
   e. $f(x) = x/2$ for $0 < x < 2$

13. a. 1.33   **b.** .222, .471   **c.** 2

15. a. $F(x) = 0$ for $x \le 0$, $= 90[\frac{x^9}{9} - \frac{x^{10}}{10}]$ for $0 < x < 1$, $=$ 1 for $x \ge 1$ **b.** .0107   **c.** .0107, .0107
   d. .9036   **e.** .818, .111   **f.** .6863

17. a. For $2 \le x \le 4$, $F(x) = .25[3x - 7 - (x - 3)^3]$
   b. 3   **c.** 3, .2

19. a. .597   **b.** .369
   c. $f(x) = .3466 - .25 \ln(x)$ for $0 < x < 4$

21. 314.79

23. 248, 3.60

25. b. 1.8(90th percentile for X) + 32
   c. $a(X$ percentile) + b

27. a. 2.14   **b.** .81   **c.** 1.17
   d. .97   **e.** 2.41

29. a. 2.54   **b.** 1.34   **c.** −.42

31. a. .9452   **b.** .0082   **c.** .9544

33. a. .3336   **b.** Approximately 0
   c. .5795   **d.** 6.524

35. a. 31.7   **b.** 17.225   **c.** 3.192

37. .002

39. 10, .2

41. a. .9938   **b.** .9876, 1   **c.** .0052

43. 15.66

45. .3174 for $k = 1$, .0456 for $k = 2$, .0026 for $k = 3$, as compared to the bounds of 1, .25, and .111, respectively.

47. a. Exact: .212, .577, .573; Approximate: .211, .567, .596   **b.** Exact: .885, .575, .017; Approximate: .885, .579, .012   **c.** Exact: .002, .029, .617; Approximate: .003, .033, .599

49. a. .9666   **b.** .0099, 0

51. b. Normal, $\mu = 239$, $\sigma^2 = 12.96$

53. a. 120   **b.** 1.329   **c.** .371
   d. .735   **e.** 0

55. a. .594   **b.** .092   **c.** .537

57. a. .424   **b.** .567; yes   **c.** 60   **d.** 66

59. a. .750, .937, .187   **b.** .050   **c.** 50.0

61. a. n/λ; 20   **b.** .930   **c.** $1 - \sum_{k=0}^{n-1} \frac{e^{-\lambda(\lambda t)^k}}{k!}$

63. $-\ln(1 - p)/\lambda$, $.693/\lambda$

67. a. .632, .632, .064   **b.** .470   **c.** 172.727

71. a. 68.0, 122.1   **b.** .3204
   c. .7257, skewness implies mean $\ne$ median

73. a. 149.157, 223.595   **b.** .9834   **c.** .0921
   d. 148.41   **e.** 9.83   **f.** 125.90

75. $\alpha = \beta$

77. b. $[\Gamma(\alpha + \beta) \cdot \Gamma(m + \beta)]/[\Gamma(\alpha + \beta + m) \cdot \Gamma(\beta)]$, $\beta/(\alpha + \beta)$

79. Yes, since the pattern in the plot is quite linear.

81. Yes

83. Yes

85. Plot $\ln(x)$ vs. z percentile. The pattern is straight, so a lognormal population distribution is plausible.

87. There is substantial curvature in the plot. λ is a scale parameter (as is σ for the normal family).

89. a. $F(y) = \frac{1}{48}(y^2 - y^3/18)$ for $0 \le y \le 12$
   b. .259, .5, .241   **c.** 6, 43.2, 7.2
   d. .518   **e.** 3.75

91. a. $f(x) = x^2$ for $0 \le x < 1$ and $= \frac{7}{4} - \frac{3}{4}x$ for $1 \le x \le 7/3$ **b.** .917   **c.** 1.213

93. a. .9162   **b.** .9549   **c.** 1.3374

95. a. .3859   **b.** .0663   **c.** (72.97, 119.03)

97. b. $F(x) = 0$ for $x < -1$, $= (4x - x^3/3)/9 + 11/27$ for $-1 \le x \le 2$, and $= 1$ for $x > 2$
   c. No. $F(0) < .5 \Rightarrow \tilde{\mu} > 0$
   d. $Y \sim \text{Bin}(10, 5/27)$

99. a. .368, .828, .460   **b.** 352.53
   c. $1/\beta \cdot \exp[-\exp(-(x - \alpha)/\beta)] \cdot \exp(-(x - \alpha)/\beta)$
   d. α **e.** $\mu = 201.95$, mode = 150, $\tilde{\mu} = 182.99$

101. a. μ **b.** No   **c.** 0
   d. $(\alpha - 1)\beta$ **e.** $\nu - 2$

103. b. $p(1 - \exp(-\lambda_1 x)) + (1 - p)(1 - \exp(-\lambda_2 x))$
   for $x \ge 0$ **c.** $p/\lambda_1 + (1 - p)/\lambda_2$
   d. $V(X) = 2p/\lambda_1^2 + 2(1 - p)/\lambda_2^2 - \mu^2$
   e. 1, $CV > 1$ **f.** $CV < 1$

105. a. Lognormal   **b.** 1   **c.** 2.72, .0185

109. a. Exponential with $\lambda = 1$
   c. Gamma with parameters α and $c\beta$

111. b. Let $u_1, u_2, u_3, \ldots$ be a sequence of observations from a Unif[0, 1] distribution (a sequence of ran-

dom numbers). Then with $x_i = (-.1)\ln(1 - u_i)$, the x_i's are observations from an exponential distribution with $\lambda = 10$.

113. $g(E(X)) \le E(g(X))$

Chapter 5

1. a. .20   **b.** .42   **c.** At least one hose is in use at each pump; .70.   **d.** $p_X(x) = .16, .34, .50$ for $x = 0, 1, 2$, respectively; $p_Y(y) = .24, .38, .38$ for $y = 0, 1, 2$, respectively; .50   **e.** No; $p(0, 0) \ne p_X(0) \cdot p_Y(0)$

3. a. .15   **b.** .40   **c.** .22   **d.** .17, .46

5. a. .054   **b.** .00018

7. a. .030   **b.** .120   **c.** .300
   d. .380   **e.** Yes.

9. a. 3/380,000   **b.** .3024   **c.** .3593
   d. $10Kx^2 + .05$ for $20 \le x \le 30$ **e.** No

11. a. $e^{-\lambda - \mu} \cdot \lambda^x \cdot \mu^y / x! y!$ **b.** $e^{-\lambda - \mu} \cdot [1 + \lambda + \mu]$
   c. $e^{-(\lambda + \mu)} \cdot (\lambda + \mu)^m / m!$; Poisson $(\lambda + \mu)$

13. a. e^{-x-y} for $x \ge 0, y \ge 0$ **b.** .400   **c.** .594
   d. .330

15. a. $F(y) = 1 - e^{-\lambda y} + (1 - e^{-\lambda y})^2 - (1 - e^{-\lambda y})^3$ for $y \ge 0$ **b.** $2/3\lambda$

17. a. .25   **b.** .318   **c.** .637
   d. $f_X(x) = 2\sqrt{R^2 - x^2}/\pi R^2$ for $-R \le x \le R$; no

19. a. $K(x^2 + y^2)/(10Kx^2 + .05)$; $K(x^2 + y^2)/(10Ky^2 + .05)$
   b. .556, .549   **c.** 25.37, 2.87

21. $= f_Y(y)$ for every x

23. .15

25. L^2

27. 1/3 hr

29. $-2/3$

31. a. $-.1082$ **b.** $-.0131$

37. a.

$\bar{x}$	25	32.5	40	45	52.5	65
$p(\bar{x})$	.04	.20	.25	.12	.30	.09

$E(\bar{X}) = \mu = 44.5$

   b.

s^2	0	112.5	312.5	800
$p(s^2)$	.38	.20	.30	.12

$E(S^2) = 212.25 = \sigma^2$

39.

Proportion	0	.1	.2	.3	.4	.5
Probability	.000	.000	.000	.001	.005	.027
Proportion	.6	.7	.8	.9	1.0	
Probability	.088	.201	.302	.269	.107	

41. a.

$\bar{x}$	1	1.5	2	2.5	3	3.5	4
$p(\bar{x})$	.16	.24	.25	.20	.10	.04	.01

   b. .85   **c.**

r	0	1	2	3
$p(r)$	.30	.40	.22	.08

47. a. .6826   **b.** .1056

49. a. .6026   **b.** .2981

51. .7720

53. a. .0062   **b.** 0

55. a. .0968   **b.** .8882

57. .9616

59. a. .9986, .9986   **b.** .9015, .3970
   c. .8357   **d.** .9525, .0003

61. a. 3.5, 2.27, 1.51   **b.** 15.4, 75.94, 8.71

63. a. .695   **b.** $4.0675 > 2.6775$

65. a. .9232   **b.** .9660

67. .1588

69. a. 2400   **b.** 1205; independence
   c. 2400, 41.77

71. a. 158, 430.25   **b.** .9788

73. a. Approximately normal with mean = 105, SD = 1.2649; Approximately normal with mean = 100, SD = 1.0142
   b. Approximately normal with mean = 5, SD = 1.6213
   c. .0068   **d.** .0010, yes

75. a. .2, .5, .3 for $x = 7, 9, 10$; .10, .35, .55 for $y = 7, 9, 10$ **b.** .25   **c.** No   **d.** 18.25   **e.** 1.05

77. a. 3/81,250   **b.** $f_X(x) = k(250x - 10x^2)$ for $0 \le x \le 20$ and $= k(450x - 30x^2 + \frac{1}{2}x^3)$ for $20 < x \le 30$; $f_Y(y)$ results from substituting y for x in $f_X(x)$. They are not independent.
   c. .355   **d.** 25.969
   e. 204.6154, $-.894$ **f.** 7.66

79. ≈ 1

81. a. 400 min   **b.** 70

83. 97

85. .9973

89. *b, c.* Chi-squared with $\nu = n$.

91. a. $\sigma_W^2/(\sigma_W^2 + \sigma_E^2)$ **b.** .9999

93. 26, 1.64

Chapter 6

1. a. $8.14, \overline{X}$ **b.** $.77, \tilde{X}$ **c.** $1.66, S$
   d. .148   **e.** $.204, S/\overline{X}$

3. a. $1.348, \overline{X}$ **b.** $1.348, \overline{X}$
   c. $1.781, \overline{X} + 1.28S$
   d. .6736   **e.** .0905

5. 1,703,000; 1,591,300; 1,601,438.281

7. a. 120.6   **b.** 1,206,000   **c.** .80   **d.** 120.0

9. a. 2.11   **b.** .119

11. b. $\left[\dfrac{p_1 q_1}{n_1} + \dfrac{p_2 q_2}{n_2}\right]^{1/2}$ **c.** Use $\hat{p}_i = x_i/n_i$ and $\hat{q}_i = 1 - \hat{p}_i$ in place of p_i and q_i in part (b) for $i = 1, 2$
   d. $-.245$ **e.** .041

15. a. $\hat{\theta} = \sum X_i^2/2n$ **b.** 74.505

17. b. .444

19. a. $\hat{p} = 2\hat{\lambda} - .30 = .20$ **b.** $\hat{p} = (100\hat{\lambda} - 9)/70$

21. b. $\hat{\alpha} = 5, \hat{\beta} = 28.0/\Gamma(1.2)$

23. $\hat{\lambda}_1 = \overline{x}, \hat{\lambda}_2 = \overline{y}$, estimate of $(\lambda_1 - \lambda_2)$ is $\overline{x} - \overline{y}$

25. a. 384.4, 18.86   **b.** 415.42

29. a. $\hat{\theta} = \min(X_i), \hat{\lambda} = n/\sum[X_i - \min(X_i)]$
   b. .64, .202

33. With x_i = time between birth $i - 1$ and birth i, $\hat{\lambda} = 6/\sum_{i=1}^{6} i x_i = .0436$.

35. 29.5

37. 1.0132

Chapter 7

1. a. 99.5%   **b.** 85%   **c.** 2.96   **d.** 1.15

3. a. Narrower   **b.** No   **c.** No   **d.** No

5. a. (4.52, 5.18)   **b.** (4.12, 5.00)
   c. .55   **d.** 94

7. By a factor of 4; the width is decreased by a factor of 5.

9. a. $(\overline{x} - 1.645\sigma/\sqrt{n}, \infty)$; (4.57, ∞)   **b.** $(\overline{x} - z_\alpha \cdot \sigma/\sqrt{n}, \infty)$ **c.** $(-\infty, \overline{x} + z_\alpha \cdot \sigma/\sqrt{n})$; $(-\infty, 59.7)$

11. 950, .8714

13. a. (.990, 1.066)   **b.** 158

15. a. 80%   **b.** 98%   **c.** 75%

17. 62.3

19. (.513, .615)

21. .218

23. a. (.438, .814)   **b.** 659

25. a. 381   **b.** 253

29. a. 2.228   **b.** 2.086   **c.** 2.845   **d.** 2.680
   e. 2.485   **f.** 2.571

31. a. 1.812   **b.** 1.753   **c.** 2.602   **d.** 3.747
   e. 2.1716 (from MINITAB)   **f.** Roughly 2.43

33. a. Reasonable amount of symmetry, no outliers
   b. Yes (based on a normal probability plot)
   c. (430.5, 446.1), yes, no

35. a. 95% CI: (2783.3, 2991.9)
   b. 95% PI: (2632.1, 3143.1), roughly 2.5 times as wide

37. a. (.888, .964)   **b.** (.752, 1.100)
   c. (.634, 1.218)

39. All 70%; (c), because it is shortest

41. a. 18.307   **b.** 3.940   **c.** .95   **d.** .10

43. (3.6, 8.1); no

45. a. 95% CI: (6.702, 9.456)   **b.** (.166, .410)

47. a. There appears to be a slight positive skew in the middle half of the sample, but the lower whisker is much longer than the upper whisker. The extent of variability is rather substantial, although there are no outliers.
   b. Yes. The pattern of points in a normal probability plot is reasonably linear.
   c. (33.53, 43.79)

49. a. (.624, .732)   **b.** 1080   **c.** No

51. $(-.84, -.16)$

53. 246

55. $(2t_r/\chi^2_{1-\alpha/2,2r}, 2t_r/\chi^2_{\alpha/2,2r}) = (65.3, 232.5)$

57. a. $(\max(x_i)/(1 - \alpha/2)^{1/n}, \max(x_i)/(\alpha/2)^{1/n})$
b. $(\max(x_i), \max(x_i)/\alpha^{1/n})$ **c.** (b); (4.2, 7.65)

59. (73.6, 78.8) versus (75.1, 79.6)

Chapter 8

1. a. Yes   **b.** No   **c.** No
d. Yes   **e.** No   **f.** Yes

5. H_0: $\sigma = .05$ versus H_a: $\sigma < .05$. I: Conclude variability in thickness is satisfactory when it isn't. II: Conclude variability in thickness isn't satisfactory when in fact it is.

7. I: concluding that the plant isn't in compliance when it is; II: concluding that the plant is in compliance when it isn't.

9. a. R_1 **b.** I: judging that one of the two companies is favored over the other when that is not the case; II: judging that neither company is favored over the other when in fact one of the two really is preferred.   **c.** .044
d. $\beta(.3) = \beta(.7) = .488$, $\beta(.4) = \beta(.6) = .845$
e. Reject H_0 in favor of H_a.

11. a. H_0: $\mu = 10$ versus H_a: $\mu \neq 10$ **b.** .01
c. .5319, .0078   **d.** 2.58
e. 10.1032 is replaced by 10.163, and 9.8968 is replaced by 9.837   **f.** $\bar{x} = 10.020$, so H_0 should not be rejected.   **g.** $z \geq 2.58$ or ≤ -2.58

13. b. .0004, 0, less than .01

15. a. .0301   **b.** .003   **c.** .004

17. a. $z = 2.56 \geq 2.33$, so reject H_0 **b.** .8413
c. 143   **d.** .0052

19. a. $z = -2.27$, so don't reject H_0.   **b.** .2266
c. 22

21. a. $t_{.025,12} = 2.179 > 1.6$, so don't reject H_0: $\mu = .5$
b. $-1.6 > -2.179$, so don't reject H_0.
c. Don't reject H_0.
d. Reject H_0 in favor of H_a: $\mu \neq .5$.

23. $t = 2.24 \geq 1.708$, so H_0 should be rejected. The data does suggest a contradiction of prior belief.

25. a. $z = -3.33 \leq -2.58$, so reject H_0.
b. .1056   **c.** 217

27. $-1.82 > -2.33$, so do not reject H_0.

29. a. $.498 < 1.895$, so do not reject H_0.   **b.** .72

31. $-1.24 > -1.397$, so prior belief does not appear to be contradicted.

35. Yes, because $2.47 \geq 1.96$.

37. $z = 5.3 \geq 2.58$, so reject H_0: $p = .40$. No.

39. -6.1 is much less than -2.33, so the data does strongly support such a conclusion.

41. a. $\{15, \ldots, 20\}$ **b.** .021, yes   **c.** .874, .196
d. Using $R = \{14, \ldots, 20\}$, no.

43. $z = -.26 > -2.33$; no

45. a. Don't reject H_0.   **b.** Don't reject H_0.
c. Don't reject H_0.   **d.** Reject H_0.
e. Don't reject H_0.   **f.** Don't reject H_0.

47. a. .0358   **b.** .0802   **c.** .5824
d. .1586   **e.** 0

49. a. P-value $= .003 < .05$, so reject H_0.
b. P-value $= .055 > .01$, so H_0 cannot be rejected.
c. P-value $> .5$, so H_0 cannot be rejected at any reasonable α.

51. P-value $< .0004 < .01$, so H_0: $\mu = 5$ should be rejected in favor of H_a: $\mu \neq 5$.

53. $t \approx 1.9$, so P-value $\approx .041$. Since P-value $\leq \alpha$, H_0: $\mu = 25$ should be rejected in favor of H_a: $\mu > 25$.

55. $t \approx 1.9$, so P-value $\approx .116$. H_0 should therefore not be rejected.

57. a. .8980, .1049, .0014   **b.** P-value ≈ 0. Yes.
c. No

59. $z = -3.12 \leq -1.96$, so H_0 should be rejected.

61. a. H_0: $\mu = .85$ versus H_a: $\mu \neq .85$
b. H_0 cannot be rejected for either α.

63. a. Yes, because $t = 12.9 \geq 2.228$
b. Normal population distribution

65. a. No; no
b. No, because $z = .44$ and P-value $= .33 > .10$

67. a. Approximately .6; approximately .2 (from Appendix Table A.17)   **b.** $n = 28$

69. a. $z = 1.64 < 1.96$, so H_0 cannot be rejected; Type II   **b.** .10. Yes.

71. Yes. $z = -3.32 \le -3.08$, so H_0 should be rejected.

73. No, since $z = 1.33 < 2.05$

75. No, since $12.11 < 21.665$

77. a. $\overline{X} + 2.33S$ **b.** $3.7145\sigma^2/n$; $1.927s/\sqrt{n}$
c. $z = -1.22$, so H_0: $\mu + 2.33\sigma = 6.75$ cannot be rejected in favor of H_a: $\mu + 2.33\sigma < 6.75$

79. b. $\Phi(z_\gamma + (\mu_0 - \mu)/(\sigma/\sqrt{n}))$
$- \Phi(-z_{\alpha-\gamma} + (\mu_0 - \mu)/(\sigma/\sqrt{n}))$ **c.** $\gamma > \alpha/2$

Chapter 9

1. a. $-.4$ hr; it doesn't   **b.** .0724, .2691   **c.** No

3. $z = 1.76 < 2.33$, so don't reject H_0.

5. a. $z = -2.90$, so reject H_0.   **b.** .0019
c. .8212   **d.** 66

7. Yes, since $z = 1.83 \ge 1.645$.

9. a. $z = 2.89$, so don't use the high purity steel.
b. .2981

11. $(-11.23, -6.31)$

13. $\hat{\theta} = -.97$, $\hat{\sigma}_{\hat{\theta}} = s_{\hat{\theta}} = .924$, $z = -1.05$, so don't reject H_0; no.

15. It increases.

17. a. 17   **b.** 21   **c.** 18   **d.** 26

19. $t = -1.20 > -t_{.01,9} = -2.821$, so do not reject H_0.

21. Yes; $-2.64 \le -2.602$, so reject H_0.

23. b. No   **c.** $t = -.38 > -t_{\alpha/2,10}$ for any reasonable α, so don't reject H_0 (P-value $\approx .7$).

25. $(.3, 6.1)$, yes, yes

27. $(6.5, 31.3)$ based on 9 df; yes, yes

29. $t = -2.10$, df $= 25$, P-value $= .023$. At significance level .05, we would conclude that cola results in a higher average strength, but not at significance level .01.

31. a. Virtually identical centers, substantially more variability in medium range observations than in higher range observations
b. $(-7.9, 9.6)$, based on 23 df; no

33. $t = -2.2$, P-value $= .022$, don't reject H_0

35. $t = -2.2$, df $= 16$, P-value $= .021 > .01 = \alpha$, so don't reject H_0.

37. a. $(-.561, -.287)$ **b.** Between -1.224 and .376

39. a. Yes
b. $t = 2.7$, P-value $= .018 < .05 = \alpha$, so H_0 should be rejected.

41. $t = 1.9$, P-value $= .047$. H_0 cannot be rejected at significance level .01, but is barely rejected at $\alpha = .05$.

43. a. No   **b.** -49.1 **c.** 49.1

45. $(-10.499, -4.589)$

47. H_0 is rejected because $-4.18 \le -2.33$

49. P-value $= .4247$, so H_0 cannot be rejected.

51. a. $z = 1.48 < 1.645$, so don't reject H_0.
b. $n = 6582$

53. a. The CI for $\ln(\theta)$ is $\ln(\hat{\theta}) \pm z_{\alpha/2}[(m - x)/(mx) + (n - y)/(ny)]^{1/2}$. Taking the antilogs of the lower and upper limits gives a CI for θ itself.
b. $(1.43, 2.31)$; aspirin appears to be beneficial.

55. $(-.35, .07)$

57. a. 3.69   **b.** 4.82   **c.** .207   **d.** .271
e. 4.30   **f.** .212   **g.** .95   **h.** .94

59. $f = 1.59$; since $.459 < 1.59 < 2.18$, don't reject H_0.

61. $f = 2.85 \ge 2.08$, so reject H_0; there does appear to be more variability in low-dose weight gain.

63. $(s_2^2 F_{1-\alpha/2}/s_1^2, s_2^2 F_{\alpha/2}/s_1^2)$; $(.023, 1.99)$

65. No. $t = 3.2$, df $= 15$, P-value $= .006$, so reject H_0: $\mu_1 - \mu_2 = 0$ using either $\alpha = .05$ or .01.

67. $z > 0 \Rightarrow$ P-value $> .5$, so H_0: $p_1 - p_2 = 0$ cannot be rejected.

69. $(-299.3, 1517.9)$

71. $(1024.0, 1336.0)$, yes

73. Yes. $t = -2.25$, df $= 57$, P-value $\approx .028$

75. a. No. $t = -2.84$, df $= 18$, P-value $\approx .012$
b. No. $t = -.56$, P-value $\approx .29$

77. Not at significance level .05. $t = -1.76 > -t_{.05,4} = -2.015$

79. No, nor should the two-sample t test be used, because a normal probability plot suggests that the good-visibility distribution is not normal.

81. a. $m = 141, n = 47$ **b.** $m = 240, n = 160$

83. a. No; $-t_{.0005,11} = -4.437 < 3.3 < 4.437$, so don't reject H_0. **b.** P-value $= .008$, so H_0 would be rejected if $\alpha = .05$ or .01.

85. .9015, .8264, .0294, .0000; true average IQs; no

87. No, because $z = .62$

89. a. Yes. $t = -6.4$, df $= 57$, and P-value ≈ 0
b. $t = 1.1$, P-value $= .14$, so don't reject H_0.

91. $(-1.29, -.59)$

Chapter 10

1. $f = 1.85 < 3.06 = F_{.05,4,15}$, so don't reject H_0.

3. $f = 1.30 < 2.57 = F_{.10,2,21}$, so P-value $> .10$. H_0 cannot be rejected at any reasonable significance level.

5. $f = 1.73 < 5.49 = F_{.01,2,27}$, so the three grades don't appear to differ.

7. $f = 1.70 < 2.46 = F_{.10,3,16}$, so P-value $> .10$. H_0 cannot be rejected at any reasonable significance level.

9. $f = 3.96$ and $F_{.05,3,20} = 3.10 < 3.96 < 4.94 = F_{.01,3,20}$, so $.01 < P$-value $< .05$. Thus H_0 can be rejected at significance level .05; there appear to be differences among the grains.

11. $w = 36.09$

3	1	4	2	5
437.5	462.0	469.3	512.8	532.1

Brands 2 and 5 don't appear to differ, nor does there appear to be any difference between brands 1, 3, and 4, but each brand in the first group appears to differ significantly from all brands in the second group.

13.

3	1	4	2	5
427.5	462.0	469.3	502.8	532.1

15. $w = 5.94$

2	1	3	4
24.69	26.08	29.95	33.84

The only significant differences are between 4 and both 1 and 2.

17. $(-.029, .379)$

19. Any value of SSE between 422.16 and 431.88 will work.

21. a. $f = 22.6$ and $F_{.01,5,78} \approx 3.3$, so reject H_0.
b. $(-99.16, -35.64), (29.34, 94.16)$

23.

	1	2	3	4
1	–	2.88 ± 5.81	7.43 ± 5.81	12.78 ± 5.48
2	–	–	4.55 ± 6.13	9.90 ± 5.81
3	–	–	–	5.35 ± 5.81
4	–	–	–	–

4	3	2	1

25. a. 1.1724, .066076, .006337
b. .029309, .000419
c. $f = 15.1$ and $F_{.001,4,70} \approx 5.2$, so P-value $< .001$. H_0 should be rejected. There appear to be differences.
d. 1 2 3 4 5

27. a. $f = 3.75 \geq 3.10 = F_{.05,3,20}$, so brands appear to differ. **b.** Normality is quite plausible (a normal probability plot of the residuals $x_{ij} - \bar{x}_{i.}$ shows a linear pattern).
c. 4 3 2 1 Only brands 1 and 4 appear to differ significantly.

31. Approximately .62

33. $\arcsin(\sqrt{x/n})$

35. a. $3.68 < 4.94$, so H_0 is not rejected.
b. $.029 > .01$, so again H_0 is not rejected.

37. $f = 8.44 > 6.49 = F_{.001}$, so P-value $< .001$ and H_0 should be rejected.

5	3	1	4	2

This underscoring pattern is a bit awkward to interpret.

39. The CI is $(-.144, .474)$, which does include 0.

41. $f = 3.96 < 4.07$, so H_0: $\sigma_A^2 = 0$ cannot be rejected.

43. $(-3.70, 1.04), (-4.83, -.33), (-3.77, 1.27), (-3.99, .15)$. Only $\mu_1 - \mu_3$ among these four contrasts appears to differ significantly from zero.

45. They are identical.

Chapter 11

1. a. $f_A = 1.55$, so don't reject H_{0A}.
b. $f_B = 2.98$, so don't reject H_{0B}.

3. a. $f_A = 12.987 \geq F_{.01,3,9}$, so conclude that there is a gas rate effect; $f_B = 105.31$, so conclude that there is a liquid rate effect. **b.** $w = 95.44$; $\underline{231.75 \ 325.25}$ $441.0 \ 613.25$, so only the lowest two rates do not differ significantly from one another.
c. $\underline{336.75 \ 382.25 \ 419.25 \ 473}$ so only the lowest and highest rates appear to differ significantly from one another.

5. $f_A = 2.56$, $F_{.01,3,12} = 5.95$, so there appears to be no effect due to angle of pull.

7. a.

Source	df	SS	MS	f
Treatments	2	28.78	14.39	1.04
Blocks	17	2977.67	175.16	12.68
Error	34	469.55	13.81	
Total	53	3476.00		

True average adaptation score does not appear to depend on which treatment is given. **b.** Yes; f_B is quite large, suggesting great variability between subjects.

9.

Source	df	SS	MS	f	$F_{.05}$
Treatments	3	81.19	27.06	22.4	3.01
Blocks	8	66.50	8.31		
Error	24	29.06	1.21		
Total	35	176.75			

1	4	3	2
8.56	9.22	10.78	12.44

11. The residuals are .0350, .0117, $-$.0750, .0283, .0875, $-$.0758, $-$.0825, .0708, $-$.1225, .0642, .1575, and $-$.0992. The pattern in the normal probability plot is quite linear.

13. b. Each SS is multiplied by c^2, but f_A and f_B are unchanged.

15. a. Approximately .20, .43 **b.** Approximately .30

17. a. $f_A = 3.76$, $f_B = 6.82$, $f_{AB} = .74$, and $F_{.05,2,9} = 4.26$, so the amount of carbon fiber addition appears significant. **b.** $f_A = 6.54$, $f_B = 5.33$, $f_{AB} = .27$

19. a.

Source	df	SS	MS	f
Coal	2	1.00241	.50121	29.49
NaOH	2	.12431	.06216	3.66
Interaction	4	.01456	.00364	.21
Error	9	.15295	.01699	
Total	17	1.29423		

Type of coal does appear to affect total acidity.
b. Coals 1 and 3 don't differ significantly from one another, but both differ significantly from coal 2.

21. a, b.

Source	df	SS	MS	f
A	2	22941.80	11470.90	22.98
B	4	22765.53	5691.38	5.60
AB	8	3993.87	499.23	.49
Error	15	15253.50	1016.90	
Total	29	64954.70		

H_{0A} and H_{0B} are both rejected.

23.

Source	df	SS	MS	f
A	2	11,573.38	5786.69	$\dfrac{\text{MSA}}{\text{MSAB}} = 26.70$
B	4	17,930.09	4482.52	$\dfrac{\text{MSB}}{\text{MSE}} = 28.51$
AB	8	1734.17	216.77	$\dfrac{\text{MSAB}}{\text{MSE}} = 1.38$
Error	30	4716.67	157.22	
Total	44	35,954.31		

Since $F_{.01,8,30} = 3.17$, $F_{.01,2,8} = 8.65$, and $F_{.01,4,30} = 4.02$, H_{0G} is not rejected but both H_{0A} and H_{0B} are rejected.

25. $(-1.39, -1.05)$

27. a.

Source	df	SS	MS	f	$F_{.05}$
A	2	14,144.44	7,072.22	61.06	3.35
B	2	5,511.27	2,755.64	23.79	3.35
C	2	244,696.39	122,348.20	1,056.27	3.35
AB	4	1,069.62	267.41	2.31	2.73
AC	4	62.67	15.67	.14	2.73
BC	4	331.67	82.92	.72	2.73
ABC	8	1,080.77	135.10	1.17	2.31
Error	27	3,127.50	115.83		
Total	53	270,024.33			

d. $Q_{.05,3,27} = 3.51$, $w = 8.90$, and all three of the levels differ significantly from one another.

29.

Source	df	SS	MS	f
A	2	12.896	6.448	1.04
B	1	100.041	100.041	16.10
C	3	393.416	131.139	21.10
AB	2	1.646	.823	<1
AC	6	71.021	11.837	1.905
BC	3	1.542	.514	<1
ABC	6	9.771	1.629	<1
Error	72	447.500	6.215	
Total	95	1037.833		

a. No interaction effects are significant.
b. Factor B and factor C main effects are significant.
c. $w = 1.89$; only machines 2 and 4 do not differ significantly from one another.

31. The P-value column shows that several interaction effects are significant at level .01.

33.

Source	df	SS	MS	f
A	6	67.32	11.02	
B	6	51.06	8.51	
C	6	5.43	.91	.61
Error	30	44.26	1.48	
Total	48	168.07		

$F_{.05,6,30} = 2.42, f_C = .61$, so H_{0C} is not rejected.

35.

Source	df	SS	MS	f
A	4	28.88	7.22	10.7
B	4	23.70	5.93	8.79
C	4	.62	.155	<1
Error	12	8.10	.675	
Total	24	61.30		

Since $F_{.05,4,12} = 3.26$, both A and B are significant.

37.

Source	df	MS	f
A	2	2207.329	2259*
B	1	47.255	48.4*
C	2	491.783	503*
D	1	.044	<1
AB	2	15.303	15.7*
AC	4	275.446	282*
AD	2	.470	<1
BC	2	2.141	2.19
BD	1	.273	<1
CD	2	.247	<1
ABC	4	3.714	3.80
ABD	2	4.072	4.17*
ACD	4	.767	<1
BCD	2	.280	<1
ABCD	4	.347	<1
Error	36	.977	
Total	71	93.621	

*Denotes a significant F ratio.

39. a. $\hat{\beta}_1 = 54.38, \hat{\gamma}_{11}^{AC} = -2.21, \hat{\gamma}_{21}^{AC} = 2.21.$
b.

Source	Effect Contrast	MS	f
A	1307	71,177.04	436.7
B	1305	70,959.34	435.4
C	529	11,660.04	71.54
AB	199	1,650.04	10.12
AC	−53	117.04	<1
BC	57	135.38	<1
ABC	27	30.38	<1
Error		162.98	

41.

Source	SS	f
A	136,640.02	1,007.6
B	139,644.19	1,029.8
C	24,616.02	181.5
D	20,377.52	150.3
AB	2,173.52	16.0
AC	2.52	<1
AD	58.52	<1
BC	165.02	1.2
BD	9.19	<1
CD	17.52	<1
ABC	42.19	<1
ABD	117.19	<1
ACD	188.02	1.4
BCD	13.02	<1
ABCD	204.19	1.5
Error	4339.33	
Total	328,607.98	

$F_{.05,1,32} \approx 4.15$, so only the four main effects and the AB interaction appear significant.

43.

Source	df	SS	f
A	1	.436	<1
B	1	.099	<1
C	1	.109	<1
D	1	414.12	851
AB	1	.003	<1
AC	1	.078	<1
AD	1	.017	<1
BC	1	1.404	3.62
BD	1	.456	<1
CD	1	2.190	4.50
Error	5	2.434	

$F_{.05,1,5} = 6.61$, so only the factor D main effect is judged significant.

45. a. 1: (1), *ab, cd, abcd*; 2: *a, b, acd, bcd*; 3: *c, d, abc, abd*; 4: *ac, bc, ad, bd.*

b.

Source	df	SS	f
A	1	14,028.13	53.89
B	1	92,235.13	345.33
C	1	3.13	<1
D	1	18.00	<1
AC	1	105.13	<1
AD	1	200.00	<1
BC	1	91.13	<1
BD	1	420.50	1.62
ABC	1	276.13	1.06
ABD	1	2.00	<1
ACD	1	450.00	1.73
BCD	1	2.00	<1
Blocks	7	898.88	<1
Error	12	3,123.72	
Total	31	111,853.88	

$F_{.01,1,12} = 9.33$, so only the A and B main effects are significant.

47. a. *ABFG*; (1), *ab, cd, ce, de, fg, acf, adf, adg, aef, acg, aeg, bcg, bcf, bdf, bdg, bef, beg, abcd, abce, abde, abfg, cdfg, cefg, defg, acdef, acdeg, bcdef, bcdeg, abcdfg, abcefg, abdefg.* {*A, BCDE, ACDEFG, BFG*}, {*B, ACDE, BCDEFG, AFG*}, {*C, ABDE, DEFG, ABCFG*}, {*D, ABCE, CEFG, ABDFG*}, {*E, ABCD, CDFG, ABEFG*}, {*F, ABCDEF, CDEG, ABG*}, {*G, ABCDEG, CDEF, ABF*}.   **b.** 1: (1), *aef, beg, abcd, abfg, cdfg, acdeg, bcdef*; 2: *ab, cd, fg, aeg, bef, acdef, bcdeg, abcdfg*; 3: *de, acg, adf, bcf, bdg, abce, cefg, abdefg*; 4: *ce, acf, adg, bcg, bdf, abde, defg, abcefg.*

49. $SSA = 2.250$, $SSB = 7.840$, $SSC = .360$, $SSD = 52.563$, $SSE = 10.240$, $SSAB = 1.563$, $SSAC = 7.563$, $SSAD = .090$, $SSAE = 4.203$, $SSBC = 2.103$, $SSBD = .010$, $SSBE = .123$, $SSCD = .010$, $SSCE = .063$, $SSDE = 4.840$. Error SS = sum of two-factor SS's = 20.568, Error MS = 2.057, $F_{.01,1,10} = 10.04$, so only the D main effect is significant.

51.

Source	df	SS	MS	f
A main effects	1	322.667	322.667	980.38
B main effects	3	35.623	11.874	36.08
Interaction	3	8.557	2.852	8.67
Error	16	5.266	0.329	
Total	23	372.113		

$F_{.05,3,16} = 3.24$, so interactions appear to be present.

53.

Source	df	SS	MS	f
A	1	30.25	30.25	6.72
B	1	144.00	144.00	32.00
C	1	12.25	12.25	2.72
AB	1	1,122.25	1,122.25	249.39
AC	1	1.00	1.00	.22
BC	1	12.25	12.25	2.72
ABC	1	16.00	16.00	3.56
Error	4	36.00	4.50	
Total	7			

Only the main effect for B and the AB interaction effect are significant at $\alpha = .01$.

55. a. $\hat{\alpha}_1 = 9.00$, $\hat{\beta}_1 = 2.25$, $\hat{\delta}_1 = 17.00$, $\hat{\gamma}_1 = 21.00$, $(\widehat{\alpha\beta})_{11} = 0$, $(\widehat{\alpha\delta})_{11} = 2.00$, $(\widehat{\alpha\gamma})_{11} = 2.75$, $(\widehat{\beta\delta})_{11} = .75$, $(\widehat{\beta\gamma})_{11} = .50$, $(\widehat{\delta\gamma})_{11} = 4.50$

b. A normal probability plot suggests that the A, C, and D main effects are quite important, and perhaps the CD interaction. In fact, pooling the 4 three-factor interaction SS's and the four-factor interaction SS to obtain an SSE based on 5 df and then constructing an ANOVA table suggests that these are the most important effects.

57.

Source	df	SS	MS	f	P
A	2	34,436	17,218	436.92	0.000
B	2	105,793	52,897	1342.30	0.000
C	2	516,398	258,199	6552.04	0.000
AB	4	6,868	1,717	43.57	0.000
AC	4	10,922	2,731	69.29	0.000
BC	4	10,178	2,545	64.57	0.000
ABC	8	6,713	839	21.30	0.000
Error	27	1,064	39		
Total	53	692,372			

All effects are significant.

59. Based on the P-values in the ANOVA table, statistically significant factors at the level $\alpha = .01$ are adhesive type and cure time. The conductor material does not have a statistically significant effect on bond strength. There are no significant interactions.

61.

Source	df	SS	MS	f
A	4	285.76	71.44	.594
B	4	227.76	56.94	.473
C	4	2867.76	716.94	5.958
D	4	5536.56	1384.14	11.502
Error	8	962.72	120.34	$F_{.05,4,8} = 3.84$
Total	24			

H_{0A} and H_{0B} cannot be rejected, while H_{0C} and H_{0D} are rejected.

Chapter 12

1. a. The accompanying displays are based on repeating each stem value five times (once for leaves 0 and 1, a second time for leaves 2 and 3, etc.).

```
17 | 0
17 | 2 3
17 | 4 4 5
17 | 6 7
17 |                    stem: hundreds and tens
18 | 0 0 0 0 1 1        leaf: ones
18 | 2 2 2 2
18 | 4 4 5
18 | 6
18 | 8
```

There are no outliers, no significant gaps, and the distribution is roughly bell-shaped with a reasonably high degree of concentration about its center at approximately 180.

```
0 | 8 8 9
1 | 0 0 0 0
1 | 3
1 | 4 4 4 4
1 | 6 6
1 | 8 8 8 9        stem: ones
2 | 1 1            leaf: tenths
2 |
2 | 5
2 | 6
2 |
3 | 0 0
```

A typical value is about 1.6, and there is a reasonable amount of dispersion about this value. The distribution is somewhat skewed toward large values, the two largest of which may be candidates for outliers.
b. No, because observations with identical x values have different y values.
c. No, because the points don't appear to fall at all close to a line or simple curve.

3. Yes. Yes.

5. b. Yes.
c. There appears to be an approximate quadratic relationship (points fall close to a parabola).

7. a. 5050   **b.** 1.3   **c.** 130   **d.** -130

9. a. .095   **b.** $-.475$ **c.** .830, 1.305
d. .4207, .3446   **e.** .0036

11. a. $-.01, -.10$ **b.** 3.00, 2.50
c. .3627   **d.** .4641

13. a. Yes, because $r^2 = .972$

15. a.
```
2 | 9
3 | 3 3 5 5 6 6 6 7 7 8 8 9
4 | 1 2 2 3 5 6 6 8 9
5 | 1
6 | 2 9
7 | 9
8 | 0
```
Typical value in low 40's, reasonable amount of variability, positive skewness, two potential outliers.
b. No
c. $y = 3.2925 + .10748x = 7.59$. No; danger of extrapolation.
d. 18.736, 71.605, .738, yes

17. a. We estimate that .144 is the expected change in calcium content associated with a 1 mg/cm² increase in the amount of dissolved material. Roughly 86% of the observed variation in calcium content can be attributed to the simple linear regression model relationship between content and the amount of dissolved material.
b. 10.88   **c.** 1.46

19. a. $y = -45.5519 + 1.7114x$ **b.** 339.51
c. -85.57 **d.** The $\hat{y}_i$'s are 125.6, 168.4, 168.4, 211.1, 211.1, 296.7, 296.7, 382.3, 382.3, 467.9, 467.9, 553.4, 639.0, 639.0; a 45° line through (0, 0).

21. b. $y = -2.182 + .660x$ **c.** 7.72   **d.** 7.72

23. a. 16,213.64; 16,205.45
b. 414,235.71; yes, since $r^2 = .961$

27. $\hat{\beta}_1 = \Sigma x_i Y_i / \Sigma x_i^2$

29.

Data set	r^2	s	Most effective: set 3
1	.43	4.03	Least effective: set 1
2	.99	4.03	
3	.99	1.90	

31. a. .001017   **b.** $(-.00956, -.00516)$

33. a. (.081, .133)
b. $H_a: \beta_1 > .1$, P-value = .277, no

35. a. $(-.16, 1.29)$, not useful
b. P-value = $.11 > \alpha$ for any reasonable α, so H_0 cannot be rejected. No.

37. a. .110, .000262   **b.** $t \approx 3.1$, P-value $\approx .014 < .10 = \alpha$, so H_0 should be rejected. The data does appear to contradict prior belief.

39. $f = 96.0$ and $F_{.01,1,9} = 10.56$, so the model is judged useful.

43. $d = 1.20$, df = 13, and $\beta \approx .1$.

45. a. (48.97, 52.63)   **b.** $t = .78$, don't reject H_0.

47. a. 95% PI is (20.21, 43.69), no
 b. (28.53, 51.92), at least 90%

49. (431.3, 628.5)

51. a. .40 is closer to $\bar{x} = .7495$ **b.** (.745, .875)
 c. (.059, .523)

53. (a) narrower than (b), (c) narrower than (d), (a) narrower than (c), (b) narrower than (d)

55. a. $x_2 = x_3 = 12$, yet $y_2 \neq y_3$.   **b.** Yes
 c. $y = -19.670 + 3.2847x$ **d.** (24.93, 67.12)

57. If, for example, 18 is the minimum age of eligibility, then for most people $y \approx x - 18$.

59. a. .966
 b. The percent dry fiber weight for the first specimen tends to be larger than for the second.
 c. No change.   **d.** 93.3%
 e. $t = 14.9$, P-value ≈ 0, so there does appear to be such a relationship.

61. $r = .9066$, $t = 7.75$, and P-value ≈ 0, so conclude that $\rho > 0$.

63. $r = .773$, yet $t = 2.44 < 2.776$; so H_0: $\rho = 0$ cannot be rejected.

65. a. The x plot is a bit curved but not disturbingly so in light of the small sample size. The y plot is quite straight.   **b.** $t = 6.3 \geq 3.355$, so there does appear to be a linear relationship.

67. $t = 2.20 \geq 1.96$, so it appears that $\rho \neq 0$. But $r = .022$ suggests that ρ is not much different from zero, indicating little practical significance.

69. a. $t = -1.24 > -2.201$, so H_0 cannot be rejected.
 b. .970

71. a. .507   **b.** .712   **c.** P-value $= .0013 < .01 = \alpha$, so reject H_0: $\beta_1 = 0$ and conclude that there is a useful linear relationship.   **d.** A 95% CI is (1.056, 1.275).   **e.** 1.0143, .2143

73. a. $y = 14.1904 - .14892x$ **b.** $t = -1.43$, so don't reject H_0: $\beta_1 = -.10$.   **c.** No; $\Sigma(x - \bar{x})^2 = 143$ here and 182 for the given data.   **d.** A 95% CI for $\mu_{Y \cdot 28}$ is (9.599, 10.443).

75. a. A substantial linear relationship
 b. $y = -.08259 + .044649x$
 c. 98.3%
 d. .7702, $-.0902$ **e.** Yes; $t = 19.96$
 f. (.0394, .0499)   **g.** (.762, .858)

79. b. .573

83. $t = -1.14$, so it is plausible that $\beta_1 = \gamma_1$.

Chapter 13

1. a. 6.32, 8.37, 8.94, 8.37, and 6.32   **b.** 7.87, 8.49, 8.83, 8.94, and 2.83   **c.** The deviation is likely to be much smaller for the x values of part (b).

3. a. Yes.   **b.** $-.75$, .31, $-.74$, 1.13, .43, $-.72$, 1.43, .93, -1.51, -1.27, .90   **c.** No.

5. $-.50$, $-.75$, $-.50$, .79, .90, .93, .19, 1.46, -1.80, -1.12; curvature; a linear pattern

7. a. No.   **b.** e_i's are -16.60, 9.70, 19.00, $-.70$, 11.40; e_i^*'s are -1.55, .68, 1.25, $-.05$, -1.06; a quadratic function.

9. For set 1, simple linear regression is appropriate. A quadratic regression is reasonable for set 2. In set 3, (13, 12.74) appears very inconsistent with the remaining data. The estimated slope for set 4 depends largely on the single observation (19, 12.5), and evidence for a linear relationship is not compelling.

11. c. $V(\hat{Y}_i)$ increases, and $V(Y_i - \hat{Y}_i)$ decreases.

13. t with $n - 2$ df; .02

15. a. A curved pattern   **b.** A linear pattern
 c. $Y = \alpha x^{\beta} \cdot \epsilon$ **d.** A 95% PI is (3.06, 6.50).
 e. One standardized residual, corresponding to the third observation, is a bit large. There are only two positive standardized residuals, but two others are essentially 0. The patterns in a standardized residual plot and normal probability plot are marginally acceptable.

17. a. $\Sigma x_i' = 15.501$, $\Sigma y_i' = 13.352$, $\Sigma(x_i')^2 = 20.228$, $\Sigma x_i' y_i' = 18.109$, $\Sigma(y_i')^2 = 16.572$, $\hat{\beta}_1 = 1.254$, $\hat{\beta}_0 = -.468$, $\hat{\alpha} = .626$, $\hat{\beta} = 1.254$ **c.** $t = -1.07$, so don't reject H_0.   **d.** H_0: $\beta = 1$, $t = -4.30$, so reject H_0.

19. a. No   **b.** $Y' = \beta_0 + \beta_1 \cdot (1/t) + \epsilon'$, where $Y' = \ln(Y)$, so $Y = \alpha e^{\beta/t} \cdot \epsilon$.   **c.** $\hat{\beta} = \hat{\beta}_1 = 3735.45$, $\hat{\beta}_0 = -10.2045$, $\hat{\alpha} = (3.70034) \cdot (10^{-5})$. $\hat{y}' = 6.7748$, $\hat{y} = 875.5$ **d.** SSE $= 1.39587$, SSPE $= 1.36594$ (using transformed values), $f = .33 < 8.68 = F_{.01,1,15}$, so don't reject H_0.

21. a. $\hat{\mu}_{Y \cdot x} = 18.14 - 1485/x$ **b.** $\hat{y} = 15.17$

23. For the exponential model, $V(Y|x) = \alpha^2 e^{2\beta x}\sigma^2$, which does depend on x. A similar result holds for the power model.

25. a. The point estimate of β_1 is $\hat{\beta}_1 = .1772$, and the estimated odds ratio is 1.194. H_0: $\beta_1 = 0$ is rejected in favor of the conclusion that experience does appear to affect the likelihood of successful task performance.

27. b. 52.88, .12　　**c.** .895　　**d.** No
　　e. (48.54, 57.22)　　**f.** (42.85, 62.91)

29. a. SSE $= 103.37$, $s^2 = 51.69$　　**b.** .961
　　c. $t = -3.83 > -4.303$, so H_0: $\beta_2 = 0$ cannot be rejected. The data does not argue strongly for inclusion of the quadratic term (perhaps a consequence of n being small; H_0 would be rejected at significance level .10).　　**d.** β_1: $(-19.87, 35.99)$ β_2: $(-5.18, 1.50)$　　**e.** (100.08, 129.34)　　**f.** 2.19

31. a. $13.636 + 11.406x - 1.7155x^2$
　　b. Yes. Yes, (6, 20)　　**c.** 2.040, .947. The model utility F test via MINITAB gives $f = 35.9$, P-value $= .003$, suggesting a useful model.　　**d.** Yes. Yes
　　e. (28.35, 35.28)

33. a. .9671, .9407
　　b. $.0000492x^3 - .000446058x^2 + .007290688x + .96034944$　　**c.** $t = 2 < 3.182 = t_{.025,3}$, so the cubic term should be deleted.　　**d.** Identical
　　e. .987, .994, yes

35. $\hat{y} = 7.6883e^{.1799x - .0022x^2}$

37. a. 4.9　　**b.** When number of deliveries is held fixed, the average change in travel time associated with a 1-mile increase in distance traveled is .060 hr. When distance traveled is held fixed, the average change in travel time associated with one extra delivery is .900 hr.　　**c.** .9861

39. a. 77.3　　**b.** 40.4

41. $f = 24.4 > 5.12 = F_{.001,6,30}$, so P-value $<< .001$. The chosen model appears to be useful.

43. a. 48.31, 3.69　　**b.** No. If x_1 increases, either x_3 or x_2 must change.　　**c.** Yes, since $f = 18.924$, P-value $= .001$.　　**d.** Yes, using $\alpha = .01$, since $t = 3.496$ and P-value $= .003$

45. a. $f = 87.6$, P-value $= 0$, there does appear to be a useful linear relationship between y and at least one of the predictors.　　**b.** .935　　**c.** (9.095, 11.087)

47. b. P-value $= .000$, so conclude that the model is useful.　　**c.** P-value $= .034 \le .05 = \alpha$, so reject H_0: $\beta_3 = 0$; % garbage does appear to provide additional useful information.　　**d.** (1479.8, 1531.1), reasonable precision　　**e.** A 95% PI is (1435.7, 1575.2).

49. a. 96.8303, -5.8303　　**b.** $f = 14.9 \ge 8.02 = F_{.05,2,9}$, so reject H_0 and conclude that the model is useful.　　**c.** (78.28, 115.38)　　**d.** (38.50, 155.16)　　**e.** (46.91, 140.66)　　**f.** No. P-value $= .208$, so H_0: $\beta_1 = 0$ cannot be rejected.

51. a. No　　**b.** $f = 5.04 \ge 3.69 = F_{.05,5,8}$. There does appear to be a useful linear relationship.　　**c.** 6.16, 3.304, (16.67, 31.91)　　**d.** $f = 3.44 < 4.07 = F_{.05,3,8}$, so H_0: $\beta_3 = \beta_4 = \beta_5 = 0$ cannot be rejected. The quadratic terms can be deleted.

53. a. The dependent variable is $\ln(q)$, and the predictors are $x_1 = \ln(a)$ and $x_2 = \ln(b)$; $\hat{\beta} = \hat{\beta}_1 = .9450$, $\hat{\gamma} = \hat{\beta}_2 = .1815$, $\hat{\alpha} = 4.7836$, $\hat{q} = 18.27$
　　b. Now regress $\ln(q)$ against $x_1 = a$ and $x_2 = b$.　　**c.** (1.24, 5.78)

55.

k	R^2	adj. R^2	C_k
1	.676	.647	138.2
2	.979	.975	2.7
3	.9819	.976	3.2
4	.9824		4

　　a. The model with $k = 2$　　**b.** No

57. The model with predictors x_1, x_3, and x_5.

59. No. All R^2 values are much less than .9.

61. a. #4　　**b.** No　　**c.** No

63. a. When gender, weight, and heart rate are held fixed, we estimate that the average change in VO$_2$ max associated with a 1-minute increase in walk time is $-.0996$.　　**b.** When weight, walk time, and heart rate are held fixed, the estimate of average difference between VO$_2$ max for males and females is .6566.　　**c.** 3.669, $-.519$　　**d.** .706
　　e. $f = 9.0 \ge 4.89 = F_{.01,4,15}$, so there does appear to be a useful relationship.

65. a. No. There is substantial curvature in the scatter plot.　　**b.** Cubic regression yields $R^2 = .998$ and a 95% PI of (261.98, 295.62), and the cubic predictor appears to be important (P-value $= .001$). A regression of y versus $\ln(x)$ has $r^2 = .991$, but there is a very large standardized residual and the standardized residual plot is not satisfactory.

67. a. $R^2 = .802$, $f = 21.03$, P-value $= .000$. pH is a candidate for deletion. Note that there is one extremely large standardized residual.
　　b. $R^2 = .920$, adjusted $R^2 = .774$, $f = 6.29$, P-value $= .002$
　　c. $f = 1.08$, P-value $> .10$, don't reject H_0: $\beta_6 = \cdots = \beta_{20} = 0$. The group of second-order predictors does not appear to be useful.

d. $R^2 = .871$, $f = 28.50$, P-value $= .000$, and now all six predictors are judged important (the largest P-value for any t-ratio is .016); the importance of pH^2 was masked in the test of (c). Note that there are two rather large standardized residuals.

69. a. $f = 1783$, so the model appears useful.
b. $t = -48.1 \leq -6.689$, so even at level .001 the quadratic predictor should be retained.
c. No   **d.** (21.07, 21.65)   **e.** (20.67, 22.05)

71. a. $f = 30.8 \geq 9.55 = F_{.01,2,7}$, so the model appears useful.   **b.** $t = -7.69$ and P-value $< .001$, so retain the quadratic predictor.   **c.** (44.01, 47.91)

73. a. 231.75   **b.** .903   **c.** $f = 41.9$, indicating a useful relationship.   **d.** (220.9, 238.1)

75. There are several reasonable choices in each case.

77. a. $f = 106$, P-value ≈ 0 **b.** (.014, .068)
c. $t = 5.9$, reject H_0: $\beta_4 = 0$, percent nonwhite appears to be important   **d.** 99.514, $y - \hat{y} = 3.486$

Chapter 14

1. a. Reject H_0.   **b.** Don't reject H_0.
c. Don't reject H_0.   **d.** Don't reject H_0.

3. No. $\chi^2 = 1.57$ and P-value $> .10$, so H_0 cannot be rejected.

5. $\chi^2 = 6.61 < 14.684 = \chi^2_{.10,9}$, so don't reject H_0.

7. $\chi^2 = 4.03$ and P-value $> .10$, so don't reject H_0.

9. a. [0, .2231), [.2231, .5108), [.5108, .9163), [.9163, 1.6094), and [1.6094, ∞)   **b.** $\chi^2 = 1.25 < \chi^2_{\alpha,4}$ for any reasonable α, so the specified exponential distribution is quite plausible.

11. a. $(-\infty, -.97)$, $[-.97, -.43)$, $[-.43, 0)$, $[0, .43)$, $[.43, .97)$, and $[.97, \infty)$ **b.** $(-\infty, .49806)$, $[.49806, .49914)$, $[.49914, .5)$, $[.5, .50086)$, $[.50086, .50194)$, and $[.50194, \infty)$ **c.** $\chi^2 = 5.53$, $\chi^2_{.10,5} = 9.236$, so P-value $> .10$ and the specified normal distribution is plausible.

13. $\hat{p} = .0843$, $\chi^2 = 280.3 > \chi^2_{\alpha,1}$ for any tabulated α, so the model gives a poor fit.

15. The likelihood is proportional to $\theta^{233}(1 - \theta)^{367}$, from which $\hat{\theta} = .3883$. The estimated expected counts are 21.00, 53.33, 50.78, 21.50, and 3.41. Combining cells 4 and 5, $\chi^2 = 1.62$, so don't reject H_0.

17. $\hat{\lambda} = 3.167$, from which $\chi^2 = 103.98 >> \chi^2_{\alpha,k-1} = \chi^2_{\alpha,7}$ for any tabulated α, so the Poisson distribution provides a very poor fit.

19. $\hat{\theta}_1 = (2n_1 + n_3 + n_5)/2n = .4275$, $\hat{\theta}_2 = .2750$, $\chi^2 = 29.1$, $\chi^2_{.01,3} = 11.344$, so reject H_0.

21. Yes. The null hypothesis of a normal population distribution cannot be rejected.

23. MINITAB gives $r = .967$, and since $c_{.10} = .9707$ and $c_{.05} = .9639$, $.05 < P$-value $< .10$. Using $\alpha = .05$, normality is judged plausible.

25. $\chi^2 = 23.18 \geq 13.277 = \chi^2_{.01,4}$, so H_0 is rejected. The proportions appear to be different.

27. Yes. $\chi^2 = 44.98$ and P-value $< .001$

29. $p_{ij} =$ proportion of jth sex combination resulting from ith genotype. $\chi^2 = 6.46$, so P-value $> .10$ and the null hypothesis of homogeneity cannot be rejected.

31. Yes. $\chi^2 = 14.15$, so $.025 < P$-value $< .03$ and H_0 should be rejected at significance level .05.

35. N_{ij}/n, $n_k N_{ij}/n$, 24

37. $\chi^2 = 3.65 < 5.992 = \chi^2_{.05,2}$, so H_0 cannot be rejected.

39. Yes. $\chi^2 = 131$ and P-value $< .001$.

41. $\chi^2 = 22.4$ and P-value $< .001$, so the null hypothesis of independence is rejected.

43. P-value $= 0$, so the null hypothesis of homogeneity is rejected.

Chapter 15

1. $s_+ = 35$ and $14 < 35 < 64$, so H_0 cannot be rejected.

3. $s_+ = 18 \leq 21$, so H_0 is rejected.

5. Reject H_0 if either $s_+ \geq 64$ or $s_+ \leq 14$. Because $s_+ = 72$, H_0 is rejected.

7. $s_+ = 442.5$, $z = 2.89 \geq 1.645$, so reject H_0.

9.

d	0	2	4	6	8	10	12	14	16	18	20
$p(d)$	$\frac{1}{24}$	$\frac{3}{24}$	$\frac{1}{24}$	$\frac{4}{24}$	$\frac{2}{24}$	$\frac{2}{24}$	$\frac{2}{24}$	$\frac{4}{24}$	$\frac{1}{24}$	$\frac{3}{24}$	$\frac{1}{24}$

11. $w = 37$ and $29 < 37 < 61$, so H_0 cannot be rejected.

13. $z = 2.27 < 2.58$, so H_0 cannot be rejected. P-value $\approx$ .023

15. $w = 39 \le 41$, so H_0 is rejected.

17. $(\bar{x}_{(5)}, \bar{x}_{(32)}) = (11.15, 23.80)$

19. $(-13.0, -6.0)$

21. $(d_{ij(5)}, d_{ij(21)}) = (16, 87)$

23. $k = 14.06 \ge 6.251$, so reject H_0.

25. $k = 9.23 \ge 5.992$, so reject H_0.

27. $f_r = 2.60 < 5.992$, so don't reject H_0.

29. $f_r = 9.62 > 7.815 = \chi^2_{.05,3}$, so reject H_0.

31. $(-5.9, -3.8)$

33. a. .021   **b.** $c = 14$ gives $\alpha = .058$; $y = 13$, so H_0 cannot be rejected.

35. $w' = 26 < 27$, so don't reject H_0.

Chapter 16

1. All points on the chart fall between the control limits.

3. .9802, .9512, 53

5. a. .0301   **b.** .2236   **c.** .6808

7. LCL = 12.20, UCL = 13.70. No.

9. LCL = 94.91, UCL = 98.17. There appears to be a problem on the 22nd day.

11. a. 200   **b.** 4.78   **c.** 384.62   (larger), 6.30 (smaller)

13. LCL = 12.37, UCL = 13.53

15. a. LCL = 0, UCL = 6.48
b. LCL = .48, UCL = 6.60

17. LCL = .045, UCL = 2.484. Yes, since all points are inside the control limits.

19. a. LCL = .105, UCL = .357
b. Yes, since .39 > UCL.

21. $\bar{p} > 3/53$

23. LCL = 0, UCL = 10.1

25. When area = .6, LCL = 0 and UCL = 14.6; when area = .8, LCL = 0 and UCL = 13.4; when area = 1.0, LCL = 0 and UCL = 12.6.

27.

l:	1	2	3	4	5	6	7	8
d_l:	0	.001	.017	0	0	.010	0	0
e_l:	0	0	0	.038	0	0	0	.054

l:	9	10	11	12	13	14	15
d_l:	0	.024	.003	0	0	0	.005
e_l:	0	0	0	.015	0	0	0

There are no out-of-control signals.

29. $n = 5$, $h = .00626$

31. Hypergeometric probabilities (calculated on an HP21S calculator) are .9919, .9317, .8182, .6775, .5343, .4047, .2964, .2110, .1464, and .0994, whereas the corresponding binomial probabilities are .9862, .9216, .8108, .6767, .5405, .4162, .3108, .2260, .1605, and .1117. The approximation is satisfactory.

33. .9206, .6767, .4198, .2321, .1183; the plan with $n = 100$, $c = 2$ is preferable.

35. .9981 .5968, and .0688

37. a. .010, .018, .024, .027, .027, .025, .022, .018, .014, .011   **b.** .0477, .0274   **c.** 77.3, 202.1, 418.6, 679.9, 945.1, 1188.8, 1393.6, 1559.3, 1686.1, 1781.6

39. $\bar{x}$ chart based on sample standard deviations: LCL = 402.42, UCL = 442.20. $\bar{x}$ chart based on sample ranges: LCL = 402.36, UCL = 442.26. S chart: LCL = .55, UCL = 30.37. R chart: LCL = 0, UCL = 82.75.

41. S chart: LCL = 0, UCL = 2.3020; because $s_{21} = 2.931 >$ UCL, the process appears to be out of control at this time. Because an assignable cause is identified, recalculate limits after deletion: for an S chart, LCL = 0 and UCL = 2.0529; for an $\bar{x}$ chart, LCL = 48.583 and UCL = 51.707. All points on both charts lie between the control limits.

43. $\bar{\bar{x}} = 430.65$, $s = 24.2905$; for an S chart, UCL = 62.43 when $n = 3$ and UCL = 55.11 when $n = 4$; for an $\bar{x}$ chart, LCL = 383.16 and UCL = 478.14 when $n = 3$, and LCL = 391.09 and UCL = 470.21 when $n = 4$.

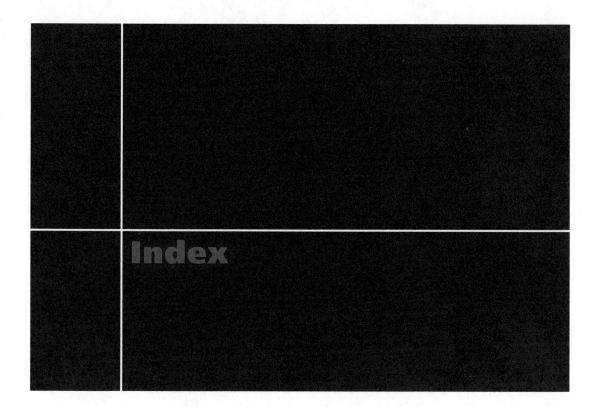

Index

Table A.5 Critical Values for t Distributions

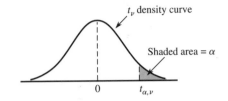

t_ν density curve

Shaded area = α

0   $t_{\alpha,\nu}$

ν	.10	.05	.025	.01	.005	.001	.0005
				α			
1	3.078	6.314	12.706	31.821	63.657	318.31	636.62
2	1.886	2.920	4.303	6.965	9.925	22.326	31.598
3	1.638	2.353	3.182	4.541	5.841	10.213	12.924
4	1.533	2.132	2.776	3.747	4.604	7.173	8.610
5	1.476	2.015	2.571	3.365	4.032	5.893	6.869
6	1.440	1.943	2.447	3.143	3.707	5.208	5.959
7	1.415	1.895	2.365	2.998	3.499	4.785	5.408
8	1.397	1.860	2.306	2.896	3.355	4.501	5.041
9	1.383	1.833	2.262	2.821	3.250	4.297	4.781
10	1.372	1.812	2.228	2.764	3.169	4.144	4.587
11	1.363	1.796	2.201	2.718	3.106	4.025	4.437
12	1.356	1.782	2.179	2.681	3.055	3.930	4.318
13	1.350	1.771	2.160	2.650	3.012	3.852	4.221
14	1.345	1.761	2.145	2.624	2.977	3.787	4.140
15	1.341	1.753	2.131	2.602	2.947	3.733	4.073
16	1.337	1.746	2.120	2.583	2.921	3.686	4.015
17	1.333	1.740	2.110	2.567	2.898	3.646	3.965
18	1.330	1.734	2.101	2.552	2.878	3.610	3.922
19	1.328	1.729	2.093	2.539	2.861	3.579	3.883
20	1.325	1.725	2.086	2.528	2.845	3.552	3.850
21	1.323	1.721	2.080	2.518	2.831	3.527	3.819
22	1.321	1.717	2.074	2.508	2.819	3.505	3.792
23	1.319	1.714	2.069	2.500	2.807	3.485	3.767
24	1.318	1.711	2.064	2.492	2.797	3.467	3.745
25	1.316	1.708	2.060	2.485	2.787	3.450	3.725
26	1.315	1.706	2.056	2.479	2.779	3.435	3.707
27	1.314	1.703	2.052	2.473	2.771	3.421	3.690
28	1.313	1.701	2.048	2.467	2.763	3.408	3.674
29	1.311	1.699	2.045	2.462	2.756	3.396	3.659
30	1.310	1.697	2.042	2.457	2.750	3.385	3.646
32	1.309	1.694	2.037	2.449	2.738	3.365	3.622
34	1.307	1.691	2.032	2.441	2.728	3.348	3.601
36	1.306	1.688	2.028	2.434	2.719	3.333	3.582
38	1.304	1.686	2.024	2.429	2.712	3.319	3.566
40	1.303	1.684	2.021	2.423	2.704	3.307	3.551
50	1.299	1.676	2.009	2.403	2.678	3.262	3.496
60	1.296	1.671	2.000	2.390	2.660	3.232	3.460
120	1.289	1.658	1.980	2.358	2.617	3.160	3.373
∞	1.282	1.645	1.960	2.326	2.576	3.090	3.291